Operational Symmetries

Heinrich Saller

Operational Symmetries

Basic Operations in Physics

Springer

Heinrich Saller
Werner-Heisenberg Institute
Max Planck Institute for Physics
Munich, Bayern
Germany

ISBN 978-3-319-58663-2 ISBN 978-3-319-58664-9 (eBook)
DOI 10.1007/978-3-319-58664-9

Library of Congress Control Number: 2017940238

Printed on acid-free paper

This Springer imprint is published by Springer Nature
The registered company is Springer International Publishing AG
The registered company address is: Gewerbestrasse 11, 6330 Cham, Switzerland

Contents

Chapter 1
Introduction and Orientation

In a first contact with nature, a child enjoys a glazed tile, a colorful pattern of a fabric, or a pebble rounded in a river by the grinding with other pebbles. When we learn at school or university the more formalized physical description of nature, such direct elementary experiences do no longer play the most important role — the remarkably possible description of the physical events by a mathematical language comes to the foreground. More with some sort of Sunday-feelings and near the end of our physical university-studies, in quantum theory and relativity, we encounter again, now with the learned tools of mathematics and physics, the childhood experiences of symmetry with the connected naive questions, often deep and difficult to grasp.

Somewhat we follow individually a development of the physical understanding: The interpretation of nature by symmetries, especially by the Pythagorean–Platonic school of the Greek natural philosophers, was replaced, after the Copernican turnaround, by a mathematical formalization with equations of motion to describe a physical dynamics. To derive these equations via extremal principles, their structures were characterized again by symmetries which had not been forgotten, but which stayed somewhat in the background. In contrast to the basic studies at universities about 50 years ago, the qualitative and quantitative possibilities, offered by symmetry considerations in physics, are of paramount importance today.

Especially quantum theory with its information theoretical background and the related newly discussed concepts "reality" and "experiment" shows the limitation and complementarity of physical concepts like "position" and "momentum," or, also, "particle" and "interaction." Operations and symmetries seem to be more fundamental tools and more adequate forms for the description of nature. Today, one has the point of view, that it is most appropriate to understand both particles and interactions and quantum theoretical wave functions — also aptly called "experience functions" or "information catalogues" — by symmetries and related properties ("quantum numbers") that characterize the physical behavior under symmetry transformations, e.g., spin and mass for the behavior under rotations and translations, respectively. Because of the many basic particle and interaction fields only a few physicists believe that the electroweak-strong standard model of elementary particles, universal and encompassing up to the quantitatively very weak, but conceptually important gravity, is

H. Saller, *Operational Symmetries*, DOI 10.1007/978-3-319-58664-9_1

the last and finally basic model of a dynamics and an end of physics. Nevertheless, this model shows a remarkable economy, not in particles, but in fundamental properties, given by mass and spin, hypercharge, isospin and color as determined by the representations of the standard symmetry operations. Particles and interactions, so far taken as basic, come in definite forms; they arise as implementations of symmetry operations, apparently better taken as fundamental concepts. All basic physical properties can be related to invariants and eigenvalues of symmetry operations.

Classical mechanics uses the concepts of "pointlike particles" and "unlocalized interactions (forces)" — originally as instantaneous actions, later with a finite velocity. In relativistic theories, spacetime-parametrized classical or operator valued fields have to be used; there remains not much of the classical particle concept. According to Wigner, a particle is described by a "full" vector space, acted on by an irreducible unitary Poincaré group representation. Rotations with spin as characteristic property may be used as an illustration: The familiar angular momentum of a rotating object can be connected with the generators of the spatial rotation group. Rotation properties are given by the "roundness" of an extended object. With the discovery of the spin — initially called "intrinsic" angular momentum, e.g., spin $\frac{1}{2}$ of the electron, a classical visualization, e.g., as a spinning top, is no longer possible: Is the electron in its rest system a sphere, or, because of nontrivial spin $\frac{1}{2}$, not spherically symmetric? Especially with the inappropriate language involving "pointlike" electrons — what is the "roundness" of a point? "Roundness" for an electron is as queer as attributing "color" or "happiness" to it. Spin $\frac{1}{2}$ is realized by compact operations on a complex two-dimensional space as used for the electron. Probably, "pointlike" is a property which makes no sense for elementary particles — similar to macroscopic color for atoms.

Another operational example may be given by the mass as invariant, characterizing spacetime translations. In this context, one may ask for the mass of quarks, e.g., in a quark propagator — are the quark mass-structures similar to the electron mass-structures? Are quarks stable or unstable, have quarks a nontrivial width? Or, is the quark mass an invariant for a non-translation operation? So far, I don't know of a satisfactory answer, especially if color confinement is postulated. Quarks and gluons seem to transcend the usual Wigner-particle concept as translation representations — a nontrivial color property for the implementation of strong interactions seems to be incompatible with a nontrivial translation property, implemented by free particles.

Are there basic operational principles which determine interactions and particles? For instance, what "causes" the simple form $\frac{1}{r}$ for the gravitational potential with the exponent for the position dependence being exactly -1? Already Newton suggested a geometrical symmetry "origin," using conic sections, with an equidistributed force on each sphere, $\frac{G}{r^2} \cdot 4\pi r^2 = 4\pi G =$ constant. Much later, in quantum field theory, the Coulomb potential, also with the $\frac{1}{r}$-form, was recognized as the position projection of the off-shell contribution in the propagator of the massless electromagnetic field, which, in turn, was connected with a symmetry operation, called gauge transformation, and embedded in nonabelian hypercharge-isospin operations. In Einstein's general relativity, Newton's gravitational potential is embedded in the metric of Schwarzschild spacetime. With those important examples in mind, could it be possible that all interactions, not only the gravitational ones, are related to nonabelian operations and their representations?

Is there a final physical theory? In this form the question is rather vague — how do we define "physics," what means "theory," what means "final"? Put otherwise, somewhat more specific: Is there a basic mathematically formalizable structure which describes all interactions and objects for space and time we are experimenting with? Quantum mechanics for the interaction of nuclei and electrons can be taken to be a final nonrelativistic theory for the states of the atoms and, in this sense, is something like a final theory for chemistry. This does not mean that quantum mechanics helps us very much to explain the structural intricacies of a DNA-molecule. Similarly, one may hope for a physical theory which is final with respect to concepts like causality with spacetime parametrized interactions and their experimental verification. However, also that hope may be presumptuous with respect to our evolutionary stage — can a chicken understand Newton's mechanics, can we understand a possibly existing final theory? Is the concept "final theory" meaningful at all? And restricted to human understanding: Has quantum epistemology and space-time relativity not taught us that we may work with fundamental prejudices we are not even aware of? After several insults concerning human egocentrism, sometimes called narcissistic prejudices, e.g., after the Copernican revolution and Darwin's insights, can we be confident that human understanding and mathematical logic is really central? And, in addition, is the historical experience of an ever simpler and more encompassing growing explanatory structure reliable also for the future?

After Galileo's (around 1610) and Newton's (1687) epochal work, there was a rather long period without equally deep progress — of course, there was no standstill and always important steps in our understanding of the physical structures. Then, starting with Faraday and Maxwell (1864, "his" equations), there came, in rather short intervals, Einstein's special relativity (1905), "his" general relativity (1915), quantum theory (around 1925 with the protagonists Bohr, Born and Schrödinger from the elder and Dirac, Heisenberg and Pauli from the younger generation), and, arguably with an equally deep character, quantum gauge field theories with quantum electrodynamics (around 1945) and the standard model of electroweak and strong interactions (around 1965). Can one expect, being aware of the two centuries between Newton and Maxwell, that we have a quantum theory-like progress all two or three decades? What remains, after ten more such revolutions, in two centuries from now? Is the treasure of physical theories inexhaustible?

Each deep theory requires and develops, slowly and sometimes even painfully, its own language which, in most cases, has to correct our anthropocentric prejudices. In the beginning, Maxwell tried an "explanation" of electromagnetic field structures in terms of mechanical ones. Later on, the field language was developed in its own right; clumsily "derived" concepts became basic and primary. Similarly, an "explanation" of quantum structures is often tried in terms of classical concepts. Also here, the language has to become "quantum-appropriate"; quantum theory is no "quantized" classical theory. "Quantization" as a procedure is a dangerous and not very well defined concept.

A visualization and explanation of quantum structures by everyday concepts may lead to a completely wrong understanding — Dirac (1930): "The new theories ...are built up from physical concepts which cannot be explained in terms of things previously known to the student, which cannot even be explained adequately in words at all." And Heisenberg (1955) in a very pointed formulation (my translation): "...the laws of nature, that we formulate in quantum theory, no longer deal with the elementary particles as such, but with our knowledge about the elementary particles ...The atoms and elementary particles constitute more a world of tendencies or possibilities than a world of things and facts." And Bohr: "There is no quantum world. There is only an abstract quantum physical description. It is wrong to think that the task of physics is to find out how nature is. Physics concerns what we can say about nature ..." Quantum theory overcomes the Cartesian distinction between "res extensa" and "res cogitans".

The validity of physical laws, apparently independent of time and space, e.g., also two million years ago in the Andromeda galaxy, hints to some kind of temporal and spatial symmetry. According to Maxwell, a basic law in physics has to be formulated as valid for all space and time. That contrasts to Aristoteles' world view with a distinction of the natural laws "under and over the moon." Could it be that a physical law without symmetry loses its meaning, that the two concepts condition each other, that physical laws determine essentially representations of symmetries, e.g., a time representation as related to a space representation? It is highly remarkable that the experimentally suggested richness of interactions is reducible, as for today, to a few fundamental ones which can be characterized by rather simple symmetry groups or basic geometrical structures.

Classical physics works with equations of motion, derivable from Lagrangians and Hamiltonians. In the characteristic case of a nonrelativistic mass point, the time derivative of the momentum conditions the position derivatives of a potential and vice versa. The equations of motion describe a causal time dependence for position and momentum as Galilei group parameters. In quantum physics, they undergo a fundamental reinterpretation. The concept of time dependent orbits for point particles makes no sense. For example, the Schrödinger equation for the states of the nonrelativistic hydrogen atom looks for space dependent wave functions related to an energy eigenvalue for a stationary time dependence — no particle orbits, no "motion." The time derivative in equations of motions is interpretable as the action of the time translations, the space dependence as the action of space related operations, e.g., of space translations and rotations. The classical equations of motion are reinterpreted and modified in quantum theory to orbit equations, expressing the action of operation groups and Lie algebras on functions which depend on the group parameters. For instance, the Schrödinger wave functions of nonrelativistic quantum mechanics are representation coefficients of the space groups under question. Or, the Klein–Gordon and Dirac equations characterize representation coefficients of the Poincaré group. The solutions of differential equations are representation coefficients of time and position operations and give the characterizing invariants. The Schrödinger equation — it is a scheme, not one special equation — is "only" an effective method for the determination of time and space representation coefficients (Schrödinger wave functions) with corresponding invariants and eigenvalues, especially for noncompact operation groups. In quantum theory, the quest for the "basic differential equation of motion" for "basic particles" seems to have to be replaced by the quest for the "basic operational structure" and "basic symmetries."

What we call "anschaulich," easily and intuitively visualizable, is founded on often experienced pictures, related to our evolution and useful for our survival. The particle-oriented mechanics of the classical world view is "anschaulich"; we also got rather well used to the field-oriented concepts of electromagnetism. Quantum theory and cosmology are not "anschaulich." Very large and very small time and space distances are rather incomprehensible to us, as exemplified for time by the unbelievably sophisticated designs of the evolution. Similarly, microphysics and macrophysics as used in the realm of particles and their interactions and of cosmological structures, respectively, elude the human intuition and are not described appropriately by the classical prejudices. Concepts like "pointlike" or "string- and membrane-like" with respect to the problem of a spatial extension have their origin in a naive extrapolation and generalization of classical, already idealizing pictures. To suggest them, without any experimental support, as basic is, therefore, a not very encouraging speculation. For physicists, it is rather dangerous that the richness of mathematical forms, even of the esthetically appealing ones only, seems to be inexhaustible. Mathematical "beauty" as property of a basic physical theory seems to be necessary, but by no means sufficient. It is good to be open for speculations — it is not good to be satisfied with speculations as "deep" as the may look at a first glance as exemplified by the use of graded Lie algebras in the "beautiful" unification of spacetime- and chargelike operations by supersymmetry without — up to now, and that lasting for almost forty years — a grain of experimental support.

The success of physics as a natural science is founded in the control by experiments where a physical theory has to have contact with. Theoretical physics is not the area of enigmatic prophesies and unsubstantiated statements as deep as they may sound. A mathematical-logical existence does not imply an experimental-physical realization. The line from theory to experiment must not be too long. Theories which cannot be falsified, e.g., worlds where we principally cannot have any measurable contact with, are physically empty. The contact between theory and experiment may be established via postdictions like the computation of Mercury's perihel rotation in general relativity and of the Balmer series of the hydrogen atom in quantum mechanics, or, even better without the danger of fudging results, in the future via yes-no or quantitative experimental predictions like the flexion of light rays at the sun and the existence of antiparticles in relativistic quantum field theories. There may be different theoretical methods to come to equal experimentally testable results, e.g., Schrödinger's differential equation or Pauli's algebraic treatment of the nonrelativistic quantum mechanical hydrogen atom — in general, the special mathematical treatment has not to be experimentally verified.

At least for me, not more phantasy and inventiveness for the creation of new structures, but a deeper understanding of the strengths and defects of basic theories looks like a more promising strategy. In this respect, this book is very conservative. Despite apparent defects and compatibility difficulties, the established theories for quantum structures and spacetime are experimentally extremely successful, and theoretically also simple, and — in some sense — beautiful. It seems useful and probably necessary to familiarize with pictures that are founded in the deeper-lying symmetry concepts whose relevance for quantum theory and relativity is supported by many experiments. To contribute to this familiarization is the goal of this book.

In the first place, it should awake the pleasure with and the love for symmetries, it is not intended primarily as a hierarchically built mathematical or physical textbook. Especially the first chapters start with simplest physical-mathematical experiences of operation and symmetry structures, follow exploringly this path, also driven by the structure itself, and reach at the end of the chapter — perhaps after a steep ascent or a daring climbing — a point where one touches actual research-fields of theoretical physics. Correspondingly, the necessary tools become more and more difficult to the end; also the mathematical requirements are more demanding from chapter to chapter. The first six chapters are a first cycle to get familiar with basic operations; they are written in a more colloquial style. The next five chapters are, in a sense, a second cycle for a deeper understanding of the structures described so far. These chapters are "very close to my heart," especially the perhaps misguided last one, where an attempt is presented to relate the particle spectrum with representations of operational electroweak spacetime, in analogy to the atomic spectrum as characterizing representations of hyperbolic space: The spectrum of hyperbolic position space explains the properties of the nonrelativistic atoms; the spectrum of electroweak spacetime is hoped to explain those of the basic interactions and elementary particles.

The last five chapters require a much deeper mathematical knowledge, and, probably, are rather hard nuts to crack. Since they are mathematically difficult, at least for me, I also have abandoned the method in the first chapters to intersperse, for a better orientation, slightly deeper mathematical remarks that come under the heading "**Good to know**." In contrast to the finite-dimensional representation structures and discrete "quantum numbers" worked with in the first six chapters that are sufficient for compact operations, e.g., for electromagnetic charge or spin, isospin and color, they deal with infinite-dimensional spaces and continuous "quantum numbers," necessary and characteristic for unitary faithful representations of noncompact groups, e.g., for the translation, Lorentz, and Poincaré groups. The appropriate analytic methods extend the algebraic ones and work especially with function spaces of the operation groups. All this remains, at least for me, rather complicated — even if one gets used to it.

Who can understand the following reflections? With a conscientiously performed structure- and application-oriented mathematical training as a physicist one can follow the first six chapters, at least by repeating the calculations, e.g., with matrices — probably a too optimistic hope. Remarkably, even an elementary mathematical-mechanical work contributes to a qualitative understanding: To detect the infinity of a divergent Feynman-integral and to attempt an ad hoc regularization is an impressive experience. This experience was so formative for Dirac that, probably for good reasons, he developed something like a "divergency-complex." It may also not hurt to go first the simple ways, to put too difficult parts to the side; in general, the next chapter starts in a simple way, some ideas and concepts are taken up again, and one may return later to the difficulties of the former chapter. Understanding is no linear procedure. One should also not underestimate the always necessary time for getting used to new concepts, to let them sink in, and to relate them to the known ones.

Difficult and incomprehensible or easily understandable — it is almost impossible to agree in attributing these properties. When one explains physical structures and their connections, one is often surprised about the elementary ignorance one encounters. On the other hand, when one listens to lectures, one is surprised about one's own ignorance, compared with the displayed virtuosity; it is almost impossible to connect it with the negative experiences. In learning and in teaching one has often the simultaneously depressing and elating experience that simplicity is difficult to realize — think of your first contact with $e^{2i\pi} = 1$. Every scientist knows the irritation coming up when a subjectively complicated and far-fetched structure is called trivial and obvious by an experienced author or discussion partner. Some supposedly trivial things have great epistemological depth.

As necessary as the experimental content, as unrenouncable is the mathematical form. Since a physical law is a structural statement, and since mathematics can be taken as the science of structures, only the mathematical form of a physical law can satisfy — perhaps after a useful phenomenological description as a rung on the ladder of understanding.

With a distinction of a "Babylonian mathematics" as application-oriented (Babylonian astronomy and, less sophisticated, Egyptian land surveying) and a "Greek mathematics" with a structural character by axioms and proofs, it may appear that, for a physicist, it is more important to familiarize oneself with a mathematical problem by working out three examples than to meditate on its provability. Learning by doing is indispensable ("it works") — that is apparent. But only with that one often fails: mathematics does not forgive any mistakes ("why does it work"). In the daily work of theorists, it is often surprising how far the "battle tested" tools in physics carry without a deeper structural understanding. The distinction, put forward by some physicists, between useful — I even heard the foolish attribute "healthy" — mathematics and esoteric formalism reflects in most cases only their biographies, sometimes even only the mathematical lectures listened to at the university. That the more abstract structural considerations of algebra and topology, of relations and categories gain the upper hand over "epsilontic" and application-oriented differential- and integral-calculus, is not surprising with respect to the structural fundamentality of today's basic physics in comparison with earlier times. If "nature knows mathematics," its capabilities should not be expected to end with elementary operations like addition and multiplication.

Since the following presentation is not systematically hierarchical, I did not include complicated proofs; I know this is a horror for a responsible scientist looking for complete deductions. Since, therefore, the presentation remains often verbal and descriptive, there may arise the mathematically oriented question what is presupposed and what is proven. If this question is really a serious one, one has to look at the relevant places in the given literature. Such a hard work cannot be avoided — also not by a thousandfold copy and repetition of the proofs given there. My floppy references may be justified with Weisskopf: "... it is no longer customary to refer to a publication by Galileo when one exploits the momentum conservation law."

By using and understanding the physical connections one should become familiar, in this book, with symmetries and the related operations. It is my experience that, in an intuitively understood structure, most proofs become almost trivial; nontrivial proofs often reflect only our not-deep-enough understanding. In addition, I have the prejudice that basic physical structure are "simple," in some sense, and with an appropriate insight almost self-evident. In this respect, the development of the theories for electrodynamics with their deeper and simpler becoming levels of understanding is an example, illustrated by the names Galvani, Volta, Coulomb, Oerstedt, Faraday, Maxwell, Hertz, Weyl, Feynman, Schwinger, Tomonaga, Glashow, Salam and Weinberg.

The concept simplicity has to be qualified: For example, Newton's gravity is described by one potential with an easy solution for the force between two mass points. In contrast, Einstein's gravity uses the ten components of the spacetime metrical tensor as gravity potentials with, sometimes, very complicated solutions for the gravitational interactions of particles or fields. However, the concepts spacetime and interaction merge and loose their independence — spacetime metric and curvature is interaction. Unexpected connections and common features lead to a deeper and qualitatively, not necessarily formally simpler understanding — by its conceptional coherence Einstein's gravity is simpler than Newton's gravity.

Kant's characterization "Gedanken ohne Inhalt sind leer, Anschauungen ohne Begriffe sind blind," (something like "ideas without content are empty, concepts without perceptions are blind") applies, with the parallelity "theories without experiments are empty" (also beautiful theories) and "experiments without theories are blind," to physics too. Although a final experimental verification is impossible, and a falsification may take centuries, as exemplified by the replacement of Newton's mechanics by special relativistic mechanics, experiments and theoretical structural insights build up some trust in the validity of theories. This is especially possible if new extended theories arise which contain the older ones as limiting cases and which tell us about our prejudices. An example is the contraction of the Lorentz group-compatible special relativistic mechanics to the Galilei group-compatible one in the limit of an infinite action velocity, or the classical commutative substructures of the noncommutative quantum structure. One may also take the "liberation" from the Euclidean geometry and the abandonment of an absolute meaning of coordinates, both successful prejudices ingrained by our evolutionary development. Or the replacement of nonrelativistic point mechanics by relativistic field theories with the prominent example of Faraday's and Maxwell's electrodynamics — mass points remain possible, but, nevertheless, they are conceptual strangers in field theories. The epistemological structure of quantum theory with information functions has to be joined with relativistic field theories: Special relativistic quantum field theory is, in some sense, one of the two deepest levels we are willing to build on. The other one is the geometrical formulation of gravity in terms of the metric induced curvature concept for spacetime. Spacetimes with group actions can be considered to be constituted by subgroup classes of the action group.

Standing on these levels, one can perceive as apparently basic, not satisfactorily solved theoretical problems:

- What is the spectrum of elementary particles, qualitatively and quantitatively — their masses, spins and charges?
- What "causes" the quarks as confined degrees of freedom — in addition to the unconfined leptons?
- What is the spectrum of the basic interactions, qualitatively and quantitatively — their coupling strengths and ranges?
- Why the dichotomy of particles and interactions?
- Are there common basic concepts for spacetime and interactions-particles?
- Why both external (spacetimelike) and internal (chargelike) degrees of freedom?

The two basic theories of today are general relativity and quantum theory. General relativity relates gravity to the metrical tensor of the spacetime manifold, quantum theory relies on Hilbert space representations of operations. There are operational spacetimes, whose defining operations are implementable, in a quantum theoretical treatment, on a Hilbert space. In such a common framework, the problems above may be considered.

For a more personal touch with respect to the mentioned philosophers, mathematicians, and physicists, I have included footnotes with their life-dates. In many cases, it is worthwhile to ponder, with the dates, over the connections of the scientific concepts of the times (Antiquity, Middle Ages, Renaissance, Enlightenment, etc.) with the general history of ideas, and how far sciences and humanities influence each other. For instance: Is it an accident that the revival of the heliocentric system by Copernicus, and Columbus' discovery of a "new world" happened in the same time? Did the explosion of the first atomic bombs facilitate the "big bang" concept? What would Newton have done with a quantum mechanical textbook? Or, Einstein in the 11 century — would he have thought about space-time relativity? Or, Einstein today, would his genius be discovered while assistant examiner at the patent office in Bern? Such reflections may also contribute to some modesty with respect to physical theories: Is it possible that, in thousand years from now, one will look back to our theories with similar qualifications as we do today viewing the pre-Copernican theories? For instance, that, at a time in the future, our initially oriented causal argumentation appears as limited as, for us today, the finally oriented causal interpretation of Aristotle? It may also be of some interest how new or how old some concepts are and which people were contemporaries. Considering the life dates of the participants in the decipherment and understanding of the fabric of the physical cosmos, there builds up a modesty with respect to one's own work and a deep reverence for the intellectual efforts and the insights, not only scientifically motivated, but also by emotions, esthetics, mystics and religions, which humanity developed over the centuries and millenia.

The following eleven chapters deal essentially with basic physical questions arising in quantum theory and relativity — all with respect to an operational interpretation in terms of groups and Lie algebras. Elementary particle physics with the standard model for the electroweak and strong interactions — a fundamental physical theory according to our present understanding and, in contrast to other speculative proposals, connected with experiments — is an often used example and goal. Much of what is given in the following can be found in other books. I also have taken over, more or less modified and, as I hope, improved at some places, contributions from my two–volume book *Operational Quantum Theory I — Nonrelativistic Structures* (2006) and *Operational Quantum Theory II — Relativistic Structures* (2006) and from my book *Operational Spacetime — Interactions and Particles* (2009), and from my contribution of the special issue *Quantum Relativity* of the International Journal of Theoretical Physics in memory of my friend David Finkelstein (1929–2016).

One should not be deluded by the representation in the sense that one reaches directly and in the only possible way, i.e., deductively a theory. In some sort of playing Easter for physicists, one finds only the colored eggs which one has hidden oneself before. Nevertheless, the possibility to formulate basic physical insights by mathematically elegant and "simple" concepts is not only fun — it is a deep scientific wonder. Sometimes, I was somewhat carried away by the attractive mathematical structures — probably more than necessary to explain and to do justice to the physical ones.

It is my aim to show again and again the limitations of the familiar concepts space and time and matter. The fascination of Kant's approach — more epistemology than traditional philosophy — to clarify the conditions which are necessary for our experience, and which, to some extent, predefine the content of our experience — like blue eyeglasses our vision, seems to me, despite all detailed criticism, basically unbroken. It is useful, for epistemological censorship, to have Kant's ideas in the back of your mind; however, even neglecting his contemporarily understandable prejudices like the necessity of the Euclidean geometry, they seem not specific enough to serve as a top-down research program for physics. Space and time and particles, or, with Bohr, possibly the whole conceptual apparatus of classical physics, are apparently necessary to talk about the experiments. However, after the labors and the birth of quantum theory with its "quantum relativity," i.e., with the different experimental projections of the operator formulation for a classical interpretation, the confidence in an all-encompassing objective fundamentality of those concepts is broken, at least in the Copenhagen interpretation. Basic interactions and particles are characterized by operation groups. Interactions and particles implement and parametrize space and time which, in this book, come as groups or equivalence classes of groups, e.g., Minkowski spacetime as Lorentz group classes of the Poincaré group. In quantum theory, the concept of free particles as building bricks for bound states becomes fuzzy. Eigenstates of time and space operations are represented on different levels, e.g., by molecules, atoms, nuclei, hadrons, quarks. For high energies, the flat spacetime implementation by particles seems to fade away as suggested by the quarks if they are really confined.

Physics started as a science with four basic concepts — forces for matter in time and space. Those concepts can be seen as arising from an operational root:

A formalization of physical laws is simultaneously a formalization of the laws of our experimental knowledge. Quantum information comes in the form of operation group coefficients. If physical dynamics is defined as the causal behavior of symmetry operations, then the concepts space and time, parameterizing these operations, and interactions and matter, implementing them, are “only” human related concepts for their description.

Chapter 2
Units and Orders of Magnitude

Especially mathematics, but also theoretical physics works basically with structures. The experimental verification or falsification of the physical structures employs units, which allow, in addition to a quantification by numbers, also a qualitative differentiation — a length, for example three meter, is different from a mass, for example three kilogram. But where is the decisive structural reason to differentiate between mass and length?

The three life-important elements *sun*, *earth*, and *water* were used for a first definition of *human or anthropomorphic* units for time, length and mass, respectively, for example in the MKS-system with, respectively, the second (s) as the 86,400th part of the averaged sun day, the meter (m) as the 10 millionth part of a quarter of the circumference of the earth, and the kilogram (kg) as mass of one liter ($10^{-3}m^3$) water. In the course of a scientific penetration of nature, such anthropocentric ("daily life") measures have been specified and replaced by *natural* measures, inherent to physical laws.

The heaven- and sky-oriented sexagesimally prone system of the Sumerians and Babylonians with about $360 = 6 \times 60 = 12 \times 30$ days a year, used for the angle number 360 of a full circle, and about $30 = \frac{1}{2} \times 60$ days a month and $86{,}400 = 2 \times 12 \times 60^2$ s a day, was replaced by the human-oriented decimal system reflecting our ten fingers. Our feeling for order of magnitudes seems to be not hexal or decimal, but logarithmic with $\log 10 \sim 2.3$, as expressed by the Weber–Fechner[1] law.

Because of our biological conditions, our physical argumentation starts from anthropomorphic concepts and measures. Empedocles[2] used a distinction between "matter," given by fire, earth, water, and air, and "interactions," presented by strife and love. The original division of physics in fields like mechanics, optics and acoustics — for working, seeing, and hearing — shows a human related conception of physics.

[1]Ernst Heinrich Weber (1795–1878), Gustav Theodor Fechner (1801–1887).

[2]Empedocles of Acragas, around -(490–430).

H. Saller, *Operational Symmetries*, DOI 10.1007/978-3-319-58664-9_2

In modern times, it was gradually given up in favor of more structural concepts like particles and interactions, and, still more abstract, symmetries and operations. Basically, the physical nature does not have to be anthropomorph; however, trivial to say, experiments are necessarily made or registered by human beings.

What is the origin of natural constants, or, as formulated in the following, of "natural units" like Planck's[3] constant $\hbar$, to express a distinction to "natural numbers" like the perimeter number π? Does it make sense to distinguish a finite number of units as basic, and, if yes, what does "basic" mean in this context? Microscopic and macroscopic fields, which differ from our mesoscopic orders of magnitude, elude our evolution-determined familiarization. The phenomena of relativity and the contact with nature in quantum theory balk at a naive understanding and a basic interpretation with anthropomorphic concepts.

Since natural units, related to deeper insights, do not necessarily directly reproduce, neither quantitatively nor qualitatively, our familiar human measures, there arise, already by a simple dimensional analysis, interesting and naively unexpected conversions. Related conclusions from units to physical processes use, in general, only simple power laws. Additional numerical factors like 4π or dimensionless ratios of dimensioned quantities, e.g., mass ratios, and, with them, natural logarithms and exponentials or other functions, cannot be conjectured so easily. In addition, special initial or boundary conditions of a specific problem are not taken into account. Dimension-analytic conclusions are rough and superficial projections like the characterization of a function by the value of its integral. They try to cherry-pick from the cake and are, on the one hand, simple and attractive, but, on the other hand, by their crudeness, unreliable or prove even wrong, if one knows, by other, in general more reliable considerations in "complete theories" with functional laws, what one has to look for. To avoid a naive numerology, a playing around with units and dimensions, as also done in the following, has to be performed with due care. It can serve only as a motivation for a further elaboration of the suggested structures and may be used as mnemonics.

2.1 External Interactions and Newton's Unit

Newton's[4] law for gravity is probably the first rather clearly formalized physical interaction with a basic universal importance. Presumably, the depth of this law is not yet exhausted: Today, after its general relativistic embedding and its interpretation in terms of spacetime metric and curvature, classical gravity waits for a connection with and an understanding in the framework of quantum theory.

Gravity is determined quantitatively by *Newton's unit G* (for *G*ravity) as given in the Newton potential of two masses $m_{1,2}$ with distance r:

---

[3] Max Planck (1858–1947).

[4] Isaac Newton (1642–1727).

$$V_{\text{Newt}}(r) = -G\frac{m_1 m_2}{r},$$
$$G \simeq 6.7 \times 10^{-11} \frac{\text{m}^3}{\text{kg s}^2} (\text{minimal(?)})$$

The Newton potential, determined up to an additive constant, contains the *heavy* masses $V_{\text{Newt}}(r) = -\frac{M_1 M_2}{r}$, characterizable as "eigenvalues" M for gravitational interactions, with the remarkable universal proportionality $(\frac{M}{m})^2 = G$ to the *inert* masses, used in the equations of motion as invariants m^2 for spacetime translations. G is no number, it has the dimension of a cubic length (volume) divided by the product of a mass with a time square, therefore Newton's "unit." Because of the rotation symmetric potential, a replacement $G = \frac{G_0}{4\pi}$ with the area $|\Omega^2| = 4\pi$ of the unit sphere would have been somewhat more appropriate.

Perhaps, the numerical value of G is minimal for all physically relevant quantities with this dimension. The soft concepts "maximal" and "minimal" will be used only up to factors like 12, $\sqrt{5}$, $\frac{1}{2\pi}, \ldots$, which do not change the order of magnitude — not however factors like $10^{\pm 4}$. The speed of light may serve as a familiar example for the concept "maximal" in the case of velocities for physical actions.

Why seems Newton's unit small for us human beings, how can it be reproduced from familiar orders of magnitude? It is related to anthropomorphic measures by the gravity acceleration $g_\bullet \simeq 9.8\frac{\text{m}}{\text{s}^2}$ (gravity field strength on the surface of the earth), the average earth density $\rho_\bullet \simeq 5.5 \times 10^3 \frac{\text{kg}}{\text{m}^3}$ (water $10^3 \frac{\text{kg}}{\text{m}^3}$, stones about $3 \times 10^3 \frac{\text{kg}}{\text{m}^3}$, iron $7.9 \times 10^3 \frac{\text{kg}}{\text{m}^3}$, platinum $21.5 \times 10^3 \frac{\text{kg}}{\text{m}^3}$) and by the circumference of the earth $2\pi r_\bullet \simeq 4 \times 10^7$ m, with $mg_\bullet = G\frac{mm_\bullet}{r_\bullet^2}$ and $m_\bullet = \frac{4\pi}{3} r_\bullet^3 \rho_\bullet \sim 6 \times 10^{24}$kg, therefore $G \simeq \frac{3g_\bullet}{4\pi r_\bullet \rho_\bullet}$. Its MKS-order of magnitude is the inverse of the number of millimeters of the earth's circumference.

The *escape velocity* from a distance r from a centrally symmetrically distributed mass m equalizes kinetic and potential energy:

$$v_{\text{esc}} = \sqrt{\frac{2Gm}{r}},$$

e.g., the escape velocity from the earth surface is $v_{\text{esc},\bullet} = \sqrt{2g_\bullet r_\bullet} \sim 11.2 \times 10^3 \frac{\text{m}}{\text{s}}$.

The remarkable universality of physical laws in time and space was realized by Newton. Newton's potential determines the fall of an apple in about one second from a five meter high tree to the earth with mass $m_\bullet \simeq 5.8 \times 10^{24}$ kg as well as the one year long orbit of this earth around the sun with the mass $m_\odot \simeq 2 \times 10^{30}$ kg.

$\frac{1}{G}$ carries the dimension of a quadratic time, multiplied with a mass density. This dimensional analysis allows a connection between a density ρ and its gravitational collapse time:

$$t_{\text{collapse}}(\rho) = \sqrt{\frac{3}{4\pi G\rho}} = \sqrt{\frac{r^3}{Gm}} \text{ for sphere } \rho = \frac{3m}{4\pi r^3}.$$

According to such a collapse time, only to be used for an order of magnitude and neglecting all other interactions, all mass, contributing to a density ρ, would be unified in one point. The collapse time of the Earth is given by $t_{\text{collapse}}(\rho_\bullet) \simeq \sqrt{\frac{r_\bullet}{g_\bullet}} \simeq 800\,s$.

Interpreting the relation between mass density and time in the inverse form as a flying apart with the daring parametrization[5] of a cosmic matter density:

$$\rho_{\cos,p}(t) = \frac{3}{4\pi G}\frac{1}{t^2},$$

thinning out with an inverse quadratic time dependence for constant G, then one obtains for the Hubble[6] time $t_{\text{Hubl}} \simeq 10^{10}$ years $\simeq 3.1 \times 10^{17}$ s, obtained from the red shifts of the spectral lines of the stars, the cosmic density $\rho_{\cos,p}(t_{\text{Hubl}}) \simeq 4 \times 10^{-26}\,\frac{\text{kg}}{\text{m}^3}$. This yields, with the proton mass $m_p \simeq 1.67 \times 10^{-27}$ kg a baryon density $n_{\cos,p}(t_{\text{Hubl}}) \simeq 24\frac{1}{\text{m}^3}$.

The Hubble time has to be used carefully, especially when it leads to the very far past, e.g., to the "first" three nanoseconds. Probably, it can give only the order of magnitude of the time span, where, backwards, the familiar pictures for space and time give physically meaningful statements, e.g., for the absolute time and the absolute space of Newton — space and time are inert boxes, wherein mass points may perform some dynamics — or for the relational space and time of Leibniz[7] — space and time are translations as operational relations of physical events.

If one calculates from the Newton force $\vec{K}_{\text{Newt}}(r) = \frac{\partial}{\partial\vec{x}}V_{\text{Newt}}(r) = G\frac{mm'}{r^3}\vec{x}$ the gravitational field strength $\frac{\vec{K}_{\text{Newt}}(r)}{m'}$ and collects the $\frac{1}{r^2}$-proportional flux on a closed surface around a mass m, e.g., for a sphere, one obtains the linearly mass-dependent[8] Kepler quotient — here, the exact $\frac{1}{r}$-behavior is important:

$$q_{\text{Kepl}}(m) = \frac{1}{4\pi}\int d\vec{F}\,\frac{\vec{K}_{\text{Newt}}(r)}{m'} = Gm = \begin{cases} Gm_\bullet \simeq 3.9 \times 10^{14}\frac{\text{m}^3}{\text{s}^2} & \text{for earth,} \\ Gm_\odot \simeq 1.5 \times 10^{20}\frac{\text{m}^3}{\text{s}^2} & \text{for sun.} \end{cases}$$

In a closed system with gravitational interaction only and one "dominating" large mass, e.g., in a quite good approximation, our planetary system or the earth–moon system, Gm is a "universal" intrinsic unit. This yields, in the example with the sun mass, the Kepler ratio $\frac{q_{\text{Kepl}}(m_\odot)}{(2\pi)^2} = \frac{R_P^3}{T_P^2} \simeq 3.8 \times 10^{18}\,\frac{\text{m}^3}{\text{s}^2}$ of the cubic averaged planetary distances R_P to the quadratic orbit times T_P for all planets. In the form

[5] Arthur Eddington (1882–1944).
[6] Edwin Hubble (1889–1953).
[7] Gottfried Wilhelm Leibniz (1646–1716).
[8] Johannes Kepler (1571–1630).

$$\frac{Gm_\odot}{R_P} = \left(\frac{2\pi R_P}{T_P}\right)^2,$$

it can be related to the virial theorem, which relates to each other the averages of potential and kinetic energy $\overline{V(x)} = \overline{E_{\text{kin}}}$.

Kepler's third law uses as intrinsic unit a ratio with orbit radius and time — not a position unit and not a time unit. It allows to estimate the moon-earth distance via $\frac{R_m^3}{T_m^2} \sim \frac{Gm_\bullet}{(2\pi)^2} = \frac{g_\bullet r_\bullet^2}{(2\pi)^2}$ and the month length $T_m \sim 28$ days, leading to $R_m \sim 3.8 \times 10^8$m and for $T_s \sim 1$ day the radius $R_s \sim 4.1 \times 10^7$m of a geostationary orbit. It is not difficult to estimate or to measure, on earth, Newton's constant G, at least the right order of magnitude, and the earth's orbit time (year). One non-terrestric measurement is necessary to quantify astronomical consequences "over the moon," e.g., the not so easy measurement of the orbit radius of the Earth which, then, allows the computation of the sun mass.

2.2 Speed of Light as Space-Time Hinge

The *speed of light c* (for *c*eleritas) as probably *maximal velocity of an action* is about seven and a half circumferences of the earth per second — "straight" light rays, of course. It defines, for an otherwise determined time unit, the associated natural length unit, so, since 1983 the meter for the second:

$$c = 299\,792\,458 \frac{\text{m}}{\text{s}} \simeq 3 \times 10^8 \frac{\text{m}}{\text{s}} \text{ (maximal)}$$

More light — away with the Parisian ur-meter! The huge ratio of the action velocity of gravity and electromagnetism to human-related velocities makes it understandable that the first physical theories assumed instantaneous long-distance interactions. Newton was not comfortable with that. The limit of nonrelativistic space–time is reached from relativistic spacetime for an infinite speed of action in the contraction:

$$\text{spacetime} \xrightarrow{\frac{1}{c}\to 0} \text{space–time}.$$

The maximal distance for the speed of light, is, for the Hubble time, about 10 billion light years, $ct_{\text{Hubl}} \simeq 0.9 \times 10^{26}$ m.

Einstein's[9] *energy equivalent of a mass m* as the trivial momentum value of the relativistic energy-momentum relation $E^2 = m^2c^4 + \vec{p}^2c^2$:

$$E_{\text{Einst}}(m) = mc^2,$$

[9] Albert Einstein (1879–1955).

gives for a hydrogen atom or the proton mass $m_p c^2 \simeq 1.5 \times 10^{-10}$ J with the energy-unit Joule[10] $\text{J} = \frac{\text{m}^2\text{kg}}{\text{s}^2}$. The first order nonrelativistic correction, for the contraction $\frac{1}{c} \to 0$, is the kinetic mass point energy:

$$E = \sqrt{m^2c^4 + \vec{p}^2c^2} = mc^2\sqrt{1 + \frac{\vec{p}^2}{m^2c^2}} = mc^2 + \frac{\vec{p}^2}{2m} + \ldots,$$

$$\lim_{c\to\infty} (E - mc^2) = \frac{\vec{p}^2}{2m}.$$

Good to know: Vectors with an arrow, here $\vec{p}$, are assumed to be from a Euclidean space with scalar product, here $\vec{p}^2 \geq 0$.

The, without concessions, maximal action velocity is, simultaneously, the *natural velocity unit* and the *natural conversion factor* between time and length. According to special relativity, there is no absolute reference frame for separate space and time translations. However, there exists an absolutely highest action velocity. Not a static position length or a time period, but the space and time connecting velocity c with dynamical implications seems to be a fundamental unit. The related lightcones in spacetime are "absolute."

The "unification" of time and space in spacetime has to be qualified: Although time and space are connected with each other by relativity to spacetime, forever and everywhere, shown by the existence of an universal velocity, both structures stay clearly separated — now in a modified form with the relativistically meaningful concepts "timelike" and "spacelike." However, the causality structure (future, past and also a generalized presence as causally not influenceable spacelike region) is no longer absolute, but observer-dependent. In a mathematical terminology, there does not exist a total order structure, but only a partial one. Completely new and relativity-characteristic is the lightlike region where nontrivial time distances coincide with nontrivial space distances, i.e., where spacetime distances vanish. The mathematical designation *singular* for nontrivial distances with vanishing "length" is, also for the physics of spacetime, very appropriate: Light is really a singular structure.

With the embedding of a three-dimensional space and a one-dimensional time into four-dimensional relativistic Minkowski[11] spacetime or, expressed otherwise, with the blowing up of a pointlike presence in time to a causally not affected and not affecting four-dimensional spacelike manifold, the classical time-dependent point-particle physics is replaced by a description with spacetime fields. Although there exist hybrid theories with relativistic point particles, which move, e.g., in gravitational and electromagnetic fields, mathematically formalizable with the eigentime, the development of physics, not at least of quantum theory with probability amplitudes and densities, shows that the concept of a point particle, perhaps even the concept of a particle, is a pragmatically useful way for the interpretation of experiments

[10] James Prescott Joule (1818–1889).

[11] Hermann Minkowski (1864–1909).

in terms of interaction-free structures, but possibly not appropriate for a fundamental formulation of dynamics and interactions.

Good to know: In mathematics, space is used as a very general concept — vector space, topological space, etc. To distinguish the physical "space" in spacetime, it will be often called "position." Inconsequentially, "spacelike" will not be changed to "positionlike." In the literature, "Minkowski spacetime" is sometimes generally used for a Riemannian manifold with $(1, 3)$-signature metric. In the following, "Minkowski spacetime" is used only for a vector space with $(1, 3)$-signature metric.

With a transition from time and space (position) to spacetime, there come some new features, often not appreciated enough. For example, it is a mortal sin against the special relativistic spirit to distinguish a basis with one time and three position vectors or also a class of bases, e.g., orthonormal bases where the Lorentz[12] metric has the matrix $\eta = \begin{pmatrix} 1 & 0 & 0 & 0 \\ 0 & -1 & 0 & 0 \\ 0 & 0 & -1 & 0 \\ 0 & 0 & 0 & -1 \end{pmatrix}$ for the spacetime "length" $x^2 = x_j \eta^{jk} x_k = x_0^2 - \vec{x}^2 = x_0^2 - x_1^2 - x_2^2 - x_3^2$, without giving a physical reason for this choice. One such reason may be a particle with nonvanishing mass, e.g., a human being, where there exist rest systems. This must not be absolutized. For example, a lightlike object, e.g., a particle with vanishing mass, distinguishes completely different spacetime bases, e.g., ligthlike tetraeder bases where the Lorentz metric has a 0-diagonal symmetric matrix $\zeta = \begin{pmatrix} 0 & 1 & 1 & 1 \\ 1 & 0 & 1 & 1 \\ 1 & 1 & 0 & 1 \\ 1 & 1 & 1 & 0 \end{pmatrix}$ for $x^2 = x_j \zeta^{jk} x_k = 2(x_0 x_1 + x_0 x_2 + x_0 x_3 + x_1 x_2 + x_1 x_3 + x_2 x_3)$ without manifest distinction for time and space.

Good to know: Nonrelativistic space-time is the direct sum of two vector spaces $\mathbb{R} \oplus \mathbb{R}^3$ for time and position. There is no vector space decomposition of special relativistic Minkowski spacetime $\mathbb{R}^4$, compatible with the Lorentz transformations. The Lorentz-compatible decomposition into two four-dimensional manifolds $\mathbb{R}^4_{\text{caus}} \uplus \mathbb{R}^4_{\text{position}}$ contains, in addition to the spacelike vectors $\mathbb{R}^4_{\text{position}}$ with $x^2 < 0$ the causal vectors $\mathbb{R}^4_{\text{caus}} = \mathbb{R}^4_{\text{time}} \uplus \{0\} \uplus \mathbb{R}^4_{\text{light}}$, with the proper timelike vectors $x^2 > 0$, the trivial vector, and the lightlike vectors $x^2 = 0$ for $x \neq 0$. The timelike vectors are the union of the future cone $\mathbb{R}^4_+$ and the past cone $\mathbb{R}^4_-$, both four-dimensional and characterized by $x^2 \geq 0$ and either $x_0 > 0$ or $x_0 < 0$. Cones are not vector spaces: They are closed under vector addition $x + y$, but under multiplication only with positive scalars, αx for $\alpha \geq 0$. The lightlike vectors $\mathbb{R}^4_{\text{light}}$ constitute a three-dimensional manifold. The order conditions $x_0 > 0$ and $x_0 < 0$ for $x^2 \geq 0$ are compatible with the orthochronous Lorentz group transformations.

Connected with all this, there is a shift of fundamental concepts: For example, the concept "velocity," dependent one a space-time decomposition, is not good for a relativistic theory. Vectors x, i.e., spacetime translations, and their dual vectors p, i.e., energy-momenta, are decomposable by $x = (x_0, \vec{x})$ and $p = (p^0, \vec{p})$ only with the distinction of a time direction. Then, the quotient yields a velocity $\frac{\vec{v}}{c} = \frac{\vec{p}}{p^0}$ in this system. It is interesting, that both relativity and also quantum theory with $[i\mathbf{p}, \mathbf{x}] = \hbar\mathbf{1}$ replace the velocity concept, familiar from daily life, by the more fundamental and somewhat more abstract momentum concept. Energy–momentum values and time-space translations are closely connected as dual partners, e.g., no time concept without the energy concept.

Starting from special relativity, characterized by the unit c, two ways open up: The macroscopic relativistic gravity road and the microscopic relativistic quantum road, which both will be pursued in the following.

---

[12] Hendrik Antoon Lorentz (1853–1928).

2.3 Gravity and Maximal Speed of Action

In Newton's gravity with the, already for Newton himself, awkward action at a distance, there is no maximal speed of action, only in Einstein's relativistic field theories. The spacetime unit c together with Newton's unit give the *natural gravity conversion factors* for time and length to mass:

$$\frac{G}{c^3} \simeq 2.5 \times 10^{-36} \frac{\text{s}}{\text{kg}} \text{ (minimal(?))},$$
$$\frac{G}{c^2} \simeq 7.4 \times 10^{-28} \frac{\text{m}}{\text{kg}} \text{ (minimal(?))}.$$

If $\frac{G}{c^3}$ would be easier experimentally accessible, the kilogram could be defined for a given time unit.

The parametrization of a mass increasing with time, called cosmic mass:

$$m_{\cos,p}(t) = \frac{c^3}{G}t,$$

(where does this mass come from?) yields for the Hubble-time $m_{\cos,p}(t_{\text{Hubl}}) \simeq 1.2 \times 10^{53}$ kg. With the proton mass, one obtains, according to Eddington, the cosmic "baryon number:"

$$N_{\cos,p}(t) = \frac{m_{\cos,p}(t)}{m_p} = \frac{c^3}{Gm_p}t,$$

e.g., $N_{\cos,p}(t_{\text{Hubl}}) \simeq 7.1 \times 10^{79}$ for the Hubble-time. Here, the speed of light is assumed to be constant in time.

The linearly mass-dependent *Schwarzschild*[13] *length:*

$$l_{\text{Schwarz}}(m) = \frac{q_{\text{Kepl}}(m)}{c^2} = \frac{G}{c^2}m$$

characterizes a black hole and leads, e.g., to the Schwarzschild radius of the sun $2l_{\text{Schwarz}}(m_\odot) \simeq 3 \times 10^3$ m. At twice the Schwarzschild length, the escape velocity is the highest action velocity $v_{\text{esc}} = c = \sqrt{\frac{2Gm}{l_{\text{Schwarz}}}}$, no action can come over the horizon of such a sperically symmetric black hole.

Reversing this relation, each length can be associated with its *Newton mass*:

$$m_{\text{Newt}}(l) = \frac{c^2}{G}l.$$

[13]Karl Schwarzschild (1873–1916).

The Newton mass characterizes the equilibrium of the gravitative energy and the mass energy of two masses with distance l:

$$m_1c^2 + m_2c^2 = G\frac{m_1m_2}{l} \Rightarrow \frac{1}{m_{\text{red}}} = \frac{1}{m_1} + \frac{1}{m_2} = \frac{G}{c^2 l}.$$

Here m_{red} is the reduced mass of the system, l its Schwarzschild length.

The ratio of the Schwarzschild length of a mass to its radius r for a rotation symmetric mass distribution gives, up to a factor, understandable in general relativity, the deflection angle for a tangent light ray:

$$\delta(m, r) = 4\frac{l_{\text{Schwarz}}(m)}{r} = 4\frac{G}{c^2}\frac{m}{r}.$$

This relation is approximately valid for a small angle. An example is a light ray from a star, tangent at the sun with radius $r_\odot \simeq 0.7 \times 10^9$ m, leading to a deflection of about 1.74 arc second. With the same general relativistic basic structures, the ratio of the Schwarzschild length of the sun to the radius of a planetary orbit R_P gives the order of magnitude for the perihel rotation of the planet for one orbit:

$$\Delta(m_\odot, R_P) = 6\pi\frac{l_{\text{Schwarz}}(m_\odot)}{R_P} = 6\pi\frac{q_{\text{Kepl}}(m_\odot)}{c^2 R_P} = 6\pi\frac{G}{c^2}\frac{m_\odot}{R_P}.$$

The radius 57.9×10^6 km of the Mercury orbit and the orbit time of 88 earth days gives as perihel rotation of the Mercury per century about 43 arc seconds.

This example shows, how a dimensional analysis is, on the one hand, inspiring, but also, on the other hand, restricted: Before Einstein, the phenomenon of the perihel rotation for Mercury was well known, as well as the values for the speed of light, the gravity constant, the Sun mass, and the Mercury orbit radius. Therefore, in principle, already in the year 1870 one could have played around with the dimensionless ratio above. For its understanding, the idea would not have been too far fetched, that the perihel rotation has to do with gravity. To have the insight and even a theory, that also the speed of light plays an essential role, needed the genius Einstein, who formulated gravity in a framework compatible with a highest speed of action and independent of a special reference frame (theory of general relativity). At this time, some theoreticians tinkered with the integer power $\frac{1}{r^{1+\epsilon}}$ in Newton's law, and some astronomers tried to understand the perihel rotation by a slowdown or by an additional planet ("Vulcanus") nearer to the Sun, which, remarkably, was observed rather often before Einstein's explanation. The modifications by Einstein's relativity lead to the effective potential with an individual correction from the planet's angular momentum:

$$V_{\text{eff}}(r) = \left(1 - \frac{\ell}{r}\right)\left(1 + \frac{\ell R}{2r^2}\right) + \cdots, \quad \ell = \frac{Gm_\odot}{c^2}, \quad R = R_P.$$

2.4 Quantum Structure and Planck's Unit

On the quantum road, the *Avogadro (Loschmidt)*[14] *number* N_{Avo}, which gives the number of molecules in one kilomol, serves as a bridge from the anthropomorphic to the atomic order of magnitudes:

$$N_{\text{Avo}} \simeq 6.03 \times 10^{26} \frac{1}{\text{kilomol}}.$$

That is about the number of poppy seeds that can be distributed on one million earth surfaces — one seed on each square millimeter. Or: To exhaust the Hubble radius $r_{\cos,p}(t_{\text{Hubl}}) \simeq 0.9 \times 10^{26}$ m, one needs about as many meter sticks as there are molecules in one kilomol.

One kilomol of one-atomic hydrogen gas has the mass of about one kilogram, i.e., one hydrogen atom has the mass $m_H \simeq 1.67 \times 10^{-27}$ kg.

To reproduce the intuitively not graspable Avogadro number, at least its order of magnitude, from daily life experiences one can measure the extension of a drop of oil on water[15]: A drop of oil (assume a cube of 1 mm^3) covers maximally, without teared up, about 3 m^2 water. Assuming the layer thickness $\frac{10^{-9}}{3}$m for a molecular cube, the oil drop contains about 2.7×10^{19} oil molecules, and 1 m^3 oil about 2.7×10^{28} molecules.

With the relative molecule mass numbers $Z(M)$ for a molecule M, normalized by carbon $Z(\text{C}) = 12$, and the density $\rho(T, P)$ for temperature T and pressure P, one obtains for the volume of a kilomol:

$$V_{\text{kilomol}}(M; T, P) = \frac{Z(M)}{\rho(T, P)} \frac{\text{kg}}{\text{kilomol}},$$

and, after division with N_{Avo}, the volume for one molecule. For example, for iron with $Z(\text{Fe}) \simeq 55.85$ and the density $7.87 \times 10^3 \ \frac{\text{kg}}{\text{m}^3}$ at normal conditions $T_0 = 273.15$ K (for Kelvin[16]) and $P_0 = 1 \text{ atm} = 1.01325 \times 10^5 \ \frac{\text{kg}}{\text{s}^2\text{m}}$, the volume of one kiloatom is around $7.1 \times 10^{-3} \ \text{m}^3$ — one iron atom has at its disposal around $(2.3 \times 10^{-10} \ \text{m})^3$. In the volume of about $22.4 \frac{\text{m}^3}{\text{kilomol}}$ at normal conditions, valid for all pure gases, a molecule has as space about $(30 \times 10^{-10} \ \text{m})^3$, thousand times as much as iron.

The macroscopic pressure of a gas is related to the density and the squared velocity average of the microscopic gas molecules by $P = \frac{\rho}{2}\langle v^2 \rangle$, where $\rho = \frac{Nm}{V}$ with the volume, the number and the mass of the molecules. The *Boltzmann*[17] *conversion factor* connects the holistic-statistical temperature concept with the energy. For the kinetic energy, the temperature is proportional to the mass:

---

[14] Amedeo Avogadro (1776–1856), Joseph Loschmidt (1821–1895).

[15] George Gamow (1904–1968).

[16] William Thomson Kelvin (1824–1907).

[17] Ludwig Boltzmann (1844–1906).

$$E = \frac{m}{2}\langle v^2\rangle = k_{\text{Boltz}} T.$$

With the kilomol volume of ideal gases, one obtains the equation for ideal gases:

$$P V_{\text{kilomol}} = N_{\text{Avo}} k_{\text{Boltz}} T \Rightarrow \begin{cases} k_{\text{Boltz}} \simeq 1.4 \times 10^{-23}\ \frac{\text{J}}{\text{K}}, \\ \frac{k_{\text{Boltz}}}{c^2} \simeq 1.6 \times 10^{-40}\ \frac{\text{kg}}{\text{K}}, \end{cases}$$

Hydrogen atoms with $m_p \simeq 1.7 \times 10^{-27}$ kg need an average speed $\sqrt{\langle v^2\rangle} \simeq 2.2 \times 10^3\ \frac{\text{m}}{\text{s}}$ for the temperature $T_0 = 273.15$ K. If there were ideal gases, the temperature unit could be defined theoretically by fixing the Boltzmann factor.

From atoms one has recognized *Planck's unit* $\hbar$ as the *minimal quantum of an action* with the dimension of a time times an energy or, what is the same, the dimension of an *angular momentum*, i.e., position times momentum:

$$\hbar = \frac{h}{2\pi} \simeq 1.05 \times 10^{-34}\ \frac{\text{m}^2\,\text{kg}}{\text{s}}\ \text{(minimal)}.$$

Perhaps qualitatively, but by no means quantitatively the word "quantum leap", loved by politicians, makes sense. There is work going on to define the kilogram, with a Watt[18] balance, via a fixed value for Planck's unit. Since 2011, the units kilogram and Kelvin are coupled to the values of the Planck and Boltzmann constant, respectively. The letter h for Planck's unit comes from the capital Greek H (eta) used for the state sum; it has nothing to do with $\hbar$eisenberg. The action quantum or angular momentum quantum is the commutator value in the commutation relations of Born[19] and Heisenberg[20] for position and momentum, i.e., a quantitative measure for the noncommutativity of the quantum structure (Heisenberg's uncertainty relation).

Without special relativity, it is not self-evident that both the product of position with momentum and the product of time with energy are equally dimensioned $\text{m} \cdot \frac{\text{kg m}}{\text{s}} = \text{s} \cdot \frac{\text{kg m}^2}{\text{s}^2}$ by an action unit. The related conclusion to postulate a nontrivial commutation for time and energy operators as for position and momentum operators $[\mathbf{x}, \mathbf{p}] = i\hbar\mathbf{1} = -[\mathbf{t}, \mathbf{E}]$ is not justified. Energy is "quantized" in the Hamilton[21] element (Hamiltonian) $\mathbf{H}(\mathbf{x}, \mathbf{p})$ via the time-dependent position and momentum $t \longmapsto \mathbf{x}(t), \mathbf{p}(t)$. The naively expected special relativistic extension $(\mathbf{x}_a)_{a=1,2,3} \hookrightarrow (\mathbf{x}_j)_{j=0,1,2,3}$ where not only position, but also time is "quantized," both as operators in $[\mathbf{x}_j, \mathbf{p}_k] = -i\hbar\eta_{jk}\mathbf{1}$ with the Lorentz metric η_{jk}, seems not to be appropriate. In canonically quantized special relativistic field theories with fields $x \longmapsto \boldsymbol{\Phi}(x)$, all spacetime coordinates, for time and also for position, are used not as operators, but as number valued parameters. The quantization related nontrivial (anti-)commutators involve the field degrees of freedom, e.g., for

[18] James Watt (1736–1819).

[19] Max Born (1882–1970).

[20] Werner Heisenberg (1901–1976).

[21] William Hamilton (1805–1865).

a scalar field $[\boldsymbol{\Phi}(x_0, \vec{x}), \frac{\partial}{\partial x_0}\boldsymbol{\Phi}(x_0, \vec{y})] = i\hbar\delta(\vec{x} - \vec{y})\mathbf{1}$ with the Dirac[22] space distribution. This sheds also some light on the attempts to "quantize spacetime."

Planck's unit with the Boltzmann factor allow the conversion of time (frequency) in temperature:

$$\frac{\hbar}{k_{\text{Boltz}}} \simeq 7.5 \times 10^{-12} \text{ s K.}$$

Avogadro's number and Planck's unit are two bridges from the anthropomorphic measures to the atoms — do they condition each other? If each molecule in one kilomol would carry an equally directed spin the total spin would be still tiny in anthropomorphic units:

$$N_{\text{Avo}}\hbar \simeq 0.6 \times 10^{-7} .$$

Therefore, this product is not appropriate to relate both bridges.

2.5 Minimal Quantum of Action and Maximal Speed of Action

The maximal action speed and the minimal action quantum define the *natural quantum conversion factors* from time and length to mass:

$$\frac{\hbar}{c^2} \simeq 1.2 \times 10^{-51} \text{ s kg (minimal),}$$
$$\frac{\hbar}{c} \simeq 3.5 \times 10^{-43} \text{ m kg (minimal).}$$

Both values should be, up to factors not changing the order of magnitude, minimal physical values with this dimensions. Also $\frac{\hbar}{c^2}$ could be used in principle, to define a mass unit for a given time unit.

The parametrization with the inverse linearly mass-dependent *Compton*[23] *length*:

$$l_{\text{Compt}}(m) = \frac{\hbar}{c}\frac{1}{m}$$

give, e.g., for the proton $l_{\text{Compt}}(m_p) \simeq 2.1 \times 10^{-16}$ m. Conversely, each length can be associated with its *Yukawa*[24] *mass*:

[22] Paul Dirac (1902–1984).

[23] Arthur Compton (1892–1962).

[24] Hideki Yukawa (1907–1981).

$$m_{\text{Yuk}}(l) = \frac{\hbar}{c} \frac{1}{l},$$

which played a role as range, e.g., for the nuclear forces in a Yukawa potential $V_{\text{Yuk}}(r) = g_0^2 \frac{e^{-\frac{r}{l}}}{r}$ with around 10^{-15} m, leading to the discovery of the π-meson, around 0.1 proton masses.

The *Lummer*[25] *conversion factor* between temperature and length:

$$\frac{\hbar c}{k_{\text{Boltz}}} \simeq 2.25 \times 10^{-3} \text{ m K},$$

allows, up to a factor of about 1.3, for the black body radiation the transition from the maximally radiated wavelength to its surface temperature by *Wien's*[26] *law*. For example, the sun light with $\lambda_{\text{max},\odot} \simeq 5 \times 10^{-7}$ m shows us the surface temperature $T_\odot \simeq 5800$ K of the sun. Here, at the temperature distribution of the black body radiation, was Planck's starting point, leading to quantum theory. The Schwarzschild length of a mass gives the *Hawking*[27] *temperature* which goes with the inverse mass:

$$8\pi T_{\text{Hawk}}(m) = \frac{\hbar c}{k_{\text{Boltz}}} \frac{1}{l_{\text{Schwarz}}(m)} = \frac{\hbar c^3}{G k_{\text{Boltz}}} \frac{1}{m},$$

e.g., $T_{\text{Hawk}}(m_\odot) \simeq 1.8 \times 10^{-7}$ K for the sun.

With Lummer's conversion factor, a photon density can be defined for the cosmic background radiation:

$$n_{\text{cos,G}}(T) = \left(\frac{k_{\text{Boltz}}}{\hbar c}\right)^3 T^3,$$

whose present order of magnitude $n_{\text{cos,G}}(T_{\text{Hubl}}) \simeq 1.7 \times 10^9 \frac{1}{\text{m}^3}$ with $T_{\text{Hubl}} \sim 2.7$ K can be compared with the present value $n_{\text{cos,p}}(t_{\text{Hubl}}) \simeq 24 \frac{1}{\text{m}^3}$ of the baryon density.

2.6 Intrinsic Units

A definition of units uses, more or less aware, some physical laws. For example, the original definition of the second relies on the regularity of the earth's motion and, with today's precision of time measurement, is no longer satisfactory. Special dynamical problems involve intrinsic units. If a nonrelativistic mechanical problem is characterized by its total energy via a Hamiltonian, i.e., by the sum of kinetic and potential energy, e.g., for a homogeneous potential $V^z(\underline{x}) = \epsilon \frac{g_0^2}{z} \underline{x}^z$ with position $\underline{x}$-dependence of order z:

---

[25]Otto Lummer (1860–1925).

[26]Wilhelm Wien (1864–1928).

[27]Stephen Hawking (1942–).

$$\underline{H} = \frac{\underline{p}^2}{2m_0} + \frac{\epsilon g_0^2}{z}\underline{x}^z, \quad m_0, g_0^2 > 0, \quad \epsilon = \pm 1, \quad z \in \mathbb{Z}, \quad z \neq 0,$$

the arising variables have the dimensions [...] of energy, length and momentum, in the MKS-system:

$$[\underline{H}] = \frac{\text{m}^2\text{kg}}{\text{s}^2}, \quad [\underline{x}] = \text{m}, \quad [\underline{p}] = \frac{\text{m kg}}{\text{s}}.$$

With that, the inert mass m_0 and the coupling constant g_0^2 can be used, obviously up to order of magnitude-compatible factors, as the intrinsic units of this dynamics with the MKS-dimension:

$$[m_0] = \text{kg}, \quad [g_0^2] = \frac{\text{m}^{2-z}\ \text{kg}}{\text{s}^2}.$$

A homogeneous potential $V^z(\underline{x})$ describes for order $z = 1$ the fall of "mass points" near the earth's surface. For order $z = -1$, it is characteristic for a central potential, e.g., the gravitational Newton potential $G\frac{m_1 m_2}{|\vec{x}_1 - \vec{x}_2|}$, and for order $z = 2$ for the oscillator potential (pendulum with small amplitudes).

Galileo[28] was the first to check the suggestion, which was later Einstein's main starting point for the foundation of his general theory of relativity, that, for all bodies, freely falling at the earth's surface, the intrinsic units g_0^2 (weight, heavy mass) and m_0 (inert mass) have the same ratio $g_\bullet$:

$$V_{\text{Newt}}(\vec{x}_\bullet + \vec{\underline{x}}) = G\frac{m_\bullet m_0}{|\vec{x}_\bullet + \vec{\underline{x}}|} = V_{\text{Gal}}(\underline{x}) + \ldots,$$
$$V_{\text{Gal}}(\underline{x}) = \frac{Gm_\bullet}{r_\bullet^2} m_0 \underline{x} = g_\bullet m_0 \underline{x} = g_0^2 \underline{x}.$$

Therefore, for the Hamiltonian with the Galileo potential, the inert mass m_0 (individual) and the acceleration $g_\bullet = \frac{Gm_\bullet}{r_\bullet^2}$, caused by the earth (universal), are the intrinsic units. If, and in an airless space, you let fall any mass from your height (about one meter), it hits the ground after about $\frac{1}{2}$ second. That the human time unit second, probably taken from such experiences, is approximately equal to our heartbeat time unit—not that of mice or elephants — is presumably a coincidence.

The Hamiltonian of a special problem effects, physically as energy as well as mathematically by implementation as transformation, the time development of all quantities, which can be built by position $\underline{x}$ und momentum $\underline{p}$. For explicitly time-independent problems, $\frac{\partial \underline{H}}{\partial t} = 0$, as exemplified earlier, the Hamiltonian gives the energy, conserved in time:

---

[28]Galileo Galilei (1564–1642).

$$\underline{H} = \frac{\underline{p}^2}{2m_0} + V(\underline{x}) \overset{\text{e.g.,}}{=} \frac{\underline{p}^2}{2m_0} + \frac{\epsilon g_0^2}{z}\underline{x}^z,$$
$$\frac{d\underline{x}}{d\underline{t}} = \frac{\partial \underline{H}}{\partial \underline{p}} = \frac{\underline{p}}{m_0}, \quad \frac{d\underline{p}}{d\underline{t}} = -\frac{\partial \underline{H}}{\partial \underline{x}} = -\frac{\partial V}{\partial \underline{x}} = -\epsilon g_0^2 \underline{x}^{z-1} \Rightarrow \frac{d\underline{H}}{d\underline{t}} = 0.$$

The equations of motion can be classically characterized by a Lagrange[29] function (Lagrangian), which connects the Hamiltonian with the time-dependence of the *canonical pair* "position-momentum" $(\underline{x}, \underline{p})$:

$$\underline{L} = \frac{1}{2}\left(\underline{p}\frac{d\underline{x}}{d\underline{t}} - \underline{x}\frac{d\underline{p}}{d\underline{t}}\right) - \underline{H}.$$

In the equation of motion of any "sufficiently smooth" position-momentum dependent function $(x, p) \longmapsto F(x, p)$, the time derivative is effected by the position and momentum derivatives via the Poisson[30] bracket with the Hamiltonian:

$$\frac{dF}{dt} = [H, F]_P = \frac{\partial H}{\partial p}\frac{\partial F}{\partial x} - \frac{\partial H}{\partial x}\frac{\partial F}{\partial p}.$$

Good to know: The bilinear Poisson bracket is a Lie[31] algebra bracket (no commutator for the pointwise product!) for such functions. It is antisymmetric and obeys Leibniz's rule, i.e., the product rule for derivations $D(f_2 f_3) = (Df_2)f_3 + f_2(Df_3)$, equivalent to the Jacobi[32] identity:

$$[F_1, F_2]_P = \frac{\partial F_1}{\partial p}\frac{\partial F_2}{\partial x} - \frac{\partial F_2}{\partial p}\frac{\partial F_1}{\partial x}: \quad \begin{cases} [F, F]_P = 0 \Rightarrow [F_1, F_2]_P = -[F_2, F_1]_P, \\ [F_1, [F_2, F_3]_P]_P = [[F_1, F_2]_P, F_3]_P + [F_2, [F_1, F_3]_P]_P. \end{cases}$$

The related Lie algebra is infinite-dimensional.

In a quantum mechanical theory, the dimension of the noncommuting canonical position–momentum pair is determined by Planck's action quantum $\hbar$:

$$[\underline{H} \cdot \underline{t}] = [\underline{x} \cdot \underline{p}] = [\hbar] = \frac{\text{m}^2\text{kg}}{\text{s}},$$
$$\underline{x}\,\underline{p} - \underline{p}\,\underline{x} = [\underline{x}, \underline{p}] = i\hbar\mathbf{1}.$$

For the Hamiltonians earlier with $z \neq -2$, the arbitrary MKS-system can be replaced by an intrinsic $(m_0, g_0^2, \hbar)$-system, specific for the given problem. It gives for the intrinsic length and time unit:

$$l_0 = \left(\frac{\hbar^2}{g_0^2 m_0}\right)^{\frac{1}{2+z}}, \quad t_0 = \left(\frac{\hbar^{2-z} m_0^z}{g_0^4}\right)^{\frac{1}{2+z}}, \quad [l_0] = \text{m}, \ [t_0] = \text{s}.$$

---

[29] Joseph-Louis Lagrange (1736–1813).

[30] Simeon Poisson (1781–1840).

[31] Sophus Lie (1842–1899).

[32] Carl Gustav Jacob Jacobi (1804–1851).

Also the intrinsic energy unit E_0 reflects the position power z in the potential:

$$E_0 = \left(\frac{\hbar^{2z} g_0^4}{m_0^z}\right)^{\frac{1}{2+z}} = \begin{cases} \frac{g_0^4 m_0}{\hbar^2}, & z = -1, \\ \hbar\sqrt{\frac{g_0^2}{m_0}}, & z = 2. \end{cases}$$

If, for the computation of the energy only natural multiples of the action quantum are admitted, i.e., $n\hbar$ with $n = 1, 2 \ldots$, then there follow the energy levels $\frac{E_0}{n^2}$ for $z = -1$, Kepler potential, e.g., the hydrogen atom, and nE_0 for $z = 2$, harmonic oscillator.

Using intrinsic units $(E_0, l_0, \hbar)$, the dynamics can be formulated in dimensionless variables:

$$\underline{H} = E_0 H, \quad \underline{x} = l_0 x, \quad \underline{p} = \frac{\hbar}{l_0} p \Rightarrow H = \frac{p^2}{2} + \epsilon \frac{x^z}{z}.$$

There remain "purely mathematical" forms with dimensionless parameters, which, in basically important cases can be characterized by symmetry considerations and representation properties of operation groups, e.g., in quantum theories, as eigenvalue equations for representation characterizing invariants. For instance, the gravitative Newton and the electromagnetic Coulomb[33] potential have equal dimensionless mathematical forms. Mathematics takes over, e.g., by a mathematically intrinsic characterization of a Hamiltonian as an invariant of an operation group.

If there occur more than three independent intrinsic units, pure numbers are physically relevant. An example is the light deflection at the sun boundary (earlier).

A analogue situation with respect to intrinsic units and canonical pairs occurs in field theories. A space or a spacetime field (for generality also a time field) is defined as a function of the corresponding time, position and spacetime parameters — in the basically important cases valued as a vector in a space with time, space and spacetime representing operations. An example in a relativistic field theory is a Lorentz-scalar spacetime field $\underline{\mathbf{\Phi}}$, where each spacetime translation $\underline{x}$ with equally dimensioned spacetime coordinates $(\underline{x}^j)_{j=0}^3 = (c\underline{t}, \underline{x}, \underline{y}, \underline{z})$ is associated with a value $\underline{\mathbf{\Phi}}(\underline{x})$, measuring, e.g., a charge, a current strength, etc. For a classical field theory, the dynamics can be given by a Lagrangian density $\underline{\mathcal{L}}$, e.g., with a power-"potential $V^n(\underline{\mathbf{\Phi}}) = g_0^2 \underline{\mathbf{\Phi}}^n$:"

$$\underline{\mathcal{L}} = \frac{1}{2} \frac{\partial \underline{\mathbf{\Phi}}}{\partial \underline{x}_j} \frac{\partial \underline{\mathbf{\Phi}}}{\partial \underline{x}^j} - g_0^2 \underline{\mathbf{\Phi}}^n, \quad n \in \mathbb{N}.$$

Good to know: Einstein made the mathematical notation more transparent by his shorthand sum convention: Doubly occurring indices involve a summation, here: $\frac{\partial \underline{\mathbf{\Phi}}}{\partial \underline{x}_j} \frac{\partial \underline{\mathbf{\Phi}}}{\partial \underline{x}^j} = \sum_{j=0}^{3} \frac{\partial \underline{\mathbf{\Phi}}}{\partial \underline{x}_j} \frac{\partial \underline{\mathbf{\Phi}}}{\partial \underline{x}^j}$.

[33]Charles Augustin de Coulomb (1736–1806).

The spacetime-integrated Lagrange density $\int d^4\underline{x}\,\underline{\mathcal{L}}(\underline{x})$ has the dimension of an action. The *canonical field pair* $(\underline{\Phi}, \frac{\partial\underline{\Phi}}{\partial \underline{x}_0})$, comprising the field and its time derivative, has the dimension of an action position density. It involves two distributions in position space, which have broken MKS-units, here square roots:

$$\left.\begin{array}{l}[\underline{x}_j] = \text{m},\\ [\underline{\mathcal{L}}\cdot \underline{x}_j] = [\underline{\Phi}\cdot \frac{\partial\underline{\Phi}}{\partial \underline{x}_j}] = \frac{[\hbar]}{\text{m}^3}\end{array}\right\} \Rightarrow, [\underline{\Phi}] = \sqrt{\frac{\text{kg}}{\text{s}}}.$$

Therefore, in addition to the universal intrinsic speed of light unit c, one has the special intrinsic unit g_0^2:

$$[c] = \frac{\text{m}}{\text{s}},\quad [g_0^4] = \frac{\text{kg}^{2-n}}{\text{m}^4\text{s}^{2-n}}.$$

In a relativistic quantum field theory with $(\hbar, c, g_0^2)$ all units are intrinsic.

2.7 The $(\hbar, c, G)$-System — Universal Units?

For a basic understanding and definition of units, it is unsatisfactory to use the intrinsic units of a special problem with a special dynamics (Hamiltonian) — one would prefer to start from a universal dynamics, or, more general, from structures that underly all physical dynamics and lead to units, valid "for all times…and cultures" (Planck 1900).

Physics describes *causal actions in space*. To quantify this, one needs three basic units: a unit for time (causality), a unit for space (position), and a unit for actions. Obviously, by phenomenological and convenience reasons, one may employ more units, e.g., in addition, an electric charge unit and a temperature unit. Interesting is the question, always again leading to long and heated discussions, if a complete dimensionalization is possible by three units and if this number makes sense as a minimal one with a deeper significance. Thus, one could think, on a first view, that the completely different and, in the historical development after gravity, new electromagnetic actions would require also a new unit, e.g., for electric charge. That, however, is not the case: Electromagnetism needs no new units. Analogously, one could think of new units for nuclear binding or decay forces (strong and weak interactions) — also that is not necessary. If one takes, in the other direction, the extreme point of view, that one considers a dynamics, formulated only in dimensionless variables — like with a dimensionless Hamiltonian H earlier, where there occur no units at all, one can ask the question, if there exist dynamical features, which can be quantified by one, two, three, or more units to introduce dimensions as quantitative markers for qualitatively different structures, e.g., for time and space and for different operation groups.

If one looks into a physics textbook with respect to the definition of units systems, two types of definitions can be distinguished: Thus, in the MKS-system (m, kg, s), the definition of meter, kilogram, and second used, originally, the simple reference to prototypes, e.g., to the ur-meter and the ur-kilogram in Paris, and to the average sun-day. Later on, the definitions were refined; e.g., the meter was defined via the speed of light and the otherwise defined second. For a "good," i.e., reliable definition of those basic units, the prototypes must have certain properties, e.g., a constant behavior in time; i.e., the validity of, possibly still unknown, physical laws is presupposed.

These three basic units are supplemented by "dynamical units," whose definition uses directly a physical law. An example is the definition of the temperature unit Kelvin via the Boltzmann conversion factor, which uses the ideal gas equation, or the definition of the charge unit Coulomb via the Coulomb law, and, then, the definition of the current strength unit Ampère.[34]

Unit systems are distinguished quantitatively and qualitatively. Instead of meter, kilogram, and second one can use handbreadth, stone, and heartbeat time — here is only a quantitative difference, effected by recalibrations (gauging, dilations). Two such unit systems are related to each other by three real dilation factors. For different dimensions of the three basic units, e.g., time, energy, and mass instead of length, momentum, and velocity, there exists a qualitative difference in addition to the three scale transformations.

A minimal *universal* unit system could be distinguished by a universal dynamics and an all physics-underlying law, whose structure determines both the number and the dimensions of the basic units. $(\hbar, c, G)$ with the quantum unit (Planck's unit) $\hbar$, the field unit (Einstein's unit) c and the gravity unit (Newton's unit) G could possibly be such a unit system. Deep structural insights are connected with the maximal Einstein unit from special relativity, as well as with the minimal Planck unit from quantum theory. It is strongly suggestive, that an understanding of the third unit — perhaps Newton's unit — will arise from a qualitative progress in the unification of the spacetime interactions, possibly by a common understanding of the geometrical structures of general relativity and the probability oriented information structures of quantum theory.

Natural equivalences for time and length, on the one side, and mass, on the other side, arise from products on the "quantum road" (Compton length), and from quotients on the "gravity road" (Schwarzschild length). If both roads merge, there arise *natural units for time and length*, perhaps minimal, if Newton's unit G is minimal:

$$t_{\min} = \sqrt{\frac{\hbar G}{c^5}} \simeq 5.5 \times 10^{-44}\ \text{s} \qquad \text{(minimal(?))},$$

$$l_{\min} = \sqrt{\frac{\hbar G}{c^3}} \simeq 1.6 \times 10^{-35}\ \text{m} \qquad \text{(minimal(?))}.$$

[34] André Marie Ampère (1775–1836).

The also arising *Planck mass* is neither minimal nor maximal:

$$M_{\text{Planck}} = \sqrt{\frac{\hbar c}{G}} \simeq 2.2 \times 10^{-8} \text{ kg}.$$

It is about 20 μg — a bacterium or almost a little flea: This magnitude is not totally exotic for us. With this mass and the easily rememberable connection of G and c to the anthropomorphic measures and the magnitude of the earth, one can compute the value of Planck's quantum unit $\hbar$ without using data books.

The real factors for the conversion of $(l_{\text{min}}, M_{\text{Planck}}, t_{\text{min}})$ to the MKS-system, basically uninteresting but characteristic for us human beings, constitute an example for the scale transformations, mentioned earlier.

It is not obvious, if the minimal(?) time and length have really an operational minimal meaning — perhaps they are nothing more than a naive combination of units. It is at least conceivable, if Newton's unit G is minimal.

The "mesoscopic" Planck mass, neither minimal nor maximal, is a borderline between macroscopic gravity and microscopic quantum structure: The product of Schwarzschild and Compton length, which are, respectively, directly and inversely directly proportional to the mass, are, for all masses, the square of the minimal(?) length. Both lengths are equal for the Planck mass:

$$\begin{aligned} l_{\text{Schwarz}}(m)\, l_{\text{Compt}}(m) &= l_{\text{min}}^2 = \frac{\hbar G}{c^3}, \\ l_{\text{Schwarz}}(M_{\text{Planck}}) &= l_{\text{Compt}}(M_{\text{Planck}}) = l_{\text{min}}, \\ \frac{l_{\text{Schwarz}}(m)}{l_{\text{min}}} &= \frac{l_{\text{min}}}{l_{\text{Compt}}(m)} = \frac{m}{M_{\text{Planck}}}. \end{aligned}$$

Then, for the minimal(?) length, the Newton and Yukawa mass are equal to Planck's mass:

$$\begin{aligned} m_{\text{Newt}}(l)\, m_{\text{Yuk}}(l) &= M_{\text{Planck}}^2 = \frac{\hbar c}{G}, \\ m_{\text{Newt}}(l_{\text{min}}) &= m_{\text{Yuk}}(l_{\text{min}}) = M_{\text{Planck}}. \end{aligned}$$

With the natural *mass logarithm*, defined by the mass, measured in Planck mass units:

$$\psi(m^2) = \log \frac{m^2}{M_{\text{Planck}}^2}, \quad m^2 = M_{\text{Planck}}^2 e^{\psi(m^2)},$$

the quantum-gravity borderline (Planck's mass) can be put to zero — smaller masses in the "quantum region" have negative, larger masses in the "gravity region" have positive mass logarithms. As an example, two masses are considered, which, with respect to the geometrical mean, have about equal distances to the human measures:

$$\text{electron: } m_e \simeq 0.91 \times 10^{-30}\text{ kg} \quad \Rightarrow \begin{cases} l_{\text{Compt}}(m_e) \simeq 3.9 \times 10^{-13}\text{ m}, \\ l_{\text{Schwarz}}(m_e) \simeq 7 \times 10^{-58}\text{ m}, \\ \psi(m_e^2) \simeq -103, \end{cases}$$

$$\text{sun: } m_\odot \simeq \ 2 \times 10^{30}\text{ kg} \quad \Rightarrow \begin{cases} l_{\text{Compt}}(m_\odot) \simeq 1.7 \times 10^{-73}\text{ m}, \\ l_{\text{Schwarz}}(m_\odot) \simeq 1.5 \times 10^{3}\text{ m}, \\ \psi(m_\odot^2) \simeq 175, \end{cases}$$

$$\sqrt{m_e m_\odot} \simeq \ 1.35\text{ kg}.$$

In a "wild argumentation" with the $(\hbar, G, c)$ units, one can conceive a science fiction scenario, not to be taken too seriously: At the minimal(?) time, there was the maximal(?) cosmic density. The cosmic mass at the minimal(?) time was the Planck mass:

$$\rho_{\cos,p}(t_{\min}) = \frac{3}{8\pi}\frac{c^5}{\hbar G^2} \simeq 6 \times 10^{95}\,\frac{\text{kg}}{\text{m}^3}, \quad m_{\cos,p}(t_{\min}) = M_{\text{Planck}}.$$

The qualitative conversion from one unit system, e.g., from the anthropomorphic MKS-system, to another one, e.g., to the possibly universally intrinsic $(\hbar, c, G)$-system, can be linearized in a dimension space — here a three-dimensional rational space $\mathbb{Q}^3$. In a *dimensional grading*, each physical quantity $a \in \mathcal{A}$ obtains its dimension powers:

$$\dim : \mathcal{A} \longrightarrow \mathbb{Q}^3, a \longmapsto \dim(a).$$

Any unit system defines a basis of the dimension space:

$$\text{grading in MKS-basis: } \begin{cases} \dim(\text{m}) \cong \begin{pmatrix}1\\0\\0\end{pmatrix}, & \dim(\text{kg}) \cong \begin{pmatrix}0\\1\\0\end{pmatrix}, & \dim(\text{s}) \cong \begin{pmatrix}0\\0\\1\end{pmatrix}, \\ \dim(\hbar) \cong \begin{pmatrix}2\\1\\-1\end{pmatrix}, & \dim(c) \cong \begin{pmatrix}1\\0\\-1\end{pmatrix}, & \dim(G) \cong \begin{pmatrix}3\\-1\\-2\end{pmatrix}, \end{cases}$$

$$\text{grading in}(\hbar, c, G) - \text{basis: } \begin{cases} \dim(\hbar) \cong \begin{pmatrix}1\\0\\0\end{pmatrix}, & \dim(c) \cong \begin{pmatrix}0\\1\\0\end{pmatrix}, & \dim(G) \cong \begin{pmatrix}0\\0\\1\end{pmatrix}, \\ \dim(\text{m}) \cong \frac{1}{2}\begin{pmatrix}1\\-3\\1\end{pmatrix}, & \dim(\text{kg}) \cong \frac{1}{2}\begin{pmatrix}1\\1\\-1\end{pmatrix}, & \dim(\text{s}) \cong \frac{1}{2}\begin{pmatrix}1\\-5\\1\end{pmatrix}. \end{cases}$$

Since the association of the dimensions powers is a grading, i.e., one has for the multiplication of physical quantities the logarithmic properties:

$$\dim(a_1 a_2) = \dim a_1 + \dim a_2, \quad \dim(a^k) = k \dim(a), \quad k \in \mathbb{Q},$$

one obtains, in this formalismus, the dimension of a product by vector addition of the dimensions of the factors. This homomorphism formalizes mathematically the dimension analysis, mentioned at the beginning of this chapter. A correct calculation with the complete physical quantities is projected to a correct calculation for the dimensions, obviously not vice versa.

Good to know: Sets S can have a structure, e.g., an order structure, an algebraic structure, or a topological structure, etc. Mappings for sets $f : S_1 \longrightarrow S_2$ with the same structure are called *morphisms*, if they are compatible with this structure. For instance, vector space morphisms $f : V_1 \longrightarrow V_2$ are compatible with the vector space defining structures — the vector addition $f(v+w) = f(v) + f(w)$ and the scalar multiplication $f(\alpha v) = \alpha f(v)$. Order morphisms are monotonous mappings, topological morphisms are continuous mappings.
The morphisms sets $\mathbf{kat}(S_1, S_2)$, e.g., all linear mappings $\mathbf{vec}_{\mathbb{R}}(V_1, V_2) = \{f : V_1 \longrightarrow V_2, \text{ real linear}\}$ for two real vector spaces, can be collected with the sets in a (little) *category* **kat**, e.g., the category $\underline{\mathbf{vec}}_{\mathbb{R}}$ for the real vector spaces and their linear mappings, or the category **top** for topological spaces with continuous mappings, etc.
Categories formalize mathematical structures. For a (little) category, e.g., complex Hilbert spaces for quantum theories, one is interested especially in its (endo-, iso, auto-)morphisms, in its product structures, and in its quotient (equivalence) structures.
Relations between different (little) categories, e.g., between topological spaces **top** and complex vector spaces $\underline{\mathbf{vec}}_{\mathbb{C}}$, are formalized by *functors*.

Complete systems with basic units are related to each other by nonsingular transformations $\mathbf{E}$ of the unit space, here 3×3-matrices with half-integer entries, where the powers of the basic units stand in the columns, e.g., the connection transformations of $(\hbar, c, G)$- and MKS-system:

$$\dim_{(\hbar,c,G)}(a) = \mathbf{E}^{(\mathrm{M,K,S})}_{(\hbar,c,G)} \dim_{(\mathrm{M,K,S})}(a), \qquad \text{with } \mathbf{E}^{(\mathrm{M,K,S})}_{(\hbar,c,G)} = \frac{1}{2}\begin{pmatrix} 1 & 1 & 1 \\ -3 & 1 & -5 \\ 1 & -1 & 1 \end{pmatrix},$$

$$\dim_{(\mathrm{M,K,S})}(a) = \mathbf{E}^{(\hbar,c,G)}_{(\mathrm{M,K,S})} \dim_{(\hbar,c,G)}(a), \qquad \text{with } \mathbf{E}^{(\hbar,c,G)}_{(\mathrm{M,K,S})} = \begin{pmatrix} 2 & 1 & 3 \\ 1 & 0 & -1 \\ 1 & -1 & 2 \end{pmatrix},$$

e.g., for energy:

$$\dim_{(\mathrm{M,K,S})}(E) = \begin{pmatrix} 2 \\ 1 \\ -2 \end{pmatrix} = \mathbf{E}^{(\hbar,c,G)}_{(\mathrm{M,K,S})} \begin{pmatrix} \frac{1}{2} \\ \frac{5}{2} \\ -\frac{1}{2} \end{pmatrix}, \quad \frac{\mathrm{m}^2\,\mathrm{kg}}{\mathrm{s}^2} \leftrightarrow \sqrt{\frac{\hbar c^5}{G}}.$$

In the opposite direction, the inverse transformation has to be taken with $\mathbf{E}^{(\mathrm{M,K,S})}_{(\hbar,c,G)} \mathbf{E}^{(\hbar,c,G)}_{(\mathrm{M,K,S})} = \mathbf{1}_3$.

Good to know: An (associative) algebra is a real or complex vector space with a bilinear (associative) product $\bullet$ of the vectors. If an associative algebra is a direct sum $\mathcal{A} = \bigoplus_{j \in H} V_j$ of vector spaces V_j, with the indices taken from an additive semigroup, and if, in addition, the multiplication is compatible with the H-addition, $V_j \bullet V_k \subseteq V_{j+k}$, and if, finally, the vector space V_0 with the neutral element $0 \in H$ is given by the scalar field $\mathbb{R}$ or $\mathbb{C}$, then $\mathcal{A}$ is *graded* by the semigroup H. The grading is the monoid morphism grad: $\mathcal{A} \longrightarrow H$. Examples are tensor, Grassmann,[35] or polynomial algebras; they are graded with the natural numbers $\mathbb{N} = \{0, 1, 2, \ldots\}$.

The dimension of the dimension-grading space, possibly exactly three, e.g., for a vector space $\mathbb{R}^3$ or for the additive groups $\mathbb{Z}^3$, $(\frac{1}{2}\mathbb{Z})^3$, and $\mathbb{Q}^3$, gives the number of the fundamental units. An integer $\mathbb{Z}^3$- and a half-integer $(\frac{1}{2}\mathbb{Z})^3$-grading uses only integer and half-integer dimensions, respectively, of exactly three basic units.

[35] Hermann Grassmann (1809–1877).

2.8 Electrodynamics and Sommerfeld's Fine-Structure Constant

The electromagnetic unit Coulomb C can be defined in a human-related way by the potential for the Coulomb force, which two resting point charges Q, Q' in the distance r exert on each other:

$$V_{\text{Coul}}(r) = \frac{1}{4\pi\epsilon_0}\frac{QQ'}{r} = \frac{\mu_0 c^2}{4\pi}\frac{QQ'}{r},$$

$$\text{definition of the unit Coulomb: } \frac{\mu_0}{4\pi} = 10^{-7}\,\frac{\text{mkg}}{\text{C}^2},$$

$$\frac{1}{c} = \sqrt{\epsilon_0\mu_0}.$$

ϵ_0 is the electric field constant (vacuum-influence constant), μ_0 is the magnetic field constant (vacuum permeability); their geometrical mean is the inverse maximal action velocity c.

Good to know: Electron is Greek for amber; Magnesia was an ancient Greek town in Minor Asia.

It is highly remarkable, that, apparently, proton and positron have precisely the same electric charge[36]:

$$e \simeq 1.6 \times 10^{-19}\ \text{C}.$$

If this charge goes through the potential difference of one Volt[37] V, it obtains with $\text{V} = \frac{\text{J}}{\text{C}}$ the energy with its mass and temperature equivalents:

$$1\ \text{eV} \simeq 1.6 \times 10^{-19}\ \text{J},\quad 1\frac{\text{eV}}{c^2} \simeq 1.8 \times 10^{-36}\ \text{kg},\quad 1\frac{\text{eV}}{k_{\text{Boltz}}} \simeq 1.15 \times 10^4\ \text{K}.$$

Then, the kinetic energy $\frac{m_e v^2}{2}$ of an electron comes with about two thousands of the speed of light $\frac{v}{c} \sim 2 \times 10^{-3}$.

The charge of particles proves to be discretized by integer multiples of this smallest *elementary charge* e, with individual *charge numbers* z:

$$Q = ez,\quad z \in \mathbb{Z}.$$

Therefore, the definition of the unit Coulomb is the arbitrary fixing of the hopefully integer $\frac{\text{C}}{e} \simeq 6 \times 10^{18}$.

---

[36] Hermann von Helmholtz (1821–1894), Robert Andrews Millikan (1868–1953).

[37] Alessandro Volta (1745–1827).

A strong hint for the discretization of charge was historically given by the common electric charge of one kilomol of simply charged ions, e.g., of hydrogen ions, determining the *Faraday*[38] *charge*:

$$Q_{\text{Faraday}} = N_{\text{Avo}}e \simeq 9.648 \times 10^7 \, \frac{\text{C}}{\text{kilomol}}.$$

One millimol has as charge about 96 C; this was the starting point for the "anthropomorphic" Coulomb definition, which comes from — not very accurate — electrolytic experiments with molecules containing silver with a relative atom mass $Z(\text{Ag}) \simeq 108$.

For atoms and elementary particles, the elementary charge serves as unit:

$$\frac{\mu_0 e^2}{4\pi} = \frac{e^2}{4\pi\epsilon_0 c^2} \simeq 2.6 \times 10^{-45} \text{ mkg}.$$

This yields the *intrinsic action unit of electrodynamics*:

$$\hbar_e = e^2\sqrt{\frac{\mu_0}{\epsilon_0}} = e^2\mu_0 c = \frac{e^2}{\epsilon_0 c} \simeq 10^{-35} \, \frac{\text{m}^2\text{kg}}{\text{s}}.$$

Here, $\sqrt{\frac{\mu_0}{\epsilon_0}}$ is the wave resistance of the vacuum, with the unit Coulomb about 377 Ω (Ohm[39] $\Omega = \frac{\text{V}}{\text{A}}$).

The electromagnetic action strength is defined as the ratio of the action unit of electrodynamics $\hbar_e$ to Planck's action unit $\hbar$, which gives up to 4π the dimensionless *Sommerfeld*[40] *number*:

$$g_e^2 = \frac{\hbar_e}{\hbar} \simeq \frac{1}{10.9}, \quad \alpha_{\text{Som}} = \frac{g_e^2}{4\pi} \simeq \frac{1}{137.04}.$$

Neither ϵ_0 nor μ_0 have a fundamental meaning, even not the elementary charge e. A basic importance has only the product and the quotient of $\mu_0 e^2$ and $\frac{\epsilon_0}{e^2}$, i.e., $\frac{1}{c^2}$ and $\hbar_e^2$, or, with Planck's unit, the pure number $g_e^2 = \alpha_{\text{Som}} 4\pi$. Electrodynamics has, in addition to c and $\hbar_e$, no third intrinsic unit, e.g., no intrinsic length.

The wave resistance of the vacuum $\frac{\hbar_e}{e^2} \simeq 377\ \Omega$ corresponds to the Hall[41] resistance $\frac{\hbar}{e^2} \simeq 4100\ \Omega$.

The electromagnetic actions are, compared with the gravitative actions, a completely new phenomenon (What was Newton's explanation of lightnings?). However, if one starts with three "mechanical" units, they do not require an extension of the dimensionality of the unit space — as intrinsic units, the maximal action

[38] Michael Faraday (1791–1867).

[39] Georg Simon Ohm (1789–1854).

[40] Arnold Sommerfeld (1868–1951).

[41] Edwin Hall (1855–1938).

velocity c and the electromagnetic action unit $\hbar_e$ have to be added, and integer individual charge numbers $z \in \mathbb{Z}$:

$$\begin{array}{ll}\text{Coulomb potential:} & V_{\text{Coul}}(r) = \frac{1}{4\pi\epsilon_0}\frac{QQ'}{r} = \frac{QQ'}{e^2} g_e^2 \frac{c\hbar}{4\pi r} = zz'\alpha_{\text{Som}}\frac{c\hbar}{r}, \\ \text{Newton potential:} & V_{\text{Newt}}(r) = -G\frac{mm'}{r} = -\frac{mm'}{M^2_{\text{Planck}}}\frac{c\hbar}{r}.\end{array}$$

Both potentials have a precise $\frac{1}{r}$-dependence — not, e.g., $\frac{1}{r^{1.0024}}$. Under the Coulomb potential for the electromagnetic "internal" interaction stands the analogously structurized Newton potential for an "external" interaction, where the masses enter with dimensionless positive real ratios $\frac{m}{M_{\text{Planck}}} \in \mathbb{R}_+$ in units of the Planck mass, in analogy to the integer charge numbers $z = \frac{Q}{e} \in \mathbb{Z}$.

Good to know: The precise $\frac{1}{r}$-dependence is required for the solution of the inhomogeneous Laplace[42] equation $-\vec{\partial}^2\frac{1}{r} = 4\pi\delta(\vec{x})$ with the rotation invariant translation action $\vec{\partial}^2 = \partial_1^2 + \partial_2^2 + \partial_3^2$ for flat 3-position and Dirac's delta-distribution. Its special relativistic embedding with Lorentz invariant spacetime translation action $\partial^2 = \partial_0^2 - \vec{\partial}^2$ leads to Feynman[43] propagators for massless fields with the lightcone-supported Dirac distribution $\frac{1}{r} = \int \frac{d^3q}{2\pi^2}\frac{1}{\vec{q}^2}e^{i\vec{q}\vec{x}} \hookrightarrow \delta(x^2)$:

$$\int \frac{d^4q}{4\pi^3}\frac{1}{-q^2 - io}e^{iqx} = \frac{i}{\pi(-x^2+io)}, \quad \partial^2\int \frac{d^4q}{4\pi^3}\frac{1}{-q^2-io}e^{iqx} = \partial^2\delta(x^2) = 4\pi\delta(x),$$

as used, for example, for the electromagnetic field. The Kepler (Newton, Coulomb) potential $\frac{1}{r}$ is embedded as off-shell (non-particle) contribution of the corresponding Feynman propagators.

The classical spacetime theory for the electrostatic Coulomb potential is Maxwell's[44] electrodynamics whereas Newton's gravitostatic potential is the entrance door to Einstein's general relativity and nonflat classical spacetime.

Maxwell's equations in the vacuum describe the spacetime source and vortex structure of the electric field strengths $\vec{\mathbf{E}}$ and the magnetic flux density $\vec{\mathbf{B}}$ with the aid of the charge density ρ and the current density $\vec{\mathbf{j}}$: The electric field swells from the charge density, it whirls around the time-dependent magnetic flux (density); the vortices of the sourcefree magnetic flux (density) are caused by the current (density) or an electric field that changes in time:

$$\begin{array}{lll}\text{homogeneous:} & \text{div}\,\vec{\mathbf{B}} = 0, & \text{rot}\,\vec{\mathbf{E}} + \frac{\partial\vec{\mathbf{B}}}{\partial t} = 0, \\ \updownarrow \text{ duality} & -\frac{\vec{\mathbf{E}}}{c} \leftrightarrow \vec{\mathbf{B}}, & \frac{1}{c^2} = \epsilon_0\mu_0, \\ \text{inhomogeneous:} & \epsilon_0\text{div}\,\vec{\mathbf{E}} = \rho, & \frac{1}{\mu_0}\text{rot}\,\vec{\mathbf{B}} - \epsilon_0\frac{\partial\vec{\mathbf{E}}}{\partial t} = \vec{\mathbf{j}}.\end{array}$$

There are additional equations of motion for the constituents of charge and current density ("matter" equations).

Introducing for the solution of the two homogeneous equations a vector potential $\vec{\mathbf{A}}$ and a scalar potential $\mathbf{U}$, determined up to gauging (ahead):

---

[42] Pierre-Simon Laplace (1749–1827).

[43] Richard Feynman (1918–1988).

[44] James Clerc Maxwell (1831–1879).

$$\vec{\mathbf{B}} = \text{rot}\,\vec{\mathbf{A}} \text{ and } \vec{\mathbf{E}} + \frac{\partial \vec{\mathbf{A}}}{\partial t} = -\text{grad}\,\mathbf{U},$$

the Maxwell equations can be written with the intrinsic units in the form

$$\left.\begin{array}{r} \partial_j \mathbf{A}_k - \partial_k \mathbf{A}_j = g_e^2 \mathbf{F}_{kj}, \\ \partial^k \mathbf{F}_{kj} = \mathbf{J}_j \end{array}\right\} \Rightarrow \partial^k(\partial_j \mathbf{A}_k - \partial_k \mathbf{A}_j) = \frac{1}{g_e^2}\mathbf{J}_j.$$

This form is manifestly compatible with the transformation of the Lorentz group (ahead), which were discovered exactly here. Either the Galileo invariance, valid for point mechanics, or the Lorentz invariance, valid for electromagnetism, can be used as basic symmetry. Point mechanics had to be reformulated in an "improved" Lorentz compatible form.

The Lorentz transformations act not only on the four dimensional spacetime translations and the spacetime derivations ("four-vectors"):

$$\partial_j \cong \left(\frac{\partial}{c\partial t}, \frac{\partial}{\partial \vec{x}}\right), \quad [\partial_j] = \frac{1}{\text{m}}.$$

Also the charge-current densities and the potentials constitute Lorentz "4-vectors"; the field strengths as a Lorentz "6-vector" arise as the nontrivial components of an antisymmetric (4×4)-matrix:

$$\mathbf{J}_k \cong \frac{\sqrt{\hbar}}{e}\left(\rho, \frac{\vec{\mathbf{j}}}{c}\right), \quad \mathbf{A}_k \cong \frac{e}{\sqrt{\hbar}}\left(\frac{\mathbf{U}}{c}, \vec{\mathbf{A}}\right),$$

$$\mathbf{F}_{kj} \cong \frac{\hbar}{\hbar_e}\frac{e}{\sqrt{\hbar}}\begin{pmatrix} 0 & -\frac{\mathbf{E}_1}{c} & -\frac{\mathbf{E}_2}{c} & -\frac{\mathbf{E}_3}{c} \\ \frac{\mathbf{E}_1}{c} & 0 & \mathbf{B}_3 & -\mathbf{B}_2 \\ \frac{\mathbf{E}_2}{c} & -\mathbf{B}_3 & 0 & \mathbf{B}_1 \\ \frac{\mathbf{E}_3}{c} & \mathbf{B}_2 & -\mathbf{B}_1 & 0 \end{pmatrix}.$$

To obtain the dual 6-vector $\epsilon^{lmkj}\mathbf{F}_{kj}$, electric and magnetic fields have to be exchanged via $-\frac{\vec{\mathbf{E}}}{c} \leftrightarrow \vec{\mathbf{B}}$.

The potentials are gauge-dependent, i.e., determined up to a spacetime dependent Lorentz-scalar function $x \longmapsto \alpha(x)$:

$$\mathbf{A}_k \longmapsto \mathbf{A}_k + \partial_k \alpha.$$

The dimensions are chosen in such a way, that the potentials $\mathbf{A}$ and the field strengths $\mathbf{F}$ constitute space densities for a canonical field pair; i.e., their product has the dimension of an action density:

$$[\mathbf{A}_k \cdot \mathbf{F}_{jk}] = \frac{[\hbar]}{\mathrm{m}^3} = \mathrm{m}^{-1}\frac{\mathrm{kg}}{\mathrm{s}} \Rightarrow \begin{cases} [\mathbf{A}_k] = \sqrt{\frac{\mathrm{kg}}{\mathrm{s}}}, \\ [\mathbf{F}_{jk}] = \mathrm{m}^{-1}\sqrt{\frac{\mathrm{kg}}{\mathrm{s}}}, \\ [\mathbf{J}_k] = \mathrm{m}^{-2}\sqrt{\frac{\mathrm{kg}}{\mathrm{s}}}. \end{cases}$$

As seen at the units, the electromagnetic vectors ($\vec{\mathbf{E}}$, $\vec{\mathbf{B}}$, $\vec{\mathbf{A}}$) are not valued like position $\vec{x}$, e.g., they do not oscillate in position. However, they transform equivalently under rotations.

Good to know: Loosely speaking, the Maxwell equations formulate the simplest case of a special relativistic theory, i.e., compatible with the Lorentz group action, for a locally conserved quantity (charge): The current $\mathbf{I} = -\frac{d\mathbf{Q}}{dt}$ as the temporal change of a charge, localized with current and charge densities $\int_{\partial V} d^2\vec{x}\ \vec{\mathbf{j}}(t, \vec{x}) = -\frac{d}{dt}\int_V d^3x\ \boldsymbol{\rho}(t, \vec{x})$ with the volume element $d^3x = dx_1dx_2dx_3$ and the three area elements $d^2\vec{x} = (dx_2dx_3, dx_3dx_1, dx_1dx_2)$, and assumed to be valid for any volume V with surface ∂V, leads to the local conservation $\dot{\boldsymbol{\rho}} + \mathrm{div}\,\vec{\mathbf{j}} = 0$. Expressing the source structure of the charge by the electric field $\boldsymbol{\rho} = \mathrm{div}\,\vec{\mathbf{D}}$, $\vec{\mathbf{D}} = \epsilon_0\vec{\mathbf{E}}$, allows the "solution" of the local conservation by the magnetic field rot $\vec{\mathbf{H}} = \partial_t\vec{\mathbf{D}} + \vec{\mathbf{j}}$, $\vec{\mathbf{B}} = \mu_0\vec{\mathbf{H}}$. The special relativistic formulation with the Lorentz 4-vector $\mathbf{J}$ and the adjoint antisymmetric Lorentz 6-vector $\mathbf{F}$ as defined above gives the inhomogeneous equations $\partial^k\mathbf{F}_{kj} = \mathbf{J}_j$. They can be solved by the "four-dimensional rotation" $\mathbf{F}_{kj} = \frac{1}{g_e^2}(\partial_j\mathbf{A}_k - \partial_k\mathbf{A}_j)$ of a Lorentz 4-vector $\mathbf{A}$ which involves the homogeneous equations for the dual adjoint Lorentz 6-vector $\partial_m\epsilon^{lmkj}\mathbf{F}_{kj} = 0$.

2.9 Binding Energies and Couplings

The Kepler dynamics with the Hamiltonian $H = \frac{\vec{p}^2}{2m} + \frac{\epsilon g_0^2}{r}$, $\epsilon = \pm 1$, has, in the classical treatment, two intrinsic units, the mass $[m] = \mathrm{kg}$ and the coupling constant $[g_0^2] = \frac{\mathrm{m^3 kg}}{\mathrm{s^2}}$, i.e., no intrinsic length. For instance, the gravitational Newton interaction can be used for apples, planets and galaxies. In the quantum framework, Planck's unit $[\hbar] = \frac{\mathrm{m^2 kg}}{\mathrm{s}}$ leads to an intrinsic length unit. In nonrelativistic atomic physics with the Coulomb potential, e.g., for the states of the hydrogen atom, there does not occur the maximal action velocity c; there arises the smaller electromagnetically normalized velocity $\alpha_{\mathrm{Som}}c \simeq 2.2 \times 10^6\ \frac{\mathrm{m}}{\mathrm{s}}$, e.g., in the *Bohr*[45] *length*:

$$l_{\mathrm{Bohr}}(m, n) = \frac{\hbar}{\alpha_{\mathrm{Som}}c}\frac{1}{nm} = \frac{\hbar^2}{4\pi c\hbar_e}\frac{1}{nm} = \frac{1}{n\alpha_{\mathrm{Som}}}l_{\mathrm{Compt}}(m),$$

which depends on a mass m and a natural number n. For $n < \frac{1}{\alpha_{\mathrm{Som}}} \sim 137$, the Bohr length is larger than the Compton length. For the Coulomb interaction of two equally charged mass points, one uses, in the related Bohr length, their reduced mass and their negative charge product:

$$\frac{1}{m} = \frac{1}{m_1} + \frac{1}{m_2}, \quad n = -z_1z_2.$$

[45]Niels Bohr (1885–1962).

For charges with equal sign $z_1 z_2 > 0$ and repulsive Coulomb interaction, the Bohr length makes no sense since a binding is impossible without additional forces. The Compton length depends linearly on the action quantum $\hbar$, the Bohr length quadratically. The intrinsic units are the reduced mass, Planck's unit and the "electromagnetically reduced" velocity, i.e., $(m, \hbar, \alpha_{\text{Som}} c)$.

The Bohr length of the hydrogen atom — in a good approximation with the electron mass as reduced mass — gives the order of magnitude for the "size" of an atom:

$$l_{\text{Bohr}}(m_e, 1) = \frac{l_{\text{Compt}}(m_e)}{\alpha_{\text{Som}}} \simeq 0.53 \times 10^{-10}\ \text{m}, \quad \frac{m_p}{m_e} \simeq 1836.$$

The product of the corresponding $\hbar^6$-proportional Bohr volume with the Avogadro number describes "densely packed" hydrogen and relates the two bridges $\hbar$ and N_{Avo} from the microscopic to the human order of magnitude phenomena:

$$\frac{4\pi}{3}[l_{\text{Bohr}}(m_e, 1)]^3 N_{\text{Avo}} \simeq 0.4 \times 10^{-3}\ \frac{\text{m}^3}{\text{kilomol}} \simeq 0.4 \frac{\text{cm}^3}{\text{Mol}}.$$

The product of the electron mass and a charge number $n = 0, 1, \ldots$ in

$$E_{\text{Ryd}}(m_e, n^2) = n^2 m_e (\alpha_{\text{Som}} c)^2 = n^2 m_e \left(\frac{e^2}{4\pi\hbar\epsilon_0}\right)^2,$$

gives, e.g., with $n = 1$ for the proton, the ionization energy $E_{\text{Ryd}}(m_e, 1) \simeq 27.2$ eV of the hydrogen atom. The binding energies contain a factor $\alpha^2_{\text{Som}} \sim 6 \times 10^{-5}$ that reduces the electron mass-energy $m_e c^2 \sim 0.5$ MeV to the Rydberg[46] energy.

To compare: The binding energy of a nucleon in an atom is about 7 bis 9 MeV, yielding as the ratio of nuclear energy to atomic energy an order of magnitude $\frac{\text{MeV}}{\text{eV}}$, i.e., about one million. If one parametrizes, in analogy to the atomic binding energy the energy for nuclear binding with a strong coupling constant α_{strong} by $E_{\text{strong}}(m) = m(\alpha_{\text{strong}} c)^2$, there follows with the nucleon and electron mass: $\frac{E_{\text{strong}}(m_p)}{E_{\text{Ryd}}(m_e, 1)} = \frac{m_p}{m_e}\left(\frac{\alpha_{\text{Som}}}{\alpha_{\text{strong}}}\right)^2$. If this ratio is assumed about 10^6, the strong coupling constant has as order of magnitude: $4\pi\alpha_{\text{strong}} = g^2_{\text{strong}} \simeq 1$.

Writing down the binding energies arising "from a fire with wood to a fire with nucleons" (nuclear reactor), one obtains the following table:

[46] Johannes Rydberg (1854–1919).

Process (matter level)	**Binding** $[\frac{\text{J}}{\text{kilomol}}]$	**Binding** $[\frac{\text{MeV}}{\text{particle}}]$	**Rest mass** [MeV]	**Coupling** k^2
Water from evaporation* (liquids)	4×10^7	4×10^{-7}	$m_{H_2O}c^2 \simeq 2 \times 10^4$	4×10^{-11}
Water binding from H_2 and O (molecules)	3×10^8	3×10^{-6}	$m_{H_2O}c^2 \simeq 2 \times 10^4$	3×10^{-10}
Hydrogen binding from p^+ and e^- (atoms)		13.6×10^{-6}	$m_e c^2 \simeq 0.5$	6×10^{-5}
Nuclear binding from nucleons (nuclei)		8	$m_p c^2 \simeq 10^3$	1.6×10^{-2}
Nucleon binding with resonances (hadrons)		5×10^2	$m_p c^2 \simeq 10^3$	1

*for normal conditions

Matter levels

As measure for the nucleon binding, the mass difference $(1440 - 940)\frac{\text{MeV}}{c^2}$ is taken between an excited state (resonance) of the nucleon and the nucleon itself. The coupling k^2 (binding coefficient) is defined as twice the ratio of binding energy and rest mass. One obtains for the hydrogen binding with the intrinsic energy unit $m_e(\alpha_{\text{Som}}c)^2 \simeq 27.2$ eV:

$$\text{hydrogen binding coefficient: } k^2 = \frac{m_e(\alpha_{\text{Som}}c)^2}{m_e c^2} = (\alpha_{\text{Som}})^2 \simeq 6 \times 10^{-5},$$

i.e., the binding coefficient is a measure for the involved coupling constant of the interaction.

Weisskopf[47] collected the matter (energy, mass) levels in a *quantum ladder* with the rungs molecules, atoms, nuclei, hadrons, quarks. For elementary particles (nucleon binding) the coupling k has 1 as its order of magnitude. To talk of "parts in a whole" makes sense only, if the binding energy is small in comparison with the energy, stored in the constituents. For the gravitationally bound sun-earth state one has the potential and mass-related energies with the ratio of Schwarzschild length to orbit radius $\frac{Gm_\bullet}{c^2 R_\bullet} \sim 10^{-12}$:

$$\overline{E}_\bullet = G\frac{m_\odot m_\bullet}{R_\bullet} \sim 10^{33}\text{J}, \quad (m_\odot + m_\bullet)c^2 \sim 2 \times 10^{45}\text{J}.$$

It looks as if for elementary particles one has reached the level, where the *particle description for an adequate formulation of a dynamics (interaction) finds its limit.* For instance, to talk about hadrons (protons, pions, etc.) as bound states of quark *particles* may be as slippery as the position concept for an "electron in an atom."

In this sense, it is satisfactory, that quarks are structures, which shape the hadronic dynamics, but which are, up to today, not observed as free particles — although a

[47] Victor F. Weisskopf (1908–2002).

mathematical proof of their confinement in a fundamental theory (quantum chromodynamics), e.g., by extending the flux-tube "arguments," is still missing. A geometric analogy to the theory of the quarks and their missing particle property: Six squares shape, as faces, a cube; however, it does not make sense to call the faces "parts of the cube." No carpenter or no goldsmith has ever manufactured an isolated square — the square lacks the property of three-dimensionality. According to the theory today, quarks shape protons: Quarks come with the homogeneous properties hypercharge, isospin, color and spin; however they lack the translation invariant mass as a free particle property.

2.10 Electroweak Interactions

The electromagnetic interaction connects with each other charged particles. It is the classical, macroscopically easily detectable "tail" of a branched charge-related structure: In addition to the long-range electromagnetic interactions, there exist *three short-ranged weak interactions* — a neutral one, which, like the electromagnetic one, does not exchange electric charges, and two oppositely charged interactions, which rearrange the electric charges, as exemplified by the radioactive decay $n \to p + e + \overline{\nu}_e$ of the neutrons (charge number $z_n = 0$)into a proton ($z_p = 1$) and an electron ($z_e = -1$) with its antineutrino ($z_{\overline{\nu}_e} = 0$), $z_n = z_p + z_e + z_{\overline{\nu}_e}$.

This charge structure with four interactions reflects four *internal (chargelike) degrees of freedom.* Analogously to *time and position*, denoting the four external (spacetimelike) coordinates, there exist four internal "coordinates," called *hypercharge and isospin.* To describe the additional strong interaction, the internal property *color* is used with eight degrees of freedom. If in a simplifying picture, time and space translations are given by one point in a corresponding Minkowski vector space, the internal properties can endow this point — think of a spacetime point with an icecream parlour — with different flavors and colors — the icecream may taste of different fruits, and may come in different colors.

The photon with charge number $z = 0$ and the three weak bosons $Z, W^{\pm}$ with charge numbers $z = 0, \pm 1$ are the particle aspect of the electroweak interactions. From this fourfold particle basis, there leads an axial rotation with the *Weinberg*[48] *angle* to an interaction basis with three isospin-vector fields and one isospin-scalar field — in the flavor picture: two mixes of cream and strawberry tastes are decomposed into the basic ingredients.

The Weinberg angle defines the rectangular *electroweak triangle*: The square of its hypotenusis can be normalized by Sommerfeld's number $\ell_e^2 = \frac{1}{g_e^2} = \frac{1}{4\pi\alpha_{\text{Som}}}$. With Pythagoras'[49] theorem, ℓ_e^2 is distributed to the orthogonal sides, which, in this normalization, give the inverse coupling constants for the isospin and hypercharge interactions. The experimental values for the electroweak length triangle are

[48] Steven Weinberg (1933–).

[49] Pythagoras of Samos, around -530.

$$\tan^2 \theta_{\text{Wein}} = \frac{\ell_W^2}{\ell_Y^2} \simeq 0.3, \quad \begin{cases} \ell_e^2 = \ell_W^2 + \ell_Y^2 \simeq 10.9, \\ \frac{1}{\ell_Z^2} = \frac{1}{\ell_W^2} + \frac{1}{\ell_Y^2} \simeq 0.5 \end{cases}$$

$$\Rightarrow \begin{cases} \text{charge hypotenusis: } \ell_e &= \frac{1}{g_e} \simeq 3.3, \\ \text{isospin side: } \ell_W &= \frac{1}{g_W} \simeq 1.6, \\ \text{hypercharge side: } \ell_Y &= \frac{1}{g_Y} \simeq 2.9, \\ \text{height: } \ell_Z &= \frac{1}{g_Z} \simeq 1.4. \end{cases}$$

If one had a symmetry argument for a Platonic[50]"beautiful" triangle, i.e., for the half of an equilateral one with the angles $(90°, 60°, \theta_{\text{Wein}} = 30°)$, the side ratios were $\ell_e : \ell_W : \ell_Y = 2 : 1 : \sqrt{3}$. In such a case, the Weinberg angle would display the distribution of the involved isospin degrees of freedom of the electroweak field:

$$\text{beautiful triangle: } \tan^2 \theta_{\text{Wein}} = \frac{1}{3} = \frac{\text{number of isospin scalars}}{\text{number of isospin vectors}}.$$

Since a closer analysis shows the angle to be dependent on the energy of the interaction, there could, in principle, arise such a "beautiful" triangle for some distinguished energy.

The experiments determine not only the form of the electroweak triangle, they give also its normalization by an intrinsic length and mass unit, respectively: The relevant unit for the electroweak interaction is the *Fermi*[51] *unit*:

$$\frac{G_{\text{Fermi}}}{(\hbar c)^3} \simeq \frac{1}{(292.8\text{GeV})^2}, \text{ with GeV } = 10^9\text{eV},$$

associated to the Fermi mass and *weak length (range)*:

$$\frac{G_{\text{Fermi}}}{(\hbar c)^3} = \frac{1}{4\sqrt{2}(M_{\text{Fermi}}c^2)^2} \Rightarrow \begin{cases} M_{\text{Fermi}} \simeq 123\frac{\text{GeV}}{c^2}, \\ l_{\text{Compt}}(M_{\text{Fermi}}) \simeq 1.6 \times 10^{-18} \text{ m}. \end{cases}$$

The electroweak length triangle has the Compton length of the charged weak bosons $W_\pm$ as isospin side, the height is related to the neutral weak boson Z:

$$\ell_W = \frac{M_{\text{Fermi}}}{m_W} \simeq 1.6 \Rightarrow l_{\text{Compt}}(m_W) = 2.6 \times 10^{-18} \text{ m},$$

$$\ell_Z = \frac{\ell_W \ell_Y}{\ell_e} = \frac{M_{\text{Fermi}}}{m_Z} \simeq 1.4 \Rightarrow l_{\text{Compt}}(m_Z) = 2.2 \times 10^{-18} \text{ m}.$$

The mass ratio of the weak bosons does not depend on the Fermi mass:

[50]Platon -(428–348).

[51]Enrico Fermi (1901–1954).

$$\frac{m_W^2}{m_Z^2} = \frac{\ell_Y^2}{\ell_e^2} = \frac{\ell_Z^2}{\ell_W^2} = \cos^2\theta_{\text{Wein}} \begin{cases} \simeq 0.77 \text{ (experiments)}, \\ = 0.75 \text{ ("beautiful" triangle.)} \end{cases}$$

The electroweak triangles for masses or coupling constants, dual to the electroweak length triangle, have the coupling constants $g^2 = \frac{1}{\ell^2}$ as squared sides, the height g_e is the elementary charge. The hypotenusis is the Z-mass, one orthogonal side the W-mass. The masses, associated with the other orthogonal side and with the height, are not particle masses:

$$g_Z^2 = g_W^2 + g_Y^2, \quad \frac{1}{g_e^2} = \frac{1}{g_W^2} + \frac{1}{g_Y^2},$$

$$(m_Z, m_W, m_Y, m_e) = (g_Z, g_W, g_Y, g_e)M_{\text{Fermi}} \simeq (91.2,\ 80.2,\ 43.4,\ 38.2)\frac{\text{GeV}}{c^2}.$$

Good to know: For a given angle $\theta < \frac{\pi}{2}$, there are two dual rectangular triangles, similar to each other, with θ as a base angle, $\tan\theta = \frac{b}{a}$. The first one with doubled area $2F = e^\psi = ab$ has sides $a^2 + b^2 = c^2$ and height in $ab = ch$. The dual partner triangle with $\frac{1}{b^2} + \frac{1}{a^2} = \frac{1}{h^2}$ and $\frac{1}{b}\frac{1}{a} = \frac{1}{h}\frac{1}{c}$ has the doubled area $e^{-\psi} = \frac{1}{ab}$. The duality relies on the two orthogonal angles — between the orthogonal sides and between hypotenusis and height. The dilation transformation for all corresponding lengths of the dual triangles multiplies by the doubled triangle area e^ψ and $e^{-\psi}$, respectively.

The symmetry (interaction-)basis for the internal isospin and hypercharge coordinates with the hypercharge gauge field $\mathbf{A}_0$ and the third component $\mathbf{A}_3$ of an isospin triplet gauge field is axially rotated to a particle basis with a photon field $\mathbf{A}$ and a neutral weak boson field $\mathbf{Z}$:

$$\begin{pmatrix} \frac{1}{g_e}\mathbf{A} \\ \frac{1}{g_Z}\mathbf{Z} \end{pmatrix} = D(\theta)\begin{pmatrix} \frac{1}{g_Y}\mathbf{A}_0 \\ \frac{1}{g_W}\mathbf{A}_3 \end{pmatrix} \Rightarrow \begin{cases} \mathbf{A} = \cos^2\theta\, \mathbf{A}_0 + \sin^2\theta\, \mathbf{A}_3, \\ \mathbf{Z} = -\mathbf{A}_0 + \mathbf{A}_3, \end{cases}$$

$$\text{with Weinberg rotation: } D(\theta) = \begin{pmatrix} \cos\theta & \sin\theta \\ -\sin\theta & \cos\theta \end{pmatrix}.$$

The electroweak rotation has a mechanical analogue in the transformations — for two mass points — from individual positions and momenta $(\mathbf{x}_i, \mathbf{p}_i)_{i=1,2}$ to center-of-mass $(\mathbf{X}, \mathbf{P})$ and relative $(\mathbf{x}, \mathbf{p})$ positions and momenta, e.g., for a sympathetic pendulum to one free component with trivial frequency Ω and one oscillator with the frequency ω, or for the gravitational attraction of two masses, as given by the Hamiltonians:

$$\begin{aligned}
\mathbf{H}_{\text{osc}} &= \frac{\mathbf{p}_1^2}{2m_1} + \frac{\mathbf{p}_2^2}{2m_2} + \frac{g_0^2}{2}(\mathbf{x}_1 - \mathbf{x}_2)^2 \\
&= \frac{\mathbf{P}^2}{2M} + \frac{\mathbf{p}^2}{2m} + \frac{g_0^2}{2}\mathbf{x}^2, \ \Omega = 0, \ \omega^2 = \frac{g_0^2}{m}, \\
\mathbf{H}_{\text{grav}} &= \frac{\mathbf{p}_1^2}{2m_1} + \frac{\mathbf{p}_2^2}{2m_2} - \frac{Gm_1m_2}{|\mathbf{x}_1 - \mathbf{x}_2|} \\
&= \frac{\mathbf{P}^2}{2M} + \frac{\mathbf{p}^2}{2m} - \frac{GMm}{|\mathbf{x}|}.
\end{aligned}$$

The associated rectangular mass triangle has the masses $m_{1,2}$ as the squares of the orthogonal sides, the mass sum $M = m_1 + m_2$ as hypotenusis square, and the reduced mass $m = \frac{m_1 m_2}{M}$ as the square of the height. The corresponding dual frequency triangle has the side squares $(\frac{g_0^2}{m_2}, \frac{g_0^2}{m_1}; \frac{g_0^2}{m})$ and $\frac{g_0^2}{M}$ as squared height. The angle θ for the axial rotation is the basis angle of the rectangular triangle; it is determined by the mass ratio:

$$\begin{aligned}
&\text{center of mass rotation: } D(\theta) = \begin{pmatrix} \cos\theta & \sin\theta \\ -\sin\theta & \cos\theta \end{pmatrix}, \\
&\quad \text{where } \tan^2\theta = \frac{m_2}{m_1}, \ \cos^2\theta = \frac{m_1}{M} = \frac{m}{m_2}, \ \sin^2\theta = \frac{m_2}{M} = \frac{m}{m_1}, \\
&\quad \text{with } m_1 + m_2 = M, \ \frac{1}{m_1} + \frac{1}{m_2} = \frac{1}{m},
\end{aligned}$$

and used for the positions in

$$\begin{pmatrix} \sqrt{M}\mathbf{X} \\ \sqrt{m}\mathbf{x} \end{pmatrix} = D(\theta)\begin{pmatrix} \sqrt{m_1}\mathbf{x}_1 \\ \sqrt{m_2}\mathbf{x}_2 \end{pmatrix} \Rightarrow \begin{cases} m_1\mathbf{x}_1^2 + m_2\mathbf{x}_2^2 = M\mathbf{X}^2 + m\mathbf{x}^2, \\ \mathbf{X} = \cos^2\theta\,\mathbf{x}_1 + \sin^2\theta\,\mathbf{x}_2, \\ \mathbf{x} = -\mathbf{x}_1 + \mathbf{x}_2, \end{cases}$$

and with inverse normalization for the momenta in

$$\begin{pmatrix} \frac{1}{\sqrt{M}}\mathbf{P} \\ \frac{1}{\sqrt{m}}\mathbf{p} \end{pmatrix} = D(\theta)\begin{pmatrix} \frac{1}{\sqrt{m_1}}\mathbf{p}_1 \\ \frac{1}{\sqrt{m_2}}\mathbf{p}_2 \end{pmatrix} \Rightarrow \begin{cases} \frac{\mathbf{p}_1^2}{m_1} + \frac{\mathbf{p}_2^2}{m_2} = \frac{\mathbf{P}^2}{M} + \frac{\mathbf{p}^2}{m}, \\ \mathbf{P} = \mathbf{p}_1 + \mathbf{p}_2, \\ \mathbf{p} = -\sin^2\theta\,\mathbf{p}_1 + \cos^2\theta\,\mathbf{p}_2. \end{cases}$$

For example, the center of mass angle in $\tan^2\theta = \frac{m_p}{m_e} \simeq 1836$, $\theta \simeq 88.7°$ for the hydrogen atom involves the electron and proton mass.

In analogy to the sympathetic pendulum in the mechanical analogue, the electromagnetic and the charge neutral weak interactions, related to the massless photon and the massive Z-boson, can be called the interaction for the *center of charge* and for the *relative charge*, respectively.

2.11 Units and Symmetry Normalizations

A dimensional grading of all physical quantities by three fundamental units, perhaps by the Planck, Einstein, and Newton unit ($\hbar$, c, G), may hint to three basic metrical structures that are connected with the normalizations and invariants of operations for time and space, and, ultimately, to the action of the dilation group. Here, the time independence of the units is assumed. Dirac has speculated about a possible time dependence — so far without any experimental substantiation.

Reversing the historical order of the arguments, one may ponder the question, which properties of perhaps existing fundamental structures in the physical description of nature could cause the introduction of units for their quantitative characterization.

First for the *field unit* c: The maximal action velocity has its operational origin in the noncompact structure of the spacetime Lorentz group $\mathbf{O}(1, 3)$, visible in the proper Lorentz transformations (boosts), which arise by an *expansion* of the Galilei transformations for finite action velocity $\frac{1}{c} > 0$. In a simplified two-dimensional spacetime, the Lorentz transformations constitute a real 1-parametric group, affecting the space translations and the time translations by hyperbolic transformations"):

$$\begin{aligned}
\begin{pmatrix} x_0 \\ x_1 \end{pmatrix} &\longmapsto \begin{pmatrix} \cosh\psi & \sinh\psi \\ \sinh\psi & \cosh\psi \end{pmatrix}\begin{pmatrix} x_0 \\ x_1 \end{pmatrix}, \quad \text{invariant: } x_0^2 - x_1^2, \\
\mathbf{SO}_0(1,1) &= \{D_2(\psi) = \begin{pmatrix} \cosh\psi & \sinh\psi \\ \sinh\psi & \cosh\psi \end{pmatrix} \mid \psi \in \mathbb{R}\}, \\
\text{group:} \quad D_2(\psi_1) &\circ D_2(\psi_2) = D_2(\psi_1 + \psi_2).
\end{aligned}$$

Only for a noncompact structure, lightlike vectors with $x_0^2 - \vec{x}^2 = 0$ can be nontrivial. Light separates time with $x_0^2 - \vec{x}^2 > 0$ from position with $x_0^2 - \vec{x}^2 < 0$ ($\vec{x}^2 = x_1^2$ for the one-dimensional space model). Only in such a case, a nontrivial future and past, compatible with orthochronous Lorentz transformations, and, therefore, a nontrivial causality can be defined.

The 2×2-matrix representation $D_2(\psi)$ of the Lorentz transformations in a spacetime basis with the diagonal matrix $g \cong \eta_2 = \begin{pmatrix} 1 & 0 \\ 0 & -1 \end{pmatrix}$ for the symmetric bilinear Lorentz metric (length squares), i.e., $g(x, x) = x_0^2 - x_1^2$, is equivalent to a diagonal matrix with two scale transformations, inverse to each other, for a light basis. In such a basis, the bilinear Lorentz metric comes in the "skew-diagonal" matrix $g \cong \zeta_2 = \begin{pmatrix} 0 & 1 \\ 1 & 0 \end{pmatrix}$, i.e., $g(x, x) = 2x_+x_-$, with $x_\pm = \frac{x_0 \pm x_1}{\sqrt{2}}$:

$$(\cosh\psi - D_\pm)^2 - \sinh^2\psi = 0 \iff D_\pm(\psi) = e^{\pm\psi}, \quad D_2(\psi) \cong \begin{pmatrix} e^{\psi} & 0 \\ 0 & e^{-\psi} \end{pmatrix}.$$

The Lorentz transformation is a self-dual dilation ("Procrustes transformation"): One of the lightlike eigenvectors $x_\pm = \frac{x_0 \pm x_1}{\sqrt{2}}$ is shortened, the other one is stretched with the inverse factor. The invariance of a hyperbola $x_+x_- = \ell^2$ under dilations

$x_\pm \longmapsto e^{\pm\psi} x_\pm$ is well known from the high-school discussions of the conic sections. The dilation structures formalize the "paradoxical" contractions of special relativity.

In the space-time basis representation, the proper Lorentz transformations display the dilation structure via the hyperbolic tangens:

$$D_2(\psi) = \frac{1}{\sqrt{2}}\begin{pmatrix}1 & 1\\ 1 & -1\end{pmatrix}\begin{pmatrix}e^\psi & 0\\ 0 & e^{-\psi}\end{pmatrix}\frac{1}{\sqrt{2}}\begin{pmatrix}1 & 1\\ 1 & -1\end{pmatrix} = \cosh\psi\begin{pmatrix}1 & \tanh\psi\\ \tanh\psi & 1\end{pmatrix},$$

$$|\tanh\psi| = |\frac{e^\psi - e^{-\psi}}{e^\psi + e^{-\psi}}| < 1.$$

Its upper limit is parametrizable by a normalization with a constant maximal parameter $c > 0$:

$$\frac{v}{c} = \tanh\psi,\ \ \psi = \frac{1}{2}\log\frac{1+\frac{v}{c}}{1-\frac{v}{c}} \ \text{ with } |v| < c \Rightarrow D_2(\psi) = \frac{1}{\sqrt{1-\frac{v^2}{c^2}}}\begin{pmatrix}1 & \frac{v}{c}\\ \frac{v}{c} & 1\end{pmatrix} = D_2\left(\frac{v}{c}\right),$$

$$\text{or } D_2(\psi) = \frac{1}{mc}\begin{pmatrix}\frac{E}{c} & p\\ p & \frac{E}{c}\end{pmatrix}, \ \text{ with energy-momentum } (E,p)$$

$$\text{and invariant } m^2c^2 = \frac{E^2}{c^2} - p^2,\ \ \det D_2(\psi) = 1.$$

The *rapidity*, as normalized velocity $\psi = \frac{v}{c} + \ldots$ for nonrelativistic small $\frac{v}{c}$, measures the length on a hyperbola, e.g.,

$$x_0^2 - x_1^2 = -1,\ \ (dx_0, dx_1) = (\cosh\psi, \sinh\psi)d\psi,$$

$$\int ds = \int\sqrt{dx_0^2 - dx_1^2} = \int dx_0\sqrt{1-\frac{v^2}{c^2}} = \int d\psi.$$

The velocity parametrization, suggested by nonrelativistic point mechanics, has some pitfalls: There does not exist a Lorentz transformation with $|v| = c$, e.g., from Einstein's rest system to a photon. In contrast to the rapidity parametrization $-\infty < \psi < \infty$, the velocity parametrization may suggest a tachyonic extension to velocities $|v| > c$, larger than the maximal value, where, with the imaginary value of $\sqrt{1-\frac{v^2}{c^2}}$, the group structure is lost. The simple linear addition of an hyperbolic "angle" ψ (rapidity) becomes the complicated combination law of its "projection" (addition theorem of the hyperbolic tangens):

$$\tanh(\psi_1+\psi_2) = \frac{\tanh\psi_1 + \tanh\psi_2}{1+\tanh\psi_1\tanh\psi_2} \iff v(\psi_1+\psi_2) = \frac{v(\psi_1)+v(\psi_2)}{1+\frac{v(\psi_1)v(\psi_2)}{c^2}}.$$

Good to know: If an axial rotation $D(\theta) = \begin{pmatrix}\cos\theta & \sin\theta\\ -\sin\theta & \cos\theta\end{pmatrix}$ from the group **SO**(2) would be parametrized not by the arc θ (rotation angle) of the unit circle, but by its projection $s = \sin\theta$ on the y-axis direction, the angle for the product of two rotation would be characterized, not by the angle sum $\theta_1 + \theta_2$ for $D(\theta_1 + \theta_2) = D(\theta_1) \circ D(\theta_2)$, but by the complicated expression $s(\theta_1 + \theta_2) = s_1\sqrt{1 - s_2^2} + s_2\sqrt{1 - s_1^2}$ for the projected arc. An s-parametrization may suggest a useless extension to the region with $s > 1$.

Because of the nontrivial signature of the Lorentz metric, for $\mathbf{O}(1,1)$ given by $g \cong \eta_2 = \begin{pmatrix} 1 & 0 \\ 0 & -1 \end{pmatrix}$ (space-time basis) or by $g \cong \zeta_2 = \begin{pmatrix} 0 & 1 \\ 1 & 0 \end{pmatrix}$ (light basis), the Lorentz group structure involves a physically relevant maximal signature parameter c, which can be taken as *relative normalization of the time and position translations*. It can be used for a renormalization of the coordinates (upper and lower components in the vectors with the action of $D_2(\psi)$):

$$\text{spacetime action: } \begin{pmatrix} x_0 \\ x_1 \end{pmatrix} \longmapsto D_2(\frac{v}{c})\begin{pmatrix} x_0 \\ x_1 \end{pmatrix} \Rightarrow \begin{cases} \frac{x_0}{c} = t \longmapsto \frac{t + \frac{v}{c^2}x_1}{\sqrt{1 - \frac{v^2}{c^2}}}, \\ x_1 \longmapsto \frac{x_1 + vt}{\sqrt{1 - \frac{v^2}{c^2}}}. \end{cases}$$

The limit of an infinite action velocity $c \to \infty$ describes the transition from the special relativistic spacetime to classical space and time, formalized by the *Inönü*[52]—Wigner[53] *contraction* from – for three position dimensions $s = 3$ — the orthochronous Lorentz group $\mathbf{SO}_0(1,s)$ to the semidirect Galileo group $\mathbf{SO}(s) \vec{\times} \mathbb{R}^s$ with position rotations $\mathbf{SO}(s)$ and velocity transformations $\vec{v} \in \mathbb{R}^s$. The limit is exemplified in a two-dimensional rotation-free spacetime, i.e., with $s = 1$:

$$\begin{pmatrix} \frac{1}{c} & 0 \\ 0 & 1 \end{pmatrix} D_2(\frac{v}{c}) \begin{pmatrix} c & 0 \\ 0 & 1 \end{pmatrix} = \frac{1}{\sqrt{1 - \frac{v^2}{c^2}}} \begin{pmatrix} 1 & \frac{v}{c^2} \\ v & 1 \end{pmatrix} \stackrel{c \to \infty}{\longrightarrow} \begin{pmatrix} 1 & 0 \\ v & 1 \end{pmatrix},$$

$$\text{i.e., } \begin{cases} t \longmapsto t, \\ x_1 \longmapsto x_1 + vt, \end{cases} \quad \text{for } s = 3: \begin{cases} t \longmapsto t, \\ \vec{x} \longmapsto \vec{x} + \vec{v}t. \end{cases}$$

If the basic law of physics — in the case it exists — is characterized by a Lorentz symmetry structure, then the unit c makes sense for the relative normalization of the positive and negative sector in the Lorentz metric $g \cong \begin{pmatrix} c^2 & 0 \\ 0 & -\mathbf{1}_3 \end{pmatrix}$.

Good to know: In general for $t, s \geq 1$, there is a contraction $\mathbf{SO}_0(t,s) \stackrel{\frac{1}{c} \to 0}{\longrightarrow} \mathbf{SO}_0(t-1,s) \vec{\times} \mathbb{R}^{t-1+s}$ from a (pseudo-)orthogonal group to a semidirect product group, formalized in the corresponding Lie algebras by $\left(\begin{array}{c|c} 0 & \frac{1}{c}\frac{v}{c} \\ \hline c\frac{v}{c} & \mathcal{L} \end{array}\right) \stackrel{\frac{1}{c} \to 0}{\longrightarrow} \left(\begin{array}{c|c} 0 & 0 \\ \hline v & \mathcal{L} \end{array}\right)$ with $\mathcal{L} \in \log \mathbf{SO}_0(t-1,s)$ and representatives $\frac{v}{c} \in \mathbb{R}^{t-1+s}$ of the cosets $\log \mathbf{SO}_0(t,s)/\log \mathbf{SO}_0(t-1,s)$.

Now for the *action unit* $\hbar$, experimentally found and interpretable as the quantum of action or as the quantum of angular momentum: As connected with the name of Noether,[54] the invariances of a dynamics and its conservation laws, i.e., its time-independent invariants, condition each other. In the quantum theoretical formalism,

[52]Erdal Inönü (1926–2007).

[53]Eugene Wigner (1902–1995).

[54]Emmy Noether (1882–1935).

this connection proves itself as a simple identity: Dynamically relevant operations, like the time translations or the "infinitesimal" rotations, are described by quantum operators — in the examples by a Hamiltonian $\mathbf{H}$ and by the three angular momenta $\vec{\mathbf{L}}$, respectively. The invariance with respect to an operation is characterized by a trivial commutator (adjoint action) with the corresponding quantum operator. For example, angular momentum is time translation invariant, i.e., conserved, for

$$\operatorname{ad}\mathbf{H}(\vec{\mathbf{L}}) = [\mathbf{H}, \vec{\mathbf{L}}] = \mathbf{H}\vec{\mathbf{L}} - \vec{\mathbf{L}}\mathbf{H} = 0 \iff \frac{d\vec{\mathbf{L}}}{dt} = 0;$$

and the time development (dynamics) is rotation invariant (with three rotation angles in $\vec{\theta}\vec{\mathbf{L}}$) with the identical condition, read in the reversed form:

$$\operatorname{ad}\vec{\mathbf{L}}(\mathbf{H}) = [\vec{\mathbf{L}}, \mathbf{H}] = \vec{\mathbf{L}}\mathbf{H} - \mathbf{H}\vec{\mathbf{L}} = 0 \iff \frac{\partial \mathbf{H}}{\partial \vec{\theta}} = 0.$$

In an operational framework, the invariants determine the representations of the corresponding action groups and Lie algebras.

The normalization of a conserved angular momentum square by a unit $\hbar^2$ can be transferred, in quantum mechanics via $\vec{\mathbf{L}} = i\vec{\mathbf{x}} \times \vec{\mathbf{p}}$ and $[i\mathbf{p}^a, \mathbf{x}^b] = \delta^{ab}\mathbf{1}$, to a normalization of the noncommutative canonical (dual) position–momentum pair, where $\hbar$ is the product of position and momentum units l_0 and π_0:

$$\underline{\vec{\mathbf{x}}} = l_0\vec{\mathbf{x}},\ \underline{\vec{\mathbf{p}}} = \pi_0\vec{\mathbf{p}},\ \text{with } l_0\pi_0 = \hbar \Rightarrow \begin{cases} [i\underline{\mathbf{p}}^a, \underline{\mathbf{x}}^b] = \delta^{ab}\hbar\mathbf{1}, \\ i\underline{\vec{\mathbf{x}}} \times \underline{\vec{\mathbf{p}}} = \hbar\vec{\mathbf{L}} = \underline{\vec{\mathbf{L}}}. \end{cases}$$

After two units c (spacetime or light unit) and $\hbar$ (dual or angular momentum unit), related rather satisfactorily with the fundamental relevance of the Lorentz group and quantum noncommutativity (duality), one has no clear qualitative understanding of a third universal intrinsic unit.

Perhaps a representation of the *causality structure* requires at least two real normalizations: Also in our daily life we describe and measure time by one always increasing number, e.g., by piling up the leaves of a tear-off calendar, and by a cyclic number, e.g., on an analogue watch with a circular dial. Take as example the imperial coronation of Charlemagne[55] in the year 800 — this is one linear noncompact date — at the 25th of December, 1 p.m. (?) — this involves three cyclic compact dates. As a mathematical model of a time, measured with a noncompact and compact invariant, one may take a screw thread: Both the noncompact and compact structure have a unit, e.g., year and day:

---

[55] Carolus Magnus (768–814).

$$\text{screw line: } (x(t), y(t), z(t)) = R_0(\cos\omega t, \sin\omega t, \Omega t),$$
$$\text{screw turning time: } \frac{2\pi}{\omega},$$
$$\text{screw raising time: } \frac{1}{\Omega}.$$

A twofold time normalization gives rise to one number like $\frac{\Omega}{\omega}$; for instance, year and day determines the number $365, 2422\ldots = 365 + \frac{1}{4} - \frac{3}{400} + \ldots$ (leap year expansion: add one day for all four years, and, subtract three days for all four hundred years, etc.). With respect to a compact and a noncompact aspect of spacetime, Newton's unit G could be related to the normalization of the external interactions (gravitative) in the noncompact real numbers $\mathbb{R}$ as time model. Its ratio to the normalization of the compact quotient group $\mathbf{U}(1) \cong \mathbb{R}/\mathbb{Z}$ could be related to the coupling constants of the internal unitary interactions (electromagnetic, weak, and strong).

All this is rather vague and, obviously, not a theory. In order not to stay completely with "empty" words, there will be given a numerical illustration — also that probably irrelevant with respect to the concrete details — which relates the Fermi and Planck mass as compact and noncompact unit, respectively, to the order of magnitude of Sommerfeld's number:

$$\frac{M_{\text{Planck}}}{M_{\text{Fermi}}} \simeq 10^{17}, \quad \log\frac{M_{\text{Planck}}}{M_{\text{Fermi}}} \simeq 39, \quad 4\frac{\hbar}{\hbar_e} = \frac{1}{\pi\alpha_{\text{Som}}} \simeq 43.6.$$

In a fundamental law of physics, if it exists, such a twofold normalization of the spacetime structure could be experimentally observable, on the one hand, in the particle and their masses, and, on the other hand, in the interactions and their normalizations (coupling constants), e.g., in the electromagnetic Coulomb or gravitative Newton interaction.

Chapter 3
How Complex Is Nature?

Numbers can be considered under the aspect of their operational properties: The abstraction of the *natural* numbers reflects experiences and activities of our daily life, e.g., the realization of different objects. It is also not difficult, e.g., by looking at the checking account or by cutting the birthday cake, to connect *negative* and *broken* numbers with our operational experiences. The extension from the rational to the *real* numbers is a much more difficult step. The real numbers are accepted as a short notation and formalization of a possibly infinite approximation procedure with rational numbers.

The *complex* or doubled real numbers — they can also be taken as vectors of a real two-dimensional plane, which can be added and multiplied — are used in physics, especially, in quantum theory — after a short flirt with the indefiniteness of the real Lorentz group. For the internal compact symmetries, i.e., for unitary symmetries like hypercharge and isospin, they are undispensable. With the real and the imaginary, the complex parametrizes noncompact and compact operations, respectively, as sketched by $z = re^{i\theta}$.

In a sense, the complex numbers seem to have a fundamental role in the connection of the concepts "time" and "object," exactly there they are used in physics. The real is embedded in the complex: The description of experiments employs real probabilities as absolute squares of complex probability amplitudes (complex truth values), getting rid of the characteristic phases with the time representation. As proposed by Wigner, the quantum implementation of the time reflection works with antilinear transformations, i.e., with the complex conjugation of the numbers $\alpha \leftrightarrow \overline{\alpha}$, and the transition from ingoing to outgoing state vectors — from bras to kets $\langle \ldots | \leftrightarrow | \ldots \rangle$.

In mathematics, the complex numbers were introduced, after some preludes, especially by Gauss,[1] who discussed and established their algebraic completeness: Each

[1]Carl Friedrich Gauss (1777–1855).

H. Saller, *Operational Symmetries*, DOI 10.1007/978-3-319-58664-9_3

complex polynomial of degree n has exactly n complex roots, possibly degenerate. This property allows the determination of eigenvalues and eigenvectors of complex linear physical operations. In the real, for example, there do not exist time translation eigenvectors for the harmonic oscillator with definite energy, or no spin eigenvectors for the position rotations, i.e., a mathematical formalization of particles by real state vectors as translation eigenvectors with a definite spin direction is impossible.

There may be a more fundamental role for the use of the complex numbers in physics. Possibly: They could constitute the smallest numerical structure for an epistemological valuation of objects in time with a distinction of the past, characterized by principally fixed facts, and the future, characterized by principally open possibilities, in a way as experiments are talked about in quantum theory — with the modality concepts certainty, probability and possibility (probability amplitudes). In quantum theory, the complex probability amplitudes are given by Hilbert space scalar products which, in turn, are determined by the unitary representations of time and space related operations.

If, starting from the naturals, the complex numbers are reached on familiar and well known ways, one can himself let be carried on by similar embedding and representation structures to the familiar spacetime structures, to Minkowski spacetime and to the Poincaré[2] group. However, one should not cheat oneself: This is no deduction, only a deductively looking and streamlined repetition of methods, obviously in hindsight, where the goal with the experimentally successful theories is well known. The original way was much more crooked. Nevertheless, it seems instructive to reverse the historical order for the understanding of the space-time structures to do justice to their relevance, e.g., with respect to a complex embedding of the real space-time structures as suggested by quantum theory.

In this chapter, let us climb this ascent, stony and with mathematical technicalities — as Sommerfeld told his students: "Soll sich dir die Wahrheit zeigen, musst du ins Komplexe steigen (Would you like to see the truth, you have to go into the complex)." We will start from numbers in countable sets:

$$\begin{array}{ccccccccccc}
\boxed{\begin{array}{c}\mathbb{P}\\ \text{prime}\end{array}} & \subset & \boxed{\begin{array}{c}\mathbb{N}\\ \text{natural}\end{array}} & \subset & \boxed{\begin{array}{c}\mathbb{Z}\\ \text{integer}\end{array}} & \subset & \boxed{\begin{array}{c}\mathbb{Q}\\ \text{rational}\end{array}} & \subset & \boxed{\begin{array}{c}\mathbb{A}\\ \text{algebraic}\end{array}} & \subset & \ldots\\
 & & & & & & & & \cap & & \\
 & & & & & & & & \boxed{\begin{array}{c}\mathbb{A}+i\mathbb{A}\\ \text{complete}\\ \text{algebraic}\end{array}} & \subset & \ldots
\end{array}$$

then proceed to numbers and operations in overcountable sets, and, finally, extend abelian structures (first column) to nonabelian ones:

[2]Henri Poincaré (1854–1912).

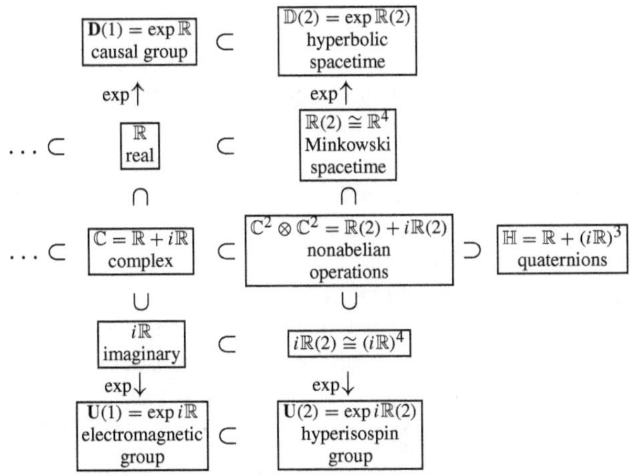

The symbols used will be explained ahead.

3.1 Numbers for Objects and Time

Since, from an early age on, we are so familiar with numbers, their construction procedures seem far-fetched and artificial. According to Kronecker[3] the natural numbers (sometimes defined without 0):

$$\mathbb{N} = \{0, 1, 2, 3, \ldots\},$$

are created by God, according to Dedekind[4] by the human mind. It is remarkable that the natural numbers carry two physically very important abstractions - the basic classical experience of separable *objects*

$$|,\ ||,\ |||,\ ||||,\ \text{etc.},$$

and, in the counting procedure, some concept of *time, orderly flowing into the future*. The time connection of the natural numbers is concretized by an hour-glass with the running sand grains showing the time flow. It is a happy coincidence that an hour-glass is formed like a spacetime-bicone where the grains come from the future cone, are forced through the small door of the presence, and end up in the wide sea of the past.

Starting with Cantor[5] the mathematicians have extended the concept of *cardinality*, familiar from the finite sets:

---

[3] Leopold Kronecker (1823–1891).

[4] Richard Dedekind (1831–1916).

[5] Georg Cantor (1845–1918).

$$\text{card}\,| = 1, \quad \text{card}\,|| = 2, \quad \text{card}\,||| = 3, \quad \text{card}\,|||| = 4, \quad \text{etc.}$$

Frege[6] and Russell[7] have taught us that, in contrast to Kronecker, the natural numbers are not something like "God-given wooden figures" (Where can I grasp you wooden figure 10^{23}?), but that they formalize already a deep abstraction. If properties are characterized by sets and classes, cardinality as a property stands for a class of equipotent (bijective) sets. For example, the number 3 is the property of the class with all sets which are bijective with $|\,|\,|$. That the classes for the natural numbers are not empty, is related to the possibility to form classes with the empty set $\emptyset$. Logicians in scholasticism had the catchphrase "ex falso quodlibet" — more prosaic in a modern set-oriented language: the empty set is a subset of any set.

Good to know: An *equivalence relation* $\sim$ for a set M is a subset $E \subseteq M \times M$ of the product set with $(x, y) \in E$ expressed by $x \sim y$: It has to be *reflexive* ($x \sim x$), *transitive* ($x \sim y$ and $y \sim z \Rightarrow x \sim z$), and *symmetric* ($x \sim y \iff y \sim x$). The subsets $[x] \subseteq M$ with equivalent elements decompose the full set M into disjoint subsets with the surjective projection $M \longrightarrow M/\sim, \quad x \longmapsto [x]$. A system $B \subseteq M$ of representatives $M = \biguplus_{x_r \in B} [x_r]$ is called an M-*base* with respect to the equivalence relation $\sim$. In general, a base is not unique.

There are several definitions of the natural numbers. Von Neumann[8] has defined the natural numbers, including zero, by using the bracing procedure $\{a\}$ for a set a. The braced $\{a\}$ differs from the unbraced a in the same way as a company, consisting of only one person, is conceptually different from this person. One starts with $\emptyset$, braces it, etc., by defining the successor of n by $n \cup \{n\}$:

$$0 = \emptyset, \quad n + 1 = n \cup \{n\} \Rightarrow \begin{cases} 1 = 0 \cup \{0\} = \{\emptyset\} = \{0\}, \\ 2 = 1 \cup \{1\} = \{\emptyset, \{\emptyset\}\} = \{0, 1\}, \\ 3 = 2 \cup \{2\} = \{\emptyset, \{\emptyset\}, \{\emptyset, \{\emptyset\}\}\} = \{0, 1, 2\}, \\ \ldots \end{cases}$$

I am never sure how much circularity is in these definitions with the multiple bracing $\{\ldots\}$ (once), then $\{\{\ldots\}\}$ (twice), etc.

Distinguished natural numbers are *prime numbers* $\mathbb{P}$, defined by the property to have exactly two divisors in $\mathbb{N}$, namely 1 and themselves, e.g., $7 = 1 \cdot 7$, $673 = 1 \cdot 673$ etc., 1 is no prime.

Good to know: To check with the sieve of Eratosthenes[9] the prime property of a natural $n \in \mathbb{N}$, one has to try as divisors the primes $2 = p_1 < p_2 \ldots < p_k$ up to p_k with $p_k^2 \geq n$, e.g., 2, 3, 5, 7, 11, 13, 17, 19 and 23 suffice for 673. Euclid's[10] remark for the proof of infinitely many primes, that with the product $n_k = p_1 \cdots p_k + 1$ of all primes up to p_k one obtains a larger prime $p > p_k$, can also lead to a factor of n_k, e.g., $p = 59$ in the smallest example $2 \cdot 3 \cdot 5 \cdot 7 \cdot 11 \cdot 13 + 1 = 30031 = 59 \cdot 509$. Are there infinitely many prime twins ($p_1, p_2 = p_1 + 2$)? One doesn't know. With the triplet definition $(p, p+2, p+4)$, there is only one prime triplet, (3, 5, 7) — that's easy to show.

---

[6]Gottlob Frege (1848–1925).

[7]Bertrand Russell (1872–1970).

[8]John (Janosz Baron) von Neumann (1903–1957).

[9]Eratosthenes of Cyrene -(276–195).

[10]Euclid of Alexandria, around -(365–275).

Euler[11] has shown how the naturals grow out of the primes. Going to larger and larger natural numbers $n \in \mathbb{N}$, the number of primes $p(n)$ smaller than n goes with $\frac{n}{\log n}$, i.e., their "density" is thinned out "more and more" as given by the inverse logarithm:

$$\text{for } n \to \infty: \quad \frac{p(n)}{n} \to \frac{1}{\log n}.$$

If primes $\mathbb{P}$ are called *simple*, then natural numbers, which are simple powers of primes, may be called *semisimple*, e.g., 13, $30 = 2 \cdot 3 \cdot 5$ or $4711 = 7 \cdot 673$, and powers of only one prime *monogeneous*, e.g., 13 or $81 = 3^4$. Each number has an — up to order — unique decomposition into monogeneous numbers:

$$\mathbb{N} \in n \geq 2: \quad n = p_1^{n_1} \cdots p_k^{n_k}, \quad p_i \in \mathbb{P}, \quad n_i \in \mathbb{N},$$

e.g., $86{,}400 = 2^7 \cdot 3^3 \cdot 5^2$. Natural numbers give the picture of a very complicated "tree," growing from 1 in the natural and divisibility order. Each prime 2, 3, 5, ..., gives rise to a new branch, all branches become multiply connected with each other, e.g., $12 = 2^2 \cdot 3$ grows from 2 and 3.

Number theory is a very fruitful "concrete" field for the abstraction of mathematical structures. There are even very elementarily sounding open problems, e.g., Goldbach's[12] conjecture, that each even natural, starting with $4 = 2 + 2$, can be written as the sum of two primes, $\mathbb{P} + \mathbb{P}$, e.g., $6 = 3 + 3, 8 = 3 + 5, 10 = 5 + 5 = 3 + 7$, etc.

Good to know: An example for the fruitfulness: The nonconstant complex polynomials $p_n(X) \in \mathbb{C}[X]$ can be decomposed into factors of degree 1 with roots $\alpha_j \in \mathbb{C}$, $p_n(X) = (X - \alpha_1) \cdots (X - \alpha_n)$. One has the *simple* polynomials $X - \alpha$, $\alpha \in \mathbb{C}$, the *semisimple* ones $(X - \alpha_1) \cdots (X - \alpha_k)$ with different roots $\alpha_n \in \mathbb{C}$, then *monogeneous* polynomials $(X - \alpha)^n$, $n \in \mathbb{N}$. Each nonconstant polynomial with highest power X^n has an — up to order - unique decomposition into monogeneous polynomials. The commutative ring $\mathbb{C}[X]$ with the complex polynomials has only principal ideals.
The irreducible real polynomials in $\mathbb{R}[X]$ are either of degree one $X - \alpha$, $\alpha \in \mathbb{R}$, or of degree two $(X - \beta)^2 + \gamma^2$ with $\beta, \gamma \in \mathbb{R}$, $\gamma \neq 0$.
A field K, whose nonconstant polynomials $p_n(X) \in K[X]$ can be decomposed into factors of degree 1 with roots $\alpha_j \in K$, is called *algebraically closed*.
Any complex $(n \times n)$-matrix f has a minimal polynomial with the matrix property $p_n(f) = 0$, in general different from the characteristic polynomial $\det(f - X\mathbf{1}_n) = 0$. The concept (semi)simple, monogeneous, etc. can be taken over for the matrices.

All the extensions of the natural numbers $\mathbb{N}$ come with denigrating names — the integers $\mathbb{Z}$ include negatives, the rationals $\mathbb{Q}$ include broken numbers, the reals $\mathbb{R}$ include irrationals and transcendentals, and the complex numbers $\mathbb{C} = \mathbb{R} + i\mathbb{R}$ include imaginaries. However, after using them long enough and after satisfactory definitions they have lost their historical stains.

Monetary transactions contributed much to a familiarization of the number extensions. With an interest $q = p\%$, "continuously" given, bankers use also irrationals, $\lim_{n\to\infty}(1 + \frac{q}{n})^n = \exp q$. It is remarkable — and, in some cases, also depressing — that, via an almost universal valuation by money, most "things," material ones (apples, etc.) and, also, immaterial ones (music, poems, etc.), are related to numbers.

---

[11] Leonhard Euler (1707–1783).

[12] Christian Goldbach (1690–1764).

3.2 Numbers as Operations — From the Naturals to the Rationals

The natural numbers are totally ordered and closed under addition and multiplication. Both operations have no inversion in $\mathbb{N}$. This defect of restricted operations is removed by $\mathbb{N}$-extensions.

The natural numbers can be generated additively by the unit 1. That natural numbers can be added (*additive semigroup*), allows to embed them in these additions, e.g., $3 \leftrightarrow (4, 7)$ or $3 \leftrightarrow (16, 19)$. In this way, each natural can be associated with a *number operation between two sets with naturals*, e.g., $3 \cong [(m, m + 3)]$, $m \in \mathbb{N}$.

Good to know: Binary relations are ordered pairs, i.e., subsets of product sets, $(m, n) \in B \subseteq M \times N$. Mappings $f : M \longrightarrow N$ are special binary relations $\{(m, f(m)) \mid m \in M\} \subseteq M \times N$; they have to be — this is included in the definition – surjective in the first place (for all $m \in N$) and injective in the second place (there is one $f(m) \in N$). For mapping values the notations $m \longmapsto f(m)$, $m \longmapsto f.m$, or $m \longmapsto f\,m$ are used. A mapping — this has to be explicitly given — is called *surjective* (in the second place) for $f(M) = N$, and *injective* (in the first place) if $f(m) = f(n)$ always entails $m = n$.

Embedding $\mathbb{N}$ in such a way in its *binary relations* $\mathbb{N} \times \mathbb{N}$, it is possible to consider the opposite relation with the *exchanged pairs*, e.g., $(7, 4)$ or, equivalent, $(19, 16)$, i.e., the class with all number pairs $[(m + 3, m)]$, $m \in \mathbb{N}$. They represent the "opposite" operation of the addition, called subtraction. A negative checking account asks for an operation — to pay off the debts.

We are used to the embedding of $\mathbb{N}$ by $n \cong [(m, m + n)] \subset \mathbb{N} \times \mathbb{N}$ and denote the class as a *negative* number $-n \cong [(m + n, m)]$, e.g., $-3 \cong [(19, 16)] = [(7, 4)]$. The classes of ordered pairs of natural numbers define the *integer numbers*:

$$\mathbb{Z} = \{[(n, m)] \mid n, m \in \mathbb{N}\} = \{\pm n \mid n \in \mathbb{N}\},$$
$$\text{with} \begin{cases} (n_1, m_1) \sim (n_2, m_2) \iff m_1 + n_2 = m_2 + n_1, \\ [(n_1, m_1)] + [(n_2, m_2)] = [(n_1 + n_2, m_1 + m_2)], \\ [(n, n)] = 0, \quad n \cong [(1, n + 1)], \quad -n \cong [(n + 1, 1)]. \end{cases}$$

In such a way, one obtains an additive group with the neutral element 0, the always experienced "now" of physics, never to be grasped.[13]

Good to know: By this universal construction, any additive semigroup with the "cross-out rule," $n + m = r + m \Rightarrow n = r$, is embedded in *its* additive group.

The numbers $\mathbb{Z}$ are interpretable as $\mathbb{N}$-operations. The extension of a set by considering its internal relations (ordered pairs) and its structure compatible embedding in "self-operations" (*adjoint realization*) is a pattern which can be often used. Here, one has the adjoint realization of the natural numbers by equivalence classes of binary relations of natural numbers:

$$\mathbb{N} \longrightarrow \mathbb{N} \times \mathbb{N}/\text{addition}, \quad n \longmapsto [(1, n + 1)]$$
$$\mathbb{Z} = \mathbb{N} \times \mathbb{N}/\text{addition}$$

[13] Augustinus of Hippo (354–430).

In such extensions by relations, the extended set may inherit some properties of the basic set. In addition to the embedding, a prescription (projection) has to be given how to recognize the basic set in the extended one.

The original abstraction of an object number by the natural numbers $\mathbb{N}$ can be refound in its binary relations $\mathbb{Z}$ (time steps) via the *absolute value* which defines *positive* integers. The integers $\mathbb{Z}$ inherit from $\mathbb{N}$ the natural order and, also, a multiplication compatible norm with triangle inequality:

$$| \ | : \mathbb{Z} \longrightarrow \mathbb{Z}_+, \ |z| = \begin{cases} z, \ z \geq 0 \iff z \in \mathbb{N} = \mathbb{Z}_+ \\ -z, \ -z \geq 0 \end{cases}$$
$$\Rightarrow |z_1 + z_2| \leq |z_1| + |z_2|, \ |z_1 z_2| = |z_1||z_2|, \ |z| > 0 \iff z \neq 0$$

Good to know: An *order relation* $\succeq$ for a set M is a subset of $M \times M$: It has to be reflexive ($x \succeq x$), transitive ($x \succeq y$ and $y \succeq z \Rightarrow x \succeq z$) as well as antisymmetric ($x \succeq y$ and $y \succeq x \Rightarrow x = y$). Each element has its past, its "behind," $[x]_\succeq = \{y \in M \mid x \succeq y\}$. For $\succeq$, there is the contra-order $x \preceq y \iff y \succeq x$ and the future, the "before," for each element $[x]_\preceq$. In an ordered additive group ($x \succeq y \Rightarrow x + z \succeq y + z$), the order can be defined with respect to the neutral element 0: $[0]_\preceq, [0]_\succeq$ are the future and the past of the presence or *positive* and *negative* elements. A *semi-order* $\succeq$ is only reflexive and transitive. One can go over to an order of the equivalence classes $[x] = \{y \mid y \succeq x$ and $x \succeq y\}$. An order with $x \succeq y$ or $y \succeq x$ for all pairs (x, y) is called *total*.
It is remarkable that also the *logical conclusion* $A \Rightarrow B$ (A involves B) is expressed by an order relation $\Rightarrow$. This hints to a connection between *causality* and *time order*. The logical *propter* is connected with the temporal *post*. An ordered set $(M, \succeq)$ can be called *causal*.

The natural numbers $\mathbb{N}$ have also a *multiplication structure* as a shorthand notation for several equal additions:

$$n \cdot m = \underbrace{m + \ldots + m}_{n\text{-times}},$$

with the unit 1 as neutral element and the prime numbers as generators.

The multiplication in $\mathbb{N}$, distributive with respect to the addition, is extended to the integers $\mathbb{Z}$ by $[(n, m)][(k, l)] = [(nk + ml, nl + mk)]$: $\mathbb{Z}$ is a commutative ring with 1. As a further property one has the freedom of zero divisors, i.e., $z_1 z_2 = 0 \iff z_1 = 0$ or $z_2 = 0$. There is a multiplication inverse only for ± 1.

The integers contain an ideal $n\mathbb{Z}$ for each natural number $n \in \mathbb{N}$ with equal cardinality as $\mathbb{Z}$, e.g., the six- or nine-ideal:

$$6\mathbb{Z} = \{0, \pm 6, \pm 12, \pm 18 \ldots\}, \ 9\mathbb{Z} = \{0, \pm 9, \pm 18, \pm 27 \ldots\}.$$

Good to know: An additive group with a distributive associative multiplication is called a *ring*. It is *unital* (involves associative) with a multiplicative unit $1_R \in R$. In a ring R, an *ideal* (invariant subring) I is an additive subgroup, stable under left- and right-multiplication with respect to R, i.e., $RI \cup IR \subseteq I$. In a commutative ring, aR with $a \in R$ is called a *principal ideal*.

All ideals of $\mathbb{Z}$ are principal ideals. Therefore, one can formulate divisor properties in terms of ideals $n\mathbb{Z}$, e.g., the least common multiple (LCM) and the greatest common divisor (GCD):

$$
\begin{aligned}
n|m, \text{ i.e., } n \text{ divides } m &\iff n\mathbb{Z} \supseteq m\mathbb{Z}, \text{ e.g., } 3\mathbb{Z} \supset 6\mathbb{Z},\\
\mathrm{LCM}(n,m)\mathbb{Z} &= n\mathbb{Z} \cap m\mathbb{Z}, \text{ e.g., } 18\mathbb{Z} = 6\mathbb{Z} \cap 9\mathbb{Z},\\
\mathrm{GCD}(n,m)\mathbb{Z} &= n\mathbb{Z} + m\mathbb{Z}, \text{ e.g., } 3\mathbb{Z} = 6\mathbb{Z} + 9\mathbb{Z},\\
\mathrm{GCD}(n,m) = 1 &\iff n\mathbb{Z} + m\mathbb{Z} = \mathbb{Z}, \text{ i.e., } n \text{ and } m \text{ are } \mathit{coprime},\\
&\text{e.g., } 10\mathbb{Z} + 21\mathbb{Z} = \mathbb{Z}.
\end{aligned}
$$

The primes $p \in \mathbb{P}$ generate the maximal proper ideals $p\mathbb{Z}$ of $\mathbb{Z}$, i.e., the extrema in the ordered set of all ideals.

Taking the naturals as an "actual infinity," it comes as a surprise that, e.g., the even naturals $2\mathbb{N}$ are bijective with all naturals $n \leftrightarrow 2n$. With Cantor an infinite set can be characterized as a set which is equipotent with a subset proper, here $k\mathbb{N} \cong \mathbb{N}$ and $k\mathbb{N} \subset \mathbb{N}$ for $2 \leq k \in \mathbb{N}$. The fact that the two properties "being a subset proper" and "having equal cardinality" are not mutually exclusive seems strange and mysterious like the concept "infinity": A set S has infinite cardinality if, and only if, it has a proper subset U with equal cardinality, i.e., with a bijective mapping $S \leftrightarrow U$.

Identifying integer numbers, which differ by a multiple of the integer $n \in \mathbb{N}$, as done, for example, in measuring time in the 24 h for one day, one obtains the *finite* quotient ring $\mathbb{Z}_n$ with the cardinality n (calculating modulo n):

$$
\mathbb{Z}_n = \mathbb{Z}/n\mathbb{Z} = \{[0], [1], \ldots, [n-1]\},
$$

$$
\text{e.g., in } \mathbb{Z}_9 = \{[0], [1], \ldots, [8]\} : \begin{cases} [1] = [10] = [19] = [28] = [1983421],\\ [-2] = [7] = [-11],\\ [4] + [7] = [2], \quad [-5] + [2] = [6],\\ [8] \cdot [2] = [7]. \end{cases}
$$

Everybody knows with $\mathbb{Z}_{24}$, that 13 h after 18 o'clock it is 7 o'clock. A representative of $\mathbb{Z}_9$ in the decimal system (in general, a representative of $\mathbb{Z}_{n-1}$ in the n-system) can be found by the iterated crossfoot (sum of the digits, "casting out nines") as surjective projection $\mathbb{Z} \longrightarrow \mathbb{Z}_9$, e.g., crossfoot(1983421) = crossfoot(28) = crossfoot(10) = $[1] \in \mathbb{Z}_9$.

Good to know: In the third year of elementary school, I learned to calculate modulo 9 (casting out nines) and the crossfoot check for addition, subtraction and multiplication of integers as a necessary, but not sufficient condition for the correctness, e.g.: $48 + 21 = 69$ with the crossfoot check $[3] + [3] = [6]$ and $48 - 21 = 27$ with $[3] - [3] = [0]$ and $48 \times 21 = 1028$ with $[3] \times [3] = [0] \neq [1028] = [2]$—here, there is a mistake! Analogously, any n allows the $\mathbb{Z}_n$-check, e.g., the $\mathbb{Z}_2$-check with even-odd.
I learned to apply the crossfoot check. In hindsight, I am not sure, if my elementary school teacher really understood the mathematical structure with the ideals involved — here the ring morphism $\mathbb{Z} \longrightarrow \mathbb{Z}/9\mathbb{Z} = \mathbb{Z}_9$.
In general: With an ideal I of a ring R, the projection $R \longrightarrow R/I$ on the classes in the quotient ring is a ring morphism.

The finite ring $\mathbb{Z}_n$ is a field, i.e., $\mathbb{Z}_n \setminus \{0\}$ is a multiplicative group, precisely if n is a prime, for example:

$$\text{in } \mathbb{Z}_7: \quad [1] \times [1] = [2] \times [4] = [3] \times [5] = [6] \times [6] = [1],$$
$$\text{in } \mathbb{Z}_{11}: \quad [1] \times [1] = [2] \times [6] = [3] \times [4] = [5] \times [9] = [7] \times [8] = [10] \times [10] = [1].$$

In $\mathbb{Z}_4$ one has $[2] \times [2] = [0]$.

$\mathbb{Z}$ itself is no field. Exactly as for the construction of $\mathbb{Z}$ above by extending the addition, one can construct the *quotient field* $\mathbb{Q}$ with the *broken numbers* by binary relations of $\mathbb{Z}$ (operation with $\mathbb{Z}$ on $\mathbb{Z}$). The division arises as inverse multiplication.

Good to know: Each commutative ring with unit and zero-divisor-freedom, $zu = 0 \Rightarrow z = 0$ or $u = 0$, is embedded, as integrity set, by this universal construction in *its* field.

The horizontal line for $[(z; 0)]$ is omitted — no division by 0:

$$\mathbb{Q} = \{[(z; u)] = \frac{z}{u} \mid z, u \in \mathbb{Z},\ u \neq 0\},$$
$$\text{with} \begin{cases} (z_1; u_1) \sim (z_2; u_2) \iff z_1u_2 = z_2u_1, \\ [(z_1; u_1)][(z_2; u_2)] = [(z_1z_2; u_1u_2)], \\ [(z_1; u_1)] + [(z_2; u_2)] = [(z_1u_2 + z_2u_1; u_1u_2)], \\ z \cong [(z; 1)],\ \frac{1}{z} = z^{-1} \cong [(1; z)], \end{cases}$$
$$\mathbb{Z} \longrightarrow \mathbb{Q} = \mathbb{Z} \times (\mathbb{Z} \setminus \{0\})/\text{multiplication}, \quad z \longmapsto [(z; 1)].$$

We are used so much to these binary relations and equivalence classes, e.g., $\frac{3}{4} = \frac{15}{20}$, and the adjoint realization of $\mathbb{Z}$ in $\mathbb{Q}$, that the denigrating, but with respect to the operational structure, well chosen attribute "broken" has been replaced by the positively sounding "rational." Life would be difficult, if we would calculate with classes of quadruples, e.g., $-\frac{3}{4} = [(19, 16; 21, 25)]$, as quartic relations of natural numbers in order to avoid all signs for subtractions and divisions. Customary usage and meaningful abbreviations have their good effects here.

With a given unit 1, the broken numbers allow to measure time more precisely. In addition, different measurements can be compared with each other by normalization to 1 or 100, one can introduce percents or arithmetic means.

$\mathbb{Q}$ inherits the ring structure of the integers $\mathbb{Z}$ and, in addition, a natural order structure and a norm by the absolute value:

$$|\ | : \mathbb{Q} \longrightarrow \mathbb{Q}_+, \quad |\frac{z}{u}| = \begin{cases} \frac{z}{u}, & zu \geq 0, \\ -\frac{z}{u}, & zu \leq 0, \end{cases}$$

which leads to the corresponding absolute value topology. With respect to the object concept, positive broken numbers $\mathbb{Q}_+$ can be related to *relative frequencies* (discrete probabilities, e.g., for playing dice).

3.3 Discrete and Continuous — From the Rationals to the Reals

The eleats, starting with Parmenides,[14] had the opinion that, ultimately, change is impossible, in contrast to Heraclitus[15] where change is central for an understanding of nature. In modern physical concepts they may coexist peacefully, Heraclitus as associated with equations of motion and Parmenides with conservation laws: constancy in change.

With the paradoxa of Zenon[16] the Greeks illustrated problems with infinities, e.g., the running contest of Achilles α with the tortoise χ (for $\chi\epsilon\lambda\omega\nu\eta$): α runs with constant velocity $v_\alpha = \frac{L}{T}$ always to the former place of χ which starts in the distance $\chi_0 = L$ with constant velocity $v_\chi = qv_\alpha$, slower than α, i.e., $q < 1$:

$$\text{after } n \text{ "runs,"} \atop n = 1, 2, \ldots : \left\{ \begin{array}{rl} \text{time: } t_n = (1 + q + \ldots + q^{n-1})T & = \frac{1-q^n}{1-q} T, \\ \text{Achilles: } \alpha_n = (1 + q + \ldots + q^{n-1})L & = \frac{1-q^n}{1-q} L, \\ \text{tortoise:} \chi_n = (1 + q + \ldots + q^{n-1} + q^n)L & = \frac{1-q^{n+1}}{1-q} L, \end{array} \right.$$

After a finite number of runs, Achilles stays behind, $\chi_n - \alpha_n = q^n L > 0$. However, he reaches the animal after infinitely many runs in a finite time:

$$t_\infty = \frac{T}{1-q}, \quad \alpha_\infty = \chi_\infty = \frac{L}{1-q}.$$

In this illustration for a distinction of potential and actual infinity, or to understand the continuum with discrete units, the number n of separated runs is a mental construction without physical relevance.

Between two different rational numbers, there is a third different one, and, therefore, infinitely many. With "broken" grains our hour glass can be rendered more and more precise — ad infinitum, at least conceptually. However, it runs never "continuously." To this end, we have to cross the rubicon from the bank "countable" to the bank "overcountable."

So far, all number extensions lead to sets with a cardinality either finite or that of the natural numbers, i.e., countably infinite:

$$\text{card}\, \emptyset = 0, \quad \text{card}\, \mathbb{N} = \text{card}\, \mathbb{Z} = \text{card}\, \mathbb{Q} = \aleph_0 : \left\{ \begin{array}{l} n + \aleph_0 = \aleph_0, \; n \in \mathbb{N}, \\ n\aleph_0 = \aleph_0, \; \aleph_0\aleph_0 = \aleph_0, \\ n < \aleph_0. \end{array} \right.$$

There are elementary concepts, for example the ratio of a diagonal in a square to a side, i.e., the square root $\sqrt{2}$, which can approximated by rational numbers, as

[14] Parmenides of Elea -(520/515–460/455).

[15] Heraclitus of Ephesus, around -(535–475).

[16] Zenon of Elea, around -(490–430).

exactly as you like. However, $\sqrt{2}$ is no quotient of integer numbers. More general: Which equations $\sqrt[k]{\frac{n}{m}} = \frac{p}{q}$ with natural $k \in \{2, 3, \dots\}$ are possible for rational numbers? With nontrivial coprime positive naturals $n, m \in \mathbb{N}$ and nontrivial coprime integers $p, q \in \mathbb{Z}$ one obtains $\frac{n}{m} = \frac{p^k}{q^k}$ wherefrom $n = p^k$ and $m = q^k$, i.e., n and m have to be k-th powers of integers. According to the historical experts, the popular belief that the discovery of this defect of the rational numbers has contributed to the dispersion of the mathematical-religious sect of the Pythagoreans[17] does not apply. Without the introduction of irrational numbers, an appropriate mathematical description of physical events would be immensely difficult, in quantum theory probably impossible.

The minimally formulated rational *approximation operations are the real numbers*: One considers series of broken numbers as the mappings:

$$\mathbb{N} \longrightarrow \mathbb{Q}, \quad n \longmapsto q_n \quad \text{series: } (q_n)_{\mathbb{N}} \in \mathbb{Q}^{\mathbb{N}}.$$

A series is a countably infinite relation for $\mathbb{Q}$, ordered by $\mathbb{N}$. The set of all series $\mathbb{Q}^{\mathbb{N}}$ with cardinality $\aleph_0^{\aleph_0}$ contains the rational numbers $\mathbb{Q}$, embedded by the constant mappings $q_n = q$ for all $n \in \mathbb{N}$, the boring melodies with one sound only. However, there are many different rational series converging to a fixed q. Series are called equivalent, if the difference series has, in the absolute value topology, the limit 0:

$$(q_n)_{\mathbb{N}} \sim (p_n)_{\mathbb{N}} \iff \lim_{n\to\infty} |q_n - p_n| = 0.$$

For example $(n)_{\mathbb{N}}$ and $(n + \frac{1}{n})_{\mathbb{N}}$ are equivalent.

All classes of equivalent series are still too many with respect to a minimal, but also complete approximation extension; there ares series which walk erratically to nowhere, e.g., $((-1)^n n)_{\mathbb{N}}$. An ingenious procedure is asked for, to choose minimally in $\mathbb{Q}^{\mathbb{N}}$ such classes of series, that they contain all limiting procedures with the rationals, and that, with them, there is nothing more left to approximate. Now, the cat bites in its tail: Each series involves, at its end, countably many tail series, where finitely many elements from the "head" are left out:

$$(q_n)_{\mathbb{N}} \in \mathbb{Q}^{\mathbb{N}}, \quad k \in \mathbb{N} \Rightarrow (q_{k+n})_{\mathbb{N}} \in \mathbb{Q}^{\mathbb{N}}: \quad \begin{cases} (q_n)_{\mathbb{N}} \quad = q_1, q_2, q_3, q_4, \dots, \\ (q_{1+n})_{\mathbb{N}} = q_2, q_3, q_4, q_5, \dots, \\ \dots \\ (q_{803+n})_{\mathbb{N}} = q_{803}, q_{804}, q_{805}, \dots \\ \dots \end{cases}$$

One distinguishes "self-equivalent" series in the following sense: A series which is *uniformly* equivalent with all its "tail series" is called a *Cauchy*[18] *series*:

[17]Pythagoras of Samos, around -(580–500).

[18]Augustin Louis Cauchy (1789–1857).

$$(q_n)_\mathbb{N} \text{ Cauchy series: } \begin{cases} \text{for all } p \in \mathbb{Q}, \text{ there is a number } n(p) \in \mathbb{N} \\ \text{with } |q_{k+n} - q_n| < |p| \text{ for al } n \geq n(p) \text{ and } k \in \mathbb{N}. \end{cases}$$

If a paper strip with any rational length $2|p|$ is put with its middle on the $n(p)$-th element of a linearly ordered Cauchy series, then almost all, i.e., up to a finite number, series members are covered by it.

Many Cauchy series have no limiting value in $\mathbb{Q}$, e.g., the decimal numbers $(q_n)_\mathbb{N}$ for $\sqrt{2} = 1,4142\ldots$ with n digits or the finite sums $q_n = \sum_{k=0}^{n} \frac{1}{k!}$. The equivalence classes of Cauchy series are the *real numbers* $\mathbb{R}$ *as Cauchy completion* of the rational numbers $\mathbb{Q}$:

$$\mathbb{R} = \{[(q_n)_{n\in\mathbb{N}}] \mid (q_n)_\mathbb{N} \text{ Cauchy}\}, \quad \begin{cases} [(q_n)_\mathbb{N}] \cdot [(p_n)_\mathbb{N})] = [(q_n p_n)_\mathbb{N}], \\ [(q_n)_\mathbb{N}] + [(p_n)_\mathbb{N})] = [(q_n + p_n)_\mathbb{N}], \end{cases}$$

$$\mathbb{Q} \longrightarrow \mathbb{R} = \mathbb{Q}^\mathbb{N}_{\text{Cauchy}}/\text{absolute value topology}, \quad q \longmapsto [(q_n = q)_\mathbb{N}].$$

Each real number α involves a rational series $\alpha = \lim_{n\to\infty} q_n$. Each irrational number, i.e., from $\mathbb{R} \setminus \mathbb{Q}$, represents necessarily an infinite series of rational numbers. We have got used also to the complicated topological relation, that real numbers formalize infinite approximation procedures, e.g., in the decimal representation:

$$\begin{aligned} \pi &= 4\left(1 - \frac{1}{3} + \frac{1}{5} - \frac{1}{7} + \ldots\right) = 4\sum_{k=0}^{\infty} \frac{(-1)^k}{2k+1} = 3.14159\ldots \\ &= \sum_{k=0}^{\infty} m_k 10^{-k} = [(\pi_n)_\mathbb{N}], \quad m_k \in \mathbb{Z}_+, \\ \pi_n &= \sum_{k=0}^{n} m_k 10^{-k} \in \mathbb{Q}, \quad \pi_1 = 3.1, \quad \pi_2 = 3.14, \ldots, \\ n \geq m &\Rightarrow \pi_n - \pi_m \leq 10^{-m}. \end{aligned}$$

On the side of an "$\mathbb{N}$-hour-glass with running sand" we can put an "$\mathbb{R}$-hour-glass with flowing water"— as far as we are ready to call a water jet "continuous."

Good to know: Any uniform, for example metrical space U can be embedded, by a universal construction, as a dense subspace in *its* Cauchy completion $\overline{U}$ which has a naturally induced uniform (metrical) structure.

The real numbers as an additive group ("translations") are isomorphic, via exponential and logarithm, to the strictly positive real numbers as a multiplicative group ("dilations"), $(\mathbb{R}, +) \cong (\exp \mathbb{R}, \cdot)$ — familiar from the sliderule in the times without pocket calculators:

$$\mathbb{R} \ni \psi = \log d \leftrightarrow d = e^\psi \in \exp \mathbb{R} = \mathbf{D}(1), \quad \begin{cases} e^{\psi_1 + \psi_2} = e^{\psi_1} \cdot e^{\psi_2}, \\ \log d_1 + \log d_2 = \log d_1 d_2. \end{cases}$$

Good to know: The letter e in the definition $e^\psi = \exp\psi = \sum_{k\geq 0}\frac{\psi^k}{k!} = 1+\psi+\frac{\psi^2}{2}+\ldots$ reminds of "exponential" and honors Euler.

Good to know: In higher dimensions, the group isomorphy $(\mathbb{R},+)\cong(\exp\mathbb{R},\cdot)$ survives as manifold isomorphy $\mathbb{R}^s\cong\mathcal{Y}^s$ of the real vector space $\mathbb{R}^s$ (additive group) and the one shell hyperboloid $\mathcal{Y}^s$ (rotated catenary $\mathcal{Y}^2$, future timelike $\mathcal{Y}^3$ in Minkowski spacetime), which is no group for $s\geq 2$. The hyperbolic function e^ψ shows up, for instance, in the hydrogen ground state function e^{-r}, or in the Yukawa potential $\frac{e^{-r}}{r}$.

The real numbers inherit the field structure of $\mathbb{Q}$ and the order and topology via the Cauchy series of the absolute value - original the positive object number $\mathbb{N}$ - as approximation operation for relative frequencies:

$$|\ |:\mathbb{R}\longrightarrow\mathbb{R}_+,\quad |[(q_n)_\mathbb{N}]| = [(|q_n|)_\mathbb{N}].$$

A real number is positive, precisely if it is a real square:

$$\alpha\in\mathbb{R},\ \ \alpha\geq 0 \iff \alpha = |\alpha| = \beta^2,\ \ \beta\in\mathbb{R}.$$

With the equivalent condition for $\alpha\geq 0$, if, and only if $\alpha = 2\beta^2$ is the square of a hypothesis of an isosceles orthogonal triangle, the Pythagorean scandal of the irrational diagonal becomes the definition of the real positivity.

The real numbers allow continuous probabilities, e.g., for the probability to hit a target, or thermodynamic probabilities.

Now, there is an end with the Cauchy extension, since all Cauchy series of real numbers have a real limit, $\mathbb{Q}\subsetneq\overline{\mathbb{Q}}=\mathbb{R}=\overline{\mathbb{R}}$.

Cantor has taught us, to consider carefully the infinite and to distinguish between different infinities. One has for the cardinalities of the continuum:

$$\text{card}\,\mathbb{R} = \aleph_0^{\aleph_0} = \aleph_1 > \aleph_0,$$

where $\mathbb{Q}$ is not only a proper subset of $\mathbb{R}$ - there is also no bijection between both sets. $\aleph_1$ is really larger than $\aleph_0$ (Cantor's diagonal slash). This game can be continued.

Good to know: One has for two sets M, N and the mappings $M^N = \{N\longrightarrow M\}$ the "triangle inequality" and the "multiplication" compatibility for the cardinalities:

$$\text{card}\,M\cup N\leq\text{card}\,M+\text{card}\,N,\quad \text{card}\,M\times N = \text{card}\,M\ \text{card}\,N,\quad \text{card}\,M^N = \text{card}\,M^{\text{card}\,N}.$$

The power set 2^S of a set S has cardinality $2^{\text{card}\,S} > \text{card}\,S$.

The question, connected with Cantor's continuum hypothesis, if there are cardinalities strictly between $\text{card}\,\mathbb{N}=\aleph_0$ and $\text{card}\,\mathbb{R}=\aleph_1$, e.g., $2^{\aleph_0}$ as the cardinality of all $\mathbb{N}$-subsets, was — after Gödel's work — answered by Cohen: In the framework of Zermelo–Fränkel's[19] axioms for set theory (more details are left out here) this well

[19]Ernst Zermelo (1871–1953), Abraham Fränkel (1891–1965), Kurt Gödel (1906–1978), Paul Cohen (1934–2007).

defined question cannot be decided: An extended axiom system can either postulate or forbid a cardinality strictly between $\aleph_0$ and $\aleph_1$.

The *finitistic prejudice* in physics assumes, based on our elementary experimental experiences, that the measurable quantities in physics can be formulated with natural numbers $\mathbb{N}$ only. It allows only for finite operations with the integers, i.e., for rational numbers $\mathbb{Q}$. With respect to operational Lie groups, rational numbers can be related to invariants and eigenvalues of compact operations, exemplified by the half-integer spins. Noncompact operations come, especially for faithful representations in Hilbert spaces, with continuous numbers and infinities. The masses of elementary particles as invariants for the noncompact spacetime translations seem to be in discordance with the finitistic prejudice: Apparently, mass ratios form a discrete set of numbers, but they seem to be "picked" from a continuous spectrum. However, with the rationals dense in the reals $\overline{\mathbb{Q}} = \mathbb{R}$, one can never be sure.

3.4 Constancy in Change — From the Real to the Complex Numbers

It is useful to consider the transition from integer discrete to real continuous from another point of view: The real numbers $\mathbb{R}$ have no continuous, additive subgroup proper. What makes the difference between the real continuum $\mathbb{R}$ and the discrete numbers $\mathbb{Z}$ - what are the integer classes $\mathbb{R}/\mathbb{Z}$ in the reals? $\mathbb{Z}$ is a discrete proper additive $\mathbb{R}$-subgroup and leads to the continuous additive quotient group $\mathbb{R}$ modulo $\mathbb{Z}$. The numbers of the half open unit interval can be taken as representatives:

$$\mathbb{R}/\mathbb{Z} \cong [0, 1[: \quad \begin{cases} [1.2] = [0.2] = [-0.8], \quad [-\frac{1}{3}] = [\frac{2}{3}], \\ [0.6] + [0.7] = [-0.7] \text{ etc.} \end{cases}$$

The interval for the additive group $\mathbb{R}/\mathbb{Z}$, where 0 is identified with 1, then with 2, etc., can be glued together at the ends to yield the unit circle in the binary real relations $\mathbb{R} \times \mathbb{R} = \mathbb{R}^2$:

$$\mathbb{R} \longrightarrow \mathbb{R}/\mathbb{Z} \longrightarrow \mathbb{R} \times \mathbb{R}, \quad t \longmapsto [t] \cong \begin{pmatrix} \cos 2\pi t \\ \sin 2\pi t \end{pmatrix}.$$

With this embedding of the additive $\mathbb{R}/\mathbb{Z}$-group structure, a commutative composition $\diamond$ is explained also in the real plane $\mathbb{R} \times \mathbb{R}$:

$$[t + s] \cong \begin{pmatrix} \cos 2\pi(t+s) \\ \sin 2\pi(t+s) \end{pmatrix} = \begin{pmatrix} \cos 2\pi t \cos 2\pi s - \sin 2\pi t \sin 2\pi \\ \cos 2\pi t \sin 2\pi s + \sin 2\pi t \cos 2\pi \end{pmatrix} \cong [t] \diamond [s].$$

The *cyclic* structure of this quotient group is exemplified, with respect to time, by the circular dial of an analogue watch which reflects the rotation of the earth around itself. All 24 h many things are repeated — not only as seen on the railway time-table. Also the calendar year with the constant return of the seasons is a physical implementation (earth orbit) of $\mathbb{R}/\mathbb{Z}$ in time. The circle is the symbol of the "finite without end" or of "constancy in change."

The composition $\diamond$ of the real pairs, defined above, can be written as commutative multiplication with a symbol i for the class $[\frac{1}{4}]$ of $t = \frac{1}{4}$, i.e., for the real pair $\binom{\cos\frac{\pi}{2}}{\sin\frac{\pi}{2}} = \binom{0}{1}$. This *imaginary unit* i has the square $[\frac{1}{2}]$ or $\binom{-1}{0}$ — because of the minus sign for the cosine of the sum of two angles:

$$\left[\frac{1}{4}\right] \cong \binom{0}{1} \cong i : \begin{cases} & [\frac{1}{4}] \diamond [\frac{1}{4}] = [\frac{1}{2}] \\ \iff & \binom{0}{1} \diamond \binom{0}{1} = \binom{-1}{0} \\ \iff & i^2 = -1. \end{cases}$$

Therefore, there is the representation:

$$[t] \cong \cos 2\pi t + i \sin 2\pi t = e^{i2\pi t} \Rightarrow [t+s] = [t] \diamond [s] \cong e^{i2\pi(t+s)}.$$

Not only the embedded cyclic group $\mathbb{R}/\mathbb{Z}$, but even the full plane $\mathbb{R}^2$ with real number pairs can be equipped with this multiplicative structure which proves distributive with respect to the separate addition of both reals in the pairs (vector addition in $\mathbb{R} \times \mathbb{R}$). Taking over the addition in $\mathbb{R}/\mathbb{Z}$ as multiplication $\diamond$ for $\mathbb{R}^2$, together with the vector addition explains the field structure of the *complex numbers* $\mathbb{C} \cong \mathbb{R}^2$:

$$\mathbb{R} \times \mathbb{R} = \{\binom{q}{p} \mid q, p \in \mathbb{R}\}, \quad \begin{cases} \binom{q}{p} \diamond \binom{r}{l} = \binom{qr - pl}{ql + pr}, \\ \binom{q}{p} + \binom{r}{l} = \binom{q+r}{p+l}, \end{cases}$$

$$\mathbb{C} = \{q + ip \mid q, p \in \mathbb{R}\}, \begin{cases} (q+ip)(r+il) = (qr - pl) + i(ql + pr), \\ (q+ip) + (r+il) = (q+r) + i(p+l). \end{cases}$$

After this adjunction of the solutions $\pm i$ of the polynomial $X^2 + 1$, irreducible (without solution) in $\mathbb{R}$, the complex numbers $\mathbb{C}$ constitute an algebraically closed field. It is remarkable, and not quite obvious, that this introduction of the complex numbers via the embedding of the cyclic group $\mathbb{R}/\mathbb{Z}$ as circle in $\mathbb{R} \times \mathbb{R}$ involves the algebraic closure of $\mathbb{C}$.

With the embedding of the cyclic group $\mathbb{R}/\mathbb{Z}$ in the complex numbers the additive inverse in $\mathbb{R}/\mathbb{Z}$ becomes the multiplicative inverse in $\mathbb{C}$:

$$\mathbb{R}/\mathbb{Z} \longrightarrow \mathbb{R}/\mathbb{Z}, \ [t] \cong e^{i2\pi t} \longmapsto [-t] \cong e^{-i2\pi t}.$$

In an interpretation with time, this inversion is the transition to a "Bavarian clock": In contrast to clocks elsewhere, the hands on a Bavarian clock turn around in the mathematically positive direction.

In the $\mathbb{R}^2$-language, the inversion in the complex representation of $\mathbb{R}/\mathbb{Z}$ by a negative second component can be extended as complex conjugation to all complex numbers, denoted by overlining:

$$\begin{aligned}
\overline{}: \mathbb{R}\times\mathbb{R} \longrightarrow \mathbb{R}\times\mathbb{R}, &\quad \overline{\begin{pmatrix} q \\ p \end{pmatrix}} = \begin{pmatrix} q \\ -p \end{pmatrix},\\
\overline{}: \mathbb{C} \longrightarrow \mathbb{C}, &\quad \overline{q+ip} = q - ip,\\
\text{with}\ &\begin{cases} \overline{\overline{\alpha}} = \alpha, & \text{involutive,}\\ \overline{\alpha+\beta} = \overline{\alpha}+\overline{\beta},\ \ \overline{\alpha\beta} = \overline{\alpha}\,\overline{\beta}, & \text{field compatible.}\end{cases}
\end{aligned}$$

The operation of time reflection is antilinear, i.e., it involves the conjugation. Complex conjugation is a nontrivial involutive field automorphism of the complex numbers. The real numbers have no nontrivial involutive field automorphism.

The canonical conjugation is not unique as an involutive $\alpha \leftrightarrow \alpha^\dagger$ field automorphism, $(\alpha+\beta)^\dagger = \alpha^\dagger + \beta^\dagger$ and $(\alpha\beta)^\dagger = \alpha^\dagger\beta^\dagger$. There exist infinitely many field automorphisms, e.g., one with $\sqrt[4]{2} \leftrightarrow i\sqrt[4]{2}$, which can be proved by general arguments, involving the axiom of choice, but whose explicit form is unknown. The canonical conjugation cannot define the reals in $\mathbb{C}$ as invariants, but it needs the reals for its definition.

The probability of quantum structures relies on the scalar product $\overline{\alpha}\alpha$ which requires the canonical conjugation. The question if quantum theory is real or complex should be answered as "quantum theory is a real theory realized in the complex."

Good to know: Holomorphic functions $\mathbb{C} \ni z \longmapsto f(z) \in \mathbb{C}$ are characterizable by their "ignorance" of the canonical conjugation: From $\frac{\partial f}{\partial \overline{z}} = 0$, there follow, for $f = u + iv$ with $z = x + iy$, the Cauchy–Riemannian[20] differential equations $\frac{\partial u}{\partial x} = \frac{\partial v}{\partial y}$ and $\frac{\partial u}{\partial y} = -\frac{\partial v}{\partial x}$.

Since only for the unit circle $\mathbb{R}/\mathbb{Z}$ in the complex the multiplicative inversion u^{-1} coincides with the conjugation $u^\star = \overline{u}$, the cyclic group is also called the *unitary group* $\mathbf{U}(1)$:

$$\mathbb{R}/\mathbb{Z} = \{[t] \mid t \in [0,1[\} \cong \mathbf{U}(1) = \{u \in \mathbb{C} \mid u^\star = u^{-1}\}.$$

In the complex formulation, the finite additive groups $\mathbb{Z}_n$ come as multiplicative *cyclotomic groups* $\mathbf{I}(n)$ (complex roots of 1) which are finite subgroups of the phase group $\mathbf{U}(1)$:

$$\mathbb{Z}_n \cong \mathbf{I}(n) = \sqrt[n]{1} = \{z \in \mathbb{C} \mid z^n = 1\} = \{e^{i2\pi\frac{k}{n}} \mid k = 0, 1, \ldots, n-1\}.$$

For primes $p \in \mathbb{P}$, $\mathbb{Z}_p$ can be generated by any nontrivial power $e^{i2\pi\frac{k}{p}}$, $k \neq 0$.

After the algebraic extensions (embeddings) $\mathbb{N} \subset \mathbb{Z} \subset \mathbb{Q}$, all countable $\aleph_0$, a topological argument (Cauchy completion) is used for the transition to $\mathbb{R}$ and $\mathbb{C} = \mathbb{R} + i\mathbb{R}$ with larger cardinality $\aleph_1$. The algebraic closure alone does not lead from $\mathbb{N}$ to $\mathbb{C}$: The complex-algebraic numbers $\mathbb{A} + i\mathbb{A}$, i.e., the roots $p(\alpha) = 0$ of integer number polynomials $p(X) \in \mathbb{Z}[X] = \mathbb{Q}[X]$, are the smallest $\mathbb{N}$ containing algebraically closed field, they are countable. The real numbers have the transcen-

[20]Bernhard Riemann (1826–1866).

dental numbers $\mathbb{T} = \mathbb{R} \setminus \mathbb{A}$ in addition to the algebraic ones. Since Euler's number $e = \sum_{k\geq 0} \frac{1}{k!} = 2.71828\ldots$ is transcendental,[21] it is impossible to define in $\mathbb{A}$ exponentials $\exp \mathbb{A}$ and logarithms $\log \mathbb{A}$. Also the embedding of the quotient group $\mathbb{A}/\mathbb{Z}$ as a "circle" $e^{i\mathbb{A}}$ in $\mathbb{A} + i\mathbb{A}$ is not possible.

Exponentiation describes the transition from "infinitesimal" to "finite," from differential equations to their solutions, from Lie algebras to Lie groups with a possible transition from "linear" to "nonlinear."

Good to know: Since $e \in \mathbb{T}$ and since complex algebraic powers of complex transcendental numbers are complex transcendental,[22] $(\mathbb{T} \oplus i\mathbb{T})^{\mathbb{A} \oplus i\mathbb{A}} = \mathbb{T} \oplus i\mathbb{T}$, π has to be transcendental because of $e^{2\pi i} = 1$. The circle cannot be squared.

The positive real cone $\mathbb{R}_+$ defines a partial *order* and the positivity for the complex numbers:

$$\begin{aligned} \alpha \in \mathbb{C}: \ \alpha \geq 0 &\iff \alpha = \overline{\delta}\delta, \ \delta \in \mathbb{C} \iff \alpha \in \mathbb{R}_+, \\ \beta, \gamma \in \mathbb{C}: \ \beta \geq \gamma &\iff \beta - \gamma \in \mathbb{R}_+, \\ \text{positive cone } \mathbb{R}_+ = |\mathbb{C}|: \ & \alpha, \beta \in \mathbb{R}_+ \Rightarrow \alpha + \beta, \alpha\beta \in \mathbb{R}_+. \end{aligned}$$

The real numbers are totally ordered, not the complex ones.

The absolute value for all the extensions

$$\mathbb{N} \subset \mathbb{Z} \subset \mathbb{Q} \subset \mathbb{R} \subset \mathbb{C},$$

defines their natural topology. Starting with the identity $|n| = n \in \mathbb{N}$, it is given, in general, by the *square root of the real square*:

$$| \ | : \mathbb{C} \longrightarrow \mathbb{R}_+, \ \ |\alpha| = \sqrt{\overline{\alpha}\alpha}, \ \ |\alpha + \beta| \leq |\alpha| + |\beta|, \ \ |\alpha\beta| = |\alpha||\beta|.$$

Good to know: Subsets $U \subseteq T$ of a topological space T inherit the *initial topology* by the continuity requirement of the injection $U \longrightarrow T$: $O_U \subseteq U$ is open $\iff O_U = O \cap U$ with O open in T.

Therefore, the norm, originating from the object number $\mathbb{N}$, can be formulated via the diagonal elements of a *scalar product* for the complex numbers, which will be of paramount importance for probability amplitudes in quantum theory:

$$\langle \ | \ \rangle : \mathbb{C} \times \mathbb{C} \longrightarrow \mathbb{C}, \quad \begin{cases} \langle \alpha | \beta \rangle = \overline{\alpha}\beta, \\ \langle \alpha | \alpha \rangle = |\alpha|^2. \end{cases}$$

The complex numbers are *polarizable* (polar coordinates), i.e., its multiplicative group $\mathbb{C}^* = \mathbb{C} \setminus \{0\} = \exp \mathbb{C} \cong \mathbf{GL}(1, \mathbb{C})$ is uniquely decomposable $e^z = e^x \cdot e^{iy}$ into the totally ordered group with the real exponentials $|\mathbb{C}^*| = \exp \mathbb{R} \cong \mathbf{D}(1)$, the "smallest" noncompact real Lie group, and the phase group with the imaginary

---

[21] Charles Hermite (1822–1901).

[22] Ferdinand Lindemann (1852–1939).

exponentials $\frac{\mathbb{C}^*}{|\mathbb{C}^*|} = \exp i\mathbb{R} \cong \mathbf{U}(1)$, the "smallest" compact real Lie group:

$$\mathbf{GL}(1, \mathbb{C}) = \mathbf{D}(1) \times \mathbf{U}(1).$$

The complex numbers are the minimal field with the natural numbers $\mathbb{N}$ which has a nontrivial field involution and which is polarizable.

With the extensions $\mathbb{N} \subset \mathbb{R} \subset \mathbb{C}$ go the extensions of the object-characterization by number, relative frequencies (probability) and quantum theoretical probability amplitudes.

3.5 Dynamics as Time Action

Since now the extension structures become much more complicated, we look back: After the familiar way from the natural numbers $\mathbb{N}$ as abstraction of the object concept to the real numbers $\mathbb{R}$ for multiple relations between objects, involving especially the time concept, the quotient group $[t] \in \mathbb{R}/\mathbb{Z}$ was represented by the unitary group $e^{2\pi it} \in \mathbf{U}(1)$ in the multiplicative group of the complex numbers $\mathbf{GL}(1, \mathbb{C}) \cong \mathbb{C} \setminus \{0\}$. With the object concept in $\mathbb{N}$ extended by the absolute value in $\mathbb{Q}_+$ and $\mathbb{R}_+$, e.g., normalized by relative frequencies and probabilities, one has, with the scalar product of the complex numbers, the tool for a formalization by complex probability amplitudes, as used in quantum theory.

The unitary group $\mathbf{U}(1)$ can be seen as defining binary complex relations $\mathbb{C} \times \mathbb{C}$, i.e., the numbers $\mathbf{U}(1)$ operate on the complex plane $\mathbb{C} \cong \mathbb{R} \times \mathbb{R}$ — exactly: as automorphisms of $\mathbb{C}$ via multiplication by $e^{i2\pi t}$:

$$D_1(t) : \mathbb{C} \longrightarrow \mathbb{C}, \quad D_1(t)(\mathrm{u}) = e^{i2\pi t}\mathrm{u} = \mathrm{u}(t), \quad D_1(t) \in \mathbf{U}(1) \subset \mathbf{GL}(1, \mathbb{C}).$$

Good to know: A *vector space* V over the scalars $\mathbb{R}$ or $\mathbb{C}$ is an additive group, closed under distributive scalar multiplication. The *morphisms* (structure compatible mappings) of a vector space V to a vector space W are the linear mappings. Bijective morphisms are called *isomorphisms*. *Endomorphisms* $\mathbf{AL}(V)$, a unital algebra, require $V = W$, the *automorphisms* $\mathbf{GL}(V)$ (general linear group) are the bijective endomorphisms.

In the real formulation, i.e., by binary $\mathbb{R}^2$-relations (quartic $\mathbb{R}$-relations), more exactly by automorphisms $\mathbb{R}^2 \longrightarrow \mathbb{R}^2$, these operations can be expressed by real (2×2)-matrices. They describe an axial rotation $\mathbf{SO}(2)$ of the vectors in the plane, parametrized by time from the unit interval $t \in]0, 1]$. Physically, this situation is realized by the time dependence of a harmonic oscillator (spring oscillations) with position and momentum as the two components in phase space vectors $\begin{pmatrix} x \\ p \end{pmatrix} \in \mathbb{R}^2$. The period of this *compact motion* gives the intrinsic time unit:

$$D_1(t) : \mathbb{R}^2 \longrightarrow \mathbb{R}^2, \quad D_1(t)\begin{pmatrix} x \\ p \end{pmatrix} = \begin{pmatrix} \cos 2\pi t & \sin 2\pi t \\ -\sin 2\pi t & \cos 2\pi t \end{pmatrix}\begin{pmatrix} x \\ p \end{pmatrix} = \begin{pmatrix} x(t) \\ p(t) \end{pmatrix},$$
$$D_1(t) \in \mathbf{SO}(2) \subset \mathbf{GL}(2, \mathbb{R}).$$

Good to know: The mass and the spring constant $m_0, g_0 > 0$ in the equations of motions $m_0 \frac{d\underline{x}}{d\underline{t}} = \underline{p}$ and $\frac{d\underline{p}}{d\underline{t}} = -g_0\underline{x}$ can be absorbed as intrinsic units $\underline{t} = 2\pi\sqrt{\frac{m_0}{g_0}}\,t$ and $\underline{x} = l_0 q$, $\underline{p} = \pi_0 p$ with $\frac{\pi_0}{l_0} = \sqrt{g_0 m_0}$ into dimensionless variables with $\frac{dx}{dt} = 2\pi p$ and $\frac{dp}{dt} = -2\pi x$. In a quantum theory all units are fixed by $l_0\pi_0 = \hbar$.

The harmonic oscillator gives a compact cyclic, i.e., not faithful, representation of the additive group time $\mathbb{R}$ by the multiplicative unitary group $\mathbf{U}(1)$. The derivative at time $t = 0$ is, up to the imaginary unit, the *angular velocity (frequency, Hamiltonian)*:

$$\frac{d}{dt}D_1(t)|_{t=0} = D_1(-t)\frac{d}{dt}D_1(t) = \mathcal{D}_1 = \begin{cases} 2\pi\begin{pmatrix} 0 & 1 \\ -1 & 0 \end{pmatrix} \text{ in } \mathbb{R}^2, \\ iH = 2\pi i \text{ in } \mathbb{C}. \end{cases}$$

The linear action of the Hamiltonian on the phase space $\mathbb{R}^2$ or on the complex numbers $\mathbb{C}$ with $\mathcal{D}_1 \cong \frac{d}{dt}$ gives the *equations of motions*, in the real and complex representation:

$$\mathcal{D}_1 : \mathbb{R}^2 \longrightarrow \mathbb{R}^2, \qquad \begin{pmatrix} x(t) \\ p(t) \end{pmatrix} \longmapsto \mathcal{D}_1\begin{pmatrix} x(t) \\ p(t) \end{pmatrix} = 2\pi\begin{pmatrix} 0 & 1 \\ -1 & 0 \end{pmatrix}\begin{pmatrix} x(t) \\ p(t) \end{pmatrix} = \frac{d}{dt}\begin{pmatrix} x(t) \\ p(t) \end{pmatrix},$$

$$\text{i.e.,}\ \frac{dx(t)}{dt} = 2\pi p(t),\quad \frac{dp(t)}{dt} = -2\pi x(t),$$

$$\mathcal{D}_1 : \mathbb{C} \longrightarrow \mathbb{C}, \qquad \mathrm{u}(t) \longmapsto \mathcal{D}_1\mathrm{u}(t) = 2\pi i\mathrm{u}(t) = \frac{d\mathrm{u}(t)}{dt}.$$

Good to know: A *representation of the group* G with neutral element e on a vector space V is a group morphism $D : G \longrightarrow \mathbf{GL}(V)$ into its automorphism group with $D(gh) = D(g) \circ D(h)$ and $D(e) = \mathbf{1}_V = \mathrm{id}_V$ (identity). The neutral group element $e \in G$ can be written as 1_G in a multiplicative notation and 0_G in an additive one. A complex linear action is a characteristic feature of quantum theory.

Good to know: The "smallest" real Lie groups are one-dimensional: $\mathbf{D}(1) = \exp\mathbb{R} \cong \mathbb{R}$ and its integer classes $\mathbf{U}(1) = \exp i\mathbb{R} \cong \mathbb{R}/\mathbb{Z}$. These two abelian groups and their representations are of paramount importance. Each dimension in a real Lie group leads to a $\mathbf{D}(1)$- or $\mathbf{U}(1)$-isomorphic subgroup.
For a "sufficiently smooth" function in the representation $\mathbb{R} \ni \psi \longmapsto D(\psi) \in \mathbb{C}$, one obtains from $D(\psi_1 + \psi_2) = D(\psi_1)D(\psi_2)$, by derivation, $D'(\psi) = D(\psi)D'(0)$, and, with the normalization $D(0) = 1$, the representation $D(\psi) = e^{D'(0)\psi}$. The cyclicity $D(\theta + 2\pi) = D(\theta)$ for $\mathbf{U}(1) \ni e^{i\theta} \longmapsto D(\theta)$ restricts the eigenvalues to integers in $D'(0) = iz \in i\mathbb{Z}$.
The irreducible complex representations of noncompact $\mathbf{D}(1)$ and compact $\mathbf{U}(1)$ act on complex one-dimensional spaces: $\mathbf{U}(1) \ni e^{i\theta} \longmapsto e^{zi\theta} \in \mathbf{U}(1)$ with integer winding numbers $z \in \mathbb{Z} = \mathbf{spec}\,\mathbf{U}(1)$ (*discrete spectrum*), and $\mathbf{D}(1) \ni e^{\psi} \longmapsto e^{iQ\psi} \in \mathbf{GL}(\mathbb{C})$ with complex invariants $iQ = iP + B \in \mathbb{C} = \mathbf{spec}\,\mathbf{D}(1)$ (*continuous spectrum*). The irreducible $\mathbf{U}(1)$-representations are unitary whereas the unitary $\mathbf{D}(1)$ representation are characterized by an imaginary invariant $iP \in i\mathbb{R} = \mathbf{spec}_+\mathbf{D}(1)$ used for energy (frequency) or momentum $E, \omega, P \in \mathbb{R}$.
The compact-noncompact transition $\mathbf{U}(1) \leftrightarrow \mathbf{D}(1)$ for the unitary representations goes with $(i\theta, z) \leftrightarrow (\psi, iP)$ for Lie parameters (imaginary-real) and invariants (discrete-continuous).
The representations are characterizable by homogeneous differential equations:

$$\mathbf{U}(1): \quad (\frac{d}{id\theta} - z)D(\theta) = 0, \quad (\frac{d^2}{d\theta^2} + z^2)D(\theta) = 0,\ D(-\theta)\frac{d}{id\theta}D(\theta) = z \in \mathbb{Z},$$

$$\mathbf{D}(1): \quad (\frac{d}{d\psi} - iQ)D(\psi) = 0, \quad (\frac{d^2}{d\psi^2} + Q^2)D(\psi) = 0,\ D(-\psi)\frac{d}{d\psi}D(\psi) = iQ = iP + B \in \mathbb{C}.$$

These differential equations arise in the time translation dependence $(\frac{d^2}{dt^2}+\omega^2)D(t)=0$ of the harmonic oscillator, in the position translation dependence of the Laplace equation $(\frac{\partial^2}{\partial\vec{x}^2}+P^2)D(\vec{x})=0$, and in the spacetime translation dependence of the Dirac equation $(\gamma^j\frac{\partial}{\partial x^j}-im\mathbf{1}_4)\mathbf{D}(x)=0$, and of the Klein–Gordon[23] equation $(\frac{\partial^2}{\partial x^2}+m^2)$ $D(x)=0$.

The irreducible complex Lie algebra action for time in $\mathbf{U}(1)$ corresponds to first order differential equations, here $\mathcal{D}_1\cong\frac{d}{dt}=2\pi i$. For a real representation with $\mathbf{SO}(2)$, two complex representations are combined with, again, first order equations, e.g., for real position and momentum, or second order time differential equations for position in nonrelativistic classical mechanics, here $\frac{d^2}{dt^2}x(t)=-(2\pi)^2x(t)$. The complete equations of motion with a potential V relate to each other a time Lie algebra action $\frac{d}{dt}$ and a position Lie algebra action $\frac{\partial}{\partial\vec{x}}$ in $\frac{d}{dt}\vec{p}(t)=-\frac{\partial}{\partial\vec{x}}V(\vec{x})$. In special relativistic point mechanics, the nonrelativistic second order structure arises as leading term only in an expansion, e.g., for a free hyperbolic mass point:

$$\begin{aligned}-mc\int ds&=-mc\int\sqrt{g^{ik}(x)dx_idx_k}\\&=-mc\int\sqrt{dx_0^2-d\vec{x}^2}=-mc\int dx_0\sqrt{1-\frac{\vec{v}^2}{c^2}}=\int dt\,\frac{m}{2}\vec{v}^2+\dots.\end{aligned}$$

The first order (space)time differential equations for canonical pairs like positions and momenta $(\vec{x},\vec{p})$, electromagnetic potentials and field strengths (A_j,F_{jk}), or conjugated pairs of Dirac fields $(\psi,\overline{\psi})$ reflect the linear Lie algebra action of the (space)time translations.

We look more general at the representations of the additive time group $\mathbb{R}$ as automorphisms of complex vector spaces V:

$$D:\mathbb{R}\longrightarrow\mathbf{GL}(V),\quad\begin{cases}D(t+s)=D(t)\circ D(s),\\ D(0)=\mathbf{1}_V.\end{cases}$$

The $\mathbb{R}$-representation (*time representation or dynamics*) D in the multiplicative automorphism group of such spaces transforms the vectors. Its action defines the *time orbit* $t\longmapsto v(t)$ for each vector $v\in V$ as an image of the represented time group $\mathbb{R}$ in the vector space V, e.g., time orbits $t\longmapsto\vec{x}(t),\vec{p}(t)$ in position and momentum space:

$$D(t):V\longrightarrow V,\quad D(t)(v)=v(t),\quad v(0)=v.$$

In mechanics, conservative time translation generators, i.e., without explicit time dependence, are build by positions and momenta. There are many $\mathbb{R}$-representations on complex spaces, any $(\vec{x},\vec{p})$-dependent Hamiltonian of a dynamics generates one. They can be built by nondecomposable, especially irreducible $\mathbb{R}$-representations. With a theorem of Schur,[24] all irreducible complex representations of the addi-

[23]Oskar Klein (1894–1977), Walter Gordon (1893–1939).

[24]Issai Schur (1885–1941).

tive time group are one-dimensional with the unitary ones given by time-dependent phases:

$$D_1(|\omega) : \mathbb{R} \longrightarrow \mathbf{U}(1) \subset \mathbf{GL}(1, \mathbb{C}), \quad D_1(t|\omega) = e^{i\omega t}.$$

In physical applications the vectors $v \in V$ with unitary time translation representations can be called *objects*. Because of their periodicity, these representations of the time group $\mathbb{R}$ are unfaithful (not injective). They are the windings of the real numbers on a circle (circular dial of an analogue watch) as used already above for the harmonic oscillator.

Good to know: If, for group or Lie algebra representation $D : G \longrightarrow \mathbf{GL}(V)$, and $\mathcal{D} : L \longrightarrow \mathbf{AL}(V)$ there exists an *invariant* proper subspace $\{0\} \neq W \subset V$, i.e., $D(g)[W] \subseteq W$ for all group elements $g \in G$, and $\mathcal{D}(l)[W] \subseteq W$ for all Lie algebra elements $l \in L$, the representation is called *reducible*, if not, *irreducible*. If the representation vector space V has no decomposition $V = W_1 \oplus W_2$ into invariant subspaces proper, D is called *nondecomposable (monogeneous)*, if not, *decomposable*. A representation may be reducible, but nondecomposable.

All nondecomposable $\mathbb{R}$-representations are characterized by a number ω and a *nilpotent* mapping N_n:

$$D_n(|\omega) : \mathbb{R} \longrightarrow \mathbf{GL}(n, \mathbb{C}), \quad D_n(t|\omega) = e^{i(\omega \mathbf{1}_n + N_n)t}, \quad n \in \mathbb{N},$$

$$(N_n)^n = 0 \Rightarrow e^{iN_n t} = \sum_{k=0}^{n-1} \frac{(iN_n t)^k}{k!}.$$

Unitary representations, definite or indefinite, need a real $\omega \in \mathbb{R}$.

Good to know: An element n of a ring with $n^{k-1} \neq 0$ und $n^k = 0$ is called *nilpotent to the power* $k \in \mathbb{N}$, 0 is nilpotent to the power 1.

Since the only nilpotent complex number is $N_1 = 0 \in \mathbb{C}$, nontrivial nilpotent structures start with the matrix $N_2 = \begin{pmatrix} 0 & 1 \\ 0 & 0 \end{pmatrix}$. The related complex representations of the real numbers (time) involve nilpotent structures, and, therefore, are not so familiar. However, they are physically not so exotic as they may look on the first sight.

The "smallest" faithful nondecomposable, but reducible representation of the real numbers (time) operates on a complex two-dimensional space:

$$D_2(|\omega) : \mathbb{R} \longrightarrow \mathbf{GL}(2, \mathbb{C}), \quad D_2(t|\omega) = e^{i\omega t} \begin{pmatrix} 1 & i\nu t \\ 0 & 1 \end{pmatrix}.$$

The number ν has to be nontrivial. It is basis dependent and has no invariant meaning. The nilquadratic contribution $i\nu t e^{i\omega t}$ draws a spiral in the complex plane. These time representations display a subtle symbiosis of the group structure of the additive group $\mathbb{R}$ with the multiplicative group $\mathbf{U}(1)$:

$$D_2(t_1|\omega) \circ D_2(t_2|\omega) = e^{i\omega t_1} e^{i\omega t_2} \begin{pmatrix} 1 & i\nu(t_1 + t_2) \\ 0 & 1 \end{pmatrix} = D_2(t_1 + t_2|\omega).$$

In the special case $\omega = 0$ without unitary circle $\mathbf{U}(1)$-contribution, they are used for the phase space with position and momentum to describe the *noncompact inertial motion* of a free mass point:

$$D_2(t|0)\begin{pmatrix} x \\ p \end{pmatrix} = \begin{pmatrix} 1 & \frac{t}{M} \\ 0 & 1 \end{pmatrix}\begin{pmatrix} x \\ p \end{pmatrix} = \begin{pmatrix} x(t) \\ p(t) \end{pmatrix} \Rightarrow \begin{cases} x(t) = x + \frac{t}{M} p, \\ p(t) = p. \end{cases}$$

Here the spiral degenerates into a line.

Good to know: The two-dimensional representations are fundamental in the following sense: The totally symmetric products $\bigvee^n$ build the representations of any dimension $n \geq 2$:

$$D_{n+1}(t|n\omega) = \bigvee^n D_2(t|\omega), \quad n \in \mathbb{N},$$

$$\text{e.g., } D_3(t|2\omega) = D_2(t|\omega) \vee D_2(t|\omega) = e^{2i\omega t}\begin{pmatrix} 1 & i\nu t & -\frac{(\nu t)^2}{2} \\ 0 & 1 & i\nu t \\ 0 & 0 & 1 \end{pmatrix} \text{ with } N_3 = \begin{pmatrix} 0 & 1 & 0 \\ 0 & 0 & 1 \\ 0 & 0 & 0 \end{pmatrix}.$$

The totally antisymmetric product gives the irreducible $\mathbb{R}$-representations $D_1(t|2\omega) = D_2(t|\omega) \wedge D_2(t|\omega)$.

The derivation of a time $\mathbb{R}$-representation at the neutral element ("now") explains, in a physical terminology, the Hamilton operator, multiplied with the imaginary unit, as a basis of the represented time translations. It leads to the equations of motion for the vectors (objects) in the space V:

$$\begin{aligned} \mathcal{D} = iH &= \frac{d}{dt} D(t)|_{t=0} = D(-t)\frac{d}{dt} D(t), \\ \mathcal{D} : V \longrightarrow V, \ \ \mathcal{D}(v(t)) &= \frac{d}{dt} v(t), \end{aligned}$$

explicitly for the nondecomposable representations:

$$\frac{d}{dt} D_n(t|\omega)|_{t=0} = \mathcal{D}_n(\omega) = i(\omega \mathbf{1}_n + N_n), \text{ e.g., } \begin{cases} \mathcal{D}_1(\omega) = i\omega, \\ \mathcal{D}_2(\omega) = i\begin{pmatrix} \omega & \nu \\ 0 & \omega \end{pmatrix}. \end{cases}$$

For the case $n = 2$ with $\omega = 0$, there arise the equations of motion for the free mass point:

$$\text{with } \omega = 0: \ \frac{d}{dt}\begin{pmatrix} x(t) \\ p(t) \end{pmatrix} = \begin{pmatrix} 0 & \frac{1}{M} \\ 0 & 0 \end{pmatrix}\begin{pmatrix} x(t) \\ p(t) \end{pmatrix} \Rightarrow \begin{cases} \frac{dx}{dt} = \frac{p}{M}, \\ \frac{dp}{dt} = 0. \end{cases}$$

In mechanics, a free motion of a mass point, i.e., with constant momentum, can be absorbed by choosing a rest system as an appropriated reference frame for space. In a field theory for four-dimensional spacetime, the lightlike structures for the massless electromagnetic field, characterized by the maximal action velocity c, cannot be "trivialized" in appropriate reference frames. There is no Lorentz transformation with $v = c$. The corresponding degrees of freedom cannot be transformed away.

The different notations for the smallest simply connected real one-dimensional Lie group can lead to confusion: The additive group $(\mathbb{R}, +)$ is isomorphic — via exponential and logarithm — to the multiplicative group $(\mathbf{D}(1), \cdot)$. The multiplicative notation is used for the dilations. The associated abelian real one-dimensional Lie algebra, of course for "both" groups $\mathbb{R} \cong \mathbf{D}(1) = \exp \mathbb{R}$ with $\log \mathbf{D}(1) = \mathbb{R}$, employs $(\mathbb{R}, +)$ as vector space with addition $x + y$ and real scalar multiplication $x \longmapsto \alpha x$. $\mathbb{R}$ is used for time and position translations. For example the time translations as group have the unitary representations $\mathbb{R} \ni t \longmapsto e^{iEt} \in \mathbf{U}(1)$ with a real energy $E \in \mathbb{R}$, the associated Lie algebra has the associated representations $\mathbb{R} \ni t \longmapsto iEt \in \mathbb{C}$.

The dilation-translation relation shows up in Einstein's equations for gravity: The metrical tensor with its four dilation degrees of freedom for the four principal axes of the metrical hyperboloid is determined by the energy-momentum tensor related to the spacetime translations.

In the classical Lagrange formalism the first order derivative and the Hamiltonian action for the time Lie algebra are equalized:

$$\text{Lagrangian: } L(\vec{x}, \vec{p}) = \vec{p}\frac{d\vec{x}}{dt} - H(\vec{x}, \vec{p}), \quad \begin{cases} \frac{d\vec{x}}{dt} = \frac{\partial H(\vec{x},\vec{p})}{\partial \vec{p}} & = [H(\vec{x}, \vec{p}), \vec{x}]_P, \\ \frac{d\vec{p}}{dt} = -\frac{\partial H(\vec{x},\vec{p})}{\partial \vec{x}} & = [H(\vec{x}, \vec{p}), \vec{p}]_P, \end{cases}$$

$$\text{Hamiltonian: } H(\vec{x}, \vec{p}) = \frac{\vec{p}^2}{2M} + V(\vec{x}) = \begin{cases} \frac{\vec{p}^2}{2M} + k\frac{\vec{x}^2}{2}, & \text{harmonic oscillator,} \\ \frac{\vec{p}^2}{2M} + \frac{\epsilon g_0^2}{|\vec{x}|}, & \text{Kepler potential,} \dots \end{cases}$$

The time derivative, topologically defined by a limiting procedure, can be algebraically expressed by the Poisson bracket $[\ ,\ ]_P$, defined via derivatives with respect to positions and momenta.

Already in classical theories, the Lagrangian and Hamiltonian have an operational character: They describe a general action on a phase space, e.g., formalized by Poisson brackets, which can be "individually" realized for special initial or boundary conditions, e.g., gravitation for Newton's apple or for the planet earth. In quantum theory, the operator action is realized on a Hilbert space. As illustrated by the states for the Hamiltonian with the Kepler potential, the classical bound state structure with elliptic orbits allows initial conditions from a continuous spectrum, the quantum structure determines a discrete spectrum for the Hamiltonian.

In quantum theory, the solution of a dynamics consists in the decomposition of the time representation by the Hamiltonian into non-decomposable complex representations with the energies as time translation invariants. Hamiltonians with potentials as Lie algebra elements can be added, e.g., $H_1 + H_2$ with $V_1 + V_2$, but not multiplied with each other as $V_1 \cdot V_2$. The divergencies in quantum field theories can be related to a pointwise multiplication of interactions.

In basic theories, the Hamiltonian as time translation generator is determined as invariant with respect to other operations, e.g., related to position, like the position translations $\mathbb{R}^3$ for free nonrelativistic scattering, or the unitary group $\mathbf{U}(3)$ for the three-dimensional isotropic quantum oscillator, or the orthogonal groups $\mathbf{O}(4)$ and $\mathbf{O}(1, 3)$ for the nonrelativistic quantum hydrogen atom.

3.6 Modalities

Already Aristoteles[25] thought — extending the "tertium non datur"— about the presentic truth value of the statement "tomorrow, there will be a sea battle." If *modality* is defined as the valuation of propositions in time, classical physics uses the time independent modality structure "false-true," quantified by the two numbers $\{0, 1\} \subset \mathbb{Z}_+$, thermostatistics embeds them as extremal values of real probabilities $[0, 1] \subset \mathbb{R}_+$, and quantum theory extends the modality structure to complex probability amplitudes $[0, 1] \times \mathbf{U}(1) \subset \mathbb{C}$ as used for Hilbert spaces.

For the time representations above, one has to look for an implementation of the object-concept which started with the natural numbers $\mathbb{N} \subset \mathbb{R}_+$ and was extended by the scalar product and the norm $|\alpha|^2 = \langle\alpha|\alpha\rangle$ for the complex numbers with the complex conjugation.

Good to know: A *conjugation* of a complex vector space $* : V \longrightarrow V$ is an involutive $v^{**} = v$ and antilinear $(\alpha v + \beta w)^* = \overline{\alpha} v^* + \overline{\beta} w^*$ mapping. A vector $v = v^*$ is called $*$-real or $*$-hermitian or $*$-symmetric. An *internal product* of a real or complex vector space $\langle\ |\ \rangle : V \times V \longrightarrow \mathbb{R}, \mathbb{C}$ is a symmetric bi- and sesqui-linear form, respectively, i.e., $\langle v|w\rangle = \overline{\langle w|v\rangle}$ and $\langle u|\alpha v + \beta w\rangle = \alpha\langle u|v\rangle + \beta\langle u|w\rangle$.

An internal product with $\langle v|v\rangle \geq 0$ for all $v \in V$ is called *positive*. The quotient space $V/V^\perp$, where the *orthogonal* $V^\perp$ contains the vectors with trivial square $\langle v|v\rangle = 0$, inherits a strictly positive internal product, called *scalar product* $\langle v||w\rangle = \langle v|w\rangle$. For the classes $|v\rangle, |w\rangle \in V/V^\perp$, one has $\langle v||v\rangle > 0 \iff |v\rangle \neq V^\perp$. Its Cauchy-completion is a[26] *Hilbert space*, abstractly defined as a complete scalar product vector space.

Quantum theory uses scalar products for the complex object Hilbert spaces with irreducible time representations and the conjugation implementing the time reflection. $\mathbf{U}(1)$ is the invariance group of the scalar product:

$$
\begin{aligned}
V \cong \mathbb{C}: \ \langle\ |\ \rangle : V \times V \longrightarrow \mathbb{C}, \quad & \begin{cases} \langle v|w\rangle = \overline{v}w \\ \langle v|v\rangle > 0 \iff v \neq 0, \end{cases} \\
D_1(-t|\omega) = e^{-i\omega t} &= D_1(t|\omega)^\star, \\
|v(t)\rangle = e^{i\omega t}|v\rangle, \quad & \langle v(t)|v(t)\rangle = \langle v|v\rangle.
\end{aligned}
$$

The scalar product defines probability amplitudes with the modality group $\mathbf{U}(1)$ for the compact time representations.

The definition of free objects in time and space requires a representation for both the time translations and the space translations, characterized by energy $E \in \mathbb{R}$ and momentum $\vec{q} \in \mathbb{R}^3$, respectively:

$$
\mathbb{R} \ni t \longmapsto e^{i\frac{Et}{\hbar}} \in \mathbf{U}(1), \quad \mathbb{R}^3 \ni \vec{x} \longmapsto e^{i\frac{\vec{q}\vec{x}}{\hbar}} \in \mathbf{U}(1).
$$

Here, the action unit $\hbar$ is used only on dimensional grounds.

For the nondecomposable time representations $D_n(t|\omega)$, $n \geq 2$, the conjugations are more complicated: The time reflection and the sesquilinear form is defined by a *non-Euclidean conjugation* $\times$ of the representation space $V \cong \mathbb{C}^n$, which, in addition

[25] Aristoteles of Stageira -(384–322).

[26] David Hilbert (1862–1943).

to the $\mathbf{U}(1)$-hermitian transposition (denoted by $\star$), involves a linear transformation of V — in the example $n = 2$:

$$D_2(-t|\omega) = D_2(t|\omega)^\times = z_2 \circ D_2(t|\omega)^\star \circ z_2^{-1} = e^{-i\omega t - iN_2^\times t},$$
$$N_2 = N_2^\times = \begin{pmatrix} 0 & 1 \\ 0 & 0 \end{pmatrix}^\times = z_2 \circ N_2^\star \circ z_2^{-1} = \begin{pmatrix} 0 & 1 \\ 1 & 0 \end{pmatrix}\begin{pmatrix} 0 & 0 \\ 1 & 0 \end{pmatrix}\begin{pmatrix} 0 & 1 \\ 1 & 0 \end{pmatrix} = \begin{pmatrix} 0 & 1 \\ 0 & 0 \end{pmatrix}.$$

The sesquilinear form, defined with z_n, is real, but, for $n \geq 2$, not positive definite, i.e., no scalar product: The representation space is no Hilbert space, e.g., for $n = 2$ with a basis $\{e^1, e^2\}$:

$$\langle\ |\ \rangle : V \times V \longrightarrow \mathbb{C}, \quad \begin{cases} \begin{pmatrix} \langle e^1|e^1\rangle & \langle e^1|e^2\rangle \\ \langle e^2|e^1\rangle & \langle e^2|e^2\rangle \end{pmatrix} = \begin{pmatrix} 0 & 1 \\ 1 & 0 \end{pmatrix} = z_2, \\ \langle e^1 \pm e^2|e^1 \pm e^2\rangle = \pm 2. \end{cases}$$

The invariance group of the sesquilinear form is the indefinite unitary group $\mathbf{U}(1, 1)$.

Good to know: In a group G with reflection $*$, i.e., $g^{**} = g$ and $(gh)^* = h^*g^*$ for all $g, h \in G$, $\mathbf{U}(G, *) = \{u \in G \mid u^* = u^{-1}\}$ is called the *$*$-unitary subgroup*. It may be definite or indefinite. The unitary groups for the nondecomposable $\mathbb{R}$-representations are $\mathbf{U}(\frac{n+1}{2}, \frac{n-1}{2})$ for $n = 1, 3, \ldots$. They are definite only for the irreducible ones $n = 1$. $\mathbf{U}(\frac{n}{2}, \frac{n}{2})$ for $n = 2, 4, \ldots$, are indefinite with neutral signature.

It is no accident, that the indefinite metric with the indefinite modality group $\mathbf{U}(1, 1)$ arises also for the Lorentz bilinear form for lightlike spacetime translations $\eta \cong \left(\begin{array}{cc|c} 0 & 1 & 0 \\ 1 & 0 & 0 \\ \hline 0 & 0 & -\mathbf{1}_2 \end{array}\right)$. Lightlike (singular) structures in quantum field theory are indefinite, their modality does not fit unitary quantum mechanics as axiomatized by von Neumann. For example, the electromagnetic field with its four components has the modality group $\mathbf{U}(1, 1) \times \mathbf{U}(1) \times \mathbf{U}(1)$: Only the two transversal components, the photons (in the explicit matrix the second and third component), have a probability and particle interpretation with modality group $\mathbf{U}(1)$. The additional two components (in the explicit matrix the zeroth and first component) describe the gauge degree of freedom and the Coulomb interaction. They are acted on by noncompact time representations in $\mathbf{U}(1, 1)$ and cannot be associated with objects (detectable particles). They are spacetime fields without particles.

3.7 Non-commutative Numbers — Quaternions

Mathematicians were not content with the complex numbers, they looked for further extensions, in a first step for fields with *noncommutative (nonabelian)*[27] numbers.

The *quaternions* $\mathbb{H}$ were introduced by Hamilton: They constitute a *nonabelian* field with one real unit $\{1\}$ and three imaginary units $\{i, j, k\}$ with the multiplication table:

[27] Niels Henrik Abel (1802–1829).

1.↓2.→	1	i	j	k
1	1	i	j	k
i	i	-1	$-k$	j
j	j	k	-1	$-i$
k	k	$-j$	i	-1

$$\text{e.g.,} \begin{cases} i^2 = j^2 = k^2 = -1, \\ ij = -k = -ji, \\ ijk = 1. \end{cases}$$

Multiplication of quaternions

Good to know: The systematic introduction of units with squares $+1$ and -1 leads to the theory of real Clifford[28] algebras.

The quaternions are a real four-dimensional algebra:

$$\mathbb{H} = \{\mathbf{q} = \sum_{k=0}^{3} \alpha_k e_k \mid \alpha_i \in \mathbb{R}\} \cong \mathbb{R}^4, \quad (e_0, e_1, e_2, e_3) = (1, i, j, k),$$

$$\mathbb{H} \times \mathbb{H} \longrightarrow \mathbb{H}, \begin{cases} e_a e_b = -\delta_{ab} 1 - \epsilon_{abc} e_c, \quad a = 1, 2, 3, \\ \epsilon_{abc} \text{ totally antisymmetric, } \epsilon_{123} = 1, \\ \mathbf{q}_1 = \alpha_0 + \vec{\alpha}\vec{e}, \quad \mathbf{q}_2 = \beta_0 + \vec{\beta}\vec{e}, \\ \mathbf{q}_1 + \mathbf{q}_2 = (\alpha_0 + \beta_0) + (\vec{\alpha} + \vec{\beta})\vec{e}, \\ \beta\mathbf{q} = \beta\alpha_0 + \beta\vec{\alpha}\vec{e}, \\ \mathbf{q}_1\mathbf{q}_2 = \alpha_0\beta_0 - \vec{\alpha}\vec{\beta} + (\alpha_0\vec{\beta} + \beta_0\vec{\alpha} - \vec{\alpha} \times \vec{\beta})\vec{e}, \\ \mathbf{q}_1(\mathbf{q}_2\mathbf{q}_3) = (\mathbf{q}_1\mathbf{q}_2)\mathbf{q}_3 \text{ (associative)}, \\ \text{e.g., } e_a(e_b e_c) = (e_a e_b) e_c, \\ \text{with } \epsilon_{abc}\epsilon_{dec} = \delta_{ad}\delta_{be} - \delta_{ae}\delta_{bd}. \end{cases}$$

They have a *conjugation*, where the order of the factors is exchanged, i.e., a *contramorphism*:

$$\mathbb{H} \overset{*}{\leftrightarrow} \mathbb{H}, \quad \mathbf{q} = \alpha_0 + \vec{\alpha}\vec{e} \leftrightarrow \overline{\mathbf{q}} = \alpha_0 - \vec{\alpha}\vec{e}$$

$$\Rightarrow \begin{cases} \overline{\mathbf{q}_1 + \mathbf{q}_2} = \overline{\mathbf{q}}_1 + \overline{\mathbf{q}}_2 \\ \overline{\mathbf{q}_1\mathbf{q}_2} = \overline{\mathbf{q}}_2\,\overline{\mathbf{q}}_1, \text{ e.g., } \overline{e_1 e_2} = -\overline{e}_3 = e_3 = e_2 e_1. \end{cases}$$

The real numbers are embedded as the conjugation invariant subfield:

$$\{\mathbf{q} \in \mathbb{H} \mid \mathbf{q} = \overline{\mathbf{q}}\} \cong \mathbb{R}.$$

There is no canonical embedding of the complex numbers in the quaternions. The many embeddings are characterized by the embeddings $\mathbf{U}(1) \hookrightarrow \mathbf{SU}(2)$ (more ahead).

[28]William Clifford (1845–1879).

The quaternions constitute a field. With the norm

$$\mathbb{H} \longrightarrow \mathbb{R}_+,\ \mathbf{q} \longmapsto |\mathbf{q}|,\ \text{with} \begin{cases} |\mathbf{q}|^2 = \mathbf{q}\overline{\mathbf{q}} = \alpha_0^2 + \vec{\alpha}^2, \\ |\mathbf{q}_1 + \mathbf{q}_2| \leq |\mathbf{q}_1| + |\mathbf{q}_2|, \\ |\mathbf{q}_1\mathbf{q}_2| = |\mathbf{q}_1||\mathbf{q}_2|, \\ |\mathbf{q}| = 0 \iff \mathbf{q} = 0, \end{cases}$$

the inverse of any nontrivial quaternion is the product of its inverse norm with the conjugated phase:

$$\mathbf{q} \neq 0 \Rightarrow \mathbf{q}^{-1} = \frac{\overline{\mathbf{q}}}{|\mathbf{q}|^2}.$$

Good to know: The three number extensions by real division algebras (with additional properties, not given here) are connected with the exactly three parallelizable spheres, i.e., with classes of rotation groups $\mathbf{SO}(1+s)/\mathbf{SO}(s) \cong \Omega^s \subset \mathbb{R}^{1+s}$, given by Ω^1, Ω^3, and Ω^7. They have continuous global vector field bases, i.e., if covered with hairs, they can be "continuously combed without bald points." Try it and comb hairs on a circle Ω^1 (continuously possible), and see the difference by combing hairs on a 2-sphere Ω^2 (there remains at least one bald point)! Those three spheres are associated with the three division $\mathbb{R}$-algebras for $1 + s = 2^n$, $n = 1, 2, 3$, which exist in addition to $\mathbb{R}$ with the "sphere" $\Omega^0 = \{\pm 1\}$: the complex numbers $\mathbb{C} \cong \mathbb{R}^2 \supset \Omega^1$, Hamilton's nonabelian quaternions $\mathbb{H} \cong \mathbb{R}^4 \supset \Omega^3$, and Cayley's[29] nonassociative and nonabelian octonions $\mathbb{O} \cong \mathbb{R}^8 \supset \Omega^7$:

$$\begin{array}{ccccccc} \mathbb{R} & \subset & \mathbb{C} & \subset & \mathbb{H} & \subset & \mathbb{O} \\ \cup & & \cup & & \cup & & \cup \\ \Omega^0 & \subset & \Omega^1 & \subset & \Omega^3 & \subset & \Omega^7 \\ \cong & & \cong & & \cong & & \cong \\ \mathbf{O}(1) & & \mathbf{SO}(2) & & \mathbf{SO}(4)/\mathbf{SO}(3) & & \mathbf{SO}(8)/\mathbf{SO}(7) \end{array}$$

3.8 Non-commutative Complex Operations

The quaternionic spread of the imaginary unit $i \hookrightarrow (i, j, k)$ has an operational formulation. The quaternions can be realized by an injective field morphism to complex (2×2)-matrices with the unit matrix $1 \longmapsto \mathbf{1}_2$ and the three hermitian Pauli[30] matrices $\vec{\sigma} = \vec{\sigma}^\star$ for the three imaginary units $\vec{e} \longmapsto i\vec{\sigma}$:

$$\mathbb{H} \longrightarrow \mathbf{AL}(2, \mathbb{C}),\ \ \mathbf{q} = \alpha_0 + \vec{\alpha}\vec{e} \longmapsto \alpha_0\mathbf{1}_2 + i\vec{\alpha}\vec{\sigma} = \begin{pmatrix} \alpha_0 + i\alpha_3 & i\alpha_1 + \alpha_2 \\ i\alpha_1 - \alpha_2 & \alpha_0 - i\alpha_3 \end{pmatrix}.$$

In general, the complex $(d \times d)$-matrices $\mathbf{AL}(d, \mathbb{C}) = \mathbb{C}^d \otimes \mathbb{C}^d$ constitute, with the matrix multiplication, an associative $*$-algebra, i.e., an algebra with conjugation. It is unital with the unit matrix $\mathbf{1}_d$. Matrices are vector space operations.

[29] Arthur Cayley (1821–1895).

[30] Wolfgang Pauli (1900–1958).

Good to know: A $*$-*algebra (involutive algebra)* A is a complex algebra with antilinear conjugation $a^{**} = a$ and $(ab)^* = b^* a^*$. It decomposes $A = A_{\mathbb{R}} \oplus iA_{\mathbb{R}}$ into real vector spaces with the symmetric elements $a = a^\star$ and the antisymmetric ones $a = -a^\star$. If A is taken, with the commutator, as real Lie[31] algebra, $iA_{\mathbb{R}}$ is a Lie subalgebra, even an ideal of A.

Both the hermitian and the antihermitian matrices in the vector space decomposition:

$$\mathbf{AL}(d, \mathbb{C}) = \mathbb{R}(d) \oplus i\mathbb{R}(d); \quad d = 1: \mathbf{AL}(1, \mathbb{C}) = \mathbb{C} = \mathbb{R} \oplus i\mathbb{R},$$

constitute a *real vector space* with dimension d^2:

$$\mathbb{R}(d) = \{\mathbf{x} \in \mathbf{AL}(d, \mathbb{C}) \mid \mathbf{x}^\star = \mathbf{x}\}, \quad \begin{cases} (\mathbf{x}_1 + \mathbf{x}_2)^\star = \mathbf{x}_1 + \mathbf{x}_2, \\ (\alpha\mathbf{x})^\star = \alpha\mathbf{x}, \ \alpha \in \mathbb{R}, \end{cases}$$

parametrizable for the simplest cases $d = 1, 2, 3$ as follows:

$$\begin{aligned} \mathbb{R}(1) &= \{\mathbf{x} = x_0 1 \mid x_0 \in \mathbb{R}\}, \\ \mathbb{R}(2) &= \{\mathbf{x} = x_0 \mathbf{1}_2 + x_a \sigma^a(2) = \begin{pmatrix} x_0 + x_3 & x_1 - ix_2 \\ x_1 + ix_2 & x_0 - x_3 \end{pmatrix} \mid x_0, x_a \in \mathbb{R}\}, \end{aligned}$$

$$\mathbb{R}(3) = \{\mathbf{x} = x_0 \mathbf{1}_3 + x_a \sigma^a(3) = \begin{pmatrix} x_0 + x_3 + \frac{x_8}{\sqrt{3}} & x_1 - ix_2 & x_4 - ix_5 \\ x_1 + ix_2 & x_0 - x_3 + \frac{x_8}{\sqrt{3}} & x_6 - ix_7 \\ x_4 + ix_5 & x_6 + ix_7 & x_0 - \frac{2x_8}{\sqrt{3}} \end{pmatrix} \mid x_0, x_a \in \mathbb{R}\}.$$

Good to know: With the given expressions all matrices $\sigma^a(2)$ and $\sigma^a(3)$ are defined, for example:

$$\sigma^2(2) = \begin{pmatrix} 0 & -i \\ i & 0 \end{pmatrix} \in \mathbb{R}(2), \quad \sigma^4(3) = \begin{pmatrix} 0 & 0 & 1 \\ 0 & 0 & 0 \\ 1 & 0 & 0 \end{pmatrix} \in \mathbb{R}(3).$$

The d^2-hermitian *generalized Weyl*[32] *matrices*, with the unit matrix $\mathbf{1}_d$ and $(d^2 - 1)$ traceless *generalized Pauli matrices* $\sigma^a(d)$ are a basis of $\mathbb{R}(d) \cong \mathbb{R}^{d^2}$:

$$\text{basis of } \mathbb{R}(d): \ \{\mathbf{1}_d, \sigma^a(d) \mid a = 1, \ldots, d^2 - 1\}, \quad \begin{cases} \sigma^a(d)^\star = \sigma^a(d), \\ \operatorname{tr} \sigma^a(d) = 0. \end{cases}$$

They involve the proper Pauli matrices $\sigma^a(2) = \sigma^a, a = 1, 2, 3$, and the Gell-Mann[33] matrices $\sigma^a(3) = \lambda^a, a = 1, 2, \ldots, 8$. The nondiagonal matrices come in pairs with entries $(1, 1)$ and $(-i, i)$, exemplified by the pair (σ^1, σ^2). The real diagonal matrices are constructed inductively, step by step for $d \geq 2$: the diagonals for $d - 1$ and the unit matrix $\mathbf{1}_{d-1}$ are embedded in traceless matrices, for example:

[31] Sophus Lie (1842–1899).

[32] Hermann Weyl (1885–1955).

[33] Murray Gell-Mann (1929–).

$$d=2:\ \sigma^3(2)=\begin{pmatrix}1&0\\0&-1\end{pmatrix},\ \text{then } d=3:\ \begin{cases}\sigma^3(3)=\begin{pmatrix}1&0&0\\0&-1&0\\0&0&0\end{pmatrix}&=\begin{pmatrix}\sigma^3(2)&0\\0&0\end{pmatrix},\\ \sigma^8(3)=\frac{1}{\sqrt{3}}\begin{pmatrix}1&0&0\\0&1&0\\0&0&-2\end{pmatrix}&=\frac{1}{\sqrt{3}}\begin{pmatrix}\mathbf{1}_2&0\\0&-2\end{pmatrix},\end{cases}$$

$$\text{then } d=4:\ \begin{cases}\sigma^3(4)=\begin{pmatrix}1&0&0&0\\0&-1&0&0\\0&0&0&0\\0&0&0&0\end{pmatrix}&=\begin{pmatrix}\sigma^3(3)&0\\0&0\end{pmatrix},\\ \sigma^8(4)=\frac{1}{\sqrt{3}}\begin{pmatrix}1&0&0&0\\0&1&0&0\\0&0&-2&0\\0&0&0&0\end{pmatrix}&=\begin{pmatrix}\sigma^8(3)&0\\0&0\end{pmatrix},\\ \sigma^{15}(4)=\frac{1}{\sqrt{6}}\begin{pmatrix}1&0&0&0\\0&1&0&0\\0&0&1&0\\0&0&0&-3\end{pmatrix}&=\frac{1}{\sqrt{6}}\begin{pmatrix}\mathbf{1}_3&0\\0&-3\end{pmatrix},\end{cases}$$

and, in general:

$$\{\sigma^{k^2-1}(d)=\begin{pmatrix}\sigma^{k^2-1}(d-1)&0\\0&0\end{pmatrix}\mid k=2,\ldots,d-1\}$$

$$\text{and } \sigma^{d^2-1}(d)=\frac{1}{\sqrt{\binom{d}{2}}}\begin{pmatrix}\mathbf{1}_{d-1}&0\\0&-(d-1)\end{pmatrix}.$$

The generalized Weyl matrices are a basis for $\mathbf{AL}(d,\mathbb{C})$ as complex d^2-dimensional vector space, and, with the antihermitian matrices in

$$\{\mathbf{1}_d, i\mathbf{1}_d, \sigma^a(d), i\sigma^a(d)\},$$

a basis for $\mathbf{AL}(d,\mathbb{C})\cong\mathbb{R}^{2d^2}$ as real $2d^2$-dimensional vector space.

The product of hermitian matrices is, in general, not hermitian. Therefore the real vector space $\mathbb{R}(d)$ for $d\geq 2$ is no algebra; it differs from the associative algebra $\mathbf{AL}(d,\mathbb{R})=\mathbb{R}^d\otimes\mathbb{R}^d$ of the real $(d\times d)$ matrices. The matrix multiplication of the hermitian Weyl matrices contains real symmetric and imaginary antisymmetric constants, arising in the anticommutators $\{a,b\}=ab+ba$ and commutators $[a,b]=ab-ba$, respectively:

$$\sigma^a(d)\sigma^b(d) = \delta_0^{ab}\mathbf{1}_d + (\delta_c^{ab} + i\epsilon_c^{ab})\sigma^c(d) \Rightarrow \begin{cases} \operatorname{tr}\sigma^a(d)\sigma^b(d) = d\delta_0^{ab}, \\ \quad \text{with } \delta_0^{ab} = \delta_0^{ba} \in \mathbb{R}, \\ \{\sigma^a(d), \sigma^b(d)\} = 2\delta_c^{ab}\sigma^c(d), \\ \quad \text{with } \delta_c^{ab} = \delta_c^{ba} \in \mathbb{R}, \\ [i\sigma^a(d), i\sigma^b(d)] = -2\epsilon_c^{ab} i\sigma^c(d), \\ \quad \text{with } \epsilon_c^{ab} = -\epsilon_c^{ba} \in \mathbb{R}. \end{cases}$$

It is possible, as done above for $d = 2$ and $d = 3$, to normalize equally the "double trace" of all traceless matrices to 2, to choose the imaginary constants *totally* antisymmetric and the real ones totally symmetric:

$$\sigma^a(d)\sigma^b(d) = \frac{2}{d}\delta^{ab}\mathbf{1}_d + (\delta^{abc} + i\epsilon^{abc})\sigma^c(d),$$

$$\text{with } \begin{cases} \operatorname{tr}\sigma^a(d)\sigma^b(d) = 2\delta^{ab}, \\ \epsilon^{abc} = \epsilon_d^{ab}\delta^{dc} = -\epsilon^{bac} = -\epsilon^{acb}, \\ \delta^{abc} = \delta_d^{ab}\delta^{dc} = \delta^{bac} = \delta^{acb}, \end{cases}$$

$$d = 2: \quad \begin{cases} \epsilon^{123} = 1, \\ \delta^{abc} = 0, \end{cases}$$

$$d = 3: \quad \begin{cases} \epsilon^{123} = 1, \quad \epsilon^{147,165,246,257,345,376} = \frac{1}{2}, \quad \epsilon^{458,678} = \frac{\sqrt{3}}{2}, \\ \delta^{118,228,338} = -\delta^{888} = \frac{1}{\sqrt{3}}, \quad \delta^{448,558,668,778} = -\frac{1}{2\sqrt{3}}, \\ \delta^{146,157,256,344,355} = -\delta^{247,366,377} = \frac{1}{2}. \end{cases}$$

The case $d = 2$ is distinguished by trivial traceless contributions in the anticommutators. i.e., $\delta^{abc} = 0$.

3.9 Causal Minkowski Spacetime

As anticipated by the notation, the additive group with the *time translations* $t = x_0 \in \mathbb{R} = \mathbb{R}(1)$ can be embedded as *trace* in the hermitian matrices $\mathbb{R}(2) \cong \mathbb{R}^4$ together with a real three-dimensional vector space of *position translations as traceless part*:

$$\mathbb{R}(2) \ni \mathbf{x} = \begin{pmatrix} x_0 + x_3 & x_1 - ix_2 \\ x_1 + ix_2 & x_0 - x_3 \end{pmatrix} = \mathbf{x}_{\text{time}} + \mathbf{x}_{\text{position}}, \quad \begin{cases} \mathbf{x}_{\text{time}} = \frac{\operatorname{tr}\mathbf{x}}{2}\mathbf{1}_2 = x_0\mathbf{1}_2, \\ \mathbf{x}_{\text{position}} = x_a\sigma^a, \\ \operatorname{tr}\mathbf{x}_{\text{position}} = 0, \end{cases}$$

$$\mathbb{R}(2) = \mathbb{R}(2)_{\text{time}} \oplus \mathbb{R}(2)_{\text{position}}, \quad \begin{cases} \mathbb{R}(2)_{\text{time}} \cong \mathbb{R}, \\ \mathbb{R}(2)_{\text{position}} \cong \mathbb{R}^3. \end{cases}$$

$\mathbb{R}(2) \subset \mathbf{AL}(2, \mathbb{C})$ is called the *Weyl or bi-spinor representation of the spacetime translations*, i.e., of *Minkowski spacetime*. It acts on a complex two-dimensional vector space, called a *Weyl spinor space*.

Good to know: The two fundamental $\mathbf{SL}(2, \mathbb{C})$-representations $(\frac{1}{2}|0)$ and $(0|\frac{1}{2})$ act on complex two-dimensional spaces with left- and right-handed Weyl spinors as bases $\mathbf{l}^A \in V_L \cong \mathbb{C}^2$, $A = 1, 2$, and $\mathbf{r}^{\dot{A}} \in V_R \cong \mathbb{C}^2$, $\dot{A} = 3, 4$, respectively. The four-dimensional Weyl spacetime representation $\mathbf{x}$ is acted on by the irreducible $\mathbf{SL}(2, \mathbb{C})$-representation $(\frac{1}{2}|\frac{1}{2})$, relating left- and right-spinors $\mathbf{x} : V_R \longrightarrow V_L$, $\mathbf{r}^{\dot{A}} \longmapsto \mathbf{x}^B_{\dot{A}} \mathbf{r}^{\dot{A}}$. This is used, for example, in the Dirac equation $(\partial - im)\Psi$=0 for a massive Dirac spinor field $\Psi \cong \mathbf{l}^A \oplus \mathbf{r}^{\dot{A}}$, comprising left and right Weyl spinors, and acted on by the decomposable four-dimensional $\mathbf{SL}(2, \mathbb{C})$-representation $(\frac{1}{2}|0) \oplus (0|\frac{1}{2})$, with $(\frac{\partial}{\partial x})^B_{\dot{A}} \mathbf{r}^{\dot{A}} = im\mathbf{l}^B$ and $(\frac{\partial}{\partial x})^{\dot{B}}_A \mathbf{l}^A = im\mathbf{r}^{\dot{B}}$.

Each translation is unitarily (by position rotation) diagonalizable with the two eigenvalues in the diagonal, i.e., $\mathbb{R}(2)$ has *rank two*:

$$\mathbf{x} = u \circ \operatorname{diag} \mathbf{x} \circ u^*, \text{ with } \begin{cases} u = \begin{pmatrix} \cos \frac{\theta}{2} & -e^{-i\phi} \sin \frac{\theta}{2} \\ e^{i\phi} \sin \frac{\theta}{2} & \cos \frac{\theta}{2} \end{pmatrix} \in \mathbf{SU}(2), \\ \operatorname{diag} \mathbf{x} = \begin{pmatrix} x_0 + |\vec{x}| & 0 \\ 0 & x_0 - |\vec{x}| \end{pmatrix}, \\ \frac{\vec{x}}{|\vec{x}|} = (\cos \phi \sin \theta, \sin \phi \sin \theta, \cos \theta). \end{cases}$$

Good to know: A real matrix $f \in \mathbf{AL}(d, \mathbb{R})$ is orthogonally diagonalizable, $f = O \circ f \circ O^T$, $O \in \mathbf{O}(d)$, if, and only if, it is *symmetric*, i.e., $f = f^T$.
A complex matrix $f \in \mathbf{AL}(d, \mathbb{C})$ is unitarily diagonalizable, $f = U \circ \operatorname{diag} f \circ U^\star$, $U \in \mathbf{U}(d)$, if, and only if, it is *normal*, i.e., $f \circ f^\star = f^\star \circ f$, or, equivalently, has trivial commutator $[f, f^\star] = 0$.

The characteristic polynomial of a spacetime matrix contains as coefficients the determinant and the trace:

$$\det(\mathbf{x} - \xi \mathbf{1}_2) = \det \mathbf{x} - \xi \operatorname{tr} \mathbf{x} + \xi^2.$$

Good to know: The development coefficients for an endomorphism $f : V \longrightarrow V$ are the traces of the Grassmann powers, $\det(f - X\mathbf{1}_d) = \sum_{k=0}^{d} t_k(f)(-\xi)^{k-d}$ with $\mathbf{t}_k(f) = \operatorname{tr} \bigwedge^k f$.

The determinant is a homogeneous form of degree two (bilinear) in the matrix elements, the trace is a form of degree one (linear). The trace is the time projection. The determinant is the indefinite *Lorentz metric of four-dimensional Minkowski spacetime* with positive and negative squares as well as nontrivial vectors with trivial square. The translations are called, respectively, *time-, space- and lightlike*:

$$\begin{aligned} \frac{1}{2} \operatorname{tr} \mathbf{x} &= x_0, \\ \det \mathbf{x} &= x_0^2 - (x_1^2 + x_2^2 + x_3^2) = x_0^2 - \vec{x}^2 = \eta(\mathbf{x}, \mathbf{x}), \\ \eta(\mathbf{x}, \mathbf{y}) &= \frac{\det(\mathbf{x} + \mathbf{y}) - \det(\mathbf{x} - \mathbf{y})}{4} = x_0 y_0 - \vec{x}\vec{y}. \end{aligned}$$

Minkowski spacetime (for $d = 2$) has a unique order which establishes *causality*. With the spectrum (eigenvalues) of a complex matrix:

$$\mathbf{z} \in \mathbf{AL}(d, \mathbb{C}), \quad \mathbf{spec}\,\mathbf{z} = \{\xi \mid \det(\mathbf{z} - \xi \mathbf{1}_2) = 0\},$$

a matrix is called *positive (futurelike)* if it is hermitian and has positive eigenvalues, or, equivalently, if it is a product of two conjugated matrices:

$$\begin{aligned}\mathbf{AL}(d, \mathbb{C}) \ni \mathbf{z} \succeq 0 &\iff \mathbf{z} = \mathbf{z}^\star \text{ and } \mathbf{spec}\,\mathbf{z} \subset \mathbb{R}_+, \\ &\iff \mathbf{z} = \mathbf{y} \circ \mathbf{y}^\star, \quad \mathbf{y} \in \mathbf{AL}(d, \mathbb{C}).\end{aligned}$$

The positive translations constitute a cone:

$$\alpha_{1,2} \geq 0, \quad \mathbf{x}_{1,2} \succeq 0 \Rightarrow \alpha_1 \mathbf{x}_1 + \alpha_2 \mathbf{x}_2 \succeq 0.$$

Good to know: A C^*-*algebra* A is a normed, complete $*$-algebra with $\| aa^\star \| = \| a \|^2$ for all elements. $\mathbf{AL}(d, \mathbb{C})$ is a C^*-algebra. A C^*-algebra is ordered: the positive cone $A_+ = \{x \in A \mid x = x^\star,\ \mathbf{spec}\, x \text{ positive}\}$ contains exactly the squared elements $A_+ = \{aa^\star \mid a \in A\}$. A_+ decomposes into the tip of the cone $\{0\}$, the open interior of the cone with strictly positive spectrum $\{gg^\star \mid g \in G(A)\}$ with the *regular group* $G(A)$ (multiplicatively invertible elements), and the tipless skin of the cone, given by the nontrivial positive elements a with $0 \in \mathbf{spec}\, a$.

Four-dimensional Minkowski spacetime $\mathbb{R}(2) \cong \mathbb{R}^4$ has the partial order (positivity), characterized by the spacetime matrices of the *future* with two positive eigenvalues:

$$\mathbf{spec}\,\mathbf{x} = \{\xi_{1,2} = x_0 \pm |\vec{x}|\} : \quad \begin{cases} \xi_1 \xi_2 = \det \mathbf{x} = x_0^2 - \vec{x}^2, \\ \xi_1 + \xi_2 = \operatorname{tr} \mathbf{x} \ = 2x_0, \end{cases}$$

$$\mathbf{x} = \mathbf{x}^\star \text{ is } \begin{cases} \text{futurelike} \iff \mathbf{spec}\,\mathbf{x} \geq 0 \iff \det \mathbf{x} \geq 0 \text{ and } \operatorname{tr} \mathbf{x} \geq 0, \\ \text{pastlike} \iff \mathbf{spec}\,\mathbf{x} \leq 0 \iff \det \mathbf{x} \geq 0 \text{ and } \operatorname{tr} \mathbf{x} \leq 0. \end{cases}$$

3.10 Hypercharge Group and Isospin Group

The real vector space $i\mathbb{R}(d) \cong (i\mathbb{R})^{d^2}$ with the antihermitian matrices in the algebra $\mathbf{AL}(d, \mathbb{C})$ constitutes, with the commutator, a real Lie algebra, the Lie algebra of the compact unitary group $\mathbf{U}(d)$, for example of the *hypercharge-isospin group* $\mathbf{U}(2)$ with $\{\frac{i\mathbf{1}_2}{2}, \frac{i\sigma^a}{2}\}$ as a Lie algebra basis:

$$\begin{aligned} i\mathbb{R}(2) = \{l &= \frac{i}{2}(\alpha_0 \mathbf{1}_2 + \alpha_a \sigma^a) = \frac{i}{2}\begin{pmatrix} \alpha_0 + \alpha_3 & \alpha_1 - i\alpha_2 \\ \alpha_1 + i\alpha_2 & \alpha_0 - \alpha_3 \end{pmatrix} \mid \alpha_0, \alpha_a \in \mathbb{R}\}, \\ l &= -l^\star, \\ [l_1, l_2] &= l_1 l_2 - l_2 l_1 = l_1^\star l_2^\star - l_2^\star l_1^\star = (l_2 l_1 - l_1 l_2)^\star = -[l_1, l_2]^\star, \\ [\frac{i\sigma^a}{2}, \frac{i\sigma^b}{2}] &= -\epsilon^{abc} \frac{i\sigma^c}{2}. \end{aligned}$$

Good to know: A *Lie algebra* is a vector space with a bilinear product (Lie bracket), which for all vectors is individually nilpotent, $[l, l] = 0$, and which fulfils, instead of the associativity, the Leibniz product rule, $[l, [m, n]] = [[l, m], n] + [m, [l, n]]$, familiar from the product rule for derivations $\partial(fg) = (\partial f)g + f(\partial g)$, and, equivalent, with the Jacobi identity. Lie algebra elements can be considered as derivations. A basis $(l^a)_{a=1}^n$ of a Lie algebra $L \cong \mathbb{K}^n$ defines basis-dependent *structure constants* $[l^a, l^b] = \epsilon_c^{ab} l^c$, valued in $\mathbb{K} \in \mathbb{R}$ or $\mathbb{K} = \mathbb{C}$ and antisymmetric in two indices $\epsilon_c^{ab} = -\epsilon_c^{ba}$. For instance, the rotation Lie algebra $\log \mathbf{SU}(2) \cong \mathbb{R}^3$ has totally antisymmetric (in all three indices pairs) real structure constants $-\epsilon^{abc}$ in the basis $(\frac{i\sigma^a}{2})_{a=1,2,3}$.

In analogy to the decomposition of the vector space $\mathbb{R}(2)$ into time and space, the Lie algebra $i\mathbb{R}(2)$ decomposes into a trace part and a traceless part:

$$i\mathbb{R}(2) = i\mathbb{R}(2)_{\mathrm{tr}} \oplus i\mathbb{R}(2)_0, \quad \begin{cases} i\mathbb{R}(2)_{\mathrm{tr}} = i\mathbb{R}\mathbf{1}_2 \cong \mathbb{R}, \\ i\mathbb{R}(2)_0 \cong \mathbb{R}^3. \end{cases}$$

$i\mathbb{R}(2)$ is the Lie algebra of the *unitary group*, i.e., its infinitesimal structure:

$$\exp i\mathbb{R}(2) = \mathbf{U}(2) \ni u^{-1} = u^\star,$$

with two Lie subalgebras: $i\mathbb{R}$ is the Lie subalgebra of the abelian unitary group $\mathbf{U}(1)$:

$$i\alpha_0 \in i\mathbb{R} \Rightarrow e^{i\alpha_0} \in \exp i\mathbb{R} = \mathbf{U}(1).$$

The group $\mathbf{U}(1)$ is realized, on the one side, as symmetry group of the *harmonic time development* (oscillator time) in the form of external (spacetime) operations, and also, on the other side, as symmetry group of the *electromagnetic charge* or of the *hypercharge* in the form of internal (chargelike) operations.

The traceless part represents the Lie algebra A_1^c of the compact special unitary nonabelian Lie group $\mathbf{SU}(2)$:

$$l = \frac{i\alpha_a\sigma^a}{2} \in i\mathbb{R}(2)_0 \Rightarrow \begin{cases} l^2 = -\frac{\alpha^2}{4}\mathbf{1}_2, \ \vec{\alpha}^2 = \alpha^2, \\ e^l = \sum_{k\geq 0} \frac{l^k}{k!} = \mathbf{1}_2 \cos\frac{\alpha}{2} + l\frac{\sin\frac{\alpha}{2}}{\frac{\alpha}{2}} \in \exp i\mathbb{R}(2)_0 = \mathbf{SU}(2). \end{cases}$$

$\mathbf{SU}(2)$ acts, on the one side, as external compact *spin group* (angular momentum group), e.g., on the space translations, and also, on the other side, as internal *isospin group* on the fields of the electroweak standard model.

3.11 Lorentz Group and Poincaré Group

Both the Minkowski spacetime translations $\mathbb{R}(2) \cong \mathbb{R}^4$ and the hyperisospin Lie algebra $i\mathbb{R}(2) \cong (i\mathbb{R})^4$ can be obtained from the algebra $\mathbf{AL}(2, \mathbb{C})$ with all complex (2×2)-matrices. The regular group of the algebra is given by the invertible matrices from the *general linear group*:

$$\mathbf{GL}(2,\mathbb{C}) = \exp \mathbf{AL}(2,\mathbb{C}) \subset \mathbf{AL}(2,\mathbb{C}), \quad \mathbf{GL}(2,\mathbb{C}) = \mathbf{GL}(\mathbf{1}_2,\mathbb{C}) \circ \mathbf{SL}(2,\mathbb{C}).$$

It contains, in addition to the abelian group $\mathbf{GL}(1,\mathbb{C}) = \mathbf{D}(1) \times \mathbf{U}(1)$ with the causal (dilation) group and the phase group, the spacetime characteristic nonabelian *special linear group* $\mathbf{SL}(2,\mathbb{C})$, i.e., the invertible matrices with unit determinant.

In analogy to the strict future and the total causal order of one-dimensional time $\mathbb{R}(1) = \mathbb{R}$:

$$\begin{aligned} \mathbb{C} \ni \tau > 0 &\iff \tau = \gamma\overline{\gamma} \text{ for } \gamma \in \mathbf{GL}(1,\mathbb{C}) \\ &\iff \tau = e^t \in \mathbf{D}(1), \end{aligned}$$

the open future cone and the partial causal order of Minkowski spacetime can be characterized by invertible operations:

$$\begin{aligned} \mathbf{AL}(2,\mathbb{C}) \ni \mathbf{z} \succ 0 &\iff \mathbf{z} = g \circ g^\star \text{ with } g \in \mathbf{GL}(2,\mathbb{C}), \\ &\iff \mathbf{z} = e^{\mathbf{x}} \in \exp \mathbb{R}(2) = \mathbb{D}(2). \end{aligned}$$

The general linear group acts on the algebra by left- and right multiplications:

$$\mathbf{GL}(2,\mathbb{C}) \circ \mathbf{AL}(2,\mathbb{C}) \circ \mathbf{GL}(2,\mathbb{C}) \ni g_1 \circ \mathbf{z} \circ g_2 \in \mathbf{AL}(2,\mathbb{C}).$$

For $g_2 = g_1^\star$, the action is compatible with (anti-)hermiticity, with time-, light- and spacelike as well as future- and pastlike:

$$\begin{aligned} \mathbf{x} = \mathbf{x}^\star &\longmapsto g \circ \mathbf{x} \circ g^\star, \\ \mathbf{x} = \mathbf{y} \circ \mathbf{y}^\star &\longmapsto g \circ \mathbf{y} \circ (g \circ \mathbf{y})^\star, \end{aligned} \quad \left\{ \begin{aligned} \eta(\mathbf{x},\mathbf{x}) = \det \mathbf{x} &\longmapsto |\det g|^2 \det \mathbf{x}, \\ \det \mathbf{x} \begin{pmatrix} >0 \\ =0 \\ <0 \end{pmatrix} &\Rightarrow |\det g|^2 \det \mathbf{x} \begin{pmatrix} >0 \\ =0 \\ <0 \end{pmatrix}, \\ \operatorname{tr} \mathbf{x} &\longmapsto \operatorname{tr} g \circ \mathbf{x} \circ g^\star \\ \operatorname{tr} \mathbf{x} \begin{pmatrix} \geq 0 \\ \leq 0 \end{pmatrix} &\Rightarrow \operatorname{tr} g \circ \mathbf{x} \circ g^\star \begin{pmatrix} \geq 0 \\ \leq 0 \end{pmatrix}. \end{aligned} \right.$$

In this left-right action $\mathbf{x} \longmapsto g \circ \mathbf{x} \circ g^\star$ the phase group $\mathbf{U}(1)$ drops out, i.e., the nontrivially acting group is given by the classes $\mathbf{GL}(2,\mathbb{C})/\mathbf{U}(1) \cong \mathbf{D}(1) \times \mathbf{SO}_0(1,3)$ of the normal subgroup $\mathbf{U}(1)$. This leaves the dilation (causal) group and the *orthochronous Lorentz group* (more below).

Good to know: In a unital algebra A, each element of the regular group $G(A)$ defines an *inner A-automorphism* $\operatorname{Ad} g : A \longrightarrow A$ by $\operatorname{Ad} g(a) = gag^{-1}$. The group center acts trivially. The *adjoint group* is the quotient group $\operatorname{Ad} G(A) = G(A)/\operatorname{centr} G(A) = \operatorname{Int} G(A)$. The adjoint regular group is represented by $\operatorname{Ad} : G(A) \longrightarrow \mathbf{GL}(A)$, $g \longmapsto \operatorname{Ad} g$. A $*$-algebra has a $*$-unitary subgroup $\mathbf{U}(A) \subseteq G(A)$ in the regular group. $G(A)$ acts on A conjugation-compatibly by hermitian internal automorphisms $\operatorname{Ad}_* g : A \longrightarrow A$ with $\operatorname{Ad}_* g(a) = gag^*$ - for unitary elements $\operatorname{Ad}_* u = \operatorname{Ad} u$.

The *eigentime* of the Minkowski spacetime translations, used for mass points, is trivial for spacelike and lightlike translations. In contrast to a time translation as a Lorentz vector component, it is invariant under the orthochronous Lorentz group. A timelike hyperboloid is projected to one real number:

$$\mathbb{R}(2) \ni \mathbf{x} \longmapsto \epsilon(x_0)\vartheta(x^2)\sqrt{x^2} \in \mathbb{R} \text{ with } x^2 = \det \mathbf{x}, \ \ \epsilon(x_0) = \frac{\operatorname{tr}\mathbf{x}}{|\operatorname{tr}\mathbf{x}|},$$
$$x^2 > 0 \Rightarrow \sqrt{x^2} \in \mathbf{D}(1),$$

where the *characteristic order functions* have been used:

$$\mathbb{R} \ni t \longmapsto \epsilon(t) = \frac{t}{|t|} \in \mathrm{i}(2) = \{1, -1\}, \ \ \vartheta(\pm t) = \frac{1 \pm \epsilon(t)}{2} = \begin{cases} 1, \ \pm t > 0, \\ 0, \ \pm t < 0, \end{cases}$$
$$\epsilon(t_1)\epsilon(t_2) = \epsilon(t_1 t_2), \ \ 1 = \vartheta(t) + \vartheta(-t),$$
$$1 = \vartheta(x^2) + \vartheta(-x^2) = [\vartheta(x_0) + \vartheta(-x_0)]\vartheta(x^2) + \vartheta(-x^2).$$

In a rest system, the eigentime invariant arises by time projection:

$$\mathbf{x} = (x_0, 0) \Rightarrow \epsilon(x_0)\vartheta(x^2)\sqrt{x^2} = x_0.$$

Good to know: In relativistic field theories, point particles can be used. They depend not of all four spacetime translations, but of relativistic eigentime only. For an eigentime dependence, the nontrivially representable group is $\mathbb{R} \cong \mathbf{I}(2) \times \exp \mathbb{R}$, not the full Poincaré group $\mathbf{O}(1,3) \vec{\times} \mathbb{R}^4$.

The Pauli matrices $\{\sigma^a\}_{a=1}^3$ are a basis of the position translations in the *Pauli or bi-spinor representation*. The double trace as bilinear form gives the position metric:

$$\mathbb{R}(2)_0 \ni \vec{x} = x_a \sigma^a, \ \ \operatorname{tr} \vec{x} = 0,$$
$$-\det \vec{x} = \frac{1}{2} \operatorname{tr} \vec{x} \circ \vec{x} = \vec{x}^2 = x_a \delta^{ab} x_b, \ \ \operatorname{tr} \sigma^a \circ \sigma^b = 2\delta^{ab}.$$

The Weyl matrices $(\sigma^j)_{j=0}^3$ are a basis for the spacetime translations with $\mathbf{1}_2$ a basis for the time translations. The dual Weyl matrices $(\check{\sigma}_j)_{j=0}^3$ are a basis for the dual Minkowski space with the energy-momenta:

$$\mathbb{R}(2) \ni \sigma^j = (\mathbf{1}_2, \sigma^a), \ \ \eta(\sigma^j, \sigma^k) = \eta^{jk} \cong \begin{pmatrix} 1 & 0 \\ 0 & -\mathbf{1}_3 \end{pmatrix},$$
$$\eta_{jk}\sigma^k = \sigma_j = \check{\sigma}^j = \eta^{jk}\check{\sigma}_k = (\mathbf{1}_2, -\sigma^a),$$
$$\operatorname{tr} \sigma^j \circ \check{\sigma}_k = 2\delta^j_k, \ \ \sigma^j \circ \check{\sigma}^k + \sigma^k \circ \check{\sigma}^j = 2\eta^{jk}\mathbf{1}_2.$$

One has to get used to a matrix representations of the position and spacetime vectors as operations on complex two-dimensional spaces, e.g., in the position reflection $\sigma^j = (\mathbf{1}_2, \sigma^a) \leftrightarrow \check{\sigma}^j = (\mathbf{1}_2, -\sigma^a)$ between left- and right-handed Weyl matrices.

The bilinear form of Minkowski spacetime is invariant under the *Lorentz cover group* $\mathbf{SL}(2, \mathbb{C})$:

$$s \in \mathbf{SL}(2, \mathbb{C}) : \begin{cases} \det \mathbf{x} \longmapsto \det s \circ \mathbf{x} \circ s^\star = \det \mathbf{x}, \\ \sigma^j \longmapsto s \circ \sigma^j \circ s^\star = \Lambda^j_k(s)\sigma^k, \\ \Lambda^j_k(s) = \frac{1}{2} \operatorname{tr} s \circ \sigma^j \circ s^\star \circ \check{\sigma}_k, \\ \mathbf{x} = x_j \sigma^j \longmapsto \Lambda^k_j(s) x_k \sigma^j. \end{cases}$$

The decomposition into time and space translations is compatible only with the unitary subgroup $\mathbf{SU}(2) \subset \mathbf{SL}(2, \mathbb{C})$, the *position rotation cover group*:

$$u \in \mathbf{SU}(2): \quad \begin{cases} \operatorname{tr} \mathbf{x} \longmapsto \operatorname{tr} u \circ \mathbf{x} \circ u^\star = \operatorname{tr} \mathbf{x}, \\ \sigma^0 = \mathbf{1}_2 \longmapsto u \circ \sigma^0 \circ u^\star = \sigma^0, \\ \sigma^a \longmapsto u \circ \sigma^a \circ u^\star = O^a_b(u)\sigma^b, \\ O^a_b(u) = \frac{1}{2} \operatorname{tr} u \circ \sigma^a \circ u^\star \circ \sigma^b, \\ x_0 \longmapsto x_0, \;\; \vec{x} = x_a\sigma^a \longmapsto O^b_a(u) x_b \sigma^a. \end{cases}$$

A characteristic explicit example is computed for the position rotations by exponentiating the Lie algebra action:

$$\begin{aligned} &\mathbf{SU}(2) \ni u = e^{i\alpha\sigma^3} = \mathbf{1}_2 + i\alpha\sigma^3 + \dots \\ &\Rightarrow O^a_b(u) = \mathbf{1}_3 + \frac{i\alpha}{2} \operatorname{tr} (\sigma^3 \circ \sigma^a \circ \sigma^b - \sigma^a \circ \sigma^3 \circ \sigma^b) + \dots = \mathbf{1}_3 - 2\alpha\epsilon^{ab3} + \dots \\ &\Rightarrow O(u) = (e^{-2\alpha\epsilon})^a_b = \begin{pmatrix} \cos 2\alpha & -\sin 2\alpha & 0 \\ \sin 2\alpha & \cos 2\alpha & 0 \\ 0 & 0 & 1 \end{pmatrix} \in \mathbf{SO}(3), \end{aligned}$$

and for the spacetime Lorentz boost transformations:

$$\begin{aligned} &\mathbf{SL}(2, \mathbb{C}) \ni s = e^{\beta\sigma^3} = \mathbf{1}_2 + \beta\sigma^3 + \dots \\ &\Rightarrow \Lambda^j_k(s) = \mathbf{1}_4 + \frac{\beta}{2} \operatorname{tr} (\sigma^3 \circ \sigma^j \circ \check{\sigma}_k + \sigma^j \circ \sigma^3 \circ \check{\sigma}_k) + \dots \\ &\text{with } \Lambda^0_b(s) = 2\beta\delta^{3b} + \dots \\ &\Rightarrow \Lambda(s) = \begin{pmatrix} \cosh 2\beta & 0 & \sinh 2\beta \\ 0 & \mathbf{1}_2 & 0 \\ \sinh 2\beta & 0 & \cosh 2\beta \end{pmatrix} \in \mathbf{SO}_0(1, 3). \end{aligned}$$

The left-right action with the negative unit matrix $-\mathbf{1}_2 \in \mathbf{SU}(2) \subset \mathbf{SL}(2, \mathbb{C})$ drops out: The special rotations $\mathbf{SO}(3)$ and the orthochronous Lorentz transformations $\mathbf{SO}_0(1, 3)$ are isomorphic to the cover groups modulo the centrum $\mathbf{I}(2) = \{\pm\mathbf{1}_2\}$, i.e., they are given by classes of this normal subgroup:

$$\begin{aligned} \mathbf{SU}(2) \ni u \longmapsto \;& O(u) \in \mathbf{SO}(3) \cong \mathbf{SU}(2)/\mathbf{I}(2), \\ & O(u_1 \circ u_2) = O(u_1) \circ O(u_2), \\ \mathbf{SL}(2, \mathbb{C}) \ni s \longmapsto \;& \Lambda(s) \in \mathbf{SO}_0(1, 3) \cong \mathbf{SL}(2, \mathbb{C})/\mathbf{I}(2), \\ & \Lambda(s_1 \circ s_2) = \Lambda(s_1) \circ \Lambda(s_2). \end{aligned}$$

The parameters are doubled to the rotation angle $\vec{\theta} = 2\vec{\alpha}$ and to the Lorentz group rapidity $\vec{\psi} = 2\vec{\beta}$. Therefore, a rotation and a Lorentz transformation look as follows as exponentials of the Lie algebra "infinitesimal" operations:

$$\mathbf{SO}(3) \ni O(\vec{\theta}) = \exp \mathcal{O}(\vec{\theta}), \qquad \mathcal{O}(\vec{\theta}) = \begin{pmatrix} 0 & -\theta_3 & \theta_2 \\ \theta_3 & 0 & -\theta_1 \\ -\theta_2 & \theta_1 & 0 \end{pmatrix},$$

$$\mathbf{SO}_0(1,3) \ni \Lambda(\vec{\theta}, \vec{\psi}) = \exp \mathcal{L}(\vec{\theta}, \vec{\psi}), \quad \mathcal{L}(\vec{\theta}, \vec{\psi}) = \left(\begin{array}{c|c} 0 & \vec{\psi}^T \\ \hline \vec{\psi} & \mathcal{O}(\vec{\theta}) \end{array}\right) = \left(\begin{array}{c|ccc} 0 & \psi_1 & \psi_2 & \psi_3 \\ \hline \psi_1 & 0 & -\theta_3 & \theta_2 \\ \psi_2 & \theta_3 & 0 & -\theta_1 \\ \psi_3 & -\theta_2 & \theta_1 & 0 \end{array}\right).$$

By $|\tanh|\vec{\psi}|| = \frac{|\vec{v}|}{c}$ and $\frac{\vec{\psi}}{|\vec{\psi}|} = \frac{\vec{v}}{|\vec{v}|}$, the rapidity is related to the velocity parametrization with the bounded hyperbolic tangent expressable by bounded velocities $\frac{|\vec{v}|}{c} < 1$.

As seen in the Pauli and Weyl representations of $\mathbf{SU}(2) \sim \mathbf{SO}(3)$ and $\mathbf{SL}(2,\mathbb{C}) \sim \mathbf{SO}_0(1,3)$, the orthochronous Lorentz group is also isomorphic to the complexified orthogonal group $\mathbf{SO}(3,\mathbb{C})$, considered as real six-dimensional group:

$$\begin{aligned} e^{\frac{i\vec{\theta}}{2}\vec{\sigma}} &\in \mathbf{SU}(2) \sim \mathbf{SO}(3), \\ e^{\frac{i\vec{\theta}+\vec{\psi}}{2}\vec{\sigma}} &\in \mathbf{SL}(2,\mathbb{C}) \sim \mathbf{SO}(3,\mathbb{C}) \cong \mathbf{SO}_0(1,3). \end{aligned}$$

The orthochronous Lorentz group and the additive translation group, together with the Lorentz group action on the translations, constitute a semidirect product group, the *Poincaré group*:

$$\mathbf{SL}(2,\mathbb{C}) \vec{\times} \mathbb{R}^4 \ni (s, \mathbf{x}) \longmapsto (\Lambda, \mathbf{x}) : \begin{cases} (\Lambda_1, \mathbf{x}_1) \bullet (\Lambda_2, \mathbf{x}_2) = (\Lambda_1 \circ \Lambda_2, \mathbf{x}_1 + \Lambda_1 . \mathbf{x}_2), \\ \Lambda . \mathbf{x} = s \circ \mathbf{x} \circ s^*. \end{cases}$$

The *Euclidean subgroup* contains the rotations of the position translations:

$$\mathbf{SU}(2) \vec{\times} \mathbb{R}^3 \ni (u, \vec{x}) \longmapsto (O, \vec{x}) : \begin{cases} (O_1, \vec{x}_1) \bullet (O_2, \vec{x}_2) = (O_1 \circ O_2, \vec{x}_1 + O_2 . \vec{x}_2), \\ O . \vec{x} = u \circ \vec{x} \circ u^\star. \end{cases}$$

The exponent of the quaternions $\mathbb{H} \cong \mathbb{R} \oplus i\mathbb{R}(2)_0$ in the $\mathbf{AL}(2,\mathbb{C})$-representation is the direct product of the noncompact causal (dilation) group and the compact spin group:

$$\begin{aligned} \exp i\mathbb{R} = \mathbf{U}(1), \quad \exp i\mathbb{R}(2) &= \mathbf{U}(2) \cong \mathbf{U}(1) \circ \mathbf{SU}(2), \\ \exp \mathbb{R} = \mathbf{D}(1), \quad \exp \mathbb{R}(2) &= \mathbb{D}(2) \cong \mathbf{GL}(2,\mathbb{C})/\mathbf{U}(2) \\ &\cong \mathbf{D}(1) \times \mathbf{SL}(2,\mathbb{C})/\mathbf{SU}(2), \\ \exp \mathbb{H} &\cong \mathbf{D}(1) \times \mathbf{SU}(2). \end{aligned}$$

The group $\mathbf{D}(1) \times \mathbf{SU}(2)$ for the quaternions is used as homogenous group for nonrelativistic space-time, where the dilations $\mathbf{D}(1) \ni e^{\psi}$ act on the time translations and the rotations $\mathbf{SU}(2) \sim \mathbf{SO}(3)$ on the position translations:

$$\begin{aligned} &[\mathbf{D}(1) \vec{\times} \mathbb{R}] \times [\mathbf{SU}(2) \vec{\times} \mathbb{R}^3] \ni (e^{\psi}, t) \times (u, \vec{x}), \\ &\text{with } (e^{\psi_1}, t_1) \bullet (e^{\psi_2}, t_2) = (e^{\psi_1+\psi_2}, t_1 + e^{\psi_1} t_2). \end{aligned}$$

Good to know: For a *semidirect* product $H \vec{\times} N$ of two groups with individual compositions $(h_1, h_2) \longmapsto h_1 h_2$ and $(n_1, n_2) \longmapsto n_1 \diamond n_2$, respectively, H has to be realized by N-automorphisms $(h, n) \longmapsto h(n)$. Then, $H \vec{\times} N$ is endowed with the group product $(h_1, n_1) \bullet (h_2, n_2) = (h_1 h_2, n_1 \diamond h_1(n_2))$. N is embedded as a normal subgroup of $H \vec{\times} N$. Is a group uniquely factorizable with $G = N \circ H$ and $N \cap H = \{1_G\}$ where N is a normal subgroup and H a subgroup, there exists the isomorphism $G \cong H \vec{\times} N$. If $H = N'$ is even a normal subgroup, the product is direct $G \cong N' \times N$.

The generators of the Lorentz group constitute the corresponding real 6-dimensional Lie algebra $A_1^c \oplus iA_1^c$. In addition to the real three-dimensional angular momentum Lie algebra A_1^c with basis $(\mathbf{J}^a)_{a=1}^3$ there is the real three-dimensional boost-vector space iA_1^c with basis $(\mathbf{B}^a)_{a=1}^3$. They have a complex two-dimensional representation by Pauli matrices:

$$A_1^c \oplus iA_1^c \ni (\mathbf{J}^a, \mathbf{B}^a) \longmapsto (\frac{i\sigma^a}{2}, \frac{\sigma^a}{2}) \in \mathbf{AL}(2, \mathbb{C})_0, \quad \begin{cases} [\mathbf{J}^a, \mathbf{J}^b] = -\epsilon^{abc}\mathbf{J}^c, \\ [\mathbf{J}^a, \mathbf{B}^b] = -\epsilon^{abc}\mathbf{B}^c, \\ [\mathbf{B}^a, \mathbf{B}^b] = \epsilon^{abc}\mathbf{J}^c, \end{cases}$$

or in a Lorentz vector notation $(\mathbf{L}^{jk})_{j,k=0}^3$:

$$\left.\begin{array}{l} \mathbf{L}^{ab} = \epsilon^{abc}\mathbf{J}^c, \\ \mathbf{L}^{0a} = -\mathbf{L}^{a0} = \mathbf{B}^a, \end{array}\right\} \text{with } [\mathbf{L}^{jk}, \mathbf{L}^{nm}] = \eta^{jn}\mathbf{L}^{km} - \eta^{kn}\mathbf{L}^{jm} - \eta^{jm}\mathbf{L}^{kn} + \eta^{km}\mathbf{L}^{jn}.$$

With the translation generators $\mathbf{P}^j = (\mathbf{P}^0, \mathbf{P}^a)$ one has the Lie brackets for the Poincaré Lie-algebra:

$$\begin{aligned} &[\mathbf{P}^j, \mathbf{P}^k] = 0, \\ &[\mathbf{L}^{jk}, \mathbf{P}^n] = \eta^{jn}\mathbf{P}^k - \eta^{kn}\mathbf{P}^j \text{ or } \begin{cases} [\mathbf{J}^a, \mathbf{P}^0] = 0, & [\mathbf{J}^a, \mathbf{P}^b] = -\epsilon^{abc}\mathbf{P}^c, \\ [\mathbf{B}^a, \mathbf{P}^0] = \mathbf{P}^a, & [\mathbf{B}^a, \mathbf{P}^b] = \delta^{ab}\mathbf{P}^0. \end{cases} \end{aligned}$$

3.12 Internal Scalar Product and External Lorentz Metric

In general, the ordered algebra $\mathbf{AL}(d, \mathbb{C}) = \mathbb{R}(d) + i\mathbb{R}(d)$ of the complex linear endomorphisms with the hermitian and antihermitian real vector subspace and the positive cone $\mathbf{GL}(d, \mathbb{C})/\mathbf{U}(d) \cong \mathbb{D}(d) \subset \mathbb{R}(d)$ is acted on by the general linear group $g \in \mathbf{GL}(d, \mathbb{C}) = \exp \mathbf{AL}(d, \mathbb{C})$:

$$(\mathbf{AL}(d, \mathbb{C}), \mathbb{R}(d), \mathbb{D}(d)) \ni \mathbf{z} \longmapsto g \circ \mathbf{z} \circ g^\star \in (\mathbf{AL}(d, \mathbb{C}), \mathbb{R}(d), \mathbb{D}(d)).$$

The hermiticity uses a $\mathbf{U}(d)$-conjugation $\mathbf{z} \leftrightarrow \mathbf{z}^\star$ with a scalar product (sesquilinear positive form) of the basic complex internal space $V \cong \mathbb{C}^d$:

$$V \times V \longrightarrow \mathbb{C}, \quad \begin{cases} \langle h|k\rangle = \overline{\langle k|h\rangle}, \\ \langle h|h\rangle \geq 0, \ \ \langle h|h\rangle = 0 \iff h = 0, \end{cases}$$
$$\text{orthonormalized basis: } h = h_\alpha e^\alpha, \ \alpha = 1, \ldots, d, \ \langle e^\alpha|e^\gamma\rangle = \delta^{\alpha\gamma}.$$

It induces an, up to overall normalization, unique scalar product ("double trace") of the algebra $\mathbf{z} = z_\alpha^\beta e^\alpha \otimes \check{e}_\beta \in \mathbf{AL}(d, \mathbb{C})$:

$$\mathbf{AL}(d, \mathbb{C}) \times \mathbf{AL}(d, \mathbb{C}) \longrightarrow \mathbb{C}, \quad \begin{cases} \langle \mathbf{y}|\mathbf{z}\rangle = \operatorname{tr} \mathbf{z} \circ \mathbf{y}^\star, \\ \langle e^\alpha \otimes \check{e}_\beta | e^\gamma \otimes \check{e}_\delta\rangle = \delta^{\alpha\gamma}\delta_{\beta\delta}, \\ \langle \mathbf{z}|\mathbf{z}\rangle = \operatorname{tr} \mathbf{z} \circ \mathbf{z}^\star = z_\alpha^\beta \overline{z_\alpha^\beta}. \end{cases}$$

In an $\mathbf{AL}(d, \mathbb{C})$-basis $\{\mathbf{1}_d, \sigma^a(d)\}$ with traceless Pauli matrices one has to use the Fierz[34] rearrangement

$$d \geq 2: \ \ \delta^{\alpha\gamma}\delta_{\beta\delta} = \frac{1}{d}\delta_\beta^\alpha \delta_\delta^\gamma + \frac{1}{d(d+1)}\sigma^a(d)_\beta^\alpha \sigma^a(d)_\delta^\gamma,$$
$$d = 2: \ \ \delta^{\alpha\gamma}\delta_{\beta\delta} = \frac{1}{2}\delta_\beta^\alpha \delta_\delta^\gamma + \frac{1}{6}\vec{\sigma}_\beta^\alpha \vec{\sigma}_\delta^\gamma.$$

The $\mathbf{U}(d)$-normalization, which can be used for the hyper-isospin group $\mathbf{U}(2)$ acting on an isospinor space $V \cong \mathbb{C}^2$, leads to the relative normalization of the abelian $\mathbf{U}(1)$ and the nonabelian $\mathbf{SU}(d)$ with Lie algebra basis $\{i\mathbf{1}_d, i\sigma^a(d)\}$ of $i\mathbb{R}(d)$:

$$\tan^2\theta(d) = \ell^2_{\mathbf{SU}(d)} : \ell^2_{\mathbf{U}(1)} = \frac{1}{d+1}, \ \ \text{e.g., } \tan^2\theta(2) = \ell^2_{\mathbf{SU}(2)} : \ell^2_{\mathbf{U}(1)} = \frac{1}{3}.$$

The related orthogonal normalization triangle has sides $(1, \sqrt{d+1}; \sqrt{d+2})$. In the case of hypercharge and isospin, there arises a "beautiful Platonic triangle" $(1, \sqrt{2}; 2)$ with angles $(\frac{\pi}{3}, \frac{\pi}{6}; \frac{\pi}{2})$.

$\mathbf{AL}(d, \mathbb{C})$ has another multilinear form: The determinant (volume element) $\mathbf{z} \longmapsto \det \mathbf{z}$ defines an $\mathbf{SL}(d, \mathbb{C})$-invariant form, multilinear of degree d. It is a "metric" (bilinear form) with a signature, relevant for the distinction of time and position, only for four-dimensional spacetime $\mathbb{R}(2)$. In this case, the bilinear volume form of the underlying two-dimensional spinor space $W \cong \mathbb{C}^2$,

$$W \times W \longmapsto \mathbb{C}, \ \ \epsilon(e^A, e^B) = \epsilon^{AB} = -\epsilon^{BA}, \ A = 1, 2, \ \epsilon^{12} = 1,$$

induces the, up to overall normalization, unique Lorentz metric η with causal signature for Minkowski spacetime $\mathbb{R}(2)$ with basis $\sigma^j = (\mathbf{1}_2, \vec{\sigma})$:

[34]Markus Fierz (1912–2006).

$$\mathbf{AL}(2,\mathbb{C}) \times \mathbf{AL}(2,\mathbb{C}) \longrightarrow \mathbb{C}, \quad \epsilon(e^A \otimes \check{e}_B, e^C \otimes \check{e}_D) = \epsilon^{AC}\epsilon_{BD},$$
$$\epsilon^{AC}\epsilon_{BD} = -\frac{1}{2}\delta^A_B\delta^C_D + \frac{1}{2}\vec{\sigma}^A_B\vec{\sigma}^C_D \cong -\frac{1}{2}\eta_{jk}\sigma^j \otimes \sigma^k, \quad \mathrm{sign}\,\eta = (1,3).$$

3.13 Operational Spacetime

Compared with the historically grown real formulation of spacetime, the complex formulation of flat Minkowski spacetime, via the Weyl representation, as the hermitian operations $\mathbb{R}(2) \subset \mathbf{AL}(2,\mathbb{C})$ in the "full" algebra opens new perspectives. It leads "naturally" to an order structure and an invariance group $\mathbf{SL}(2,\mathbb{C})$, covering the orthochronous Lorentz group $\mathbf{SO}_0(1,3) \cong \mathbf{SL}(2,\mathbb{C})/\mathbf{I}(2)$, as simple subgroup in the full linear group with additional dilation $\mathbf{D}(1)$- and phase $\mathbf{U}(1)$-operations, in this connection called hyper-Lorentz group:

$$\exp \mathbf{AL}(2,\mathbb{C}) = \mathbf{GL}(2,\mathbb{C}) = \mathbf{D}(1) \times \mathbf{U}(1) \circ \mathbf{SL}(2,\mathbb{C}).$$

In this approach, the $\mathbf{SL}(2,\mathbb{C})$-invariant volume elements induce the Lorentz metric, i.e., there is no choice with respect to the possible signatures of the metric for real four-dimensional spacetime—$\mathbf{O}(4)$ for definite signature, or $\mathbf{O}(1,3)$ for causal signature, or $\mathbf{O}(2,2)$ for neutral signature. With $\mathbf{SL}(2,\mathbb{C})$, there is no basic implementation of the discrete time and position reflection groups $\mathbf{O}(1,3)/\mathbf{SO}_0(1,3) \cong \mathbf{I}(2) \times \mathbf{I}(2)$ which, as seen in experiments, proved to be broken. The strict spacetime future cone (causality) is given by the group operations representing the unitary classes, $\mathbb{R}(2)_+ \cong \exp \mathbb{R}(2)$; flat Minkowski spacetime $\mathbb{R}(2)$ can be interpreted as the corresponding tangent Lie algebra classes

$$\log \mathbf{GL}(2,\mathbb{C})/\log \mathbf{U}(2) \cong \mathbb{R}(2) \ni \mathbf{x} = x_0\,\mathbf{1}_2 + \vec{x}\vec{\sigma},$$
$$\mathbf{GL}(2,\mathbb{C})/\mathbf{U}(2) \cong \exp \mathbb{R}(2) \ni e^{\mathbf{x}} = e^{x_0}(\cosh r\,\mathbf{1}_2 + \frac{\sinh r}{r}\vec{x}\vec{\sigma}), \quad r = |\vec{x}|.$$

The complex formulation also connects closely the spacetime structure with Pauli, Weyl and Dirac spinors as fundamental and experimentally important implementations of the spacetime groups, as used, e.g., by lepton and quark fields.

The ontology of position (space) and time is a difficult question: Position and time do not "exist by their own", they are not empty boxes or stages for the interactions of mass points and particles. To characterize spacetime as "that arena within which all the phenomena of the physical universe appear to take place" may be misleading. In this book: Time and position parametrize, or are defined by physically relevant operations, or classes of operations, as will be illustrated in many examples. For example, the Euclidean plane is given by the rotation classes $\mathbb{E}^2 = \mathbf{SO}(2) \vec{\times} \mathbb{R}^2/\mathbf{SO}(2)$ of the Euclidean group, as manifold isomorphic to $\mathbb{R}^2$. A rather complicated example is real four-dimensional Galileo space-time $\mathbb{G}^4 = \frac{[\mathbf{O}(1)\times(\mathbf{O}(3)\vec{\times}\mathbb{R}^3)]\vec{\times}[\mathbb{R}\oplus\mathbb{R}^3]}{\mathbf{O}(1)\times(\mathbf{O}(3)\vec{\times}\mathbb{R}^3)}$, only as manifold isomorphic to $\mathbb{R}^4$, parametrizing the equivalence classes with respect to, i.e., defined up to real six-dimensional rotation and velocity transformations in the also time involving real ten-dimensional rotation-translation group, i.e., space-times related to each other by rotations and velocity transformations are equivalent and identifiable.

Interactions with particles as one concept and position with time as another concept cannot be separated; they cannot be thought of without each other. Einstein (1950): "According to general relativity, the concept of space detached from any physical content does not exist. The physical reality is represented by a field." And in 1954: "Space as opposed to 'what fills space' which is dependent on the coordinates has no separate existence...There is no such thing as an empty space, i.e., a space without field." Or Born (1962): "It is not space that is there and that impresses its forms on things, but the things and their physical laws that determine space" — no spacetime without interactions and matter.

Important dynamical structures in physics show the connection between spacetime concepts, coming as group related manifolds, and operations: Time operations, parametrized by the real group $\mathbb{R}$, are implemented by the harmonic oscillator. Flat position, parametrizing structures of the Euclidean group $\mathbf{O}(3) \vec{\times} \mathbb{R}^3$ with rotations and translations, is implemented by the asymptotic structures of the nonrelativistic scattering theory (potential scattering). The atomic spectrum, characterized by the states of the nonrelativistic hydrogen atom, is an implementation of the Lorentz group as motion group of the 3-hyperboloid $\mathcal{Y}^3 \cong \mathbf{SO}_0(1,3)/\mathbf{SO}(3)$ as a maximally symmetric curved position, in this context called Kepler position. Flat Minkowski spacetime, given by the equivalence classes[35] $\mathbb{R}^{(1,3)} = \mathbf{O}(1,3) \vec{\times} \mathbb{R}^4/\mathbf{O}(1,3)$ of the Poincaré group $\mathbf{O}(1,3) \vec{\times} \mathbb{R}^4$ with respect to the Lorentz group $\mathbf{O}(1,3)$, is implemented by free particles. The inhomogeneous tangent groups:

$$\left.\begin{array}{c}\mathbf{O}(1) \vec{\times} \mathbb{R}(\text{time}) \\ \mathbf{O}(3) \vec{\times} \mathbb{R}^3 \text{ (flat position)}\end{array}\right\} \subset \mathbf{O}(1,3) \vec{\times} \mathbb{R}^4 \text{ (flat spacetime)},$$

can be related to operation groups or to manifolds of operation group classes:

$$\left.\begin{array}{c}\mathbf{D}(1) \text{ (eigentime)} \\ \mathcal{Y}^3 \cong \mathbf{SL}(2,\mathbb{C})/\mathbf{SU}(2) \\ \text{(hyperbolic position)}\end{array}\right\} \subset \begin{array}{c}\mathbf{D}(1) \times \mathcal{Y}^3 \cong \mathbf{GL}(2,\mathbb{C})/\mathbf{U}(2) \\ \text{(electroweak spacetime).}\end{array}$$

Possibly, and this attitude is assumed in this book, physics should not be characterized as the theory of particles and interactions "in" time and position and spacetime, but — especially quantum theory — as theory "of" time and position and spacetime, as parametrized by operation groups and represented by interactions and particles. A distinguished operation group for spacetime, here $\mathbf{GL}(2,\mathbb{C})$ with its four noncompact degrees of freedom, its unitary classes $\mathbf{GL}(2,\mathbb{C})/\mathbf{U}(2)$, as spacetime model asks for its origin and motivation – what is so special about $\mathbf{GL}(2,\mathbb{C})$? Perhaps:

[35] Albert Michelson (1852–1931), Edward Morley (1838–1923).

In an operational interpretation, the complex numbers with canonical conjugation $\mathbb{C} = \mathbb{R} + i\mathbb{R}$ have to be used to parametrize noncompact and compact operations by real $\mathbb{R}$ and imaginary $i\mathbb{R}$, respectively. Then, the minimal complex-linear *nonabelian* structures $\mathbf{GL}(2, \mathbb{C})$ formalize the basic properties for quantum theories with interaction. Is there a basic connection with the mathematical formalization of the human epistemology?

Chapter 4
Plato's Beautiful Symmetry

In the history of physics, the regularity, symmetry and beauty of the five Platonic solids in three-dimensional Euclidean space induced philosophers and scientists to use them for a description of nature — by Plato directly, by Kepler in his mathematical-religious hymns concerning our solar system, today for elementary particles. In the course of deeper experimental and theoretical insights, those attempts proved mostly wrong with respect to their original meaning. Nevertheless, they initiated often methods in physics which were strongly influenced by similar symmetry considerations.

If we interpret Plato's or Kepler's symmetry ideas with the concepts, methods and experiences of today, we run the danger either to admire them, in displaying a superhuman in- and foresight, or to denigrate them as lofty speculations — both attitudes do them no justice. On the one hand, we can talk today, with the many experimental results and the sharper mathematical tools, much more distinctly about the physical structures. On the other hand, I do not believe that Plato and Kepler did not know about the antinomies of the infinitely small, e.g., "all particles have particle constituents" versus "there exist last indivisible particles," as clearly formulated later by Kant,[1] which arise from a radical atomistic point of view, advocated by Demokrit and Leukipp[2] and revived by Gassendi.[3] That they thought about basic concepts and replaced objects and particles (basic states) by symmetries (basic operations), also motivated by esthetical and religious attitudes and considerations, is admirable.

To clarify my attitude with respect to the description of basic physical structures by symmetries, I quote from the last public talk (1975) of Heisenberg *Was ist ein Elementarteilchen* (my translation):

---

[1]Immanuel Kant (1724–1804).

[2]Democritos, around -(460–370), Leukippos (?), around -440.

[3]Pierre Gassendi (1592–1655).

H. Saller, *Operational Symmetries*, DOI 10.1007/978-3-319-58664-9_4

> It is unavoidable that we use a language originating from classical philosophy. We ask: What does the proton consist of? Is the quantum of light elementary or composite?, etc. However, all these questions are wrongly asked since the words "divide" and "consist of" have lost almost all their meaning. Therefore, it would be our task to adjust our language, our thinking, i.e., our scientific philosophy, to this new situation that has been created by the experiments. Unfortunately, that is very difficult. Therefore, there creep into particle physics, always again, wrong questions and wrong conceptions, ...
>
> We have to come to terms with the fact that the experimental experiences for very small and for very large distances no longer provide us with an *anschauliches Bild* and we have to learn to live there without *Anschauung* (something like "without familiar everyday pictures"). In this case we realize that the antinomy of the infinitely small for the elementary particles is resolved in a very subtle way — in a way that neither Immanuel Kant nor the Greek philosophers could think of — the words "to divide" lose their sense.
>
> If one wants to compare the insights of today's particle physics with any earlier philosophy, it could only be the philosophy of Plato, since the particles of today's physics are representations of symmetry groups — that's what quantum theory teaches us — and, therefore, the particles resemble the symmetric Platonic solids.

Perhaps one can mitigate the resignation to live without "Anschauung" in the following sense: We have to get used to a new kind of "Anschauung," oriented at symmetries and related concepts like operations, invariants, eigenvalues, etc.

Today's standard theory of particles essentially assumes that all matter is made up of two families of particles, quarks and leptons, all of which are "pointlike" concentrations of quantum fields. In contrast to the composite "brick" and bound state picture, quantum structure suggests a "quantum ladder" with different energy (matter) levels where energy and mass related operations are implemented by eigenstates, e.g., by molecules, atoms, nuclei, hadrons, quarks, leptons. "Elementarity" is a question of the available energy. In this context also Dirac may be quoted from his Nobel lecture (1933):

> The distinction between elementary particles and composite particles now becomes a matter of convenience. This reason alone is sufficient to compel one to give up the attractive philosophical idea that all matter is made up of one kind, or perhaps two kinds of bricks.

4.1 Regular Polygons and Platonic Triangles

The mathematical content of Plato's work starts with polygons: A polygon in the real two-dimensional plane $\mathbb{R}^2$ has the same number of corners n_0 and sides n_1:

$$\text{in } \mathbb{R}^2 : \quad n_0 - n_1 = 0.$$

This trivial relation is an "academic" example for the general Euler relation which connects with each other the numbers of subpolytopes in a given polytope - here the number of corners n_0 and sides n_1 as zero- and one-dimensional "subpolytopes," respectively, of a two-dimensional polygon (more ahead).

Lengths and angles make sense in the Euclidean plane $\mathbb{E}^2$ (more ahead): For each natural number $n_0 \geq 3$, there exists a *regular*, i.e., an equilateral and equiangular polygon $\{n_0\}$: From the sum of all angles $(n_0 - 2)\pi$, there follows the angle $\theta(n_0)$ at each corner:

$$\text{in } \mathbb{E}^2: \; n_0 = n_1 \geq 3, \; \theta(n_0) = \frac{n_0 - 2}{n_0}\pi.$$

The "smallest" regular polygon is the fundamental simplex $\{3\}$ of $\mathbb{E}^2$. Only the regular hexagon $\{6\}$ can be composed by such regular triangles:

$$\theta(n_0) = 2\theta(3) = 2\frac{\pi}{3} \iff n_0 = 6.$$

In the complex plane $\mathbb{C} \cong \mathbb{R}^2$, and with appropriate units, the corners of a regular polygon on its out-circle (circumcircle) or the central side points on the tangent in-circle realize the elements of the cyclic group:

$$\{n_0\} \cong \mathbf{I}(n_0) = \{z \in \mathbb{C} \,\big|\, z^{n_0} = 1\} = \{e^{i\frac{2\pi}{n_0}k} \,\big|\, k = 0, 1, \ldots, n-1\}.$$

For $n_0 \geq 3$, the cyclic groups $\mathbf{I}(n_0)$ define stars:

Cyclic group	Star
$\mathbf{I}(1)$	trivial
$\mathbf{I}(2)$	reflection
$\mathbf{I}(3)$	Mercedes
$\mathbf{I}(4)$	NATO
$\mathbf{I}(5)$	Texaco
$\mathbf{I}(6)$	David

For the hexagon, the David[4] star group $\mathbf{I}(6) = \mathbf{I}(2) \times \mathbf{I}(3)$ involves, in the exponent, the classes $[\frac{k}{6}] = [\frac{1}{6}], [\frac{1}{3}], [-\frac{1}{2}], [-\frac{2}{3}], [-1]$. Those numbers are relevant for the lepton and quark hypercharges in the electroweak-strong standard model where the cyclic factor groups $\mathbf{I}(2)$ and $\mathbf{I}(3)$ are the centers (abelian normal subgroups) of the isospin and color operations, respectively.

Good to know: Each group G of order n is isomorphic to a subgroup of the permutation group of n elements, i.e., of the bijections $\mathbf{G}(n) = \{\{1, 2, \ldots, n\} \leftrightarrow \{1, 2, \ldots, n\}\}$ with order $n!$.
The group law of G can be given in the group table where both each column and each line contains all group elements,

$g_1 = 1_G$	g_2	$\cdots$	g_n
g_2	g_2^2	$\cdots$	$g_2 g_n$
$\cdots$	$\cdots$	$\cdots$	$\cdots$
g_n	$g_n g_2$	$\cdots$	g_n^2

In a finite group, the order of each subgroup $H \subseteq G$ divides the group order, $\text{card}\, G/\text{card}\, H \in \mathbb{N}$. Each group element defines, by its powers $\{g^k\}_{k=0}^{n-1}$, a cyclic subgroup. Therefore: Each group of prime order p is cyclic, i.e., isomorphic to $\mathbf{I}(p)$. The isomorphic additive groups $\mathbb{Z}_p$ are even fields. Each group of squared prime order p^2 is abelian and isomorphic to either $\mathbf{I}(p^2)$ or $\mathbf{I}(p) \times \mathbf{I}(p)$. From order p^3 on, it becomes more complicated: At least, all groups of order p^k are solvable (meta-abelian).

---

[4]David -(1004–965).

If, for a finite group G, the prime number decomposition of the group order is given by $n = p_1^{k_1} \cdots p_r^{k_r}$, a subgroup $U \subseteq G$ of order $p_j^{k_j}$ is called a *Sylow*[5] *subgroup*. For each monogeneous factor p^k of n there exists a Sylow subgroup in G. All finite groups are "known" and classified.

Regular polygons, especially those which are constructable with ruler and compasses alone played a role in Kepler's 'Harmonice Mundi (1619)', a mystically motivated synopsis of nature. Gauss has shown which regular polygons can be constructed with ruler and compasses alone: The number of corners n_0 has to be the product of a power of 2 and pairwise different Fermat[6] prime numbers:

$$n_0 = \begin{cases} 2^n, \quad n = 2, 3, \ldots, \\ 2^n \times p_1 \times \cdots \times p_k, \quad n = 0, 1, 2, \ldots \\ \text{with } p_s \in \mathbb{P}_{\text{Fermat}}, \quad p_i \neq p_j \text{ for } i \neq j. \end{cases}$$

The Fermat numbers involve special powers of 2:

$$\mathbb{N}_{\text{Fermat}} = \{n_s = 2^{2^s} + 1 \mid s = 0, 1, \ldots\}$$

s	0	1	2	3	4	5
n_s	3	5	17	257	65 537	4 294 967 297

They have the product property:

$$n_{k+1} - 1 = \prod_{s=0}^{k} (n_s + 1) = (n_k - 1)^2, \text{ i.e., } 2^{2^{k+1}} - 1 = \prod_{s=0}^{k} (2^{2^s} + 1).$$

The first five Fermat numbers are primes, the sixth one not:

$$\begin{aligned} \mathbb{P}_{\text{Fermat}} &= \mathbb{N}_{\text{Fermat}} \cap \mathbb{P} = \{3, 5, 17, 257, 65\,537, ???\}, \\ n_5 &= 2^{2^5} + 1 = 641 \times 6\,700\,417. \end{aligned}$$

Up to today, one does not know if there exist more than five Fermat primes. If the explicitly given Fermat primes are really the only ones (Can one really believe, with respect to all the computer power already applied, that one will find an additional one, e.g., near $2^{2^{137}}$?), all regular polygons, constructable with ruler and compasses, can be obtained from the monster-polygon with $2^{32} - 1 = 4\,294\,967\,295$ corners, almost a table for mankind. The triangle construction is easy, the pentagon construction uses the regular 10-gon, constructable with the golden ratio. From a regular n_1- and n_2-gon in a circle with central angles $\frac{2\pi}{n_{1,2}}$, one can construct, for coprime n_1 and n_2, the regular $n_1 n_2$-gon since $n_1\mathbb{Z} + n_2\mathbb{Z} = \mathbb{Z}$, e.g., from the triangle and pentagon the

[5] Ludwig Sylow (1832–1918).

[6] Pierre de Fermat (1601–1665).

15-gon with the central angle $\frac{2\pi}{15} = 2\frac{2\pi}{5} - \frac{2\pi}{3}$. A bisection of all sides doubles the corner number.

Good to know: It is amusing to look for the "largest number ever to have arisen naturally in mathematics." For example, looking for solutions of $m_1^n + \cdots + m_{n-1}^n = m_n^n$ with integer powers of integers, generalizing Fermat's question for $m_1^3 + m_2^3 = m_3^3$, the smallest solution for $m_1^4 + m_2^4 + m_3^4 = m_4^4$ is given by $(m_1, m_2, m_3, m_4) = (95\,800, 217\,519, 414\,510, 422\,481)$ — and that are by far not the "largest numbers ever to have arisen naturally in mathematics."

A regular polygon $\{n_0\}$ consists of triangles (the wedges of the cake) with two equal sides and $\frac{2\pi}{n_0}$ as angle in the center. Half of such a triangle is orthogonal with the angles:

$$\frac{\pi}{2}, \frac{\pi}{n_0}, \frac{\pi}{2} - \frac{\pi}{n_0} = \frac{(n_0 - 2)\pi}{2n_0}.$$

This gives the ratio of the radii of the tangent in-circle to the corners containing out-circle:

$$\frac{r_{\{n_0\}}}{R_{\{n_0\}}} = \cos\frac{\pi}{n_0}.$$

Some regular polygons are distinguished by the property that they can be used for a complete tiling of the Euclidean plane or for the walls of your bathroom or as chicken wire. If, in such a tiling, m_0 regular polygons of type $\{n_0\}$ come together in one point, their angles have to add up to 2π:

$$\text{in } \mathbb{E}^2: \quad \begin{cases} n_0, m_0 \geq 3, \;\; m_0\theta(n_0) = 2\pi \iff \frac{1}{n_0} + \frac{1}{m_0} - \frac{1}{2} = 0. \\ \textit{Diophantic equality}: D = 2(n_0 + m_0) - n_0 m_0 = 0. \end{cases}$$

This allows only the *Platonic mosaics* with regular triangles, squares and hexagons:

Polygon	n_0	m_0	D	$\cos^2\frac{\pi}{n_0}$
Regular triangle	3	6	0	$\frac{1}{4}$
Square	4	4	0	$\frac{1}{2}$
Regular hexagon	6	3	0	$\frac{3}{4}$

Platonic mosaics in the plane $\mathbb{E}^2$

Equivalent to the Diophantic[7] equality number $D = 0$ is the cosine relation for mosaic angles:

$$\frac{\pi}{2} - \frac{\pi}{n_0} = \frac{\pi}{m_0} \Rightarrow 1 - \cos^2\frac{\pi}{n_0} - \cos^2\frac{\pi}{m_0} = 0.$$

---

[7]Diophantus of Alexandria, around (200–284).

Triangle and hexagon mosaic with exchanged numbers for corners and sides $n_0 \leftrightarrow m_0$ are *dual* to each other in the following sense: If the central points of neighbouring triangles in the triangle mosaic are connected with each other over the centers of the sides, one obtains the hexagon lattice, and vice versa. The sides of both lattices are orthogonal with the ratio of their lengths:

$$l_3 : l_6 = \sqrt{3} : 1.$$

For the square lattice, there arises the dual square lattice with sides of equal length. The triangle is the *vertex polygon* of the hexagon mosaic and vice versa; the square is its own vertex polygon.

The bisected equilateral triangle with angles 30, 60 and 90° and ratio $1 : 2 : \sqrt{3}$ for the side lengths and the diagonally bisected square with corresponding values $45, 45, 90°$ and $1 : 1 : \sqrt{2}$ are the two *Platonic triangles*. They can be also constructed by one quarter of the square and one sixth of the regular triangle. The two Platonic triangles are the two bisected orthogonal triangles of the three mosaic polygons:

$$\left(\frac{\pi}{n_0}, \frac{\pi}{2} - \frac{\pi}{n_0}, \frac{\pi}{2}\right) = \begin{cases} (\frac{\pi}{3}, \frac{\pi}{6}, \frac{\pi}{2}), \; n_0 = 3, \text{ with } 1^2 + \sqrt{3}^2 &= 2^2, \\ (\frac{\pi}{6}, \frac{\pi}{3}, \frac{\pi}{2}), \; n_0 = 6, \text{ with } \sqrt{3}^2 + 1^2 &= 2^2, \\ (\frac{\pi}{4}, \frac{\pi}{4}, \frac{\pi}{2}), \; n_0 = 4, \text{ with } 1^2 + 1^2 &= \sqrt{2}^2. \end{cases}$$

Platonic triangles

The Pythagorean number mysticism for 1, 2 and 3 played an essential role in choosing these triangles.

4.2 Platonic Solids

In Plato's geometry one goes from the plane with polygons to the space with polyhedra — plane and space assumed, unconsciously by Plato, as Euclidean. That it is possible to define a non-Euclidean "plane" and space had to wait for over two thousand years till Bolyai and Lobachevski.[8]

Platonic solids are regular convex polyhedra in Euclidean space $\mathbb{E}^3$, they have as faces equal regular polygons and equal solid angles at each corner (vertex). They are "honest" solids with an equipartition of the surface for casting them and playing "dice," generalizing the cubus (hexahedron).

If at each vertex of a Platonic solid there meet m_0 regular polygons $\{n_0\}$, the sum of the m_0 equal polygon angles $\theta(n_0)$ has to be smaller than 2π to form a "convex corner":

[8]Nikolai Lobachevsky (1792–1856), Janos Bolyai (1802–1860).

$$\text{in } \mathbb{E}^3: \quad \begin{cases} n_0 \geq 3, \; m_0 \geq 3, \; m_0\theta(n_0) < 2\pi, \\ \Rightarrow \frac{1}{n_0} + \frac{1}{m_0} - \frac{1}{2} > 0 \iff D > 0 \; (\textit{Diophantic inequality}). \end{cases}$$

This necessary inequality, symmetric in n_0 and m_0, has five solutions (table ahead) which can be realized with the explicit coordinates given ahead.

The number N_0 of the vertices of a simply connecting polyhedron minus the number N_1 of its edges plus the number N_2 of its faces fulfils the general

$$\textit{Euler equation for polyhedra in } \mathbb{R}^3: N_0 - N_1 + N_2 = 2.$$

Therefore the numbers $N_{0,1,2}$ for the Platonic solids are given as follows: The polygons with n_0 corners are glued together, joining two edges, and at each vertex there join m_0 polygons:

$$\text{in } \mathbb{E}^3: \quad \begin{cases} N_1 = \frac{N_2 n_0}{2}, \; N_0 = \frac{N_2 n_0}{m_0}, \quad D = 2(n_0 + m_0) - n_0 m_0 \\ \Rightarrow N_0 = \frac{4n_0}{D}, \; N_1 = \frac{2n_0 m_0}{D}, \; N_2 = \frac{4m_0}{D}. \end{cases}$$

In the resulting table, the number N_2 of the regular faces gives the solids their names:

Polyhedron	n_0	m_0	D	N_0	N_1	N_2	$\cos^2 \frac{\pi}{n_0}$	$\tan^2 \frac{\pi}{n_0}$
Tetrahedron	3	3	3	4	6	4	$\frac{1}{4}$	3
Octahedron	3	4	2	6	12	8	$\frac{1}{4}$	3
Icosahedron	3	5	1	12	30	20	$\frac{1}{4}$	3
Hexahedron	4	3	2	8	12	6	$\frac{1}{2}$	1
Dodecahedron	5	3	1	20	30	12	$\frac{3+\sqrt{5}}{8}$	$5 - 2\sqrt{5}$

Vertices, edges and polygons of the Platonic solids

All characterizing integers $N_{0,1,2}$ are even. As faces, there occur three types of regular polygons:

triangles (tetrahedron, octahedron, icosahedron),
squares (hexahedron),
pentagons (dodecahedron).

4.3 Plato's Immaterial Basic Physics

For the Greeks as a seafaring people, a composition of their surroundings was suggested as given by basic elements earth, water, air and fire (lightnings, sun, stars). Plato's natural philosophy looked for ideal ("beautiful") forms in the "world of mathematics," realized by these four elements and by the whole cosmos. As a solution he found the distinguished symmetry structure of the exactly five — therefore so called — Platonic solids, mathematically developed by his disciple Theaitetos.[9]

Plato judged an explanation of matter by "last matter" as unsatisfactory, he proposed as a deepest origin immaterial symmetries, whose holistic aspects contain also the property "matter." In his *Timaios* dialogue, the four solids with regular triangles and squares — both constructable from his two triangles — are associated with the four elements, the fifth solid with pentagons serves as "ground state" model for space, i.e., for the cosmic quintessence (ether):

fire(F)	air(A)	earth(E)	water(W)	quintessence
tetrahedron	octahedron	hexahedron	icosahedron	dodecahedron

Association of the elements with Platonic solids

Simplest dynamical considerations, similar to earlier ideas of the Pythagoreans, illustrated the transitions of the elements (without earth) into each other by rearranging the regular polygons which were, in a sense, the "first" structures, i.e., the "protons." Interaction is interpreted as symmetry rearrangement. Denoting the three Platonic solids with triangles by a capital letter for the associated element with an upper index for the number N_2 of triangles:

$$\{3,3\} = F^4, \quad \{3,4\} = A^8, \quad \{3,5\} = W^{20},$$

one could formulate a few reactions by a rearrangement of the triplet-protons (immaterial triangles) to form different material solids, e.g., water decomposes into fire and air:

$$W^{20} \longrightarrow F^4 + 2A^8, \quad A^8 \longrightarrow 2F^4, \text{ etc.}$$

Obviously, a detailed comparison with chemical or elementary particle reactions, possibly even with the protons of today composed by three quarks, would immensely overstretch and, also, misinterpret Plato's ideas. Nevertheless, his consequent immaterial interpretation remains impressive.

Plato's idealistic and deductive epistemology was, to some degree, replaced and modified by Aristotle's nominalistic and inductive one. These two philosophical approaches were characterized, in scholastic times, by the catchphrases "universalia ante res" and "universalia post res," respectively. Physics today, with theory and experiment, employs both points of view — "universalia in rebus."

---

[9]Theaitetos of Athens, around -(417–369).

4.4 Dual Platonic Solids

The five Platonic solids are related to each other. For example, at each vertex of the hexahedron, there meet *three* regular polygons with *four* vertices (squares)—at each vertex of the octahedron, there meet—with an exchange of the two numbers—*four* regular polygons with *three* vertices (triangles). With an equal sum $n_0 + m_0$ of the number of polygon corners n_0 and meeting number m_0, the exchange $n_0 \leftrightarrow m_0$ and, therefore, the exchange of the numbers of vertices and faces $N_0 \leftrightarrow N_2$ determines *dual (reciprocal) Platonic solids* with an equal number N_1 of edges:

$$\begin{array}{rcl} (N_0, N_1, N_2) & \leftrightarrow & (N_2, N_1, N_0), \\ (n_0, m_0) & \leftrightarrow & (m_0, n_0), \end{array} \left\{ \begin{array}{lll} N_1 = 6 & \text{edges:} & \text{tetrahedron is self-dual,} \\ N_1 = 12 & \text{edges:} & \text{hexahedron and} \\ & & \text{octahedron is a dual pair,} \\ N_1 = 30 & \text{edges:} & \text{dodecahedron and} \\ & & \text{icosahedron is a dual pair.} \end{array} \right.$$

The dual relationship can be visualized as follows: If, for a Platonic $\{n_0, m_0\}$-solid (take a cube), the centers of the $\{n_0\}$-polygons (squares), meeting in one vertex (three squares), are connected by lines, one obtains the regular *vertex polygon* $\{m_0\}$ (a triangle). N_0 vertices lead to N_0 vertex polygons (eight triangles). Inside the original Platonic solid with N_0 vertices there arises a Platonic N_0-solid:

$$1 - \cos^2 \frac{\pi}{n_0} - \cos^2 \frac{\pi}{m_0} > 0.$$

Repeating this procedure, one obtains an N_2-solid, inside the N_0-solid and not identical with the original N_2-solid, i.e., this procedure is not involutive.

Good to know: The linear forms of a finite-dimensional vector space V constitute its *dual vector space* V^T. There exist pairs with dual bases, $\{e^j\}_{j=1}^n$ for V and $\{\check{e}_j\}_{j=1}^n$ for V^T with the dual product $\langle \check{e}_k, e^j \rangle = \delta_k^j$. A nondegenerate bilinear form defines an isomorphism of a vector space and its dual $V^T \longrightarrow V$. For definite positive forms, duality, e.g., dual bases, can be re-interpreted in terms of orthogonality, e.g., orthogonal bases via $\check{e}_j \longmapsto e^j$.

For the regular polygons in two dimensions with $n_0 = n_1$, there exists two characteristic circles: the out-circle with the n_0 corners and the in-circle, tangent to the n_1 sides. For the regular polyhedra in three dimensions, there are three characteristic spheres: the *vertex sphere* (out-sphere, circumsphere) with the N_0 vertices, the *polygon sphere* (in-sphere), tangent to the N_2 regular polygons in their central points, and the *edge sphere*, tangent to the N_1 edges in their central points. To have an involutive dual relation with $N_0 \leftrightarrow N_2$, i.e., to come back to the original Platonic solid one has to use the edge sphere: In the center of all N_1 edges, one constructs orthogonal lines, tangent to the edge sphere. In such a way, there arise two dual solids with equal edge sphere. They penetrate each other, all $N_0 + N_2$ vertices are on the common out-sphere.

In this way, the self-dual tetrahedron $\{3, 3\}$ is reproduced in a rotated form which intersects with the original tetrahedron in an *octahedron*. The intersection of octahedron and dual hexahedron, penetrating each other, is a *hexoctahedron (cuboctahedron)* with twelve vertices. It has as faces six squares and eight regular triangles. In two different ways, one can recognize three regular polygons, slicing the hexoctahedron in three parallel planes: In one view $3+6+3=12$ vertices for triangle, hexagon, and triangle — in the other view $4+4+4=12$ vertices for little square, big square, and little square. In addition to this *Archimedean*[10] *solid* as intersection of $\{3,4\}$ and $\{4,3\}$ there arises the *icosidodecahedron* as intersection of $\{3,5\}$ and $\{5,3\}$. Its faces are 20 regular triangles and 12 pentagons with 30 vertices which can be seen in 5 parallel planes with $5+5+10+5+5$ vertices for four pentagons and one decagon. The intersection-polyhedra of the three dual classes of Platonic solids are called *quasiregular*.

With $r_{\{n_0,m_0\}}$ the radius of the in-sphere of a Platonic solid, and $R_{\{n_0,m_0\}}$ the radius for the out-sphere, their ratio coincides for dual solids:

$$\frac{R_{\{n_0,m_0\}}}{r_{\{n_0,m_0\}}}=\tan\frac{\pi}{n_0}\tan\frac{\pi}{m_0}=\frac{R_{\{m_0,n_0\}}}{r_{\{m_0,n_0\}}},$$

n_0, m_0	$\tan^2\frac{\pi}{n_0}\tan^2\frac{\pi}{m_0}$
3, 3	9
3, 4	3
3, 5	$15-6\sqrt{5}$

Each dual class has one member with meeting number $m_0=3$: These three Platonic solids with the regular triangle as vertex polygon can be nested into each other - the tetrahedron in the hexahedron, the latter one in the dodecahedron:

$$\text{tetrahedron}(6=3+3)\subset\text{hexahedron}(7=3+4)\subset\text{dodecahedron}(8=3+5).$$

4.5 Soccer Molecules

For the Platonic solids, the "cornerness" δ (angle defect to 2π at each vertex) is inverse proportional to the vertex number — the more vertices, the "rounder." The solid angles ω at each vertex distribute 4π to the faces:

$$\{n_0,m_0\}:\quad\begin{cases}\text{"cornerness"}: 2\pi-m_0\theta(n_0)=\delta=\frac{\pi}{n_0}D=\frac{4\pi}{N_0},\\ \text{solid angle:}\qquad\qquad\qquad\quad \omega=\frac{\pi}{m_0}D=\frac{4\pi}{N_2}.\end{cases}$$

In the same way as all central angles of a polygon have the sum 2π, i.e., the length of the unit circle, all solid angles of a convex polyhedron give the full space angle 4π, i.e., the area of the unit sphere — here $N_0\delta=N_2\omega=4\pi$. This was already known to Descartes.[11] An intuition for this fact can be obtained by the area relation of spherical trigonometry: The area of a triangle with great circles on the unit sphere as its sides is the spherical excess, which gives the difference of the angle sum to π, i.e., to the angle sum of the triangle in the Euclidean plane. Take as examples

[10] Archimedes of Syracuse, around -(287–212).

[11] René Descartes (1596–1650).

triangles with two great circles from the north pole to the equator where the area is the "north pole angle."

There are many possibilities, to modify the somewhat boring regularity of the Platonic solids and to construct less regular polyhedra. For example: For any simply connected polyhedron where there meet m_0 possibly different regular polygons in all vertices (constant meeting number) and which has as polygon-faces k_3 regular triangles, k_4 squares, etc., one has for the numbers of vertices, edges and faces:

$$[k_3, k_4, \ldots, k_m] : \quad (N_0, N_1, N_2) = \sum_{n_0=3}^{m} k_{n_0}(\frac{n_0}{m_0}, \frac{n_0}{2}, 1),$$

and, therefore, Euler's relation:

$$2 = N_0 - N_1 + N_2 = \sum_{n_0=3}^{m} (\frac{n_0}{m_0} - \frac{n_0}{2} + 1)k_{n_0}$$

Two more restricting conditions may be convexity and a constant meeting pattern at each of the N_0 vertices involving m_3 regular triangles, m_4 squares, etc. This gives as additional conditions:

$$m_0 = \sum_{n_0=3}^{m} m_{n_0}, \quad N_0 = \frac{n_0}{m_{n_0}} k_{n_0}, \quad n_0 = 3, \ldots, m.$$

In this case, one realizes that, because of the convex vertices with the angles, there can arise only three types of regular polygons: regular triangles, squares, and pentagons. By a careful discussion, all possibilities can be enumerated. For example, there exist 18 convex polyhedra with a constant meeting pattern.

As an explicit example the "soccer molecules" are considered: N_0 carbon atoms build molecules $[k_5, k_6]$ with pentagon- and hexagon-faces, the *fullerenes*[12] C_{N_0}. An application of the relations above for convex polyhedra with constant meeting pattern, there follows the meeting number $m_0 = 3$ and, therefore, the conditions:

$$\text{fullerenes: } C_{N_0} = [k_5, k_6] \Rightarrow \begin{cases} \text{vertices: } N_0 = \frac{5k_5+6k_6}{3}, \\ \text{edges: } N_1 = \frac{5k_5+6k_6}{2}, \\ \text{faces: } N_2 = k_5 + k_6, \\ 2 = N_0 - N_1 + N_2 = \frac{k_5}{6} \\ \Rightarrow k_5 = 12, \quad k_6 = \frac{N_0}{2} - 10. \end{cases}$$

[12]Richard Buckminster Fuller (1895–1983).

Fullerenes have always 12 pentagons. Since with the exception of the dodecahedron [12, 0], the meeting pattern at each vertex has to be either $(m_5, m_6) = (1, 2)$ or $(2, 1)$, one has necessarily:

$$N_0 = \frac{6k_6}{m_6} = \frac{5k_5}{m_5} \Rightarrow \begin{cases} (m_5, m_6) = (1, 2), & [k_5, k_6] = [12, 20], \; N_0 = 60, \\ (m_5, m_6) = (2, 1), & [k_5, k_6] = [12, 5], \quad N_0 = 30. \end{cases}$$

Only for the Buckminster-fulluren ("buckyball") C_{60} this condition is also sufficient, a polyhedron [12, 5] cannot be constructed.

4.6 Coordinates for Platonic Solids

It is quite useful to have *standard Platonic solids* with explicit Cartesian coordinates in Euclidean space $\mathbb{E}^3$ with orthonormal basic vectors:

$$e^1 = (1, 0, 0), \quad e^2 = (0, 1, 0), \quad e^3 = (0, 0, 1).$$

One should realize the difficulties for Plato–Theaitetos and Kepler without that tool, dreamed of by Descartes in Neuburg (Bavaria) during the 30-years war.

In the following, all Platonic solids $\{n_0, m_0\} \cong N_2$ are coordinatized with their center (of mass, if they are made of homogeneous matter) in the origin $(0, 0, 0)$ and nested in each other according to their duality and other relationships described above. In addition to the coordinates of all vertices, there will be given a *characteristic regular polygon* with its central point. Calculations with these standard solids yield, e.g., coordinate independent angles, lengths ratios involving the radii of in- and out-sphere and the edge length $(r_{N_2}, R_{N_2}, l_{N_2})$.

Good to know: The central point of a general simplex with n vertex vectors $\{a_j\}_{j=1}^n$ is the arithmetical mean $\langle a \rangle = \frac{a_1 + \cdots + a_n}{n}$. The center of mass of mass points is $\langle a \rangle = \frac{\sum_i m_i a_i}{m}$, $m = \sum_i m_i$, and, in the continuous case, $\langle \vec{a} \rangle = \frac{\int d^3x \, \mu(\vec{x}) \vec{x}}{m}$, $m = \int d^3x \, \mu(\vec{x})$ with a mass density μ.

One starts with the familiar *hexahedron* (all sign combinations):

$$\text{hexahedron:} \begin{cases} \text{8 vertices: } (\pm 1, \pm 1, \pm 1), \\ \text{characteristic square: } (1, 1, 1), \; (1, -1, 1), \; (1, -1, -1), \; (1, 1, -1), \\ \text{square center: } r_6 = (1, 0, 0), \; \text{edge length: } l_6^2 = 4, \\ \text{out-radius: } R_6^2 = 3, \; \text{ in-radius: } r_6^2 = 1. \end{cases}$$

The *octahedron* with its vertices at the central points of the squares of the hexahedron has the coordinates:

$$\text{octahedron:}\begin{cases}\text{6 vertices: } (\pm1,0,0),\ \ (0,\pm1,0),\ \ (0,0,\pm1),\\ \text{characteristic triangle: } (1,0,0),\ (0,1,0),\ (0,0,1),\\ \text{triangle center: } r_8=\frac{1}{3}(1,1,1),\ \text{edge length: } l_8^2=2,\\ \text{out-radius: } R_8^2=1,\ \ \text{in-radius: } r_8^2=\frac{1}{3}.\end{cases}$$

Here, one has the coordinate independent data:

$$\frac{R_6^2}{r_6^2}=\frac{R_8^2}{r_8^2}=3,\quad \tan^2\frac{\pi}{4}=1,\quad \tan^2\frac{\pi}{3}=3.$$

There are two ways to use the $8 = 4 + 4$ vertices of the hexahedron for the four vertices of a *tetrahedron*. They are related to each other by central reflection at the origin. One possibility is given as follows:

$$\text{tetrahedron:}\begin{cases}\text{4 vertices: } (1,1,1),\ (1,-1,-1),\ (-1,1,-1),\ (-1,-1,1),\\ \text{characteristic triangle: } (1,1,1),\ (1,-1,-1),\ (-1,1,-1),\\ \text{triangle center: } r_4=\frac{1}{3}(1,1,-1),\ \text{edge length: } l_4^2=8,\\ \text{out-radius: } R_4^2=3,\ \text{in-radius: } r_4^2=\frac{1}{3},\\ \frac{R_4^2}{r_4^2}=9,\ \ \tan^2\frac{\pi}{3}=3.\end{cases}$$

Now, there come the two more complicated dual Platonic solids involving regular pentagons: The $20 = 8 + 12$ vertices of the *dodecahedron* with their coordinates to be determined allow five possibilities for the eight vertices of a hexahedron, for example:

$$\text{dodecahedron:}\begin{cases}\text{20 vertices:}\begin{cases}(\pm1,\pm1,\pm1)\ \text{(hexahedron vertices)},\\ (\pm\alpha,\pm\beta,0),\ (0,\pm\alpha,\pm\beta),\ (\pm\beta,0,\pm\alpha),\\ \beta,\alpha>0,\end{cases}\\ \text{characteristic pentagon:}\begin{cases}v_1=(1,1,1),\quad v_2=(\alpha,\beta,0),\\ v_3=(\alpha,-\beta,0),\ v_4=(1,-1,1),\\ v_5=(\beta,0,\alpha).\end{cases}\end{cases}$$

To determine β and α: The vectors of corresponding vertices of the characteristic pentagon have equal scalar product for equal angle:

$$\langle v_1|v_3\rangle=\langle v_2|v_4\rangle=\alpha-\beta=\langle v_3|v_5\rangle=\langle v_5|v_2\rangle=\alpha\beta=\langle v_4|v_1\rangle=1$$
$$\Rightarrow \alpha=\frac{1}{2}+\frac{\sqrt{5}}{2},\ \beta=-\frac{1}{2}+\frac{\sqrt{5}}{2}.$$

This allows the complete determination:

$$\text{dodecahedron:}\begin{cases}\text{pentagon center: } r_{12}=(\gamma,0,\delta)=\left(\frac{1}{2}+\frac{3}{2\sqrt{5}},0,\frac{1}{2}+\frac{1}{2\sqrt{5}}\right),\\ \text{edge length: } l_{12}^2=4\beta^2=6-2\sqrt{5},\\ \text{out-radius: } R_{12}^2=3,\ \text{in-radius: } r_{12}^2=\gamma^2+\delta^2=1+\frac{2}{\sqrt{5}}.\end{cases}$$

There occur the two lengths:

$$\gamma + \delta = \gamma^2 + \delta^2 = 1 + \frac{2}{\sqrt{5}}, \quad \gamma = \frac{1}{2} + \frac{3}{2\sqrt{5}}, \quad \delta = \frac{1}{2} + \frac{1}{2\sqrt{5}}.$$

The *icosahedron* inside with its vertices at the central points of the pentagons of the dodecaedron is given as follows:

$$\text{icosahedron:} \begin{cases} \text{12 vertices: } (\pm\delta, \pm\gamma, 0), \ (0, \pm\delta, \pm\gamma), \ (\pm\gamma, 0, \pm\delta), \\ \text{characteristic triangle: } (\delta, -\gamma, 0), \ (0, -\delta, \gamma), \ (\gamma, 0, \delta), \\ \text{triangle center: } r_{20} = \frac{\gamma+\delta}{3}(1, -1, 1), \ \text{edge length: } l_{20}^2 = \frac{6}{5} + \frac{2}{\sqrt{5}}, \\ \text{out-radius: } R_{20}^2 = r_{12}^2, \ \text{in-radius: } r_{20}^2 = \frac{r_{12}^4}{R_{12}^2}, \end{cases}$$

with the ratio of the radii:

$$\frac{R_{12}^2}{r_{12}^2} = \frac{R_{20}^2}{r_{20}^2} = 3(5 - 2\sqrt{5}) = 1.5835\ldots, \quad \tan^2\frac{\pi}{3} = 3, \quad \tan^2\frac{\pi}{5} = 5 - 2\sqrt{5}.$$

For dodecahedron and icosahedron, the *golden ratio* plays a role: There, a length l has to be cut into two parts b and $l - b$ in such a way that b is the geometrical mean of l and $l - b$:

$$b = \sqrt{l(l-b)} \Rightarrow \frac{b}{l} = \beta = \frac{\sqrt{5} - 1}{2}.$$

b is the side of a regular 10-gon with out-circle radius l.

4.7 Kepler's *Mysterium Cosmographicum*

In Kepler's time, one was convinced that there exist exactly six planets, Mercury, Venus, Earth, Mars, Jupiter and Saturn, and, after Copernicus, that they move on perfect circles. The sun, moon, and the five planets from the geocentric point of view remain in our 2 + 5-weekdays (easy to recognize in French and English): Sunday, Monday, mardi, mercredi, jeudi, vendredi, Saturday. After long investigations of the geometry of regular polyhedra, Kepler tried, in his *Mysterium cosmographicum* (1595), to interpret the planetary orbits by great circles on the out- and in-spheres of the exactly five Platonic solids, nested "correctly" in each other. The out-sphere of a Platonic solid should be the in-sphere of the next Platonic solid. In such a way the radii of the planetary orbits would reflect "harmonious" ratios.

Today, one has the following observations for the averaged radii of the planetary orbits, together with the orbit times and the masses (earth mass $\simeq 5.98 \times 10^{24}$ kg):

Name	Orbit radius $\frac{R_{max}+R_{min}}{2}$ (10^6 km)	Orbit time (earth days)	Excentricity $\frac{R_{max}-R_{min}}{R_{max}+R_{min}}$	Mass (earth masses)
Sun	0	0	–	33 000
Mercury	57.9	88	0.206	0.054
Venus	108.2	225	0.007	0.814
Earth	149.6	365.25	0.017	1
Mars	227.9	687	0.093	0.107
Jupiter	778.0	4 332	0.048	317.45
Saturn	1 427.0	10 760	0.056	95.06
Uranus	2 872.0	30 681	0.047	14.5
Neptun	4 496.6	60 267	0.009	17.6
Pluto	5 900.0	90 730	0.247	0.05?

Planets of our sun

Kepler could explain rather well the square of the five dimensionless ratios of the six orbit radii $\frac{R^2}{r^2}$ for neighboring planets, known at his time, by the radii of the in- and out-spheres of the five Platonic solids $\{n_0, m_0\}$, appropriately nested in each other:

Planet	Orbit radius ($\frac{1}{4}$Mercury)	$\frac{R^2}{r^2}$ (observed)	Platonic solids $\{n_0, m_0\}$	$\frac{R^2}{r^2}$ $\tan^2\frac{\pi}{n_0}\tan^2\frac{\pi}{m_0}$	N	Titius–Bode $4+3\cdot 2^N$
Mercury	4				$-\infty$	4
		3.5	octahedron $\{3, 4\}$	3		
Venus	7.5				0	7
		1.9	icosahedron $\{3, 5\}$	1.58		
Earth	10.3				1	10
		2.3	dodecahedron $\{5, 3\}$	1.58		
Mars	15.7				2	16
		11.9	tetrahedron $\{3, 3\}$	9		
Jupiter	53.7				4	52
		3.4	hexahedron $\{4, 3\}$	3		
Saturn	98.5				5	100
		4.1	−−	−		
Uranus	198.4				6	196
		2.4	−−	−		
Neptun	310.6				?	?
		1.7	−−	−		
Pluto	407.6				7	388

Kepler's Platonic planet orbits

For example, the Earth orbits on the out-sphere of the cosmic icosahedron, which is simultaneously the in-sphere of the cosmic dodecahedron. Kepler could have exchanged with each other dual solids, i.e., octahedron with hexahedron or dodecahedron with icosahedron. Kepler had to try, with respect to the ratios, altogether $30 = \frac{5!}{2\cdot 2}$ different arrangements, and with the selfdual tetrahedron's large $\frac{R^2}{r^2}$ between Jupiter and Mars, only $6 = \frac{4!}{2\cdot 2}$ possibilities.

In the last two columns of the table one finds the *Titius-Bode*[13] *parametrization* of the planetary orbit radii, which fits so lala without a convincing physical explanation (perhaps turbulence properties of a planetary ur-vortex):

[13] Johann Daniel Titius (1729–1796), Johann Elert Bode (1747–1826).

$$\frac{\text{planetary orbit radius}(N)}{\text{Mercury orbit radius}} = \frac{4 + 3 \cdot 2^N}{4}.$$

The gap at $N = 3$, where Kepler used the big ratio for the tetrahedron, led to the detection of the planetoid belt (Ceres, Pallas, etc.) between Mars and Jupiter. The two external planets, Neptun and probably also Pluto (perhaps not from the ur-vortex, but captured later), do not fit the Titius-Bode power law.

Probably, Kepler's belief in the divine harmony and symmetry of the cosmos and his "sleepwalker-like" playing around with ratios of orbit radii was the decisive experience to find, in his'Harmonice Mundi' his third law — the constancy, for all planets, of the ratio of the third power of the big half-axis R_P to the square of the orbit time T_P. With Newton, it is given by the product of the mass $m_\odot$ of the central body and Newton's unit G:

$$q_{\text{Kepl}}(m_\odot) = \frac{R_P^3}{T_P^2} = \frac{G}{(2\pi)^2} m_\odot \simeq 3.8 \times 10^{18} \, \frac{\text{m}^3}{\text{s}^2}.$$

More exactly, the quotient is proportional to the mass sum $m_\odot + m_P$, therefore it is not constant for all planets. In addition, the planetary system is no two-body system, there are relativistic corrections, etc. Since all these influences are small in our sun system, Kepler had the luck of the brave.

Kepler's proposal to the duke Friedrich of Württemberg, to manufacture a drinking vessel in the form of Platonic solids, nested in each other — by independent silversmiths to keep the cosmic mystery — was rejected by the duke. He invited Kepler to produce first a copper model. By monetary reasons, it remained with a paper model and, finally, the whole affair petered out.

Up to the end of his life, Kepler thought of his *Mysterium cosmographicum* as of his greatest achievement. From the present-day point of view, the application of the Platonic solids to the planetary system is completely unfounded, if not to say, with the hybris of today, nonsense. Especially, classical physics does not know a discrete manifold of solutions for the orbits. Kepler's symmetry ideas for the central potential with orbits characterized by natural numbers $\{n_0, m_0\}$ for regular face and vertex polygon was resuscitated around 300 years later. An apparently more fundamental and brilliantly successful application in the structure of the atoms started with the discrete orbits of the Bohr-Sommerfeld pre-quantum theory, characterized by the integers $\{2j, l, m\}$ for principal and angular momentum quantum numbers. Kepler's law for the orbit areas was extended by Bohr's conditions for discrete angular momenta ("quantization"). The virial relation $(\frac{2\pi R_P}{T_P})^2 = \frac{G m_\odot}{R_P}$ for averaged kinetic and potential energy is extended to the quantized energy levels $E_{2j} = -\frac{1}{(2j+1)^2} \frac{m_e(\alpha_{\text{Som}} c)^2}{2}$ with $2j = 0, 1, 2, \ldots$.

The concrete application of the Platonic solids for an understanding of our planetary system was unfounded. Nevertheless, the intuition of basic symmetry structures for nature was, perhaps really, Kepler's greatest achievement.

4.8 Polytopes and Mosaics

Generalizing the considerations in the Euclidean plane and space to higher-dimensional Euclidean spaces $\mathbb{E}^n$, Plato's geometry can be carried on: Which are the regular four-dimensional polytopes with Platonic solids as side volumes (cells), etc.? With which regular polyhedra can the Euclidean space $\mathbb{E}^3$ be completely filled up, etc.? All those questions, only superficially touched upon in the following, are beautifully treated and extensively answered by the work of Schläfli and Coxeter.[14]

If the regular $\mathbb{E}^3$-polyhedra are denoted by the *Schläfli symbol*:

$$\{n_0, m_0\} = \{p_1, p_2\} = \{\text{vertex numbers of the face and vertex polygon}\},$$

there are the five Platonic solids (dual partners with exchanged numbers):

$$\begin{array}{c}\text{regular } \mathbb{E}^3\text{-polyhedra :}\\ \{p_1, p_2\} \overset{\text{dual}}{\leftrightarrow} \{p_2, p_1\}\end{array} \left\{\begin{array}{l} \alpha_3 = \{3, 3\} \text{ (4 faces)},\\ \beta_3 = \{3, 4\} \text{ (8 faces)},\\ \gamma_3 = \{4, 3\} \text{ (6 faces)},\\ \iota_3 = \{3, 5\} \text{ (20 faces)},\\ \kappa_3 = \{5, 3\} \text{ (12 faces)}.\end{array}\right.$$

The name is given by the number of the regular two-dimensional faces, e.g., the dodecahedron has 12 faces.

With the Platonic solids as regular side volumes (three-dimensional cells) in four-dimensional Euclidean space, there are even six regular polytopes:

$$\begin{array}{c}\text{regular } \mathbb{E}^4\text{-polytopes :}\\ \{p_1, p_2, p_3\} \overset{\text{dual}}{\leftrightarrow} \{p_3, p_2, p_1\}\end{array} \left\{\begin{array}{l} \alpha_4 = \{3, 3, 3\} \text{ (5 cells)},\\ \beta_4 = \{3, 3, 4\} \text{ (16 cells)},\\ \gamma_4 = \{4, 3, 3\} \text{ (8 cells)},\\ \epsilon_4 = \{3, 4, 3\} \text{ (24 cells)},\\ \iota_4 = \{3, 3, 5\} \text{ (600 cells)},\\ \kappa_4 = \{5, 3, 3\} \text{ (120 cells)}.\end{array}\right.$$

The first two numbers give the Platonic $\mathbb{E}^3$-polyhedron as side volume, the last number the local meeting number — how man $\mathbb{E}^3$-polyhedra meet in one line. It is possible to find in $\{p_1, p_2, p_3\}$ three regular polygons $\{p_1\}$, $\{p_2\}$ and $\{p_3\}$. From dimension five on, there are only three possibilities, which lead to three main series:

$$\begin{array}{c}\text{regular } \mathbb{E}^n\text{-polytopes :}\\ n \geq 5,\end{array} \left\{\begin{array}{ll} \alpha_n = \{3^{n-1}\} & \text{(fundamental simplex)},\\ \beta_n = \{3^{n-2}, 4\} & \text{(cross polytop)},\\ \gamma_n = \{4, 3^{n-2}\} & \text{(measure polytop)},\end{array}\right.$$

$$\{p_1, p_2, \ldots, p_{n-2}, p_{n-1}\} \overset{\text{dual}}{\leftrightarrow} \{p_{n-1}, p_{n-2}, \ldots, p_2, p_1\} : \ \alpha_n \overset{\text{dual}}{\leftrightarrow} \alpha_n, \ \beta_n \overset{\text{dual}}{\leftrightarrow} \gamma_n,$$

[14]Ludwig Schläfli (1814–1895), Harald Scott McDonald Coxeter (1907–2003).

as continuations of the two- and three-dimensional polytopes in the three series:

$$\begin{aligned}\text{fundamental simplices} &: \text{regular triangle, tetrahedron, } \ldots,\\ \text{cross polytopes} &: \text{square, octahedron, } \ldots,\\ \text{measure polytopes} &: \text{square, hexahedron, } \ldots\end{aligned}$$

Pentagon structures are not continued, only regular triangles and squares.

The mosaic property in the plane is generalized as follows:

$$\mathbb{E}^1\text{-mosaic: } \{1\} \text{ (distance)}, \quad \mathbb{E}^2\text{-mosaics: } \begin{cases}\{3,6\} \text{ (3 corners)},\\ \{6,3\} \text{ (6 corners)},\\ \{4,4\} \text{ (4 corners)},\end{cases}$$

$$\mathbb{E}^3\text{-mosaic: } \{4,3,4\} \text{ (6 faces)}, \ \mathbb{E}^4\text{-mosaics: } \begin{cases}\{3,3,4,3\} \text{ (16 cells)},\\ \{3,4,3,3\} \text{ (24 cells)},\\ \{4,3,3,4\} \text{ (8 cells)}.\end{cases}$$

The first $n-1$ numbers give the building block, the last one the meeting number. In space $\mathbb{E}^3$, there are only hexahedric building blocks (cubes), for $n=4$ three different mosaics, and then only one main series with the measure polytopes, starting with square and hexahedron (cube):

$$n \geq 5, \ \mathbb{E}^n\text{-mosaic:}\{4, 3^{n-2}, 4\} \text{ (measure polytop)}.$$

The $\mathbb{E}^n$-mosaics can be looked at to be degenerate $\mathbb{E}^{n+1}$-polytopes.

With Schläfli and Coxeter, the regular polytopes and mosaics are obtained from the Diophantic (in)equality $D > 0$ and $D = 0$, respectively, written with the related cosines:

$$\begin{aligned}\text{regular } \mathbb{E}^3\text{-polyhedra } \{p_1, p_2\} &: 1 - \cos^2\tfrac{\pi}{p_1} - \cos^2\tfrac{\pi}{p_2} > 0,\\ \text{regular } \mathbb{E}^2\text{-mosaics } \{p_1, p_2\} &: 1 - \cos^2\tfrac{\pi}{p_1} - \cos^2\tfrac{\pi}{p_2} = 0.\end{aligned}$$

The generalization uses the $n-1$ characteristic integers of a regular $\mathbb{E}^n$-polytope and an $\mathbb{E}^{n-1}$-mosaic in the Schläfli symbol for the Diophantic (in)equality — starting from the matrix:

$$D(p_1, \ldots, p_{n-1}) = \begin{pmatrix} 1 & \cos\frac{\pi}{p_1} & 0 & 0 & 0 \ldots & 0 & 0 & 0 \\ \cos\frac{\pi}{p_1} & 1 & \cos\frac{\pi}{p_2} & 0 & 0 \ldots & 0 & 0 & 0 \\ 0 & \cos\frac{\pi}{p_2} & 1 & \cos\frac{\pi}{p_3} & 0 \ldots & 0 & 0 & 0 \\ & \ldots & & & \ldots\ldots & & \ldots & \\ 0 & 0 & 0 & 0 & 0 \ldots & \cos\frac{\pi}{p_{n-2}} & 1 & \cos\frac{\pi}{p_{n-1}} \\ 0 & 0 & 0 & 0 & 0 \ldots & 0 & \cos\frac{\pi}{p_{n-1}} & 1 \end{pmatrix}$$

and its determinant:

$$\det D(p_1, \ldots, p_{n-1}) \begin{cases} > 0 \text{ regular } \mathbb{E}^n\text{-polytope}, \\ = 0 \text{ regular } \mathbb{E}^{n-1}\text{-mosaic}. \end{cases}$$

The slightly academic case $n = 2$ determines the regular polygons and the mosaic for a line, given by the distance:

$$\det D(p_1) = 1 - \cos^2 \frac{\pi}{p_1} \begin{cases} > 0 \Rightarrow p_1 = 2, 3, 4, \ldots; & \text{regular } \mathbb{E}^2\text{-polygons}, \\ = 0 \Rightarrow p_1 = 1; & \mathbb{E}^1\text{-mosaic with a distance}. \end{cases}$$

Euler's equation relating to each other the number of vertices N_0, edges N_1, faces N_2, volumes N_3, etc., for simply connecting polytopes, more generally proved by Poincaré, is (in)homogeneous for (odd)even space dimension:

$$\mathbb{R}^n\text{-polytopes: } N_0 - N_1 + N_2 - \cdots + (-1)^{n-1} N_{n-1} = \begin{cases} 2, & n = 1, 3, \ldots, \\ 0, & n = 2, 4, \ldots. \end{cases}$$

If one takes into account also the full polytope, one is talking about, with $N_n = 1$, the relation does no longer look differently for even and odd dimensions:

$$\text{with } N_n = 1: \quad \sum_{k=0}^{n} (-1)^k N_k = 1.$$

Euler's equation can be easily understood for the important special case of the fundamental simplex in $\mathbb{R}^n$ with $n + 1$-vertices, i.e., the distance for the line, the triangle for the plane, the tetrahedron for the space. Each pair of vertices defines an edge, each triplet a triangular face, each quartet a tetrahedric volume, etc.:

$$\mathbb{R}^n\text{-fundamental simplex: } N_k = \binom{n+1}{k+1}, \quad k = 0, \ldots, n.$$

The sum of all $(-1)^k N_k$ of the $\mathbb{R}^n$-fundamental simplex is the binomial expansion of $-(1-1)^{n+1}$ without the first term:

$$N_k = \binom{n+1}{k+1} \Rightarrow \sum_{k=0}^{n} (-1)^k N_k = 1 - (1-1)^{n+1} = 1.$$

If one discusses, in addition, the decomposition of general polytopes into fundamental simplices, one has understood a lot about Euler's relation.

4.9 Orthogonal Symmetry Groups

All Platonic geometry — in the Euclidean plane and space — describes properties that are connected with the corresponding orthogonal group. The discrete Platonic symmetries are embedded in and are distinguished as subsymmetries of continuous orthogonal ones.

All symmetry of the former sections depends on the possibility to define lengths and angles, i.e., that there exists a *scalar product* for the vector space $\mathbb{R}^n$ which is used as the decisive structure: The Euclidean space $\mathbb{E}^n$ is a real n-dimensional vector space endowed by a symmetric, bilinear, positive definite product:

$$x, y \in \mathbb{R}^n : \ \langle x|y\rangle = \langle y|x\rangle \in \mathbb{R}, \quad \| x \|^2 = \langle x|x\rangle > 0 \iff x \neq 0.$$

Two nontrivial vectors with trivial scalar product are called orthogonal. This product allows, via the Cauchy–Schwarz[15]-inequality, the definition of an angle between two nontrivial vectors:

$$\begin{aligned}
&\| x + \alpha y \|^2 = \langle x|x\rangle + 2\alpha\langle x|y\rangle + \alpha^2\langle y|y\rangle \geq 0 \text{ for all } \alpha \in \mathbb{R},\\
&\Rightarrow \text{discriminant: } \det \left| \begin{pmatrix} \langle x|x\rangle & \langle x|y\rangle \\ \langle y|x\rangle & \langle y|y\rangle \end{pmatrix} \right| \\
&\qquad\qquad = \langle x|x\rangle\langle y|y\rangle - \langle x|y\rangle^2 \geq 0 \text{ (Cauchy–Schwarz)},\\
&\text{if } x, y \neq 0 : \ \frac{\langle x|y\rangle}{\| x \| \ \| y \|} = \cos\theta(x, y),\\
&\langle x|y\rangle = \| x \| \| y \| \cos\theta(x, y), \ \ \cos^2\theta(x, y) \leq 1.
\end{aligned}$$

From here on, the familiar flat trigonometry in $\mathbb{E}^2$ can be developed, e.g., for a triangle $x + y = z$, one obtains Pythagoras' theorem $\| x \|^2 + \| y \|^2 = \| z \|^2$ for $\langle x|y\rangle = 0$ (orthogonal triangle), or the theorem of Thales,[16] concerning the orthogonality of a triangle over a circle diameter, i.e., $\langle x|y\rangle = 0$ for $x = \frac{z}{2} + r$ and $y = \frac{z}{2} - r$ with $\| r \| = \| \frac{z}{2} \|$.

Good to know: A very prominent example of the Cauchy–Schwarz inequality in a complex Hilbert space, i.e., with a symmetric sesquilinear positive product $\langle x|y\rangle = \overline{\langle y|x\rangle} \in \mathbb{C}$ leading to $\langle x|x\rangle\langle y|y\rangle \geq (\frac{\langle x|y\rangle + \langle y|x\rangle}{2})^2$, is Heisenberg's uncertainty relation $(\Delta\mathbf{x})^2(\Delta\mathbf{p})^2 \geq |\frac{[\mathbf{x},\mathbf{p}]}{2}|^2 = \frac{\hbar^2}{4}$ for the measurement of position $\mathbf{x}$ and momentum $\mathbf{p}$.

Good to know: A bilinear form $V \times V \ni (v, w) \longmapsto \zeta(v, w) \in \mathbb{K} = \mathbb{R}, \mathbb{C}$ or a sesquilinear form for $\mathbb{K} = \mathbb{C}$ of a vector space $V \cong \mathbb{K}^n$ is *nondegenerate*, if, and only if, in any basis, the discriminant $\det \zeta(e^a, e^b)^n_{a,b=1}$ does not vanish. A symmetric bilinear form $\zeta(v, w) = \zeta(w, v)$ for $\mathbb{K} = \mathbb{R}, \mathbb{C}$ or a symmetric sesquilinear form $\zeta(v, w) = \overline{\zeta(w, v)}$ for $\mathbb{K} = \mathbb{C}$ is positive, $\zeta \succeq 0 \iff \zeta(v, v) \geq 0$ for all $v \in V$, if, and only if, in any basis, the discriminant and all $d \times d$-subdiscriminants for all vector subspaces $\mathbb{K}^d \cong W \subseteq V$ with $d = 1, \ldots, n$ are positive, $\det \zeta(e^a, e^b)|_{d\times d} \geq 0$. Then there exists a basis with positive diagonal form $\zeta \cong \operatorname{diag}(|\xi_1|^2, \ldots, |\xi_n|^2)$.
Positivity of a hermitian $n \times n$-matrix is also characterized by the factorization possibility $\zeta = \xi \circ \xi^\star$ with a complex $n \times n$-matrix and its transposed conjugated one.

---

[15]Hermann Amandus Schwarz (1843–1921).

[16]Thales of Miletus -(624–546).

With linear operations $\mathbb{R}^n \ni x \longmapsto O(x) \in \mathbb{R}^n$, the invariance group of a scalar product and, therefore, also of lengths and angles, e.g., orthogonal angles, is called *orthogonal group* and denoted by $\mathbf{O}(n)$:

$$O \in \mathbf{O}(n) \iff \langle O(x)|O(y)\rangle = \langle x|y\rangle \text{ for all } x, y \in \mathbb{R}^n.$$

It is a subgroup of all automorphisms, i.e., of the invertible linear transformations $\mathbf{GL}(n, \mathbb{R})$ of the vector space $\mathbb{R}^n$. Euclidean spaces have orthonormal bases:

$$\{e^a\}_{a=1}^n : \ \langle e^a|e^b\rangle = \delta^{ab}, \ \mathbb{R}^n \ni x = x_a e^a, \ \ \langle x|y\rangle = x_a y_a,$$

used in the following.

The nontrivial *orbits* of an orthogonal group $\mathbf{O}(n)$ are *spheres*, all similar (equiformal) to the *unit sphere* Ω^{n-1}:

$$n = 0, 1, 2, \ldots : \ \Omega^n = \{x \in \mathbb{R}^{1+n} \bigm| \langle x|x\rangle = 1\},$$

starting with two points $\Omega^0 = \{-1, 1\}$, the circle Ω^1, and the proper sphere Ω^2.

The orthogonal group $\mathbf{O}(n)$ is abelian (commutative) only for $n = 1$. There exist orthogonal transformations which change the orientation of an object. The normal subgroup with the orientiation compatible orthogonal transformations, i.e., with $\det O = 1$, is the *special* orthogonal or *rotation* group $\mathbf{SO}(n)$. The rotation group classes, i.e., the quotient group $\mathbf{O}(n)/\mathbf{SO}(n)$ is isomorphic to a 2-elementic reflection group $\mathbf{I}(2) = \{\pm 1\}$. In general, there are many $\{\pm 1\}$-isomorphic subgroups $\{\epsilon, \epsilon^2 = \mathbf{1}_n\} \subseteq \mathbf{O}(n)$, $\det \epsilon = -1$, which allow the isomorphy to the semidirect product:

$$\mathbf{O}(n) \cong \{\pm 1\} \vec{\times} \mathbf{SO}(n) \ni (\epsilon, O) : \ (\epsilon_1, O_1) \bullet (\epsilon_2, O_2) = (\epsilon_1\epsilon_2, O_1 \circ \epsilon_1 \circ O_2 \circ \epsilon_1).$$

For odd dimensions, a normal subgroup $\{\pm \mathbf{1}_{1+2n}\}$ with the identity and its inverse, the central reflection $-\mathbf{1}_{1+2n}$, can represent the classes $\mathbf{O}(1+2n)/\mathbf{SO}(1+2n)$. The corresponding group decomposition with the central reflection is even direct:

$$\mathbf{O}(1+2n) = \{\pm \mathbf{1}_{1+2n}\} \times \mathbf{SO}(1+2n).$$

For even dimension $\{\pm \mathbf{1}_{2n}\}$ is a normal subgroup of the special group $\mathbf{SO}(2n)$. The quotient group $\mathbf{SO}(2n)/\{\pm \mathbf{1}_{2n}\}$ is not representable by a subgroup of $\mathbf{SO}(2n)$. That is different for the related noncompact groups. For example, the corresponding classes (quotient group) of the special Lorentz group are isomorphic to the orthochochronous Lorentz group $\mathbf{SO}_0(1, 3) \cong \mathbf{SO}(1, 3)/\{\pm \mathbf{1}_4\}$ and $\mathbf{SO}(1, 2n-1) = \{\pm \mathbf{1}_{2n}\} \times \mathbf{SO}_0(1, 2n-1)$.

Good to know: The classes $G/N = \{[g] = gN \bigm| g \in G\}$ of a group G with respect to a normal subgroup N, i.e., $gN = Ng$ for all $g \in G$, carry a natural group (quotient group) structure via $[g_1][g_2] = [g_1 g_2]$ and the neutral element $[1_G] = N$. Any subgroup $H \subseteq G$ of an abelian group leads to a group G/H.

The infinitesimal structure of the orthogonal group $\mathbf{O}(n)$, i.e., its generators in all linear $\mathbb{R}^n$-transformations (endomorphisms) $\mathbf{AL}(n, \mathbb{R})$ is the Lie algebra $\log \mathbf{O}(n)$:

$$\mathcal{O} \in \log \mathbf{O}(n) \iff \langle \mathcal{O}(x)|y\rangle + \langle x|\mathcal{O}(y)\rangle = 0 \text{ for all } x, y \in \mathbb{R}^n.$$

$\mathbf{O}(n)$ and $\mathbf{SO}(n)$ have the same Lie algebra.

Good to know: The notations with a logarithm for the transition from a Lie group to its Lie algebra $G \longmapsto \log G$ and with an exponent for the transition from a Lie algebra to its simply connected universal covering Lie group $L \longmapsto \exp L$ reflect corresponding matrix operations.
Two Lie groups (real or complex) with isomorphic Lie algebras $\log G_1 \cong \log G_2 \cong L$ are called *locally isomorphic*, $G_1 \sim G_2$, e.g., $\log \mathbb{R} \cong \log \mathbf{U}(1) \cong \mathbb{R}$ and $\log \mathbf{O}(3) \cong \log \mathbf{SU}(2) \cong \mathbb{R}^3$. That is the case, if they differ only by discrete subgroups: For a Lie group with Lie algebra L, there is exactly one locally isomorphic simply connected Lie group, the *universal covering group* $G = \exp L$. Its discrete normal subgroups $N_1, N_2, \ldots, N_k \subset G$ give the quotient groups G/N_j which constitute all Lie groups with Lie algebra L.
The Lie algebra of a finite or a discrete group is trivial, e.g., $\log \mathbb{Z}_n \cong \log \mathbb{Q} \cong \{0\}$.

Now, more about the individual orthogonal groups:

The orthogonal group $\mathbf{O}(1)$, acting on the line $\mathbb{E}^1$, is the discrete 2-elementic abelian *reflection or orientation group*, isomorphic to the *permutation (transposition) group* $\mathbf{G}(2)$ of two elements:

$$\mathbf{O}(1) \cong \{\pm 1\} \cong \mathbf{G}(2) \cong \mathbf{I}(2).$$

The group $\mathbf{SO}(1) \cong \{1\}$ is trivial, likewise the Lie algebra $\log \mathbf{O}(1) = \{0\}$.

The nonabelian orthogonal group $\mathbf{O}(2)$, acting on the plane $\mathbb{E}^2$, inherits for each line $\mathbb{E}^1 \subset \mathbb{E}^2$ a reflection subgroup $\mathbf{I}(2)$, with the re-orientation R_2, e.g., for the 2nd axis as $\mathbb{E}^1$-line:

$$\mathrm{R}_2 \cong \begin{pmatrix} 1 & 0 \\ 0 & -1 \end{pmatrix}, \quad \mathbf{I}(2) \cong \{\mathrm{R}_2, \mathrm{R}_2^2 = \mathbf{1}_2\}.$$

The *axial rotation group* $\mathbf{SO}(2)$ is abelian and parametrizable by one rotation angle:

$$O \in \mathbf{SO}(2) \Rightarrow O = O(\theta) = \begin{pmatrix} \cos\theta & \sin\theta \\ -\sin\theta & \cos\theta \end{pmatrix}, \quad \theta \in \mathbb{R}.$$

The finite subgroups of $\mathbf{SO}(2)$ are the cyclic groups $\mathbf{I}(n)$ with order n. They leave invariant a regular n-gon with its orientation:

$$\mathbf{SO}(2) \supset \mathbf{I}(n) \cong \{O(\frac{2\pi}{n})^k \bigm| k = 0, \ldots, n-1\}, \quad n = 1, 2, \ldots$$

Therefore, all finite $\mathbf{O}(2)$-subgroups are given by $\mathbf{I}(n)$ and the reflection including *dihedral groups* $\mathbf{I}^\pm(n)$ with order $2n$: $\mathbf{I}^\pm(n)$ is the semidirect product of the reflection group $\mathbf{I}(2) = \{\pm 1\}$ with $\mathbf{I}(n) = \{i_n^k \bigm| k = 0, \ldots, n-1\}$:

$$\mathbf{O}(2) \cong \mathbf{I}(2) \vec{\times} \mathbf{SO}(2) \supset \mathbf{I}(2) \vec{\times} \mathbf{I}(n) \cong \mathbf{I}^\pm(n), \quad \mathrm{R}_2 \circ O(\theta) \circ \mathrm{R}_2 = O(-\theta),$$

$$\text{with } \begin{cases} \mathbf{I}^\pm(2) \cong \mathbf{I}(2) \times \mathbf{I}(2) \cong \mathbf{K}(4), \\ \mathbf{I}^\pm(3) \cong \mathbf{I}(2) \vec{\times} \mathbf{I}(3) \cong \mathbf{G}(3), \\ \mathbf{I}^\pm(4) \cong \mathbf{I}(2) \vec{\times} \mathbf{I}(4) \\ \ldots, \end{cases}$$

$$\text{product, } \mathbf{I}^\pm(n) \ni (\epsilon, i_n^k) : \ (\epsilon_1, i_n^{k_1}) \circ (\epsilon_2, i_n^{k_2}) = (\epsilon_1\epsilon_2, i_n^{k_1+\epsilon_1 k_2}).$$

The dihedral groups are abelian for $n = 1, 2$. $\mathbf{I}^{\pm}(2) = \mathbf{I}(2) \times \mathbf{I}(2) = \mathbf{K}(4)$ is *Klein's*[17] *four-group*. $\mathbf{I}^{\pm}(3)$ is isomorphic to the permutation group $\mathbf{G}(3)$.

The real one-dimensional Lie algebra of $\mathbf{O}(2)$ consists of antisymmetric 2×2-matrices:

$$\mathcal{O} \in \log \mathbf{O}(2) \iff \mathcal{O} = \theta \begin{pmatrix} 0 & 1 \\ -1 & 0 \end{pmatrix}, \quad \theta \in \mathbb{R}.$$

The finite subgroups of $\mathbf{O}(2)$ are also subgroups of the orthogonal group $\mathbf{O}(3)$, i.e., as cyclic groups $\mathbf{I}(n)$ and dihedral groups $\mathbf{I}^{\pm}(n)$ acting on any two-dimensional hyperplane $\mathbb{E}^2 \subset \mathbb{E}^3$. The genuinely new finite subgroups of the proper *rotation group* $\mathbf{SO}(3)$ are associated with the three dual pairs of Platonic solids:

$$\begin{array}{r} \mathbf{SO}(3) \supset \mathbf{I}(n_0, m_0) \\ = \mathbf{I}(m_0, n_0): \end{array} \left\{ \begin{array}{l} \text{tetrahedron group: } \mathbf{I}(3,3) \cong \mathbf{S}(4), \text{ order } 12, \\ \text{octahedron group: } \mathbf{I}(3,4) \cong \mathbf{G}(4), \text{ order } 24, \\ \text{icosahedron group: } \mathbf{I}(3,5) \cong \mathbf{S}(5), \text{ order } 60. \end{array} \right.$$

These three groups are isomorphic to *permutation groups* $\mathbf{G}(n)$ with order $n!$ or their normal subgroups, the *alternating (special permutation) groups* $\mathbf{S}(n)$ with the positive signature permutations and order $\frac{n!}{2}$. Perhaps, it would have been cumbersome, but more appropriate to have named these three finite groups $\mathbf{I}(n_0, m_0) = \mathbf{I}(m_0, n_0)$ for the self-dual Archimedean intersection polyhedra, i.e., octahedron, hexoctaedron and icosidodecahedron group with the number of vertices 6, 12 and 30, respectively, half the group order.

Good to know: The kernel $\text{kern} f = \{g \in G \,|\, f(g) = 1 \in G'\}$ of a group morphism $f : G \longrightarrow G'$ is a normal subgroup with the quotient group isomorphic to the image: $G/\text{kern} f \cong f[G]$.
The signature mapping of the permutation group $\mathbf{G}(n) \ni \pi \longmapsto \text{sign}\pi \in \mathbf{I}(2)$ is a group morphism with the alternating group $\mathbf{S}(n)$ (special permutation group) as kernel. It is the discrete precursor of the determinant mapping of vector space automorphisms (general linear group) $\mathbf{GL}(n, \mathbb{K}) \ni g \longmapsto \det g \in \mathbf{GL}(1, \mathbb{K})$ where the kernel $\mathbf{SL}(n, \mathbb{K})$ (special linear group) contains the matrices with unit determinant. The signature group $\mathbf{I}(2) \subset \mathbf{U}(1)$ is embedded in $\mathbf{GL}(1, \mathbb{R}) = \mathbf{I}(2) \times \mathbf{D}(1)$ and $\mathbf{GL}(1, \mathbb{C}) = \mathbf{U}(1) \times \mathbf{D}(1)$.

Each rotation in Euclidean $\mathbb{E}^3$ is characterized by an invariant rotation axis with the direction $\vec{n}$ and unit length $\vec{n}^2 = 1$ and a rotation angle θ. The trace of the matrix leads to the angle:

$$O \in \mathbf{SO}(3) \Rightarrow \left\{ \begin{array}{l} O = O(\theta, \vec{n}) \cong O_{ab} = \delta_{ab} \cos\theta + n_a n_b (1 - \cos\theta) - \epsilon_{abc} n_c \sin\theta \\ \text{tr}\, O(\theta, \vec{n}) = 1 + 2\cos\theta, \quad \frac{\epsilon_{abc}(O_{ab} - O_{ba})}{12 \sin\theta} = -n_c. \end{array} \right.$$

There exists always a basis with the rotation axis as basic vector, for example the third unit vector:

$$O(\theta, \vec{n}_3) = \begin{pmatrix} \cos\theta & \sin\theta & 0 \\ -\sin\theta & \cos\theta & 0 \\ 0 & 0 & 1 \end{pmatrix}.$$

Two rotations commute if, and only if, their rotation axes are parallel, i.e., if they belong to the same subgroup $\mathbf{SO}(2) \subset \mathbf{SO}(3)$.

---

[17] Felix Klein (1849–1925).

If the rotation axis is combined with the rotation angle $\vec{\theta} = \theta\vec{n}$, all rotations can be parametrized by the points of a ball with radius π and antipodally identified points.

The tetra-, octa- and icosahedron group can be generated by the cyclic groups for the face and vertex polygon in a Platonic solid $\{n_0, m_0\}$: If the corners for two cyclic groups are on the orbits of axial rotations $\mathbf{I}(n_0) = O(\frac{2\pi}{n_0}, \vec{n})$ and $\mathbf{I}(m_0) = O(\frac{2\pi}{m_0}, \vec{m})$ with non-parallel rotation axes $\vec{n}, \vec{m}$, then those groups do not commute. All finite products of the elements of these two finite groups generate a subgroup of $\mathbf{SO}(3)$. The condition for the finiteness of this subgroup leads to a relation for n_0 and m_0, equivalent to the Diophantic inequality:

$$\begin{array}{rcl} \mathbf{I}(n_0, m_0) \text{ generated by } \mathbf{I}(n_0) \text{ and } \mathbf{I}(m_0) \text{ for } (n_0, m_0) &=& (3,3),\ (3,4),\ (3,5), \\ \text{order: } 2N_1 &=& 12, \quad 24, \quad 60. \end{array}$$

The group order $2N_1$ sums up 1 for the identity, the order of the cyclic groups for N_2 regular n_0-faces and for N_0 regular m_0-vertex polygons, and, in addition, the reflections at the N_1 edges — because of antipodic pairs, always one half only:

$$\text{group order: } 1 + \frac{N_2(n_0 - 1) + N_0(m_0 - 1) + N_1}{2} = 1 + \frac{N_2 n_0 + N_0 m_0 - 2}{2} = 2N_1.$$

For all finite subgroups of $\mathbf{O}(3) \cong \mathbf{I}(2) \times \mathbf{SO}(3)$ one has to take semidirect products with the reflection group which yields groups of order $4N_1$:

$$\mathbf{O}(3) \supset \left\{ \begin{array}{ll} \mathbf{I}^{\pm}(3,3) \cong \mathbf{I}(2) \vec{\times} \mathbf{S}(4) \cong \mathbf{G}(4), & \text{order } 24, \\ \mathbf{I}^{\pm}(3,4) \cong \mathbf{I}(2) \vec{\times} \mathbf{G}(4), & \text{order } 48, \\ \mathbf{I}^{\pm}(3,5) \cong \mathbf{I}(2) \vec{\times} \mathbf{S}(5) \cong \mathbf{G}(5), & \text{order } 120. \end{array} \right.$$

The nonabelian permutation groups $\mathbf{G}(n)$, $n \geq 3$, have the alternating (special) groups $\mathbf{S}(n)$ as a normal subgroups proper. Only $\mathbf{G}(4)$ has a second normal subgroup proper, the Klein four-group $\mathbf{K}(4) = \mathbf{I}(2) \times \mathbf{I}(2)$. Therefore, $\mathbf{G}(3)$ and $\mathbf{G}(4)$ have a semidirect factorization into abelian groups:

$$\begin{array}{rl} \mathbf{G}(2) & \cong \mathbf{I}(2), \\ \mathbf{G}(3) & \cong \mathbf{I}(2) \vec{\times} \mathbf{S}(3) \quad \cong \mathbf{I}(2) \vec{\times} \mathbf{I}(3), \\ \mathbf{G}(4) & \cong \mathbf{I}(2) \vec{\times} \mathbf{S}(4) \\ & \cong \mathbf{G}(3) \vec{\times} \mathbf{K}(4) \cong [\mathbf{I}(2) \vec{\times} \mathbf{I}(3)] \vec{\times} [\mathbf{I}(2) \times \mathbf{I}(2)], \\ \text{for } n \geq 5 \text{ only: } \mathbf{G}(n) & \cong \mathbf{I}(2) \vec{\times} \mathbf{S}(n). \end{array}$$

The decomposition of the permutation groups $\mathbf{G}(n)$ into abelian factors, direct or semidirect, up to $\mathbf{G}(4)$ is related with the possibility to solve polynomial equations $\sum_{k=0}^{n} \alpha_{n-k} x^k = 0$ up to order $n = 4$ by radicals, i.e., $\sqrt[n]{\alpha}$. Those factors are cyclic groups with the solutions of "basic" order n equations $\sqrt[n]{1} = \mathbf{I}(n) = \{x \in \mathbb{C} \,\big|\, x^n - 1 = 0\}$ for $n = 2, 3$.

Good to know: Order 2 equations $x^2+\alpha_1 x+\alpha_2=0$ can be brought with $x\longmapsto x-\frac{\alpha_1}{2}$ to the *depressed* form $x^2=q$ with two solutions $x=\pm\sqrt{q}$ where $\mathbf{G}(2)\cong\mathbf{I}(2)=\{\pm 1\}$. Similarily, and following Cardano,[18] order 3 equations $x^3+\alpha_1 x^2+\alpha_2 x+\alpha_3=0$ with $x\longmapsto x-\frac{\alpha_1}{3}$ give the depressed form $x^3=2q-3px$. With the ansatz $x=\sqrt[3]{q+r}+\sqrt[3]{q-r}$ the absolute term vanishes, leaving the condition $\sqrt[3]{r^2-q^2}=p$, i.e., $r=\sqrt{q^2+p^3}$. What about factors from $\mathbf{I}(3)=\sqrt[3]{1}$?
Now, following Lagrange, with permutation groups: The coefficients of an order n equation are symmetric with respect to the permutation group $\mathbf{G}(n)$ of the roots, e.g., for $n=3$:

$$(x-x_1)(x-x_2)(x-x_n)=x^3+\alpha_1x^2+\alpha_2x+\alpha_3,\quad \begin{cases}-\alpha_1 &= x_1+x_2+x_3,\\ \alpha_2 &= x_1x_2+x_2x_3+x_3x_1,\\ -\alpha_3 &= x_1x_2x_3.\end{cases}$$

The *discrete Fourier*[19] *transform* $\{x_i\}\longrightarrow\{p_i\}$ of the three roots and its inverse connects with the cyclic group $\mathbf{I}(3)=\{i_3^k=\exp\frac{2i\pi k}{3}\mid k=0,1,2\}\cong\mathbf{S}(3)$ with $i_3=\frac{-1+i\sqrt{3}}{2}$, $i_3+i_3^2=-1$:

$$\begin{aligned}p_1&=x_1+x_2+x_3,\\ p_2&=x_1+x_2i_3+x_3i_3^2,\\ p_3&=x_1+x_3i_3+x_2i_3^2,\end{aligned}\quad \begin{pmatrix}p_1\\p_2\\p_3\end{pmatrix}=\begin{pmatrix}1&1&1\\1&i_3&i_3^2\\1&i_3^2&i_3\end{pmatrix}\begin{pmatrix}x_1\\x_2\\x_3\end{pmatrix},\quad \begin{pmatrix}x_1\\x_2\\x_3\end{pmatrix}=\frac{1}{3}\begin{pmatrix}1&1&1\\1&i_3^2&i_3\\1&i_3&i_3^2\end{pmatrix}\begin{pmatrix}p_1\\p_2\\p_3\end{pmatrix}.$$

$p_1=-\alpha_1$ is symmetric under all root permutations. The cyclic permutations $\mathbf{S}(3)$, acting on $p_{2,3}$, give $\mathbf{I}(3)$-factors. Therefore, the cubes $p_{2,3}^3$ are $\mathbf{S}(3)$-invariant, e.g.,

$$\begin{aligned}&x_1\longmapsto x_2\longmapsto x_3\longmapsto x_1 \text{ gives } p_2\longmapsto i_3^2p_2,\ p_3\longmapsto i_3p_3,\\ &p_2^3=x_1^3+x_2^3+x_3^3+3(x_1x_3^2+x_2x_1^2+x_3x_2^2)i_3+3(x_1x_2^2+x_2x_3^2+x_3x_1^2)i_3^2+6x_1x_2x_3.\end{aligned}$$

The transposition $x_2\leftrightarrow x_3$ leads to the transposition $p_2\leftrightarrow p_3$. Therefore, the order 2 polynomial $(z-p_2^3)(z-p_3^3)$ is invariant under all permutations $\mathbf{G}(3)=\mathbf{G}(2)\vec{\times}\mathbf{S}(3)$. Its coefficients can be combined by the coefficients of the original polynomial.
From the depressed cubic polynomial $x^3+3px-2q=0$, one obtains the polynomial for the cubes of the Fourier transforms and its two solutions

$$(z-p_2^3)(z-p_3^3)=z^2-54qz-27^2p^3=0\Rightarrow z_{1,2}=p_{2,3}^3=27(q\pm\sqrt{q^2+p^3}),$$

$\{p_1=0,p_2,p_3\}$ are Fourier transformed back to the roots $\{x_1,x_2,x_3\}$ which involve square and cubic roots $\mathbf{G}(3)\cong\mathbf{I}(2)\vec{\times}\mathbf{I}(3)$:

$$\begin{pmatrix}x_1\\x_2\\x_3\end{pmatrix}=\begin{pmatrix}1&1&1\\1&i_3^2&i_3\\1&i_3&i_3^2\end{pmatrix}\begin{pmatrix}0\\ \sqrt[3]{q+\sqrt{q^2+p^3}}\\ \sqrt[3]{q-\sqrt{q^2+p^3}}\end{pmatrix}.$$

The angular momentum Lie algebra $\log\mathbf{O}(3)$ has the familiar Cartesian bases, which exhibit the Lie subalgebras $\log\mathbf{O}(2)$ of the three hyperplanes:

$$\begin{aligned}&\mathcal{O}\in\log\mathbf{O}(3)\iff\mathcal{O}=\theta_a\mathcal{O}^a=\begin{pmatrix}0&\theta_3&-\theta_2\\-\theta_3&0&\theta_1\\\theta_2&-\theta_1&0\end{pmatrix},\\ &\mathcal{O}^a\circ\mathcal{O}^b-\mathcal{O}^b\circ\mathcal{O}^a=[\mathcal{O}^a,\mathcal{O}^b]=-\epsilon^{abc}\mathcal{O}^c.\end{aligned}$$

[18]Gerolamo Cardano (1501–1576).
[19]Jean Baptiste Joseph Fourier (1768–1830).

The angular momentum Lie algebra is the "smallest" compact Lie algebra, as abstract Lie algebra denoted by the symbol A_1^c. Similar to the generation of the discrete group $\mathbf{I}(n_0, m_0)$ by the cyclic groups of its two characteristic polygons, the rotations can be generated by two abelian axial rotations, e.g., the Lie algebra by two generators $[\mathcal{O}^1, \mathcal{O}^2] = -\mathcal{O}^3$.

The orthogonal group $\mathbf{O}(4) \cong \mathbf{I}(2) \vec{\times} \mathbf{SO}(4)$ has the local isomorphy to a direct product of two rotation groups:

$$\mathbf{SO}(4) \sim \mathbf{SO}(3) \times \mathbf{SO}(3).$$

The universal covering group is the doubled spin group $\mathbf{SU}(2) \times \mathbf{SU}(2)$ where $\mathbf{SO}(4)$ and $\mathbf{SO}(3) \times \mathbf{SO}(3)$ are related to the classes with respect to the discrete normal subgroups $\mathbf{I}(2)$ and $\mathbf{I}(2) \times \mathbf{I}(2)$, respectively (more ahead).

In general, the symmetry group of the regular fundamental simplex α_n as a finite subgroup of the orthogonal group $\mathbf{O}(n) \cong \mathbf{I}(2) \vec{\times} \mathbf{SO}(n)$, $n \geq 3$, is isomorphic to the permutation group $\mathbf{G}(n+1) = \mathbf{I}(2) \vec{\times} \mathbf{S}(n+1)$ of its vertices. The symmetry group of the cross polytope and the measure polytope $\beta_n \overset{\text{dual}}{\leftrightarrow} \gamma_n$ involves, in a semidirect product, the 2^n reflections in the n-fold direct product $\mathbf{I}(2)^n$ of the n axes of an orthogonal basis and, in addition, their permutations $\mathbf{G}(n) = \mathbf{I}(2) \vec{\times} \mathbf{S}(n)$:

$$\begin{array}{l} \text{in } \mathbf{O}(n): \\ n \geq 3 \end{array} \left\{ \begin{array}{rl} \text{fundamental simplex:} & \mathbf{I}^{\pm}(3^{n-1}) \cong \mathbf{G}(n+1), \\ & \text{order } (n+1)!, \\ \text{measure and cross polytop:} & \mathbf{I}^{\pm}(3^{n-2}, 4) \cong \mathbf{G}(n) \vec{\times} \mathbf{I}(2)^n, \\ & \text{order } n!\, 2^n. \end{array} \right.$$

4.10 Erlangen Program of Geometry

According to the *Erlangen Program* (1872) of Felix Klein, to conduct *geometry* boils down to investigate the action $\bullet$ of a *Lie group G on a manifold $\mathcal{M}$*, with $G \bullet \mathcal{M}$ called a Klein space, and to classify group compatible stuctures, especially invariant ones. A *linear geometry* is characterizable by a vector space, acted on by a group of linear transformations.

Some examples for linear geometries of a real vector space $V \cong \mathbb{R}^n$ with automorphism groups and meaningful concepts:

General linear	$\mathbf{GL}(n, \mathbb{R}) = \{g \mid \det g \neq 0\}$	parallelity, simplex
Orientiation linear	$\mathbf{GL}_+(n, \mathbb{R}) = \{g \mid \det g > 0\}$	n-simplex orientation
Special linear	$\mathbf{SL}(n, \mathbb{R}) = \{g \mid \det g = 1\}$	n-simplex volume
Equiformal	$\mathbf{DO}(n) = \{g \mid \langle g(x)\vert g(y)\rangle = e^{\lambda}\langle x\vert y\rangle,\ \lambda \in \mathbb{R}\}$	angle
Euclidean	$\mathbf{O}(n) = \{g \mid \langle g(x)\vert g(y)\rangle = \langle x\vert y\rangle\}$	angle and length
Special Euclidean	$\mathbf{SO}(n) = \{g \mid g \in \mathbf{O}(n), \det g = 1\}$	all that

Geometries, acting groups and invariant concepts

For example, the n-dimensional special Euclidean geometry is defined by a vector space $\mathbb{R}^n$ with the action of the rotations $\mathbf{SO}(n)$, for an equiformal geometry, there act rotations $\mathbf{SO}(n)$, dilations $\mathbf{D}(1) = \exp\mathbb{R}$ and reorientations (reflections) $\mathbf{I}(2)$. Meaningful concepts are inherited by subgroups, but not the other way — the larger the group, the less restrictive. In the special Euclidean geometry all the concepts are meaningful:

$$\mathbf{GL}(n, \mathbb{R}) \supset \left\{ \begin{matrix} \mathbf{GL}_+(n, \mathbb{R}) \supset \mathbf{SL}(n, \mathbb{R}) \\ \mathbf{DO}(n) \supset \mathbf{O}(n) \end{matrix} \right\} \supset \mathbf{SO}(n),$$

$$\begin{aligned} \mathbf{GL}(n, \mathbb{R})/\mathbf{GL}_+(n, \mathbb{R}) &\cong \mathbf{O}(n)/\mathbf{SO}(n) \cong \mathbf{I}(2) = \{1, -1\}, \\ \mathbf{GL}_+(n, \mathbb{R})/\mathbf{SL}(n, \mathbb{R}) &\cong \mathbf{DO}(n)/\mathbf{O}(n) \cong \mathbf{D}(1) = e^{\mathbb{R}}. \end{aligned}$$

With the distinction of a subgroup $G \subseteq \mathbf{GL}(V)$ acting on a vector space — as $V \cong \mathbb{R}^n$ above, one can explain the associated *affine group* $G \vec{\times} V$, consisting of the homogeneous group G and the translations (vectors) V as inhomogeneous transformations — they constitute an abelian normal subgroup in the semidirect product:

$$\text{affine group: } G \vec{\times} V = \{(g, v) \mid g \in G \subseteq \mathbf{GL}(V),\ v \in V\},$$

$$\text{group multiplication: } \left\{ \begin{aligned} \text{composition: } & (g_1, v_1) \bullet (g_2, v_2) \\ & = (g_1 \circ g_2, v_1 + g_1.v_2), \\ \text{neutral element: } & (\mathbf{1}_V, 0) \bullet (g, v) = (g, v), \\ \text{inversion: } & (g, v)^{-1} = (g^{-1}, -g^{-1}.v), \\ \text{factorization: } & G \vec{\times} V = V \bullet G, \\ & (g, v) = (\mathbf{1}_V, v) \bullet (g, 0). \end{aligned} \right.$$

Therefore, a linear geometry comes with an affine group.

4.11 Erlangen Program of Flat Space-Time

It is instructive, to track, with the Erlangen Program as a structurizing figure of thought, the development of the linear space-time concepts with their symmetry aspects.

A static world view, neglecting time and dynamics, can be related to an Euclidean geometry with a $3 + 3$ real parametric affine group:

$$\text{Euclidean space with } \mathbf{O}(3) \vec{\times} \mathbb{R}^3.$$

If time, better time translations, is formalized as a real one-dimensional vector space $\mathbb{R}$ in a direct sum with the space translations $\mathbb{R}^3$, one has *absolute* space-time:

$$\text{Aristotle–Newton space-time with } [\mathbf{O}(1) \vec{\times} \mathbb{R}] \times [\mathbf{O}(3) \vec{\times} \mathbb{R}^3].$$

This group has $1+3+3$ real parameters. The time factor $\mathbf{O}(1) \vec{\times} \mathbb{R}$ describes the time translations with the time reflection group $\mathbf{O}(1) \cong \mathbf{I}(2)$, the space factor $\mathbf{O}(3) \vec{\times} \mathbb{R}^3$ the space (position) translations with the rotations and the space reflections $\mathbf{O}(3) \cong \mathbf{I}(2) \times \mathbf{SO}(3)$.

The Galileo group is peculiar: On the one hand, it respects a space-time decomposition $(t, \vec{x}) \in \mathbb{R} \oplus \mathbb{R}^3$, on the other hand, the proper Galileo transformations connect space and time via velocities $\vec{v} \in \mathbb{R}^3$:

$$\begin{aligned} &\text{Galileo space-time with } [\mathbf{O}(3) \vec{\times} \mathbb{R}^3] \vec{\times} [\mathbb{R} \oplus \mathbb{R}^3], \\ &\text{composition: } \begin{cases} (O_1, \vec{v}_1; t_1, \vec{x}_1) \bullet (O_2, \vec{v}_2; t_2, \vec{x}_2) \\ = (O_1 \circ O_2, \vec{v}_1 + O_1.\vec{v}_2; t_1 + t_2, \vec{x}_1 + \vec{v}_1 t_2 + O_1.\vec{x}_2). \end{cases} \end{aligned}$$

This $3+3+4$-parametric space-time group with the two semidirect factors displays a rather complicated structure.

Good to know: Aristotle space $\mathbb{A}^3$ distinguishes absolute rest whereas Galileo space $\mathbb{G}^3$ is determined "up to a constant velocity." In a group theoretical language this can be expressed by classes with respect to the homogeneous group of the corresponding inhomogeneous Euclid and Galileo group, i.e., $\mathbb{A}^3 = \mathbf{O}(3) \vec{\times} \mathbb{R}^3/\mathbf{O}(3)$ and $\mathbb{G}^3 = [\mathbf{O}(3) \vec{\times} \mathbb{R}^3] \vec{\times} \mathbb{R}^3/[\mathbf{O}(3) \vec{\times} \mathbb{R}^3]$, respectively which both, as manifolds, are isomorphic to the additive translation group $\mathbb{R}^3$.

In general, the multiplication law can be obtained by a multiplicative matrix representation of the additive group $\mathbb{R}^n$:

$$\begin{aligned} &\mathbb{R}^n \ni v \longmapsto \begin{pmatrix} \mathbf{1}_n & v \\ 0 & 1 \end{pmatrix} \in \mathbf{GL}(n+1, \mathbb{R}), \\ &\begin{pmatrix} \mathbf{1}_n & v_1 \\ 0 & 1 \end{pmatrix} \circ \begin{pmatrix} \mathbf{1}_n & v_2 \\ 0 & 1 \end{pmatrix} = \begin{pmatrix} \mathbf{1}_n & v_1 + v_2 \\ 0 & 1 \end{pmatrix}, \end{aligned}$$

which can be extended to include a nontrivial homogeneous group:

$$\begin{aligned} &G \subseteq \mathbf{GL}(n, \mathbb{R}), \quad G \vec{\times} \mathbb{R}^n \ni (g, v) \longmapsto \left(\begin{array}{c|c} g & v \\ \hline 0 & 1 \end{array}\right) \in \mathbf{GL}(n+1, \mathbb{R}), \\ &\begin{pmatrix} g_1 & v_1 \\ 0 & 1 \end{pmatrix} \circ \begin{pmatrix} g_2 & v_2 \\ 0 & 1 \end{pmatrix} = \begin{pmatrix} g_1 \circ g_2 & v_1 + g_1.v_2 \\ 0 & 1 \end{pmatrix}. \end{aligned}$$

A twofold application leads to the multiplication law of the Galileo group:

$$\begin{aligned} &[\mathbf{O}(3) \vec{\times} \mathbb{R}^3] \vec{\times} [\mathbb{R} \oplus \mathbb{R}^3] \ni (O, \vec{v}; t, \vec{x}) \longmapsto \left(\begin{array}{cc|c} O & \vec{v} & \vec{x} \\ 0 & 1 & t \\ \hline 0 & 0 & 1 \end{array}\right) \in \mathbf{GL}(5, \mathbb{R}), \\ &\begin{pmatrix} O_1 & \vec{v}_1 & \vec{x}_1 \\ 0 & 1 & t_1 \\ 0 & 0 & 1 \end{pmatrix} \circ \begin{pmatrix} O_2 & \vec{v}_2 & \vec{x}_2 \\ 0 & 1 & t_2 \\ 0 & 0 & 1 \end{pmatrix} = \begin{pmatrix} O_1 \circ O_2 & \vec{v}_1 + O_1.\vec{v}_2 & \vec{x}_1 + \vec{v}_1 t_2 + O_1.\vec{x}_2 \\ 0 & 1 & t_1 + t_2 \\ 0 & 0 & 1 \end{pmatrix}. \end{aligned}$$

Special relativity simplifies the geometry for space-time: The proper Galileo transformations with the velocities $\vec{v} \in \mathbb{R}^3$ are "expanded," as homogeneous transformations, to the level of the rotations $\mathbf{SO}(3)$. The $4+6$-parametric Poincaré group consists of the spacetime translations $\mathbb{R}^4$ and the Lorentz group $\mathbf{O}(1, 3)$:

$$\begin{aligned} &\text{Minkowski spacetime with } \mathbf{O}(1,3) \vec{\times} \mathbb{R}^4, \\ &(\Lambda_1, x_1) \bullet (\Lambda_2, x_2) = (\Lambda_1 \circ \Lambda_2, x_1 + \Lambda_1.x_2). \end{aligned}$$

The homogeneous Lorentz group is the semidirect product of the special Lorentz group with a not Lorentz compatible reflection group which can be built by a space or a time reflection. The special group itself is the direct product of the orthochronous subgroup and the Lorentz compatible central reflection. The quotient group with the orthochronous Lorentz group classes characterizes all reflections of Minkowski spacetime and is isomorphic to the Klein four-group:

$$\begin{aligned} \mathbf{O}(1,3) &\cong \mathbf{I}(2) \vec{\times} \mathbf{SO}(1,3) \cong \mathbf{I}(2) \vec{\times} [\mathbf{I}(2) \times \mathbf{SO}_0(1,3)]), \\ \mathbf{O}(1,3)/\mathbf{SO}_0(1,3) &\cong \mathbf{I}(2) \times \mathbf{I}(2) \cong \mathbf{K}(4). \end{aligned}$$

Good to know: General relativity uses, for spacetime, a real four-dimensional differential manifold $\mathbb{M}^{(1,3)}$, endowed, at each spacetime point, with a tangent space metric with causal signature. Special spacetimes can be distinguished by characteristic action groups for Klein spaces $G \bullet \mathbb{M}^{(1,3)}$. For example, the three maximally symmetric de Sitter-,[20] Minkowski and anti-de Sitter-spacetimes are characterized by the real 10-dimensional action groups $\mathbf{O}(1,4)$, $\mathbf{O}(1,3) \vec{\times} \mathbb{R}^4$, and $\mathbf{O}(2,3)$, respectively. In these cases, spacetime $\mathbb{M}^{(1,3)}$ parametrizes the corresponding classes of the Lorentz group, i.e., $\mathbf{O}(1,4)/\mathbf{O}(1,3)$, $[\mathbf{O}(1,3) \vec{\times} \mathbb{R}^4]/\mathbf{O}(1,3)$, and $\mathbf{O}(2,3)/\mathbf{O}(1,3)$.

4.12 Simple Lie Operations

In atomic and elementary particle physics, one tries to describe the observed symmetry in the space-time and internal charge behavior by Lie groups and by Lie algebras as their infinitesimal and generating structures, the internal unitary symmetry by compact (Euclidean) Lie groups. All these structures have their origin in, are connected with, and generalize Plato's beautiful symmetry.

Linearization by infinitesimalization was twice an essential tool for physical progress: the description of a physical dynamics by differential equations in time (Newton, Leibniz) and the differential geometrical description of symmetry structures (Riemann, Lie, Einstein). Both cases are connected with a transition from a differentiable manifold, e.g., a Lie group, to the tangent space structure, in the example to the Lie algebra, and, therefore, mathematically-structurally closely related to each other.

Given the general, almost amorphous structure of a group, it is surprising that it is possible to enumerate exhaustively and very explicitly symmetry structures of an important type of Lie groups and their Lie algebras. First Killing[21] and then, more complete, Cartan[22] classified all *simple complex finite-dimensional Lie algebras*, and later Cartan all *simple real finite-dimensional Lie algebras*.

[20]Willem de Sitter (1872–1934).

[21]Wilhelm Killing (1847–1923).

[22]Élie Cartan (1869–1951).

Good to know: Lie algebras and Lie groups are defined without a characterization as linear transformations. However, many Lie algebras can be thought of consisting of matrices: The theorem of Ado[23] ensures the possibility to represent all real or complex finite-dimensional Lie algebras injectively (*faithfully*) by endomorphisms (square matrices) of correspondingly real or complex finite-dimensional vector spaces with the Lie bracket as commutator. The largest nilpotent ideal is represented by nilpotent matrices. A reductive Lie algebra L, i.e., with semisimple commutator ideal $[L, L]$, has a faithful representation by conjugation stable matrices, i.e., stable under transposition $l \leftrightarrow l^T$ for $\mathbb{R}$ and hermitian transposition $l \leftrightarrow l^\star$ for $\mathbb{C}$.

There exist four countably infinite *main series* of simple complex finite-dimensional Lie algebras:

$$\begin{aligned} &{}_{r(r+2)}A_r \ (r \geq 1), \ {}_{r(2r+1)}C_r \ (r \geq 1), \\ &{}_{r(2r+1)}B_r \ (r \geq 1), \ {}_{r(2r-1)}D_r \ (r \geq 3), \end{aligned}$$

and five *exceptional simple Lie algebras*:

$$\begin{aligned} &{}_{14}G_2, \ {}_{52}F_4, \\ &{}_{78}E_6, \ {}_{133}E_7, \ {}_{248}E_8. \end{aligned}$$

The left index in ${}_dL_r$ gives the vector space dimension d of the Lie algebra, i.e., the number of linearly independent operations, the right one the rank r — the maximal dimension number of a subvector space with simultaneously diagonalizable operations (Cartan elements), in physics "simultaneously measurable." r is related to the number of the independent "quantum numbers." It divides the dimension $\frac{d}{r} \in \mathbb{N}$.

Good to know: An *ideal* I of a Lie algebra L is a Lie subalgebra, stable under adjoint action, i.e., $[L, I] \subseteq I$. The classes L/I inherit a natural Lie algebra structure. A nonabelian Lie algebra L without proper ideal I, i.e., $[L, I] \subseteq I \Rightarrow I = L$ oder $\{0\}$, is called *simple*.

The missing members of the D-series are not simple: The complex numbers ${}_1D_1 \cong \mathbb{C}$ are abelian and ${}_6D_2 \cong {}_3A_1 \oplus {}_3A_1$ is semisimple. The four main series unfold fully and independently after some isomorphies for rank 1, 2, and 3:

$$\begin{aligned} &\text{rank 1: } A_1 \cong C_1 \cong B_1, \ D_1 \text{ (abelian)}, \\ &\text{rank 2: } D_2 \cong A_1 \oplus A_1, \ C_2 \cong B_2, \\ &\text{rank 3: } D_3 \cong A_3. \end{aligned}$$

The main series Lie algebras have the inclusion relations:

$$A_r \subset D_{1+r} \subset B_{1+r} \cap D_{1+r}.$$

The Lie algebras of the main series are the infinitesimal or generating operations (tangent or derivative structures) of the complex special-linear, symplectic and orthogonal groups. The Lie algebra exponents give the universal covering groups:

$$\begin{aligned} &\exp A_r \cong \mathbf{SL}(r+1, \mathbb{C}), \quad \exp C_r \cong \mathbf{Sp}(2r, \mathbb{C}), \\ &\exp B_r \cong \mathbf{SO}(2r+1, \mathbb{C}), \ \exp D_r \cong \mathbf{SO}(2r, \mathbb{C}). \end{aligned}$$

---

[23]Igor Dmitrievich Ado (1910–1983).

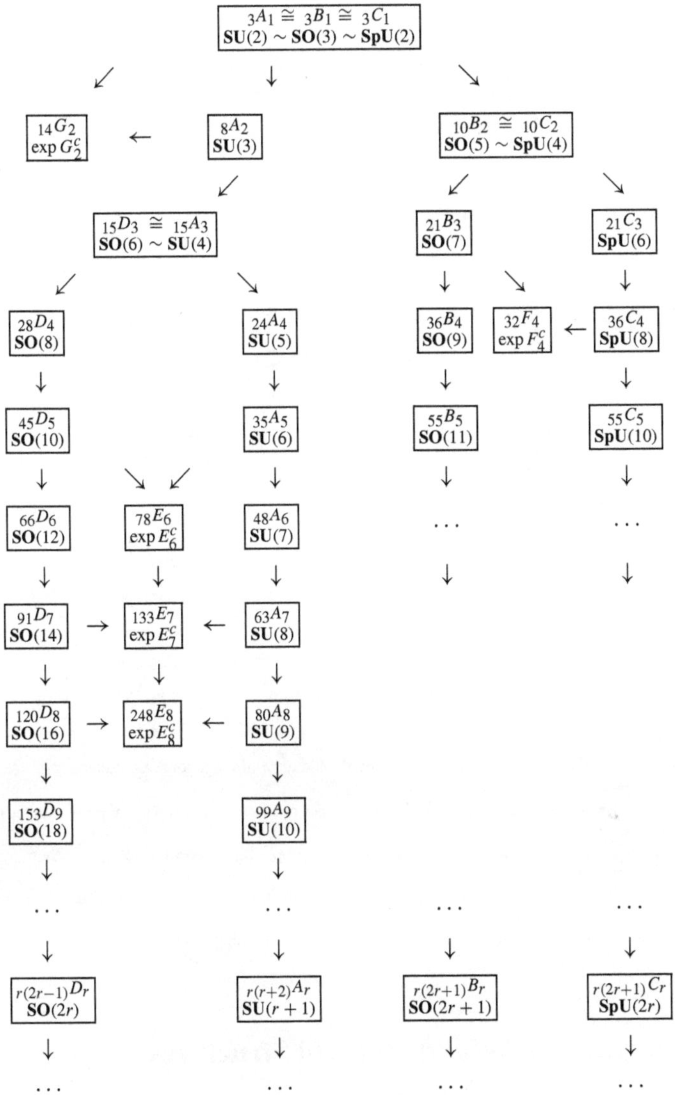

Cartan Classification of the Simple Lie Algebras $_{\text{dim}}\text{L}_{\text{rang}}$ and their Compact Groups $\exp \mathbf{L}^c$

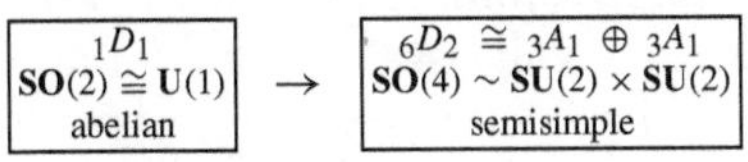

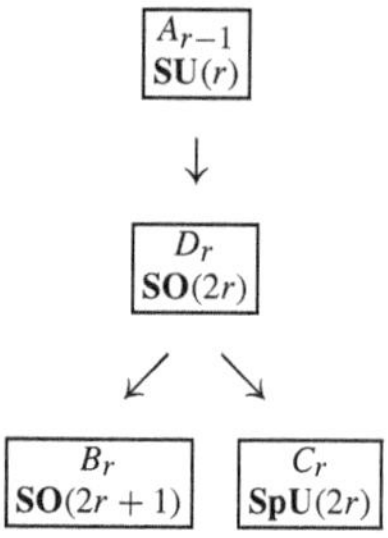

Inclusions for the Main Series Simple Lie Algebras (with $\rightarrow$ for $\subset$)

Good to know: A *representation* of a Lie algebra $L \ni l \longmapsto \mathcal{D}(l) \in \mathbf{AL}(V)$ in the endomorphisms of a vector space V is a Lie algebra morphism with the Lie bracket in the endomorphisms defined by the commutator $\mathcal{D}([l,k]) = [\mathcal{D}(l), \mathcal{D}(k)] = \mathcal{D}(l) \circ \mathcal{D}(k) - \mathcal{D}(k) \circ \mathcal{D}(l)$. The Lie algebra acts on the vector space $L \times V \longrightarrow V, l \bullet v = \mathcal{D}(l).v$. The representation morphisms are the *intertwiners*, defined as the action compatible linear mappings f with a commutative

diagram f $\begin{array}{ccc} V_1 & \overset{\mathcal{D}_1(l)}{\longrightarrow} & V_1 \\ \downarrow & & \downarrow \\ V_2 & \underset{\mathcal{D}_2(l)}{\longrightarrow} & V_2 \end{array}$ f for all $l \in L$. With an isomorphism f, the two representations are *equivalent*.

The *adjoint representation* $L \ni l \longmapsto \mathrm{ad}\, l \in \mathbf{AL}(L)$ of a Lie algebra on itself is given by the commutator of the Lie bracket associated endomorphisms $\mathrm{ad}\, l(m) = [l, m]$. In a basis $[l^a, l^b] = \epsilon_c^{ab} l^c$, the structure constants yield the matrices of the adjoint representation $(\mathrm{ad}\, l^a)_c^b = \epsilon_c^{ab}$.
The symmetric and invariant bilinear *Killing form* is the double trace of the adjoint representation $L \times L \ni (l, m) \longmapsto \kappa(l, m) = \mathrm{tr}\ \mathrm{ad}\, l \circ \mathrm{ad}\, m \in \mathbb{R}$ or $\mathbb{C}$ — in a basis $\kappa^{ab} = \epsilon_d^{ac} \epsilon_c^{bd}$.
A representation is simple (irreducible), if the vector space does not contain an L-invariant proper vector subspace, i.e., with the exception of V and $\{0\}$.
A Lie algebra is called *semisimple* as a direct sum-product of simple Lie algebras, or, equivalently, if, and only if the Killing form is nondegenerate.
A real Lie algebra with negativ definite Killing form is called *compact*. For example, $\log \mathbf{SU}(2)$ carries a negative definite Killing form $-\delta \cong -\mathbf{1}_3$ with signature $(0, 3)$.
The exponent of a complex linear endomorphism is a complex linear automorphism, $\exp \mathbf{AL}(n, \mathbb{C}) = \mathbf{GL}(n, \mathbb{C})$.

4.13 Lie Symmetries and Metrical Structures

A Lie structure is defined purely algebraically, e.g., by real (or complex) structure constants $[l^a, l^b] = \epsilon_c^{ab} l^c$ with respect to a vector space basis $\{l^a\}_{a=1}^d$ of a real (or complex) Lie algebra ${}_dL$. Therefore, it is somewhat surprising that algebraic structures involve metrical ones, e.g., the bilinear Killing form — for a basis $\kappa(l^a, l^b) = \epsilon_d^{ac} \epsilon_c^{bd} \in \mathbb{R}$ or $\mathbb{C}$. Starting, vice versa, from a metrical structure of a vector space, a Lie symmetry can be introduced via invariance operations, as familiar for the definition of the orthogonal $\mathbf{O}(3)$-symmetry in space $\mathbb{R}^3$ as invariance structure of the lengths $\vec{x}^2 = \ell^2$, or of the Lorentz $\mathbf{O}(1, 3)$-symmetry in energy-momentum space $\mathbb{R}^4$ as invariance structure of the mass hyperboloids $q_0^2 - \vec{q}^2 = m^2$.

Good to know: A multilinear form $\kappa(v_1, \ldots, v_k)$ of a vector space V is called *invariant* for an automorphism $g \in \mathbf{GL}(V)$, if $\kappa(v_1, \ldots, v_k) = \kappa(g.v_1, \ldots, g.v_k)$. All such g define a $\mathbf{GL}(V)$-subgroup. The endomorphisms $l \in \mathbf{AL}(V)$ with the property $\kappa(l.v_1, v_2, \ldots, v_k) + \ldots + \kappa(v_1, \ldots, v_{k-1}, l.v_k) = 0$ constitute an $\mathbf{AL}(V)$-Lie subalgebra.
The Killing form of a semisimple Lie algebra $L \cong \mathbb{R}^n$ has diagonal form $(\mathbf{1}_t, -\mathbf{1}_s)$, $t + s = n$, with t noncompact and s compact degrees of freedom, and invariance Lie algebra $\log \mathbf{O}(t, s) \supseteq L$. For example, $\log \mathbf{SL}(2, \mathbb{C})$ has the indefinite Killing form $\eta \wedge \eta \cong \begin{pmatrix} \mathbf{1}_3 & 0 \\ 0 & -\mathbf{1}_3 \end{pmatrix}$ with neutral signature (3, 3).

The Lie algebras and Lie groups of the four main series are characterizable by metrical structures: The special linear group $\mathbf{SL}(r+1, \mathbb{C})$ is the invariance group of the totally antisymmetric multilinear forms of degree $r+1$, i.e., of the *volume elements* of the vector space $\mathbb{C}^{r+1}$, given by the determinant of $r+1$ vectors $v_j \in V \cong \mathbb{C}^{r+1}$. It characterizes the special complex $r+1$-dimensional geometry. The symplectic group $\mathbf{Sp}(2n, \mathbb{C})$ keeps the nondegenerate *antisymmetric bilinear forms* of $\mathbb{C}^{2n}$ invariant and stands for the symplectic complex $2n$-dimensional geometry. The special orthogonal group $\mathbf{SO}(n, \mathbb{C})$ as invariance group for the nondegenerate *symmetric bilinear forms* of $\mathbb{C}^n$ is connected with the orthogonal complex n-dimensional geometry.

The transition from the complex to the real requires a conjugation and unitary groups, definite and indefinite ones. Simple real Lie algebras can be defined, via conjugations, as *real forms* of complex ones. For each simple complex Lie algebra above ${}_dL_r \cong \mathbb{C}^d$ there is an, up to isomorphy, unique *compact real form* ${}_dL_r^c \cong \mathbb{R}^d$ (Euclidean form) with an invariant *scalar product* given by the negative definite Killing form. It has a compact covering group.

In the transition to the compact forms of the orthogonal symmetries (invariant symmetric bilinear forms), there is the replacement of the complex vector space by a real one with equal dimension, $\mathbb{C}^n \longrightarrow \mathbb{R}^n$, which has an invariant scalar product:

$$\begin{array}{c} \exp B_r^c \cong \overline{\mathbf{SO}(2r+1)}, \quad \exp D_r^c \cong \overline{\mathbf{SO}(2r)} \\ \text{with } \mathbf{SO}(n, \mathbb{R}) = \mathbf{SO}(n) \cong \overline{\mathbf{SO}(n)}/\mathbf{I}(2). \end{array}$$

The rotation groups $\mathbf{SO}(n)$ arise from the simply connected covering groups $\overline{\mathbf{SO}(n)}$ as classes with respect to a normal subgroup $\mathbf{I}(2)$. For $n \leq 6$, the covering groups are isomorphic to linear groups:

$$\begin{array}{l} \overline{\mathbf{SO}(3)} \cong \mathbf{SU}(2), \quad \overline{\mathbf{SO}(4)} \cong \mathbf{SU}(2) \times \mathbf{SU}(2), \\ \overline{\mathbf{SO}(5)} \cong \mathbf{SpU}(4), \ \overline{\mathbf{SO}(6)} \cong \mathbf{SU}(4). \end{array}$$

In the transition to the compact forms of the special linear and the symplectic symmetries (invariant antisymmetric multi- and bilinear forms, respectively) the vector space remains complex and obtains an additional positive sesquilinear form with positive unitary invariance group $\mathbf{U}(n, \mathbb{C}) = \mathbf{U}(n)$:

$$\begin{array}{llll} \exp A_r^c & \cong \mathbf{SL}(r+1, \mathbb{C}) \cap \mathbf{U}(r+1) & = \mathbf{SU}(r+1, \mathbb{C}) & = \mathbf{SU}(r+1), \\ \exp C_r^c & \cong \mathbf{Sp}(2r, \mathbb{C}) \cap \mathbf{U}(2r) & = \mathbf{SpU}(2r, \mathbb{C}) & = \mathbf{SpU}(2r). \end{array}$$

The complex vector space has two invariant forms, the antisymmetric multi- and bilinear one and a positive sesquilinear one, i.e., a scalar product.

Good to know: Unitary Lie groups have a **U** in their symbols. This indicates the existence of a sesquilinear form or a conjugate linear isomorphism to the dual vector space, i.e., a conjugation. Unitary Lie groups are *real* Lie groups, acting on complex vector spaces, e.g., $\mathbf{U}(1)$ and $\mathbf{SU}(2)$.

If for a complex Lie algebra group, the characterizing complex vector spaces is replaced by a real one, ${}_dL_r^{\mathbb{R}} \cong \mathbb{R}^d$, one ends up with the noncompact groups $\mathbf{SL}(r+1,\mathbb{R}) = \mathbf{SL}(r+1)$ (normal forms) for the complex special linear group $\mathbf{SL}(r+1,\mathbb{C})$ and with the noncompact groups $\mathbf{Sp}(2r,\mathbb{R}) = \mathbf{Sp}(2r)$ in the case of the complex symplectic group $\mathbf{Sp}(2r,\mathbb{C})$.

All simple real Lie algebras have been classified by Cartan.

Good to know: The complex Lie algebra ${}_{15}D_3 \cong {}_{15}A_3$ has as complex and real Lie groups:

$$\mathbf{SO}(6,\mathbb{C}) \cong \mathbf{SL}(4,\mathbb{C}) \text{ and } \begin{cases} \mathbf{SO}(6) & \sim \mathbf{SU}(4) \text{ (compact form)}, \\ \mathbf{SO}(1,5) & \sim \mathbf{SU}^*(4), \\ \mathbf{SO}^*(6) & \sim \mathbf{SU}(1,3), \\ \mathbf{SO}(2,4) & \sim \mathbf{SU}(2,2), \\ \mathbf{SO}(3,3) & \sim \mathbf{SL}(4,\mathbb{R}) \text{ (normal form)}. \end{cases}$$

The real 15-dimensional groups have physical applications: $\mathbf{SU}(4)$ can be used as color $\mathbf{SU}(3)$-extension with the leptons supplementing the quarks. $\mathbf{SU}(1,3)$ is the invariance group of a symmetric sesquilinear product which extends the Lorentz $\mathbf{SO}(1,3)$-metric into a complex four-dimensional space. $\mathbf{SO}(2,4)$ is used as conformal group, e.g., for massless structures. $\mathbf{SO}(3,3)$ is the invariance group of the Lorentz $\mathbf{SO}(1,3)$-Killing form, and, in the $\mathbf{SL}(4,\mathbb{R})$-form, a subgroup of the structure group $\mathbf{GL}(4,\mathbb{R})$ of the general relativistic tangent spacetime bases. The contraction of $\mathbf{SO}(1,5)$ can be used to lead to the real 14-dimensional extended Poincaré group $\mathbf{SO}(1,3) \vec{\times} [\mathbb{R}^4 \oplus \mathbb{R}^4]$ for spacetime translations and energy-momenta, which are noncommuting in the real nine-dimensional Heisenberg group $\mathbf{H}(4)$. $\mathbf{SO}(1,3)$ gives the remaining six parameters.

4.14 The Simplest Simple Symmetry

Lie symmetry starts with rank 1, i.e., with one independent invariant. There is only one simple complex Lie algebra with rank 1. Its compact form is locally isomorphic to the rotation group $\mathbf{SO}(3)$, characterizing Plato's symmetry.

If the Lie algebras ${}_dL_r$ are ordered with respect to their vector space dimension $d \geq r \geq 1$, they start as follows:

The one-dimensional Lie algebras are abelian, denoted by ${}_1D_1 \cong \mathbb{C}$ with ${}_1D_1^{\mathbb{R}} \cong \mathbb{R}$ and ${}_1D_1^c \cong i\mathbb{R}$. Their invariant linear forms are their algebraic dual spaces. In physical applications, $D_1^{\mathbb{R}}$ is used as time translations $t \in \mathbb{R}$ with the frequencies $\omega \in \mathbb{R}$ as invariants.

A nonabelian Lie algebra ${}_2L$ with dimension two has a basis with $[l^1, l^2] = l^2$. Therefore, l^2 spans a proper abelian ideal i.e., ${}_2L$ is not semisimple.

The lowest-dimensional semisimple Lie algebras have dimension three and rank one. Semisimple Lie algebras have only trivial invariant linear forms, they start nontrivially with invariant nondegenerate bilinear forms. Each invariant multilinear form of a Lie algebra allows, via its normalization, the introduction of, in physical terms, a scale (unit). The rank r of a Lie algebra L_r gives the number of the independent scales (units) for a physical theory, characterized by L_r.

Good to know: The rank r of a semisimple Lie algebra gives the number of the generators for the ring with the invariants in the enveloping algebra.

There is — up to isomorphy — exactly one complex semisimple Lie algebra with rank 1, the simple Lie algebra of the *binary alternative* as basic building block of all semisimple Lie algebras:

$$_3A_1 \cong {}_3C_1 \cong {}_3B_1.$$

It has dimension 3 and is the infinitesimal structure of the complex group $\mathbf{SL}(2, \mathbb{C})$ (volume elements of $\mathbb{C}^2$), of the complex group $\mathbf{Sp}(2, \mathbb{C})$ (antisymmetric bilinear forms of $\mathbb{C}^2$) as well as of the complex group $\mathbf{SO}(3, \mathbb{C})$ (symmetric bilinear forms of $\mathbb{C}^3$). The notation A_1 of Cartan — it starts with A in the alphabet and with 1 in the naturals — can also be read as *one alternative*.

The symmetry of the binary alternative plays a fundamental role also in elementary particle physics: for the spin group, for the Lorentz transformations and, also, for isospin. The square of Planck's unit $\hbar^2$ can be chosen as normalization of the A_1-related Killing form, and $\hbar$ can be used as a scale for angular momenta.

From a spherical basis of A_1, also called canonical, normal or Weyl basis, with the *raising and lowering element* $l_\pm$ and their Lie bracket, the *Cartan-element* l_0:

$$\text{spherical basis } \{l_\pm, l_0\}: \;\; [l_-, l_+] = l_0, \;\; [l_0, l_\pm] = \pm 2 l_\pm,$$

one obtains, by complex linear transformations, a Cartesian basis, also called compact or Euclidean basis:

$$\text{Cartesian basis } \{l^{1,2,3}\}: \;\; [l^a, l^b] = -\epsilon^{abc} l^c.$$

This is a basis of the real three-dimensional compact form $A_1^c = \{\alpha_a l^a \,\big|\, \alpha_a \in \mathbb{R}\}$ which is the infinitesimal or generating structure of the unitary group $\mathbf{SU}(2)$, the rotation group $\mathbf{SO}(3)$ with the classes of $\mathbf{SU}(2)$ with respect to the discrete normal subgroup $\mathbf{I}(2) \cong \{\pm \mathbf{1}_2\}$, and of the symplectic group $\mathbf{SpU}(2)$:

$$\exp A_1^c \cong \mathbf{SU}(2) \cong \mathbf{SpU}(2) \sim \mathbf{SO}(3) \cong \mathbf{SU}(2)/\mathbf{I}(2).$$

The real-linear combinations of a Weyl basis define the Lie algebra $A_1^{\mathbb{R}} = \{\alpha_\pm l_\pm + \alpha_0 l_0 \,\big|\, \alpha_\pm, \alpha_0 \in \mathbb{R}\}$ of noncompact groups as invariance groups of indefinite bilinear products:

$$\exp A_1^{\mathbb{R}} \sim \mathbf{SL}(2, \mathbb{R}) \sim \mathbf{SU}(1, 1) \cong \mathbf{SpU}(1, 1) \sim \mathbf{SO}(1, 2) \cong \mathbf{SU}(1, 1)/\mathbf{I}(2).$$

The three-dimensional Lie symmetry A_1 has as its fundamental representation, denoted by [1], the complex *Pauli-representation* acting on two-dimensional *spinors*. The representation space can be constructed from a two elementic set $\{\circ, \bullet\}$, called *binary alternative* ("yin and yang" or "open and closed" or "up and down"), as basis of the complex mappings $\{\circ, \bullet\} \longrightarrow \mathbb{C}$, i.e., of the free vector space $\mathbb{C}^{\{\bullet,\circ\}} \cong \mathbb{C}^2$:

$$\bullet = e^\uparrow \cong \begin{pmatrix} 1 \\ 0 \end{pmatrix}, \quad \circ = e^\downarrow \cong \begin{pmatrix} 0 \\ 1 \end{pmatrix}, \quad \alpha e^\uparrow + \beta e^\downarrow \cong \begin{pmatrix} \alpha \\ \beta \end{pmatrix} \in \mathbb{C}^2.$$

Good to know: The real (or complex) functions of a set $\{f : I \longrightarrow \mathbb{R}\}$, $\iota \longmapsto f_\iota$, are given by the product subset $\mathbb{R}^I = \{(\iota, f_\iota) \in I \times \mathbb{R}\}$ which is called the *free vector space* with vectors $f \cong (f_\iota)_{\iota \in I}$ and pointwise addition and scalar multiplication. For a finite set, card $I = n$, the finite dimensional vector space $\mathbb{R}^I \cong \mathbb{R}^n$ contains the linear combinations of the function values $f = f_\iota|\iota\rangle$ with the I-elements $\{|\iota\rangle\}_{\iota \in I}$ as natural basis.

The operations of the Lie algebra A_1 act on the vector space for the complex alternative by flip-operations and their commutator σ_0:

$$\begin{aligned} l^{\pm,0} \longmapsto \sigma_{\pm,0} : \sigma_+ = \begin{pmatrix} 0 & 1 \\ 0 & 0 \end{pmatrix}, \quad \sigma_- &= \begin{pmatrix} 0 & 0 \\ 1 & 0 \end{pmatrix}, \quad \sigma_0 = \begin{pmatrix} 1 & 0 \\ 0 & -1 \end{pmatrix}, \\ [\sigma_+, \sigma_-] &= \sigma_0, \ [\sigma_0, \sigma_\pm] = \pm 2\sigma_\pm, \\ \sigma_-(e^\uparrow) &= e^\downarrow, \quad \sigma_+(e^\uparrow) = 0, \qquad \sigma_0(e^\uparrow) = e^\uparrow, \\ \sigma_-(e^\downarrow) &= 0, \quad \sigma_+(e^\downarrow) = e^\uparrow, \qquad \sigma_0(e^\downarrow) = -e^\downarrow. \end{aligned}$$

A $\mathbf{U}(2)$-invariant scalar product of the compact binary alternative with Pauli representation $l^{1,2,3} \longmapsto \frac{i}{2}\sigma^{1,2,3}$ is $\langle e^A | e^B \rangle = \delta^{AB}$. The invariant bilinear form of the Lie algebra A_1 in the Pauli representation is the double trace, $\operatorname{tr} \frac{i}{2}\sigma^a \frac{i}{2}\sigma^b = -\frac{1}{2}\delta^{ab}$ which can be spin-normalized by $\hbar^2$ for a basis $\{\hbar \frac{i}{2}\sigma^a\}_{a=1}^3$.

The Pauli spinors are fundamental in the following sense: They build, up to isomorphy, all finite-dimensional representations of the Lie algebra A_1. By totally symmetric products (tensor products) of Pauli spinors each natural number $n \in \mathbb{N}$ characterizes the representation space of an irreducible A_1-representation $[n]$ whose basis may be called an $(n+1)$-fold alternative. More familiar is the characterization by the *spin* $j = \frac{n}{2} \in \frac{\mathbb{N}}{2}$, integer or half-integer. With the "loss" of the faithful $\mathbf{I}(2)$- representation, the rotation group $\mathbf{SO}(3) \cong \mathbf{SU}(2)/\mathbf{I}(2)$ has only representations for integer spin. The l_0-eigenvalues in the $(n+1)$-dimensional representation $[n]$ occupy its *weight diagram* with $(n+1)$ equidistant points on the Euclidean line $\mathbb{E}^1$, situated symmetrically on both sides of the origin—in spin normalization $m \in \{-j, -j+1, \ldots, j-1, j\}$.

With his solids, Plato investigated properties of the simplest simple Lie operations $\exp A_1^c \sim \mathbf{SO}(3)$. The simplest simple Lie algebra A_1 occurs everywhere: As shown by Cartan, Weyl and Jacobson,[24] it governs the structure of all semisimple Lie algebras ${}_dL_r$ of rank r. All representations of all semisimple Lie algebras can be decomposed into simple A_1-representations - for ${}_dL_r$ in $\frac{d-r}{2}$ different ways: The semisimple Lie algebra ${}_dL_r$ (take as example the Lie algebra ${}_{15}A_3$ of $\mathbf{SL}(4, \mathbb{C})$ or its compact form ${}_{15}A_3^c$ of $\mathbf{SU}(4)$) contains $\frac{d-r}{2}$ Lie subalgebras ${}_3A_1$ (in the example 6) with their $3\frac{d-r}{2}$ generators in a spherical basis $\{l_\pm^k, l_0^k\}_{k=1}^{\frac{d-r}{2}}$ (in the example 18). However, only d generators are linearly independent (in the example 15); therefore, there exist $\frac{d-3r}{2}$ nontrivial linear relations (in the example 3) between the simultaneously diagonalizable operations, the Cartan elements $\{l_0^k\}_{k=1}^{\frac{d-r}{2}}$.

[24]Nathan Jacobson (1910–1999).

4.15 Quantum Numbers and Weights of Simple Lie Algebras

If the abstract concept of a symmetry is compared with Plato's concept of an idea, a representation of the symmetry-related operations may be associated with a shadow on the wall of the cave (Plato's allegory of the cave in his *Politeia*). Or: A general physical law, e.g., a dynamics, can be associated with Plato's idea, its application for special boundary or initial conditions with the wall-images.

Cartan has given all representations of simple complex and real Lie algebras by linear operations on complex finite-dimensional vector spaces.

Properties of interactions or particles are informations with respect to the actions of the represented symmetry operations. For example, spin characterizes a rotation property (roundness), mass a spacetime translation property (inertia), the electric charge can be connected with a winding property.

Actions of transformations on vectors are — in complex spaces — characterizable by *eigenvalues*. For semisimple operations there exists even an eigenvector basis for the complex representation space. Several transformations can have common eigenvectors. This is the case for semisimple (diagonalizable) complex transformations, if, and only if they commute with each other. Eigenvalues of simultaneously diagonalizable transformations constitute *weights*, more precise weight vectors. Semisimple Lie algebras are characterizable by Lie subalgebras with simultaneously diagonalizable operations — the Cartan algebras with the rank as dimension - and by the weights for their representations.

Good to know: Any complex matrix $f \in \mathbf{AL}(n, \mathbb{C})$ is similar $g \circ f \circ g^{-1}$ with $g \in \mathbf{GL}(n, \mathbb{C})$ to a triangular Jordan[25] form. The eigenvalues are in the diagonal. The Jordan form can be written, i.e., there exist appropriate bases, with block matrices, which are either one-dimensional (α) (eigenvalue) or the sum of an eigenvalue-diagonal matrix and a nilcyclic matrix $\begin{pmatrix} \alpha & 1 & 0 & \dots & 0 \\ 0 & \alpha & 1 & \dots & 0 \\ & \dots & \dots & & \\ 0 & 0 & \dots & \alpha & 1 \\ 0 & 0 & \dots & 0 & \alpha \end{pmatrix}$. A semisimple matrix allows a diagonal form with the eigenvalues and without nilpotent contributions.

In physics, the *discrete weights* of finite-dimensional representations of semisimple Lie algebras were the prototype for *quantum numbers*, e.g., the angular momentum operator $\mathcal{D}^{2j}(l_0) \cong \begin{pmatrix} -j & 0 & \dots & 0 \\ 0 & -j+1 & \dots & 0 \\ \dots & \dots & \dots & \dots \\ 0 & \dots & j-1 & 0 \\ 0 & \dots & 0 & j \end{pmatrix}$, $2j \in \mathbb{N}$, for spin $\mathbf{SU}(2)$. Since compact operations are characterized by integers, it would be misleading to declare those discrete representation eigenvalues ("natura facit saltus") as the essential difference of the quantum description in contrast to the classical description ("natura non facit saltus"). With such an emphasis, the much more revolutionary quantum modality structure via probability amplitudes is not appreciated enough. Also noncompact operations with continuous "quantum numbers" are representable on Hilbert spaces

[25] Camille Jordan (1838–1922).

with scalar product-induced probability amplitudes. With this point of view, the probably unalterable name "quantum theory" is too narrow.

In the following, a very explicit Euclidean geometrical visualization will be given for Lie algebras and their representations, explicitly associated with the simple geometry of Plato: All finite-dimensional simple, i.e. irreducible representations of a simple rank r Lie algebra L_r by linear mappings ($n \times n$-matrices) of a complex vector space $V \cong \mathbb{C}^n$ can be illustrated by *weight diagrams* in an r-dimensional Euclidean space $\mathbb{E}^r$. Here, the number of the weights — exactly: of the weight vectors in $\mathbb{E}^r$ — coincides with the dimension n of the representation. The r components of the n weight vectors are the eigenvalues ("quantum numbers") of the simultaneously diagonalizable operations. It is useful to use a Weyl basis of the Lie algebra with a generating system involving $\frac{d-r}{2}$ A_1-triplets $\{l_\pm^k, l_0^k\}_{k=1}^{\frac{d-r}{2}}$. The weights collect the eigenvalues of the representation of r linearly independent Cartan operations $\{l_0^k\}_{k=1}^r$.

4.16 Weights, Roots, and Weyl Mirrors

Each semisimple Lie algebra L_r of rank r has r *fundamental representations*. Each L_r-representation is characterized by a *fundamental weight* $\{\gamma_k\}_{k=1,\dots,r}$ which, to remember, is a vector with r components in a Euclidean space $\gamma_k \in \mathbb{E}^r$.

The integer linear combinations $\sum_{k=1}^{r} z_k\gamma_k \in \mathbb{E}^r, z_k \in \mathbb{Z}$, of the fundamental weights define the $\mathbb{Z}^r$-isomorphic *weight lattice* as the lattice with the possible "quantum numbers" of the Lie symmetry L_r. The non-negative combinations $\sum_{k=1}^{r} n_k\gamma_k, n_k \geq 0$, of the fundamental weights characterize the simple (irreducible) representations by the r non-negative numbers $[n_1, \dots, n_r]$ for the highest weight of the representation.

Good to know: Only for the rank 1 Lie algebra A_1 the dimension $n+1$ suffices to characterize a simple representation $[n]$. Already for the $\mathbf{SL}(3, \mathbb{C})$-Lie algebra A_2 and the color symmetry $A_2^{\mathbb{C}} = \log \mathbf{SU}(3)$, there are non-equivalent simple representations with equal dimension, e.g., two 15-dimensional simple representations, characterized by [2, 1] and [4, 0]. For A_2, the representation $[n_1, n_2]$ has dimension $\frac{(n_1+1)(n_2+1)(n_1+n_2+2)}{2}$.

The simple representations come in dual pairs $[n_1, \dots, n_r] \overset{\text{dual}}{\leftrightarrow} [n_r, \dots, n_1]$ whose partners are reflected to each other at the origin of the weight space. Reflection symmetric weight diagrams characterize *self-dual representations*. Their representation vector spaces have an invariant bilinear form. Not self-dual representations, i.e., with centrally asymmetric weight diagrams, do not have an invariant bilinear form.

Good to know: *Dual representations* of a group $G \ni g$ are related to each other by inverse transposition, $D(g) \leftrightarrow D(g^{-1})^T$ and, for a Lie algebra, $L \ni l$ by negative transposition, $\mathcal{D}(l) \leftrightarrow -\mathcal{D}(l)^T$. A representation, equivalent to its dual one, is called *self-dual*.

The adjoint representation of ${}_dL_r$ acts on the vector space structure of the Lie algebra itself. The weight diagram of the d-dimensional adjoint representation is called *root diagram*: It has r trivial weights in the origin. The $d-r$ nontrivial weights, which come in reflection related pairs, are called *roots* $\{\pm\omega_k\}_{k=1}^{\frac{d-r}{2}}$, $\omega_k \in \mathbb{E}^r$. The root diagram is central reflection symmetric (self-dual) with the Killing-Form as invariant symmetric bilinear form. Each adjoint weight can be associated with a basic element of the Lie algebra — the roots with raising and lowering operators $\pm\omega_k \leftrightarrow l_\pm^k$, the trivial weights with r linear independent combinations of the $\frac{d-r}{2}$ associated Lie brackets, the Cartan operators $0 \leftrightarrow l_0^k = [l_-^k, l_+^k]$. The adjoint representation can, but does not have to be fundamental.

The $\frac{d-r}{2}$ roots of the adjoint representation involve r *fundamental roots* $\{\omega_k\}_{k=1}^r$ with associated *raising operators* $\{l_+^k\}_{k=1}^r$. Their linear combinations with only positive (only negative) numbers give half of all roots.

The r fundamental weights $\{\gamma_k\}_{k=1}^r$ for the fundamental representations and the r fundamental roots $\{\omega_k\}_{k=1}^r$ for the adjoint representation are Euclidean orthogonal as follows: There exist a scalar product in the weight space $\mathbb{E}^r$ with

$$\langle\gamma_k|\omega_j\rangle = \frac{\delta_{kj}}{2}\langle\omega_k|\omega_j\rangle .$$

Fundamental roots and fundamental weights for different indices are orthogonal. For equal index k, they do not have to be parallel.

The hyperplanes $s_{\omega_k} \cong \mathbb{E}^{r-1}$ in the Euclidean weight space $\mathbb{E}^r$, orthogonal to the roots ω_k, i.e., $s_{\omega_k} = \{v \in \mathbb{E}^r \bigm| \langle v|\omega_k\rangle = 0\}$, are the *Weyl mirrors*. The Weyl mirrors for the fundamental roots $\{\omega_k\}_{k=1}^r$ constitute the walls of the *Weyl chamber* W_K. All weight diagrams are symmetric and complete with respect to a reflection at the Weyl mirrors. These reflections determine the discrete *Weyl reflection group* of the Lie algebra, given by products of permutation groups $\mathbf{G}(n)$:

Lie algebra	${}^{r(r+2)}A_r$	${}^{r(2r+1)}C_r$	${}^{r(2r+1)}B_r$	${}^{r(2r-1)}D_r$
Weyl reflections	$\mathbf{G}(r+1)$	$\mathbf{G}(r) \vec{\times} \mathbf{I}(2)^r$	$\mathbf{G}(r) \vec{\times} \mathbf{I}(2)^r$	$\mathbf{G}(r) \vec{\times} \mathbf{I}(2)^{r-1}$

The weights can be ordered lexicographically. From the highest weight of an irreducible representation one reaches, with the negative fundamental roots (*lowering operators* $\{l_-^k\}_{k=1}^r$), all weights, i.e., the highest weight determines all the other weights.

4.17 Decomposition of Product Representations

A physical dynamics, characterized by a symmetry group, determines and rearranges the related representations: Physical objects, acted on by representations, especially by irreducible ones, e.g., the rotation group for the spin of particles, may lead to composite structures with possibly decomposable product representations.

The tensor product of two irreducible representations of a simple Lie algebra L_r, characterized by their highest weights $[n] = [n_1, \ldots, n_r]$ for an n-dimensional vector space and $[m] = [m_1, \ldots, m_r]$ for an m-dimensional one, is again an

L_r-representation. In general, its $n \cdot m$-dimensional representation space is decomposable into subspaces acted on by irreducible representations. For a graphical decomposition: First, the origin of the weight diagram of the representation $[n]$ is put on all weights of the representation $[m]$ — only by translations without rotations. By the mutual *vector addition of all weights* there arise the $n \cdot m$ weights of the product representation in the Euclidean weight space $\mathbb{E}^r$, sometimes multiply occupied. To decompose these $n \cdot m$ weights into irreducible weight diagrams, one starts with the simply occuring highest weight and completes it to the uniquely defined weight diagram of the representation $[n_1 + m_1, \ldots, n_r + m_r]$. After taking away all the weights of this representation, one repeats this procedure with the remaining highest weight — simply or multiply occuring. In such a way one strips away one irreducible "layer" after the other of the product representation.

As a first example for such a reduction, the familiar rules of the spin "vector addition" for the angular momentum Lie algebra $A_1^{\mathbb{C}}$ are reproduced: The product of two irreducible representations $[2j] = (j)$ with spin $j = 0, \frac{1}{2}, 1, \ldots$, and dimension $2j + 1$ and one-dimensional weight diagrams are decomposed as follows:

$$\text{for } A_1 \text{ and } \mathbf{SU}(2): \begin{cases} (j_1) \otimes (j_2) \cong (j_1 + j_2) \oplus (j_1 + j_2 - 1) \oplus \ldots \oplus (|j_1 - j_2|), \\ \text{dimension: } (2j_1 + 1)(2j_2 + 1) = \sum\limits_{J=|j_1-j_2|}^{j_1+j_2} (2J + 1). \end{cases}$$

This can also be obtained by projecting the $(2j_1 + 1)(2j_2 + 1)$ points of a rectangular grid in the Euclidean weight space $\mathbb{E}^1 \oplus \mathbb{E}^1 \cong \mathbb{E}^2$ on the diagonal in the first and third quadrant and collecting them in $(2J + 1)$-multiplets, starting with the highest weight $J = j_1 + j_2$.

Good to know: For the spin $\mathbf{SU}(2)$-representation, the angular momentum notation is used, $[2j] \cong (j)$ or $[n] \cong (\frac{n}{2})$.

A more complicated example is the reduction of the product of two two-dimensional weight diagrams, e.g., of two adjoint real eight-dimensional representations $[1, 1]$ of $A_2^{\mathbb{C}}$, the Lie algebra of $\mathbf{SU}(3)$. Here, the doubly occupied origin of a hexagon (octet representation) is put on all the eight weights of the other octet. The 64 product weights are decomposed into weights for irreducible $\mathbf{SU}(3)$-representations:

$$\text{for } A_2 \text{ and } \mathbf{SU}(3): \begin{cases} [1,1] \otimes [1,1] \cong [2,2] \oplus [2,0] \oplus [0,2] \oplus 2 \times [1,1] \oplus [0,0], \\ \text{dimensions: } 8 \cdot 8 = 64 = 27 + 10 + \overline{10} + 2 \times 8 + 1. \end{cases}$$

4.18 Squares and the Physics of Atoms

The theoretical description of the atoms as the characteristic state examples of nonrelativistic quantum mechanics allows a symmetry interpretation of the arising states and its wave functions (probability amplitudes) — the information for the atoms is given by representations of a Lie group. This insight may give some credit to Kepler's attempt to interpret the discrete planetary orbits with the aid of Plato's symmetry.

The Kepler potential $\frac{\epsilon}{r}$ with $\epsilon = \pm 1$ for repulsion and attraction, respectively, e.g., the Newton potential for sun and planets or the Coulomb potential for nucleus and electron, displays rotational symmetry with respect to the angular momentum group $\mathbf{SO}(3)$ with Lie algebra A_1^c. This leads, in classical mechanics, to planar orbits and Kepler's area theorem. The naive suggestion, that the rotation symmetry of dynamics involves circular orbits as solutions, seems obvious. However it is superficial, even wrong: A planar circular or elliptic orbit is an $\mathbf{SO}(2)$-orbit, not an orbit of the full spatial rotation group $\mathbf{SO}(3)$. In a classical dynamics, the rotation symmetry of the dynamics involves that, given one orbit as a solution, e.g., a planar ellipse, also the spatially rotated ellipse is a solution. The invariance group of the dynamics is the action group for the set with of its solutions.

In general, the individual states of a dynamics — here the solution orbits — do not have to display the invariance of the dynamics — here rotation symmetry. The states for the general law of motion are representations of its symmetry group, and, in general, there exists not only the trivial representation. The totality of all states for a dynamics — here of all possible solution orbits — is closed (invariant) under the action of the symmetry operations of the dynamics.

The Kepler dynamics with rotation invariant Hamiltonian $H = \frac{\vec{p}^2}{2} + \frac{\epsilon}{r}$ (intrinsic units) has, in addition to the conserved angular momentum vector $\vec{L}$, related to Kepler's second law, three conservation laws. For classical bound states like planetary orbits, they are recognizable in the conservation of the perihelion vector (nonrelativistically there are no rosette orbits), formalized by the Lenz–Runge[26] vector $\vec{R}$ from one focus of the ellipse to its perihelion. The Lenz–Runge-vector is in the orbit plane. Therefore, it is orthogonal to the angular momentum. The six conserved quantities, related to the angular momentum and Lenz–Runge vectors, allow, in the case of an attractive Kepler potential $\epsilon = -1$ and for negative energy states two combinations (sum and difference for appropriate normalization) as basis of the real six-dimensional Lie algebra $D_2^c \cong A_1^c \oplus A_1^c \cong \mathbb{R}^6$ whose universal covering group is the unitary product group $\mathbf{SU}(2) \times \mathbf{SU}(2) \cong \overline{\mathbf{SO}(4)}$.

In the classical description, the energy can be obtained by the invariant values of angular momentum $L \geq 0$ and Lenz–Runge-vector $R \geq 0$:

$$\left.\begin{aligned} H &= \tfrac{\vec{p}^2}{2} + \tfrac{\epsilon}{r}, \\ \vec{L} &= \vec{x} \times \vec{p}, \\ \vec{R} &= \vec{p} \times \vec{L} + \epsilon \tfrac{\vec{x}}{r}, \end{aligned}\right\} \Rightarrow \begin{aligned} \vec{R}^2 &= 1 + 2H\vec{L}^2, \\ E &= \tfrac{R^2 - 1}{2L^2}. \end{aligned}$$

For nontrivial $L, R > 0$, the classical orbits are given, in polar coordinates, by an ellipse (Kepler's first law), or a hyperbola (one branch), or a parabola for energies $E < 0, E > 0$, and $E = 0$, respectively:

[26]Heinrich Lenz (1804–1865), Carl Runge (1856–1927).

$$\begin{aligned}
\vec{R}\vec{x} = Rr\cos\varphi = L^2 + \epsilon r \Rightarrow r(\varphi) = \tfrac{L^2}{R\cos\varphi - \epsilon},\\
E < 0 \text{ for } R < 1: r(\varphi) = \tfrac{L^2}{R\cos\varphi+1}, &\text{ attraction } \epsilon = -1 \text{ (ellipse)},\\
E > 0 \text{ for } R > 1: r(\varphi) = \tfrac{L^2}{R\cos\varphi-\epsilon}, &\text{ attraction, repulsion } \epsilon = \mp 1 \text{ (hyperbola)},\\
E = 0 \text{ for } R = 1: r(\varphi) = \tfrac{L^2}{\cos\varphi+1}, &\text{ attraction } \epsilon = -1 \text{ (parabola)}.
\end{aligned}$$

As possible for an electromagnetic Coulomb potential, the hyberbolic orbit of one charged mass point (with the reduced mass) "includes" ("excludes") for attraction (repulsion) $\epsilon = \mp 1$ the other mass point (with the sum mass). The hyperbolic polar angle for infinite distance $r \to \infty$:

$$\cos^2\varphi_\infty = \frac{1}{R^2} = \frac{1}{1+2EL^2} \Rightarrow 2EL^2 = \tan^2\varphi_\infty,$$

is related to the scattering angle by $\theta = \pi - 2\varphi_\infty$. The conserved energy and angular momentum are expressable by asymptotic momentum p_∞ and impact length ρ:

$$\text{with } E = \frac{p_\infty^2}{2},\quad L = \rho p_\infty:\quad \rho^2 p_\infty^4 = \cot^2\frac{\theta}{2}.$$

The relation between scattering angle and impact length leads to Rutherford's[27] formula for the scattering cross section $d\sigma = \pi d\rho^2 = 2\pi\rho d\rho$ with:

$$\begin{aligned}
\text{in intrinsic units: } d\rho^2 &= \tfrac{1}{2p_\infty^4}\tfrac{d\cos\theta}{\sin^4\frac{\theta}{2}},\\
\text{for } \underline{H} = \frac{\underline{\vec{p}}^2}{2m} + \frac{\epsilon g^2}{\underline{r}}:\ d\underline{\rho}^2 &= \tfrac{m^2 g^4}{2\underline{p}_\infty^4}\tfrac{d\cos\theta}{\sin^4\frac{\theta}{2}}.
\end{aligned}$$

The states with negative energies of the nonrelativistic quantum mechanical atoms can be classified by complex representations of the "double spin" group $\mathbf{SU}(2) \times \mathbf{SU}(2)$. The weight diagrams of the irreducible A_1^c-representations contain $2j+1$ equidistant $\mathbb{E}^1$-points with the values $m = -j, -j+1, \ldots, j-1, j$ ($\mathbb{E}^1$-lattice), where $2j+1$ is the dimension of the $\mathbf{SU}(2)$-representation and $j = 0, \frac{1}{2}, 1, \frac{3}{2}, \ldots$. Therefore, the weight diagrams of $A_1^c \oplus A_1^c$ are $(2j_+ + 1)(2j_- + 1)$-rectangular lattices in $\mathbb{E}^2$. They constitute the weight lattice with the weights $(m_+, m_-) \in \frac{\mathbb{Z}}{2} \times \frac{\mathbb{Z}}{2}$. The two fundamental representations of $\mathbf{SU}(2) \times \mathbf{SU}(2)$ have the dominant weights $(\frac{1}{2}, 0)$ and $(0, \frac{1}{2})$, both have complex dimension 2.

Good to know: The complex group $\mathbf{SO}(4,\mathbb{C})$ has the semisimple Lie algebra $D_2 \cong A_1 \oplus A_1 \cong \mathbb{C}^6$. Its compact form $D_2^c \cong \mathbb{R}^6$ has the covering group $\mathbf{SU}(2) \times \mathbf{SU}(2)$ with the proper discrete normal subgroups for the associated $\mathbf{SU}(2) \times \mathbf{SU}(2)$-locally isomorphic quotient groups: $\mathbf{I}(2) \times \{1\}$ for $\mathbf{SO}(3) \times \mathbf{SU}(2)$, $\mathbf{I}(2) \times \mathbf{I}(2)$ for $\mathbf{SO}(3) \times \mathbf{SO}(3)$ and $\mathbf{I}(2)$ (diagonal) for $\frac{\mathbf{SU}(2)\times\mathbf{SU}(2)}{\mathbf{I}(2)} \cong \mathbf{SO}(4)$. The noncompact partner, the simple Lie algebra $A_1^c \oplus iA_1^c \cong \mathbb{R}^6$ belongs to the Lorentz group $\mathbf{SO}_0(1,3) \cong \mathbf{SL}(2,\mathbb{C})/\mathbf{I}(2)$ as real 6-dimensional group. The real three-dimensional Lenz-Runge vector for $\mathbf{SO}(4)$ in the Kepler problem and the boost vector for the Lorentz group $\mathbf{SO}_0(1,3)$ in special relativity are compact-noncompact partners.

[27]Ernest Rutherford (1871–1937).

Not the simply connected group $\mathbf{SU}(2) \times \mathbf{SU}(2)$ is the invariance group, but only the quotient group $\frac{\mathbf{SU}(2)\times\mathbf{SU}(2)}{\mathbf{I}(2)} \cong \mathbf{SO}(4)$ with the discrete, both factors connecting normal subgroup in the "diagonal" group $\mathbf{I}(2) \subset \mathbf{SU}(2) \ni (u, u) \in \mathbf{SU}(2) \times \mathbf{SU}(2)$:

$$\mathbf{I}(2) \cong \{\pm\mathbf{1}_2\} \subset \mathbf{SU}(2), \quad \mathbf{SU}(2) \times \mathbf{SU}(2) \supset \{(\mathbf{1}_2, \mathbf{1}_2), (-\mathbf{1}_2, -\mathbf{1}_2)\} \cong \mathbf{I}(2).$$

The weight lattice of $\mathbf{SO}(4)$ is a sublattice of the weight lattice of $\mathbf{SU}(2) \times \mathbf{SU}(2)$. It contains the points $(m_+, m_-) \in \frac{\mathbb{Z}}{2} \times \frac{\mathbb{Z}}{2}$ with integer sum $m_+ + m_- \in \mathbb{Z}$. The representations $(\frac{1}{2}, \frac{1}{2})$, faithful for $\mathbf{SO}(4)$ and $(1, 0)$, $(0, 1)$, faithful for $\mathbf{SO}(3)$ only, have the dimensions 4 and 3, respectively.

Good to know: Angular momentum $\vec{L} = i\vec{x} \times \vec{p}$ and Lenz–Runge vector $\vec{R} = i[\vec{p}, \frac{\vec{L}^2}{2} + \epsilon r]$ are conserved $[\vec{L}, H] = 0 = [\vec{R}, H]$ and orthogonal to each other $\vec{L}\vec{R} = 0$. With the sum and difference $\vec{J}_\pm = \frac{1}{2}(\vec{L} \pm \frac{\vec{R}}{\sqrt{-2H}})$ for negative energies the Hamiltonian can be written as $2(\vec{J}_+^2 + \vec{J}_-^2) = 1 + \frac{1}{2H}$. The orthogonality of angular momentum and Lenz–Runge vectors $\vec{L}\vec{R} = 0$ involves the equality of the invariants $\vec{J}_+^2 = \vec{J}_-^2 = -J(J+1)$ in the energy $-4J(J+1) = 1 + \frac{1}{2E_{2J}}$.

As a consequence of the orthogonality of angular momentum and Lenz–Runge vector for the $A_1^c \oplus A_1^c$-invariant dynamics of the atomic negative energy states, there arises the additional condition for the Kepler potential related weight diagrams, that the dimensions of both A_1^c-representations have to coincide. This condition, $J_+ = J_- = J$, determines a sublattice of the $\mathbf{SO}(4)$-sublattice of the $\mathbf{SU}(2) \times \mathbf{SU}(2)$-lattice, associated with the representations, generated only by the fundamental representation $(\frac{1}{2}, \frac{1}{2})$ with dimension 4, called *Kepler representations* or *harmonic* $\mathbf{SO}(4)$*-representations*. Similar to the harmonic oscillator, where the energy levels $E_n = n\hbar\omega$ characterize the n-fold product $n = 0, 1, \ldots$ of the basic nontrivial representation with energy $E = \hbar\omega$, the atomic states (J, J) are acted on by totally symmetric products of the fundamental representation $(\frac{1}{2}, \frac{1}{2})$. The characterizing weight diagrams are squared lattices with $(2J+1)^2$ weights. $2J + 1$ is the principal quantum number, where J can take all integer and halfinteger values. The totally symmetric n-fold products of the four-dimensional fundamental representation with power $n = 2J$ have the dimension $\binom{n+3}{n}$, i.e., 1 for $n = 0$, then 4 for $n = 1$, then $10 = 9 + 1$ for $n = 2$, then $20 = 16 + 4$ for $n = 3$, etc. In general, there arises the leading irreducible representation in the n-fold product:

$$\text{dimension of } \left(\frac{n}{2}, \frac{n}{2}\right) : \binom{n+3}{n} - \binom{n+1}{n-2} = (n+1)^2 = (2J+1)^2.$$

None of the two factors in $\mathbf{SU}(2) \times \mathbf{SU}(2)$ is the angular momentum group. The $A_1^c \oplus A_1^c$-representation are decomposable with respect to angular momentum properties by reduction of the product of the two A_1^c-representations with equal J (projection of the square lattice points to the diagonal):

$$(J) \otimes (J) \cong (2J) \oplus (2J-1) \oplus \ldots \oplus (1) \oplus (0),$$

$$\text{dimension } (2J+1)^2 = \sum_{L=0}^{2J} (2L+1),$$

$$2J = L + N \Rightarrow (L, N) = (2J, 0), (2J-1, 1), \ldots, (0, 2J).$$

N characterizes the rotation group classes $\mathbf{SO}(4)/\mathbf{SO}(3)$. With respect to the radial part in the associated Schrödinger wave functions, k is called knot number. Because of $J_+ = J_- = J$, i.e., $J_\pm$ are either both integer or both halfinteger, there occur only integer orbit angular momenta l, i.e., only representations of the rotation group $\mathbf{SO}(3) \cong \mathbf{SU}(2)/\mathbf{I}(2)$.

Only the lowest energy state with $J = 0$, i.e., with trivial rotation properties ($l = 0$, spherical symmetric) and without knots, has the full symmetry of the Kepler dynamics — the ground state is acted on by the trivial representation of $\mathbf{SO}(4)$.

As important especially in quantum field theory, the ground state does not have to display the full symmetry of the dynamics. Such a spontaneous symmetry breakdown characterizes a degenerate ground state: there exists a manifold of symmetry equivalent ground states. It allows the coexistence of a symmetric dynamics and "asymmetric" states. The rearrangement of basic symmetries, related to a nontrivial symmetry direction of the ground state, plays an important role in the standard model of elementary particles.

If one takes into consideration also the spin property of the electron – without dynamical action, only for counting the energy eigenstates — the nonrelativistic hydrogen atom is characterized by representations of the product group $\mathbf{SO}(4) \times \mathbf{SU}(2)$ with the Kepler group and the spin group as direct factors. Since the electron has spin $\frac{1}{2}$ with two directions ("up, down"), spin comes only with the two-dimensional A_1^c-representation. Therefore, the $A_1^c \oplus A_1^c \oplus A_1^c$-diagrams of the atoms in Euclidean space $\mathbb{E}^3$ with $2(2J+1)^2$ weights are quadratic columns with always two vertical points. One uses the symbols $s, p, d, \ldots$ for angular momenta $l = 0, 1, 2, \ldots$. Therefore, the two $J = 0$-states are s-states with spin up and down, the eight $J = \frac{1}{2}$ states comprise 2×1 s-states and 2×3 p-states, etc., with the 2×9 states for $J = 1$, etc.

The binding energies of the related atomic states are $\mathbf{SO}(4)$-invariants. They are degenerate for the weights of an irreducible representation:

$$2E_{2J} = -\frac{1}{(2J+1)^2}, \quad \begin{cases} J = 0, \frac{1}{2}, 1, \frac{3}{2}, \ldots \\ m_\pm = -J, -J+1, \ldots, J-1, J \end{cases}$$

The binding energies of all $2(2J+1)^2$ energy-degenerate electrons sum up to the binding energy of a full shell, identical for all shells:

$$-(2J+1)^2 2E_{2J} = 1.$$

The same "energy chunk" is equipartitioned to all states with equal principal quantum number $2J+1$.

The intrinsic energy unit for a n-fold positively charged nucleus is $n^2 m(\alpha_{\text{Som}}c)^2$:

$$2\underline{E}_{2J}(n) = -\frac{n^2 m(\alpha_{\text{Som}}c)^2}{(2J+1)^2}.$$

The Rydberg energy involves the relativistic energy equivalent of the reduced mass mc^2 (with m almost the electron mass), multiplied with the square of Sommerfeld's fine structure constant:

$$\begin{aligned} m(\alpha_{\text{Som}}c)^2 &\simeq 27.2\ \text{eV},\quad m_e c^2 \simeq 0.5\ \text{MeV} \\ (\alpha_{\text{Som}})^2 &= (\tfrac{e^2}{4\pi\epsilon_0 c\hbar})^2 \simeq \tfrac{1}{137^2} \simeq 5.3\times 10^{-5}, \\ M &= m_e + m_{\text{nucleus}},\quad \tfrac{1}{m} = \tfrac{1}{m_e} + \tfrac{1}{m_{\text{nucleus}}}. \end{aligned}$$

The Bohr length $\frac{\hbar}{m_e\alpha_{\text{Som}}c} \sim 0.5\times 10^{-10}$ m and the Rydberg time $\frac{\hbar}{m_e(\alpha_{\text{Som}}c)^2} \sim 2\times 10^{-17}$ s, related to the characteristic binding energy, have no meaning as orbit radius and circulation time — there are no electron orbits.

4.19 Plato's Polytopes and Cartan's Weight Diagrams

Simple Lie algebras have a hierarchical structure like Euclidean polytopes: Similar to the shaping of polygons by distances (sides), of polyhedra by polygons, etc. — a polygon does not "consist" of its sides, a polytope not of its polygons —, the Lie symmetry A_1 of the binary alternative "shapes" the Lie symmetries A_2, B_2, C_2, D_2, G_2 for rank two, these, in turn, "shape" those for rank three, etc. What the distance is for Euclidean geometry, that is the Lie algebra A_1 for semisimple symmetry. This suggests a comparison of Plato's mosaics with Cartan's root diagrams, especially for compact Lie algebras. Both structures involve Euclidean geometry.

For the Euclidean line, there is only one "mosaic" with the distance $\{2\}$ as building "block," which can be associated with the symmetry of the simple alternative as the origin of all simple symmetry:

$$\mathbb{E}^1: \begin{cases} \text{Plato:} & \{2\} & \text{(distance)}, \\ \text{Cartan:} & A_1 \cong B_1 \cong C_1 & \text{(distance)}. \end{cases}$$

In the Euclidean plane, there are two dual classes of Platonic mosaics, with triangles (dual to hexagons) and squares. They can be associated with the root diagrams of the three rank 2 Lie algebras:

$$\mathbb{E}^2: \begin{cases} \text{Plato:} \quad \{3,6\},\ \{6,3\}, \quad \{4,4\} \quad \text{(triangle, hexagon, square)}, \\ \text{Cartan:} \ A_2,\ G_2\ \text{(hexagon)},\ B_2 \cong C_2\ \text{(square)}. \end{cases}$$

The root diagram of A_2 is a hexagon, that of G_2 has the form of a David star, i.e., two reflected triangles with the convex corners on a hexagon.

In three Euclidean dimensions, there is only one mosaic, the hexahedron (cube), and three rank 3 Lie algebras:

$$\mathbb{E}^3: \begin{cases} \text{Plato:} & \{3,4,3\} \quad \text{(hexahedron)}, \\ \text{Cartan:} & A_3 \cong D_3,\ B_3,\ C_3 \ \text{(hexoctahedron)}. \end{cases}$$

The root diagrams for all three Lie algebras have the form of an Archimedean hexoctahedron, the self-dual intersection of hexahedron and octahedron. The difference in the actual root diagrams comes from the additional roots that do not coincide with vertices, e.g., at central points of the edges.

In the following, the simple Lie algebras and their irreducible representations will be considered in some detail with respect to their Platonic symmetry.

The Main Series A_r

The fundamental representations of the Lie algebra A_r for the invariance group $\mathbf{SL}(r+1,\mathbb{C})$ of the volume elements have the complex dimension $\binom{r+1}{k}$, $k = 1,\dots,r$. The weight diagrams come in dual pairs, reflected to each other at the origin of the weight space. For odd rank A_{1+2n}, $n = 0, 1, \dots$, there is one fundamental self-dual weight diagram, symmetric under central reflection. It characterizes a representation with dimension $\binom{2n+2}{n+1}$, e.g., the complex two-dimensional Pauli spinor representation for $\mathbf{SU}(2) = \exp A_1^c$. Dual representations of the compact $\mathbf{SU}(r+1)$-Lie algebra A_r^c, i.e., $[n_1,\dots,n_r] \leftrightarrow [n_r,\dots,n_1]$, are also called *antirepresentations*, e.g., quark and antiquark representations $[1,0] \leftrightarrow [0,1]$ for $\mathbf{SU}(3) = \exp A_2^c$.

The weight diagram for the $(r+1)$-dimensional *defining* A_r-representation $[1,0,\dots,0]$ is given by the vertices of the fundamental regular simplex. A_r may be called the *simplex series* with distance, triangle, tetrahedron, etc. A_r has the permutation group $\mathbf{G}(r+1)$ with order $(r+1)!$ as Weyl-reflection group.

The product of the two reflected defining representations of A_r, $[1,0,\dots,0]$ and $[0,\dots,0,1]$, decomposes into the trivial and the adjoint representation, $[2]$ for A_1 and $[1,0,\dots,0,1]$ for $r \geq 2$ — in dimensions:

$$(r+1)^2 = 1 + r(r+2).$$

The adjoint representation of A_{r+1} for $r \geq 1$ decomposes with respect to its Lie subalgebra A_r into the adjoint representation, and the dual pair of the defining representations and a trivial one:

$$A_{r+1} \supset A_r: \ (r+1)(r+3) = r(r+2) + 2(r+1) + 1.$$

Now more details for rank 1,2, and 3:

A_1^c, the Lie algebra of $\mathbf{SU}(2)$, used for spin and isospin, has one fundamental representation $[1]$, the two-dimensional Pauli–representation. Its one-dimensional weight diagram consists of the two "corners" of the *distance (edge)* in $\mathbb{E}^1$. The weight diagram of the three-dimensional adjoint representation [2] occupies the two endpoints and the central point of a distance:

$$\begin{aligned} A_1\text{-dublet in } \mathbb{E}^1 &: \{g_2, g_1\} = \{-\tfrac{1}{2}, +\tfrac{1}{2}\},\\ A_1\text{-triplet in } \mathbb{E}^1 &: \{-1, 0, +1\}. \end{aligned}$$

A_2^c, the Lie algebra of $\mathbf{SU}(3)$, used for flavor and color, has two centrally reflected defining representations, $[1, 0]$ and $[0, 1]$, with their dimensions called triplet 3 and antitriplet $\overline{3}$, respectively, and used for quarks and anti-quarks. The *regular triangle* as weight diagram in $\mathbb{E}^2$ for the triplet- (quark-) representation is either parametrized with barycentric coordinates or, as ususal in physics, with the parameters related to a decomposition with respect to an (iso)spin Lie subalgebra A_1 with dublet and singlet. Here, the origin is chosen as the center of the diagram. The weights of the adjoint octet-representation $[1, 1]$ are the vertices of a *regular hexagon* with two additional weights in the origin:

$$\begin{aligned} &\begin{matrix} A_2\text{-triplet in } \mathbb{E}^2 : \\ \text{(triangle)} \end{matrix} \left\{ \begin{aligned} g_{1,2} &= (\pm\tfrac{1}{2}, \tfrac{1}{2\sqrt{3}}),\\ g_3 &= (0, -\tfrac{1}{\sqrt{3}}), \end{aligned} \right.\\ &\begin{matrix} A_2\text{-octet in } \mathbb{E}^2 : \\ \text{(hexagon)} \end{matrix} \left\{ \begin{aligned} &A_1\text{-dublets: } \pm(\pm\tfrac{1}{2}, \tfrac{\sqrt{3}}{2}),\\ &A_1\text{-triplet: } (\pm 1, 0), \quad (0, 0),\\ &A_1\text{-singlet: } (0, 0). \end{aligned} \right. \end{aligned}$$

A_3^c for $\mathbf{SU}(4)$, used for higher flavor symmetries, has as four-dimensional fundamental defining representations $[1, 0, 0]$ and $[0, 0, 1]$, or 4 and $\overline{4}$, dual to each other, i.e., reflected at the origin, with the tetrahedron vertices as weight diagrams in Euclidean space:

$$\begin{matrix} A_3\text{-quartet in } \mathbb{E}^3 : \\ \text{(tetrahedron)} \end{matrix} \left\{ \begin{aligned} g_{1,2} = (\pm\tfrac{1}{2}, \tfrac{1}{2\sqrt{3}}, \tfrac{1}{2\sqrt{6}}),\ g_3 &= (0, -\tfrac{1}{\sqrt{3}}, \tfrac{1}{2\sqrt{6}}),\\ g_4 &= (0, 0, -\tfrac{3}{2\sqrt{6}}). \end{aligned} \right.$$

With respect to the Lie subalgebra A_2 the representations of A_3 have horizontal layers. The product of two tetrahedra ($4 \otimes 4 \cong 6 \oplus 10$) decomposes into the six-dimensional reflection symmetric fundamental representation $[0, 1, 0]$ with two layers with an A_2-triplet and an antitriplet as the vertices of an *octahedron*:

$$\begin{matrix} A_3\text{-sextet in } \mathbb{E}^3 : \\ \text{(octahedron)} \end{matrix} \left\{ \begin{aligned} A_2\text{- triplet: } &(\pm\tfrac{1}{2}, \tfrac{1}{2\sqrt{3}}, \tfrac{1}{\sqrt{6}}), && (0, -\tfrac{1}{\sqrt{3}}, \tfrac{1}{\sqrt{6}}),\\ A_2\text{-antitriplet: } &(\mp\tfrac{1}{2}, -\tfrac{1}{2\sqrt{3}}, -\tfrac{1}{\sqrt{6}}), && (0, \tfrac{1}{\sqrt{3}}, -\tfrac{1}{\sqrt{6}}). \end{aligned} \right.$$

The product of a tetrahedron and its dual ($4 \otimes \overline{4} \cong 1 \oplus 15$) decomposes into the trivial and the adjoint 15-dimensional A_3-representation $[1, 0, 1]$ with 12 roots (nontrivial weights) and three trivial weights in the origin, horizontally layered with respect to A_2-representations in triplet, octet and singlet, and antitriplet:

$$A_3 - 15 - \text{plet in } \mathbb{E}^3 : \atop \text{(hexoctahedron)} \left\{ \begin{array}{c} A_2\text{-triplet: } (\pm\frac{1}{2}, \frac{1}{2\sqrt{3}}, \frac{2}{\sqrt{6}}),\ (0, -\frac{1}{\sqrt{3}}, \frac{2}{\sqrt{6}}), \\ A_2\text{-octet and -singlet}, \\ A_2\text{-antitriplet: } (\mp\frac{1}{2}, -\frac{1}{2\sqrt{3}}, -\frac{4}{\sqrt{6}}),\ (0, \frac{1}{\sqrt{3}}, -\frac{2}{\sqrt{6}}). \end{array} \right.$$

This 12-cornered polyhedron with the 15 weights for $A_3^{\mathbb{C}} = \log \mathbf{SU}(4)$ is a *hexoctaeder* (cuboctaeder) with the three layers $3 + (6 + 3) + 3 = 15$ for $A_2^{\mathbb{C}} = \log \mathbf{SU}(3)$.

The Main Series D_r

The main series $D_r (r \geq 3)$ for the invariance group $\mathbf{SO}(2r, \mathbb{C})$ of the symmetric nondegenerate bilinear forms of even-dimensional $\mathbb{C}^{2r}$, starts independently at $r = 4$ after the isomorphies $D_2 \cong A_1 \oplus A_1$ and $\mathbf{SO}(4) \sim \mathbf{SU}(2) \times \mathbf{SU}(2)$ (symmetry of the negative energy states of the nonrelativistic atoms) and $D_3 \cong A_3$ and $\mathbf{SO}(6) \sim \mathbf{SU}(4)$. The fundamental D_r-representations have the dimension $\binom{2r}{k}$ with $k = 1, \ldots r - 2$ and twice 2^{r-1} for the two spinor representations (starting with $\mathbf{SU}(2) \times \mathbf{SU}(2) \cong \overline{\mathbf{SO}(4)}$). The spinor representations are dual to each other, the other fundamental representations have a symmetric invariant bilinear form. The Weyl reflection group of D_r is $\mathbf{G}(r) \vec{\times} \mathbf{I}(2)^{r-1}$ with order $r!\, 2^{r-1}$.

The Main Series B_r

All representations of B_r for the invariance group $\mathbf{SO}(2r + 1, \mathbb{C})$ of the symmetric nondegenerate bilinear forms of odd-dimensional $\mathbb{C}^{2r+1}$ have reflection symmetric weight diagrams (self-dual representations). The weight diagram of the one fundamental complex 2^r-dimensional B_r-*spinor representation* 2^{r-1} (starting with $\mathbf{SU}(2) \cong \overline{\mathbf{SO}(3)}$) occupies the vertices of the r-dimensional regular measure polytopes — distance for B_1, square $\{4\}$ for B_2, and $\{4, 3^{r-2}, 4\}$ for B_r, $r = 3, 4, \ldots$. For $r \geq 2$, there are other fundamental representations with dimension $\binom{2r+1}{k}$, $k = 1, \ldots, r - 1$, which includes the defining representation of dimension $2r + 1$. The Weyl-reflection group of B_r is $\mathbf{G}(r) \vec{\times} \mathbf{I}(2)^r$ with order $r!\, 2^r$. For $r \geq 3$, the adjoint representation with dimension $\binom{2r+1}{2}$ is fundamental.

The defining B_{r+1} representation ($r \geq 2$) has dimension $2r + 3$ and decomposes with respect to the Lie subalgebra B_r into a $2r+1$-dimensional defining representation and two trivial ones. The adjoint representation of B_{r+1} decomposes into the adjoint B_r-representation, two defining representations and a trivial one:

$$(r + 1)(2r + 3) = r(2r + 1) + 2(2r + 1) + 1.$$

$B_2^{\mathbb{C}}$ for the group $\mathbf{SO}(5)$ has the fundamental spinor representation 4 with the vertices of a *square* as weight diagram. The defining representation 5 has as weights the four square vertices and the central point. The adjoint representation 10 occupies the four vertices of a square, the four central points of its sides, and twice the central square point (two trivial weights):

$$B_2 \text{ in } \mathbb{E}^2: \begin{cases} \text{quartet with two } A_1\text{-dublets } \pm(-\frac{1}{2}, \frac{1}{2}) \text{ (square)}, \\ \text{quintet with} \begin{cases} A_1\text{-triplet } (-1, 0, 1), \\ \text{two } A_1\text{-singlets } (-1, 0),\ (0, 1), \end{cases} \\ \text{decuplet with} \begin{cases} \text{three } A_1\text{-singlets } (1, 0),\ (0, 0),\ (0, 1), \\ A_1\text{-triplet } (-1, 0, 1), \\ \text{two } A_1\text{-dublets } \pm(-\frac{1}{2}, \frac{1}{2}). \end{cases} \end{cases}$$

B_3^c for the group **SO**(7) can be horizontally layered with respect to B_2^c for **SO**(5). It has one fundamental spinor representation on the vertices of a *hexahedron* as weight diagram, The defining representation 7 has a centrally dotted *octahedron* as weight diagram. The 21-dimensional adjoint representation has as root diagram a *hexoctahedron* (cuboctahedron). In contrast to A_3^c where it is layered, with respect to A_2^c, by triangle, hexagon and (anti)triangle, it is rotated for B_3^c and layered with respect to B_2^c by a (small) square, a $\frac{\pi}{4}$-rotated (large) square, and a (small) square again. The small squares have each one central point weight, the large square has as additional weights the four central side points and three central points in the origin of the weight space — altogether $2 \times (4 + 1) + (4 + 4 + 3) = 21$ weights:

$$B_3 \text{ in } \mathbb{E}^3: \begin{cases} \text{octet with two } B_2\text{-quartets on a hexahedron}, \\ \text{septet with } B_2\text{-quintet and two } B_2\text{-singlets}, \\ \text{21-plet with two } B_2\text{-quintets} \\ \quad \text{and, between, } B_2\text{-decuplet and } B_2\text{-singlet}. \end{cases}$$

The Main Series C_r

The representations of C_r for the invariance group $\mathbf{Sp}(2r, \mathbb{C})$ of the antisymmetric nondegenerate bilinear forms of even-dimensional complex vector spaces have reflection symmetric weight diagrams (self-dual representations). The fundamental representations have dimensions $\binom{2r}{k} - \binom{2r}{k-2}$, $k = 1, \ldots, r$, with the defining representation of dimension $2r$. B_r and C_r have isomorphic Weyl reflection groups $\mathbf{G}(r) \vec{\times} \mathbf{I}(2)^r$ with order $r!\, 2^r$.

The defining C_{r+1}-representation with dimension $2r+2$ decomposes with respect to C_r into a $2r$-dimensional defining representation and two trivial ones.

C_3^c for **Sp**(6) has horizontal layers for $C_2^c \cong B_2^c$. Its fundamental representations are the 6-dimensional defining C_3-representation with octahedric weight diagram, the 15-plet and 14-plet, which, also, are on a *hexoctahedron*:

$$C_3 \text{ in } \mathbb{E}^3: \begin{cases} \text{sextet on an octahedron,} \\ \text{15-plet with two } C_2\text{-quartets, } C_2\text{-quintet and two } C_2\text{-singlets,} \\ \text{14-plet with two } C_2\text{-Quartets, } C_2\text{-quintet and } C_2\text{-singlet.} \end{cases}$$

The Exceptional Algebras $G_2,\ F_4,\ E_{6,7,8}$

The exceptional Lie algebra G_2 has dimension 14 with A_2 as Lie subalgebra. Its adjoint 14-dimensional representation is fundamental and decomposes into an A_2-octet, triplet and antitriplet. The diagram with 12 roots is a David star with two reflected regular triangles. The second fundamental representation is seven-dimensional with a centrally dotted *regular hexagon* as weight diagram. G_2 has the dihedral group $\mathbf{I}^{\pm}(6)$ as Weyl reflection group.

The four fundamental representations of the exceptional Lie algebra F_4 have the dimensions with the adjoint underlined:

$$\text{for } F_4: \ \ 26,\ 273,\ 1\,274,\ \underline{52}.$$

The Weyl reflection group is given by $\mathbf{G}(3) \vec{\times} [\mathbf{G}(4) \vec{\times} \mathbf{I}(2)^3]$.

Also for the little series $E_{6,7,8}$ the adjoint representations are fundamental. The six, seven, and eight fundamental representations have the following dimensions:

$$\begin{aligned} &\text{for } E_6: 27,\ 351,\ 2\,925,\ 351,\ 27,\ \underline{78}, \\ &\text{for } E_7: 56,\ 1\,539,\ 27\,664,\ 365\,750,\ 8\,645,\ \underline{133},\ 912, \\ &\text{for } E_8: \underline{248},\ 30\,380,\ 2\,450\,240,\ 146\,325\,270, \\ &\qquad\quad 6\,899\,079\,264,\ 6\,696\,000,\ 3\,875. \end{aligned}$$

Their Weyl reflection groups have the orders:

Lie algebra	$_{14}G_2$	$_{52}F_4$	$_{78}E_6$	$_{133}E_7$	$_{248}E_8$
Weyl group order	$2^2 \cdot 3$	$2^7 \cdot 3^2$	$2^7 \cdot 3^4 \cdot 5$	$2^{10} \cdot 3^4 \cdot 5 \cdot 7$	$2^{14} \cdot 3^5 \cdot 5^2 \cdot 7$

4.20 Quark Flavors

In the theory of quarks, the classification with *unitary flavor symmetries* played an important role. Starting in the sixties of the last century, the rational electrical charge number q of the quarks in units of the positron charge:

$$q(\mathbf{u}\text{p}, \mathbf{d}\text{own}, \mathbf{s}\text{trange}, \mathbf{c}\text{harm}, \mathbf{b}\text{ottom}, \mathbf{t}\text{op}) = \left(\frac{2}{3}, -\frac{1}{3}, -\frac{1}{3}, \frac{2}{3}, -\frac{1}{3}, \frac{2}{3}\right),$$

was parametrized by the *baryon number* N (all quarks have $N = \frac{1}{3}$), the *third strong isospin direction* T^3, *strangeness* S, *charm* C, *bottomness* B and *topness* T:

$$q = \frac{2}{3}: \begin{pmatrix} \mathbf{u}, & T^3 = \frac{1}{2} \\ \mathbf{c}, & C = 1 \\ \mathbf{t}, & T = 1 \end{pmatrix}, \quad q = -\frac{1}{3}: \begin{pmatrix} \mathbf{d}, & T^3 = -\frac{1}{2} \\ \mathbf{s}, & S = -1 \\ \mathbf{b}, & B = -1 \end{pmatrix}$$

$$\text{charge operator: } Q = \frac{N}{2} + T^3 + \frac{S + C + B + T}{2}$$

Only the explicitly given quantum numbers are nontrivial, e.g., $T^3(\mathbf{c}) = 0, B(\mathbf{d}) = 0$, etc.

The charge numbers can be explained as eigenvalues of linear combinations of Cartan operators for the unitary groups $\mathbf{U}(1)$ and $\mathbf{SU}(r+1)$ and their Lie algebras $D_1^c \cong i\mathbb{R}$ and A_r^c. The quarks and anti-quarks are acted on by the dual pair of the defining $\mathbf{SU}(r+1)$-representations.

First for up and down quark in an A_1^c-dublet and the quantum numbers baryon number and strong isospin, represented by Pauli matrices:

$$\begin{pmatrix} \mathbf{u} \\ \mathbf{d} \end{pmatrix}, \ \mathbf{SU}(2), \ A_1^c: \begin{cases} \mathbf{SU}(2)\text{-dublet weights: } \begin{pmatrix} g_1 \\ g_2 \end{pmatrix} = \begin{pmatrix} \frac{1}{2} \\ -\frac{1}{2} \end{pmatrix}, \\ \text{charge operator: } Q = \begin{cases} \frac{N}{2} + T^3, \\ \frac{\mathbf{1}_2}{6} + \frac{\sigma^3(2)}{2} \end{cases} \\ \text{charge numbers: } q = \frac{2}{3}, -\frac{1}{3}, \\ \text{charge number sum: } q_u + q_d = 2 \cdot \frac{1}{6}. \end{cases}$$

By including the strange quark in an A_2^c quark triplet (triangle), the scalar $\mathbf{U}(1)$-contribution $\frac{\mathbf{1}_2}{6}$ is embedded in an $\mathbf{SU}(3)$-generator, represented by Gell-Mann matrices:

$$\begin{pmatrix} \mathbf{u} \\ \mathbf{d} \\ \mathbf{s} \end{pmatrix}, \ \mathbf{SU}(3), \ A_2^c: \begin{cases} \mathbf{SU}(3)\text{-triplet weights: } \begin{pmatrix} g_1 \\ g_2 \\ g_3 \end{pmatrix} = \begin{pmatrix} \frac{1}{2} & \frac{1}{2\sqrt{3}} \\ -\frac{1}{2} & \frac{1}{2\sqrt{3}} \\ 0 & -\frac{1}{\sqrt{3}} \end{pmatrix}, \\ \text{charge operator: } Q = \begin{cases} T^3 + \frac{N+S}{2}, \\ \frac{\sigma^3(3)}{2} + \frac{\sigma^8(3)}{2\sqrt{3}}, \end{cases} \\ \text{charge numbers: } q = \frac{2}{3}, -\frac{1}{3}, -\frac{1}{3}, \\ \text{charge number sum: } q_u + q_d + q_s = 0. \end{cases}$$

The charge numbers of the quarks can be connected with an element of a simple Lie algebra A_r^c only if the charge number sum vanishes. With the advent of charm as fourth quark flavor, the charge number had an $\mathbf{SU}(4)$-scalar contribution in the A_3^c-quark-quartet (tetrahedron):

$$\begin{pmatrix}\mathbf{u}\\ \mathbf{d}\\ \mathbf{c}\\ \mathbf{s}\end{pmatrix},\ \mathbf{SU}(4),\ A_3^c:\begin{cases}\mathbf{SU}(4)\text{-quartet weights: } \begin{pmatrix}g_1\\ g_2\\ g_3\\ g_4\end{pmatrix} = \begin{pmatrix}\frac{1}{2} & \frac{1}{2\sqrt{3}} & \frac{1}{2\sqrt{6}}\\ -\frac{1}{2} & \frac{1}{2\sqrt{3}} & \frac{1}{2\sqrt{6}}\\ 0 & -\frac{1}{\sqrt{3}} & \frac{1}{2\sqrt{6}}\\ 0 & 0 & -\frac{3}{2\sqrt{6}}\end{pmatrix}\\ \text{charge operator: } Q = \begin{cases}\frac{N}{2} + T^3 + \frac{S+C}{2},\\ \frac{\mathbf{1}_4}{6} + \frac{\sigma^3(4)}{2} + \frac{-\sigma^8(4)+\sqrt{2}\sigma^{15}(4)}{2\sqrt{3}},\end{cases}\\ \text{charge numbers: } q = \frac{2}{3}, -\frac{1}{3}, \frac{2}{3}, -\frac{1}{3},\\ \text{charge number sum: } q_u + q_d + q_c + q_s = 4\cdot\frac{1}{6}.\end{cases}$$

After the advent of the bottom and top quarks, there was some playing around with the groups $\mathbf{SU}(5)$ and $\mathbf{SU}(6)$. Summarizing the history of the unitary flavor classification:

$$\underbrace{\underbrace{\underbrace{\underbrace{\mathbf{U}(1)_N \times \mathbf{SU}(2) \times \mathbf{U}(1)_S}_{\mathbf{U}(1)_N\times\mathbf{SU}(3)} \times\mathbf{U}(1)_C}_{\mathbf{U}(1)_N\times\mathbf{SU}(4)} \times\mathbf{U}(1)_B}_{\mathbf{U}(1)_N\times\mathbf{SU}(5)} \times\mathbf{U}(1)_T}_{\mathbf{U}(1)_N\times\mathbf{SU}(6)}$$

Finally, the higher $\mathbf{SU}(n)$-flavor symmetries have been qualified as useful for a phenomenological classification, not, however, as fundamental symmetries of a basic dynamics. Going back to field one in this game, the flavor structure of the quarks is looked at to be given by three dublets for *weak* isospin $\mathbf{SU}(2)$, i.e., (up, down), (charm, strange), (top, bottom), coming with the hypercharge group in the product group $\mathbf{U}(1) \circ \mathbf{SU}(2)$ with charge numbers $q = \frac{1}{6} \pm \frac{1}{2}$:

$$\begin{pmatrix}\mathbf{u}\\ \mathbf{d}\end{pmatrix},\ \begin{pmatrix}\mathbf{c}\\ \mathbf{s}\end{pmatrix},\ \begin{pmatrix}\mathbf{t}\\ \mathbf{b}\end{pmatrix}:\ Q = Y + T_3 \longmapsto \frac{\mathbf{1}_2}{6} + \frac{\sigma^3}{2} = \begin{pmatrix}\frac{2}{3} & 0\\ 0 & -\frac{1}{3}\end{pmatrix}.$$

The appearance of three families remains, up to today, a theoretically unsolved enigma.

4.21 Plato's Triangles and Dynkin's Diagrams

Probably the most direct connection between Plato's geometry and semisimple Lie symmetry starts with the three mosaics (triangle, square and hexagon) in the Euclidean plane $\mathbb{E}^2$. The equilateral triangle and the square can be cut into halves, the *three beautiful isosceles triangles* with the mosaic angles $\frac{\pi}{n_0}$ for $n_0 = 3, 4, 6$ as base angle:

$$\begin{aligned}\text{base angle: } & \Phi(\omega_1,\omega_2) = \frac{\pi}{3}, \frac{\pi}{4}, \frac{\pi}{6},\\ \text{length ratio: } & \left(\frac{\text{base}}{\text{side}}\right)^2 = 4\cos^2\Phi(\omega_1,\omega_2) = \frac{\|\omega_1\|^2}{\|\omega_2\|^2} = 1, 2, 3.\end{aligned}$$

Those three triangles are symbolized, as a shorthand and as a construction description, by their Dynkin[28] diagrams with two little circles (vertices) connected by one, two, and three lines, respectively. The two vertices display the dimension two of the Euclidean plane, the line numbers $\{1, 2, 3\}$ reflect the values of $4\cos^2\Phi$. The Dynkin diagram of a one-dimensional distance (edge) is simply one vertex.

If one takes one of the three beautiful isosceles triangles, nails it down, with one corner of the base, at the origin of a Euclidean plane $\mathbb{E}^2$ and reflects it as long over its sides (Weyl reflections) until the arising polygon is complete — here the mosaic property of square and equilateral triangle is used — one obtains polygons with $2n_0$ corners and $\frac{\pi}{n_0}$-triangles: A hexagon (six corners and $\frac{\pi}{3}$-triangles), a square, dotted at the central side points (eight corners and $\frac{\pi}{4}$-triangles or four squares) and a 12-cornered David-star (12 corners and $\frac{\pi}{6}$-triangles). The vectors from the origin to the corners are the roots of the three simple rank-2 Lie algebras ${}_{2n_0+2}L_2$, i.e., of ${}_8A_2$, ${}_{10}B_2$, ${}_{14}G_2$. One has to add two trivial weights in the origin.

The roots of ${}_3A_1$ are the two endpoints of a distance in the Euclidean line $\mathbb{E}^1$, its central point is the trivial weight.

With the two isosceles $\frac{\pi}{3}$- and $\frac{\pi}{4}$-triangles, the equilateral one and the diagonally bisected square, the Dynkin diagrams of all simple Lie algebras of rank $r \geq 2$ can be obtained by joining the Dynkin diagrams of these triangles. The threefold connection line of G_2 ($\frac{\pi}{6}$-triangle) does not occur for more than two vertices. In such a way, the equilateral triangle and the square (with the $\frac{\pi}{4}$-triangle as its bisection) determine all simple Lie algebras.

A Dynkin diagram with r connected vertices prescribes the construction of the root diagram of the associated simple Lie algebra L_r with rank r in a Euclidean space $\mathbb{E}^r$. Each vertex stands for a fundamental root. The direct connection of two vertices gives, by the number $1, 2, 3$ of the connecting lines, the angle between these roots — the obtuse outside angle of the corresponding beautiful triangle is taken:

$$4\cos^2\Phi = 0, 1, 2, 3: \quad \Phi = \frac{\pi}{2}, \frac{2\pi}{3}, \frac{3\pi}{4}, \frac{5\pi}{6}.$$

Therefore, each subdiagram with two connected vertices shows a Lie subalgebra A_2 (simple line as connection, hexagon) or B_2 (double line as connection, square). Not directly connected vertices describe orthogonal roots, $\cos^2\Phi = 0$. The number over an vertex gives the squared length ratio of the root lengths. It is redundant up to the Lie algebras B_r, C_r — there the order matters. With the r fundamental roots, constructed from the Dynkin diagram, the full root system arises by Weyl-reflections: For each fundamental root one constructs the $(r-1)$-dimensional orthogonal hyperplane $\mathbb{E}^{r-1} \subset \mathbb{E}^r$ as mirrors for the fundamental roots, which are reflected thereon as long as possible.

There are rules for joining Dynkin diagrams, which result from the condition that, after finitely many reflections, the root diagram has to close without "gaps."

[28] Eugene Dynkin (1924–2014).

(1) r vertices have at most $r - 1$ connection, where a connection can be a simple, a double or a triple line.

(2) There are no closed lines.

(3) A vertex can be the source of at most three lines. Therefore, only the exceptional Lie algebra G_2 has a connection by three lines.

(4) If two vertices, connected by a simple line, are shrinked to one vertex, there has to arise an admissable Dynkin diagram. Therefore, with rule (3), there is at most one double line, i.e., one square contribution. In addition to that, all root diagrams are built by equilateral triangles. In addition, there is at most one branching.

(5) There is only one diagram F_4 with a double line not at the fringe.

(6) There are only three diagrams $E_{6,7,8}$ with a branching not at the next to the last vertex.

The four main series are characterizable as follows: A_r and D_r arise only from equilateral triangles, D_r has one branching at the next to the last vertex. B_r and C_r use, in addition to triangles, one square at the fringe — their Dynkin diagrams are equal up to the ratio of the side length for square and triangle:

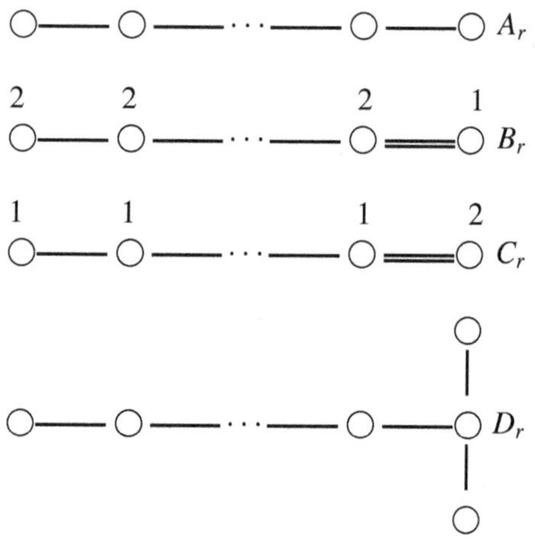

The little series with the three exceptional Lie algebras $E_{6,7,8}$ are characterized by (n_0, m_0)-vertices on both side of the branching, where (n_0, m_0) belongs to the three dual pairs of the five Platonic solids with their Archimedean polyhedra:

$$E_{n_0+m_0} \leftrightarrow \{n_0, m_0\}, \{m_0, n_0\} : \begin{cases} E_6 \leftrightarrow \{3,3\} & \text{tetrahedron,} \\ E_7 \leftrightarrow \{4,3\}, \{3,4\} & \text{hexoctahedron,} \\ E_8 \leftrightarrow \{5,3\}, \{3,5\} & \text{icosidodecahedron.} \end{cases}$$

The exceptional Lie algebras have the Dynkin diagrams:

 G_2,  F_4

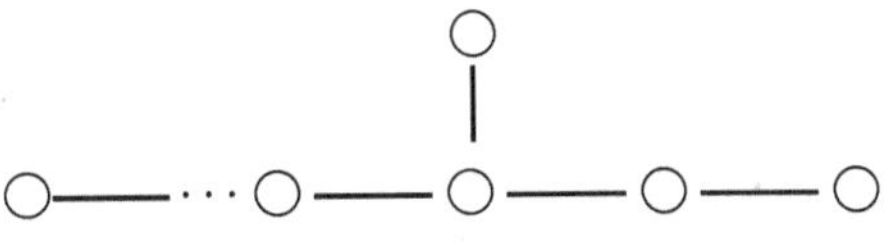 $E_{6,7,8}$.

4.22 Fundamental Representations

For a simple Lie algebra L_r with rank r, each of the r fundamental roots ω_k can be uniquely associated with that fundamental weight γ_k, which is the only non-orthogonal weight. The projection of γ_k on ω_k is half of the fundamental root:

$$k = 1, \ldots, r: \quad \omega_k \leftrightarrow \gamma_k \text{ with } \langle \omega_j | \gamma_k \rangle = \frac{\delta_{jk}}{2} \parallel \omega_j \parallel^2 .$$

Therefore, the r vertices of a Dynkin diagram can be labelled not only by the fundamental roots $\{\omega_k\}_{k=1}^r$, but also by the associated fundamental representations $\{\gamma_k\}_{k=1}^r$ with their dimensions.

The weights in the weight space $\mathbb{E}^r$ are integer linear combinations of the fundamental weights, which, therefore, are a a $\mathbb{Z}$-basis of the weight lattice. The r fundamental weights are an $\mathbb{N}$-basis of the cone with the irreducible representations: Via natural number addition of the fundamental weights $\sum_{k=1}^{r} n_k \gamma_k$ one obtains the highest weight of the irreducible representation $[n_1, \ldots, n_r]$.

The multiplication of the fundamental weights with natural numbers is related to a tensor product for a product representation: The representation space of $[n, 0, \ldots, 0]$ is isomorphic to a vector subspace of the *totally symmetric* n-fold product of the representation space of the fundamental representation $[1, 0, \ldots, 0]$ — analogously for $[0, n, 0, \ldots, 0]$ etc.

There are not only totally symmetric contributions in tensor products: The n-fold tensor product of a representation is decomposable with respect to symmetry classes for the permutation group $\mathbf{G}(n)$, totally symmetric as above, then totally antisymmetric, and also "mixed symmetric." It can be seen without difficulties, that the permutation of the vector space factors is compatible with the action of the represented Lie algebra — linear transformations and permutations commute.

Taking into account permutations, the r fundamental representations of the simple Lie algebra L_r can be generated, up to equivalence, by at most three irreducible representations — those at the maximally three ends of the Dynkin diagram ("outside"). Their *totally antisymmetric* products can be decomposed, containing all fundamental representations "inside."

The dimensions of antisymmetric product vector spaces (Grassmann powers) are given by $\binom{d}{n}$ for a d-dimensional vector space. If an "outside" representation has dimension d, the distance n to a vertex, which can be reached over simple line connections, determines the antisymmetric tensor product dimension $\binom{d}{n}$. Its decomposition into irreducible L_r-representation contains exactly once the fundamental representation associated with this vertex. In such a way, the diagram can be "gobbled up" from outside up to the vertices with a branching or with a double and triple line connection. Fundamental representations at branching points can be reached on different ways from outside. All this becomes clearer by the following examples.

The fundamental representations of the Lie algebras A_r have the following dimensions for their complex representation spaces:

$$\begin{array}{ccccccccc} \gamma_1 & — & \gamma_2 & — & \cdots & — & \gamma_{r-1} & — & \gamma_r \\ \binom{r+1}{1} & — & \binom{r+1}{2} & — & \cdots & — & \binom{r+1}{r-1} & — & \binom{r+1}{r} \end{array}$$

$$\gamma_k \overset{\text{dual}}{\leftrightarrow} \gamma_{r-k}$$

Fundamental representations of A_r and SL$(r+1, \mathbb{C})$, $r \geq 1$,
Generating representation: $r+1$.

It is possible to obtain all fundamental representations by totally antisymmetric products of the lowest-dimensional defining representation (dimension $r+1$) at the fringe. A self-dual representation has an invariant bilinear form — in the case of A_r only for odd rank $r = 2l+1$: For A_{2l+1}, the representation with fundamental weight γ_{l+1} has an invariant bilinear form with symmetry $(-1)^{l+1}$, where $\pm$ stands for symmetric and antisymmetric, respectively — e.g., A_1 an antisymmetric one, A_2 none, A_3 a symmetric one, etc.

For the Lie algebras C_r, the products of the defining representation with dimension $2r$ yield all the other fundamental ones by appropriate subspaces of the antisymmetric products. The product spaces have to be decomposed into irreducible spaces — $\binom{2r}{0} = 1$ is the dimension of the trivial representation:

$$\binom{2r}{1} — \binom{2r}{2} - \binom{2r}{0} — \binom{2r}{3} - \binom{2r}{1} — \cdots — \binom{2r}{r-1} - \binom{2r}{r-3} = \binom{2r}{r} - \binom{2r}{r-2}$$

$$\left[\binom{2r}{k} - \binom{2r}{k-2}\right]_{\epsilon_k} \quad \text{self-dual with } \epsilon_k = (-1)^k$$

Fundamental representations of C_r and Sp$(2r, \mathbb{C})$, $r \geq 2$,
Generating representation: $2r$.

The subindex $\epsilon_k = \pm 1$ denotes the existence of a symmetric and antisymmetric invariant bilinear form, respectively, of the representation space.

For the Lie algebras B_r, $r \geq 3$ ($B_2 \cong C_2$), a multiplicative buildup of the fundamental representations requires the defining representation with dimension $2r+1$. In addition, there is the complex 2^r-dimensional spinor representation:

$$\binom{2r+1}{1}_+ \text{———} \binom{2r+1}{2}_+ \text{———} \cdots\cdots \text{———} \binom{2r+1}{r-2}_+ \text{———} \binom{2r+1}{r-1}_+ \Longrightarrow (2^r)_\epsilon$$

dual with $\epsilon = (-1)^{\frac{r(1+r)}{2}}$

Fundamental representations for B_r **and** $\mathbf{SO}(2r+1,\mathbb{C}),\ r \geq 2,$
Generating representations: $2r+1,\ 2^r.$

For the Lie algebras D_r, $r \geq 3$, there exist two 2^{r-1}-dimensional spinor representations. The $2r$-dimensional defining representation generates the other ones:

$$\begin{array}{ccccccccc} & & & & & & & & (2^{r-1})_\epsilon \\ & & & & & & & & | \\ \binom{2r}{1}_+ & \text{———} & \binom{2r}{2}_+ & \text{———} & \cdots\cdots & \text{———} & \binom{2r}{r-3}_+ & \text{———} & \binom{2r}{r-2}_+ \\ & & & & & & & & | \\ & & & & & & & & (2^{r-1})_\epsilon \end{array}$$

self-dual with $\epsilon = (-1)^{\frac{r}{2}}$ for $r \in 2\mathbb{N}$ and $2^{r-1} \overset{\text{dual}}{\leftrightarrow} 2^{r-1}$ for $r \in 2\mathbb{N}+1$

Fundamental representations of D_r **and** $\mathbf{SO}(2r,\mathbb{C}),\ r \geq 3,$
Generating representations: $2r,\ 2^{r-1},\ 2^{r-1}.$

The two spinor representations are dual to each other for odd rank and self-dual for even rank.

For the exceptional Lie algebra G_2, the seven-dimensional fundamental representation is generating:

$$\binom{7}{1} \equiv \binom{7}{2} - \binom{7}{1}$$

$$7_+ \equiv \underline{14}_+$$

Fundamental representations of G_2
Generating representation: 7.

The adjoint representation is underlined.

The exceptional Lie algebra F_4 needs two fundamental representations with dimensions 26 and 52 at both ends:

$$\binom{26}{1} \text{———} \binom{26}{2} - \binom{52}{1} \Longrightarrow \binom{52}{2} - \binom{52}{1} \text{———} \binom{52}{1}$$

$$26_+ \text{———} 273_+ \Longrightarrow 1274_+ \text{———} \underline{52}_+$$

Fundamental representations for F_4
Generating representations: 26, 52.

For the little series $E_{6,7,8}$ all three fundamental representations at the three ends are neccessary to generate the remaining ones. The representation at the branching point can be obtained from all three ends.

One has for E_6:

$$\binom{78}{1}$$

$$\binom{27}{1} — \binom{27}{2} — \binom{27}{3} = \binom{78}{2} - \binom{78}{1} — \binom{27}{2} — \binom{27}{1}$$

$$\underline{78}_+$$

$$27 — 351 — 2925_+ — 351 — 27$$

$$(27, 351) \overset{\text{dual}}{\leftrightarrow} (27, 351)$$

Fundamental representations for E_6
Generating representations: 78, 27, 27.

Continuing for E_7:

$$\binom{912}{1}$$

$$\binom{56}{1} — \binom{56}{2} - \binom{56}{0} — \binom{56}{3} - \binom{56}{1} — 365\,750 — \binom{133}{2} - \binom{133}{1} — \binom{133}{1}$$

$$\binom{56}{4} - \binom{56}{2} = 365\,750 = \binom{133}{3} - 2\binom{133}{1}$$

$$912_-$$

$$56_- — 1\,539_+ — 27\,664_- — 365\,750_+ — 8\,645_+ — \underline{133}_+$$

Fundamental representations for E_7
Generating representations: 912, 56, 133.

For E_8, finally, there arise astronomical dimensions:

$$147\,250_+$$

$$\underline{248}_+ — 30\,380_+ — 2\,450\,240_+ — 146\,325\,270_+ — 6\,899\,079\,264_+ — 6\,696\,000_+ — 3\,875_+$$

Fundamental representations for E_8
Generating representations: 147250, 248, 3875.

The totally antisymmetric products of the generating representations at the three ends decompose into irreducible representations, where the highest-dimensional one is the fundamental representation for the vertex. In the reduction, there occur also non-fundamental irreducible representations (in parentheses):

$$\binom{248}{2} = 30\,380 \oplus 248$$

$$\binom{248}{3} = 2\,450\,240 \oplus 2 \times 30\,380 \oplus 2 \times 248$$

$$\binom{248}{4} = 146\,325\,270 \oplus (4\,096\,000) \oplus 2\,450\,240 \oplus (779\,247) \oplus 147\,250 \oplus (27\,000) \oplus 3\,875 \oplus 248$$

$$\binom{248}{5} = 6\,899\,079\,264 \oplus (344\,452\,500) \oplus 146\,325\,270 \oplus (76\,271\,625) \oplus (26\,411\,008) \oplus 6\,696\,000 \oplus (4\,096\,000) \oplus (1\,763\,125) \oplus 2 \times (779\,247) \oplus 147\,250 \oplus 2 \times 30\,380 \oplus 248$$

$$\binom{147\,250}{2} = 6\,899\,079\,264 \oplus (2\,275\,896\,000) \oplus \ldots \oplus 248$$

$$\binom{3\,875}{2} = 6\,696\,000 \oplus (779\,247) \oplus 30\,380 \oplus 248$$

$$\binom{3\,875}{3} = 6\,899\,079\,264 \oplus \ldots$$

Despite these deterrent numbers ("naturally arising"), the E_8-symmetry plays a not unimportant role in theoretical-physical speculations.

4.23 Quantum Theory and Symmetry

The still "anschauliche" Bohr–Sommerfeld model of the atoms with the electron orbits, ad hoc discretized by the "quantization" of angular momentum, and the electrons jumping from orbit to orbit, suggests an individuality of the constituting electrons and nuclei, also in the atoms. The experiments seem to show clearly, that the atoms do not deserve their name since they consist of electrons and nuclei.

The situation is not so simple. Especially Heisenberg himself has emphasized indefatigably, up to his last public talk (1975), that the fully developed quantum theory has taught us to take seriously, not only philosophically, but also experimentally, limitations and complementarities of macroscopically successful concepts. e.g., of "orbits," of "position" and "momentum," or, also, of "parts in a whole" and "to consist of."

A visualization of the atoms, more appropriate than with electron orbits in space, uses the complex probability amplitudes — the "information clouds" as given by Schrödinger's[29] wave functions. Those "information catalogues" associate a modulus and a phase with each position point. They are used for probabilities of experimental results, e.g., in the measurement of a position, of a momentum, of a spin, etc. The wave functions oscillate via their amplitudes in a complex information space — by no means in the phase space or even in position space.

For example, the "information clouds" for the atomic states can be seen as oscillating structures, fixed by the condition of absolute normalizability. The space part, not

[29]Erwin Schrödinger (1887–1961).

oscillating in time, is characterized by knot-surfaces: The principal quantum number $2j + 1$, characterizing symmetry properties with respect to the **SO**(4)-invariance of the states for the Kepler potential, indicates altogether $2j$ knot-surfaces in the "clouds." The **SO**(3)-angular momentum decomposition specifies $k = 2j - l$ surfaces in the form of 2-spheres. The remaining l noncompact knot-surfaces are m planes for **SO**(2)-axial rotations in **SO**(3) and $l - |m|$ bicone-surfaces:

$$2j = k + (l - |m|) + |m|.$$

It is meaningful only up to a certain point to look at an atom as to consist of a nucleus and electrons. The limitation to talk about "parts of a whole" becomes more clear for the nucleus itself where an individuality of protons and neutrons does not make much sense. It becomes complete nonsense, if one tries, for example, to characterize a neutron as consisting of a proton, an electron and an anti-neutrino since it can decay into those particles, or a photon as consisting of an electron and a positron with respect to the pair creation.

The atom as a state is a whole. In this sense, it is undivisible and has no parts. It is really an $\alpha\tau o\mu o\nu$ with respect to its character as energy eigenstate. The binding energy (mass defect), compared with the energy of the constituents gives a quantitative measure for the limitation to talk of parts in a whole. With the quantum structure it is really possible — for the atom with their small binding energy strongly exaggerated — to reverse the argumentation of divisibility: quantum theory has not shown, that the atom can be divided, but, on the contrary, that electrons, protons, nuclei, atoms, molecules are indivisible as energy eigenstates. Probably, as often in quantum theory, one should not argue too much about classical properties - here about "divisibility." The main point is to develop an openness for the revaluation, relativization and limitation of old classical concepts.

We are informed about interactions and particles by their symmetry forms. If a photon with its state-form goes over the form of an atom, both can combine to a new form, the excited atom, similar to the pouring of the content of two vessels with each a special form into another one with a different form. What governs the physical processes as structurizing features are the symmetries, which are multiply rearranged — Heraclitus: all is in flux — but obey conservation laws — Parmenides: nothing changes.

To define fundamentality with Democritus via the particle concept as energy eigenstate, is not in the spirit of quantum theory, to try it with Plato via forms, i.e., with the operational symmetry concept, seems to be more appropriate. Therefore, the question, not for fundamental particles, but for fundamental symmetries seems to be the right one. It is a deep insight of today's basic physics that the possible information forms of particles and interactions are, apparently, characterizable by a few simple symmetries.

4.24 Basic Internal Symmetry Program

The Erlangen program of Felix Klein led to a structural clarification of the, at this time, sometimes rather confusing geometrical investigations and discussions, mixing up algebraic and topological methods. A similarly clearing effect had the standard model of elementary particles with respect to the description of the electromagnetic, weak and strong interactions, their carriers and the objects they are acting on. There may even be found a content-related analogy: In the same way as a geometry can be characterized by symmetry operations acting on a manifold, basic quantum physics is characterized by symmetry-related interactions, acting on vector spaces. For example, Euclidean geometry with $\mathbf{SO}(3) \vec{\times} \mathbb{R}^3$ for the Erlangen program can be seen in analogy to $\mathbf{U}(1) \circ [\mathbf{SU}(2) \times \mathbf{SU}(3)] \vec{\times} \mathbb{C}^6$ as some sort of unitary geometry for the left handed quark isospin dublet fields in the standard model of particle physics.

With the present interpretation of the experiments, the *internal (chargelike) interactions* are determined by *compact* symmetry groups and Lie algebras with concepts, meaningful for these symmetries:

Electromagnetic (weak)	$\mathbf{U}(1)$	$D_1^c \cong \mathbb{R}$	hypercharge
Weak (electromagnetic)	$\mathbf{SU}(2)$	$A_1^c \cong \mathbb{R}^3$	isospin
Strong	$\mathbf{SU}(3)$	$A_2^c \cong \mathbb{R}^8$	color

Interactions, invariance groups, Lie algebras and meaningful concepts

These internal interactions with the real 12-parametric three factor *standard symmetry*:

$$\text{internal symmetry,} \begin{cases} \text{covering group: } \mathbf{U}(1) \times \mathbf{SU}(2) \times \mathbf{SU}(3), \\ \text{Lie algebra: } D_1^c \oplus A_1^c \oplus A_2^c \cong \mathbb{R}^{12}, \end{cases}$$

are interlaced: There exists a central correlation of hypercharge $\mathbf{U}(1)$, isospin $\mathbf{SU}(2)$ and color $\mathbf{SU}(3)$, leading to the observed electromagnetic charge, also with a symmetry group $\mathbf{U}(1)$. The operation group, faithfully realized in the standard model, is a quotient group of the covering group above.

Representations of compact groups are characterizable by integers, or, possibly after a renormalization, by rational numbers, as exemplified by hypercharge, isospin and the electrical charge number.

As in the Erlangen program, after the information about the operation group, the vector spaces have to be given for its action. Experimentally, this information comes from the fundamental fields for leptons and quarks and from the gauge fields. For the simple symmetries $\mathbf{SU}(2)$ and $\mathbf{SU}(3)$, there arise, in addition to trivial one-dimensional representations, also the fundamental representations on complex two- and three-dimensional vector spaces for leptons and quarks as well as the adjoint ones on real three- and eight-dimensional vector spaces for the gauge fields:

Field	Charge number $\mathbf{U}(1)$-weight	Hypercharge $\mathbf{U}(1)$-weight, y	Isospin $\mathbf{SU}(2)$ weights, d_I	Color $\mathbf{SU}(3)$ representation, d_C	$\mathbb{C}^{d_I \cdot d_C}$ $d_I \cdot d_C$
Right leptons **e**	-1	-1	0 singlet, 1	$[0, 0]$ singlet, 1	1
Left leptons **l**	$-1, 0$	$-\frac{1}{2}$	$(-\frac{1}{2}, \frac{1}{2})$ dublet, 2	$[0, 0]$ singlet, 1	2
Right quarks **d**, **u**	$-\frac{1}{3}, \frac{2}{3}$	$-\frac{1}{3}, \frac{2}{3}$	0 singlet, 1	$[1, 0]$ triplet, 3	3
Left quarks **q**	$-\frac{1}{3}, \frac{2}{3}$	$\frac{1}{6}$	$(-\frac{1}{2}, \frac{1}{2})$ dublet, 2	$[1, 0]$ triplet, 3	6
Electroweak gauge $\mathbf{A}_0, \vec{\mathbf{A}}$	$-1, 0, 0, 1$	0	$0, (-1, 0, 1)$ singlet, 1 triplet, 3	$[0, 0]$ singlet, 1	1, 3
Color gauge **C**	0	0	0 singlet, 1	$[1, 1]$ octet, 8	8
Higgs $\mathbf{\Phi}$	0, 1	$\frac{1}{2}$	$(-\frac{1}{2}, \frac{1}{2})$ dublet, 2	$[0, 0]$ singlet, 1	2

$\mathbf{U}(1) \times \mathbf{SU}(2) \times \mathbf{SU}(3)$ **-representations for standard model fields**

The electromagnetic charge number, i.e., the $\mathbf{U}(1)$-weight, is given by the sum of hypercharge weight and isospin weight. The last column displays the *internal multiplicity* $d_I \cdot d_C$, which for the leptons and quarks, and also for the Higgs field, coincides, for $y \neq 0$, with the *fractionality* of the conventionally normalized rational hypercharge number:

$$y \in \mathbb{Q}, \quad d_I, d_C \in \mathbb{N}, \quad y \cdot d_I \cdot d_C = \begin{cases} -1 \text{ for right leptons,} \\ -1 \text{ for left leptons,} \\ -1, 2 \text{ for right quarks,} \\ 1 \text{ for left quarks.} \end{cases}$$

Good to know: The fractionality of a nontrivial rational number q is the smallest natural number n that makes the product $n \cdot q$ an integer, i.e., $|q| = \frac{m}{n}$ with coprime m, n.

Moreover, the sum of all numbers $y \cdot d_I \cdot d_C$ vanishes. This is true also for some subsets. The latter insight motivated theories with larger embedding compact internal Lie groups like $\mathbf{SU}(5)$ or $\mathbf{SO}(10)$, so far without convincing experimental predictions.

Firmly established by experiments, but so far without convincing theoretical reasons, the leptons and quarks come both in three families, equal with respect to their homogeneous properties (internal and Lorentz group), and with different translation properties and different invariant masses — given for the leptons by the electron, muon and tauon family, and for the quarks by the (up-down), (charm-strange) and (top-bottom) family:

Right leptons **e**	e_R	μ_R	τ_R
Left leptons **l**	(ν_e, e_L)	(ν_μ, μ_L)	(ν_τ, τ_L)
Right quarks **d**, **u**	d_R, u_R	s_R, c_R	b_R, t_R
Left quarks **q**	(d_L, u_L)	(s_L, c_L)	(b_L, t_L)

Fermion families

4.25 Basic External Symmetry Program

Via the translations containing gauge interactions (covariant derivatives), the compact symmetry $\mathbf{U}(1) \circ [\mathbf{SU}(2) \times \mathbf{SU}(3)]$ of the internal interactions is interlaced with the *external (spacetimelike) noncompact symmetry*. In special relativity, the external operations are characterized by a Klein space $\mathbf{O}(1,3) \vec{\times} \mathbb{R}^4$ as semidirect product of the Lorentz group with the spacetime translations.

After the Stern–Gerlach[30] experiment with the discovery of the spin-$\frac{1}{2}$ property, established later for protons, neutrons, electrons, etc., the Lorentz group $\mathbf{O}(1,3)$ is replaced, with respect to its fundamentality, by the complex formulated group $\mathbf{SL}(2,\mathbb{C})$ as real 6-dimensional Lie group. The spacetime translations are acted on by a non-faithful representation of $\mathbf{SL}(2,\mathbb{C})$, in contrast, for example, to an electron field which is acted on by a faithful one.

If one starts from the Lorentz group $\mathbf{O}(1,3)$, i.e., the invariance group for the translation metric, as in some sense, "fundamental," the half-integer spin representations are called "doubled valued" representations. With respect to the experimentally established violation of the position and time reflection symmetry, the $\mathbf{O}(1,3)$-restriction to the orthochronous subgroup $\mathbf{SO}_0(1,3)$ as "fundamental" is suggested. This, together with the complex formulation of quantum structures, are arguments for the "fundamentality" of $\mathbf{SL}(2,\mathbb{C})$.

The two groups $\mathbf{O}(1,3)$ and $\mathbf{SL}(2,\mathbb{C})$ are different with respect to discrete subgroups. Therefore, they have isomorphic Lie algebras:

$$\mathbf{SL}(2,\mathbb{C})/\mathbf{I}(2) \cong \mathbf{SO}_0(1,3), \quad \begin{cases} \mathbf{SO}(1,3) \cong \mathbf{I}(2) \times \mathbf{SO}_0(1,3), \\ \mathbf{O}(1,3) \cong \mathbf{I}(2) \vec{\times} \mathbf{SO}(1,3). \end{cases}$$

The three arising discrete subgroups are isomorphic to the reflection group $\mathbf{I}(2) = \{\pm 1\}$, but operationally different: The orthochronous Lorentz group $\mathbf{SO}_0(1,3)$ parametrizes the classes of the simply connected group $\mathbf{SL}(2,\mathbb{C})$ with respect to $\mathbf{I}(2) \cong \{\pm \mathbf{1}_2\}$. Its direct product with the spacetime central reflection $\mathbf{I}(2) \cong \{\pm \mathbf{1}_4\}$ and the semidirect product with $\mathbf{I}(2) \cong \{\mathrm{P}, \mathbf{1}_4\}$ with, for example, a position reflection $\mathrm{P} = \begin{pmatrix} 1 & 0 \\ 0 & -\mathbf{1}_3 \end{pmatrix}$, gives the full Lorentz group $\mathbf{O}(1,3)$.

The Lorentz Lie algebra can be seen — in analogy to the complex numbers $\mathbb{C} \cong \mathbb{R} \oplus i\mathbb{R}$ as real two-dimensional vector space — as a complexification of the angular momentum Lie algebra A_1^c, leading to the — in the real — doubled Lie algebra $A_1^c \oplus iA_1^c \cong \mathbb{R}^6$ with the vector space $iA_1^c \cong \mathbb{R}^3$ for the proper Lorentz transformations (boosts):

$$\begin{aligned} \exp A_1^c &\cong \mathbf{SU}(2), & \mathbf{SU}(2)/\mathbf{I}(2) &\cong \mathbf{SO}(3), \\ \exp[A_1^c \oplus iA_1^c] &\cong \mathbf{SL}(2,\mathbb{C}), & \mathbf{SL}(2,\mathbb{C})/\mathbf{I}(2) &\cong \mathbf{SO}(3,\mathbb{C}) \cong \mathbf{SO}_0(1,3). \end{aligned}$$

---

[30] Walther Gerlach (1889–1979), Otto Stern (1888–1969).

Here, $\mathbf{SO}(3, \mathbb{C})$ has to be considered as real 6-dimensional, not as complex three-dimensional Lie group. It is remarkable, that, starting with the rotation group $\mathbf{SO}(3)$ of the Euclidean space with Lie algebra A_1^c, one ends up, by doubling in the complex, with the Lie algebra $A_1^c \oplus iA_1^c$ of the Lorentz group $\mathbf{SO}_0(1,3)$ without ever mentioning the concept "time."

The analogue to the Pauli spinors as fundamental representation $(\frac{1}{2})$ (spin notation) of $\mathbf{SU}(2)$ and the Lie algebra A_1^c with a basis $\{\mathbf{J}^a\}_{a=1}^3$ for rotations of position are the two *Weyl spinor representations* $(\frac{1}{2}|0)$ and $(0|\frac{1}{2})$ for $\mathbf{SL}(2, \mathbb{C})$ and its Lie algebra $A_1^c \oplus iA_1^c$ with a basis $\{\mathbf{J}^a, \mathbf{B}^a\}_{a=1}^3$ for rotations and boosts of spacetime:

$$\begin{array}{rcl} A_1^c \ni \vec{\mathbf{J}} & \longmapsto \frac{i}{2}\vec{\sigma} & \in \mathbf{AL}(2, \mathbb{C}), \\ A_1^c \oplus iA_1^c \ni (\vec{\mathbf{J}}, \vec{\mathbf{B}}) & \longmapsto (\frac{i}{2}\vec{\sigma}, \pm\frac{1}{2}\vec{\sigma}) & \in \mathbf{AL}(2, \mathbb{C}). \end{array}$$

The two Weyl representations are conjugated to each other $(0|\frac{1}{2}) \overset{\times}{\leftrightarrow} (\frac{1}{2}|0)$ in analogy to the conjugation of the complex numbers $\alpha + i\beta \overset{\times}{\leftrightarrow} \alpha - i\beta$.

The products of the two fundamental complex two-dimensional Weyl representations as building blocks give — up to isomorphy — all finite-dimensional representations of the Lorentz symmetry. The important four-dimensional Minkowski translations are acted on by the smallest *real* nontrivial representation of the Lorentz group which is irreducible, but not fundamental, but the product of the two fundamental Weyl representations:

$$\left(\frac{1}{2}|0\right) \otimes \left(0|\frac{1}{2}\right) \cong \left(\frac{1}{2}|\frac{1}{2}\right) = \left(\frac{1}{2}|\frac{1}{2}\right)^{\times}.$$

Also the spinors, found by Dirac in an admirably intuitive way in his famous equation for the electrons, are not acted on by a fundamental representation of the Lorentz group. The Dirac spinors are even not irreducible; their complex four-dimensional representation space decomposes into two two-dimensional spaces for the two fundamental Weyl spinors. The Dirac equation with the electromagnetic interaction of the electrons proved as a part of the deeper equations of the electroweak standard model with the irreducible left- and right-handed Weyl spinor fields for leptons and quarks.

All finite-dimensional irreducible representations of the Lorentz Lie algebra are characterized by rectangular weight diagrams in a two-dimensional weight space $\mathbb{E}^2$:

$$\text{weight diagram of } (j_+|j_-): \ (m_+, m_-) \text{ with } \begin{cases} m_+ = -j_+, j_+ + 1, \ldots, j_+, \\ m_- = -j_-, j_- + 1, \ldots, j_-. \end{cases}$$

In contrast to its compact partner, the semisimple, not simple rank 2 Lie algebra $D_2^c \cong A_1^c \oplus A_1^c$, used for the atomic states with two spin-like quantum number $m_\pm = \frac{m \pm n}{2}$ and their sum $m = m_+ + m_-$ as angular momentum quantum number, the representations of the noncompact simple rank-2 Lorentz Lie algebra $A_1^c \oplus iA_1^c$ have complex quantum numbers $m_\pm = \frac{m \pm in}{2}$. The real part $m = m_+ + m_-$ is a quantum number for a compact group (angular momentum); the imaginary part $in = m_+ - m_-$ characterizes the noncompact boosts with the hyperbolic transformations $\mathbf{SO}_0(1,1)$. This abelian Lorentz group is a dually represented dilation group $\mathbf{D}(1)$. The abelian groups, associated with the Cartan subalgebras are in the compact case (atomic states):

$$\exp[i\mathbb{R} \oplus i\mathbb{R}] \cong \mathbf{SO}(2) \times \mathbf{SO}(2) \subset \mathbf{SU}(2) \times \mathbf{SU}(2),$$

in contrast to

$$\exp[i\mathbb{R} \oplus \mathbb{R}] \cong \mathbf{SO}(2) \times \mathbf{SO}_0(1,1) \subset \mathbf{SL}(2,\mathbb{C})$$

for the Lorentz cover group.

For example, the electromagnetic field with values in a four-dimensional representation space of the Lorentz symmetry (representation $(\frac{1}{2}|\frac{1}{2})$) has two transversal components, for the left- and right-polarized photons as particles with a nontrivial $\mathbf{SO}(2)$-behavior and a trivial $\mathbf{SO}_0(1,1)$-behavior. The remaining two components involve the contribution for the Coulomb-potential and have trivial $\mathbf{SO}(2)$-polarization and nontrivial $\mathbf{SO}_0(1,1)$-behavior. They cannot be associated with real particles.

The finite-dimensional irreducible representations of the Lorentz Lie algebra $A_1^c \oplus iA_1^c$ come in conjugated pairs $(j_+|j_-) \overset{\times}{\leftrightarrow} (j_-|j_+)$ with opposite *chirality* (handedness). The experimentally well established standard model implements a connection between the external chirality and the internal unitary multiplicity properties: In the conventional assignment, the left-handed leptons and left-handed quarks are isospin dublets, in contrast to the right-handed leptons and right-handed quarks that are isospin singlets:

Field	Lorentz symmetry $A_1^c \oplus iA_1^c$-representation	Isospin symmetry A_1^c-representation
Right spinors $\mathbf{e}, \mathbf{u}, \mathbf{d}$	$(0\|\frac{1}{2})$	singlet (0)
Left spinors $\mathbf{l}, \mathbf{q}$	$(\frac{1}{2}\|0)$	dublet $(\frac{1}{2})$
Higgs scalar $\mathbf{\Phi}$	$(0\|0)$	dublet $(\frac{1}{2})$

Chirality and isospin in the standard model

It is impossible to remove the chirality-isospin connection, e.g., by transition to the left-handed antirepresentations of the right-handed spinors. The connection remains in the parametrization of the mass structure as given by a Yukawa coupling between left-handed and right-handed representations with a spinless Higgs field that is an isospin dublet.

The full Poincaré operations:

$$\text{external symmetries:} \left\{ \begin{array}{r} \text{noncompact group: } \mathbf{SO}_0(1,3) \vec{\times} \mathbb{R}^4, \\ \text{Lie algebra: } [A_1^c \oplus iA_1^c] \vec{\oplus} \mathbb{R}^4 \cong \mathbb{R}^{10}. \end{array} \right.$$

can be obtained in a chain of affine tangent and complexification structures: The compact rotation group $\mathbf{SO}(3) \cong \mathbf{SU}(2)/\mathbf{I}(2)$ with Lie algebra $A_1^c \cong \mathbb{R}^3$ has as tangent structure the semidirect Euclidean group $\mathbf{SO}(3) \vec{\times} \mathbb{R}^3$ and Lie algebra $A_1^c \vec{\oplus} \mathbb{R}^3$ with the position translations:

$$\begin{array}{l} A_1^c : [\mathbf{J}^a, \mathbf{J}^b] = -\epsilon^{abc}\mathbf{J}^c, \\ \mathbb{R}^3 : [\mathbf{P}^a, \mathbf{P}^b] = 0, \end{array} \quad A_1^c \vec{\oplus} \mathbb{R}^3 : [\mathbf{J}^a, \mathbf{P}^b] = -\epsilon^{abc}\mathbf{P}^c.$$

The complexification of the rotation group with the — in the real — independent generators $\{i\mathbf{J}^a\}_{a=1}^3 = \{\mathbf{B}^a\}_{a=1}^3$ (boosts) leads to the homogeneous Lorentz group with Lie algebra $A_1^c \oplus iA_1^c$ (multiplication of $[\mathbf{J}^a, \mathbf{J}^b] = -\epsilon^{abc}\mathbf{J}^c$ with i and $i \times i$):

$$[\mathbf{J}^a, \mathbf{B}^b] = -\epsilon^{abc}\mathbf{B}^c, \quad [\mathbf{B}^a, \mathbf{B}^b] = \epsilon^{abc}\mathbf{J}^c.$$

To close the Lie algebra, which includes the position translations, one has to give the Lie bracket of the boosts iA_1^c with basis $\{\mathbf{B}^a\}_{a=1}^3$ and the position translations $\mathbb{R}^3$ with basis $\{\mathbf{P}^a\}_{a=1}^3$. There suffices one additional rotation trivial element $\mathbf{P}^0$ for a rotation scalar, defined by the Lie bracket of boosts and position translations:

$$[\mathbf{B}^a, \mathbf{P}^b] = \delta^{ab}\mathbf{P}^0 \Rightarrow \mathbf{P}^0 = \frac{1}{3}[\mathbf{B}^a, \mathbf{P}^a].$$

There follows, with the Jacobi identity, the Lie brackets for $\mathbf{P}^0$:

$$[\mathbf{B}^a, \mathbf{P}^0] = \mathbf{P}^a, \quad [\mathbf{J}^a, \mathbf{P}^0] = 0, \quad [\mathbf{P}^a, \mathbf{P}^0] = 0.$$

Therefore, the semidirect Lie algebra closes by adding the rotation scalar element $\mathbf{P}^0$ for the time translations, resulting in the Poincaré Lie algebra. Summarizing the chain of the symmetry extensions for position and time:

$$A_1^c \subset \left\{ \begin{array}{l} A_1^c \vec{\oplus} \mathbb{R}^3 \text{ (Euclid)} \\ A_1^c \oplus iA_1^c \text{ (Lorentz)} \end{array} \right\} \subset [A_1^c \oplus iA_1^c] \vec{\oplus} \mathbb{R}^4 \text{ (Poincare)},$$

$$\mathbf{SO}(3) \subset \left\{ \begin{array}{l} \mathbf{SO}(3) \vec{\times} \mathbb{R}^3 \text{ (Euclid)} \\ \mathbf{SO}_0(1,3) \text{ (Lorentz)} \end{array} \right\} \subset \mathbf{SO}_0(1,3) \vec{\times} \mathbb{R}^4 \text{ (Poincare)}.$$

Good to know: For general n, the unitary $\mathbf{SU}(n)$-Lie algebra $A_{n-1}^c \cong \mathbb{R}^{n^2-1}$ has the semidirect tangent Lie algebra $A_{n-1}^c \oplus \mathbb{R}^{n^2-1}$ (Euclid) and the real doubling in the complex $A_{n-1}^c \oplus iA_{n-1}^c$ (Lorentz). The associated tangent Lie algebra $[A_{n-1}^c \oplus iA_{n-1}^c] \vec{\oplus} \mathbb{R}^{n^2} \cong \mathbb{R}^{3n^2-2}$ (Poincaré) is a Lie subalgebra of the $(4n^2-1)$-dimensional Lie algebra of $\mathbf{SU}(n,n)$. Only for $n = 2$, there exists a local isomorphy to a real orthogonal (conformal) group $\mathbf{SO}(2,4)$.

Chapter 5
Circles and Winding Numbers

The circle as an "everywhere equal," unbounded, but, nevertheless, finite line seems to be perfect — these properties contributed in the past, and also today, to look for its realizations in the basic structures of physics and, apparently, to find it there quite often. A nearly religious fixation on the circle as the "handwriting of God" caused also sometimes a stagnation in the physical understanding, e.g., in the epicycles, especially of Ptolemy,[1] which retarded a deeper understanding of the planetary orbits almost for one and a half millenium. By Copernicus, after some preludes by Cusanus,[2] and by Kepler, the heliocentric interpretation of our planetary system by Aristarch[3] was taken up again. In today's theory for the planets, the basic rotational symmetry structure does not show up in individual orbits, i.e., not in invariance of states (solutions) of a dynamics, but in the symmetry operations of the dynamical law. States (solutions) of a dynamics are acted on by representations of the symmetry which, in general, do not have to be trivial — even not in the case of ground states.

Also today the circle and its possible realizations play an important role in the physical description of natural phenomena. The circle can be associated with the axial rotations $\mathbf{SO}(2)$, or, in a complex framework, with the unitary symmetry structure $\mathbf{U}(1)$. The phase group $\mathbf{U}(1)$ is basic for all unitary groups like the spin group $\mathbf{SU}(2)$. All discrete quantum numbers, especially those for the unitary internal groups of particle physics, can be related, via their representation weights, to winding numbers of the circle, characterizing the mappings of the circle on itself.

In addition, the unitary symmetry $\mathbf{U}(1) \cong \mathbb{R}/\mathbb{Z}$ realizes, as quotient group, the real classes of the integers. This is of importance for the probability interpretation of quantum theory and characterizes the time representations by the phase property of particles and their scalar products.

---

[1]Claudius Ptolemaios, around (90–163).

[2]Nikolaus Cryfftz of Kues (1401–1464), Niclas Koppernigk (1473–1543).

[3]Aristarchos of Samos, around -(310–230).

H. Saller, *Operational Symmetries*, DOI 10.1007/978-3-319-58664-9_5

Since the unitary Lie group $\mathbf{SU}(1+r)$ is the compact form of the complex linear Lie operations $\mathbf{SL}(1+r, \mathbb{C})$, and since all simple Lie algebras are subalgebras of the $\mathbf{SL}(1+r, \mathbb{C})$-generating Lie algebras A_r, the structure of $\mathbf{SU}(1+r)$ with its winding numbers casts a light on all semisimple Lie symmetry.

5.1 Circles on the Earth and in the Sky

The Eleat Parmenides, around the year -500, represented the being, in its unchangeable and unlimited form, as a sphere.

In Egypt, around the year -250, Eratosthenes suggested that the earth is not a flat plane, but a sphere. For a proof of his suggestion, also quantitatively, he measured the length and the radius of great circles. To this end, he looked at the shadow-length s of a rectangularly standing stick with length l, and put the ratio of both lengths in relation to the distance d of a point where a stick throws no shadow. Apparently, he even knew, e.g., via the pole star, that the direction of the chosen distance along the Nile was almost parallel to the earth's rotation axis.

With elementary trigonometry, in the language of today, and without qualifying details: The quotient of the shadow-length to the stick-length gives the angle $\theta \simeq \frac{s}{l}$ between sun rays and the stick direction, which, with $\theta \simeq \frac{d}{R_\bullet}$ for an rectangularly standing stick, leads to the radius of the earth $R_\bullet \simeq \frac{ld}{s}$.

Of course, it is possible to compare the shadow-lengths at two different points with equal longitude:

$$\tan \theta_k = \frac{s_k}{l} \Rightarrow \tan(\theta_1 - \theta_2) = \frac{\frac{s_1}{l} - \frac{s_2}{l}}{1 + \frac{s_1 s_2}{l^2}}.$$

For example, and with some numerical glossing over: If you drive by car from Rome to Munich — that is almost in a northern direction, you can measure, for a distance of about 800 km, a shadow difference of about 12.5 cm for a 1m-stick. This gives you a latitude difference for the two cities of $\theta \simeq \frac{1}{8} \simeq 7°$ with $2\pi \cong 360°$ and an earth radius $R_\bullet \simeq 6400$ km — from the books: Rome has (latitude, longitude) = $(\theta, \phi) = (41°53', 12°29')$, and Munich $(48°08', 11°34')$. Obviously, instead of using a stick, the geographical latitude can be measured directly via the horizontal angle of the pole star.

Good to know: The addition theorem for the spherical tangent, used for a spherical earth, has the hyperbolic tangent analogue for hyperbolic spacetime in special relativity. It is used for the addition of velocities v with c the speed of light:

$$\tanh \psi_k = \frac{v_k}{c} \Rightarrow \tanh(\psi_1 + \psi_2) = \frac{\frac{v_1}{c} + \frac{v_2}{c}}{1 + \frac{v_1 v_2}{c^2}}.$$

Circles were suspected in the sky too: If, with the formalization of today, the planetary orbits are approximated by circles with frequency $\nu = \frac{\omega}{2\pi}$ in a Euclidean plane $\mathbb{E}^2 \cong \mathbb{C}$ around the Sun in the center:

$$\text{orbit of planet } P\text{:} \quad P(t) = P_0 \begin{pmatrix} \cos \omega_P t \\ \sin \omega_P t \end{pmatrix} \cong P_0 e^{i\omega_P t},$$
$$\text{orbit of Sun } S\text{:} \quad S(t) = 0,$$

the orbit-coordinates are given for the Earth E as center by

$$P_E(t) = P(t) - E(t) \cong P_0 e^{i\omega_P t} - E_0 e^{i\omega_E t},$$
$$S_E(t) = -E(t) \cong -E_0 e^{i\omega_E t}.$$

The time $t = 0$ marks the constellation with the planet P, the Earth E and the Sun S on one line. The Earth-centered orbit of an outside planet, e.g., Saturn, looks like the Sun-centered orbit of Earth (epicycle) rolling with its center on the Sun-centered orbit of the Saturn (deferent). This approximation with circles gives for all non-Earth planets only one epicycle with the frequency of the Earth. A circle-analysis of elliptic orbits needs a discrete infinity of epicycles.

Good to know: A periodic function $f(t+1) = f(t)$, square integrable $f \in L^2[0,1]$, has a Fourier series (harmonic analysis) $f(t) = \sum_{z=-\infty}^{\infty} \alpha_z e^{z2\pi i t}$ with product representations of $\mathbb{R} \ni t \longmapsto e^{2\pi i t} \in \mathbf{U}(1)$ and coefficients $\alpha_z \in \mathbb{C}$. The Fourier components for the Kepler ellipses are $\alpha_z = \int_0^{2\pi} \frac{d\phi}{2\pi} e^{-iz\phi} \frac{L^2}{1+R\cos\phi}$.

5.2 Winding Numbers and Unit Roots

The circle in the Euclidean plane as a multiplicative group is parametrizable by the unitary phase group $\mathbf{U}(1)$, a real one-dimensional Lie group:

$$\mathbf{U}(1) = \{e^{2\pi i\alpha} \mid \alpha \in \mathbb{R}\}.$$

$\mathbf{U}(1)$ is the invariance group of the scalar product $\langle z_1|z_2\rangle = \overline{z}_1 z_2$ and the norm $\| z \|$ of the complex numbers $z \in \mathbb{C}$.

The circle $\mathbf{U}(1)$ can be wound on itself with an integer winding number, with equal or opposite orientation, by the group endomorphisms:

$$z \in \mathbb{Z}: \quad D_z : \mathbf{U}(1) \longrightarrow \mathbf{U}(1), \quad e^{2\pi i\alpha} \longmapsto D_z(\alpha) = e^{z2\pi i\alpha} = D_z(\alpha+1),$$
$$D_z(\alpha_1) \circ D_z(\alpha_2) = D_z(\alpha_1 + \alpha_2).$$

These are all continuous, irreducible complex representations of the Lie group $\mathbf{U}(1)$. The $\mathbf{U}(1)$-operations act on a complex one-dimensional vector space, $\mathbf{U}(1) \subset \mathbf{GL}(1, \mathbb{C})$. Integers are necessary to account for the cyclicity $e^{2i\pi} = 1$ and the uniqueness. To call the compact group-characteristic integer winding number condition $z \in \mathbb{Z}$ "quantization" is inappropriate.

The weight diagram with the $\mathbf{U}(1)$-winding numbers of a representation D_z consists of exactly one point $z \in \mathbb{Z}$ on the real axis. Opposite windings D_z and D_{-z} are dual to each other (antirepresentations). Only the trivial representation D_0 is self-dual. In the product of two windings, their numbers are added, $D_z \otimes D_{z'} \cong D_{z+z'}$. The irreducible continuous representations occupy all integers. $(\mathbb{Z}, +)$ is called the *dual group* of $\mathbf{U}(1)$, the $\mathbf{U}(1)$-weight lattice $\mathbb{Z}$ contains the $\mathbf{U}(1)$-*characters*.

Good to know: All group automorphisms $\{f : G \longrightarrow G\}$ of an abelian group G constitute themselves an abelian group with the composition $f \circ g$.

The $\mathbf{U}(1)$-representation D_z winds the $|z|$-th part of a circle on the full circle. D_z is faithful (injective) only for the sign group $z \in \mathbf{I}(2) = \{\pm 1\}$. The two representations $D_{\pm 1}$ are fundamental in so far that their tensor products give, up to equivalence, all irreducible $\mathbf{U}(1)$-representations, $D_{\pm n} \cong \bigotimes^n D_{\pm 1}$.

The *cyclotomic group* $\mathbf{I}(z)$ for $|z| \in \mathbb{N}$ is the kernel of the $\mathbf{U}(1)$-representation D_z. Therefore, the image of the representation is isomorphic to the corresponding quotient group, which itself is isomorphic to $\mathbf{U}(1)$ for $z \neq 0$:

$$z \neq 0,\ k \in \mathbb{Z} \Rightarrow D_z(\frac{k}{z}) = 1 \Rightarrow \begin{cases} \text{kern } D_z & = \mathbf{I}(z) \subset \mathbf{U}(1) \\ D_z(\mathbf{U}(1)) & \cong \mathbf{U}(1)/\mathbf{I}(z) \cong \mathbf{U}(1), \\ \text{with} & e^{\frac{2\pi i \alpha}{z}} \mathbf{I}(z) \leftrightarrow e^{2\pi i \alpha}. \end{cases}$$

The abstract $\mathbf{U}(1)$ can be visualized by a closed rubber band. Its windings around the wrist characterize representations with integer winding numbers. This picture is not quite right, since the winding orientation is reversed in the crossing points. To change that, the band has to be cut, to be wound z-times around the wrist and to be glued together again. The $|z| - 1$ crossovers are related to the nontrivial elements of the group $\mathbf{I}(z)$.

The cyclic group $\mathbf{I}(z) \subset \mathbf{U}(1)$ contains the complex solutions of the corresponding cyclotomic equation:

$$z \in \mathbb{Z},\ z \neq 0: \ \mathbf{I}(z) = \{u \in \mathbb{C} \mid u^z = 1\} = \mathbf{I}(-z) = \{e^{2\pi i \frac{k}{z}} \mid k = 0, \ldots, |z| - 1\}.$$

One can restrict the $\mathbf{I}(z)$-considerations to the natural numbers $z = n \in \mathbb{N}$. Only the trivial group $\mathbf{I}(1) = \{1\}$ and the sign group $\mathbf{I}(2) = \{\pm 1\}$ are real. $\mathbf{I}(n)$ as multiplicative group is isomorphic to the additive group $\mathbb{Z}_n$:

$$\mathbf{I}(n) \cong \mathbb{Z}_n = \mathbb{Z}/n\mathbb{Z} = \{[0], [1], \ldots, [n-1]\},\ \ \mathbf{I}(n) = e^{\frac{2\pi i}{n}\mathbb{Z}_n}.$$

It is useful to have a multiplicative notation $\mathbf{I}(n)$ and an additive one $\mathbb{Z}_n$.

$\mathbf{I}(n)$ occupies the corners of a regular polygon in the unit circle with the Gauss-Fermat possibilities for a construction with compass and ruler. For a regular n-gon and m-gon, in the same circle, one constructs, for coprime n and m, a regular $n \cdot m$-gon as follows: Starting from each of the n corners, one marks with the compass all $k = 0, \ldots, m - 1$ arcs of the m-gon — this yields $n \cdot m$ points on the circle. The doubling of the circle points $2 \cdot n$ can be easily done by orthogonally halfing the n sides.

The cyclotomic group $\mathbf{I}(n)$ with the unit roots can be generated by the element i_n for the smallest angle, which, for the Gauss-Fermat polygons can be expressed by radicals. The circumference of a regular n-polygon in and out of the unit circle is

$$I(n) = 2n \sin\frac{\pi}{n} < 2\pi < 2n \tan\frac{\pi}{n} = O(n),$$

which allows the Archimedean exhaustion method for the determination of π. The doubling of the number of corners uses the relations $\cos^2\frac{\theta}{2} = \frac{1+\cos\theta}{2}$ and $\sin^2\frac{\theta}{2} = \frac{1-\cos\theta}{2}$:

$$i_n = e^{\frac{2\pi i}{n}} \Rightarrow \mathbf{I}(n) = \{i_n^k \mid k = 0, \ldots, n-1\},$$

$$i_n = \cos\frac{2\pi}{n} + i \sin\frac{2\pi}{n} : \begin{cases} n = 2^N : & i_1 = 1,\ i_2 = -1, \\ & i_4 = i,\ i_8 = \frac{1+i}{\sqrt{2}}, \ldots, \\ n = 2^N \cdot 3 : & i_3 = \frac{-1+i\sqrt{3}}{2},\ i_6 = \frac{\sqrt{3}+i}{2}, \ldots, \\ n = 2^N \cdot 5 : & i_5 = \frac{1+\sqrt{5}}{4} + i\frac{\sqrt{10-2\sqrt{5}}}{4}, \ldots, \\ n = 2^N \cdot 17 : & \begin{cases} \cos\frac{2\pi}{17} = \frac{-1+\sqrt{17}+\sqrt{2(17-\sqrt{17})}}{16} \\ + \frac{\sqrt{68+12\sqrt{17}-4\sqrt{2(17-\sqrt{17})}-8\sqrt{2(17+\sqrt{17})}}}{16}, \end{cases} \\ n = 2^N \cdot 257 : & \ldots, \\ n = 2^N \cdot 65\,537 : & \ldots \end{cases}$$

Also the power i_n^k can be used as a generator, if k and n are coprime. For p prime, $\mathbf{I}(p)$ can be generated by each nontrivial power of i_p — the additive group $\mathbb{Z}_p$ is even a field.

The possibility of the construction of regular polygons with compass and ruler, of the solution of polynomials by radicals, of the trisection of an angle, of the doubling of the cube (Delian problem), etc., have been cleared up satisfactorily by the theory of algebraic field extensions, of solvable groups, etc., initiated by the work of Galois.[4]

Good to know: To construct, with compass and ruler, two squares with area ratio 2 : 1, i.e., to construct $\sqrt{2}$, is easy — to construct in this way two cubes with volume ratio 2 : 1, i.e., $\sqrt[3]{2}$, is impossible.

$\mathbf{I}(n)$ has as subgroups $\mathbf{I}(m)$ for all n-divisors m. It is the direct product of factors corresponding to a pairwise coprime decomposition of n:

$$m \text{ divides } n \Rightarrow \mathbf{I}(m) \subseteq \mathbf{I}(n),$$
$$n = m_1 \cdots m_r \text{ pairwise coprime} \Rightarrow \mathbf{I}(n) \cong \mathbf{I}(m_1) \times \cdots \times \mathbf{I}(m_r).$$

For example, $\mathbf{I}(6) \cong \mathbf{I}(2) \times \mathbf{I}(3)$ and $\mathbf{I}(12) \cong \mathbf{I}(4) \times \mathbf{I}(3)$, but $\mathbf{I}(4)$ is not isomorphic to Klein's four-group $\mathbf{K}(4) \cong \mathbf{I}(2) \times \mathbf{I}(2)$, a bicycle. The cyclic group $\mathbf{I}(p^k)$ with primes p are the *simple* cyclic groups.

Looking for subgroups of the direct product $\mathbf{I}(n) \times \mathbf{I}(m)$ of two cyclic groups, one can restrict to the direct product of simple group factors $\mathbf{I}(p^k) \times \mathbf{I}(p^l)$ with a

[4]Evariste Galois (1811–1832).

prime p. In addition to the obvious subgroups $\mathbf{I}(p^r) \times \mathbf{I}(p^s)$ with $0 < r \leq k$ and $0 < s \leq l$, there are cyclic subgroups in the form of *correlated bicycles* where the two cycles "turn around synchronized" — like a real bicycle running on the ground:

$$\mathbf{I}(p^k) \times \mathbf{I}(p^l) \supset \mathbf{I}(p^{\max(r,s)}) \text{ for all } 0 \leq r \leq k, \;\; 0 \leq s \leq l,$$
$$\text{generated by } (i_{p^r}, i_{p^s}) \in \mathbf{I}(p^r) \times \mathbf{I}(p^s).$$

For example, $\mathbf{I}(2) \times \mathbf{I}(8)$ has the nontrivially correlated cyclic subgroups with two cycles turning:

$$\begin{aligned}
&\mathbf{I}(2) \cong \{(-1,-1), (1,1)\} \text{ with } (-1,-1)^2 = (1,1),\\
&\mathbf{I}(4) \cong \{(-1,i), (1,-1), (-1,-i), (1,1)\},\\
&\mathbf{I}(8) \cong \{(-1,i_8), (1,i), (-1,i_8^3), (1,-1), (-1,-i_8), (1,-i), (-1,-i_8^3), (1,1)\}.
\end{aligned}$$

Good to know: The direct product group $G \times G'$ has the composition $(g, g') \bullet (h, h') = (gh, g'h')$.

An N-fold product ("wheelwork") of prime power groups $\mathbf{I}(p^{k_i})$, $i = 1, \ldots, N$, has cyclic subgroups with none, one, etc., $(N-1)$ and N correlated (synchronized) cycles.

5.3 Electromagnetic Winding Numbers

With the circle $\mathbf{U}(1)$ as symmetry group of the *electromagnetic interactions*, the integer multiples of the elementary electron charge e as particle charges $Q = ze$ are oriented winding numbers of the circle with representations D_z, $z \in \mathbb{Z}$. The property "charge" describes the behavior under $\mathbf{U}(1)$-operations acting on a complex one-dimensional vector space. Particle and antiparticle with antirepresentations have opposite charge number $\pm z$. If a particle is its own antiparticle, its charge has to vanish.

That the rational, but non-integer quark charge numbers $q = \pm\frac{1}{3}, \pm\frac{2}{3} \in \mathbb{Q}$ do not fit in such a picture, may have different reasons: Either their broken charge number can be understood only in connection with additional operations, e.g., with color $\mathbf{SU}(3)$. This possibility wil be considered below in more detail. Or: the elementary charge $\pm e$ (electron, positron) is not really elementary, but describes a multiple $\mathbf{U}(1)$-winding, e.g., with $z = \pm 3$, if quarks carry the really elementary charge, e.g., $\tilde{e} = \frac{e}{3}$. Finally, it may be possible, as Heisenberg insisted stubbornly, that the phenomenologically successful quark parametrization of strong interactions is an aberration what concerns a basic understanding. He used, as an example for an illustration, the epicycle theory, which was successfully used by the antique seafarers for one and a half millenium without yielding a deeper understanding of the planetary dynamics as given later by the Kepler–Newton theory.

With the $\mathbf{U}(1)$-interpretation of the electromagnetic phenomena, charge conservation becomes winding number conservation. A system with a trivial charge, e.g., an electron with a positron, visualized by two rubber bands with opposite winding numbers, can be collected in a product $\mathbf{U}(1)$-representation $D_0(\alpha) = 1$ with trivial winding number. For example, a pair annihilation is visualizable by the two oppositely oriented rubber bands around your wrist connecting, shrinking to one point, and giving you free.

Sometimes one talks about the "quantization of charge." This is not an appropriate expression since the representations of a compact group, here of $\mathbf{U}(1)$, are characterized by integers — without any quantum theory.

5.4 Energies and Time-Winding Numbers

Any dynamics represents time and involves energy. Time and energy constitute a dual pair with the energies the eigenvalues and invariants of time representations.

A harmonic oscillator (spring oscillations, pendulum with small amplitude) with Hamiltonian $\underline{\mathbf{H}} = \frac{\underline{\mathbf{p}}^2}{2\underline{m}_0} + \frac{\underline{g}_0}{2}\underline{\mathbf{x}}^2$, involving the mass $\underline{m}_0$ and the spring constant $\underline{g}_0$, has frequency $\underline{\omega}_0 = \sqrt{\frac{\underline{g}_0}{\underline{m}_0}}$. In its periodic motion, it returns always to its starting position $\underline{\mathbf{x}}(0)$ with, there, equal momentum $\underline{\mathbf{p}}(0)$:

$$\underline{\mathbf{x}}(\underline{t}) = \underline{\mathbf{x}}(0) \cos \underline{\omega}_0 \underline{t} + \frac{\underline{\mathbf{p}}(0)}{\sqrt{\underline{g}_0 \underline{m}_0}} \sin \underline{\omega}_0 \underline{t}.$$

After some time, the observer comes to the idea to use the oscillator as a clock with the oscillator period the intrinsic time unit $\frac{1}{\underline{\omega}_0}$. If also momentum and position is measured in intrinsic units $(\underline{\pi}_0, \underline{l}_0)$ with the ratio $\frac{\underline{\pi}_0}{\underline{l}_0} = \sqrt{\underline{g}_0 \underline{m}_0}$, there remains a purely mathematical structure:

$$2\pi t = \underline{\omega}_0 \underline{t} \Rightarrow \begin{cases} \mathbf{x}(t) = \frac{\underline{\mathbf{x}}(\underline{t})}{\underline{l}_0} = \mathbf{x} \cos 2\pi t + \mathbf{p} \sin 2\pi t, \\ \mathbf{p}(t) = \frac{\underline{\mathbf{p}}(\underline{t})}{\underline{\pi}_0} = -\mathbf{x} \sin 2\pi t + \mathbf{p} \cos 2\pi t, \end{cases}$$
$$\text{with } \mathbf{x} = \mathbf{x}(0), \quad \mathbf{p} = \mathbf{p}(0).$$

Underlined symbols have a dimension, non-underlined ones are number-valued. For the harmonic oscillator, the time, from the infinite past to the infinite future and formalized by the additive group with the reals, is wound on a circle, determined by its period, e.g., on the dial of an analogue watch whose cyclic form reflects the compact time representations $\mathbb{R} \longrightarrow \mathbf{U}(1)$.

The purely imaginary eigenvalues in the spectrum of the Hamilton matrix $\mathbf{h}$ require the use of the complex numbers:

$$\frac{d}{dt}\begin{pmatrix}\mathbf{x}\\\mathbf{p}\end{pmatrix} = \mathbf{h}\begin{pmatrix}\mathbf{x}\\\mathbf{p}\end{pmatrix}, \quad \mathbf{h} = 2\pi\begin{pmatrix}0 & 1\\-1 & 0\end{pmatrix},$$
$$\textbf{spec}\,\mathbf{h} = \{\omega |\det(\mathbf{h} - \omega\mathbf{1}_2) = 0\} = \{\pm 2\pi i\}.$$

Only in a complex vector space $\mathbf{V} \cong \mathbb{C}^2$, the eigenvectors of the time development can be given:

$$\mathrm{u}(t) = \frac{\mathbf{x}(t) - i\mathbf{p}(t)}{\sqrt{2}} = e^{2\pi i t}\mathrm{u}, \quad \mathrm{u}^\star(t) = \frac{\mathbf{x}(t) + i\mathbf{p}(t)}{\sqrt{2}} = e^{-2\pi i t}\mathrm{u}^\star.$$

In a real vector space $V \cong \mathbb{R}^2$, built by position and momentum, there are no time translation eigenvectors. Time eigenvectors have unitary phases.

All polynomials in position and momentum of the oscillator can be formulated as polynomials in the eigenvectors. The time dependence of the generating eigenvector monomials is characterized by *time winding numbers*:

$$[\mathrm{u}^\star(t)]^n[\mathrm{u}(t)]^m = e^{2\pi i z t}\mathrm{u}^{\star n}\mathrm{u}^m, \quad z = m - n \in \mathbb{Z}.$$

Their unitary time representations involve two steps:

$$\mathbb{R} \longrightarrow \mathbf{U}(1) \longrightarrow \mathbf{U}(1), \quad \underline{t} \longmapsto e^{i\omega_0 \underline{t}} \longmapsto e^{z i\omega_0 \underline{t}},$$
$$i\omega_0 \in i\mathbb{R}, \quad z \in \mathbb{Z}.$$

The first step defines the frequency ω_0 in the imaginary dual group $i\mathbb{R}$ with the irreducible unitary representations of the time translations. The second step involves the winding numbers z from the dual group $\mathbb{Z}$, characterizing the irreducible $\mathbf{U}(1)$-representations:

$$\text{dual groups} \begin{cases} \mathbb{R} \leftrightarrow i\mathbb{R}, \\ \mathbf{U}(1) \leftrightarrow \mathbb{Z}. \end{cases}$$

Good to know: For abelian groups, the duality relation is reflexive, e.g., the irreducible representations of $\mathbb{Z}$ constitute a $\mathbf{U}(1)$-isomorphic group.

We experience time by its representations via objects and interactions. If you were a harmonic oscillator, your time model would be $\mathbf{U}(1)$, not $\mathbb{R}$. The irreducible unitary time representations $\mathbb{R} \longrightarrow \mathbf{U}(1)$ are not faithful. That most people have the impression of a non-cyclic time, may have different reasons: Perhaps, time is cyclic and we live in a pulsating cosmos without remnants of former periods and with, for human time units, extremely long periods. Or time is really not cyclic and there exist, in addition to the unitary $\mathbf{U}(1)$-representations, also time representations with noncompact contributions from "eternity."

In a classical theory, the conserved energy:

$$\underline{E} = \frac{\underline{\mathbf{p}}^2}{2m_0} + g_0\frac{\underline{\mathbf{x}}^2}{2} = \pi_0 l_0 \omega_0 \frac{\mathbf{p}^2 + \mathbf{x}^2}{2} = \pi_0 l_0 \omega_0 \frac{\mathrm{u}\mathrm{u}^\star + \mathrm{u}^\star\mathrm{u}}{2},$$

is continuous, i.e., it can take any real value. In quantum theory with Planck's constant as intrinsic unit $\pi_0 l_0 = \hbar$ and a noncommuting position-momentum operator pair:

$$[i\underline{\mathbf{p}}, \underline{\mathbf{x}}] = \hbar\mathbf{1}, \quad [i\mathbf{p}, \mathbf{x}] = \mathbf{1} = [\mathrm{u}^\star, \mathrm{u}],$$

the energy eigenvalues are restricted to time winding numbers $\mathbb{Z}$ for the $(\mathrm{u}, \mathrm{u}^\star)$-monomials from the $\mathbf{U}(1)$-dual group $(\mathbb{Z}, +)$, and the measurable energy differences to the ground state are restricted to natural winding numbers:

$$\text{measurable values of } E = \frac{\underline{E}}{\hbar\omega_0} = \frac{\{\mathrm{u}, \mathrm{u}^\star\}}{2} \in \mathbb{N} = \{0, 1, 2, \dots\}.$$

5.5 Spin Winding Numbers

The two endpoints $z = \pm 1$ of a distance on the real axis $\mathbb{R}$ constitute the weight diagram of the fundamental Pauli representation of the complex Lie group $\mathbf{SL}(2, \mathbb{C})$. Its Lie algebra A_1 acts on a binary alternative, given by a basis of a complex two-dimensional vector space:

$$e^+ = e^1 = \begin{pmatrix} 1 \\ 0 \end{pmatrix}, \quad e^- = e^2 = \begin{pmatrix} 0 \\ 1 \end{pmatrix},$$

and, analogously, for its compact form $\mathbf{SU}(2)$ with Lie algebra A_1^c. The two vectors of an alternative carry opposite winding numbers for the dual representations $D_{\pm 1}$ of an abelian Cartan subgroup $\mathbf{U}(1) \cong \mathbf{SO}(2) \subset \mathbf{SU}(2)$, e.g., in the third direction $\mathbf{U}(1)_3$:

$$D^{[1]}(\beta) = \begin{pmatrix} D_1(\beta) & 0 \\ 0 & D_{-1}(\beta) \end{pmatrix} = \begin{pmatrix} e^{2\pi i\beta} & 0 \\ 0 & e^{-2\pi i\beta} \end{pmatrix} : \quad \begin{cases} D^{[1]}(\beta)(e^+) = e^{2\pi i\beta} e^+, \\ D^{[1]}(\beta)(e^-) = e^{-2\pi i\beta} e^-. \end{cases}$$

Its Lie algebra $i\mathbb{R}$ involves the $\mathbf{SU}(2)$-*winding number matrix* w_2:

$$\frac{d}{d\beta}|_{\beta=0} D^{[1]}(\beta) = 2\pi i w_2, \quad w_2 = \begin{pmatrix} 1 & 0 \\ 0 & -1 \end{pmatrix} = \sigma_3, \quad e^{\pi i w_2} = -\begin{pmatrix} 1 & 0 \\ 0 & 1 \end{pmatrix} = -\mathbf{1}_2.$$

The fundamental $\mathbf{SU}(2)$-representation with two opposite fundamental $\mathbf{U}(1)$-winding numbers $z = \pm 1$ is denoted by $D^{[1]}$.

Good to know: The Lie algebra of a special group $G \subseteq \mathbf{SL}(n, \mathbb{C})$ has traceless matrices, $\operatorname{tr} \log G = \{0\}$.

There are many $\mathbf{SO}(2)$-subgroups in the real or complex automorphisms $\mathbf{GL}(2, \mathbb{R})$ and $\mathbf{GL}(2, \mathbb{C})$. Usually, the orthogonal group $\mathbf{SO}(2)$ in the real is defined by the invariance of the unit metric $\mathbf{1}_2$:

$$\mathbf{SO}(2) = \{O \in \mathbf{GL}(2, \mathbb{R}) \mid O \circ \mathbf{1}_2 \circ O^T = \mathbf{1}_2, \quad \det O = 1\}.$$

In the complex with $O \circ O^\star = \mathbf{1}_2$, $O \in \mathbf{GL}(2, \mathbb{C})$, one obtains with the Pauli matrices $\vec{\sigma}$ and a unit vector in the Euclidean space $\vec{n} \in \mathbb{E}^3$, $\vec{n}^2 = 1$:

$$O = e^{i\vec{\sigma}\vec{n}2\pi\beta} = \mathbf{1}_2 \cos 2\pi\beta + i\vec{\sigma}\vec{n} \sin 2\pi\beta = \begin{cases} \begin{pmatrix} \cos 2\pi\beta & i\sin 2\pi\beta \\ i\sin 2\pi\beta & \cos 2\pi\beta \end{pmatrix}, & \vec{\sigma}\vec{n} = \sigma_1, \\ \begin{pmatrix} \cos 2\pi\beta & \sin 2\pi\beta \\ -\sin 2\pi\beta & \cos 2\pi\beta \end{pmatrix}, & \vec{\sigma}\vec{n} = \sigma_2, \\ \begin{pmatrix} e^{i2\pi\beta} & 0 \\ 0 & e^{-i2\pi\beta} \end{pmatrix}, & \vec{\sigma}\vec{n} = \sigma_3. \end{cases}$$

Here, the expression with $\vec{\sigma}\vec{n} = \sigma_2$ stays in the real, the expression with $\vec{\sigma}\vec{n} = \sigma_3$ is diagonal. In the real vector space $\mathbb{R}^2$, the orbits of $\mathbf{SO}(2) \ni O = e^{i\sigma_2 2\pi\beta}$ are circles with metric $a^2\mathbf{1}_2$. However, also ellipses with metric E are $\mathbf{SO}(2)$-orbits. The related groups arise by similarity transformations:

$$\begin{aligned} &E = \begin{pmatrix} a^2 & 0 \\ 0 & b^2 \end{pmatrix} = A \circ A^T, \quad A = \begin{pmatrix} |a| & 0 \\ 0 & |b| \end{pmatrix}, \quad O \circ \mathbf{1}_2 \circ O^T = O_A \circ E \circ O_A^T, \\ &\text{with } O_A = A \circ O \circ A^{-1} \in A \circ \mathbf{SO}(2) \circ A^{-1} \cong \mathbf{SO}(2), \\ &\text{e.g., } O = \begin{pmatrix} \cos 2\pi\beta & \sin 2\pi\beta \\ -\sin 2\pi\beta & \cos 2\pi\beta \end{pmatrix} \Rightarrow O_A = \begin{pmatrix} \cos 2\pi\beta & e^{\lambda} \sin 2\pi\beta \\ -e^{-\lambda} \sin 2\pi\beta & \cos 2\pi\beta \end{pmatrix}, \quad e^{2\lambda} = \frac{a^2}{b^2}. \end{aligned}$$

The representations of $\mathbf{U}(1)_3 \cong \mathbf{SO}(2)$ and, therefore, also the $\mathbf{SU}(2)$-representations contain, with each winding number of $\mathbf{U}(1)$, also the opposite one. The weight diagram of a totally symmetrized product of n Pauli-representations, each with winding numbers $\{-1, 1\}$, characterizes the irreducible $\mathbf{SU}(2)$-representation $D^{[n]}$. It uses all $n + 1$ $\mathbf{U}(1)_3$-winding numbers $z \in \{-n, -n + 2, \ldots, n - 2, n\}$, each exactly once. This is reflected by dimension $n + 1$ of the $D^{[n]}$-representation space.

The sum of all these winding numbers vanishes. For an (iso)spin interpretation of $\mathbf{SU}(2)$ with $j = \frac{n}{2}$, the winding numbers give twice the usually normalized third component of the (iso)spin $j_3 = \frac{n}{2}$. The even dimensional representations (half-integer spin) contain always two faithful $\mathbf{U}(1)_3$-representations with winding numbers $z = 2j_3 = \pm 1$, the odd dimensional ones (integer spin) have one trivial $\mathbf{U}(1)_3$-representation (no winding) and start nontrivially with a twofold winding $z = 2j_3 = \pm 2$. Therefore, they represent faithfully only the $\mathbf{SO}(3)$-isomorphic quotient group (equivalence classes) $\mathbf{SU}(2)/\mathbf{I}(2)$. The group duals with the representation characterizing natural numbers (invariants) are no longer groups, but cones:

$$\text{group dual of } \begin{cases} \mathbf{SU}(2) : \mathbb{N} = \{n = 2j = 0, 1, 2, \ldots\}, \\ \mathbf{SO}(3) : 2\mathbb{N} = \{n = 2j = 0, 2, 4, \ldots\}. \end{cases}$$

Good to know: Since for nonabelian groups, the dual does not constitute a group, it is called "group dual" — not "dual group" as possible in the abelian case.

Extending the addition of the winding numbers in the product of $\mathbf{U}(1)$-representations, $D_z \otimes D_{z'} \cong D_{z+z'}$, the product of irreducible $\mathbf{SU}(2)$-representations $D^{[n]} \otimes D^{[n']}$ with positive maximal winding number n, n' of the factors:

$$D^{[n]} \otimes D^{[n']} \cong D^{[n+n']} \oplus D^{[n+n'-2]} \oplus \cdots \oplus D^{[|n-n'|]},$$

decomposes into irreducible $\mathbf{SU}(2)$-representations, each with multiplicity one, starting with $D^{[n+n']}$ and maximal $\mathbf{U}(1)_3$-winding numbers $\pm(n+n')$.

5.6 Windings of the Binary Alternative

The winding numbers of $\mathbf{U}(1)$ are given by the integers z in the representations D_z, those of $\mathbf{SU}(2)$ by natural numbers $n = 2j$, which give the largest winding numbers $\pm n$ in the representation $D^{[n]}$. What happens if windings of both groups $\mathbf{U}(1)$ and $\mathbf{SU}(2)$ are combined with a correlation (synchronization) of both circle groups?

For a binary alternative $\{e^+, e^-\}$ as basis of a complex two-dimensional vector space, a winding in the unitary group $\mathbf{U}(2)$ can employ two abelian $\mathbf{U}(1)$-subgroups, the phase group $e^{2\pi i\alpha_0 \mathbf{1}_2} \in \mathbf{U}(\mathbf{1}_2)$, equally oriented for both components, and the subgroup $e^{2\pi i\alpha_3\sigma_3} \in \mathbf{U}(1)_3 \subset \mathbf{SU}(2)$ with opposite windings for both components. The double circle group can be visualized in the multiplicative form $\mathbf{U}(1) \times \mathbf{U}(1)$ as 2-torus or, in the additive form, as square $\mathbb{R}/\mathbb{Z} \times \mathbb{R}/\mathbb{Z}$ with periodic boundary conditions.

A winding of a circle $\mathbf{U}(1)_0 = \{e^{2\pi i\alpha} \mid \alpha \in \mathbb{R}\}$ on the subgroup $\mathbf{U}(\mathbf{1}_2) \circ \mathbf{U}(1)_3$, similar to the winding of a ribbon around a torus, is given by a group morphism:

$$\begin{aligned} R: \ \mathbf{U}(1)_0 &\longrightarrow \mathbf{U}(\mathbf{1}_2) \circ \mathbf{U}(1)_3 \subset \mathbf{U}(2) \\ e^{2\pi i\alpha} &\longmapsto R(\alpha) = e^{2\pi i\gamma}\begin{pmatrix} e^{2\pi i\beta} & 0 \\ 0 & e^{-2\pi i\beta} \end{pmatrix}, \ \begin{cases} R(\alpha+\alpha') = R(\alpha) \circ R(\alpha'), \\ R(0) = R(1) = \mathbf{1}_2. \end{cases} \end{aligned}$$

The parameters β, γ depend linearly on α (more below). Combining the purely vertical and horizontal windings with $\beta = 0$ and $\gamma = 0$, i.e., $\mathbf{U}(1)_0 \longrightarrow \mathbf{U}(\mathbf{1}_2)$ and $\mathbf{U}(1)_0 \longrightarrow \mathbf{U}(1)_3$, respectively, a winding may employ both directions. To find group morphisms, involving both subgroups, is in analogy to give correlated bicycles, i.e., bicycles with two synchronized wheels.

For ribbons nontrivial in both directions, three circle groups $\mathbf{U}(1)_0, \mathbf{U}(\mathbf{1}_2) \subset \mathbf{U}(2)$ and $\mathbf{U}(1)_3 \subset \mathbf{SU}(2) \subset \mathbf{U}(2)$ with corresponding parameters α, γ and β have to be taken into account. For a compatible winding of three circles, there are several integer number conditions to be fulfilled — the possibilities for ribbons are restricted. $m \in \mathbb{N}$ horizontal windings for $\mathbf{U}(\mathbf{1}_2)$ can go with $n \in \mathbb{N}$ vertical windings for $\mathbf{U}(1)_3 \subset \mathbf{SU}(2)$ — always in two orientations. A realization of $\mathbf{U}(1)_0$, nontrivial in both directions, requires a nontrivial rational number (Lissajous[5] number) as ratio of vertical to horizontal windings:

$$\frac{\gamma}{\beta} = \pm\frac{m}{n}: \ R(\alpha| \pm \frac{m}{n}) = e^{\pm 2\pi i \frac{m}{n}\beta}\begin{pmatrix} e^{2\pi i\beta} & 0 \\ 0 & e^{-2\pi i\beta} \end{pmatrix}.$$

[5]Jules Antoine Lissajous (1822–1880).

The multiplicative group of the nontrivial rational numbers $\mathbb{Q}^*$ is the direct product of three multiplicative monoids $\mathbf{I}(2)$, $\mathbb{N}$ and $\frac{1}{\mathbb{N}}$, i.e., each nontrivial rational number has a corresponding unique decomposition:

$$\begin{array}{c}\mathbb{Q}^* = \mathbf{I}(2) \times \mathbb{N} \times \frac{1}{\mathbb{N}},\\ q = \epsilon \times n \times \frac{1}{m},\\ \text{with } n, m \text{ coprime},\end{array} \left\{\begin{array}{rl}\mathbf{I}(2) = \{\pm 1\} & \text{(group)},\\ \mathbb{N} = \{1, 2, 3, \dots\} & \text{(monoid)},\\ \frac{1}{\mathbb{N}} = \{1, \frac{1}{2}, \frac{1}{3}, \dots\} & \text{(monoid)},\\ \mathbf{I}(2) \cap \mathbb{N} = \mathbf{I}(2) \cap \frac{1}{\mathbb{N}} = \mathbb{N} \cap \frac{1}{\mathbb{N}} = \{1\}, & \end{array}\right.$$

$$\text{e.g., } -\frac{3}{5} = (-1) \times 3 \times \frac{1}{5},\ \frac{3}{9} = (+1) \times (+1) \times \frac{1}{3}.$$

Now, "proper" ribbons will be constructed as $\mathbf{U}(1)_0$-realizations with $\frac{1}{n}$ for the monoid $\frac{1}{\mathbb{N}}$, i.e., for one horizontal winding for $\mathbf{U}(\mathbf{1}_2)$, $m = 1$, there are one, two, etc. vertical windings for the subgroup $\mathbf{U}(1)_3 \subset \mathbf{SU}(2)$. For the two basic vectors of the complex two-dimensional vector space, there arise weight dublets $(\underline{q}_-, \underline{q}_+)$ for the $\mathbf{U}(1)_0$-realization where the components have the difference 2β. The phase groups of the basic vectors $e^{\pm}$ is denoted by $\mathbf{U}(1)_\pm \subset \mathbf{U}(2)$:

$$\frac{\gamma}{\beta} = \pm\frac{1}{n} \Rightarrow \left\{\begin{array}{l}R\left(|\frac{1}{n}\right) : \mathbf{U}(1)_0 \longrightarrow \mathbf{U}(1)_+ \times \mathbf{U}(1)_-,\\ R\left(\alpha| \pm \frac{1}{n}\right) = \begin{pmatrix} e^{2\pi i \underline{q}_+ \beta} & 0\\ 0 & e^{2\pi i \underline{q}_- \beta}\end{pmatrix},\ \underline{q}_+ - \underline{q}_- = 2,\end{array}\right.$$

$$(\underline{q}_+, \underline{q}_-) = \left(1 \pm \frac{1}{n}, -1 \pm \frac{1}{n}\right) = \left\{\begin{array}{l}(2, 0), (0, -2),\ n = 1,\\ (\frac{3}{2}, -\frac{1}{2}), (\frac{1}{2}, -\frac{3}{2}),\ n = 2,\\ (\frac{4}{3}, -\frac{2}{3}), (\frac{2}{3}, -\frac{4}{3}),\ n = 3,\\ \dots.\end{array}\right.$$

The second winding numbers pair arises from the first one by negative transposition (dual or anti-realization).

If one of the $\mathbf{U}(1)_0$-winding numbers $(\underline{q}_+, \underline{q}_-)$ contains the defining $\mathbf{U}(1)_0$-winding number $|z| = 1$, the other has to be an integer multiple. This integer number condition, arising by the embedding $\mathbf{U}(1)_3 \subset \mathbf{SU}(2)$, allows only three possibilities to wind the circle $\mathbf{U}(1)_0$ nontrivially on the torus $\mathbf{U}(1)_+ \times \mathbf{U}(1)_- = \mathbf{U}(\mathbf{1}_2) \circ \mathbf{U}(1)_3 \subset \mathbf{U}(2)$:

$$\left.\begin{array}{l}n \in \mathbb{N} \text{ and}\\ \frac{\underline{q}_+}{\underline{q}_-} \text{ or } \frac{\underline{q}_-}{\underline{q}_+} \in \mathbb{Z}\end{array}\right\} \iff \left\{\begin{array}{l}n \in \mathbb{N} \text{ and}\\ \frac{2}{n-1} \text{ or } \frac{2}{n+1} \in \mathbb{Z}\end{array}\right. \iff n = 1, 2, 3$$

The realizations of a circle with $\frac{\gamma}{\beta} = \pm\frac{1}{n}$ in the windings of a binary alternative allow at most the 3×2 explicitly given possibilities for the winding numbers $(\underline{q}_+, \underline{q}_-)$.

One can normalize the parameter ratio of the circles $\mathbf{U}(1)_0$ (parameter α) and $\mathbf{U}(1)_3$ (parameter β). For the case $n = 1$, where one basic vector carries a trivial winding number, the $\mathbf{U}(1)$-property of the other vector is equalized with the represented circle $\mathbf{U}(1)_0$. Then, there are two $\mathbf{U}(1)_3$-windings for one $\mathbf{U}(1)_0$-winding:

$$\begin{aligned}
\tfrac{\alpha}{\beta} &= 2, \qquad (\underline{q}_+, \underline{q}_-) = 2(q_+, q_-) \Rightarrow R(\alpha| \pm \tfrac{1}{n}) = \begin{pmatrix} e^{2\pi i q_+ \alpha} & 0 \\ 0 & e^{2\pi i q_- \alpha} \end{pmatrix}, \\
\tfrac{\gamma}{\alpha} &= \pm \tfrac{1}{2n} = \pm y, \ (q_+, q_-) = (\tfrac{1}{2} \pm y, -\tfrac{1}{2} \pm y) \\
&= \begin{cases} (1,0), (0,-1), \ n = 1, \ y = \tfrac{1}{2}, \\ (\tfrac{3}{4}, -\tfrac{1}{4}), (\tfrac{1}{4}, -\tfrac{3}{4}), \ n = 2, \ y = \tfrac{1}{4}, \\ (\tfrac{2}{3}, -\tfrac{1}{3}), (\tfrac{1}{3}, -\tfrac{2}{3}), \ n = 3, \ y = \tfrac{1}{6}. \end{cases}
\end{aligned}$$

With this normalization, the cases $n = 2, 3$ have *broken winding numbers*. The winding numbers for $n = 1$ and $n = 3$ can be related to the charge numbers of leptons and quarks. Since broken winding numbers are not allowed for a $\mathbf{U}(1)$-representation, the representations with $n = 3$ have to be modified. This can be done by the introduction of color $\mathbf{SU}(3)$ as an additional operation group where the center $\mathbf{I}(3) \ni e^{2\pi i \frac{k}{3}}$ with $\frac{k}{3} \in \{\frac{1}{3}, \frac{2}{3}, 1\}$ reflects the broken winding winding numbers (more ahead).

Good to know: As I use it, a group representation as a mapping $D : G \longrightarrow \mathbf{GL}(V)$ has to be unique — for each $g \in G$ there has to exist exactly one $D(g) \in \mathbf{GL}(V)$. Multiple valued "representations" are considered below with another definition.

With an appropriate normalization, one can impose a unique $\mathbf{U}(1)_0$-representation for each case $n = 1, 2, 3$, e.g., for $n = 3$ with $(q_+, q_-) = (2, -1)$ and $(1, -2)$.

5.7 Charge Numbers of Leptons and Quarks

With the normalization above of the circle $\mathbf{U}(1)_0$, realized in the automorphisms $\mathbf{U}(2)$ of a binary alternative, the winding numbers (q_+, q_-) can be taken as electromagnetic charge numbers. The electromagnetic group $\mathbf{U}(1)_0$, wound on the torus, is thus embedded in the windings $\mathbf{U}(1)_+ \times \mathbf{U}(1)_- = \mathbf{U}(\mathbf{1}_2) \circ \mathbf{U}(1)_3$ of the electroweak symmetry $\mathbf{U}(2)$ (hypercharge, isospin) of the electroweak standard model. The binary alternative $(e^+, e^-) = (e^1, e^2)$ as basis of a complex two-dimensional vector space can be related to isospin-dublet fields (lepton, quark and Higgs fields). Then, half of the winding number ratio $y = \frac{1}{2n}$ is the hypercharge number.

For $n = 1$, the *lepton dublets* with antiparticles can be taken:

$$n = 1 \Rightarrow y = \frac{1}{2}, \ (q_+, q_-) = \begin{cases} (0, -1) \text{ for } (\boldsymbol{\nu}_e, \mathbf{e}^-), (\boldsymbol{\nu}_\mu, \boldsymbol{\mu}^-), (\boldsymbol{\nu}_\tau, \boldsymbol{\tau}^-), \\ (1, 0) \text{ for } (\mathbf{e}^+, \boldsymbol{\nu}_e^\star), (\boldsymbol{\mu}^+, \boldsymbol{\nu}_\mu^\star), (\boldsymbol{\tau}^+, \boldsymbol{\nu}_\tau^\star), \end{cases}$$

and also the Higgs fields. $n = 3$ is appropriate for the *quark dublets* with antiquarks:

$$n = 3 \Rightarrow y = \frac{1}{6}, \ (q_+, q_-) = \begin{cases} (\frac{2}{3}, -\frac{1}{3}) \text{ for } (\mathbf{u}, \mathbf{d}), \quad (\mathbf{c}, \mathbf{s}), \quad (\mathbf{t}, \mathbf{b}), \\ (\frac{1}{3}, -\frac{2}{3}) \text{ for } (\mathbf{d}^\star, \mathbf{u}^\star), (\mathbf{s}^\star, \mathbf{c}^\star), (\mathbf{b}^\star, \mathbf{t}^\star). \end{cases}$$

For the even number case $n = 2$, there is no association with standard model dublets.

Normalizing the electromagnetic circle $\mathbf{U}(1)_0$ for the leptons with $n = 1$, i.e., electron-positrons carry the elementary charge, the charge numbers for the hypercharges $\frac{1}{2n} = \frac{1}{4}$ and $\frac{1}{n} = \frac{1}{6}$ cannot be used without additional structures - there are no broken winding numbers for linear representations. There is the hope for the standard model, that the color $\mathbf{SU}(3)$-symmetry for the strong interaction has the confinement property, i.e., there exist only color-singlet *particles* — in contrast to color nontrivial fields for *interactions*. Experimentally so far, quarks have not be observed as free particles. Since color singlets arise as combinations of three quarks or three antiquarks or of quark with antiquark, and since all these combinations have integer $\mathbf{U}(1)_0$-winding numbers (charge numbers), a color confinement would also avoid broken charge numbers. One can even reverse this argument and argue that the integer charge number condition may be a reason for the introduction a "custodian" color group $\mathbf{SU}(3)$: The *third* integer charge and the *three* color degrees of freedom of the quarks fit together. More of that ahead in connection of the central correlation of the hypercharge $\mathbf{U}(1)$ with isospin-color $\mathbf{SU}(2) \times \mathbf{SU}(3)$.

5.8 Winding Number Matrices and Representation Weights

Opposite winding numbers for $\mathbf{U}(1)_3 \cong \mathbf{SO}(2) \subset \mathbf{SU}(2)$ in the *winding number matrix* $w_2 = \begin{pmatrix} 1 & 0 \\ 0 & -1 \end{pmatrix} = \sigma^3(2)$ for the simple symmetry $\mathbf{SU}(2)$ of a binary alternative can be found in all unitary groups $\mathbf{SU}(n)$. Depending on the place where the positive winding number is located, there are $n-1$ independent $\mathbf{SO}(2)$-winding number matrices $\{w_2(k) \mid k = 1, \ldots, n-1\}$, e.g., for $\mathbf{SU}(3)$:

$$w_2(1) = \begin{pmatrix} 1 & 0 & 0 \\ 0 & 0 & 0 \\ 0 & 0 & -1 \end{pmatrix},\quad w_2(2) = \begin{pmatrix} 0 & 0 & 0 \\ 0 & 1 & 0 \\ 0 & 0 & -1 \end{pmatrix},\quad e^1 = \begin{pmatrix} 1 \\ 0 \\ 0 \end{pmatrix},\ e^2 = \begin{pmatrix} 0 \\ 1 \\ 0 \end{pmatrix},\ e^3 = \begin{pmatrix} 0 \\ 0 \\ 1 \end{pmatrix}.$$

The sum of all these winding number matrices displays "interlaced" winding numbers for n basis vectors $\{e^1, \ldots, e^n\}$ in the following sense: For one distinguished vector e^n, the $\mathbf{U}(1)$-circle is wound $(n-1)$-times in the negative direction, whereas one "counteracting" positive winding for all the remaining $n-1$ vectors establishes a "winding-equilibrium." This is formalized by the traceless winding number matrix of the group $\mathbf{SU}(n)$:

$$n = 2, 3, \ldots: \quad w_n = \sum_{j=1}^{n-1} w_2(j) = \begin{pmatrix} 1 & 0 & 0 & \ldots & 0 & 0 \\ 0 & 1 & 0 & \ldots & 0 & 0 \\ & & \cdots & & & \cdots \\ 0 & 0 & \ldots & 0 & 1 & 0 \\ 0 & 0 & \ldots & 0 & 0 & -(n-1) \end{pmatrix}$$

$$= \begin{pmatrix} \mathbf{1}_{n-1} & 0 \\ 0 & -(n-1) \end{pmatrix} = \sqrt{\binom{n}{2}}\, \sigma^{n^2-1}(n),$$

$$\operatorname{tr} w_n = 0, \quad w_n = w_n^\star.$$

After multiplication with the imaginary unit i, the traceless hermitian winding number matrix is an element of the fundamental n-dimensional representation of the $\mathbf{SU}(n)$-Lie algebra A^c_{n-1}.

By successive embedding, "opening" one dimension after the other:

$$\mathbf{SU}(2) \subset \mathbf{SU}(3) \subset \mathbf{SU}(4) \subset \ldots,$$

one obtains, via the winding number matrices, the n weights g_k of the defining $\mathbf{SU}(n)$-representation. Each weight vector g_k has $n-1$ components:

$$k = 1, 2, \ldots, n: \quad g_k = (g_k^{a^2-1})_{a=2}^n = (g_k^3, g_k^8, g_k^{15} \ldots, g_k^{n^2-1}) \in \mathbb{E}^{n-1}.$$

Starting from $\mathbf{SU}(2)$ and normalized with equal Euclidean length, the explicit components of the weights are as follows:

$$\mathbf{SU}(2): \begin{cases} g_1 = 1, \\ g_2 = -1, \end{cases}$$

$$\mathbf{SU}(3): \begin{cases} g_1 = (1, g_1^8), \\ g_2 = (-1, g_1^8), \\ g_3 = (0, -2g_1^8), \end{cases} \qquad 4(g_1^8)^2 = 1 + (g_1^8)^2 \Rightarrow g_1^8 = \frac{1}{\sqrt{3}},$$

$$\mathbf{SU}(4): \begin{cases} g_1 = (1, \frac{1}{\sqrt{3}}, g_1^{15}), \\ g_2 = (-1, \frac{1}{\sqrt{3}}, g_1^{15}), \\ g_3 = (0, -\frac{2}{\sqrt{3}}, g_1^{15}), \\ g_4 = (0, 0, -3g_1^{15}), \end{cases} \qquad 9(g_1^{15})^2 = \frac{4}{3} + (g_1^{15})^2 \Rightarrow g_1^{15} = \frac{1}{\sqrt{6}},$$

etc.

The weight components can be collected in $(n-1)$ diagonal matrices $\sigma^{a^2-1}(n)$, which can be built successively by the $\mathbf{U}(1)$-winding matrices w_n:

$$\mathbf{SU}(2): \quad \sigma^3(2) = \begin{pmatrix} 1 & 0 \\ 0 & -1 \end{pmatrix} = w_2,$$

$$\mathbf{SU}(3): \begin{cases} \sigma^3(3) = \begin{pmatrix} 1 & 0 & 0 \\ 0 & -1 & 0 \\ 0 & 0 & 0 \end{pmatrix} = \begin{pmatrix} \sigma^3(2) & 0 \\ 0 & 0 \end{pmatrix}, \\[2ex] \sigma^8(3) = \frac{1}{\sqrt{3}} \begin{pmatrix} 1 & 0 & 0 \\ 0 & 1 & 0 \\ 0 & 0 & -2 \end{pmatrix} = \frac{1}{\sqrt{3}} \begin{pmatrix} \mathbf{1}_2 & 0 \\ 0 & -2 \end{pmatrix} = \frac{1}{\sqrt{3}} w_3, \end{cases}$$

$$\mathbf{SU}(4): \begin{cases} \sigma^3(4) = \begin{pmatrix} 1 & 0 & 0 & 0 \\ 0 & -1 & 0 & 0 \\ 0 & 0 & 0 & 0 \\ 0 & 0 & 0 & 0 \end{pmatrix} = \begin{pmatrix} \sigma^3(3) & 0 \\ 0 & 0 \end{pmatrix}, \\ \sigma^8(4) = \frac{1}{\sqrt{3}} \begin{pmatrix} 1 & 0 & 0 & 0 \\ 0 & 1 & 0 & 0 \\ 0 & 0 & -2 & 0 \\ 0 & 0 & 0 & 0 \end{pmatrix} = \begin{pmatrix} \sigma^8(3) & 0 \\ 0 & 0 \end{pmatrix}, \\ \sigma^{15}(4) = \frac{1}{\sqrt{6}} \begin{pmatrix} 1 & 0 & 0 & 0 \\ 0 & 1 & 0 & 0 \\ 0 & 0 & 1 & 0 \\ 0 & 0 & 0 & -3 \end{pmatrix} = \frac{1}{\sqrt{6}} \begin{pmatrix} \mathbf{1}_3 & 0 \\ 0 & -3 \end{pmatrix} = \frac{1}{\sqrt{6}} w_4, \end{cases}$$

and in general:

$$\mathbf{SU}(n): \begin{cases} \{\sigma^{a^2-1}(n) = \begin{pmatrix} \sigma^{a^2-1}(n-1) & 0 \\ 0 & 0 \end{pmatrix} \mid a = 2, \ldots, n-1\}, \\ \sigma^{n^2-1}(n) = \frac{1}{\sqrt{\binom{n}{2}}} \begin{pmatrix} \mathbf{1}_{n-1} & 0 \\ 0 & -(n-1) \end{pmatrix} = \frac{1}{\sqrt{\binom{n}{2}}} w_n. \end{cases}$$

The weights of the defining $\mathbf{SU}(n)$-representation occupy the n corners of the Euclidean fundamental simplex in the weight space $\mathbb{E}^{n-1}$ with edge length 2 and the center of mass in the origin. There is the distance for $\mathbf{SU}(2)$, the equilateral triangle for $\mathbf{SU}(3)$, the regular tetrahedron for $\mathbf{SU}(4)$, etc., in the matrices with n rows and $(n-1)$ columns:

$\mathbb{E}^{n-1}$-coordinates of a Euclidean fundamental simplex for $\mathbf{SU}(n)$, $n \geq 2$

$$\text{simpl}\,(n) = \begin{pmatrix} g_1 \\ g_2 \\ g_3 \\ g_4 \\ \cdots \\ \cdots \\ g_n \end{pmatrix} = \begin{pmatrix} 1, & \frac{1}{\sqrt{3}}, & \frac{1}{\sqrt{6}}, & \cdots & \frac{1}{\sqrt{\binom{n}{2}}} \\ -1, & \frac{1}{\sqrt{3}}, & \frac{1}{\sqrt{6}}, & \cdots & \frac{1}{\sqrt{\binom{n}{2}}} \\ 0, & -\frac{2}{\sqrt{3}}, & \frac{1}{\sqrt{6}}, & \cdots & \frac{1}{\sqrt{\binom{n}{2}}} \\ 0, & 0, & -\frac{3}{\sqrt{6}}, & \cdots & \frac{1}{\sqrt{\binom{n}{2}}} \\ & & \cdots & \cdots & \cdots \\ & & \cdots & \cdots & \cdots \\ 0, & 0, & 0, & \cdots & -\frac{n-1}{\sqrt{\binom{n}{2}}} \end{pmatrix},$$

$$|g_k - g_j| = 2(1 - \delta_{kj}), \quad |g_k| = \sqrt{\frac{2(n-1)}{n}}.$$

The exponent of the $\mathbf{SU}(n)$-winding number matrix w_n generates the cyclotomic group $\mathbf{I}(n)$, faithfully represented in the group $\mathbf{SU}(n)$:

$$\mathbf{I}(n) \longrightarrow \mathbf{SU}(n), \quad i_n \longmapsto \mathbf{i}_n = e^{\frac{2\pi i}{n} w_n} = \begin{pmatrix} e^{\frac{2\pi i}{n}} \mathbf{1}_{n-1} & 0 \\ 0 & e^{-\frac{2\pi i(n-1)}{n}} \end{pmatrix} = i_n \mathbf{1}_n.$$

The powers $\mathbf{i}_n^k$ of $\mathbf{i}_n$ realize the *"broken winding numbers"* $\frac{k}{n}$ with $k = 0, \ldots, n-1$. The isomorphic image of $\mathbf{I}(n)$ in $\mathbf{SU}(n)$ is the discrete center of the unitary group, centr $\mathbf{SU}(n) \cong \mathbf{I}(n)$. The traceless matrix w_n or an equivalent matrix $u \circ w_n \circ u^{-1}$ with $u \in \mathbf{SU}(n)$ allows to embed the broken winding number of the discrete group $\mathbf{I}(n)$ in the Lie algebra of $\mathbf{SU}(n)$, i.e., into the continuous $\mathbf{SU}(n)$-transformations:

$$\log \mathbf{i}_n = \frac{2\pi i w_n}{n} \in A_{n-1}^c.$$

Good to know: The *center of a group* consists of those elements which commute with all group elements centr $G = \{c \in G \mid cg = gc \text{ for all } g \in G\}$. It is an abelian normal subgroup.

5.9 Fractionality of Unitary Group Representations

The endomorphisms of the cyclic group $\mathbf{I}(n)$ are characterized by an integer $z \in \mathbb{Z}$ — in analogy to the circle winding number. The generating element $i_n = e^{\frac{2\pi i}{n}} \in \mathbf{I}(n)$ is mapped to its z-th power:

$$R_z : \mathbf{I}(n) \longrightarrow \mathbf{I}(n), \quad i_n \longmapsto i_n^z, \quad z \in \mathbb{Z}.$$

The *fractionality* of a nontrivial rational number $q \in \mathbb{Q}^*$ is defined as the smallest natural number $F(q)$ leading to an integer $q \cdot F(q) \in \mathbb{Z}$ i.e., its $\frac{1}{\mathbb{N}}$-factor in the coprime factorization $\mathbb{Q}^* = \mathbf{I}(2) \times \mathbb{N} \times \frac{1}{\mathbb{N}}$ with three monoids $\{\mathbf{I}(2), \frac{1}{\mathbb{N}}, \mathbb{N}\}$. It can be expressed with the greatest common divisor:

$$\begin{aligned}
\mathbb{Q}^* \longrightarrow \mathbb{N}, \quad q \longmapsto F(q), \quad & F(q) = \min_{N \in \mathbb{N}}\{Nq \in \mathbb{Z}\}, \\
n, k \in \mathbb{N} \Rightarrow F\left(\pm\frac{k}{n}\right) &= \frac{n}{\mathrm{GCD}(n, k)}, \\
\text{e.g., } F(z) = 1 \text{ for } |z| \in \mathbb{N}, \quad & F(\pm\frac{1}{3}, \pm\frac{2}{3}, \pm\frac{4}{6}) = 3.
\end{aligned}$$

Then, the kernel and the image of the $\mathbf{I}(n)$-endomorphism R_z is given as follows:

$$\text{kern } R_z = \mathbf{I}(\mathrm{GCD}(n, |z|), \quad R_z(\mathbf{I}(n)) \cong \mathbf{I}(n)/\mathbf{I}(\mathrm{GCD}(n, |z|)) \cong \mathbf{I}(F(\frac{z}{n})).$$

$F(\frac{z}{n})$ is a divisor of n. Only for coprime $|z|$ and n, i.e., $\mathrm{GCD}(n, |z|) = 1$, the cyclic group is represented faithfully. The different divisors $|k|$ of n characterize the different endomorphisms of $\mathbf{I}(n)$ and, therefore, all subgroups, e.g., for $\mathbf{I}(6)$:

$$R_z : \mathbf{I}(6) \ni e^{\pi i \frac{1}{3}} \longmapsto e^{\pi i \frac{z}{3}} \in \mathbf{I}(6), \quad \text{kern } R_z \cong \begin{cases} \mathbf{I}(1) \text{ for } z = \pm 1 \bmod 6, \\ \mathbf{I}(2) \text{ for } z = \pm 2 \bmod 6, \\ \mathbf{I}(3) \text{ for } z = \pm 3 \bmod 6, \\ \mathbf{I}(6) \text{ for } z = 0 \bmod 6. \end{cases}$$

The unitary automorphism group $\mathbf{U}(n)$ of a complex n-dimensional vector space $\mathbb{C}^n$ contains the phase group $\mathbf{U}(\mathbf{1}_n)$ as center and the special unitary automorphisms $\mathbf{SU}(n)$. Both are normal subgroups:

$$\mathbf{U}(n) = \{u \in \mathbf{GL}(n, \mathbb{C}) \mid u^\star = u^{-1}\}, \quad \begin{cases} \mathbf{U}(\mathbf{1}_n) = \{e^{i\alpha}\mathbf{1}_n \mid \alpha \in \mathbb{R}\} \cong \mathbf{U}(1), \\ \mathbf{SU}(n) = \{u_0 \in \mathbf{U}(n) \mid \det u_0 = 1\}. \end{cases}$$

For $n \geq 2$, $\mathbf{U}(n)$ is not the direct product $\mathbf{U}(\mathbf{1}_n) \times \mathbf{SU}(n)$ of both normal subgroups: A corresponding decomposition is not unique, it allows the freedom of the cyclic group $\mathbf{I}(n)$. The intersection group $\mathbf{I}(n)\mathbf{1}_n$ is the center of $\mathbf{SU}(n)$:

$$\left.\begin{aligned} \mathbf{U}(n) = \mathbf{U}(\mathbf{1}_n) \circ \mathbf{SU}(n), \\ \mathbf{U}(\mathbf{1}_n) \cap \mathbf{SU}(n) \cong \mathbf{I}(n), \end{aligned}\right\} \Rightarrow \mathbf{U}(n)/\mathbf{SU}(n) \cong \mathbf{U}(\mathbf{1}_n)/\mathbf{I}(n) \cong \mathbf{U}(1),$$

$$u = e^{i\alpha(k)} u_0(k), \quad \begin{cases} e^{i\alpha(k)} = i_n^k |\det u|^{\frac{1}{n}}, \\ u_0(k) = e^{-i\alpha(k)} u, \end{cases} \quad k = 0, \ldots, n-1.$$

Good to know: With U a subgroup and N a normal subgroup of a group G, there holds the group isomorphism $UN/N \cong U/N \cap U$. Here with $G = UN = \mathbf{U}(n)$ for $(U, N) = (\mathbf{U}(\mathbf{1}_n), \mathbf{SU}(n))$ or $(U, N) = (\mathbf{SU}(n), \mathbf{U}(\mathbf{1}_n))$, i.e., $\mathbf{U}(n)/\mathbf{U}(\mathbf{1}_n) \cong \mathbf{SU}(n)/\mathbf{I}(n)$ and $\mathbf{U}(n)/\mathbf{SU}(n) \cong \mathbf{U}(\mathbf{1}_n)/\mathbf{I}(n) \cong \mathbf{U}(1)$.

Since the intersection $\{\mathbf{i}_n^k \mid k = 0, \ldots, n-1\} \cong \mathbf{I}(n)$ is discrete, $\mathbf{U}(n)$ has the same Lie algebra $i\mathbb{R} \oplus A_{n-1}^c$ as the direct product group $\mathbf{U}(\mathbf{1}_n) \times \mathbf{SU}(n)$.

The irreducible representations of the $\mathbf{U}(n)$-center $\mathbf{U}(\mathbf{1}_n)$ are characterized by integer winding numbers $z \in \mathbb{Z}$. The irreducible representations of the real (n^2-1)-dimensional special unitary Lie group $\mathbf{SU}(n)$, $n \geq 2$, and its Lie algebra A_{n-1}^c are characterized by $n-1$ non-negative integers $(m_j)_{j=1}^{n-1}$, given by the highest weight:

$$\begin{aligned} \text{irreducible } \mathbf{SU}(n) \text{ -representations:} & [m_1, m_2, \ldots, m_{n-1}], \\ \text{group dual of } \mathbf{SU}(n) \text{ :} & \mathbb{N}^{n-1}. \end{aligned}$$

The group dual is a monoid, not a group — in contrast to the group duals $(\mathbb{Z}, i\mathbb{R})$ of $(\mathbf{U}(1), \mathbb{R})$.

The defining representation is the group $\mathbf{SU}(n)$ itself, denoted with 1 at the first index-place:

$$D^{[1,0,\ldots,0]}(\mathbf{SU}(n)) = \mathbf{SU}(n) \text{ acting on } \mathbb{C}^n.$$

The cyclotomic group $\mathbf{I}(n)$ is realized by the $\mathbf{SU}(n)$-center, with the $\mathbf{SU}(n)$-winding number matrix w_n, $i_n \longmapsto \mathbf{i}_n = e^{2\pi i \frac{w_n}{n}} = i_n \mathbf{1}_n$.

The k-fold *totally antisymmetrized tensor products* of the defining representations give all the other fundamental representations of $\mathbf{SU}(n)$ — for the tensor power k the 1 is at the k-th index-place:

$$D^{[0,\ldots,0,1,0,\ldots,0]} \cong \bigwedge^{k} D^{[1,0,\ldots,0]}, \quad k = 1, \ldots, n-1,$$

$$\text{dimension of the representation: } d_k = \binom{n}{k}.$$

In this product, the generating element of $\mathbf{I}(n)$ is represented by the corresponding power:

$$i_n \longmapsto i_n^k \mathbf{1}_{d_k}.$$

Therefore, one has a representation of the cyclic group for the *fractionality* $F(\frac{k}{n})$ *of the* $\mathbf{SU}(n)$*-representation*:

$$D^{[0,\ldots,0,1,0,\ldots,0]}(\mathbf{SU}(n)) : \begin{cases} D^{[0,\ldots,0,1,0,\ldots,0]}(\mathbf{I}(n)) \cong \mathbf{I}(F(\frac{k}{n})), \\ \text{kern } D^{[0,\ldots,0,1,0,\ldots,0]} \cong \mathbf{I}(\text{GCD}(n,k)), \\ D^{[0,\ldots,0,1,0,\ldots,0]}(\mathbf{SU}(n)) \cong \mathbf{SU}(n)/\mathbf{I}(\text{GCD}(n,k)), \end{cases}$$

with the examples:

$$\begin{aligned} D^{[1]}(\mathbf{SU}(2)) &\cong \mathbf{SU}(2), \\ D^{[1,0]}(\mathbf{SU}(3)), D^{[0,1]}(\mathbf{SU}(3)) &\cong \mathbf{SU}(3), \\ D^{[1,0,0]}(\mathbf{SU}(4)), D^{[0,0,1]}(\mathbf{SU}(4)) &\cong \mathbf{SU}(4), \\ D^{[0,1,0]}(\mathbf{SU}(4)) &\cong \mathbf{SU}(4)/\mathbf{I}(2), \\ &\text{etc.} \end{aligned}$$

The m-fold *totally symmetric tensor product* of the k-th fundamental representation contains another distinguished irreducible $\mathbf{SU}(n)$-representation — the m is at the k-th index-place

$$D^{[0,\ldots,0,m,0,\ldots,0]} \subseteq \bigvee^{m} D^{[0,\ldots,0,1,0,\ldots,0]}, \quad m \in \mathbb{N}.$$

The generating $\mathbf{I}(n)$-element i_n is represented, up to the identity in the representation space, by $i_n^{m\cdot k}$ with corresponding consequences for the images of $\mathbf{I}(n)$ and $\mathbf{SU}(n)$.

Each of the $n-1$ fundamental representations $D^{[0,\ldots,0,1,0,\ldots,0]}$ of $\mathbf{SU}(n)$ is characterized by a fundamental weight γ_k, a vector in the Euclidean space $\mathbb{E}^{n-1}$, $D^{[0,\ldots,0,m,0,\ldots,0]}$ is characterized by the m-fold vector $m\gamma_k$ as highest weight. In the irreducible $\mathbf{SU}(n)$-representation $D^{[m_1,m_2,\ldots,m_{n-1}]}$ the highest weight is the vector sum $\sum_{k=1}^{n-1} m_k\gamma_k$ of the m_k-fold fundamental weights. The dimension $d(m_1, m_2, \ldots, m_{n-1})$ of the representation space is rather complicated, given recursively:

$$\begin{aligned}
\mathbf{SU}(2): &\quad d(m_1) = m_1+1,\\
\mathbf{SU}(3): &\quad d(m_1,m_2) = \tfrac{(m_1+1)(m_2+1)(m_2+m_1+2)}{2},\\
\mathbf{SU}(4): &\quad d(m_1,m_2,m_3) = \tfrac{(m_1+1)(m_2+1)(m_2+m_1+2)(m_3+1)(m_3+m_2+2)(m_3+m_2+m_1+3)}{2!\,3!},\\
&\quad \ldots\\
\mathbf{SU}(n+1): &\quad d(m_1,\ldots,m_n) = \tfrac{d(m_1,\ldots,m_{n-1})}{2!3!\cdots n!}\,(m_n+1)(m_n+m_{n-1}+2)\cdots(m_n+m_{n-1}+\cdots+m_1+n).
\end{aligned}$$

The generating $\mathbf{I}(n)$-element is, with the identity $\mathbf{1}_d$ in the representation space, given as follows:

$$i_n \longmapsto i_n^{m_1+2m_2+\cdots+(n-1)m_{n-1}}\mathbf{1}_d.$$

For nontrivial irreducible representations the fractionality is given as follows:

$$D^{[m_1,m_2,\ldots,m_{n-1}]}: \begin{cases}
\text{fractionality: } F = F(\frac{m_1+2m_2+\cdots+(n-1)m_{n-1}}{n}),\\
\text{divisor: } N = \mathrm{GCD}(n, m_1+2m_2+\ldots\\
\qquad\qquad +(n-1)m_{n-1}),\\
D^{[m_1,m_2,\ldots,m_{n-1}]}(\mathbf{I}(n)) \cong \mathbf{I}(F),\\
\text{kern } D^{[m_1,m_2,\ldots,m_{n-1}]} \cong \mathbf{I}(N),\\
D^{[m_1,m_2,\ldots,m_{n-1}]}(\mathbf{SU}(n)) \cong \mathbf{SU}(n)/\mathbf{I}(N),
\end{cases}$$

with the examples for nontrivial representations:

$$D^{[m]}(\mathbf{SU}(2)) \cong \begin{cases} \mathbf{SU}(2), & m \in 2\mathbb{Z}+1,\\ \mathbf{SU}(2)/\mathbf{I}(2), & m \in 2\mathbb{Z}, \end{cases}$$

$$D^{[m_1,m_2]}(\mathbf{SU}(3)) \cong \begin{cases} \mathbf{SU}(3), & m_1-m_2 \in 3\mathbb{Z}\pm 1,\\ \mathbf{SU}(3)/\mathbf{I}(3), & m_1-m_2 \in 3\mathbb{Z}, \end{cases}$$

$$D^{[m_1,m_2,m_3]}(\mathbf{SU}(4)) \cong \begin{cases} \mathbf{SU}(4), & m_1+2m_2-m_3 \in 4\mathbb{Z}\pm 1,\\ \mathbf{SU}(4)/\mathbf{I}(2), & m_1+2m_2-m_3 \in 4\mathbb{Z}+2,\\ \mathbf{SU}(4)/\mathbf{I}(4), & m_1+2m_2-m_3 \in 4\mathbb{Z}, \end{cases}$$

etc.

The fractionalities for $\mathbf{SU}(2)$ and $\mathbf{SU}(3)$ are called two-ality and tri-ality, respectively.

The $n-1$ fundamental representations of the Lie algebra A_{n-1} are also the fundamental representations of the simply connecting covering group $\mathbf{SU}(n)$. As representation images of the group, there arise all quotient groups $\mathbf{SU}(n)/\mathbf{I}(k)$ with the n-divisors $k \in \mathbb{N}$ and subgroups $\mathbf{I}(k) \subseteq \mathbf{I}(n)$ of the $\mathbf{SU}(n)$-center. The weight lattices of the irreducible representations of the quotient groups $\mathbf{SU}(n)/\mathbf{I}(k)$ are sublattices of the weight lattice of $\mathbf{SU}(n)$. They are defined by the given relations for the components of $(m_1,\ldots,m_{n-1})$. For nontrivial discrete normal subgroup $\mathbf{I}(k)$, the proper sublattice is spanned by $n-1$ different representations, which may be called fundamental with respect to the quotient group $\mathbf{SU}(n)/\mathbf{I}(k)$.

5.10 Adjoint Winding Numbers and Adjutopes

The adjoint representation of the Lie algebra $A_r^c = \log \mathbf{SU}(1+r)$ gives the group representation $D^{[2]}$ in the case of $\mathbf{SU}(2)$ and $D^{[1,0,\ldots,0,1]}$ for $\mathbf{SU}(1+r)$, $r \geq 2$. Because of the fractionality $F(1) = 1$ the $\mathbf{SU}(1+r)$-center is trivially represented, i.e., the adjoint image of $\mathbf{SU}(1+r)$ is $\mathbf{SU}(1+r)/\mathbf{I}(1+r)$.

The Lie algebra A_r has rank r and dimension $r(r+2)$ and, therefore, r trivial weights and $\binom{1+r}{2}$ root pairs $\pm\omega_j$, with their lengths normalized to 2 (winding number). The weights of the adjoint representation of A_1 and A_2 are as follows:

A_1 and $\mathbf{SL}(2,\mathbb{C})$, endpoints ("corners") of the distance and one central point:

$$\mathrm{adj}(2) = \begin{pmatrix} \omega \\ 0 \\ -\omega \end{pmatrix} = \begin{pmatrix} 2 \\ 0 \\ -2 \end{pmatrix} \subset \mathbb{E}^1,$$

A_2 and $\mathbf{SL}(3,\mathbb{C})$, corners of the hexagon and two central points:

$$\mathrm{adj}(3) = \begin{pmatrix} \omega_3 \\ \omega_2 \\ \omega_1 \\ 0 \\ 0 \\ -\omega_1 \\ -\omega_2 \\ -\omega_3 \end{pmatrix} = \begin{pmatrix} 1 & \sqrt{3} \\ -1 & \sqrt{3} \\ 2 & 0 \\ 0 & 0 \\ 0 & 0 \\ -2 & 0 \\ 1 & -\sqrt{3} \\ -1 & -\sqrt{3} \end{pmatrix} \subset \mathbb{E}^2$$

The octet of the pseudoscalar bosons $(K, \pi, \eta, \overline{K})$ is an example for the adjoint representation of the flavor group $\mathbf{SU}(3)$. Here, the units for strong isospin and hypercharge have to be chosen according to the physical convention, e.g., for the kaons $(\pm 1, \pm\sqrt{3}) \longmapsto (\pm\frac{1}{2}, \pm 1)$. The gluon octet for the color group $\mathbf{SU}(3)$ is another example.

The general principle for the adjoint weights can be easily recognized in the example with $r = 2$ for embedding A_1 into A_2: The weights $\mathrm{adj}(1+r)$ of the adjoint representation of A_r are on three hyperplanes in Euclidean $\mathbb{E}^r$, symmetric with respect to central reflection. The hyperplane $\mathbb{E}^{r-1}$ through the origin contains the weights $\mathrm{adj}(r)$ of A_{r-1} (dimension $(r-1)(r+1)$) and one additional trivial weight. The two "sandwiching" parallel hyperplanes $\mathbb{E}^{r-1}$ contain the weights $\mathrm{simpl}(r)$ of the fundamental simplex for the defining A_{r-1}-representation (dimension r) and its dual. The lengths of the roots, i.e., of the $r(r+1)$ nontrivial weights, are normalized to 2:

$$A_r \text{ and } \mathbf{SL}(1+r,\mathbb{C}): \ \mathrm{adj}(1+r) = \begin{pmatrix} \omega_{\binom{1+r}{2}} \\ \cdots \\ \omega_2 \\ \omega_1 \\ 0(r\,\text{zeros}) \\ -\omega_1 \\ -\omega_2 \\ \cdots \\ -\omega_{\binom{1+r}{2}} \end{pmatrix} = \left(\begin{array}{cc} \mathrm{simpl}(r) & \sqrt{\frac{2(1+r)}{r}} \\ \hline \mathrm{adj}(r) & 0 \\ 0 & 0 \\ \hline -\mathrm{simpl}(r) & -\sqrt{\frac{2(1+r)}{r}} \end{array}\right) \subset \mathbb{E}^r$$

The rank r gives the number of columns, the dimension $r(r+2)$ the number of rows. The decomposition with respect to representations of A_{r-1} contains a trivial representation scal:

$$\mathbf{SL}(1+r,\mathbb{C}) \cong \bigoplus \mathbf{SL}(r,\mathbb{C}),$$

$$\begin{aligned} \mathrm{adj}(1+r) &\cong \mathrm{simpl}(r) \oplus [\mathrm{adj}(r) \oplus \mathrm{scal}] \oplus \mathrm{simpl}(r)^{\star}, \\ \text{dimensions: } r(r+2) &= r + [(r-1)(r+1)+1] + r. \end{aligned}$$

The Archimedean hexoctahedron as a nice example in three-dimensional space for the adjoint representation for $\mathbf{SL}(4,\mathbb{C})$ and $\mathbf{SU}(4)$, with, in three parallel planes, a regular triangle, a regular hexagon with two central points, and a regular "antitriangle" for the triplet (quarks), the octet (gluons), and the antitriplet (antiquarks) $\mathbf{SU}(3)$-representations, respectively:

A_3 and $\mathbf{SL}(4,\mathbb{C})$, corners of the hexoctahedron and three central points:

$$\mathrm{adj}(4) = \begin{pmatrix} \omega_6 \\ \omega_5 \\ \omega_4 \\ \hline \omega_3 \\ \omega_2 \\ \omega_1 \\ 0 \\ 0 \\ 0 \\ -\omega_1 \\ -\omega_2 \\ -\omega_3 \\ \hline -\omega_4 \\ -\omega_5 \\ -\omega_6 \end{pmatrix} = \left(\begin{array}{ccc} 1 & \frac{1}{\sqrt{3}} & \sqrt{\frac{8}{3}} \\ -1 & \frac{1}{\sqrt{3}} & \sqrt{\frac{8}{3}} \\ 0 & -\frac{2}{\sqrt{3}} & \sqrt{\frac{8}{3}} \\ \hline 1 & \sqrt{3} & 0 \\ -1 & \sqrt{3} & 0 \\ 2 & 0 & 0 \\ 0 & 0 & 0 \\ 0 & 0 & 0 \\ 0 & 0 & 0 \\ -2 & 0 & 0 \\ 1 & -\sqrt{3} & 0 \\ -1 & -\sqrt{3} & 0 \\ \hline 0 & \frac{2}{\sqrt{3}} & -\sqrt{\frac{8}{3}} \\ 1 & -\frac{1}{\sqrt{3}} & -\sqrt{\frac{8}{3}} \\ -1 & -\frac{1}{\sqrt{3}} & -\sqrt{\frac{8}{3}} \end{array}\right) \subset \mathbb{E}^3$$

The general geometrical procedure: One has a series of nested, dimensionality-increasing, convex polytopes with their corners (roots) on spheres with radius 2: For A_1 a distance with length 4 and one central point, for A_2 a regular hexagon in a circle with radius 2 and two central points, for A_3 the 12 corners of the Archimedean hexoctahedron on a 2-sphere with radius 2 and 3 central points, in general for A_r an *adjutope* (for adjoint polytope) with $r(1+r)$ corners on an $(r-1)$-sphere with radius 2 and r central points in Euclidean space $\mathbb{E}^r$.

5.11 Additive Unification of Unitary Symmetries

In the electroweak standard model of the elementary particles with the representation of the internal unitary symmetries $\mathbf{U}(1)$, $\mathbf{SU}(2)$ and $\mathbf{SU}(3)$, the hypercharge numbers are winding numbers for the group $\mathbf{U}(1)$. Furthermore, there are winding numbers for the isospin and color group **SU**(2) and **SU**(3) with the centers **I**(2) and **I**(3), respectively, which are "synchronized" with the **U**(1)-winding numbers.

The quark and lepton fields with their antifields display the following hypercharges y, isospin- and color-multiplicities d_I and d_C, respectively:

Field	Hypercharge y	d_I, d_C	yd_Id_C	Handedness	**SU**(5)-multiplet
Right lepton **e**	-1	1, 1	-1	r	10
Right anti-lepton **e***	$+1$	1, 1	$+1$	l	$\overline{10}$
Left quark **q**	$+\frac{1}{6}$	2, 3	$+1$	l	$\overline{10}$
Left anti-quark **q***	$-\frac{1}{6}$	2, 3	-1	r	10
Right up quark **u**	$+\frac{2}{3}$	1, 3	$+2$	r	10
Right anti-up quark **u***	$-\frac{2}{3}$	1, 3	-2	l	$\overline{10}$
Right down quark **d**	$-\frac{1}{3}$	1, 3	-1	r	$\overline{5}$
Right anti-down quark **d***	$+\frac{1}{3}$	1, 3	$+1$	l	5
Left lepton **l**	$-\frac{1}{2}$	2, 1	-1	l	5
Left anti-lepton **l***	$+\frac{1}{2}$	2, 1	$+1$	r	$\overline{5}$

Hypercharges of leptons and quarks

Antifields have conjugated representations of the Lorentz symmetry — here: left antifields are right-handed (chirality r) and right antifields are left-handed (chirality l).

At first and, also, at second view attractive unifications embed the 3-factor group of the internal symmetry in a larger symmetry group G with the direct isospin-color product group as direct sum by block matrices:

$$\mathbf{SU}(2) \times \mathbf{SU}(3) \longrightarrow \begin{pmatrix} \mathbf{SU}(2) & 0 \\ 0 & \mathbf{SU}(3) \end{pmatrix} \subset G.$$

If **U**(1)-hypercharge is represented in the special linear group of the sum vector space, the represented **U**(1)-Lie algebra $i\mathbb{R}$ has a traceless matrix Y as basis. The collection of quark and lepton fields in representations of a larger special group requires a vanishing sum of the hypercharges of all fields — taken with equal handedness, e.g., left-handed $\sum yd_Id_C = 0$. Here, the internal multiplicity d_Id_C of isospin and color has to be taken into account, e.g., for left quarks $\frac{1}{6} \cdot 2 \cdot 3 = 1$, for right antileptons $1 \cdot 1 \cdot 1 = 1$. Looking at the table above, not only the hypercharge sum of all left-handed fields vanishes, there are subsets with vanishing sum, e.g., left-handed leptons and right-handed anti-downquarks.

As compact simple group with rank $1 + 1 + 2$ for **U**(1), **SU**(2) and **SU**(3) as subgroups, there is only **SU**(5). This minimal case of an additive unification embeds the fundamental representations $D^{[1]}$ of **SU**(2) (two-dimensional $d_I = 2$) and $D^{[0,1]}$ of **SU**(3) (three-dimensional $d_C = 3$) in an action of the special unitary group **SU**(5) on a complex five-dimensional vector space.

SU(5) has four fundamental representations, denoted by their dimensions $\binom{5}{k}$, $k = 1, 2, 3, 4$:

$$D^{[1,0,0,0]} \cong 5,\ D^{[0,1,0,0]} \cong 10,\ D^{[0,0,1,0]} \cong \overline{10}, \quad D^{[0,0,0,1]} \cong \overline{5},$$
$$5 \wedge 5 \cong 10, \quad 5 \wedge 5 \wedge 5 \cong \overline{10},\ 5 \wedge 5 \wedge 5 \wedge 5 \cong \overline{5}.$$

Because of the prime number 5, the **SU**(5)-center **I**(5) is represented like the full group — either faithfully or trivially. The four fundamental **SU**(5)-representations have fractionality 5, they are all faithful.

In the defining representation 5, hypercharge **U**(1) is represented as follows:

$$\mathbf{U}(1) \longrightarrow \mathbf{SU}(5),\ e^{2\pi i\alpha} \longmapsto e^{2\pi i Y\alpha},$$
$$Y = \begin{pmatrix} y_1\mathbf{1}_2 & 0 \\ 0 & y_2\mathbf{1}_3 \end{pmatrix} \text{ with } 2y_1 + 3y_2 = 0.$$

With the conventional normalization $y_1 = -\frac{1}{2}$, there follows $y_2 = \frac{1}{3}$ and one obtains the following **U**(1) and **SU**(2) × **SU**(3)-decompositions of the four **SU**(5)-representations, denoted by the hypercharge number and the multiplicities (y, d_I, d_C):

$$\begin{aligned}
\mathbf{SU}(5) &\cong \bigoplus \mathbf{U}(1) \times \mathbf{SU}(2) \times \mathbf{SU}(3),\\
5 &\cong \left(-\tfrac{1}{2}, 2, 1\right) \oplus \left(\tfrac{1}{3}, 1, \overline{3}\right),\\
10 &\cong (-1, 1, 1) \oplus \left(\tfrac{2}{3}, 1, 3\right) \oplus \left(-\tfrac{1}{6}, 2, \overline{3}\right),\\
\overline{10} &\cong (1, 1, 1) \oplus \left(-\tfrac{2}{3}, 1, \overline{3}\right) \oplus \left(\tfrac{1}{6}, 2, 3\right),\\
\overline{5} &\cong \left(\tfrac{1}{2}, 2, 1\right) \oplus \left(-\tfrac{1}{3}, 1, 3\right).
\end{aligned}$$

The 15 left-handed lepton and quark fields fit precisely into the two fundamental **SU**(5)-representations 5 and $\overline{10}$ — that is not selfevident, and, then, the right-handed antifields into $\overline{5}$ and 10:

$$\begin{aligned}
5 &\Rightarrow \mathcal{L} \cong \mathbf{l} \oplus \mathbf{d}^\star,\\
\overline{10} &\Rightarrow \mathcal{E}^\star \cong \mathbf{e}^\star \oplus \mathbf{u}^\star \oplus \mathbf{q},\\
\overline{5} &\Rightarrow \mathcal{L}^\star \cong \mathbf{l}^\star \oplus \mathbf{d},\\
10 &\Rightarrow \mathcal{E} \cong \mathbf{e} \oplus \mathbf{u} \oplus \mathbf{q}^\star.
\end{aligned}$$

So far, only quarks and leptons are accommodated in representations of a larger symmetry. Physically interesting are the dynamical consequences of the operations of the larger symmetry, i.e., the additional gauge interactions for the standard model which come in the adjoint 24-dimensional **SU**(5)-representation $D^{[1,0,0,1]}$ with the following **U**(1) and **SU**(2) × **SU**(3)-decomposition:

$$\begin{aligned}
24 &\cong (0,1,1) \oplus (0,3,1) \oplus (0,1,8) \oplus (0,2,3) \oplus (0,2,\overline{3}),\\
\mathcal{A} &\cong \mathbf{A}_0 \quad \oplus \mathbf{A} \quad\ \ \oplus \mathbf{G} \quad\ \ \oplus \mathbf{X} \quad\ \ \oplus \mathbf{X}^\star.
\end{aligned}$$

In addition to the electroweak and strong interactions, implemented by the twelve gauge fields $\mathbf{A}_0$, $\mathbf{A}$ and $\mathbf{G}$ for $\mathbf{U}(1)$, $\mathbf{SU}(2)$ and $\mathbf{SU}(3)$, respectively, there arise interactions via the isospin-color connecting twice $2 \cdot 3$ gauge fields $\mathbf{X}, \mathbf{X}^\star$ as isospin dublets and color triplets. They mediate between quarks and leptons.

Unfortunately, the exciting consequences of an $\mathbf{SU}(5)$-unified theory, e.g., decaying protons ("diamonds are not forever"), did not come to pass experimentally. They should arise with a spontaneously, i.e., by the ground state broken $\mathbf{SU}(5)$-symmetry with a remaining $\mathbf{U}(1) \times \mathbf{SU}(3)$-symmetry for electromagnetism and color.

In addition to this experimental disappointment, such an additive unification is somewhat unimaginative in the following respect: The rational, non-integer hypercharge numbers for the quark and lepton fields, $\frac{1}{6}, \frac{2}{3}, -\frac{1}{3}, -\frac{1}{2}$ with normalization -1 for the right-handed lepton, have nothing to do with the fractionality of the $\mathbf{SU}(5)$-representations, i.e., its winding number group $\mathbf{I}(5)$ and fractions $\frac{1}{5}$ do not play a role.

5.12 Fractionality Correlations for Unitary Groups

The color trivial fields of the standard model for the elementary particles display a correlation between hypercharge and isospin. Conventionally normalized hypercharge and isospin are both either integer or half-integer: The left-handed lepton or the Higgs field are isodublets with $|y| = \frac{1}{2}$, the right-handed leptons are isosinglets with $|y| = 1$, the electroweak gauge fields with trivial hypercharge constitute an isotriplet and isosinglet.

The hypercharge cycle $\mathbf{U}(1)$ and the isospin cycle $\mathbf{U}(1)_3 \subset \mathbf{SU}(2)$ appear "synchronized" — in analogy to the cyclic subgroup $\mathbf{I}(2)$ in the Klein four-group $\mathbf{K}(4)$ as a bicycle:

$$\{(1, 1), (-1, -1)\} \cong \mathbf{I}(2) \subset \mathbf{I}(2) \times \mathbf{I}(2) \cong \mathbf{K}(4).$$

This correlation suggests the basic importance of the *hyperisospin group* $\mathbf{U}(2) = \mathbf{U}(\mathbf{1}_2) \circ \mathbf{SU}(2)$ instead of the direct product group $\mathbf{U}(1) \times \mathbf{SU}(2)$. The intersection group $\mathbf{U}(\mathbf{1}_2) \cap \mathbf{SU}(2) \cong \mathbf{I}(2)$ as *fractionality correlation* is discrete, therefore the Lie algebras of $\mathbf{U}(2)$ and $\mathbf{U}(1) \times \mathbf{SU}(2)$ coincide.

In general the locally isomorphic groups $\mathbf{U}(n) = \mathbf{U}(\mathbf{1}_n) \circ \mathbf{SU}(n)$ and $\mathbf{U}(1) \times \mathbf{SU}(n)$ (simply connected) are related to each other via classes with respect to the "synchronizing" cycle $\mathbf{I}(n)$:

$$\mathbf{U}(n) \cong \frac{\mathbf{U}(1) \times \mathbf{SU}(n)}{\mathbf{I}(n)}, \text{ with } \mathbf{I}(n) \cong \{(i_n^k, \mathbf{i}_n^k) \mid k = 0, \ldots, n-1\}$$
$$\subset \mathbf{I}(n) \times \mathbf{I}(n)\mathbf{1}_n \subset \mathbf{U}(1) \times \mathbf{SU}(n).$$

The dual defining representations $D_{\pm 1}^{[1]}$ of $\mathbf{U}(2)$ with the upper index for the dominant weight of the $\mathbf{SU}(2)$-representation and the lower one for the weight of the

$\mathbf{U}(1)$-representation, have as product the $\mathbf{SU}(2)$-trivial $\mathbf{U}(2)$-representation $D_{\pm 2}^{[0]}$. The tensor products of two

$$\text{fundamental } \mathbf{U}(2) - \text{ representations: } D_1^{[1]}, \;\; D_{-2}^{[0]},$$

give all irreducible $\mathbf{U}(2)$-representations:

$$\mathbf{U}(2) - \text{ representation } D_z^{[m]} : \begin{cases} m = 1, 3, 5, \ldots \text{ with all } z \in 2\mathbb{Z} + 1, \\ m = 0, 2, 4, \ldots \text{ with all } z \in 2\mathbb{Z}. \end{cases}$$

The intersection group $\mathbf{I}(2) \cong \mathbf{U}(\mathbf{1}_2) \cap \mathbf{SU}(2)$ imposes the correlation above between isospin $j = \frac{m}{2}$ and hypercharge $y = \frac{z}{2}$.

The two-dimensional weight lattice for all irreducible representations of $\mathbf{U}(1) \times \mathbf{SU}(2)$ is $\mathbb{Z} \times \mathbb{Z}$, the dominant weights $D_z^{[m]}$, in spin notation $(\frac{z}{2}, \frac{m}{2})$, of all representations can be combined, with natural number coefficients, by $D_0^{[1]}$ and $D_{\pm 1}^{[0]}$. The $\mathbf{I}(2)$-correlation leads to the sublattice $\{D_{2y}^{[2j]} \cong (y, j) \mid y - j \in \mathbb{Z}\}$ for $\mathbf{U}(2)$: All representations can be combined, with non-negative coefficients, by $D_1^{[1]} \cong (\frac{1}{2}, \frac{1}{2})$ and $D_{-2}^{[0]} \cong (-1, 0)$.

In the general case $\mathbf{U}(n) = \mathbf{U}(\mathbf{1}_n) \circ \mathbf{SU}(n)$ with fractionality correlation $\mathbf{I}(n)$, it is not difficult to see that the products of the dual defining representations $D_1^{[1,0,\ldots,0]}$ and $D_{-1}^{[0,\ldots,0,1]}$ and their n-fold $\mathbf{SU}(n)$-trivial products $D_{\pm n}^{[0,0,\ldots,0]}$ lead to the following correlation between $\mathbf{U}(1)$- and $\mathbf{SU}(n)$-representations:

$$\text{fundamental } \mathbf{U}(n) \text{ representations: } D_1^{[1,0,\ldots,0]}, \ldots, D_{n-1}^{[0,0,\ldots,1]}, \;\; D_{-n}^{[0,0,\ldots,0]},$$
$$\mathbf{U}(n) \text{ -representations } D_z^{[m_1,\ldots,m_{n-1}]} \text{ with } z \in n\mathbb{Z} + \sum_{i=1}^{n-1} m_i,$$

with the example of the $\mathbf{SU}(3)$-irreducible $\mathbf{U}(3) = \mathbf{U}(\mathbf{1}_3) \circ \mathbf{SU}(3)$-representations:

$$\text{fundamental } \mathbf{U}(3) \text{ representations: } D_1^{[1,0]}, \;\; D_2^{[0,1]}, \;\; D_{-3}^{[0,0]},$$
$$\mathbf{U}(3)\text{-representations } D_z^{[m_1,m_2]} \text{ with } \begin{cases} z \in 3\mathbb{Z} \text{ for } m_1 - m_2 \in 3\mathbb{Z}, \\ z \in 3\mathbb{Z} + 1 \text{ for } m_1 - m_2 \in 3\mathbb{Z} + 1, \\ z \in 3\mathbb{Z} - 1 \text{ for } m_1 - m_2 \in 3\mathbb{Z} - 1. \end{cases}$$

The weight lattice for $\mathbf{U}(n)$ is a proper sublattice of the product weight lattice for $\mathbf{U}(1) \times \mathbf{SU}(n)$. In contrast to $\mathbf{SU}(n)$ or $\mathbf{U}(1)$, the n fundamental representations of $\mathbf{U}(n)$ are not unique. The $n - 1$ fundamental representations of $\mathbf{SU}(n)$ can be extended to $\mathbf{U}(n)$-representations with the following $\mathbf{U}(1)$-representation types:

$$D_{1+n\mathbb{Z}}^{[1,0,\ldots,0]}, \ldots, D_{n-1+n\mathbb{Z}}^{[0,0,\ldots,1]}, \;\; D_{n\mathbb{Z}}^{[0,0,\ldots,0]}.$$

With $\mathbf{I}(n) \cong \mathbb{Z}_n$, the n representation types of $\mathbf{U}(n)$ can be associated with the corners of an $\mathbf{I}(n)$-polygon.

Generalizing the $\mathbf{I}(n)$-correlation of the groups $\mathbf{U}(1)$ and $\mathbf{SU}(n)$ in $\mathbf{U}(n)$, one can look for fractionality correlations via discrete normal subgroups of products $\mathbf{SU}(n) \times \mathbf{SU}(m)$ with two special groups. An example is the symmetry group $\frac{\mathbf{SU}(2)\times\mathbf{SU}(2)}{\mathbf{I}(2)} \cong \mathbf{SO}(4)$ for the states of the Kepler potential as seen in the atoms. Here, there is a "synchronization" in the Klein bicycle $\mathbf{K}(4) \subset \mathbf{SU}(2) \times \mathbf{SU}(2)$ via the subgroup $\mathbf{I}(2) \cong \{(1, 1), (-1, -1)\}$. More complicated are quotient groups of products $\mathbf{U}(1) \times \mathbf{SU}(n) \times \mathbf{SU}(m)$ with three "wheels" in the center $\mathbf{U}(1) \times \mathbf{I}(n) \times \mathbf{I}(m)$, e.g., for the group $\mathbf{U}(1) \times \mathbf{SU}(2) \times \mathbf{SU}(3)$ in the electroweak-strong standard model. More of that ahead.

5.13 Broken Winding Numbers

If a circle $\mathbf{U}(1)$ is wound on an inverse integer part of another circle, one has compatibility with the multiplicative $\mathbf{U}(1)$-group structure:

$$n \in \mathbb{N}, \ d_{\frac{1}{n}} : \mathbf{U}(1) \ni e^{2\pi i\alpha} \longmapsto e^{\frac{2\pi i}{n}\alpha} \in \mathbf{U}(1),$$
$$d_{\frac{1}{n}}(e^{2\pi i\alpha})d_{\frac{1}{n}}(e^{2\pi i\beta}) = d_{\frac{1}{n}}(e^{2\pi i(\alpha+\beta)}).$$

However, for $n \geq 2$, the "broken" representation is multiply valued with multiplicity n, i.e., with the order of the cyclic group $\mathbf{I}(n)$:

$$k \in \mathbb{Z}, \ \frac{d_{\frac{1}{n}}(e^{2\pi i(\alpha+k)})}{d_{\frac{1}{n}}(e^{2\pi i\alpha})} = e^{\frac{2\pi i}{n}k} = i_n^k \in \mathbf{I}(n),$$

and similarily for all "broken" representations d_y with rational numbers $y = \frac{z}{n}$ and coprime $z \in \mathbb{Z}$ and $n \in \mathbb{N}$.

This multi-valuedness is formalizable as a *realization* of $\mathbf{U}(1)$ in the $\mathbf{I}(n)$-classes of $\mathbf{U}(1)$:

$$R_{\frac{1}{n}} : \mathbf{U}(1) \longrightarrow \mathbf{U}(1)/\mathbf{I}(n), \ R_{\frac{1}{n}}(e^{2\pi i\alpha}) = e^{\frac{2\pi i}{n}\alpha}\mathbf{I}(n) = \{e^{\frac{2\pi i(\alpha+k)}{n}} \mid k = 0, \dots n-1\}.$$

The realizations $R_{\frac{1}{n}}$ are even isomorphisms between the group $\mathbf{U}(1)$ and the *fractionality groups* $\mathbf{U}(1)/\mathbf{I}(n)$. For the isomorphism $R_{\frac{1}{n}}$, a $\mathbf{U}(1)$-element on the circle is mapped to the n associated points on this circle, i.e., it may be compared to a fork with n points.

Good to know: A subgroup $H \subseteq G$ of an abelian group is a normal subgroup and defines classes G/H with a natural abelian group structure.
A G-representation is a group morphism $G \longrightarrow \mathbf{GL}(V)$ into a linear group. More general, a G-*realization* is a group morphism $G \longrightarrow G'$ into any group G'.

For $n = 1$, R_1 is the $\mathbf{U}(1)$-identity, for $n = 2$ with the sign group $\mathbf{I}(2)$, the dublet isomorphism $R_{\frac{1}{2}}$, the Mercedes-star group $\mathbf{I}(3)$ leads to the Mercedes-fork isomorphism $R_{\frac{1}{3}}$, then one has the NATO-fork for $\mathbf{I}(4)$, the Texaco-fork for $\mathbf{I}(5)$, etc. For the standard model of the elementary particles, the David-star group $\mathbf{I}(6) \cong \mathbf{I}(2) \times \mathbf{I}(3)$ is relevant with the sextet isomorphism.

Good to know: The daughter of the Austrian-Hungarian consul general Emil Jellinek in Nizza, who, around the turn of the century 1900, was responsible for the marketing of Daimler's motorcars, especially in France, had the name Mercedes.

Because of the fractionality correlation $\mathbf{SU}(n) \cap \mathbf{U}(\mathbf{1}_n) \cong \mathbf{I}(n)$ for both normal subgroups $\mathbf{SU}(n)$ and $\mathbf{U}(\mathbf{1}_n)$ in $\mathbf{U}(n)$ and the isomorphy of the fractionality group to the classes of $\mathbf{SU}(n)$:

$$\mathbf{U}(1) \cong \mathbf{U}(1)/\mathbf{I}(n) \cong \mathbf{U}(n)/\mathbf{SU}(n),$$

it is possible, to implement the multi-valuedness in the discrete normal subgroup $\mathbf{I}(n)$ in a Lie algebra with the $\mathbf{SU}(n)$-realization of $\mathbf{I}(n)$:

$$\mathbf{I}(n) \longrightarrow \mathbf{SU}(n), \quad i_n \longmapsto \mathbf{i}_n.$$

Then, $\mathbf{U}(1)$ is realized modulo $\mathbf{SU}(n)$-transformations for the group, and modulo $\log \mathbf{SU}(n)$-transformations for the Lie algebra:

$$\begin{aligned} R_{\frac{1}{n}}^{[1,0,\ldots,0]} &: \mathbf{U}(1) \longrightarrow \mathbf{U}(n)/\mathbf{SU}(n), & e^{2\pi i\alpha} &\longmapsto e^{\frac{2\pi i}{n}\alpha}\mathbf{SU}(n), \\ \mathcal{R}_{\frac{1}{n}}^{[1,0,\ldots,0]} &: i\mathbb{R} \longrightarrow i\mathbb{R} \oplus A_{n-1}^c, & i\alpha &\longmapsto \frac{i}{n}\alpha\mathbf{1}_n + A_{n-1}^c. \end{aligned}$$

Only for $n = 1$ and for monogeneous numbers $n = p^k$ with p prime, $\mathbf{SU}(n)$ is the "minimal" unitary group for the Lie algebra implementation of $\mathbf{I}(n)$. In general, one obtains for y-broken $\mathbf{U}(1)$-realizations with

$$y = \frac{z}{n}, \text{ with } \begin{cases} n \in \mathbb{N}, \ z \in \mathbb{Z} \text{ coprime} \\ n = m_1 \cdots m_r \text{ pairwise coprime decomposition,} \end{cases}$$

and the Sylow isomorphy $\mathbf{I}(n) \cong \mathbf{I}(m_1) \times \cdots \times \mathbf{I}(m_r)$ as "minimal" special unitary *multi-valuedness group* $\mathbf{SU}(m_1) \times \cdots \times \mathbf{SU}(m_r)$. Therefore, the *fundamental y-broken* $\mathbf{U}(1)$-*realization* in a $\mathbf{U}(n)$-quotient group is given as follows:

$$\begin{aligned} R_y^{[1,0,\ldots,0]\times\cdots\times[1,0,\ldots,0]} &: \mathbf{U}(1) \longrightarrow \mathbf{U}(\mathbf{1}_n) \circ [\mathbf{SU}(m_1) \times \cdots \times \mathbf{SU}(m_r)]_n / [\mathbf{SU}(m_1) \times \cdots \times \mathbf{SU}(m_r)]_n, \\ e^{2\pi i\alpha} &\longmapsto e^{2\pi i y\alpha}\mathbf{SU}(m_1) \times \cdots \times \mathbf{SU}(m_r). \end{aligned}$$

The fundamental realizations act on a vector space $V \cong \mathbb{C}^n$ with complex dimension $n = m_1 \cdots m_r$. As a representative one can choose an appropriate $(n \times n)$-matrix in a *multiplicative unification*:

$$e^{2\pi i\alpha} \longmapsto e^{2\pi i y\alpha} u(m_1) \otimes u(m_2) \otimes \cdots \otimes u(m_r) \in e^{2\pi i y\alpha}\mathbf{SU}(m_1) \otimes \mathbf{SU}(m_2) \otimes \cdots \otimes \mathbf{SU}(m_r) \subset \mathbf{U}(n).$$

Good to know: A direct product group $G_1 \times G_2$ can be represented either *additively* on the direct sum of two vector spaces, $G_1 \times G_2 \ni (g_1, g_2) \longmapsto D_1(g_1) \oplus D_2(g_2) \in \mathbf{GL}(V_1 \oplus V_2)$, or *multiplicatively* on a tensor product $G_1 \times G_2 \ni (g_1, g_2) \longmapsto D_1(g_1) \otimes D_2(g_2) \in \mathbf{GL}(V_1 \otimes V_2)$. For $V_1 \oplus V_2$ the dimensions are added, for $V_1 \otimes V_2$ they are multiplied with $\dim_{\mathbb{K}} V_1 \otimes V_2 \geq \dim_{\mathbb{K}} V_1 \oplus V_2$ for $\dim_{\mathbb{K}} V_i \geq 2$.

5.14 Broken Hypercharges and Nonabelian Internal Symmetry

The ratios of the hypercharges of the standard model fields do not involve irrational numbers like $\frac{1}{\sqrt{2}}$ or $\frac{1}{\pi}$. All hypercharges can be related to the cyclotomic David-star group:

$$e^{2\pi i y} \in \mathbf{I}(6), \text{ with } y \in \{\pm\frac{k}{6} \mid k = 0, 1, 2, 3, 4, 5\}.$$

There do not occur other fractions.

The simply connected compact internal symmetry group for the Lie algebra of the standard model is the direct product group:

$$\mathbf{U}(1) \times \mathbf{SU}(2) \times \mathbf{SU}(3).$$

However, not this direct product group seems to be faithfully represented: Those standard model fields, which have a non-integer rational hypercharge number (in the normalization $y = 1$ for the positron charge number), also have always a nontrivial property with respect to the nonabelian internal symmetry, isospin $\mathbf{SU}(2)$ or color $\mathbf{SU}(3)$:

Field	Hypercharge y	$6y$ mod 2	$6y$ mod 3	$6y$ mod 6
Lepton **e**	-1	0	0	0
Anti-lepton **e***	$+1$	0	0	0
Quark **q**	$+\frac{1}{6}$	1	$+1$	1
Anti-quark **q***	$-\frac{1}{6}$	1	-1	5
Up quark **u**	$+\frac{2}{3}$	0	$+1$	4
Anti-up quark **u***	$-\frac{2}{3}$	0	-1	2
Down quark **d**	$-\frac{1}{3}$	0	$+1$	4
Anti-down quark **d***	$+\frac{1}{3}$	0	-1	2
Lepton **l**	$-\frac{1}{2}$	1	0	3
Anti-lepton **l***	$+\frac{1}{2}$	1	0	3

Hypercharge fractionalities of leptons and quarks

"Anti" goes with the reflection left-handed $[1|0] \leftrightarrow [0|1]$ right-handed. The fractionality n of the hypercharge numbers $y = \frac{z}{n}$ (z and n coprime) coincide with the multiplicity of their respective nonabelian unitary representation. The maximal fractionality $F(\frac{1}{6}) = 6$, occurs for the left quarks.

The hypercharge circle group $\mathbf{U}(1)$ is correlated with the nonabelian isospin-color group $\mathbf{SU}(2) \times \mathbf{SU}(3)$ via its center $\mathbf{I}(2) \times \mathbf{I}(3)$. The center of the group $\mathbf{U}(1) \times \mathbf{SU}(2) \times \mathbf{SU}(3)$ is $\mathbf{U}(1) \times \mathbf{I}(2) \times \mathbf{I}(3)$ with the discrete subgroups $\mathbf{I}(n) \times \mathbf{I}(2) \times \mathbf{I}(3)$ for all $n \geq 1$. Here, those subgroups are "synchronizing" which relate to each other at least two cycles. A central correlation of two factors is obtained by the $\mathbf{I}(2)$- and $\mathbf{I}(3)$-classes:

$$\frac{\mathbf{U}(1) \times \mathbf{SU}(2)}{\mathbf{I}(2)} \times \mathbf{SU}(3) \cong \mathbf{U}(2) \times \mathbf{SU}(3),$$
$$\frac{\mathbf{U}(1) \times \mathbf{SU}(3)}{\mathbf{I}(3)} \times \mathbf{SU}(2) \cong \mathbf{SU}(2) \times \mathbf{U}(3).$$

$\mathbf{I}(2)$ and $\mathbf{I}(3)$ with coprime group order cannot be correlated.

Finally, there is a cyclic subgroup which correlates all three cycles:

$$\mathbf{I}(6) \cong \mathbf{I}(2) \times \mathbf{I}(3) \cong \{(\pm 1, \pm \mathbf{1}_2)\} \times \{(i_3^k, \mathbf{i}_3^k) \mid k = 0, 1, 2\},$$

leading to the quotient group:

$$\mathbf{U}(2 \times 3) = \frac{\mathbf{U}(1) \times \mathbf{SU}(2) \times \mathbf{SU}(3)}{\mathbf{I}(6)},$$

also denoted by $\mathbf{S}[\mathbf{U}(2) \times \mathbf{U}(3)]$. Apparently, these classes constitute the maximal symmetry group, faithfully represented by the fields of the standard model.

The classes of $6y$ in $\mathbf{I}(2) \cong \mathbb{Z}_2 = \{[0], [1]\}$ and in $\mathbf{I}(3) \cong \mathbb{Z}_3 = \{[0], [\pm 1]\}$, i.e., $6y\text{mod}2$ and $6y\text{mod}3$, respectively, reflect the isospin and color representations:

$$\begin{array}{r} 6y\text{mod}2 \leftrightarrow \mathbf{SU}(2): \\ m \in 2\mathbb{Z} + 6y\text{mod}2 \end{array} \begin{cases} [0] \leftrightarrow D^{[m]}, \; m \in 2\mathbb{Z}, \\ [1] \leftrightarrow D^{[m]}, \; m \in 2\mathbb{Z}+1, \end{cases}$$
$$\begin{array}{r} 6y\text{mod}3 \leftrightarrow \mathbf{SU}(3): \\ m_1 - m_2 \in 3\mathbb{Z} + 6y\text{mod}3 \end{array} \begin{cases} [0] \leftrightarrow D^{[m_1,m_2]}, \; m_1 - m_2 \in 3\mathbb{Z}, \\ [+1] \leftrightarrow D^{[m_1,m_2]}, \; m_1 - m_2 \in 3\mathbb{Z}+1, \\ [-1] \leftrightarrow D^{[m_1,m_2]}, \; m_1 - m_2 \in 3\mathbb{Z}-1. \end{cases}$$

All representations of $\mathbf{U}(1) \times \mathbf{SU}(2) \times \mathbf{SU}(3)$ are characterized by the four-dimensional product lattice of the weight lattices of the direct factors which are one-, one-, and two-dimensional. The weights of the irreducible $\mathbf{U}(2 \times 3)$-representations occupy a proper sublattice for the representations:

$$D_{6y}^{[m]\times[m_1,m_2]} \text{ with } \begin{cases} [m] & \text{for isospin } \mathbf{SU}(2), \\ [m_1, m_2] & \text{for color } \mathbf{SU}(3), \\ y & \text{for hypercharge } \mathbf{U}(1), \\ \text{where } y \bmod \mathbb{Z} = \frac{m}{2} - \frac{m_1 - m_2}{3}. \end{cases}$$

The following six examples as representation types contain the fundamental representations [1] for isospin and [1, 0] and [0, 1] for color:

	[0]	[1]
[0, 0]	$D_{6\mathbb{Z}}^{[0]\times[0,0]}$	$D_{3+6\mathbb{Z}}^{[1]\times[0,0]}$
[1, 0]	$D_{4+6\mathbb{Z}}^{[0]\times[1,0]}$	$D_{1+6\mathbb{Z}}^{[1]\times[1,0]}$
[0, 1]	$D_{2+6\mathbb{Z}}^{[0]\times[0,1]}$	$D_{5+6\mathbb{Z}}^{[1]\times[0,1]}$

Isospin $\mathbf{SU}(2)$ and color $\mathbf{SU}(3)$ need their $(1 + 2 = 3)$ fundamental representations to combine, by their dominant weights with natural numbers as coefficients, the dominant weights of all irreducible $\mathbf{SU}(2) \times \mathbf{SU}(3)$-representations in their three-dimensional weight lattice. As fundamental $\mathbf{U}(2 \times 3)$-representations four left-handed fermion fields of the standard model of electroweak and strong interactions can be taken — either $\{\mathbf{e}^\star, \mathbf{l}, \mathbf{q}, \mathbf{d}^\star\}$ or $\{\mathbf{e}^\star, \mathbf{l}, \mathbf{q}, \mathbf{u}^\star\}$:

	[0]	[1]
[0, 0]	$D_{6}^{[0]\times[0,0]}$	$D_{-3}^{[1]\times[0,0]}$
[1, 0]	–	$D_{1}^{[1]\times[1,0]}$
[0, 1]	$D_{2}^{[0]\times[1,0]}, D_{-4}^{[0]\times[1,0]}$	–

for

	[0]	[1]
[0, 0]	$\mathbf{e}^\star$	$\mathbf{l}$
[1, 0]	–	$\mathbf{q}$
[0, 1]	$\mathbf{d}^\star, \mathbf{u}^\star$	–

The isosinglet quark field $\mathbf{u}$ and $\mathbf{d}$ have equal $\mathbf{SU}(2) \times \mathbf{SU}(3)$-representations. With all five representations, one has a $\mathbf{U}(1)$-dual ensemble, i.e., the sum of all hypercharges of the five types of standard model fermion fields vanishes:

$$\text{for } (\mathbf{e}^\star, \mathbf{l}, \mathbf{q}, \mathbf{d}^\star, \mathbf{u}^\star): \ 1 \times 1 + 2 \times \left(-\frac{1}{2}\right) + 6 \times \frac{1}{6} + 3 \times \frac{1}{3} + 3 \times \left(-\frac{2}{3}\right) = 0.$$

The ten types of fermion fields in the standard model as given in the table above occupy the six elements of the David-star group $\mathbf{I}(6) \cong \mathbb{Z}_6$ as correlating normal subgroup of the unitary David group $\mathbf{U}(2 \times 3)$. The isosinglets occupy the Mercedes-star $\mathbf{I}(3)$, the isodoublets the reflected Mercedes-star $-\mathbf{I}(3)$, possible with the sign group $\mathbf{I}(2)$. The lepton fields occupy $\mathbf{I}(2)$, the quark fields the complement $\mathbf{I}(6) \setminus \mathbf{I}(2)$. Both the five left-handed and the five right-handed fields constitute each an ensemble of fundamental $\mathbf{U}(2 \times 3)$-representation with trivial hypercharge sum.

The representations $D_{6y}^{[m]\times[m_1,m_2]}$ of the group $\mathbf{U}(2 \times 3)$ lead to $\mathbf{U}(1)$-realizations $R_y^{[m]\times[m_1,m_2]}$ with broken winding numbers in the quotient group $\mathbf{U}(2 \times 3)/[\mathbf{SU}(2) \times \mathbf{SU}(3)]$, i.e., for the $\mathbf{SU}(2) \times \mathbf{SU}(3)$-classes. The Weyl spinor fields of the standard symmetry come with the following $\mathbf{U}(1)$-realizations:

$$\text{right lepton } \mathbf{e}: \begin{cases} y = -1, \ D_{-6}^{[0]\times[0,0]}, \\ R_{-1} = D_{-1} : \mathbf{U}(1) \longrightarrow \mathbf{U}(1), \\ \text{representation } D_{-1}(e^{2\pi i \alpha}) = e^{-2\pi i \alpha}, \\ \text{acts on } V \cong \mathbb{C} \text{ by } \mathbf{U}(1), \end{cases}$$

$$\text{left lepton } \mathbf{l}: \begin{cases} y = -\frac{1}{2}, \ D_{-3}^{[1]\times[0,0]}, \\ R_{-\frac{1}{2}}^{[1]} : \mathbf{U}(1) \longrightarrow \mathbf{U}(2)/\mathbf{SU}(2), \\ \text{realization } R_{-\frac{1}{2}}^{[1]}(e^{2\pi i \alpha}) = e^{-\pi i \alpha}\mathbf{SU}(2), \\ \text{acts on } V \cong \mathbb{C}^2 \text{ by } \mathbf{U}(2), \end{cases}$$

$$\text{right quarks } \mathbf{u}, \mathbf{d}: \begin{cases} y = -\frac{1}{3}, \frac{2}{3}, \ D_{-2,4}^{[0]\times[1,0]}, \\ R_y^{[1,0]} : \mathbf{U}(1) \longrightarrow \mathbf{U}(3)/\mathbf{SU}(3), \\ \text{realization } R_y^{[1,0]}(e^{2\pi i \alpha}) = e^{2\pi i y \alpha}\mathbf{SU}(3), \\ \text{acts on } V \cong \mathbb{C}^3 \text{ by } \mathbf{U}(3), \end{cases}$$

$$\text{left quark } \mathbf{q}: \begin{cases} y = \frac{1}{6}, \ D_1^{[1]\times[1,0]}, \\ R_{\frac{1}{6}}^{[1]\times[1,0]} : \mathbf{U}(1) \longrightarrow \mathbf{U}(2\times 3)/\mathbf{SU}(2)\times\mathbf{SU}(3), \\ \text{realization } R_{\frac{1}{6}}^{[1]\times[1,0]}(e^{2\pi i \alpha}) = e^{\frac{\pi i \alpha}{3}}\mathbf{SU}(2)\times\mathbf{SU}(3), \\ \text{acts on } V \cong \mathbb{C}^6 \text{ by } \mathbf{U}(2\times 3). \end{cases}$$

In the standard model, the $\mathbf{SU}(2) \times \mathbf{SU}(3)$-classes are taken into account by the gauge fields, i.e., by the three electroweak $\mathbf{SU}(2)$-gauge fields and by the eight color $\mathbf{SU}(3)$-gauge fields.

It is easy to realize that irreducible realization and antirealization come with negatively reflected hypercharge numbers and dual unitary representations:

$$\text{antirealizations: } R_y^{[m_1,\ldots,m_{m_1}-1]\times\cdots\times[k_1,\ldots,k_{m_r}-1]} \leftrightarrow R_{-y}^{[m_{m_1}-1,\ldots,m_1]\times\cdots\times[k_{m_r}-1,\ldots,k_1]},$$
$$\text{e.g., } R_{\frac{1}{6}}^{[1]\times[1,0]} \leftrightarrow R_{-\frac{1}{6}}^{[1]\times[0,1]}.$$

In product realizations, the $\mathbf{U}(1)$-winding numbers, broken or not, are added. Only products with trivial realizations of the nonabelian factor $\mathbf{SU}(m_1) \times \cdots \times \mathbf{SU}(m_r)$ are $\mathbf{U}(1)$-*representations*, i.e., uniquely valued group endomorphism $\mathbf{U}(1) \longrightarrow \mathbf{U}(1)$ with integer winding numbers. In the standard model, such representations of the electromagnetic group $\mathbf{U}(1)$ are proposed or expected to arise by two mechanisms — by the "bleaching" mechanism of spontaneous symmetry breakdown (ground state degeneracy) in the case of isospin $\mathbf{SU}(2)$ and by the, not yet proven, but hoped for "confinement" mechanism in the case of color $\mathbf{SU}(3)$.

Representations without class-related multiple-valuedness, i.e., with trivial non-abelian unitary internal property and integer charge number, can be obtained, e.g., for the fundamental realizations with a noninteger hypercharge $y = \frac{z}{n}$ and prime n by the totally antisymmetric product of power n:

$$\bigwedge^{n} R^{[1,0,\dots,0]}_{\frac{z}{n}} \cong D_z,$$

$$\text{e.g. } \bigwedge^{3} R^{[1,0]}_{\frac{z}{3}} \cong D_z \text{ for right-handed quark } n = 3,\ z = -1, 2.$$

This example plays a role in the standard model in the combination of three quarks as color triplets with hypercharge fractionality three to a hadronic fermion as color-trivial particle with integer charge number, e.g., to a proton or to a neutron.

Obvioulsy, it is also possible to combine different realizations with equal fractionality representation to integer winding number representations. For example in the standard model, both the left-handed lepton field and the scalar symmetry breakdown effecting field (Higgs[6] field for the definition of the ground state) have fractionality two, i.e., hypercharge number $y = \pm\frac{1}{2}$:

$$\text{Higgs field, } y = \pm\frac{1}{2}: \quad R^{[1]}_{y} : \mathbf{U}(1) \longrightarrow \mathbf{U}(2)/\mathbf{SU}(2).$$

The product of the broken hypercharge $\mathbf{U}(1)$-realizations of the Higgs field and the left-handed lepton field has integer winding numbers $(-1, 0)$ as charge numbers for the electromagnetic $\mathbf{U}(1)$-symmetry, e.g., for the electron particle and its neutrino:

$$\text{left-handed lepton with Higgs field: } \begin{cases} R^{[1]}_{-\frac{1}{2}} \wedge R^{[1]}_{-\frac{1}{2}} \cong D_{-1}, \\ R^{[1]}_{-\frac{1}{2}} \wedge R^{[1]}_{\frac{1}{2}} \cong D_0. \end{cases}$$

For the left-handed quark, this product is a $\mathbf{U}(1)$-realization with fractionality three:

$$\text{left-handed quark with Higgs field : } \begin{cases} R^{[1]\times[1,0]}_{\frac{1}{6}} \wedge R^{[1]}_{\frac{1}{2}} \cong R^{[1,0]}_{\frac{2}{3}}, \\ R^{[1]\times[1,0]}_{\frac{1}{6}} \wedge R^{[1]}_{-\frac{1}{2}} \cong R^{[1,0]}_{-\frac{1}{3}}. \end{cases}$$

With the antisymmetric power three product, described above, one obtains a $\mathbf{U}(1)$-representation.

5.15 Young Frames for $\mathbf{SU}(n)$-Representations

The fractionality of an $\mathbf{SU}(n)$-representation describes the representation of its center $\mathbf{I}(n)$. It can be visualized as follows: The *Young*[7] *frame* for the fundamental $\mathbf{SU}(n)$-representation $D^{[0,\dots,1,\dots,0]}$ (the 1 at the k-th place, $k = 1, \dots, n-1$), which is isomorphic to the totally antisymmetrized k-fold tensor product of the defining

[6]Peter Higgs (1929–).

[7]Alfred Young (1873–1940).

representation $D^{[1,0,\ldots,0]}$, is defined as a vertical scheme, consisting of k squares (boxes):

$$\mathbf{SU}(n): \quad \underbrace{D^{[0,\ldots,1,\ldots,0]}}_{k\text{th place}} \sim k \text{ boxes}$$

$$, \quad k = 1, \ldots, n-1.$$

For a given n, the number k of boxes determines the fractionality $F(\frac{k}{n})$.

The Young frame for a representation $D^{[0,\ldots,m,\ldots,0]}$ (the m at the k-th place), which occurs in the totally symmetrized tensor product of $D^{[0,\ldots,1,\ldots,0]}$, consists of m horizontally connected columns, each with k boxes — it is a $k \times m$-rectangle, divided into $N = k \cdot m$ boxes:

$$\mathbf{SU}(n): \quad \underbrace{D^{[0,\ldots,m,\ldots,0]}}_{k\text{th place}} \sim k \text{ boxes}$$

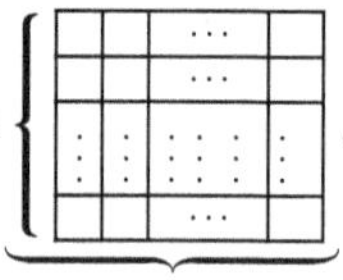

$$, \quad \begin{array}{l} k = 1, 2, \ldots, n-1, \\ m = 1, 2, \ldots \end{array}$$

m columns

Finally, an irreducible $\mathbf{SU}(n)$-representation $D^{[m_1,\ldots,m_{n-1}]}$ has the Young frame, which arises by drawing first the $(n-1) \times m_{n-1}$-Young frame for $D^{[0,\ldots,0,m_{n-1}]}$, then, to the right with the same upper line, the Young frame for $D^{[0,\ldots,0,m_{n-2},0]}$, etc. One obtains a Young frame, where, going to the right, from column to column, the number of squares does not increase. Such a Young frame with $\sum_{k=1}^{n-1} km_k$ boxes looks like an upside-down staircase with possibly unequal widths and heights of the steps,

$$\text{e.g., for } \mathbf{SU}(5): \quad D^{[2,1,0,3]} \sim$$

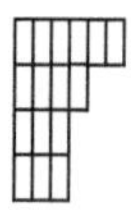

The number of boxes modulo n in an $\mathbf{SU}(n)$-Young frame gives the fractionality of the representation, in the $\mathbf{SU}(5)$-example $16 \bmod 5 = 1 \bmod 5$.

Reflecting the totally antisymmetric or the totally symmetric products of fundamental $\mathbf{SU}(n)$-representations, the vertical structure of a Young frame has to do with total antisymmetrization, the horizontal one with total symmetrization. A Young frame describes a representation of the permutation group $\mathbf{G}(N)$ for the number of their boxes N (more ahead).

The dimension $d(m_1, \ldots, m_{n-1})$ of an $\mathbf{SU}(n)$-representation, associated to a Young frame, is given by the number of possibilities to fill in numbers from the set $\{1, 2, \ldots, n\}$, possibly also repeatedly, with the following prescription: The numbers have to increase strictly in each column, (vertical total antisymmetrization), they must not decrease in each line (horizontal total symmetrization).

5.16 Young Tableaux for Young Frames

Young frames characterize irreducible representations of the permutation group and, therefore, can be investigated independent of their importance for the representations of unitary groups.

For each natural number n, there are Young frames with n boxes:

$$Y(s_n, s_{n-1}, \dots, s_1), \quad s_k \geq 0, \quad \sum_{k=1}^{n} k s_k = n,$$

with s_n columns of length n, then, to the right, with equal upper line, s_{n-1} columns of length $n-1$ etc., and, finally, s_1 single boxes. The general number $z(n)$ of Young n-frames which are possible for a box number n, starting with

$$z(1) = 1, \quad z(2) = 2, \quad z(3) = 3, \quad z(4) = 5, \quad z(5) = 7, \quad \dots,$$

is given by a rather complicated combinatorial expression.

With respect to the upside-down staircase, a Young frame can be described by coordinates for the staircase corners with natural number pairs (y_k, x_k) in the fourth quadrant of a Euclidean plane: y_k is the height of the step and x_k the horizontal distance to the origin, both numbers are taken positive:

$$\mathcal{Y}(y_1, x_1; y_2, x_2; \dots; y_r, x_r) \text{ with } \begin{cases} 0 \leq y_k \leq n, \quad 0 \leq x_k \leq n, \\ x_k = \sum_{l=1}^{k-1} x'_l, \quad \sum_{k=1}^{r} y_k x'_k = n, \\ y_k = \sum_{l=1}^{k-1} y'_l, \quad \sum_{k=1}^{r} x_k y'_k = n. \end{cases}$$

As examples, all Young frames are given up to $n = 4$ in the *column notation* Y, and all up to $n = 5$ in the *staircase notation* $\mathcal{Y}$

$$\begin{aligned}
n = 1: &\quad Y(1) = \mathcal{Y}(1, 1), \\
n = 2: &\quad Y(1, 0) = \mathcal{Y}(2, 1), \; Y(0, 2) = \mathcal{Y}(1, 2), \\
n = 3: &\quad Y(1, 0, 0) = \mathcal{Y}(3, 1), \; Y(0, 1, 1) = \mathcal{Y}(2, 1; 1, 2), \; Y(0, 0, 3) = \mathcal{Y}(1, 3), \\
n = 4: &\quad \begin{cases} Y(1, 0, 0, 0) = \mathcal{Y}(4, 1), \; Y(0, 0, 0, 4) = \mathcal{Y}(1, 4), \; Y(0, 0, 2, 0) = \mathcal{Y}(2, 2), \\ Y(0, 1, 0, 1) = \mathcal{Y}(3, 1; 1, 2), \; Y(0, 0, 1, 2) = \mathcal{Y}(2, 1; 1, 3), \end{cases} \\
n = 5: &\quad \begin{cases} \mathcal{Y}(5, 1), \; \mathcal{Y}(4, 1; 1, 2), \; \mathcal{Y}(3, 1; 2, 2), \; \mathcal{Y}(3, 1; 1, 3), \\ \mathcal{Y}(1, 5), \; \mathcal{Y}(2, 1; 1, 4), \; \mathcal{Y}(2, 2; 1, 3). \end{cases}
\end{aligned}$$

Young frames can be reflected at the diagonal of the fourth quadrant—in the staircase notation:

$$\text{reflection: } \mathcal{Y}(y_1, x_1; \dots; y_r, x_r) \leftrightarrow \mathcal{Y}(x_r, y_r; \dots; x_1, y_1).$$

If a Young frame with n boxes is filled up with all numbers $\{1, \dots, n\}$ according to the rule that they increase both rightwards and downwards, i.e., 1 is always left on top, there arises a *Young tableau* for a Young frame $Y(n)$. The number of possible tableaux for one frame is its dimension $d(Y(n)) = d(y_1, x_1; \dots; y_r, x_r)$, e.g.:

$$\mathcal{Y}(2,2) = \begin{array}{|c|c|}\hline \ & \ \\ \hline \ & \ \\ \hline \end{array} \qquad \text{with } \begin{array}{|c|c|}\hline 1 & 2 \\ \hline 3 & 4 \\ \hline \end{array}\,,\ \begin{array}{|c|c|}\hline 1 & 3 \\ \hline 2 & 4 \\ \hline \end{array} \Rightarrow d(2,2) = 2,$$

$$\mathcal{Y}(3,1;1,3) = \begin{array}{|c|c|c|}\hline \ & \ & \ \\ \hline \ \\ \cline{1-1} \ \\ \cline{1-1} \end{array} \qquad \text{with } \begin{array}{|c|c|c|}\hline 1 & 2 & 3 \\ \hline 4 \\ \cline{1-1} 5 \\ \cline{1-1} \end{array}\,,\ \begin{array}{|c|c|c|}\hline 1 & 2 & 4 \\ \hline 3 \\ \cline{1-1} 5 \\ \cline{1-1} \end{array}\,,\ \begin{array}{|c|c|c|}\hline 1 & 2 & 5 \\ \hline 3 \\ \cline{1-1} 4 \\ \cline{1-1} \end{array}\,,$$

$$\begin{array}{|c|c|c|}\hline 1 & 3 & 4 \\ \hline 2 \\ \cline{1-1} 5 \\ \cline{1-1} \end{array}\,,\ \begin{array}{|c|c|c|}\hline 1 & 3 & 5 \\ \hline 2 \\ \cline{1-1} 4 \\ \cline{1-1} \end{array}\,,\ \begin{array}{|c|c|c|}\hline 1 & 4 & 5 \\ \hline 2 \\ \cline{1-1} 3 \\ \cline{1-1} \end{array}$$

$$\Rightarrow d(3,1;1,3) = 6.$$

There exists a complicated combinatorial formula for the general case - the examples above give the following dimensions:

$$\begin{aligned}
&n = 1 : d(1,1) = 1,\\
&n = 2 : d(2,1) = 1 = d(1,2),\\
&n = 3 : d(3,1) = 1 = d(1,3),\ d(2,1;1,2) = 2,\\
&n = 4 : d(4,1) = 1 = d(1,4),\ d(2,2) = 2,\ d(3,1;1,2) = 3 = d(2,1;1,3)\\
&n = 5 : d(5,1) = 1 = d(1,5),\ d(4,1;1,2) = 4 = d(2,1;1,4),\\
&\qquad\quad d(3,1;2,2) = 5 = d(2,2;1,3),\ d(3,1;1,3) = 6.
\end{aligned}$$

Reflected frames have equal dimensions, some dimensions are easily computed and especially simple:

$$\begin{aligned}
&d(y_1, x_1; \dots, y_r, x_r) = d(x_r, y_r; \dots, x_1, y_1),\\
&d(n,1) = 1,\ \ d(n-k, 1; 1, k+1) = \binom{n-1}{k}.
\end{aligned}$$

The square of the dimensions of all Young frames for fixed n sum up to the order $n!$ of the permutation group $\mathbf{G}(n)$:

$$\sum_{Y(n)} d(Y(n))^2 = n! : \begin{cases} n = 1: \ 1^2 = 1,\\ n = 2: \ 1^2 + 1^2 = 2,\\ n = 3: \ 1^2 + 2^2 + 1^2 = 6,\\ n = 4: \ 1^2 + 3^2 + 2^2 + 3^2 + 1^2 = 4!,\\ n = 5: \ 1^2 + 4^2 + 5^2 + 6^2 + 5^2 + 4^2 + 1^2 = 5!,\\ \text{etc.} \end{cases}$$

This, on the first sight surprising number mysticism is understandable by the structures of complex box-matrix algebras.

5.17 Box-Matrix Algebras

The symbols $\mathbf{AL}(n, \mathbb{K})$ for real or complex numbers and more general:

$$\mathbf{AL}(n_1, \mathbb{K}) \oplus \ \dots \ \oplus \mathbf{AL}(n_k, \mathbb{K}), \ \ n_j \in \mathbb{N}, \ \ \mathbb{K} = \mathbb{R}, \mathbb{C},$$

denote real or complex *box-matrix algebras*:

$$\text{e.g.,} \quad \mathbf{AL}(2, \mathbb{K}) = \begin{pmatrix} \bullet & \bullet \\ \bullet & \bullet \end{pmatrix},$$

$$\mathbf{AL}(2, \mathbb{K}) \oplus \mathbf{AL}(2, \mathbb{K}) \oplus \mathbf{AL}(3, \mathbb{K}) = \begin{pmatrix} \bullet & \bullet & 0 & 0 & 0 & 0 & 0 \\ \bullet & \bullet & 0 & 0 & 0 & 0 & 0 \\ 0 & 0 & \bullet & \bullet & 0 & 0 & 0 \\ 0 & 0 & \bullet & \bullet & 0 & 0 & 0 \\ 0 & 0 & 0 & 0 & \bullet & \bullet & \bullet \\ 0 & 0 & 0 & 0 & \bullet & \bullet & \bullet \\ 0 & 0 & 0 & 0 & \bullet & \bullet & \bullet \end{pmatrix}.$$

$\bullet$ can be any number from the field $\mathbb{K} = \mathbf{AL}(1, \mathbb{K})$. There may occur equal subsquares with $n_j = n_k$. $\mathbf{AL}(n, \mathbb{K}) \cong \mathbb{K}^n \otimes \mathbb{K}^n \cong \mathbb{K}^{n^2}$ is a *full matrix algebra*. With matrix multiplication, $\mathbf{AL}(n_1, \mathbb{K}) \oplus \ \dots \ \oplus \mathbf{AL}(n_k, \mathbb{K})$ is a unital $\mathbb{K}$-algebra with $\mathbb{K}$-dimension $\sum_{i=1}^{k} n_k^2$. The algebra unit has the subunits $\mathbf{1}_{n_j} \in \mathbf{AL}(n_j, \mathbb{K})$ in the diagonal and zeros elsewhere.

If only some columns are nontrivial in a box-matrix algebra, e.g., in the example above:

$$L_1 = \begin{pmatrix} \bullet & 0 & 0 & 0 & 0 & 0 & 0 \\ \bullet & 0 & 0 & 0 & 0 & 0 & 0 \\ 0 & 0 & 0 & 0 & 0 & 0 & 0 \\ 0 & 0 & 0 & 0 & 0 & 0 & 0 \\ 0 & 0 & 0 & 0 & 0 & 0 & 0 \\ 0 & 0 & 0 & 0 & 0 & 0 & 0 \\ 0 & 0 & 0 & 0 & 0 & 0 & 0 \end{pmatrix}, \quad L_7 = \begin{pmatrix} 0 & 0 & 0 & 0 & 0 & 0 & 0 \\ 0 & 0 & 0 & 0 & 0 & 0 & 0 \\ 0 & 0 & 0 & 0 & 0 & 0 & 0 \\ 0 & 0 & 0 & 0 & 0 & 0 & 0 \\ 0 & 0 & 0 & 0 & 0 & 0 & \bullet \\ 0 & 0 & 0 & 0 & 0 & 0 & \bullet \\ 0 & 0 & 0 & 0 & 0 & 0 & \bullet \end{pmatrix},$$

$$L_{13457} = \begin{pmatrix} \bullet & 0 & 0 & 0 & 0 & 0 & 0 \\ \bullet & 0 & 0 & 0 & 0 & 0 & 0 \\ 0 & 0 & \bullet & \bullet & 0 & 0 & 0 \\ 0 & 0 & \bullet & \bullet & 0 & 0 & 0 \\ 0 & 0 & 0 & 0 & \bullet & 0 & \bullet \\ 0 & 0 & 0 & 0 & \bullet & 0 & \bullet \\ 0 & 0 & 0 & 0 & \bullet & 0 & \bullet \end{pmatrix},$$

one has a subalgebra, closed under left multiplication with the whole box-matrix algebra. An analogue situation occurs for the right multiplication, if one restricts to subalgebras with nontrivial rows.

Good to know: A *left (or right) ideal* of an associative algebra A is defined as a subalgebra L (or R), closed under left (or right) multiplication with the algebra A:

$$AL \subseteq L, \quad RA \subseteq R.$$

The sum and the intersection of left (or right) ideals are left (or right) ideals. Each algebra element $a \in A$ defines by the left (or right) products Aa (or aA) a left (or right) *principal ideal.* A nontrivial left (or right) ideal of A is *minimal*, if it contains no proper ideal of A. A subalgebra of A which is both left and right ideal, is an *ideal* I, proper for $I \neq \{0\}, A$.

A nontrivial number at one place in $A = \mathbf{AL}(n_1, \mathbb{K}) \oplus \cdots \oplus \mathbf{AL}(n_k, \mathbb{K})$ leads — by left (or right) multiplication with the algebra A to the left (or right) principal ideal with the full column and full line, respectively. Left (or right) multiplication of such an element with A can shift it to each place of its column (or its line). By linear combinations, the column (or line) of $a \in A$ can be filled up completely.

For a box-matrix algebra, the minimal left (or right) ideals are those with one nontrivial column (or line), L_1 or L_7 in the example above. Each left (or right) ideal is a direct sum of left (or right) minimal ideals. A minimal left (or right) ideal is the left (or right) principal ideal of all its nontrivial elements — otherwise it would not be minimal:

$$e_L \in L^{\min} \subseteq A, \; e_L \neq 0 \Rightarrow L^{\min} = Ae_L.$$

Good to know: A unital algebra $A \ni 1_A$ that is the direct sum of its minimal left ideals, is called *semisimple*:

$$A = \bigoplus_{j \in J} L_j^{\min} = \bigoplus_{j \in J} Ae_j = \bigoplus_{j \in J} R_j^{\min} = \bigoplus_{j \in J} e'_j A.$$

Then it is also the direct sum of its minimal right ideals. It is possible to choose the generating elements e_j of the minimal left (or right) ideals from a decomposition of the algebra unit 1_A into an orthonormal system of projectors:

$$1_A = \sum_{j=1}^{N} e_j = \sum_{j=1}^{N} e'_j, \quad \begin{cases} e_j e_k = \delta_{jk} e_j, & e'_j e'_k = \delta_{jk} e'_j \\ L_j^{\min} L_k^{\min} = \delta_{jk} L_j^{\min}, & R_j^{\min} R_k^{\min} = \delta_{jk} R_j^{\min}. \end{cases}$$

In a semisimple algebra A, the direct sum of isomorphic left (or right) ideals constitute an ideal $I \subseteq A$, an *isotypical component* of A. An isotypical component is a direct sum of minimal ideals:

$$A = \bigoplus_{m \in M} I_m^{\text{isotyp}} = \bigoplus_{n \in N} I_n^{\min}.$$

A nontrivial semisimple algebra without proper ideal is called *simple*.

Each matrix in the box-matrix algebra $\mathbf{AL}(n_1, \mathbb{K}) \oplus \ldots \oplus \mathbf{AL}(n_k, \mathbb{K})$ is a unique sum of column matrices (or line matrices). Here, the projectors have a number 1 in the diagonal. The proper ideals of are given by the nontrivial box-matrix subalgebras, where at least one, but not all full matrix algebras $\mathbf{AL}(n_j, \mathbb{K})$ are occupied.

In the example above, there are two isotypical components, isomorphic to $\mathbf{AL}(2, \mathbb{K}) \oplus \mathbf{AL}(2, \mathbb{K})$ and $\mathbf{AL}(3, \mathbb{K})$, and three minimal ideals, isomorphic to $\mathbf{AL}(2, \mathbb{K})$, $\mathbf{AL}(2, \mathbb{K})$ and $\mathbf{AL}(3, \mathbb{K})$:

$$\begin{pmatrix} \bullet&\bullet&0&0&0&0&0\\ \bullet&\bullet&0&0&0&0&0\\ 0&0&\bullet&\bullet&0&0&0\\ 0&0&\bullet&\bullet&0&0&0\\ 0&0&0&0&\bullet&\bullet&\bullet\\ 0&0&0&0&\bullet&\bullet&\bullet\\ 0&0&0&0&\bullet&\bullet&\bullet \end{pmatrix} = \begin{pmatrix} \bullet&\bullet&0&0&0&0&0\\ \bullet&\bullet&0&0&0&0&0\\ 0&0&\bullet&\bullet&0&0&0\\ 0&0&\bullet&\bullet&0&0&0\\ 0&0&0&0&0&0&0\\ 0&0&0&0&0&0&0\\ 0&0&0&0&0&0&0 \end{pmatrix} \oplus \begin{pmatrix} 0&0&0&0&0&0&0\\ 0&0&0&0&0&0&0\\ 0&0&0&0&0&0&0\\ 0&0&0&0&0&0&0\\ 0&0&0&0&\bullet&\bullet&\bullet\\ 0&0&0&0&\bullet&\bullet&\bullet\\ 0&0&0&0&\bullet&\bullet&\bullet \end{pmatrix}$$

$$= \begin{pmatrix} \bullet&\bullet&0&0&0&0&0\\ \bullet&\bullet&0&0&0&0&0\\ 0&0&0&0&0&0&0\\ 0&0&0&0&0&0&0\\ 0&0&0&0&0&0&0\\ 0&0&0&0&0&0&0\\ 0&0&0&0&0&0&0 \end{pmatrix} \oplus \begin{pmatrix} 0&0&0&0&0&0&0\\ 0&0&0&0&0&0&0\\ 0&0&\bullet&\bullet&0&0&0\\ 0&0&\bullet&\bullet&0&0&0\\ 0&0&0&0&0&0&0\\ 0&0&0&0&0&0&0\\ 0&0&0&0&0&0&0 \end{pmatrix} \oplus \begin{pmatrix} 0&0&0&0&0&0&0\\ 0&0&0&0&0&0&0\\ 0&0&0&0&0&0&0\\ 0&0&0&0&0&0&0\\ 0&0&0&0&\bullet&\bullet&\bullet\\ 0&0&0&0&\bullet&\bullet&\bullet\\ 0&0&0&0&\bullet&\bullet&\bullet \end{pmatrix}.$$

For the box-matrix algebras, the full ones $\mathbf{AL}(n, \mathbb{K})$ are the simple algebras.

The theorem of Wedderburn[8] implies that the box-matrix algebras, used so far for illustrations, are characteristic for complex numbers: Each *complex finite-dimensional* semisimple associative algebra is isomorphic to a box-matrix algebra $\mathbf{AL}(n_1, \mathbb{C}) \oplus \cdots \oplus \mathbf{AL}(n_k, \mathbb{C})$, the simple ones to the full matrix algebras $\mathbf{AL}(n, \mathbb{C}) \cong \mathbb{C}^{n^2}$.

5.18 Algebras for Finite Groups

To combine Young frames and Wedderburn's theorem, box-matrix algebras and permutation groups have to be connected.

Good to know: The bijections of a set S constitute a group $\mathbf{G}(S)$ — for a set with cardinality n isomorphic to the permutation group $\mathbf{G}(n)$ with $n!$ elements. Any group G can be faithfully realized in its bijections by the left multiplications L_g:

$$\begin{aligned} &g \in G,\ L_g : G \longrightarrow G,\quad L_g(h) = gh,\\ &\qquad G \longrightarrow \mathbf{G}(G),\quad g \longmapsto L_g,\ \text{with } L_{g_1 g_2} = L_{g_1} \circ L_{g_2}. \end{aligned}$$

Therefore, each group of order n is isomorphic to a subgroup of $\mathbf{G}(n)$.

There are many notations for the elements of the permutation groups. The notation $(k_1, \ldots, k_n) \in \mathbf{G}(n)$ with n different numbers $\{k_1, \ldots, k_n\} = \{1, \ldots, n\}$, e.g., $(2, 1, 3) \in \mathbf{G}(3)$, denotes the permutation, where the j-th element is brought to the k_j-th position, $j \longmapsto k_j$, in the example $1 \longmapsto 2$, $2 \longmapsto 1$ and $3 \longmapsto 3$. Each permutation can be generated by transpositions of two elements $\mathbf{G}(2) = \{(12), (21)\}$. The odd or even number of the generating transposition factors defines the signature ± 1 of a permutation. The corresponding mapping

$$\epsilon : \mathbf{G}(n) \longrightarrow \{\pm 1\} \cong \mathbf{G}(2),$$

[8]Joseph Wedderburn (1882–1948).

is a group morphism with kernel $\mathbf{S}(n)$, the alternating or special permutation group, a normal subgroup of $\mathbf{G}(n)$.

According to Maschke's[9] theorem the representations of a finite group G in the automorphisms of a finite-dimensional vector space are decomposable into irreducible representations on vector subspaces, i.e., there exists a basis, which leads to box-matrices for the representation:

$$\text{finite } G \ni g \longmapsto D(g) \in \mathbf{AL}(N_1, \mathbb{K}) \oplus \ \dots \ \oplus \mathbf{AL}(N_k, \mathbb{K}) \subseteq \mathbf{GL}(V),$$
$$V = \bigoplus_{j=1}^{k} W_j \cong \mathbb{K}^N, \quad W_j \cong \mathbb{K}^{N_j}.$$

A finite group G of order N defines an N-dimensional unital *group* $\mathbb{K}$*-algebra* by the "formal" sums with $\mathbb{K}$-coefficients:

$$\mathbb{K}^G = \{\sum_{i=1}^{N} \alpha_i g_i \mid \alpha_i \in \mathbb{K}, \ g_i \in G\} \cong \mathbb{K}^N.$$

The elements can be considered as mappings from the groups into the numbers:

$$\alpha : G \longrightarrow \mathbb{K}, \quad \alpha(g_i) = \alpha_i, \quad \alpha = \sum_{i=1}^{N} \alpha_i g_i \in \mathbb{K}^G.$$

The group elements constitute a vector space basis of the group algebra. They can be added and multiplied with a $\mathbb{K}$-number. They have the associative *convolution product*, distributive over the addition:

$$\alpha * \beta = \sum_{i=1}^{N} \alpha_i g_i \sum_{j=1}^{N} \beta_j g_j = \sum_{i,j=1}^{N} \alpha_i \beta_j g_i g_j = \sum_{k=1}^{N} \gamma_k g_k \in \mathbb{K}^G,$$
$$\text{with } \gamma_k = \sum_{i,j=1}^{N} \delta^{g_k}_{g_i g_j} \alpha_i \beta_j.$$

The *permutation algebra* $\mathbb{K}^{\mathbf{G}(n)} = \mathbf{AL}(n!, \mathbb{K})$ is a vector space with dimension $n!$.

Each unital algebra is faithfully represented in its endomorphisms $a \longmapsto L_a$ by left multiplications $L_a(b) = ab$, called adjoint representation. The irreducible subspaces $AL \subseteq L$ are the minimal left ideals. For the algebra of a finite group, the group algebra $\mathbb{K}^G$ as representation space decomposes (with Maschke) into irreducible subspaces, the minimal left ideals. Therefore, $\mathbb{K}^G$ is a semisimple finite-dimensional algebra. The complex group algebra $\mathbb{C}^G$ is (with Wedderburn) isomorphic to a direct sum of full matrix algebras, which are isomorphic to the minimal ideals.

[9]Heinrich Maschke (1853–1908).

Especially: The complex permutation algebra is semisimple and decomposes into minimal Young ideals, isomorphic to full matrix algebras:

$$\mathbb{C}^{\mathbf{G}(n)} = \bigoplus_{Y(n)} I^{\min}(Y(n)) = \mathbf{AL}(n!, \mathbb{C}), \quad I^{\min}(Y(n)) \cong \mathbf{AL}(d(Y(n)), \mathbb{C}).$$

A minimal ideal $I^{\min}(Y(n))$ of the permutation algebra is characterized by a Young frame $Y(n)$ with n boxes. The number $z(n)$ of minimal ideals is the number of all possible Young frames. It gives the number of the different (inequivalent) irreducible representations of the permutation algebra and permutation group. In the example $n = 5$, the seven irreducible $\mathbf{G}(5)$-representations on $(1, 4, 5, 6, 5, 4, 1)$-dimensional vector spaces can be given as block-matrices in (26×26)-matrices with $26 = 1 + 4 + 5 + 6 + 5 + 4 + 1$:

$$\mathbb{C}^{\mathbf{G}(5)} \cong \mathbb{C}^{5!}, \quad 5! = 1^2 + 4^2 + 5^2 + 6^2 + 5^2 + 4^2 + 1^2, \quad z(5) = 7.$$

A minimal ideal of the permutation algebra for a Young frame $Y(n)$ is isomorphic to a full matrix algebra. Therefore, it is the direct sum of isomorphic minimal left ideals, the columns in the full matrix algebra. These minimal left ideals are characterized by the $d(Y(n))$ Young tableaux (columns) for this Young frame $Y(n)$ (full matrix):

$$I^{\min}(Y(n)) = \bigoplus_{j=1}^{d(Y(n))} L_j^{\min}(Y(n)), \quad L_j^{\min}(Y(n)) \cong \mathbb{C}^{d(Y(n))}.$$

The number of isomorphic left ideals is the dimension $d(Y(n))$ of a Young frame, therefore $n! = \sum_{Y(n)} d(Y(n))^2$ with $d(Y(n))$ the complex dimension of the corresponding irreducible representation of the permutation group.

The horizontal Young tableau with n boxes in one line characterizes the trivial representation of the permutation group by $\mathbf{G}(1) \cong \{1\}$, the vertical one with n boxes in one column the representation by the signature $\mathbf{G}(2) \cong \{\pm 1\}$. Both representations are one-dimensional.

Each element of a minimal left ideal can be used for a principal ideal form:

$$L_j^{\min}(Y(n)) = \mathbb{C}^{\mathbf{G}(n)} e_j(Y(n)).$$

The numbers in a Young tableau (increasing rightwards and downwards) define an element of the permutation algebra according to the following rule: Starting with $(12 \ldots n)$ one constructs, by total antisymmetrization over all columns with the signature, the associated *vertical operator* $V_j(Y(n))$ *of the Young tableau*:

$$V_j(Y(n)) = \sum_{\pi} \epsilon(\pi)\pi,$$

$$\text{e.g.,} \begin{cases} \begin{array}{|c|}\hline 1 \\ \hline 2 \\ \hline \vdots \\ \hline n \\ \hline \end{array} \Rightarrow V_A = (12\ldots n) - (21\ldots n) + \cdots \pm (n\,n-1\ldots 21) \\[2ex] \begin{array}{|c|c|c|c|}\hline 1 & 2 & \ldots & n \\ \hline \end{array} \Rightarrow V_S = 1_{\mathbf{G}(n)} = (12\ldots n) \\[2ex] \begin{array}{|c|c|}\hline 1 & 2 \\ \hline 3 & 4 \\ \hline \end{array} \Rightarrow V = (1234) - (3214) - (1432) + (3412) \end{cases}$$

Then one builds from $(12\ldots n)$, by total symmetrization over all lines and their sum, the associated *horizontal operator* $H_j(Y(n))$ *of the Young tableau*:

$$H_j(Y(n)) = \sum_{\pi} \pi,$$

$$\text{e.g.,} \begin{cases} \begin{array}{|c|}\hline 1 \\ \hline 2 \\ \hline \vdots \\ \hline n \\ \hline \end{array} \Rightarrow H_A = 1_{\mathbf{G}(n)} = (12\ldots n) \\[2ex] \begin{array}{|c|c|c|c|}\hline 1 & 2 & \ldots & n \\ \hline \end{array} \Rightarrow H_S = (12\ldots n) + (21\ldots n) + \cdots + (n\,n-1\ldots 21) \\[2ex] \begin{array}{|c|c|}\hline 1 & 2 \\ \hline 3 & 4 \\ \hline \end{array} \Rightarrow H = (1234) + (2134) + (1243) + (2143) \end{cases}$$

The product of both in the correct order:

$$e_j(Y(n)) = H_j(Y(n))V_j(Y(n)),$$

yields a generating element for the associated left ideal $L_j^{\min}(Y(n))$. The elements, associated to all Young tableaux for all Young frames, constitute an orthogonal system, i.e., they are, up to normalizations, projectors.

Chapter 6
The Hall of Mirrors

With the Stern–Gerlach experiment 1921 that showed a twofold split of a ray with silver atoms in an inhomogeneous magnetic field ("direction quantization"), and with the introduction of the concept *spin* 1925 by Goudsmit and Uhlenbeck[1] the rotation group $\mathbf{SO}(3)$ was dethroned in its fundamentality by the unitary spin group $\mathbf{SU}(2)$, and, equally, the orthochronous Lorentz group $\mathbf{SO}_0(1,3)$ by the group $\mathbf{SL}(2,\mathbb{C})$. $\mathbf{SO}(3)$ and $\mathbf{SO}_0(1,3)$ and their covering groups $\mathbf{SU}(2)$ for rotations and $\mathbf{SL}(2,\mathbb{C})$ for Lorentz transformations are real three- and six-dimensional Lie groups, respectively.

Irreducible real representations of the rotations $\mathbf{SO}(3)$ act on vector spaces with odd dimensions $1+2l$ for angular momentum $l \in \{0,1,2,\ldots\}$, e.g., on three-dimensional Euclidean position $\mathbb{E}^3$ with $l=1$. Those of the group $\mathbf{SU}(2)$ act on complex spaces with dimensions $1+2j$ for spin $j \in \{0,\frac{1}{2},1,\frac{3}{2},\ldots\}$, i.e., also on even-dimensional spaces, e.g., on Pauli spinors from $\mathbb{C}^2$, or as seen in the *two*fold split of the silver atom rays.

The reflection groups for position and time translations, both isomorphic to $\mathbf{I}(2) = \{\pm 1\}$, are subgroups of the full orthogonal group $\mathbf{O}(3) \cong \mathbf{I}(2) \times \mathbf{SO}(3)$ and of the full Lorentz group $\mathbf{O}(1,3) \cong \mathbf{I}(2) \vec{\times} [\mathbf{I}(2) \times \mathbf{SO}_0(1,3)]$. The two-ality group $\mathbf{I}(2)$ for the cover group-classes $\mathbf{SO}(3) \cong \mathbf{SU}(2)/\mathbf{I}(2)$ and $\mathbf{SO}_0(1,3) \cong \mathbf{SL}(2,\mathbb{C})/\mathbf{I}(2)$ has nothing to do with these reflections. How are the time and position reflections implemented if one starts from the reflection-free groups $\mathbf{SU}(2)$ and $\mathbf{SL}(2,\mathbb{C})$ as fundamental groups?

With the embedding of the orthogonal group $\mathbf{O}(3)$ in the Lorentz group $\mathbf{O}(1,3)$ the reflection of position translations is not a relativistically invariant operation. In special relativity, even more after the introduction of the spin, the invariance of a fundamental dynamics under position and time reflection has to be questioned. Are the reflections part of the fundamental symmetry of a dynamics or are they only additional symmetric "accidents" of special representations of the symmetry operations?

[1] Samuel Goudsmit (1902–1978), George Uhlenbeck (1900–1988).

H. Saller, *Operational Symmetries*, DOI 10.1007/978-3-319-58664-9_6

With the realization of real operations on complex vector spaces, reflections are implementable by conjugations too. For an operation group, there may exist nondegenerate invariant bilinear or sesquilinear forms. Such forms are an important tool for the implementation of invariant reflections — both for the representation spaces and for the operations.

Starting from $\mathbf{SL}(2, \mathbb{C})$, Weyl gave, already in the thirties, irreducible half-integer spin representations of the Lorentz group, which, in contrast to the decomposable Dirac representations, have not enough degrees of freedom to implement the position reflection by a chirality reflection (right-handed $\leftrightarrow$ left-handed). That these Weyl representations of the real Lorentz symmetry on complex vector spaces are not only a mathematical peculiarity without physical application, but that the reflections of the position and time translations are really not respected by a fundamental dynamics, was shown experimentally for the position reflection by Wu 1957 after a theoretical analysis of Lee and Yang[2] (P-violation), and for the time reflection — under the assumption of CPT-invariance — 1964 by Cronin and Fitch[3] (CP-violation).

When the physicists announced the discovery of the broken time reflection invariance, the journalists were not really impressed: "Some years ago, these guys told us that left is different from right. Today they sell us the difference of yesterday and tomorrow as a big sensation. What is all the fuss about?"

Apparently, the neutrinos with their weak interactions play a distinguished role in the breakdown of position and chirality reflection invariance. Here and also for the violation of CP-invariance, observed at the neutral kaons that have also strong interactions, the internal degrees of freedom also play a role.

To get used to a mirror, mathematically to the reflection group $\mathbf{I}(2) = \{\pm 1\}$, is easy. It is remarkable how difficult the orientation becomes with each additional mirror. Similar to how one can get easily lost in a hall with differently directed mirrors, the many reflections in basic physics are difficult to work with. In addition to the mirrors in real position and time, there act complex conjugation mirrors. Of basic importance is the mirror-symmety of particles and antiparticles, and of creation and annihilation operators.

6.1 Reflection Group and Orthogonality

The sign group $\mathbf{I}(2) = \{\pm 1\}$, the simplest nontrivial group, arises physically as *reflection group* $\{\mathbf{1}, \mathrm{R}\}$ with $\mathrm{R}^2 = \mathbf{1}$. It is isomorphic to the *orthogonal* group of the Euclidean line, $\mathbf{I}(2) \cong \mathbf{O}(1)$ — rotations in one dimension are reflections. Its action reflects the points $\mathbb{R} \ni x \stackrel{\mathrm{R}}{\leftrightarrow} -x \in \mathbb{R}$ at the origin as a zero-dimensional mirror, which is the sole unchanged point of the line. The length is unchanged $x^2 \stackrel{\mathrm{R}}{\leftrightarrow} x^2$.

---

[2]Chien-Shiung Wu (1912–1997), Tsung-Dao Lee (1926–), Chen Ning (Frank) Yang (1922–).
[3]James Cronin (1931–), Val Fitch (1923–).

In the Euclidean plane the reflection of all points on a line through the origin generates a reflection group in the nonabelian orthogonal group $\mathbf{O}(2)$, e.g., with the 1-axis as mirror:

$$\mathbf{I}(2) \cong \{\mathbf{1}, \mathrm{R}\} = \{\mathbf{1}_2, \begin{pmatrix} 1 & 0 \\ 0 & -1 \end{pmatrix}\} \subset \mathbf{O}(2).$$

A one-dimensional mirror in the plane is not unique, it is not rotation invariant. Any line through the origin is a possible mirror: The decomposition of the nonabelian orthogonal group $\mathbf{O}(2) \cong \mathbf{I}(2) \vec{\times} \mathbf{SO}(2)$ contains the reflection group as semidirect, not as direct factor. The unique *central reflection* $-\mathbf{1}_2$ as rotation $\begin{pmatrix} \cos\alpha & \sin\alpha \\ -\sin\alpha & \cos\alpha \end{pmatrix}$ for $\alpha = \pi$ with the origin as zero-dimensional mirror generates the normal subgroup $\{\pm\mathbf{1}_2\}$ in the abelian rotation group $\mathbf{SO}(2)$. The rotations $\mathbf{SO}(2)$ have no $\mathbf{SO}(2)/\mathbf{I}(2)$-isomorphic subgroup. This is different for the hyperbolic dilations $\mathbf{SO}(1,1)$ (Lorentz group for two-dimensional spacetime) with the orthochronous subgroup $\mathbf{SO}_0(1,1) \cong \mathbf{SO}(1,1)/\mathbf{I}(2)$.

The central reflection $-\mathbf{1}_2$ and a one-dimensional reflection $\begin{pmatrix} 1 & 0 \\ 0 & -1 \end{pmatrix}$ generate a *Klein four-group* $\mathbf{K}(4)$ in $\mathbf{O}(2)$ and in $\mathbf{O}(1,1)$:

$$\mathbf{K}(4) \cong \{\pm\mathbf{1}_2, \ \pm \begin{pmatrix} 1 & 0 \\ 0 & -1 \end{pmatrix}\}.$$

The multiplication table of the Klein four-group looks as follows:

$$\mathbf{K}(4) = \{\mathbf{1}, \mathrm{R}_1, \mathrm{R}_2, \mathrm{R}_3\} \cong \mathbf{I}(2) \times \mathbf{I}(2) \begin{cases} \cong \{\mathbf{1}, \mathrm{R}_1\} \times \{\mathbf{1}, \mathrm{R}_2\} \\ \cong \{\mathbf{1}, \mathrm{R}_2\} \times \{\mathbf{1}, \mathrm{R}_3\} \\ \cong \{\mathbf{1}, \mathrm{R}_3\} \times \{\mathbf{1}, \mathrm{R}_1\}, \end{cases}$$

1	R1	R2	R3
R1	**1**	R3	R2
R2	R3	**1**	R3
R3	R2	R1	**1**

All that can be easily generalized from one and two dimensions to reflections of n-dimensional vector spaces with the orthogonal action groups $\mathbf{O}(n_+, n_-)$: For each vector subspace $\mathbb{R}^k$ as a *k-dimensional mirror*, $k = 0, 1, \ldots, n-1$, there is an *$(n-k)$-dimensional reflection* that defines an $\mathbf{I}(2)$-subgroup of the orthogonal group. The reflection group for an odd-dimensional mirror leads to a semidirect factor $\mathbf{O}(n_+, n_-) \cong \mathbf{I}(2) \vec{\times} \mathbf{SO}(n_+, n_-)$. For odd dimension n, the central reflection group is a direct factor $\mathbf{O}(n_+, n_-) \cong \mathbf{I}(2) \times \mathbf{SO}(n_+, n_-)$, e.g., the reflection group of the position translations $\mathbf{O}(3) \cong \mathbf{I}(2) \times \mathbf{SO}(3)$.

The vectors of a mirror $\mathbb{R}^k \subset \mathbb{R}^n$ are reflection invariant (fixed), they have positive *parity* $+1$ as eigenvalue of the reflection; the orthogonal complement $\mathbb{R}^{n-k}$ has negative parity -1.

For nontrivial signature $\mathbf{O}(n_+, n_-)$, the reflections on n_-- and n_+-dimensional mirrors with vectors of negative and positive squares, respectively, are of interest — in adapted bases:

$$\mathbf{O}(n_+, n_-) : \mathtt{T} = \begin{pmatrix} -\mathbf{1}_{n_+} & 0 \\ 0 & \mathbf{1}_{n_-} \end{pmatrix}, \quad \mathtt{P} = \begin{pmatrix} \mathbf{1}_{n_+} & 0 \\ 0 & -\mathbf{1}_{n_-} \end{pmatrix},$$
$$\mathtt{T} \circ \mathtt{P} = \mathtt{Z} = -\mathbf{1}_n, \quad \mathtt{T} = -\mathtt{P}.$$

Their product is the central reflection $\mathtt{Z}$, which completes a Klein four-group in the orthogonal group,

$$\mathbf{O}(n_+, n_-) \supset \{\mathbf{1}_n, \mathtt{T}, \mathtt{P}, \mathtt{Z}\} \cong \mathbf{K}(4).$$

For even dimensions $n = 2, 4, \ldots$, the unique central reflection is in the normal subgroup $\{\pm\mathbf{1}_n\} \subset \mathbf{SO}(n_+, n_-)$. For $n_+ = 1$, there exists even a subgroup $\mathbf{SO}_0(1, n-1) \cong \mathbf{SO}(1, n-1)/\{\pm\mathbf{1}_n\}$ for the classes. This *orthochronous group* $\mathbf{SO}_0(1, n-1)$ is compatible with a nontrivial order relation (causality), familiar from Minkowski spacetime. It is a factor in the semidirect-direct product decomposition:

$$\mathbf{O}(1, n-1) \cong \mathbf{I}(2) \vec{\times} [\mathbf{I}(2) \times \mathbf{SO}_0(1, n-1)],$$
$$\mathbf{O}(1, n-1)/\mathbf{SO}_0(1, n-1) \cong \mathbf{K}(4) \cong \mathbf{I}(2) \times \mathbf{I}(2).$$

A one-dimensional reflection $\mathtt{R}$ in $\mathbb{R}^n$ is not a rotation. There exist orthogonal transformations which change the orientation of an object. By embedding $\mathbb{R}^n$ in a vector space $\mathbb{R}^{n+1}$ with one additional dimension, the reflection can be effected by a rotation $\mathbf{SO}(n+1)$ in $\mathbb{R}^{n+1}$:

$$\text{for } \mathbb{R}^n,\ n \geq 2: \ \mathtt{R} = \begin{pmatrix} \mathbf{1}_{n-1} & 0 \\ 0 & -1 \end{pmatrix} \in \mathbf{O}(n), \quad \det \mathtt{R} \circ \mathbf{SO}(n) = \{-1\},$$

$$\begin{matrix} \text{for } \mathbb{R}^n \subset \mathbb{R}^{n+1}, \\ \\ \mathbf{O}(n) \subset \mathbf{SO}(n+1): \end{matrix} \left\{ \begin{array}{l} \mathbf{SO}(n) \hookrightarrow \begin{pmatrix} \mathbf{SO}(n) & 0 \\ 0 & 1 \end{pmatrix}, \\ \mathtt{R} \circ \mathbf{SO}(n) \hookrightarrow \begin{pmatrix} \mathtt{R} \circ \mathbf{SO}(n) & 0 \\ 0 & -1 \end{pmatrix}, \\ \mathtt{R} \hookrightarrow \begin{pmatrix} \mathbf{1}_{n-1} & 0 & 0 \\ 0 & -1 & 0 \\ 0 & 0 & -1 \end{pmatrix} \in \mathbf{SO}(n+1). \end{array} \right.$$

Two reflected noodles on a tabletop, e.g., formed as a letter N, cannot be aligned by rotation or translation — however with the aid of a rotation in three-dimensional position. As parents of small children know, a left-handed glove cannot be rotated into a right-handed one. If one could use rotations $\mathbf{SO}(4)$ in a four-dimensional Euclidean space, glove factories would manufacture only, e.g., left-handed gloves.

6.2 Reflections of Time and Position

Time and position translations have linear reflections by orthogonal subgroups:

$$\mathbf{O}(1) \cong \mathbf{I}(2), \mathbf{O}(3) \cong \mathbf{I}(2) \times \mathbf{SO}(3) \qquad \mathbf{I}(2) \cong \{\pm \mathbf{1}_3\}.$$
$$t \overset{\mathtt{T}}{\leftrightarrow} -t \qquad \vec{x} \overset{\mathtt{P}}{\leftrightarrow} -\vec{x}$$

They are embedded in a spacetime reflection Klein group in the Lorentz group acting on spacetime translations $\mathbb{R}^{1+3}$ (Minkowski spacetime):

$$\mathbf{O}(1,3) \cong \mathbf{I}(2) \vec{\times} \mathbf{SO}(1,3) \cong \mathbf{I}(2) \vec{\times} [\mathbf{I}(2) \times \mathbf{SO}_0(1,3)].$$

A *time reflection* $\mathtt{T}$ employs a position mirror $\mathbb{E}^3 \cong \mathbb{R}^3$ with $(t, \vec{x}) \overset{\mathtt{T}}{\leftrightarrow} (-t, \vec{x})$ and a *position reflection* $\mathtt{P}$ a time mirror $\mathbb{R}$ with $(t, \vec{x}) \overset{\mathtt{P}}{\leftrightarrow} (t, -\vec{x})$. There are as many position reflections $\mathtt{P}$ with associated time reflection $\mathtt{T}$ as there are decompositions into time and position translations. $\{\mathtt{P}, \mathtt{T}\}$ are only rotation, not boost invariant. Their product, the central spacetime reflection $x \overset{-\mathbf{1}_4}{\leftrightarrow} -x$, is invariant under the full Lorentz group $\mathbf{O}(1,3)$:

$$\mathbf{O}(1,3)/\mathbf{SO}_0(1,3) \cong \mathbf{I}(2) \times \mathbf{I}(2) \cong \mathbf{K}(4) \cong \{\mathbf{1}_4, \mathtt{P}\} \times \{\pm \mathbf{1}_4\} \cong \{\pm \mathbf{1}_4, \mathtt{P}, \mathtt{T}\},$$
$$\mathtt{P} \in \mathbf{O}(1,3)/\mathbf{SO}(1,3),$$
$$\mathtt{T} = -\mathtt{P}, \quad \det \mathtt{P} = -1 = \det \mathtt{T},$$
$$[\mathtt{P}, \mathbf{SO}_0(1,3)] \neq \{0\}, \quad [\mathtt{P}, \mathbf{O}(3)] = \{0\}.$$

$\mathbf{1}_4$	$-\mathbf{1}_4$	T	P
$-\mathbf{1}_4$	$\mathbf{1}_4$	P	T
T	P	$\mathbf{1}_4$	$-\mathbf{1}_4$
P	T	$-\mathbf{1}_4$	$\mathbf{1}_4$

Good to know: Minkowski spacetime has bases where a position reflection $\mathtt{P}$ and the Lorentz metric η have the same matrix $\begin{pmatrix} 1 & 0 \\ 0 & -\mathbf{1}_3 \end{pmatrix}$. Nevertheless, they are, as an automorphism $\mathtt{P}$ and as a bilinear form η, conceptually completely different. A similarity transformation f, e.g., for a basis change, uses the inverse transformation for vector space transformations, e.g., for the position reflection $\mathtt{P} \longmapsto f \circ \mathtt{P} \circ f^{-1}$, and the transposed transformation for bilinear vector space forms, e.g., for the Lorentz metric $\eta \longmapsto f \circ \eta \circ f^T$.

Position rotations with reflection $\mathbf{O}(3) \cong \mathbf{I}(2) \times \mathbf{SO}(3)$ can be embedded in the special Lorentz group, $\mathbf{O}(3) \subset \mathbf{SO}(1,3)$, where the reflection of the position translations is part of the central reflection of all spacetime translations $-\mathbf{1}_3 \hookrightarrow -\mathbf{1}_4$. Invariance under $\mathbf{O}(3)$ must not lead to invariance under the full Lorentz group $\mathbf{O}(1,3)$ — analogously for the embedding of the time reflection $\mathbf{O}(1) \cong \mathbf{I}(2) \subset \mathbf{SO}(1,3)$. Already the special Lorentz group $\mathbf{SO}(1,3)$ contains $-\mathbf{1}_4$ as a simultaneous position- and time-translation reflection: Projections of all translations on position and time translations reduce the central spacetime reflection $-\mathbf{1}_4$ to a position reflection $\mathtt{P}$ and a time reflection $\mathtt{T}$, respectively.

6.3 Reflection by Conjugation

If a real two-dimensional Euclidean space is considered as a complex plane $\mathbb{C} \cong \mathbb{R}^2$ — how do the real reflections look like? The canonical conjugation $z \leftrightarrow \overline{z}$ is a reflection on the real axis:

$$\star : \mathbb{C} \longrightarrow \mathbb{C},\ z^\star = \overline{z},\ \ \mathbb{C} \ni z \cong \begin{pmatrix} x \\ y \end{pmatrix} \in \mathbb{R}^2,\ \ \overline{z} \cong \begin{pmatrix} x \\ -y \end{pmatrix},\ \star \cong \begin{pmatrix} 1 & 0 \\ 0 & -1 \end{pmatrix}.$$

The conjugation $\star$ is an antilinear bijection of $\mathbb{C}$: It is additive and involutive, $z^{\star\star} = z$, however no complex-linear automorphism $\mathbf{GL}(1, \mathbb{C})$, i.e., $(\alpha z)^\star \neq \alpha z^\star$ for $\alpha \notin \mathbb{R}$.

Good to know: An *antilinear* mapping of complex vector spaces $f : V \longrightarrow W$ is additive $f(v+w) = f(v) + f(w)$ with conjugation for the scalars $f(\alpha v) = \overline{\alpha} f(v)$. It is a real linear mapping.

The canonical conjugation of the complex numbers comes with the scalar product $(z, z') \longmapsto \langle z|z'\rangle = \overline{z}z'$, strictly positive $\overline{z}z > 0 \iff z \neq 0$. Its invariance group is the *unitary* group $\mathbf{U}(1) \subset \mathbf{GL}(1, \mathbb{C})$. The conjugation inverts the unitary group:

$$\star : \mathbf{U}(1) \longrightarrow \mathbf{U}(1),\ \ u = e^{i\alpha} \longmapsto u^\star = u^{-1} = e^{-i\alpha},$$

It reflects its Lie algebra $\mathbb{R}$ via the exponent $i\alpha \leftrightarrow -i\alpha$ in the Lie group $\mathbf{U}(1)$.

The conjugation $\star$ generates a reflection group $\mathbf{I}(2) \cong \{\mathbf{1}, \star\}$, which supplements the circle group $\mathbf{U}(1)$, arising as rotation group $\mathbf{SO}(2)$ (abelian) in the real plane, to the full orthogonal group $\mathbf{O}(2)$ (nonabelian). The semidirect group $\mathbf{I}(2) \vec{\times} \mathbf{U}(1)$ uses the action of the conjugation $\star$ on $\mathbf{U}(1)$:

$$\begin{aligned}
\mathbf{O}(2) \cong \mathbf{I}(2) \vec{\times} \mathbf{SO}(2) &\cong \{\mathbf{1}, \star\} \vec{\times} \mathbf{U}(1) \cong \mathbf{U}(1, \star),\\
\begin{pmatrix} \cos\alpha & \sin\alpha \\ -\sin\alpha & \cos\alpha \end{pmatrix} &\cong (1, e^{i\alpha}),\ \ \begin{pmatrix} 1 & 0 \\ 0 & -1 \end{pmatrix} \cong (\star, 1),\\
\begin{pmatrix} \cos\alpha & -\sin\alpha \\ -\sin\alpha & -\cos\alpha \end{pmatrix} &\cong (\star, e^{i\alpha}) = (1, e^{i\alpha}) \bullet (\star, 1),\\
\text{in } \mathbf{U}(1, \star) : &\begin{cases} (1, e^{i\alpha}) \bullet (1, e^{i\beta}) = (1, e^{i(\alpha+\beta)}),\\ (\star, e^{i\alpha}) \bullet (1, e^{i\beta}) = (\star, e^{i(\alpha-\beta)}),\\ (1, e^{i\alpha}) \bullet (\star, e^{i\beta}) = (\star, e^{i(\alpha+\beta)}),\\ (\star, e^{i\alpha}) \bullet (\star, e^{i\beta}) = (1, e^{i(\alpha-\beta)}). \end{cases}
\end{aligned}$$

A conjugation is a reflection; one has to get used to this connection. That Wigner connected the reversal of the time translations, parametrized by the real numbers, with an antilinear transformation in the complex-formulated quantum structure with the time representations $\mathbb{R} \longrightarrow \mathbf{U}(1)$, should — post festum — not be too surprising.

Good to know: The Fourier components of the characteristic functions for future and past

$$\vartheta(t) = \int \frac{dE}{2i\pi} \frac{1}{E - io} e^{iEt},\ \ \vartheta(-t) = -\int \frac{dE}{2i\pi} \frac{1}{E + io} e^{iEt},$$

(with distributive "small" $o > 0$ and limit $o \to 0$ after integration) are related to each other by conjugation, especially for the simple poles $E = \pm io$ in the upper (for future) and lower (for past) complex half-plane.

6.4 Dual Involutors and Conjugations

For an orientation, the most important general abstract structures of reflections are given in this section, with many examples below.

A bijection of a set that is its own inverse is called *involutive* or a reflection:

$$\mathrm{R} : M \longrightarrow M, \quad x \longmapsto x^{\mathrm{R}}, \quad \mathrm{R} \circ \mathrm{R} = \mathbf{1}_M, \quad x^{\mathrm{RR}} = x,$$

trivial for $\mathrm{R} = \mathbf{1}_M$. The mappings $\{\mathbf{1}_M, \mathrm{R}\}$ realize (implement) the reflection group $\mathbf{I}(2)$ in the bijection (permutation) group $\mathbf{I}(2) \longrightarrow \mathbf{G}(M)$, either trivial or faithful. Reflections as discrete groups have trivial Lie algebras.

A reflection is called an *involutor* in the case of a linear isomorphism of a vector space V — then one has a group representation $\mathbf{I}(2) \longrightarrow \mathbf{GL}(V)$, and a *conjugation* in the case of an antilinear bijection of a complex vector space. In the following, conjugations (antilinear bijections) are written with different "stars," $*, \star, \times, +$, etc. Only linear mappings should be written as matrices for given bases, matrices are not useful for antilinear mappings, e.g., a composition of two antilinear mappings is linear.

A vector space reflection, $\mathbf{I}(2) \times V \longrightarrow V$, $-1 \bullet v = v^{\mathrm{R}}$, which can be linear or antilinear for complex spaces, decomposes V into the reflection invariants V_+, the eigenspace with R-*eigenvalue* (R-*parity*) $+1$ for a linear reflection and with the $*$-*hermitian* elements for an antilinear $\mathrm{R} = *$, and the eigenvectors V_- with R-eigenvalue (R-parity) -1 and the $*$-*antihermitian* vectors $V_- = iV_+$ for $\mathrm{R} = *$:

$$V = V_+ \oplus V_-, \quad V_\pm = \{v = \pm v^{\mathrm{R}} \mid v \in V\} = \{v \pm v^{\mathrm{R}} \mid v \in V\}.$$

The implementation of reflection groups by bijections on sets leads to the implementation on product sets, e.g., on the tensor product of vector spaces $V_1 \otimes V_2 \ni v_1 \otimes v_2 \longmapsto \mathrm{R}_1(v_1) \otimes \mathrm{R}_2(v_2) \in V_1 \otimes V_2$. The eigenvalues are multiplied for products of linear reflection groups.

If a reflection $\mathrm{R} \in \mathbf{I}(2)$ acts on two sets, related by a mapping $\varphi : M \longrightarrow V$, e.g., the time reflection for a time orbit $\mathbb{R} \longrightarrow V$, or time and position reflections for a relativistic field $\mathbb{R}^{1+3} \longrightarrow V$, the reflection is induced on the mapping as defined by the commutativity of the following diagram:

$$\begin{array}{ccc} M & \xrightarrow{\mathrm{R}\bullet} & M \\ {\scriptstyle\varphi}\downarrow & & \downarrow{\scriptstyle\varphi^{\mathrm{R}}} \\ V & \xrightarrow[\mathrm{R}\bullet]{} & V \end{array} \qquad \begin{array}{l} \varphi \overset{\mathrm{R}}{\leftrightarrow} \varphi^{\mathrm{R}}, \quad \varphi^{\mathrm{RR}} = \varphi, \\ \varphi^{\mathrm{R}}(x) = \mathrm{R} \bullet \varphi(\mathrm{R} \bullet x). \end{array}$$

In the case of two vector spaces $\varphi : V_1 \longrightarrow V_2$, the (anti)linearity of the induced action is the "product" of the (anti)linearity of the individual actions, e.g., two antilinear actions induce a linear one.

A bijection of two sets $\mathtt{I} : M_1 \longrightarrow M_2$ defines an involutive bijection of the disjoint union, $\mathtt{R} = \mathtt{I} \uplus \mathtt{I}^{-1} : M_1 \uplus M_2 \longrightarrow M_1 \uplus M_2$, especially of the direct sum $V_1 \oplus V_2$ for an (anti)linear bijection $\mathtt{I} : V_1 \longrightarrow V_2$ of two vector spaces.

Bilinear or sesquilinear vector space forms provide *dual reflections*: A nondegenerate bilinear form of a finite-dimensional vector space $V \times V \ni (v, w) \longmapsto \mathtt{E}(v, w) \in \mathbb{C}$, e.g. the Lorentz metric of the Minkowski translations $\mathbb{R}^{1+3} \times \mathbb{R}^{1+3} \ni (x, y) \longmapsto \eta(x, y) \in \mathbb{R}$, defines an isomorphism between dual spaces $V \ni v \longmapsto \mathtt{E}(v, \ ) \in V^T$, and vice versa, illustrated by the isomorphy of Minkowski spacetime translations and energy-momenta and formalized by the raising and lowering of Lorentz vector indices. For an (anti-)symmetric form $\mathtt{E}(v, w) = \pm\mathtt{E}(w, v)$, the dual reflection $\mathtt{E}(v, \ ) = \pm\mathtt{E}(\ , v)$ is unique up to a sign. $\mathtt{E} \oplus \mathtt{E}^{-1}$ defines an involution of the selfdual space $\mathbf{V} = V \oplus V^T$, being the direct sum of vectors and their linear forms. For a sesquilinear nondegenerate form, there arises an antilinear bijection. An antilinear bijection $\mathtt{Z} : V \longrightarrow V^T$ and its inverse $\mathtt{Z}^{-1} : V^T \longrightarrow V$ are collected in the conjugation $* : \mathbf{V} \longrightarrow \mathbf{V}$ with $* = \mathtt{Z} \oplus \mathtt{Z}^{-1}$. (Anti)symmetry $\mathtt{Z}(v, w) = \pm\overline{\mathtt{Z}(w, v)}$ gives uniqueness — up to a sign. An example is the bra-ket conjugation $|v\rangle \overset{\star}{\leftrightarrow} \langle v|$, induced by the scalar product $\langle v|w\rangle$ of a Hilbert space.

A dual reflection $\mathtt{I}$ induces an operator reflection, i.e., a reflection of vector space endomorphisms as defined by the commutativity of the following diagram, which involves the transposed endomorphism f^T. The reflected endomorphism is linear:

$$\begin{array}{ccc} V & \xrightarrow{\mathtt{I}} & V^T \\ {\scriptstyle f^{\mathtt{I}}}\downarrow & & \downarrow{\scriptstyle f^T} \\ V & \xrightarrow[\mathtt{I}]{} & V^T \end{array}, \text{ with } \begin{cases} f^{\mathtt{I}} = \mathtt{I}^{-1} \circ f^T \circ \mathtt{I}, & (f^T)^{\mathtt{I}} = \mathtt{I} \circ f \circ \mathtt{I}^{-1}, \\ f \overset{\mathtt{I}}{\leftrightarrow} f^{\mathtt{I}}, \ f^{\mathtt{II}} = f, & (f \circ g)^{\mathtt{I}} = g^{\mathtt{I}} \circ f^{\mathtt{I}}. \end{cases}$$

There exist distinguished V-bases for nondegenerate forms with the parity eigenvalues $\epsilon \in \{\pm 1\}$ as nontrivial matrix-elements, leading to $\mathtt{I} \cong \begin{pmatrix} \mathbf{1}_{n_+} & 0 \\ 0 & -\mathbf{1}_{n_-} \end{pmatrix}$ in the nondegenerate symmetric bilinear (orthogonal) and symmetric and antisymmetric sesquilinear (unitary) case, and $\mathtt{I} \cong \begin{pmatrix} \mathbf{0} & \mathbf{1}_{\frac{n}{2}} \\ -\mathbf{1}_{\frac{n}{2}} & 0 \end{pmatrix}$ in the nondegenerate bilinear antisymmetric (symplectic) case with, necessarily, even-dimensional vector spaces V.

A bi- or sesquilinear nondegenerate form (metric) $\mathtt{I}$, invariant under an action group G or Lie algebra L, defines an invariant reflection:

$$\mathtt{I} : V \times V \longrightarrow \mathbb{K}, \quad \begin{cases} \mathtt{I}(g \bullet v, g \bullet w) = \mathtt{I}(v, w), & \text{for all } g \in G, \\ \mathtt{I}(l \bullet v, w) = -\mathtt{I}(v, l \bullet w), & \text{for all } l \in L. \end{cases}$$

6.5 Creation-Annihilation Conjugation

The unitary group $\mathbf{U}(1)$ with its conjugation $\mathbf{U}(1, \star) \cong \mathbf{O}(2)$ is used in quantum theory, e.g., for the representation of translations in the description of *particles*. The related dual conjugation $\star$ reflects *creation operators* u with *annihilation opera-*

tors $\mathrm{u}^\star$, e.g., the time translation eigenoperators of the one-dimensional harmonic quantum mechanical oscillator:

$$\text{creation operator: } \mathrm{u} = \frac{\mathbf{x} - i\mathbf{p}}{\sqrt{2}}, \quad \text{annihilation operator: } \mathrm{u}^\star = \frac{\mathbf{x} + i\mathbf{p}}{\sqrt{2}}.$$

The reflection invariant combinations for Bose operators are position and momentum:

$$\mathbf{x} = \frac{\mathrm{u} + \mathrm{u}^\star}{\sqrt{2}} = \mathbf{x}^\star, \quad \mathbf{p} = i\frac{\mathrm{u} - \mathrm{u}^\star}{\sqrt{2}} = \mathbf{p}^\star.$$

Good to know: The star $\star$ is used for the annihilation operator. The quantum structure uses products of u and $\mathrm{u}^\star$ modulo the commutator $[\mathrm{u}^\star, \mathrm{u}] = 1$ for bosons and the anticommutators $\{\mathrm{u}^\star, \mathrm{u}\} = 1$, $\{\mathrm{u}, \mathrm{u}\} = 0$, $\{\mathrm{u}^\star, \mathrm{u}^\star\} = 0$ for fermions.

The unitary phase group $\mathbf{U}(1)$ represents the time behavior of a free particle with the frequency (energy, mass) m as intrinsic unit:

$$\begin{aligned} \mathbf{U}(1):\quad \mathrm{u} \longmapsto \mathrm{u}(t) &= e^{imt}\mathrm{u}, & \mathrm{u}^\star \longmapsto \mathrm{u}^\star(t) &= e^{-imt}\mathrm{u}^\star, \\ d_t\mathrm{u}(t) &= im\mathrm{u}(t), & d_t\mathrm{u}^\star(t) &= -im\mathrm{u}^\star(t), \\ d_t\mathbf{x}(t) &= m\mathbf{p}(t), & d_t\mathbf{p}(t) &= -m\mathbf{x}(t), \\ (d_t^2 + m^2)(\mathrm{u}, \mathrm{u}^\star, \mathbf{x}, \mathbf{p}) &= 0. \end{aligned}$$

The $\mathbf{U}(1)$-scalar product leads to the quantum theoretical *probability interpretation*, normalized for the creation operator by $\parallel \mathrm{u} \parallel^2 = \langle \mathrm{u}|\mathrm{u}\rangle = 1$. The annihilation operator has trivial norm $\parallel \mathrm{u}^\star \parallel^2 = \langle \mathrm{u}^\star|\mathrm{u}^\star\rangle = 0$ and is orthogonal to the creation operator $\langle \mathrm{u}^\star|\mathrm{u}\rangle = 0$.

The representation of the infinitesimal time operations (time translations, time Lie algebra) is given by the i-multiplied Hamiltonian, reflected by the $\mathbf{U}(1)$-conjugation $(i\mathbf{H})^\star = -i\mathbf{H}$ (antihermitian),

$$\mathbf{H} = m\frac{\mathbf{p}^2 + \mathbf{x}^2}{2} = \frac{m}{2}\{\mathrm{u}, \mathrm{u}^\star\} = \mathbf{H}^\star.$$

For elementary particles, the unitary group with reflection arises in the simplest case for a charge-neutral spinless particle with trivial position, e.g., for a neutral pion, assumed as stable:

$$\pi_0(t) = \frac{e^{imt}\mathrm{u} + e^{-imt}\mathrm{u}^\star}{\sqrt{2}}.$$

Good to know: For Minkowski spacetime, a spinless particle field $\mathbf{\Phi}$ without chargelike (internal) properties has the harmonic analysis by the direct integral $\mathbf{\Phi}(x) = {}^{\oplus}\!\int \frac{d^3q}{2q_0} \frac{e^{iqx}\mathrm{u}(\vec{q}) + e^{-iqx}\mathrm{u}^\star(\vec{q})}{\sqrt{2}}$ with $qx = q_0x_0 - \vec{q}\vec{x}$, the momentum $\vec{q}$, the energy $q_0 = \sqrt{m^2 + \vec{q}^2}$, and the momentum-dependent creation and annihilation operators, $\mathrm{u}(\vec{q})$ and $\mathrm{u}^\star(\vec{q})$, respectively, and the Lorentz invariant measure $\frac{d^3q}{2\sqrt{m^2+\vec{q}^2}}$ of the momentum hyperboloid. The quantum structure of spacetime fields uses the Dirac momentum-distribution $[\mathrm{u}^\star(\vec{q}), \mathrm{u}(\vec{p})] = 2q_0\delta(\vec{p} - \vec{q})$ and the Hilbert metric-distribution $\langle \mathrm{u}(\vec{q})|\mathrm{u}(\vec{p})\rangle = 2q_0\delta(\vec{p} - \vec{q})$.

One should realize that the $\mathbf{U}(1)$-conjugation, here $\mathrm{u} \overset{\star}{\leftrightarrow} \mathrm{u}^\star$, has to do with the time translation action $\mathbb{R} \longrightarrow \mathbf{U}(1)$ — it has nothing to do with a nontrivial electric charge — chargeless particles also have a creation-annihilation conjugation.

6.6 The Dual Reflection Klein Group

If the dual $\mathbf{U}(1)$-conjugation $\star$ of the complex self-dual two-dimensional vector space

$$\mathbf{V} = \mathbb{C}\mathrm{u} \oplus \mathbb{C}\mathrm{u}^\star \cong \mathbb{C}^2,$$

with a basis of a creation and an annihilation operator is supplemented with a linear *dual involutor* E, which exchanges creation and annihilation operator, one generates four bijections of the self-dual vector space $\mathbf{V} \cong \mathbb{C}^2$; two are linear $\mathbf{I}(2) \cong \{\mathbf{1}, \mathrm{E}\}$, and two antilinear $\{\star, \mathrm{T}\}$, as seen in the behavior of the scalars $\alpha \in \mathbb{C}$:

$\mathrm{u} \overset{\star,\mathrm{E}}{\leftrightarrow} \mathrm{u}^\star$

Reflection		Scalars	Vectors
Identity	1	$\alpha \leftrightarrow \alpha$	$\mathrm{u} \leftrightarrow \mathrm{u}, \quad \mathrm{u}^\star \leftrightarrow \mathrm{u}^\star$
U(1)-conjugation	$\star$	$\alpha \leftrightarrow \overline{\alpha}$	$\mathrm{u} \leftrightarrow \mathrm{u}^\star$
Dual involutor	E	$\alpha \leftrightarrow \alpha,$	$\mathrm{u} \leftrightarrow \epsilon\mathrm{u}^\star,$
Scalar conjugation	$\mathrm{T} = \mathrm{E} \circ \star$	$\alpha \leftrightarrow \overline{\alpha}$	$\mathrm{u} \leftrightarrow \epsilon\mathrm{u}, \quad \mathrm{u}^\star \leftrightarrow \epsilon\mathrm{u}^\star$

The involutor E is normalized by the E-*parity* $\epsilon \in \{\pm 1\}$. It commutes with the $\mathbf{U}(1)$-conjugation, i.e., $\mathrm{E} \circ \star = \star \circ \mathrm{E}$. The four bijections implement Klein's four-group $\mathbf{K}(4)$ on the self-dual vector space $\mathbf{V}$,

$$\textit{dual reflection group: } \mathbf{K}(4) \cong \mathrm{grp}(\mathrm{E}, \star) = \{\mathbf{1}, \star, \mathrm{E}, \mathrm{T}\},$$
$$\mathbf{I}(2) \cong \{\mathbf{1}, \mathrm{E}\} \overset{\star}{\leftrightarrow} \{\star, \mathrm{T}\}.$$

Good to know: In a monoid M with associative product $M \times M \longrightarrow M$ and neutral element $1_M x = x 1_M = x \in M$, a subset T of invertible elements generates, by all finite products with elements of $T \cup T^{-1}$, a group, denoted by $\mathrm{grp}\, T \subseteq M$.

The scalar conjugation T is used for the implementation of the *time reflection*. It multiplies creation and annihilation operator with the E-parity ϵ and conjugates the scalars, e.g., for the neutral pion:

$$\text{time reflection: } \begin{array}{ccc} \mathbb{R} & \overset{-1}{\longrightarrow} & \mathbb{R} \\ f \downarrow & & \downarrow f^{\mathrm{T}} \\ \mathbf{V} & \underset{\mathrm{T}}{\longrightarrow} & \mathbf{V} \end{array} \quad \begin{array}{l} \mathrm{u}(t) \overset{\mathrm{T}}{\leftrightarrow} \epsilon\mathrm{u}(-t), \\ \mathrm{u}^\star(t) \overset{\mathrm{T}}{\leftrightarrow} \epsilon\mathrm{u}^\star(-t), \end{array}$$

$$\text{e.g.,} \boldsymbol{\pi}_0(t) = \frac{e^{imt}\mathrm{u} + e^{-imt}\mathrm{u}^\star}{\sqrt{2}} \overset{\mathrm{T}}{\leftrightarrow} \epsilon\boldsymbol{\pi}_0(-t).$$

In contrast to the conjugation $\star$ with the $\mathbf{U}(1)$-invariant scalar product:

$$\mathrm{u} \overset{\star}{\longmapsto} \mathrm{u}^\star \overset{\mathbf{U}(1)}{\longmapsto} e^{-imt}\mathrm{u}^\star, \quad \mathrm{u} \overset{\mathbf{U}(1)}{\longmapsto} e^{imt}\mathrm{u} \overset{\star}{\longmapsto} e^{-imt}\mathrm{u}^\star,$$

the time reflection it is not compatible, $[\mathrm{T}, \mathbf{U}(1)] \neq 0$, with the $\mathbf{U}(1)$-time expansion of a translation eigenvector (free particle):

$$\mathrm{u} \stackrel{\mathrm{T}}{\longmapsto} \epsilon \mathrm{u} \stackrel{\mathbf{U}(1)}{\longmapsto} e^{imt} \epsilon \mathrm{u}, \quad \mathrm{u} \stackrel{\mathbf{U}(1)}{\longmapsto} e^{imt} \mathrm{u} \stackrel{\mathrm{T}}{\longmapsto} e^{-imt} \epsilon \mathrm{u}.$$

6.7 (Iso)Spin Reflections

A reflection may have to be compatible with other operations, e.g., with external or internal symmetry operations for elementary particles and their fields. For example, spin and isospin are properties of the unitary nonabelian real three-dimensional Lie group $\mathbf{SU}(2)$, represented by fields and particles. To see spin and isospin in a mirror, the implemented reflection group $\mathbf{I}(2)$ has to be compatible with the represented $\mathbf{SU}(2)$-operations. The $\mathbf{SU}(2)$-representations have two dual reflections, an involutor and a conjugation.

Good to know: A complex finite-dimensional irreducible representation of a simple Lie algebra has at most, up to a nontrivial scalar, one invariant bilinear form and one invariant sesquilinear form — sometimes both, sometime only one of them, sometimes none. This is a simple consequence of the lemma of Schur and Burnside.[4] If, and only if the weight diagram is central reflection-symmetric, there exists one invariant bilinear form. In this case, the irreducible representation is self-dual.

(Iso)spin $\mathbf{SU}(2)$ is irreducibly represented as a unitary subgroup of a complex $(1+2j)$-dimensional vector space. The $\mathbf{SU}(2)$-Lie algebra $A_1^c \ni \vec{\alpha}\vec{\Sigma}$ with three basic vectors $(\Sigma^1, \Sigma^2, \Sigma^3)$ and real components $\alpha_a \in \mathbb{R}$ comes in the corresponding antihermitian linear transformations:

$$\begin{aligned}
\text{trivial representation}\, j = 0: &\quad \vec{\alpha}\vec{\Sigma} \longmapsto 0,\\
\text{Pauli (iso)spinor representation}\, j = \frac{1}{2}: &\quad \vec{\alpha}\vec{\Sigma} \longmapsto \vec{\alpha}\frac{i\vec{\sigma}}{2} \cong \frac{i}{2}\begin{pmatrix} \alpha_3 & \alpha_1 - i\alpha_2 \\ \alpha_1 + i\alpha_2 & -\alpha_3 \end{pmatrix},\\
\text{angular momentum representation}\, j = 1: &\quad \vec{\alpha}\vec{\Sigma} \longmapsto \vec{\alpha}\vec{L} \cong \begin{pmatrix} 0 & \alpha_3 & -\alpha_2 \\ -\alpha_3 & 0 & \alpha_1 \\ \alpha_2 & -\alpha_1 & 0 \end{pmatrix}.
\end{aligned}$$

The (iso)spin Lie algebra is centrally reflected by the unitary conjugation:

$$\begin{array}{ccc} V & \stackrel{\star}{\longrightarrow} & V^T \\ \vec{\Sigma}\downarrow & & \downarrow -\vec{\Sigma}^\star \\ V & \underset{\star}{\longrightarrow} & V^T \end{array} \quad \text{e.g.,} \quad \begin{cases} j = 0: 0^\star = 0, \\ j = \frac{1}{2}: (i\vec{\sigma})^\star = -i\vec{\sigma} \iff \delta_{AC}\overline{\vec{\sigma}_D^C}\delta^{DB} = \vec{\sigma}_A^B, \\ \qquad\qquad A, B = 1, 2, \\ j = 1: \vec{L}^\star = \vec{L}^T = -\vec{L} \iff \epsilon^{abc} = -\epsilon^{bac}, \\ \qquad\qquad a, b = 1, 2, 3. \end{cases}$$

Obviously, the analogous conjugation structures apply for all compact groups $\mathbf{SU}(1+r)$ and their Lie algebras $A_r^c \cong \mathbb{R}^{r(r+2)}$, $r \geq 1$.

[4]Issai Schur (1875–1941), William Burnside (1852–1927).

All representation spaces of the group **SU**(2) (spin and isospin) have, in addition to the unitary conjugation, a *linear* dual involution via an **SU**(2)-invariant bilinear form that also gives a central (iso)spin reflection. The defining complex two-dimensional Pauli (iso)spinor representation starts with the **SU**(2)-invariant antisymmetric *(iso)spinor metric*:

$$\text{Pauli (iso)spinors:} \begin{cases} \epsilon : V \times V \longrightarrow \mathbb{C}, \ \ \epsilon(v,w) = -\epsilon(w,v), \\ \{\mathrm{u}^1, \mathrm{u}^2\} \text{ basis von } V \cong \mathbb{C}^2 \Rightarrow \epsilon(\mathrm{u}^A, \mathrm{u}^B) = \epsilon^{AB} \cong \begin{pmatrix} 0 & 1 \\ -1 & 0 \end{pmatrix}. \end{cases}$$

It induces the real (iso)spin involution:

$$\begin{aligned} &\begin{pmatrix} 0 & -1 \\ 1 & 0 \end{pmatrix}\begin{pmatrix} \alpha_3 & \alpha_1 - i\alpha_2 \\ \alpha_1 + i\alpha_2 & -\alpha_3 \end{pmatrix}^T\begin{pmatrix} 0 & 1 \\ -1 & 0 \end{pmatrix} = -\begin{pmatrix} \alpha_3 & \alpha_1 - i\alpha_2 \\ \alpha_1 + i\alpha_2 & -\alpha_3 \end{pmatrix}, \\ &\epsilon^{-1} \circ \vec{\sigma}^T \circ \epsilon = -\vec{\sigma} \iff \epsilon_{AC}\vec{\sigma}^C_D \epsilon^{DB} = -\vec{\sigma}^B_A. \end{aligned}$$

As a (2×2)-matrix $\epsilon \cong \begin{pmatrix} 0 & 1 \\ -1 & 0 \end{pmatrix}$ has the form of $i\sigma_2$ (spin) or $i\tau_2$ (isospin) — a possibly misleading identification, since $\epsilon_\sigma, \epsilon_\tau : V \times V \longrightarrow \mathbb{C}$ are bilinear forms in contrast to the endomorphisms $i\sigma_2, i\tau_2 : V \longrightarrow V$ for $V \cong \mathbb{C}^2$.

In contrast to that: Not all representation spaces for $\mathbf{SU}(1+r)$ with rank $r \geq 2$, e.g., not the quark spaces for color **SU**(3), have an invariant bilinear form.

Good to know: The totally antisymmetric multilinear $\mathbf{SL}(1+r, \mathbb{C})$-invariant volume elements of the vector space $\mathbb{C}^{1+r}$ are bilinear forms exactly for $r = 1$. A fortiori, the volume elements are $\mathbf{SU}(1+r)$-invariant.

In parallel to the definition of all irreducible (iso)spin representations by totally symmetric products of the defining complex two-dimensional Pauli representation, the (iso)spinor bilinear form gives, via totally symmetric powers, invariant nondegenerate bilinear forms E for the representation vector spaces $V \cong \mathbb{C}^{1+2j}$ and, with them, invariant linear (iso)spin reflections. Here, a *connection between symmetry properties and (iso)spin* shows up: The representations with half-integer (iso)spin, i.e., $j = \frac{1}{2}, \frac{3}{2}, \ldots$, have an antisymmetric (symplectic) metric, those with integer (iso)spin, i.e., $j = 0, 1, \ldots$, a symmetric (orthogonal) one:

$$\mathrm{E} \cong \epsilon_j \bigvee^{2j} \epsilon, \mathrm{E}(v,w) = (-1)^{2j}\mathrm{E}(w,v), \ \mathrm{E}-\text{ parity } \epsilon_j \in \{\pm 1\},$$

$$\mathrm{E} : V \longrightarrow V^T, \begin{cases} \langle \mathrm{E}(v), w \rangle = \mathrm{E}(v,w), \\ \mathrm{u}^a \leftrightarrow \mathrm{E}^{ab}\mathrm{u}^\star_b, \ \ \mathrm{E}^{ab}\mathrm{E}_{bc} = \delta^a_c, \end{cases} \quad \mathrm{E}^{-1} \circ \vec{\Sigma} \circ \mathrm{E} = -\vec{\Sigma},$$

$$\begin{array}{ccc} V & \xrightarrow{\mathrm{E}} & V^T \\ \vec{\Sigma}\downarrow & & \downarrow -\vec{\Sigma}^T \\ V & \xrightarrow[\mathrm{E}]{} & V^T \end{array} \text{ e.g., } \begin{cases} j = 0; \ \ V = \mathbb{C}\mathrm{u}, \ \mathrm{E} \cong \epsilon_0 \, 1, \\ \qquad \mathrm{u} \leftrightarrow \epsilon_0 \mathrm{u}^\star, \ \ 1 \cdot 0 \cdot 1 = 0, \\ j = \frac{1}{2} : V = \mathbb{C}^2\mathrm{u}^A, \ \mathrm{E} \cong \epsilon_{\frac{1}{2}}\epsilon, \ \epsilon \cong \epsilon^{AB} = -\epsilon^{BA}, \\ \qquad \begin{cases} \mathrm{u}^1 \leftrightarrow \epsilon_{\frac{1}{2}}\mathrm{u}^\star_2, \\ \mathrm{u}^2 \leftrightarrow -\epsilon_{\frac{1}{2}}\mathrm{u}^\star_1, \end{cases} \quad \epsilon^{-1} \circ \vec{\sigma}^T \circ \epsilon = -\vec{\sigma}, \\ j = 1 : \ V = \mathbb{C}^3\mathrm{u}^a, \ \mathrm{E} \cong \epsilon_1 \mathbf{1}_3, \ \mathbf{1}_3 \cong \delta^{ab} = +\delta^{ba}, \\ \qquad \mathrm{u}^a \leftrightarrow \epsilon_1 \delta^{ab}\mathrm{u}^\star_b, \ \ \delta^{-1} \circ \vec{L}^T \circ \delta = -\vec{L}. \end{cases}$$

Good to know: The vector space notation $V = \mathbb{K}^n \mathrm{u}^A$ gives the scalars $\mathbb{K} \in \{\mathbb{R}, \mathbb{C}\}$, the dimension n, and a basis $\{\mathrm{u}^A\}_{a=1}^n$ with index symbols A.

The representation spaces for integer (iso)spin $j = 0, 1, \ldots$, e.g., Euclidean position space with $j = 1$, can be chosen as real spaces. For $j \neq 0$, they are acted on faithfully only by the two-ality classes, i.e., by the (iso)rotation group $\mathbf{SO}(3) \cong \mathbf{SU}(2)/\mathbf{I}(2)$. For the real representation spaces, the metric from the dual isomorphism can be identified with the scalar product from the unitary conjugation $\star$, i.e., the conjugation looses an independent meaning. Complex representation spaces for half-integer (iso)spin have two nontrivial invariant reflections with Klein-reflection group, real representation spaces for integer (iso)spin have only one nontrivial invariant reflection:

$$\mathbf{SU}(2) \longrightarrow \mathbf{SU}(1+2j), j = \frac{1}{2}, \frac{3}{3}, \cdots : \quad \begin{cases} \mathbf{K}(4) \cong (\mathbf{1}_{1+2j}, \star, \mathrm{E}, \star \circ \mathrm{E}), \\ \mathrm{E} \cong \epsilon_j \begin{pmatrix} 0 & \mathbf{1}_{\frac{1+2j}{2}} \\ -\mathbf{1}_{\frac{1+2j}{2}} & 0 \end{pmatrix}, \end{cases}$$

$$\mathbf{SO}(3) \longrightarrow \mathbf{SO}(1+2j), j = 0, 1, \ldots : \quad \begin{cases} \mathbf{I}(2) \cong (\mathbf{1}_{1+2j}, \mathrm{E}), \\ \mathrm{E} \cong \epsilon_j \mathbf{1}_{1+2j}. \end{cases}$$

The spin-related linear dual reflection will be used for the representation of the position reflection $\mathtt{P}$. As seen in the electroweak standard model, electromagnetic $\mathbf{U}(1)$-actions are embedded in hyperisospin $\mathbf{U}(2)$-actions. Therefore, the isospin-related linear dual reflection will be used for the electromagnetic charge reflection $\mathtt{C}$:

$$\mathtt{P} \longmapsto \mathrm{E}(\text{spin}), \quad \mathtt{C} \longmapsto \mathrm{E}(\text{isospin}).$$

6.8 Euclidean Conjugation and Anticonjugation

Real orthogonal structures are embeddable in complex unitary ones. In analogy to the different involutors in the definite and indefinite orthogonal groups $\mathbf{O}(n_+, n_-)$, there also are different conjugations, definite and indefinite ones, $\mathbf{U}(n_+, n_-)$.

A complex vector space with two dimensions, $V \cong \mathbb{C}^2$ will be looked at in more detail: A definite $\mathbf{U}(2)$-conjugation $\star$ is associated with a scalar product and $\mathbf{U}(2)$ as invariance group:

$$\begin{aligned} &\star : V \longrightarrow V^T, \quad v \longmapsto v^\star, \quad \begin{pmatrix} v_1 \\ v_2 \end{pmatrix} \overset{\star}{\leftrightarrow} (\overline{v_1}, \overline{v_2}), \\ &V \times V \longrightarrow \mathbb{C}, \quad \langle v|w\rangle = \langle v^\star, w\rangle = \overline{v_1} w_1 + \overline{v_2} w_2 = \langle u.v|u.w\rangle, \quad u \in \mathbf{U}(2). \end{aligned}$$

The unitary group $\mathbf{U}(2)$ is a real four-dimensional subgroup of the real eight-dimensional group $\mathbf{GL}(2, \mathbb{C})$. The antilinear conjugation $\star$ is not an element of the unitary group $\mathbf{U}(2)$, even not an element of all linear automorphisms $\mathbf{GL}(2, \mathbb{C})$.

A second indefinite conjugation of $V \cong \mathbb{C}^2$ exchanges the two components:

$$\times : V \longrightarrow V^T, \quad v \longmapsto v^\times, \quad \begin{pmatrix} v_1 \\ v_2 \end{pmatrix} \overset{\times}{\leftrightarrow} (\overline{v_2}, \overline{v_1}),$$
$$V \times V \longrightarrow \mathbb{C}, \quad \langle v||w\rangle = \langle v^\times, w\rangle = \overline{v_2}w_1 + \overline{v_1}w_2 = \langle u.v||u.w\rangle, \quad u \in \mathbf{U}(1,1).$$

$\times$ is called the *anticonjugation* of $\mathbb{C}^2$. The associated $\mathbf{U}(1,1)$-invariant inner product is indefinite; there are vectors with negative square, e.g., $\begin{pmatrix} 1 \\ -1 \end{pmatrix}$.

The five-corners star $\star$ is reserved for the definite $\mathbf{U}(n)$-conjugations, called *Euclidean conjugations*, the four-corners star $\times$ for the neutral signature indefinite $\mathbf{U}(n,n)$-conjugations, called *anticonjugations*, of complex even-dimensional vector spaces.

Good to know: Positivity $\langle v|v\rangle \geq 0$ for all vectors v entails symmetry $\langle v|w\rangle = \overline{\langle w|v\rangle}$.

The product (composition) of two conjugations, more general of two antilinear mappings, is a linear mapping, here a linear automorphism — in the example with the Euclidean $\mathbf{U}(2)$-conjugation and the $\mathbf{U}(1,1)$-anticonjugation:

$$\mathsf{C} = \times \circ \star \cong \begin{pmatrix} 0 & 1 \\ 1 & 0 \end{pmatrix} \in \mathbf{GL}(2, \mathbb{C}),$$
$$V \overset{\mathsf{C}}{\leftrightarrow} V, \quad \begin{pmatrix} v_1 \\ v_2 \end{pmatrix} \leftrightarrow \begin{pmatrix} v_2 \\ v_1 \end{pmatrix}.$$

With one conjugation fixed, e.g., a Euclidean $\mathbf{U}(2)$-conjugation $\star$, any other conjugation is determined by an automorphism of the complex vector space.

Two different commuting conjugations generate a Klein four-group — in the example:

$$\mathrm{grp}(\star, \times) = \{\mathbf{1}, \star, \times, \mathsf{C}\} \cong \mathbf{K}(4).$$

6.9 Particle-Antiparticle Quartets

The two unitary groups $\mathbf{U}(2)$ and $\mathbf{U}(1,1)$ are the invariance groups of the two conjugation types $\star$ and $\times$ of a complex two-dimensional vector space $V_{\text{doub}} = \mathbb{C}\mathrm{u} \oplus \mathbb{C}\mathrm{a}$ with a basis $\{\mathrm{u}, \mathrm{a}\}$ and its linear forms $V^T_{\text{doub}} = \mathbb{C}\mathrm{u}^\star \oplus \mathbb{C}\mathrm{a}^\star$ with dual basis $\{\mathrm{u}^\star, \mathrm{a}^\star\}$. In quantum theories, the definite $\mathbf{U}(2)$-scalar product is used for the probability interpretation, $\langle \mathrm{u}|\mathrm{u}\rangle = 1 = \langle \mathrm{a}|\mathrm{a}\rangle$.

This conjugation structure is realized on the *complex quartets of particle and antiparticle creation and annihilation operators*:

$$\begin{aligned} \text{particle creation operator:} &\quad \mathrm{u}, \\ \text{particle annihilation operator:} &\quad \mathrm{u}^\star = \mathrm{a}^\times, \\ \text{antiparticle creation operator:} &\quad \mathrm{a}, \\ \text{antiparticle annihilation operator:} &\quad \mathrm{a}^\star = \mathrm{u}^\times. \end{aligned}$$

An example is given by the charged pions (more ahead).

The action of the phase group $\mathbf{U}(\mathbf{1}_2)$, equal for the two basic vectors, realizes the unitary *time expansion* of free particle and antiparticle:

$$\text{time } \mathbb{R} \longrightarrow \mathbf{U}(\mathbf{1}_2) : \begin{cases} \mathrm{u} \longmapsto e^{imt}\mathrm{u} &= \mathrm{u}(t), \\ \mathrm{a} \longmapsto e^{imt}\mathrm{a} &= \mathrm{a}(t), \\ \mathrm{u}^\star \longmapsto e^{-imt}\mathrm{u}^\star &= \mathrm{u}^\star(t), \\ \mathrm{a}^\star \longmapsto e^{-imt}\mathrm{a}^\star &= \mathrm{a}^\star(t). \end{cases}$$

The associated Hamiltonian $\mathbf{H}$ generates via $i\mathbf{H}$ the Lie algebra of $\mathbf{U}(\mathbf{1}_2)$. It is hermitian for both conjugations,

$$\mathbf{H} = m(\mathrm{uu}^\star + \mathrm{aa}^\star) = m(\mathrm{ua}^\times + \mathrm{au}^\times) = \mathbf{H}^\star = \mathbf{H}^\times \cong m\begin{pmatrix} 1 & 0 \\ 0 & 1 \end{pmatrix},$$
$$\mathbf{H}(\mathrm{u}) = m\mathrm{u}, \quad \mathbf{H}(\mathrm{a}) = m\mathrm{a}.$$

The (2×2)-matrix form for the Hamiltonian as multiple of the identity is related to the space $V_{\text{doub}} \cong \mathbb{C}^2$.

Good to know: The conjugation of a product reverses the order of the factors, in general $(ab)^* = b^*a^*$.

The phase group $\mathbf{U}(1)_3$, inverse for the two basic vectors, can represent an *internal transformation*, e.g., the electromagnetic $\mathbf{U}(1)$:

$$\text{charge } \mathbf{U}(1) \longrightarrow \mathbf{U}(1)_3, \quad z \in \mathbb{Z} : \begin{cases} \mathrm{u} \longmapsto e^{zi\alpha}\mathrm{u}, \\ \mathrm{a} \longmapsto e^{-zi\alpha}\mathrm{a}, \\ \mathrm{u}^\star \longmapsto e^{-zi\alpha}\mathrm{u}^\star, \\ \mathrm{a}^\star \longmapsto e^{zi\alpha}\mathrm{a}^\star. \end{cases}$$

The Lie algebra of $\mathbf{U}(1)_3$ is generated by $i\mathbf{Q}$ with a charge $\mathbf{Q}$, e.g., an electromagnetic charge. The charge is invariant (hermitian) for the Euclidean conjugation, however antihermitian for the anticonjugation. It commutes with the Hamiltonian, i.e., $\mathbf{Q}$ is a time invariant:

$$\mathbf{Q} = z(\mathrm{uu}^\star - \mathrm{aa}^\star) = z(\mathrm{ua}^\times - \mathrm{au}^\times) = \mathbf{Q}^\star = -\mathbf{Q}^\times \cong \begin{pmatrix} 1 & 0 \\ 0 & -1 \end{pmatrix},$$
$$\mathbf{Q}(\mathrm{u}) = z\mathrm{u}, \quad \mathbf{Q}(\mathrm{a}) = -z\mathrm{a}, \quad [\mathbf{H}, \mathbf{Q}] = 0.$$

Particle and antiparticle have opposite $\mathbf{Q}$-properties, e.g., opposite electromagnetic charge for charged pions, or opposite hypercharge for neutral kaons, or opposite fermion numbers for neutrons.

The product $\mathsf{C} = \star \circ \times$ of Euclidean conjugation and anticonjugation is a linear reflection. It is conserved (time independent) and anticommutes with the charge $\mathbf{Q}$. It will be used to implement a *charge reflection* (since linear, the usual charge "conjugation" is avoided):

$$\mathsf{C} = \mathrm{ua}^\star + \mathrm{au}^\star = \mathrm{uu}^\times + \mathrm{aa}^\times = \mathsf{C}^\star = \mathsf{C}^\times \cong \begin{pmatrix} 0 & 1 \\ 1 & 0 \end{pmatrix},$$

$$\mathrm{u} \overset{\mathsf{C}}{\leftrightarrow} \mathrm{a}, \quad \mathrm{u}^\star \overset{\mathsf{C}}{\leftrightarrow} \mathrm{a}^\star, \quad [\mathbf{H}, \mathsf{C}] = 0, \quad \{\mathbf{Q}, \mathsf{C}\} = 0.$$

The product group of the two abelian groups $\mathbf{U}(\mathbf{1}_2)$ and $\mathbf{U}(1)_3$ for time and charge action,

$$\mathbf{U}(\mathbf{1}_2) \times \mathbf{U}(1)_3 = \{\begin{pmatrix} e^{i(t+\alpha)} & 0 \\ 0 & e^{i(t-\alpha)} \end{pmatrix} \mid \alpha, t \in \mathbb{R}\},$$

is acted on by the Klein four-group $\mathbf{K}(4) \cong \{\mathbf{1}, \star, \times, \mathsf{C}\}$ as follows:

$$\mathbf{U}(\mathbf{1}_2) \times \mathbf{U}(1)_3 \xrightarrow{\mathbf{K}(4)} \mathbf{U}(\mathbf{1}_2) \times \mathbf{U}(1)_3, \begin{cases} \begin{pmatrix} e^{i(t+\alpha)} & 0 \\ 0 & e^{i(t-\alpha)} \end{pmatrix}^\star = \begin{pmatrix} e^{-i(t+\alpha)} & 0 \\ 0 & e^{-i(t-\alpha)} \end{pmatrix}, \\ \begin{pmatrix} e^{i(t+\alpha)} & 0 \\ 0 & e^{i(t-\alpha)} \end{pmatrix}^\times = \begin{pmatrix} e^{-i(t-\alpha)} & 0 \\ 0 & e^{-i(t+\alpha)} \end{pmatrix}, \\ \begin{pmatrix} e^{i(t+\alpha)} & 0 \\ 0 & e^{i(t-\alpha)} \end{pmatrix}^\mathsf{C} = \begin{pmatrix} e^{i(t-\alpha)} & 0 \\ 0 & e^{i(t+\alpha)} \end{pmatrix}. \end{cases}$$

The simplest examples for complex quartets of elementary particles involve spinless particles with one nontrivial abelian "internal" property, e.g., the electromagnetically charged pions $\pi_\pm$ or neutral kaons with opposite hypercharge $(\mathrm{K}_0, \overline{\mathrm{K}}_0)$ — if assumed as stable with equal masses for the pair members. The fields for trivial position use the following combinations:

$$\boldsymbol{\Phi}(t) = \frac{e^{imt}\mathrm{u} + e^{-imt}\mathrm{a}^\star}{\sqrt{2}} \overset{\mathsf{C}}{\leftrightarrow} \boldsymbol{\Phi}^\star(t) = \frac{e^{imt}\mathrm{a} + e^{-imt}\mathrm{u}^\star}{\sqrt{2}},$$

$$\text{e.g., } \boldsymbol{\pi}(t) = \frac{e^{imt}\pi_+ + e^{-imt}\pi_-^\star}{\sqrt{2}} \overset{\mathsf{C}}{\leftrightarrow} \boldsymbol{\pi}^\star(t) = \frac{e^{imt}\pi_- + e^{-imt}\pi_+^\star}{\sqrt{2}},$$

where, for the pions, π_+ creates and $\pi_+^\star$ annihilates a positively charged pion — analogously $(\pi_-, \pi_+^\star)$ for the negatively charged pion. For a full Minkowski spacetime translation analysis, the complex quartet operators are momentum-dependent.

6.10 Spinless Particles with Isospin

The $\mathbf{SU}(2)$-invariant bilinear form for the reflection of (iso)spin plays an important role for elementary particles and their fields. To separate its use for spin as external spacetime-related property and for isospin as internal chargelike property, the reflection for spinless bosons is considered first.

In a symmetric idealization for strong isospin with stable particles one has for the eight lowest-mass pseudoscalar bosons as mass-degenerate isospin multiplets:

$$\begin{aligned}\pi-\text{meson triplet (strong isospin1)}:&\quad m\simeq 140\,\frac{\text{MeV}}{c^2},\\ \text{two K}-\text{meson dublets (strong isospin}\frac{1}{2}):&\quad M\simeq 495\,\frac{\text{MeV}}{c^2},\\ \eta-\text{meson singlet (strong isospin0)}:&\quad \mu\simeq 550\,\frac{\text{MeV}}{c^2}.\end{aligned}$$

In the general case with isospin $I = 0, \frac{1}{2}, 1, \ldots$, the complex quartets with particle and antiparticle creation and annihilation operators span a complex $4(1+2I)$-dimensional vector space:

$$\mathbf{V}_{\text{doub}} = \mathbb{C}^{1+2I}\mathrm{u}_a \oplus \mathbb{C}^{1+2I}\mathrm{a}^a \oplus \mathbb{C}^{1+2I}\mathrm{u}^{\star a} \oplus \mathbb{C}^{1+2I}\mathrm{a}^\star_a.$$

The location of the indices $a = 1, \ldots, 1+2I$ at the bases indicates isospin representations for the particle creation operators u_a and antiparticle annihilation operators $\mathrm{a}^\star_a$ which are dual to those of the particle annihilation operators $\mathrm{u}^{\star a}$ and antiparticle creation operators a^a.

The eight-elementic quartet reflection group $\mathrm{grp}(\mathtt{E}, \star, +) \cong \mathbf{I}(2)^3$ is generated by three isospin-compatible reflections, two conjugations, called dual conjugation and antidual conjugation, and one bilinear form-induced involutor:

$$\{\mathrm{u}_a, \mathrm{u}^{\star a}, \mathrm{a}^a, \mathrm{a}^\star_a \mid a = 1, \ldots, 1+2I\}$$

$$\begin{array}{ccccc} & \mathrm{u}_a & \overset{\star,\mathtt{E}}{\leftrightarrow} & \mathrm{u}^{\star a} & \\ + & \updownarrow & & \updownarrow & + \\ & \mathrm{a}^a & \underset{\star,\mathtt{E}}{\leftrightarrow} & \mathrm{a}^\star_a & \end{array}$$

Reflection		Scalars	Vectors
Identity	$\mathbf{1}$	$\alpha \leftrightarrow \alpha$	$(\mathrm{u}_a, \mathrm{u}^{\star b}, \mathrm{a}^a, \mathrm{a}^\star_a) \leftrightarrow (\mathrm{u}_a, \mathrm{u}^{\star b}, \mathrm{a}^a, \mathrm{a}^\star_a)$
Dual conjugation	$\star$	$\alpha \leftrightarrow \overline{\alpha}$	$\mathrm{u}_a \leftrightarrow \delta_{ab}\mathrm{u}^{\star b},\ \mathrm{a}^a \leftrightarrow \delta^{ab}\mathrm{a}^\star_b$
Dual involutor	$\mathtt{E}$	$\alpha \leftrightarrow \alpha$	$\mathrm{u}_a \leftrightarrow \mathtt{E}_{ab}\mathrm{u}^{\star b},\ \mathrm{a}^a \leftrightarrow \mathtt{E}^{ab}\mathrm{a}^\star_b$
Scalar conjugation	$\mathtt{T} = \mathtt{E} \circ \star$	$\alpha \leftrightarrow \overline{\alpha}$	$\mathrm{u}_a \leftrightarrow \mathtt{E}_{ab}\delta^{bc}\mathrm{u}_c,\ \mathrm{u}^{\star a} \leftrightarrow \mathtt{E}^{ab}\delta_{bc}\mathrm{u}^{\star c}$ $\mathrm{a}^\star_a \leftrightarrow \mathtt{E}_{ab}\delta^{bc}\mathrm{a}^\star_c,\ \mathrm{a}^a \leftrightarrow \mathtt{E}^{ab}\delta_{bc}\mathrm{a}^c$
Antidual conjugation	$+$	$\alpha \leftrightarrow \overline{\alpha}$	$\mathrm{u}_a \leftrightarrow \delta_{ab}\mathrm{a}^b,\ \mathrm{u}^{\star a} \leftrightarrow \delta^{ab}\mathrm{a}^\star_b$
Antidual involutor	$\mathtt{I} = \star \circ +$	$\alpha \leftrightarrow \alpha$	$\mathrm{u}_a \leftrightarrow \mathrm{a}^\star_a,\ \mathrm{a}^a \leftrightarrow \mathrm{u}^{\star a}$
Anticonjugation	$\times = \mathtt{E} \circ +$	$\alpha \leftrightarrow \overline{\alpha}$	$\mathrm{u}_a \leftrightarrow \mathtt{E}_{ab}\delta^{bc}\mathrm{a}^\star_c,\ \mathrm{a}^a \leftrightarrow \mathtt{E}^{ab}\delta_{bc}\mathrm{u}^{\star c}$
Antiinvolutor	$\mathtt{C} = \mathtt{E} \circ \star \circ +$	$\alpha \leftrightarrow \alpha$	$\mathrm{u}_a \leftrightarrow \mathtt{E}_{ab}\mathrm{a}^b,\ \mathrm{a}^\star_a \leftrightarrow \mathtt{E}_{ab}\mathrm{u}^{\star b}$

Isospin $\mathbf{SU}(2)$ – compatible reflections

Via the four linear reflections in a Klein group, and the four antilinear reflections,

$$\mathbf{K}(4) \cong \{\mathbf{1}, \mathtt{E}, \mathtt{I}, \mathtt{C}\} \overset{\mathtt{T}}{\leftrightarrow} \{\mathtt{T}, \star, \times, +\},$$

each of the four subspaces, e.g., particle creators u_a, is bijectively connected with itself and with the three other subspaces, both linearly and antilinearly, in the example with $\mathrm{u}^{\star a}, \mathrm{a}^a, \mathrm{a}^\star_a$.

The Euclidean reflection relates with each other dual spaces. The dual involutor $\mathtt{E}$ has its origin in the invariant metrics of the isospin structure. All isospin representations on the four subspaces are equivalent. $\mathtt{E}$ is normalized with an $\mathtt{E}$-parity

$\epsilon_I \in \{\pm 1\}$ in such a way that the matrix E^{ab} has only ± 1 as nontrivial entries — in the examples:

$$\begin{aligned} I = 0 &: \mathrm{E} \cong \epsilon_0 \cong \pm 1, \\ I = \tfrac{1}{2} &: \mathrm{E}^{ab} \cong \epsilon_{\frac{1}{2}} \epsilon^{ab} \cong \pm \begin{pmatrix} 0 & 1 \\ -1 & 0 \end{pmatrix}, \\ I = 1 &: \mathrm{E}^{ab} \cong \epsilon_1 \delta^{ab} \cong \pm \begin{pmatrix} 1 & 0 & 0 \\ 0 & 1 & 0 \\ 0 & 0 & 1 \end{pmatrix}. \end{aligned}$$

For a particle and its antiparticle with isospin and trivial position,

$$\boldsymbol{\Phi}_a(t) = \frac{e^{imt}\mathrm{u}_a + e^{-imt}\mathrm{a}^\star_a}{\sqrt{2}}, \quad \boldsymbol{\Phi}^{\star a}(t) = \frac{e^{imt}\mathrm{a}^a + e^{-imt}\mathrm{u}^{\star a}}{\sqrt{2}},$$

the "scalar" conjugation for the antilinear time reflection involves, in addition to the canonical conjugation for the complex numbers, also the isospin reflection:

Time reflection $\mathrm{T} = \mathrm{E} \circ \star$	$\mathrm{u}_a \leftrightarrow \mathrm{E}_{ab}\delta^{bc}\mathrm{u}_c,\ \mathrm{u}^{\star a} \leftrightarrow \mathrm{E}^{ab}\delta_{bc}\mathrm{u}^{\star c}$ $\mathrm{a}^\star_a \leftrightarrow \mathrm{E}_{ab}\delta^{bc}\mathrm{a}^\star_c,\ \mathrm{a}^a \leftrightarrow \mathrm{E}^{ab}\delta_{bc}\mathrm{a}^c$

$$\boldsymbol{\Phi}_a(t) \leftrightarrow \mathrm{E}_{ab}\delta^{bc}\boldsymbol{\Phi}_c(-t), \quad \boldsymbol{\Phi}^{\star a}(t) \leftrightarrow \mathrm{E}^{ab}\delta_{bc}\boldsymbol{\Phi}^{\star c}(-t).$$

For isomultiplets with integer isospin $I = 0, 1, 2, \ldots$, the $\mathbf{SO}(3)$-isospin degrees of freedom can be identified with the antiparticle ones. One obtains for nontrivial isospin particle-antiparticle pairs with opposite isospin winding numbers:

$$I = 0,\ \mathrm{E} \cong \epsilon_0 : \begin{cases} \boldsymbol{\eta}(t) = \frac{e^{i\mu t}\mathrm{u} + e^{-i\mu t}\mathrm{u}^\star}{\sqrt{2}}, \\ \text{with } \mathrm{a} = \mathrm{u},\ \ \mathrm{a}^\star = \mathrm{u}^\star, \\ \boldsymbol{\eta}(t) \overset{\mathrm{T}}{\leftrightarrow} \epsilon_0 \boldsymbol{\eta}(-t). \end{cases}$$

$$\begin{matrix} I = 1,\ \mathrm{E} \cong \epsilon_1 \mathbf{1}_3 : \\ a = 1, 2, 3, \end{matrix} \begin{cases} \boldsymbol{\pi}_a(t) = \frac{e^{imt}\mathrm{u}_a + e^{-imt}\mathrm{a}^\star_a}{\sqrt{2}}, \quad \boldsymbol{\pi}^{\star a}(t) = \frac{e^{imt}\mathrm{a}^a + e^{-imt}\mathrm{u}^{\star a}}{\sqrt{2}}, \\ \text{with } \mathrm{u}_1 \pm i\mathrm{u}_2 = \mathrm{a}^1 \mp i\mathrm{a}^2_2,\ \mathrm{u}_3 = \mathrm{a}^3, \\ \boldsymbol{\pi}_a(t) \overset{\mathrm{T}}{\leftrightarrow} \epsilon_1 \boldsymbol{\pi}_a(-t), \quad \boldsymbol{\pi}^{\star a}(t) \overset{\mathrm{T}}{\leftrightarrow} \epsilon_1 \boldsymbol{\pi}^{\star a}(-t). \end{cases}$$

For example, in an isotriplet $(\pi_1, \pi_2, \pi_3) \sim (\pi_-, \pi_0, \pi_+)$ the charged pions (π_+, π_-) are antiparticles, the neutral pion π_0 is its own antiparticle. The pion creation operators u_a as a Cartesian basis of the isospin representation space are connected with the charge eigenvectors in a spherical basis $\mathrm{u}_\pm$ by complex linear transformations $\mathrm{u}_\pm = \mathrm{u}_1 \pm i\mathrm{u}_2$, $\mathrm{u}_0 = \mathrm{u}_3$. If particles are identified with antiparticles, the complex quartet reflection group is reduced to the dual reflection group $\mathrm{grp}(\mathrm{E}, \star)$ for $\mathrm{u}_a \overset{\star,\mathrm{E}}{\leftrightarrow} \mathrm{u}^{\star a}$.

For halfinteger isospin, i.e., for necessarily complex $\mathbf{SU}(2)$-representation spaces, each particle comes with a different antiparticle, even for a trivial electromagnetic charge number. The simplest realization are isospin dublet kaons $(\mathrm{K}_0, \mathrm{K}_+)$, idealized as stable particles, created by $(\mathrm{u}_\alpha)_{\alpha=1,2}$, with its antiparticle dublet $(\mathrm{K}_-, \overline{\mathrm{K}}_0)$ created by $(\mathrm{a}^\alpha)_{\alpha=1,2}$:

$$I = \frac{1}{2},\ \mathrm{E} \cong \epsilon_{\frac{1}{2}}\epsilon : \quad a = 1, 2, \quad \begin{cases} \mathbf{K}_a(t) = \frac{e^{iMt}\mathrm{u}_a + e^{-iMt}\mathrm{a}_a^\star}{\sqrt{2}}, & \mathbf{K}^{\star a}(t) = \frac{e^{iMt}\mathrm{a}^a + e^{-iMt}\mathrm{u}^{\star a}}{\sqrt{2}}, \\ \mathbf{K}_a(t) \overset{\mathrm{T}}{\leftrightarrow} \epsilon_{\frac{1}{2}}\epsilon_{ab}\delta^{bc}\mathbf{K}_c(-t), & \mathbf{K}^{\star a}(t) \overset{\mathrm{T}}{\leftrightarrow} \epsilon_{\frac{1}{2}}\epsilon^{ab}\delta_{bc}\mathbf{K}^{\star c}(-t). \end{cases}$$

The linear charge reflection (conjugation) involves the isospin reflection (dual isomorphism):

Charge reflection $\mathrm{C} = \mathrm{E} \circ \star \circ +$	$\mathrm{u}_a \leftrightarrow \mathrm{E}_{ab}\mathrm{a}^b,\ \ \mathrm{a}_a^\star \leftrightarrow \mathrm{E}_{ab}\mathrm{u}^{\star b}$

$$\mathbf{\Phi}_a(t) \overset{\mathrm{C}}{\leftrightarrow} \mathrm{E}_{ab}\mathbf{\Phi}^{\star b}(t) : \quad \begin{cases} \boldsymbol{\eta} \overset{\mathrm{C}}{\leftrightarrow} \epsilon_0 \boldsymbol{\eta}, \\ \mathbf{K}_a \overset{\mathrm{C}}{\leftrightarrow} \epsilon_{\frac{1}{2}}\epsilon_{ab}\mathbf{K}^{\star b}, \\ \boldsymbol{\pi}_a \overset{\mathrm{C}}{\leftrightarrow} \epsilon_1 \delta_{ab}\boldsymbol{\pi}^{\star b}, \end{cases}$$

e.g., for the kaons $(\mathrm{K}_0, \mathrm{K}_+) \overset{\mathrm{C}}{\leftrightarrow} \epsilon_{\frac{1}{2}}(-\overline{\mathrm{K}}_0, \mathrm{K}_-)$.

6.11 Quartets of Complex Representations

A real operation structure, e.g., a real Lie group or a real Lie algebra, has to be recognizable as real, also in a complex representation. This requires a conjugation of the complex vector space acted on, and embeds a real symmetry as a substructure of a unitary group and an antisymmetric Lie algebra, e.g., $\mathbf{U}(1)$ and $\mathbf{SU}(n)$ are complex represented real Lie groups.

A conjugation of a complex vector space induces a conjugation of the $\mathbb{C}$-linear mappings. The Euclidean conjugation of $V \cong \mathbb{C}^2$ leads to the familiar hermitian transposition:

$$\begin{pmatrix} \alpha & \beta \\ \gamma & \delta \end{pmatrix}^\star = \star \circ \begin{pmatrix} \alpha & \beta \\ \gamma & \delta \end{pmatrix}^T \circ \star = \star \circ \begin{pmatrix} \alpha & \gamma \\ \beta & \delta \end{pmatrix} \circ \star = \begin{pmatrix} \overline{\alpha} & \overline{\gamma} \\ \overline{\beta} & \overline{\delta} \end{pmatrix},$$
e.g., for Pauli matrices: $(\sigma^a)^\star = \sigma^a, \quad a = 1, 2, 3.$

Other conjugations use, in addition to the hermitean transposition, a similarity transformation (inner automorphism) with the conjugation-characterizing vector space automorphism, e.g., $f^\times = \mathrm{C} \circ f^\star \circ \mathrm{C}^{-1}$ for the anticonjugation $\times$ on $V \cong \mathbb{C}^2$ with $\mathrm{C} = \times \circ \star$:

$$\begin{pmatrix} \alpha & \beta \\ \gamma & \delta \end{pmatrix}^\times = \begin{pmatrix} 0 & 1 \\ 1 & 0 \end{pmatrix} \begin{pmatrix} \alpha & \beta \\ \gamma & \delta \end{pmatrix}^\star \begin{pmatrix} 0 & 1 \\ 1 & 0 \end{pmatrix} = \begin{pmatrix} \overline{\delta} & \overline{\beta} \\ \overline{\gamma} & \overline{\alpha} \end{pmatrix},$$
e.g., for Pauli matrices: $(\sigma^1)^\times = \sigma^1, \quad (\sigma^2)^\times = -\sigma^2, \quad (\sigma^3)^\times = -\sigma^3.$

In general: A conjugation $V \overset{*}{\leftrightarrow} V^T$ of a complex vector space with its linear forms $V \cong \mathbb{C}^n \cong V^T$ and the related $\mathbf{U}(n_+, n_-)$-invariant sesquilinear form are the two sides of one coin. The conjugation also acts on all linear mappings $f : V \longrightarrow V$

by using the definite unitarity $\mathbf{U}(n)$ for the Euclidean conjugation $\star$ and a signature diagonal matrix as V-automorphism $\begin{pmatrix} \mathbf{1}_{n_+} & \\ 0 & -\mathbf{1}_{n_-} \end{pmatrix}$:

$$f^* = \begin{pmatrix} \mathbf{1}_{n_+} & \\ 0 & -\mathbf{1}_{n_-} \end{pmatrix} \circ f^\star \circ \begin{pmatrix} \mathbf{1}_{n_+} & \\ 0 & -\mathbf{1}_{n_-} \end{pmatrix}, \quad (f \circ g)^* = g^* \circ f^*$$

Both f and the conjugated f^* are $\mathbb{C}$-linear mappings.

Different conjugations of the vector space $V \cong \mathbb{C}^n$ are characterized by the signature (n_+, n_-) of the associated unitary invariance groups. A subgroup $\mathbf{U}(n_+, n_-) \subset \mathbf{GL}(n, \mathbb{C})$, defined by its $*$-unitarity, has a real Lie algebra, $\dim_{\mathbb{R}} \log \mathbf{U}(n_+, n_-) = n^2$, characterized by the corresponding $*$-antisymmetry:

$$\begin{aligned} \mathbf{U}(n_+, n_-) &= \{u \in \mathbf{GL}(n, \mathbb{C}) \mid u^* = u^{-1}\}, \\ \log \mathbf{U}(n_+, n_-) &= \{l \in \mathbf{AL}(n, \mathbb{C}) \mid l^* = -l\}, \end{aligned}$$

i.e., the $*$-conjugation of the real n^2-dimensional Lie algebra $\log \mathbf{U}(n_+, n_-)$ is the central reflection.

For example, the real three-dimensional Lie algebra of $\mathbf{SU}(2)$ can be spanned by the $\star$-antisymmetric basis $\{i\vec{\sigma}\}$ with the three Pauli matrices, the Lie algebra of $\mathbf{SU}(1,1)$ with equal real dimension by the $\times$-antisymmetric basis $\{i\sigma^1, \sigma^2, \sigma^3\}$ where two generators come, in comparison with those of $\mathbf{SU}(2)$, multiplied by an i, e.g., $\{\sigma^2, \sigma^3\}$:

$$\begin{aligned} \log \mathbf{SU}(2) &: (\vec{\alpha} i \vec{\sigma})^\star = -(\vec{\alpha} i \vec{\sigma}), \quad \alpha_a \in \mathbb{R}, \\ \log \mathbf{SU}(1,1) &: (\alpha_1 i\sigma^1 + \alpha_2\sigma^2 + \alpha_3\sigma^3)^\times = -(\alpha_1 i\sigma^1 + \alpha_2\sigma^2 + \alpha_3\sigma^3). \end{aligned}$$

Good to know: In physics, the commutators for the Pauli-matrices are familiar in the form $[\sigma^a, \sigma^b] = 2i\epsilon^{abc}\sigma^c$. Mathematically more appropriate for the real Lie algebra $\log \mathbf{SU}(2)$ is the pedantic notation $[i\sigma^a, i\sigma^b] = -2\epsilon^{abc}(i\sigma^c)$, since the generators of a unitary Lie group, here $\{i\sigma^a\}_{a=1}^3$, are antihermitian and the structure constants of a real Lie group are real, here $-2\epsilon^{abc}$.

With the quartet of complex vector spaces — space, dual space, antispace, dual antispace — there goes a quartet structure for representations of real Lie groups and Lie algebras. For special groups, like for $\mathbf{SU}(2)$ and $\mathbf{SL}(2, \mathbb{C})$, there may exist additional isomorphies which narrow down the quartet multiplicity.

Good to know: A real or complex finite-dimensional vector space V has its dual vector space V^T with the linear forms. A complex vector space $V \cong \mathbb{C}^n$ has, in addition to its dual $V^T \cong \mathbb{C}^n$, its *antispace* or *conjugated space* $\overline{V} \cong \mathbb{C}^n$. Together with the antidual space $\overline{V}^T$, there arises the complex quartet $(V, V^T, \overline{V}, \overline{V}^T)$. *Antispaces* have an isomorphic additive group-structure for their vectors, $v + w$, however an antistructure with respect to the complex scalar multiplication $\mathbb{C} \times \overline{V} \ni (\alpha, v) \longmapsto \overline{\alpha} v \ni \overline{V}$. With the set identity as *anticonjugation*, antivector spaces are isomorphic as real vector spaces and antilinearly bijective.

For a representation of a Lie group G and its Lie algebra L on a complex vector space $V \cong \mathbb{C}^n$, the equal-dimensional *dual representation* on the linear forms $V^T \cong \mathbb{C}^n$ is defined by inverse and negative transposition for group and Lie algebra, respectively. Both are involutive operations:

$$\text{representation on } V: \begin{cases} G \ni g \longmapsto D(g) \in \mathbf{GL}(V), \\ L \ni l \longmapsto \mathcal{D}(l) \in \mathbf{AL}(V), \end{cases}$$

$$\text{dual representation on } V^T: \begin{cases} G \ni g \longmapsto \check{D}(g) = D(g^{-1})^T \in \mathbf{GL}(V^T), \\ L \ni l \longmapsto \check{\mathcal{D}}(l) = -\mathcal{D}(l)^T \in \mathbf{AL}(V^T), \end{cases}$$

$$\text{involutive:} \check{\check{D}} = D, \ \check{\check{\mathcal{D}}} = \mathcal{D},$$

with the two-dimensional Pauli representation of (iso)spin $\mathbf{SU}(2)$ and $A_1^c \cong \mathbb{R}^3$ as an example:

$$\text{for } \mathbf{SU}(2) \text{ and } A_1^c: \begin{cases} D(u) \cong e^{i\vec{\alpha}\vec{\sigma}}, & \mathcal{D}(\vec{\alpha}\vec{l}) \cong i\begin{pmatrix} \alpha_3 & \alpha_1 - i\alpha_2 \\ \alpha_1 + i\alpha_2 & -\alpha_3 \end{pmatrix}, \\ \check{D}(u) \cong (e^{-i\vec{\alpha}\vec{\sigma}})^T, & \check{\mathcal{D}}(\vec{\alpha}\vec{l}) \cong i\begin{pmatrix} -\alpha_3 & -\alpha_1 - i\alpha_2 \\ -\alpha_1 + i\alpha_2 & \alpha_3 \end{pmatrix}. \end{cases}$$

With an invariant nondegenerate bilinear form, dual representations are equivalent, e.g., for $\mathbf{SU}(2)$ with the bilinear (iso)spinor metric, $\epsilon \circ \vec{\sigma} \circ \epsilon^{-1} = -\vec{\sigma}^T$. The dual quark and antiquark representations of $\mathbf{SU}(3)$ with Lie algebra bases $i\vec{\lambda}$ and $-i\vec{\lambda}^T$, respectively, involving the eight Euclidean hermitean Gell-Mann matrices $\vec{\lambda} = \vec{\lambda}^\star$, are not equivalent, since there does not exist an invariant bilinear quark metric.

Good to know: The existence of an invariant bilinear form is representation-dependent. The adjoint representation of a semisimple Lie algebra has the Killing-Form as an invariant nondegenerate symmetric bilinear form.

With the canonical conjugation of the complex numbers there exists for the representation D on $V \cong \mathbb{C}^n$ the equal-dimensional *antirepresentation* $\overline{D}$ on the conjugated vector space (antispace) $\overline{V} \cong \mathbb{C}^n$, and, correspondingly, the *dual antirepresentation* $\check{\overline{D}}$ on $\overline{V}^T \cong \mathbb{C}^n$ for the dual representation $\check{D}$ on $V^T \cong \mathbb{C}^n$. The antirepresentation $\overline{D}$ arises from the representation D by the canonical conjugation of only the numbers — in a matrix form without additional transposition. The dual antirepresentation uses, in addition, the transposition:

$$\text{antirepresentation on } \overline{V}: \begin{cases} G \ni g \longmapsto \overline{D}(g) \in \mathbf{GL}(\overline{V}), \\ L \ni l \longmapsto \overline{\mathcal{D}}(l) \in \mathbf{AL}(\overline{V}), \end{cases}$$

$$\text{dual antirepresentation on } \overline{V}^T: \begin{cases} G \ni g \longmapsto \check{\overline{D}}(g) = \overline{D}(g^{-1})^T \in \mathbf{GL}(\overline{V}^T), \\ L \ni l \longmapsto \check{\overline{\mathcal{D}}}(l) = -\overline{\mathcal{D}}(l)^T \in \mathbf{AL}(\overline{V}^T), \end{cases}$$

$$\text{involutive:} \overline{\overline{D}} = D, \ \overline{\overline{\mathcal{D}}} = \mathcal{D},$$

in the example for (iso)spin:

$$\text{for } \mathbf{SU}(2) \text{ and } A_1^c: \begin{cases} \overline{D}(u) \cong e^{-i\alpha_a \overline{\sigma}^a}, \quad \overline{\mathcal{D}}(\vec{\alpha}\vec{l}) \cong i\begin{pmatrix} -\alpha_3 & -\alpha_1 - i\alpha_2 \\ -\alpha_1 + i\alpha_2 & \alpha_3 \end{pmatrix}, \\ \check{\overline{D}}(u) \cong (e^{i\alpha_a \overline{\sigma}^a})^T, \check{\overline{\mathcal{D}}}(\vec{\alpha}\vec{l}) \cong i\begin{pmatrix} \alpha_3 & \alpha_1 - i\alpha_2 \\ \alpha_1 + i\alpha_2 & -\alpha_3 \end{pmatrix}. \end{cases}$$

Again, there are cases where antirepresentations are equivalent, e.g., for $\mathbf{SU}(2)$ where there exists, up to equivalence, only one complex two-dimensional representation, the Pauli spinor representation:

$$\text{for } A_1^c = \log \mathbf{SU}(2) : D \cong \overline{D}$$
$$\text{with}(i\sigma^1, i\sigma^2, i\sigma^3) = \sigma^2 \circ (-i\sigma^1, i\sigma^2, -i\sigma^3) \circ (\sigma^2)^{-1}.$$

Unitary groups $\mathbf{SU}(n_+, n_-)$ have an invariant conjugation. Therefore, in these cases, the defining n-dimensional anti- and dual representations and all their products are equivalent, $D \cong \check{\overline{D}}$ and $\overline{D} \cong \check{D}$.

In general, there exist four possibly inequivalent equal-dimensional representations of a real group G and of a real Lie algebra L on the complex quartet of vector spaces:

$$G \ni g \longmapsto (D, \check{D}, \overline{D}, \check{\overline{D}})(g) = (g, g^{-1T}, \overline{g}, \overline{g}^{-1T}) \in \mathbf{GL}(V, V^T, \overline{V}, \overline{V}^T),$$
$$L \ni l \longmapsto (\mathcal{D}, \check{\mathcal{D}}, \overline{\mathcal{D}}, \check{\overline{\mathcal{D}}})(l) = (l, -l^T, \overline{l}, -\overline{l}^T) \quad \in \mathbf{AL}(V, V^T, \overline{V}, \overline{V}^T).$$

6.12 The Weyl Spinor Quartet

According to the remarks above concerning the representations of real Lie groups on complex vector spaces as, necessarily, subgroups of unitary groups, also the real Lie group $\mathbf{SL}(2, \mathbb{C})$ has to be represented in a unitary group. Which unitary group and which related conjugation belong to the Lorentz group?

The Lorentz (covering) group $\mathbf{SL}(2, \mathbb{C})$ is noncompact, its Lie algebra is the direct vector space sum $A_1^c \oplus iA_1^c \cong \mathbb{R}^6$ of the spin (rotation) Lie algebra A_1^c for the compact (positive unitary) group $\mathbf{SU}(2)$ and the boost vector space iA_1^c. The four representations of the Lie algebra $A_1^c \oplus iA_1^c$ on the complex quartet of two-dimensional spaces are called *Weyl representations*:

Representation	Spinor space $\mathbb{C}^2$	Lie algebra $A_1^c \oplus iA_1^c$	Group $\mathbf{SL}(2, \mathbb{C})$
$\mathcal{D}, D$	$V = \mathbb{C}^2 \mathbf{l}^A$	$(i\vec{\sigma}, \vec{\sigma})$	$s(\vec{\alpha}, \vec{\beta}) = e^{(i\vec{\alpha}+\vec{\beta})\vec{\sigma}}$
$\overline{\mathcal{D}}, \overline{D}$	$\overline{V} = \mathbb{C}^2 \mathbf{l}^{\times}_{\dot{A}}$	$(-i\vec{\sigma}^T, \vec{\sigma}^T)$	$\overline{s}(\vec{\alpha}, \vec{\beta}) = (e^{-(i\vec{\alpha}-\vec{\beta})\vec{\sigma}})^T$
$\check{\overline{\mathcal{D}}}, \check{\overline{D}}$	$\overline{V}^T = \mathbb{C}^2 \mathbf{r}^{\dot{A}}$	$(i\vec{\sigma}, -\vec{\sigma})$	$\check{\overline{s}}(\vec{\alpha}, \vec{\beta}) = e^{(i\vec{\alpha}-\vec{\beta})\vec{\sigma}}$
$\check{\mathcal{D}}, \check{D}$	$V^T = \mathbb{C}^2 \mathbf{r}^{\times}_{A}$	$(-i\vec{\sigma}^T, -\vec{\sigma}^T)$	$\check{s}(\vec{\alpha}, \vec{\beta}) = (e^{-(i\vec{\alpha}+\vec{\beta})\vec{\sigma}})^T$

Weyl spinor quartet for the Lorentz group $\mathbf{SL}(2, \mathbb{C})$

An $\mathbf{SO}_0(1,3)$-related notation uses four left- and four right-handed Weyl matrices, $j = 0, 1, 2, 3$:

$$\sigma^j = (\mathbf{1}_2, \vec{\sigma}), \quad \check{\sigma}^j = (\mathbf{1}_2, -\vec{\sigma}), \quad \begin{cases} \sigma^{jk} = \frac{\sigma^k \check{\sigma}^j - \sigma^j \check{\sigma}^k}{2} \cong (i\vec{\sigma}, \vec{\sigma}), \\ \hat{\sigma}^{jk} = \frac{\check{\sigma}^k \sigma^j - \check{\sigma}^j \sigma^k}{2} \cong (i\vec{\sigma}, -\vec{\sigma}). \end{cases}$$

Via the spinor metric with $\epsilon^{-1} \circ \vec{\sigma}^T \circ \epsilon = -\vec{\sigma}$ dual Weyl representations are equivalent. The *left-handed Weyl spinors* $\mathbf{l}^A \in V = V_L \cong \mathbb{C}^2$ and the *dual right-handed Weyl spinors* $\mathbf{l}^\times_{\dot{A}} \in \overline{V} = V_R^T \cong \mathbb{C}^2$ realize the two nonequivalent fundamental representations of the Lorentz group. Each of the pairs $(\mathbf{l}^A, \mathbf{l}^\times_{\dot{A}})$ and $(\mathbf{r}^{\dot{A}}, \mathbf{r}^\times_A)$ carries both fundamental representations of the real Lorentz group $\mathbf{SL}(2,\mathbb{C})$ in the unitary group $\mathbf{SU}(2,2)$. Weyl has introduced the useful index notation with dotted and undotted indices for the two nonequivalent fundamental $\mathbf{SL}(2,\mathbb{C})$-representations and their products with $\hat{s} = \check{\bar{s}} = s^{-1\star}$:

$$\begin{aligned} s &= e^{(i\vec{\alpha}+\vec{\beta})\vec{\sigma}} \in \mathbf{SL}(2,\mathbb{C}), & \hat{s} &= e^{(i\vec{\alpha}-\vec{\beta})\vec{\sigma}} \in \mathbf{SL}(2,\mathbb{C}), \\ s &: V_L \longrightarrow V_L,\ \mathbf{l}^A \longmapsto s^A_B\, \mathbf{l}^B, & \hat{s} &: V_R \longrightarrow V_R,\ \mathbf{r}^{\dot{A}} \longmapsto s^{\dot{A}}_{\dot{B}}\, \mathbf{r}^{\dot{B}}, \\ s &= s^A_B\, \mathbf{l}^B \otimes \mathbf{r}^\times_A \in V_L \otimes V_L^T, & \hat{s} &= \hat{s}^{\dot{A}}_{\dot{B}}\, \mathbf{r}^{\dot{B}} \otimes \mathbf{l}^\times_{\dot{A}} \in V_R \otimes V_R^T. \end{aligned}$$

The product of the two fundamental $\mathbf{SL}(2,\mathbb{C})$-representations is the Weyl bi-spinor representation acting upon the Minkowski translations and relating left- and right-handed spinors. It is used in the Dirac equation for spacetime translation derivatives $(\frac{\partial}{\partial \mathbf{x}})^{\dot{B}}_A$:

$$\begin{aligned} \mathbf{x} &: V_L \longrightarrow V_R, \quad \mathbf{x} = \mathbf{x}^A_{\dot{B}}\, \mathbf{r}^{\dot{B}} \otimes \mathbf{r}^\times_A \in V_R \otimes V_L^T, \\ \mathbf{x} &= \sigma^j x_j \cong \begin{pmatrix} x_0 + x_3 & x_1 - ix_2 \\ x_1 + ix_2 & x_0 - x_3 \end{pmatrix} = \mathbf{x}^\star \in \mathbf{AL}(2,\mathbb{C}). \end{aligned}$$

The $\mathbf{SL}(2,\mathbb{C})$-compatible $\mathbf{U}(2,2)$-anticonjugation for the Weyl representations is denoted by $\times$. The representation of the Lorentz group as $\mathbf{SU}(2,2)$-subgroup is explicitly seen by the related antisymmetry of the Lie algebra in the Weyl-decomposable *Dirac representation*:

$$\begin{aligned} &\mathbf{\Psi} = \begin{pmatrix} \mathbf{l} \\ \mathbf{r} \end{pmatrix} \in V_L \oplus V_R \cong \mathbb{C}^4, \\ &\begin{pmatrix} (i\vec{\sigma},\vec{\sigma}) & 0 \\ 0 & (i\vec{\sigma},-\vec{\sigma}) \end{pmatrix}^\times = \begin{pmatrix} 0 & \mathbf{1}_2 \\ \mathbf{1}_2 & 0 \end{pmatrix} \begin{pmatrix} (i\vec{\sigma},\vec{\sigma}) & 0 \\ 0 & (i\vec{\sigma},-\vec{\sigma}) \end{pmatrix}^\star \begin{pmatrix} 0 & \mathbf{1}_2 \\ \mathbf{1}_2 & 0 \end{pmatrix} = -\begin{pmatrix} (i\vec{\sigma},\vec{\sigma}) & 0 \\ 0 & (i\vec{\sigma},-\vec{\sigma}) \end{pmatrix}. \end{aligned}$$

For an $\mathbf{SL}(2,\mathbb{C})$-invariant dynamics, e.g., a special relativistic field theory, the unitary group $\mathbf{U}(2,2)$ is the invariance group of the related anticonjugation $\times$, not, however, the invariance group of the dynamics. The unitary group $\mathbf{SU}(2,2)$ is locally isomorphic to the orthogonal group $\mathbf{SO}(2,4)$.

For general dimension, the compact unitary symmetry has the complex embedding $\mathbf{SU}(1+r) \hookrightarrow \mathbf{SL}(1+r, \mathbb{C})$ with $A_r^c \hookrightarrow A_r^c \oplus iA_r^c \cong \mathbb{R}^{2r(r+2)}$ in a real rank $2r$ Lie algebra, e.g., for the color-group $\mathbf{SU}(3)$. For $1+r \geq 3$, all four $\mathbf{SL}(1+r, \mathbb{C})$-representations on complex n-dimensional spaces are inequivalent, e.g., the complex three-dimensional representations of the real 16-dimensional Lie group $\mathbf{SL}(3, \mathbb{C})$ with the eight Gell-Mann matrices $\vec{\lambda}$:

$$\begin{array}{c}\textbf{for } \mathbf{SL}(3,\mathbb{C}) \textbf{ and}\\ A_2^c \oplus iA_2^c \cong \mathbb{R}^{16}:\end{array} \left\{ \begin{array}{r} \text{representation } \mathcal{D}: (i\vec{\lambda}, \vec{\lambda}),\\ \text{dual representation } \check{\mathcal{D}}: (-i\vec{\lambda}^T, -\vec{\lambda}^T),\\ \text{antirepresentation } \overline{\mathcal{D}}: (i\vec{\lambda}, -\vec{\lambda}),\\ \text{dual antirepresentation } \check{\overline{\mathcal{D}}}: (-i\vec{\lambda}^T, \vec{\lambda}^T). \end{array} \right.$$

There is no similarity transformation (quark metric) for the Gell-Mann matrices and their negative transposed — in contrast to $\epsilon \circ \vec{\sigma} \circ \epsilon^{-1} = -\vec{\sigma}^T$ for the Pauli matrices with the spinor metric. The dual triplet representations of $\mathbf{SU}(3)$, i.e., $(1,0) \cong 3 \longrightarrow i\vec{\lambda}$ and $(0,1) \cong \overline{3} \longrightarrow -i\vec{\lambda}^T$, and those of $\mathbf{SL}(3, \mathbb{C})$ — $(3|1)$, $(\overline{3}|1)$, $(1|3)$, $(1|\overline{3})$ — are inequivalent.

6.13 Interlocking Field and Particle Quartets

What is relevant with respect to the reflection implementation of isospin $\mathbf{SU}(2)$, holds equally for spin $\mathbf{SU}(2)$. However, there are important differences: In contrast to isospin, spin as external property is also connected with the rotations of the Euclidean position translations and momenta. In addition, the real three-dimensional spin operations $\mathbf{SU}(2)$ for massive particles are embedded — in the procedure of induced representations — into the real six-dimensional Lorentz operations $\mathbf{SL}(2, \mathbb{C})$ for fields.

The particle expansion (harmonic analysis) of fields interlocks two complex quartets — for the particles with spin and translation representations, and for the fields with Lorentz group representations. The members of both quartets are connected by conjugations and involutors.

The $\mathbf{SU}(2)$-compatible spin metric $\mathtt{E}$ for the complex quartets, acting on particle creation operators u, particle annihilation operators u*, antiparticle creation operators a, and antiparticle annihilation operator a*, all with spin j,

$$\mathbf{V}_{\text{doub}} = \mathbb{C}^{1+2j}\mathrm{u}^{\alpha} \oplus \mathbb{C}^{1+2j}\mathrm{a}_{\alpha} \oplus \mathbb{C}^{1+2j}\mathrm{u}_{\alpha}^{\star} \oplus \mathbb{C}^{1+2j}\mathrm{a}^{\star\alpha},$$

generate, together with the conjugations $\star$ and $+$, in analogy to the isospin earlier, a reflection group $\mathbf{I}(2)^3$:

$$\{\mathrm{u}^{\alpha}, \mathrm{u}_{\alpha}^{\star}, \mathrm{a}_{\alpha}, \mathrm{a}^{\star\alpha} \mid \alpha = 1, \ldots, 1+2j\}$$

$$\begin{array}{ccc} \mathrm{u}^\alpha & \overset{\star,\,\mathtt{E}}{\leftrightarrow} & \mathrm{u}^\star_\alpha \\ +\updownarrow & & \updownarrow + \\ \mathrm{a}_\alpha & \underset{\star,\,\mathtt{E}}{\leftrightarrow} & \mathrm{a}^{\star\alpha} \end{array}$$

Dual conjugation	$\star$	$\mathrm{u}^\alpha \leftrightarrow \delta^{\alpha\beta}\mathrm{u}^\star_\beta,\ \mathrm{a}_\alpha \leftrightarrow \delta_{\alpha\beta}\mathrm{a}^{\star\beta}$
Dual involutor	$\mathtt{E}$	$\mathrm{u}^\alpha \leftrightarrow \mathtt{E}^{\alpha\beta}\mathrm{u}^\star_\beta,\ \mathrm{a}_\alpha \leftrightarrow \mathtt{E}_{\alpha\beta}\mathrm{a}^{\star\beta}$
Antidual conjugation	$+$	$\mathrm{u}^\alpha \leftrightarrow \delta^{\alpha\beta}\mathrm{a}_\beta,\ \mathrm{u}^\star_\alpha \leftrightarrow \delta_{\alpha\beta}\mathrm{a}^{\star\beta}$

Spin $\mathbf{SU}(2)$ — compatible reflections

With the time representation $e^{\pm imt} \in \mathbf{U}(1)$ for particle and antiparticle and trivial position (harmonic oscillator with spin and charge):

$$\mathbf{\Phi}^\alpha(t) = \frac{e^{imt}\mathrm{u}^\alpha + e^{-imt}\mathrm{a}^{\star\alpha}}{\sqrt{2}}, \quad \mathbf{\Phi}^\star_\alpha(t) = \frac{e^{imt}\mathrm{a}_\alpha + e^{-imt}\mathrm{u}^\star_\alpha}{\sqrt{2}},$$

the eight-elementic reflection group $\mathbf{I}(2)^3$ with four linear and four antilinear elements involves seven nontrivial actions:

Dual conjugation	$\star$	$\mathbf{\Phi}^\alpha(t) \leftrightarrow \delta^{\alpha\beta}\mathbf{\Phi}^\star_\beta(t)$
Dual involutor for spin	$\mathtt{E}$	$\mathbf{\Phi}^\alpha(t) \leftrightarrow \mathtt{E}^{\alpha\beta}\mathbf{\Phi}^\star_\beta(-t)$
Scalar conjugation	$\mathtt{T} = \mathtt{E} \circ \star$	$\mathbf{\Phi}^\alpha(t) \leftrightarrow \mathtt{E}^{\alpha\beta}\delta_{\beta\gamma}\mathbf{\Phi}^\gamma(-t)$, $\mathbf{\Phi}^\star_\alpha(t) \leftrightarrow \mathtt{E}_{\alpha\beta}\delta^{\beta\gamma}\mathbf{\Phi}^\star_\gamma(-t)$
Antidual conjugation	$+$	$\mathbf{\Phi}^\alpha(t) \leftrightarrow \delta^{\alpha\beta}\mathbf{\Phi}^\star_\beta(-t)$
Anticonjugation	$\times = \mathtt{E} \circ +$	$\mathbf{\Phi}^\alpha(t) \leftrightarrow \mathtt{E}^{\alpha\beta}\delta_{\beta\gamma}\mathbf{\Phi}^\gamma(t)$, $\mathbf{\Phi}^\star_\alpha(t) \leftrightarrow \mathtt{E}_{\alpha\beta}\delta^{\beta\gamma}\mathbf{\Phi}^\star_\gamma(t)$
Antidual involutor	$\mathtt{I} = \star \circ +$	$\mathbf{\Phi}^\alpha(t) \leftrightarrow \mathbf{\Phi}^\alpha(-t)$, $\mathbf{\Phi}^\star_\alpha(t) \leftrightarrow \mathbf{\Phi}^\star_\alpha(-t)$
Antiinvolutor	$\mathtt{CP} = \mathtt{E} \circ \star \circ +$	$\mathbf{\Phi}^\alpha(t) \leftrightarrow \mathtt{E}^{\alpha\beta}\mathbf{\Phi}^\star_\beta(t)$

The notation $\mathtt{CP}$ for the external antiinvolutor without product symbol $\circ$ for the charge reflection, so far $\mathtt{C}$ for the internal antiinvolutor, should indicate that a factorization into two reflections $\mathtt{C}$ and $\mathtt{P}$ is not possible yet at this stage. The antidual conjugation is the product of CP-reflection and time conjugation $+ = \mathtt{CP} \circ \mathtt{T}$.

Now the transition from "time-fields" with trivial position to Minkowski spacetime fields: To embed particles in spacetime fields, a transmutation from the particle-characterizing rotation property (spin) to the Lorentz property for a field is necessary. For massive particles, the embedding $\mathbf{SU}(2) \hookrightarrow \mathbf{SL}(2, \mathbb{C})$ from a rest system of the particle is effected by the boosts with momenta $\vec{q}$. All transmutators are products of the dual fundamental Weyl representations of the boosts $s(\frac{\vec{q}}{m}), \hat{s}(\frac{\vec{q}}{m}) \in \mathbf{SL}(2, \mathbb{C})$:

$$\begin{array}{l} (m, 0) \longmapsto (q_0, \vec{q}), \\ s(\frac{\vec{q}}{m}) \circ m\mathbf{1}_2 \circ s^\star(\frac{\vec{q}}{m}) = \sigma^j q_j, \end{array} \text{ by } \begin{cases} s(\frac{\vec{q}}{m}) = e^{\vec{\sigma}\frac{\vec{q}}{|\vec{q}|}\beta(\frac{\vec{q}}{m})} &= \sqrt{\frac{q_0+m}{2m}}(\mathbf{1}_2 + \frac{\vec{\sigma}\vec{q}}{q_0+m}), \\ \hat{s}(\frac{\vec{q}}{m}) = e^{-\vec{\sigma}\frac{\vec{q}}{|\vec{q}|}\beta(\frac{\vec{q}}{m})} &= \sqrt{\frac{q_0+m}{2m}}(\mathbf{1}_2 - \frac{\vec{\sigma}\vec{q}}{q_0+m}), \end{cases}$$

$$\text{with } \tanh 2\beta(\frac{\vec{q}}{m}) = \frac{|\vec{q}|}{q_0}, \quad \frac{q_0}{m} = \sqrt{1 + \frac{\vec{q}^2}{m^2}},$$

for example, the Lorentz boosts:

$$\Lambda_j^k(\frac{\vec{q}}{m}) = \frac{1}{2}\operatorname{tr}\check{\sigma}^k \circ s(\frac{\vec{q}}{m}) \circ \sigma_j \circ s^\star(\frac{\vec{q}}{m}) \cong \frac{1}{m}\left(\begin{array}{c|c} q_0 & \vec{q} \\ \hline \vec{q} & m\mathbf{1}_3 + \frac{\vec{q}\otimes\vec{q}}{q_0+m}\end{array}\right).$$

The left and right Weyl transmutators with *position-reflected momentum (boost)* $\vec{q} \leftrightarrow -\vec{q}$ are related to each other by the symplectic metrics of Pauli and Weyl spinors:

$$\epsilon_{AB}s^B_\beta(\frac{\vec{q}}{m})\epsilon^{\beta\alpha} = \hat{s}^\alpha_A(\frac{\vec{q}}{m}) = s^\alpha_A(-\frac{\vec{q}}{m}),\ \ \eta_{jl}\Lambda^l_m(\frac{\vec{q}}{m})\eta^{mk} = [\Lambda^{-1}(\frac{\vec{q}}{m})]^k_j = \Lambda^k_j(-\frac{\vec{q}}{m}).$$

The direct sum Dirac transmutator solves the Dirac equation in energy-momentum space:

$$\left.\begin{array}{l}\sigma^j q_j \hat{s}(\frac{\vec{q}}{m}) = m s(\frac{\vec{q}}{m}),\\ \check{\sigma}^j q_j s(\frac{\vec{q}}{m}) = m\hat{s}(\frac{\vec{q}}{m})\end{array}\right\} \iff (\gamma^j q_j - m\mathbf{1}_4)\mathbf{s}(\frac{\vec{q}}{m}) = 0,$$

$$\text{with } \gamma^j = \begin{pmatrix} 0 & \sigma^j \\ \check{\sigma}^j & 0\end{pmatrix},\ \mathbf{s}(\frac{\vec{q}}{m}) = \begin{pmatrix} s(\frac{\vec{q}}{m}) & 0 \\ 0 & \hat{s}(\frac{\vec{q}}{m})\end{pmatrix}$$

Good to know: In the $\mathbf{SU}(2)$-spin decomposition of an $\mathbf{SL}(2,\mathbb{C})$-Lorentz group representation $[j|j'] \cong \bigoplus_{J=|j-j'|}^{j+j'} [J]$, all $\mathbf{SU}(2)$-contributions $[J]$ have either half-integer or integer spin J. The spin representation $[J]$ can be found in many Lorentz group representations.

Massless particles, e.g., photons, have no spin $\mathbf{SU}(2)$, they have polarization or helicity for the axial rotations $\mathbf{SO}(2)$. Here, one uses the helicity projectors. e.g., in the Weyl representations:

$$\lim_{m\to 0}\sqrt{\frac{m}{2q_0}}s(\frac{\vec{q}}{m}) = \frac{\mathbf{1}_2 + \frac{\vec{\sigma}\vec{q}}{|\vec{q}|}}{2},\quad \lim_{m\to 0}\sqrt{\frac{m}{2q_0}}\hat{s}(\frac{\vec{q}}{m}) = \frac{\mathbf{1}_2 - \frac{\vec{\sigma}\vec{q}}{|\vec{q}|}}{2},\quad q_0 = |\vec{q}|.$$

Because of the complex representation spaces, half-integer spin particles can never be identical with their antiparticles. For spin-$\frac{1}{2}$ particles with Pauli $\mathbf{SU}(2)$-representations, the two-dimensional Weyl $\mathbf{SL}(2,\mathbb{C})$-representations are the "smallest" extensions. Particle and antiparticle creation operators $\mathrm{u}^\alpha(\vec{q})$ and $\mathrm{a}_\alpha(\vec{q})$, respectively, and the corresponding particle and antiparticle annihilation operators $\mathrm{u}^\star_\alpha(\vec{q})$ and $\mathrm{a}^{\star\alpha}(\vec{q})$ for all momenta $\vec{q}$ and both spin directions $\pm\frac{1}{2}$ are embedded in Weyl fields, e.g., in an electron-positron field:

$$\begin{array}{l}\mathbf{l}^A(x) = \int \frac{d^3q}{2q_0}\, s^A_\alpha(\frac{\vec{q}}{m})\, \frac{e^{iqx}\mathrm{u}^\alpha(\vec{q}) + e^{-iqx}\mathrm{a}^{\star\alpha}(\vec{q})}{\sqrt{2}},\ \mathbf{r}^{\dot{A}}(x) = \int \frac{d^3q}{2q_0}\, s^{\dot{A}}_\alpha(-\frac{\vec{q}}{m})\, \frac{e^{iqx}\mathrm{u}^\alpha(\vec{q}) - e^{-iqx}\mathrm{a}^{\star\alpha}(\vec{q})}{\sqrt{2}},\\ \mathbf{l}^\star_{\dot{A}}(x) = \int \frac{d^3q}{2q_0}\, s^\alpha_{\dot{A}}(\frac{\vec{q}}{m})\, \frac{e^{-iqx}\mathrm{u}^\star_\alpha(\vec{q}) + e^{iqx}\mathrm{a}_\alpha(\vec{q})}{\sqrt{2}},\ \mathbf{r}^\star_A(x) = \int \frac{d^3q}{2q_0}\, s^\alpha_A(-\frac{\vec{q}}{m})\, \frac{e^{-iqx}\mathrm{u}^\star_\alpha(\vec{q}) - e^{iqx}\mathrm{a}_\alpha(\vec{q})}{\sqrt{2}},\\ \quad\text{with } q_0 = \sqrt{m^2 + \vec{q}^2}.\end{array}$$

The definition with sums and differences of momentum creation and annihilation operators is — up to an imaginary i — in parallel with the position-momentum definition for the time-dependent harmonic oscillator as real and imginary combination $\mathbf{x}(t) = \frac{e^{imt}\mathrm{u}+e^{-imt}\mathrm{u}^\star}{\sqrt{2}}$, $\mathbf{p}(t) = i\frac{e^{imt}\mathrm{u}-e^{-imt}\mathrm{u}^\star}{\sqrt{2}}$. The field quartet contains two pairs, $(\mathbf{l}^A, \mathbf{l}^\star_{\dot{A}})_{A,\dot{A}=1,2}$ and $(\mathbf{r}^{\dot{A}}, \mathbf{r}^\star_A)_{A,\dot{A}=1,2}$, both with the two inequivalent Weyl spinor representations, and related to each other — as spacetime translation orbits — by the Weyl-Dirac equations — in parallel to the oscillator equations $d_t\mathbf{x}(t) = m\mathbf{p}(t)$, $d_t\mathbf{p}(t) = -m\mathbf{x}(t)$:

$$\left.\begin{array}{l} \sigma^j\partial_j\mathbf{l}(x) = im\mathbf{r}(x), \\ \check{\sigma}^j\partial_j\mathbf{r}(x) = im\mathbf{l}(x) \end{array}\right\} \iff (\gamma^j\partial_j - im\mathbf{1}_4)\mathbf{\Psi}(x) = 0, \text{ with } \mathbf{\Psi} = \begin{pmatrix}\mathbf{l}\\ \mathbf{r}\end{pmatrix},$$
$$(\partial^2 + m^2)(\mathbf{l}, \mathbf{r}, \mathbf{\Psi}) = 0.$$

Already the particle analysis of only two $\mathbf{SL}(2,\mathbb{C})$-nonequivalent Weyl spinor fields, e.g., $\mathbf{l} \sim \frac{\mathrm{u}+\mathrm{a}^\star}{\sqrt{2}}$ and $\mathbf{l}^\star \sim \frac{\mathrm{a}+\mathrm{u}^\star}{\sqrt{2}}$, needs the full particle quartet. This minimal representation structure with respect to the Lorentz group is used for massless spinor fields. Massive charged particles, e.g., electron-positrons, need all four Weyl fields as seen in the Lorentz scalar Lagrangian mass term $m(\mathbf{l}^\star\mathbf{r} + \mathbf{r}^\star\mathbf{l})$.

The particle quartet contains equivalent $\mathbf{SU}(2)$-spin dublets for all momenta, acted on by dual translation representations, $\mathbb{R}^{1+3} \ni x \longmapsto e^{\pm iqx} \in \mathbf{U}(1)$ with $q^2 = m^2$. The definition of the external dual involutor has to take into account also the dual reflection of the momenta $\vec{q} \leftrightarrow -\vec{q}$:

$$\{\mathrm{u}^\alpha(\vec{q}), \mathrm{a}_\alpha(\vec{q}), \mathrm{u}^\star_\alpha(\vec{q}), \mathrm{a}^{\star\alpha}(\vec{q}) \mid \alpha = 1,\dots,1+2j,\ \ \vec{q} \in \mathbb{R}^3\},$$

Dual conjugation	$\star$	$\mathrm{u}^\alpha(\vec{q}) \leftrightarrow \delta^{\alpha\beta}\mathrm{u}^\star_\beta(\vec{q})$,	$\mathrm{a}_\alpha(\vec{q}) \leftrightarrow \delta_{\alpha\beta}\mathrm{a}^{\star\beta}(\vec{q})$
Dual involutor for spin	$\mathtt{E}$	$\mathrm{u}^\alpha(\vec{q}) \leftrightarrow \mathtt{E}^{\alpha\beta}\mathrm{u}^\star_\beta(-\vec{q})$,	$\mathrm{a}_\alpha(\vec{q}) \leftrightarrow \mathtt{E}_{\alpha\beta}\mathrm{a}^{\star\beta}(-\vec{q})$
Scalar conjugation	$\mathtt{T} = \mathtt{E}\circ\star$	$\mathrm{u}^\alpha(\vec{q}) \leftrightarrow \mathtt{E}^{\alpha\beta}\delta_{\beta\gamma}\mathrm{u}^\gamma(-\vec{q})$,	$\mathrm{u}^\star_\alpha(\vec{q}) \leftrightarrow \mathtt{E}_{\alpha\beta}\delta^{\beta\gamma}\mathrm{u}^\star_\gamma(-\vec{q})$
		$\mathrm{a}^{\star\alpha}(\vec{q}) \leftrightarrow \mathtt{E}^{\alpha\beta}\delta_{\beta\gamma}\mathrm{a}^{\star\gamma}(-\vec{q})$,	$\mathrm{a}_\alpha(\vec{q}) \leftrightarrow \mathtt{E}_{\alpha\beta}\delta^{\beta\gamma}\mathrm{a}_\gamma(-\vec{q})$
Antidual conjugation	$+$	$\mathrm{u}^\alpha(\vec{q}) \leftrightarrow \delta^{\alpha\beta}\mathrm{a}_\beta(\vec{q})$,	$\mathrm{u}^\star_\alpha(\vec{q}) \leftrightarrow \delta_{\alpha\beta}\mathrm{a}^{\star\beta}(\vec{q})$
Anticonjugation	$\times = \mathtt{E}\circ +$	$\mathrm{u}^\alpha(\vec{q}) \leftrightarrow \mathtt{E}^{\alpha\beta}\delta_{\beta\gamma}\mathrm{a}^{\star\gamma}(-\vec{q})$,	$\mathrm{u}^\star_\alpha(\vec{q}) \leftrightarrow \mathtt{E}_{\alpha\beta}\delta^{\beta\gamma}\mathrm{a}_\gamma(-\vec{q})$
Antidual involutor	$\mathtt{I} = \star\circ +$	$\mathrm{u}^\alpha(\vec{q}) \leftrightarrow \mathrm{a}^{\star\alpha}(\vec{q})$,	$\mathrm{a}_\alpha(\vec{q}) \leftrightarrow \mathrm{u}^\star_\alpha(\vec{q})$
Antiinvolutor	$\mathtt{CP} = \mathtt{E}\circ\star\circ +$	$\mathrm{u}^\alpha(\vec{q}) \leftrightarrow \mathtt{E}^{\alpha\beta}\mathrm{a}_\beta(-\vec{q})$,	$\mathrm{u}^\star_\alpha(\vec{q}) \leftrightarrow \mathtt{E}_{\alpha\beta}\mathrm{a}^{\star\beta}(-\vec{q})$

Spin $\mathbf{SU}(2)$ – compatible reflections
Particle reflections $\mathrm{grp}(\mathtt{E}, \star, +) \cong \mathbf{I}(2) \times \mathbf{I}(2) \times \mathbf{I}(2)$

For example, the time reflected electron creator $\mathrm{u}^\alpha(\vec{q}) \leftrightarrow \epsilon^{\alpha\beta}\delta_{\beta\gamma}\mathrm{u}^\gamma(-\vec{q})$ has the same energy $q_0 = \sqrt{m^2 + \vec{q}^2}$, reflected momentum $\vec{q} \leftrightarrow -\vec{q}$ and spin $1 \leftrightarrow 2$.

The four spaces are related to each other by four linear and four antilinear reflections:

$$\mathbf{K}(4) \cong \{\mathbf{1}, \mathtt{E}, \mathtt{I}, \mathtt{E}\circ\mathtt{I} = \mathtt{CP}\} \overset{\star}{\leftrightarrow} \{\star, \mathtt{T}, +, \times\}.$$

The reflections of the particle quartet induce field reflections. The action of the seven nontrivial elements of the particle quartet reflection group $\mathbf{I}(2)^3$ on a particle analyzable charged spinless field, e.g., (π_+, π_-), with parity $\epsilon \in \{\pm 1\}$ looks as follows:

$$\mathbf{\Phi}(x) = \int \frac{d^3q}{2q_0} \frac{e^{iqx}\mathrm{u}(\vec{q}) + e^{-iqx}\mathrm{a}^\star(\vec{q})}{\sqrt{2}}, \quad \mathbf{\Phi}^\star(x) = \int \frac{d^3q}{2q_0} \frac{e^{iqx}\mathrm{a}(\vec{q}) + e^{-iqx}\mathrm{u}^\star(\vec{q})}{\sqrt{2}},$$

Dual conjugation	$\star$	$\mathbf{\Phi}(x) \leftrightarrow \mathbf{\Phi}^\star(x)$
Dual involutor for spin	$\mathtt{E}$	$\mathbf{\Phi}(t, \vec{x}) \leftrightarrow \epsilon\, \mathbf{\Phi}^\star(-t, \vec{x})$
Scalar conjugation	$\mathtt{T} = \mathtt{E} \circ \star$	$\mathbf{\Phi}(t, \vec{x}) \leftrightarrow \epsilon\, \mathbf{\Phi}(-t, \vec{x})$ $\mathbf{\Phi}^\star(t, \vec{x}) \leftrightarrow \epsilon\, \mathbf{\Phi}^\star(-t, \vec{x})$
Antidual conjugation	$+ = \mathtt{CP} \circ \mathtt{T}$	$\mathbf{\Phi}(x) \leftrightarrow \mathbf{\Phi}^\star(-x)$
Anticonjugation	$\times = \mathtt{E} \circ +$	$\mathbf{\Phi}(t, \vec{x}) \leftrightarrow \epsilon\, \mathbf{\Phi}(t, -\vec{x})$ $\mathbf{\Phi}^\star(t, \vec{x}) \leftrightarrow \epsilon\, \mathbf{\Phi}^\star(t, -\vec{x})$
Antidual involutor	$\mathtt{I} = \star \circ +$	$\mathbf{\Phi}(x) \leftrightarrow \mathbf{\Phi}(-x)$ $\mathbf{\Phi}^\star(x) \leftrightarrow \mathbf{\Phi}^\star(-x)$
Antiinvolutor	$\mathtt{CP} = \mathtt{E} \circ \star \circ +$	$\mathbf{\Phi}(t, \vec{x}) \leftrightarrow \epsilon\, \mathbf{\Phi}^\star(t, -\vec{x})$

and for one pair $(\mathbf{l}, \mathbf{l}^\star)$ of particle analyzable Weyl fields:

$$\mathbf{l}^A(x) = \int \frac{d^3q}{2q_0}\, s^A_\alpha\left(\frac{\vec{q}}{m}\right) \frac{e^{iqx}\mathrm{u}^\alpha(\vec{q}) + e^{-iqx}\mathrm{a}^{\star\alpha}(\vec{q})}{\sqrt{2}}, \quad \mathbf{l}^\star_{\dot{A}}(x) = \int \frac{d^3q}{2q_0}\, s^\alpha_{\dot{A}}\left(\frac{\vec{q}}{m}\right) \frac{e^{-iqx}\mathrm{u}^\star_\alpha(\vec{q}) + e^{iqx}\mathrm{a}_\alpha(\vec{q})}{\sqrt{2}},$$

Dual conjugation	$\star$	$\mathbf{l}^A(x) \leftrightarrow \delta^{A\dot{A}}\mathbf{l}^\star_{\dot{A}}(x)$
Dual involutor for spin	$\mathtt{E}$	$\mathbf{l}^A(t, \vec{x}) \leftrightarrow \epsilon^{AB}\delta^{\dot{B}}_{B}\mathbf{l}^\star_{\dot{B}}(-t, \vec{x})$
Scalar conjugation	$\mathtt{T} = \mathtt{E} \circ \star$	$\mathbf{l}^A(t, \vec{x}) \leftrightarrow \epsilon^{AB}\delta_{B\dot{C}}\mathbf{l}^C(-t, \vec{x})$ $\mathbf{l}^\star_{\dot{A}}(t, \vec{x}) \leftrightarrow \epsilon_{\dot{A}\dot{B}}\delta^{\dot{B}\dot{C}}\mathbf{l}^\star_{\dot{C}}(-t, \vec{x})$
Antidual conjugation	$+ = \mathtt{CP} \circ \mathtt{T}$	$\mathbf{l}^A(x) \leftrightarrow \delta^{A\dot{B}}\mathbf{l}^\star_{\dot{B}}(-x)$
Anticonjugation	$\times = \mathtt{E} \circ +$	$\mathbf{l}^A(t, \vec{x}) \leftrightarrow \epsilon^{AB}\delta_{B\dot{C}}\mathbf{l}^C(t, -\vec{x})$ $\mathbf{l}^\star_{\dot{A}}(t, \vec{x}) \leftrightarrow \epsilon_{\dot{A}\dot{B}}\delta^{\dot{B}\dot{C}}\mathbf{l}^\star_{\dot{C}}(t, -\vec{x})$
Antidual involutor	$\mathtt{I} = \star \circ +$	$\mathbf{l}^A(x) \leftrightarrow \mathbf{l}^A(-x)$ $\mathbf{l}^\star_{\dot{A}}(x) \leftrightarrow \mathbf{l}^\star_{\dot{A}}(-x)$
Antiinvolutor	$\mathtt{CP} = \mathtt{E} \circ \star \circ +$	$\mathbf{l}^A(t, \vec{x}) \leftrightarrow \epsilon^{AB}\delta^{\dot{B}}_{B}\mathbf{l}^\star_{\dot{B}}(t, -\vec{x})$

A Euclidean star $(\mathbf{l}, \mathbf{l}^\star)$ for the $\mathbf{SL}(2, \mathbb{C})$-antirepresentation can be justified by the Euclidean conjugation of the particle quartet.

The antilinear scalar conjugation $\mathtt{T}$ implements the *time reflection*, the linear antiinvolutor $\mathtt{CP}$ the *position reflection*, the product is the antilinear *antidual conjugation* $+ = \mathtt{CP} \circ \mathtt{T}$ which implements the central Minkowski spacetime reflection $-\mathbf{1}_4$. It is compatible with the Lorentz group; the reflections $\mathtt{T}$ (scalar conjugation, time reflection) and $\mathtt{CP}$ (antiinvolutor, position reflection) are compatible with spin $\mathbf{SU}(2)$-transformations only:

$$\mathbf{K}(4) \cong \{\mathbf{1}, \mathtt{T}, \mathtt{CP}, \mathtt{CP} \circ \mathtt{T} = +\},$$

$$\begin{array}{ccc} \mathbb{R} & \stackrel{-1}{\longrightarrow} & \mathbb{R} \\ f \downarrow & & \downarrow f^{\mathrm{T}} \\ \mathbf{V}_{\mathrm{doub}} & \underset{\mathrm{T}}{\longrightarrow} & \mathbf{V}_{\mathrm{doub}} \end{array}, \quad \begin{array}{ccc} \mathbb{E}^3 & \stackrel{-\mathbf{1}_3}{\longrightarrow} & \mathbb{E}^3 \\ f \downarrow & & \downarrow f^{\mathrm{CP}} \\ \mathbf{V}_{\mathrm{doub}} & \underset{\mathrm{CP}}{\longrightarrow} & \mathbf{V}_{\mathrm{doub}} \end{array}, \quad \begin{array}{ccc} \mathbb{R}^{(1,3)} & \stackrel{-\mathbf{1}_4}{\longrightarrow} & \mathbb{R}^{(1,3)} \\ f \downarrow & & \downarrow f^{+} \\ \mathbf{V}_{\mathrm{doub}} & \underset{\mathrm{CP\circ T}}{\longrightarrow} & \mathbf{V}_{\mathrm{doub}} \end{array},$$

T-reflection	T	$\mathbf{l}^A(t,\vec{x}) \leftrightarrow \epsilon^{AB}\delta_{BC}\mathbf{l}^C(-t,\vec{x}),\ \mathbf{l}^\star_{\dot{A}}(t,\vec{x}) \leftrightarrow \epsilon_{\dot{A}\dot{B}}\delta^{\dot{B}\dot{C}}\mathbf{l}^\star_{\dot{C}}(-t,\vec{x})$ $\mathbf{r}^{\dot{A}}(t,\vec{x}) \leftrightarrow \epsilon^{\dot{A}\dot{B}}\delta_{\dot{B}\dot{C}}\mathbf{r}^{\dot{C}}(-t,\vec{x}),\ \mathbf{r}^\star_A(t,\vec{x}) \leftrightarrow \epsilon_{AB}\delta^{BC}\mathbf{r}^\star_C(-t,\vec{x})$
CP-reflection	CP	$\mathbf{l}^A(t,\vec{x}) \leftrightarrow \epsilon^{AB}\delta_B^{\dot{B}}\mathbf{l}^\star_{\dot{B}}(t,-\vec{x})$ $\mathbf{r}^{\dot{A}}(t,\vec{x}) \leftrightarrow -\epsilon^{\dot{A}\dot{B}}\delta_{\dot{B}}^{B}\mathbf{r}^\star_B(t,-\vec{x})$
CPT-reflection	$\mathrm{CP}\circ\mathrm{T}$	$\mathbf{l}^A(x) \leftrightarrow \delta^{A\dot{B}}\mathbf{l}^\star_{\dot{B}}(-x)$ $\mathbf{r}^{\dot{A}}(x) \leftrightarrow -\delta^{\dot{A}A}\mathbf{r}^\star_A(-x)$

The induced reflection for the pair $(\mathbf{r}, \mathbf{r}^\star)$ display — because of the particle momentum operator differences involved — a minus sign:

$$\mathbf{r}^{\dot{A}}(x) = \int \frac{d^3q}{2q_0}\, s_\alpha^{\dot{A}}\big(-\frac{\vec{q}}{m}\big)\, \frac{e^{iqx}\mathrm{u}^\alpha(\vec{q}) - e^{-iqx}\mathrm{a}^{\star\alpha}(\vec{q})}{\sqrt{2}},\quad \mathbf{r}^\star_A(x) = \int \frac{d^3q}{2q_0}\, s_A^\alpha\big(-\frac{\vec{q}}{m}\big)\, \frac{e^{-iqx}\mathrm{u}^\star_\alpha(\vec{q}) - e^{iqx}\mathrm{a}_\alpha(\vec{q})}{\sqrt{2}}.$$

6.14 Left-Right (Chiral) Reflection

The external dual involutor (spinor metric) of the Weyl quartet with two irreducible spacetime field pairs $(\mathbf{l}, \mathbf{l}^\star)$ and $(\mathbf{r}, \mathbf{r}^\star)$, which come with sum $\mathbf{l} \sim \frac{\mathrm{u}+\mathrm{a}^\star}{\sqrt{2}}$ and difference $\mathbf{r}^\star \sim \frac{-\mathrm{a}+\mathrm{u}^\star}{\sqrt{2}}$ of particle momentum operators, respectively, can be realized on the particle quartet in a second form — with a relative sign in the dual involution of particle and antiparticle:

Lorentz dual involutor $\tilde{\mathrm{E}}$	$\mathrm{u}^\alpha(\vec{q}) \leftrightarrow \epsilon^{\alpha\beta}\mathrm{u}^\star_\beta(\vec{q}),\ \mathrm{a}_\alpha(\vec{q}) \leftrightarrow -\epsilon_{\alpha\beta}\mathrm{a}^{\star\beta}(\vec{q})$ $\mathbf{l}^A(x) \leftrightarrow \epsilon^{AB}\mathbf{r}^\star_B(x),\ \mathbf{l}^\star_{\dot{A}}(x) \leftrightarrow \epsilon_{\dot{A}\dot{B}}\mathbf{r}^{\dot{B}}(x)$

Majorana[5] fields identify reflected partners, $\mathbf{l}^A = \epsilon^{AB}\mathbf{r}^\star_B$, $\mathbf{l}^\star_{\dot{A}} = \epsilon_{\dot{A}\dot{B}}\mathbf{r}^{\dot{B}}$.

The product of both external dual involutors (metric for Pauli and Weyl spinors) — E for spin group $\mathbf{SU}(2)$ and $\tilde{\mathrm{E}}$ for Lorentz group $\mathbf{SL}(2,\mathbb{C})$ — defines the *chiral involutor* which reflects the momenta and, because of the sum and difference of the momentum operators in the fields, displays a relative particle-antiparticle sign:

Chiral involutor $\mathrm{P} = \mathrm{E}\circ\tilde{\mathrm{E}}$	$\mathrm{u}^\alpha(\vec{q}) \leftrightarrow \mathrm{u}^\alpha(-\vec{q}),$ $\mathrm{a}_\alpha(\vec{q}) \leftrightarrow -\mathrm{a}_\alpha(-\vec{q}),$ $\mathbf{l}^A(t,\vec{x}) \leftrightarrow \delta^A_{\dot{A}}\mathbf{r}^{\dot{A}}(t,-\vec{x}),$	$\mathrm{u}^\star_\alpha(\vec{q}) \leftrightarrow \mathrm{u}^\star_\alpha(-\vec{q}),$ $\mathrm{a}^{\star\alpha}(\vec{q}) \leftrightarrow -\mathrm{a}^{\star\alpha}(-\vec{q}),$ $\mathbf{r}^\star_A(t,\vec{x}) \leftrightarrow \delta^{\dot{A}}_A\mathbf{l}^\star_{\dot{A}}(t,-\vec{x})$

[5] Ettore Majorana (1906–1938(?)).

In the field of a massive spin-$\frac{1}{2}$ particle, e.g., in the electron-positron field $\boldsymbol{\Psi}(x) \cong \begin{pmatrix} \mathrm{e}^A \\ \mathrm{e}^{\dot{A}} \end{pmatrix}(x)$, the creation operators for particle $\mathrm{u}^\alpha(\vec{q})$ (electron) and antiparticle $\mathrm{a}_\alpha(\vec{q})$ (positron) have opposite parity with respect to the chiral reflection. The chiral involutor reflects left-handed Weyl representations $(\mathbf{l}, \mathbf{r}^\star)$ with right-handed ones $(\mathbf{r}, \mathbf{l}^\star)$, as used in the left-right direct sum for a Dirac field $\boldsymbol{\Psi} \cong \begin{pmatrix} \mathbf{l}^A \\ \mathbf{r}^{\dot{A}} \end{pmatrix}$, $\boldsymbol{\Psi}^\star \cong (\mathbf{l}^\star_{\dot{A}}, \mathbf{r}^\star_A)$.

That the chiral involutor $\mathtt{P}$ is taken to implement the reflection of the position translations reflection, which is already possible for a minimal pair $(\mathbf{l}, \mathbf{l}^\star)$ with dual Lorentz properties, is a historically motivated misnomer. The earlier introduced involutor $\mathtt{CP} : \mathbf{l}(t, \vec{x}) \leftrightarrow \epsilon \mathbf{l}^\star(t, -\vec{x})$ effects the minimal implementation of the position translations reflection, i.e., for one Weyl spinor pair, e.g., $(\mathbf{l}, \mathbf{l}^\star)$, without chiral partners.

The large reflection group $\mathrm{grp}(\mathtt{E}, \star, +, \mathtt{P})$ supplements the quartet reflection group $\mathbf{I}(2)^3$ by the chiral involutor $\mathtt{P}$. The C-transformation is defined as product $\mathtt{CP} \circ \mathtt{P}$:

C-transformation $\mathtt{C} = \mathtt{CP} \circ \mathtt{P} = -\mathtt{P} \circ \mathtt{CP}$	$\mathrm{u}^\alpha(\vec{q}) \longmapsto \epsilon^{\alpha\beta}\mathrm{a}_\beta(\vec{q}),\quad \mathrm{a}_\alpha(\vec{q}) \longmapsto -\epsilon_{\alpha\beta}\mathrm{u}^\beta(\vec{q})$ $\mathrm{u}^\star_\alpha(\vec{q}) \longmapsto \epsilon_{\alpha\beta}\mathrm{a}^{\star\beta}(\vec{q}),\quad \mathrm{a}^{\star\alpha}(\vec{q}) \longmapsto -\epsilon^{\alpha\beta}\mathrm{u}^\star_\beta(\vec{q})$ $\mathbf{l}^A(t, \vec{x}) \longmapsto -\epsilon^{AB}\mathbf{r}^\star_B(t, -\vec{x}),\quad \mathbf{l}^\star_{\dot{A}}(t, \vec{x}) \longmapsto -\epsilon_{\dot{A}\dot{B}}\mathbf{r}^{\dot{B}}(t, -\vec{x})$ $\mathbf{r}^{\dot{A}}(t, \vec{x}) \longmapsto \epsilon^{\dot{A}\dot{B}}\mathbf{l}^\star_{\dot{B}}(t, -\vec{x}),\quad \mathbf{r}^\star_A(t, \vec{x}) \longmapsto \epsilon_{AB}\mathbf{l}^B(t, -\vec{x})$

It anticommutes with the P-transformation $\{\mathtt{P}, \mathtt{C}\} = 0$ and is an order four element, $\{\mathbf{1}, \mathtt{C}, \mathtt{C}^2 = -\mathbf{1}, \mathtt{C}^3 = -\mathtt{C}\} \cong \mathbf{I}(4)$. This complicated mirror hall — if I am not lost therein — has 32 elements, since each element arises with its negative, e.g., $\pm\mathtt{P}$. The group is a semidirect product $\mathrm{grp}(\mathtt{E}, \star, +, \mathtt{P}) \cong \mathbf{I}(4) \vec{\times} \mathbf{I}(2)^3$.

6.15 Fields with Spin and Isospin

The reflection structure is even more complicated: In a particle analysis of fields with spin and isospin the reflections of external and internal $\mathbf{SU}(2)$-structures have to be taken into account, i.e., a spin-reflection with $\mathtt{E}^{\alpha\beta}$ (Greek indices) and an isospin-reflection with $\mathtt{E}_{ab}$ (Latin indices). The $\mathbf{I}(2)^3$-isomorphic mirror group $\mathrm{grp}(\mathtt{E}, \star, +)$ of the complex quartet reads for an idealized mass degenerate particle isomultiplet:

$$\{\mathrm{u}^\alpha_a(\vec{q}), \mathrm{a}^a_\alpha(\vec{q}), \mathrm{u}^{\star a}_\alpha(\vec{q}), \mathrm{a}^{\star\alpha}_a(\vec{q}) \mid \alpha = 1, \ldots, 1 + 2j,\ a = 1, \ldots, 1 + 2I,\ \vec{q} \in \mathbb{R}^3\},$$

Dual conjugation	$\star$	$\mathrm{u}^\alpha_a(\vec{q}) \leftrightarrow \delta^{\alpha\beta}\delta_{ab}\mathrm{u}^{\star b}_\beta(\vec{q}),\quad \mathrm{a}^a_\alpha(\vec{q}) \leftrightarrow \delta_{\alpha\beta}\delta^{ab}\mathrm{a}^{\star\beta}_b(\vec{q})$
Dual involutor for spin and isospin	$\mathtt{E}$	$\mathrm{u}^\alpha_a(\vec{q}) \leftrightarrow \mathtt{E}^{\alpha\beta}\mathtt{E}_{ab}\mathrm{u}^{\star b}_\beta(-\vec{q}),\quad \mathrm{a}^a_\alpha(\vec{q}) \leftrightarrow \mathtt{E}_{\alpha\beta}\mathtt{E}^{ab}\mathrm{a}^{\star\beta}_b(-\vec{q})$
Antidual conjugation	$+$	$\mathrm{u}^\alpha_a(\vec{q}) \leftrightarrow \delta^{\alpha\beta}\delta_{ab}\mathrm{a}^b_\beta(\vec{q})\quad \mathrm{u}^{\star a}_\alpha(\vec{q}) \leftrightarrow \delta_{\alpha\beta}\delta^{ab}\mathrm{a}^{\star\beta}_b(\vec{q})$

Spin – isospin $\mathbf{SU}(2) \times \mathbf{SU}(2)$ – compatible reflections

The dual involutor $\mathtt{E}$ reflects both spin and isospin.

For Weyl-representations of the Lorentz-group, e.g., for a left-handed field $\mathbf{l}_\alpha^A(x)$ with spin $\frac{1}{2}$ and isospin $\frac{1}{2}$, one has for the time and position translation reflections:

Scalar conjugation	$\mathtt{T} = \mathtt{E} \circ \star$	$\mathbf{l}_a^A(t,\vec{x}) \leftrightarrow \epsilon^{AB}\delta_{B\dot{C}}\epsilon_{ab}\delta^{bc}\mathbf{l}_c^{\star\dot{C}}(-t,\vec{x})$ $\mathbf{l}^{\star a}_{\dot{A}}(t,\vec{x}) \leftrightarrow \epsilon_{\dot{A}\dot{B}}\delta^{\dot{B}C}\epsilon^{ab}\delta_{bc}\mathbf{l}^{\star c}_{\dot{C}}(-t,\vec{x})$
Antiinvolutor	$\mathtt{GP} = \mathtt{E} \circ \star \circ +$	$\mathbf{l}_a^A(t,\vec{x}) \leftrightarrow \epsilon^{AB}\delta_B^{\dot{B}}\epsilon_{ab}\mathbf{l}^{\star b}_{\dot{B}}(t,-\vec{x})$
Antidual conjugation	$\mathtt{GP} \circ \mathtt{T} = +$	$\mathbf{l}_a^A(x) \leftrightarrow \delta^{A\dot{A}}\delta_{ab}\mathbf{l}^{\star b}_{\dot{A}}(-x)$

If a nontrivial isospin multiplet is involved, the CP-reflection is called GP-reflection. Examples for such a transformation behavior are the lepton fields in the standard model:

$$\text{left-handed lepton fields: } \mathbf{l}_\alpha^A \cong (\boldsymbol{\nu}^A, \mathbf{e}^A).$$

6.16 Quark Fields Without Quark Particles

The antilinear GPT-reflection, e.g., for a left-handed Weyl-field with isospin dublet-transformation properties like the left-handed lepton field $(\mathbf{l}_a^A)_{a=1,2}^{A=1,2}$ of the standard model:

GPT-reflection	$\mathbf{l}_a^A(x) \leftrightarrow \delta^{A\dot{A}}\delta_{ab}\mathbf{l}^{\star b}_{\dot{A}}(-x)$

can be factorized, because of the self-duality of the isospin $\mathbf{SU}(2)$-representation and its isospin metric into a linear GP- and an antilinear T-reflection — for an isodublet:

T-reflection	$\mathbf{l}_a^A(t,\vec{x}) \leftrightarrow \epsilon^{AB}\delta_{B\dot{C}}\epsilon_{ab}\delta^{bc}\mathbf{l}_c^{\star\dot{C}}(-t,\vec{x})$ $\mathbf{l}^{\star a}_{\dot{A}}(t,\vec{x}) \leftrightarrow \epsilon_{\dot{A}\dot{B}}\delta^{\dot{B}C}\epsilon^{ab}\delta_{bc}\mathbf{l}^{\star c}_{\dot{C}}(-t,\vec{x})$
GP-reflection	$\mathbf{l}_a^A(t,\vec{x}) \leftrightarrow \epsilon^{AB}\delta_B^{\dot{B}}\epsilon_{ab}\mathbf{l}^{\star b}_{\dot{B}}(t,-\vec{x})$

The reflection GP cannot be factorized into G and P for one pair $(\mathbf{l}, \mathbf{l}^\star)$ of Weyl fields.

The standard model is lopsided with respect to left- and right-handed Weyl fields — they do not come in pairs with equal isospin properties, e.g., the left-handed electron field is a member of an isodublet, the right-handed one is an isosinglet. Therefore, the chirality reflection P (left-handed $\leftrightarrow$ right-handed) is not implementable. The question if the standard model dynamics is P-invariant or not is meaningless — the P-reflection cannot even be formulated. Only with a ground state and particle states, there arise sectors with P-representations and a possible P-violation, e.g., for the weak neutral interactions.

Since there is an equivalence of all dual representations for isospin $\mathbf{SU}(2)$, however not for $\mathbf{SU}(n)$ with $n \geq 3$, it is possible to give a Lorentz- and color-compatible GPT-reflection for quark fields, e.g., for the right-handed down-quark color triplets:

$$\mathbf{d}_c^{\dot{A}}(x),\ \ \mathbf{d}_A^{\star c}(x), \quad \begin{cases} A, \dot{A} = 1, 2 \text{ for Lorentz-group } \mathbf{SL}(2,\mathbb{C}), \\ c = 1, 2, 3 \text{ for color group } \mathbf{SU}(3), \end{cases}$$

$$\text{GPT-reflection: } \mathbf{d}_c^{\dot{A}}(x) \leftrightarrow -\delta^{\dot{A}A}\delta_{cd}\mathbf{d}_A^{\star d}(-x),$$

not, however, a factorization into GP- and T-reflections. In contrast to isospin $\mathbf{SU}(2)$ with the compatible metric ϵ^{ab} (volume form in two dimensions), there does not exist a color $\mathbf{SU}(3)$-compatible linear dual isomorphism between triplet-representation $(1, 0)$ for quark fields $\mathbf{d}$ and antitriplet-representation $(0, 1)$ for antiquark fields $\mathbf{d}^\star$. Quark triplets do not have an invariant bilinear form; there exists no quark metric.

The existence of an invariant bilinear form is representation-dependent. Therefore, a color-compatible non-implementability of the GP-reflection for the quarks cannot be generalized: Color octets and $\mathbf{SU}(3)$-representations with trivial triality $\mathbf{I}(3) \subset \mathbf{SU}(3)$, e.g., the gluon fields or quark-parametrized color-currents, have an invariant bilinear form and, therefore, an $\mathbf{SU}(3)$-compatible GP-reflection.

In analogy to the P-transformation one can say with respect to the GP-reflection in the standard model: The model displays neither GP-invariance nor GP-violation — it is impossible to implement the GP-reflection for all field of the standard dynamics, especially for quark field color-triplets.

There is the related difficulty for a harmonic analysis of quark fields in terms of quark particle operators with a free $\mathbf{U}(1)$-translation behavior:

$$\mathbf{d}_c^{\dot{A}}(x) = \int \frac{d^3q}{2q_0}\, s_\alpha^{\dot{A}}\Big(-\frac{\vec{q}}{m}\Big)\, \frac{e^{iqx}\mathrm{u}_c^\alpha(\vec{q}) - e^{-iqx}\mathrm{a}_c^{\star\alpha}(\vec{q})}{\sqrt{2}},$$
$$\mathbf{d}_A^{\star c}(x) = \int \frac{d^3q}{2q_0}\, s_A^\alpha\Big(-\frac{\vec{q}}{m}\Big)\, \frac{e^{-iqx}\mathrm{u}_\alpha^{\star c}(\vec{q}) - e^{iqx}\mathrm{a}_\alpha^c(\vec{q})}{\sqrt{2}}.$$

What is the quark mass, what its rest system? Are quarks stable, or do they have a width? It is impossible to formulate, in a color-compatible form, a time and a position reflection (T- and GP-reflection, respectively) for triplets since such reflections rely on the existence of a linear dual isomorphism $\mathrm{u} \overset{\mathrm{E}}{\leftrightarrow} \mathrm{u}^\star$.

These difficulties fit together with the experimental fact, that, so far, quarks have never been detected as free particles, and that the CP-violation has been observed in experiments with quark-parameterizable strongly interacting particles. A derivation of the CP-violation starting from such symmetry- and representation-oriented ideas, does, as far as I know, not exits. Today, the parametrization of the CP-violation uses essentially a free phase in the mixing of three quark families whose flavor triality has nothing to do with the color triality.

Chapter 7
Telescopes for Symmetries

A telescope allows us to see more exactly things that are far away. The telescopes, attributed to the pirates in those nice picture books, have nested tubes becoming smaller and smaller. Nested symmetry structures, transitions between groups or Lie algebras, projections to subgroups or their embedding in larger groups — these are concepts for the realization and closer investigation of a physical symmetry group or Lie algebra.

If a symmetry of a dynamics is distinguished as "fundamental," a "smaller" ("less") symmetry may be related to a special coordinate system or basis. For example, a measuring apparatus in a rest system of an observer in spacetime is not symmetric under all Lorentz transformations, a rest system is characterized by a subgroup $\mathbf{O}(3) \subset \mathbf{O}(1,3)$. Or, a ground state of lowest energy, less symmetric than the dynamics, e.g., for a ferromagnet with a fixed magnetization direction with $\mathbf{O}(2) \subset \mathbf{O}(3)$ or in superconductivity without definite charges with $\{1\} \subset \mathbf{U}(1)$ or in the standard model of the electroweak interactions with the distinction of electromagnetism $\mathbf{U}(1) \subset \mathbf{U}(2)$, leads to a breakdown and a rearrangement of the interaction symmetry for the energy eigenstates, e.g., for the asymptotically observed particles, characterized by the smaller symmetry. The introduction of a "larger "("more") symmetry may be suggested by a convenient parametrization, e.g., the introduction of separate angular momentum- and spin-symmetry for a weak spin-orbit coupling with $\mathbf{SU}(2) \times \mathbf{SO}(3) \supset \mathbf{SU}(2)$, or a spectrum classifying operation group for a many particle system, or the use of complex structures as an algebraically complete framework for real structures, e.g., the "color" symmetry of the isotropic quantum oscillator extending the rotation symmetry in three real position dimensions with $\mathbf{SU}(3) \supset \mathbf{SO}(3)$.

Embedding manifolds of subgroups define homogeneous spaces. Physically important examples are the mass hyperboloid with the momenta of an elementary particle as embedding manifold of (the compact rotations for) Euclidean position

H. Saller, *Operational Symmetries*, DOI 10.1007/978-3-319-58664-9_7

in (the Lorentz group for) Minkowskian spacetime, or the Goldstone[1] manifold for the degenerate ground state of the electroweak interactions which characterizes the embedding of the electromagnetic phase group in the hyperisospin group.

7.1 Symmetry Telescopes

A *group or a Lie algebra telescope* is a finite chain of nested groups or Lie algebras:

$$G_1 \subseteq G_2 \subseteq \ldots \subseteq G_k,$$
$$L_1 \subseteq L_2 \subseteq \ldots \subseteq L_k.$$

One endpoint G_1 or L_1 denotes the *smallest (coarsest) symmetry*, the other one G_k or L_k the *largest (finest) symmetry* of the telescope. A Lie group telescope has a corresponding Lie algebra telescope and vice versa. One can look into a telescope from both sides — looking from the large symmetries, it may be interesting how one can find or define the smaller ones. Looking in the other direction, the embedding manifold of the coarse symmetry in the finer one may be considered. There is a minimal initial symmetry: the initial (trivial) group $\{1\}$ and Lie algebra $\{0\}$. However, not each law for building a symmetry telescope involves the minimal initial symmetry.

Some telescope examples from basic physics: A homogeneous space-time telescope is characterized by the linear groups:

$$\mathbf{O}(3) \text{ (Euclid)} \subset \mathbf{O}(1,3) \text{ (Lorentz)} \subset \mathbf{GL}(4,\mathbb{R}) \text{ (Einstein)},$$

where, for general relativity, the general linear group $\mathbf{GL}(4,\mathbb{R})$ is the structural group of all bases for tangent Minkowski translations, or in a complex formulation with associated real Lie algebras:

Group	$\mathbf{SU}(2)$	$\subset$	$\mathbf{SL}(2,\mathbb{C})$	$\subset$	$\mathbf{GL}(4,\mathbb{C})$
for Lie algebra	$\mathbb{R}^3$	$\subset$	$\mathbb{R}^6$	$\subset$	$\mathbb{R}^{32}$

An inhomogeneous space-time telescope with translations involving semidirect groups is

$$\mathbf{O}(3) \vec{\times} \mathbb{R}^3 \text{ (Euclid)} \subset [\mathbf{O}(3) \vec{\times} \mathbb{R}^3] \vec{\times} \mathbb{R}^3 \text{ (Galileo)} \subset \mathbf{O}(1,3) \vec{\times} \mathbb{R}^4 \text{ (Poincaré)}.$$

Here, $\mathbb{R}^3$ is the vector space with the position translations or with the velocity transformations; $\mathbb{R}^4$ denotes the spacetime translations.

[1]Jeffrey Goldstone (1933–).

A unitary telescope with internal operations is

$$\begin{array}{ccccc} \mathbf{U}(1) & \subset & \mathbf{U}(2) & \subset & \mathbf{U}(2\times 3) \\ & & \cong \frac{\mathbf{U}(1)\times\mathbf{SU}(2)}{\mathbf{I}(2)} & & \cong \frac{\mathbf{U}(1)\times\mathbf{SU}(2)\times\mathbf{SU}(3)}{\mathbf{I}(6)} \\ \text{(charge)} & & \text{(hypercharge–isospin)} & & \text{(hyperisospin–color).} \end{array}$$

An internal-external telescope embeds compact unitary operations in general linear ones:

$$\text{(hyperisospin) } \mathbf{U}(2) \subset \mathbf{GL}(2, \mathbb{C}) \text{ (hyper-Lorentz group).}$$

Here, all symmetries are to be understood as real Lie symmetries.

7.2 Subgroup Equivalences and Embedding Manifolds

A group pair $G_j \subseteq G_{j+1}$ or a Lie algebra pair $L_j \subseteq L_{j+1}$ in a symmetry telescope designates a *group or a Lie algebra transition (embedding)*. It is trivial for the initial structure $\{1_G\} \subseteq G$ or $\{0\} \subseteq L$. A proper symmetry transition is a little telescope with two different tubes.

For a subgroup $H \subseteq G$ one looks for the "remainder" in G — what is G modulo H? A transition is associated with H-embeddings as G-subsets using a left multiplication (also "left translation") with any group element $g \in G$:

$$L_g : H \longrightarrow G, \quad L_g(h) = gh, \quad L_g(H) = gH.$$

A subgroup $H \subseteq G$ gives rise to an *equivalence relation* for the elements of the full group: Two G-elements are H-equivalent, called H-*conjugated*, if they are connected with each other by a H-multiplication:

$$g \overset{H}{\approx} k \iff g^{-1}k \in H \iff k \in gH.$$

The three defining properties of a subgroup equivalence are related to three defining group properties:

$$\begin{array}{rl} g \overset{H}{\approx} g & \iff g^{-1}g = 1_G \in H, \\ \text{(reflexive)} & \quad \text{(neutral element),} \\ g \overset{H}{\approx} k \Rightarrow k \overset{H}{\approx} g & \iff g^{-1}k \in H \Rightarrow k^{-1}g \in H, \\ \text{(symmetric)} & \quad \text{(inverse),} \\ g \overset{H}{\approx} k,\ k \overset{H}{\approx} l \Rightarrow g \overset{H}{\approx} l & \iff g^{-1}k \in H,\ k^{-1}l \in H \Rightarrow g^{-1}l \in H, \\ \text{(transitive)} & \quad \text{(group composition).} \end{array}$$

Subgroup equivalences are relevant for positions and spacetimes, e.g., as Galileo and Lorentz equivalences of reference systems.

A *H-right orbit in G* is defined by the right-multiplication of a G-element with all H-elements. This gives the disjoint decomposition of the *group G into H-orbits*, which contain H-conjugated elements:

$$\begin{aligned} gH &= \{gh \,\big|\, h \in H\} = \{k \in G \,\big|\, k \overset{H}{\approx} g\}, \\ G &= \biguplus_{\text{repr } g_r \in B} g_r H, \quad G/H = \{gH \,\big|\, g \in G\}, \end{aligned}$$

with the choice of a representative g_r for each class. The classes g_rH are equipotent. For a finite group G, each class has order H elements. Therefore, the order of G, devided by the order of H gives the number of the finitely many H-orbits and class representatives in a *G-base* $B \subseteq G$, card $G/H =$ card B. The projection of the group to the H-orbits maps group elements to group subsets with representatives:

$$G \longrightarrow G/H \longrightarrow (G/H)_r = B, \quad g \longmapsto gH \longmapsto g_r.$$

Analoguous are the left-orbits of a subgroup $H \setminus G = \{Hg \,\big|\, g \in G\}$. Left- and right-structures are isomorphic. Finally, one can define, with two subgroups $H_{1,2} \subseteq G$, the left-right-orbits $H_1 \setminus G/H_2 = \{H_1 g H_2 \,\big|\, g \in G\}$.

The H-orbits G/H are also called the quotient of G modulo H, or H-cosets, or H-classes, or — especially for Lie groups — the *embedding manifold of the subgroup H in G*.

In general, the subgroup orbits G/H do not constitute a group: For instance, the rotation $\mathbf{SO}(s)$-orbits in $\mathbf{SO}(1+s)$ are representable and parametrizable by an s-sphere Ω^s — a group only for $s = 1$ with $\mathbf{SO}(1) \cong \{1\}$. For $s = 2$, the embedding manifold $\mathbf{SO}(3)/\mathbf{SO}(2)$ for the axial rotation group orbits is parametrizable by a spherical surface: Each rotation $\mathbf{SO}(3)$ is an axial rotation $\vec{\theta}$ around the axis $\frac{\vec{\theta}}{|\theta|}$ for $\vec{\theta} \neq 0$ with rotation angle between π and $-\pi$. Therefore, each point of the solid sphere with radius π in the Euclidean space $\mathbb{E}^3$ can be associated with an $\mathbf{SO}(3)$-rotation — unique up to the diametral point pairs on the surface that parametrize the same rotation. Axial rotations $\mathbf{SO}(2)$ are parametrized by the points on a diameter of the sphere in the direction of the rotation axis. All $\mathbf{SO}(2)$-orbits in $\mathbf{SO}(3)$ are parametrizable by all possible sphere-diameters, i.e., by the points of the surface, $\Omega^2 \cong \mathbf{SO}(3)/\mathbf{SO}(2)$.

Good to know: In the complex, an orthogonal transformation $O \in \mathbf{O}(s)$ is unitarily diagonalizable with the eigenvalues coming in pairs $(e^{i\theta}, \epsilon e^{-i\theta})$, $\epsilon = \pm 1$, and one additional eigenvalue (± 1) for odd dimension n. In the real, the eigenvalue-pairs determine $\mathbf{O}(2)$ transformations, the lonely eigenvalue (± 1) displays an $\mathbf{SO}(s)$-invariant one-dimensional subspace (rotation axis).
With the isomorphy $\mathbf{SO}(3) \cong \mathbf{SU}(2)/\mathbf{I}(2)$, the spin group $\mathbf{SU}(2)$ is parametrizable by all points of the solid sphere — without diametral identification, expressed by the two-elementic $\mathbb{I}(2)$-classes.

For a Lie algebra transition $M \subseteq L$ the full Lie algebra L is decomposable into vector subspaces, *hyperplanes in L*, that all are isomorphic to M as vector space:

$$l_1 \overset{M}{\approx} l_2 \iff l_1 - l_2 \in M, \quad L/M = \{l + M \,\big|\, l \in L\},$$
$$L = \bigoplus_{\text{repr } l_r \in B} (l_r + M).$$

For Lie algebras, and for abelian groups, a left-right distinction does not make sense. All M-parallel hyperplanes constitute a vector space L/M, in general not a Lie algebra. For instance, the angular momentum Lie algebra $\log \mathbf{O}(3)$ in the Lie algebra of the Lorentz group $\log \mathbf{O}(1,3)$ leads to the real three-dimensional vector space with the boosts $\log \mathbf{O}(1,3)/\log \mathbf{O}(3) \cong \mathbb{R}^3$. It consists of three-dimensional hyperplanes in the six-dimensional Lorentz-Lie algebra.

In a symmetry telescope the cosets G_{j+1}/G_j and covector spaces L_{j+1}/L_j are something like the seams, where there meet adjacent tubes:

$$\begin{array}{ccccccccc} G_k & \supseteq & G_{k-1} & \supseteq & \cdots & \supseteq & G_2 & \supseteq & G_1, \\ & G_k/G_{k-1} & & G_{k-1}/G_{k-2} & \cdots & G_3/G_2 & & G_2/G_1 & \end{array}$$

$$\begin{array}{ccccccccc} L_k & \supseteq & L_{k-1} & \supseteq & \cdots & \supseteq & L_2 & \supseteq & L_1. \\ & L_k/L_{k-1} & & L_{k-1}/L_{k-2} & \cdots & L_3/L_2 & & L_2/L_1 & \end{array}$$

7.3 Self-Operations

Groups and Lie algebras have actions on themselves, induced by their characterizing structures.

A group G acts on itself by *left- and right-multiplications* (translations), which are group realizations in the group bijections:

$$L_g, R_g : G \longrightarrow G, \quad L_g(k) = gk, \quad R_g(k) = kg^{-1},$$
$$L_{g_1} \circ L_{g_2} = L_{g_1 g_2}, \quad R_{g_1} \circ R_{g_2} = R_{g_1 g_2}.$$

A rather trivial example are the rotations $\mathbf{SO}(2)$, acting on the circle $\Omega^1 \cong \mathbf{SO}(2)$.

Left-right-multiplications $L \times R$ constitute the *biregular realization* of the doubled group $G \times G$, acting on the group $G \ni k \longmapsto (L_{g_1} \times R_{g_2})(k) = g_1 k g_2^{-1}$. The "diagonal" action defines the *inner group automorphisms*, trivial for the elements from the centrum:

$$\operatorname{Int} g = L_g \times R_g : G \longrightarrow G, \quad \operatorname{Int} g(k) = gkg^{-1}, \quad \operatorname{Int} g_1 \circ \operatorname{Int} g_2 = \operatorname{Int} g_1 g_2.$$

The left-multiplications can be defined as group realization on left cosets with respect to a subgroup $H \subseteq G$ (and, analogously, right on right):

$$G \times G/H \longrightarrow G/H, \quad (g, kH) \longmapsto gkH.$$

Examples are rotations $\mathbf{SO}(3)$, acting on the 2-sphere $\Omega^2 \cong \mathbf{SO}(3)/\mathbf{SO}(2)$, or Lorentz transformations $\mathbf{SO}_0(1,3)$, acting on a momentum 3-hyperboloid $\mathcal{Y}^3 \cong \mathbf{SO}_0(1,3)/\mathbf{SO}(3)$.

The left multiplication with a group element $g \in G$ on a base with representatives gives the shifted representative up to the right multiplication with a subgroup element (*Wigner element*):

$$G = \biguplus_{k_r \in B} k_r H, \quad B = (G/H)_r \subseteq G,$$

$$L_g : B \longrightarrow B, \quad k_r \longmapsto (gk)_r, \quad \text{with} \begin{cases} gk_r = (gk)_r h(g, k_r) \in G, \\ (gk)_r = (gkH)_r \in B, \\ h(g, k_r) \in H. \end{cases}$$

Rotate as example a convex potato peel, not a 2-sphere, with the representatives for the axial rotation classes $\mathbf{SO}(3)/\mathbf{SO}(2)$.

A Lie group G has the *adjoint representation* on its Lie algebra $\log G$, i.e., it is realized in the general linear group of the vector space $\log G$:

$$\begin{aligned} &G \times \log G \longrightarrow \log G, \quad g \bullet l = \operatorname{Ad} g(l) \cong g \circ l \circ g^{-1} \text{ for matrices,} \\ &\text{for } g = 1_G + m + \ldots \Rightarrow \operatorname{Ad} g(l) = l + [m, l] + \ldots, \\ &\operatorname{Ad} g_1 \circ \operatorname{Ad} g_2 = \operatorname{Ad} g_1 g_2, \end{aligned}$$

e.g., the spin group $u \in \mathbf{SU}(2)$ acting on the angular momentum Lie algebra $i\sigma^a \in \log \mathbf{SU}(2) \cong \mathbb{R}^3$ by the familiar Euclidean rotations:

$$\begin{aligned} &i\sigma^a \longmapsto u \circ i\sigma^a \circ u^\star = O^a_b(u) i\sigma^b, \\ &\text{with } O \in \mathbf{SO}(3): \quad O^a_b(u) = \frac{1}{2} \operatorname{tr} u \circ i\sigma^a \circ u^\star \circ \sigma^b. \end{aligned}$$

Therefore, each Lie group defines its associated semidirect tangent group or *adjoint affine Lie group* $G \vec{\times} \log G$. Nontrivial examples are the Euclidean group $\mathbf{SU}(2) \vec{\times} \mathbb{R}^3 = \mathbf{SO}(3) \vec{\times} \mathbb{R}^3$ for the spin group, or $\mathbf{SL}(2,\mathbb{C}) \vec{\times} \mathbb{R}^6 = \mathbf{SO}_0(1,3) \vec{\times} \mathbb{R}^6$ — not the Poincaré group — for the Lorentz group with Lie algebra $\log \mathbf{SL}(2,\mathbb{C}) \cong \mathbb{R}^6$:

$$\begin{aligned} s &\in \mathbf{SL}(2,\mathbb{C}), \\ \sigma^{jk} &= \tfrac{\sigma^k \check{\sigma}^j - \sigma^j \check{\sigma}^k}{4} \in \log \mathbf{SL}(2,\mathbb{C}): \end{aligned} \begin{cases} \sigma^{jk} \longmapsto s \circ \sigma^{jk} \circ s^{-1} = \Lambda^j_m(s) \Lambda^k_n(s) \sigma^{mn}, \\ \text{with } \Lambda \in \mathbf{SO}_0(1,3). \end{cases}$$

The restriction of the group action to cosets (irreducible G-realizations) gives, for Lie groups, the semidirect products with Lie algebra classes:

$$\begin{aligned} &G \times G/H \longrightarrow G/H, \quad g \bullet kH = gkH, \\ &G \times \log G/\log H \longrightarrow \log G/\log H, \quad g \bullet (l + \log H) = \operatorname{Ad} g(l) + \log H. \end{aligned}$$

An example is the action $\mathbf{GL}(2,\mathbb{C}) \times \mathbb{R}^4_+ \longrightarrow \mathbb{R}^4_+$ of the full group $\mathbf{GL}(2,\mathbb{C}) \supset \mathbf{SL}(2,\mathbb{C})$ with the Lorentz group on the strict future $\mathbb{R}^4_+ \cong \mathbf{GL}(2,\mathbb{C})/\mathbf{U}(2)$ and on the Lie algebra classes $\mathbf{GL}(2,\mathbb{C}) \vec{\times} \mathbb{R}^4 \supset \mathbf{SO}_0(1,3) \vec{\times} \mathbb{R}^4$ with the Poincaré group for full Minkowski spacetime $\log \mathbf{GL}(2,\mathbb{C})/\log \mathbf{U}(2) \cong \mathbb{R}^4$.

In these semidirect products, only the linear vector space structure of the Lie algebra $\log G$ or its classes $\log G/\log H$ is relevant — not the Lie bracket.

The *adjoint representation* of a Lie algebra uses the Lie bracket:

$$\begin{aligned}
&\log G \times \log G \longrightarrow \log G,\quad (l,m) \longmapsto \operatorname{ad} l(m) = [l,m],\\
&\operatorname{ad} l \circ \operatorname{ad} m = \operatorname{ad}[l,m],\\
&\text{for } g = \exp l = 1_G + l + \ldots \Rightarrow \operatorname{Ad} g = \exp \operatorname{ad} l = \mathbf{1}_{\log G} + \operatorname{ad} l + \ldots,\\
&\text{e.g. } i\sigma^a \in \log \mathbf{SU}(2) \Rightarrow \operatorname{ad} i\sigma^a(i\sigma^b) = -2\epsilon^{abc} i\sigma^c.
\end{aligned}$$

Obviously, all self-operations can be also performed in representations,

$$\begin{aligned}
g, l &\longmapsto D(g), \mathcal{D}(l),\\
\text{e.g., } \operatorname{Int} D(g)(D(k)) &= D(g) \circ D(k) \circ D(g^{-1}) = D(gkg^{-1}),\\
\operatorname{ad} \mathcal{D}(l)(\mathcal{D}(m)) &= [\mathcal{D}(l), \mathcal{D}(m)].
\end{aligned}$$

To summarize the self-operations of groups and Lie algebras:

$G \times G \longrightarrow G$	left: $k \longmapsto L_g(k) = gk$, right: $k \longmapsto R_g(k) = kg^{-1}$, inner: $k \longmapsto \operatorname{Int} g(k) = gkg^{-1}$,
$G \times L \longrightarrow L$	adjoint: $l \longmapsto \operatorname{Ad} g(l) = g \circ l \circ g^{-1}$,
$L \times L \longrightarrow L$	adjoint: $m \longmapsto \operatorname{ad} l(m) = [l, m]$.

Self – Operations

7.4 Normal and Ideal Transitions, Simple Symmetries

The left multiplication gH of a subgroup $H \subseteq G$ with a group element gives, in general, not a group. This is in contrast to the inner group automorphisms: gHg^{-1} is an H-isomorphic subgroup. The stability group of H consists of all group elements $g \in G$ with $gHg^{-1} = H$.

Precisely for a normal subgroup $N \subseteq G$, i.e., for *normal transition*, the left- and right-orbits coincide $gN = Ng$ for all $g \in G$: The stability group of N is the full group G. In this case, the orbits G/N are endowed with a natural, i.e., representative-independent, group structure with composition $g_1N \bullet g_2N = g_1g_2N$.

Analogously: If the Lie subalgebra $M \subseteq L$ is an ideal, i.e., for an *ideal transition* with $[L,M] \subseteq M$, the vector space L/M carries a natural Lie algebra structure by $[l_1 + M, l_2 + M] = [l_1, l_2] + M$.

If one considers in the Poincaré group $\mathbf{O}(1,3) \vec{\times} \mathbb{R}^4$ the embeddings of the Minkowski translations $\mathbb{R}^4$, a normal subgroup, the corresponding classes inherit the group structure of the Lorentz group $\mathbf{O}(1,3)$, in contrast to Minkowski spacetime as the manifold of the Lorentz group embeddings, which are isomorphic to the spacetime translations $\mathbf{O}(1,3) \vec{\times} \mathbb{R}^4/\mathbf{O}(1,3) \cong \mathbb{R}^{(1,3)} \cong \mathbb{R}^4$ only as manifold — Minkowski spacetime (the Lorentz group classes) can be parametrized by four real parameters, however, the classes have no naturally inherited additive group structure,

$$\begin{aligned} \text{natural group structure:} \quad & \mathbf{O}(1,3) \vec{\times} \mathbb{R}^4/\mathbb{R}^4 \cong \mathbf{O}(1,3), \\ \text{no natural group structure:} \quad & \mathbf{O}(1,3) \vec{\times} \mathbb{R}^4/\mathbf{O}(1,3) \cong \mathbb{R}^4. \end{aligned}$$

This can be visualized by the $\mathbf{O}(s)$-orbits $\mathbb{E}^s \cong \mathbf{O}(s) \vec{\times} \mathbb{R}^s/\mathbf{O}(s)$ for the Euclidean plane and space, $s = 2, 3$.

Normal series are group telescopes with normal transitions, where, in addition, the smallest group is a normal subgroup of the largest one, and, therefore, of all groups in the telescope:

$$G_1 \subseteq \ldots \subseteq G_k, \text{ normal: } G_j \subseteq G_{j+1} \text{ and } G_1 \subseteq G_k.$$

All embeddings G_{j+1}/G_j are groups, called the *quotients of the normal series* from G_k to G_1. From the neighborhood property to be a normal subgroup there does not follow that G_j is a normal subgroup of G_{j+2}.

Each group and Lie algebra has the improper normal subgroups G, $\{1_G\}$ and ideals L, $\{0\}$, respectively. In addition to those boring structures, there is the abelian *center of the symmetry*, whose elements "slide" through the composition, i.e., they commute with all group and Lie algebra elements:

$$\begin{aligned} \operatorname{centr} G &= \{z \in G \mid z = gz \text{ for all } g \in G\}, \\ \operatorname{centr} L &= \{c \in L \mid [c, l] = 0 \text{ for all } l \in L\}. \end{aligned}$$

The centers are the kernels of the inner automorphisms of G and the adjoint representation of L. The symmetry modulo the center is called the *adjoint group* of the group G and the *adjoint Lie algebra* of the Lie algebra L:

$$\operatorname{Int} G = G/\operatorname{centr} G, \quad \operatorname{ad} L = L/\operatorname{centr} L.$$

Examples are the special unitary groups with the centers defining the n-ality, $\operatorname{centr} \mathbf{SU}(n) \cong \mathbf{I}(n)$, e.g., two-ality with the classes $\mathbf{SU}(2)/\mathbf{I}(2) \cong \mathbf{SO}(3)$ and triality $\mathbf{I}(3)\mathbf{1}_3 \subset \mathbf{SU}(3)$, and, analogously the full linear groups:

$$\begin{aligned} \operatorname{Int} \mathbf{U}(n) &\cong \mathbf{U}(n)/\mathbf{U}(\mathbf{1}_n) && \cong \mathbf{SU}(n)/\mathbf{I}(n) && \cong \operatorname{Int} \mathbf{SU}(n). \\ \operatorname{Int} \mathbf{GL}(n, \mathbb{C}) &\cong \mathbf{GL}(n, \mathbb{C})/\mathbf{GL}(1, \mathbb{C}) && \cong \mathbf{SL}(n, \mathbb{C})/\mathbf{I}(n) && \cong \operatorname{Int} \mathbf{SL}(n, \mathbb{C}). \end{aligned}$$

with the orthochronous Lorentz group $\mathbf{SL}(2,\mathbb{C})/\mathbf{I}(2) \cong \mathbf{SO}_0(1,3)$ as example. The adoint group of an abelian group A is trivial $\mathrm{Int}\,A \cong \{1\}$.

In a transition from a Lie group G with the product as internal composition $(g,h) \longmapsto gh$ to its Lie algebra $\log G$, one has, in addition to the Lie bracket, the additive group structure of the vector space $\log G$, i.e., *two internal* compositions $(l,m) \longmapsto [l,m]$ and $(l,m) \longmapsto l+m$.

For an abelian group, e.g., $\mathbf{U}(1) = \exp i\mathbb{R}$ or $\mathbf{D}(1) = \exp\mathbb{R}$, the Lie bracket of $\log G$ is trivial. Here, only the additive composition of $\log G$ defines, after exponentiation, the group product:

$$l, m \in \log G \text{ (abelian)} \Rightarrow e^{l+m} = e^l e^m.$$

Good to know: The regular group of the complex $(n \times n)$-matrices consists of their exponents, the general linear group $\mathbf{GL}(n,\mathbb{C}) = \exp\mathbf{AL}(n,\mathbb{C})$ with $\log\mathbf{GL}(n,\mathbb{C}) = \mathbf{AL}(n,\mathbb{C})$. Each endomorphism $f \in \mathbf{AL}(n,\mathbb{C})$ has a minimal polynomial of finite degree $\sum_{k=0}^{d}\alpha_k f^k = 0$, e.g., $\mathbf{1}_2 - \sigma_3^2 = 0$ for the Pauli matrix σ_3, or $\lambda_3 - \lambda_3^3 = 0$ and $2\mathbf{1}_3 - \sqrt{3}\lambda_8 - 3\lambda_8^2 = 0$ for the Gell-Mann matrices λ_3 and λ_8.

In addition to the center, there are other natural normal subgroups or ideals (more ahead). If a group does not have a proper normal subgroup, e.g., the cyclic groups $\mathbf{I}(p)$ for a prime number p, or if a nonabelian Lie algebra has no proper ideal, e.g., the $\mathbf{SU}(n+1)$-Lie algebras A_n^c for $n \geq 1$, group or Lie algebra are called *simple*.

7.5 Relativities and Homogeneous Spaces

An embedding of a subgroup leads to the concept of its *relativity*, parametrizable by the corresponding coset manifold. In the following, five physically relevant relativities are described: "up-down" or *perpendicular relativity* as realized after discovering the earth's surface to be spherical; then rotation or space and time relativity, as used in what we call *special relativity*; then Lorentz group or Minkowski spacetime relativity or, also, with Wigner's definition, *particle relativity* as an important ingredient (local inertial systems) of general relativity; then *electromagnetic relativity* as used for the ground state and the particle definition in the standard model of electroweak interactions; and, finally, *unitary relativity* with "spacetimelike" and "chargelike" operations.

Relativity will be defined by operation group classes, e.g., in special relativity, the distinction of your rest system determines a decomposition of spacetime translations into time and position translations. Compatible with this decomposition is your position rotation group $\mathbf{SO}(3)$ as a subgroup of the orthochronous Lorentz group $\mathbf{SO}_0(1,3)$. There are as many decompositions of spacetime into time and position as there are rotation groups in a Lorentz group. The rotation group classes are parametrizable by the points of a one-shell three-dimensional hyperboloid $\mathcal{Y}^3 \cong \mathbf{SO}_0(1,3)/\mathbf{SO}(3)$ that give the momenta (velocities) for all your possible motions, i.e., by your mass hyperboloid $q_0^2 - \vec{q}^2 = m^2$. Another example: The perpendicularities of mankind, if earthbound, are characterized by the axial rotation groups in a rotation group and parametrizable by the two coordinates of the earth's surface $\Omega^2 \cong \mathbf{SO}(3)/\mathbf{SO}(2)$. Or: There are as many Wigner particle definitions for spacetime as there are Lorentz groups $\mathbf{O}(1,3)$ in the general linear group $\mathbf{GL}(4,\mathbb{R})$.

Now in general: The distinction of an "idolized" operation subgroup[2] H in a "general" operation group G picks one element in the G-symmetric space G/H, which stands for the relativity of the "idolized" group, called H-*relativity*. A decomposition of a Lie group with G/H and H can be formalized by fiber bundles (more ahead). Relativity groups H, i.e., the local fixgroups in G/H, are candidates for gauge groups. An "idolization" goes, negatively, with the "narrow-minded" assumption of an absolute point of view, e.g., absolute up-down, absolute time, absolute Minkowski spacetime (particle universality), or, positively, with the distinction of a smaller operation symmetry, enforced, e.g., by initial or boundary conditions. Important examples are degenerate ground states ("spontaneous symmetry breakdown"), where, by distinguishing one ground state from the symmetric space G/H, i.e., from the H-degeneracy manifold, an "interaction-symmetry" G is reduced to a "particle-symmetry" H. Examples are the ground states of superconductivity, superfluidity, ferromagnetism, and the electroweak standard model. The degeneracy degrees of freedom give rise to long range interactions.

All of this gives the first four columns of the following table, which together with the last one will be discussed with their representations in more detail ahead

Relativity	"General" group G (r_G, r)	"Idolized" subgroup H (fiber)	Homogeneous space$\mathbb{M} \cong G/H$ (base manifold)	Relativity manifold parameters
Axial rotation (perpendicular) relativity	$\mathbf{SO}(3)$ $\sim \mathbf{SU}(2)$ $(1,0)$	$\mathbf{SO}(2)$	2-sphere $\Omega^2 \cong \mathbf{SO}(3)/\mathbf{SO}(2)$ $\cong \mathbf{SU}(2)/\mathbf{SO}(2)$	two spherical coordinates
Rotation (special) relativity	$\mathbf{SO}_0(1,3)$ $\sim \mathbf{SL}(2,\mathbb{C})$ $(2,1)$	$\mathbf{SO}(3)$ $\sim \mathbf{SU}(2)$	3-hyperboloid $\mathcal{Y}^3 \cong \mathbf{SO}_0(1,3)/\mathbf{SO}(3)$ $\cong \mathbf{SL}(2,\mathbb{C})/\mathbf{SU}(2)$	three momenta
Lorentz group (particle) relativity	$\mathbf{GL}(4,\mathbb{R})$ $(4,4)$	$\mathbf{O}(1,3)$	tetrad or metric manifold $\mathcal{M}^{10} \cong \mathbf{GL}(4,\mathbb{R})/\mathbf{O}(1,3)$ $\cong \mathbf{D}(1) \times \mathbf{SO}_0(3,3)/\mathbf{SO}_0(1,3)$	10 components for metrical tensor
Electro-magnetic relativity	$\mathbf{U}(2)$ $(2,0)$	$\mathbf{U}(1)_+$	Goldstone manifold $\mathcal{G}^3 \cong \mathbf{U}(2)/\mathbf{U}(1)_+$	three weak coordinates
Unitary (electroweak) relativity	$\mathbf{GL}(2,\mathbb{C})$ $(4,2)$	$\mathbf{U}(2)$	spacetime future 4-cone $\mathbb{D}(2) \cong \mathbf{GL}(2,\mathbb{C})/\mathbf{U}(2)$ $\cong \mathbf{D}(1) \times \mathcal{Y}^3$	four spacetime coordinates

Somewhat in accordance with the historical development, the "general" operations of one relativity can constitute the "idolized" group of the next relativity as seen in the two telescopes ending in full general linear groups, a real one for spacetime concepts, from flat to spherical earth to special and general relativity, and a complex

[2]In general, H is assumed to be no normal subgroup. Otherwise, it "disappears" in the quotient group $G_0 = G/H$.

one for interactions, from electromagnetic to electroweak transformations and their spacetime (gauge) dependence:

$$\begin{array}{l} \mathbf{SO}(2) \subset \mathbf{SO}(3) \subset \mathbf{SO}_0(1,3) \subset \mathbf{GL}(4,\mathbb{R}), \\ \mathbf{U}(1)_+ \subset \mathbf{U}(2) \subset \mathbf{GL}(2,\mathbb{C}). \end{array}$$

Starting from the larger groups, the transitions to the smaller ones are related for $\mathbf{SO}_0(1,3) \subset \mathbf{GL}(4,\mathbb{R})$ to the distinction of a local inertial system in general relativistic spacetime, for $\mathbf{SO}(3) \subset \mathbf{SO}_0(1,3)$ to the distinction of a rest system in special relativistic spacetime, and for $\mathbf{SO}(2) \subset \mathbf{SO}(3)$ to the distinction of an axial system in Euclidean position.

There is a connection of the two telescopes on the level of the adjoint groups — the rotations $\mathbf{SO}(3)$ and its cover group $\mathbf{SU}(2)$ and the orthochronous Lorentz group $\mathbf{SO}_0(1,3)$ and its cover group $\mathbf{SL}(2,\mathbb{C})$:

$$\begin{array}{rl} \mathbf{SO}(3) \cong & \mathbf{U}(2)/\mathbf{U}(\mathbf{1}_2) \cong \mathbf{SU}(2)/\{\pm\mathbf{1}_2\}, \\ \mathbf{SO}_0(1,3) \cong & \mathbf{GL}(2,\mathbb{C})/\mathbf{GL}(\mathbf{1}_2,\mathbb{C}) \cong \mathbf{SL}(2,\mathbb{C})/\{\pm\mathbf{1}_2\}. \end{array}$$

All groups in the five relativities considered are real Lie groups. All "general" groups have reductive Lie algebras; for perpendicular and rotation relativity they are even simple. Perpendicular and electromagnetic relativity have a compact "general" group. With the exception of Lorentz group relativity, all "idolized" groups are compact subgroups. The second column contains the dimension r_G of maximal abelian subgroups, which is the rank of the group G generating Lie algebra $L = \log G$, and the dimension $r \le r_G$ of maximal noncompact abelian subgroups. The rank r_G gives the number of independent invariants, rational or continuous, that characterize a G-representation. r is the maximal number of independent continuous invariants.

7.6 Products and Factorizations of Groups

In a normal transition $N \subset G$ (analogously an ideal transition for Lie algebras) the classes G/N constitute a group. It may occur, that the group G contains a G/N-isomorphic subgroup H. In such a case, there is the natural question if the group G is a product of the normal subgroup N and a G/N-isomorphic subgroup H. That there does not have to exist a G/N-isomorphic subgroup, can be seen already at the integers $\mathbb{Z} \subset \mathbb{R}$. The classes $\mathbb{R}/\mathbb{Z}$ have a group structure — however, there does not exist a subgroup of the reals isomorphic to the 1-torus $\mathbb{R}/\mathbb{Z} \cong \mathbf{U}(1)$. Or, the rotation group $\mathbf{SO}(3) \cong \mathbf{SU}(2)/\{\pm\mathbf{1}_2\}$ is no $\mathbf{SU}(2)$-subgroup.

In a factorization $G = H \diamond N$ for the case $G/N \cong H$, if possible, three product types $\diamond$ are important: the group composition product, the semidirect product and the direct product.

First, for the group product: With a normal subgroup $N \subseteq G$ and a subgroup $H \subseteq G$, also $NH = HN \subseteq G$ with the group multiplication is a subgroup. In general, for a normal transition $G/N \cong H \subseteq G$, the product HN does not factorize the full group. For instance, the cyclic group $\mathbf{I}(4)$ has only one normal proper subgroup, given by $\mathbf{I}(2)$, and $\mathbf{I}(4)/\mathbf{I}(2) \cong \mathbf{I}(2)$, however $\mathbf{I}(4) \neq \mathbf{I}(2) \times \mathbf{I}(2)$. Is a group the product of a normal subgroup and a subgroup, $G = NH$, then each group element is factorizable into an N- and a H-element. Examples are the unitary groups as products of two normal subgroups $\mathbf{U}(n) = \mathbf{U}(\mathbf{1}_n) \circ \mathbf{SU}(n)$. Physically relevant is the hypercharge-isospin group $\mathbf{U}(2) = \mathbf{U}(\mathbf{1}_2) \circ \mathbf{SU}(2)$.

A second "more separating" product: A factorization of the group elements for $HN = G$ does not have to be unique—it has the multi-valuedness of the intersecting normal subgroup $H \cap N$. The factorization of the unitary groups with the phase group is determined up to the cyclotomic group:

$$\mathbf{U}(n) = \mathbf{U}(\mathbf{1}_n) \circ \mathbf{SU}(n), \text{ with } \mathbf{SU}(n) \cap \mathbf{U}(\mathbf{1}_n) \cong \mathbf{I}(n),$$

i.e., ambiguous for $n \geq 2$, e.g., $\mathbf{U}(2) \not\cong \mathbf{U}(1) \times \mathbf{SU}(2)$.

If a factorization $G = NH$ is possible with the additional property $H \cap N = \{1_G\}$, i.e., for a *unique factorization*, one has a *semidirect product group*:

$$G \cong H \vec{\times} N = NH \ni (h, n) = (1_H, n) \circ (h, 1_N).$$

The factorization type, not the subgroup factor H is unique. For example, the factorization of the orthogonal group $\mathbf{O}(n_+, n_-)$ into the special group $\mathbf{SO}(n_+, n_-)$ (normal subgroup) and a reflection group $\mathbf{I}(2) \subset \mathbf{O}(n_+, n_-)$ is unique,

$$\begin{aligned} \mathbf{SO}(n_+, n_-) &\subset \mathbf{O}(n_+, n_-), \quad \mathbf{O}(n_+, n_-)/\mathbf{SO}(n_+, n_-) \cong \mathbf{I}(2) = \{\pm 1\}, \\ \mathbf{O}(n_+, n_-) &= \mathbf{I}(2) \vec{\times} \mathbf{SO}(n_+, n_-), \quad \mathbf{I}(2) \cap \mathbf{SO}(n_+, n_-) \cong \{1\}. \end{aligned}$$

For the implementation of $-1 \in \mathbf{I}(2)$ one can take a diagonal matrix in $\mathbf{O}(n_+, n_-)$ with an odd number of -1's, e.g., the central position reflection $-\mathbf{1}_3$ in the orthogonal group $\mathbf{O}(3)$. In the full Lorentz group $\mathbf{O}(1, 3)$ with special Lorentz group $\mathbf{SO}(1, 3)$, one may take a time reflection $\begin{pmatrix} -1 & 0 \\ 0 & \mathbf{1}_3 \end{pmatrix}$ or a position reflection $\begin{pmatrix} 1 & 0 \\ 0 & -\mathbf{1}_3 \end{pmatrix}$:

$$\mathbf{O}(1, 3) \cong \mathbf{I}(2) \vec{\times} \mathbf{SO}(1, 3).$$

For a semidirect group $H \vec{\times} N$, the subgroup H is realized in the automorphisms of the normal subgroup N. Prominent examples are the affine groups $H \vec{\times} V$, e.g., for any group representation with a subgroup of the automorphism $H \subseteq \mathbf{GL}(V)$ of a vector space V with translations (additive group), e.g., the Euclidean group $\mathbf{SO}(3) \vec{\times} \mathbb{R}^3$, the Poincaré group $\mathbf{O}(1, 3) \vec{\times} \mathbb{R}^4$, or the Higgs group $\mathbf{U}(2) \vec{\times} \mathbb{C}^2$ with the hyperisospin group acting on a complex two-dimensional vector space. An affine group for a finite-dimensional vector space $V \cong \mathbb{K}^n$ has a representation in the linear group of a vector space with one more dimension $V' \cong \mathbb{K}^{1+n}$:

$$\begin{aligned}\mathbf{GL}(n,\mathbb{K}) \vec{\times} \mathbb{K}^n \supseteq H \vec{\times} \mathbb{K}^n &\longrightarrow \mathbf{GL}(1+n,\mathbb{K}),\\ (h,v) &\longmapsto \begin{pmatrix} h & v\\ 0 & 1\end{pmatrix} = \begin{pmatrix} \mathbf{1}_n & v\\ 0 & 1\end{pmatrix}\begin{pmatrix} h & 0\\ 0 & 1\end{pmatrix},\\ (h,v) \bullet (k,w) &= (h \circ k, v + h.w).\end{aligned}$$

It can be derived by a contraction of a $\mathbf{GL}(1+n,\mathbb{K})$-subgroup, e.g., the Euclidean group by nonrelativistic contraction $\mathbf{SO}_0(1,3) \overset{\frac{1}{c}\to 0}{\longrightarrow} \mathbf{SO}(3) \vec{\times} \mathbb{R}^3$ of the Lorentz group.

Finally, the third complete separation of two unique factor subgroups: If in a semidirect product $G \cong H \vec{\times} N$ with $H \cap N = \{1_G\}$ the N-orbits $G/N \cong H \subseteq G$ are even isomorphic to a normal subgroup $H = N' \subseteq G$, one has a *direct product group* $G \cong N \times N'$, e.g., the orthogonal transformations for odd dimensions, e.g., $\mathbf{O}(3)$ for Euclidean position:

$$\begin{aligned}\mathbf{O}(n_+, n_-) &\cong \mathbf{I}(2) \times \mathbf{SO}(n_+, n_-) \quad \text{for } n_+ + n_- = 1, 3, 5, \ldots,\\ \mathbf{O}(n_+, n_-) &\cong \mathbf{I}(2) \vec{\times} \mathbf{SO}(n_+, n_-) \quad \text{for } n_+ + n_- = 2, 4, \ldots\end{aligned}$$

Two groups $H_{1,2}$ can be directly multiplied $(h_1, h_2) \in H_1 \times H_2$ with the product a separate composition for both factors, which are normal subgroups.

The standard model for electroweak interactions is invariant under the direct product $\mathbf{SL}(2,\mathbb{C}) \times \mathbf{U}(2)$ of the Lorentz group and the hypercharge-isospin group.

There exist some systematic methods, to find normal subgroups N of a group G and ideals I of a Lie algebra L, which give rise to factorizations $G = H \diamond N$ and $L = M \oplus I$ with subgroups H and Lie subalgebras M, respectively.

7.7 Abel Telescopes

Characteristic for a group G is its invertible and associative product, and for a Lie algebra L its antisymmetric Lie bracket with Jacobi identity. Related to these compositions, one can look for natural normal subgroups and ideals, i.e., corresponding substructures, which can be always constructed, independently of the specific symmetry. This amounts to an investigation of the natural properties of the group product and the Lie bracket.

To look into the abelian structure of the Lie-bracket, one considers the finite linear combinations of all Lie-brackets:

$$L^2 = [L, L] = \{\textstyle\sum_i [l_i, m_i] \bigm| l_i, m_i \in L\}.$$

L is abelian, if, and only if, $L^2 = \{0\}$. L^2 is the *commutator ideal* of L, also called its derived Lie algebra $L^2 = \partial L$. The ideal-property $[L, L^2] \subseteq L^2$ can be easily verified via the Jacobi identity. It also is easy to check, that the quotient Lie algebra L/L^2 is abelian, $[l + L^2, k + L^2] \subseteq L^2$.

The analogue construction for a group G involves the finite products of all "commutators" $\mathrm{com}(g, h) = ghg^{-1}h^{-1}$ and is called its *commutator group* $G^2 = \mathrm{com}(G, G)$, a normal subgroup. It consists precisely of the neutral element $\{1_G\}$, if G is abelian. The quotient group G/G^2 is abelian.

If a group G has a factorization with its commutator subgroup, or a Lie algebra with its commutator ideal, then G has to contain as factor an abelian subgroup H and L an abelian Lie subalgebra:

$$\text{if } G = H \diamond G^2, \quad L = M \diamond L^2, \quad \text{then abelian } H, M.$$

The iteration of the commutator-procedure, defines the *Abel telescopes* of a symmetry with normal and ideal transitions and abelian quotients; all members are even normal subgroups and ideals of the largest symmetry:

$$\begin{array}{ll} G = G^{(0)} \supseteq G^2 = G^{(2)} \supseteq G^{(4)} \supseteq \ldots, & G^{(2n+2)} = \mathrm{com}(G^{(2n)}, G^{(2n)}) = \partial^{n+1}G \\ L = L^{(0)} \supseteq L^2 = L^{(2)} \supseteq L^{(4)} \supseteq \ldots, & L^{(2n+2)} = [L^{(2n)}, L^{(2n)}] = \partial^{n+1}L \end{array}$$

For a finite group and a finite-dimensional Lie algebra there arises, with the proper transitions, a finite normal series and ideal series.

The permutation groups $\mathbf{G}(n)$ for $n = 2, 3, 4$ have famous Abel telescopes, which lead to the trivial group. They involve the alternating groups (special permutation groups) as commutator since the signature of $\mathbf{G}(n)^2$ is positive. The Klein four-group is a commutator group:

$$\begin{array}{lll} n \geq 2: & \mathbf{G}(n) \supset \mathbf{S}(n) = \mathbf{G}(n)^2, & \\ & \mathbf{G}(2) \supset \mathbf{S}(2) \supset \{1\}, & \mathbf{G}(2) \cong \mathbf{I}(2), \\ & \mathbf{G}(3) \supset \mathbf{S}(3) \supset \{1\}, & \mathbf{S}(3) \cong \mathbf{I}(3), \\ & \mathbf{G}(4) \supset \mathbf{S}(4) \supset \mathbf{K}(4) \supset \{1\}, & \mathbf{K}(4) \cong \mathbf{S}(4)^2 \cong \mathbf{I}(2) \times \mathbf{I}(2). \end{array}$$

The abelian quotients allow semidirect factorizations:

$$\begin{array}{lr} n \geq 2: & \mathbf{I}(2) \Rightarrow \mathbf{G}(n) \cong \mathbf{I}(2) \vec{\times} \mathbf{S}(n), \\ n = 3: & \mathbf{I}(2),\ \mathbf{I}(3) \Rightarrow \mathbf{G}(3) \cong \mathbf{I}(2) \vec{\times} \mathbf{I}(3), \\ n = 4: & \mathbf{I}(2),\ \mathbf{I}(3),\ \mathbf{I}(2) \times \mathbf{I}(2) \Rightarrow \mathbf{G}(4) \cong \mathbf{I}(2) \vec{\times} [\mathbf{I}(3) \vec{\times} \mathbf{K}(4)]. \end{array}$$

If an Abel telescope leads to the initial symmetry, i.e., to the trivial group $\{1_G\}$ or Lie algebra $\{0\}$, the group or Lie algebra is called *solvable*, or *meta-abelian of degree k*, if the initial symmetry is reached exactly for $\partial^k G = G^{(2k)} = \{1_G\}$ or $\partial^k L = L^{(2k)} = \{0\}$. $\{1_G\}$ and $\{0\}$ have the Abel degree 0, nontrivial abelian symmetries have Abel degree 1, etc. The structure before the initial symmetry is abelian. The predicate "solvable" originates from the permutation groups $\mathbf{G}(n)$ with the related property of complex polynomials $\sum_{k=0}^{n} \alpha_k x^k = 0$ with n solutions, which are solvable by radicals for $n \leq 4$.

Good to know: The simply expressable theorems of Burnside, Feit, and Thompson,[3] that all finite groups of odd order, card $G \in 2\mathbb{N}+1$, and in addition to the groups of "simple-prime generated order" (monogenous), card $G = p^n$, also groups with "double-prime generated order" (bigeneous), card $G = (p_1)^{n_1} (p_2)^{n_2}$ are solvable, are by no means simply to prove.

In some sense the opposite of an abelian symmetry is the case where the commutator reproduces the symmetry, $L^2 = L \neq \{0\}$ or $G^2 = G \neq \{1_G\}$. Such a nonabelian group and Lie algebra may be called *idempotent*. Examples are simple Lie algebras or simple nonabelian groups. All simple Lie algebras of the Cartan classification, especially all traceless matrix algebras have "boring" Abel telescopes, e.g., the Lie algebra $A_1 \cong \mathbb{C}^3$ or its compact form $A_1^c \cong \mathbb{R}^3$:

$$A_1 = [A_1, A_1], \text{ since } i\sigma^a = \epsilon_{abc}\left[\frac{i\sigma^b}{2}, \frac{i\sigma^c}{2}\right].$$

Not only semisimple Lie algebras are idempotent. A non-semisimple example is the Poincaré group $(\mathbf{SO}_0(1,3) \vec{\times} \mathbb{R}^4)^2 = \mathbf{SO}_0(1,3) \vec{\times} \mathbb{R}^4$.

A Lie algebra L with a semisimple commutator algebra $[L, L]$ is called *reductive* — then $L^4 = L^2$. It can be equivalently characterized by being a direct sum of a semisimple Lie subalgebra H and an abelian ideal A:

$$\begin{aligned} \text{reductive } L = A \oplus H: \quad & [A, A] = \{0\} = [H, A], \\ & H \cong [L, L] \cong L/A, \quad A \cong L/[L, L]. \end{aligned}$$

Examples are the full matrix Lie algebras $\log \mathbf{GL}(n, \mathbb{K}) = \mathbf{AL}(n, \mathbb{K})$, $\mathbb{K} = \mathbb{R}, \mathbb{C}$, of the full matrix groups with the commutator ideal $[\mathbf{AL}(n, \mathbb{K})]^2 = \log \mathbf{SL}(n, \mathbb{K})$ the traceless matrices. For the unitary groups $\mathbf{U}(n)$ one has $[\log \mathbf{U}(n)]^2 = \log \mathbf{SU}(n)$.

The commutator subgroup $\mathrm{com}(G, G)$ of a compact, connected Lie group is closed and leads to a factorization with a normal subgroup $C_0 \subseteq \mathrm{centr}\, G$ of the abelian center, unique up to a finite normal subgroup,

$$G = C_0 \circ G^2, \quad \text{finite } C_0 \cap G^2.$$

C_0 is the center contribution of the connected component G_0 of the neutral group element, $C_0 = \mathrm{centr}\, G \cap G_0$. Examples for such products are the unitary groups $\mathbf{U}(n)$ or $\mathbf{U}(1) \times \mathbf{SU}(n_1) \times \ldots \times \mathbf{SU}(n_k)$ and locally isomorphic groups.

The universal covering group $\overline{G} = \exp L$ of a reductive Lie algebra L is a product with the commutator group as direct factor:

$$\begin{aligned} \overline{G} &= \overline{C}_0 \times \overline{G}^2, \quad \text{connection component of } 1_G : \overline{C}_0 \subseteq \mathrm{centr}\, \overline{G}, \\ \mathrm{centr}\, \overline{G} &\cong \overline{C}_0 \times \overline{C}_2, \quad \text{finite } \overline{C}_2 = \mathrm{centr}\, \overline{G}^2. \end{aligned}$$

Examples are the $\mathbf{U}(n)$-Lie algebras $\log \mathbf{U}(n) \cong \mathbb{R} \oplus \log \mathbf{SU}(n)$ with the universal covering group $\mathbf{D}(1) \times \mathbf{SU}(n)$ containing the dilations $\mathbf{D}(1) = \exp \mathbb{R}$.

[3] Walter Feit (1930–2004), John Thompson (1932–).

7.8 Power Telescopes

Abel telescopes are defined by group or Lie algebra products with an even number of elements. All powers are used in the sharper *power telescopes.*

In general, the finite linear combinations of the Lie brackets of two ideals $I, I' \subseteq L$ give an ideal in L — a fortiori an ideal in I and I':

$$\text{ideal } [I, I'] = \left\{ \sum_i [l_i, l_i'] \,\Big|\, l_i \in I, l_i' \in I' \right\}.$$

Therefore, the inductively defined *powers of a Lie algebra or a group,*

$$\begin{aligned} L^1 &= L, \quad L^{n+1} = [L, L^n], \\ G^1 &= G, \quad G^{n+1} = \text{com}(G, G^n), \end{aligned}$$

lead to the *power telescopes* with L-ideals and normal G-subgroups:

$$\begin{aligned} L &\supseteq L^2 \supseteq L^3 \supseteq \ldots, \\ G &\supseteq G^2 \supseteq G^3 \supseteq \ldots. \end{aligned}$$

If a power telescope ends with the trivial symmetry, the symmetry is *nilpotent with power k*, if triviality is reached exactly for $L^k = \{0\}$ or $G^k = \{1_G\}$. Initial symmetries are nilpotent with power 1, nontrivial abelian symmetries with power 2.

The $(n \times n)$-triagonal matrices, i.e., with zeros in the left lower part, have a commutator ideal that is strictly triagonal, i.e., with trivial diagonal. In the Lie algebra powers, the diagonal-parallel lines become trivial, power by power — one has a power n nilpotent Lie algebra:

$$\begin{pmatrix} \times & \times & \times & \ldots & \times \\ 0 & \times & \times & \ldots & \times \\ & \ldots & & \ldots & \\ 0 & 0 & 0 & \ldots & \times \end{pmatrix} \in \nabla(n, \mathbb{K}) \subseteq \mathbf{AL}(n, \mathbb{K}), \quad n = 1, 2, \ldots,$$

$$\begin{aligned} \nabla(n, \mathbb{K})^2 &= [\nabla(n, \mathbb{K}), \nabla(n, \mathbb{K})] \ni \begin{pmatrix} 0 & \times & \times & \ldots & \times \\ 0 & 0 & \times & \ldots & \times \\ & \ldots & & \ldots & \\ 0 & 0 & 0 & \ldots & 0 \end{pmatrix}, \\ \nabla(n, \mathbb{K})^n &= [\nabla(n, \mathbb{K}), \nabla(n, \mathbb{K})^{n-1}] = \{\mathbf{0}_n\}. \end{aligned}$$

The Abel telescopes with even powers are subtelescopes of the power telescopes with all powers — nilpotent symmetries are a subset of solvable symmetries.

Good to know: One has $[L^k, L^j] \subseteq L^{k+j}$ for the powers, and the relation $L^{2k} \subseteq L^{2^k}$ for the commutator ideals and the powers.

Summarizing, there are the relations for the properties connected with the group product or the Lie bracket:

$$\begin{aligned} \{\text{initial}\} \subset \{\text{abelian}\} \subset \{\text{nilpotent}\} \subset \{\text{solvable}\}, \\ \{\text{solvable}\} \cap \{\text{nonabelian, simple}\} = \emptyset. \end{aligned}$$

7.9 The Heisenberg Lie Algebra and Group

It is possible to relate the quantum-mechanical Born–Heisenberg commutation relations to a Lie algebra structure: In nonrelativistic quantum mechanics the position-momentum operators $(\vec{\mathbf{x}}, \vec{\mathbf{p}})$ come with the nontrivial commutator $\mathbf{I}$ — think of the *Heisenberg operator* $\mathbf{I}$ as an abstract vector, not as the imaginary unit i! $\{\vec{\mathbf{x}}, \vec{\mathbf{p}}, \mathbf{I}\}$ span as basis the power three nilpotent real *Heisenberg Lie algebra* $\log \mathbf{H}(s) \cong \mathbb{R}^{1+2s}$, characterized by $s = 1, 2, \dots$, nontrivial brackets for $\vec{\mathbf{x}} = (\mathbf{x}^1, \dots, \mathbf{x}^s)$ and $\vec{\mathbf{p}}$:

$$[\vec{\mathbf{x}}, \vec{\mathbf{p}}] = \mathbf{1}_s\mathbf{I}, \quad [\vec{\mathbf{x}}, \mathbf{I}] = 0 = [\vec{\mathbf{p}}, \mathbf{I}].$$

The commutator $\mathbf{I}$ spans the real one-dimensional center:

$$\log \mathbf{H}(s)^2 = \text{ centr } \log \mathbf{H}(s) = \mathbb{R}\mathbf{I}.$$

The adjoint Heisenberg Lie algebra, i.e., the classes of the center, is abelian and can be spanned by the classes (underlined) of position and momentum, commuting as in the classical framework:

$$\text{ad } \log \mathbf{H}(s) \cong \log \mathbf{H}(s) / \text{ centr } \log \mathbf{H}(s) \cong \mathbb{R}^{2s}, \quad \text{with } [\underline{\vec{\mathbf{x}}}, \underline{\vec{\mathbf{p}}}] = 0.$$

The Heisenberg Lie algebra has a *faithful nonunitary* real $(2+s)$-dimensional representation where the Heisenberg operator $\mathbf{I}$ is *not* proportional to the unit matrix:

$$\begin{aligned}
\log \mathbf{H}(s) \ &\ni \vec{\pi}\vec{\mathbf{x}} + \vec{\xi}\vec{\mathbf{p}} + \tau\mathbf{I} \longmapsto \left(\begin{array}{cc|c} 0 & \vec{\pi} & \tau \\ 0 & \mathbf{0}_s & \vec{\xi} \\ \hline 0 & 0 & 0 \end{array}\right) \in \mathbf{AL}(2+s, \mathbb{R}), \\
\log \mathbf{H}(s)^2 \ &\ni [\vec{\pi}_1\vec{\mathbf{x}} + \vec{\xi}_1\vec{\mathbf{p}} + \tau_1\mathbf{I}, \vec{\pi}_2\vec{\mathbf{x}} + \vec{\xi}_2\vec{\mathbf{p}} + \tau_2\mathbf{I}] = (\vec{\pi}_1\vec{\xi}_2 - \vec{\xi}_1\vec{\pi}_2)\mathbf{I}, \\
\log \mathbf{H}(s)^3 \ &= \{0\}.
\end{aligned}$$

All operators — positions, momenta and the Heisenberg operator — are representable by nilpotent matrices, e.g., for $s = 1$:

$$\mathbf{x} \longmapsto \begin{pmatrix} 0 & 1 & 0 \\ 0 & 0 & 0 \\ 0 & 0 & 0 \end{pmatrix}, \ \mathbf{x} \circ \mathbf{x} = 0,$$

$$\mathbf{p} \longmapsto \begin{pmatrix} 0 & 0 & 0 \\ 0 & 0 & 1 \\ 0 & 0 & 0 \end{pmatrix}, \ \mathbf{p} \circ \mathbf{p} = 0,$$

$$\mathbf{I} \longmapsto \begin{pmatrix} 0 & 0 & 1 \\ 0 & 0 & 0 \\ 0 & 0 & 0 \end{pmatrix}, \ \mathbf{I} \circ \mathbf{I} = 0, \quad [\mathbf{x}, \mathbf{p}] = \mathbf{I}, \quad [\mathbf{x}, \mathbf{I}] = 0 = [\mathbf{p}, \mathbf{I}].$$

Additive translations as vectors of a finite-dimensional space $V \cong \mathbb{K}^n$ have a representation as multiplicative group in $\mathbf{GL}(1+n, \mathbb{K}) \supset \mathbf{GL}(n, \mathbb{K}) \vec{\times} \mathbb{K}^n$, generated by nilquadratic matrices:

$$\mathbb{K}^n \ni \vec{v}\vec{\mathbf{e}} = v_a \mathbf{e}^a \longmapsto \left(\begin{array}{c|c} \mathbf{1}_n & \vec{v} \\ \hline 0 & 1 \end{array}\right) = \exp\left(\begin{array}{c|c} \mathbf{0}_n & \vec{v} \\ \hline 0 & 0 \end{array}\right)$$
$$= \left(\begin{array}{cccc|c} 1 & 0 & \ldots & 0 & v_1 \\ 0 & 1 & \ldots & 0 & v_2 \\ \cdot & \cdot & \cdots & \cdot & \cdot \\ 0 & 0 & \ldots & 1 & v_n \\ \hline 0 & 0 & \ldots & 0 & 1 \end{array}\right) \in e^{\mathbb{K}^n} \subset \mathbf{GL}(1+n, \mathbb{K}),$$
$$\left(\begin{array}{c|c} \mathbf{0}_n & \vec{v} \\ \hline 0 & 0 \end{array}\right) \circ \left(\begin{array}{c|c} \mathbf{0}_n & \vec{w} \\ \hline 0 & 0 \end{array}\right) = 0, \quad \left(\begin{array}{c|c} \mathbf{1}_n & \vec{v} \\ \hline 0 & 1 \end{array}\right) \circ \left(\begin{array}{c|c} \mathbf{1}_n & \vec{w} \\ \hline 0 & 1 \end{array}\right) = \left(\begin{array}{c|c} \mathbf{1}_n & \vec{v}+\vec{w} \\ \hline 0 & 1 \end{array}\right).$$

Exponentiation of the Heisenberg Lie algebra gives the noncompact real $(1+2s)$-dimensional *Heisenberg group* $\mathbf{H}(s)$, for illustration again in a nonunitary faithful real $(2+s)$-dimensional representation:

$$\mathbf{H}(s) \longrightarrow \mathbf{GL}(2+s, \mathbb{R}),$$
$$\exp(\vec{\pi}\vec{\mathbf{x}} + \vec{\xi}\vec{\mathbf{p}} + \tau\mathbf{I}) \longmapsto \exp\left(\begin{array}{cc|c} 0 & \vec{\pi} & \tau \\ 0 & \mathbf{0}_s & \vec{\xi} \\ \hline 0 & 0 & 0 \end{array}\right) = \left(\begin{array}{cc|c} 1 & \vec{\pi} & \tau+\vec{\pi}\vec{\xi} \\ 0 & \mathbf{1}_s & \vec{\xi} \\ \hline 0 & 0 & 1 \end{array}\right),$$
$$e^{\vec{\pi}_1\vec{\mathbf{x}}+\vec{\xi}_1\vec{\mathbf{p}}+\tau_1\mathbf{I}} \circ e^{\vec{\pi}_2\vec{\mathbf{x}}+\vec{\xi}_2\vec{\mathbf{p}}+\tau_2\mathbf{I}} = e^{(\vec{\pi}_1+\vec{\pi}_2)\vec{\mathbf{x}}+(\vec{\xi}_1+\vec{\xi}_2)\vec{\mathbf{p}}+(\tau_1+\tau_2-\vec{\pi}_2\vec{\xi}_1)\mathbf{I}}.$$

It is a semidirect product with semidirect Lie algebra with the position translations as homogeneous contribution $e^{\mathbb{R}^s}$ acting on $(1+s)$-dimensional translations spanned by the Heisenberg operator $\mathbf{I}$ and the momenta $\vec{\mathbf{p}}$:

$$\log \mathbf{H}(s) \cong \mathbb{R}^s \vec{\oplus} \mathbb{R}^{1+s},$$
$$\mathbf{H}(s) \cong e^{\mathbb{R}^s} \vec{\times} \mathbb{R}^{1+s} \ni e^{\vec{\pi}\vec{\mathbf{x}}+\vec{\xi}\vec{\mathbf{p}}+\tau\mathbf{I}} = e^{\vec{\xi}\vec{\mathbf{p}}+\tau\mathbf{I}} \circ e^{\vec{\pi}\vec{\mathbf{x}}},$$
$$e^{\mathbb{R}^s} \times \mathbb{R}^{1+s} \longrightarrow \mathbb{R}^{1+s}, \quad \begin{pmatrix} 1 & \vec{\pi} \\ 0 & \mathbf{1}_s \end{pmatrix}\begin{pmatrix} \tau \\ \vec{\xi} \end{pmatrix} = \begin{pmatrix} \tau+\vec{\pi}\vec{\xi} \\ \vec{\xi} \end{pmatrix}.$$

For higher-dimensional positions, it is possible to include homogeneous transformations $\mathbf{L} \in \mathbf{AL}(s, \mathbb{R})$ and $\exp \mathbf{L} \in \mathbf{GL}(s, \mathbb{R})$, e.g., rotations for $\mathbf{H}(3)$, with dual action on positions and momenta in the double semidirect *homogeneously extended Heisenberg Lie algebra and Lie group* — illustrated in the nonunitary faithful real $(2+s)$-dimensional representation:

$$\mathbf{AL}(s, \mathbb{R}) \vec{\oplus} \log \mathbf{H}(s) \ni \theta\mathbf{L} + \vec{\pi}\vec{\mathbf{x}} + \vec{\xi}\vec{\mathbf{p}} + \tau\mathbf{I}$$
$$\longmapsto \left(\begin{array}{cc|c} 0 & \vec{\pi} & \tau \\ 0 & \theta & \vec{\xi} \\ \hline 0 & 0 & 0 \end{array}\right) \in \mathbf{AL}(2+s, \mathbb{R}),$$
$$\left[\left(\begin{array}{cc|c} 0 & \vec{\pi}_1 & \tau_1 \\ 0 & \theta_1 & \vec{\xi}_1 \\ \hline 0 & 0 & 0 \end{array}\right), \left(\begin{array}{cc|c} 0 & \vec{\pi}_2 & \tau_2 \\ 0 & \theta_2 & \vec{\xi}_2 \\ \hline 0 & 0 & 0 \end{array}\right)\right] = \left(\begin{array}{cc|c} 0 & -\theta_1^T(\vec{\pi}_2)+\theta_2^T(\vec{\pi}_1) & \vec{\pi}_1\vec{\xi}_2 - \vec{\pi}_2\vec{\xi}_1 \\ 0 & [\theta_1, \theta_2] & \theta_1(\vec{\xi}_2) - \theta_2(\vec{\xi}_1) \\ \hline 0 & 0 & 0 \end{array}\right),$$

$$\mathbf{GL}(s, \mathbb{R}) \vec{\times} \mathbf{H}(s) \ni \exp(\theta\mathbf{L} + \vec{\pi}\vec{\mathbf{x}} + \vec{\xi}\vec{\mathbf{p}} + \tau\mathbf{I})$$
$$\longmapsto \left(\begin{array}{cc|c} 1 & \vec{\pi} & \tau+\vec{\pi}\vec{\xi} \\ 0 & \exp\theta & \vec{\xi} \\ \hline 0 & 0 & 1 \end{array}\right) \in \mathbf{GL}(2+s, \mathbb{R}).$$

The semidirect Heisenberg groups and Lie algebras can be obtained as *nonrelativistic Inönü–Wigner contractions* of the corresponding semidirect Poincaré groups and Lie algebras, starting with the minimal rotation free case $\mathbf{H}(1)$ and the 1-position Poincaré group $\mathbf{O}(1,1)\vec{\times}\mathbb{R}^2$ with Lie algebra for boost and translations:

$$\log\mathbf{O}(1,1)\ \vec{\oplus}\ \mathbb{R}^2 : [\mathbf{B},\mathbf{p}^3]=\mathbf{p}^0,\quad [\mathbf{B},\mathbf{p}^0]=\mathbf{p}^3,\quad [\mathbf{p}^0,\mathbf{p}^3]=0,$$

$$\text{with } (\mathbf{B},\mathbf{p}^0)=c(\mathbf{x}^3,\mathbf{I})\ \Big\downarrow\ \tfrac{1}{c}\to 0\ \text{(contraction)}$$

$$\log\mathbf{H}(1)\cong\mathbb{R}\ \vec{\oplus}\ \mathbb{R}^2 : [\mathbf{x}^3,\mathbf{p}^3]=\mathbf{I},\quad [\mathbf{x}^3,\mathbf{I}]=0,\quad [\mathbf{I},\mathbf{p}^3]=0.$$

The boosts contract to the position operators, the time translation to the Heisenberg operator — in general for $s=1,2,\ldots$ with s boosts $\vec{\mathbf{B}}$ and momenta $\vec{\mathbf{p}}$, $\binom{s}{2}$ rotations $\mathbf{J}$, one time translation $\mathbf{p}^0$, and the nontrivial Lie brackets:

$$\log\mathbf{O}(1,s)\ \vec{\oplus}\ \mathbb{R}^{1+s} : \begin{cases}[\mathbf{J},\mathbf{J}]=\mathbf{J}, & [\mathbf{J},\vec{\mathbf{B}}]=\vec{\mathbf{B}}, \quad [\vec{\mathbf{B}},\vec{\mathbf{B}}]=\mathbf{J},\\ [\mathbf{J},\vec{\mathbf{p}}]=\vec{\mathbf{p}}, & [\vec{\mathbf{B}},\mathbf{p}^0]=\vec{\mathbf{p}}, \quad [\vec{\mathbf{B}},\vec{\mathbf{p}}]=\mathbf{p}^0,\end{cases}$$

$$\text{with } (\vec{\mathbf{B}},\mathbf{p}^0)=c(\vec{\mathbf{x}},\mathbf{I})\ \Big\downarrow\ \left(\begin{array}{cc|c}0 & \vec{\mathbf{B}} & \mathbf{p}^0\\ \vec{\mathbf{B}} & \mathbf{L} & \vec{\mathbf{p}}\\ \hline 0 & 0 & 0\end{array}\right) \xrightarrow{\frac{1}{c}\to 0} \left(\begin{array}{cc|c}0 & \vec{\mathbf{x}} & \mathbf{I}\\ 0 & \mathbf{L} & \vec{\mathbf{p}}\\ \hline 0 & 0 & 0\end{array}\right)\text{(contraction)}$$

$$\begin{array}{r}\log\mathbf{O}(s)\ \vec{\oplus}\ \log\mathbf{H}(s) :\\ \log\mathbf{H}(s)\cong\mathbb{R}^s\ \vec{\oplus}\ \mathbb{R}^{1+s}\end{array} \begin{cases}[\mathbf{J},\mathbf{J}]=\mathbf{J}, & [\mathbf{J},\vec{\mathbf{x}}]=\vec{\mathbf{x}}, \quad [\vec{\mathbf{x}},\vec{\mathbf{x}}]=0,\\ [\mathbf{J},\vec{\mathbf{p}}]=\vec{\mathbf{p}}, & [\vec{\mathbf{x}},\mathbf{I}]=0, \quad [\vec{\mathbf{x}},\vec{\mathbf{p}}]=\mathbf{I}.\end{cases}$$

The Inönü–Wigner contraction was invented with a different interpretation: There, the boosts $\mathbf{B}$ contract to the special Galileo transformations $\mathbf{G}(s)\cong\mathbf{H}(s)$.

The special relativistic semidirect Poincaré groups themselves can be obtained by contraction from indefinite orthogonal groups with simple Lie algebras, starting from the real three-dimensional case with $\mathbf{O}(2)$ as maximal compact subgroup:

$$\log\mathbf{O}(1,2) : [\mathbf{J}^1,\mathbf{\Sigma}^2]=\mathbf{\Sigma}^3,\quad [\mathbf{\Sigma}^3,\mathbf{J}^1]=\mathbf{\Sigma}^2,\quad [\mathbf{\Sigma}^2,\mathbf{\Sigma}^3]=-\mathbf{J}^1,$$

$$\text{with } (\mathbf{J}^1,\mathbf{\Sigma}^2)=\frac{1}{\xi}(\mathbf{p}^0,\mathbf{p}^3),\ \mathbf{\Sigma}^3=\mathbf{B}\ \Big\downarrow\ \xi\to 0\ \text{(contraction)}$$

$$\log\mathbf{O}(1,1)\ \vec{\oplus}\ \mathbb{R}^2 : [\mathbf{p}^0,\mathbf{p}^3]=0,\quad [\mathbf{B},\mathbf{p}^0]=\mathbf{p}^3,\quad [\mathbf{B},\mathbf{p}^3]=\mathbf{p}^0,$$

$$\begin{array}{r}\text{with } (\mathbf{B},\mathbf{p}^0)=c(\mathbf{x}^3,\mathbf{I}),\\ \text{or } \mathbf{J}^1=\frac{c}{\xi}\mathbf{I},\ \mathbf{\Sigma}^2=\frac{1}{\xi}\mathbf{p}^3,\ \mathbf{\Sigma}^3=\mathbf{B}=c\mathbf{x}^3\end{array}\ \Big\downarrow\ \xi,\frac{1}{c}\to 0\ \text{(contractions)}$$

$$\log\mathbf{H}(1)\cong\mathbb{R}\ \vec{\oplus}\ \mathbb{R}^2 : [\mathbf{I},\mathbf{p}^3]=0,\quad [\mathbf{x}^3,\mathbf{I}]=0,\quad [\mathbf{x}^3,\mathbf{p}^3]=\mathbf{I},$$

and, in general, the contraction chain for the real $\binom{2+s}{s}$-dimensional groups:

$$s = 1, 2, \ldots : \quad \mathbf{O}(1, 1+s) \overset{\xi \to 0}{\longrightarrow} \mathbf{O}(1, s) \,\vec{\times}\, \mathbb{R}^{1+s} \overset{\frac{1}{c} \to 0}{\longrightarrow} \mathbf{O}(s) \,\vec{\times}\, \mathbf{H}(s),$$

e.g., starting from the de Sitter group $\mathbf{O}(1, 4)$.

There are also expansions of *relativistically extended Heisenberg groups* $\mathbf{H}(1, s)$ which include, from the beginning, the time translations in a special relativistic generalization with the Lorentz metric:

$$\log \mathbf{H}(1, s): \quad [\mathbf{x}^j, \mathbf{p}^k] = -\eta^{jk}\mathbf{I}, \quad [\mathbf{x}^j, \mathbf{I}] = 0 = [\mathbf{p}^j, \mathbf{I}], \quad j = 0, a; \quad a = 1, \ldots, s.$$

There exists a non-unitary real $3+s$-dimensional faithful representation of the corresponding Poincaré-Heisenberg Lie algebra which has $1+s$ spacetime translations and momenta $\{\mathbf{x}^j, \mathbf{p}^j\}$, s boosts $\{\mathbf{B}^a\}$, $\binom{s}{2}$ rotations $\{\mathbf{J}^{ab} = -\mathbf{J}^{ba}\}$, and one Heisenberg operator $\mathbf{I}$, sketched by

$$\log \mathbf{O}(1, s) \,\vec{\oplus}\, \log \mathbf{H}(1, s) \ni \begin{pmatrix} 0 & \mathbf{x}^0 & \mathbf{x}^a & \mathbf{I} \\ 0 & 0 & \mathbf{B}^a & \mathbf{p}^0 \\ 0 & \mathbf{B}^b & \mathbf{J}^{ab} & \mathbf{p}^b \\ 0 & 0 & 0 & 0 \end{pmatrix} \in \mathbf{AL}(3+s, \mathbb{R}).$$

The Poincaré–Heisenberg group $\mathbf{O}(1, s) \,\vec{\times}\, \mathbf{H}(1, s)$ can be expanded to orthogonal groups $\mathbf{O}(n, m)$ with real dimensions $\binom{n+m}{2} = 2(1+s) + s + \binom{s}{2} + 1 = \binom{3+s}{2}$, e.g., to the real 15-dimensional group $\mathbf{O}(2, 4)$ for $s = 3$.

The main difference to the minimal expansion scheme, given above, is that, in the extended scheme, the time translations $\mathbf{p}^0$ are not contracted to the Heisenberg operator $\mathbf{I}$ which remains as independent operator also in the expanded group. In the minimal scheme, there is no relativistic Heisenberg group and the Heisenberg operator is a contraction of the time translations $\mathbf{I} = \frac{1}{c}\mathbf{p}^0$.

The structures, e.g., the representations of the noncompact groups $\mathbf{O}(1, 1+s)$ and of the Poincaré groups $\mathbf{O}(1, s) \,\vec{\times}\, \mathbb{R}^{1+s}$, non-unitary finite-dimensional and unitary infinite-dimensional as induced from subgroups and classified by Wigner, can be contracted to the corresponding structures of the Heisenberg groups.

In the contraction procedure the Killing invariant morphs into the translation invariant, and, finally, into the invariant Heisenberg operator $\mathbf{I}$:

$$\begin{array}{ccccc} \log \mathbf{O}(1, 2) & \overset{\xi \to 0}{\longrightarrow} & \log \mathbf{O}(1, 1) \,\vec{\oplus}\, \mathbb{R}^2 & \overset{\frac{1}{c} \to 0}{\longrightarrow} & \log \mathbf{H}(1), \\ -(\mathbf{J}^1)^2 + (\mathbf{\Sigma}^2)^2 + (\mathbf{\Sigma}^3)^2 & & -(\mathbf{p}^0)^2 + (\mathbf{p}^3)^2 & & -\mathbf{I}^2. \end{array}$$

The spectrum of the translation invariant with nontrivial Wigner particle masses morphs into the spectrum of the Heisenberg invariant $\mathbf{I} \longmapsto \hbar i \mathbf{1}$ with nontrivial Planck constants.

Is quantum mechanics a real or complex structure? The Heisenberg group (Lie algebra) consists of a real operations, implemented on complex Hilbert spaces. The familiar Hilbert space with the square-integrable function classes $L^2(\mathbb{R}^s)$ of position translations $\vec{x} \in \mathbb{R}^s$, $s = 1, 2, \ldots$, as eigenvalues of $\vec{\mathbf{x}}$ and the momentum representation $\vec{\mathbf{p}} \longmapsto -i\hbar\vec{\partial}$ is acted on by these *faithful unitary* representations. According to a theorem of Stone[4] and von Neumann, all faithful unitary representations of the Heisenberg Lie algebra are of this type. This Hilbert space is related to the Hilbert space for the unitary representations of the Poincaré group which includes the Wigner particle representation spaces. In the complex infinite-dimensional Hilbert representations of the Heisenberg group, the invariant Heisenberg operator $\mathbf{I} \longmapsto i\hbar\mathbf{1}$ is represented by the unit multiplied with one nontrivial spectral value $i\hbar \in \textbf{spec}\,\mathbf{I} = i\mathbb{R}$ as action unit (Planck's constant). With the trace property of $[\mathbf{x}, \mathbf{p}] = \mathbf{I}$ there is no finite-dimensional representation with the Heisenberg operator as unit $\mathbf{I} \longmapsto i\mathbf{1}_n$.

The Heisenberg Lie algebra and group can be related to a duality structure with pairs of creation-annihilation operators $[\mathrm{u}^*, \mathrm{u}] = \mathbf{1}$ for $\mathbf{x} = \frac{\mathrm{u}+\mathrm{u}^*}{\sqrt{2}}$ and $\mathbf{p} = i\frac{\mathrm{u}-\mathrm{u}^*}{\sqrt{2}}$ with $[i\mathbf{p}, \mathbf{x}] = \mathbf{1}$. Generalizing from their origin from the harmonic oscillator in position, they describe the *quantization of n degrees of freedom* by $[\mathrm{u}_a^*, \mathrm{u}^b]_\epsilon = \delta_a^b\mathbf{1}$ for $a, b = 1, \ldots n$, bosonic for commutators $\epsilon = -1$ and fermionic for anticommutators $\epsilon = +1$. Here $\{\mathrm{u}^a\}_{a=1}^n$ is a basis of a complex vector space $V \cong \mathbb{C}^n$ and $\{\mathrm{u}_a^*\}_{a=1}^n$ the dual basis for the dual space $V^T \cong \mathbb{C}^n$. For quantum fields, the degrees of freedom may become continuous (distributional), e.g., indexed by the momenta of a particle $[\mathrm{u}(\vec{q})^*, \mathrm{u}(\vec{p})]_\epsilon \sim \delta(\vec{q} - \vec{p})$ for $\vec{q}, \vec{p} \in \mathbb{R}^3$.

If the Born–Heisenberg relations are induced by the duality of complex representation vector spaces, their interpretation as Heisenberg action groups and Lie algebras as above is mathematically possible, but inappropriate with respect to their origin.

7.10 Orbits and Fixgroups

Realizations of groups on sets and representations of Lie algebras on vector spaces distinguish subgroups and Lie subalgebras. This is important, e.g., for the embedding of massive and massless particles in special relativistic fields.

A group acting on a set by bijections, e.g., the rotation group acting on the Euclidean space,

$$\mathbf{SO}(3) \times \mathbb{R}^3 \longrightarrow \mathbb{R}^3, \quad O \bullet \vec{x} = O(\vec{x}),$$

leads to orbits, in the example to 2-spheres as orthogonal orbits. Or in classical mechanics, if one considers the action of the additive group with the time translations $(\mathbb{R}, +)$ on the phase space (position-momentum space):

$$\mathbb{R} \times \mathcal{R}(q, p) \longrightarrow \mathcal{R}(q, p), \quad t \bullet (q, p) = (q(t), p(t)),$$

the phase space curves are *time orbits*. Some time orbits in the phase space remain in the finite — periodic or not; examples are the time orbits $\frac{q^2+p^2}{2} = E$ of the harmonic

[4] Marshall Harvey Stone (1903–1989).

oscillator, characterized by the energy E, or the orbits of the planets. Other time orbits do not stay in the finite, e.g., the motion of a free Newton mass point, considered not in his rest system, or the hyperbolic orbits of never returning comets.

A group G-realization on a set S via a morphism ρ into the bijection (permutation) group $\mathbf{G}(S)$:

$$G \times S \longrightarrow S, \quad (g, x) \longmapsto g \bullet x = \rho(g)(x),$$

defines an *orbit equivalence* relation, which leads to a disjoint decomposition of the set S into G-orbits with representatives $\{x_r\}_{r \in B}$:

$$\begin{aligned} &x \overset{G}{\approx} y \iff \text{There exists a } g \in G \text{ with } g \bullet x = y, \\ &G \bullet x = \{y \,|\, y \overset{G}{\approx} x\}, \quad S/G = \{G \bullet x \,|\, x \in S\}, \quad S = \biguplus_{x_r \in B} G \bullet x_r. \end{aligned}$$

The orbits are, as equivalence classes, either disjoint $G \bullet x \cap G \bullet y = \emptyset$ for $y \notin G \bullet x$, or identical $G \bullet x = G \bullet y$ for $y \in G \bullet x$. Examples are the s-dimensional spheres for the action of the rotation group $\mathbf{SO}(1+s)$ on the Euclidean space $\mathbb{E}^{1+s}$ or phase space curves for a given dynamics — not position space curves — as time orbits; they are either identical or they do not cross. Another example is given by *projective spaces*: The one-dimensional subspaces $\mathbf{GL}(1, \mathbb{K})v \in V$ of a vector space $\mathbb{K}^n \cong V \ni v \neq 0$ define its projective space $\mathbf{p}(V) = V \setminus \{0\}/\mathbf{GL}(\mathbb{K})$.

The action of a group may be trivial: A set element $x \in S$ is called *invariant* for the group G actions if $G \bullet x = \{x\}$. It may be only a subgroup that leaves an element $x \in S$ invariant, e.g., for a nontrivial point in Euclidean space $\mathbb{E}^3$ with respect to $\mathbf{O}(3)$-action, where the invariance subgroup is given by the axial rotations $\mathbf{O}(2)$ around the direction of this point to the origin. In classical mechanics, a periodic orbit in the phase space has equal values for the position-momentum pair after each period, i.e., for the times $\{0, \pm T, \pm 2T, \dots\}$, which gives $\mathbb{Z} \subset \mathbb{R}$ as the time translation fixgroup of a phase space point. The general definition of the *fixgroup of a point $x \in S$ with respect to G-action* is:

$$G_x = \{g \in G \,|\, g \bullet x = x\}.$$

With a fixgroup one has the transition $G_x \subseteq G$ and the cosets (manifold) G/G_x.

Since the action of the fixgroup $G_x \subseteq G$ leaves the element x invariant, the full orbit $G \bullet x$ is reached already by the action of the group elements modulo the fixgroup. Therefore, there exists a bijection between orbit points $G \bullet x$ in the set S and G_x-orbits in the group G:

$$G \bullet x \cong G/G_x, \quad g \bullet x \leftrightarrow gG_x,$$

and for the disjoint union:

$$S = \biguplus_{x_r \in B} G \bullet x_r \cong \biguplus_{x_r \in B} G/G_{x_r}.$$

The classes G/H of a subgroup $H \subseteq G$ with the H-right-orbits kH as elements constitute a set with an action of the group G — for the right orbits the action is defined by left-multiplications. The fixgroup of the left action of G on the cosets G/H is, for each element $kH \in G/H$, isomorphic to the class defining subgroup H:

$$\begin{aligned} G \times G/H &\longrightarrow G/H, \;\; g \bullet kH = gkH, \\ G_{kH} &= \{g \in G \,\big|\, gkH = kH\} = kHk^{-1} \cong H. \end{aligned}$$

Therefore, one obtains the remarkably simple result, that *all orbits, possible for the action of a group G, are of the type G/H with a subgroup $H \subseteq G$.* Up to isomorphy, the set $\{G/H \,\big|\, H \text{ subgroup}\}$ characterizes all possible irreducible G-sets, i.e., all irreducible G-realizations. G/H is one G-orbit.

An example are the s-spheres as $\mathbf{O}(1+s)$-orbits of nontrivial vectors of the Euclidean space $\mathbb{E}^{1+s}$: they are isomorphic to the quotients $\mathbf{O}(1+s)/\mathbf{O}(s)$. The origin does not move, it is isomorphic to the trivial group $\{0\} \cong \{1\} \cong \mathbf{O}(1+s)/\mathbf{O}(1+s)$. Other examples are the time orbits in the phase space with the action of the additive time group $\mathbb{R}$: A time action with trivial fixgroup $\{1\}$ describes orbits with the cardinality of $\mathbb{R}$, realized, e.g., by hyperbolic orbits of comets or by orbits of a two-dimensional Lissajous-oscillator in the real four-dimensional phase space

$$\begin{pmatrix} q_1 + ip_1 \\ q_2 + ip_2 \end{pmatrix}(t) = \begin{pmatrix} e^{i\omega_1 t}(q_1 + ip_1) \\ e^{i\omega_2 t}(q_2 + ip_2) \end{pmatrix} \in \mathbb{C}^2 \cong \mathbb{R}^4,$$

with an irrational frequency ratio $\frac{\omega_1}{\omega_2} \notin \mathbb{Q}$. The full group $\mathbb{R}$ as fixgroup characterizes an orbit with only one element, e.g., a free mass point in a rest system or the limit of an infinitely heavy sun in our solar system. The periodic motions with fixgroup $\mathbb{Z}$ realize the orbits that are equipotent with the circle $\mathbf{U}(1) \cong \mathbb{R}/\mathbb{Z}$, e.g., Lissajous orbits for $\frac{\omega_1}{\omega_2} \in \mathbb{Q}$.

A group action on a set S gives rise to a coarser equivalence for elements with isomorphic fixgroups. The associate *orbit types (strata)* $[G \bullet x]$ are equivalence classes which may include disjoint G-orbits:

$$\begin{aligned} &x \overset{G}{\sim} y \iff G_x \cong G_y, \;\; [G \bullet x] = \{y \in S \,\big|\, G_y \cong G_x\}, \\ &S/[G] = \{[G \bullet x] \,\big|\, x \in S\}, \;\; S = \biguplus_{[x_r] \in [B]} [G \bullet x_r], \\ &x \overset{G}{\approx} y \Rightarrow x \overset{G}{\sim} y. \end{aligned}$$

A stratum for a group G may be an orbit $G' \bullet x$ for a larger group $G' \supseteq G$. For example, an $\mathbf{O}(3)$-strata decomposition of the three-dimensional Euclidean space consists of the origin and the union of all nontrivial spheres as disjoint $\mathbf{SO}(3)$-orbits:

$$x \neq 0: \quad \mathbf{O}(3) \bullet x \cong \mathbf{O}(3)/\mathbf{O}(2) \cong \Omega^2,$$
$$\mathbb{E}^3 = \{1\} \uplus [\mathbf{O}(3) \bullet x] \cong \{1\} \uplus \frac{\mathbf{D}(1) \times \mathbf{O}(3)}{\mathbf{O}(2)}.$$

The two $\mathbf{O}(3)$-strata are orbits of the equiformal group $\mathbf{D}(\mathbf{1}_3) \times \mathbf{O}(3)$ that contains also the dilations $\mathbf{D}(\mathbf{1}_3) \cong \exp \mathbb{R}$.

A Lie group, represented on a vector space $G \times V \longrightarrow V$, can be investigated by its "infinitesimal" actions, i.e., by the actions of its Lie algebra:

$$L \times V \longrightarrow V, \quad l \bullet v = \mathcal{D}(l)(v).$$

Those actions are realizable in equations of motion by differentiation with respect to the Lie-parameters, e.g., by time or spacetime derivatives $\frac{d}{dt}$, ∂^j. The vector subspaces $L \bullet x$ for the Lie algebra action on a vector $x \in V$ give rise to an equivalence relation in V and a V-decomposition into vector subspaces:

$$v \overset{L}{\approx} w \iff L \bullet v = L \bullet w, \quad V/L = \bigoplus_{v_r \in B} L \bullet v_r.$$

In contrast to the group orbit $G \bullet x$, the Lie algebra orbit $L \bullet v$ does not have to contain the start-vector v.

A vector is *invariant* under Lie algebra-action for $L \bullet v = \{0\}$. The *fix-Lie algebra* of a vector is the Lie subalgebra

$$L_v = \{l \in L \,\big|\, l \bullet v = 0\}.$$

For Lie algebras, there is no analogue to the group orbit-isomorphy $G \bullet x \cong G/G_x$.

7.11 Orbits of Minkowski Spacetime

To characterize the fixgroups of the orthochronous Lorentz group $\mathbf{SO}_0(1,3)$, acting on the real four-dimensional vector space with Minkowski spacetime translations $\mathbb{R}^4$, and, similarly, for the dual space with the energy-momenta $\check{\mathbb{R}}^4 \cong \mathbb{R}^4$, one can go over to the corresponding fix-Lie algebras:

$$\mathbf{SO}_0(1,3) \times \mathbb{R}^4 \longrightarrow \mathbb{R}^4, \quad \text{fix: } \Lambda \bullet x = \Lambda(x) = x,$$
$$\log \mathbf{O}(1,3) \times \mathbb{R}^4 \longrightarrow \mathbb{R}^4, \quad \text{fix: } \mathbf{L} \bullet x \cong \begin{pmatrix} 0 & \psi_1 & \psi_2 & \psi_3 \\ \psi_1 & 0 & \theta_3 & -\theta_2 \\ \psi_2 & -\theta_3 & 0 & \theta_1 \\ \psi_3 & \theta_2 & -\theta_1 & 0 \end{pmatrix} \begin{pmatrix} x^0 \\ x^1 \\ x^2 \\ x^3 \end{pmatrix} = \begin{pmatrix} 0 \\ 0 \\ 0 \\ 0 \end{pmatrix}.$$

It is easy to see that there are four orbit types (strata) — for the trivial translation, and for the nontrivial timelike, spacelike, and lightlike translations. Obviously, the fixgroup of the trivial translation is the full Lorentz group $\mathbf{SO}_0(1,3)$, the fixgroup of a time translation, e.g., with $x^j = \delta^j_0$, is the rotation group $\mathbf{SO}(3)$, and that of a position translation, e.g., with $x^j = \delta^j_3$, is the "slimmed down" Lorentz group $\mathbf{SO}_0(1,2)$.

It is more difficult for the lightlike translations $x \neq 0$ with trivial Lorentz square $x^2 = 0$. In a two-dimensional model $\mathbf{SO}_0(1,1)$ with one position direction, the fixgroup of the nontrivial lightlike translations $\begin{pmatrix} 0 & \psi \\ \psi & 0 \end{pmatrix}\begin{pmatrix} 1 \\ -1 \end{pmatrix} = \begin{pmatrix} 0 \\ 0 \end{pmatrix}$ is trivial, $\psi = 0$, i.e., $\mathbf{SO}_0(1,1)_x \cong \{1\}$. For four spacetime dimensions, the additional two-dimensional position translations allow an axial rotation, but not only that: The elements of the fix-Lie algebra of the light translation $\begin{pmatrix} 1 \\ 0 \\ 0 \\ -1 \end{pmatrix}$ look as follows:

$$\begin{pmatrix} 0 & \psi_1 & \psi_2 & 0 \\ \psi_1 & 0 & \theta_3 & \psi_1 \\ \psi_2 & -\theta_3 & 0 & \psi_2 \\ 0 & -\psi_1 & -\psi_2 & 0 \end{pmatrix}\begin{pmatrix} 1 \\ 0 \\ 0 \\ -1 \end{pmatrix} = \begin{pmatrix} 0 \\ 0 \\ 0 \\ 0 \end{pmatrix}.$$

$\psi_1, \psi_2 = 0$ characterize the afore-mentioned axial rotations in the $(1,2)$-plane. For the cases $\psi_1, \psi_2 \neq 0$, one can construct a basis with two lightlike vectors as combinations of a time- and spacelike basic vector:

$$\begin{pmatrix} -\frac{1}{\sqrt{2}} & 0 & 0 & \frac{1}{\sqrt{2}} \\ 0 & 1 & 0 & 0 \\ 0 & 0 & 1 & 0 \\ \frac{1}{\sqrt{2}} & 0 & 0 & \frac{1}{\sqrt{2}} \end{pmatrix}\begin{pmatrix} 0 & \psi_1 & \psi_2 & 0 \\ \psi_1 & 0 & \theta_3 & \psi_1 \\ \psi_2 & -\theta_3 & 0 & \psi_2 \\ 0 & -\psi_1 & -\psi_2 & 0 \end{pmatrix}\begin{pmatrix} -\frac{1}{\sqrt{2}} & 0 & 0 & \frac{1}{\sqrt{2}} \\ 0 & 1 & 0 & 0 \\ 0 & 0 & 1 & 0 \\ \frac{1}{\sqrt{2}} & 0 & 0 & \frac{1}{\sqrt{2}} \end{pmatrix}$$

$$= \begin{pmatrix} 0 & -\psi_1\sqrt{2} & -\psi_2\sqrt{2} & 0 \\ 0 & 0 & \theta_3 & \psi_1\sqrt{2} \\ 0 & -\theta_3 & 0 & \psi_2\sqrt{2} \\ 0 & 0 & 0 & 0 \end{pmatrix} = \mathbf{O}(\theta_3) + \mathbf{p}(\psi_1, \psi_2).$$

The fix-Lie algebra involves the infinitesimal axial rotations $\mathbf{O}(\theta_3)$ (one real parameter) and, from the boosts, real two-dimensional lightlike "translations" $\mathbf{p}(\psi_1, \psi_2)$ with the brackets:

$$\begin{aligned} [\mathbf{O}(\theta_3), \mathbf{O}(\theta_3')] &= 0, \\ [\mathbf{p}(\psi_1, \psi_2), \mathbf{p}(\psi_1', \psi_2')] &= 0, \end{aligned} \quad [\mathbf{O}(\theta_3), \mathbf{p}(\psi_1, \psi_2)] = \theta_3 \mathbf{p}(-\psi_2, \psi_1).$$

Therefore, the fixgroup of a lightlike translation is isomorphic to the Euclidean group for the lightlike planes in the Lorentz group $\mathbf{SO}(2) \vec{\times} \mathbb{R}^2 \subset \mathbf{SO}_0(1,3)$.

With nontrivial timelike, lightlike and spacelike translations (e^0, e^+, e^3), respectively, the $\mathbf{SO}_0(1,3)$-orbit type decomposition of the Minkowski translations uses

$$\begin{aligned} \text{timelike:} \quad & \mathbf{SO}_0(1,3) \bullet e^0 \cong \mathbf{SO}_0(1,3)/\mathbf{SO}(3), \\ \text{lightlike:} \quad & \mathbf{SO}_0(1,3) \bullet e^+ \cong \mathbf{SO}_0(1,3)/\mathbf{SO}(2) \vec{\times} \mathbb{R}^2, \\ \text{spacelike:} \quad & \mathbf{SO}_0(1,3) \bullet e^3 \cong \mathbf{SO}_0(1,3)/\mathbf{SO}_0(1,2). \end{aligned}$$

All proper fix-symmetries are three-dimensional.

For the real seven-dimensional equiformal Lorentz group $\mathbf{D}(\mathbf{1}_4) \times \mathbf{SO}_0(1,3)$ with the dilations $\mathbf{D}(\mathbf{1}_4) \cong \exp \mathbb{R}$ of the translations, $x \longmapsto e^{\psi_0}x$, there are, in addition to the trivial orbit, three four-dimensional orbits, the future and past timelike ones, and the spacelike orbit. The future and past lightlike vectors are real three-dimensional manifolds:

$$\begin{aligned}\mathbb{R}^{1+3} &\cong \{1\} \uplus 2 \cdot \frac{\mathbf{D}(1)\times\mathbf{SO}_0(1,3)}{\mathbf{SO}(3)} \uplus 2 \cdot \frac{\mathbf{SO}_0(1,3)}{\mathbf{SO}(2) \vec{\times} \mathbb{R}^2} \uplus \frac{\mathbf{D}(1)\times\mathbf{SO}_0(1,3)}{\mathbf{SO}_0(1,2)}\\ &\cong \{1\} \uplus 2 \cdot \mathbf{D}(1) \times \mathcal{Y}^3 \uplus 2 \cdot \mathcal{C}^3 \uplus \mathbf{D}(1) \times \mathcal{Y}^{(1,2)}.\end{aligned}$$

Fixgroups and orbits lead to *orbit coordinates*: The analogue to the polar coordinates for nontrivial vectors $\vec{x}$ in Euclidean space with the unit sphere Ω^2 as $\mathbf{SO}(3)$-orbit:

$$\begin{aligned}\vec{x}^2 > 0: \ \vec{x} = e^{\psi_0}\vec{\omega}_2, \ \vec{\omega}_2 &= \begin{pmatrix}\sin\theta\ \cos\varphi\\ \sin\theta\ \sin\varphi\\ \cos\theta\end{pmatrix} \in \Omega^2 \cong \mathbf{SO}(3)/\mathbf{SO}(2),\\ e^{\psi_0} &= |\vec{x}| \in \mathbf{D}(1), \ \ \vec{\omega}_2^2 = 1.\end{aligned}$$

are the hyperbolic coordinates for the three types of nontrivial vectors $x = \begin{pmatrix}x_0\\ \vec{x}\end{pmatrix}$ in Minkowskian spacetime with the time- and spacelike unit hyperboloids and the lightcone without $x = 0$ as Lorentz group $\mathbf{SO}_0(1,3)$-orbits:

$$\begin{aligned}x^2 > 0: x = \pm e^{\psi_0}\mathrm{y}_3, \qquad \mathrm{y}_3 &= \begin{pmatrix}\cosh\psi_3\\ \sinh\psi_3\ \vec{\omega}_2\end{pmatrix} \in \mathcal{Y}^3 \cong \mathbf{SO}_0(1,3)/\mathbf{SO}(3),\\ e^{\psi_0} &= \sqrt{x^2} \in \mathbf{D}(1), \qquad \mathrm{y}_3^2 = 1,\\ x^2 < 0: x = e^{\psi_0}\mathrm{y}_{(1,2)}, \qquad \mathrm{y}_{(1,2)} &= \begin{pmatrix}\sinh\psi_3\\ \cosh\psi_3\ \vec{\omega}_2\end{pmatrix} \in \mathcal{Y}^{(1,2)} \cong \mathbf{SO}_0(1,3)/\mathbf{SO}_0(1,2),\\ e^{\psi_0} &= \sqrt{-x^2} \in \mathbf{D}(1), \ \ \mathrm{y}_{(1,2)}^2 = -1,\\ x^2 = 0,\ x \neq 0: \ x = \pm \mathrm{c}_3, \qquad \mathrm{c}_3 &= e^{\psi_3}\begin{pmatrix}1\\ \vec{\omega}_2\end{pmatrix} \in \mathcal{C}^3 \cong \mathbf{SO}_0(1,3)/\mathbf{SO}(2) \vec{\times} \mathbb{R}^2,\\ &\qquad \mathrm{c}_3^2 = 0.\end{aligned}$$

The determination of fixgroups and orbits is technically much easier in a complex formulation with the Lorentz group covering group. Spacetime translations in the Weyl representation are hermitian (2×2)-matrices. Their Lorentz transformation is effected by a left-right $\mathbf{SL}(2,\mathbb{C})$-action. The fixgroups are characterizable as invariance groups of diagonal matrices:

$$\begin{aligned} x = \begin{pmatrix}x_0 + x_3 & x_1 - ix_2\\ x_1 + ix_2 & x_0 - x_3\end{pmatrix} &: \mathbf{SL}(2,\mathbb{C})_x = \{s \in \mathbf{SL}(2,\mathbb{C}) \,\big|\, s \circ x \circ s^\star = x\},\\ \text{timelike} \quad e^0 \cong \mathbf{1}_2 = \begin{pmatrix}1 & 0\\ 0 & 1\end{pmatrix} &: \mathbf{SL}(2,\mathbb{C})_{\mathbf{1}_2} = \mathbf{SU}(2) \sim \mathbf{SO}(3),\\ \text{spacelike} \quad e^3 \cong \sigma_3 = \begin{pmatrix}1 & 0\\ 0 & -1\end{pmatrix} &: \mathbf{SL}(2,\mathbb{C})_{\sigma_3} = \mathbf{SU}(1,1) \sim \mathbf{SO}_0(1,2),\\ \text{lightlike } e^+ \cong \tfrac{\mathbf{1}_2+\sigma_3}{2} = \begin{pmatrix}1 & 0\\ 0 & 0\end{pmatrix} &: \mathbf{SL}(2,\mathbb{C})_{\frac{\mathbf{1}_2+\sigma_3}{2}} = \mathbf{SO}(2) \vec{\times} \mathbb{R}^2.\end{aligned}$$

With the group parametrization $s = \exp \frac{(i\vec{\theta}+\vec{\psi})\vec{\sigma}}{2} \in \mathbf{SL}(2,\mathbb{C})$, the Lie algebra condition for the fixgroup involves commutators and anticommutators, e.g., for lightlike translations:

$$\begin{aligned}
\left(\mathbf{1}_2 + \tfrac{(i\vec{\theta}+\vec{\psi})\vec{\sigma}}{2} + \dots\right) \circ x \circ \left(\mathbf{1}_2 - \tfrac{(i\vec{\theta}-\vec{\psi})\vec{\sigma}}{2} + \dots\right) &= x \\
\iff [i\vec{\theta}\vec{\sigma}, x] + \{\vec{\psi}\vec{\sigma}, x\} &= 0, \\
\text{for } x = \tfrac{\mathbf{1}_2+\sigma_3}{2} : \ [i\vec{\theta}\vec{\sigma}, \sigma_3] + \{\vec{\psi}\vec{\sigma}, \mathbf{1}_2 + \sigma_3\} &= 0, \\
\Rightarrow \tfrac{(i\vec{\theta}+\vec{\psi})\vec{\sigma}}{2} = \tfrac{i\theta_3\sigma_3+(\psi_1-i\psi_2)(\sigma_1+i\sigma_2)}{2} &= i\begin{pmatrix} \frac{\theta_3}{2} & \psi_1 - i\psi_2 \\ 0 & -\frac{\theta_3}{2} \end{pmatrix}.
\end{aligned}$$

The future and past timelike and the spacelike manifolds are isomorphic to classes with the two types of unitary groups, $\mathbf{U}(2)$ (definite) and $\mathbf{U}(1,1)$ (indefinite), in the full linear group $\mathbf{GL}(2,\mathbb{C})$ as real eight-dimensional Lie group:

$$\begin{aligned}
\text{timelike (future): } \mathbf{D}(1) \times \mathbf{SO}_0(1,3)/\mathbf{SO}(3) &\cong \mathbf{GL}(2,\mathbb{C})/\mathbf{U}(2), \\
\text{spacelike: } \mathbf{D}(1) \times \mathbf{SO}_0(1,3)/\mathbf{SO}_0(1,2) &\cong \mathbf{GL}(2,\mathbb{C})/\mathbf{U}(1,1).
\end{aligned}$$

7.12 Orbits of the Heisenberg Group

The semidirect affine Heisenberg group $\mathbf{H}(s) \cong e^{\mathbb{R}^s} \vec{\times} \mathbb{R}^{1+s}$, $s \geq 1$, for the Heisenberg Lie algebra with the characterizing Lie bracket $[\mathbf{x}, \mathbf{p}] = \mathbf{1}_s\mathbf{I}$ has two fixgroup types in the homogeneous group $e^{\mathbb{R}^s}$ — the full group and the group with the ξ-dual action:

$$\begin{aligned}
&\mathbb{R}^{1+s} \ni \begin{pmatrix} \tau \\ \vec{\xi} \end{pmatrix} \longmapsto \begin{pmatrix} 1 & \vec{\pi} \\ 0 & \mathbf{1}_s \end{pmatrix}\begin{pmatrix} \tau \\ \vec{\xi} \end{pmatrix} = \begin{pmatrix} \tau + \vec{\pi}\vec{\xi} \\ \vec{\xi} \end{pmatrix} \\
&\Rightarrow \mathrm{FIX}_{\begin{pmatrix} \tau \\ \vec{\xi} \end{pmatrix}} e^{\mathbb{R}^s} = \begin{cases} e^{\mathbb{R}^s}, & \vec{\xi} = 0, \\ e^{\mathbb{R}^{s-1}} = \left\{\begin{pmatrix} 1 & \vec{\pi} \\ 0 & \mathbf{1}_s \end{pmatrix} \mid \vec{\pi}\vec{\xi} = 0\right\}, & \vec{\xi} \neq 0. \end{cases}
\end{aligned}$$

The corresponding orbits in the vector space $\mathbb{R}^{1+s}$ consist of the start-vector only (zero-dimensional) or are given by a line (one-dimensional), respectively. They are isomorphic to the fixgroup classes in the homogeneous group:

$$e^{\mathbb{R}^s}\text{-orbits in } \mathbb{R}^{1+s} : \begin{cases} \left\{\begin{pmatrix} \tau \\ 0 \end{pmatrix}\right\} \cong e^{\mathbb{R}^s}/e^{\mathbb{R}^s} \cong \{1\}, & \vec{\xi} = 0, \\ \left\{\begin{pmatrix} \tau + \vec{\pi}\vec{\xi} \\ \vec{\xi} \end{pmatrix} \mid \vec{\pi} \in \mathbb{R}^s\right\} \cong e^{\mathbb{R}^s}/e^{\mathbb{R}^{s-1}} \cong \mathbb{R}, & \vec{\xi} \neq 0. \end{cases}$$

For example, the two-dimensional vector space $\mathbb{R}^2 \ni \begin{pmatrix} \tau \\ \xi \end{pmatrix}$ for $s = 1$ has the disjoint orbit-decomposition into the points on the first axis as the one-elementic orbits, and the first axis-parallel lines — with the exception of the first axis — which are the line orbits.

Space(-time) translations and their dual (energy-)momenta in the cases of the Euclidean or the Poincaré group have isomorphic fixgroups and orbits. That is not the case for the Heisenberg group. The dual "translations" $\check{\mathbb{R}}^{1+s}$ for the Heisenberg group:

$$\begin{pmatrix}\tau\\\vec{\xi}\end{pmatrix} \in \mathbb{R}^{1+s}, \quad (\hbar, \vec{p}) \in \check{\mathbb{R}}^{1+s}, \quad (\hbar, \vec{p}) \begin{pmatrix}\tau\\\vec{\xi}\end{pmatrix} = \hbar\tau + \vec{p}\vec{\xi} \in \mathbb{R},$$

have different fixgroups and orbits under the dual action

$$(\hbar, \vec{p}) \begin{pmatrix}1 & -\vec{\pi}\\0 & \mathbf{1}_s\end{pmatrix} = (\hbar, \vec{p} - \hbar\vec{\pi})$$

The full and the trivial homogeneous group are the fixgroups for trivial and nontrivial first component $\hbar \in \mathbb{R}$, respectively:

$$\text{FIX}_{(\hbar,\vec{p})} e^{\mathbb{R}^s} = \begin{cases} e^{\mathbb{R}^s}, \ \hbar = 0, \\ \{\mathbf{1}_{1+s}\}, \ \hbar \neq 0. \end{cases}$$

The corresponding orbits in $\check{\mathbb{R}}^{1+s}$ are points (one vector, zero-dimensional) and hyperplanes (s-dimensional):

$$e^{\mathbb{R}^s}\text{-orbits in } \check{\mathbb{R}}^{1+s}: \begin{cases} \{(0, \vec{p})\} \cong \{1\}, & \hbar = 0, \\ \{(\hbar, \vec{p} - \hbar\vec{\pi}) \bigm| \vec{\pi} \in \mathbb{R}^s\} \cong \mathbb{R}^s, & \hbar \neq 0. \end{cases}$$

7.13 Squares and Their Symmetries

Some linear groups $G \subseteq \mathbf{GL}(V)$ for a vector space V can be characterized by metrical structures. For instance, the Lorentz transformations $\mathbf{O}(1, 3)$ of the Minkowski translations are definable by the invariance of a symmetric bilinear form $g(x, y) = g(y, x)$ with signature $(1, 3)$, or the unitary hyperisospin group $\mathbf{U}(2)$ by the invariance of a scalar product of a complex two-dimensional Higgs vector space with $\langle x|x\rangle > 0$ for all nontrivial vectors x. For a structural understanding of units and scales in physical theories, one has to investigate the dilation structures, especially of linear, bilinear or sesquilinear forms, perhaps also of higher degree multilinear forms, in general related to operational invariants.

Nondegenerate symmetric bilinear forms of real vector spaces $\mathbb{K} = \mathbb{R}$ and hermitian sesquilinear forms of complex ones $\mathbb{K} = \mathbb{C}$ are called *squares*:

$$\zeta : V \times V \longrightarrow \mathbb{K}, \quad \begin{cases} \zeta(x, \alpha y + \beta z) &= \alpha\zeta(x, y) + \beta\zeta(x, z), \\ \zeta(x, y) &= \overline{\zeta(y, x)}. \end{cases}$$

Each vector has a real square:

$$\zeta^2 : V \longrightarrow \mathbb{R},\ \ \zeta^2(x) = \zeta(x,x),\ \ \begin{cases} \zeta^2(\alpha x) = |\alpha|^2\zeta^2(x),\ \ \alpha \in \mathbb{K}, \\ \zeta^2(x+y) + \zeta^2(x-y) = 2\zeta^2(x) + 2\zeta^2(y) \\ \text{(parallelogram equality)}. \end{cases}$$

For a real semisimple Lie algebra, e.g., for the angular momenta, the symmetric nondegenerate Killing form is a square.

Nondegenerate forms define (anti)linear dual isomorphisms and a square of the linear forms:

$$\begin{aligned} &V^T \ni \theta \leftrightarrow \theta^\zeta \in V, \text{ with } \zeta(\theta^\zeta, x) = \langle \theta, x\rangle, \\ &V^T \times V^T \longrightarrow \mathbb{K}, \text{ with } \zeta^{-1}(\theta, \omega) = \zeta(\theta^\zeta, \omega^\zeta). \end{aligned}$$

For instance, the Lorentz metric defines an isomorphism between spacetime translations and energy-momenta, $x \longmapsto g(x,\ \ )$ and $p \longmapsto g^{-1}(p,\ \ )$, used for the "raising and lowering of the indices":

$$\begin{aligned} g(x,y) &= x^a g_{ab} y^b &&= x_b y^b,\ x_b = x^a g_{ab}, \\ g^{-1}(p,q) &= p_c g^{cd} q_d &&= p^d q_d,\ p^d = p_c g^{cd}. \end{aligned}$$

Another example is Dirac's bra-ket notation, which makes sense with a Hilbert space product. It associates, in an antilinear dual isomorphism of complex spaces, a unique linear form $\langle w|$ with each vector $|w\rangle$:

$$\begin{aligned} &V \times V \ni (v,w) \longmapsto \langle v|w\rangle \in \mathbb{C}, \\ &V \ni |w\rangle \leftrightarrow \langle w| \in V^T, \text{ with } V \ni |v\rangle \overset{\langle w|}{\longmapsto} \langle w|v\rangle \in \mathbb{C}. \end{aligned}$$

Squares are real, but not necessarily definite. The invariance group of a square ζ, called *square group* $\mathbf{Q}(n_+, n_-)$, is an *orthogonal group* $\mathbf{O}(n_+, n_-)$ for the real cases $V \cong \mathbb{R}^n$, and a *unitary group* $\mathbf{U}(n_+, n_-)$ for the complex case $V \cong \mathbb{C}^n$, in both cases with a signature (n_+, n_-), $n = n_+ + n_-$. An indefinite square group has, with regard to the signature of the Killing form for the related Lie algebra, compact and noncompact dimensions, the latter ones being $d_{nc}^{\mathbf{O}} = n_+n_-$ and $d_{nc}^{\mathbf{U}} = 2n_+n_-$:

$$\dim_{\mathbb{R}} \mathbf{Q}(n_+, n_-) = d_c + d_{nc} : \begin{cases} \dim_{\mathbb{R}} \mathbf{O}(n_+, n_-) = \binom{n}{2} = \binom{n_+}{2} + \binom{n_-}{2} + n_+n_-, \\ \dim_{\mathbb{R}} \mathbf{U}(n_+, n_-) = n^2 = n_+^2 + n_-^2 + 2n_+n_-. \end{cases}$$

For a nontrivial *index* $I = \min(n_+, n_-) > 0$ of a square, there exist nontrivial positive, negative, and singular vectors with $\zeta^2(x) > 0$, $\zeta^2(x) < 0$, and $\zeta^2(x) = 0$, respectively. For the Minkowski translations they are called nontrivial timelike, spacelike, and lightlike, respectively.

The values of a square coincide for all points of a square group orbit $\mathbf{Q}(n_+, n_-) \bullet x \in V/\mathbf{Q}(n_+, n_-)$:

$$g \in \mathbf{Q}(n_+, n_-) \iff \begin{cases} \zeta(g \bullet x, g \bullet y) = \zeta(x, y) \\ \zeta^2(g \bullet x) = \zeta^2(x) \text{ for all } x, y \in V \end{cases}$$

Vectors, related to each other by a nontrivial scalar multiplication, have isomorphic fixgroup and, therefore, equal orbit type:

$$\mathbf{Q}(n_+, n_-)_x \cong \mathbf{Q}(n_+, n_-)_{\alpha x} \text{ for } \alpha \neq 0.$$

The invariance property for the invariance-group corresponding Lie algebra $\log \mathbf{Q}(n_+, n_-)$ is:

$$l \in \log \mathbf{Q}(n_+, n_-) \iff \zeta(l \bullet x, y) + \zeta(x, l \bullet y) = 0 \text{ for all } x, y \in V.$$

Each vector $x \in V$ defines a bilinear form of the invariance Lie algebra of a square. It is trivial for the fix-Lie algebra $\log \mathbf{Q}(n_+, n_-)_x$:

$$\begin{aligned} \zeta_x : \log \mathbf{Q}(n_+, n_-) \times \log \mathbf{Q}(n_+, n_-) \longrightarrow \mathbb{R}, \quad \zeta_x(l, m) &= \zeta(l \bullet x, m \bullet x), \\ \zeta_x(l, m) &= 0 \text{ for } l \text{ or } m \in \log \mathbf{Q}(n_+, n_-)_x. \end{aligned}$$

This structure is important in the electroweak standard model for the association of particle masses to the gauge fields of the spontaneously broken symmetries. The related transition from the hyperisospin transformations $\mathbf{U}(2)$ (square group) for fields to the electromagnetic charge transformations $\mathbf{U}(1)_+ \cong \mathbf{U}(1)$ (fixgroup) for particles employs the ground state-induced metric for the real three-dimensional Goldstone or weak hyperplanes $\log \mathbf{U}(2)/\log \mathbf{U}(1)$ (more ahead).

For a definite square with compact square group $\mathbf{Q}(n)$, there occurs as proper fixgroup only $\mathbf{Q}(n-1) \subset \mathbf{Q}(n)$, $n \geq 2$, leading to the orbit type decomposition of the vector space acted on:

$$\mathbf{Q}(n) \vec{\times} \mathbb{K}^n : \ \mathbb{K}^n \cong \{1\} \uplus \mathbf{D}(1) \times \frac{\mathbf{Q}(n)}{\mathbf{Q}(n-1)}.$$

An indefinite square with noncompact square group $\mathbf{Q}(n_+, n_-)$, $n_\pm \geq 1$, the vector space $\mathbb{K}^n$ for $n \geq 3$ is the disjoint union of four orbit types: the trivial one, and the positive, negative and singular one:

$$\begin{aligned} \mathbf{Q}(n_+, n_-) \vec{\times} \mathbb{K}^n,\ n \geq 3 :\ \mathbb{K}^n \cong \{1\} \uplus \mathbf{D}(1) \times \frac{\mathbf{Q}(n_+, n_-)}{\mathbf{Q}(n_+ - 1, n_-)} \\ \uplus \mathbf{D}(1) \times \frac{\mathbf{Q}(n_+, n_-)}{\mathbf{Q}(n_+, n_- - 1)} \uplus \frac{\mathbf{Q}(n_+, n_-)}{\mathbf{Q}(n_+ - 1, n_- - 1) \vec{\times} \mathbb{K}^{n-2}}. \end{aligned}$$

For $n = 2$, positive and negative type are equal:

$$\mathbf{Q}(1,1) \vec{\times} \mathbb{K}^2 : \ \mathbb{K}^2 \cong \{1\} \uplus 2 \cdot \mathbf{D}(1) \times \frac{\mathbf{Q}(1,1)}{\mathbf{Q}(1)} \uplus \mathbf{Q}(1,1).$$

An example are the generalized Minkowski translations $\mathbb{R}^{(1,s)}$ with indefinite orthogonal $\mathbf{O}(1,s)$-squares where one obtains, as orbit types for a position dimension $s \geq 2$, two-shell hyperboloids for the timelike vectors and one-shell hyperboloids for the spacelike ones. For one position dimension, all nonsingular orbit types are hyperbolas with two branches. The singular orbits are two tip-punctured cone surfaces for $s \geq 2$ and four lines for $s = 1$:

$$\mathbf{O}(1,s) \vec{\times} \mathbb{R}^{1+s} : \begin{cases} s \geq 2 : \begin{cases} \mathbb{R}^{1+s} \cong \{1\} \uplus \mathbf{D}(1) \times \frac{\mathbf{O}(1,s)}{\mathbf{O}(s)} \\ \qquad \uplus \mathbf{D}(1) \times \frac{\mathbf{O}(1,s)}{\mathbf{O}(1,s-1)} \uplus \frac{\mathbf{O}(1,s)}{\mathbf{O}(s-1) \vec{\times} \mathbb{R}^{s-1}}, \\ \frac{\mathbf{O}(1,s)}{\mathbf{O}(s)} \cong \mathcal{Y}^s_+ \uplus \mathcal{Y}^s_- \quad \text{(two shells, time)}, \\ \frac{\mathbf{O}(1,s)}{\mathbf{O}(1,s-1)} \cong \mathcal{Y}^{(1,s-1)} \quad \text{(one shell, position)}, \\ \frac{\mathbf{O}(1,s)}{\mathbf{O}(s-1) \vec{\times} \mathbb{R}^{s-1}} \cong \mathcal{C}^s_+ \uplus \mathcal{C}^s_- \quad \text{(two shells, light)}, \end{cases} \\ s = 1 : \begin{cases} \mathbb{R}^{1+1} \cong \{1\} \uplus \mathbf{D}(1) \times \mathbf{O}(1,1) \uplus 4 \times \mathbb{R}, \\ \frac{\mathbf{O}(1,1)}{\mathbf{O}(1)} = \frac{\mathbf{O}(1,1)}{\mathbf{I}(2)} \cong \mathbb{R} \uplus \mathbb{R} \text{ (two branches)}, \\ \mathbf{O}(1,1) \cong \mathbb{R} \uplus \mathbb{R} \uplus \mathbb{R} \uplus \mathbb{R} \text{ (four branches)}. \end{cases} \end{cases}$$

7.14 Stability Groups of Vector Space Decompositions

A decomposition of a vector space by orbits of an acting group involves, in general, not only vector subspaces. For a linear group $G \subseteq \mathbf{GL}(V)$ of a finite-dimensional $\mathbb{K}$-vector space ($\mathbb{K} = \mathbb{R}, \mathbb{C}$), the fixgroup G_x of a vector $x \in V$, $x \neq 0$, keeps fixed the one-dimensional vector space $\mathbb{K}x$. No $\mathbb{K}x$-complement $U \cong V/\mathbb{K}x$ has to be invariant, and there does not have to exist a fixgroup G_x-compatible vector space decomposition $V \cong \mathbb{K}x \oplus U$.

In the case of the Minkowski translations and a fixed time translation, the $\mathbf{SO}(3)$-action is compatible with a *Sylvester*[5] *decomposition* into time and position translations. Analogously, the fixgroup $\mathbf{SO}_0(1,2)$ of a position translation is compatible with a decomposition which involves a one-dimensional vector subspace of position translations.

For a fixed light translation $l = \begin{pmatrix} 1 \\ 0 \\ 0 \\ -1 \end{pmatrix}$ the spanned one-dimensional subspace $\mathbb{L}^1 = \mathbb{R}l$ is fixed — however, there is no direct complement that is stable under the

[5] James Joseph Sylvester (1814–1897).

action of this fixgroup. In general, the fixgroup action on the position translations produces a lightlike component in the direction of $\mathbb{L}^1$:

$$\mathbf{SO}(2) \vec{\times} \mathbb{R}^2 : \begin{pmatrix} 0 & \psi_1 & \psi_2 & 0 \\ \psi_1 & 0 & \theta_3 & \psi_1 \\ \psi_2 & -\theta_3 & 0 & \psi_2 \\ 0 & -\psi_1 & -\psi_2 & 0 \end{pmatrix} \begin{pmatrix} 0 \\ 1 \\ 0 \\ 0 \end{pmatrix} = \begin{pmatrix} \psi_1 \\ 0 \\ -\theta_3 \\ -\psi_1 \end{pmatrix}.$$

A group compatible with the vector space decomposition $V \cong U_1 \oplus U_2$, where the subspaces — not necessarily the vectors — remain stable, is called its *stability group* $G \subseteq \mathbf{GL}(V)$:

$$\mathrm{STAB}_{U_1 \oplus U_2} G = \{g \in G \mid g(U_{1,2}) = U_{1,2}\}.$$

Good to know: A stability group is a fixgroup $G_{\{U_1,U_2\}}$ for the G-action on vector subspaces.

The stability groups of decompositions of the Minkowski translations $\mathbb{R}^{(1,s)}$ with time translations $\mathbb{T} \cong \mathbb{R}$ and position translations $\mathbb{E}^s \cong \mathbb{R}^s$ are in the case of stable time translations:

$$\begin{array}{llll} \mathbb{R}^{(1,3)} & \cong \mathbb{T} \oplus \mathbb{E}^3 \cong & \mathbb{T} \oplus \mathbb{E}^2 \oplus \mathbb{E}^1 \cong & \mathbb{T} \oplus \mathbb{E}^1 \oplus \mathbb{E}^1 \oplus \mathbb{E}^1, \\ \mathbf{SO}_0(1,3) \supset & \mathbf{SO}(3) \quad \supset & \mathbf{SO}(2) \quad \supset & \{1\}, \end{array}$$

and for stable position translations:

$$\begin{array}{lllll} \mathbb{R}^{(1,3)} & \cong \mathbb{R}^{(1,2)} \oplus \mathbb{E}^1 \cong & \left\{ \begin{array}{c} \mathbb{T} \oplus \mathbb{E}^2 \oplus \mathbb{E}^1 \\ \mathbb{R}^{(1,1)} \oplus \mathbb{E}^1 \oplus \mathbb{E}^1 \end{array} \right\} & \cong \mathbb{T} \oplus \mathbb{E}^1 \oplus \mathbb{E}^1 \oplus \mathbb{E}^1, \\ \mathbf{SO}_0(1,3) \supset & \mathbf{SO}_0(1,2) \quad \supset & \left\{ \begin{array}{c} \mathbf{SO}(2) \\ \mathbf{SO}_0(1,1) \end{array} \right\} & \supset \quad \{1\}. \end{array}$$

For stable light translations, a *Witt*[6] *decomposition* keeps fixed two one-dimensional light translation-spaces $\mathbb{L}^1 \cong \mathbb{R}$:

$$\begin{array}{lll} \mathbb{R}^{(1,3)} & \cong \mathbb{L}^1 \oplus \mathbb{E}^2 \oplus \mathbb{L}^1 \cong & \mathbb{L}^1 \oplus \mathbb{E}^1 \oplus \mathbb{E}^1 \oplus \mathbb{L}^1, \\ \mathbf{SO}_0(1,3) \supset & \mathbf{SO}(2) \quad \supset & \{1\}. \end{array}$$

7.15 Polar Decompositions

In the case of a definite square with invariance group $\mathbf{Q}(1+s) = \mathbf{O}(1+s)$ (orthogonal) or $\mathbf{Q}(1+s) = \mathbf{U}(1+s)$ (unitary) for a vector space $\mathbb{R}^{1+s}$ or $\mathbb{C}^{1+s}$, $s \geq 1$, respectively, there exists only the trivial and the positive orbit type. The fixgroup $\mathbf{Q}(1+s)_v \cong \mathbf{Q}(s)$ of a nontrivial vector v, which can be taken as a basis vector

[6] Ernst Witt (1911–1991).

$v = e^{1+s} \cong \begin{pmatrix} 0 \\ \vdots \\ 0 \\ 1 \end{pmatrix}$, is the stability group of a corresponding Euclidean decomposition of the vector space $\mathbb{K}^{1+s}$ with $\mathbb{K}v$ as direct summand:

$$\textit{Euclidean transition}: \mathbf{Q}(1+s) \supset \mathbf{Q}(s) \text{ with } \mathbb{K}^{1+s} \cong \mathbb{K} \oplus \mathbb{K}^s.$$

A definite square $\zeta : \mathbb{K} \times \mathbb{K} \longrightarrow \mathbb{K}$ can be diagonalized $\zeta \cong M^2 \mathbf{1}_{1+s}$, orthogonally or unitarily. The *Euclid unit* $M^2 > 0$ gives the square of all vectors of an orthogonal, equally normalized *Euclidean basis*. It determines the scale of the Euclidean transition $\mathbf{Q}(s) \subset \mathbf{Q}(1+s)$.

For the rotation group $\mathbf{O}(3)$ acting on a three-dimensional space:

$$\mathbf{O}(3) \vec{\times} \mathbb{R}^3 : \ \mathbb{R}^3 \cong \{1\} \uplus \mathbf{D}(1) \times \Omega^2,$$

the Euclid unit M characterizes the radius of a 2-sphere $\mathbf{O}(3)/\mathbf{O}(2) \cong \Omega^2$ and the length of orthogonal basic vectors $\langle e^a | e^b \rangle = \delta^{ab} M^2$. The fix-Lie algebra of the vector $e^3 = \begin{pmatrix} 0 \\ 0 \\ 1 \end{pmatrix}$ is spanned by the angular momentum L^3, the corresponding classes of the angular momenta $L^{1,2}$ for the first and second direction are a basis of the orthogonal hyperplane $\log \mathbf{O}(3)/\log \mathbf{O}(2)_3 \cong \mathbb{R}^2$:

$$\begin{aligned} L(\theta) &= \theta_1 L^1 + \theta_2 L^2 + \theta_3 L^3 = \begin{pmatrix} 0 & \theta_3 & -\theta_2 \\ -\theta_3 & 0 & \theta_1 \\ \theta_2 & -\theta_1 & 0 \end{pmatrix}, \\ L(\theta) e^3 &= \begin{pmatrix} -\theta_2 \\ \theta_1 \\ 0 \end{pmatrix}, \quad \mathbf{O}(3)_{e^3} \cong \mathbf{O}(2). \end{aligned}$$

With the Killing form, the lenghts of the angular momenta are determined up to a factor g_0^2:

$$\begin{aligned} \langle L(\theta) | L(\psi) \rangle_{e^3} &= \langle -\theta_2 e^1 + \theta_1 e^2 | -\psi_2 e^1 + \psi_1 e^2 \rangle = g_0^2 M^2 (\theta_1 \psi_1 + \theta_2 \psi_2), \\ \langle L^a | L^b \rangle_{e^3} &= \begin{cases} g_0^2 M^2 \delta^{ab}, & a, b = 1, 2, \\ 0, & a = 3 \text{ or } b = 3. \end{cases} \end{aligned}$$

In general, the *orthogonal orbit (polar) decomposition* for real vector spaces involves spheres with fixgroup $\mathbf{O}(s)$, $s \geq 1$:

$$\mathbf{O}(1+s) \vec{\times} \mathbb{R}^{1+s}: \mathbb{R}^{1+s} \cong \{1\} \uplus \mathbf{D}(1) \times \Omega^s, \quad \Omega^s \cong \mathbf{O}(1+s)/\mathbf{O}(s),$$
$$\mathbf{O}(1+s)_{e^{1+s}} = \begin{pmatrix} \mathbf{O}(s) & 0 \\ 0 & 1 \end{pmatrix} \cong \mathbf{O}(s).$$

In the electroweak standard model, the gauge interactions for leptons and quarks are particle-analyzed with the *electromagnetic Euclidean transition*. The hyperisospin group $\mathbf{U}(2)$ is the invariance group for the scalar product of the Higgs Hilbert space:

$$\boldsymbol{\Phi} = h \in \mathbb{H} \cong \mathbb{C}^2, \quad ||h||^2 = \langle h|h\rangle > 0 \iff h \neq 0,$$

The normalization of the electromagnetic stability group $\mathbf{U}(1)_+$ in the hyperisospin group $\mathbf{U}(2)$ is determined by the ground state expectation value of the Higgs-vector, e.g., with the Fermi mass as Euclid unit:

$$\langle \boldsymbol{\Phi} \rangle = c^2 = \begin{pmatrix} 0 \\ M \end{pmatrix} \in \mathbb{H}, \quad M = M_{\text{Fermi}} \simeq 123 \frac{\text{GeV}}{c^2}.$$

The $\mathbf{U}(2)$-orbit decomposition of the Higgs space contains the real three-dimensional Goldstone or weak manifold $\mathcal{G}^3 \cong \mathbf{U}(2)/\mathbf{U}(1)_+$:

$$\mathbf{U}(2) \vec{\times} \mathbb{C}^2: \quad \mathbb{C}^2 \cong \{1\} \uplus \mathbf{D}(1) \times \mathcal{G}^3.$$

The Higgs vector space is acted on by the $\mathbf{U}(2)$-Lie algebra with basis $\{i\frac{\tau^j}{2}\}^{j=0,1,2,3} = (i\frac{\mathbf{1}_2}{2}, i\frac{\vec{\tau}}{2})$:

$$i\frac{\tau}{2}(\alpha) = \alpha_0 i\frac{\mathbf{1}_2}{2} + \vec{\alpha} i\frac{\vec{\tau}}{2} = i\begin{pmatrix} \alpha_0+\alpha_3 & \alpha_1 - i\alpha_2 \\ \alpha_1 + i\alpha_2 & \alpha_0 - \alpha_3 \end{pmatrix} \in \log \mathbf{U}(2) \cong \mathbb{R}^4.$$

By the ground state square, the real three-dimensional Goldstone or weak hyperplanes $\log \mathbf{U}(2)/\log \mathbf{U}(1)_+ \cong \mathbb{R}^3$ are endowed with a nontrivial bilinear form (metric) — with a free factor g_0^2:

$$i\frac{\tau}{2}(\alpha).c^2 = i\frac{M}{2}\begin{pmatrix} \alpha_1 - i\alpha_2 \\ \alpha_0 - \alpha_3 \end{pmatrix}, \quad \mathbf{U}(2)_{c^2} = \mathbf{U}(1)_+ \cong \mathbf{U}(1),$$
$$\Rightarrow \langle i\frac{\tau}{2}(\alpha)|i\frac{\tau}{2}(\beta)\rangle_{c^2} = \frac{g_0^2 M^2}{4}[(\alpha_1 + i\alpha_2)(\beta_1 - i\beta_2) + (\alpha_0 - \alpha_3)(\beta_0 - \beta_3)],$$
$$\Rightarrow \begin{cases} \langle i\frac{\tau^\pm}{2}|i\frac{\tau^\pm}{2}\rangle_{c^2} = g_0^2 M^2, & \langle i\frac{\tau^+}{2}|i\frac{\tau^-}{2}\rangle_{c^2} = 0, \quad \frac{\tau^\pm}{2} = \frac{\tau^1}{2} \pm i\frac{\tau^2}{2}, \\ \langle i\frac{\tau^0-\tau^3}{2}|i\frac{\tau^0-\tau^3}{2}\rangle_{c^2} = \frac{g_0^2 M^2}{4}, & \langle i\frac{\tau^0 \pm \tau^3}{2}|i\frac{\tau^0+\tau^3}{2}\rangle_{c^2} = 0. \end{cases}$$

The invariance Lie algebra $\mathbb{R}(i\frac{\tau^0+\tau^3}{2}) = \log \mathbf{U}(1)_+$ with trivial square is related to the massless electromagnetic field for the remaining invariance group $\mathbf{U}(1)_+$ (more ahead).

In general, the *unitary orbit (polar) decomposition* for complex vector spaces involves the generalized compact Goldstone manifolds with fixgroup $\mathbf{U}(s)$, $s \geq 1$:

$$\mathbf{U}(1+s) \vec{\times} \mathbb{C}^{1+s} : \mathbb{C}^{1+s} \cong \{1\} \uplus \mathbf{D}(1) \times \mathcal{G}^{1+2s}, \quad \mathcal{G}^{1+2s} \cong \mathbf{U}(1+s)/\mathbf{U}(s)_+,$$
$$\mathbf{U}(1+s)_{e^{1+s}} = \mathbf{U}(s)_+ = \begin{pmatrix} \mathbf{U}(s) & 0 \\ 0 & 1 \end{pmatrix} \cong \mathbf{U}(s).$$

7.16 Sylvester and Witt Transitions

The signature decomposition of the dimensions in a square ζ can be replaced by a triplet which decomposes the dimensions with its index I and its *character* C:

$$I = \min(n_+, n_-), \quad C = |n_+ - n_-|, \quad \epsilon = \text{ sign } (n_+ - n_-) = \pm 1,$$
$$n = n_+ + n_- = 2I + C, \quad (n_+, n_-) \sim (I, \epsilon C, I),$$
$$\text{e.g., for } \begin{cases} \mathbf{O}(1,3) : (1,3) \sim (1,-2,1), \\ \mathbf{U}(1,1) : (1,1) \sim (1,0,1), \\ \mathbf{Q}(n) : (n,0) \sim (0,n,0). \end{cases}$$

With ζ also $-\zeta$ is a square with the interchange $n_\pm \leftrightarrow n_\mp$ and $\epsilon \leftrightarrow -\epsilon$.

For signature (n_+, n_-) and index-character $(I, \epsilon C, I)$, the *Sylvester* and *Witt transitions* from a square characterizing group $\mathbf{Q}(n_+, n_-)$ to the stability group of the corresponding vector space decomposition with two and three direct summands, respectively, are

$$\textit{Sylvester transition}: \mathbf{Q}(n_+, n_-) \supset \mathbf{Q}(n_+) \times \mathbf{Q}(n_-),$$
$$\text{with } \mathbb{K}^n \cong \mathbb{K}^{n_+} \oplus \mathbb{K}^{n_-},$$
$$\textit{Witt transition}: \mathbf{Q}(n_+, n_-) \supset \mathbf{Q}(I) \times \mathbf{Q}(C),$$
$$\text{with } \mathbb{K}^n \cong \mathbb{K}^I \oplus \mathbb{K}^C \oplus \mathbb{K}^I.$$

The vector subspaces $\mathbb{K}^{n_\pm}$ in a Sylvester decomposition are related to the positive and negative squares $\pm M_\pm^2$ in the diagonal matrix for a Sylvester basis:

$$\text{Sylvester transition:} \begin{cases} \zeta^2(x_+) > 0, \quad \zeta^2(x_-) < 0 \text{ for } x_\pm \in \mathbb{K}^{n_\pm}, \quad x_\pm \neq 0, \\ \mathbf{1} \cong \begin{pmatrix} M_+^2 \mathbf{1}_{n_+} & 0 \\ 0 & -M_-^2 \mathbf{1}_{n_-} \end{pmatrix}, \quad M_\pm^2 > 0. \end{cases}$$

The squares $M_\pm^2$ or the *Sylvester unit* M_s^2 and the *signature factor* $e^{2\psi}$:

$$M_s^4 = M_+^2 M_-^2 > 0,\ \ e^{2\psi} = \frac{M_+^2}{M_-^2} > 0,\ \ \mathbf{1} \cong M_s^2 \begin{pmatrix} e^{\psi}\mathbf{1}_{n_+} & 0 \\ 0 & -e^{-\psi}\mathbf{1}_{n_-} \end{pmatrix},$$

determine the scales of the stability group $\mathbf{Q}(n_+) \times \mathbf{Q}(n_-)$ of the Sylvester decomposition.

In the case of Minkowski translations $x = (t, \vec{x})$, the Sylvester transition $\mathbf{O}(1, s) \supset \mathbf{O}(1) \times \mathbf{O}(s)$ has as signature unit the maximal action velocity (speed of light) c with the dimension $\frac{\text{m}}{\text{s}}$. A Sylvester unit M^2 has the dimension ms:

$$\text{translations } \mathbb{R}^4 : \begin{cases} g \cong M^2 \begin{pmatrix} \frac{1}{c} & 0 \\ 0 & -c\mathbf{1}_3 \end{pmatrix}, \\ g(t,t) \sim \frac{M^2}{c} \sim \text{s}^2, \\ \qquad [\text{time}] = \text{second} = \text{s}, \\ -g(\vec{x},\vec{x}) \sim M^2 c \sim \text{m}^2, \\ \qquad [\text{position}] = \text{meter} = \text{m}. \end{cases}$$

The linear forms of Minkowski spacetime constitute the energy-momentum space $q = (E, \vec{q})$ with the inverse units[7]:

$$\text{energy-momenta } \check{\mathbb{R}}^4 \cong \mathbb{R}^4 : \begin{cases} g^{-1} \cong \frac{1}{M^2} \begin{pmatrix} c & 0 \\ 0 & -\frac{1}{c}\mathbf{1}_3 \end{pmatrix}, \\ g^{-1}(E,E) \sim \frac{c}{M^2} \sim \text{s}^{-2}, \\ \qquad [\text{energy}] \text{ with Hertz} = \frac{1}{\text{s}}, \\ -g^{-1}(\vec{q},\vec{q}) \sim \frac{1}{M^2 c} \sim \text{m}^{-2}, \\ \qquad [\text{momentum}] \text{ with dioptry} = \frac{1}{\text{m}}. \end{cases}$$

The ususal energy-momentum units $(\frac{[\hbar]}{\text{s}}, \frac{[\hbar]}{\text{m}})$ can be obtained by multiplication with the dimension of Planck's unit $[\hbar] = \frac{\text{kg m}^2}{\text{s}}$.

In a Witt decomposition the subspace $\mathbb{K}^C$ with the character C as dimension has either positive or negative squares, $\epsilon = \pm 1$, for all nontrivial vectors. The two subspaces $\mathbb{K}^I$ contain vectors with trivial squares. These two spaces are not orthogonal with respect to an indefinite square, as seen in the skew-diagonal Witt matrix for a Witt basis:

$$\text{Witt transition:} \begin{cases} \epsilon\zeta^2(x) > 0 \text{ for } x \in \mathbb{K}^C,\ \ x \neq 0, \\ \zeta^2(x) = 0 \text{ for } x \in \mathbb{K}^I \oplus \mathbb{K}^I, \\ \zeta \cong \epsilon M_w^2 \begin{pmatrix} 0 & 0 & \nu\mathbf{1}_I \\ 0 & \mathbf{1}_C & 0 \\ \nu\mathbf{1}_I & 0 & 0 \end{pmatrix},\ \ \nu \neq 0. \end{cases}$$

---

[7] Heinrich Hertz (1857–1894).

The *Witt unit* M_w^2 as square of a definite orbit $\epsilon\zeta^2(x) = M_w^2$ determines the scale of the stability group $\mathbf{Q}(C)$ of the Witt decomposition.

In the case of Minkowski translations, the Witt transition $\mathbf{O}(1,3) \supset \mathbf{O}(1) \times \mathbf{O}(2)$ corresponds to the Witt decomposition $\mathbb{R}^{(1,3)} \cong \mathbb{L}^1 \oplus \mathbb{E}^2 \oplus \mathbb{L}^1$ into two one-dimensional nonorthogonal light-spaces $\mathbb{L}^1$ and a two-dimensional space $\mathbb{E}^2$ with position translations. To characterize the action of the orthochronous group $\mathbf{SO}_0(1,3)$ on $\mathbb{R}^{(1,3)}$ by eigenvalues with their eigenvectors, one goes over to a Cartan group $\mathbf{SO}_0(1,1) \times \mathbf{SO}(2)$ as maximal abelian subgroup. The two lightlike spaces $\mathbb{L}^1 \oplus \mathbb{L}^1$ are spanned by two eigenvectors of the real two-dimensional Cartan subgroup with trivial action of the axial rotations $\mathbf{SO}(2)$; the dilations $\mathbf{SO}_0(1,1)$ act by shrinking or stretching. The dilations $\mathbf{SO}_0(1,1)$ act trivially on the two-dimensional position translations $\mathbb{E}^2$, the axial rotations $\mathbf{SO}(2)$ nontrivially. $\mathbf{SO}(2)$-eigenvectors as left- and right-handed screws can be constructed only in a complex embedding.

7.17 Bases of Minkowski Translations

In a Sylvester decomposition, the Minkowski translations can be spanned by an orthogonal basis $\{s^i\}_{i=0}^3$ with one time and three position translations:

$$\text{Sylvester bases:} \begin{cases} g(s^j, s^k) \cong M^2\begin{pmatrix} \frac{1}{c^2} & 0 \\ 0 & -\mathbf{1}_3 \end{pmatrix}, \\ \mathbb{R}^{(1,3)} \cong \mathbb{T} \oplus \mathbb{E}^3, \quad \mathbf{O}(1,3) \supset \mathbf{O}(1) \times \mathbf{O}(3); \end{cases}$$

whereas for a Witt decomposition bases $\{w^i\}_{i=0}^3$ with two nonorthogonal light translations and two orthogonal position translations are used:

$$\text{Witt bases:} \begin{cases} g(w^j, w^k) \cong -m^2\begin{pmatrix} 0 & 0 & \nu \\ 0 & \mathbf{1}_2 & 0 \\ \nu & 0 & 0 \end{pmatrix}, \\ \mathbb{R}^{(1,3)} \cong \mathbb{L}^1 \oplus \mathbb{E}^2 \oplus \mathbb{L}^1, \quad \mathbf{O}(1,3) \supset \mathbf{O}(2). \end{cases}$$

Finkelstein[8] *bases* $\{f^i\}_{i=0}^3$ contain four light translations. The corresponding stability group is the permutation group $\mathbf{G}(4)$; the "democratic light matrix" of the bilinear form does not distinguish any basis direction:

$$\text{Finkelstein bases:} \begin{cases} g(f^j, f^k) \cong \mu^2\begin{pmatrix} 0 & 1 & 1 & 1 \\ 1 & 0 & 1 & 1 \\ 1 & 1 & 0 & 1 \\ 1 & 1 & 1 & 0 \end{pmatrix}, \\ \mathbb{R}^{(1,3)} \cong \mathbb{L}^1 \oplus \mathbb{L}^1 \oplus \mathbb{L}^1 \oplus \mathbb{L}^1, \quad \mathbf{O}(1,3) \supset \mathbf{G}(4). \end{cases}$$

The masses $M, m, \mu > 0$ and the factors $c > 0$ or $\nu \neq 0$ determine the scales of the relevant stability groups in the Lorentz group.

[8] David Finkelstein (1929–2016).

Each of these basis types b defines, by the invariance of the associated Lorentz metric, a Lorentz group $\mathbf{O}(1,3)(b)$ in the automorphism group $\mathbf{GL}(4,\mathbb{R})$ of the Minkowski translations:

$$\{b^i\}_{i=0}^3: \quad \Lambda_j^l(b)g(b^j,b^k)\Lambda_k^m(b) = g(b^l,b^m).$$

The three classes with Sylvester, Witt and Finkelstein bases, $\{s^i\}_{i=0}^3$, $\{w^i\}_{i=0}^3$ and $\{f^i\}_{i=0}^3$, are connected with each other by automorphisms $e \in \mathbf{GL}(4,\mathbb{R})$:

$$e_j^l(w,s)g(s^j,s^k)e_k^m(w,s) = g(w^l,w^m), \quad e_j^l(f,s)g(s^j,s^k)e_k^m(f,s) = g(f^l,f^m).$$

The automorphisms between the classes are determined up to invariance groups $\mathbf{O}(1,3)$ — with possible representatives:

$$e(w,s) \cong \frac{m}{M}\begin{pmatrix} c\frac{1}{\sqrt{2}} & 0 & 0 & \frac{1}{\sqrt{2}} \\ 0 & 1 & 0 & 0 \\ 0 & 0 & 1 & 0 \\ c\frac{1}{\sqrt{2}} & 0 & 0 & -\frac{1}{\sqrt{2}} \end{pmatrix}, \quad e(f,s) \cong \frac{\mu}{M\sqrt{2}}\begin{pmatrix} c\sqrt{\frac{3}{2}} & 1 & \frac{1}{\sqrt{3}} & \frac{1}{\sqrt{6}} \\ c\sqrt{\frac{3}{2}} & -1 & \frac{1}{\sqrt{3}} & \frac{1}{\sqrt{6}} \\ c\sqrt{\frac{3}{2}} & 0 & -\frac{2}{\sqrt{3}} & \frac{1}{\sqrt{6}} \\ c\sqrt{\frac{3}{2}} & 0 & 0 & -\frac{3}{\sqrt{6}} \end{pmatrix}.$$

It is remarkable that, in general, a square of real $(1+s)$-dimensional[9] spacetime translations, $s \geq 1$, with singular, nonorthogonal basis vectors (Finkelstein basis) $\{f^i\}_{i=0}^s$ and permutation-invariance $\mathbf{G}(1+s)$:

$$g(f^j,f^k) \cong \mu^2 \begin{pmatrix} 0 & 1 & 1 & \dots & 1 \\ 1 & 0 & 1 & \dots & 1 \\ & \dots & & \dots & \\ 1 & \dots & 1 & 0 & 1 \\ 1 & \dots & 1 & 1 & 0 \end{pmatrix},$$

defines a Lorentz group $\mathbf{O}(1,s)$: There exist Sylvester bases $\{s^i\}_{i=0}^s$ with one time translation $\{s^0\}$ and s position translations $\{s^i\}_{i=1}^s$:

$$g(s^j,s^k) = M^2\begin{pmatrix} \frac{1}{c^2} & 0 \\ 0 & -1_s \end{pmatrix}, \quad e_j^l(f,s)g(s^j,s^k)e_k^m(f,s) = g(f^l,f^m).$$

This allows the introduction of an $\mathbf{SO}_0(1,s)$-compatible order structure, i.e., of causality. The bases-connecting automorphism contains, in the rows, from the second ($i=1$) to the last ($i=s$) column the $(1+s)$ vectors $\{\vec{e}^j\}_{j=0}^s$ of the vertices of a regular fundamental simplex in Euclidean space $\mathbb{E}^s$ with the center in the origin:

[9] Do not mix up the symbol s used both for position dimension and Sylvester bases!

$$e(f,s) = \frac{\mu}{M\sqrt{2}}\begin{pmatrix} cN & \vec{e}^0 \\ cN & \vec{e}^1 \\ \dots & \dots \\ cN & \vec{e}^s \end{pmatrix}, \text{ with } \begin{cases} (\vec{e}^j)^2 &= N^2, \ j = 0, \dots, s, \\ \vec{e}^j\vec{e}^k &= N^2 \cos\varphi, \ j \neq k, \\ \sum_{j=0}^{s} \vec{e}^j &= 0, \end{cases}$$

i.e., a distance for $s = 1$, a regular triangle for $s = 2$, a regular tetrahedron for $s = 3$, in general:

$$e(f,s) \cong \frac{\mu}{M\sqrt{2}}\left(\begin{array}{c|cccccc} c\sqrt{\frac{2s}{1+s}} & 1 & \frac{1}{\sqrt{3}} & \frac{1}{\sqrt{6}} & \cdots\cdots & & \frac{1}{\sqrt{\binom{1+s}{2}}} \\ c\sqrt{\frac{2s}{1+s}} & -1 & \frac{1}{\sqrt{3}} & \frac{1}{\sqrt{6}} & \cdots\cdots & & \frac{1}{\sqrt{\binom{1+s}{2}}} \\ c\sqrt{\frac{2s}{1+s}} & 0 & -\frac{2}{\sqrt{3}} & \frac{1}{\sqrt{6}} & \cdots\cdots & & \frac{1}{\sqrt{\binom{1+s}{2}}} \\ c\sqrt{\frac{2s}{1+s}} & 0 & 0 & -\frac{3}{\sqrt{6}} & \cdots\cdots & & \frac{1}{\sqrt{\binom{1+s}{2}}} \\ \cdots & & \cdots & & \cdots & & \\ \cdots & & \cdots & & \cdots & & \\ c\sqrt{\frac{2s}{1+s}} & 0 & 0 & \cdots & \cdots & 0 & -\frac{s}{\sqrt{\binom{1+s}{2}}} \end{array}\right),$$

$$N^2 = \frac{2s}{1+s}, \quad \cos\varphi = -\frac{1}{s}.$$

As coordinates of the regular fundamental simplex, determined up to $\mathbf{O}(s)$-rotations, the weights of the defining $(1+s)$-dimensional representation of the unitary group $\mathbf{SU}(1+s)$ are chosen, i.e., for $s = 1$ the two weights $\begin{pmatrix} 1 \\ -1 \end{pmatrix}$ of $\mathbf{SU}(2)$, for $s = 2$ the three weights $\begin{pmatrix} 1 & \frac{1}{\sqrt{3}} \\ -1 & \frac{1}{\sqrt{3}} \\ 0 & -\frac{2}{\sqrt{3}} \end{pmatrix}$ of $\mathbf{SU}(3)$, etc. If you switch on, in your rest system, a pocket lamp, the light rays to the four vertices of a spatial tetrahedron constitute a Finkelstein basis with the pocket lamp in the center (origin).

Simplex bases describe a transition from the automorphism group of the spacetime translations via the Lorentz group to the permutation group, which can be supplemented by the solvable telescopes of the permutation groups. All that together yields telescopes that end, up to three position dimensions, with the trivial symmetry:

$$\begin{array}{rl} s = 1: & \mathbf{GL}(2,\mathbb{R}) \supset \mathbf{O}(1,1) \supset \mathbf{G}(2) \supset \{1\}, \\ s = 2: & \mathbf{GL}(3,\mathbb{R}) \supset \mathbf{O}(1,2) \supset \mathbf{G}(3) \supset \mathbf{S}(3) \supset \{1\}, \\ s = 3: & \mathbf{GL}(4,\mathbb{R}) \supset \mathbf{O}(1,3) \supset \mathbf{G}(4) \supset \mathbf{S}(4) \supset \mathbf{K}(4) \supset \{1\}, \\ s \geq 4: & \mathbf{GL}(1+s,\mathbb{R}) \supset \mathbf{O}(1,s) \supset \mathbf{G}(1+s) \supset \mathbf{S}(1+s). \end{array}$$

The permutation group $\mathbf{G}(1+s)$ is the Weyl group for the Lie algebra A_s of $\mathbf{SL}(1+s,\mathbb{C})$ and of its real forms, e.g., of $\mathbf{SU}(1+s)$. Only the permutations $\mathbf{S}(1+s)$ with positive signature are a subgroup of the special group $\mathbf{SL}(1+s,\mathbb{C})$. The negative signature transposition $\begin{pmatrix} 0 & 1 \\ 1 & 0 \end{pmatrix}$ of two basic vectors has determinant -1.

7.18 The Manifold of the Lorentz Tetrads

The linear group $\mathbf{GL}(V)$ of a finite-dimensional vector space $V \cong \mathbb{K}^n$ acts on the set with all bases:

$$\mathcal{B}(V) = \{b = \{b^i\}_{i=1}^n \mid \text{basis of } V\}.$$

The columns of a nonsingular matrix give the components of a basis with the components of the associated dual basis in the rows of the inverse matrix,

$$\mathbf{GL}(\mathbb{K}^n) \ni b \cong (b^1, b^2, \dots, b^n), \quad b^{-1} \cong \begin{pmatrix} \check{b}_1 \\ \check{b}_2 \\ \vdots \\ \check{b}_n \end{pmatrix}, \quad \langle \check{b}_j, b^i \rangle = \delta^i_j.$$

The manifold of all bases $\mathcal{B}(V)$ is, as $\mathbf{GL}(V)$-set, isomorphic to the group $\mathbf{GL}(V)$ — there is only one orbit $\mathcal{B}(V) \cong \mathbf{GL}(V) \bullet \mathrm{id}_V$. Any two bases are connected by an automorphism:

$$c^i = e(c, b)(b^i), \quad e(c, b) = c^i \otimes \check{b}_i \in \mathbf{GL}(V), \quad \begin{cases} e(b, b) &= \mathbf{1}_V \\ e(b, c) &= e^{-1}(c, b) \\ e(b, c) \circ e(c, d) &= e(b, d). \end{cases}$$

The automorphism $e(c, b)$, connecting two bases, can be written as tensor $c^i \otimes \check{b}_i$ (sum over i), using the dual basis $\{\check{b}_i\}_{i=1}^n$ of the space V^T with the linear forms of V with dual products $\langle \check{b}_j, b^i \rangle = \delta^i_j$.

Good to know: The endomorphisms of a vector space $V \cong \mathbb{K}^n$ are isomorphic to the tensor product $f \in \mathbf{AL}(V) \cong V \otimes V^T$ with the dual vector space V^T. In components with dual bases, $\{b^i\}$ for V and $\{\check{b}_i\}$ for V^T, one has $f = f_i^j b^i \otimes \check{b}_j$, e.g., for two dimensions $v = v_i b^i \cong \begin{pmatrix} v_1 \\ v_2 \end{pmatrix} \in V \cong \mathbb{K}^2$ and $\omega = \omega^i \check{b}_i \cong (\omega^1, \omega^2) \in V^T \cong \mathbb{K}^2$, the matrix for the tensor product reads $v \otimes \omega \cong \begin{pmatrix} v_1\omega^1 & v_1\omega^2 \\ v_2\omega^1 & v_2\omega^2 \end{pmatrix} \in V \otimes V^T \cong \mathbb{K}^4$. The identity is given by $\mathrm{id}_V = b^i \otimes \check{b}_i$. The components in different dual bases (for distinction also different index types) $f = f_i^j b^i \otimes \check{b}_j = f_\mu^\nu c^\mu \otimes \check{c}_\nu$ are related to each other by automorphisms $e(c, b) = e^i_\mu c^\mu \otimes \check{b}_i$ with $e^i_\mu = \langle \check{c}_\mu, b^i \rangle$ — for the bases $b^i = e^i_\mu c^\mu$, and for the endomorphisms $e^i_\mu f_i^j e^\nu_j = f_\mu^\nu$ (similarity transformation).

Two bases of a vector space V are equivalent with respect to a linear group $H \subseteq \mathbf{GL}(V)$, if they are related to each other by an automorphism $h \in H$. Then, the corresponding tensor is in H:

$$\begin{aligned} \{b^i\}_{i=1}^n \overset{H}{\approx} \{c^i\}_{i=1}^n &\iff \text{There exists } h \in H \text{ with } c^i = h(b^i) \text{ for } i = 1, \dots n, \\ &\iff c^i \otimes \check{b}_i \in H. \end{aligned}$$

If the group H is defined as the invariance group of a bilinear or a sesquilinear form ζ of the vector space V, the equivalence of the bases can be defined equivalently by equal forms of all basic vector pairs:

$$H \bullet \zeta = \zeta : \ \{b^i\}_{i=1}^n \overset{H}{\approx} \{c^i\}_{i=1}^n \iff \zeta(b^i, b^j) = \zeta(c^i, c^j) \text{ for all } i, j = 1, \ldots n.$$

The invariance group $\mathbf{Q}(n_+, n_-)$ of a square ζ is, e.g., for the real definite cases $\mathbf{O}(n)$, the invariance group of ellipses, ellipsoids (elliptic surfaces), etc., e.g., for $n = 2$:

$$\zeta \cong \begin{pmatrix} \alpha & \gamma \\ \gamma & \beta \end{pmatrix}, \quad \text{sign } \zeta = 2.$$

The $\mathbf{O}(2)$-orbits do not have to be circles or to have the coordinate axes as principal axes. For any square group $\mathbf{Q}(n_+, n_-)$ there exist orthogonal bases, where the metrical matrix has diagonal elements only, $\zeta = u \circ \text{diag}\, \zeta \circ u^*$, with orthogonal u for $\mathbb{K} = \mathbb{R}$ and unitary u for $\mathbb{K} = \mathbb{C}$ — then normalizable to ± 1. Orthogonal basic vectors coincide with the principal axes of ellipses, ellipsoids, etc.; for orthonormal bases one has circles, spheres, etc.

The freedom in the choice of equivalent squares (metrics) can be also formulated in terms of automorphism classes for the transformation to a standard metric, e.g. to a standard unit circle $x^2 + y^2 = 1$ or to a standard unit hyperbola $xy = 1$, etc. For example, a spacetime metric g with signature $(1, 3)$ can always be transformed, via a tetrad transformation, to an orthonormal basis class, $g_{jk}(x) = e_j^a(x)\eta_{ab}e_k^b(x)$.

The elements of the embedding manifold $\mathbf{GL}(V)/H \cong \mathcal{B}(V)/H$ for an invariance group H are the H-inequivalent basis classes, i.e., the H-orbits, starting from representative bases.

Examples for the Minkowski translations are the Sylvester, Witt and Finkelstein classes as three inequivalent elements of $\mathbf{GL}(4, \mathbb{R})/\mathbf{O}(1, 3)$. All bases of the Minkowski translations can be reached from a fixed basis, e.g., a Sylvester basis $\{s^i\}_{i=0}^3$. The related automorphisms are the *Lorentz tetrads* for the basis $\{s^i\}_{i=0}^3$:

$$e(b, s)(s^i) = b^i, \quad e(b, s) = b^i \otimes \check{s}_i, \quad i = 0, 1, 2, 3,$$

with the matrix form in corresponding dual bases $\{s^i, \check{s}_i\}_{i=0}^3$:

$$e(b, s) = e_j^i(b, s)\, s^j \otimes \check{s}_i.$$

In the rows of the matrix $e_j^i(b, s)$ are the components of the goal-basis $\{b^i\}_{i=0}^3$ in the start-basis $\{s^i\}_{i=0}^3$, expressable as dual products or as traces of the linear transformations $\{b^i \otimes \check{s}_j\}_{i,j=0}^3$:

$$\text{with respect to } (s^i, \check{s}_i) : \ e(b, s) \cong e_j^i(b, s) = \langle \check{s}_j, b^i \rangle = \text{tr}\, b^i \otimes \check{s}_j.$$

Examples are the Lorentz tetrads from Sylvester to Witt or Finkelstein bases:

$$e(w,s) = w^i \otimes \check{s}_i \cong e^i_j(w,s) \cong \begin{pmatrix} w^0 \\ w^1 \\ w^2 \\ w^3 \end{pmatrix} \cong \frac{m}{M} \begin{pmatrix} c\frac{1}{\sqrt{2}} & 0 & 0 & \frac{1}{\sqrt{2}} \\ 0 & 1 & 0 & 0 \\ 0 & 0 & 1 & 0 \\ c\frac{1}{\sqrt{2}} & 0 & 0 & -\frac{1}{\sqrt{2}} \end{pmatrix},$$

$$e(f,s) = f^i \otimes \check{s}_i \cong e^i_j(f,s) \cong \begin{pmatrix} f^0 \\ f^1 \\ f^2 \\ f^3 \end{pmatrix} \cong \frac{\mu}{M\sqrt{2}} \begin{pmatrix} c\sqrt{\frac{3}{2}} & 1 & \frac{1}{\sqrt{3}} & \frac{1}{\sqrt{6}} \\ c\sqrt{\frac{3}{2}} & -1 & \frac{1}{\sqrt{3}} & \frac{1}{\sqrt{6}} \\ c\sqrt{\frac{3}{2}} & 0 & -\frac{2}{\sqrt{3}} & \frac{1}{\sqrt{6}} \\ c\sqrt{\frac{3}{2}} & 0 & 0 & -\frac{3}{\sqrt{6}} \end{pmatrix}.$$

All Lorentz tetrads $\mathbf{GL}(4,\mathbb{R})/\mathbf{O}(1,3)$ as orthogonal classes in the structural group of Minkowski spacetime as tangent spacetime constitute a real 10-dimensional manifold. The symmetric $\mathbf{O}(1,3)$-metric $g(b^j, b^k)$ as metrical tensor for general relativity is determined by $\binom{4+1}{2} = 10$ parameters: maximally six compact parameters from the orthogonal group $\mathbf{O}(4)$ are necessary for a diagonalization, the remaining four noncompact parameters in the diagonal describe the gravity characteristic dilations. The metrical hyperboloid, determined by ten real parameters — six for its orientation and four dilations for its main axes, contains the metrical ellipsoid for position, determined by six real parameters — three for its orientation and three dilations for its main axes. The structure of the metrical position ellipsoid is familiar from the inertia ellipsoid with its three main axes.

The Lorentz tetrads, as automorphisms, allow the transformation of all Lorentz structures in a basis $\{b^i\}_{i=0}^3$ into the corresponding structures in another basis $\{s^i\}_{i=0}^3$, e.g., of the Lorentz group $\Lambda(b) \in \mathbf{O}(1,3)(b) \subset \mathbf{GL}(4,\mathbb{R})$, of the Lorentz bilinear form $g(b)$, etc., as shown in the following commutative diagram:

$$\begin{array}{ccc} \mathbb{R}^{(1,3)}(s) & \xrightarrow{\Lambda(s)} & \mathbb{R}^{(1,3)}(s) \\ {\scriptstyle e(b,s)}\downarrow & & \downarrow{\scriptstyle e(b,s)} \\ \mathbb{R}^{(1,3)}(b) & \xrightarrow[\Lambda(b)]{} & \mathbb{R}^{(1,3)}(b) \end{array}, \text{ with } \left\{ \begin{array}{l} e(b,s) = b^i \otimes \check{s}_i \in \mathbf{GL}(4,\mathbb{R}), \\ \Lambda(s) \in \mathbf{O}(1,3)(s), \\ \Lambda(b) \in \mathbf{O}(1,3)(b), \end{array} \right.$$

$$\begin{aligned} \mathbf{O}(1,3)(b) &= e(b,s) \circ \mathbf{O}(1,3)(s) \circ e^{-1}(b,s), \\ g(b) &= e(b,s) \circ g(s) \circ e^T(b,s). \end{aligned}$$

$\mathbb{R}^{(1,3)}(b)$ denotes the Minkowski translations with the basis $\{b^i\}_{i=0}^3$. If the bases are from the same class, then $e(b,s) \in \mathbf{O}(1,3)(s)$ is a Lorentz transformation.

All bases of the Minkowski translations are connected by $\mathbf{GL}(4,\mathbb{R})$-automorphisms $e(c,b)$. If they are all related to a fixed basis as representative of a Lorentz group basis class, e.g., to a Sylvester basis $\{s^i\}_{i=0}^3$ as necessary for the definition of particles, one uses often different index types, e.g., Latin indices for the fixed basis and Greek indices for general bases $\{b^\mu\}_{\mu=0}^3$. The tensors $b^\mu \otimes \check{s}_i$ for the Lorentz tetrads $e(b,s) = \delta^i_\mu b^\mu \otimes \check{s}_i$ as representatives of $\mathbf{GL}(4,\mathbb{R})/\mathbf{O}(1,3)$ have a hybrid transformation behavior — from left with the full general linear group $\mathbf{GL}(4,\mathbb{R})$, and from right with a subgroup $\mathbf{O}(1,3)(s)$:

$$\begin{array}{ccc} \mathbb{R}^{(1,3)}(s) & \xrightarrow{\Lambda(s)} & \mathbb{R}^{(1,3)}(s) \\ {\scriptstyle b^\mu\otimes\check{s}_i}\downarrow & & \downarrow{\scriptstyle c^\mu\otimes\check{s}_i(\Lambda)} \\ \mathbb{R}^{(1,3)}(b) & \xrightarrow[e(c,b)]{} & \mathbb{R}^{(1,3)}(c) \end{array}, \text{ with } \left\{ \begin{aligned} c^\mu &= e^\mu_\nu(c,b)\, b^\nu, \\ \check{s}_i(\Lambda) &= \Lambda^{-1j}{}_i(s)\check{s}_j, \end{aligned} \right.$$

$$\Lambda(s) \in \mathbf{O}(1,3)(s), \quad e(c,b) = c^\mu \otimes \check{b}_\mu \in \mathbf{GL}(4,\mathbb{R}).$$

7.19 External Transitions in the Electroweak Standard Model

The standard model of the electroweak and strong interactions is formulated as a dynamics of special relativistic *quantum fields*, characterized by their representation properties with respect to the external (spacetimelike) Poincaré transformations and the internal (chargelike) unitary operations. The *particle properties* of the fields require a rest system (laboratory) and a "projection" of the fields on a ground state with lowest energy.

The transition from the Poincaré symmetry-characterized field dynamics to the particles needs the *field-particle telescope* with the homogeneous and inhomogeneous fixgroups:

$$\begin{array}{ccccc} \mathbb{R}^{(1,3)} & \cong & \mathbb{T}\oplus\mathbb{E}^3 & \cong & \mathbb{T}\oplus\mathbb{E}^2\oplus\mathbb{E}^1, \\ \mathbf{SL}(2,\mathbb{C}) & \supset & \mathbf{SU}(2) & \supset & \mathbf{SO}(2), \\ \mathbf{SL}(2,\mathbb{C})\ \vec{\times}\ \mathbb{R}^4 & \supset & \mathbf{SU}(2)\times\mathbb{R} & \supset & \mathbf{SO}(2)\times\mathbb{R}. \end{array}$$

The first transition leads from the semidirect product of the noncompact Lorentz group and the Minkowski translations $\mathbb{R}^4$ to the direct product of a compact spin group $\mathbf{SU}(2)$ or rotation group $\mathbf{SO}(3)$, and time translations $\mathbb{R}$. For the definition of the associated Sylvester decomposition a time axis has to be distinguished. From a rest system, defined up to $\mathbf{SO}(3)$-position rotations, a second transition leads to a circularity (polarization) group $\mathbf{SO}(2)$. In a given Sylvester decomposition, the distinction of a position direction e.g., the thumb of the right or left hand or the momentum of a massless particle, allows the definition of the associated Witt decomposition. Rest systems with a fixed position direction are defined up to $\mathbf{SO}(2)$-axial rotations and called axial systems.

For particles, the time translation group $\mathbb{T} \cong \mathbb{R}$ is represented by the phase $e^{i\omega t} \in \mathbf{U}(1)$ with the frequency $\omega = \frac{cq_0}{\hbar}$ related via the energy $cq_0 = c\sqrt{\vec{q}^2 + m^2c^2}$ to the particle mass m. For a massive particle $m^2 > 0$, e.g., for the weak bosons or the electrons, the representation of the spin group $\mathbf{SU}(2)$ is compatible with a Sylvester decomposition of the Minkowski translations. A massless particle $m^2 = 0$ with $q_0 = |\vec{q}| > 0$, e.g., the photons or a massless neutrino, does not define a time axis (rest system); one needs a Witt decomposition of the Minkowski translations $\mathbb{R}^4$ with fixed light translations $\mathbb{L}^1 \oplus \mathbb{L}^1 \cong \mathbb{T} \oplus \mathbb{E}^1 \cong \mathbb{R}^2$. Its stability group $\mathbf{SO}(2)$ leaves invariant time translations $\mathbb{T}$ and a position direction $\mathbb{E}^1$. The representation of the axial rotations $\mathbf{SO}(2)$ for a massless particle involves an integer as the winding number of the circularity (polarization, helicity).

The *neutral massive and massless vector fields* in the electroweak standard model are particle-analyzed with a Sylvester and Witt transition, respectively. Vector fields carry a Lorentz invariant bilinear form as metric in their complex four-dimensional representation space $V \cong \mathbb{C}^4$:

$$-\eta \cong \begin{pmatrix} -1 & 0 \\ 0 & \mathbf{1}_3 \end{pmatrix} \text{ with } (n_+, n_-) = (3, 1), \quad (I, C, I) = (1, 2, 1), \quad \epsilon = +1.$$

For the massive neutral vector field, the three Z-bosons with spin 1 for the group $\mathbf{O}(3)$ display the particle content; the positive definiteness and the embedding of the orthogonal group in a unitary group $\mathbf{O}(3) \subset \mathbf{U}(3)$ determine the Hilbert metric of the particles in the complex representation space $V \cong \mathbb{C}^4$. Such an orthogonal-unitary embedding is characteristic for a quantum framework; it is known from the isotropic harmonic oscillator in s position dimensions with the rotations-encompassing unitary symmetry group $\mathbf{U}(s) \supset \mathbf{O}(s)$. In the case of relativistic vector fields, the unitary group $\mathbf{U}(3)$ does not describe a symmetry of the dynamics, it is the conjugation group for the unitarily implemented time translations. The mass of the Z-particle can be used as Sylvester unit for the spin group $\mathbf{O}(3)$ in the Lorentz group $\mathbf{O}(1, 3)$:

$$\mathbf{O}(1, 3) \supset \mathbf{O}(1) \times \mathbf{O}(3), \quad \zeta \cong m_Z^2 \begin{pmatrix} -e^{-2\psi} & 0 \\ 0 & \mathbf{1}_3 \end{pmatrix}, \quad m_Z \simeq 91.2 \frac{\text{GeV}}{c^2}.$$

The signature factor $e^{-2\psi}$ for the negatively normalized direction does not play a role — the corresponding degree of freedom is projected out for a spin-1 particle which needs only the three degrees of freedom in $\mathbf{1}_3$.

For a massless neutral vector field, e.g. the electromagnetic field, the two polarized photons, acted on by the axial rotation group $\mathbf{O}(2)$, give the particle content. Their Hilbert space-structure is determined by the definite unitary group $\mathbf{U}(2) \supset \mathbf{O}(2)$. As related Witt unit for the polarization group $\mathbf{O}(2)$ in the Lorentz group $\mathbf{O}(1, 3)$ the electromagnetic mass from the standard model electroweak mass orthogonal triangle can be taken — it is not a particle mass:

$$\mathbf{O}(1,3) \supset \mathbf{O}(1,1) \times \mathbf{O}(2), \quad \zeta \cong m_e^2 \begin{pmatrix} 0 & 0 & \nu \\ 0 & 1_2 & 0 \\ \nu & 0 & 0 \end{pmatrix}, \quad m_e \simeq 38.2 \frac{\mathrm{GeV}}{c^2}.$$

The nondiagonal contributions for the indefinite subgroup $\mathbf{O}(1,1) \subset \mathbf{U}(1,1)$ are connected with the Coulomb interaction and the gauge degree of freedom. In this connection, the factor $\nu \in \mathbb{R}$ is called *gauge fixing factor*. It can be chosen arbitrarily nontrivial $\nu \neq 0$.

Associating in such a way the external groups with the Euclid, Sylvester, and Witt mass units, used for the Higgs field, the massive and the massless vector fields, respectively, their ratios yield the coupling constants of the neutral vector fields and the Weinberg angle:

$$\begin{aligned} g_Z^2 &= \frac{m_Z^2}{M_{\mathrm{Fermi}}^2} \sim \frac{1}{2}, \\ g_e^2 &= \frac{m_e^2}{M_{\mathrm{Fermi}}^2} \sim \frac{1}{10} \sim \frac{4\pi}{137}, \end{aligned} \qquad \sin^2\theta_{\mathrm{Wein}} \cos^2\theta_{\mathrm{Wein}} = \frac{m_e^2}{m_Z^2} \sim 0.2.$$

7.20 Fields with Group Actions

The action of a group G on two sets M, S, e.g., G represented on two vector spaces $V_{1,2}$, is used to define the group action on the connecting mappings $S^M = \{f : M \longrightarrow S\}$ between these two sets, e.g., on the linear mappings $\{f : V_1 \longrightarrow V_2\}$, by the commutativity of the following diagram:

$$\begin{array}{ccc} M & \xrightarrow{g\bullet} & M \\ f \downarrow & & \downarrow {}_g f \\ S & \xrightarrow[g\bullet]{} & S \end{array}, \quad {}_g f(x) = g \bullet f(g^{-1} \bullet x)$$

$$\Rightarrow \begin{cases} {}_e f &= f \text{ for neutral element } e \in G, \\ {}_{gh} f &= {}_g({}_h f) \text{ for } g, h \in G. \end{cases}$$

The group action is denoted, also for different sets, by the general symbol $\bullet$ for "acts on". The G-action on the mapping $f \in S^M$ gives another mapping ${}_g f$:

$$G \times S^M \longrightarrow S^M, \quad (g, f) \longmapsto {}_g f.$$

If a mapping $f \in S^M$ is invariant under G-action, f is compatible with the group action and is called a G-*intertwiner*:

$${}_g f = f \iff f(g \bullet x) = g \bullet f(x).$$

Such G-invariant mappings are the morphisms for G-sets. The whole graph of f, not the individual mapping values must be G-invariant, in general $g \bullet f(x) \neq f(x)$.

Since a group acts on itself, also the orbit mappings $\varphi \in S^G$ with starting point $\varphi(1_G) \in S$ have a G-action:

$$\varphi : G \longrightarrow S, \ \ \varphi(g) = g \bullet \varphi(1_G).$$

These orbit mappings are invariants, i.e., group intertwiners, compatible with the left multiplication on the group:

$$\begin{array}{ccc} G & \xrightarrow{g\bullet} & G \\ \varphi \downarrow & & \downarrow {}_g\varphi = \varphi, \\ S & \xrightarrow[g\bullet]{} & S \end{array} \qquad {}_g\varphi(h) = g \bullet \varphi(g^{-1}h) = \varphi(h).$$

If, for instance, the additive time translation group $\mathbb{R}$ acts, in a $\mathbf{U}(1)$-representation, on a complex one-dimensional space $V \cong \mathbb{C}$, the time orbits $\varphi \in V^{\mathbb{T}}$ are $t \longmapsto \varphi(t) = e^{i\omega t}\varphi(0)$ with a frequency ω. The $\mathbf{U}(1)$-orbits $\varphi \in V^{\mathbb{R}^4}$ of the Minkowski translations $\mathbb{R}^4$ look analogously $\varphi(x) = e^{iqx}\varphi(0)$ with an energy-momentum $q \in \check{\mathbb{R}}^4$.

If the values of a mapping $f : M \longrightarrow V$ with G-action are in a finite-dimensional vector space $V \cong \mathbb{K}^n$, the vector space with all mappings $f \in V^M$ contains *the fields of M with values in V and the action of the group G*. V^M can be spanned by *basis fields* $\{f^k\}_{k=1}^n$ which, at each point $x \in M$, constitute a V-basis, $\{f^k(x)\}_{k=1}^n \in \mathcal{B}(V)$.

Good to know: Vector space mappings inherit pointwise the vector space structure $(f+h)(x) = f(x) + h(x)$ and $(\alpha f)(x) = \alpha f(x)$. The $\mathbb{K}$-dimension of all mappings V^M for $V \cong \mathbb{K}^n$ is the product $n \cdot \text{Card}M$ with the cardinality of M.

Special relativistic fields $\boldsymbol{\varphi} \in V^{\mathbb{R}^4}$ associate with each spacetime translation $x \in \mathbb{R}^4$ a field value in a vector space V with a representation of the Lorentz group $\mathbf{SL}(2, \mathbb{C})$, i.e., the fields $\boldsymbol{\varphi} \in V^{\mathbb{R}^4}$ carry the induced action of the Lorentz group:

$$\begin{array}{ccc} \mathbb{R}^4 & \xrightarrow{\Lambda} & \mathbb{R}^4 \\ \varphi \downarrow & & \downarrow {}_s\varphi, \\ V & \xrightarrow[D(s)]{} & V \end{array} \qquad \begin{array}{l} \mathbf{SL}(2, \mathbb{C}) \ni s \longmapsto \begin{cases} \Lambda \in \mathbf{SO}_0(1,3), \\ D(s) \in \mathbf{SL}(V), \end{cases} \\ {}_s\boldsymbol{\varphi}(x) = s \bullet \boldsymbol{\varphi}(s^{-1} \bullet x) = D(s)(\boldsymbol{\varphi}(\Lambda^{-1}.x)). \end{array}$$

$\Lambda = \Lambda(s)$ is the real four-dimensional representation of the Lorentz symmetry as orthochronous group $\mathbf{SO}_0(1,3) \cong \mathbf{SL}(2, \mathbb{C})/\mathbf{I}(2)$, acting on the spacetime translations, $D(s)$ is the representation on the value-vector space V. In general, special relativistic fields are not Lorentz group intertwiners.

An example is a scalar field with $V \cong \mathbb{C}$, e.g., the standard model Higgs field $\boldsymbol{\Phi}$, with a nontrivial Lorentz transformation behavior only via the translations:

$$\mathbf{\Phi} : \mathbb{R}^4 \longrightarrow V, \ {}_s\mathbf{\Phi}(x) = \mathbf{\Phi}(\Lambda^{-1}.x).$$

A vector field with $V \cong \mathbb{C}^4$, e.g., the electromagnetic field $\mathbf{A}$, shows, via its four basic fields $\{\mathbf{A}^j\}_{j=0}^3$, the Lorentz transformation behavior:

$$\mathbf{A} : \mathbb{R}^4 \longrightarrow V, \ {}_s\mathbf{A}^j(x) = \Lambda^j_k \mathbf{A}^k(\Lambda^{-1}.x), \ j, k = 0, 1, 2, 3.$$

Another example is a left-handed Weyl-spinor field with value space $V \cong \mathbb{C}^2$ and two basic fields $\{\mathbf{l}^A\}_{A=1}^2$, e.g., a left-handed neutrino field:

$$\mathbf{l} : \mathbb{R}^4 \longrightarrow V, \ {}_s\mathbf{l}^A(x) = s^A_B \mathbf{l}^B(\Lambda^{-1}.x), \ A, B = 1, 2.$$

As mappings between sets with a Lorentz group action, special relativistic fields are, in general, not linear in the translations, i.e., $\varphi(x + y) \neq \varphi(x) + \varphi(y)$. They contain orbits $x \longmapsto e^{iqx}\mathrm{u}(0)$ of the translations $\mathbb{R}^4$ in the value space. Their orbit structure with respect to the translations is investigated by their harmonic analysis, or rest system analysis, that displays the particle content.

7.21 Symmetry Transitions by Fields

Embedding manifolds G/H for a subgroup like the 2-sphere $\Omega^2 \cong \mathbf{O}(3)/\mathbf{O}(2)$, the Sylvester boost hyperboloid $\mathcal{Y}^3 \cong \mathbf{SO}_0(1, 3)/\mathbf{SO}(3)$, and the Goldstone manifold $\mathcal{G}^3 \cong \mathbf{U}(2)/\mathbf{U}(1)_+$ are related to symmetry transitions $H \subseteq G$. In general, they are not vector spaces. Especially for the quantum structure, transitions between Lie groups have to be represented by linear transformations of vector spaces.

G/H, the set of the H-right orbits in G, feels the action of the product group $G \times H$: The embedding manifold is acted from left by the full group G with fixgroup H. It has a trivial H-action from right:

$$\begin{aligned} k \in G, \ L_k : G/H &\longrightarrow G/H, \ gH \cong x \longmapsto k \bullet x \cong kgH, \\ h \in H, \ R_h : G/H &\longrightarrow G/H, \ gH \cong x \longmapsto x \bullet h \cong gHh = gH. \end{aligned}$$

$gH \cong x \cong (x^k)_{k=1}^{d_G - d_H}$ may be a parametrization of the differentiable manifold G/H by real coordinates for $d_{G,H}$ the real dimensions of the Lie groups G, H, e.g., by two spherical coordinates (φ, θ) for $\mathbf{O}(3)/\mathbf{O}(2)$.

Transmutators are defined as fields of a coset manifold with values in a vector space of linear transformations: Each orbit gH of the "little" group H in the "large" group G is associated with $n \cdot m$ linear mappings $\{\mathbf{G}^\mu_i\}_{i=1,\dots,n}^{\mu=1,\dots,m}$ (no numbers!) from an n-dimensional vector space V with H-representation to an m-dimensional vector space W with G-representation:

$$\begin{aligned}
&\mathbf{G} : G/H \longrightarrow W \otimes V^T, \;\; gH \cong x \longmapsto \mathbf{G}(x) = \{\mathbf{G}_i^\mu(x)\}_{i=1,\dots,n}^{\mu=1,\dots,m}, \\
&\text{with } \mathbf{G}_i^\mu(x) : V \longrightarrow W, \;\; \mathbf{G}_i^\mu(x) \in W \otimes V^T.
\end{aligned}$$

For each orbit, the transmutators constitute a basis of the linear mappings $W \otimes V^T$, i.e., $\{\mathbf{G}_i^\mu(x)\}_{i=1,\dots,n}^{\mu=1,\dots,n} \in \mathcal{B}(W \otimes V^T)$.

Good to know: The linear mappings $\{f : V \longrightarrow W\}$ for finite-dimensional vector spaces are naturally isomorphic to the tensor product $W \otimes V^T$ with the dual space V^T — familiar from the $(n \times m)$-matrix form of linear mappings $f : \mathbb{K}^n \longrightarrow \mathbb{K}^m$.

The groups G and H act on the vector spaces W and V, respectively, by linear representations:

$$\begin{aligned}
k \in G, \;\; D(k) : W &\longrightarrow W, \; w \longmapsto D(k)(w), \\
h \in H, \;\; D(h) : V &\longrightarrow V, \;\; v \longmapsto D(h)(v).
\end{aligned}$$

If $\{b^\mu\}_{\mu=1}^m$ and $\{c^i\}_{i=1}^n$ are bases of W and V, respectively, then $\{b^\mu \otimes \check{c}_i\}_{i=1,\dots n}^{\mu=1,\dots,m}$ is a basis of the linear mappings $W \otimes V^T$. The transmutators are transformed like bases:

$$\mathcal{B}(W \otimes V^T) \longrightarrow \mathcal{B}(W \otimes V^T), \;\; \begin{cases} b^\mu \otimes \check{c}_i \longmapsto D(k)^\mu_\nu \, b^\nu \otimes \check{c}_j \, D(h^{-1})^j_i, \\ \mathbf{G}_i^\mu(x) \longmapsto D(k)^\mu_\nu \, \mathbf{G}(x)^\nu_j \, D(h^{-1})^j_i. \end{cases}$$

The field structure of a transmutator $(W \otimes V^T)^{G/H}$ on the manifold G/H with values in $W \otimes V^T$ and $G \times H$-action is given in the following diagram:

$$\begin{array}{ccc}
G/H & \stackrel{k\bullet}{\longrightarrow} & G/H \\
\mathbf{G}\downarrow & & \downarrow {}_k\mathbf{G}_h, \\
W \otimes V^T & \underset{D(k)\otimes D(h^{-1})}{\longrightarrow} & W \otimes V^T
\end{array} \quad k \in G, \; h \in H,$$

$${}_k\mathbf{G}_h(k \bullet x) = D(k) \circ \mathbf{G}(x) \circ D(h^{-1}) \iff \begin{cases} {}_k\mathbf{G}(x) &= D(k) \circ \mathbf{G}(k^{-1} \bullet x), \\ \mathbf{G}_h(x) &= \mathbf{G}(x) \circ D(h^{-1}). \end{cases}$$

The subgroup H acts nontrivially only on the value space $W \otimes V^T$ — in contrast to its trivial action on the manifold G/H with the subgroup classes.

The structure of transmutators is rather complicated and needs some familiarization by examples.

7.22 Decomposition-Transmutators

A group G-representation on a vector space $W \cong \mathbb{K}^n$ may be decomposable $W = V_1 \oplus V_2 \cong \mathbb{K}^n$ with respect to a stability subgroup $H \subset G$,

$$G \longrightarrow \mathbf{GL}(W), \;\; H \longrightarrow \mathbf{GL}(V_1 \oplus V_2).$$

Decomposition-transmutators associate each orbit $x \cong gH \in G/H$ with an isomorphism between W and its decomposition $V_1 \oplus V_2$:

$$\mathbf{G} : (G/H)_{\text{rep}} \longrightarrow W \otimes (V_1 \oplus V_2)^T, \;\; x \longmapsto \{\mathbf{G}_i^\mu(x)\}_{i=1,\dots,n}^{\mu=1,\dots,n}.$$

Euclidean, Sylvester, and Witt transitions have corresponding *decomposition transmutators* $\mathbf{G}(x)$, which linearize the group transitions by x-dependent $(n \times n)$-matrices.

With a constant start-basis $\{c^i\}_{i=1}^{n_1+n_2}$ for the decomposition $V_1 \oplus V_2$ and an x-dependent W-basis $\{b^\mu(x)\}_{\mu=1}^n$, decomposition transmutators are given by tensors $b^\mu(x) \otimes \check{c}_i$, where the W-basis is the G-transformed start-basis:

$$\begin{aligned} \mathbf{G}_i^\mu(x) &= b^\mu(x) \otimes \check{c}_i \Rightarrow b^\mu(x) = c^i \operatorname{tr} \mathbf{G}_i^\mu(x), \\ \text{start-basis: } b^\mu(x_0) &= c^i \delta_i^\mu \Rightarrow \operatorname{tr} \mathbf{G}_i^\mu(x_0) = \delta_i^\mu. \end{aligned}$$

The transmutators constitute, at each manifold point x, the isomorphism $\delta^i_\mu b^\mu(x) \otimes \check{c}_i$ for the vector space decomposition $W \cong V_1 \oplus V_2$. The traces $\operatorname{tr} \mathbf{G}_i^\mu(x)$ are the elements of the $(n \times n)$-transmutator matrix for the G-transformation in the start-basis, i.e., the start-basis-components of the W-basis give the matrix rows.

As an explicit example, the Euclidean decomposition $\mathbb{E}^2 \oplus \mathbb{E}^1$ of the three-dimensional space $\mathbb{E}^3$ is considered with the associated real two-parametric embedding manifold $\mathbf{O}(3)/\mathbf{O}(2)$ of the axial rotations as stability group in the rotation group, parametrizable by a 2-sphere with radius M:

$$q = \frac{\vec{q}}{M} \in \Omega^2 \cong \mathbf{O}(3)/\mathbf{O}(2) \text{ with } \vec{q}^{\,2} = M^2 > 0.$$

The transmutators map the 2-sphere to decomposition isomorphisms of $\mathbb{E}^3$:

$$\begin{aligned} \mathbf{O} : \Omega^2 &\longrightarrow \mathbb{E}^3 \otimes (\mathbb{E}^2 \oplus \mathbb{E}^1)^T, \\ q &\longmapsto \mathbf{O}(q)_i^\mu = \vec{b}^\mu(q) \otimes \vec{\check{c}}_i, \;\; \mu = 1,2,3, \;\; i = 1,2,3. \end{aligned}$$

They are fields of the 2-sphere with values in the isomorphisms of the Euclidean space $\mathbb{E}^3$. To obtain their matrix representations, a vector on the sphere with radius M is fixed as basis $\vec{c}^{\,3} \cong \begin{pmatrix} 0 \\ 0 \\ M \end{pmatrix}$ of the one-dimensional subspace $\mathbb{E}^1 = \mathbb{R}\vec{c}^{\,3}$. Two orthogonal basic vectors $\{\vec{c}^{\,1}, \vec{c}^{\,2}\}$ with equal length for the complement $\mathbb{E}^2$ are on the same 2-sphere; they are determined up to axial rotations $\mathbf{O}(2)$ around $\mathbb{E}^1$. Each vector on the sphere $\vec{q}$ can be obtained from the fixed vector $\vec{c}^{\,3}$ by an $\mathbf{O}(3)$-rotation:

$$\vec{q} = \mathbf{O}(q)\vec{c}^{\,3}, \quad \begin{pmatrix} q_1 \\ q_2 \\ q_3 \end{pmatrix} = \begin{pmatrix} 1 - \frac{q_1^2}{M(M+q_3)} & -\frac{q_1 q_2}{M(M+q_3)} & \frac{q_1}{M} \\ -\frac{q_1 q_2}{M(M+q_3)} & 1 - \frac{q_2^2}{M(M+q_3)} & \frac{q_2}{M} \\ -\frac{q_1}{M} & -\frac{q_2}{M} & \frac{q_3}{M} \end{pmatrix} \begin{pmatrix} 0 \\ 0 \\ M \end{pmatrix}.$$

The $\mathbf{O}(3)$-rotation $\mathbf{O}(q)$ is determined up to the $\vec{c}^{\,3}$-fixgroup $\mathbf{O}(2)$; the normalization $\mathbf{O}(0, 0, M) = \mathbf{1}_3$ is chosen. The (3×3)-matrix is the transmutator matrix for $\mathbf{O}(3)/\mathbf{O}(2)$. The three rows display general basic vectors in the components with respect to the start-basis $\{c^i\}_{i=1}^{2+1}$.

7.23 Field-Particle Transmutators in the Standard Model

The standard model of electroweak interactions is analyzed with decomposition-transmutators for Euclidean, Sylvester, and Witt transitions from field-relevant operations to particle-relevant ones.

The transition from the hyperisospin symmetry $\mathbf{U}(2)$ of the interactions to the ground state determined electromagnetic $\mathbf{U}(1)$-symmetry of the particles defines the real three-parametric compact Goldstone manifold with the two complex components of the Higgs field $\boldsymbol{\Phi} = h$:

$$\begin{aligned} h &\cong \begin{pmatrix} h_1 \\ h_2 \end{pmatrix} \in \mathbb{H} \cong \mathbb{C}^2, \text{ with } \langle h|h\rangle = \| h \|^2 = |h_1|^2 + |h_2|^2 = M^2 > 0, \\ \frac{h_{1,2}}{\| h \|} &\in \mathcal{G}^3 \cong \mathbf{U}(2)/\mathbf{U}(1)_+. \end{aligned}$$

This leaves three independent real parameters in the two complex components.

The transmutator describes the decomposition of the Higgs Hilbert space $\mathbb{H}$ with a $\mathbf{U}(2)$-representation into two one-dimensional subspaces $V_{1,2} \cong \mathbb{C}$ with $\mathbf{U}(1)_+$-representations:

$$\begin{aligned} \mathbf{U} : \mathcal{G}^3 &\longrightarrow \mathbb{H} \otimes (V_1 \oplus V_2)^T, \\ h &\longmapsto \mathbf{U}_i^\alpha(h) = b^\alpha(h) \otimes \check{c}_i, \quad \alpha = 1, 2, \quad i = 1, 2. \end{aligned}$$

One vector of a start-basis $\{c^i\}_{i=1,2}$ of $V_1 \oplus V_2$ is fixed with the Fermi mass as Euclid unit M:

$$\langle c^2|c^2\rangle = M^2, \quad c^2 = \begin{pmatrix} 0 \\ M \end{pmatrix}.$$

The orthogonal complement of $V_2 = \mathbb{C}c^2$ is spanned by an equally normalized second basic vector $c^1 = \binom{M}{0}$. The decomposition-transmutator matrix contains, as columns, the start-basis components of orthogonal $\mathbb{H}$-bases:

$$\mathbf{U}_i^\alpha(h) \cong (b^1(h), b^2(h)) \cong \frac{1}{M}\begin{pmatrix} \overline{h}_2 & h_1 \\ -\overline{h}_1 & h_2 \end{pmatrix}.$$

For the field-particle telescope of the massive Z-fields, assumed as stable, the momentum hyperboloid as Sylvester manifold,

$$q \cong \frac{q_j}{m_Z} \in \mathcal{Y}^3 \cong \mathbf{SO}_0(1,3)/\mathbf{SO}(3) \text{ with } q_0^2 - \vec{q}^{\,2} = m_Z^2 > 0,$$

is represented by transmutators for a Sylvester decomposition of the Minkowski translations and energy-momenta:

$$\begin{aligned} \mathbf{\Lambda} : \mathcal{Y}^3 &\longrightarrow \mathbb{R}^4 \otimes (\mathbb{T} \oplus \mathbb{E}^3)^T, \\ q &\longmapsto \mathbf{\Lambda}_{0,a}^j(q) = s^j(q) \otimes \check{b}_{0,a}, \quad j = 0,1,2,3, \quad a = 1,2,3. \end{aligned}$$

The transmutators for the Sylvester transition $\mathbf{SO}_0(1,3) \supset \mathbf{SO}(3)$ are parametrizable by the three real momenta $\frac{\vec{q}}{m_Z}$ of an hyperbolic orbit $q^2 = m_Z^2$. A rest system fixes an energy axis $b^0 = \binom{m_Z}{\vec{0}}$ with an orthonormal basis $\{b^j\}^{j=0,1,2,3}$. The four rows of the transmutator matrix display the rest system-coordinates of the four basic vectors $\{s^j(q)\}_{j=0}^3$ for a Sylvester decomposition associated with $q^2 = m_Z^2$:

$$\mathbf{\Lambda}_{0,a}^j(q) \cong \frac{1}{m_Z}\begin{pmatrix} q_0 & \vec{q} \\ \vec{q} & \delta_{bc} m_Z + \frac{q_b q_c}{q_0 + m_Z} \end{pmatrix}.$$

The Witt transition $\mathbf{SO}_0(1,3) \supset \mathbf{SO}(2)$, e.g., for the photon analysis of the massless electromagnetic field, can be obtained, after a Sylvester transition $\mathbf{SO}_0(1,3) \supset \mathbf{SO}(3)$ with a distinguished time direction by a Euclidean transition from the rotation group to the axial rotation group around a distinguished momentum direction:

$$q \cong \frac{\vec{q}}{q_0} \in \Omega^2 \cong \mathbf{SO}(3)/\mathbf{SO}(2), \text{ with } \vec{q}^{\,2} = q_0^2 > 0,$$

with the transmutator:

$$\begin{aligned} \mathbf{O} : \Omega^2 &\longrightarrow (\mathbb{T} \oplus \mathbb{E}^3) \otimes (\mathbb{T} \oplus \mathbb{E}^2 \oplus \mathbb{E}^1)^T, \\ q &\longmapsto \mathbf{O}_{0,\alpha,3}^{0,a}(q) = b^{0,a}(q) \otimes \check{c}_{0,\alpha,3}, \quad a = 1,2,3, \quad \alpha = 1,2. \end{aligned}$$

The manifold $\mathbf{SO}(3)/\mathbf{SO}(2)$ and the transmutator are parametrizable by two real numbers from $\frac{\vec{q}}{q_0}$ for a lightlike nontrivial orbit $q^2 = q_0^2 - \vec{q}^{\,2} = 0$, $q \neq 0$:

$$\mathbf{O}_{0,\alpha,3}^{0,a}(q) \cong \frac{1}{q_0}\left(\begin{array}{c|cc|c} q_0 & 0 & 0 & 0 \\ \hline 0 & q_0 - \frac{q_1^2}{q_0+q_3} & -\frac{q_1 q_2}{q_0+q_3} & q_1 \\ 0 & -\frac{q_1 q_2}{q_0+q_3} & q_0 - \frac{q_2^2}{q_0+q_3} & q_2 \\ 0 & -q_1 & -q_2 & q_3 \end{array}\right).$$

The decomposition transmutator of the Witt transition is the product of Sylvester and Euclid transmutators:

$$\begin{aligned} (\mathbf{SO}_0(1,3)/\mathbf{SO}(2))_{\text{rep}} &\longrightarrow \mathbb{R}^4 \otimes (\mathbb{T} \oplus \mathbb{E}^2 \oplus \mathbb{E}^1)^T \\ (p,q) &\longmapsto \mathbf{L}_{0,\alpha,3}^j(p,q) = \mathbf{\Lambda}_{0,a}^j(p) \circ \mathbf{O}_{0,\alpha,3}^{0,a}(q). \end{aligned}$$

$p, q \neq 0$ with $p^2 = M_s^2 > 0$ and $q^2 = 0,\ q \neq 0$ give five real parameters for the Witt manifold $\mathbf{SO}_0(1,3)/\mathbf{SO}(2)$.

7.24 The Four Symmetry Regimes of the Standard Model

The symmetry operations that are important in the standard model of the electroweak and strong interactions, can be grouped into two times two, i.e., into four sectors: There are the homogeneous symmetry operations for interaction fields and for particles, which both can be separated into external (spacetimelike) and internal (chargelike) operations:

	External	Internal
Interactions	$\mathbf{SL}(2,\mathbb{C})$	$\mathbf{U}(2\times 3) \cong \frac{\mathbf{U}(1)\times\mathbf{SU}(2)\times\mathbf{SU}(3)}{\mathbf{I}(2)\times\mathbf{I}(3)}$
	$\cup$	$\cup$
Particles	$\mathbf{SU}(2) \supset \mathbf{SO}(2)$	$\mathbf{U}(1)$

The external homogeneous interaction symmetries of a special relativistic dynamics is the Lorentz group, which leaves, in a rest system, spin $\mathbf{SU}(2) \subset \mathbf{SL}(2,\mathbb{C})$ as massive particle symmetry, and, in an axial system, polarization $\mathbf{SO}(2) \subset \mathbf{SL}(2,\mathbb{C})$ for the massless case. The internal hyperisospin $\mathbf{U}(2)$- and color $\mathbf{SU}(3)$-symmetry in the centrally correlated product $\mathbf{U}(2\times 3)$ for the interactions are slimmed down to an electromagnetic $\mathbf{U}(1)$-symmetry for the particles, via the hyperisospin bleaching $\mathbf{U}(2) \to \mathbf{U}(1)_+$ in an electromagnetic system from the degenerated ground state, and the suspected color confinement $\mathbf{SU}(3) \to \{1\}$.

The transitions from the interaction symmetry to the particle symmetry is displayed by the harmonic analysis: The analysis with respect to representations of spacetime translations $\mathbb{R}^4$ employs, for massive particles with a rest system, Sylvester transitions $\mathbf{SL}(2,\mathbb{C})/\mathbf{SU}(2) \cong \mathbf{SO}_0(1,3)/\mathbf{SO}(3)$ and, for massless particles with an axial system, Witt transitions $\mathbf{SL}(2,\mathbb{C})/\mathbf{SO}(2) \cong \mathbf{SO}_0(1,3)/\mathbf{SO}(2)$, characterized by two masses, e.g., the particle Z-mass m_Z and the electromagnetic mass m_e. The analysis with respect to representations of the electromagnetic group $\mathbf{U}(1)$ employs the Goldstone transition $\mathbf{U}(2)/\mathbf{U}(1)_+$, characterized by the Fermi mass M_{Fermi}.

Essential structures of the standard model of the electroweak interactions are interpretable with fields of the embedding manifold $\mathbb{D}(2) = \mathbf{GL}(2, \mathbb{C})/\mathbf{U}(2)$ for the unitary group $\mathbf{U}(2)$, a compact real four-dimensional Lie group, in the full complex linear group $\mathbf{GL}(2, \mathbb{C})$, a real eight-dimensional Lie group. The real four-dimensional noncompact manifold $\mathbb{D}(2)$ has as elements the $\mathbf{U}(2)$-classes of $\mathbf{GL}(2, \mathbb{C})$. With such an interpretation, the standard fields implement the transition in the *external-internal telescope*:

$$\text{(hyperisospin) } \mathbf{U}(2) \subset \mathbf{GL}(2, \mathbb{C}) \text{ (hyper-Lorentz group).}$$

The external *hyper-Lorentz group* (*extended Lorentz group*) $\mathbf{GL}(2, \mathbb{C})$ is the direct product of the totally ordered *causal group* $\mathbf{D}(1) = \exp \mathbb{R}$ and the *unimodular group* $\mathbf{UL}(2)$ which, itself, is the product of the *phase group* $\mathbf{U}(1)$ and the simply connected Lorentz group $\mathbf{SL}(2, \mathbb{C})$. The internal hyperisospin group is the product of the *hypercharge group* $\mathbf{U}(\mathbf{1}_2)$ and the *isospin group* $\mathbf{SU}(2)$:

$$\begin{aligned} \mathbf{GL}(2, \mathbb{C}) &= \mathbf{D}(\mathbf{1}_2) \times \mathbf{UL}(2) = \mathbf{D}(\mathbf{1}_2) \times \mathbf{U}(\mathbf{1}_2) \circ \mathbf{SL}(2, \mathbb{C}), \\ \mathbf{U}(2) &= \mathbf{U}(\mathbf{1}_2) \circ \mathbf{SU}(2). \end{aligned}$$

The adjoint groups, i.e., the centrum classes, are the really representable external Lorentz group $\mathbf{SO}_0(1, 3)$ and the internal isospin group $\mathbf{SO}(3)$:

$$\begin{aligned} \mathbf{GL}(2, \mathbb{C})/\mathbf{GL}(\mathbf{1}_2, \mathbb{C}) &\cong \mathbf{SL}(2, \mathbb{C})/\mathbf{I}(2) \cong \mathbf{SO}_0(1, 3), \\ \mathbf{U}(2)/\mathbf{U}(\mathbf{1}_2) &\cong \ \mathbf{SU}(2)/\mathbf{I}(2) \quad \cong \mathbf{SO}(3). \end{aligned}$$

The embedding manifold of the compact group is the manifold product of the real one-dimensional causal group and the real three-dimensional Sylvester or boost hyperboloid $\mathcal{Y}^3$:

$$\begin{aligned} \mathbb{D}(2) &= \mathbf{GL}(2, \mathbb{C})/\mathbf{U}(2) \cong \mathbf{D}(1) \times \mathcal{Y}^3, \\ \mathcal{Y}^3 &= \quad \mathbf{UL}(2)/\mathbf{U}(2) \quad \cong \mathbf{SL}(2, \mathbb{C})/\mathbf{SU}(2) \cong \mathbf{SO}_0(1, 3)/\mathbf{SO}(3). \end{aligned}$$

The tangent space at each point of the homogeneous manifold $\mathbb{D}(2)$ is a real four-dimensional vector space $\mathbb{R}(2)$ — isomorphic to the classes of the Lie algebra $\log \mathbf{U}(2) \cong \mathbb{R}^4$ in the Lie algebra $\log \mathbf{GL}(2, \mathbb{C}) \cong \mathbb{R}^8$. It is isomorphic to the Minkowski translations $\mathbb{R}(2) \cong \mathbb{R}^4$. The action of the external group $\mathbf{GL}(2, \mathbb{C})$ on the tangent space $\mathbb{R}(2)$ is faithful for the adjoint group, i.e., for the orthochronous Lorentz group $\mathbf{SO}_0(1, 3)$. In addition, the additive group $\mathbb{R}(2)$ acts on itself:

$$\mathbf{GL}(2, \mathbb{C}) \times \mathbb{D}(2) \longrightarrow \mathbb{D}(2) \Rightarrow \begin{cases} \mathbf{SO}_0(1, 3) \times \mathbb{R}(2) \longrightarrow \mathbb{R}(2), \\ \qquad \mathbb{R}(2) \times \mathbb{R}(2) \longrightarrow \mathbb{R}(2). \end{cases}$$

The hyperisospin group $\mathbf{U}(2)$ acts trivially on the manifold $\mathbb{D}(2)$.

The causal manifold $\mathbb{D}(2)$ is, as manifold, isomorphic to the strictly future translations of the Minkowski translations, i.e., to the open forward cone.

Those standard model fields, which have a trivial action of the color group $\mathbf{SU}(3)$, can be related to transmutators φ for the transition $\mathbf{U}(2) \subset \mathbf{GL}(2,\mathbb{C})$ from the internal group to the external group. They associate each hyperisospin orbit in the hyper-Lorentz group, dependent on four real parameters $g\mathbf{U}(2) \cong x \cong (x^\mu)_{\mu=0}^3 \in \mathbb{D}(2)$, with field values that are linear mappings $\{V \longrightarrow W\}$ between a vector space V with hyperisospin $\mathbf{U}(2)$-action and a vector space W with hyper-Lorentz $\mathbf{GL}(2,\mathbb{C})$-action:

$$\begin{array}{ccc} \mathbb{D}(2) & \stackrel{g\bullet}{\longrightarrow} & \mathbb{D}(2) \\ \varphi \downarrow & & \downarrow {}_g\varphi_u, \\ W \otimes V^T & \underset{D(g)\otimes d(u^{-1})}{\longrightarrow} & W \otimes V^T \end{array} \quad \begin{array}{l} \mathbb{D}(2) = \mathbf{GL}(2,\mathbb{C})/\mathbf{U}(2), \\ g \in \mathbf{GL}(2,\mathbb{C}), \quad u \in \mathbf{U}(2), \\ W \cong \mathbb{C}^n, \; D(g) \in \mathbf{GL}(W), \\ V \cong \mathbb{C}^n, \; d(u) \in \mathbf{GL}(V), \end{array}$$

$$x \longmapsto \varphi(x) \text{ with } {}_g\varphi_u(x) = D(g) \circ \varphi(g^{-1} \bullet x) \circ d(u^{-1}).$$

The classes $\mathbb{D}(2) = \mathbf{GL}(2,\mathbb{C})/\mathbf{U}(2)$, i.e., the real four-dimensional $\mathbf{U}(2)$-orbits, can (and will) be taken as a model of operational spacetime; they do not constitute a vector space. Therefore, the transmutators φ with causal support $\mathbb{D}(2)$ are not immediately connected with particle-analyzable fields of flat spacetime $\mathbb{R}(2)$.

The transmutator interpretation for the color trivial fields of the standard model is looked at in more detail: The left-handed lepton fields, e.g., with the electron and its neutrino, implement fundamentally the hyperisospin group:

$$\begin{array}{c} \text{Lorentz spinor} \\ \text{isodoublet:} \\ x \longmapsto \mathbf{l}^{A=1,2}_{\alpha=1,2}(x), \end{array} \quad \text{with} \quad \left\{ \begin{array}{l} \mathbf{l}^A_\alpha(x) \longmapsto s^A_B\, \mathbf{l}^B_\beta(s^{-1} \bullet x)\, (u^\star)^\beta_\alpha, \\ W \cong \mathbb{C}^2, \quad s \in \mathbf{SL}(2,\mathbb{C}), \\ V \cong \mathbb{C}^2, \quad u \in \mathbf{U}(2). \end{array} \right.$$

The right-handed lepton fields, e.g., with an electron, represent the internal group faithfully only for the hypercharge group $\mathbf{U}(1)$:

$$\begin{array}{c} \text{Lorentz spinor} \\ \text{isosinglet:} \\ x \longmapsto \mathbf{r}^{\dot{A}=1,2}(x), \end{array} \quad \text{with} \quad \left\{ \begin{array}{l} \mathbf{r}^{\dot{A}}(x) \longmapsto \hat{s}^{\dot{A}}_{\dot{B}}\, \mathbf{r}^{\dot{B}}(s^{-1} \bullet x)\, p^\star, \\ W \cong \mathbb{C}^2, \quad \hat{s} \in \mathbf{SL}(2,\mathbb{C}), \\ V \cong \mathbb{C}, \quad p \in \mathbf{U}(1) \subset \mathbf{U}(2). \end{array} \right.$$

The $\mathbf{SL}(2,\mathbb{C})$-transformations s and $\hat{s}$ for the left- and right-handed fields are related by inverse conjugation, $\hat{s} = s^{-1\star}$ (dual antirepresentations.)

The electroweak gauge fields with the photons and the weak bosons represent faithfully the adjoint groups:

Lorentz vector isosinglet: $x \longmapsto \mathbf{A}_0^{j=0,1,2,3}(x)$, with $\begin{cases} \mathbf{A}_0^j(x) \longmapsto \Lambda_k^j \, \mathbf{A}_0^k(s^{-1} \bullet x), \\ W \cong \mathbb{C}^4, \ \Lambda \in \mathbf{SL}(2,\mathbb{C})/\mathbf{I}(2) \cong \mathbf{SO}_0(1,3). \\ V \cong \mathbb{C}, \end{cases}$

Lorentz vector isotriplet: $x \longmapsto \mathbf{A}_{a=1,2,3}^{j=0,1,2,3}(x)$, with $\begin{cases} \mathbf{A}_a^j(x) \longmapsto \Lambda_k^j \, \mathbf{A}_b^k(s^{-1} \bullet x) \, (O^{-1})_a^b, \\ W \cong \mathbb{C}^4, \ \Lambda \in \mathbf{SL}(2,\mathbb{C})/\mathbf{I}(2) \cong \mathbf{SO}_0(1,3), \\ V \cong \mathbb{C}^3, \ O \in \mathbf{U}(2)/\mathbf{U}(\mathbf{1}_2) \cong \mathbf{SO}(3). \end{cases}$

7.25 Fiber Bundles and Spacetime Bundles

In general, a proper subgroup $H \subset G$ and its cosets G/H do not factorize the group G. However, the two separate structures, the subgroup H and the equivalence classes G/H with H-conjugated elements, can be joined in a coset fiber bundle for the full group with the subgroup H as fiber over the classes G/H as base.

A bundle is the union of the local fibers with the base manifold, the prickles and skin of a hedgehog. Properties related to the base $\mathbb{M}$ are called *horizontal*, whereas properties related to the fiber F are called *vertical*. More detailed: A *fiber bundle* $F(\mathbb{M})$ has a *base* $\mathbb{M}$ where each base point "has" its *local fiber*, isomorphic to a *typical fiber* F, e.g., a typical vector space for a vector space bundle, a typical Lie group for a Lie group bundle, etc. Both base and fiber are assumed as differentiable manifolds which are, with the mappings considered, "sufficiently smooth." A fiber bundle is characterized by a projection $\pi : F(\mathbb{M}) \longrightarrow \mathbb{M}$ from bundle to base manifold whose local inverses are the local fibers $\pi^{-1}(x) = F_x \cong F$. With the equivalence relation to belong to the same fiber, the base manifold characterizes the disjoint equivalence classes:

$$g, g' \in F(\mathbb{M}) : \ g\pi g' \iff \pi(g) = \pi(g') \Rightarrow F(\mathbb{M})/\pi \cong \mathbb{M},$$
$$F(\mathbb{M}) = \biguplus_{x \in \mathbb{M}} (x, F_x).$$

A *section* $\sigma : \mathbb{M} \longrightarrow F(\mathbb{M})$ is a projection compatible mapping, i.e., $\pi \circ \sigma = \mathrm{id}_{\mathbb{M}}$, e.g., the mappings $F^{\mathbb{M}} = \{f : \mathbb{M} \longrightarrow F\}$.

The *trivial F-bundle for* $\mathbb{M}$ is the set product $F \times \mathbb{M}$. A bundle is required to be *locally trivial* $F \times U_\iota$ for the sets $U_\iota \subseteq \mathbb{M}$ of an $\mathbb{M}$-covering $\{U_\iota\}_{\iota \in I}$. The diffeomorphisms group connecting all local trivializations with their possibly different $\mathbb{M}$-coverings is required to be equal for all manifold points and called the *structural group of the bundle*. The fiber F with the structural group H is a *Klein space* $H \bullet F$, i.e., a differential manifold with Lie group action.

The definition of a fiber bundle $H \bullet F(\mathbb{M})$ has to specify the structural group. For example, the trivial bundle $F \times \mathbb{M}$ with trivialization $(\mathbb{M}, \mathrm{id}_F \times \mathrm{id}_{\mathbb{M}})$ has as structural group only the fiber identity $\{\mathrm{id}_F\}$. It is also possible to impose the structural group via fiber properties, e.g., maximality, a fiber metric, etc., with a correspondingly chosen trivialization. For example, the maximal structural group of a tangent bundle

$\mathbf{T}(\mathbb{M}) = \mathbb{R}^n(\mathbb{M})$ with the local tangent vector spaces $\mathbf{T}_x \cong \mathbb{R}^n$ is the full linear group $\mathbf{GL}(n, \mathbb{R})$ of the typical fiber. For a Riemannian manifold (more ahead), the structural group can be restricted via the tangent space metric and orthonormal bases to the tangent Lorentz group $\mathbf{SO}_0(t, s)$. In any case, the structural group is realized faithfully in the diffeomorphism group of the fiber, $H \subseteq \mathbf{G}(F)$. A *$H$-representation bundle* $H \bullet W(\mathbb{M})$ carries vector space fibers W_x with an H-representation, e.g., for the adjoint group representation on the Lie algebra $H \bullet \log H(\mathbb{M})$.

A *principal bundle* is a Lie group bundle $H(\mathbb{M})$, where the Lie group H as typical fiber coincides with its structural group. The structural group has to act on itself either by left or by right translations. An example is the frame bundle of a real n-dimensional manifold $\mathbb{M}$ with $H = \mathbf{GL}(n, \mathbb{R})$.

A Lie group with a closed subgroup $H \subset G$ defines the associated *coset bundle* $H(\mathbb{M})$, a principal bundle with the subgroup as structural group. The base manifold is constituted by representatives for the cosets (H-equivalence classes):

$$\pi : G \longrightarrow G/H \longrightarrow \mathbb{M} = (G/H)_r \subseteq G, \;\; g \longmapsto gH \longmapsto g_r = (gH)_r, \quad \text{with } g_r H = gH.$$

The local fiber is the fixgroup $G_{g_r} \cong H$ of the representative. As manifold, a coset bundle is isomorphic to the full group:

$$G \cong H(\mathbb{M}) = \biguplus_{g_r \in \mathbb{M}} (g_r, G_{g_r}), \;\; G_{g_r} \cong H, \;\; \mathbb{M} = (G/H)_r.$$

The coset bundle notation of a direct product group is $H_1 \times H_2 \cong H_1(H_2)$. It is a trivial bundle.

The left multiplication with a group element $k \in G$ on the base manifold with the representatives gives the shifted representative up to the right multiplication with a subgroup Wigner element. These right multiplications define isomorphisms R of the structural group:

$$\begin{array}{ccc} H(\mathbb{M}) & \stackrel{R}{\longrightarrow} & H(\mathbb{M}) \\ \pi \downarrow & & \downarrow \pi \\ \mathbb{M} & \underset{L_k}{\longrightarrow} & \mathbb{M} \end{array}, \quad \begin{array}{l} L_k : \mathbb{M} \longrightarrow \mathbb{M}, \;\; g_r \longmapsto (kg)_r \\ \text{with} \begin{cases} kg_r = (kg)_r h(k, g_r) \in G, \\ (kg)_r = (kgH)_r \in \mathbb{M}, \\ h(k, g_r) \in H, \end{cases} \\ R|_{G_{g_r}} : H \cong G_{g_r} \longrightarrow G_{(kg)_r} \cong H, \\ \qquad h_r \longmapsto h_r h(k, g_r). \end{array}$$

$$(G/H)_r = \mathbb{M},$$

A Lie group coset bundle $H(\mathbb{M})$ with $\mathbb{M} = (G/H)_r$ has its adjoint Lie algebra bundle $\log H(\mathbb{M})$, also with the subgroup H as structural group.

Examples for Lie group coset bundles are the spheres and hyperboloids with the rotation group $\mathbf{SO}(s)$ as typical fiber. They can be visualized for $s = 2$, e.g., the antipodally identified 2-sphere with all diameters, where a diameter parametrizes the axial rotations around itself,

$$\mathbf{SO}(1+s) \cong \mathbf{SO}(s)(\Omega^s), \text{ with } \Omega^s \cong \mathbf{SO}(1+s)/\mathbf{SO}(s),$$
$$\mathbf{SO}_0(1,s) \cong \mathbf{SO}(s)(\mathcal{Y}^s), \text{ with } \mathcal{Y}^s \cong \mathbf{SO}_0(1,s)/\mathbf{SO}(s).$$

An affine group $G = H \vec{\times} V$ with homogeneous subgroup $H \subseteq \mathbf{GL}(n, \mathbb{K})$ and the translations $V \cong \mathbb{K}^n$ as normal subgroup leads to the homogeneous group for the translation equivalence classes:

$$[h] = \begin{pmatrix} h & V \\ 0 & 1 \end{pmatrix} \in H \vec{\times} V/V \cong H: \ [h] \circ [k] = [hk].$$

In general, the cosets with respect to the homogeneous group, i.e., the classes with H-conjugated elements, do not inherit a natural group multiplication. They constitute a translation V-isomorphic manifold:

$$\begin{pmatrix} H & v \\ 0 & 1 \end{pmatrix} \in H \vec{\times} V/H \cong V \text{ (as manifold)}: \begin{pmatrix} H & v \\ 0 & 1 \end{pmatrix} \circ \begin{pmatrix} H & w \\ 0 & 1 \end{pmatrix} = \begin{pmatrix} H & Hw+v \\ 0 & 1 \end{pmatrix}.$$

One can chose representatives $v_r = \begin{pmatrix} 1_H & v \\ 0 & 1 \end{pmatrix} \in \begin{pmatrix} H & v \\ 0 & 1 \end{pmatrix}$ with the group law of of the translations. However, with respect to the action of the homogenous group, it is more appropriate to consider the affine groups as coset bundles $H \vec{\times} V \cong H(\mathbb{M})$ with the homogeneous group H as typical fiber and coset representatives $\mathbb{M} = (H \vec{\times} V/H)_r$ as an V-isomorphic base.

Important examples for the distinction of the cosets $H \vec{\times} \mathbb{R}^n/H$ and the translations $\mathbb{R}^n$ are the physical concepts for position (space) and spacetimes, which, in an operational language, are all cosets of operation groups, i.e., groups with a subgroup H-induced equivalence relation. The Euclidean spaces $\mathbb{E}^s \cong \mathbf{O}(s) \vec{\times} \mathbb{R}^s/\mathbf{O}(s)$ consist of rotation group classes; they are isomorphic as manifold to, but conceptually different from the translation groups $\mathbb{R}^s$ — similarly the affine groups with the bundles for spacetime which, as manifolds, are all isomorphic to $\mathbb{R}^4$, but different as Klein spaces, i.e., as manifolds with a group action:

$$\begin{aligned} \text{Aristotle–Newton space-time: } & \mathbb{E}^1 \oplus \mathbb{E}^3 \cong \frac{\mathbf{O}(1) \vec{\times} \mathbb{R}}{\mathbf{O}(1)} \oplus \frac{\mathbf{O}(3) \vec{\times} \mathbb{R}^3}{\mathbf{O}(3)}, \\ \text{Galileo space-time: } & \mathbb{G}^4 \cong \frac{[\mathbf{O}(1) \times (\mathbf{O}(3) \vec{\times} \mathbb{R}^3)] \vec{\times} [\mathbb{R} \oplus \mathbb{R}^3]}{\mathbf{O}(1) \times (\mathbf{O}(3) \vec{\times} \mathbb{R}^3)}, \\ \text{Minkowski spacetime: } & \mathbb{R}^{(1,3)} \cong \frac{\mathbf{O}(1,3) \vec{\times} \mathbb{R}^4}{\mathbf{O}(1,3)}. \end{aligned}$$

Galileo space-time contains classes with respect to velocity transformations, i.e., space-times related to each other by velocity transformations are equivalent. Minkowski spacetime parametrizes the Lorentz group classes of the Poincaré group where the class property shows up in the Michelson–Morley experiment. All these space(times) are base manifolds for coset bundles, e.g.:

$$\mathbf{O}(s) \vec{\times} \mathbb{R}^s \cong \mathbf{O}(s)(\mathbb{E}^s), \ \ \mathbf{O}(1,s) \vec{\times} \mathbb{R}^{1+s} \cong \mathbf{O}(1,s)(\mathbb{R}^{(1,s)}).$$

7.26 Riemannian Manifolds

Manifolds ("sufficiently smooth" and differentiable parametrizations) as such are rather amorphous structures. Physically important and manifold-characterizing are the operation groups, they are connected with. Spheres and hyperboloids as important examples will be treated in later sections in more detail.

A manifold is endowed with a *Riemannian structure* $(\mathbb{M}, \mathbf{g})$ by a tensor field $\mathbf{g}$ for symmetric nondegenerate bilinear forms as *metrics of the tangent spaces* $\mathbf{T}_x \cong \mathbb{R}^n$ for $x \in \mathbb{M}$:

$$\mathbf{g}(x) : \mathbf{T}_x \vee \mathbf{T}_x \longrightarrow \mathbb{R}, \;\; \mathbf{g}(e^l(x), e^i(x)) = \mathbf{g}^{li}(x),$$
$$\mathbf{g} = \mathbf{g}^{li}\check{e}_l \otimes \check{e}_i = ds^2 \in \mathbf{T}^T \vee \mathbf{T}^T(\mathbb{M}), \;\; \mathbf{g}(v, w) = \mathbf{g}(w, v).$$

$\mathbf{T}^T(\mathbb{M})$ is the cotangent bundle.

Vector spaces (V, η) with a metric are linear Riemannian manifolds with $V = \mathbb{M} = \mathbf{T}_x$ and $\mathbf{g}(x) = \eta$. A nondegenerate bilinear real metric is characterized by an orthogonal invariance group $\mathbf{O}(t, s)$. Manifolds related to Lie groups with semi-simple Lie algebras, e.g., coset manifolds G/H, inherit, for the tangent space, the nondegenerate symmetric Killing form for a Riemannian structure, invariant for $\log \mathbf{SO}_0(t, s) \supseteq \log G$, e.g., $\log \mathbf{SO}(n^2 - 1)$ for unitary groups $\mathbf{SU}(n)$, or $\log \mathbf{SO}_0(3, 3)$ for the Lorentz group $\mathbf{SL}(2, \mathbb{C}) \sim \mathbf{SO}_0(1, 3)$.

The morphisms for Riemannian manifolds $\varphi : \mathbb{M}_1 \longrightarrow \mathbb{M}_2$ are compatible with the metrics $\mathbf{g}_2(\varphi.x, \varphi.y) = \mathbf{g}_1(x, y)$. The metric-associated *global invariance group* $\mathbf{G}_\mathbf{g}(\mathbb{M})$ consists of the metric-compatible diffeomorphisms $\varphi : \mathbb{M} \longrightarrow \mathbb{M}$ with $\mathbf{g} \circ (\varphi \times \varphi) = \mathbf{g}$ (isometries, Riemannian automorphisms).

The metrical structure of a Riemannian manifold defines *three operation groups*: its characterizing global and local invariance group and the not-so-specific tangent Poincaré or Euclidean group. For example: The global groups (motion groups) for the sphere Ω^2 and the hyperboloid $\mathcal{Y}^2$ are given by the rotations $\mathbf{SO}(3)$ and Lorentz transformations $\mathbf{SO}_0(1, 2)$, respectively; both have the axial rotations $\mathbf{SO}(2)$ as the local group and the Euclidean group $\mathbf{SO}(2) \,\vec{\times}\, \mathbb{R}^2$ as the tangent group. The Einstein cosmos $\mathbb{R} \times \Omega^3$ has $\mathbb{R} \times \mathbf{SO}(4)$ as the global group, $\mathbf{SO}(3)$ as the local group, and, with ad-hoc definitions because of the direct product structure, either $\mathbf{SO}_0(1, 3) \,\vec{\times}\, \mathbb{R}^4$ (causal signature, Poincaré group) or $\mathbf{SO}(4) \,\vec{\times}\, \mathbb{R}^4$ (definite signature, Euclidean group) as tangent groups.

Two metrics of a Riemannian manifold define isomorphisms of the tangent spaces:

$$\mathbf{g}' \circ \mathbf{g}^{-1} : \mathbf{T}(\mathbb{M}) \longrightarrow \mathbf{T}(\mathbb{M}), \;\; e^j \longmapsto \mathbf{g}^{jk}\mathbf{g}'_{kl}e^l, \;\; \mathbf{g}'(x) \circ \mathbf{g}^{-1}(x) \in \mathbf{GL}(n, \mathbb{R}).$$

With a subgroup $H \subseteq \mathbf{GL}(n, \mathbb{R})$, they are *H-equivalent* if $\mathbf{g}'(x) \circ \mathbf{g}^{-1}(x) \in H$ for all $x \in \mathbb{M}$. This equivalence relation collects the metrics into disjoint classes $\mathbf{GL}(n, \mathbb{R})/H$, the *n-bein manifold modulo H*. The local transformation from one metric to the other gives representatives of the equivalence classes, defined up to local H-transformations, $\mathbf{g} = e^T \circ \mathbf{g}' \circ e$, $e(x) \in H$.

A special case is given by the center $\mathbf{GL}(\mathbb{R})\mathbf{1}_n \subseteq \mathbf{GL}(n, \mathbb{R})$ as direct factor for the volume invariance group $\mathbf{GL}(n, \mathbb{R}) \cong \mathbf{GL}(1, \mathbb{R}) \times \mathbf{SL}(n, \mathbb{R})$. Two metrics are *conformal* to each other (dilation-equivalent) if the isomorphism above is the identity up to a scalar factor, i.e., a dilation from the unit connection subgroup $\mathbf{D}(1) \subset \mathbf{GL}(1, \mathbb{R})$:

$$\mathbf{g} \overset{\mathbf{D}(1)}{\sim} \mathbf{g}' \iff \mathbf{g}' \circ \mathbf{g}^{-1} = e^{2\psi}\, \mathrm{id}_{\mathbf{T}(\mathbb{M})}, \quad \mathbf{g}_{jk}(x)\mathbf{g}'^{kl}(x) = e^{2\psi(x)}\delta_j^l.$$

The local orthonormalization of a Riemannian metric displays a manifold-characteristic *signature*[10] (t, s) and allows the definition of a local n-bein field (boldface $\mathbf{e}$) with respect to a fixed constant metric η, e.g., diagonally orthonormalized:

$$\mathbf{g} = ds^2 = \eta^{ab}\check{\mathbf{e}}_a \otimes \check{\mathbf{e}}_b, \quad \mathbf{g}^{jk}(x) = \mathbf{e}_a^j(x)\eta^{ab}\mathbf{e}_b^k(x), \quad \eta = \left(\begin{array}{c|c} \mathbf{1}_t & 0 \\ \hline 0 & -\mathbf{1}_s \end{array}\right).$$

This leaves an orthogonal Lorentz invariance group $\mathbf{O}(t, s)$ with the unit connection subgroup $\mathbf{SO}_0(t, s)$. It is a rotation group for definite signature $ts = 0$ and a proper Lorentz group for causal signature $(t, s) = (1, s)$. Its Lie algebra $\log \mathbf{SO}_0(t, s)$ is simple for $n = t + s \geq 3$ with the exception of the two semisimple Lie algebras $\log \mathbf{SO}(4) = \log \mathbf{SO}(3) \oplus \log \mathbf{SO}(3)$ and $\log \mathbf{SO}_0(2, 2) = \log \mathbf{SO}_0(1, 2) \oplus \log \mathbf{SO}_0(1, 2)$.

In contrast to the Lorentz group, the overall dilations $\mathbf{D}(\mathbf{1}_n)$ cannot be characterized as the invariance group of a metric. A manifold measure $d^n x$ is $\mathbf{SL}(n, \mathbb{R}) \vec{\times} \mathbb{R}^n$-invariant in any coordinates. It becomes the up-to-a-constant unique, affine-invariant, i.e., $\mathbf{GL}(n, \mathbb{R}) \vec{\times} \mathbb{R}^n$-invariant measure (volume element) of the manifold $\mathbb{M}$ by multiplication with the absolute value of the square root of the metric determinant for the dilations $\mathbf{D}(1) \cong \mathbf{GL}(n, \mathbb{R})/\mathbf{SL}(n, \mathbb{R})$:

$$\sqrt{|\det \mathbf{g}(x)|}d^n x = |\det \mathbf{e}(x)|d^n x.$$

The symmetric tensor $\mathbf{g}$ has $\binom{1+n}{2}$ real parameters, which are taken over completely by n-beins from the equal-dimensional manifold $\mathbf{GL}(n, \mathbb{R})/\mathbf{O}(t, s)$ as parameters for the local orthonormalization by $\binom{n}{2}$ rotations $\mathbf{O}(n)$, for the rotation to principal axes of the metrical ellipsoid or hyperboloid, and by n dilations for choosing n units. An n-bein representative is defined up to local $\mathbf{O}(t, s)$-transformations:

$$\mathbf{g} = \mathbf{e}^T \circ \eta \circ \mathbf{e}, \quad \mathbf{e}(x) \in \mathbf{O}(n) \times \mathbf{D}(1) \times \mathbf{SO}_0(1, 1)^{n-1} \in \mathbf{GL}(n, \mathbb{R})/\mathbf{O}(t, s),$$
$$\binom{1+n}{2} = \binom{n}{2} + n = n^2 - \binom{n}{2}.$$

Together with the tangent translations $\mathbb{R}^n$, one obtains the $\binom{1+n}{2}$ parametric *tangent Poincaré group with homogeneous Lorentz group*, sometimes called pseudo-Euclidean and pseudo-Riemannian for $ts \neq 0$,

[10] Obviously, the natural numbers (t, s) should remind us of time and space. In general, it will be difficult to confuse the dimension $t \in \mathbb{N}$ with a coordinate $t \in \mathbb{R}$.

$$\mathbf{SO}_0(t,s) \vec{\times} \mathbb{R}^n, \ t+s=n \geq 1.$$

It is a proper Euclidean group with metrical ellipsoids for $ts = 0$, e.g., for the spheres Ω^s and the hyperboloids $\mathcal{Y}^t$, and a proper Poincaré group with metrical "timelike" hyperboloids for $(t,s) = (1,s)$. There are characteristically different Riemannian manifolds $\mathbb{M}^{(t,s)}$ with isomorphic tangent groups, e.g., the compact sphere Ω^s and the noncompact hyperboloid $\mathcal{Y}^s$, both with Euclidean tangent group $\mathbf{SO}(s) \vec{\times} \mathbb{R}^s$.

A semisimple Lie algebra $\log G \cong \mathbb{R}^n$ has a nondegenerate Killing form which is invariant for the larger Lie algebra $\log \mathbf{SO}_0(t,s) \supseteq \log G$. The maximal invariance arises for the orthogonal groups.

A metrical tensor field defines its *Killing vector fields* $v = v_k \partial^k$ by the Lie derivations (more ahead) from its invariance Lie algebra, related to the global motion group (more ahead). This associates a real Lie algebra to a real Riemannian manifold:

$$\mathbf{T}_{\mathbf{g}}(\mathbb{M}) = \{v \in \mathbf{T}(\mathbb{M}) \,\big|\, \mathcal{L}_v(\mathbf{g}) = [v, \mathbf{g}] = 0\}.$$

With the $\binom{1+n}{2}$ metric components, one obtains, with $[v_k\partial^k, dx_i] = (\partial^k v_i) dx_k$, a system of $\binom{1+n}{2}$ differential equations for the n functions $\{v_k\}_{k=1}^n$:

$$\begin{aligned}
[v_k\partial^k, \mathbf{g}^{ij} dx_i \otimes dx_j] &= [v_k(\partial^k \mathbf{g}^{ij}) + \mathbf{g}^{kj}(\partial^i v_k) + \mathbf{g}^{ki}(\partial^j v_k)] dx_i \otimes dx_j,\\
v_k\partial^k \in \mathbf{T}_{\mathbf{g}}(\mathbb{M}) &\iff v_k(\partial^k \mathbf{g}^{ij}) + \mathbf{g}^{kj}(\partial^i v_k) + \mathbf{g}^{ki}(\partial^j v_k) = 0\\
&\quad\ \text{for all } i,j \in \{1,\ldots,n\}.
\end{aligned}$$

Each one-dimensional manifold has one Killing field and $\mathbb{R}$ as invariance Lie algebra:

$$\mathbf{g} = b^2(\tau) d\tau^2 : \ \ v\partial_\tau b^2 + 2b^2 \partial_\tau v = 2b\partial_\tau(vb) = 0 \Rightarrow v\partial_\tau = \frac{1}{b}\partial_\tau.$$

For two dimensions, there are three equations for the Killing fields $v = v_0\partial_\tau + v_1\partial_\rho$:

$$\mathbf{g} = b^2 d\tau^2 \pm a^2 d\rho^2 : \ \ \begin{cases} (v_0\partial_\tau + v_1\partial_\rho) b^2 + 2b^2\partial_\tau v_0 &= 0,\\ (v_0\partial_\tau + v_1\partial_\rho) a^2 + 2a^2 \partial_\rho v_1 &= 0,\\ b^2\partial_\rho v_0 \pm a^2 \partial_\tau v_1 &= 0. \end{cases}$$

For the metrical tensor $\eta = \eta_{ab} dx^a \otimes dx^b \cong \left(\begin{array}{c|c} \mathbf{1}_t & 0 \\ \hline 0 & -\mathbf{1}_s \end{array}\right)$ of the vector space $\mathbb{R}^{(t,s)}$ as Riemannian manifold, the Killing fields constitute a Poincaré Lie algebra:

$$\begin{aligned}
\mathbf{T}_\eta(\mathbb{R}^{(t,s)}) &= \log \mathbf{SO}_0(t,s) \ \vec{\oplus}\ \mathbb{R}^n,\\
v_a\partial^a &\in \mathbf{T}_\eta(\mathbb{R}^{(t,s)}) \iff \partial^a v^b + \partial^b v^a = 0 \Rightarrow v^a(x) = (\lambda^{ab} - \lambda^{ba}) x_b + \xi^a,\\
v_a\partial^a &= \lambda_{ab}(\eta^{ac} x_c \partial^b - \eta^{bc} x_c \partial^a) + \xi_a \partial^a.
\end{aligned}$$

The Poincaré Lie algebra is represented by Lie derivations,

$$\mathbf{L}^{ab} \longmapsto \eta^{ac}x_c\partial^b - \eta^{bc}x_c\partial^a, \ \mathbf{p}^a \longmapsto \partial^a.$$

For a metric without Killing vectors, the manifold has no symmetry, $\mathbf{T_g}(\mathbb{M}) = \{0\}$. If a metric does not depend, in a parametrization, on a coordinate, $\frac{\partial}{\partial x_k}\mathbf{g}^{ij}(x) = 0$ for all $i, j \in \{1, \ldots, n\}$, the corresponding derivative $v = \partial^k$ characterizes a one-dimensional invariance Lie algebra. More generally, for each Killing field $v \in \mathbf{T_g}(\mathbb{M})$, there exists a parametrization with $v = \partial^k$. Such a one-dimensional Lie algebra $\mathbb{R}$ leads by integration locally to the noncompact covering Lie group $e^{x_k} \in \mathbf{D}(1) = e^{\mathbb{R}} \cong \mathbf{SO}_0(1,1) \ni e^{\sigma_3 x_k}$ or to its compact quotient group $e^{i\sigma_2 x_k} \in \mathbf{SO}(2) \cong \mathbb{R}/\mathbb{Z}$.

Linearly independent Killing vectors $\{v^K\}_{K=1}^d$ of $(\mathbb{M}, \mathbf{g})$ define a d-dimensional invariance Lie algebra $\mathbf{T_g}(\mathbb{M})$ of the metric with cover group $\exp \mathbf{T_g}(\mathbb{M})$. Under appropriate smoothness and connectivity conditions, at least locally in a chart, a locally isomorphic d-dimensional quotient Lie group $\mathbf{G_g}(\mathbb{M})$ (classes of $\exp \mathbf{T_g}(\mathbb{M})$) is the *global symmetry or motion or isometry group* with diffeomorphisms of the Riemannian manifold:

$$\exp \mathbf{T_g}(\mathbb{M}) \sim \mathbf{G_g}(\mathbb{M}) = \{\varphi : \mathbb{M} \longrightarrow \mathbb{M} \,\big|\, \mathbf{g}(\varphi.x, \varphi.y) = \mathbf{g}(x, y)\}.$$

Its maximal dimension is $\binom{1+n}{2}$: The *maximal global symmetry group* is either an orthogonal group, e.g., for spheres and hyperboloids (ahead), or a Poincaré group for a flat vector space,

$$\begin{aligned} \dim_{\mathbb{R}} \mathbf{G_g}(\mathbb{M}) &\le \binom{1+n}{2} = \dim_{\mathbb{R}} \mathbf{G}^{\max}(t,s), \\ \mathbf{G}^{\max}(t,s) &\in \{\mathbf{SO}_0(t, 1+s),\ \mathbf{SO}_0(1+t, s),\ \mathbf{SO}_0(t,s) \vec{\times} \mathbb{R}^n\}. \end{aligned}$$

The actual motion group $\mathbf{G_g}(\mathbb{M})$ must not be a subgroup of $\mathbf{G}^{\max}(t,s)$. It is a subgroup of an $\binom{1+n}{2}$-dimensional product of maximal motion groups — possibly also from smaller-dimensional manifolds.

$\mathbf{G_g}(\mathbb{M})$ acts on the manifold $\mathbb{M}$, which, therefore, is the disjoint union of $\mathbf{G_g}(\mathbb{M})$-orbits with fixgroups (isotropy subgroup) $\mathbf{H}^\iota_{\mathbf{g}}(\mathbb{M}) \subseteq \mathbf{G_g}(\mathbb{M})$ for orbit representatives,

$$\mathbb{M} \cong \biguplus_{x^\iota \in \mathbb{M}} \mathbf{G_g}(\mathbb{M}) \bullet x^\iota \cong \biguplus_{\iota} \mathbf{G_g}(\mathbb{M})/\mathbf{H}^\iota_{\mathbf{g}}(\mathbb{M}).$$

The fixgroups of the points of one orbit, $g_{1,2} \bullet x \in \mathbf{G_g}(\mathbb{M}) \bullet x$, are conjugate to each other, i.e., related by inner $\mathbf{G_g}(\mathbb{M})$-automorphisms with $g_1 g_2^{-1}$. There may be disjoint orbits with isomorphic fixgroup, collectable into strata.

The *local invariance group* of a Riemannian manifold is defined by the fixgroup of all manifold points, i.e., by the intersection of appropriate representative fixgroups for all orbit types (examples ahead):

$$\mathbf{H_g}(\mathbb{M}) = \bigcap_\kappa \mathbf{H_g^\kappa}(\mathbb{M}) \subseteq \mathbf{G_g}(\mathbb{M}).$$

For a maximal global symmetry group, the local invariance group is maximal and isomorphic to the tangent Lorentz group. In general, the local invariance group is a Lorentz subgroup:

$$\mathbf{H_g}(\mathbb{M}) \subseteq \mathbf{H}^{\max}(t,s) = \mathbf{SO}_0(t,s).$$

The *metrical coefficients are representation matrix elements* of the global symmetry group $\mathbf{G_g}(\mathbb{M})$, invariant under the action of the local group $\mathbf{H_g}(\mathbb{M})$:

$$\biguplus_\iota \mathbf{G_g}(\mathbb{M})/\mathbf{H_g^\iota}(\mathbb{M}) \cong \mathbb{M} \ni x \longmapsto \mathbf{g}^{jk}(x),$$

$$\mathbf{G_g}(\mathbb{M})/\mathbf{H_g^\iota}(\mathbb{M}) \subseteq \mathbf{G_g}(\mathbb{M})/\mathbf{H_g}(\mathbb{M}).$$

Some examples: A manifold with trivial global symmetry group $\{1\}$ is the union of its points $\mathbb{M} = \biguplus_{x\in\mathbb{M}} \{x\}$. A one-dimensional manifold is isomorphic to noncompact $\mathbb{R} \cong \mathbf{SO}_0(1,1) \cong \mathcal{Y}^1$ or to compact $\mathbf{SO}(2) \cong \Omega^1$. A torus is a product of axial rotation groups $\mathbb{T}^n = \mathbf{SO}(2) \times \cdots \times \mathbf{SO}(2)$.

Spheres and hyperboloids are one orbit. The metrical coefficients are spherical harmonics $\mathrm{Y} : \mathbf{SO}(1+s)/\mathbf{SO}(s) \cong \Omega^s \longrightarrow \mathbb{R}$ and their hyperbolic counterparts $\mathbf{SO}_0(1,s)/\mathbf{SO}(s) \cong \mathcal{Y}^s \longrightarrow \mathbb{R}$. For example, the Ω^2-metric involves the spherical harmonics $\sqrt{4\pi}\mathrm{Y}_0^0(\varphi,\theta) = 1$ and $\sqrt{\frac{4\pi}{5}}\mathrm{Y}_0^2(\varphi,\theta) = 1 - \frac{3}{2}\sin^2\theta$. Its three Killing fields are given by the angular momenta, generating the rotation group $\mathbf{SO}(3)$ — in polar coordinates,

$$\Omega^2 : \ d\omega_2^2 = d\theta^2 + \sin^2\theta d\varphi^2, \ \ \vec{\mathbf{J}} \cong \begin{pmatrix} \cos\varphi\cot\theta\frac{\partial}{\partial\varphi} + \sin\varphi\frac{\partial}{\partial\theta} \\ \sin\varphi\cot\theta\frac{\partial}{\partial\varphi} - \cos\varphi\frac{\partial}{\partial\theta} \\ \frac{\partial}{\partial\varphi} \end{pmatrix}, \ \ [\mathbf{J}^a, \mathbf{J}^b] = -\epsilon^{abc}\mathbf{J}^c.$$

The local invariance group $\mathbf{SO}(2) \subset \mathbf{SO}(3)$ is parametrizable by a third coordinate in (χ,φ,θ). This coordinate does not show up — neither in the metrical coefficients, $\frac{\partial \mathbf{g}^{jk}}{\partial\chi} = 0$, nor in the basis as $d\chi^2$.

In contrast to the metrical coefficients for the compact spheres, which are definite-unitary matrix elements, i.e., of Hilbert representations, those of the noncompact hyperboloids are not Hilbert representation coefficients; e.g.,

$$\mathcal{Y}^2 : \ dy_2^2 = d\psi^2 + \sinh^2\psi d\varphi^2.$$

Minkowski spacetime $\mathbb{R}^{(1,s)}$, $s \geq 2$, has metrical coefficients $dx^2 = dx_0^2 - d\vec{x}^2$ with a trivial representation of the Poincaré group. The action of the Lorentz group $\mathbf{SO}_0(1,s)$, $s \geq 2$, on the translations has four fixgroup types for the "many" orbits:

$\mathbf{SO}_0(1, s)$ for the trivial translation, $\mathbf{SO}(s)$ for timelike translations, $\mathbf{SO}_0(1, s-1)$ for spacelike translations, and $\mathbf{SO}(s-1) \vec{\times} \mathbb{R}^{s-1}$ for lightlike translations. The common fixgroup for all translations is $\mathbf{SO}(s-1)$, i.e., the axial rotations $\mathbf{SO}(2)$ for four-dimensional spacetime. Vector spaces $\mathbb{R}^n$ are one orbit with respect to their own action as translations. Vector spaces $\mathbb{R}^{(t,s)}$ as Riemannian manifolds have the Poincaré group as both the tangent and motion groups; the local invariance group is the Lorentz group $\mathbf{SO}_0(t, s) \vec{\times} \mathbb{R}^n / \mathbf{SO}_0(t, s) \cong \mathbb{R}^{(t,s)}$.

Summarizing: The three Riemannian operation groups for a Riemannian manifold $(\mathbb{M}, \mathbf{g})$ are

$\mathbb{M}^{(t,s)}$:

Tangent Poincare group
$\mathbf{SO}_0(t, s) \times \mathbb{R}^n$

Global symmetry (motion) group $\mathbf{G}_\mathbf{g}(\mathbb{M})$
$\mathbf{G}^{\max}(t, s) \in \{\mathbf{SO}_0(t, 1+s),\ \mathbf{SO}_0(1+t, s),\ \mathbf{SO}_0(t, s) \vec{\times} \mathbb{R}^n\}$

Local invariance group
$\mathbf{H}_\mathbf{g}(\mathbb{M}) \subseteq \mathbf{G}_\mathbf{g}(\mathbb{M}) \cap \mathbf{SO}_0(t, s)$

In a physical context, the global group $\mathbf{G}_\mathbf{g}(\mathbb{M})$ and its generators give the symmetries and conserved quantities of a dynamics, like energy for time translations and static spacetimes or angular momenta for rotation-symmetric spacetimes, whereas the local subgroup $\mathbf{H}_\mathbf{g}(\mathbb{M})$ will be related to gauge transformations, e.g., to electromagnetic and isospin transformations.

Functions of a Riemannian manifold with metric for tangent and cotangent spaces in local coordinates:

$$\mathbf{g} = ds^2 = \mathbf{g}^{ij}(x) dx_i \otimes dx_j, \quad \mathbf{g}^{-1} = \mathbf{g}_{ij}(x) \partial^i \otimes \partial^j,$$

have an invariant derivative by the *Laplace-Beltrami*[11] *operator* as the divergence of the gradient, both metric-appropriate:

$$\begin{aligned} \mathrm{grad}_\mathbf{g} &: \mathcal{C}(\mathbb{M}) \longrightarrow \mathbf{T}(\mathbb{M}),\ f \longmapsto v = \mathbf{g}_{ij}(\partial^j f)\partial^i = v_i \partial^i, \\ div_\mathbf{g} &: \mathbf{T}(\mathbb{M}) \longrightarrow \mathcal{C}(\mathbb{M}),\ v \longmapsto \tfrac{1}{\sqrt{|\det \mathbf{g}|}} \partial^i \sqrt{|\det \mathbf{g}|}\, v_i, \\ \partial^2_\mathbf{g} &: \mathcal{C}(\mathbb{M}) \longrightarrow \mathcal{C}(\mathbb{M}),\ f \longmapsto \tfrac{1}{\sqrt{|\det \mathbf{g}|}} \partial^i \sqrt{|\det \mathbf{g}|}\, \mathbf{g}_{ij}\, \partial^j f, \end{aligned}$$

$$\partial^2_\mathbf{g} = div_\mathbf{g} \circ \mathrm{grad}_\mathbf{g} = \tfrac{1}{\sqrt{|\det \mathbf{g}|}} \partial^i \sqrt{|\det \mathbf{g}|}\, \mathbf{g}_{ij}\, \partial^j.$$

7.27 Spheres and Hyperboloids

The embeddings of the rotation group $\mathbf{SO}(s)$, $s = 1, 2, \ldots$, in the rotation group $\mathbf{SO}(1+s)$ with one more dimension, or into "its" causal orthochronous Lorentz group $\mathbf{SO}_0(1, s)$, or into the associated Euclidean group $\mathbf{SO}(s) \vec{\times} \mathbb{R}^s$ with flat space translations are parametrizable by, or define, maximally symmetric simply connected

[11] Eugenio Beltrami (1835–1900).

complete Riemannian manifolds with definite metric and constant curvature (more ahead). The embedding manifolds are isomorphic, respectively, to a compact *sphere*, to a noncompact one-shell "timelike" *hyperboloid*, and to a noncompact *Euclidean space*:

$$s = 1, 2, \ldots : \quad \begin{cases} \Omega^s \cong \mathbf{SO}(1+s)/\mathbf{SO}(s), \\ \mathcal{Y}^s \cong \mathbf{SO}_0(1,s)/\mathbf{SO}(s), \\ \mathbb{E}^s \cong \mathbf{SO}(s) \vec{\times} \mathbb{R}^s/\mathbf{SO}(s). \end{cases}$$

The tangent spaces of all three manifolds are characterized by the Euclidean group $\mathbf{SO}(s) \vec{\times} \mathbb{R}^s$. One can define for $s = 0$ the discrete groups $\Omega^0 = \{\pm 1\}$, $\mathcal{Y}^0 = \{1\}$, and $\mathbb{R}^0 = \{0\}$. The structures for $s = 1$, circle, hyperbola (catenary), and line, with $\mathbf{SO}(1) = \{1\}$ are abelian groups. The sphere Ω^2 can be visualized in Euclidean space $\mathbb{E}^3$ by an orange skin, breaking if pressed flat, the hyperboloid $\mathcal{Y}^2$ by a rotated catenary $\mathcal{Y}^1$, crumpling if pressed flat.

As manifolds, not as symmetric spaces, hyperboloids and flat spaces are isomorphic, $\mathcal{Y}^s \cong \mathbb{E}^s$. Using Lie algebras, an isomorphy for symmetric spaces can be expressed with the exponential $\mathcal{Y}^s \cong \exp \mathbb{R}^s$.

The Killing forms κ for the global and local groups have the signatures:

$$\begin{aligned} \text{for } \Omega^s : \log \mathbf{SO}(1+s) &\text{ with } \kappa \sim -\mathbf{1}_{\binom{1+s}{2}}, \\ \log \mathbf{SO}(s) \quad &\text{ with } \kappa \sim -\mathbf{1}_{\binom{s}{2}}, \\ \text{for } \mathcal{Y}^s : \log \mathbf{SO}_0(1,s) &\text{ with } \kappa \sim \begin{pmatrix} -\mathbf{1}_{\binom{s}{2}} & 0 \\ 0 & \mathbf{1}_s \end{pmatrix}, \end{aligned}$$

which, for the cosets, leaves the definite metrics $k\mathbf{1}_s$. The *definiteness (parameter) k* indicates the positivity or negativity of the relevant Killing metric, translations have a trivial Killing metric:

$$k\begin{pmatrix} \Omega^s \\ \mathcal{Y}^s \\ \mathbb{E}^s \end{pmatrix} = \begin{pmatrix} -1 \\ 1 \\ 0 \end{pmatrix} = \iota^2, \quad \iota\begin{pmatrix} \Omega^s \\ \mathcal{Y}^s \\ \mathbb{E}^s \end{pmatrix} = \begin{pmatrix} i \\ 1 \\ 0 \end{pmatrix}.$$

The definiteness parameter is the square of the *compactness (parameter)* ι.

The tangent metrics, measures (volume elements) and Laplace derivatives for the Euclidean spaces $\mathbb{E}^s$ are:

$$\mathbb{E}^s \ni \vec{x} = \begin{pmatrix} x_1 \\ \vdots \\ x_s \end{pmatrix} \in \mathbb{R}^s, \quad \begin{cases} d\vec{x}_s^2 = dx_1^2 + \cdots + dx_s^2, \\ d^s x = dx_1 \cdots dx_s, \\ \vec{\partial}_s^2 = \frac{\partial^2}{\partial x_1^2} + \cdots + \frac{\partial^2}{\partial x_s^2}. \end{cases}$$

Unit spheres and hyperboloids have the *Eulerian parametrization* by polar coordinates from their global symmetry groups with nonabelian degrees of freedom for $s \geq 2$:

$$s \geq 1: \quad \Omega^s \ni \vec{\omega}_s = \begin{pmatrix} \cos\theta \\ \sin\theta \ \vec{\omega}_{s-1} \end{pmatrix} \in \mathbb{R}^{1+s}, \quad \vec{\omega}_0 = 1,$$
$$\mathcal{Y}^s \ni \mathrm{y}_s = \begin{pmatrix} \cosh\psi \\ \sinh\psi \ \vec{\omega}_{s-1} \end{pmatrix} \in \mathbb{R}^{1+s}.$$

Therefrom, one obtains, with the compact-noncompact transition $i\theta \leftrightarrow \psi$ for the characteristic "leading" coordinate, the metrics $\mathbf{g} = \mathbf{g}_{ij}(x)dx^i dx^j$ and measures (volume elements) $\sqrt{|\det \mathbf{g}(x)|}d^s x$:

$$d\omega_s^2 = \begin{cases} d\theta^2, & s = 1, \\ d\theta^2 + \sin^2\theta \, d\varphi^2, & s = 2, \\ d\theta^2 + \sin^2\theta \, d\omega_{s-1}^2, & s \geq 2, \end{cases} \qquad d\mathrm{y}_s^2 = \begin{cases} d\psi^2, \\ d\psi^2 + \sinh^2\psi \, d\varphi^2, \\ d\psi^2 + \sinh^2\psi \, d\omega_{s-1}^2, \end{cases}$$

$$\mathbf{g}_{\Omega^s} \cong \begin{cases} 1, \\ \begin{pmatrix} 1 & 0 \\ 0 & \sin^2\theta \end{pmatrix}, \\ \begin{pmatrix} 1 & 0 \\ 0 & \sin^2\theta \, \mathbf{g}_{\Omega^{s-1}} \end{pmatrix}, \end{cases} \qquad \mathbf{g}_{\mathcal{Y}^s} \cong \begin{cases} 1, \\ \begin{pmatrix} 1 & 0 \\ 0 & \sinh^2\psi \end{pmatrix}, \\ \begin{pmatrix} 1 & 0 \\ 0 & \sinh^2\psi \, \mathbf{g}_{\Omega^{s-1}} \end{pmatrix}, \end{cases}$$

$$d^s\omega = \begin{cases} d\theta, \\ \sin\theta \, d\theta \, d^1\omega, \\ \sin^{s-1}\theta \, d\theta \, d^{s-1}\omega, \end{cases} \qquad d^s\mathrm{y} = \begin{cases} d\psi, \\ \sinh\psi \, d\psi \, d^1\omega, \\ \sinh^{s-1}\psi \, d\psi \, d^{s-1}\omega. \end{cases}$$

The measures involve $\sin\theta$ and $\sinh\psi$ as radii of "subspheres" (circles of latitudes for Ω^2).

The Laplace-Beltrami derivatives $\partial_{\mathbf{g}}^2 = \frac{1}{\sqrt{|\det \mathbf{g}|}}\partial^i \sqrt{|\det \mathbf{g}|} \, \mathbf{g}_{ij} \, \partial^j$ for spheres and hyperboloids are in the Eulerian parametrization:

$$\partial_{\Omega^s}^2 = \begin{cases} \partial_\theta^2, \\ \frac{1}{\sin\theta}\partial_\theta \sin\theta \, \partial_\theta + \frac{1}{\sin^2\theta}\partial_\varphi^2 & = \partial_\theta^2 + \cot\theta \, \partial_\theta + \frac{1}{\sin^2\theta}\partial_\varphi^2, \\ \frac{1}{\sin^{s-1}\theta}\partial_\theta \sin^{s-1}\theta \, \partial_\theta + \frac{1}{\sin^2\theta}\partial_{\Omega^{s-1}}^2 & = \partial_\theta^2 + (s-1)\cot\theta \, \partial_\theta + \frac{1}{\sin^2\theta}\partial_{\Omega^{s-1}}^2, \end{cases}$$

$$\partial_{\mathcal{Y}^s}^2 = \begin{cases} \partial_\psi^2, \\ \frac{1}{\sinh\psi}\partial_\psi \sinh\psi \, \partial_\psi + \frac{1}{\sinh^2\psi}\partial_\varphi^2 & = \partial_\psi^2 + \coth\psi \, \partial_\psi + \frac{1}{\sinh^2\psi}\partial_\varphi^2, \\ \frac{1}{\sinh^{s-1}\psi}\partial_\psi \sinh^{s-1}\psi \, \partial_\psi + \frac{1}{\sinh^2\psi}\partial_{\Omega^{s-1}}^2 & = \partial_\psi^2 + (s-1)\coth\psi \, \partial_\psi + \frac{1}{\sinh^2\psi}\partial_{\Omega^{s-1}}^2. \end{cases}$$

Geodesic polar coordinates are obtained by a transformation of the "leading angle" to a length coordinate:

$$\begin{array}{l}\Omega^s: \theta \in [-\frac{\pi}{2}, \frac{\pi}{2}], \qquad \sin\theta = \rho \in [-1, 1],\\ \mathcal{Y}^s: \psi \in]-\infty, \infty[, \ \sinh\psi = \rho \in]-\infty, \infty[,\end{array}$$

$$k\begin{pmatrix}\Omega^s\\ \mathcal{Y}^s\\ \mathbb{E}^s\end{pmatrix} = \begin{pmatrix}-1\\ 1\\ 0\end{pmatrix}: \left\{\begin{array}{rl}\begin{pmatrix}d\omega_s^2\\ dy_s^2\\ dx_s^2\end{pmatrix} &= \frac{d\rho^2}{1+k\rho^2} + \rho^2 d\omega_{s-1}^2, \quad \begin{pmatrix}\mathbf{g}_{\Omega^s}\\ \mathbf{g}_{\mathcal{Y}^s}\\ \mathbf{g}_{\mathbb{E}^s}\end{pmatrix} \cong \begin{pmatrix}\frac{1}{1+k\rho^2} & 0\\ 0 & \rho^2 \mathbf{g}_{\Omega^{s-1}}\end{pmatrix},\\ \begin{pmatrix}d^s\omega\\ d^s y\\ d^s x\end{pmatrix} &= \frac{\rho^{s-1}}{\sqrt{1+k\rho^2}} d\rho\, d^{s-1}\omega,\\ \begin{pmatrix}\partial^2_{\Omega^s}\\ \partial^2_{\mathcal{Y}^s}\\ \partial^2_s\end{pmatrix} &= \frac{\sqrt{1+k\rho^2}}{\rho^{s-1}}\partial_\rho\, \rho^{s-1}\sqrt{1+k\rho^2}\,\partial_\rho + \frac{1}{\rho^2}\partial^2_{\Omega^{s-1}}\\ &= (1+k\rho^2)\partial^2_\rho + \frac{s(1+k\rho^2)-1}{\rho}\partial_\rho + \frac{1}{\rho^2}\partial^2_{\Omega^{s-1}}.\end{array}\right.$$

The compact spheres have a coordinate singularity at $\rho^2 = 1$. For the momentum hyperboloid one writes $d^s y = \frac{|\vec{q}|^{s-1}d|\vec{q}|}{\sqrt{1+\vec{q}^2}}\, d^{s-1}\omega = \frac{d^s q}{\sqrt{1+\vec{q}^2}}$.

The s-dimensional subvolumes $|\omega^s| \leq |\Omega^s|$ for the compact classes of orthogonal groups $\mathbf{SO}(1+s)/\mathbf{SO}(s) \cong \Omega^s$ are angles $0 \leq |\omega^1| \leq 2\pi$ for $s = 1$ and areas (solid angles) $0 \leq |\omega^2| \leq 4\pi$ for $s = 2$. In the volumes of unit spheres, computed by using momentum measures and the parametrizations $\sin\theta = q = \frac{p}{\sqrt{1+p^2}}$, the powers of π increase by one in two-dimensional steps:

$$\begin{array}{l}|\Omega^s| = \int d^s\omega = \int_0^\pi d\theta\ \sin^{s-1}\theta \int d^{s-1}\omega = \frac{\Gamma(\frac{s}{2})\Gamma(\frac{1}{2})}{\Gamma(\frac{1+s}{2})}|\Omega^{s-1}|\\ \qquad = \int d^{1+s}q\ 2\delta(q_0^2 + \vec{q}^2 - 1) = \int_{\vec{q}^2\leq 1}\frac{d^s q}{\sqrt{1-\vec{q}^2}} = \int \frac{2d^s p}{(1+\vec{p}^2)^{\frac{1+s}{2}}},\end{array}$$

polar decomposition: $q = |q|\vec{\omega}_s$ with $|q|^2 = q_0^2 + \vec{q}^2$, $\vec{\omega}_s \in \Omega^s$,

$$\int d^{1+s}q = \int_0^\infty |q|^s d|q| \int d^s\omega,$$

$$|\Omega^s| = \frac{2\pi^{\frac{1+s}{2}}}{\Gamma(\frac{1+s}{2})}: \left\{\begin{array}{rcl}\frac{|\Omega^{2+s}|}{|\Omega^s|} &=& \frac{2\pi}{1+s}, \quad \frac{\Gamma(\frac{1+s}{2})}{\Gamma(\frac{1}{2})} = \frac{\Gamma(s)}{2^{s-1}\Gamma(\frac{s}{2})},\\ |\Omega^{2R}| &=& \frac{2\pi^R\Gamma(\frac{1}{2})}{\Gamma(R+\frac{1}{2})} = 2, 4\pi, \frac{8\pi^2}{3}, \ldots,\\ |\Omega^{2R-1}| &=& \frac{2\pi^R}{\Gamma(R)} = 2\pi, 2\pi^2, \ldots\end{array}\right.$$

Both the motion groups $\mathbf{SO}(2R+1)$ for Ω^{2R} and $\mathbf{SO}(2R)$ for Ω^{2R-1} have Lie algebras with rank R which is also the power of π in the related sphere volumes.

The volume of the "full" s-dimensional ball, $s \geq 1$, with radius r is $\mathcal{V}^s(r) = \frac{|\Omega^{s-1}|}{s} r^s$ with $\frac{d\mathcal{V}^s(r)}{dr} = |\Omega^{s-1}|$.

The spherical–hyperbolic, i.e., compact–noncompact transition, gives the parametrizations and volume elements of the unit hyperboloids, parametrizing noncompact classes of orthogonal groups — using energy-momentum measures and the parametrizations $\sinh\psi = q = \frac{p}{\sqrt{1-p^2}}$:

$$\begin{array}{l}\int d^s y = \int_0^\infty d\psi\ \sinh^{s-1}\psi \int d^{s-1}\omega = \int_0^\infty d\psi\ \sinh^{s-1}\psi|\Omega^{s-1}|\\ \qquad = \int \vartheta(q_0)\, d^{1+s}q\ 2\delta(q_0^2 - \vec{q}^2 - 1) = \int \frac{d^s q}{\sqrt{1+\vec{q}^2}} = \int_{\vec{p}^2\leq 1}\frac{2d^s p}{(1-\vec{p}^2)^{\frac{1+s}{2}}},\end{array}$$

"polar" decomposition: $q = |q|y_s$ with $|q|^2 = q_0^2 - \vec{q}^2$, $y_s \in \mathcal{Y}^s$,

$$\int \vartheta(q_0)\vartheta(q^2)d^{1+s}q = \int_0^\infty |q|^s d|q| \int d^s y.$$

Spheres and hyperboloids are conformally, i.e., "angle-faithfully" Euclidean, as seen explicitly in a *Cartesian parametrization* with $e^{\psi_0(x)} = \frac{1}{1-k\frac{\vec{x}^2}{4}}$ in the dilation factors:

$$k\begin{pmatrix}\Omega^s\\ \mathcal{Y}^s\\ \mathbb{E}^s\end{pmatrix} = \begin{pmatrix}-1\\ 1\\ 0\end{pmatrix}: \begin{cases} \begin{pmatrix}d\omega_s^2\\ dy_s^2\\ d\vec{x}_s^2\end{pmatrix} = \frac{d\vec{x}_s^2}{(1-k\frac{\vec{x}^2}{4})^2}, \quad \begin{pmatrix}\mathbf{g}_{\Omega^s}\\ \mathbf{g}_{\mathcal{Y}^s}\\ \mathbf{g}_{\mathbb{E}^s}\end{pmatrix} = \frac{1}{(1-k\frac{\vec{x}^2}{4})^2}\mathbf{1}_s,\\ \begin{pmatrix}d^s\omega\\ d^s y\\ d^s x\end{pmatrix} = \frac{1}{(1-k\frac{\vec{x}^2}{4})^s}\, d^s x,\\ \begin{pmatrix}\partial^2_{\Omega^s}\\ \partial^2_{\mathcal{Y}^s}\\ \vec{\partial}^2_s\end{pmatrix} = (1-k\frac{\vec{x}^2}{4})^s\vec{\partial}(1-k\frac{\vec{x}^2}{4})^{2-s}\vec{\partial}\\ \qquad = (1-k\frac{\vec{x}^2}{4})^2\vec{\partial}^2 + (s-2)(1-k\frac{\vec{x}^2}{4})k\frac{\vec{x}^2}{4}\frac{\partial}{\partial\frac{\vec{x}^2}{4}}. \end{cases}$$

The open disk $r < 2$ for the 2-hyperboloid $\mathcal{Y}^2$ (Poincaré's model) has the non-Euclidean (Lobachevski) geometry $\frac{d\vec{x}_2^2}{(1-\frac{r^2}{4})^2}$ where the conformal $\vec{x}^2$-dependent dilation factor for the length expansion goes to infinity at the boundary $r \to 2$. For half of the 2-sphere Ω^2, projected on the disk $r \le 2$, the non-Euclidean geometry $\frac{d\vec{x}_2^2}{(1+\frac{r^2}{4})^2}$ arises with the conformal factor for the length contraction. The finite distances on the nonflat geodesics are given by the "leading angle":

$$d(0,r) = \int_0^r \frac{dR}{1-k\frac{R^2}{4}} = \begin{cases} 4\arctan\frac{r}{2} = 2\theta, & k = -1,\\ 2\log\frac{2+r}{2-r} = 4\,\mathrm{artanh}\,\frac{r}{2} = 2\psi, & k = 1,\end{cases}$$

$$\begin{pmatrix}\tan^2\theta\\ \tanh^2\psi\end{pmatrix} = r^2 = \frac{\rho^2}{1+k\rho^2}, \quad \begin{pmatrix}\sin^2\theta\\ \sinh^2\psi\end{pmatrix} = \rho^2 = \frac{r^2}{1-kr^2}.$$

7.28 Maximally Symmetric Manifolds

The foregoing structures are characteristic for the three maximally symmetric signature (t, s)-manifolds with dimension $n = t + s = 1, 2, \ldots$. Precisely for a constant curvature (more ahead), a Riemannian manifold is a general hyperboloid or a flat vector space,

$$\begin{aligned} \mathcal{Y}^{(t,s)} &\cong \mathbf{SO}_0(t, 1+s)/\mathbf{SO}_0(t,s), & k &= -1,\\ \mathbb{R}^{(t,s)} &\cong \mathbf{SO}_0(t,s)\,\vec{\times}\,\mathbb{R}^n/\mathbf{SO}_0(t,s), & k &= 0,\\ \mathcal{Y}^{(s,t)} &\cong \mathbf{SO}_0(1+t, s)/\mathbf{SO}_0(t,s), & k &= 1,\end{aligned}$$

with the spheres $\mathcal{Y}^{(0,s)} = \Omega^s$ and the proper hyperboloids $\mathcal{Y}^{(s,0)} = \mathcal{Y}^s$. The numbers (t, s) and (s, t) denote the (noncompact, compact) dimensions in the hyperboloids, e.g., noncompact $t(1+s) - ts = t$ for $\mathcal{Y}^{(t,s)}$. Examples are the (anti-)de Sitter universes $\mathcal{Y}^{(1,3)}$ and $\mathcal{Y}^{(3,1)}$. The flat spaces $\mathbb{R}^{(t,s)} = \mathbf{SO}_0(t,s)\,\vec{\times}\,\mathbb{R}^n/\mathbf{SO}_0(t,s) \cong \mathbb{R}^n$

have to be distinguished, not as manifolds, but as symmetric spaces, from the abelian groups $\mathbb{R}^n$ for $n \geq 2$.

The tangent Poincaré groups, isomorphic for these three manifolds, arise as Inönü–Wigner contractions of the global symmetry groups,

$$\begin{array}{c} \\ \left.\begin{array}{rl} (t,s) &= (1,0), \\ (\theta,\psi) &= \frac{x}{\ell}, \\ \ell &\to \infty \end{array}\right\} \end{array} \begin{array}{ccccc} \mathbf{SO}_0(1+t,s) & \rightarrow & \mathbf{SO}_0(t,s) \vec{\times} \mathbb{R}^{t+s} & \leftarrow & \mathbf{SO}_0(t,1+s), \\ \begin{pmatrix} \cos\theta & \ell\sin\theta \\ -\frac{1}{\ell}\sin\theta & \cos\theta \end{pmatrix} & \rightarrow & \begin{pmatrix} 1 & x \\ 0 & 1 \end{pmatrix} & \leftarrow & \begin{pmatrix} \cosh\psi & \ell\sinh\psi \\ \frac{1}{\ell}\sinh\psi & \cosh\psi \end{pmatrix}. \end{array}$$

All three maximal global symmetry groups are subgroups of $\mathbf{SO}_0(1+t,1+s)$. They are characterizable as fixgroups of nontrivial vectors $x \in \mathbb{R}^{2+t+s}$ with negative, trivial, and positive square x^2 as familiar from the $\mathbf{SO}_0(1,3)$-subgroups $\mathbf{SO}_0(1,2)$, $\mathbf{SO}(2) \vec{\times} \mathbb{R}^2$, and $\mathbf{SO}(3)$ as fixgroups for spacelike, nontrivial lightlike, and timelike $x \in \mathbb{R}^4$, respectively.

From the cases with definite metric $\mathcal{Y}^{(0,s)} = \Omega^s$ and $\mathcal{Y}^{(s,0)} = \mathcal{Y}^s$, one generalizes to hyperboloids $\mathcal{Y}^{(t,s)}$ and $\mathcal{Y}^{(s,t)}$ with $\vec{\omega}_0 = 1$, $d\omega_0^2 = 0$, and $d^0\omega = 0$ (sloppy ω-notation):

$$\begin{aligned} t \geq 1,\ \mathbf{SO}_0(t,1+s)/\mathbf{SO}_0(t,s) \cong \mathcal{Y}^{(t,s)} \ni \mathrm{y}_{(t,s)} &= \begin{pmatrix} \sinh\psi\,\vec{\omega}_{t-1} \\ \cosh\psi\,\vec{\omega}_s \end{pmatrix} \in \mathbb{R}^{1+t+s}, \\ d\mathrm{y}_{(t,s)}^2 &= d\psi^2 + \sinh^2\psi d\omega_{t-1}^2 - \cosh^2\psi\, d\omega_s^2, \\ d^{(t,s)}\mathrm{y} &= \sinh^{t-1}\psi\, d^{t-1}\omega\ \cosh^s\psi\, d^s\omega\, d\psi, \\ s \geq 1,\ \mathbf{SO}_0(1+t,s)/\mathbf{SO}_0(t,s) \cong \mathcal{Y}^{(s,t)} \ni \mathrm{y}_{(s,t)} &= \begin{pmatrix} \cosh\psi\,\vec{\omega}_t \\ \sinh\psi\,\vec{\omega}_{s-1} \end{pmatrix} \in \mathbb{R}^{1+t+s}, \\ d\mathrm{y}_{(s,t)}^2 &= d\psi^2 - \cosh^2\psi d\omega_t^2 + \sinh^2\psi\, d\omega_{s-1}^2, \\ d^{(s,t)}\mathrm{y} &= \cosh^t\psi\, d^t\omega\ \sinh^{s-1}\psi\, d\psi d^{s-1}\omega, \end{aligned}$$

with the lowest-dimensional nontrivial examples:

$$\begin{aligned} \mathbf{SO}(3)/\mathbf{SO}(2) \cong \Omega^2 \ni \vec{\omega}_2 &= \begin{pmatrix} \cos\theta \\ \sin\theta\ \cos\varphi \\ \sin\theta\ \sin\varphi \end{pmatrix} \in \mathbb{R}^3, \\ d\omega_2^2 &= d\theta^2 + \sin^2\theta\, d\varphi^2,\ \ d^2\omega = \sin\theta\, d\theta\, d\varphi, \\ \mathbf{SO}_0(1,2)/\mathbf{SO}(2) \cong \mathcal{Y}^2 \ni \mathrm{y}_2 &= \begin{pmatrix} \cosh\psi \\ \sinh\psi\ \cos\varphi \\ \sinh\psi\ \sin\varphi \end{pmatrix} \in \mathbb{R}^3, \\ d\mathrm{y}_2^2 &= d\psi^2 + \sinh^2\psi\, d\varphi^2,\ \ d^2\mathrm{y} = \sinh\psi\, d\psi\, d\varphi, \\ \mathbf{SO}_0(1,2)/\mathbf{SO}_0(1,1) \cong \mathcal{Y}^{(1,1)} \ni \mathrm{y}_{(1,1)} &= \begin{pmatrix} \sinh\psi \\ \cosh\psi\ \cos\varphi \\ \cosh\psi\ \sin\varphi \end{pmatrix} \in \mathbb{R}^3, \\ d\mathrm{y}_{(1,1)}^2 &= d\psi^2 - \cosh^2\psi\, d\varphi^2,\ \ d^{(1,1)}\mathrm{y} = \cosh\psi\, d\psi\, d\varphi. \end{aligned}$$

Sphere Ω^2 and "timelike" hyperboloid $\mathcal{Y}^2$ have a definite metric, the "spacelike" hyperboloid $\mathcal{Y}^{(1,1)}$ has an indefinite one. The compact-noncompact transition $\psi \leftrightarrow i\theta$ relates both hyperboloids to the sphere Ω^2:

$$\psi \leftrightarrow i\theta, \quad \begin{cases} \mathcal{Y}^2 : d\psi^2 + \sinh^2\psi\, d\varphi^2 \leftrightarrow -[d\theta^2 + \sin^2\theta\, d\varphi^2] : \Omega^2, \\ \mathcal{Y}^{(1,1)} : d\psi^2 - \cosh^2\psi\, d\varphi^2 \leftrightarrow -[d\theta^2 + \cos^2\theta\, d\varphi^2] : \Omega^2. \end{cases}$$

The metrical tensor is invariant under the maximal local group, i.e., it is conformal to the corresponding flat bilinear $\mathbf{SO}_0(t,s)$-form. In the Cartesian parametrization, the Lorentz invariant dilation factor for the metrical tensor depends on the invariant x^2:

$$\begin{aligned} \mathbf{g}^{li}(x) &= e^{2\psi_0(x)}\eta^{li}, \quad e^{\psi_0(x)} = \frac{1}{1-k\frac{x^2}{4}} \in \mathbf{D}(1) \text{ with } x^2 = x^r\eta_{rm}x^m = \vec{x}_t^2 - \vec{x}_s^2, \\ \psi_0(x) &= -\log(1-k\tfrac{x^2}{4}) = -k\tfrac{x^2}{4} + \dots. \end{aligned}$$

7.29 Derivations and Gauge Fields

For spacetime, global group actions, e.g., internal electromagnetic $\mathbf{U}(1)$-transformations of an electron-positron field $\boldsymbol{\Psi} \longmapsto e^{i\alpha}\boldsymbol{\Psi}$, can be extended to local or gauge transformations, e.g., $\boldsymbol{\Psi}(x) \longmapsto e^{i\alpha(x)}\boldsymbol{\Psi}(x)$. In a spacetime dynamics, the locally possibly different transformations have to be made compatible with appropriately modified derivations ∂^j, connecting "infinitesimally different" points of the manifold, e.g., compatible with the translations for flat space-times. The appropriate mathematical formalism for classical gauge transformations uses fiber bundle connections.

Good to know: A tangent bundle $\mathbf{T}(\mathbb{M})$ of a manifold is an example for a *module* over a ring (scalars), here over the manifold functions $f \in \mathcal{C}(\mathbb{M})$. A module generalizes the concept of a vector space over a field (scalars), e.g., the local tangent spaces $\mathbf{T}_x \cong \mathbb{R}^n, x \in \mathbb{M}$. The vector space corresponding definitions and properties for modules, especially those related to the scalar multiplication, have to be reconsidered. For instance, linearity may be restricted to additivity. Or: In contrast to a vector space, a module must not have a basis. Or: The restriction of the Lie algebra structure leads to the concept of a *Lie ring*, defined as an additive group with Lie bracket which has to be additive, antisymmetric and with Jacobi identity — scalar multiplication compatibility like $[fv, w] = f[v, w]$ is not required. If a ring contains a field, e.g., the constant functions $\mathcal{C}(\mathbb{M}) \supseteq \mathbb{R}$, the module is also a corresponding vector space which requires a careful distinction.

Derivations are characterized by additivity and the Leibniz property. The derivations of the commutative unital ring with the "sufficiently smooth" functions $\mathcal{C}(\mathbb{M}) = \{\mathbb{M} \ni x \longmapsto f(x) \in \mathbb{R}\}$ of a real n-dimensional manifold $\mathbb{M}$, especially a spacetime,

$$v : \mathcal{C}(\mathbb{M}) \longrightarrow \mathcal{C}(\mathbb{M}), \quad \text{with} \begin{cases} \text{additivity: } v(f+g) &= v(f) + v(g), \\ \text{Leibniz property: } v(fg) &= v(f)g + fv(g), \\ \text{locally in a basis: } v(x) &= v_k(x)e^k(x), \\ \text{holonomic bases: } e^k(x) &= \frac{\partial}{\partial x_k} = \partial^k, \end{cases}$$

constitute the tangent bundle $\mathbf{T}(\mathbb{M})$, a module for the ring $\mathcal{C}(\mathbb{M})$, with the tangent vector spaces $\mathbf{T}_x \cong \mathbb{R}^n$ as fibers. The derivations of these derivations, i.e., of the tangent fields, in directions $v \in \mathbf{T}(\mathbb{M})$:

$$\nabla_v : \mathbf{T}(\mathbb{M}) \longrightarrow \mathbf{T}(\mathbb{M}), \quad \text{with} \begin{cases} \nabla_v(u+w) &= \nabla_v u + \nabla_v w, \\ \nabla_v(fw) &= v(f)w + f\nabla_v w, \\ \nabla_{e^k}(e^j) &= \nabla^k e^j = \boldsymbol{\Gamma}^{kj}_i e^i. \end{cases}$$

constitute a *connection*, a morphism of the tangent bundle $\mathbf{T}(\mathbb{M})$ into its endomorphisms:

$$\begin{aligned} &\nabla : \mathbf{T}(\mathbb{M}) \longrightarrow \mathbf{T} \otimes \mathbf{T}^T(\mathbb{M}), \quad v \longmapsto \nabla_v, \\ &\mathcal{C}(\mathbb{M})\text{-linear: } \nabla_{fv+gw} = f\nabla_v + g\nabla_w. \end{aligned}$$

$\mathbf{T}^T(\mathbb{M})$ is the dual $\mathcal{C}(\mathbb{M})$-module, the cotangent bundle.

A derivation ∇_v with $\nabla_v(f) = v(f)$ on the scalar ring is extendable, by the characteristic derivation properties, to all tangent and cotangent fields, e.g., to tangent 1-form fields by $v(\langle \omega, u \rangle) = \langle \nabla_v \omega, u \rangle + \langle \omega, \nabla_v u \rangle$, e.g., for a dual basis $\nabla^k \check{e}_i = -\boldsymbol{\Gamma}^{kj}_i \check{e}_j$, or to 2-forms like metrics $\nabla_v(\omega \otimes \theta) = \nabla_v(\omega) \otimes \theta + \omega \otimes \nabla_v(\theta)$.

Good to know: A *derivation* D of a ring (an algebra) is an additive (linear) mapping with Leibniz rule, i.e., $D : A \longrightarrow A$ with $D(ab) = D(a)b + aD(b)$ for an associative and $D : L \longrightarrow L$ with $D([l, m]) = [D(l), m] + [l, D(m)]$ for a Lie ring (algebra). The commutator $[D_1, D_2]$ defines a Lie bracket for the derivations. A derivation is *inner*, if it acts, for an associative ring (associative algebra), as commutator $A \ni a \longmapsto \text{ad}\, b(a) = [b, a] = ba - ab$ with $b \in A$, and, for a Lie ring (Lie algebra), as Lie bracket $L \ni l \longmapsto \text{ad}\, m(l) = [m, l]$ with $m \in L$.

The tangent fields have a Lie bracket $[v, w]$, the tangent spaces $\mathbf{T}_x \cong \mathbb{R}^n$ are Lie algebras. The Lie brackets are trivial for a holonomic basis, $[\partial^j, \partial^k] = 0$.

A *Lie derivation* $\mathcal{L}_v$ is also additive with Leibniz property:

$$\mathcal{L}_v : \mathbf{T}(\mathbb{M}) \longrightarrow \mathbf{T}(\mathbb{M}) \text{ with } \mathcal{L}_v(w) = \text{ ad}\, v(w) = [v, w].$$

However, in general, $\mathcal{L}$ is no affine connection since it is not $\mathcal{C}(\mathbb{M})$-linear, $\mathcal{L}_{fv} \neq f\mathcal{L}_v$, with $[fv, w] = f[v, w] - w(f)v$. $\mathcal{L}_v$ with $\mathcal{L}_v(f) = v(f)$ is also extendable, by the characteristic derivation properties, to all tangent and cotangent fields.

The *invariance Lie algebra of a tensor field* (invariance under Lie derivation) is a Lie subalgebra of $\mathbf{T}$:

$$\mathbf{T}_a(\mathbb{M}) = \{v \in \mathbf{T}(\mathbb{M}) \,\big|\, \mathcal{L}_v(a) = [v, a] = 0\}.$$

By the definitions above, a local dual frame is invariant under all Lie derivations $[v, e^j \otimes \check{e}_j] = 0$.

The conditions for the Killing fields (invariances of the metric) of a Riemannian manifold,

$$(\mathbb{M}, \mathbf{g}) : \;\; v_k \partial^k \in \mathbf{T}_\mathbf{g}(\mathbb{M}) \iff \mathbf{g}^{kj}(\partial^i v_k) + \mathbf{g}^{ki}(\partial^j v_k) + v_k(\partial^k \mathbf{g}^{ij}) = 0,$$

can be written in analogy to the simple derivative form of "flat" vector spaces,

$$(\mathbb{R}^{(t,s)}, \eta): \ v_a\partial^a \in \mathbf{T}_\eta(\mathbb{R}^{(t,s)}) \iff \partial^a v^b + \partial^b v^a = 0, \ v^a = \eta^{ab} v_b,$$

by introducing the *Riemannian connection*,

$$(\mathbb{M}, \mathbf{g}): \ v_k\partial^k \in \mathbf{T}_\mathbf{g}(\mathbb{M}) \iff \nabla^i v^j + \nabla^j v^i = 0, \ v^i = \mathbf{g}^{ij} v_j, \\ \text{with} \nabla^i v^j = \partial^i v^j - \boldsymbol{\Gamma}^{ij}_{\ k} v^k.$$

It takes into account a possible "nonflatness" by including local changes via derivatives of the metrical tensor via the connection coefficients,

$$\boldsymbol{\Gamma}^{ij}_{\ k} = \frac{1}{2}\mathbf{g}_{kl}(\partial^i \mathbf{g}^{lj} + \partial^j \mathbf{g}^{li} - \partial^l \mathbf{g}^{ij}).$$

The Killing field condition for the metric motivates the introduction of a second derivative on a Riemannian manifold in addition to and constructed with the Lie derivative.

In general, a *connection for a representation bundle* $H \bullet W(\mathbb{M})$ with local bases $\{\mathbf{e}^\gamma(x)\}_{\gamma=1}^N$ of the fiber $W_x \cong \mathbb{K}^N$ with structural group H is called a *gauge field*. It is a section of the adjoint Lie algebra bundle,

$$\mathfrak{D}: \mathbf{T}(\mathbb{M}) \longrightarrow W \otimes W^T(\mathbb{M}), \ v \longmapsto \mathfrak{D}_v, \ \mathfrak{D}_{fv+gw} = f\mathfrak{D}_v + g\mathfrak{D}_w,$$

$$\mathfrak{D}_v: W(\mathbb{M}) \longrightarrow W(\mathbb{M}), \ \text{ with } \begin{cases} \mathfrak{D}_v(\boldsymbol{\Psi} + \boldsymbol{\Phi}) &= \mathfrak{D}_v\boldsymbol{\Psi} + \mathfrak{D}_v\boldsymbol{\Phi}, \\ \mathfrak{D}_v(f\boldsymbol{\Psi}) &= v(f)\boldsymbol{\Psi} + f\mathfrak{D}_v\boldsymbol{\Psi}, \\ \mathfrak{D}_{e^k}(\mathbf{e}^\gamma) &= \mathfrak{D}^k\mathbf{e}^\gamma = \mathbf{A}^{k\gamma}_{\ \ \beta}\mathbf{e}^\beta. \end{cases}$$

$$\mathbb{M} \ni x \longmapsto \mathbf{A}^k(x) \in (\log H)_x \subseteq (W \otimes W^T)_x.$$

As before, $(v, \mathfrak{D}_v)$ is extendable to all tensor powers of W and its dual W^T in the tensor algebra bundle $\bigotimes \mathbf{W}(\mathbb{M})$ with $\mathbf{W} = W \oplus W^T$ with the induced tensor representations of the group H and its Lie algebra $\log H \cong \mathbb{R}^d$. $\mathfrak{D}$ defines the *covariant derivatives* for the H-representations bundles on flat spacetime with $\mathfrak{D}^k = \partial^k - \mathbf{A}^k$ (more ahead).

7.30 Curvatures and Field Strengths

Similar to the tangent fields with the Lie bracket $[v, w]$ for the local Lie algebra $\mathbf{T}_x \cong \mathbb{R}^n$, also a connection has a Lie bracket $[\nabla_v, \nabla_w]$ for the local Lie algebra $(\mathbf{T} \otimes \mathbf{T}^T)_x \cong \mathbf{AL}(n, \mathbb{R})$ with the tangent space endomorphisms, and, more general,

$[\mathfrak{D}_v, \mathfrak{D}_w]$ for the local Lie algebra $(W \otimes W^T)_x \cong \mathbf{AL}(N, \mathbb{C})$. In general, a connection $v \longmapsto \nabla_v$ is not a representation of the tangent field Lie algebra in the tangent space endomorphisms, which would require $\nabla_{[v,w]} = [\nabla_v, \nabla_w]$, and, more general, $\mathfrak{D}_{[v,w]} = [\mathfrak{D}_v, \mathfrak{D}_w]$ for $v \longmapsto \mathfrak{D}_v$.

The *curvature* for a manifold connection is the "improved Lie algebra representation" as the difference of the two Lie brackets for two tangent fields. It is a module morphism, linear and antisymmetric:

$$\mathbf{R} : \mathbf{T} \wedge \mathbf{T}(\mathbb{M}) \longrightarrow \mathbf{T} \otimes \mathbf{T}^T(\mathbb{M}), \quad \begin{cases} \mathbf{R}(v \wedge w) &= [\nabla_v, \nabla_w] - \nabla_{[v,w]} \\ &= -\mathbf{R}(w \wedge v), \\ \mathbf{R}(fv \wedge w) &= f\mathbf{R}(v \wedge w),\ f \in \mathcal{C}(\mathbb{M}), \\ \mathbf{R}(e^k \wedge e^j) &= \mathbf{R}^{kj} = -\mathbf{R}^{jk} = \mathbf{R}_i^{lkj} e^i \otimes \check{e}_l. \end{cases}$$

The first *Bianchi*[12] *identity* $\mathbf{R}_i^{lkj} + \mathbf{R}_i^{kjl} + \mathbf{R}_i^{jlk} = 0$ corresponds to the Jacobi identity.

The curvature for a Riemannian connection with *Ricci*[13] *tensor* $\mathbf{R}_\bullet$ and curvature scalar $\mathbf{R}_\bullet^\bullet$,

$$\begin{aligned} \mathbf{R}_i^{lkj} &= \partial^k \boldsymbol{\Gamma}_i^{jl} - \partial^j \boldsymbol{\Gamma}_i^{kl} - \boldsymbol{\Gamma}_m^{kl} \boldsymbol{\Gamma}_i^{jm} + \boldsymbol{\Gamma}_m^{jl} \boldsymbol{\Gamma}_i^{km}, \\ \boldsymbol{\Gamma}_k^{ij} &= \frac{1}{2} \mathbf{g}_{kl} (\partial^i \mathbf{g}^{lj} + \partial^j \mathbf{g}^{li} - \partial^l \mathbf{g}^{ij}) \text{ (holonomic bases)}, \\ \mathbf{R}_\bullet^\bullet &= \mathbf{g}_{lk} \mathbf{R}_\bullet^{lk} = \mathbf{g}_{lk} \mathbf{R}_j^{lkj}, \end{aligned}$$

multiplied with the manifold metric, is, in the index pairs $(kj) \leftrightarrow (il)$, a symmetric bilinear form ($\binom{n}{2} \times \binom{n}{2}$-metric) for the two-dimensional tangent space areas $e^k \wedge e^j = -e^j \wedge e^k$:

$$\begin{aligned} \mathbf{g} \bullet \mathbf{R} : \mathbf{T} \wedge \mathbf{T}(\mathbb{M}) \longrightarrow \mathbf{T}^T \wedge \mathbf{T}^T(\mathbb{M}),\ \mathbf{g} \bullet \mathbf{R}(e^k \wedge e^j) &= \mathbf{R}^{ilkj} \check{e}_i \wedge \check{e}_l, \\ \mathbf{g} \bullet \mathbf{R}(e^i \wedge e^l, e^k \wedge e^j) = \mathbf{g}^{im} \mathbf{R}_m^{lkj} = \mathbf{R}^{ilkj} &= -\mathbf{R}^{iljk} = \mathbf{R}^{kjil}. \end{aligned}$$

In the index pair-diagonal form, $\mathbf{g} \bullet \mathbf{R}$ displays the dilation factors for the $\binom{t+s}{2}$ areas.

Einstein's equations relate the curvature of a Riemannian spacetime with the energy-momentum tensor $\mathbf{T}$, and, possibly, a cosmological term with cosmological constant Λ:

$$\mathbf{R}_\bullet - \frac{\mathbf{g}}{2} \mathbf{R}_\bullet^\bullet + \Lambda \mathbf{g} = -\frac{8\pi G}{c^2} \mathbf{T}.$$

With the curvature determined by the second order derivatives of the metric, linear and nonlinear, there is, with $\Lambda = 0$, the schematic form $\partial \boldsymbol{\Gamma} + \boldsymbol{\Gamma\Gamma} = \partial\partial \mathbf{g} + \partial \mathbf{g} \partial \mathbf{g} = \mathbf{T}$.

---

[12] Luigi Bianchi (1856–1928).

[13] Gregorio Ricci-Curbastro (1853–1925).

Einstein's theory extends Newton's nonrelativistic gravity for a mass density $\boldsymbol{\rho}_m$ with the special case of Newton's potential for a mass point:

$$\begin{aligned} \vec{\partial}^2 \mathbf{V}_m(\vec{x}) &= 4\pi G \boldsymbol{\rho}_m(\vec{x}) \Rightarrow \mathbf{V}_m(\vec{x}) = -\int d^3y \frac{G\boldsymbol{\rho}_m(\vec{y})}{|\vec{x}-\vec{y}|}, \\ \text{with } -\vec{\partial}^2 \frac{Gm}{|\vec{x}|} &= 4\pi G m \delta(\vec{x}). \end{aligned}$$

Mass is considered as distributed by its energy-momentum vector $m \hookrightarrow p^j, p^2 = m^2$, with the mass density embedded in the energy-momentum density $\boldsymbol{\rho}_m(\vec{x}) \hookrightarrow \mathbf{T}^j_k(x)$. In parallel, Newton's potential has to be embedded in the metrical tensor $\mathbf{V}_m(\vec{x}) \hookrightarrow \mathbf{g}^{jk}(x)$ as illustrated, in two parametrizations, by the Schwarzschild metric outside the horizon $r > \ell$:

$$\begin{aligned} ds^2 &= \tfrac{(1-\frac{\ell}{r})^2}{(1+\frac{\ell}{r})^2} dt^2 - (1+\tfrac{\ell}{r})^4 d\vec{x}^2, \qquad r = |\vec{x}|, \\ &= (1-\tfrac{4\ell}{\underline{r}})dt^2 - \tfrac{d\vec{\underline{x}}^2}{1-\frac{4\ell}{\underline{r}}} - \underline{r}^2 d\omega_2^2,\ \underline{r} = |\vec{\underline{x}}|. \end{aligned}$$

The corresponding procedure for electrodynamics uses the embedding of the charge density in the electromagnetic vector current $\boldsymbol{\rho}_z(\vec{x}) \hookrightarrow \mathbf{J}_k(x)$. The equation for the Coulomb potential of a point charge $\vec{\partial}^2 \frac{Q}{4\pi\epsilon_0|\vec{x}|} = -\frac{Q}{\epsilon_0}\delta(\vec{x})$ and, in parallel, a corresponding charge density has to be embedded in the inhomogeneous Maxwell equations $\partial^j \mathbf{F}_{jk} = \mathbf{J}_k$, or the second order equations for the vector potential $\partial\mathbf{F} = \partial\partial\mathbf{A} = \mathbf{J}$ with the embedding of the potential $\mathbf{V}_z(\vec{x}) = \int d^3y \frac{\boldsymbol{\rho}_z(\vec{y})}{4\pi\epsilon_0|\vec{x}-\vec{y}|} \hookrightarrow \mathbf{A}^j(x)$.

The dynamics of a charged mass point ($Q = ze$, charge number $z \in \mathbb{Z}$) in flat and a mass point in Riemannian spacetime arises from the classical actions with eigentime τ and point orbit coordinates (geodesics) $\tau = \epsilon(x_0)\vartheta(x^2)\sqrt{x^2} \longmapsto \mathbf{x}_i(\tau)$:

$$\begin{aligned} &W_z^{\text{em}} = \tfrac{\hbar}{4\pi\alpha_{\text{Som}}} \int d^4x (\mathbf{F}_{ij} \tfrac{\partial^i \mathbf{A}^j - \partial^j \mathbf{A}^i}{2} + \tfrac{\mathbf{F}_{ij}\mathbf{F}^{ij}}{4}) + \tfrac{m}{2} \int d\tau d_\tau \mathbf{x}_i d_\tau \mathbf{x}^i - \hbar z \int d\tau \mathbf{A}^i(\mathbf{x}) d_\tau \mathbf{x}_i, \\ &\text{current vector: } \mathbf{J}_i(x) = z \int d\tau\ d_\tau \mathbf{x}_i\ \delta(x - \mathbf{x}), \end{aligned}$$

$$\begin{aligned} &W_m^{\text{grav}} = \tfrac{c^3}{8\pi G} \int d^4x \sqrt{|\det \mathbf{g}|} \mathbf{R}^{\bullet}_{\bullet} + \tfrac{m}{2} \int d\tau d_\tau \mathbf{x}_i \mathbf{g}^{ij}(\mathbf{x}) d_\tau \mathbf{x}_j, \\ &\text{energy-momentum tensor: } \sqrt{|\det \mathbf{g}|} c \mathbf{T}_{ij}(x) = m \int d\tau\ d_\tau \mathbf{x}_i d_\tau \mathbf{x}_j\ \delta(x - \mathbf{x}). \end{aligned}$$

In a harmonic rotation analysis, nonrelativistic potentials for a mass and charge density can be expanded by multipoles, involving Legendre[14] polynomials P^L for the irreducible $\mathbf{SO}(3)$-representations with their factorization in spherical harmonics Y^L_m and harmonic $\mathbf{SO}(3)$-polynomials $\mathrm{Y}^L_m(\frac{\vec{y}}{|\vec{y}|})|\vec{y}|^L = (\vec{y})^L_m$:

$$\mathrm{P}^L(\cos\theta) = \sqrt{\frac{4\pi}{1+2L}} \mathrm{Y}^L_0(\varphi, \theta),$$

[14] Adrien-Marie Legendre (1752–1833).

$$\frac{1}{|\vec{x}-\vec{y}|} = \frac{1}{|\vec{x}|}\sum_{L=0}^{\infty} \mathrm{P}^L(\cos\theta)\left(\frac{|\vec{y}|}{|\vec{x}|}\right)^L \text{ for } |\vec{y}| \leq |\vec{x}|,$$

$$\cos\theta = \frac{\vec{x}}{|\vec{x}|}\frac{\vec{y}}{|\vec{y}|} : \mathrm{P}^L\left(\frac{\vec{x}\vec{y}}{|\vec{x}||\vec{y}|}\right) = \frac{4\pi}{1+2L}\sum_{m=-L}^{L} \overline{\mathrm{Y}}^L_m\left(\frac{\vec{x}}{|\vec{x}|}\right)\mathrm{Y}^L_m\left(\frac{\vec{y}}{|\vec{y}|}\right),$$

$$\Rightarrow \int d^3y\, \frac{\rho(\vec{y})}{|\vec{x}-\vec{y}|} = \sum_{L=0}^{\infty}\frac{4\pi}{(1+2L)|\vec{x}|^{1+L}}\sum_{m=-L}^{L}\overline{\mathrm{Y}}^L_m\left(\frac{\vec{x}}{|\vec{x}|}\right)\rho^L_m,$$

$$\text{with } \rho^L_m = \int d^3y\, \rho(\vec{y})\, \mathrm{Y}^L_m\left(\frac{\vec{y}}{|\vec{y}|}\right)|\vec{y}|^L = \int d^3y\, \rho(\vec{y})\, (\vec{y})^L_m.$$

For one-dimensional manifolds (abelian groups) $(\Omega^1, \mathbb{R}, \mathcal{Y}^1)$, e.g., real one-dimensional time, the curvature concept does not apply — the naively supposed "curvature" of a circle is not "intrinsic"; it can be "understood" only with a surrounding 2-space. A curvature as metric of two-dimensional tangent areas $\{e^k \wedge e^j \,\big|\, j, k = 1, \ldots, n\}$ needs at least two dimensions — starting for the maximally symmetric manifolds with $(\Omega^2, \mathcal{Y}^2)$ (positive curvature) and $\mathcal{Y}^{(1,1)}$ (negative curvature) which can be visualized in $\mathbb{E}^3$. For maximally symmetric Riemannian manifolds the curvature is the antisymmetric square of the metrical tensor, i.e., of the metric for tangent space vectors $\{e^k \,\big|\, j = 1, \ldots, n\}$. Via the tangent space metric, the curvature has a signature. The Ricci tensor can be expressed by a flat space metric with the conformal dilation factor:

$$\mathbf{R} = -k\,\mathbf{g}\wedge\mathbf{g}, \qquad \mathbf{R}^{ilkj}(x) = -\frac{k}{(1-k\frac{x^2}{4})^4}(\eta^{ik}\eta^{lj} - \eta^{ij}\eta^{lk}),$$

$$\mathbf{R}_\bullet = (n-1)k\,\mathbf{g}, \qquad \mathbf{R}^{lij}_j(x) = (n-1)\frac{k}{(1-k\frac{x^2}{4})^2}\eta^{li},$$

$$\mathbf{R}^\bullet_\bullet = 2\binom{n}{2}k, \qquad \mathbf{R}^{ij}_{ij}(x) = 2\binom{n}{2}k,$$

$$\mathbf{R}_\bullet - \frac{1}{2}\mathbf{R}^\bullet_\bullet\mathbf{g} = -\binom{n-1}{2}k\,\mathbf{g}, \qquad \text{with } \mathbf{g} = e^{2\psi_0}\eta,\ e^{\psi_0(x)} = \frac{1}{1-k\frac{x^2}{4}}.$$

For $n \geq 3$, the curvature is related to the Killing form of the nonabelian local Lie algebra $\log \mathbf{SO}_0(t,s)$ with dimension $\binom{n}{2}$, composed by ts noncompact and $\binom{t}{2} + \binom{s}{2}$ compact dimensions:

$$\text{for } \log \mathbf{SO}_0(t,s) : \eta \cong \begin{pmatrix} \mathbf{1}_t & 0 \\ 0 & -\mathbf{1}_s \end{pmatrix}, \quad \eta\wedge\eta \cong \begin{pmatrix} \mathbf{1}_{ts} & 0 \\ 0 & -\mathbf{1}_{\binom{t}{2}+\binom{s}{2}} \end{pmatrix},$$

$$\text{e.g., } \log \mathbf{SO}(s) : \eta = -\delta \cong -\mathbf{1}_s,\ \delta\wedge\delta \cong -\mathbf{1}_{\binom{s}{2}},$$

$$\text{e.g., } \log \mathbf{SO}_0(1,s) : \eta \cong \begin{pmatrix} 1 & 0 \\ 0 & -\mathbf{1}_s \end{pmatrix}, \quad \eta\wedge\eta \cong \begin{pmatrix} \mathbf{1}_s & 0 \\ 0 & -\mathbf{1}_{\binom{s}{2}} \end{pmatrix}.$$

More general, a curvature can be defined for representation bundles: The *field strengths* (*curvature*) of an H-connection on $\mathbb{M}$ define $\log H$-Lie algebra operations in the endomorphisms of the vector space fibers:

$$\mathbf{F} : \mathbf{T} \wedge \mathbf{T}(\mathbb{M}) \longrightarrow W \otimes W^T(\mathbb{M}), \begin{cases} \mathbf{F}(v \wedge w) &= [\mathfrak{D}_v, \mathfrak{D}_w] - \mathfrak{D}_{[v,w]} \\ &= -\mathbf{F}(w \wedge v), \\ \mathbf{F}(e^k \wedge e^j) &= \mathbf{F}^{kj} = -\mathbf{F}^{jk} = \mathbf{F}^{\alpha kj}_{\beta} \mathbf{e}^{\beta} \otimes \check{\mathbf{e}}_{\alpha} \\ &= \partial^k \mathbf{A}^j - \partial^j \mathbf{A}^k - [\mathbf{A}^k, \mathbf{A}^j], \\ &\text{(holonomic bases)}, \end{cases}$$

$$\mathbf{F}(v \wedge w) : W(\mathbb{M}) \longrightarrow W(\mathbb{M}), \begin{cases} \mathbf{F}^{\alpha kj}_{\beta} &= \partial^k \mathbf{A}^{j\alpha}_{\ \ \beta} - \partial^j \mathbf{A}^{k\alpha}_{\ \ \beta} \\ &\quad - \mathbf{A}^{k\gamma}_{\ \ \beta} \mathbf{A}^{j\alpha}_{\ \ \gamma} + \mathbf{A}^{j\gamma}_{\ \ \beta} \mathbf{A}^{k\alpha}_{\ \ \gamma}, \\ &\text{(holonomic bases)}, \end{cases}$$

$$\mathbb{M} \ni x \longmapsto \mathbf{F}^{kj}(x) \in (\log H)_x \subseteq (W \otimes W^T)_x.$$

For a Riemannian manifold, the field strengths (curvature) map the Lorentz Lie algebra, acting on the tangent spaces, into the structural Lie algebra $\mathbf{F}^{\alpha kj}_{\beta}$: $\log \mathbf{SO}_0(t, s) \longrightarrow \log H$, acting on the fiber (no Lie algebra morphism). The geometrical meaning of "curvature" (local area change) applies only for the tangent bundle.

The electroweak-strong standard model is formalized with representation bundles $\mathbf{U}(2 \times 3) \bullet W(\mathbb{R}^{(1,3)})$ and the adjoint Lie algebra bundle $\log \mathbf{U}(2 \times 3)(\mathbb{R}^{(1,3)})$. The representation properties of fields, gauge potentials, and field strengths are characterized as follows:

$$\mathbf{U}(2 \times 3) \ni U \longmapsto u \in \mathbf{U}(W) : \begin{cases} \mathbf{\Psi}(x) \longmapsto u(x)\mathbf{\Psi}(x), \\ \mathbf{A}^k(x) \longmapsto u(x)\mathbf{A}^k(x)u^\star(x) + (\partial^k u)(x)u^\star(x), \\ \mathbf{F}^{kj}(x) \longmapsto u(x)\mathbf{F}^{kj}(x)u^\star(x). \end{cases}$$

7.31 Electroweak and Strong Gauge Interactions

The *standard model* of the elementary interactions in Minkowski spacetime is a theory of compatibly represented external and internal operations. It embeds the historically first formulated quantum field theory for the electromagnetic interaction of an electron Dirac field (quantum electrodynamics) into the electroweak and strong gauge interactions of quark and lepton Weyl fields. The fields involved are acted on by irreducible representations $(L|R)$ of the Lorentz group $\mathbf{SL}(2, \mathbb{C})$ and irreducible representations of the hypercharge group $\mathbf{U}(1)$ (rational hypercharge number y), of the isospin group $\mathbf{SU}(2)$ (integer or halfinteger isospin in (T)), and the color group $\mathbf{SU}(3)$ as given in the following table:

Field	Symbol	$\mathbf{SL}(2,\mathbb{C})$ $(L\|R)$	$\mathbf{U}(1)$ y	$\mathbf{SU}(2)$ (T)	$\mathbf{SU}(3)$ $[C_1,C_2]$	$\mathbf{U}(1)_+$ $q=y+T_3$
Left lepton	$\mathbf{l}$	$(\frac{1}{2}\|0)$	$-\frac{1}{2}$	$(\frac{1}{2})$	$[0,0]$	$0,-1$
"Mute" neutrino	$\boldsymbol{\nu}$	$(0\|\frac{1}{2})$	0	(0)	$[0,0]$	0
Right lepton	$\mathbf{e}$	$(0\|\frac{1}{2})$	-1	(0)	$[0,0]$	-1
Left quark	$\mathbf{q}$	$(\frac{1}{2}\|0)$	$\frac{1}{6}$	$(\frac{1}{2})$	$[\frac{1}{2},0]$	$\frac{2}{3},-\frac{1}{3}$
Right up quark	$\mathbf{u}$	$(0\|\frac{1}{2})$	$\frac{2}{3}$	(0)	$[\frac{1}{2},0]$	$\frac{2}{3}$
Right down quark	$\mathbf{d}$	$(0\|\frac{1}{2})$	$-\frac{1}{3}$	(0)	$[\frac{1}{2},0]$	$-\frac{1}{3}$
Hypercharge gauge	$\mathbf{A}_0$	$(\frac{1}{2}\|\frac{1}{2})$	0	(0)	$[0,0]$	0
Isospin gauge	$\vec{\mathbf{A}}$	$(\frac{1}{2}\|\frac{1}{2})$	0	(1)	$[0,0]$	$0,\pm 1$
Color gauge	$\mathbf{C}$	$(\frac{1}{2}\|\frac{1}{2})$	0	(0)	$[\frac{1}{2},\frac{1}{2}]$	0
Higgs	$\boldsymbol{\Phi}$	$(0\|0)$	$\frac{1}{2}$	$(\frac{1}{2})$	$[0,0]$	$1,0$

The fields of the minimal standard model

abelian hypercharge $\mathbf{U}(1)$ and the nonabelian isospin-color group $\mathbf{SU}(2)\times\mathbf{SU}(3)$,

$$\mathbf{U}(1)\hookrightarrow\mathbf{U}(2\times 3)=\mathbf{U}(\mathbf{1}_6)\circ[\mathbf{SU}(2)\times\mathbf{SU}(3)]=\frac{\mathbf{U}(1)\times\mathbf{SU}(2)\times\mathbf{SU}(3)}{\mathbf{I}(2)\times\mathbf{I}(3)}.$$

A degenerate (asymmetric) ground state (more ahead) distinguishes an electromagnetic phase group with the charge as sum of hypercharge $y\in\frac{\mathbb{Z}}{6}$ and third isospin component $T_3\in\frac{\mathbb{Z}}{2}$:

$$\mathbf{U}(1)\subset\mathbf{U}(2):\ \ q=y+T_3\in\frac{\mathbb{Z}}{3}.$$

The fields are acted on homogeneously by a direct product of Lorentz and internal transformations. With the exception of the Higgs field, the isospin $\mathbf{SU}(2)$-representation is a subrepresentation of the Lorentz group $\mathbf{SL}(2,\mathbb{C})$-representation. This is a characteristic structure of induced representations that start with the two-sided regular representation and the doubled group.

In Weinberg's original model of leptons with no experimental indication for a massive neutrino, there was no right-handed neutrino, $\boldsymbol{\nu}=0$. For a massive neutrino, an internally scalar ("mute") neutrino $\boldsymbol{\nu}$ can be added.

Both factors in the internal group $\mathbf{U}(\mathbf{1}_6)\circ[\mathbf{SU}(2)\times\mathbf{SU}(3)]$ are centrally correlated, i.e., the representations of hypercharge $\mathbf{U}(1)$ are related to the representations of the $\mathbf{SU}(2)\times\mathbf{SU}(3)$-center, the cyclotomic group $\mathbf{I}(2)\times\mathbf{I}(3)=\mathbf{I}(6)$ (hexality = two-triality, "star of David"). All $\mathbf{U}(2\times 3)$-representations $[y\|T;2C_1,2C_2]$ (not the Lorentz properties!) carried by the standard model fields with the isospin and color multiplicities:

$$d_{\mathbf{SU}(2)}=1+2T,\ \ d_{\mathbf{SU}(3)}=(1+2C_1)(1+2C_2)(1+C_1+C_2),$$

can be generated by the dual defining representations of $\mathbf{U}(2 \times 3)$,

$$u = \left[\frac{1}{6}||\frac{1}{2}; 1, 0\right], \quad \check{u} = \left[-\frac{1}{6}||\frac{1}{2}; 0, 1\right],$$

as seen in the powers $\bigwedge^n u \otimes \bigwedge^m \check{u}$ (all fermion fields are taken as left-handed),

Field	$\mathbf{U}(2\times 3)$ $[y\|\|T; 2C_1, 2C_2]$	(n, m)	$n - m = 6y$	$6y$ mod 2	$6y$ mod 3
$\mathbf{l}$	$[-\frac{1}{2}\|\|\frac{1}{2}; 0, 0]$	$(0, 3)$	-3	1	0
$\boldsymbol{\nu}^\star$	$[0\|\|0; 0, 0]$	$(0, 0)$	0	0	0
$\mathbf{e}^\star$	$[1\|\|0; 0, 0]$	$(6, 0)$	6	0	0
$\mathbf{q}$	$[\frac{1}{6}\|\|\frac{1}{2}; 1, 0]$	$(1, 0)$	1	1	1
$\mathbf{u}^\star$	$[-\frac{2}{3}\|\|0; 0, 1]$	$(0, 4)$	-4	0	-1
$\mathbf{d}^\star$	$[\frac{1}{3}\|\|0; 0, 1]$	$(2, 0)$	2	0	-1
$\mathbf{A}_0$	$[0\|\|0; 0, 0]$	$(0, 0)$	0	0	0
$\vec{\mathbf{A}}$	$[0\|\|1; 0, 0]$	$(1, 1)$	0	0	0
$\mathbf{C}$	$[0\|\|0; 1, 1]$	$(1, 1)$	0	0	0
$\boldsymbol{\Phi}$	$[\frac{1}{2}\|\|\frac{1}{2}; 0, 0]$	$(3, 0)$	3	1	0

The central correlations of the internal symmetries are expressed by the modulo relations:

$$\begin{aligned} &6y \text{ mod } 2 = 2T \text{ mod } 2, \quad 6y \text{ mod } 3 = 2(C_1 - C_2) \text{ mod } 3, \\ &y \cdot d_{\mathbf{SU}(2)} \cdot d_{\mathbf{SU}(3)} \in \mathbb{Z}. \end{aligned}$$

The non-gauge fields with the free Lagrangians:

$$\begin{aligned} \text{left fermions: } \mathbf{L}(\mathbf{l}) &= i\mathbf{l}^\star_\alpha \partial^k \check{\sigma}_k \mathbf{l}^\alpha + i\mathbf{q}^\star_{\alpha c} \partial^k \check{\sigma}_k \mathbf{q}^{\alpha c}, \quad \alpha = 1, 2; \ c = 1, 2, 3, \\ \text{right fermions: } \mathbf{L}(\mathbf{r}) &= i\mathbf{e}^\star \partial^k \sigma_k \mathbf{e} + +i\boldsymbol{\nu}^\star \partial^k \sigma_k \boldsymbol{\nu} + i\mathbf{u}^\star_c \partial^k \sigma_k \mathbf{u}^c + i\mathbf{d}^\star_c \partial^k \sigma_k \mathbf{d}^c, \\ \text{Higgs: } \mathbf{L}(\boldsymbol{\Phi}) &= \boldsymbol{\Phi}^\star_{k\alpha} \partial^k \boldsymbol{\Phi}^\alpha + \boldsymbol{\Phi}^\alpha_k \partial^k \boldsymbol{\Phi}^\star_\alpha - \boldsymbol{\Phi}^\alpha_k \boldsymbol{\Phi}^{k\star}_\alpha \cong (\partial^k \boldsymbol{\Phi}^\star_\alpha)(\partial_k \boldsymbol{\Phi}^\alpha), \end{aligned}$$

interact with the four gauge fields $\mathbf{A}_0$ and $\vec{\mathbf{A}}$ for the electroweak interactions and the eight color gauge fields $\mathbf{C}$ (gluons) for the strong interactions, with the local transformation properties:

$$\begin{aligned} \mathbf{U} = \exp L: \quad & \mathbf{A} \longmapsto u \circ (\mathbf{A} - \mathbf{1}\partial) \circ u^\star = \mathbf{A} + [l, \mathbf{A}] + \partial l + \ldots \text{for } u = \exp l, \\ \mathbf{U}(1): \quad i\mathbf{A}_0^k & \longmapsto e^{i\alpha_0}(i\mathbf{A}_0^k - \partial^k)e^{-i\alpha_0}, \\ \mathbf{SU}(2): \quad \tfrac{i\vec{\tau}}{2}\vec{\mathbf{A}}^k & \longmapsto e^{\frac{i\vec{\tau}}{2}\vec{\alpha}}(\tfrac{i\vec{\tau}}{2}\vec{\mathbf{A}}^k - \partial^k)e^{-\frac{i\vec{\tau}}{2}\vec{\alpha}}, \\ \mathbf{SU}(3): \quad \tfrac{i\lambda^A}{2}\mathbf{C}^k_A & \longmapsto e^{\frac{i\lambda^B}{2}\gamma_B}(\tfrac{i\lambda^A}{2}\mathbf{C}^k_A - \partial^k)e^{-\frac{i\lambda^B}{2}\gamma_B}. \end{aligned}$$

and the Lagrangians:

$$\begin{aligned}
\mathbf{L}(\mathbf{A}_0) &= \mathbf{F}_{kj}\frac{\partial^k \mathbf{A}_0^j - \partial^j \mathbf{A}_0^k}{2} + g_1^2\frac{\mathbf{F}_{kj}\mathbf{F}^{kj}}{4},\\
\mathbf{L}(\vec{\mathbf{A}}) &= \mathbf{F}_{kj}^c\frac{\partial^k \mathbf{A}_c^j - \partial^j \mathbf{A}_c^k - \epsilon_c^{ab}\mathbf{A}_a^k\mathbf{A}_b^j}{2} + g_2^2\frac{\mathbf{F}_{kj}^b\mathbf{F}_b^{kj}}{4},\\
\mathbf{L}(\mathbf{C}) &= \mathbf{F}_{kj}^C\frac{\partial^k \mathbf{C}_C^j - \partial^j \mathbf{C}_C^k - \epsilon_C^{AB}\mathbf{C}_A^k\mathbf{C}_B^j}{2} + g_3^2\frac{\mathbf{F}_{kj}^B\mathbf{F}_B^{kj}}{4}.
\end{aligned}$$

The indices differentiate between the different Lie algebras in the case of the field strengths $\mathbf{F}$. The structure constants are taken in a Pauli and Gell-Mann basis:

$$\log \mathbf{SU}(2) : \left\{\frac{i}{2}\tau^{a\beta}_{\ \ \gamma} | a = 1,2,3,\ \ \beta = 1,2\right\}, \qquad \left[\frac{i}{2}\tau^a, \frac{i}{2}\tau^b\right] = \epsilon_c^{ab}\frac{i}{2}\tau^c,$$

$$\log \mathbf{SU}(3) : \left\{\frac{i}{2}\lambda^{Ab}_{\ \ c} | A = 1,\ldots,8;\ \ b = 1,2,3\right\}, \quad \left[\frac{i}{2}\lambda^A, \frac{i}{2}\lambda^B\right] = \epsilon_C^{AB}\frac{i}{2}\lambda^C.$$

The gauge field coupling constants are the normalization ratios of the internal Lie algebras and the Lorentz Lie algebra:

$$\left(\frac{1}{g_1^2}, \frac{1}{g_2^2}, \frac{1}{g_3^2}\right) = \frac{(\kappa^2_{\mathbf{U}(1)}, \kappa^2_{\mathbf{SU}(2)}, \kappa^2_{\mathbf{SU}(3)})}{\kappa^2_{\mathbf{SO}_0(1,3)}}.$$

With the local transformation behavior of the matter and Higgs fields:

$$\mathbf{U}(2\times 3) : \begin{cases}
\mathbf{l} \longmapsto e^{-\frac{i\mathbf{1}_2}{2}\alpha_0 + \frac{i\vec{\tau}}{2}\vec{\alpha}}\,\mathbf{l}, & \mathbf{q} \longmapsto e^{\frac{i\mathbf{1}_6}{6}\alpha_0 + \frac{i\vec{\tau}\otimes\mathbf{1}_3}{2}\vec{\alpha} + \frac{i\mathbf{1}_2\otimes\lambda^A}{2}\gamma_A}\,\mathbf{q},\\
\boldsymbol{\nu} \longmapsto \boldsymbol{\nu}, & \mathbf{u} \longmapsto e^{\frac{2i\mathbf{1}_3}{3}\alpha_0 + \frac{i\lambda^A}{2}\gamma_A}\,\mathbf{u},\\
\mathbf{e} \longmapsto e^{-i\alpha_0}\,\mathbf{e}, & \mathbf{d} \longmapsto e^{-\frac{i\mathbf{1}_3}{3}\alpha_0 + \frac{i\lambda^A}{2}\gamma_A}\,\mathbf{d},\\
\boldsymbol{\Phi} \longmapsto e^{i\frac{\mathbf{1}_2\alpha_0 + \vec{\tau}\vec{\alpha}}{2}}\,\boldsymbol{\Phi}, &
\end{cases}$$

their gauge interactions in

$$\mathbf{L}(\mathbf{A}_0) + \mathbf{L}(\vec{\mathbf{A}}) + \mathbf{L}(\mathbf{C}) + \mathbf{L}(\mathbf{l}) + \mathbf{L}(\mathbf{r}) + \mathbf{L}(\boldsymbol{\Phi}) - (\mathbf{A}_0^k\mathbf{J}_k + \mathbf{A}_a^k\mathbf{J}_k^a + \mathbf{C}_A^k\mathbf{J}_k^A),$$

involve the currents for the non-gauge fields

$$\begin{aligned}
\log \mathbf{U}(1)\colon\ \mathbf{J}_k &= -\tfrac{1}{2}\mathbf{l}^*\check{\sigma}_k\mathbf{l} - \mathbf{e}^*\sigma_k\mathbf{e} + \tfrac{1}{6}\mathbf{q}^*\check{\sigma}_k\mathbf{q} + \tfrac{2}{3}\mathbf{u}^*\sigma_k\mathbf{u} - \tfrac{1}{3}\mathbf{d}^*\sigma_k\mathbf{d}\\
&\quad - \tfrac{i}{2}\left(\boldsymbol{\Phi}^\star\boldsymbol{\Phi}_k - \boldsymbol{\Phi}\boldsymbol{\Phi}_k^\star\right),
\end{aligned}$$

$$\begin{aligned}
\log \mathbf{SU}(2)\colon\ \mathbf{J}_k^a &= \mathbf{l}^*\check{\sigma}_k\tfrac{\tau^a}{2}\mathbf{l} + \mathbf{q}^*\check{\sigma}_k\tfrac{\tau^a}{2}\mathbf{q}\\
&\quad + \left(i\boldsymbol{\Phi}\tfrac{\tau^a}{2}\boldsymbol{\Phi}_k^\star - i\boldsymbol{\Phi}_k\tfrac{\tau^a}{2}\boldsymbol{\Phi}^\star\right),
\end{aligned}$$

$$\log \mathbf{SU}(3)\colon\ \mathbf{J}_k^A = \mathbf{q}^*\check{\sigma}_k\tfrac{\lambda^A}{2}\mathbf{q} + \mathbf{u}^*\sigma_k\tfrac{\lambda^A}{2}\mathbf{u} + \mathbf{d}^*\sigma_k\tfrac{\lambda^A}{2}\mathbf{d}.$$

The gauge field strengths equations are given with the corresponding currents:

$$\partial^j \mathbf{F}_{kj} + \mathbf{F}_{kj} \times \mathbf{A}^j = \mathbf{J}_k.$$

The spacetime translations of leptons and quarks come with minimally coupled internal gauge field actions in the covariant derivatives:

$$\begin{aligned}
\left(\partial^k + \tfrac{i}{2}\mathbf{A}_0^k - i\tfrac{\vec{\tau}}{2}\vec{\mathbf{A}}^k\right)\check{\sigma}_k \mathbf{l} &= 0, & \left(\partial^k - \tfrac{i}{6}\mathbf{A}_0^k - i\tfrac{\vec{\tau}}{2}\vec{\mathbf{A}}^k - i\tfrac{\lambda^A}{2}\mathbf{C}_A^k\right)\check{\sigma}_k \mathbf{q} &= 0,\\
\partial^k \sigma_k \boldsymbol{\nu} &= 0, & \left(\partial^k - \tfrac{2i}{3}\mathbf{A}_0^k - i\tfrac{\lambda^A}{2}\mathbf{C}_A^k\right)\sigma_k \mathbf{u} &= 0,\\
(\partial^k + i\mathbf{A}_0^k)\sigma_k \mathbf{e} &= 0, & \left(\partial^k + \tfrac{i}{3}\mathbf{A}_0^k - i\tfrac{\lambda^A}{2}\mathbf{C}_A^k\right)\sigma_k \mathbf{d} &= 0,
\end{aligned}$$

as well as for the scalar Higgs field:

$$\left(\partial^k - i\frac{\mathbf{1}_2\mathbf{A}_0^k + \vec{\tau}\vec{\mathbf{A}}^k}{2}\right)\boldsymbol{\Phi} = \boldsymbol{\Phi}^k, \quad \left(\partial^k - i\frac{\mathbf{1}_2\mathbf{A}_0^k + \vec{\tau}\vec{\mathbf{A}}^k}{2}\right)\boldsymbol{\Phi}_k = 0,$$

or, with second order Lagrangian:

$$\mathbf{L}(\boldsymbol{\Phi}, \mathbf{A}) = \left[\left(\partial^k - i\frac{\mathbf{1}_2\mathbf{A}_0^k + \vec{\tau}\vec{\mathbf{A}}^k}{2}\right)\boldsymbol{\Phi}\right]\left[\left(\partial_k + i\frac{\mathbf{1}_2\mathbf{A}_{0k} + \vec{\tau}\vec{\mathbf{A}}_k}{2}\right)\boldsymbol{\Phi}^\star\right].$$

7.32 Representation Currents and Gauge Vertices

Gauge theories connect spacetime translations with internal Lie algebra operations. All standard model basic interactions are described by gauge vertices. The currents are position distributions of the represented Lie algebra. A Lie algebra L representation is characterized by a power three tensor $L^T \otimes W \otimes W^T$, involving the dual Lie algebra L^T and Lie algebra representation, formalized by a power two tensor $W \otimes W^T$ of representation space and its dual. Therefore, via the currents, all gauge vertices involve in addition to the gauge field two fields, e.g., spinor fields in the electromagnetic vertex $\mathbf{A}^k\overline{\boldsymbol{\Psi}}\gamma_k\boldsymbol{\Psi}$ for half-integer spin electron-positrons. For integer spin fields, e.g., charged pions or the gauge field itself in the nonabelian case, the corresponding current may be re-written in a first order derivative formulation which leads to power three tensors in the current and power four tensors in the gauge vertices. All Feynman diagrams for the standard electroweak-strong interactions are composed by such three-prongs (or integer spin four-prongs) vertices.

With dual bases for a Lie algebra L and its linear forms L^T and Lie bracket with structure constants $\epsilon_c^{ab} = -\epsilon_c^{ba}$:

$$L \cong \mathbb{R}^d : \ \langle \check{l}_a, l^b \rangle = \delta_a^b, \ \ [l^a, l^b] = \epsilon_c^{ab} l^c,$$

the internal Lie algebra is represented by endomorphisms of a vector space and its dual W, $W^T \cong \mathbb{C}^N$ with dual bases $\langle \check{\mathbf{e}}_\beta, \mathbf{e}^\gamma \rangle = \delta^\gamma_\beta = \epsilon \langle \mathbf{e}^\gamma, \check{\mathbf{e}}_\beta \rangle$, $\epsilon = \pm 1$ (Fermi and Bose):

$$\begin{array}{ll} \mathcal{D} : L \longrightarrow \mathbf{AL}(W), & l^a \longmapsto \mathcal{D}(l^a) = \mathcal{D}^{a\beta}_{\ \ \gamma} \mathbf{e}^\gamma \otimes \check{\mathbf{e}}_\beta, \\ L \times W \longrightarrow W, & l^a \bullet \mathbf{e}^\beta = \mathcal{D}^{a\beta}_{\ \ \gamma} \mathbf{e}^\gamma, \\ \check{\mathcal{D}} : L \longrightarrow \mathbf{AL}(W^T), & l^a \longmapsto -\mathcal{D}(l^a)^T = -\mathcal{D}^{a\beta}_{\ \ \gamma} \epsilon \check{\mathbf{e}}_\beta \otimes \mathbf{e}^\gamma, \\ L \times W^T \longrightarrow W^T, & l^a \bullet \check{\mathbf{e}}_\gamma = -\mathcal{D}^{a\beta}_{\ \ \gamma} \check{\mathbf{e}}_\beta, \end{array}$$

e.g., the Lie algebras of $\mathbf{U}(1)$ and $\mathbf{SU}(N)$ with the irreducible and defining representations in a Pauli basis:

$$\begin{array}{lll} \log \mathbf{U}(1) & \longrightarrow \mathbf{AL}(1, \mathbb{C}) \cong \mathbb{C}, & \mathcal{D}(l^0) \cong iz, \ \ z \in \mathbb{Z}, \\ \log \mathbf{SU}(N) & \longrightarrow \mathbf{AL}(N, \mathbb{C}), & \mathcal{D}(l^a) \cong \frac{i}{2} \tau^a(N)^\beta_\gamma, \\ \multicolumn{3}{l}{\text{Pauli matrices: } \{\tau^a(N) \,\big|\, a = 1, \ldots, N^2 - 1\},} \end{array}$$

with proper Pauli matrices $\tau(2) = \tau$ and Gell-Mann matrices $\tau(3) = \lambda$.

For spacetime fields the Lie algebra representation is given by the charges:

$$l^a \longmapsto iQ^a = i \int d^3x \, \mathbf{J}^a_0(x), \ \ [iQ^a, iQ^b] = \epsilon^{ab}_c iQ^c.$$

They are position integrals over the currents, which are defined with the quantization opposite (anti)commutators — exemplified by

Field	Quantization	Current
Scalar (Hermitian)	$[i\mathbf{\Phi}_{k\beta}, \mathbf{\Phi}^\gamma](\vec{x}) = \delta^\gamma_\beta \delta^0_k \delta(\vec{x})$	$i\mathbf{J}^a_k = \mathcal{D}^{a\beta}_{\ \ \gamma} \frac{\{\mathbf{\Phi}^\gamma, i\mathbf{\Phi}_{k\beta}\}}{2}$
Scalar (complex)	$[i\mathbf{\Phi}^\star_{k\beta}, \mathbf{\Phi}^\gamma](\vec{x}) = [i\mathbf{\Phi}^\gamma_k, \mathbf{\Phi}^\star_\beta](\vec{x}) = \delta^\gamma_\beta \delta^0_k \delta(\vec{x})$	$i\mathbf{J}^a_k = \mathcal{D}^{a\beta}_{\ \ \gamma} \frac{\{\mathbf{\Phi}^\gamma, i\mathbf{\Phi}^\star_{k\beta}\} + \{-i\mathbf{\Phi}^\gamma_k, \mathbf{\Phi}^\star_\beta\}}{2}$
Vector (Hermitian)	$[i\mathbf{F}^\gamma_{kj}, \mathbf{A}^l_\beta](\vec{x}) = \delta^\gamma_\beta \delta^0_k \delta^a_j \delta^l_b \delta^b_a \delta(\vec{x})$	$i\mathbf{J}^a_k = \mathcal{D}^{a\beta}_{\ \ \gamma} \frac{\{\mathbf{A}^j_\beta, i\mathbf{F}^\gamma_{kj}\}}{2}$
Weyl (left)	$\{\mathbf{l}^\star_\beta, \mathbf{l}^\gamma\}(\vec{x}) = \delta^\gamma_\beta \sigma^0 \delta(\vec{x})$	$i\mathbf{J}^a_k = \mathcal{D}^{a\beta}_{\ \ \gamma} \frac{[\mathbf{l}^\gamma \check{\sigma}_k, \mathbf{l}^\star_\beta]}{2}$
Weyl (right)	$\{\mathbf{r}^\star_\beta, \mathbf{r}^\gamma\}(\vec{x}) = \delta^\gamma_\beta \check{\sigma}^0 \delta(\vec{x})$	$i\mathbf{J}^a_k = \mathcal{D}^{a\beta}_{\ \ \gamma} \frac{[\mathbf{r}^\gamma \sigma_k, \mathbf{r}^\star_\beta]}{2}$
Dirac	$\{\overline{\mathbf{\Psi}}_\beta, \mathbf{\Psi}^\gamma\}(\vec{x}) = \delta^\gamma_\beta \gamma^0 \delta(\vec{x})$	$i\mathbf{J}^a_k = \mathcal{D}^{a\beta}_{\ \ \gamma} \frac{[\mathbf{\Psi}^\gamma \gamma_k, \overline{\mathbf{\Psi}}_\beta]}{2}$

The Lie algebra action on the fields uses the commutator:

$$[iQ^a, \mathbf{\Phi}^\beta] = \mathcal{D}^{a\beta}_{\ \ \gamma} \mathbf{\Phi}^\gamma, \ \ [iQ^a, \mathbf{\Phi}^\star_\gamma] = -\mathcal{D}^{a\beta}_{\ \ \gamma} \mathbf{\Phi}^\star_\beta.$$

The simultaneous external-internal action is expressed by covariant derivatives:

$$(\partial^k \delta^\beta_\gamma - \mathcal{D}^{a\beta}_{\ \ \gamma} \mathbf{A}^k_a) \mathbf{\Phi}^\gamma, \ \ (\partial^k \delta^\beta_\gamma + \mathcal{D}^{a\beta}_{\ \ \gamma} \mathbf{A}^k_a) \mathbf{\Phi}^\star_\beta.$$

A Lie algebra L representation in the endomorphisms $W \otimes W^T$ of a finite-dimensional vector space is a power three tensor, for physics called a *gauge vertex*:

$$\mathcal{D} : L \longrightarrow W \otimes W^T, \quad \mathcal{D} = \check{l}_a \otimes \mathcal{D}(l^a) = \check{l}_a \otimes \mathcal{D}^{a\beta}_{\ \ \gamma} \mathbf{e}^\gamma \otimes \check{\mathbf{e}}_\beta \in L^T \otimes W \otimes W^T.$$

For gauge interactions in Minkowski spacetime, the gauge vertices as tensor products of three vectors are implemented by three quantum fields and Lorentz compatible spacetime distributions, e.g., for a Dirac field $\mathbf{\Psi}$:

$$\begin{aligned} &\check{l}_a \otimes \mathcal{D}(l^a) \hookrightarrow \mathbf{A}^k_a \mathbf{J}^a_k \overset{\text{e.g.}}{=} \mathbf{A}^k_a \mathcal{D}^{a\beta}_{\ \ \gamma} \frac{[\mathbf{\Psi}^\gamma \gamma_k, \overline{\mathbf{\Psi}}_\beta]}{2}, \\ &\text{with } (\check{l}_a, \mathbf{e}^\gamma, \check{\mathbf{e}}_\beta) \hookrightarrow (\mathbf{A}_a, \mathbf{\Psi}^\gamma, \overline{\mathbf{\Psi}}_\beta). \end{aligned}$$

The gauge fields themselves go with the Lie algebra and its currents: Their number is given by the Lie algebra dimension $L \cong \mathbb{R}^d$, they transform under the adjoint Lie algebra representation (for field strength $\mathbf{F}^b$ and the currents $\mathbf{j}^b, \mathbf{J}^b$) and its dual co-adjoint representation (for gauge fields $\mathbf{A}_c$):

$$\begin{aligned} \text{ad} : L &\longrightarrow \mathbf{AL}(L), & \text{ad}\, l^a &= \epsilon^{ab}_c l^c \otimes \check{l}_b, & l^b &\longmapsto \epsilon^{ab}_c l^c, \\ \check{\text{ad}} : L &\longrightarrow \mathbf{AL}(L^T), & \check{\text{ad}}\, l^a &= -\epsilon^{ab}_c \check{l}_b \otimes l^c, & \check{l}_c &\longmapsto \epsilon^{ab}_c \check{l}_b. \end{aligned}$$

The gauge field currents are products of the dual pairs $(\mathbf{A}_a, \mathbf{F}^a)$:

$$\begin{aligned} \mathbf{j}_k{}^a &= \mathbf{A}^j{}_b \epsilon^{ab}{}_c \mathbf{F}_{kj}{}^c, \\ Q^a &= \int d^3x (\mathbf{j}_0{}^a + \mathbf{J}_0{}^a) \end{aligned} \Rightarrow \begin{cases} [iQ^a, (\mathbf{F}_{kj}{}^b, \mathbf{j}^b, \mathbf{J}^b)] = \epsilon^{ab}{}_c (\mathbf{F}_{kj}{}^c, \mathbf{j}^c, \mathbf{J}^c), \\ \qquad [iQ^a, \mathbf{A}^k{}_c] = -\epsilon^{ab}{}_c \mathbf{A}^k{}_b. \end{cases}$$

With the adjoint Lie algebra representation the gauge field self-coupling is nontrivial only for a nonabelian Lie algebra.

In the Lagrangian for the gauge field sector,

$$\mathbf{L}(\mathbf{A}, \mathbf{F}) = \mathbf{F}^c_{kj} \frac{\partial^k \mathbf{A}^j_c - \partial^j \mathbf{A}^k_c}{2} + g_{L,cb} \frac{\mathbf{F}^c_{kj} \mathbf{F}^{kjb}}{4} - \frac{1}{2} \mathbf{A}^k_a \mathbf{j}^a_k,$$

the statistical factor $\frac{1}{2}$ in $\frac{1}{2}\mathbf{A}^k_a \mathbf{j}^a_k$ takes into account the tensor power 2 of the gauge field $\mathbf{A} \vee \mathbf{A}$ in the interaction. The current arises by derivation with respect to the gauge field:

$$\frac{1}{2} \mathbf{A}^k_a \mathbf{j}^a_k = \epsilon^{ab}_c \frac{\mathbf{A}^k_a \mathbf{A}^j_b}{2} \mathbf{F}^c_{kj}, \quad \frac{\partial \frac{1}{2} \mathbf{A}^k_a \mathbf{j}^a_k}{\partial \mathbf{A}^k_a} = \epsilon^{ab}_c \mathbf{A}^j_b \mathbf{F}^c_{kj}.$$

In a second order derivative formulation, there occur gauge field derivatives and cubic gauge field products in the current

$$\mathbf{j}_k^a = \mathbf{A}_b^j \epsilon_c^{ab}(\partial_k \mathbf{A}_j^c - \partial_j \mathbf{A}_k^c + \epsilon^{cde}\mathbf{A}_k^d \mathbf{A}_j^e + \delta_{kj}\partial^l \mathbf{A}_l^c).$$

For the *gauge field normalization* there has to exist an invariant nondegenerate symmetric bilinear form of the Lie algebra and its dual, diagonalizable in a Sylvester basis:

$$L^T \times L^T \longrightarrow \mathbb{R}, \ \langle \check{l}_a | \check{l}_b \rangle = \kappa_{ab} = \kappa_{ba} = \pm \frac{1}{\kappa_L^2}\delta_{ab},$$

e.g., the Killing form for a semisimple Lie algebra like $\mathbf{SU}(N)$, $N \geq 2$, or a squared linear form for an abelian Lie algebra like $\mathbf{U}(1)$.

In the following with compact gauge group $\mathbf{U}$, the normalization $\kappa_{U,ab} = \frac{1}{\kappa_U^2}\delta_{ab}$ is assumed for Lie algebra bases with totally antisymmetric structure constants $\epsilon_d^{ab}\delta^{dc} = -\epsilon^{abc}$. The gauge field coupling constant in the field strength square is the normalization ratio of the represented internal Lie algebra $L = \log \mathbf{U}$ and the external Lorentz Lie algebra $\log \mathbf{SO}_0(1,3)$ with its Killing form $\eta \wedge \eta$, normalized by $\kappa^2_{\mathbf{SO}_0(1,3)}$:

$$\begin{aligned} \langle \mathbf{F} | \mathbf{F} \rangle &= g_U^2 \, \delta_{ab}\eta^{kl}\eta^{jm}\mathbf{F}_{kj}^a \mathbf{F}_{lm}^b = g_U^2 \mathbf{F}_{kj}^a \mathbf{F}_a^{kj}, \\ g_U^2 &= \frac{\kappa^2_{\mathbf{SO}_0(1,3)}}{\kappa_U^2}. \end{aligned}$$

The Lagrangian for the gauge field sector,

$$\mathbf{L}(\mathbf{A},\mathbf{F}) = \mathbf{F}_{kj}^c \frac{\partial^k \mathbf{A}_c^j - \partial^j \mathbf{A}_c^k - \epsilon_c^{ab}\mathbf{A}_a^k \mathbf{A}_b^j}{2} + g_U^2 \frac{\mathbf{F}_{kj}^c \mathbf{F}_c^{kj}}{4},$$

gives the field equations:

$$\partial^k \mathbf{A}_c^j - \partial^j \mathbf{A}_c^k - \epsilon_c^{ab}\mathbf{A}_a^k \mathbf{A}_b^j = g_U^2 \mathbf{F}_c^{jk}, \ \ \partial^j \mathbf{F}_{kj}^b + \epsilon_c^{ab}\mathbf{A}_a^j \mathbf{F}_{kj}^c = 0.$$

7.33 Electroweak Ground State and Masses

An asymmetric ground state (ground state degeneracy) for the standard model leads to a rearrangement of the interaction $\mathbf{U}(2)$-hyperisospin symmetry properties with respect to the particle related electromagnetic $\mathbf{U}(1)$-symmetry. The interaction fields can be rearranged in terms of the asymptotic particle (mass and electromagnetic eigenstate) fields — for the gauge vector fields, and for the lepton and quark spinor fields.

The Higgs field serves the purpose to distinguish the electromagnetic $\mathbf{U}(1)$-interaction from the weak interactions, related to the three real degrees of freedom of the compact degeneracy *Goldstone manifold* $\mathcal{G}^3 \cong \mathbf{U}(2)/\mathbf{U}(1)_+$ with the remaining electromagnetic $\mathbf{U}(1)_+$-invariance from the Cartan subgroup $\mathbf{U}(1)_+ \times \mathbf{U}(1)_- \subset \mathbf{U}(2)$. The hyperisospin dublet Higgs field characterizes the ground state $|\mathrm{U}\rangle$ by the Fermi mass $M \sim 123\frac{\mathrm{GeV}}{c^2}$ as nontrivial expectation value for the chargeless component:

$$\begin{aligned}
&\langle \mathrm{U}|\mathbf{\Phi}^\star\mathbf{\Phi}|\mathrm{U}\rangle = M^2 \neq 0, \quad \mathbf{\Phi} \overset{\mathbf{SU}(2)}{=} \begin{pmatrix}\Phi_+\\ \Phi_0\end{pmatrix},\\
&\langle \mathrm{U}|\mathbf{\Phi}|\mathrm{U}\rangle = \begin{pmatrix}\langle \mathrm{U}|\mathbf{\Phi}_+|\mathrm{U}\rangle\\ \langle \mathrm{U}|\mathbf{\Phi}_0|\mathrm{U}\rangle\end{pmatrix} = \begin{pmatrix}0\\ M\end{pmatrix}, \quad \langle \mathrm{U}|\mathbf{\Phi}\mathbf{\Phi}^\star|\mathrm{U}\rangle = \frac{\mathbf{1}_2 - \tau_3}{2} M^2.
\end{aligned}$$

Such a symmetry reduction can be implemented by a potential with the corresponding minimum in the Lagrangian:

$$\mathbf{L} = \ldots + g_\Phi(\mathbf{\Phi}^*\mathbf{\Phi} - M^2)^2 + \ldots$$

Taking into account only the constant term $\mathbf{\Phi} = \begin{pmatrix}0\\ M\end{pmatrix} + \ldots$ for the Higgs field, the Higgs field-gauge fields coupling produces mass terms for the three vector fields related to the Goldstone degrees of freedom:

$$\begin{aligned}
\mathbf{L}(\mathbf{\Phi}, \mathbf{A}) &= \left|\partial^k - i\frac{\mathbf{1}_2\mathbf{A}_0^k + \tau^a\mathbf{A}_a^k}{2})\mathbf{\Phi}\right|^2 = \frac{M^2}{4}\,\mathrm{tr}\left|\begin{pmatrix}\mathbf{A}_0^k + \mathbf{A}_3^k & \mathbf{A}_1^k - i\mathbf{A}_2^k\\ \mathbf{A}_1^k + i\mathbf{A}_2^k & \mathbf{A}_0^k - \mathbf{A}_3^k\end{pmatrix}\frac{\mathbf{1}_2 - \tau_3}{2}\right|^2 + \ldots\\
&= \frac{M^2}{4}\,\mathrm{tr}\left|\begin{pmatrix}0 & \mathbf{A}_1^k - i\mathbf{A}_2^k\\ 0 & \mathbf{A}_0^k - \mathbf{A}_3^k\end{pmatrix}\right|^2 + \ldots = \frac{M^2}{4}(\mathbf{Z}^k\mathbf{Z}_k + \mathbf{A}_+^k\mathbf{A}_{-k}) + \ldots,\\
\mathbf{Z} &= -\mathbf{A}_0 + \mathbf{A}_3, \quad \mathbf{A}_\pm = \mathbf{A}_1 \pm i\mathbf{A}_2.
\end{aligned}$$

The interaction-basis for the internal isospin and hypercharge coordinates with the hypercharge gauge field $\mathbf{A}_0$ and the third component $\mathbf{A}_3$ of an isospin triplet gauge field, both chargeless, is *Weinberg-rotated* by an orthogonal matrix $\mathbf{SO}(2)$ to a mass eigenfield particle basis with a massless electromagnetic field $\mathbf{A}$ and a massive neutral weak boson field $\mathbf{Z}$:

$$\begin{aligned}
&\mathbf{L}(\mathbf{A}) = \frac{1}{g_1^2}\left(\frac{\partial^k\mathbf{A}_0^j - \partial^j\mathbf{A}_0^k}{2}\right)^2 + \frac{1}{g_2^2}\left(\frac{\partial^k\mathbf{A}_c^j - \partial^j\mathbf{A}_c^k - \epsilon_c^{ab}\mathbf{A}_a^k\mathbf{A}_b^j}{2}\right)^2\\
&= \frac{1}{g_e^2}\left(\frac{\partial^k\mathbf{A}^j - \partial^j\mathbf{A}^k}{2}\right)^2 + \frac{1}{g_Z^2}\left(\frac{\partial^k\mathbf{Z}^j - \partial^j\mathbf{Z}^k}{2}\right)^2 + \ldots,\\
&\begin{pmatrix}\frac{1}{g_e}\mathbf{A}\\ \frac{1}{g_Z}\mathbf{Z}\end{pmatrix} = \begin{pmatrix}\cos\theta_W & \sin\theta_W\\ -\sin\theta_W & \cos\theta_W\end{pmatrix}\begin{pmatrix}\frac{1}{g_1}\mathbf{A}_0\\ \frac{1}{g_2}\mathbf{A}_3\end{pmatrix} \Rightarrow \begin{cases}\frac{g_Z}{g_1}\sin\theta_W = 1 = \frac{g_Z}{g_2}\cos\theta_W,\\ g_Z^2 = g_2^2 + g_1^2, \ \frac{1}{g_e^2} = \frac{1}{g_2^2} + \frac{1}{g_1^2},\\ \mathbf{A} = \cos^2\theta_W\,\mathbf{A}_0 + \sin^2\theta_W\,\mathbf{A}_3,\end{cases}\\
&\text{with } \mathbf{Z} = -\mathbf{A}_0 + \mathbf{A}_3,
\end{aligned}$$

$$\begin{aligned}
\text{experimental input: } &\sin^2\theta_W \sim 0.23, \ \frac{g_e^2}{4\pi} = \alpha_{\mathrm{Som}} \sim \frac{1}{137}, \ \frac{m_W}{g_2} = M \sim 123\frac{\mathrm{GeV}}{c^2},\\
&\Rightarrow (m_Z^2, m_W^2) = (g_Z^2, g_2^2)M^2 = \left(\frac{1}{\cos^2\theta_W}, 1\right)\frac{4\pi\alpha_{\mathrm{Som}}}{\sin^2\theta_W}M^2,\\
&(m_Z, m_W) \simeq (91.2,\ 80.2)\frac{\mathrm{GeV}}{c^2}.
\end{aligned}$$

Three of the four real components of the complex Higgs isodoublet, i.e., the Goldstone degrees of freedom, serve as spinless components of the massive vector fields $\mathbf{Z}, \mathbf{A}_\pm$. The radial fourth Higgs field component $\sqrt{\mathbf{\Phi}^*\mathbf{\Phi}} - M$ gives rise to a spin- and chargeless Higgs particle with mass $m_\Phi^2 = 2g_\Phi M^2$.

In general, the fermions come in triplets ("families") encompassing a left-handed isodoublet and two right-handed isosinglets with hypercharges $(y, y+\frac{1}{2}, y-\frac{1}{2})$, e.g., the first quark family $(\mathbf{q}, \mathbf{u}_R, \mathbf{d}_R)$ with hypercharges $(\frac{1}{6}, \frac{2}{3}, -\frac{1}{3})$ or the electron family $(\mathbf{l}, \boldsymbol{\nu}_R, \mathbf{e}_R)$ with hypercharges $(-\frac{1}{2}, 0, -1)$.

With respect to the asymptotic Weinberg-rotated neutral vector fields,

$$\mathbf{A}_0 = \mathbf{A} - s^2\mathbf{Z}, \quad \mathbf{A}_3 = \mathbf{A} + c^2\mathbf{Z}, \text{ with } (\sin^2\theta_W, \cos^2\theta_W) = (s^2, c^2),$$

the current-gauge fields couplings are rearranged to the electromagnetic and neutral weak interactions (quark notation):

$$\begin{aligned}
\mathbf{JA} &= \left[y\mathbf{q}^*\check{\sigma}\mathbf{1}_2\mathbf{q} + \left(y+\frac{1}{2}\right)\mathbf{u}_R^*\sigma\mathbf{u}_R + \left(y-\frac{1}{2}\right)\mathbf{d}_R^*\sigma\mathbf{d}_R \right]\mathbf{A}_0 + \mathbf{q}^*\check{\sigma}\frac{\tau^a}{2}\mathbf{q}\mathbf{A}_a \\
&= [\mathbf{q}^*\check{\sigma}\frac{2y\mathbf{1}_2+\tau^3}{2}\mathbf{q} + \frac{2y+1}{2}\mathbf{u}_R^*\sigma\mathbf{u}_R + \frac{2y-1}{2}\mathbf{d}_R^*\sigma\mathbf{d}_R]\mathbf{A} \\
&\quad + \left[\mathbf{q}^*\frac{-2ys^2\mathbf{1}_2 + c^2\tau^3}{2}\check{\sigma}\mathbf{q} - \frac{2y+1}{2}s^2\mathbf{u}_R^*\sigma\mathbf{u}_R - \frac{2y-1}{2}s^2\mathbf{d}_R^*\sigma\mathbf{d}_R\right]\mathbf{Z} + \ldots \\
&= \left[\frac{2y+1}{2}\overline{\mathbf{\Psi}}_u\gamma\mathbf{\Psi}_u + \frac{2y-1}{2}\overline{\mathbf{\Psi}}_d\gamma\mathbf{\Psi}_d\right]\mathbf{A} \\
&\quad + \left\{\overline{\mathbf{\Psi}}_u\frac{[1-(2y+1)2s^2]\mathbf{1}_4 + i\gamma_5}{4}\gamma\mathbf{\Psi}_u - \overline{\mathbf{\Psi}}_d\frac{[1+(2y-1)2s^2]\mathbf{1}_4 + i\gamma_5}{4}\gamma\mathbf{\Psi}_d\right\}\mathbf{Z} + \ldots
\end{aligned}$$

$$\text{with } \mathbf{q} \overset{\mathbf{SU}(2)}{=} \begin{pmatrix}\mathbf{u}_L\\ \mathbf{d}_L\end{pmatrix}, \quad \mathbf{\Psi}_u, \mathbf{\Psi}_d \overset{\text{chiral}}{=} \begin{pmatrix}\mathbf{u}_L\\ \mathbf{u}_R\end{pmatrix}, \begin{pmatrix}\mathbf{d}_L\\ \mathbf{d}_R\end{pmatrix}, \quad \frac{\mathbf{1}_4 \pm i\gamma_5}{2}\gamma = \begin{pmatrix}\check{\sigma}\\ \sigma\end{pmatrix}, \quad \gamma_5^2 = -\mathbf{1}_4.$$

The neutral interactions for the lepton case $(\mathbf{l}, \boldsymbol{\nu}_R, \mathbf{e}_R)$ are especially simple:

$$\begin{aligned}
y = -\frac{1}{2}: \quad \mathbf{JA} &= -\overline{\mathbf{\Psi}}_e\gamma\mathbf{\Psi}_e\mathbf{A} + \left\{\frac{1}{2}\boldsymbol{\nu}_L^*\check{\sigma}\boldsymbol{\nu}_L - \overline{\mathbf{\Psi}}_e\frac{[1-4s^2]\mathbf{1}_4 + i\gamma_5}{4}\gamma\mathbf{\Psi}_e\right\}\mathbf{Z} + \ldots, \\
\text{with } \mathbf{l} &\overset{\mathbf{SU}(2)}{=} \begin{pmatrix}\nu_L\\ \mathbf{e}_L\end{pmatrix}, \quad \mathbf{\Psi}_e \overset{\text{chiral}}{=} \begin{pmatrix}\mathbf{e}_L\\ \mathbf{e}_R\end{pmatrix}.
\end{aligned}$$

In the minimal lepton case, the Higgs field can induce, via a Yukawa interaction, the electron mass:

$$\begin{aligned}
L(\mathbf{l}, \mathbf{e}_R, \mathbf{\Phi}) &= g_e(\mathbf{e}_R^*\mathbf{l}\mathbf{\Phi}^* + \mathbf{l}^*\mathbf{e}_R\mathbf{\Phi}) = m_e(\mathbf{e}_R^*\mathbf{e}_L + \mathbf{e}_L^*\mathbf{e}_R) + \ldots = m_e\overline{\mathbf{\Psi}}_e\mathbf{\Psi}_e + \ldots, \\
\text{with } m_e &= g_e M.
\end{aligned}$$

The value for the electron mass goes with the value of the Yukawa coupling constant, both basically unknown.

If the quarks are confined, the definition of a translation related mass is impossible. If they can be treated as possibly free particles, a mass inducing Yukawa interaction looks as follows (generalizable to any left-right triplet):

$$\begin{aligned} L(\mathbf{q}, \mathbf{u}_R, \mathbf{d}_R, \boldsymbol{\Phi}) &= g_u \mathbf{u}_R^* \mathbf{q}\ \boldsymbol{\Phi} + g_d \mathbf{d}_R^* \mathbf{q}\ \boldsymbol{\Phi}^* + \text{h.c.} \\ &= m_u(\mathbf{u}_R^* \mathbf{u}_L + \mathbf{u}_L^* \mathbf{u}_R) + m_d(\mathbf{d}_R^* \mathbf{d}_L + \mathbf{d}_L^* \mathbf{d}_R) + \dots \\ &= m_u \overline{\boldsymbol{\Psi}}_u \boldsymbol{\Psi}_u + m_d \overline{\boldsymbol{\Psi}}_d \boldsymbol{\Psi}_d + \dots, \\ \text{with } m_u &= g_u M, \quad m_d = g_d M. \end{aligned}$$

For leptons and quarks, each family has equal internal invariants for hyperisospin and color, e.g., electron, muon and tau leptons $\mathbf{Q}_{1,2,3} = \begin{pmatrix} \nu_e \\ \mathrm{e} \end{pmatrix}, \begin{pmatrix} \nu_\mu \\ \mu \end{pmatrix}, \begin{pmatrix} \nu_\tau \\ \tau \end{pmatrix}$, or the three quark families, $\mathbf{Q}_{1,2,3} = \begin{pmatrix} \mathrm{u} \\ \mathrm{d} \end{pmatrix}, \begin{pmatrix} \mathrm{c} \\ \mathrm{s} \end{pmatrix}, \begin{pmatrix} \mathrm{t} \\ \mathrm{b} \end{pmatrix}$. The experiments show a mixing of the mass eigenfields for the weak interactions, effected by the symmetry rearrangement. This can be parametrized in a complex flavor space $W \cong \mathbb{C}^f$, e.g., with $f = 3$, by complex Yukawa coupling matrices $G_U, G_D \in \mathbb{C}^f \otimes \mathbb{C}^f$:

$$L(\mathbf{Q}, \mathbf{U}_R, \mathbf{D}_R, \boldsymbol{\Phi}) = \mathbf{U}_R^* G_U \mathbf{Q}\ \boldsymbol{\Phi} + \mathbf{D}_R^* G_D \mathbf{Q}\ \boldsymbol{\Phi}^* + \text{h.c.}$$

Good to know: Any real (complex) matrix G has a polar decomposition into a positive symmetric (hermitian) and an orthogonal (unitary) matrix, $G = HU_G$, unique for invertible G. Any symmetric (hermitian) matrix is orthogonally (unitarily) diagonalizable, $H = U\Delta U^*$. Therefore $G = U\Delta U^* U_G$.

The Yukawa coupling matrices (G_U, G_D) can be diagonalized, each by two unitary matrices (one from left, one from right), altogether four unitary matrices. The two unitary matrices, transforming the isosinglets $(\mathbf{U}_R, \mathbf{D}_R)$ can be absorbed in a redefinition, irrelevant for their gauge interaction. That is different for the isodoublets. The Yukawa coupling can be reformulated with two unitary matrices $(U_U, U_D) \in \mathbf{U}(f)$, defining the mass eigenfields, and two diagonal matrices (Δ_U, Δ_D) with related diagonal mass matrices (m_U, m_D) as follows:

$$\begin{aligned} L(\mathbf{Q}, \mathbf{U}_R, \mathbf{D}_R, \boldsymbol{\Phi}) &= \mathbf{U}_R^* \Delta_U U_U \mathbf{Q}\ \boldsymbol{\Phi} + \mathbf{D}_R^* \Delta_D U_D \mathbf{Q}\ \boldsymbol{\Phi}^* + \text{h.c.} \\ &= \mathbf{U}_R^* m_U \mathbf{U}_L' + \mathbf{D}_R^* m_D \mathbf{D}_L' + \text{h.c.} + \dots, \\ \text{with } \mathbf{Q} &\overset{\mathbf{SU}(2)}{=} \begin{pmatrix} \mathbf{U}_L \\ \mathbf{D}_L \end{pmatrix}, \\ (m_U, m_D) = (\Delta_U, \Delta_D)M &\overset{\text{flavor}}{=} \begin{pmatrix} (m_u, m_d) & 0 & 0 \\ 0 & (m_c, m_s) & 0 \\ 0 & 0 & (m_t, m_b) \end{pmatrix}, \\ \text{mass eigenfields: } \mathbf{U}_L' &= U_U \mathbf{U}_L, \ \ \mathbf{D}_L' = U_D \mathbf{D}_L. \end{aligned}$$

The remaining two unitary matrices are relevant in the isospin changing weak interaction with the currents written with the mass eigenfields:

$$\begin{aligned}\mathbf{Q}^*\check{\sigma}\vec{\tau}\mathbf{Q}\vec{\mathbf{A}} &= \mathbf{U}_L^*\check{\sigma}\mathbf{D}_L\mathbf{A}_+ + \ldots = \mathbf{U}_L'^* U_U\check{\sigma}U_D^*\mathbf{D}_L'\mathbf{A}_+ + \ldots\\ &= \mathbf{U}_L'^*\check{\sigma}V_{UD}\mathbf{D}_L'\mathbf{A}_+ + \ldots,\\ V_{UD} &= U_U U_D^* \in \mathbf{U}(f).\end{aligned}$$

The unitary matrix V_{UD} with maximally f^2 real parameters allows a left-right factorization $V_{UD} = u_U^* v_{UD} u_D$ with two diagonal phase matrices $u_{U,D}$ whose maximally $2f-1$ phases can be absorbed by an additional physically irrelevant redefinition of the left-handed mass eigenfields. The special unitary *Cabibbo–Kobayashi–Maskawa*[15] *matrices* $v_{UD} \in \mathbf{SU}(f)$ with maximally $f^2-(2f-1)=(f-1)^2$ physically relevant real parameters have an orthogonal $\mathbf{SO}(f)$-contribution with $\binom{f}{2}$ families-mixing angles which leaves $\binom{f-1}{2}$ phases:

$$\begin{aligned}V_{UD} = u_U^* v_{UD} u_D \in \mathbf{U}(f) : f^2 &= (2f-1) + \tbinom{f}{2} + \tbinom{f-1}{2},\\ f=2:4 &= 3+1+0,\\ f=3:9 &= 5+3+1.\end{aligned}$$

For two quark families, one has the Cabibbo rotation $v_{UD} \in \mathbf{SO}(2)$ without relevant phases. The charge changing weak interaction in the mass eigenfields looks as follows:

$$\begin{aligned}f=2:\ v_{UD}\mathbf{D}_L' &= \begin{pmatrix}\cos\theta_C & \sin\theta_C\\ -\sin\theta_C & \cos\theta_C\end{pmatrix}\begin{pmatrix}\mathbf{d}_L'\\ \mathbf{c}_L'\end{pmatrix} = \begin{pmatrix}\cos\theta_C\mathbf{d}_L' + \sin\theta_C\mathbf{c}_L'\\ -\sin\theta_C\mathbf{d}_L' + \cos\theta_C\mathbf{c}_L'\end{pmatrix},\\ \mathbf{Q}^*\check{\sigma}\vec{\tau}\mathbf{Q}\vec{\mathbf{A}} &= [\mathbf{u}_L'^*\check{\sigma}(\cos\theta_C\mathbf{d}_L' + \sin\theta_C\mathbf{c}_L')\\ &\quad +\mathbf{c}_L'^*\check{\sigma}(-\sin\theta_C\mathbf{d}_L' + \cos\theta_C\mathbf{c}_L')]\mathbf{A}_+ + \ldots,\end{aligned}$$

$$\text{experiments: } \theta_C \sim 13.04^0.$$

A standard parametrization of the proper Kobayashi–Maskawa matrix $v_{UD} \in \mathbf{SU}(3)$ for three quark families, transforming from weak interaction eigenfields to mass eigenfields, with three Cabibbo-like angles in $c_{ij} = \cos\theta_{ij}$ etc. and experimental values reads:

$$v_{UD} = \begin{pmatrix}1 & 0 & 0\\ 0 & c_{23} & s_{23}\\ 0 & -s_{23} & c_{23}\end{pmatrix}\begin{pmatrix}c_{13} & 0 & s_{13}e^{-i\delta_{13}}\\ 0 & 1 & 0\\ -s_{13}e^{i\delta_{13}} & 0 & c_{13}\end{pmatrix}\begin{pmatrix}c_{12} & s_{12} & 0\\ -s_{12} & c_{12} & 0\\ 0 & 0 & 1\end{pmatrix},$$

$$\text{experiments: } \theta_{12} = \theta_C \sim 13.04^o,\ \theta_{13} \sim 0.20^o,\ \theta_{23} \sim 2.38^o,\ \delta_{13} \sim 1.2.$$

The one physically relevant phase is used for the parametrization of the *CP*-violation structure.

The expansion of the Higgs field around the ground state value $\mathbf{\Phi}_0 = M + \ldots$ is the leading asymmetric part of the full rearrangement with the three Goldstone degrees of freedom in the $\mathbf{U}(2)/\mathbf{U}(1)_+$-transmutator field, normalized by the radial field:

---

[15]Nicola Cabibbo (1935–2010), Makoto Kobayashi (1944–), Toshihide Maskawa (1940–).

$$\mathbf{U} \cong \frac{1}{\mathbf{R}}(\mathbf{\Phi}^\star_\beta \epsilon^{\beta\alpha}, \mathbf{\Phi}^\alpha) = \frac{1}{\mathbf{R}}\begin{pmatrix} \mathbf{\Phi}^\star_0 & \mathbf{\Phi}_+ \\ -\mathbf{\Phi}^\star_+ & \mathbf{\Phi}_0 \end{pmatrix}, \quad \mathbf{R}^2 = \mathbf{\Phi}^\star\mathbf{\Phi} = \mathbf{\Phi}^\star_0\mathbf{\Phi}_0 + \mathbf{\Phi}^\star_+\mathbf{\Phi}_+.$$

With the projective property of the ground state, the $\mathbf{U}(2)/\mathbf{U}(1)_+$-transmutator contains only three Goldstone fields which reflect the massive vector fields for a broken gauge symmetry:

$$\frac{\mathbf{\Phi}}{\mathbf{R}} = \exp[i\frac{\mathbf{1}_2\varphi_0 + \vec{\tau}\vec{\varphi}}{2}]\begin{pmatrix} 0 \\ 1 \end{pmatrix} = \exp[i\frac{(\mathbf{1}_2 - \tau_3)(\varphi_0 - \varphi_3)}{4} + i\frac{\tau_1\varphi_1 + \tau_2\varphi_2}{2}]\begin{pmatrix} 0 \\ 1 \end{pmatrix},$$

Goldstone fields: $\{\varphi_0 - \varphi_3, \varphi_{1,2}\}$.

By their couplings with the Higgs field, the fermion and vector fields are undressed ("frozen" from interaction to particle fields), corresponding to their transformation behavior, with the Goldstone transmutators, rearranging the hyperisospin $\mathbf{U}(2)$-properties into electromagnetic $\mathbf{U}(1)_+$-properties, e.g., for the electroweak gauge fields and the left-handed quark fields :

$$\mathbf{A} \longmapsto \mathbf{U}^\star \circ (\mathbf{A} - \mathbf{1}\partial) \circ \mathbf{U},$$
$$\mathbf{q} \longmapsto \mathbf{U}^\star\mathbf{q} = \begin{pmatrix} \mathbf{u}_L \\ \mathbf{d}_L \end{pmatrix}.$$

Chapter 8
Classical and Quantum Logics

It seems obvious that physics, mathematics, and logics reflect — in this order — three steps, upwards or downwards, of empirical connections or abstractions: Physics is closely connected with experiments, mathematics abstracts the structures involved, and logics, finally, is the theory of the rules for the formalization of such structures — without a detailed specification of their content. There are people who thought, like Leibniz or Carnap,[1] that there may exist "worlds with a different physics," e.g., with a fine-structure constant $\frac{1}{8\pi}$, but that our logic has to be valid in all "possible" worlds. Is that really obvious? Can one not take logic as the most empirical of all sciences since it condenses all our experiences? Does logic not consist of human prejudices which are successful in their applications? After the "discovery" of quantum theory, why does one think anew about logic and its foundation?

It has some justification to separate mathematics in the time before and the time after Cantor's pioneering work on sets. After Cantor and Frege: Are mathematics and logic separable? Where belongs Russell's paradox, where Gödel's theorem? Similarly — are mathematics and physics, or physics and logic (epistemology) separable? Why – with Wigner – "the unreasonable effectiveness of mathematics in the natural sciences"? A mathematician — I think it was Bolyai, familiar from the investigation of non-Euclidean geometries — had the opinion that each mathematical structure has a physical relevance. Obviously, all mathematical structures are, in a sense, connected, e.g., if based on a common class- and set-related foundation with order structures, topological structures and algebraic structures. But with respect to specific details — where in nature arise the complex zeros of Riemann's zeta-function, or where does nature need p-adic topologies or ultrametric fields?

There is much work to characterize mathematics as growing from logic. That is possible with respect to the formal logical structure; it is less convincing with

[1] Rudolf Carnap (1891–1970).

H. Saller, *Operational Symmetries*, DOI 10.1007/978-3-319-58664-9_8

respect to the content of the mathematical concepts — what is the logical motivation for manifolds, for distributions, etc.? There are also attempts to "derive" physics as an extension of logic, e.g., Kant's attempts to relate basic physics with the conditions for the possibility of experience.

What is reality, what is structure? Is the ontology of the electron in its abstract definition as irreducible unitary Poincaré group representation so different from the ontology of a regular triangle definable in a Euclidean geometry? All these epistemological problems are millenia old — starting with Platon's ideas, taken up, in the Middle ages, in the universalia controversy, more recently reformulated in the analytical philosophy, and so on. The discussion of such questions gives us an awareness of the problems involved; philosophical analyses and definitions alone seem not to solve them. Connected with the phenomena of relativity and quantum theory we got epistemological suggestions which, probably, no deep philosophical thinking could have provided us with. When Berkeley's[2] empiricism was formulated by the catchword "esse est percipi," it was easy to ridicule him. With the knowledge of the quantum structure his ideas have to be judged more favorably, and one can confront his critics, for example, with the interpretation of the quantum-mechanical two-slits experiment. However, without the help of experiments, the Irish bishop was lost in a scientific discussion.

The history of science has shown that nature is more inventive than we are — it offers structures, unexpected for our philosphical inventiveness. Sometimes, a better knowledge of the scientific concepts and of related experiments could help or could have helped the subtle philosophical investigations, e.g., a knowledge of relativity in the context of causality, and a knowledge of quantum theory in the discussion of the subject-object dichotomy.

Remarkably, the basic structures are not really complicated, but, in a deep sense simple, even self-evident. However, it may be very difficult to find the simple basis. Is it not self-evident, post festum in quantum theory, that the experimental apparatus cannot be left out and that all experience involves relations which connect and even define the properties of what we easily and superficially call "object" and "subject"?

To think about logic is self-referential; one can never be sure to what extent one already uses logic — what is a deduction, what a consistency reflection, or what is syntax, what is semantics? In addition, one enters the foundations of mathematics where with the concepts of sets and classes and the problematic compression (comprehension, forming new classes, "bracing" {....}) there open up abysses, both in depth and triviality. To avoid the related problems, the definitions in the following are formulated mostly for sets S, i.e., for classes that are elements of a class, denoted by $S \in \underline{\mathbf{set}}$, although they can be generalized in many cases to "un-sets," i.e., to classes that cannot be elements.

A logic of quantum structure becomes especially awkward if it is described in the language of a different classical logic. Does that make sense? How far influences

---

[2]George Berkeley (1685–1735).

the form the content, or the syntax the semantics? Two attitudes offer themselves: First, it may exist an encompassing "super-logic," or, second, more simple, one talks about a quantum logic with a classical logic, i.e., one has some sort of logical typing. The methods of physics may be encouraging in this respect: Our experiments and the measuring apparatus, also for quantum phenomena, are talked about with the classical physical concepts. Applied to logic, such a point of view requires an understanding of the classical logic as some projection of the quantum logic.

A first characterization describes the classical logic of Boole[3] as set-oriented — as logics of power set lattices with a true-false and a probability quantification. In contrast, the quantum logic is operation-oriented with complex-linear connections. In classical logic, the experiments are talked about with a unique negation which fulfills the "tertium non datur" and the "principium contradictionis." The complex-linear formalism of quantum physics with its superposition principle does not display the Boolean structures.

First in this chapter, a short review is given of the Aristotelian syllogisms, then, after a foray into elementary logic and the foundations of mathematics, a short characterization of Boole–Stone logics, and, in the end, of the unitary logics of quantum theory as formalized by Birkhoff[4] and von Neumann. Quantum logics work with lattices of vector subspaces of complex Hilbert spaces with state vectors. "Quantum relativity" is seen in the basis dependence of the experimental verification of the quantum structure.

Probably for most physicists, familiar with the natural phenomena by experiments, the structural logical formalization looks like "abstract nonsense," and for most philosophers, defining their subtle concepts by deeply reflected well formulated cautious words, a reckless concretization of logical structures, e.g., by the famous two-slits experiment, will be unintelligible. Thus is the situation in border areas one has to live with; they are laden with conflicts, dangerous and fruitful.

8.1 Predicates and Relations

The earliest ansatz for logic was put forward by Aristotle. With the knowledge and formalization of today, not in a historically contemporary language, his proposals will be looked at. They were so good and knowledgeable that nobody dared to correct and improve them for a long time — an example for the paralyzing influence of an overwhelming intelligence. The logical work of the elder Stoa (about -(300–150)) was not surpassed until Leibniz, about two thousand years later. However, as we will see ahead in the context of a set- and class-oriented buildup of mathematics, the Aristotelian syllogistic, restricted to elementary object-predicate propositions,

[3]George Boole (1815–1864).

[4]Garrett Birkhoff (1884–1944).

scratches only the surface of the logical structures and problems. Especially Boole, Frege and Russell opened up new vistas and broke new ground. With some arrogant exaggeration and with the hubris of today, the Aristotelean syllogisms have only a historical interest, and can be considered to describe not much more than an exercise in naive set theory, easily visualizable by "Euler circles" as sets for propositions.

"All Athenians are Greeks" and "all Greeks are humans", both assumed as "true" propositions, allows the conclusion "all Athenians are humans" as "true":

$$\mathcal{A}a\mathcal{G} \text{ and } \mathcal{G}a\mathcal{M} \Rightarrow \mathcal{A}a\mathcal{M}.$$

With such syllogisms, Aristotle started the *formal logic*, i.e., a logic which is independent of the content of the propositions, but looks only for the form of the logical conclusions. One has formally the *syllogism*:

$$\begin{aligned} &PaQ \text{ denotes "all } P \text{ are } Q\text{"},\\ &PaM \text{ and } MaQ \Rightarrow PaQ. \end{aligned}$$

(P, Q) is an ordered pair of properties, standing in the *relation* "all," as given in the example $(\mathcal{A}, \mathcal{G})$ and $(\mathcal{G}, \mathcal{M})$:

$$\mathcal{A}a\mathcal{G}, \mathcal{G}a\mathcal{M}, PaQ.$$

Immediately, there arise different questions: Why do we trust in such a syllogism? Which relations like "all" should we consider? And given the relations — which syllogisms are possible?

Aristotle considered the following four ($4 = 2 \times 2$) *binary relations* for *predicates* (properties):

	affirmative	negated
generally	a, PaQ "all P are Q"	e, PeQ "no P is Q"
particularly	i, PiQ "some P are Q"	o, PoQ "not all P are Q"

e.g., a for *generally affirmative*: PaQ, "all P are Q."

Scholastics brought all this in an easily memoralizable "school form": The letters a and i are from "affirmo," the letters e and o from "nego."

In the set-oriented language of today with elements $x \in S$, predicates (properties) are formalizable by subsets: The *power set* $\mathbf{2}^S = \{X \subseteq S\}$ with S-subsets denotes all possible predicates (properties) the S-elements can have. The negation of the property X is explained by its *complement* $\complement X = \complement_S X$ with respect to the "full" set in question.

A binary relation ρ between two sets is a subset of the product set:

$$x\rho y \iff (x, y) \in \rho \subseteq S \times U.$$

A binary relation which is surjective in the first place (domain) and injective in the second place (range) is called a mapping:

$$\begin{aligned} &\rho : S \longrightarrow U \text{ with } x\rho y \iff x \longmapsto \rho(x) = y, \\ &(x, y) = (x, \rho(x)) \in S \times \rho[S] \subseteq S \times U. \end{aligned}$$

Each binary relation has its *converse (transposition)*:

$$y\rho^T x \iff x\rho y, \ \ \rho^T \subseteq U \times S, \ \ \rho^{TT} = \rho,$$

and its negation:

$$x\rho^c y \iff \text{not } x\rho y, \ \ \rho^c = \complement_{S\times U}\rho \subseteq S \times U, \ \ \rho^{cc} = \rho.$$

The Aristotelean relations have the following set-theoretical formalization which exhaust the four intersection possibilities of a set P with another set Q and its complement (negation) $\complement Q$:

$P, \ Q, \ \complement P, \ \complement Q \subseteq S$:

	affirmative	negated
generally	PaQ "all P are Q" $P \subseteq Q$ $P \cap \complement Q = \emptyset$	PeQ "no P is Q" $P \subseteq \complement Q$ $P \cap Q = \emptyset$
particularly	PiQ "some P are Q" $P \not\subseteq \complement Q$ $P \cap Q \neq \emptyset$	PoQ "not all P are Q" $P \not\subseteq Q$ $P \cap \complement Q \neq \emptyset$

8.2 Aristotelian Syllogisms

The task of the syllogistic consists in giving, for the four relations above, how one can draw conclusions from two premisses. If the conclusion involves the predicates P and Q, then the two premisses contain a joining predicate M. With that, the following four syllogism-patterns for relations ρ, σ, τ from a, i, e, o can be differentiated:

$$\begin{aligned} &I. \ \ \ \ M\rho Q \text{ and } P\sigma M \Rightarrow P\tau Q, \\ &II. \ \ Q\rho M \text{ and } P\sigma M \Rightarrow P\tau Q, \\ &III. \ M\rho Q \text{ and } M\sigma P \Rightarrow P\tau Q, \\ &IV. \ \ Q\rho M \text{ and } M\sigma P \Rightarrow P\tau Q. \end{aligned}$$

The task can be facilitated by using the converse (transposed) relations which allows the reduction to one pattern only:

$$P\rho M \text{ and } M\sigma Q \Rightarrow P\tau Q,$$

for the resulting six different relations a, a^T, i, e, o, o^T. "Some" and "no" are *symmetric*:

$$i = i^T, \quad e = e^T.$$

An example for the transposition of relations: "Teacher" is converse to "pupil." In monogameous cultures, the relation "spouse" is symmetric.

The particular relations are the negations of the general ones:

$$\begin{array}{lll} PeQ \iff Pi^cQ \iff \text{"not some } P \text{ are } Q\text{,"} \\ PoQ \iff Pa^cQ \iff \text{"not all } P \text{ are } Q\text{."} \end{array}$$

One works with relations (predicates) for subsets $\mathbf{2}^S$ of objects S:

$$\begin{array}{l} P, \; Q, \; \complement P, \; \complement Q \in \mathbf{2}^S, \\ a, a^T, i, e, o, o^T \in \mathbf{2}^S \times \mathbf{2}^S. \end{array}$$

Take for S "all humans of this earth" with subsets like "Athenian, Greek, tall, stupid, uncle, spouse," etc. Another example, used by Aristotle, is the predicate set, composed by the four predicates "human" $\mathcal{M}$, "stone" $\mathcal{S}$, "living object" $\mathcal{L}$ and "horse" $\mathcal{H}$. The four disjoint elements $\{\mathcal{M}, \; \mathcal{H}, \; \complement_{\mathcal{L}}(\mathcal{M} \cup \mathcal{H}), \; \mathcal{S}\}$, assumed to have no subsets proper, e.g., centaurs do not exist, combine the power set of the full set $\mathcal{L} \cup \mathcal{S}$ (living objects and stones) with $2^4 = 16$ elements:

$$\begin{aligned} \mathcal{L} \cup \mathcal{S} &= \mathcal{M} \cup \mathcal{H} \cup \complement_{\mathcal{L}}(\mathcal{M} \cup \mathcal{H}) \cup \mathcal{S} \text{ (disjoint)}, \\ \mathbf{2}^{\mathcal{L} \cup \mathcal{S}} &= \{\emptyset, \\ &\quad \mathcal{M}, \; \mathcal{H}, \; \complement_{\mathcal{L}}(\mathcal{M} \cup \mathcal{H}), \; \mathcal{S}, \\ &\quad \mathcal{M} \cup \mathcal{H}, \; \complement_{\mathcal{L}}\mathcal{H}, \; \mathcal{M} \cup \mathcal{S}, \; \complement_{\mathcal{L}}\mathcal{M}, \; \mathcal{H} \cup \mathcal{S}, \; \complement_{\mathcal{L}}(\mathcal{M} \cup \mathcal{H}) \cup \mathcal{S}, \\ &\quad \complement_{\mathcal{L}}\mathcal{H} \cup \mathcal{S}, \; \complement_{\mathcal{L}}\mathcal{M} \cup \mathcal{S}, \; \mathcal{M} \cup \mathcal{H} \cup \mathcal{S}, \; \mathcal{L}, \\ &\quad \mathcal{L} \cup \mathcal{S}\}. \end{aligned}$$

The original properties $A = \{\mathcal{M}, \mathcal{H}, \mathcal{L}, \mathcal{S}\} \subset \mathbf{2}^{\mathcal{L} \cup \mathcal{S}}$ constitute a subset of the power set. The 4^2 relations in the product $\rho \subseteq A \times A$ with $1\rho 2$ are

$2 \backslash 1$	$\mathcal{M}$ human	$\mathcal{H}$ horse	$\mathcal{L}$ living	$\mathcal{S}$ stone
$\mathcal{M}$	a	e	i, o	e
$\mathcal{H}$	e	a	i, o	e
$\mathcal{L}$	a	a	a	e
$\mathcal{S}$	e	e	e	a

e.g., $\mathcal{L}i\mathcal{H}$ and $\mathcal{L}o\mathcal{H}$.

8.2.1 Strong Syllogisms

To solve the task of syllogistic, one looks first for all *equivalences* (strong syllogisms), i.e., for the cases where a syllogism is equivalent with its premisses:

$$P\rho M \text{ and } M\sigma Q \text{ for an } M \iff P\tau Q.$$

For such an equivalence, τ is called a *relation product* of ρ and σ:

$$\sigma \circ \rho = \tau.$$

An example: If ρ is the relation "son" and σ the relation "sibling," then τ is the relation "nephew":

$$\text{nephew} = \text{sibling} \circ \text{son}.$$

Therefore, as done first by De Morgan,[5] the syllogisms can be formulated in a multiplication table displaying which products of the six relations give one of these relations.

Syllogistics uses two axioms for "all" — reflexivity and transitivity:

$$\begin{aligned}&(\text{IR}) : PaP \text{ for all } P,\\ &(\text{IT}) : PaM \text{ and } MaQ \text{ for an } M \Rightarrow PaQ.\end{aligned}$$

With the validity of PaP, there is the equivalence:

$$PaQ \iff PaM \text{ and } MaQ \text{ for an } M.$$

The particular relations "some" and "not all" are definable by the general relation "all" and "no" if another predicate is used to connect them:

$$\begin{aligned}PiQ &\iff MaP \text{ and } MaQ \text{ for an } M,\\ PoQ &\iff MaP \text{ and } MeQ \text{ for an } M.\end{aligned}$$

Therefore, three equivalences (syllogisms) are obtained by the axiom and the definitions above:

$$a \circ a = a, \quad a \circ a^T = i, \quad e \circ a^T = o.$$

Using the *associativity* of the relation product:

$$\tau \circ (\sigma \circ \rho) = (\tau \circ \sigma) \circ \rho,$$

there follows:

$$a \circ i = a \circ a \circ a^T = a \circ a^T = i.$$

The next equivalence

$$PeM \text{ and } QaM \Rightarrow PeQ, \text{ i.e., } e \circ a = e = e^T,$$

[5] Augustus De Morgan (1806–1871).

is proved in the set-theoretical formulation:

$$P \cap M = \emptyset \text{ and } Q \subseteq M \iff P \cap Q = \emptyset.$$

Using the product property of the conversion:

$$\sigma \circ \rho = \tau \iff \rho^T \circ \sigma^T = \tau^T,$$

one obtains from the equivalences above with $e^T = e, i^T = i$:

$$a^T \circ a^T = a^T, \quad a \circ e = o^T, \quad i \circ a^T = i, \quad a^T \circ e = e.$$

Associativity implies:

$$\begin{aligned} o \circ a^T &= o \circ a^T \circ a^T &= e \circ a^T &= o, \\ a^T \circ o &= a^T \circ e \circ a^T &= e \circ a^T &= o, \\ i \circ e &= a \circ a^T \circ e &= a \circ e &= o^T, \end{aligned}$$

and application of the conversion:

$$a \circ o^T = o^T, \quad o^T \circ a = o^T, \quad e \circ i = o.$$

Altogether, there are 15 *strong syllogisms* in the (6×6)-matrix:

$2 \setminus 1$	a	a^T	e	i	o	o^T
a^T	i	a^T	o	i	o	$-$
a	a	$-$	e	$-$	$-$	o^T
e	o^T	e	$-$	o^T	$-$	$-$
i	i	$-$	o	$-$	$-$	$-$
o^T	o^T	$-$	$-$	$-$	$-$	$-$
o	$-$	o	$-$	$-$	$-$	$-$

e.g., $a \circ i = i$.

It is useful to concretize the meaning of the entries by examples, and to realize that their derivations did not use concrete examples. We trust more a "mechanical calculation" than a content-related argumentation.

In the matrix above *conversion* is *transposition*. With the definition of the particular relations "some" $i = a \circ a^T$ and "not all" $o = e \circ a^T$ only the (3×3)-left upper half of the table for the products of the three general relations a, a^T, e is of interest, the remainder is obtained by transposition and multiplication:

$2 \setminus 1$	a	a^T	e
a^T	i (definition)	a^T (transposition)	o (definition)
a	a (axiom)	$-$	e
e	o^T (transposition)	e (transposition)	$-$

8.2.2 *Weak Syllogisms*

In addition to the strong syllogisms there are *weak syllogisms* (implications) only in one direction:

$$\begin{array}{llll} PaQ & \Rightarrow PiQ & \text{for sets: } P \subseteq Q & \Rightarrow P \not\subseteq \complement Q, \\ Pa^T Q & \Rightarrow PiQ & \text{for sets: } Q \subseteq P & \Rightarrow P \not\subseteq \complement Q, \\ PeQ & \Rightarrow PoQ & \text{for sets: } P \subseteq \complement Q & \Rightarrow P \not\subseteq Q, \\ PeQ & \Rightarrow Po^T Q & \text{for sets: } P \subseteq Q & \Rightarrow Q \not\subseteq P. \end{array}$$

A conclusion with a or a^T can be attenuated to i, one with e to o or o^T. Therefore, one obtains from the strong syllogisms above six weak syllogisms:

$$\text{with } a \subseteq i, \quad e \subseteq o, \quad e \subseteq o^T: \qquad \begin{cases} a \circ a \subseteq i, & a^T \circ a^T \subseteq i, \\ e \circ a \subseteq o, & a^T \circ e \subseteq o, \\ e \circ a \subseteq o^T, & a^T \circ e \subseteq o^T. \end{cases}$$

Altogether there are $15 + 6 = 21$ *normalized syllogisms*. To finish the task of syllogisms, one has to give counterexamples for the empty 21 places in the relation-multiplication 6×6-matrix above. Five empty places are in the diagonal, the 16 not-diagonal ones come in pairs, i.e., one has to give only eight counterexamples. This was done already by Aristotle.

8.2.3 *Traditional Syllogisms*

Scholastics introduced phantasy words whose vowels give the syllogisms. Since, at this time, one did not use the converse relations, but stayed with the four patterns above for a, i, e, o, the 21 normalized syllogisms lead to 24 *traditional syllogisms*:

	I	II	III	IV
$a \circ a = a$	barbara	–	–	–
$e \circ a = e$	celarent	cesare	–	–
$a^T \circ e = e$	–	camestres	–	calemnes
$a \circ i = i$	darii	–	datisi	–
$i \circ a^T = i$	–	–	disamis	dimatis
$e \circ i = o$	ferio	festino	ferison	fresison
$a^T \circ o = o$	–	baroco	–	–
$o \circ a^T = o$	–	–	bocardo	–
$a \circ a^T = i$	–	–	darapti	–
$e \circ a^T = o$	–	–	felapton	fesapo
$a \circ a \subseteq i$	barbari	–	–	–
$e \circ a \subseteq o$	celaront	cesaro	–	–
$a^T \circ e \subseteq o$	–	camestros	–	calemos
$a^T \circ a^T \subseteq i$	–	–	–	bamalip

8.3 Conjunction, Negation, and Adjunction

For the Aristotelean syllogisms, logic has been applied naively. This exemplifies the difficulty to describe logic with logic. Especially, the particles "and" and "or" have been used thoughtlessly. That these logical particles, called *junctors*, and the negation are so important should not surprise with the acquaintance of quantum theory: Has the electron in an atom the sharp properties "position" and "momentum," or "position" or "momentum," or one of these properties only, or even none? Precisely, at this point with the junctor "or," one can characterize the distinction of classical and quantum logic.

Historically, the logic of the junctors has been founded by Euclid,[6] a pupil of Socrates,[7] and his school, and by Chrysippos[8] of the elder Stoa. It has been later developed by Boole and Grassmann. Already Leibniz tried some sort of logical pangloss (a language for everything) for a formalization and a quasi-mechanical proof-method of logical statements, in analogy to Descartes' analytic geometry.

Starting from primitive propositions $A, B, C, \ldots$ new propositions can be composed by conjunction and negation, formalized ($\doteq$) as follows:

$$\begin{array}{lrl} \text{(C)} & \text{and, et} & \doteq \wedge, \text{ e.g., } A \wedge B, \\ \text{(N)} & \text{not } A, \text{ non } A & \doteq A^c. \end{array}$$

The implication is denoted by a doubled arrow:

$$\text{if } A, \text{ then } B \doteq A \Rightarrow B.$$

The *power set* $\mathbf{2}^S$ of a set S provides a good realization of these structures:

$$A, B \subseteq S \text{ then } \begin{cases} A \wedge B \sim A \cap B, \\ A^c \sim \complement A, \\ A \Rightarrow B \sim A \subseteq B. \end{cases}$$

It will play a characteristic role in Boole–Stone logics (ahead).

An implication calculus uses reflexivity and transitivity:

$$\begin{array}{ll} \text{(IR)} & \text{then } A \Rightarrow A, \\ \text{(IT)} & \text{if } (A \Rightarrow B) \text{ and } (B \Rightarrow C) \text{ then } (A \Rightarrow C). \end{array}$$

It allows further implications with the associativity axiom and the following rules for conjunction and negation:

---

[6] Euclid of Megara, around -(435–365).

[7] Socrates -(469–399).

[8] Chrysippos of Soli, around -(280–207).

$$
\begin{array}{lll}
\text{(C0)} & (A \wedge B) \wedge C \iff A \wedge (B \wedge C) \iff A \wedge B \wedge C, & \\
\text{(C1)} & & \text{then } A \wedge B \Rightarrow A, \\
\text{(C2)} & & \text{then } A \wedge B \Rightarrow B, \\
\text{(C3)} & \text{if } (C \Rightarrow A) \text{ and } (C \Rightarrow B) & \text{then } (C \Rightarrow A \wedge B), \\
\text{(NC)} & \text{if } (A \wedge B^c \Rightarrow C^c) & \text{then } (A \wedge C \Rightarrow B).
\end{array}
$$

(C1), (C2) are the basic implications that can be posited without any preconditions. (C3) is an infimum-condition for the conjunction. Only (NC) contains the negation, it is a contraposition and not so easy to accept immediately.

Does this axiomatization allow the independence of the elementary logic of a meta-logic? What is the ultimate foundation of these prescriptions?

With these rules, there arise equivalences (idempotency and commutativity) with the implication $\iff$ in both directions, $\Rightarrow$ and $\Leftarrow$:

$$
\begin{aligned}
A \wedge A &\iff A \text{ with (C1) and (C3)}, \\
A \wedge B &\iff B \wedge A \text{ with (C1), (C2), (C3)}.
\end{aligned}
$$

Furthermore, the equivalence of a proposition with its doubled negation can be derived:

$$A \iff A^{cc}.$$

The contraposition rule (NC) allows the derivation of the conversion for the implication of the negations:

$$\text{if } A \Rightarrow B \text{ then } B^c \Rightarrow A^c.$$

If all Athenians are Greeks, a non-Greek cannot be an Athenian. The negation of "if you cut your finger it bleeds" is not "if you don't cut your finger it doesn't bleed"— cut a toe! In general, such derivations are elementary, but sometimes tedious.

The *adjunction* $\vee$ (or, aut) is the most important junctor besides the conjunction $\wedge$ (and, et). It can be defined by the doubled negation in De Morgan's equivalences:

$$
\begin{aligned}
A \vee B &\iff (A^c \wedge B^c)^c, \\
A \wedge B &\iff (A^c \vee B^c)^c.
\end{aligned}
$$

Dualization comes by the following exchanges:

$$(\wedge, \Rightarrow, \Leftarrow) \leftrightarrow (\vee, \Leftarrow, \Rightarrow).$$

A dualized logical implication always leads to a logical implication.

The prescriptions for the conjunction imply the dual prescriptions for the adjunction:

$$\begin{array}{lrl} (A0): & (A \vee B) \vee C \iff A \vee (B \vee C) & \iff A \vee B \vee C, \\ (A1): & & \text{then } A \Rightarrow A \vee B, \\ (A2): & & \text{then } B \Rightarrow A \vee B, \\ (A3): & \text{if } (A \Rightarrow C) \text{ and } (B \Rightarrow A) & \text{then } (A \vee B \Rightarrow C), \\ (NA): & \text{if } (A^c \Rightarrow B^c \vee C) & \text{then } (B \Rightarrow A \vee C). \end{array}$$

In elementary logic, there holds the distributivity of "et" over "aut" and, dually, of "aut" over "et":

$$\begin{aligned} (A \vee B) \wedge C &\iff (A \wedge C) \vee (B \wedge C), \\ (A \wedge B) \vee C &\iff (A \vee C) \wedge (B \vee C). \end{aligned}$$

For the proof, one needs, remarkably, the negation with the prescriptions (NC), (NA). Such a distributivity does not hold in a quantum logic of Birkhoff and von Neumann (more ahead).

8.4 The System of the Junctors

Starting from the conjunction "et" ($\wedge$) and the negation "non" (c) more junctors can be defined. The ten junctors, defined in the following, constitute, in a certain sense, a complete system. The table gives the name, the definition, a verbalization, and a set-theoretical realization:

conjunction	$A \wedge B$ A and B	$A \cap B$
adjunction	$A \vee B \doteq (A^c \wedge B^c)^c$ A or (aut)B)	$A \cup B$
subtraction	$A \setminus B \doteq A \wedge B^c$ A without B	$A \cap \complement B$
subjunction	$A \setminus^c B \doteq A^c \vee B$ (not A) or B	$\complement A \cup B$
converse subtraction	$A \setminus^T B \doteq A^c \wedge B$ B without A	$\complement A \cap B$
converse subjunction	$A \setminus^{Tc} B \doteq A \vee B^c$ A or not B	$A \cup \complement B$
negated adjunction	$A \vee^c B \doteq A^c \wedge B^c$ neither A nor B	$\complement(A \cup B) = \complement A \cap \complement B$
negated conjunction	$A \wedge^c B \doteq (A \wedge B)^c$ not (A and B)	$\complement(A \cap B) = \complement A \cup \complement B$
disjunction	$A \dot{+} B \doteq (A \vee B) \wedge (A^c \vee B^c)$ either A or B	$(A \cup B) \cap \complement(A \cap B)$
negated disjunction	$A \dot{+}^c B \doteq (A \wedge B) \vee (A^c \wedge B^c)$ both or none (A and B)	$(A \cap B) \cup \complement(A \cup B)$

Adjunctions and conjunctions are binary connections, disjunctions are quaternary connections involving two propositions and their negations.

The system of the ten junctors can be arranged in two rows with each five junctors:

conjunction (et) $\wedge$		$\vee$ adjunction (aut)
subtraction (sine) $\setminus$		$\setminus^T$ converse subtraction
disjunction (vel) $\dot{+}$	$\bullet$	$\dot{+}^c$ negated disjunction
converse subjunction $\setminus^{Tc}$		$\setminus^c$ subjunction
negated adjunction $\vee^c$		$\wedge^c$ negated conjunction

Duality uses the right-left reflection $\leftrightarrow$. Negation is obtained by central $\bullet$ reflection.

8.5 Falsum and Verum

In the classical implication calculus above for conjunction and negation, the presence of a proposition and its negation implies any proposition:

$$\begin{array}{rll} A^c \wedge B^c & \Rightarrow A^c & \text{with (C1)}, \\ \text{then } A^c \wedge A & \Rightarrow B & \text{with (NC)}. \end{array}$$

A proposition, which implies all propositions is called *logically false*, used with the variable $\bigwedge$, called *falsum*:

$$\bigwedge \Rightarrow A \qquad \text{“ex falso quodlibet.”}$$

Already Aristotle knew the principle of contradiction, given by the presence of a proposition and its negation:

$$A \wedge A^c \iff \bigwedge \qquad \text{“principium contradictionis.”}$$

In a set-theoretical model, falsum is realized by the empty set:

$$\bigwedge \sim \emptyset \subseteq A$$

By contraposition, a proposition implied by any proposition, e.g., the negation $(A \wedge A^c)^c$ of the logically false, is called *logically true* with the variable $\bigvee$, called *verum*:

$$\bigwedge \Rightarrow A^c, \text{ then } A \Rightarrow \bigwedge^c = \bigvee \qquad \text{“ex quolibet verum.”}$$

In the Euler-circle visualization, “verum” is the full sheet of paper. Either a proposition or its negation has to be logically true. That constitutes the principle of the excluded third, already formulated by Aristotle:

$$A \vee A^c \iff \bigvee \qquad \text{“tertium non datur.”}$$

One has the equivalences:

$$\bigwedge^c \iff \bigvee, \; \bigvee^c \iff \bigwedge.$$

In a set-theoretical model, verum is realized by the full set under consideration:

$$\bigvee \sim S \supseteq A.$$

8.6 Classes and Sets

Logic and the foundation of mathematics are closely intertwined. The history of mathematics can be separated into the time before and the time after Cantor (1845–1918). Poor Cantor was partly ostracized, especially by the influential Kronecker, and driven into a mental hospital for a while. After Cantor's epochal work, also strongly theologically motivated, mathematics works with sets.

Frege, who may be called the founder of analytic philosophy, working in painful isolation, complained of the neglection of his publications: "Mathematicians say 'Philosophica sunt, non leguntur', and philosophers say 'Mathematica sunt, non leguntur'." He almost despaired in 1902, shortly before the print of his book *Grundgesetze der Arithmetik*, as Russell mailed him a paradox, a logically sharpened modernization of the ancient truth-puzzle: When a Cretan says: "All Cretans always lie" — is he lying or not? The Russell construction R comprehends all sets x that do not contain themselves as an element:

$$R = \{x \mid x \notin x\}.$$

An analysis of this construction as a set leads to the self-contradictory statement:

$$R \in R \iff R \notin R.$$

Russell's construction illustrates the *compression-problem* (comprehension, "bracing"): How can objects be collected into sets or classes? Russell himself proposed a theory of types — there are less restricting systems of axioms, especially such which use an *all-class*. In the following, a sketch of a set theory of Bernays[9] and von Neumann is given.

One starts from a "field" with "objects" $x, y, \ldots$ which will be called *classes*. In addition, there is "given" a *binary relation* $\in$ which, for any two classes, applies or not:

[9] Paul Bernays (1888–1977).

$$x \in y \top x \notin y$$
$$\text{with } x \notin y \iff (x \in y)^c.$$

The last paragraph contains words which make it difficult, at least for me and my superficial analysis, to feel on a safe foundation.

$\in$ is the problematic undefined abstract concept to be elaborated in set theory. It is translated in the everyday language by its original naive wording of membership or "is element of." The symbol $\in = \epsilon$ ($\epsilon\sigma\tau\iota$) was introduced by Peano.[10]

A class x is called a *set*, if it is an element of a class:

$$x \text{ is a set } \iff \mathbf{set}(x) \doteq \text{There exists a class } y \text{ with } x \in y.$$

If there is no such class, the class x may be called an *"un-set."* In German, there exists the appropriate word "Unmenge" — a set over all sizes. Also this approach distinguishes two types, sets and classes.

8.6.1 Compression of Sets

It is assumed that the "field" of classes is complete with respect to the *operation of compression (comprehension)*: All sets (no un-sets!) that fulfill a *logically pertinent condition* **log** (more ahead) are collected in a class of the "field":

$$\text{(COMPRESS)} \qquad \bigvee_{\text{class } c} \bigwedge_{x} (x \in c \iff \mathbf{set}(x) \wedge \mathbf{log}(x)).$$

Here $\bigvee_c$ has the meaning: "there exists a c" (existence quantor), and the dual $\bigwedge_x$ "for all x" (all-quantor). This class c is denoted with the bracing $\{\ldots\}$:

$$c = \{x \mid \mathbf{set}(x) \wedge \mathbf{log}(x)\} = \{x \mid \mathbf{log}(x)\}.$$

For convenience, in most cases, the explicit notation of the important set property $\mathbf{set}(x)$ is omitted: Only sets can be compressed into a class. The quant(ificat)ors $\bigvee_c$ and $\bigwedge_x$ are often symbolized by $\exists c$ and $\forall x$.

The compression of all sets is called the *all-class*:

$$\underline{\mathbf{set}} \doteq \{x \mid \mathbf{set}(x)\}, \quad x \in \underline{\mathbf{set}} \iff \mathbf{set}(x).$$

[10]Giuseppe Peano (1858–1932).

How looks Russell's construction in this framework?

$$R = \{x \mid x \in \underline{\textbf{set}} \wedge (x \notin x)\}.$$

As a set, R leads to a contradiction, the Russell class R is an un-set:

$$R \notin \underline{\textbf{set}} \text{ and } R \notin R.$$

There exist trivial models for "fields" of "objects" that fulfill the compression axiom (COMPRESS) of Bernays and von Neumann. For instance, take objects where all pairs fulfill $x \notin y$, i.e., only un-sets. The existence problem arises by adding more axioms, especially if all natural numbers are included.

The *logically pertinent propositions* are defined with Skolem[11] as composed by relations $x \in y$ with implication, negation, conjunction, adjunction, and quantors:

(I)	implication "then" $\Rightarrow$, "if, and only if" $\Longleftrightarrow$,	e.g., $(x \in y) \Rightarrow (x \in z)$,
(N)	negation "non" ^{c},	e.g., $(x \in y)^c \Longleftrightarrow x \notin y$,
(C)	conjunction "et" $\wedge$,	e.g., $(x \in y) \wedge (x \in z)$,
(A)	adjunction "aut" $\vee$,	e.g., $(x \in y) \vee (x \in z)$,
(QC)	all-quantor "for all x" $\bigwedge\limits_x$,	e.g., $\bigwedge\limits_{x \in y}$,
(QA)	existence quantor "there exists an x" $\bigvee\limits_x$,	e.g., $\bigvee\limits_{x \in y}$.

The implication calculus used is the self-dual calculus of classical logic with the rules given above, i.e., (IR) and (IT), and (C0)–(C3), (NC) for the conjunction, and (A0)–(A3), (NA) for the adjunction. The rules for the adjunction, defined by $A \vee B \Longleftrightarrow (A^c \wedge B^c)^c$, can be obtained from the rules for the conjunction. The rules for the quantors are not discussed here.

8.6.2 Subclasses and Junctor Classes

Set theory is classical logic with the inclusion of the characteristic mathematical structure, the membership relation. All axioms of set theory concern the binary element relation $\in$. Especially, it has to be determined which classes are sets.

"Verum" and "falsum" correspond to the *all-class* and the *empty class*:

$$\underline{\textbf{set}} \doteq \{x \mid \textbf{set}(x)\}, \qquad \emptyset \doteq \{x \mid \textbf{set}(x) \wedge \neg\textbf{set}(x)\}.$$

[11] Thoralf Skolem (1887–1963).

The empty class is postulated as a set:

$$\text{(EMPTY)} \quad \mathbf{set}(\emptyset), \text{ i.e., } \emptyset \in \underline{\mathbf{set}}.$$

Therefore, one "has" at least one set. The minimal "field" of "objects" contains the all-class **set** (underlined) and the empty set $\emptyset$, nothing more.

The implication defines a subclass:

$$a \subseteq b \doteq \bigwedge_x (x \in a \Rightarrow x \in b).$$

The subset axiom is postulated for sets:

$$\text{(SUBSET)} \quad (b \in \underline{\mathbf{set}}) \wedge (a \subseteq b) \Rightarrow a \in \underline{\mathbf{set}}.$$

Using the opposite direction and a "subclass," the equality is defined via equal elements:

$$a \supseteq b \doteq b \subseteq a$$
$$a = b \doteq (a \subseteq b) \wedge (a \supseteq b) \iff \bigwedge_x (x \in a \iff x \in b).$$

The extensionality axiom for the equality postulates that equal sets are elements of the same classes:

$$\text{(EQUAL)} \quad (a = b) \wedge (a \in z) \Rightarrow b \in z.$$

Each class a consists precisely of its elements:

$$a = \{x \mid x \in a\},$$

and is "between" the all-class and the empty set:

$$\emptyset \subseteq a \subseteq \underline{\mathbf{set}},$$

This applies for each set:

$$a \in \underline{\mathbf{set}} \Rightarrow a \subset \underline{\mathbf{set}}.$$

Especially the Russell un-set $R \subseteq \underline{\mathbf{set}}$ is a subclass, and the all-class is an un-set.

The conjunction and adjunction define the intersection class and union class, respectively, of two classes:

$$a \cap b \doteq \{x \mid (x \in a) \wedge (x \in b)\},$$
$$a \cup b \doteq \{x \mid (x \in a) \vee (x \in b)\}.$$

The union set axiom postulates a set for two sets:

$$\text{(UNION)} \quad a, b \in \underline{\mathbf{set}} \Rightarrow a \cup b \in \underline{\mathbf{set}}.$$

With $a \cap b \subseteq a$, the intersection of two sets is a set.

Each class has its complementary class, defined by the negation, with the involution:

$$\complement a \doteq \{x \mid x \notin a\}, \quad \complement\complement a = a.$$

The complement of a set is an un-set, and vice versa:

$$a \in \underline{\mathbf{set}} \iff \complement a \notin \underline{\mathbf{set}}.$$

De Morgan's rules give:

$$a \cup b = \complement(\complement a \cap \complement b), \quad a \cap b = \complement(\complement a \cup \complement b),$$

and, furthermore, the "tertium non datur" and the "principium contradictionis":

$$a \cap \complement a = \emptyset = \complement\underline{\mathbf{set}}, \quad a \cup \complement a = \underline{\mathbf{set}} = \complement\emptyset.$$

There exist finite "fields" where the axioms (COMPRESS) and (EQUAL) are valid, and exactly one of the axioms (EMPTY), (SUBSET) and (UNION) does not hold — therefore, they are independent. Furthermore, there exist finite models with the validity of all five axioms — e.g., comprising two classes $\emptyset \in \underline{\mathbf{set}}$. So far, one is on a safe, but rather boring ground.

8.6.3 Atom Classes and Power Classes

In this section, set theory is postulated in such a way that no longer simple models apply.

Each class a has its *unit class* in the Peano compression:

$$\{x \mid x \in \underline{\mathbf{set}} \wedge (a \in \underline{\mathbf{set}} \Rightarrow x = a)\}.$$

The shortened symbol $\{a\}$ is used only for sets:

$$\begin{aligned} &a \in \mathbf{set} \Rightarrow \{x \mid x = a\} = \{a\} \\ &a \notin \underline{\mathbf{set}} \Rightarrow \{x \mid x \in \underline{\mathbf{set}} \wedge (a \in \underline{\mathbf{set}} \Rightarrow x = a)\} = \underline{\mathbf{set}}. \end{aligned}$$

A unit set is postulated as a set:

$$\text{(ATOM)} \quad a \in \underline{\mathbf{set}} \Rightarrow \{a\} \in \underline{\mathbf{set}}.$$

There holds the membership:

$$a \in \underline{\mathbf{set}} \iff a \in \{a\}.$$

With the axiom (ATOM) there is no finite "field" of objects fulfilling all axioms so far.

Atom sets are defined by the property to have no subset proper. The unit sets coincide with the atom sets:

$$\emptyset \subseteq b \subseteq \{a\} \Rightarrow (b = \emptyset) \vee (b = \{a\})$$

Each class a allows the construction of its *power class* comprising its subsets:

$$\{x \mid x \in \underline{\mathbf{set}} \wedge x \subseteq a\}$$

The subsets of a set a come with the notation:

$$a \in \underline{\mathbf{set}} \Rightarrow \{x \mid x \subseteq a\} = \mathbf{2}^a,$$

with the two examples:

$$\begin{aligned} \mathbf{2}^{\emptyset} &= \{\emptyset\} \\ a \in \underline{\mathbf{set}} \Rightarrow \mathbf{2}^{\{a\}} &= \{\emptyset, \{a\}\} \end{aligned}$$

The power class of the all-class is the all-class itself:

$$\{x \mid x \in \underline{\mathbf{set}} \wedge x \subseteq \underline{\mathbf{set}}\} = \underline{\mathbf{set}}.$$

The *intersection and union class of the elements of a class a* are

$$\bigcap_{x \in a} x, \quad \bigcup_{x \in a} x.$$

For a set a, the union class of the elements of the power class gives the set a; the intersection class is the empty set:

$$a \in \underline{\mathbf{set}} \Rightarrow \bigcup_{x \in \mathbf{2}^a} x = a, \quad \bigcap_{x \in \mathbf{2}^a} x = \emptyset.$$

In general, one obtains for a nonempty class:

$$a \neq \emptyset \Rightarrow \bigcap_{x \in a} x \subseteq \bigcup_{x \in a} x.$$

The "anomaly of the empty set" looks as follows:

$$\text{for } \emptyset : \bigcap_{x \in \emptyset} x = \underline{\mathbf{set}} \supseteq \bigcup_{x \in \emptyset} x = \emptyset.$$

Set–faithfulness is postulated by the *strong union axioms* — in both directions:

$$\text{(QUNION)} \qquad a \in \underline{\mathbf{set}} \iff \bigcup_{x \in a} x \in \underline{\mathbf{set}}.$$

With these axioms, there follows set-faithfulness of the power operation:

$$a \in \underline{\mathbf{set}} \Rightarrow \mathbf{2}^a \in \underline{\mathbf{set}}.$$

8.7 The Natural Numbers

The *successor* a^+ of a set a is defined by its union with its unit-set:

$$a \in \underline{\mathbf{set}} \Rightarrow a^+ = a \cup \{a\} \in \underline{\mathbf{set}}.$$

With von Neumann: Starting from the empty set, all natural numbers are generated as successors:

$$\begin{array}{lll} 0 & = \emptyset, & \\ 1 & = 0 \cup \{0\} = \{0\} & = \{\emptyset\}, \\ 2 & = 1 \cup \{1\} = \{0, 1\} & = \{\emptyset, \{\emptyset\}\} = \{0\} \cup \{1\}, \\ 3 & = 2 \cup \{2\} = \{0, 1, 2\} & = \{\, \emptyset\, ,\ \{\emptyset\}\, ,\ \{\emptyset, \{\emptyset\}\}\, \}, \\ & \ldots & \\ n+1 & = n \cup \{n\} = \{0, 1, \ldots, n\}. & \end{array}$$

Following Dedekind, the natural numbers $\mathbb{N} = \{0, 1, 2, \ldots\}$ are defined by the intersection of all classes c which contain $0 = \emptyset$ as an element, and which are closed (complete) with respect to the formation of successors:

$$\mathbb{N} = \bigcap_{(\emptyset \in c) \wedge \bigwedge_x (x \in c \Rightarrow x^+ \in c)} c.$$

The *axiom of the natural numbers* postulates them as a set:

$$\text{(NUMBER)} \qquad \mathbb{N} \in \underline{\mathbf{set}}.$$

This requires an infinite set and opens up, for Hilbert at least, the "paradise of the analysis." Against the establishment of "actual infinities" serious objections have been raised, e.g., by Kronecker, Brouwer,[12] Poincaré, and Weyl.

8.8 Axioms of Set Theory

Summarizing all axioms of Bernays–von Neumannn's set theory which tries to harmonize the elementary logic with the membership relation, involving the empty set $\emptyset$ and the all-class $\underline{\mathbf{set}}$:

$$\text{(COMPRESS)} \qquad \bigvee_{\text{class } c} \bigwedge_{x} \Big(x \in c \iff (x \in \underline{\mathbf{set}}) \wedge \mathbf{log}(x) \Big),$$

$$\text{notation (bracing): } c = \{x \in \underline{\mathbf{set}} \mid \mathbf{log}(x)\},$$

$$\text{(EMPTY)} \qquad \emptyset \in \underline{\mathbf{set}},$$

$$\text{(ATOM)} \qquad a \in \underline{\mathbf{set}} \Rightarrow \{a\} \in \underline{\mathbf{set}},$$

$$\text{(SUBSET)} \qquad (a \in \underline{\mathbf{set}}) \wedge (b \subseteq a) \Rightarrow b \in \underline{\mathbf{set}},$$

$$\text{(EQUAL)} \qquad (a = b) \wedge (a \in z) \Rightarrow b \in z,$$

$$\text{(UNION)} \qquad a, b \in \underline{\mathbf{set}} \Rightarrow a \cup b \in \underline{\mathbf{set}},$$

$$\text{(QUNION)} \qquad a \in \underline{\mathbf{set}} \iff \bigcup_{x \in a} x \in \underline{\mathbf{set}},$$

$$\text{(NUMBER)} \qquad \mathbb{N} = \bigcap_{(\emptyset \in c) \wedge \bigwedge_x (x \in c \Rightarrow x^+ \in c)} c, \qquad \mathbb{N} \in \underline{\mathbf{set}},$$

$$\text{(MAPPING)} \qquad f : a \longrightarrow b, \;\; a \in \underline{\mathbf{set}} \Rightarrow f\,[a] \in \underline{\mathbf{set}}.,$$

The last axiom (MAPPING) postulates sets as mapping-images of sets.

This set theory is essentially equivalent to the set theory established by Zermelo, Fraenkel, and Skolem. Does there "exist" a "field" with such "objects"? Has such a "field" something to do with the physical nature?

According to Gödel, any consistent axiom system that involves the definition of the natural numbers is incomplete: It allows the formulation of undecidable theorems, i.e., the validity of neither the theorem nor its negation can be proved. An example in Bernays–von Neumann's set theory was given by Cohen in the continuum hypothesis: Between the cardinalities of the natural numbers and its power set there does not exist a different cardinality:

$$\begin{aligned} &\text{Card } \mathbb{N} \leq \text{Card } a \leq \text{Card } \mathbf{2}^{\mathbb{N}} \\ &\Rightarrow \text{either } (\text{Card } a = \text{Card } \mathbb{N}) \text{ or } (\text{Card } a = \text{Card } \mathbf{2}^{\mathbb{N}}). \end{aligned}$$

[12]Luitzen Egbertus Jan Brouwer (1881–1966).

With the axioms above, this statement can neither be proved nor disproved. The system can be extended with the validity or the negation of the continuum hypothesis. Again, the extended system allows the formulation of undecidable theorems.

The openness of an axiom system — is it connected with the principally open future of physics as formulated in quantum theory? Is that some sort of temporal property of an axiom system? Is it related to the classical incompleteness of quantum structures?

8.9 Implication (Order) and Equivalence

The essential content of physics is given by dynamics, defined as realizations of causality by interactions and objects. Probably not accidentally, time parametrizes causality: The logical implication and the order of time have similar structures.

A set S — more general for classes too — is called *semi-implicative* (*semi-ordered*) via a binary inner relation with the following two properties:

$$a, b, c \in S : \begin{cases} \textit{reflexive} : a \sqsubseteq a, \\ \textit{transitive} : a \sqsubseteq b \text{ and } b \sqsubseteq c \Rightarrow a \sqsubseteq c. \end{cases}$$

Such relations are used in different contexts:

$$a \sqsubseteq b \cong \begin{cases} a \Rightarrow b : a \text{ implies } b, \text{ or } b \text{ follows from } a, \\ a \preceq b : \ a \text{ is earlier than } b, \text{ or } a \text{ is smaller than } b. \end{cases}$$

A semi-implication (–order) $\sqsubseteq$ comes with the *converse (contrary, transposed) semi-implication (-order)*:

$$a \sqsupseteq b \iff b \sqsubseteq a.$$

A semi-implication (-order) is *total on* S, if the converse pair fills up the product set $S \times S$:

$$(a, b) \in S \times S \iff a \sqsubseteq b \text{ or } a \sqsupseteq b.$$

A semi-implicative (-ordered) set allows important concepts which are worth to be formulated in the logical and the temporal jargon, with $a \in S$:

$$\begin{aligned} &\textit{future of } a : [a]_{\preceq} = \{b \mid a \preceq b\}, && \textit{implications of } a : [a]_{\Rightarrow} = \{b \mid a \Rightarrow b\} \\ &\textit{past of } a : [a]_{\succeq} = \{b \mid a \succeq b\}, && \textit{conditions for } a : [a]_{\Leftarrow} = \{b \mid a \Leftarrow b\} \\ &\textit{presence of } a : [a]_{\preceq} \cap [a]_{\succeq}, && \textit{equivalences of } a : [a]_{\Leftarrow} \cap [a]_{\Rightarrow}, \end{aligned}$$

and for both languages:

$$\begin{aligned}
&\textit{causally connected set of } a: && [a]_{\text{caus}} = [a]_{\preceq} \cup [a]_{\succeq} = [a]_{\Leftarrow} \cup [a]_{\Rightarrow}\,,\\
& && (= S, \text{ for a total order,})\\
&\textit{causally disconnected set of } a: && \complement\,[a]_{\text{caus}} = \{x \notin [a]_{\text{caus}}\},\\
& && (= \emptyset, \text{ for a total order.})
\end{aligned}$$

With an additional symmetry property, a semi-implication (-order) becomes either an equivalence or an implication (order):

semi-implication (-order) (reflexive, transitive) ↗ equivalence (symmetric)
semi-implication (-order) (reflexive, transitive) ↘ implication (order) (antisymmetric)

An *equivalence relation* $\sim$ of a set is a *symmetric* semi-implication (-order):

$$a, b \in S: \quad \textit{symmetric}: a \sim b \iff b \sim a.$$

It defines equivalence subsets which connect the equivalence with the equality as the finest equivalence relation:

$$[a]_{\sim} = \{b \in S \mid b \sim a\}, \quad a \sim b \iff [a]_{\sim} = [b]_{\sim}\,,$$

leading to a disjoint decomposition with a system $R \subseteq S$ of representatives:

$$S = \biguplus_{r \in R} [a_r]_{\sim}\,, \qquad S/\sim \; = \{[a_r]_{\sim} \mid r \in R\}.$$

$S/\sim$ is also called the *fibration* of S with the equivalence relation $\sim$.

A semi-implication (-order) $\sqsubseteq$ defines "its" equivalence relation $\simeq$:

$$a \simeq b \iff a \sqsubseteq b \text{ and } a \sqsupseteq b.$$

The equivalence sets $S/\simeq$ carry a well defined, i.e., representative-independent *induced implication (order)*:

$$[a]_{\simeq} \sqsubseteq [b]_{\simeq} \iff a \sqsubseteq b.$$

It is *antisymmetric*:

$$[a]_{\simeq} \sqsubseteq [b]_{\simeq} \text{ and } [b]_{\simeq} \sqsubseteq [a]_{\simeq} \iff [a]_{\simeq} = [b]_{\simeq}\,.$$

In general, an antisymmetric semi-implication (-order) is called an *implication (order)*:

$$(S, \sqsubseteq) \in \underline{\textbf{oset}} : \quad a, b, c \in S, \quad \begin{cases} \text{reflexive: } a \sqsubseteq a, \\ \text{transitive: } a \sqsubseteq b \text{ and } b \sqsubseteq c \Rightarrow a \sqsubseteq c, \\ \text{antisymmetric: } a \sqsubseteq b \text{ and } b \sqsubseteq a \iff a = b. \end{cases}$$

The induced equivalence coincides with the equality:

$$\begin{aligned} \textit{presence of } a : \quad & [a]_\preceq \cap [a]_\succeq = \{a\}, \\ \textit{equivalences of } a : \quad & [a]_\Leftarrow \cap [a]_\Rightarrow = \{a\}. \end{aligned}$$

An ordered set S is *initialized* with an initial (earliest, empty) element (origin) $\square$, and *finalized* with a final (latest, full) element (end) $\blacksquare$ with the properties:

$$a \in S : \ \square \subseteq a, \ \ a \subseteq \blacksquare .$$

If there exist such elements, they are unique.

8.9.1 Examples for Order and Implication

The inclusion $X \subseteq Y$ orders the power set $\mathbf{2}^S$ of a set S.

The integer, rational, and real numbers, $\mathbb{N}$, $\mathbb{Z}$, $\mathbb{Q}$ and $\mathbb{R}$, carry the natural total order; for the complex numbers it is no longer total, $\mathbb{R}_+ \subset \mathbb{C}$:

$$\begin{aligned} &\alpha, \beta \in \mathbb{C} : \alpha \succeq \beta \iff \alpha - \beta \succeq 0 \\ &\gamma \in \mathbb{C} : \quad \gamma \succeq 0 \iff \text{There exists } \delta \in \mathbb{C} \text{ with } \gamma = \overline{\delta}\delta. \end{aligned}$$

In general, an addition-compatible order of a vector space V with ordered scalars K can be characterized at the trivial translation 0:

$$\begin{aligned} \text{additive group:} \quad & x, y \in V, \ x \succeq y \iff x - y \succeq 0, \\ \text{reflexive:} \quad & 0 \succeq 0, \\ \text{transitive:} \quad & x, y \succeq 0 \Rightarrow x + y \succeq 0, \\ \text{antisymmetric:} \quad & x, -x \succeq 0 \Rightarrow x = 0, \\ \text{scalar multiplication:} \quad & x \succeq 0, \ \alpha \in K, \ \alpha \geq 0 \Rightarrow \alpha x \succeq 0. \end{aligned}$$

The future of the neutral element (presence) is a *cone*:

$$x, y \in [0]_\preceq , \quad \alpha, \beta \geq 0 \Rightarrow \alpha x + \beta y \in [0]_\preceq .$$

The $\mathbb{C}$-order is a special case of the unique order structure of $C*$-algebras: There, an element x is positive if it is symmetric, $x = x^*$, and if it has a positive spectrum, or, equivalently, if it allows a product form $x = y^\star \circ y$. Another special case is given by the translations $x, y, \ldots \in \mathbb{R}^{1+3}$ of Minkowski spacetime with $\mathbf{O}(1,3)$-invariant Lorentz metric in the Weyl-representation where Minkowski spacetime is embedded in the $C*$-algebra of the complex 2×2-matrices:

$$\begin{aligned}
&x = \begin{pmatrix} x_0 + x_3 & x_1 - ix_2 \\ x_1 + ix_2 & x_0 - x_3 \end{pmatrix}, \quad x^2 = \det x, \quad x_0 = \tfrac{1}{2} \operatorname{tr} x \\
&\tau \in \mathbf{spec}\, x \iff \det x - \tau \operatorname{tr} x + \tau^2 = 0 \\
&x \succeq 0 \iff \tau_\pm \geq 0 \iff \det x \geq 0, \ \operatorname{tr} x \geq 0.
\end{aligned}$$

The equivalence sets $g\mathbf{U}(n) \in \mathbf{GL}(n, \mathbb{C})/\mathbf{U}(n)$, i.e., the orbits of the unitary group in the general complex-linear group $\mathbf{GL}(n, \mathbb{C})$ as real $2n^2$-dimensional Lie group, carry a total semi-order by the modulus of the determinant:

$$g_1\mathbf{U}(n) \succeq g_2\mathbf{U}(n) \iff |\det g_1| \geq |\det g_2|.$$

The induced equivalence relation:

$$g_1\mathbf{U}(n) \simeq g_2\mathbf{U}(n) \iff |\det g_1| = |\det g_2|,$$

defines the equivalence sets $[g\mathbf{U}(n)]_\simeq \in \mathbf{GL}(n, \mathbb{C})/\mathbf{U}(1) \circ \mathbf{SL}(n, \mathbb{C})$, i.e., the orbits of the product of the phase group $\mathbf{U}(1)$ with the special linear group $\mathbf{SL}(n, \mathbb{C})$ in the general linear group. These totally order equivalence sets are isomorphic to the dilation group $\mathbf{GL}(n, \mathbb{C})/\mathbf{U}(1) \circ \mathbf{SL}(n, \mathbb{C}) \cong \mathbf{D}(1)$ and the real numbers $\mathbf{D}(1) = \exp \mathbb{R} \cong \mathbb{R}$. The cardinality of the equivalence class is given by the real (n^2-1)-dimensional Sylvester manifold $\mathbf{SL}(n, \mathbb{C})/\mathbf{SU}(n)$.

For $n = 1$, the moduli $\mathbf{GL}(1, \mathbb{C})/\mathbf{U}(1) = \mathbf{D}(1) \cong \mathbb{R}$ constitute a model for time. For $n = 2$, the totally semi-ordered real four-dimensional manifold $\mathbf{GL}(2, \mathbb{C})/\mathbf{U}(2)$ is isomorphic to the strict future cone $\mathbb{R}^{1+3}_+$ of Minkowski space with the timelike vectors $\{x \in \mathbb{R}^{1+3} \mid x^2 > 0 \text{ and } x_0 > 0\}$. $\mathbf{GL}(2, \mathbb{C})/\mathbf{U}(1) \circ \mathbf{SL}(2, \mathbb{C}) \cong \mathbf{D}(1)$ is isomorphic to the totally ordered set with the strictly future orbits of the Lorentz group in Minkowski spacetime. The real three-dimensional Sylvester manifold is the "boost" manifold $\mathbf{SL}(2, \mathbb{C})/\mathbf{SU}(2) \cong \mathbf{SO}_0(1,3)/\mathbf{SO}(3)$.

A mapping of two sets defines, by equal images, a *mapping equivalence relation* for S:

$$f : S \longrightarrow T : \ x \overset{f}{\sim} y \iff f(x) = f(y),$$

with the equivalence sets and the fibration of S:

$$[x] = \{y \mid f(y) = f(x)\} \in S/f,$$

and the bijection to the mapping image:

$$S/f \cong f[S] \subseteq T \text{ with } [x] \leftrightarrow \{z \in T \mid z = f(x)\}.$$

If additional structures are involved, the mapping has to be compatible with them as exemplified by a group morphism:

$$f : G \longrightarrow H : \ f(1_G) = 1_H, \ f(gh) = f(g)f(h).$$

In this case, the quotient with the kernel of the mapping, a normal subgroup of G, contains the equivalence sets

$$g \overset{f}{\sim} h \iff f(g^{-1}h) = 1_H \iff g^{-1}h \in \text{kern } f \iff h \in g \,\text{kern } f.$$

The bijection above is a group isomorphy:

$$G/f = G/\text{kern } f \cong f[G] \subseteq H.$$

8.10 Lattices and Logics

The definitions of sets and classes and the operations with them use the classical logic. In this framework, classical logic itself can be refound. It is remarkable, that, there, also new different logics can be formulated — also those where the linear structure of quantum theory becomes important.

A set with two associative and commutative inner compositions, called *join* $\sqcup$ and *meet* $\sqcap$, is a *lattice* if both compositions have an *absorptive* relationship to each other:

$$(L, \sqcup, \sqcap) \in \underline{\textbf{latt}} : \begin{cases} \text{associative:} \begin{cases} (a \sqcup b) \sqcup c = a \sqcup (b \sqcup c), \\ (a \sqcap b) \sqcap c = a \sqcap (b \sqcap c), \end{cases} \\ \text{commutative:} \begin{cases} a \sqcup b = b \sqcup a \\ a \sqcap b = b \sqcap a, \end{cases} \\ \text{absorptive: } (a \sqcap b) \sqcup a = a = (a \sqcup b) \sqcap a. \end{cases}$$

Both compositions are equivalent structures with the duality $\sqcap \leftrightarrow \sqcup$. They are connected by the absorptivity which leads to their idempotency:

$$a \sqcup a = a = a \sqcap a.$$

A lattice carries its *natural order* by the definition:

$$a \sqsubseteq b \iff a \sqcap b = a.$$

With respect to the order, join and meet characterize the least upper bound $a \sqcup b = \sup(a, b)$ and the greatest lower bound $a \sqcap b = \inf(a, b)$.

Lattices with the intersection $\cap$ as meet have the inclusion $a \subseteq b \iff b \cap a = a$ as their natural order.

Ordered sets $(S, \preceq)$ with supremum and infimum for all pairs, e.g., totally ordered sets, are lattices:

$$a, b \in S : a \sqsubseteq b \iff a \preceq b, \ a \sqcup b = \sup(a, b), \ a \sqcap b = \inf(a, b).$$

A subset $I \subseteq \mathbb{R}$ of the naturally ordered reals is a lattice:

$$a \geq b \in I \subseteq \mathbb{R} : a \sqcup b = \sup(a, b) = a, \ a \sqcap b = \inf(a, b) = b,$$
$$(a \sqcup b) + (a \sqcap b) = a + b.$$

The causal order of two-dimensional Minkowski spacetime $\mathbb{R}^{1+1}$ with only one position direction and the Lorentz bilinear form $xy = x_0 y_0 - x_1 y_1$ defines a lattice structure: For example, the position translations $\pm x = (0, \pm a)$ with $a > 0$ has as meet and join time translations $\sup(x, -x) = (a, 0) = -\inf(x, -x)$. Ordered Minkowski space $\mathbb{R}^{1+s}$ with $s \geq 2$ is no lattice.

Divisibility as order relation defines a lattice structure for the nontrivial natural numbers. The greatest common divisor and the least common multiple constitute the infimum and the supremum with respect to this order:

$$n, m \in \mathbb{N} = \{1, 2, \dots\} : \begin{cases} n \sqsubseteq m \iff n|m (n \text{ divides } m), \\ n \sqcap m = \text{GCD}(n, m) = \inf(n, m), \\ n \sqcup m = \text{LCM}(n, m) = \sup(n, m). \end{cases}$$

The principal ideals $n\mathbb{Z}$ of the integers constitute a lattice:

$$(\mathbb{N}, \text{GCD}, \text{LCM}) \cong (\{n\mathbb{Z}\}_{n \in \mathbb{N}}, +, \cap) \in \underline{\textbf{latt}},$$
$$n, m \in \mathbb{N} : \ n|m \iff n\mathbb{Z} \supseteq m\mathbb{Z},$$
$$\text{GCD}(n, m)\mathbb{Z} = n\mathbb{Z} + m\mathbb{Z}, \ \ \text{LCM}(n, m)\mathbb{Z} = n\mathbb{Z} \cap m\mathbb{Z}.$$

The prime numbers give the maximal ideals $p\mathbb{Z}$.

An *initialized* lattice has an origin:

$$\Box \in L \text{ with } \Box \sqsubseteq a, \ \text{ i.e., } \Box = \Box \sqcap a \text{ for all } a \in L.$$

It allows the definition of

$$a \text{ and } b \text{ are disjoint elements} : \ a \sqcap b = \Box,$$
$$a \text{ is elementary} : \ a \neq \Box, \ \ b \sqsubseteq a \Rightarrow b = \Box.$$

A real subset I with a lower bound $\iota \in I$ is initialized. The nontrivial naturals $(\mathbb{N}, |)$ with the divisibility order are initialized with $1 \in \mathbb{N}$. Disjoint elements are coprime, the primes are elementary.

A *complementary* lattice has an involutive contra-automorphism relating meet and join with the origin as meet for each lattice element and its *complement*:

$$L \longrightarrow L,\ a \longmapsto a^c,\ a^{cc} = a, \quad \begin{cases} (a \sqcup b)^c = a^c \sqcap b^c, \\ a \sqcap a^c = \square \text{ for all } a \in L. \end{cases}$$

The complement of the origin defines the end:

$$\square^c = \blacksquare \sqsupseteq a,\ a \sqcup a^c = \blacksquare \text{ for all } a \in L.$$

The contra-automorphism structure is equivalent to the contra-montony of the natural order (implication) — the De Morgan rule:

$$(a \sqcup b)^c = b^c \sqcap a^c \iff (b \sqsubseteq a \Rightarrow a^c \sqsubseteq b^c).$$

A complementary real subset has to consist of two numbers $I = \{\iota = \square, \sigma = \blacksquare\} \subset \mathbb{R}$, e.g., $I = \{0, 1\}$. The nontrivial naturals $(\mathbb{N}, |)$ are not complementary.

A complementary lattice is called a *logic*. It allows an appropriate language for the logical concepts:

$$(L, \sqcup, \sqcap, \square, c) \in \underline{\textbf{logic}} : \quad \begin{cases} a \in L : \textit{proposition}, \\ \sqcap : \textit{conjunction (and, et)}, \\ \sqcup : \textit{adjunction (or, aut)}, \\ \sqsubseteq : \textit{implication (then, ergo)}, \\ \square : \textit{absurd proposition (falsehood, falsum)}, \\ a^c \in L : \textit{negation (not, non)}, \\ \blacksquare : \textit{self-evidentproposition (truth, verum)}. \end{cases}$$

One has in the logical language:

$$\text{principium contradictionis: } a \sqcap a^c = \square,$$
$$\text{tertium non datur: } a \sqcup a^c = \blacksquare.$$

An *elementary proposition* is nontrivial minimal. One obtains for the *subtraction (without, sine)* and the *disjunction (exclusive or, vel)* $\dot{+}$

$$a \text{ and not} b : a \setminus b = a \sqcap b^c,$$
$$\text{either } a \text{ or} b : a \dot{+} b = (a \sqcap b^c) \sqcup (b \sqcap a^c).$$

One proposition gives the trivial logic $\{\square = \blacksquare\}$. Two different propositions define the atomic logic $\{\square, \blacksquare\}$.

One may try to define a composition $\sqcup$ by starting with $\sqcap$ and the contra-automorphism c via $a \sqcup b = (a^c \sqcap b^c)^c$. However, this definition does not entail absorptivity. What about the contraposition rule

$$a \sqcap b^c \sqsubseteq c^c \overset{?}{\iff} a \sqcap c \sqsubseteq b.$$

Self-duality entails $b^c \sqsubseteq c^c \iff c \sqsubseteq b$. However, it is not possible to join a general element. As in the classical logic one has (C0) – (C3) and (A0) – (A3), not however, in general, the full contrapositions (NC) and (NA).

8.11 Classical (Boole–Stone) Logics

Lattices in a power set are relevant for *classical logics*.

Somewhat unexpected, the following property formalizes the decisive difference of classical and quantum logic: A lattice is *distributive* for

$$\begin{aligned} a \sqcup (b \sqcap c) &= (a \sqcup b) \sqcap (a \sqcup c), \\ a \sqcap (b \sqcup c) &= (a \sqcap b) \sqcup (a \sqcap c). \end{aligned}$$

The two dual conditions are equivalent in a logic. A distributive logic is called a *Boole–Stone logic*; one has for the disjunction:

$$L \in \underline{\textbf{logic}}(\text{Boole}-\text{Stone}) \Rightarrow a \dot{+} b = (a \sqcup b) \sqcap (a \sqcap b)^c.$$

Weaker than distributivity is *modularity* for a lattice, a partial associativity for meet and join:

$$a \sqsubseteq c \Rightarrow a \sqcup (b \sqcap c) = (a \sqcup b) \sqcap c.$$

A distributive lattice has at most one complementarity, i.e., the negation $a \longmapsto a^c$ in a Boole–Stone logic is unique. With the distributivity, two complements a^c and a' have to coincide:

$$\begin{aligned} a' \quad &= a' \sqcap \blacksquare = a' \sqcap (a \sqcup a^c) = (a' \sqcap a) \sqcup (a' \sqcap a^c) \\ &= \square \sqcup (a' \sqcap a^c) = a' \sqcap a^c \\ &\iff a' \sqsubseteq a^c, \text{ similarly } a^c \sqsubseteq a' \Rightarrow a' = a^c. \end{aligned}$$

The power set $(\mathbf{2}^S, \cup, \cap)$ of a set S with union and intersection is a distributive lattice. With $(\emptyset, S) = (\square, \blacksquare)$ and the complement $X \leftrightarrow X^c = \complement X = \complement_S X$ a power set constitutes a Boole–Stone logic:

$$(\mathbf{2}^S, \cup, \cap, \emptyset, \complement) \in \underline{\textbf{logic}}(\text{Boole}-\text{Stone}).$$

A subset family $\mathcal{S} \subseteq \mathbf{2}^S$ having the structure of a lattice with origin is called a *measure ring*, its elements *measurable sets*, the pair $(S, \mathcal{S})$ a *measurable space*. For the measure ring property of a subset family it is enough that it includes the empty set and is stable with respect to differences and unions;

$$\emptyset \in \mathcal{S},\ \ X, Y \in \mathcal{S} \Rightarrow \begin{cases} \complement_X Y = X \setminus Y \in \mathcal{S}, \\ X \cup Y \in \mathcal{S}, \end{cases}$$

which entails the measurability of the intersections $X \cap Y = X \setminus (X \setminus Y) \in \mathcal{S}$. The algebraic ring property uses the intersection as multiplication, $X \cdot Y = X \cap Y$, the disjoint union as addition, $X + Y = (X \cup Y) \setminus (X \cap Y)$, and the empty set as neutral element $0 = \emptyset$.

The coarsest (smallest with respect to the inclusion) measure ring is $\{\emptyset\}$, the coarsest full one is $\{\emptyset, S\}$, the finest one is $\mathbf{2}^S$. A σ-measure ring contains also the union of countably many measurable sets.

If a mapping f between two measurable spaces $(S, \mathcal{S})$ and $(T, \mathcal{T})$ gives measurable inverse images of measurable sets, i.e., $f^{-1}[\mathcal{T}] \subseteq \mathcal{S}$, then it is called a *measure morphism or measurable*. In analogy to topological spaces, the condition of measurability of mappings defines *initial (coarsest) and final (finest) measure rings*, e.g., the measure ring $\mathcal{X}$ of a subset by measurability of the injection $X \hookrightarrow S$ or the finite *product of measure rings* $\mathcal{S}_1 \otimes \mathcal{S}_2$ by the measurability of the projections $S_1 \times S_2 \longrightarrow S_{1,2}$. In general, the measure ring $\mathcal{S}_1 \otimes \mathcal{S}_2$ contains not only the "rectangular products" $X_1 \times X_2$.

A *full measure ring* $\mathcal{S}$ of a set S is a complementary measure ring:

$$\mathcal{S} \subseteq \mathbf{2}^S \text{ with } \begin{cases} X \in \mathcal{S} \Rightarrow \complement X \in \mathcal{S}, \\ \emptyset, S \in \mathcal{S}, \\ X, Y \in \mathcal{S} \Rightarrow X \cap Y,\ X \cup Y \in \mathcal{S}. \end{cases}$$

$(\mathcal{S}, \cup, \cap, \complement, \emptyset)$ is a Boole–Stone logic.

Each subset family $\mathcal{X}$ in $S \in$ **<u>set</u>** has a unique embracing coarsest (full) measure ring on S.

According to *Stone's theorem* each Boole–Stone logic is isomorphic to a power set lattice $\mathbf{2}^S$ of a set S. The trivial logic is Stone-isomorphic to the power set of the empty set $\mathbf{2}^\emptyset = \{\emptyset\} \cong \{\square\}$. The atomic logic is Stone-isomorphic to the power set of a set I with precisely one element $\circ$:

$$I = \{\circ\} : \ \mathbf{2}^I = \{\emptyset, I\} \cong \{\square, \blacksquare\}$$

$\{\emptyset\}$ is such a 1-elementic set.

A *topology* $\underline{S}$ of a set S is a subset of its power set with the full and the empty set, stable for pairwise intersections and arbitrary unions:

$$\underline{S} \subseteq \mathbf{2}^S \text{ with } \begin{cases} \emptyset, S \in \underline{S}, \\ X_i \in \underline{S},\ i \in I \Rightarrow X_k \cap X_j,\ \bigcup_{i \in I} X_i \in \underline{S}. \end{cases}$$

A topology is a distributive lattice, not necessarily complementary. The sets which are both open and closed, i.e., $X, \complement X \in \underline{S}$, constitute a sublattice that is a Boole–Stone logic.

For a topological space S with the open sets $\underline{S}$ the *Borel*[13] *ring* with the *Borel sets* is the coarsest $\underline{S}$ embracing σ-measure ring $\mathcal{S} \supseteq \underline{S}$. Hence also the point sets $\{x\}$ for $x \in S$ are Borel sets. If not stated otherwise, a topological space as measurable space carries always its full Borel measure ring (*Borel space*).

A continuous mapping $f : S \longrightarrow T$, $f^{-1}\left[\underline{T}\right] \subseteq \underline{S}$, is Borel measurable (*Borel mappings*).

8.12 Quantum (Birkhoff–von Neumann) Logics

Quantum theory is formulated with complex linear operators. Quantum propositions are Hilbert spaces, i.e., sets with a linear structure, determined by operator representations and their invariants, not individual vectors. Examples are irreducible $(1 + 2j)$-dimensional spaces for $\mathbf{SU}(2)$-spin j, or, for elementary particles, infinite-dimensional spaces acted on by irreducible unitary Poincaré group representations. There is an appropriate elucidation of *quantum relativity*, describing the basis-dependence of measurements: "God made the invariants, each experimenter measures his own eigenvalues." Quantum measurement has an non-objectifiable part.

If a linear operator f, acting on, for simplicity, a complex finite-dimensional vector space $V \cong \mathbb{C}^n$ is diagonalizable, it can be written $f = \alpha_j e^j \otimes \check{e}_j$ (sum over j) with its eigenvectors $f(e^j) = \alpha_j e^j$ (no sum over j) as a V-basis and the associate dual basis, $\langle \check{e}_j, e^k \rangle = \delta_j^k$. $f = \alpha_j \mathcal{P}_j$ is a sum (over j) with elementary projectors $\mathcal{P}_j = e^j \otimes \check{e}_j$ (no sum over j) onto one-dimensional subspaces $\mathbb{C}e^j$.

Good to know: For a Hilbert space, f is unitarily diagonalizable, $u \circ f \circ u^\star = \text{diag}\, f$, if, and only if, it is normal $f \circ f^\star = f^\star \circ f$, i.e., it commutes with its adjoint partner $[f, f^\star] = 0$.
Two diagonalizable operators f, g are simultaneously diagonalizable, i.e., there exists a common eigenvector basis, if, and only if, they commute with each other $[f, g] = 0$.

Lattices with Hilbert spaces as propositions are relevant for quantum logics. Vector subspaces define *linear lattices* with, as in classical logics, the intersection for the meet (the logical "et"), and, different from classical logics, the span for the join (the logical "aut") and the trivial space as origin (the logical "falsum"):

$$(\sqcap, \sqcup, \square) \sim (\cap, +, \{0\}) : \begin{cases} 2^V = \{W \subseteq V \mid \text{subspace}\} \in \underline{\mathbf{latt}}, \\ 2^{(V)} = \{W \subseteq V \mid \text{finite dimension}\} \in \underline{\mathbf{latt}}. \end{cases}$$

---

[13] Émile Borel (1871–1956).

A lattice with finite-dimensional subspaces is modular.

For dimension $n \geq 2$, where the vector space endomorphisms $\mathbf{AL}(n, \mathbb{C})$ are nonabelian, these vector space lattices are *not distributive*, e.g., a complex two-dimensional vector space with position-momentum basis $\{e^1, e^2\} = \{\mathbf{x}, \mathbf{p}\}$:

$$\begin{aligned} W_j = \mathbb{C}e^j \cong \mathbb{C},\ \ j = 1, 2,\ \ &\text{full space: } V = W_1 + W_2 \cong \mathbb{C}^2,\\ &\text{diagonal space: } \Delta = \mathbb{C}(e^1 + e^2) \cong \mathbb{C},\\ (W_1 + W_2) \cap \Delta = \Delta \neq (W_1 \cap \Delta) + (W_2 \cap \Delta) &= \{0\} + \{0\} = \{0\} = W_1 \cap W_2. \end{aligned}$$

Complementarity (the logical "negation") is defined by orthogonality: The dually orthogonal partner of a subspace $W \subseteq V$ is defined by the orthogonal linear forms $W_\perp^T \subseteq V^T$:

$$W_\perp^T = \{\omega \in V^T \mid \langle \omega, W\rangle = \{0\}\}.$$

With a nondegenerate square (inner product, symmetric bilinear, or sesquilinear form)

$$\begin{aligned} \langle\,|\,\rangle : V \times V \longrightarrow \mathbb{K},\ \zeta(v, w) = \langle v|w\rangle = \overline{\langle v|w\rangle},\\ \langle v|w + u\rangle = \langle v|w\rangle + \langle v|u\rangle,\ \ \langle v|\alpha w\rangle = \alpha\langle v|w\rangle, \end{aligned}$$

there is an *orthogonal subspace in* V. In the case of a finite-dimensional space V orthogonality defines an involution:

$$\begin{aligned} \perp : 2^V \longrightarrow 2^V,\ W \longmapsto W_\perp = \{v \in V \mid \langle W|v\rangle = \{0\},\\ V_\perp = \{0\},\\ W \subseteq W_{\perp\perp},\ W_\perp = W_{\perp\perp\perp},\ \text{for } V \cong \mathbb{K}^n:\ \ W = W_{\perp\perp}. \end{aligned}$$

Orthogonality is not complementary for an indefinite nondegenerate square. For example, in a two-dimensional Minkowski space with Lorentz metric $\begin{pmatrix} 1 & 0 \\ 0 & -1 \end{pmatrix}$ for a basis $\{e^0, e^3\}$, time and position translations $\mathbb{T}$ and $\mathbb{E}$ respectively are orthogonal to each other, whereas the isotropic lightlike subspaces $\mathbb{L}_\pm$ are self-orthogonal:

$$\begin{aligned} \mathbb{T}_\perp &= (\mathbb{R}e^0)_\perp = \mathbb{R}e^3 = \mathbb{E},\\ \mathbb{L}_\pm &= \mathbb{R}(e^0 \pm e^3) = (\mathbb{L}_\pm)_\perp. \end{aligned}$$

A *complementary* linear lattice requires a definite square, i.e., a scalar product:

$$\langle v|v\rangle = 0 \iff v = 0 \Rightarrow \mathbb{K}^n \cong V = W + W_\perp = W \oplus W_\perp = W \perp W_\perp.$$

The vector subspaces of a finite-dimensional Hilbert space constitute a complementary lattice, called a *Birkhoff–von Neumann logic* (linear logic):

$$V \cong \mathbb{K}^d,\ \zeta \succeq 0 \Rightarrow (2^V, \cap, +, \{0\}, \perp) \in \underline{\mathbf{logic}}(\text{Birkhoff}-\text{von Neumann}),$$

as well as the closed subspaces of any Hilbert space, i.e., the Hilbert subspaces, possibly also infinite-dimensional:

$$\text{Hilbert space } V \Rightarrow 2^{\overline{V}} = \{W \subseteq V \mid \text{closed}\} \in \underline{\mathbf{logic}}(\text{Birkhoff}-\text{von Neumann}).$$

Chapter 9
Classical and Quantum Probability

What is truth? In philosophy, there are, over the millenia, subtle definitions, discussions and differentiations of the truth concept, starting with Aristotle's "adequatio intellectus et rei," then "dialectical," "factual," "pragmatic," "empirical," and "logical" truth and falsehood, even sociological definitions, and so on, which, obviously, are not taken up here.

Truth in science is quantifiable by numbers — integer, real and complex numbers. Truth in physics cannot be talked about without experiments. A "classical physicist" makes an experimental prediction for a physical quantity, a real number with a unit, e.g., six Volt, and, then, reads off the result of one corresponding event on the scale of the experimental apparatus — the proposition is true or false, allowing some error for accuracy. In quantum theory it becomes more complicated; the prediction is a probability proposition, calculated from complex amplitudes, and approachable by the measurement of "many" corresponding events. Since founded on a conceptually changed theory, quantum probabilities cannot be sharpened by higher accuracy. *Quantum uncertainty* is not measurement inaccuracy or unsufficient knowledge. Its origin is also not the perturbing influence of the measurement apparatus. In contrast to a classical uncertainty for one property, e.g., for the position of a mass point, quantum uncertainty involves and relates to each other two properties. The best known example: Pressing the position accuracy squeezes out the momentum values and vice versa. There remains a "basic" uncertainty for complementary properties — not experimentally caused. Quantum theory leads to an epistemological extension of the simplest true-false structures. In the language of individual events as registered by the experimental apparatus, quantum properties like momentum or spin direction of an individual particle make sense only as collective properties, like the concept of the thermostatistical temperature which makes no sense for an individual molecule.

A probability is a numerical valuation of a logic by a positive normalized measure, i.e., an association of the logical structure of numbers to a Boole–Stone logic with sets or to a Birkhoff–von Neumann logic with Hilbert spaces. Measures can be written in

H. Saller, *Operational Symmetries*, DOI 10.1007/978-3-319-58664-9_9

local forms, e.g., with distributions (generalized functions) or probability densities of event spaces, e.g., of position in quantum mechanics.

Probability as such is not the distinguishing characteristics of quantum structures. Probability is used also in classical thermostatistics where it arises from measures on phase space subsets, i.e., from Boole-Stone lattices with a classical logic. Probability in quantum theories, introduced by Born, arises from the quantum characteristic concept of complex additive probability *amplitudes*,

$$\text{valuation:} \quad \underset{\text{classical mechanics}}{\{0, 1\}} \quad \subset \quad \underset{\text{thermostatistics}}{[0, 1]} \quad \subset \underset{\text{quantum theory}}{\mathbf{U}(1) \times [0, 1]}$$

given by scalar products of Hilbert space vectors and scalar product densities (distributions), e.g., with position-dependent Schrödinger functions. Probabilities and probability amplitudes have no dimensional units, information is measured in pure numbers, positive and complex, respectively, possibly distributed, e.g., as probability densities on position. The valuation chain above can be looked at to be a quantitative numerical extension of the original truth concept.

Uncertainty relations, the basis of Bohr's *complementarity* concept, are quantum characteristic. They are scalar product related binary relations involving two operators and are unknown to classical probabilities. Product Hilbert spaces, e.g., for two and more particles, carry product Hilbert products. In these cases, the probability interpretation gives rise to the also quantum characteristic phenomenon of entanglement.

It is remarkable, and, in a sense, unexpected, that the set-oriented structures of probability, as formalized by Kolmogorow,[1] can be founded on complex linear structures. Here, there exist metrics based on scalar products. The complex Hilbert subspaces constitute nondistributive linear lattices for a Birkhoff–von Neumann logic. Quantum theories can be mathematically characterized by complex Hilbert space representations of nonabelian operations, historically starting with the Heisenberg Lie algebra $[\mathbf{x}, \mathbf{p}] = i\mathbf{1}$ for noncommuting position-momentum operations. Like the position-momentum pairs, all noncommuting operator pairs give rise to uncertainty relations.

With the distinguished Haar[2] measure of a locally compact group, positive and unique up to normalization, one is lead to probabilities of groups as event spaces, i.e., to probabilities for the logic of group operations. The complex algebra with the absolute integrable functions of a locally compact group is ordered with the positive cone of scalar product inducing functions which determine the Hilbert spaces for the cyclic representations of the group, and, therefore, the related quantum logics with probability amplitudes.

The positive type functions of time, positions and spacetimes are of paramount importance as ground-states for the implementation of the corresponding external and internal operations.

---

[1]Andrei Kolmogorow (1903–1987).

[2]Alfred Haar (1885–1933).

9.1 Truth Tables and Truth Valuations

The joining of two propositions A and B, either verum $v = \bigvee$ or falsum $f = \bigwedge$ gives four possibilities for v or f, altogether $2^4 = 16$ arrangements, collected in *truth tables*.

Eight truth columns arise from the eight binary junctors involving the four junctors $\wedge$ (et), $\vee$ (aut), $\setminus$ (sine), T (transposed), and their negations c (non) with the $1+3$ and $3+1$ distributions for true and false:

A	B	$A \wedge B$	$A \wedge^c B$	$A \vee B$	$A \vee^c B$	$A \setminus B$	$A \setminus^c B$	$A \setminus^T B$	$A \setminus^{Tc} B$
v	v	v	f	v	f	f	v	f	v
v	f	f	v	v	f	v	f	f	v
f	v	f	v	v	f	f	v	v	f
f	f	f	v	f	v	f	v	f	v

The negations and the quaternary junctors $\dot{+}$ (vel) and $\dot{+}^c$ give four truth columns for the $2+2$ true-false distributions. The remaining two arrangements are verum and falsum themselves, always true or false:

A	B	A^c	B^c	$A \dot{+} B$	$A \dot{+}^c B$
v	v	f	f	f	v
v	f	f	v	v	f
f	v	v	f	v	f
f	f	v	v	f	v

$\bigvee$	$\bigwedge$
v	f
v	f
v	f
v	f

With two propositions A, B, their two negations, the ten junctors above, and the two extremal propositions $\bigwedge$, $\bigvee$ the 16 truth arrangements are exhausted — in this sense the ten junctors are complete.

False and true are formalizable by a Chrysippean mapping of propositions into the numbers $\{0, 1\}$, called truth values:

$$A \longmapsto \mu(A) \in \{0, 1\},$$

with the properties for implication, negation, and conjunction:

$$\begin{aligned}
&\text{(I)}\ A \Rightarrow B \text{ then } \mu(A) \leq \mu(B),\\
&\text{(N)}\ \mu(A) + \mu(A^c) = 1,\\
&\text{(C)}\ \mu(A \wedge B) = \mu(A)\mu(B).
\end{aligned}$$

The truth values of other junctors follow from their definitions, e.g., for the adjunction:

$$\begin{aligned}
A \vee B = (A^c \wedge B^c)^c \Rightarrow \mu(A \vee B) &= 1 - \mu(A^c \wedge B^c)\\
&= 1 - \mu(A^c)\mu(B^c)\\
&= 1 - (1 - \mu(A))(1 - \mu(B))\\
&= \mu(A) + \mu(B) - \mu(A)\mu(B),
\end{aligned}$$

which leads to the following sum equality for two propositions and their "aut" and "et"-junctions, called *disjoint additivity*:

$$\mu(A \vee B) + \mu(A \wedge B) = \mu(A) + \mu(B).$$

For an atomic logic with two propositions $\{\emptyset, I\}$, there is only one Chrysippean truth mapping:

$$(\mu(\emptyset), \mu(I)) = (0, 1).$$

A Boole–Stone logic with more than two propositions has more than one Chrysippean mapping:

$$\emptyset \subset X \subset S \Rightarrow (\mu(X), \mu(X^c)) = \begin{cases} \text{either } (0, 1), \\ \text{or } (1, 0). \end{cases}$$

For example, a binary set II with two elements, isomorphic to the power set of an atomic set I has two Chrysippean truth valuations ($i = 1, 2$):

$$\begin{aligned} II &= \{\circ, \bullet\} \cong \mathbf{2}^I = \{\emptyset, I\},\ \mathbf{2}^{II} = \{\emptyset, \{\circ\}, \{\bullet\}, II\} \cong \mathbf{2}^{(\mathbf{2}^I)}, \\ \mu_i(\emptyset) &= 0,\ \mu_i(II) = 1,\ \text{and} \begin{cases} \text{either } \mu_1(\{\bullet\}) &= 1,\ \mu_1(\{\circ\}) = 0, \\ \quad\text{or } \mu_2(\{\bullet\}) &= 0,\ \mu_2(\{\circ\}) = 1. \end{cases} \end{aligned}$$

9.2 Measures of Lattices and Logics

Measures algebraize the commutative lattice (logic) operations join (or) and meet (and) by addition and multiplication, the origin (false) and end (truth) elements by the corresponding neutral elements.

A commutative ring *R-valued measure* of an initialized lattice $(L, \sqcup, \sqcap, \Box)$ is defined by disjoint-additivity and origin-triviality:

$$\mu : L \longrightarrow R, \quad \begin{cases} \mu(a \sqcup b) = \mu(a) + \mu(b) \text{ for } a \sqcap b = \Box, \\ \mu(\Box) = 0. \end{cases}$$

A *σ-additive* measure has this property also for countably many disjoint elements.

Lattice measures inherit properties from the valuation ring R, e.g., reflection (conjugation), order, etc. R can be an algebra, e.g., an abelian $C*$-algebra with complex Hilbert space operators. For an algebra R, measures can constitute vector spaces, i.e., $\alpha_1\mu_1 + \alpha_2\mu_2$ with $\alpha_{1,2} \in \mathbb{K}$ is a lattice measure.

Two elements are *μ-independent* for

$$\mu(a \sqcap b) = \mu(a)\mu(b).$$

A *unital measure* of a complementary lattice $(L, \sqcup, \sqcap, \square, c)$, i.e., of a logic, is valued in a unital ring $1_R \in R$ and normalized:

$$\mu(\blacksquare) = \mu(a) + \mu(a^c) = 1_R.$$

A unital σ-additive measure $\mathcal{P}$ of a complementary lattice (logic) with $\mathcal{P}$-independent elements is *R-projector valued*:

$$\mathcal{P} : L \longrightarrow R, \quad \begin{cases} \mathcal{P}(\square) = 0, \ \mathcal{P}(\blacksquare) = 1_R, \\ \mathcal{P}(a \sqcup b) = \mathcal{P}(a) + \mathcal{P}(b) \text{ for } a \sqcap b = \square, \\ \mathcal{P}(a \sqcap b) = \mathcal{P}(a)\mathcal{P}(b) \Rightarrow \mathcal{P}(a) = \mathcal{P}(a)\mathcal{P}(a). \end{cases}$$

Complex or real measures are valued in a subset $M \subseteq \hat{\mathbb{C}}$ of the complex numbers $\hat{\mathbb{C}} = \mathbb{C} \cup \{\infty\}$, additively extended by infinity, with the rules $\alpha + \infty = \infty$ for $\alpha \in \hat{\mathbb{C}}$ and $\overline{\infty} = \infty$: With the $\mathbb{C}$-injection topology $\hat{\mathbb{C}}$ is a topological space with Borel ring. $\hat{\mathbb{C}}$ has many relevant subsets, especially $\mathbb{C}$ (finite), $\hat{\mathbb{R}} = \mathbb{R} \cup \{\infty\}$ (real), $\mathbb{R}$ (real and finite), $\mathbb{R}_+$ (positive and finite), giving the name for measures valued therein.

An order compatible complex measure is positive:

$$\begin{aligned} a \sqsubseteq b &\Rightarrow \mu(a) \leq \mu(b), \\ \square \sqsubseteq a &\Rightarrow 0 \leq \mu(a) \in \hat{\mathbb{R}}_+. \end{aligned}$$

The positive lattice measures $\mu \succeq 0$ constitute a cone, i.e., $\alpha_1\mu_1 + \alpha_2\mu_2 \succeq 0$ for $\alpha_{1,2} \geq 0$.

An order compatible real mapping of a real subset with lower bound $\iota \in I \subseteq \mathbb{R}$ is a positive lattice measure:

$$\mu : I \longrightarrow \mathbb{R}, \quad \iota \leq \alpha \leq \beta \Rightarrow 0 = \mu(\iota) \leq \mu(\alpha) \leq \mu(\beta).$$

A nontrivial order (implication) compatible finite measure of a logic can be normalized:

$$(L, \sqcup, \sqcap, \square, c) : \quad \begin{cases} 0 < \mu(\blacksquare) \in \mathbb{R}_+, & \mu(a^c) = \mu(\blacksquare) - \mu(a), \\ \underline{\mu} = \frac{\mu}{\mu(\blacksquare)}, \ \underline{\mu}(\blacksquare) = 1, & \underline{\mu}(a^c) = 1 - \underline{\mu}(a) \in [0, 1]. \end{cases}$$

An order compatible, therefore positive, and normalized measure is called a *probability (measure)* of the logic, as first explicitly axiomatized by Kolmogorow. The endowment of a logic with a probability can be compared with the endowment of a vector space with a norm, $V \ni v \longmapsto \| v \| \in \mathbb{R}_+$ with $\| v + w \| \leq \| v \| + \| w \|$ and $\| \alpha v \| = |\alpha| \, \| v \|$.

9.3 Probabilities of Boole-Stone Logics

Probabilities for classical theories start from subset-lattices of a basic set S which, for physics, may be called an *event space* (also sample space), e.g., a position or a spacetime or a phase space.

A disjoint-additive mapping from the measurable space $(S, \mathcal{S})$ into numbers $M \subseteq \hat{\mathbb{C}}$ is called a complex M*-valued measure* for the measurable sets (events) $X \subseteq S$:

$$\mu : \mathcal{S} \longrightarrow M, \quad \begin{cases} \mu(\emptyset) = 0, \\ \mu(X \cup Y) = \mu(X) + \mu(Y) \text{ for } X \cap Y = \emptyset, \\ \quad \Longleftrightarrow \mu(X \cup Y) + \mu(X \cap Y) = \mu(X) + \mu(Y) \text{ in general.} \end{cases}$$

A property for a space with a measure is valid *μ-almost everywhere* in the following situation: There exists a set with trivial μ-value, $\mu(N) = 0$, where the property in question is valid on the complementary set $\complement_S N$ (it may be valid even in N). For example, one works with μ-almost everywhere defined measurable mappings $f : S \longrightarrow T$.

For the finest measure ring $\mathbf{2}^S$ a *discrete measure* is given by a complex number for each element:

$$\mu : \mathbf{2}^S \longrightarrow \hat{\mathbb{C}}, \ \{x\} \longmapsto \mu(x).$$

A probability of a Boole-Stone logic in the power set $\mathbf{2}^S$ is implication (order) compatible, therefore positive, and normalized:

$$\mu : \mathcal{S} \longrightarrow [0, 1], \quad \begin{cases} \mu(\emptyset) = 0, \ \mu(S) = 1, \\ X \subseteq Y \Rightarrow \mu(X) \le \mu(Y), \\ \mu(X \cup Y) + \mu(X \cap Y) = \mu(X) + \mu(Y). \end{cases}$$

The *U-conditioned probability* for a subset (event) $U \subseteq S$ arises by renormalization, $\mu_U(A) = \frac{\mu(A)}{\mu(U)}$ for $\mathcal{S} \ni A \subseteq U$.

Neglecting the intersection probability, a positive measure of subsets $\mathcal{S}$ obeys *Bell's*[3] *inequality*:

$$\mu \succeq 0 \Rightarrow \mu(X \cup Y) \le \mu(X) + \mu(Y).$$

In a Birkhoff–von Neumann logic for quantum theory Bell's inequality with $\mu(X+Y)$ for subspaces can be violated.

Probability values of the coarsest logics are $\{0, 1\} = [0, 1] \cap \mathbb{N}$. The *Dirac probability* for any element $x \in S$ is projector-valued:

[3] John Bell (1928–1990).

$$\vartheta_x : \mathcal{S} \longrightarrow [0, 1],\ \vartheta_x(X) = \begin{cases} 1, & x \in X, \\ 0, & x \notin X. \end{cases}$$
$$\vartheta_x(X \cap Y) = \vartheta_x(X)\vartheta_x(Y).$$

Discrete probabilities (relative frequencies) have rational values in $[0, 1] \cap \mathbb{Q}$.

The cardinality provides an order compatible measure of the discrete measure ring $\mathbf{2}^S$, called *counting measure*:

$$\text{card} : \mathbf{2}^S \longrightarrow \hat{\mathbb{N}} = \mathbb{N} \cup \{\infty\},\ X \longmapsto \text{card}\, X.$$

A finite set S allows *counting probabilities*:

$$\underline{\text{card}} : \mathbf{2}^S \longrightarrow [0, 1] \cap \mathbb{Q},\ \underline{\text{card}} = \frac{\text{card}}{\text{card}\, S}.$$

An example: Throwing "dice" with a Platonic solid, e.g., with a hexaeder (a die proper), gives the results (events) $S = \{1, 2, \ldots, N_2\}$, e.g., $N_2 = 6$. The Platonic solid leads to the discrete measure ring $\mathbf{2}^S$ with the 2^{N_2} subsets of S, e.g., $2^6 = 64$ for the die. The cardinality $\text{card}\,\{n\} = 1, n \in S$, gives an equipartition for the results and a probability:

$$\underline{\text{card}}\,\{n\} = \frac{\text{card}\,\{n\}}{\text{card}\, S} = \frac{1}{N_2}.$$

Throwing two Platonic solids, e.g., two dice, is described with the product ring $\mathbf{2}^S \otimes \mathbf{2}^S$, etc.

Throwing darts on a circular target with the numbers (events) $\{1, 2, \ldots, N\}$, the circle $\{1\}$ as bull's eye with radius ℓ and the surrounding $(N-1)$ rings of equal width ℓ may be valued with the positive finite measures given by the areas:

$$\mu(\{n\}) = \pi(2n-1)\ell^2,\ n = 1, 2, \ldots, N,\ \mu(\{1, 2, \ldots, N\}) = \pi N^2 \ell^2.$$

They are normalizable as probabilities:

$$\underline{\mu}(\{n\}) = \frac{2n-1}{N^2},\ \ \underline{\mu}(\{1, 2, \ldots, N\}) = 1.$$

In quantum theory, Bose–Einstein and Fermi–Dirac statistics are relevant, connected with integer and half-integer spin, correspondingly. The relevant mulitplicity factors w^k, $k = 1, 2, \ldots$, make use of the dimensions of the tensor products of vector spaces $V \cong \mathbb{K}^n$, $\mathbb{K} = \mathbb{R}, \mathbb{C}$, and their totally (anti-)symmetric subspaces:

$$\begin{array}{lll} \text{Boltzmannn, tensor product:} & \dim_{\mathbb{K}} \bigotimes^k V = n^k, & w_B^k = \frac{n^k}{k!}, \\ \text{Bose–Einstein, totally symmetric:} & \dim_{\mathbb{K}} \bigvee^k V = \binom{n+k-1}{k} & = w_{BE}^k, \\ \text{Fermi–Dirac, totally antisymmetric:} & \dim_{\mathbb{K}} \bigwedge^k V = \binom{n}{k} & = w_{FD}^k. \end{array}$$

9.4 Probability Densities and Distributions

For an event space with positive σ-additive measure (S, μ), *integrals* can be defined of μ-almost everywhere defined measurable mappings on a measurable set $X \in \mathcal{S}$. The definition procedure is not given here. The measure can be localized $\mu \longmapsto d\mu$ with the integral notation:

$$\mu(X) = \int_X d\mu.$$

A measurable mapping $f : S \longrightarrow \mathbb{K}$ is called *μ-integrable* if its μ-integral is finite for all measurable sets:

$$f \in \mathcal{F}_\mu(S, \mathbb{K}) \iff \int_X d\mu\, f \in \mathbb{K} \text{ for all } X \in \mathcal{S}.$$

All μ-integrable mappings constitute a vector space, with conjugation for $\mathbb{K} = \mathbb{C}$. The integration over the set $X \in \mathcal{S}$ can be written as integration over the full measure space by using the *characteristic function* χ_X:

$$\int_X d\mu\, f = \int d\mu\, \chi_X\, f,$$

with the short notation $\int_S d\mu\, f = \int d\mu\, f$ for the full space.

The integral defines a linear form $\mathcal{F}_\mu(S, \mathbb{K})^T$ of the μ-integrable functions:

$$\mathcal{F}_\mu(S, \mathbb{K}) \ni f \longmapsto \langle \mu, f \rangle = \int d\mu\, f \in \mathbb{K}.$$

Such a linear form on S-functions is an example for a distribution, in general defined by a continuous linear form of a topological vector space with continuous test functions. Since distributions $f \longmapsto \mu(f) = \langle \mu, f \rangle$ act on functions $x \longmapsto f(x)$, they have, in general, no "point values" $x \longmapsto \mu(x)$, and no pointwise products $\mu_1(x) \cdot \mu_2(x)$ as familiar from the properties of the Dirac "function." A probability of the event space (S, μ) with probability density $d\mu$ is, in general, not expressible by a positive function, but by a positive distribution (generalized function) with the notation $d\mu = \mu(x)dx$.

Dirac probabilities can be written with Dirac distributions, e.g., for intervals on the reals

$$\begin{aligned} c \in \mathbb{R},\ \mathbb{R} \supset [a, b] \longmapsto \vartheta_c(a, b) &= \int_a^b dx\, \delta(x - c) = \vartheta(a - c)\vartheta(c - b) \\ &= \begin{cases} 1, & c \in [a, b], \\ 0, & c \notin [a, b], \end{cases} \\ \vartheta(a - c) &= \vartheta_c(-\infty, a) = \vartheta_a(c, \infty). \end{aligned}$$

Each μ-integrable mapping h defines a *measure* μ_h *on the basis* μ, i.e., μ explains a linear mapping from the corresponding μ-integrable mappings into the *measures on the basis* μ:

$$\mathcal{F}_\mu(S,\mathbb{K}) \ni h \longmapsto \mu_h \in \mathcal{F}_\mu(S,\mathbb{K})^T \text{ with } \langle \mu_h, f\rangle = \langle \mu, hf\rangle = \int d\mu\, hf.$$

An example is the Gauss or normal probability density $\frac{1}{\sigma\sqrt{2\pi}}e^{-\frac{1}{2}(\frac{x-\ell}{\sigma})^2}dx, x \in \mathbb{R}$, with the mean (location) $\ell \in \mathbb{R}$ and the variance $\sigma^2 > 0$. It plays a role as ground-state coefficient of the quantum harmonic oscillator in the position representation.

In classical statistical mechanics, the phase space $\mathbb{P}^{2N}$ as event space for n mass points in s-dimensional position, e.g., $s = 3$, is a real $2N$-dimensional manifold with $N = sn$. A positive phase space distribution (density) for a fixed energy $E > 0$, position-momentum-dependent via a Hamiltonian $H(q, p)$,

$$d^N q d^N p\, \mu(\tfrac{H(q,p)}{E}),\ state\, sum\colon \mathcal{P}_H(E) = \int d^N q d^N p\, \mu(\tfrac{H(q,p)}{E})$$

characterizes an *ensemble*. A distribution with sharp energy (energy width $\Delta \to 0$) is used for a *microcanonical ensemble*:

$$\begin{aligned}&d^N q d^N p\, \vartheta(E + \Delta - H)\vartheta(H - E),\\ &\text{state density: } \rho_H(E) = \int d^N q d^N p\, \delta(\tfrac{H(q,p)}{E} - 1).\end{aligned}$$

An example is given by the circles $H = \frac{\vec{q}^2+\vec{p}^2}{2}$ for the harmonic oscillator with state density $\rho_H(E) = (2\pi E)^N$.

A *canonical or Gibbs*[4] *ensemble* is defined by an exponential phase space density with Boltzmann's unit k and the temperature T:

$$\begin{aligned}&Boltzmann - Gibbs\, measure\colon d^N q d^N p\, e^{-\frac{H(q,p)}{kT}} \text{ with } E = kT,\\ &\text{e.g., } H = \tfrac{\vec{q}^2+\vec{p}^2}{2} \Rightarrow \mathcal{P}_H(E) = \int d^N q d^N p\, e^{-\frac{\vec{q}^2+\vec{p}^2}{2E}} = (2\pi E)^N.\end{aligned}$$

The "averaged" exponent in the state sum is the *entropy* $S_H(E)$, the logarithm of the state sum:

$$\mathcal{P}_H(E) = \int d^N q d^N p\, \mu(\tfrac{H(q,p)}{kT}) = \text{const. } e^{\frac{S_H(E)}{k}},\ S_H(E) = k\log \mathcal{P}_H(E) + \text{const.}$$

For a probability μ, the positive linear form

$$f \longmapsto \langle f\rangle_\mu = \langle \mu, f\rangle = \int d\mu f = \int \mu(x)dx f(x),$$

gives the *μ-expectation value* of a μ-integrable function f of the event space. For a discrete probability, the expectation value sums up the weighted function values for the events,

[4]Josiah Willard Gibbs (1839–1903).

$$\langle \mu, f \rangle = \sum_{n \in S} \mu\{n\} f(n).$$

Examples are the expectation value for throwing Platonic solids, e.g., a die with $N_2 = 6$, or for throwing darts on a target as given above:

$$f(n) = n : \begin{cases} \mu\{n\} = \frac{1}{N_2}, & \langle \mu, f \rangle = \sum_{n=1}^{N_2} \frac{n}{N_2} = \frac{N_2+1}{2}, \\ \mu\{n\} = \frac{2n-1}{N^2}, & \langle \mu, f \rangle = \sum_{n=1}^{N} \frac{n(2n-1)}{N^2} = \frac{(N+1)(4N-1)}{6N}. \end{cases}$$

Planck's black body radiation law gives the energy expectation value for integer spaced levels E_n (harmonic oscillator with n quanta - here it started all) with measures (Boltzmann factors) w^n and summation $\sum_{n=0}^{\infty}$:

$$\mathbb{N} \ni n \longmapsto w^n, E_n, \begin{cases} \text{with } w = e^{-\frac{h\nu}{kT}} = e^{-\omega\tau}, \; E_n = n\omega, \\ \langle E \rangle = \frac{\sum E_n w^n}{\sum w^n} = \frac{\sum n\omega e^{-n\omega\tau}}{\sum e^{-n\omega\tau}} \\ \quad = -\frac{\partial}{\partial\tau} \log \sum e^{-n\omega\tau} = -\frac{\partial}{\partial\tau} \log \frac{1}{1-e^{-\omega\tau}} \\ \quad = \frac{\omega}{e^{\omega\tau}-1} = \frac{h\nu}{e^{\frac{h\nu}{kT}}-1}. \end{cases}$$

The μ-centralized function $f_\mu = f - \langle f \rangle_\mu 1$ has trivial μ-expectation value. It can be used in the positive *μ-variance* $\text{var}_\mu f$ of the function, the square of the *standard μ-deviation* $\Delta_\mu f$:

$$\text{var}_\mu f = (\Delta_\mu f)^2 = \langle f^2 \rangle_\mu - \langle f \rangle_\mu^2 = \langle f_\mu^2 \rangle_\mu.$$

9.5 Lebesgue Function Spaces

An event space S with a positive σ-additive measure $\mu : \mathcal{S} \longrightarrow \mathbb{R}_+$ gives rise to prenorms on the μ-almost everywhere defined measurable $\mathbb{K}$-valued functions

$$f : (S, \mu) \longrightarrow \mathbb{K}, \; 1 \leq p < \infty : \| f \|_p = \left(\int d\mu \, |f|^p \right)^{\frac{1}{p}} \in \hat{\mathbb{R}}.$$

The additive *Minkowski inequality* for $f + g$ (if defined) follows from the multiplicative *Hölder*[5] *inequality* (Cauchy–Schwarz for different indices) for fg:

[5]Otto Hölder (1859–1937).

$$\begin{aligned} 1 \le p, r < \infty : & \parallel f + g \parallel_p \le \parallel f \parallel_p + \parallel g \parallel_p, \\ & \text{(subadditivity)}, \\ \tfrac{1}{p} + \tfrac{1}{r} = 1 : & \parallel fg \parallel_1 \le \parallel f \parallel_p \parallel g \parallel_r, \\ & \text{(submultiplicity)}. \end{aligned}$$

Hence one can define *vector spaces with functions, absolutely integrable to the power* p, i.e., with seminorm $\parallel f \parallel_p$,

$$1 \le p < \infty : \mathcal{L}^p_\mu(S, \mathbb{K}) = \{f : (S, \mu) \longrightarrow \mathbb{K} \,\big|\, \parallel f \parallel_p \in \mathbb{R}\},$$

and, for $p = \infty$, the seminorm space with the *essentially bounded functions*

$$\begin{aligned} \mathcal{L}^\infty_\mu(S, \mathbb{K}) &= \{f : (S, \mu) \longrightarrow \mathbb{K} \,\big|\, |f(x)| \in \mathbb{R}\mu\text{-almost everywhere}\}, \\ \parallel f \parallel_\infty &= \inf\{\alpha \,\big|\, |f(x)| \le \alpha, \mu\text{-almost everywhere}\}. \end{aligned}$$

Hölder's inequality entails the *duality pairing* with *conjugated powers* (p, r) on a hyperbola $(p - 1)(r - 1) = 1$:

$$\begin{aligned} 1 \le p, r \le \infty, \tfrac{1}{p} + \tfrac{1}{r} = 1 : \mathcal{L}^p_\mu(S, \mathbb{K}) \times \mathcal{L}^r_\mu(S, \mathbb{K}) &\longrightarrow \mathcal{L}^1_\mu(S, \mathbb{K}), \\ (f, g) &\longmapsto fg. \end{aligned}$$

For all these function spaces, the functions that vanish μ-almost everywhere constitute the subspace $\mathcal{L}^p_\mu(S, 0)$ with trivial prenorm. The normed quotients, called *Lebesgue*[6] *spaces*, are even Banach[7] spaces:

$$1 \le p \le \infty : \ L^p_\mu(S, \mathbb{K}) = \mathcal{L}^p_\mu(S, \mathbb{K}) / \mathcal{L}^p_\mu(S, 0).$$

The duality pairing leads to a bilinear product:

$$\tfrac{1}{p} + \tfrac{1}{r} = 1 : \ L^p_\mu(S, \mathbb{K}) \times L^r_\mu(S, \mathbb{K}) \longrightarrow \mathbb{K}, \ \langle h, f \rangle = \int d\mu\, hf,$$

whence the topological vector space isomorphism $L^p_\mu(S, \mathbb{K}) \cong L^r_\mu(S, \mathbb{K})'$ with the strong dual can be derived.

The absolutely square integrable functions $\mathcal{L}^2_\mu(S, \mathbb{K})$ constitute a pre-Hilbert space leading to the *Hilbert space of the square* μ*-integrable functions*:

$$L^2_\mu(S, \mathbb{K}) \times L^2_\mu(S, \mathbb{K}) \longrightarrow \mathbb{K}, \ \langle h | f \rangle_\mu = \int d\mu\, \overline{h} f.$$

It is separable (countable Hilbert space basis) for a separable Borel space S (countable basis for the topology).

[6] Henri Lebesgue (1875–1941).

[7] Stefan Banach (1892–1945).

$L^2_\mu(S,\mathbb{K})$ with countable Hilbert basis $\{|e^j\rangle\}_{j\in\mathbb{N}}$ has as orthogonality:

$$|e^j\rangle : S \longrightarrow \mathbb{K},\ \delta^{jk} = \langle e^j|e^k\rangle_\mu = \int d\mu\, \overline{e^j} e^k.$$

The completeness

$$L^2_\mu(S,\mathbb{K}) \ni |f\rangle = \sum_j f^j|e^j\rangle,\quad f^j = \langle e^j|f\rangle_\mu = \int d\mu\, \overline{e^j} f$$

can be expressed with the measure-associated Dirac distribution:

$$\begin{aligned} f(x) &= \sum_j \int d\mu(x')\, e^j(x)\overline{e^j(x')} f(x') = \int d\mu(x')\, \delta(\mu(x,x')) f(x'),\\ \mathbf{1}_{L^2_\mu(S,\mathbb{K})} &\cong \sum_j |e^j\rangle\langle e^j| \iff \delta(\mu(x,x')) = \sum_j e^j(x)\overline{e^j(x')}, \end{aligned}$$

e.g., with a Lebesgue-based measure involving a positive function μ:

$$\begin{aligned} L^2_\mu(\mathbb{R}^n,\mathbb{K}) \ni |f\rangle,\ f(x) &= \int d\mu(x')\, \delta(\mu(x,x')) f(x')\\ &= \int d^n x'\, \delta(x-x') f(x'),\\ \text{with} &\begin{cases} d\mu(x) &= \mu(x) d^n x,\\ \delta(\mu(x,x')) &= \frac{1}{\mu(x)}\delta(x-x'). \end{cases} \end{aligned}$$

9.6 Haar Measures of Operation Groups

Operation groups are event spaces with distinguished positive measures and probabilities: For a *locally compact group* G with Borel measure ring $\mathcal{G}$, there exists a positive σ-additive measure $dg \in \mathcal{M}(G)_+$, called *Haar measure*, invariant with respect to all left translations (multiplications):

$$\mu^G : \mathcal{G} \longmapsto \hat{\mathbb{R}}_+,\ \mu^G(X) = \int_X dg = \mu^G(kX) \text{ for all } k \in G.$$

It is unique up to a multiplicative constant $\mu \longmapsto e^\lambda \mu$, $\lambda \in \mathbb{R}$, and finite on all compact sets: The analogous statement holds with respect to right translations.

A Haar measure is an important tool to investigate group structures and related scalar products and probabilities (Schur product).

Time, positions, and spacetimes, defined by locally compact operation groups, come with characteristic Haar measures.

With a left Haar measure μ, also μ_g with right action $\mu_g(X) = \mu(Xg^{-1}) = \Delta(g)\mu(X)$ for all $g \in G$ is a left Haar measure. The left-right connecting scalar factors define the *modular function* as group realization:

$$\Delta : G \longrightarrow \mathbb{R}_+,\ \Delta(g_1 g_2) = \Delta(g_1)\Delta(g_2).$$

The modular function is a dilation representation $G \ni g \longmapsto \Delta(g) \in \mathbf{D}(1)$. A nonunimodular group has to have a normal subgroup with dilation classes $G/N \cong \mathbf{D}(1)$, $N \cong \text{kern}\, \Delta$. If a group is unimodular, i.e., $\Delta(g) = 1$ for all g, then left Haar measures are right Haar measures. Finite, discrete, abelian, compact, and semisimple groups are unimodular.

For a compact group G, the normalized Haar measure with $\mu^G(G) = \int dg = 1$ is a probability.

A Haar measure for a group G with discrete topology uses the counting measure, finite and normalizable for a finite group by $\underline{\mu}^G = \frac{\mu^G}{\text{card}\, G}$:

$$\begin{aligned} &\mu^G : \mathbf{2}^G \longrightarrow \hat{\mathbb{N}},\ \mu^G(X) = \text{card}\, X, \\ &f : G \longrightarrow \mathbb{K},\ \int_G dg\, f(g) = \sum_{g \in G} f(g). \end{aligned}$$

Haar integrable functions $h \in \mathcal{F}(G, \mathbb{K})$ define measures $dg\, h(g)$ with a Haar measure as basis.

A Haar measure on the additive Lie group $\mathbb{R}^n$ with $\int_a^b dx = b - a$ is called a *Lebesgue measure*:

$$\lambda^n : \mathcal{R}^n \longrightarrow \mathbb{R}_+,\ X \longmapsto \lambda^n(X) = \int_X d^n x.$$

A Lebesgue measure is $\mathbf{SL}(n, \mathbb{R})\, \vec{\times}\, \mathbb{R}^n$ invariant, $d^n x \longmapsto d^n x$ for $x \longmapsto s.x + a$. With an additional appropriate dilation factor for $\mathbf{D}(1) \cong \mathbf{GL}_0(n, \mathbb{R})/\mathbf{SL}(n, \mathbb{R})$, e.g., $e^\lambda = \sqrt{|\det g|}$ from a metric, or $e^\lambda = \langle \psi | \psi \rangle$ from a scalar product, it can be used in the form $e^{\lambda(x)} d^n x$ for any parametrization of a real n-dimensional Lie group, e.g. $d^3 u = \sin\theta\, d\chi d\varphi d\theta$ for $\mathbf{SU}(2)$.

For locally compact groups with a Haar measure one uses the measure free notation $L(G, \mathbb{K}) = L_\mu(G, \mathbb{K})$ for the Lebesgue spaces, and $L(G) = L(G, \mathbb{C})$ for complex functions $\mathbb{K} = \mathbb{C}$.

9.7 Operators and Matrices

The formulation of the basic mathematical quantum structures helps to distinguish quantum and classical physics and to come to grips with the related difficult epistemological problems.

Basic physics formalizes operations, given — in the framework considered here — by a real finite-dimensional Lie group G and its Lie algebra L. Characteristic examples for operator groups are the abelian compact group $\mathbf{U}(1)$, e.g., for electromagnetism, the nonabelian compact group $\mathbf{SU}(2)$, e.g., for rotations and isospin operations, the abelian noncompact groups $\mathbb{R}^n$, e.g., for time and position translations, the Euclidean group $\mathbf{SO}(3)\, \vec{\times}\, \mathbb{R}^3$ for nonrelativistic scattering and the Poincaré group $\mathbf{SL}(2, \mathbb{C}) \vec{\times} \mathbb{R}^4$ with the homogeneous Lorentz cover group for free particles and relativistic scattering, the two latter ones both nonabelian noncompact groups.

The historically first investigated quantum problems worked with the nonabelian noncompact Heisenberg Lie algebra for positions-momenta with $[\mathbf{x}, \mathbf{p}] = i\mathbf{1}_s$, acting faithfully on the Hilbert space $L^2(\mathbb{R}^s)$. Group or Lie algebra representations act on vector spaces, e.g., spin $\mathbf{SU}(2)$ on complex $1 + 2j$-dimensional Hilbert spaces. In this example, a spin $\frac{1}{2}$-particle "is," with respect to spin, a full two-dimensional space which can be spanned by not less than two linearly independent vectors. Therefore, a measurement enforced restriction to one state vector, is, in general, incomplete.

In quantum theory, the group operations are implemented by unitary operations $\mathbf{U}(V)$ in the endomorphism algebra $\mathbf{AL}(V)$ of a complex Hilbert space with conjugation and scalar product $\langle v|w\rangle = \langle v^\star, w\rangle$, given via the Schur product by Haar integration over the Lie group. The endomorphisms have, via the commutator, a natural Lie algebra structure with the antihermitian elements as real Lie subalgebra $\{f = -f^\star \in \mathbf{AL}(V)\}$ for the implementation of L. Faithful irreducible Hilbert representations of (non)compact groups are (in)finite-dimensional. Physical properties are given by quantum numbers, i.e., invariants and eigenvalues for the represented operations, which come from a discrete spectrum for a compact group, and from a continuous spectrum for a noncompact group.

For a complex finite-dimensional space $V \cong \mathbb{C}^n$, a Lie algebra representation $L \hookrightarrow \mathbf{AL}(V)$ can be exponentiated to a group, $\exp L \hookrightarrow \mathbf{GL}(V)$. The free algebra (enveloping algebra) for the Lie algebra with the operation invariants is also represented by endomorphisms, $\mathbf{E}(L) \hookrightarrow \mathbf{AL}(V)$.

With a group or Lie algebra representation on a vector space V, one has the dual representation (inverse and negative transposed, respectively) on the dual space (linear V-forms) V^T, then product representations on the tensor products $\bigotimes^k V$, and, collecting all multilinear structures, on the tensor algebra $\bigotimes \mathbf{V} = \bigoplus_{k=0}^{\infty} \bigotimes^k \mathbf{V}$ of the selfdual direct sum space $\mathbf{V} = V \oplus V^T$.

For a complex finite-dimensional space V, an operation $f \in \mathbf{AL}(V)$ is unitarily diagonalizable with an eigenvector basis $\{e^j\}_{j=1}^n$ precisely if it is normal (if it commutes with its adjoint), i.e., $[f, f^\star] = 0$. Examples are (anti)hermitian operations $f = \pm f^\star$ with the *spectral decomposition*:

$$f = f_j^k b^j \otimes \check{b}_k = \sum_{j=1}^{n} f(j) e^j \otimes e_j^\star = \sum_{f(j) \in \mathbf{spec}\, f} f(j) \mathcal{P}_{f(j)},$$

with eigenvalues $f(j) \in \mathbb{C}$ (all real or all imaginary) for $\operatorname{diag} f = u \circ f \circ u^\star$, $u \in \mathbf{SU}(n)$, and the eigenspaces related orthogonal projectors $\mathcal{P}_{f(j)}$ which, for degenerate values, can be sums of nondecomposable projectors $\mathcal{P}_j$:

$$j = 1, \dots, n: \ \mathcal{P}_j = e^j \otimes e_j^\star, \ \mathcal{P}_j \circ \mathcal{P}_k = \delta_{jk}\mathcal{P}_j, \ \sum_{j=1}^{n} \mathcal{P}_j = \mathbf{1}_V.$$

A diagonalizable operator with eigenvalues defines a mapping from its spectrum to an abelian endomorphisms subalgebra with eigenspace projectors:

$$\mathbf{spec}\, f \ni f(j) \longmapsto \mathcal{P}_{f(j)} \in \mathbf{AL}(V).$$

Two unitarily diagonalizable operations are simultaneously unitarily diagonalizable, if, and only if, they commute $[f_1, f_2] = 0$. Therefore, an endomorphism algebra $A \subseteq \mathbf{AL}(V)$ contains maximal simultaneously unitarily diagonalizable abelian subalgebras $H \subseteq A \subseteq \mathbf{AL}(V)$, i.e., with a common eigenvector basis, for the full algebra given by the diagonal matrices.

For a complex infinite-dimensional Hilbert space, the corresponding statements need some modifications in a more subtle discussion. The following superficial formulation gives a flavor of the rather difficult general *spectral theorem* which brings together infinite-dimensional operators, e.g., for noncompact group representations, and measure theory: For H an abelian unital $C*$-subalgebra $\mathbf{1}_V \in H \subseteq \mathbf{B}(V) \subseteq \mathbf{AL}(V)$ of the bounded linear operators of a Hilbert space, there exists a regular H-projection valued measure $\mathcal{P}_H$ of the spectrum $\mathbf{spec}\, H \subseteq \mathbb{C}$ (with appropriate Borel ring)

$$\mathbf{spec}\, H \supseteq X \longmapsto \mathcal{P}_H(X) \in H, \quad \begin{cases} \mathcal{P}_H(\emptyset) = 0, \ \mathcal{P}_H(\mathbf{spec}\, H) = \mathbf{1}_V, \\ \text{disjoint } \sigma\text{-additivity, starting with} \\ \mathcal{P}_H(X \cup Y) = \mathcal{P}_H(X) + \mathcal{P}_H(Y) \\ \qquad\qquad \text{for } X \cap Y = \emptyset, \\ \mathcal{P}_H(X \cap Y) = \mathcal{P}_H(X)\mathcal{P}_H(Y). \end{cases}$$

The *spectral measure* can be used to define a H-valued integral:

$$\mathcal{P}_H(X) = \int_{q \in X} d\mathcal{P}_H(q),$$

and, the spectral decomposition $f = \int d\mathcal{P}_H(q)\, \hat{f}(q)$ for operators $f \in H$ with eigenvalue related functions $\hat{f} : \mathbf{spec}\, H \longrightarrow \mathbb{C}$.

A suggestive formalization for infinite-dimensional operators uses complex measures of the spectrum and distributive bases (more ahead), e.g., for the particle representations of the Poincaré group with creation and annihilation operators on a momentum hyperboloid with the translations spectral values:

$$\mathcal{P}_e = e \otimes e^\star \to d\mathcal{P}_{\mathcal{Y}^3}(\vec{q}) = \frac{d^3q}{\sqrt{\vec{q}^2 + m^2}} \mathrm{u}(\vec{q}) \otimes \mathrm{u}^\star(\vec{q}), \ \vec{q} \in \mathbb{R}^3 \cong \mathcal{Y}^3.$$

Quantum measurements are operator measurements: The measurement of an operator involves its expectation values (matrix elements, coefficients) $\langle v|f|w\rangle \in \mathbb{C}$

for normalized vectors $\langle v|v\rangle = \langle w|w\rangle = 1$. The projector $|w\rangle\langle w| = \mathcal{P}_w = \mathcal{P}_w^2$, associated to a normalized vector, has the distinguished positive expectation values $\langle v|\mathcal{P}_w|v\rangle = |\langle v|w\rangle|^2 \in [0, 1]$, called transition probabilities as absolute square of the transition probability amplitudes $\langle v|w\rangle \in \mathbf{U}(1) \times [0, 1]$. The Cauchy–Schwarz inequality leads to the uncertainty relations for the measurement of noncommuting operators.

In an experimental setup that reflects an eigenvector basis of a Cartan subalgebra $H \subset A$, e.g., an appropriately directed magnetic field for the spin operator $\mathbf{J}^3$, e.g., in a spin-$\frac{1}{2}$ representation,

$$\mathbf{J}^3 \longmapsto \mathcal{D}^{\frac{1}{2}}(\mathbf{J}^3) = \tfrac{i}{2}|\uparrow\rangle\langle\uparrow| - \tfrac{i}{2}|\downarrow\rangle\langle\downarrow|,$$

the related diagonalized operators can be treated as classical ones. The nondiagonalizable operators in A have no classical correspondence — in this sense quantum theory with diagonalized ("classized") operators H is not complete for nonabelian operators A.

Learning vector space structures, physicists are inclined to denigrate as mathematically pedantic the conceptual distinction between an abstract vector or operator — an element of a vector space — and its number oriented representation in a basis. With two dual bases an endomorphism $f = f_j^k b^j \otimes \check{b}_k = f_\alpha^\beta e^\alpha \otimes \check{e}_\beta$ has, in general, different matrices, $f_j^k \cong f_\alpha^\beta$. For example, with any direction on the 2-sphere,

$$\begin{aligned}
&\text{for } \Omega^2 \ni \vec{\omega} \cong (\theta, \varphi) : \ \vec{\sigma}\vec{\omega} = \begin{pmatrix} \cos\theta & e^{-i\varphi}\sin\theta \\ e^{i\varphi}\sin\theta & -\cos\theta \end{pmatrix},\\
&\text{e.g. } \vec{\omega} = \Big((0,0),\ (\tfrac{\pi}{2}, 0),\ (\tfrac{\pi}{2}, \tfrac{\pi}{2})\Big) \Rightarrow \vec{\sigma}\vec{\omega} = (\sigma^3, \sigma^1, \sigma^2)
\end{aligned}$$

the spin $\frac{1}{2}$-represented third component operator can be written in any orthonormal basis $\{|\vec{\omega}\rangle, |\overleftarrow{\omega}\rangle\}$ of $V \cong \mathbb{C}^2$:

$$\begin{aligned}
\mathcal{D}^{\frac{1}{2}}(\mathbf{J}^3) &= \tfrac{i}{2}\left[\cos\theta|\vec{\omega}\rangle\langle\vec{\omega}| + e^{-i\varphi}\sin\theta|\vec{\omega}\rangle\langle\overleftarrow{\omega}| + e^{i\varphi}\sin\theta|\overleftarrow{\omega}\rangle\langle\vec{\omega}| - \cos\theta|\overleftarrow{\omega}\rangle\langle\overleftarrow{\omega}|\right]\\
&= \tfrac{i}{2}|\uparrow\rangle\langle\uparrow| - \tfrac{i}{2}|\downarrow\rangle\langle\downarrow| = \tfrac{i}{2}|\leftarrow\rangle\langle\rightarrow| + \tfrac{i}{2}|\rightarrow\rangle\langle\leftarrow| = \tfrac{1}{2}|\leftarrow\rangle\langle\rightarrow| - \tfrac{1}{2}|\rightarrow\rangle\langle\leftarrow|.
\end{aligned}$$

The bases in the example are related to each other by an $\mathbf{SU}(2)$-rotation, the (2×2)-componentic matrix representations are isomorphic:

$$\begin{aligned}
\begin{pmatrix} |\vec{\omega}\rangle \\ |\overleftarrow{\omega}\rangle \end{pmatrix} &= u(\vec{\omega})\begin{pmatrix} |\uparrow\rangle \\ |\downarrow\rangle \end{pmatrix},\ \vec{\sigma}\vec{\omega} = u(\vec{\omega}) \circ \sigma^3 \circ u^\star(\vec{\omega}) \Rightarrow \sigma^3 \cong \sigma^1 \cong \sigma^2,\\
\text{with } u(\vec{\omega}) &= \begin{pmatrix} \cos\frac{\theta}{2} & -e^{-i\varphi}\sin\frac{\theta}{2} \\ e^{i\varphi}\sin\frac{\theta}{2} & \cos\frac{\theta}{2} \end{pmatrix} \in \mathbf{SU}(2),\ \text{e.g.,} \begin{pmatrix} |\rightarrow\rangle \\ |\leftarrow\rangle \end{pmatrix} = \tfrac{1}{\sqrt{2}}\begin{pmatrix} 1 & -1 \\ 1 & 1 \end{pmatrix}\begin{pmatrix} |\uparrow\rangle \\ |\downarrow\rangle \end{pmatrix}.
\end{aligned}$$

Quantum relativity relies on the conceptual difference between an operator, on the one side, and, on the other side, its many matrices as possible measurement values given by component representations in experimentally distinguished bases.

9.8 Probability Amplitudes and Uncertainties

For metrical structures there is the specialization from norm to scalar product with, as familiar from Euclidean geometry, the concept of angles. A scalar product, defined as a binary vector space relation, sketched by $\langle v|w\rangle$, leads to a norm as an unary relation, $\| v \| = \sqrt{\langle v|v\rangle}$; for a norm to be based an a scalar product, additional conditions have to be fulfilled, e.g., the parallelogram equation $\| v + w \|^2 + \| v - w \|^2 = 2 \| v \|^2 + 2 \| w \|^2$. In parallel is the specialization from probability (positive normalized measure) as the general structure to probability amplitudes as the more restricted one: A logic with a probability amplitude is implemented by a complex Hilbert space, i.e., a complex vector space with a scalar product. There are norms and probabilities, e.g., thermostatistical ones, which are not based on scalar products. The geometry of complex Hilbert spaces is interpretable with probability concepts (or vice versa) — vectors as probability amplitudes, scalar products with cosines as transition amplitudes and uncertainties, etc.

Probabilities and probability amplitudes for quantum theories arise on complex Hilbert spaces which have their origin in *operator algebras* A. In the following, A is assumed to be an associative algebra, real or complex $\mathbb{K} \in \{\mathbb{R}, \mathbb{C}\}$ with unit $1_A \in A$ and equipped with a (conjugate) linear reflection, $A \ni a \leftrightarrow a^\star$, e.g., in quantum mechanics, a complex operator algebra with positions $\mathbf{x} = \mathbf{x}^\star$, momenta $\mathbf{p} = \mathbf{p}^\star$, angular momenta $\mathbf{J} = i\mathbf{L} = -\mathbf{J}^\star$, creation and annihilation operators $\mathrm{u} \leftrightarrow \mathrm{u}^\star$, etc. Bose and Fermi quantum algebras as the most important examples will be discussed in the next section, and, in later sections, convolution algebras with complex group functions.

9.8.1 States of Operator Algebras

A linear form $d \in A^T$ of the algebra A is called *positive*, $d \succeq 0$, for positive values of the hermitian squares $a^\star a = (a^\star a)^\star$:

$$d : A \longrightarrow \mathbb{K},\ d(a^\star a) = \| a \|_d^2 \geq 0 \text{ for all } a \in A.$$

The condition generalizes the positivity condition $b = a^\star a$ for C*-algebra elements. However, A must not be a C*-algebra. If the algebra has topological properties, the topological dual $A' \subseteq A^T$ with the continuous linear A-forms has to be considered. Algebra forms can be reflected $d \leftrightarrow d^\star$ with $d^\star(a) = \overline{d(a^\star)}$. They are symmetric for $d = d^\star$. A positive unital form, i.e., with $d(1_A) = 1$, is called a *state of the algebra*.

Given a (symmetric and positive) form $d \in A^T$, each algebra element gives rise to a (symmetric and positive) form d_a:

$$A \ni a \longmapsto d_a \in A^T, \text{ with } d_a(b) = d(a^\star ba), \begin{cases} d_{1_A} = d, \\ d = d^\star \Rightarrow d_a = d_a^\star, \\ d \succeq 0 \Rightarrow d_a \succeq 0, \end{cases}$$

normalizable for d-positive elements $a \in A$, i.e., with $d(a^\star a) > 0$.

The dual vector space $A' \subseteq A^T$ with the linear algebra forms carries, in addition to the reflection, an order $d \succeq 0$ that defines the positive cone A^T_+ wherein the states are a convex "polytope." An algebra state is called *pure* (nondecomposable, irreducible) if there do not exist states for its nontrivial combination $d = \alpha_1 d_1 + \alpha_2 d_2$ with positive scalars. The pure states are the "corners" of the convex "polytope." They lead to irreducible representations of the algebra A (more ahead).

The Cauchy–Schwarz inequality for a positive form is relevant for Heisenberg's uncertainty relation: It gives a lower bound for the prenorm product of two operators $a, b \in A$:

$$\begin{aligned}
&\text{for all } \alpha \in \mathbb{R} \text{ and } e^{i\beta} \in \mathbf{U}(1) \cap \mathbb{K} = \begin{cases} \{\pm 1\}, \; \mathbb{K} = \mathbb{R}, \\ \mathbf{U}(1), \; \mathbb{K} = \mathbb{C}, \end{cases} \\
&d((a^\star + \alpha e^{-i\beta} b^\star)(a + \alpha e^{i\beta} b)) \geq 0 \Rightarrow \| a \|_d^2 \| b \|_d^2 \geq [d(\tfrac{e^{i\beta} a^\star b + e^{-i\beta} b^\star a}{2})]^2,
\end{aligned}$$

If the algebra elements are both symmetric or both antisymmetric, the product of their prenorms is bounded from below by the absolute d-value of a combination of their anticommutator and commutator:

$$(a, b) = \pm(a^\star, b^\star) \Rightarrow \| a \|_d \| b \|_d \geq |d(\tfrac{\{a,b\}}{2} \cos\beta + \tfrac{[ia,b]}{2} \sin\beta)|.$$

For special values of the phase, there remains only the commutator or only the anticommutator:

$$\begin{aligned}
\beta = 0 \; &\Rightarrow \| a \|_d \| b \|_d \geq |d(\tfrac{\{a,b\}}{2})|, \\
\mathbb{K} = \mathbb{C} : \; \beta = \tfrac{\pi}{2} \; &\Rightarrow \| a \|_d \| b \|_d \geq |d(\tfrac{[ia,b]}{2})|.
\end{aligned}$$

Here, the commutator, e.g., for position-momentum $[i\mathbf{p}, \mathbf{x}] = \hbar 1_A$, occurs only for complex algebras.

The standard deviation (square root of the variance) $\Delta a = \sqrt{\text{var}\, a}$ of a symmetric algebra element in a state d,

$$a = a^\star, \; a_d = a - d(a)1_A = a_d^\star \Rightarrow \begin{cases} d(a_d) &= 0, \\ (\Delta a)^2 &= d(a_d^2) = d(a^2) - d(a)^2 \geq 0, \end{cases}$$

is trivial for an a-eigenform d, defined by $d(a^2) = d(a)^2$. For any pair of (anti)symmetric elements in a complex algebra (with an eventual i-multiplication both can be assumed to be symmetric) the *uncertainty relation* — a binary relation arising from the Cauchy–Schwarz inequality above — bounds the product of their standard deviations from below by half the absolute state value of their commutator:

$$(a, b) = (a^\star, b^\star) \Rightarrow \Delta a\, \Delta b = \| a_d \| \| b_d \| \geq \tfrac{1}{2}|d([a_d, b_d])| = \tfrac{1}{2}|d([a, b])|.$$

The most famous example is Heisenberg's position-momentum uncertainty where the commutator involves the unit operator:

$$[\mathbf{x}, \mathbf{p}] = \hbar i \mathbf{1} : \ \Delta\mathbf{x}\, \Delta\mathbf{p} \geq \tfrac{1}{2}|d([\mathbf{x}, \mathbf{p}])| = \tfrac{\hbar}{2}.$$

An immediate analogy for a time-energy uncertainty (complementarity) requires two corresponding operators: The Hamiltonian can be taken for energy, however, there is no time operator in conventional quantum mechanics.

In general, every Hilbert-representation of a nonabelian real Lie-algebra by anti-hermitian operators gives rise — via the commutators – to nontrivial uncertainties:

$$\begin{array}{l} L \ni l \longmapsto \mathbf{L} \in \log \mathbf{U}(V), \text{ with } [\mathbf{L}^a, \mathbf{L}^b] = \epsilon_c^{ab} \mathbf{L}^c, \\ \text{e.g., angular momenta: } \Delta\mathbf{L}^1\, \Delta\mathbf{L}^2 \geq \tfrac{1}{2}|d(\mathbf{L}^3)|. \end{array}$$

The operators of an abelian Lie algebra $\{[l, m] = 0 \,\big|\, l, m \in L\}$ can be simultaneously measured without uncertainties (classical subalgebra). An invariant operator I has no uncertainty relations with the acting operators $\{a \,\big|\, [a, I] = 0\}$, $\Delta a\, \Delta I = 0$. For instance, the rotation invariant angular momentum operator $\vec{\mathbf{L}}^2$ can be measured without uncertainty $\Delta\mathbf{L}^a\, \Delta\vec{\mathbf{L}}^2 = 0$.

9.8.2 Cyclic Hilbert Spaces for Algebra States

A positive form gives the algebra the structure of a *pre-Hilbert space* with prescalar product, prenorm, and topology (not necessarily Hausdorff[8]):

$$\begin{aligned} \langle\ ||\ \rangle_d : A \times A \longrightarrow \mathbb{C},\ \langle a||b\rangle_d &= d(a^\star b), \\ \| a \|_d^2 &= \langle a||a\rangle_d = d(a^\star a) \geq 0. \end{aligned}$$

For a positive form, the d-trivial elements constitute the orthogonal $A_d^\perp$, a left ideal:

$$d \succeq 0 \Rightarrow A_d^\perp = \{n \in A \,\big|\, d(n^\star n) = 0\}.$$

The classes $|A\rangle_d = A/A_d^\perp$ with respect to the orthogonal have a definite scalar product:

$$\begin{array}{c} \langle\ |\ \rangle_d : |A\rangle_d \times |A\rangle_d \longrightarrow \mathbb{K},\ \langle a|b\rangle_d = d(a^\star b),\ \langle a|a\rangle_d \geq 0, \\ \langle n|n\rangle_d = 0 \iff n \in A_d^\perp \iff |n\rangle_d = 0. \end{array}$$

[8]Felix Hausdorff (1868–1942).

The equivalence classes $|a\rangle_d = a + A_d^\perp$ are called *state vectors*. d-strictly positive operators $a \in A$ define nontrivial state vectors. The class $|1_A\rangle_d$ of the algebra unit is a *cyclic state vector*, i.e., $A|1_A\rangle_d = |A\rangle_d$. The state values can be written as expectation values of this cyclic state vector:

$$d : A \longrightarrow \mathbb{K},\ d(a) = \langle 1_A|a|1_A\rangle_d.$$

The Cauchy completion of the scalar product space is the *d-associate Hilbert space* $A \xrightarrow{d} A/A_d^\perp \xrightarrow{\text{Cauchy}} \overline{|A\rangle_d} = V_d$. The algebra is represented on $|A\rangle_d$, dense in V_d. The norm-bounded left multiplications are extendable to the Hilbert space:

$$\begin{array}{l} \text{if } \ \| ba \|_d \leq k \ \| a \|_d,\ k \in \mathbb{R},\ \text{for all } a \in |A\rangle_d, \\ b : V_d \longrightarrow V_d,\ b|a\rangle_d = |ba\rangle_d. \end{array}$$

Pure states lead to irreducible algebra representations.

For a Hilbert space basis B, one has Parseval's[9] equation and the decomposition of the Hilbert space identity that represents the algebra unit 1_A:

$$\sum_{|e\rangle_d \in B} \langle a|e\rangle_d \langle e|a\rangle_d = \langle a|a\rangle_d,\ \sum_{|e\rangle_d \in B} |e\rangle_d \langle e| \cong 1_A.$$

For complex Hilbert spaces, the decomposition of the bilinear identity 1_A is sesquilinear, therefore the isomorphism notation $\cong$.

There is a unique correspondence between Hilbert subspaces $W \subseteq V_d$, used for Birkhoff–von Neumann logics, and projectors $\mathcal{P}_W$ which can be written as sum of primitive (nondecomposable) projectors for an orthonormal basis. The subspace dimension, if finite, is given by the trace of the projector (summation over κ):

$$V_d \supseteq W \leftrightarrow \mathcal{P}_W = \mathcal{P}_W^\star = \mathcal{P}_W^2,\ \left\{ \begin{array}{rl} \mathcal{P}_W & \cong |e^\kappa\rangle_d \langle e^\kappa|, \\ \mathrm{tr}\mathcal{P}_W & = \dim_{\mathbb{K}} W = d(W). \end{array} \right.$$

Each nontrivial finite-dimensional subspace W with the coarsest unital ring gives rise to the *yes-no W-probability measure* normalized with the discriminant of the scalar product:

$$\mu : \{\emptyset, W\} \longrightarrow \mathbb{R}_+,\ \mu(\emptyset) = 0,\ \mu(W) = \det \langle e^\kappa|e^{\kappa'}\rangle_d = 1.$$

Each nontrivial state vector $|a\rangle_d \neq 0$ defines the primitive projector for the Hilbert ray $\mathbb{C}|a\rangle_d$:

$$|e^a\rangle_d = \tfrac{|a\rangle_d}{\|a\|_d},\ \mathcal{P}_a \cong |e^a\rangle_d \langle e^a|,\ \langle e^a|e^a\rangle_d = 1,\ \text{ e.g., } \mathcal{P}_{1_A} \cong |1_A\rangle_d \langle 1_A|.$$

[9]Marc-Antoine Parseval (1755–1836).

The *expectation value* of an algebra operator in a subspace W (nontrivial, finite-dimensional) is given by the trace:

$$A \ni b \longmapsto d_W(b) = \mathrm{tr}\mathcal{P}_W \circ b = \langle e^\kappa|b|e^\kappa\rangle_d = d(e^{\kappa\star}be^\kappa),$$
$$\text{e.g.,} \begin{cases} d_a(b) &= \mathrm{tr}\mathcal{P}_a \circ b = \langle e^a|b|e^a\rangle_d = \frac{\langle a|b|a\rangle_d}{\langle a|a\rangle_d} = \frac{d(a^\star ba)}{d(a^\star a)}, \\ d_{1_A}(b) &= \mathrm{tr}\mathcal{P}_{1_A} \circ b = \langle 1_A|b|1_A\rangle_d = d(b). \end{cases}$$

Characteristic for quantum theory are the *transition probabilities* between two subspaces, computable with the corresponding projectors:

$$\begin{aligned} d_U(\mathcal{P}_W) &= d_W(\mathcal{P}_U) &= \mathrm{tr}\mathcal{P}_W \circ \mathcal{P}_U &= \langle e^\kappa|e^\lambda\rangle_d\langle e^\lambda|e^\kappa\rangle_d \in [0, d(W)d(U)], \\ p_{a\to b} = p_{b\to a} &= d_a(\mathcal{P}_b) &= \mathrm{tr}\mathcal{P}_a \circ \mathcal{P}_b &= \frac{|\langle a|b\rangle_d|^2}{\|a\|_d^2\,\|b\|_d^2} = |\langle e^a|e^b\rangle_d|^2 \in [0, 1]. \end{aligned}$$

The transition probabilities involve the *transition amplitude* between two state vectors as scalar product of the normalized vectors with values in the complex unit disk:

$$\frac{\langle a|b\rangle_d}{\|a\|_d\,\|b\|_d} = \langle e^a|e^b\rangle_d \in \mathbf{U}(1) \times [0, 1].$$

It is the complex generalization of the cosinus value for the angle between two Euclidean vectors:

$$\vec{a}, \vec{b} \in \mathbb{E}^n : \ \cos\vec{a}\vec{b} = \frac{\vec{a}\vec{b}}{|\vec{a}||\vec{b}|} \in \mathbf{I}(2) \times [0, 1].$$

9.9 Bose and Fermi Quantum Algebras

Important examples of operator algebras are the *Bose*[10] *and Fermi quantum algebras*, familiar from the harmonic oscillators, and used, with distributions, for all free particle fields. They are characterized by (anti-)commutators, $[a, b]_+ = \{a, b\} = ab + ba$, $[a, b]_- = [a, b] = ab - ba$, for creation and annihilation operators $\mathrm{u} \overset{\star}{\leftrightarrow} \mathrm{u}^\star$:

$$\left.\begin{aligned} [\mathrm{u}^\star, \mathrm{u}]_\epsilon &= 1, \\ [\mathrm{u}, \mathrm{u}]_\epsilon = 0,\ [\mathrm{u}^\star, \mathrm{u}^\star]_\epsilon &= 0, \end{aligned}\right\} \text{ for Bose } \epsilon = -1, \text{ and Fermi } \epsilon = +1.$$

The corresponding Fermi and Bose quantum algebras contain all finite complex linear combinations of all finite products of $(\mathrm{u}, \mathrm{u}^\star)$, modulo the (anti-)commutators:

$$\text{in } \mathbf{Q}_\epsilon(\mathbb{C}^2) : [\mathrm{u}^\star, \mathrm{u}]_\epsilon - 1 = 0,\ [\mathrm{u}, \mathrm{u}]_\epsilon = 0,\ [\mathrm{u}^\star, \mathrm{u}^\star]_\epsilon = 0.$$

[10] Satyendra Nath Bose (1894–1974).

The monomials $\{(\mathrm{u})^k(\mathrm{u}^\star)^l \mid k,l=0,1,2,\ldots\}$ lead to a basis: The finite-dimensional Fermi quantum algebra $\mathbf{Q}_+(\mathbb{C}^2) \cong \mathbb{C}^4$ has basis $\{1, \mathrm{u}, \mathrm{u}^\star, \mathrm{u}\mathrm{u}^\star\}$; the nilpotency $(\mathrm{u})^2 = 0$, $(\mathrm{u}^\star)^2 = 0$ is the origin of the *Pauli exclusion principle* for fermionic degrees of freedom. The Bose quantum algebra $\mathbf{Q}_-(\mathbb{C}^2) \cong \mathbb{C}^{\aleph_0}$ has countably infinite dimension with all the monomials as a basis. For the Bose quantum algebra, it makes sense to define the hermitian combinations $\mathbf{x} = \frac{\mathrm{u}+\mathrm{u}^\star}{\sqrt{2}}$ and $\mathbf{p} = i\frac{\mathrm{u}-\mathrm{u}^\star}{\sqrt{2}}$ as position and momentum, respectively, with $[i\mathbf{p}, \mathbf{x}] = \mathbf{1}$.

To obtain, in general, the quantum algebras $\mathbf{Q}_\epsilon(\mathbb{C}^{2s})$ for a complex finite-dimensional vector space $V \cong \mathbb{C}^s$ with dual space V^T and dual bases $\langle \mathrm{u}^\star_a, \mathrm{u}^b\rangle = \delta^b_a$, the tensor algebra $\bigotimes \mathbf{V}$ of the selfdual direct sum $\mathbf{V} = V \oplus V^T \cong \mathbb{C}^{2s}$ is factorized by the minimal ideal which identifies the bilinear dual products (not the sesquilinear Hilbert space scalar products) and corresponding tensor (anti-)commutators:

$$\mathbf{Q}_\epsilon(\mathbb{C}^{2s}) = \bigotimes \mathbf{V} \Big/ \text{minimal ideal} \begin{pmatrix} [\mathrm{u}^\star_a, \mathrm{u}^b]_\epsilon - \langle \mathrm{u}^\star_a, \mathrm{u}^b\rangle, \\ [\mathrm{u}^a, \mathrm{u}^b]_\epsilon,\ [\mathrm{u}^\star_a, \mathrm{u}^\star_b]_\epsilon \end{pmatrix}.$$

The construction of the quantum algebras for a vector space is similar to the construction of the enveloping algebra for a Lie algebra with the identification of Lie bracket and corresponding tensor commutator.

For $V \cong \mathbb{C}^s$, the complex 2^{2s}-dimensional Fermi quantum algebra and the complex $\aleph_0$-dimensional Bose quantum algebra are characterized by the duality-induced basic (anti-)commutators

$$\begin{aligned}
&\text{in Fermi } \mathbf{Q}_+(\mathbb{C}^{2s}) \cong \bigotimes^s \mathbf{Q}_+(\mathbb{C}^2) : \begin{cases} \{\mathrm{u}^\star_a, \mathrm{u}^b\} &= \delta^b_a, \\ \{\mathrm{u}^a, \mathrm{u}^b\} &= 0,\ \{\mathrm{u}^\star_a, \mathrm{u}^\star_b\} = 0, \end{cases}\\
&\text{in Bose } \mathbf{Q}_-(\mathbb{C}^{2s}) \cong \bigotimes^s \mathbf{Q}_-(\mathbb{C}^2) : \begin{cases} [\mathrm{u}^\star_a, \mathrm{u}^b] &= \delta^b_a, \\ [\mathrm{u}^a, \mathrm{u}^b] &= 0,\ [\mathrm{u}^\star_a, \mathrm{u}^\star_b] = 0. \end{cases}
\end{aligned}$$

Historically, they were used first for s bosonic degrees of freedom of harmonic oscillators, e.g., the Bose case $\mathbf{Q}_-(\mathbb{C}^3)$ for 3-position $\mathbf{x}^a = \frac{\mathrm{u}^a+\mathrm{u}^\star_a}{\sqrt{2}}$ with momenta $\mathbf{p}^a = i\frac{\mathrm{u}^a-\mathrm{u}^\star_a}{\sqrt{2}}$ and $[i\mathbf{p}^a, \mathbf{x}^b] = \delta^a_b$, and angular momenta $\mathbf{J}^a = -i\epsilon^{abc}\mathbf{x}^b \otimes \mathbf{p}^c$. Their structure is much more general: Quantum algebras $\mathbf{Q}_\epsilon(\mathbb{C}^{2s})$ give the *quantization of s fermionic or bosonic degrees of freedom* in the algebra $\mathbf{AL}(s,\mathbb{C})$.

The quantum algebra has representatives for the basic space endomorphisms $\mathbf{AL}(V) \ni f = f^b_a \mathrm{u}^a \otimes \mathrm{u}^\star_b$ whose endomorphism commutator is in the same class (simply denoted by $=$) as the quantum product commutator:

$$\begin{aligned}
f, g \in \mathbf{AL}(V) &: [f,g]_\circ = f \circ g - g \circ f,\\
\text{in } \mathbf{Q}_\epsilon(\mathbb{C}^{2s}) &: [f,g]_\otimes = f \otimes g - g \otimes f,\ [f,g]_\circ = [f,g]_\otimes,\\
&\quad [f, \mathrm{u}^c] = f^c_a \mathrm{u}^a,\ [f, \mathrm{u}^\star_d] = -f^b_d \mathrm{u}^\star_b.
\end{aligned}$$

In such a way, quantum algebras are representation spaces of the basic space Lie algebra $\mathbf{AL}(V)$. They contain product representations of the basic space V and its dual V^T. For instance in the quantum algebras $\mathbf{Q}_\epsilon(\mathbb{C}^3)$, the angular momenta for the infinitesimal rotations obey the familiar Lie brackets:

$$\mathbf{J}^a = -\epsilon^{abc}\mathrm{u}^b \otimes \mathrm{u}_c^\star : \mathbf{J}^a \circ \mathbf{J}^b - \mathbf{J}^b \circ \mathbf{J}^b = -\epsilon^{abc}\mathbf{J}^c = \mathbf{J}^a \otimes \mathbf{J}^b - \mathbf{J}^b \otimes \mathbf{J}^b = [\mathbf{J}^a, \mathbf{J}^b].$$

Usually, the tensor quantum product sign $\otimes$ is omitted, e.g., $\mathbf{x}^b\mathbf{p}^a - \mathbf{p}^a\mathbf{x}^b = i\delta_b^a$ and $\mathbf{J}^a = -i\epsilon^{abc}\mathbf{x}^b\mathbf{p}^c$. The class property, e.g., $\mathrm{u} = \mathrm{u}\mathrm{u}^\star\mathrm{u} + \epsilon\mathrm{u}^2\mathrm{u}^\star$, is taken into account by "getting used to it" — similar as for the rationals, e.g., $\frac{3}{4} = \frac{9}{12}$.

The *Fock*[11] *state* $\langle \ldots \rangle$ of quantum algebras is defined by nontrivial values for the abelian subalgebra with the $\mathbf{U}(1)$-invariants $\{(\mathrm{u}^\star\mathrm{u})^k \mid k = 0, 1, 2, \ldots\}$:

$$\begin{aligned} \mathbf{Q}_\epsilon(\mathbb{C}^2) \longrightarrow \mathbb{C},\ &\langle (\mathrm{u}^\star\mathrm{u})^k \rangle = 1,\ k = 0, 1, 2, \ldots, \\ &\langle [\mathrm{u}^\star\mathrm{u}, a] \rangle = 0,\ a \in \mathbf{Q}_\epsilon(\mathbb{C}^2). \end{aligned}$$

It is a unital form, $\langle 1 \rangle = 1$, and positive:

$$\begin{aligned} &\Rightarrow \langle a\mathrm{u}^\star \rangle = 0,\ \langle (\mathrm{u}^\star)^k(\mathrm{u})^l \rangle = \delta^{kl}k!,\ \langle a^\star a \rangle \geq 0, \\ \text{e.g., } &\langle \mathrm{u}^\star\mathrm{u} \rangle = \parallel \mathrm{u} \parallel^2 = 1, \quad \text{therefore } creation \text{ operator u}, \\ &\langle \mathrm{u}\mathrm{u}^\star \rangle = \parallel \mathrm{u}^\star \parallel^2 = 0, \quad \text{therefore } annihilation \text{ operator u}^\star. \end{aligned}$$

The quantum algebras are no C*-algebras where $\parallel a \parallel = \parallel a^\star \parallel$ for all elements.

The orthogonals for the quantum algebras above are given by the left ideal $\mathbf{Q}_\epsilon(\mathbb{C}^2)\mathrm{u}^\star$ with the annihilation operator. The classes $\mathbf{Q}_\epsilon(\mathbb{C}^2)/\mathbf{Q}_\epsilon(\mathbb{C}^2)\mathrm{u}^\star$ with the Fock state vectors can be spanned by the monomials $\{(\mathrm{u})^k \mid k = 0, 1, 2, \ldots\}$ of the creation operator, for Fermi by two elements $\{1, \mathrm{u}\}$. Usually one uses the occupation number notation for the normalized k-quantum state vectors:

$$\begin{aligned} |k\rangle &= \tfrac{1}{\sqrt{k!}}(\mathrm{u})^k + \mathbf{Q}_\epsilon(\mathbb{C}^2)\mathrm{u}^\star,\ k = \begin{cases} 0, 1, 2, & \text{for Fermi } \epsilon = +1, \\ 0, 1, 2, \ldots, & \text{for Bose } \epsilon = -1, \end{cases} \\ \langle l|k\rangle &= \delta^{lk},\ \langle l|\mathrm{u}|k\rangle = \delta_l^{k+1}\sqrt{k+1} = \langle k|\mathrm{u}^\star|l\rangle. \end{aligned}$$

The class of the algebra unit $|0\rangle = 1 + \mathbf{Q}_\epsilon(\mathbb{C}^2)\mathrm{u}^\star$ is the zero-quantum state vector, used as ground-state of the harmonic oscillators. The annihilation ideal in the definition of the ground-state $|0\rangle$ may remind of Dirac's attempts to postulate a populated ground-state (Dirac sea). The Fermi–Fock Hilbert space $V_+^{\text{Fock}} \cong \mathbb{C}^2$ has basis $\{|0\rangle, |1\rangle\}$, the Bose-Fock Hilbert space with Hilbert basis $\{|k\rangle \mid k = 0, 1, 2, \ldots\}$ is isomorphic to the square integrable position functions $V_-^{\text{Fock}} \cong L^2(\mathbb{R})$.

With the "vacuum" $|0\rangle$, all occupation number vectors:

$$|k\rangle = \tfrac{(\mathrm{u})^k}{\sqrt{k!}}|0\rangle,\ k = 0, 1, 2, \ldots,$$

define quantum algebra states $b \longmapsto \langle k|b|k\rangle = \frac{1}{k!}\langle (\mathrm{u}^\star)^k b(\mathrm{u})^k \rangle$. The Fock state values are the "vacuum" expectation values:

[11] Vladimir Fock (1898–1974).

$$\langle b \rangle = \langle 0|b|0 \rangle.$$

With occupation number bases for the two-dimensional Fermi Hilbert space,

$$\text{Fermi:} \begin{cases} |0\rangle = \binom{1}{0},\ |1\rangle = \binom{0}{1}, \\ \mathrm{u} = \begin{pmatrix} 0 & 0 \\ 1 & 0 \end{pmatrix},\ \mathrm{u}^\star = \begin{pmatrix} 0 & 1 \\ 0 & 0 \end{pmatrix}, \\ \mathrm{u}^\star\mathrm{u} = \begin{pmatrix} 1 & 0 \\ 0 & 0 \end{pmatrix},\ \mathrm{u}\mathrm{u}^\star = \begin{pmatrix} 0 & 0 \\ 0 & 1 \end{pmatrix}, \\ \{\mathrm{u}^\star, \mathrm{u}\} = \begin{pmatrix} 1 & 0 \\ 0 & 1 \end{pmatrix}, \end{cases}$$

and the countably infinite-dimensional Bose Hilbert space,

$$\text{Bose:} \begin{cases} |0\rangle = \begin{pmatrix} 1 \\ 0 \\ 0 \\ \vdots \end{pmatrix},\ |1\rangle = \begin{pmatrix} 0 \\ 1 \\ 0 \\ \vdots \end{pmatrix},\ |2\rangle = \begin{pmatrix} 0 \\ 0 \\ 1 \\ \vdots \end{pmatrix}, \dots, \\ \mathrm{u} = \left(\begin{array}{ccc|c} 0 & 0 & 0 & 0 \dots \\ \sqrt{1} & 0 & 0 & 0 \cdots \\ 0 & \sqrt{2} & 0 & 0 \cdots \\ \hline 0 & 0 & \sqrt{3} & 0 \cdots \\ & \dots & & \dots \end{array}\right),\ \mathrm{u}^\star = \left(\begin{array}{ccc|c} 0 & \sqrt{1} & 0 & 0 \dots \\ 0 & 0 & \sqrt{2} & 0 \cdots \\ 0 & 0 & 0 & \sqrt{3} \cdots \\ \hline 0 & 0 & 0 & 0 \cdots \\ & \dots & & \dots \end{array}\right), \\ \mathrm{u}^\star\mathrm{u} = \left(\begin{array}{ccc|c} 1 & 0 & 0 & 0 \dots \\ 0 & 2 & 0 & 0 \cdots \\ 0 & 0 & 3 & 0 \cdots \\ \hline 0 & 0 & 0 & 4 \cdots \\ & \dots & & \dots \end{array}\right),\quad \mathrm{u}\mathrm{u}^\star = \left(\begin{array}{ccc|c} 0 & 0 & 0 & 0 \dots \\ 0 & 1 & 0 & 0 \cdots \\ 0 & 0 & 2 & 0 \cdots \\ \hline 0 & 0 & 0 & 3 \cdots \\ & \dots & & \dots \end{array}\right), \\ [\mathrm{u}^\star, \mathrm{u}] = \left(\begin{array}{ccc|c} 1 & 0 & 0 & 0 \dots \\ 0 & 1 & 0 & 0 \cdots \\ 0 & 0 & 1 & 0 \cdots \\ \hline 0 & 0 & 0 & 1 \cdots \\ & \dots & & \dots \end{array}\right), \end{cases}$$

the Bose case illustrates the circumvention of the second historical obstacle in the buildup of quantum mechanics: The first obstacle was the non-number nature of the non-commuting observables in $[i\mathbf{p}, \mathbf{x}] = \mathbf{1}$ or, equivalently for the harmonic oscillator, in $[\mathrm{u}^\star, \mathrm{u}] = \mathbf{1}$. A realization of the operators by matrices seems to lead to a contradiction for the traces $\mathrm{tr}(\mathrm{u}^\star\mathrm{u} - \mathrm{u}\mathrm{u}^\star) = 0 \neq \mathrm{tr}\mathbf{1}_n = n$, as illustrated by the three-dimensional substructure $|_3$ of the Bose case:

$$\begin{aligned} \mathrm{u}^\star|_3\mathrm{u}|_3 &= \begin{pmatrix} 1 & 0 & 0 \\ 0 & 2 & 0 \\ 0 & 0 & 0 \end{pmatrix},\ \mathrm{u}|_3\mathrm{u}^\star|_3 = \begin{pmatrix} 0 & 0 & 0 \\ 0 & 1 & 0 \\ 0 & 0 & 2 \end{pmatrix}, \\ [\mathrm{u}^\star|_3, \mathrm{u}|_3] &= \begin{pmatrix} 1 & 0 & 0 \\ 0 & 1 & 0 \\ 0 & 0 & -2 \end{pmatrix} \neq [\mathrm{u}^\star, \mathrm{u}]|_3 = \mathbf{1}_3. \end{aligned}$$

For an infinite-dimensional Hilbert space the trace argument is no longer valid.

The Fermi and Bose Fock-Hilbert spaces are used for the representations of time, position and spacetime translations, especially for free particles. In the massive particle case, the creation and annihilation operators are distributions of the energy-momentum hyperboloid $\vec{q} \in \mathcal{Y}^3$ with $q_0^2 - \vec{q}^2 = m^2$ (the discrete dimension $\mathrm{u}^a \in V \cong \mathbb{C}^s$ is replaced by a "continuous" one $\mathrm{u}(\vec{q}) \in V \cong \mathbb{C}^{\mathcal{Y}^3}$), possibly with additional discrete indices for the eigenvalues of compact homogeneous groups ((iso)spin, color etc.):

$$\begin{array}{rl}[\mathrm{u}^\star_\alpha(\vec{p}), \mathrm{u}^\beta(\vec{q})]_\epsilon &= 2q_0\delta(\frac{\vec{q}-\vec{p}}{2\pi})\delta^\beta_\alpha, \text{ with } q_0 = \sqrt{\vec{q}^2 + m^2}\\ \langle 0|\mathrm{u}^\star_\alpha(\vec{p})\mathrm{u}^\beta(\vec{q})|0\rangle &= \langle m^2; \vec{p}, \alpha|m^2; \vec{q}, \beta\rangle = 2q_0\delta(\frac{\vec{q}-\vec{p}}{2\pi})\delta^\beta_\alpha\\ \text{distributive Hilbert basis: } &\{|m^2; \vec{q}, \alpha\rangle = \mathrm{u}^\alpha(\vec{q})|0\rangle \,\big|\, \vec{q} \in \mathbb{R}^3, \alpha = 1, \ldots, d.\}.\end{array}$$

Product states use appropriately (anti)symmetrized tensor product distributions, e.g.,

$$|\vec{q}_1, \ldots, \vec{q}_k, \alpha_1, \ldots, \alpha_k\rangle \sim \mathrm{u}^{\alpha_1}(\vec{q}_1) \otimes \cdots \otimes \mathrm{u}^{\alpha_k}(\vec{q}_k)|0\rangle.$$

Lie algebra elements have to integrate the momentum operators on the mass hyperboloid, e.g., for the rotation (spin) Lie algebra:

$$\begin{array}{l}\underline{\mathbf{O}} = {}^\oplus\!\int \frac{d^3q}{(2\pi)^3 2q_0}\, \mathrm{u}^b(\vec{q})\mathbf{O}^c_b\mathrm{u}^\star_c(\vec{q}), \quad [\underline{\mathbf{O}}_1, \underline{\mathbf{O}}_2] = \underline{[\mathbf{O}_1, \mathbf{O}_2]},\\ \mathbf{J}^a = {}^\oplus\!\int \frac{d^3q}{(2\pi)^3 2q_0}\mathbf{J}^a(\vec{q}),\ \mathbf{J}^a(\vec{q}) = -\epsilon^{abc}\mathrm{u}^b(\vec{q}) \otimes \mathrm{u}^\star_c(\vec{q}),\\ [\mathbf{J}^a(\vec{p}), \mathbf{J}^b(\vec{q})] = -2q_0\delta(\frac{\vec{q}-\vec{p}}{2\pi})\epsilon^{abc}\mathbf{J}^c(\vec{q}),\ [\mathbf{J}^a, \mathbf{J}^b] = -\epsilon^{abc}\mathbf{J}^c.\end{array}$$

The momentum integrals become, in a formulation with spacetime fields, position integrals over currents $(\mathbf{J}^a_k(\vec{x}))_{k=0,1,2,3}$, e.g., $\mathbf{J}^a = \int d^3x\, \mathbf{J}^a_0(\vec{x})$.

For the quantum algebras considered, $\mathbf{Q}_\epsilon(\mathbf{V})$ may be based on infinite-dimensional representation spaces, e.g., as subspaces of a convolution algebra with group G functions $V \subseteq \mathbb{C}^G$.

9.10 Hilbert Spaces for U(1) and D(1)

The "smallest" real Lie groups are one-dimensional $\mathbf{D}(1) \cong \mathbb{R}$ (simply connected and noncompact) and $\mathbf{U}(1) \cong [-\pi, \pi] \cong \mathbb{R}/\mathbb{Z}$ (infinitely connected and compact). Taken as subgroups, their structures are basically characteristic for noncompact and compact Lie groups. Their Haar measures with $d\psi = e^{-\psi}de^\psi$ and $id\theta = e^{-i\theta}de^{i\theta}$, both Lebesgue measures, and their $\mathbf{U}(1)$-characters (irreducible Hilbert representations), i.e., the dual groups $i\mathbb{R}$ for $\mathbf{D}(1)$, continuous with Lebesgue Plancherel[12] measure, and $\mathbb{Z}$ for $\mathbf{U}(1)$, discrete with counting Plancherel measure, are given by

$$\begin{array}{ll}\int_\mathbb{R} d\psi = \int d\psi, & D^{iP} : \mathbf{D}(1) \ni e^\psi \longmapsto e^{iP\psi} \in \mathbf{U}(1),\ iP \in i\mathbb{R} \cong \mathbf{irrep}_+\mathbf{D}(1),\\ \int_{-\pi}^{\pi} \frac{d\theta}{2\pi} = 1, & D^z : \mathbf{U}(1) \ni e^{i\theta} \longmapsto e^{zi\theta} \in \mathbf{U}(1), \quad z \in \mathbb{Z} \cong \mathbf{irrep}\,\mathbf{U}(1).\end{array}$$

[12] Michel Plancherel (1885–1967).

In the separable Hilbert spaces

$$\begin{array}{l} L^2(\mathbf{D}(1)), \text{ with } \langle f|g\rangle = \int d\psi\, \overline{f(\psi)}g(\psi),\\ L^2(\mathbf{U}(1)), \text{ with } \langle f|g\rangle = \int_{-\pi}^{\pi} \frac{d\theta}{2\pi}\, \overline{f(\theta)}g(\theta), \end{array}$$

the irreducible $\mathbf{U}(1)$-representations constitute an orthonormal Hilbert space basis:

$$L^2(\mathbf{U}(1))\text{-basis: } \{D^z \,\big|\, z \in \mathbb{Z}\}, \quad \begin{cases} \int_{-\pi}^{\pi} \frac{d\theta}{2\pi}\, e^{-zi\theta} e^{z'i\theta} &= \delta_{zz'},\\ \sum_{z\in\mathbb{Z}} e^{-zi\theta} e^{zi\theta'} &= \delta(\frac{\theta-\theta'}{2\pi}), \end{cases}$$

whereas the irreducible $\mathbf{D}(1)$-representations in $\mathbf{U}(1)$ are no Hilbert space elements; they constitute an orthogonal *distributive basis*:

$$D^{iP} \notin L^2(\mathbf{D}(1)) : \quad \begin{cases} \int d\psi\, e^{-iP\psi} e^{iP'\psi} &= \delta(\frac{P-P'}{2\pi}),\\ \int \frac{dP}{2\pi}\, e^{-iP\psi} e^{iP\psi'} &= \delta(\psi - \psi'). \end{cases}$$

The $\mathbf{U}(1)$-representations of $\mathbf{D}(1) \cong \mathbb{R}$ define the Fourier transformation, i.e., the harmonic analysis of $\mathbf{D}(1)$-functions and relate to each other the isomorphic Hilbert spaces with the square integrable functions on the eigenvalues $iP \in \mathbf{irrep}_+\mathbf{D}(1) \cong i\mathbb{R}$ on the one hand and the square integrable $\mathbf{D}(1)$-representation coefficients on the other hand:

$$L^2_{dP}(\mathbb{R}) \cong L^2_{d\psi}(\mathbb{R}), \quad \begin{cases} f(\psi) &= \int \frac{dP}{2\pi}\, \tilde{f}(P) e^{iP\psi},\\ \langle f|g\rangle = \int d\psi\, \overline{f(\psi)}g(\psi) &= \int \frac{dP}{2\pi}\, \overline{\tilde{f}(P)}\tilde{g}(P) = \langle \tilde{f}|\tilde{g}\rangle. \end{cases}$$

9.11 Group Algebras and Representation Spaces

Each group determines, via group-induced algebras, its Hilbert spaces, finite- or infinite-dimensional, where its action can be represented by definite unitary automorphisms, e.g., the complex $(1+2J)$-dimensional Hilbert spaces for the irreducible $\mathbf{SU}(2)$-representations. A Hilbert space, e.g., the Fock space for translations and free particles, may not be appropriate for another group, e.g., as Hilbert space for bound states or for the implementation of interactions.

Each representation of a *compact Lie group* K, maybe a finite group, is equivalent to a representation $D : K \longrightarrow \mathbf{U}(V)$ in the unitary group of a Hilbert space V. In generalization of Maschke's theorem for finite groups, it is decomposable into irreducible finite-dimensional ones, $V \cong \mathbb{C}^d$. The invariants and eigenvalues, such as quantum numbers for charge $\mathbf{U}(1)$, isospin $\mathbf{SU}(2)$, color $\mathbf{SU}(3)$, etc., are rational and ultimately connected to *integer winding numbers* $z \in \mathbb{Z}$ from the dual group of $\mathbf{U}(1)$, which characterize the irreducible representations $\mathbf{U}(1) \ni e^{i\theta} \longmapsto e^{zi\theta} \in \mathbf{U}(1)$. A maximal abelian compact subgroup with direct factors $\mathbf{U}(1) \cong \mathbf{SO}(2) \cong \exp i\mathbb{R}$ is called a *Cartan torus*, e.g., $\mathbf{SO}(2)^r \subset \mathbf{SU}(1+r)$ or, for hyperisospin,

$\mathbf{U}(1)_+ \times \mathbf{U}(1)_- \subset \mathbf{U}(2)$. The representations (characters) of the Cartan tori are basic for the representations of compact groups.

Compared with the compact group structures, the theory for *noncompact Lie groups*, especially for the nonabelian ones, e.g., for the Lorentz group, is more complicated and difficult. In addition to the finite-dimensional non-Hilbert representations, there are Hilbert representations, necessarily infinite-dimensional, if faithful. A noncompact semisimple group has the *Iwasawa*[13] *decomposition (factorization)* $G = K \circ A \circ \exp N$ into maximal compact subgroup K, maximal noncompact abelian subgroup A, and a subgroup $\exp N$ with nilpotent Lie algebra, e.g.,

$$\begin{aligned} 1 \le t \le s : \mathbf{SO}_0(t,s) &= [\mathbf{SO}(t) \times \mathbf{SO}(s)] \circ \mathbf{SO}_0(1,1)^t \circ \exp \mathbb{R}^{t(s-1)}, \\ \mathbf{SL}(1+r,\mathbb{R}) &= \mathbf{SO}(1+r) \circ \mathbf{SO}_0(1,1)^r \circ \exp \mathbb{R}^{\binom{1+r}{2}}, \\ \mathbf{SL}(1+r,\mathbb{C}) &= \mathbf{SU}(1+r) \circ \mathbf{SO}_0(1,1)^r \circ \exp \mathbb{R}^{(1+r)r}. \end{aligned}$$

A *minimal parabolic subgroup* $P = K_0 \circ A \circ \exp N$ extends $A \circ \exp N$ by the A-centralizer $K_0 = \{k \in K \,|\, [k, A] = \{0\}\}$ in K. A maximal noncompact abelian group A is isomorphic to a translation (dilation) group $A \cong \mathbb{R}^r$ with factors $\mathbf{D}(1) \cong \mathbf{SO}_0(1,1) \cong \exp \mathbb{R} \cong \mathbb{R}$, and called a *Cartan plane*. Its dimension r gives the maximal number of representation characterizing invariants from a continuous spectrum, called *continuous invariants*. In physics, they can be used as basic units for the related operations.

The *Cartan decomposition* $G = K \circ \exp R$ of a noncompact group with semisimple Lie algebra displays the maximal compact group and the subset of noncompact elements. The Lie algebra $\log G = \log K \oplus R$ has the brackets:

$$[\log K, \log K] \subseteq \log K,\ [\log K, R] \subseteq R,\ [R, R] \subseteq \log K.$$

The *Cartan factorization* $G = K \circ A \circ K$ implies the form $g = k_1 a k_2$ with $a \in A$ and $k_{1,2} \in K$ for each group element, for the examples above:

$$\begin{aligned} 1 \le t \le s : \mathbf{SO}_0(t,s) &= [\mathbf{SO}(t) \times \mathbf{SO}(s)] \circ \mathbf{SO}_0(1,1)^t \circ [\mathbf{SO}(t) \times \mathbf{SO}(s)], \\ \mathbf{SL}(1+r,\mathbb{R}) &= \mathbf{SO}(1+r) \circ \mathbf{SO}_0(1,1)^r \circ \mathbf{SO}(1+r), \\ \mathbf{SL}(1+r,\mathbb{C}) &= \mathbf{SU}(1+r) \circ \mathbf{SO}_0(1,1)^r \circ \mathbf{SU}(1+r). \end{aligned}$$

Hilbert representations of a Lie group contain, as basic substructures, Hilbert representations of its Cartan tori and planes, which determine, respectively, the rational and continuous representation invariants and eigenvalues, e.g., one rational and one continuous invariant for a Cartan cylinder with axial rotations and dilations $\mathbf{SO}(2) \times \mathbf{SO}_0(1,1) \subset \mathbf{SO}_0(1,3)$ in the Lorentz group, and either one rational $\mathbf{SO}(2) \subset \mathbf{SO}_0(1,2)$ or one continuous invariant $\mathbf{SO}_0(1,1) \subset \mathbf{SO}_0(1,2)$ in the Lorentz group for two position dimensions.

The cyclic Hilbert representations $\mathbf{cycrep}_+\mathbb{R}^r$ (definition ahead) of a Cartan plane with translations (dilations) are characterized by essentially bounded *positive-type*

[13] Kenkichi Iwasawa (1917–1998).

functions (*scalar product or Hilbert metric-inducing functions*) $\mathbb{R}^r \ni x \longmapsto d(x)$, $d \in L^\infty(\mathbb{R}^r)_+$. Scalar product–inducing functions are Fourier-transformed positive Radon[14] measures (distributions, densities) $\mathcal{M}(\check{\mathbb{R}}^r)_+$ of the group dual $\check{\mathbb{R}}^r$, i.e., of the irreducible, but not faithful translation representations $\mathbb{R} \ni x \longrightarrow e^{iqx} \in \mathbf{U}(1)$ and dilation representations like $\mathbf{SO}_0(1,1) \ni \begin{pmatrix} \cosh\psi & \sinh\psi \\ \sinh\psi & \cosh\psi \end{pmatrix} \longmapsto \begin{pmatrix} \cos P\psi & i\sin P\psi \\ i\sin P\psi & \cos P\psi \end{pmatrix} \in \mathbf{SO}(2)$ with a continuous eigenvalue iq for, respectively, energy-momenta and a dilation unit $P \in \mathbb{R}$.

The equivalence classes of the irreducible complex Hilbert representations of a group $G \longrightarrow \mathbf{U}(V)$ constitute the *group dual* $\check{G} = \mathbf{irrep}_+ G$. Its explicit determination, especially for noncompact nonabelian groups, may be a difficult task. The group dual is induced by the group duals of the real one-dimensional "continuous" groups which are discrete $\check{\mathbf{U}}(1) = \mathbf{irrep}\,\mathbf{U}(1) \cong \mathbb{Z}$ and continuous $\check{\mathbf{D}}(1) = \mathbf{irrep}_+\mathbf{D}(1) \cong \mathbf{irrep}_+\mathbb{R} \cong i\mathbb{R}$. The example $\mathbf{U}(1)$ shows the compatibility of continuous and discrete structures — for group parameters and group invariants, respectively — which may be of relevance with respect to the general question of "discretization" and "quantization" of operations, e.g., for "quantized" spacetime.

In general, the eigenvalues for irreducible representations (not only Hilbert representations) are characterized by the examples in the following table displaying the twofold dichotomy compact–noncompact and abelian–nonabelian for the represented groups:

	Compact	Noncompact
Abelian	$\mathbf{U}(1) \longrightarrow \mathbf{U}(1)$ $e^{i\theta} \longmapsto e^{zi\theta}$ $z \in \mathbb{Z} \cong \mathbf{irrep}\,\mathbf{U}(1)$	$\mathbf{D}(1) \longrightarrow \mathbf{U}(1) \times \mathbf{D}(1)$ $e^{\psi} \longmapsto e^{iQ\psi}$ $iQ = iP + B \in i\mathbb{R} + \mathbb{R} \cong \mathbf{irrep}\,\mathbf{D}(1)$
Nonabelian $n \geq 2$	$\mathbf{SU}(2) \longrightarrow \mathbf{SU}(1+2J)$ $u \longmapsto \bigvee^{2J} u,\ 2J = \lvert z\rvert \in \mathbb{N}$ $z \in \mathbb{Z}$	$\begin{cases} \mathbf{SL}(2,\mathbb{C}) \longrightarrow \text{compact} \\ \mathbf{SL}(2,\mathbb{C}) \longrightarrow \text{noncompact} \end{cases}$ $\mathbb{Z} \times [i\mathbb{R} + \mathbb{R}]$

Irreducible group representations with example weights

Compact groups have a discrete spectrum; noncompact groups have a continuous one with, possibly, discrete contributions. The noncompact nonabelian case — that is really difficult.

Group representation matrix elements are group functions, called *representation coefficients*. In physics, operator groups as event spaces are experimentally valued by numbers. Therefore, it is appropriate to join the complex numbers $\mathbb{C}$ with the group operations G in *complex group algebras* whose (minimal) ideals lead to the (irreducible) representation algebras for the group algebra, and, in those ideals, to the representations of the group.

For an orientation, it is useful to recapitulate in this section, in general, the relevant vector spaces and algebras for group functions with their group induced convolution product and their number induced pointwise product.

[14] Johann Radon (1887–1956).

9.11.1 Bi-Regular and Induced Representations

The group orbit of any vector $G \bullet v \subseteq V$ in a representation space – irreducible or not, unitary or not — spans a representation space which may be V or a subspace proper. Especially, for a finite dimensional representation $V \cong \mathbb{C}^n$ with matrices in any vector space basis $G \ni g \longmapsto D^k_m(g)$, the orbits of the columns for any fixed $k_0 \in \{1, \dots, n\}$ (similarly for the rows) span n representation spaces. These orbits are n group functions:

$$m = 1, \dots, n : v_m : G \longrightarrow \mathbb{C},\ v_m(g) = D^{k_0}_m(g) \Rightarrow D^m_l(g_2)v_m(g_1) = v_l(g_2 g_1).$$

Examples are the $(1 + 2L)$ columns in irreducible $\mathbf{SO}(3)$-representation matrices, e.g., the central column which contains – up to normalization – the $(1+2L)$ spherical harmonics $D^{L0}_m(g) \sim \mathrm{Y}^L_m(\theta, \varphi)$, and energy-momentum vectors in dilation-Lorentz transformations:

$$D^1(g) = \left(\begin{array}{c|c|c} e^{i(\chi+\varphi)}\cos^2\frac{\theta}{2} & ie^{i\varphi}\frac{\sin\theta}{\sqrt{2}} & -e^{-i(\chi-\varphi)}\sin^2\frac{\theta}{2} \\ ie^{i\chi}\frac{\sin\theta}{\sqrt{2}} & \cos\theta & ie^{-i\chi}\frac{\sin\theta}{\sqrt{2}} \\ -e^{i(\chi-\varphi)}\sin^2\frac{\theta}{2} & ie^{-i\varphi}\frac{\sin\theta}{\sqrt{2}} & e^{-i(\chi+\varphi)}\cos^2\frac{\theta}{2} \end{array}\right) \in \mathbf{SO}(3),$$

$$\begin{aligned} D(\psi) &= e^{\psi_0}\left(\begin{array}{c|c} \cosh\psi & \sinh\psi \\ \sinh\psi & \cosh\psi \end{array}\right) = \left(\begin{array}{c|c} E & q \\ q & E \end{array}\right) \in \mathbf{D}(1) \times \mathbf{SO}_0(1, 1), \\ &\text{with } E^2 - q^2 = m^2 = e^{2\psi_0}. \end{aligned}$$

For a group G, all representation matrix elements are complex group functions. For a real finite-dimensional Lie group, complex-valued mappings (functions, distributions, measures) of the group are considered. Summation over group elements is expressed by invariant integration with a *Haar measure* $dg \in \mathcal{M}(G)_+$, $\int_G dg = \int dg$, a counting measure for finite and discrete groups. With a Haar measure basis, all group measures can be characterized by (generalized) functions $\mu(g)dg$.

Complex functions $f = {}^{\oplus}\!\!\int dg\, g f(g) \in \mathcal{F}(G)$ with a direct integral Haar measure constitute group representation spaces, in general highly reducible. They inherit the group multiplication in the *left-right regular (bi-regular) representation of* G, which is actually a *representation of the doubled group* $G \times G$,

$$\begin{array}{ccc} G & \overset{L_k \times R_l}{\longrightarrow} & G \\ f \big\downarrow & & \big\downarrow {}_k f_l, \\ \mathbb{C} & \underset{1_{\mathbb{C}}}{\longrightarrow} & \mathbb{C} \end{array} \qquad \begin{aligned} (G \times G) \bullet G &\longrightarrow G, \\ (k, l) \bullet g &= kgl^{-1}, \\ (G \times G) \bullet \mathcal{F}(G) &\longrightarrow \mathcal{F}(G), \\ (k, l) \bullet f &= {}_k f_l = {}^{\oplus}\!\!\int dg\, g f(k^{-1}gl). \end{aligned}$$

Distributions $\mathcal{F}'(G)$ have, via duality to their test functions $\mathcal{F}(G)$, the dual action:

$$(G \times G) \bullet \mathcal{F}'(G) \longrightarrow \mathcal{F}'(G), \qquad \begin{aligned} \langle (k, l) \bullet \mu, f \rangle &= \langle \mu, (k, l) \bullet f \rangle, \\ \int dg\, \mu(kgl^{-1}) f(g) &= \int dg\, \mu(g) f(k^{-1}gl). \end{aligned}$$

It is important to realize that in a bi-regular representation of $G \times H \subseteq G \times G$ the subgroup H acts independently twice, e.g., $G \ni g \longmapsto L_{h_1} \times R_{h_2}(g) = h_1 g h_2^{-1}$ for $h_{1,2} \in H$, as familiar from the independently acting pair spin-isospin with isomorphic groups $\mathbf{SU}(2)$.

A G-representation $D^G = \mathrm{ind}_H^G(D)$, *induced from a subgroup H-representation*, is a bi-regular subrepresentation for $G \times H \subseteq G \times G$. It collects complex functions into H-intertwiners, valued in a complex vector space $\begin{pmatrix} w_1 \\ \vdots \\ w_N \end{pmatrix} \cong w \in W$ with H-representation, $D : H \longrightarrow \mathbf{GL}(W)$:

$$\begin{array}{ccc} G & \xrightarrow{L_k \times R_h} & G \\ {\scriptstyle w}\downarrow & & \downarrow{\scriptstyle {}_k w}, \\ W & \xrightarrow[D(h)]{} & W \end{array} \qquad \begin{array}{rrl} H\text{-intertwiner: } & w(gh^{-1}) & = D(h).w(g), \\ G\text{-action: } & w & \longmapsto D^G(k).w = {}_k w, \\ \text{with} & {}_k w(g) & = w(k^{-1}g). \end{array}$$

An H-intertwiner on the group maps H-cosets of the group into H-orbits in the Hilbert space W. It is defined by its values on representatives $g_r \in \mathbb{M} = (G/H)_r \subseteq G$ of the coset manifold. The G-action comes with the representative-dependent H-action of the related Wigner element $h(k, g_r) \in H$:

$$\begin{array}{ccc} \mathbb{M} & \xrightarrow{L_k} & \mathbb{M} \\ {\scriptstyle w}\downarrow & & \downarrow{\scriptstyle {}_k w}, \\ W & \xrightarrow[D(h^{-1})]{} & W \end{array} \qquad \begin{array}{l} \mathbb{M} = (G/H)_r \subseteq G,\ G \bullet W^{\mathbb{M}} \longrightarrow W^{\mathbb{M}}, \\ {}_{k^{-1}}w(g_r) = w(kg_r) = D(h^{-1}(k, g_r)).w((kg)_r), \\ \text{with } kg_r = (kg)_r h(k, g_r). \end{array}$$

Hilbert representations of noncompact groups can be induced by well known irreducible finite-dimensional Hilbert representations of compact and abelian subgroups, e.g., by direct product or Cartan subgroups.

A G-representation D^G on a Hilbert space V, induced from a subgroup H-representation D has a *system of imprimitivity* $(D^G, G/H, \mathcal{P})$ with a G-action-compatible bounded projection valued measure $\mathcal{P}$ on the H-classes:

$$G/H \supseteq X \longmapsto \mathcal{P}(X) \in \mathbf{B}(V), \text{ with } D^G(k) \circ \mathcal{P}(X) \circ D^G(k^{-1}) = \mathcal{P}(kX).$$

9.11.2 Finite Groups and Convolution Algebras

For a finite group, e.g., for the cyclic and permutation groups $\mathbf{I}(n) \subseteq \mathbf{G}(n)$, compact with the discrete topology, the complex linear combinations of group elements $f = \bigoplus_{g \in G} g f(g) \in \mathbb{C}^G$ realize the group functions $g \longmapsto f(g)$. They constitute the complex *group algebra* $\mathbb{C}^G = \{G \longrightarrow \mathbb{C}\}$ with the group elements as natural basis $\mathbb{C}^G \cong \mathbb{C}^{\operatorname{card} G}$. The group product induces a unital algebra structure with *convolution product*:

$$\begin{aligned} f_1 * f_2 &= \bigoplus_{g \in G} g\,(f_1 * f_2)(g) \in \mathbb{C}^G, \\ f_1 * f_2(g) &= \sum_{g_1, g_2 \in G} f_1(g_1)\delta^g_{g_1 g_2} f_2(g_2) \text{ with } \delta^g_{g_1 g_2} = \begin{cases} 1 \text{ for } g_1 g_2 = g, \\ 0 \text{ for } g_1 g_2 \neq g. \end{cases} \end{aligned}$$

The theory of group representations is embedded in the theory of algebra representations.

The *minimal two-sided ideals of the group algebra*, $I = \mathbb{C}^G * I * \mathbb{C}^G$, are generated by the conjugacy classes $G(k) = \{gkg^{-1} \mid g \in G\}$ of the group. They characterize *the irreducible representations of the group algebra* $\mathbb{C}^G$ by full matrix algebras $I \cong \mathbf{AL}(d_Z, \mathbb{C}) \cong \mathbb{C}^{d_Z} \otimes \mathbb{C}^{d_Z}$ with dimensions d_Z^2 and left-right action $G \bullet \mathbf{AL}(d_Z, \mathbb{C}) \bullet G$,

$$\text{finite } G: \quad \mathbb{C}^G \cong \bigoplus_{Z=1}^{n} \mathbb{C}^{d_Z} \otimes \mathbb{C}^{d_Z}, \ \operatorname{card} G = \sum_{Z=1}^{n} d_Z^2.$$

Via the columns or rows as d_Z-dimensional vector spaces in these algebras, one obtains *the irreducible representations of the group G* by finite unitary groups in $\mathbf{U}(d_Z) \subseteq \mathbf{AL}(d_Z, \mathbb{C})$.

All finite groups are permutation subgroups $G \subseteq \mathbf{G}(\operatorname{card} G)$. Therefore, a characteristic example is given by the familiar Young tableaux–related irreducible representations of the permutation group $\mathbf{G}(n)$ with cardinality $n!$ and its algebra $\mathbb{C}^{n!}$.

9.11.3 Algebras and Vector Spaces for Locally Compact Groups

The structures and concepts for finite groups are generalizable to countably infinite groups, e.g., to the lattices $\mathbb{Z}^n$, and to finite-dimensional Lie groups, e.g., to the general groups $\mathbf{GL}(n, \mathbb{K})$. As exemplified by the finite groups with the convolution algebra $\mathbb{C}^G$, all group representation spaces come from a convolution algebra with group functions.

Two products are important for (generalized) group functions $\mu = {}^{\oplus}\!\int dg\, g\mu(g)$ of a locally compact group (unimodular $\Delta(g) = 1$ if necessary). The group functions inherit the group composition $g_1 g_2 = g \in G$ as *convolution product* (where defined),

$$\begin{aligned} \mu_1 * \mu_2 &= [{}^{\oplus}\!\textstyle\int dg\, g\mu_1(g)] * [{}^{\oplus}\!\int dg\, g\mu_2(g)] &&= {}^{\oplus}\!\textstyle\int dg\, g(\mu_1 * \mu_2)(g), \\ \mu_1 * \mu_2(g) &= \textstyle\int dg_1 dg_2\, \mu_1(g_1)\, \delta(g_1 g_2 g^{-1})\mu_2(g_2) &&= \textstyle\int dg_1\, \mu_1(g_1)\mu_2(g_1^{-1} g). \end{aligned}$$

The associative convolution is abelian if, and only if, the group multiplication is abelian. As a second product, the group functions inherit from the complex numbers the abelian *pointwise multiplication* (where defined), important for product representations,

$$\mu_1 \cdot \mu_2(g) = \mu_1(g)\mu_2(g).$$

With respect to the convolution and pointwise product, the *Lebesgue Banach spaces* $L^p(G)$, $1 \leq p \leq \infty$, with the classes of Haar measure almost everywhere defined absolute p-integrable group functions, i.e., $(\| f \|_p)^p = \int dg\,|f(g)|^p < \infty$, are connected with each other as follows:

$$1 \leq p, r, s \leq \infty : \begin{cases} L^p(G) * L^r(G) \subseteq L^s(G), \text{ with } \frac{1}{p} + \frac{1}{r} - \frac{1}{s} = 1, \\ L^p(G) \cdot L^r(G) \subseteq L^s(G), \text{ with } \frac{1}{p} + \frac{1}{r} - \frac{1}{s} = 0. \end{cases}$$

They are left-right modules, i.e., stable under the corresponding multiplications, respectively, for the absolute integrable group function classes $L^1(G)$, a *convolution algebra*, and for the essentially bounded group functions $L^\infty(G)$, i.e., $|f(g)| < \infty$ almost everywhere, a *pointwise product algebra* with unit,

$$\begin{aligned} L^1(G) * L^r(G) &\subseteq L^r(G), \quad L^1(G) * L^1(G) \subseteq L^1(G), \\ L^\infty(G) \cdot L^r(G) &= L^r(G), \; L^\infty(G) \cdot L^\infty(G) = L^\infty(G). \end{aligned}$$

They are left-right convolution modules even for the *Radon distributions* $\mathcal{M}(G)$ of the group (definition ahead), a unital convolution Banach algebra,

$$\begin{aligned} 1 \leq p \leq \infty : \; \mathcal{M}(G) * L^p(G) * \mathcal{M}(G) &= L^p(G), \\ \mathcal{M}(G) * \mathcal{M}(G) &= \mathcal{M}(G). \end{aligned}$$

The Radon measures, in the form of Radon distributions with a Haar measure $\omega(g)dg$, embed the group by Dirac measures:

$$\begin{aligned} G \ni k \longmapsto \delta_k \in \mathcal{M}(G) \text{ with } \langle \delta_k, f \rangle &= \int dg\, \delta_k(g) f(g) = f(k), \\ \delta_k * \delta_l &= \delta_{kl}, \; \delta_{1_G} = \delta, \\ \delta_k(g) &= \delta(gk^{-1}). \end{aligned}$$

Dirac distributions and Haar measures are defined as "inverse to each other," especially their normalizations:

$$\langle \delta_k, 1 \rangle = \int \delta_k(g) dg = 1 = \int \delta(g) dg.$$

The left-right action of a group on itself $g \longmapsto g_1 g g_2^{-1}$ is embedded in the left-right convolution module property of the function spaces for the Radon group distributions,

$$\begin{aligned} \mathcal{M}(G) * L^p(G) * \mathcal{M}(G) &\longrightarrow L^p(G),\ f \longmapsto \mu_1 * f * \mu_2, \\ \mu_1 * f * \mu_2(g) &= \int dg_1 \int dg_2\, \mu_1(g_1) f(g_1^{-1} g g_2) \mu_2(g_2^{-1}), \\ \text{e.g., } \delta_k * f * \delta_l(g) &= f(k^{-1} g l^{-1}). \end{aligned}$$

The Radon distributions are defined as the dual $\mathcal{M}(G) = \mathcal{C}_c(G)'$ of the compactly supported continuous functions $\mathcal{C}_c(G)$, their test functions; these functions are dense in all $L^p(G)$, $1 \leq p < \infty$. $\mathcal{M}(G)$ contains the function algebra $L^1(G)$ as two-sided ideal. The involutive convolution algebra $\mathcal{C}_c(G)$ is a subspace of the bounded continuous functions $\mathcal{C}_b(G)$, which, in their turn, can be considered as a closed, in general, proper subspace of the essentially bounded functions,

$$\begin{aligned} L^1(G) &\subseteq \mathcal{M}(G) \supset G, \\ \mathcal{C}_c(G) &\subseteq \mathcal{C}_b(G) \ \subseteq L^\infty(G). \end{aligned}$$

As suggested by the dimension of the measure $d^n x$ in the integration with a distribution — start with $\int_a^b dx\, \frac{df(x)}{dx} = f(a) - f(b)$ — every Radon distribution of an open real set $T \subseteq \mathbb{R}^n$ is a finite sum of derivatives up to order n of locally essentially bounded functions,

$$T \subseteq \mathbb{R}^n: \ \mathcal{M}(T) \subseteq \left\{ \sum_{N=0}^{n} \alpha_N \partial^N L^\infty(T) \right\},$$

e.g., the Dirac distribution as derivation of the step and sign functions $L^\infty(\mathbb{R}) \ni \vartheta, \epsilon \notin \mathcal{M}(\mathbb{R})$, or the Yukawa potential:

$$\begin{aligned} \mathbb{R} \text{ with } dx: \quad \tfrac{d}{dx}\vartheta(x) &= \tfrac{d}{dx}\tfrac{\epsilon(x)}{2} = \delta(x), \\ \tfrac{\mathbf{SO}(3) \vec{\times} \mathbb{R}^3}{\mathbf{SO}(3)} \cong \mathbb{R}^3 \text{ with } d^3x: \quad \tfrac{d}{dr^2} e^{-r} &= -\tfrac{e^{-r}}{2r}, \\ \tfrac{\mathbf{SO}_0(1,3) \vec{\times} \mathbb{R}^4}{\mathbf{SO}_0(1,3)} \cong \mathbb{R}^4 \text{ with } d^4x: \ (\tfrac{d}{dx^2})^N \vartheta(x^2) &= \delta^{(N-1)}(x^2),\ N = 1, 2. \end{aligned}$$

All the (generalized) function vector spaces and algebras considered have an *involution* with group inverse and conjugation:

$$\mu \leftrightarrow \hat{\mu} \text{ with } \hat{\mu}(g) = \overline{\mu(g^{-1})} \text{ for unimodular } G.$$

With a group representation $D: G \longrightarrow \mathbf{GL}(V)$, there may go a representation of the three convolution group algebras $A(G) \in \{\mathcal{C}_c(G),\ L^1(G),\ \mathcal{M}(G)\}$ in the endomorphism algebra $\mathbf{AL}(V)$ — for a function or a Radon distribution μ:

$$\begin{aligned} D: A(G) \longrightarrow \mathbf{AL}(V),\ \mu \longmapsto D(\mu) &= \int dg\, D(g)\mu(g), \\ D(\mu_1 * \mu_2) &= D(\mu_1) \circ D(\mu_2), \\ D(\delta_k) &= D(k). \end{aligned}$$

In the following, the unital convolution algebra $\mathcal{M}(G) \supset G$ with the Radon distributions (measures) and the embedded group operations, and its ideal the Lebesgue functions $L^1(G)$, and the unital abelian pointwise product algebra $L^\infty(G)$ with the essentially bounded functions play the most important roles. With the two-sided convolutive action of $\mathcal{M}(G)$ and $L^1(G)$ and the pointwise action of $L^\infty(G)$, these spaces will be used for the group G and the group dual $\check{G}$,

$*$	$\mathcal{M}(G)$	$L^1(G)$	$L^\infty(G)$
$\mathcal{M}(G)$	$\mathcal{M}(G)$	$L^1(G)$	$L^\infty(G)$
$L^1(G)$	$L^1(G)$	$L^1(G)$	$L^\infty(G)$
$L^\infty(G)$	$L^\infty(G)$	$L^\infty(G)$	–

$\cdot$	$L^\infty(G)$	$L^1(G)$	$\mathcal{M}(G)$
$L^\infty(G)$	$L^\infty(G)$	$L^1(G)$	$\mathcal{M}(G)$
$L^1(G)$	$L^1(G)$	–	–
$\mathcal{M}(G)$	$\mathcal{M}(G)$	–	–

Convolution product	Pointwise product
$\mu_1 * \mu_2(g)$	$\mu_1 \cdot \mu_2(g)$
from group product	for product representations
$G \times G \longrightarrow G$	$\mathbf{rep}G \times \mathbf{rep}G \longrightarrow \mathbf{rep}G$
$g_1 g_2 = g_{12}$	$D_1 \otimes D_2 = D_{12}$

The "divergences" of quantum field theory have their origin in undefined pointwise multiplications of Radon distributions (measures) $\mathcal{M}(\mathbb{R}^4) \cdot \mathcal{M}(\mathbb{R}^4)$ of spacetime translations.

9.12 Schur Product and Hilbert Spaces for Groups

For a vector space with group functions, the value at the neutral element, if defined, is a linear form (distribution):

$$\mathcal{F}(G) \ni f \longmapsto f(1_G) = \int dg\, \delta(g) f(g) = \langle \delta_{1_G}, f \rangle \in \mathbb{C},$$

normalized for $f(1_G) = 1$.

Two vector spaces are *put in duality by a bilinear form* $W \times V \ni (w, v) \longmapsto d(w, v) \in \mathbb{K}$. It defines a mapping to the linear forms $W \ni w \longmapsto w_d = d(w, \ ) \in V^T$ with $\langle w_d, v \rangle = d(w, v)$. For finite dimension, all duality structures are described by the algebraic dual V^T and the dual product $V^T \times V \ni (\theta, v) \longmapsto \langle \theta, v \rangle \in \mathbb{K}$, for dual bases $\langle \check{e}^a, e_b \rangle = \delta^a_b$. This is used also in the trace of an endomorphism $f : V \longrightarrow V$, $f = f^b_a e_b \otimes \check{e}^a$, $\mathrm{tr} f = f^a_a$. That is more complicated for infinite dimensions and continuous linear forms from the topological dual $V' \subseteq V^T$.

For group functions, the discrete summation over the components, e.g., in a dual product $\langle \theta, v \rangle = \sum_a \theta_a v^a$, is generalized to the Haar-integration over the group. The trace comes as integration of the group functions $\int dg\, \mu(g)$. The bilinear and sesquilinear products are generalized by the *Schur product*, which is the *convolution product $\mathcal{F}_1(G) \times \mathcal{F}_2(G)$ at the neutral element*, if defined,

$$\begin{aligned} \mathcal{F}_1(G) \times \mathcal{F}_2(G) \ni (\mu_1, \mu_2) \longmapsto \langle \mu_1, \mu_2 \rangle &= \mu_1 * \mu_2(1_G) = \int dg\, \mu_1(g^{-1}) \mu_2(g), \\ \langle \mu_1 | \mu_2 \rangle &= \hat{\mu}_1 * \mu_2(1_G) = \int dg\, \overline{\mu_1(g)} \mu_2(g). \end{aligned}$$

The Schur product, if defined, gives a positive product for function spaces:

$$\mathcal{F}(G) \times \mathcal{F}(G) \ni (\mu, \mu) \longmapsto \langle \mu | \mu \rangle = \int dg \, |\mu(g)|^2.$$

The Schur product with the group integration does not have to be a complex number as it is the case, e.g., for essentially bounded functions $L^\infty(G) \cdot L^\infty(G) = L^\infty(G)$. It may be a distribution, e.g., of the group dual $\check{G}$ with the Plancherel measure, as illustrated for $L^\infty(\mathbb{R}^n)$ by $\int d^n x \, e^{-iq_1 x} e^{iq_2 x} = \delta(\frac{q_1 - q_2}{2\pi})$ (more ahead).

9.12.1 Duality for Group Function Spaces

With the convolution of dual Lebesgue spaces valued in the essentially bounded functions, the dual product is given by the Schur product,

$$\begin{array}{l} \frac{1}{p} + \frac{1}{r} = 1, \; 1 < p, r < \infty : \; L^p(G)' = L^r(G), \text{ with } L^p(G) * L^r(G) \subseteq L^\infty(G) \\ L^p(G) \times L^r(G) \longrightarrow \mathbb{C}, \; \langle f_p, f_r \rangle = f_p * f_r(1_G) = \int dg \, f_p(g^{-1}) f_r(g). \end{array}$$

The space $L^2(G)$ with the square-integrable functions is self-dual.

The essentially bounded functions $L^\infty(G)$ constitute the dual space for the Lebesgue convolution algebra $L^1(G)$,

$$L^1(G)' = L^\infty(G), \text{ with } L^1(G) * L^\infty(G) \subseteq L^\infty(G).$$

The Radon distributions are dual to the compactly supported functions — their test functions. They are also in duality with the essentially bounded functions by $\langle \mu, d \rangle = \mu * d(1_G)$:

$$L^1(G) \subseteq \mathcal{M}(G) = \mathcal{C}_c(G)' \subseteq L^\infty(G)', \text{ with } \mathcal{M}(G) * L^\infty(G) \subseteq L^\infty(G).$$

9.12.2 Hilbert Metrics of Cyclic Representation Spaces

For a *cyclic* ("one orbit-based") *Hilbert representation* of a locally compact group G, the representation Hilbert space is the closure of the $\mathbb{C}$-span of the G-orbit $V = \overline{\mathbb{C}^{(G \bullet |\mathcal{C}\rangle)}}$ of a vector $|\mathcal{C}\rangle$, which is called a *cyclic vector*. With the fixgroup H of a cyclic vector, $G \bullet |\mathcal{C}\rangle \cong G/H$, one has $V \cong \overline{\mathbb{C}^{(G/H)}}$. All vectors $|\mathcal{C}'\rangle \in G \bullet |\mathcal{C}\rangle$ of the orbit are cyclic. An invariant vector $G \bullet |\mathcal{C}\rangle = |\mathcal{C}\rangle$ gives a one-dimensional trivial representation space $\mathbb{C}|\mathcal{C}\rangle \cong \mathbb{C}$. For example, the Fock state vector, cyclic for translations, is nondegenerate $\mathbb{R}^n \bullet |0\rangle = |0\rangle$. Or, each nontrivial vector $|\mathcal{C}\rangle \in \mathbb{R}^s$ with $\mathbf{SO}(s)$-action is cyclic with an orthogonal basis on the spherical ("cyclic") orbit $\mathbf{SO}(s)|\mathcal{C}\rangle \cong \mathbf{SO}(s)/\mathbf{SO}(s-1) \cong \Omega^{s-1}$.

Ground-state vectors in physics are cyclic; they determine cyclic G-representations. With a proper subgroup $H \neq G$ as fixgroup, e.g., $\mathbf{U}(1)_+ \subset \mathbf{U}(2)$ in the electroweak standard model or $\{1\} \subset \mathbf{U}(1)$ in superconductivity, the ground-state is *degenerate*; all vectors of the ground-state orbit, called *degeneracy manifold*, $|\mathcal{C}'\rangle \in G \bullet |\mathcal{C}\rangle \cong G/H$, are possible ground-state vectors.

The wave function conditions in quantum mechanics with their behavior for infinity and near by singular points are more generally considered in the framework of Lebesgue function spaces. All cyclic Hilbert representations spaces of a locally compact group can be constructed from the Lebesgue function algebra $L^1(G)$ with its dual (continuous linear forms) $L^\infty(G)$: A *positive-type function* or *metric-(scalar product-)inducing function* is an essentially bounded function $d \in L^\infty(G)$, which defines a positive product of the convolution algebra $L^1(G)$ with the absolute integrable function classes:

$$d \in L^\infty(G)_+ \iff \langle f|f\rangle_d = \int dg_1 dg_2\, \overline{f(g_1)} d(g_1^{-1}g_2) f(g_2) \geq 0 \quad \text{for all } f \in L^1(G).$$

A positive-type function is locally almost everywhere a continuous bounded function,

$$L^\infty(G)_+ \overset{dg}{=} \mathcal{C}_b(G)_+.$$

Connected to each positive-type function $d \in L^\infty(G)_+$ is a cyclic Hilbert space: The induced scalar product is the value of the convolution $L^1(G)*L^\infty(G)*L^1(G) \subseteq L^\infty(G)$ at the neutral group element $1_G \in G$:

$$L^1(G) \times L^1(G) \longrightarrow \mathbb{C},\ \langle f|f'\rangle_d = \langle \hat{f} * f'\rangle_d = \hat{f} * d * f'(1_G), \quad \text{with } \hat{f}(g) = \overline{f(g^{-1})}.$$

With this product, the algebra functions define a pre-Hilbert space and, by canonical Cauchy completion of the nontrivial norm classes $|f\rangle_d$, a Hilbert space $\overline{|L^1(G)\rangle_d}$. There exists a cyclic vector $|\mathcal{C}\rangle \in \overline{|L^1(G)\rangle_d}$ whose positive-type function is the expectation value $d(g) = \langle \mathcal{C}|g \bullet |\mathcal{C}\rangle$.

All diagonal matrix elements of a Hilbert representation $D: G \longrightarrow \mathbf{U}(V)$ define continuous positive-type functions,

$$0 \neq |v\rangle \in V,\ G \ni g \longmapsto d_v(g) = \langle v|D(g)|v\rangle,\ d_v \in \mathcal{C}_b(G)_+,$$

e.g., $\mathbb{R} \ni \psi \longmapsto d^{iq}(\psi) = e^{iq\psi} \in \mathbf{U}(1)$. Nontrivial positive-type functions are not invariant under group action.

Compact group examples are the diagonal elements in the $\mathbf{SU}(2)$- and $\mathbf{SO}(3)$-matrices:

$$D^{\frac{1}{2}}(u) = u = \begin{pmatrix} e^{i\frac{\chi+\varphi}{2}} \cos\frac{\theta}{2} & ie^{-i\frac{\chi-\varphi}{2}} \sin\frac{\theta}{2} \\ ie^{i\frac{\chi-\varphi}{2}} \sin\frac{\theta}{2} & e^{-i\frac{\chi+\varphi}{2}} \cos\frac{\theta}{2} \end{pmatrix} \in \mathbf{SU}(2),$$

$$D^1(u) = u \vee u = \begin{pmatrix} e^{i(\chi+\varphi)} \cos^2\frac{\theta}{2} & ie^{i\varphi}\frac{\sin\theta}{\sqrt{2}} & -e^{-i(\chi-\varphi)} \sin^2\frac{\theta}{2} \\ ie^{i\chi}\frac{\sin\theta}{\sqrt{2}} & \cos\theta & ie^{-i\chi}\frac{\sin\theta}{\sqrt{2}} \\ -e^{i(\chi-\varphi)} \sin^2\frac{\theta}{2} & ie^{-i\varphi}\frac{\sin\theta}{\sqrt{2}} & e^{-i(\chi+\varphi)} \cos^2\frac{\theta}{2} \end{pmatrix} \in \mathbf{SO}(3).$$

For $\mathbf{U}(1)$ as a simple explicit example, the irreducible representations give the positive type functions $d_z(\theta) = e^{zi\theta}$, $d_z \in L^\infty(\mathbf{U}(1))_+ \subseteq L^2(\mathbf{U}(1))$, $z \in \mathbb{Z}$. The scalar product is reduced to the corresponding harmonic components:

$$\begin{aligned} \langle f|f'\rangle_z &= \int_{-\pi}^{\pi} \int_{-\pi}^{\pi} \frac{d\theta_1}{2\pi}\frac{d\theta_2}{2\pi}\, \overline{f(\theta_1)} e^{zi(\theta_2-\theta_1)} f(\theta_2) = \overline{\tilde{f}}_z \tilde{f}'_z, \\ \text{with } \tilde{f}_z &= \int_{-\pi}^{\pi} \frac{d\theta}{2\pi}\, e^{zi\theta} f(\theta) \in \mathbb{C} \end{aligned}$$

as coefficients in the complex one-dimensional representation spaces $V_z \cong \mathbb{C}$.

According to Gelfand and Raikov,[15] there is *a surjection from the positive-type functions (representation metrics) to the equivalence classes of cyclic Hilbert representations*:

$$L^\infty(G)_+ \longrightarrow \mathbf{cycrep}_+ G \supseteq \check{G}.$$

In general, there exist different positive type functions for a cyclic representation. For an abelian group G, one obtains a bijection. The characterization of a cyclic Hilbert representation, finite or infinite dimensional, by only one function is a very strong and fruitful tool for the investigation of its structure.

Via the Schur product, the positive-type functions of a group $d \in L^\infty(G)_+$, carry a positive product $\langle d_1|d_2\rangle = \int dg\, \overline{d_1(g)} d_2(g)$, which is not number-valued for noncompact groups.

In general, an essentially bounded function $d \in L^\infty(G)$ defines a bilinear form on the Radon distributions, $\mathcal{M}(G) * L^\infty(G) * \mathcal{M}(G) = L^\infty(G)$:

$$\mathcal{M}(G) \times \mathcal{M}(G) \longrightarrow \mathbb{C},\ \omega * d * \omega'(1_G) = \int dg_1 dg_2\, \omega(g_1^{-1}) d(g_1^{-1} g_2) \omega'(g_2).$$

The essentially bounded functions $L^\infty(G)$ are ordered. The *cone* $L^\infty(G)_+$ with the positive-type functions is convex:

$$d, d' \in L^\infty(G)_+,\ \alpha, \alpha' \geq 0 \Rightarrow \alpha d + \alpha' d' \in L^\infty(G)_+.$$

The *representation normalization* of a positive-type functions at the neutral element, then called a *state*, is related to a quantum theoretical probability normalization,

$$G \ni 1_G \longmapsto d(1_G) = 1 = \langle \mathcal{C}|D(1_G)|\mathcal{C}\rangle = \langle \mathcal{C}|\mathcal{C}\rangle.$$

[15] Izrail Gelfand (1913–2009), Dmitrii Raikov (1905–).

The extremal continuous states, i.e., there exist only trivial cone combinations with states, $d = \alpha d' + (1-\alpha)d''$, $\alpha \geq 0$, with $\alpha \in \{0, 1\}$, are called *pure states*. They give the equivalence classes of the irreducible Hilbert representations. For an irreducible Hilbert representation, every nontrivial vector $|v\rangle \neq 0$ is cyclic.

A positive-type function is *reflection-symmetric*, i.e., *unitary*, and bounded by the neutral element value:

$$d \in L^\infty(G)_+ \Rightarrow \begin{cases} \quad d &= \hat{d} \iff \overline{d} = d^-, \text{ with } d^-(g) = d(g^{-1}),\ \overline{d}(g) = \overline{d(g)}, \\ |d(g)| &\leq d(1_G). \end{cases}$$

The unit function characterizes the trivial group representation. Functions of the inverse, i.e., conjugate functions, are for *dual (inverse-transposed) representations*. The pointwise product of two positive-type functions gives a positive-type function for the product representation:

$$L^\infty(G)_+ : \begin{cases} \text{trivial representation} & D = \mathbf{1}: & d^1 : G \ni g \longmapsto d^1(g) = 1, \\ \text{dual representation} & D \leftrightarrow \check{D}: & d \leftrightarrow d^- = \overline{d}, \\ & & \text{self-dual: } d = d^- = \overline{d}, \\ \text{sum representation} & D_1 \oplus D_2 : & (d_1 + d_2)(g) = d_1(g) + d_2(g), \\ \text{product representation} & D_1 \otimes D_2 : & (d_1 \cdot d_2)(g) = d_1(g) d_2(g). \end{cases}$$

Positive-type functions are the "continuous" extension of positive sesquilinear forms (scalar products) of finite-dimensional vector spaces:

$$\begin{aligned} d : V \times V \longrightarrow \mathbb{C}, \qquad\qquad & \langle v|w\rangle_d = \sum_{i_1,i_2} \overline{v}^{i_1} d_{i_1 i_2} w^{i_2}, \\ d : L^1(G) \times L^1(G) \longrightarrow \mathbb{C},\ & \langle f|f'\rangle_d = \int dg_1 dg_2\, \overline{f(g_1)} d(g_1^{-1} g_2) f'(g_2). \end{aligned}$$

Finite-dimensional scalar products can be unitarily diagonalized $d = u^\star \circ \operatorname{diag} d \circ u = \xi^\star \circ \xi$ with $u \in \mathbf{U}(n)$.

The familiar Hilbert spaces with square-integrable functions are included as follows: An absolute square of an L^2-function is a positive-type function:

$$L^2(G) * L^2(G) \subseteq L^\infty(G),\ L^2(G) \ni \xi \longmapsto d = \hat{\xi} * \xi \in L^\infty(G)_+.$$

If, and only if, a positive-type function is the absolute square of a square-integrable one, the Hilbert space can be constructed with square-integrable group functions:

$$\langle f|f'\rangle_d = \hat{f} * \hat{\xi} * \xi * f'(1_G) = \langle \xi * f|\xi * f'\rangle_\delta \text{ with } \xi * L^1(G) \subseteq L^2(G).$$

In general, the mapping above $L^2(G) \longrightarrow L^\infty(G)_+$ is not surjective; i.e., not all cyclic representations d can be characterized by a square-integrable G-function ξ_d.

9.13 Harmonic Analysis of Group Functions

A set with group action is the disjoint union of group orbits, $\biguplus_r G/H_r$, each characterized by a fixgroup $G \bullet x_r \cong G/H_r$. The linear extension: A Hilbert representation space is a direct sum $\bigoplus_r \overline{\mathbb{C}^{(G/H_r)}}$ of cyclic ones. Any irreducible representation is cyclic, but the converse is not true:

$$\underset{\bigoplus}{\mathbf{rep}G \supset \mathbf{cycrep}G} \underset{\oplus\int}{\supset} \mathbf{irrep}\,G \supseteq \mathbf{irrep}_+G = \check{G}.$$

A cyclic representation and, therefore, any Hilbert representation of a locally compact group is a direct integral ${}^{\oplus}\!\int d\check{g}$ of irreducible ones — even a direct sum for a compact group. Corresponding to a Haar measure of a locally compact group G, unique up to a constant factor, there is a positive *Plancherel measure* $d\check{g} \in \mathcal{M}(\check{G})_+$, ${}^{\oplus}\!\int_{\check{G}} d\check{g} = {}^{\oplus}\!\int d\check{g}$, of the group dual $\check{G} = \mathbf{irrep}_+G \subseteq L^\infty(G)_+$ with the equivalence classes of the irreducible Hilbert representations. It is defined by the biregular representation and unique for a given Haar measure, e.g., for the normalized one of a compact group. For a unimodular group, the measure renormalizations are inverse to each other:

$$\mathcal{M}(G)_+ \times \mathcal{M}(\check{G})_+ \ni (dg, d\check{g}) \to (e^\lambda dg, e^{-\lambda} d\check{g}).$$

The Plancherel measure is a measure of the group invariants, a discrete (continuous) set for a (non)compact group, which, for a Lie group, are generated by multilinear Lie algebra forms. Topologies of the group dual $\check{G}$ are not discussed here. The inverse Plancherel measure is the *Plancherel density (distribution)*. Group integration with Haar measure is used for the orthogonality of representations, group dual integration with Plancherel measure is used for the completeness of representations.

Familiar examples are the compact spin group $\mathbf{SU}(2)$ with normalized Haar measure $d^3u \in \mathcal{M}(\mathbf{SU}(2))_+$ and the associated Plancherel measure $d_J \in \mathcal{M}(\check{\mathbf{SU}}(2))_+$, a counting measure of the group dual with the representation dimensions:

$$\mathbf{irrep}\,\mathbf{SU}(2) = \{(J) \mid J = 0, \tfrac{1}{2}, 1 \ldots\} \ni (J) \longmapsto d_J = 1 + 2J \in \mathbb{N},$$

and the noncompact translation group with Haar measure $d^n x \in \mathcal{M}(\mathbb{R}^n)_+$, the energy-momenta as dual group, and associated Plancherel measure $d^n \frac{q}{2\pi} \in \mathcal{M}(\check{\mathbb{R}}^n)_+$ with Plancherel distribution $\delta(\frac{q}{2\pi})$. The familiar Fourier transformation relates to each other translation functions $L^1(\mathbb{R}^n)$ and energy-momentum functions $\mathcal{C}_0(\check{\mathbb{R}}^n)$, vanishing at infinity (more ahead):

$$\tilde{f}(q) = \int d^n x\, e^{-iqx}\, f(x) \text{ and } f(x) = \int d^n \tfrac{q}{2\pi}\, e^{iqx}\, \tilde{f}(q).$$

In general, the unitary *Fourier transformation* or *harmonic analysis* associates functions of the group dual, called *harmonic components*, with group function algebras $\mathbf{A}(G)$, acted on by the bi-regular representation:

$$\mathbf{F}: \mathbf{A}(G) \ni f \longmapsto \mathbf{F}f = \tilde{f} \in {}^{\oplus}\!\!\int d\check{g}\, \mathbf{AL}(V_{\check{g}}).$$

The Fourier transform is a conjugation-compatible algebra morphism, relating to each other convolution and pointwise product:

$$\begin{aligned} \mathbf{F}(f+f') &= \mathbf{F}f + \mathbf{F}f',\ \mathbf{F}(\alpha f) &&= \alpha\mathbf{F}f, \\ \mathbf{F}(f * f') &= \mathbf{F}f \cdot \mathbf{F}f', \quad \mathbf{F}(\overline{f}) &&= \overline{\mathbf{F}f}. \end{aligned}$$

It is injective, i.e., invertible on the image $\mathbf{F}(\mathbf{A}(G))$.

For compact groups, the full algebra $\mathbf{A}(G) = L^1(G)$ with the representation coefficients consists of Fourier transformable functions. For noncompact groups, which are second countable, unimodular and type I, a restriction to square integrable functions $\mathbf{A}(G) = L^1(G) \cap L^2(G)$ is necessary.

$\mathbf{F}f$ is valued in the direct integral (sum) of the endomorphism algebras $\mathbf{AL}(V_{\check{g}})$ with *unitary irreducible G-representations*, $G \ni g \longmapsto D^{\check{g}}(g) \in \mathbf{U}(V_{\check{g}})$ with $D^{\check{g}} \cong \check{g} \in \check{G}$ acting on the Hilbert space $V_{\check{g}}$. The Fourier transform $\tilde{f}(\check{g})$ is the coefficient (harmonic component) of this representation:

$$\check{G} \ni \check{g} \longmapsto \tilde{f}(\check{g}) = \int dg\, D^{\check{g}}(g^{-1}) f(g) = \langle \check{g} | f \rangle = D^{\check{g}} * f(1_G).$$

The group function f is decomposed with its Fourier integral. Its unit value is the integral with a trace over the irreducible components:

$$G \ni g \longmapsto f(g) = \int d\check{g}\ \mathrm{tr} D^{\check{g}}(g) \circ \tilde{f}(\check{g}),\ f(1_G) = \int d\check{g}\ \mathrm{tr}\tilde{f}(\check{g}).$$

This allows the formal decomposition of the Dirac distribution of the group:

$$\delta(g) = \int d\check{g}\ \mathrm{tr} D^{\check{g}}(g), \text{ e.g., } x \in \mathbb{R}^n : \delta(x) = \int \tfrac{d^n q}{(2\pi)^n} e^{iqx}.$$

The use of representation equivalence classes makes all this rather troublesome for nonabelian groups.

The Plancherel measure of a group dual is defined by and allows an orthogonal direct integral decomposition of the both-sided regular $G \times G$-representation, and the left- and right-regular G-representation,

$$\begin{aligned} L^2(G) &\cong {}^{\oplus}\!\!\int d\check{g}\, \mathbf{AL}(V_{\check{g}}),\ \mathbf{1}_{L^2(G)} \cong {}^{\oplus}\!\!\int d\check{g}\, D^{\check{g}} \otimes \overline{D}^{\check{g}}, \\ f = {}^{\oplus}\!\!\int dg\, g f(g) &= \tilde{f} = {}^{\oplus}\!\!\int d\check{g}\, D^{\check{g}} \otimes \tilde{f}(\check{g}) \text{ (where defined)}, \\ \text{right-, left-regular: } R &\cong {}^{\oplus}\!\!\int d\check{g}\, D^{\check{g}} \otimes \mathbf{1},\ \ L \cong {}^{\oplus}\!\!\int d\check{g}\, \mathbf{1} \otimes \overline{D}^{\check{g}}. \end{aligned}$$

The *Parseval formulas* for the scalar product and the *Fourier inversion* are

$$f_{1,2} \in \mathbf{A}(G): \quad \int dg\, \overline{f_1(g)} f_2(g) = \int d\check{g}\ \mathrm{tr}\overline{\tilde{f}_1(\check{g})} \circ \tilde{f}_2(\check{g}) = \langle f_1 | f_2 \rangle,$$
$$f(g) = \int d\check{g}\ \mathrm{tr} D^{\check{g}}(g) \circ \tilde{f}(\check{g}).$$

For noncompact groups, the inversion holds only for a subspace of functions.

A direct decomposition of a Hilbert representation D displays the *normalizations* (cardinal multiplicities for compact groups) of the irreducible components $D^{\check{g}}$ by a positive *spectral distribution* ρ_D *of the group dual* $\check{G}$:

$$D = {}^{\oplus}\!\int d\check{g}\, \rho_D(\check{g})\, D^{\check{g}},\ d\check{g}\, \rho_D(\check{g}) \in \mathcal{M}(\check{G})_+,\ D(1_G) = {}^{\oplus}\!\int d\check{g}\, \rho_D(\check{g}) = 1.$$

A positive-type function d for a cyclic representation is an integral of positive-type functions $d_{\check{g}}$ for its irreducible components, i.e., matrix elements for cyclic vectors $d_{\check{g}}(g) = \langle C | D^{\check{g}}(g) | C \rangle$, with a positive distribution $\tilde{d}$ of the group dual (Plancherel multiplicities for compact groups),

$$d(g) = \int d\check{g}\, \tilde{d}(\check{g})\, d_{\check{g}}(g),\ d(1_G) = \int d\check{g}\, \tilde{d}(\check{g})\, \langle C | C \rangle.$$

The Schur product for cyclic representations integrates their positive-type functions over the group:

$$L^\infty(G)_+ \times L^\infty(G)_+ \longrightarrow \mathcal{M}(\check{G}),$$
$$\{d_2 | d_1\} = \hat{d}_2 * d_1(1_G) = \int dg\, \overline{d_2(g)} d_1(g) = \int d\check{g}\ \mathrm{tr}\overline{\tilde{d}_2(\check{g})} \circ \tilde{d}_1(\check{g}).$$

For a compact group K with normalized Haar measure, the Plancherel measure is the counting measure $\bigoplus_{Z \in \check{K}} d_Z$ with the dimensions $d_Z = \dim_{\mathbb{C}} V_Z$ of the irreducible representation spaces, i.e., the number of columns or rows in the full matrix algebra $\mathbb{C}^{d_Z} \otimes \mathbb{C}^{d_Z}$. Plancherel integration is representation summation $\int d\check{g} = \sum_{Z \in \check{K}} d_Z$ (more ahead).

The Plancherel measure for a noncompact group has continuous support. For noncompact nonabelian groups, its determination is difficult. Its support is the reduced dual — it must not be the full group dual. If not, the group is called nonamenable; in such a case there exist irreducible Hilbert representations, not in the support of the Plancherel measure. With the continuous $\check{G}$-support of the Plancherel measure, orthogonality and completeness of Hilbert representations of a noncompact group use distributions of the group dual $\check{G}$.

9.14 Schur Orthogonality and Frobenius Reciprocity for Compact Groups

The Lebesgue function spaces for a compact group K are all subspaces of the convolution algebra with the absolute integrable function classes,

$$L^\infty(K) \subseteq L^p(K) \subseteq L^q(K) \subseteq L^1(K) \text{ for } \infty \geq p \geq q \geq 1.$$

All representations can be formulated as acting on square-integrable functions $L^2(K) \subseteq L^1(K)$. The positive-type functions are square-integrable $L^\infty(K) \subseteq L^2(K)$. There is the *Peter*[16]*–Weyl decomposition* of the square-integrable functions into the countably many finite-dimensional full $(d_Z \times d_Z)$-matrix algebras as irreducible group algebra representations with the columns (or rows) the irreducible Hilbert spaces $V_Z \cong \mathbb{C}^{d_Z}$:

$$\text{compact } K\colon \ L^2(K) \overset{\text{dense}}{=} \bigoplus_{Z \in \check{K}} \mathbf{AL}(d_Z, \mathbb{C}), \ \ \mathbf{AL}(d_Z, \mathbb{C}) \cong \mathbb{C}^{d_Z} \otimes \mathbb{C}^{d_Z}.$$

This is a special case of *Frobenius'*[17] *reciprocity* for the K-representation induced by the trivial representation of the trivial subgroup,

$$L^2(K) = L^2(K/\{1\}, \mathbb{C}) \overset{\text{dense}}{\cong} \bigoplus_{Z \in \check{K}} d_Z \mathbb{C}^{d_Z}.$$

Frobenius' reciprocity is a consequences of the squared matrix structure for the irreducible group algebra representations with biregular left-right group action $K \bullet \mathbf{AL}(d_Z, \mathbb{C}) \bullet K$. It makes use of rectangular submatrices: The K-representation on $L^2(K/H, V)$, induced by an irreducible subgroup $H \subseteq K$-representation D on a complex vector space V, decomposes into irreducible ones D^I. There arise all K-representations D^I which are suprepresentations of D. Their multiplicity n_I coincides with the multiplicity of the inducing H-representation D in D^I:

$$\begin{aligned} D_{\text{ind}}(K) &= \bigoplus_I n_I D^I(K), \\ D^I(H) &= n_I D(H) \oplus \ldots \Rightarrow D_{\text{ind}}(H) = \bigoplus_I [n_I^2 D(H) \oplus \ldots]. \end{aligned}$$

An example: The $\mathbf{SU}(3)$-representation on the functions $L^2(\mathbf{SU}(3)/\mathbf{SU}(2), \mathbb{C}^2)$, induced by a Pauli spinor $\mathbf{SU}(2)$-representation, decomposes into all nontrivial irreducible $\mathbf{SU}(3)$-representations $[n_1, n_2] \in \mathbb{N} \times \mathbb{N}$, $n_1 + n_2 = 1, 2, \ldots$ which contain Pauli spinor subrepresentations, with, e.g., no $\mathbf{SU}(3)$-singlet, one triplet, one

[16]Fritz Peter (1899–1949).

[17]Ferdinand Georg Frobenius (1849–1917).

antitriplet, one sextet, one antisextet, twice the octet-representation [1, 1], since an octet contains two Pauli spinor **SU**(2)-representations [1] on $\mathbb{C}^2$, etc.

The algebras $\mathbf{AL}(d_Z, \mathbb{C}) \cong \mathbb{C}^{d_Z} \otimes \mathbb{C}^{d_Z}$ are *Schur-orthogonal* to each other with respect to the *Schur scalar product*, defined with normalized Haar group measure $du \in \mathcal{M}(K)_+$,

$$\begin{aligned} D^Z : K &\longrightarrow \mathbf{U}(d_Z), \\ \{D^{Z'}|D^Z\} &= \hat{D}^{Z'} \circledast D^Z(1_K) = \int_K du\, \overline{D^{Z'}(u)} \otimes D^Z(u) = \delta_{ZZ'} \tfrac{1}{d_Z} \mathbf{1}_{d_Z} \otimes \mathbf{1}_{d_Z}. \end{aligned}$$

More explicit: Two representation matrix elements of inequivalent irreducible Hilbert representations are orthogonal, with $v_Z, w_Z \in V_Z$:

$$\int_K du\, \overline{\langle v_{Z'}|D^{Z'}(u)|w_{Z'}\rangle} \langle v_Z|D^Z(u)|w_Z\rangle = \begin{cases} 0, & D^Z \not\cong D^{Z'}, \\ \frac{1}{d_Z} \langle v_Z|v_{Z'}\rangle\langle w_{Z'}|w_Z\rangle, & D^Z = D^{Z'}. \end{cases}$$

The Plancherel measure $d_Z \in \mathcal{M}(\check{K})_+$ of an irreducible representation D^Z of a compact group with normalized Haar measure is a counting measure, given by the number (Plancherel multiplicity) d_Z of equivalent irreducible representations $\mathbb{C}^{d_Z}$ in one full matrix algebra $\mathbb{C}^{d_Z} \otimes \mathbb{C}^{d_Z}$, which coincides with its dimension. The Schur orthogonality involves the Plancherel density (inverse Plancherel measure).

The Schur orthogonality for representations is conceptually different from a possible orthonormality of the vectors in one representation space, e.g., for a Hilbert basis $\{e^j \mid j = 1, \ldots, d_Z\}$,

$$\langle e^i|e^j\rangle = \delta^{ij} \Rightarrow \int_K du\, \overline{D^{Z'}(u)^{j'}_{i'}}\, D^Z(u)^j_i = \delta_{ZZ'} \tfrac{1}{d_Z} \delta_{ii'}\delta^{jj'}.$$

The Schur orthogonality is exemplified for **SU**(2) with Plancherel measure $d_J = 1 + 2J$ by the positive-type square-integrable functions arising as diagonal matrix elements of the two- and three-dimensional representation:

$$\int d^3u = \int_{-2\pi}^{2\pi} \tfrac{d\chi}{4\pi} \int_0^{2\pi} \tfrac{d\varphi}{2\pi} \int_{-1}^{1} \tfrac{d\cos\theta}{2} = 1, \quad \begin{cases} \int d^3u\, |e^{i\frac{\chi+\varphi}{2}} \cos\frac{\theta}{2}|^2 = \frac{1}{2}, \\ \int d^3u\, |e^{i(\chi+\varphi)} \cos^2\frac{\theta}{2}|^2 = \frac{1}{3}, \\ \int d^3u\, e^{-i(\chi+\varphi)} \cos^2\frac{\theta}{2}\, e^{i\frac{\chi+\varphi}{2}} \cos\frac{\theta}{2} = 0. \end{cases}$$

Positive-type functions, normalized as states do not have to be Schur-normalized. *Representation normalization*, used as $\langle v|D(1_G)|v\rangle = \langle v|v\rangle$ for state vectors $v \in V$, differs from *Schur normalization*, used as $\{D'|D\} = \int dg\, \overline{D'}(g) \otimes D(g)$ for operations $D(g) \in V \otimes V^T$.

A familiar example is given by the irreducible $(1 + 2L)$-dimensional representations (L) of the rotations **SO**(3), with the Rodrigues[18] formula for the matrix elements:

---

[18]Olinde Rodrigues (1795–1851).

$$\langle L; k|D^L(\chi, \varphi, \theta)|L; m\rangle = \frac{i^{m-k}e^{i(k\chi+m\varphi)}}{2^L}\sqrt{\frac{(L+m)!}{(L-k)!(L+k)!(L-m)!}}\frac{(1-\zeta)^{\frac{k-m}{2}}}{(1+\zeta)^{\frac{k+m}{2}}}(-\frac{d}{d\zeta})^{L-m}\frac{(1+\zeta)^{k+L}}{(1-\zeta)^{k-L}}\Big|_{\zeta=\cos\theta}.$$

The $(m, k) = (0, 0)$-components, i.e., the central elements in the $(1+2L)\times(1+2L)$-matrices, e.g., in the $\mathbf{SO}(3)$-matrix $D^1(u)$ above, are the *Legendre polynomials*. They are representation-normalized positive-type functions, $\mathrm{P}^L \in L^\infty(\mathbf{SO}(3))_+$:

$$\begin{aligned}\langle L; 0|D^L(\chi, \varphi, \theta)|L; 0\rangle = \mathrm{P}^L(\cos\theta) &= \tfrac{1}{2^L L!}(-\tfrac{\partial}{\partial\cos\theta})^L \sin^{2L}\theta \\ = \sum_{k=0}^{L}\frac{(L+k)!}{(L-k)!}\frac{(-\sin^2\frac{\theta}{2})^k}{(k!)^2} &= \begin{cases} 1, & L=0,\\ 1-2\sin^2\frac{\theta}{2} = \cos\theta, & L=1,\\ 1-6\sin^2\frac{\theta}{2}+6\sin^4\frac{\theta}{2} & \\ \quad = \frac{3}{2}(\cos^2\theta-\frac{1}{3}), & L=2,\\ \frac{5}{2}(\cos^3\theta-\frac{3}{5}\cos\theta), & L=3,\\ \ldots. & \end{cases}\end{aligned}$$

and define the scalar product for the Hilbert spaces $\overline{|L^1(\mathbf{SO}(3))\rangle}_{\mathrm{P}^L} \cong \mathbb{C}^{1+2L}$. The Legendre polynomials are Schur-orthogonal for different angular momenta L and have as Schur norm the inverse $\frac{1}{1+2L}$ of the dimension of the representation space, i.e., the Plancherel density:

$$\begin{aligned}\text{positive-type function: } |L\}(\zeta) &= \mathrm{P}^L(\zeta),\ \zeta = \cos\theta,\\ \text{representation-normalized: } \mathrm{P}^L(1) &= 1,\\ \text{Schur product: } \{L'|L\} &= \int_{-1}^{1}\tfrac{d\zeta}{2}\,\mathrm{P}^{L'}(\zeta)\mathrm{P}^L(\zeta) = \delta^{LL'}\tfrac{1}{1+2L},\\ \text{completeness: } &\sum_{L=0}^{\infty}(1+2L)\mathrm{P}^L(\zeta)\mathrm{P}^L(\zeta') = \delta(\tfrac{\zeta-\zeta'}{2}).\end{aligned}$$

The transition from the central column (or row) of the $(1+2L)\times(1+2L)$-matrices to the spherical harmonics involves a renormalization with the square root of the quotient of sphere surface and Plancherel measure:

$$\begin{aligned}\langle L; 0|D^L(\chi, \varphi, \theta)|L; m\rangle &= \tfrac{i^m e^{im\varphi}}{2^L L!}\sqrt{\tfrac{(L+m)!}{(L-m)!}}\tfrac{1}{\sin^m\theta}(-\tfrac{d}{d\cos\theta})^{L-m}\sin^{2L}\theta\\ &= \sqrt{\tfrac{|\Omega^2|}{1+2L}}\,\mathrm{Y}^L_m(\varphi,\theta) \overset{m=0}{=} \mathrm{P}^L(\cos\theta),\\ \int d^2\omega\, \overline{\mathrm{Y}^{L'}_{m'}(\varphi,\theta)}\mathrm{Y}^L_m(\varphi,\theta) &= \delta^{LL'}\delta_{mm'},\ \int d^2\omega = \int_0^{2\pi}d\varphi\int_{-1}^{1}d\cos\theta,\\ \sum_{L=0}^{\infty}\sum_{m=-L}^{L}\overline{\mathrm{Y}^L_m(\varphi',\theta')}\mathrm{Y}^L_m(\varphi,\theta) &= \delta(\vec{\omega}-\vec{\omega}') = \delta(\cos\theta-\cos\theta')\delta(\varphi-\varphi').\end{aligned}$$

The spherical harmonics $\{\mathrm{Y}^L_m \,\big|\, m = -L, \ldots, L\}$ are an orthonormal basis for the irreducible $\mathbf{SO}(3)$-representation space with angular momentum L. The representation properties of the spherical harmonics and the Legendre polynomials give the factorizations as used in multipole expansions with the cosine definition for $L = 1$:

$$\frac{\vec{x}}{|\vec{x}|}\frac{\vec{y}}{|\vec{y}|} = \cos\theta,$$

$$\frac{4\pi}{1+2L}\sum_{m=-L}^{L}\overline{\mathrm{Y}^L_m(\frac{\vec{x}}{|\vec{x}|})\mathrm{Y}^L_m(\frac{\vec{y}}{|\vec{y}|})} = \sum_{m=-L}^{L} D^L_{0m}(\frac{\vec{x}}{|\vec{x}|})D^L_{m0}(\frac{\vec{y}}{|\vec{y}|}) = D^L_{00}(\frac{\vec{x}\vec{y}}{|\vec{x}||\vec{y}|}) = \mathrm{P}^L(\cos\theta).$$

A notational remark: A positive type function for a representation will be written as $|I\}$ with the invariants, e.g., $|J\}$ where $J = 0, \frac{1}{2}, 1, \ldots$ for the group $\mathbf{SU}(2)$, and Schur product $\{I'|I\}$ for representations. The eigenvectors $\{|I; w\rangle \bigm| w\}$ which span the corresponding representation space, display the representation characterizing invariants I before the semicolon, the eigenvalues w after it, e.g., the eigenvectors $\{|J; m\rangle \bigm| J = -m, \ldots, m\}$ for each invariant J, with Hilbert product $\langle I; w'|I; w\rangle$ for the eigenvectors of this representation.

For finite groups, both kinds of Hilbert bases of the group algebra, the canonical one and the harmonic ones, are finite. For compact groups K in general, e.g., for $\mathbf{U}(n)$, the group elements are a Haar-measure-related distributive basis; the harmonic Hilbert bases remain discrete. The Fourier expansion (decomposition of the two-sided regular $K \times K$-representation, harmonic analysis) with a discrete harmonic basis $\{\sqrt{d_Z}|Z^a_b\} \bigm| a, b = 1, \ldots, d_Z,\ Z \in \check{K}\}$ is described by the orthogonal Peter-Weyl decomposition of a square integrable function in $L^2(K) = \overline{+}\!\!\int du\ |u\}\,\mathbb{C}$:

$$\begin{aligned}
|f\} = {}^{\oplus}\!\!\int du\ |u\}f(u) &= \bigoplus_{Z\in\check{K}} d_Z|\tilde{f}(Z)\} = \bigoplus_{Z\in\check{K}} d_Z|Z^a_b\}\tilde{f}(Z)^b_a,\\
\text{function values: } f(u) &= \{u|f\} \qquad = \sum_{Z\in\check{K}} d_Z Z(u)^a_b \tilde{f}(Z)^b_a,\\
\text{harmonic coefficients: } \tilde{f}(Z)^b_a &= \{Z^b_a|f\} \qquad = \int du\ \overline{Z(u)^a_b} f(u),\\
\text{Plancherel unitarity: } \{f|f\} &= \int_K du\ \overline{f(u)}f(u) = \sum_{Z\in\check{K}_+} d_Z \overline{\tilde{f}(Z)^a_b}\tilde{f}(Z)^a_b.
\end{aligned}$$

The canonical distributive basis $\{|u\} \bigm| u \in K\}$, which is a proper Hilbert basis only for a finite group, has a harmonic expansion and vice versa:

$$\mathbf{1}_{L^2(K)} \cong {}^{\oplus}\!\!\int du\ |u\}\{u| = \bigoplus_{Z\in\check{K}} d_Z|Z^b_a\}\{Z^b_a| \Rightarrow \begin{cases} |u\} = \bigoplus_{Z\in\check{K}} d_Z|Z^b_a\}\overline{Z(u)^a_b},\\ |Z^a_b\} = {}^{\oplus}\!\!\int du\ |u\}Z(u)^a_b,\\ Z(u)^a_b = \{u|Z^b_a\} = \langle a|Z(u)b\rangle. \end{cases}$$

It involves the normalized Haar measure of the group and the Plancherel measure of the group dual (invariants of the irreducible representations), which counts the multiplicity of irreducible group K-representation vector spaces, with either left or right K-action, in an irreducible algebra $L^2(K)$-representation with left-right $K \times K$-action.

Examples are the Fourier series for square integrable $\mathbf{U}(1)$- and $\mathbf{SU}(2)$-functions:

$$L^2(\mathbf{U}(1)) \overset{\text{dense}}{\cong} \bigoplus_{Z\in\mathbb{Z}} \mathbf{A}_Z, \quad \mathbf{A}_Z \cong \mathbb{C}, \quad \text{with } |u\} = |\theta\} : \begin{cases} |f\} &= {}^{\oplus}\int_{-\pi}^{\pi} \frac{d\theta}{2\pi} |\theta\} f(\theta) = \bigoplus_{Z\in\mathbb{Z}} |Z\} \tilde{f}(Z), \\ e^{Zi\theta} &= \{\theta|Z\},\ e^{-Zi\theta} = \{Z|\theta\}, \\ f(\theta) &= \{\theta|f\} = \sum_{Z\in\mathbb{Z}} e^{Zi\theta} \tilde{f}(Z), \\ \tilde{f}(Z) &= \{Z|f\} = \int_{-\pi}^{\pi} \frac{d\theta}{2\pi} e^{-Zi\theta} f(\theta), \end{cases}$$

$$L^2(\mathbf{SU}(2)) \overset{\text{dense}}{\cong} \bigoplus_{2J\in\mathbb{N}} \mathbf{A}_{1+2J}, \quad \mathbf{A}_{1+2J} \cong \mathbf{AL}(1+2J,\mathbb{C}), \quad \text{e.g., with } |u\} = |\chi,\varphi,\cos\theta\} : \begin{cases} |f\} &= {}^{\oplus}\int d^3u\, |u\} f(u) \\ &= \bigoplus_{2J\in\mathbb{N}} (1+2J)|J^a_b\} \tilde{f}(J)^b_a, \\ J(u)^a_b &= \{u|J\}^a_b \in \mathbf{U}(1+2J), \\ & \quad \text{(representation matrix)}, \\ f(u) &= \{u|f\} = \sum_{2J\in\mathbb{N}} (1+2J)(J)(u)^a_b \tilde{f}(J)^b_a, \\ \tilde{f}(J)^b_a &= \{J^b_a|f\} = \int d^3u\, \overline{(J)(u)^a_b} f(u). \end{cases}$$

The irreducible representations decompose the Haar measure associate Dirac measure of the group:

$$\begin{aligned} u, v \in K: \quad f(u) &= \int dv\, \{u|v\} f(v) = \sum_{Z\in\check{K}} d_Z \int dv\ \mathrm{tr}[Z(u) \circ Z(v^{-1})] f(v), \\ \{u|v\} = \delta(vu^{-1}) &= \sum_{Z\in\check{K}} d_Z\, \mathrm{tr}Z(v) \circ Z(u^{-1}), \\ \delta(u) &= \sum_{Z\in\check{K}} d_Z\, \mathrm{tr}Z(u),\ \delta(1_K) = \sum_{Z\in\check{K}} d_Z^2 \text{ (if defined)}, \end{aligned}$$

with the examples

$$\mathbf{U}(1): \delta(\tfrac{\theta-\varphi}{2\pi}) = \sum_{Z\in\mathbb{Z}} e^{Zi\theta} e^{-Zi\varphi},\ \delta(\tfrac{\theta}{2\pi}) = \sum_{Z\in\mathbb{Z}} e^{Zi\theta},$$

$$\mathbf{SU}(2): \begin{cases} \delta(vu^{-1}) &= \sum_{2J\in\mathbb{N}} (1+2J)\, \mathrm{tr}(J)(v) \circ (J)(u^{-1}), \\ \delta(\frac{\chi}{4\pi})\delta(\frac{\varphi}{2\pi})\delta(\frac{\cos\theta}{2}) &= \sum_{2J\in\mathbb{N}} (1+2J)\, \mathrm{tr}(J)(\chi,\varphi,\theta). \end{cases}$$

The convolution of square integrable functions is the multiplication of the harmonic matrices in the Fourier algebra automorphism:

$$\begin{aligned} L^2(K) \leftrightarrow L^2(K),\ f \leftrightarrow \tilde{f} \text{ with } \widetilde{f_1 * f_2} &= \tilde{f}_1 \circ \tilde{f}_2, \\ \text{for all } Z\colon \{Z|f_1 * f_2\} &= \{Z|f_1\} \circ \{Z|f_2\}, \\ \text{e.g., } \mathbf{U}(1): (f_1 * f_2)(\theta) &= \sum_{Z\in\mathbb{Z}} \tilde{f}_1(Z) \tilde{f}_2(Z) e^{Zi\theta}. \end{aligned}$$

9.15 Translation Representations

In the following, the structures for locally compact groups, as given above, are specialized to the noncompact abelian translation groups $\mathbb{R}^n$, e.g., to a Cartan plane A in a semisimple Lie group $G = K \circ A \circ N$ or to translations in an affine group $G \vec{\times} \mathbb{R}^n$. The (energy-)momenta constitute the group dual with the equivalence classes of the irreducible Hilbert representations $x \longmapsto e^{iqx} = \langle x|q\rangle \in \mathbf{U}(1)$. The translation group has the distinction of being isomorphic to its group dual $\check{\mathbb{R}}^n$, an abelian group,

$$x \in \mathbb{R}^n \cong \check{\mathbb{R}}^n = (i\mathbb{R})^n \ni iq.$$

All faithful Hilbert representations of the translations $\mathbb{R}$ are infinite-dimensional.

9.15.1 *Fourier Transformation*

The harmonic components of a translation distribution, its *Fourier components*, are given by the Fourier transform:

$$\begin{aligned}\langle q|\mu\rangle = \tilde{\mu}(q) = \int \tfrac{d^n x}{N}\, e^{-iqx}\mu(x) = \int \tfrac{d^n x}{N}\, \langle q|x\rangle\langle x|\mu\rangle,\\ \int \tfrac{Nd^n q}{(2\pi)^n}\, |\tilde{\mu}(q)|^2 = \int \tfrac{d^n x}{N}\, |\mu(x)|^2, \text{ where defined.}\end{aligned}$$

If the Haar measure of the translations $\mathbb{R}^n$ is normalized as $\frac{d^n x}{N}$ with a free normalization factor $N > 0$, e.g., $N = 1$ or $N = (2\pi)^n$ or $N = (2\pi)^{\frac{n}{2}}$, the corresponding Plancherel measure of the (energy-)momenta $\check{\mathbb{R}}^n$ is $\frac{Nd^n q}{(2\pi)^n}$.

The Fourier transforms of the three relevant spaces $L^1(\mathbb{R}^n)$, $\mathcal{M}(\mathbb{R}^n)$, and $L^\infty(\mathbb{R}^n)$ are as follows: According to a theorem of Lebesgue, the Fourier transformation of the convolution algebra $L^1(\mathbb{R}^n)$ is an injective algebra morphism, with a dense range, but not surjective, into the continuous functions $\mathcal{C}_0(\check{\mathbb{R}}^n)$, vanishing at infinity. The Fourier transformation can be extended to the Radon distribution convolution algebra $\mathcal{M}(\mathbb{R}^n)$ with values in the bounded continuous functions $\mathcal{C}_b(\check{\mathbb{R}}^n)$. Positive Radon measures with spectral energy-momentum distributions and the continuous positive-type functions are bijective (*Bochner's*[19] *theorem*):

$$\text{Fourier:} \begin{cases} \widetilde{L^p(\mathbb{R}^n)} \subseteq L^r(\check{\mathbb{R}}^n),\ \frac{1}{p} + \frac{1}{r} = 1,\ 1 \le p \le 2,\ \infty \ge r \ge 2,\\ \widetilde{L^1(\mathbb{R}^n)} = \dot{\mathcal{C}}_0(\check{\mathbb{R}}^n) \text{ dense in } \mathcal{C}_0(\check{\mathbb{R}}^n) \text{ (Lebesgue)},\\ \widetilde{\mathcal{M}(\mathbb{R}^n)} = \dot{\mathcal{C}}_b(\check{\mathbb{R}}^n) = \text{complex span of} \mathcal{C}_b(\check{\mathbb{R}}^n)_+,\\ \widetilde{\mathcal{M}(\mathbb{R}^n)_+} = \mathcal{C}_b(\check{\mathbb{R}}^n)_+ \overset{d^n q}{=} L^\infty(\check{\mathbb{R}}^n)_+ \quad \text{(Bochner)}.\end{cases}$$

The convolution and pointwise product are Fourier-compatible (with convenient measure normalization $N = 1$),

[19]Salomon Bochner (1899–1982).

$$\mu(x) = \int \frac{d^n q}{(2\pi)^n}\, \tilde{\mu}(q) e^{iqx},\ \tilde{\mu}(q) = \int d^n x\, \mu(x) e^{-iqx},\ \delta(x) = \int \frac{d^n q}{(2\pi)^n} e^{iqx},$$

$$\tilde{\tilde{\mu}} = \mu,\ \widetilde{\mu_1 \cdot \mu_2} = \tilde{\mu}_1 * \tilde{\mu}_2,\ \begin{cases} \mu_1 \cdot \mu_2(x) = \int \frac{d^n q}{(2\pi)^n}\, \tilde{\mu}_1 * \tilde{\mu}_2(q) e^{iqx}, \\ \tilde{\mu}_1 * \tilde{\mu}_2(q) = \int \frac{d^n p}{(2\pi)^n}\, \tilde{\mu}_1(p) \tilde{\mu}_2(q-p). \end{cases}$$

For the representation relevant spaces, the products are exchanged in the spaces with the harmonic components:

$*$	$\mathcal{M}(\mathbb{R}^n)$	$L^1(\mathbb{R}^n)$	$\dot{\mathcal{C}}_b(\mathbb{R}^n)$
$\mathcal{M}(\mathbb{R}^n)$	$\mathcal{M}(\mathbb{R}^n)$	$L^1(\mathbb{R}^n)$	$\dot{\mathcal{C}}_b(\mathbb{R}^n)$
$L^1(\mathbb{R}^n)$	$L^1(\mathbb{R}^n)$	$L^1(\mathbb{R}^n)$	$\dot{\mathcal{C}}_b(\mathbb{R}^n)$
$\dot{\mathcal{C}}_b(\mathbb{R}^n)$	$\dot{\mathcal{C}}_b(\mathbb{R}^n)$	$\dot{\mathcal{C}}_b(\mathbb{R}^n)$	–

$G = \mathbb{R}^n$

$\cdot$	$\dot{\mathcal{C}}_b(\mathbb{R}^n)$	$L^1(\mathbb{R}^n)$	$\mathcal{M}(\mathbb{R}^n)$
$\dot{\mathcal{C}}_b(\mathbb{R}^n)$	$\dot{\mathcal{C}}_b(\mathbb{R}^n)$	$L^1(\mathbb{R}^n)$	$\mathcal{M}(\mathbb{R}^n)$
$L^1(\mathbb{R}^n)$	$L^1(\mathbb{R}^n)$	–	–
$\mathcal{M}(\mathbb{R}^n)$	$\mathcal{M}(\mathbb{R}^n)$	–	–

$\mu_1 * \mu_2(x)$ **F** $\mu_1 \cdot \mu_2(x)$

$\updownarrow$ Fourier $\updownarrow$

$\cdot$	$\dot{\mathcal{C}}_b(\check{\mathbb{R}}^n)$	$\dot{\mathcal{C}}_0(\check{\mathbb{R}}^n)$	$\mathcal{M}(\check{\mathbb{R}}^n)$
$\dot{\mathcal{C}}_b(\check{\mathbb{R}}^n)$	$\dot{\mathcal{C}}_b(\check{\mathbb{R}}^n)$	$\dot{\mathcal{C}}_0(\check{\mathbb{R}}^n)$	$\mathcal{M}(\check{\mathbb{R}}^n)$
$\dot{\mathcal{C}}_0(\check{\mathbb{R}}^n)$	$\dot{\mathcal{C}}_0(\check{\mathbb{R}}^n)$	$\dot{\mathcal{C}}_0(\check{\mathbb{R}}^n)$	$\mathcal{M}(\check{\mathbb{R}}^n)$
$\mathcal{M}(\check{\mathbb{R}}^n)$	$\mathcal{M}(\check{\mathbb{R}}^n)$	$\mathcal{M}(\check{\mathbb{R}}^n)$	–

$\check{G} = \check{\mathbb{R}}^n$

$*$	$\mathcal{M}(\check{\mathbb{R}}^n)$	$\dot{\mathcal{C}}_0(\check{\mathbb{R}}^n)$	$\dot{\mathcal{C}}_b(\check{\mathbb{R}}^n)$
$\mathcal{M}(\check{\mathbb{R}}^n)$	$\mathcal{M}(\check{\mathbb{R}}^n)$	$\dot{\mathcal{C}}_0(\check{\mathbb{R}}^n)$	$\dot{\mathcal{C}}_b(\check{\mathbb{R}}^n)$
$\dot{\mathcal{C}}_0(\check{\mathbb{R}}^n)$	$\dot{\mathcal{C}}_0(\check{\mathbb{R}}^n)$	–	–
$\dot{\mathcal{C}}_b(\check{\mathbb{R}}^n)$	$\dot{\mathcal{C}}_b(\check{\mathbb{R}}^n)$	–	–

$\tilde{\mu}_1 \cdot \tilde{\mu}_2(q)$ from group product

$\tilde{\mu}_1 * \tilde{\mu}_2(q)$ for product representations

The Fourier transformation defines algebra isomorphisms:

$$\mathcal{M}(\mathbb{R}^n) \overset{\mathbf{F}}{\cong} \dot{\mathcal{C}}_b(\check{\mathbb{R}}^n) \overset{d^n q}{=} L^\infty(\check{\mathbb{R}}^n),\ L^1(\mathbb{R}^n) \overset{\mathbf{F}}{\cong} \dot{\mathcal{C}}_0(\check{\mathbb{R}}^n).$$

Dual and sesquilinear products use the convolution product at the trivial translation or at the trivial (energy-)momentum as integrals over all translations or over all (energy-)momenta. For a sesquilinear form, the conjugated (generalized) function is used:

$$\begin{aligned} \langle \mu_1, \mu_2 \rangle &= \int d^n x\, \mu_1(-x) \mu_2(x) = \mu_1 * \mu_2(0) \\ &= \int \frac{d^n q}{(2\pi)^n}\, \tilde{\mu}_1(q) \tilde{\mu}_2(q) \ = \tilde{\mu}_1^- \underset{(2\pi)^n}{*} \tilde{\mu}_2(0), \\ \langle \mu_1 | \mu_2 \rangle &= \langle \overline{\mu}_1, \mu_2 \rangle,\ \text{with } \hat{\mu}(x) = \overline{\mu^-(x)} = \overline{\mu(-x)} = \int \frac{d^n q}{(2\pi)^n}\, \overline{\tilde{\mu}(q)} e^{iqx}. \end{aligned}$$

9.15.2 Cyclic Translation Representations

The continuous translation positive-type functions, isomorphic to the positive (energy-)momentum Radon distributions (densities), characterize all cyclic translation representations in the bijection

$$\mathcal{C}_b(\mathbb{R}^n)_+ \stackrel{d^nq}{=} L^\infty(\mathbb{R}^n)_+ \cong \mathcal{M}(\check{\mathbb{R}}^n)_+ \leftrightarrow \mathbf{cycrep}_+\mathbb{R}^n.$$

The Hilbert product induced by a function positive-type function,

$$\mathcal{C}_b(\mathbb{R}^n)_+ \ni d \leftrightarrow \tilde{d} \in \mathcal{M}(\check{\mathbb{R}}^n)_+, \; d(x) = \int \frac{d^nq}{(2\pi)^n} \tilde{d}(q)\, e^{iqx},$$

can be transformed into an integration of the pointwise product of the harmonic components with a representation characteristic positive (energy-)momentum spectral measure $\tilde{d}(q)\frac{d^nq}{(2\pi)^n}$, which involves a positive Radon distribution,

$$\begin{aligned} L^1(\mathbb{R}^n) \underset{d}{*} L^1(\mathbb{R}^n) &\longrightarrow \mathbb{C} \stackrel{\mathbf{F}}{\leftrightarrow} \dot{\mathcal{C}}_0(\check{\mathbb{R}}^n) \underset{\tilde{d}}{\cdot} \dot{\mathcal{C}}_0(\check{\mathbb{R}}^n) \longrightarrow \mathbb{C}, \\ \langle f|f'\rangle_d &= \int d^nx_1 d^nx_2\, \overline{f(x_1)} d(x_2 - x_1) f'(x_2) = \hat{f} * d * f'(0) \\ &= \int \frac{d^nq}{(2\pi)^n} \overline{\tilde{f}(q)}\, \tilde{d}(q)\, \tilde{f}'(q) = \overline{\tilde{f}} \cdot \tilde{d} \cdot \tilde{f}'(\check{\mathbb{R}}^n). \end{aligned}$$

A finite product gives a finite Radon distribution $\overline{\tilde{f}} \cdot \tilde{d} \cdot \tilde{f}'$ of the (energy-)momenta.

The product can be extended to the Radon distributions and their Fourier-transformed bounded (energy-)momentum functions

$$\begin{aligned} \mathcal{M}(\mathbb{R}^n) \underset{d}{*} \mathcal{M}(\mathbb{R}^n) &\longrightarrow \mathbb{C} \stackrel{\mathbf{F}}{\leftrightarrow} \dot{\mathcal{C}}_b(\check{\mathbb{R}}^n) \underset{\tilde{d}}{\cdot} \dot{\mathcal{C}}_b(\check{\mathbb{R}}^n) \longrightarrow \mathbb{C}, \\ \langle \omega|\omega'\rangle_d &= \int \frac{d^nq}{(2\pi)^n} \overline{\tilde{\omega}(q)}\, \tilde{d}(q)\, \tilde{\omega}'(q). \end{aligned}$$

The transition to a Hilbert space with square-integrable $\mathbb{R}^n$-functions requires the positive-type function to be an absolute square of a square-integrable function:

$$\begin{aligned} d = \hat{\xi} * \xi \iff \tilde{d} = |\tilde{\xi}|^2 \text{ with } L^2(\mathbb{R}^n) \ni \xi \leftrightarrow \tilde{\xi} \in L^2(\check{\mathbb{R}}^n), \\ \xi * L^1(\mathbb{R}^n) \subseteq L^2(\mathbb{R}^n) \leftrightarrow \tilde{\xi} \cdot \dot{\mathcal{C}}_0(\check{\mathbb{R}}^n) \subseteq L^2(\check{\mathbb{R}}^n). \end{aligned}$$

The Hilbert space $V_d = \overline{|L^1(\mathbb{R}^n)\rangle_d}$ with the cyclic translation representation has a *distributive basis*, labeled by the eigenvalues (energy-momenta),

$$\{|q\rangle_d = |d; q\rangle \,\big|\, q \in \mathbb{R}^n\} : \begin{cases} \langle q'|q\rangle_d = \langle d; q'|d; q\rangle = \tilde{d}(q)\delta(\frac{q-q'}{2\pi}), \\ {}^{\oplus}\!\!\int \frac{d^nq}{(2\pi)^n \tilde{d}(q)} |q\rangle_d \langle q|_d \cong \mathbf{1}_{V_d}, \\ D(x)|q\rangle_d = e^{iqx}|q\rangle_d, \\ V_d \ni |f\rangle_d = |d; \tilde{f}\rangle = {}^{\oplus}\!\!\int \frac{d^nq}{(2\pi)^n} \tilde{f}(q)|q\rangle_d, \\ \langle f|f'\rangle_d = \int \frac{d^nq}{(2\pi)^n} \overline{\tilde{f}}(q)\tilde{d}(q)\tilde{f}'(q). \end{cases}$$

The normalizations of the product and the integration are inverse to each other, $\tilde{d}(q)\delta(q) \leftrightarrow \frac{d^nq}{\tilde{d}(q)}$. The distributive eigenvectors $|q\rangle_d = |d; q\rangle$ are no Hilbert space vectors. If normalizable, the integration of a distributive basis over the eigenvalues gives a cyclic vector:

$$\begin{aligned}|1\rangle_d = |d;1\rangle = {}^{\oplus}\!\!\int \tfrac{d^nq}{(2\pi)^n}|q\rangle_d : \ d(x) = \langle 1|D(x)|1\rangle_d &= \int \tfrac{d^nq}{(2\pi)^n}\tfrac{d^nq'}{(2\pi)^n} e^{iqx}\langle q'|q\rangle_d\\ &= \int \tfrac{d^nq}{(2\pi)^n}\tilde d(q)\, e^{iqx},\\ \text{representation normalization: } d(0) = \langle 1|D(0)|1\rangle_d &= \int \tfrac{d^nq}{(2\pi)^n}\tilde d(q) = 1.\end{aligned}$$

The Schur product of two translation representations, characterized by positive-type functions, integrates over the translations

$$\begin{aligned}\{d_2|d_1\} = \int d^nx\,\overline{d_2(x)}d_1(x) &= \int \tfrac{d^nq}{(2\pi)^n}\tilde d_1(q)\tilde d_2(q)\\ = \quad \hat d_2 * d_1(0) \quad &= \tilde d_1^{-} \tfrac{*}{(2\pi)^n} \tilde d_2(0).\end{aligned}$$

It is a Plancherel distribution of the (energy-)momenta.

9.16 Product Representations on Polynomials

Finite-dimensional representations of a group G and a Lie algebra L can be formulated as actions on polynomials $P(e^1,\dots,e^n) \in \mathbb{K}[e^1,\dots,e^n]$ whose variables are a basis $(e^j)_{j=1}^n$ of a vector space $V \cong \mathbb{K}^n$ with a corresponding representation, e.g., positions $e^j = x^j$ for $\mathbf{SO}(n)$:

$$\begin{aligned}G \ni g &\longmapsto D(g) \in \mathbf{GL}(n,\mathbb{K}),\ G \times V \longrightarrow V,\ g \bullet e^j = D(g)^j_k e^k,\\ L \ni l &\longmapsto \mathcal{D}(l) \in \mathbf{AL}(n,\mathbb{K}),\ L \times V \longrightarrow V,\ l \bullet e^j = \mathcal{D}(l)^j_k e^k.\end{aligned}$$

The linear forms $V^T \cong \mathbb{K}^n$ carry the dual representations, i.e., the inverse transposed one for the group G and the negative transposed one for the Lie algebra L. They can be formulated with dual bases, $(\check e_j)_{j=1}^n$, e.g., with momenta $\check e_j = q_j$ for $\mathbf{SO}(n)$:

$$\begin{aligned}G \ni g &\longmapsto \check D(g) = D(g^{-1})^T,\ G \times V^T \longrightarrow V^T,\ g \bullet \check e_k = D(g^{-1})^j_k \check e_j,\\ L \ni l &\longmapsto \check{\mathcal{D}}(l) = -\mathcal{D}(l)^T,\ L \times V^T \longrightarrow V^T,\ l \bullet \check e_k = -\mathcal{D}(l)^j_k \check e_j.\end{aligned}$$

The basic space may be acted on by a fundamental representation, e.g., a spinor representation on $V \cong \mathbb{C}^2$ for $\mathbf{SU}(2)$ (Pauli spinors) or $\mathbf{SL}(2,\mathbb{C})$ (Weyl spinors), or by an adjoint representation on the Lie algebra, e.g., on $V \cong \mathbb{R}^3$ for $\mathbf{SU}(2)$ (like angular momenta) and on $V \cong \mathbb{R}^6$ for $\mathbf{SL}(2,\mathbb{C})$ (like the electromagnetic field tensor), or a defining representation, e.g., on $V \cong \mathbb{R}^{t+s}$ for $\mathbf{O}(t,s)$.

The representations on the direct sum $V_1 \oplus V_2 \cong \mathbb{K}^{n_1+n_2}$ and on the tensor product $V_1 \otimes V_2 \cong \mathbb{K}^{n_1 n_2}$ of acted-on vector spaces are:

$$\begin{aligned}G \times (V_1 \oplus V_2) &\longrightarrow V_1 \oplus V_2,\ g \bullet (v_1 + v_2) = g \bullet v_1 + g \bullet v_2,\\ L \times (V_1 \oplus V_2) &\longrightarrow V_1 \oplus V_2,\ l \bullet (v_1 + v_2) = l \bullet v_1 + l \bullet v_2,\\ G \times (V_1 \otimes V_2) &\longrightarrow V_1 \otimes V_2,\ g \bullet (v_1 \otimes v_2) = (g \bullet v_1) \otimes (g \bullet v_2),\\ L \times (V_1 \otimes V_2) &\longrightarrow V_1 \otimes V_2,\ l \bullet (v_1 \otimes v_2) = (l \bullet v_1) \otimes v_2 + v_1 \otimes (l \bullet v_2).\end{aligned}$$

For irreducible representations on basic spaces, the representations on the tensor powers may be decomposable. The action on the scalars $\bigotimes^{0} V = \mathbb{K} \ni \alpha$ is defined as trivial: $g \bullet \alpha = \alpha$ and $l \bullet \alpha = 0$.

Therefore, with a vector space V action, the corresponding action is defined on the corresponding tensor algebra $\bigotimes V = \bigoplus_{k=0}^{\infty} \bigotimes^{k} V$ as direct sum of the finite tensor powers and on all its vector sub- and quotient spaces and dual structures. Important examples are the finite-dimensional Grassmann algebra $\bigwedge V \cong \mathbb{K}^{2^n}$ as direct sum of the totally antisymmetric powers $\bigwedge^{k} V \cong \mathbb{K}^{\binom{n}{k}}$ and the infinite-dimensional polynomial algebra $\bigvee V = \mathbb{K}[e^1, \dots, e^n]$ as direct sum of the totally symmetric powers $\bigvee^{k} V \cong \mathbb{K}^{\binom{n+k-1}{k}}$ with the homogeneous polynomials $\mathbb{K}^k[e^1, \dots, e^n]$ of degree k in n variables.

For dual polynomial algebras $\mathbb{K}[e^j]$ and $\mathbb{K}[\check{e}_j]$ of dual basic spaces with actions:

$$\begin{aligned} D(g) &= D(g)^j_k e^k \otimes \check{e}_j \ \in \mathbf{GL}(n, \mathbb{K}) = \exp \mathbf{AL}(n, \mathbb{K}), \\ \mathcal{D}(l) &= \mathcal{D}(l)^j_k e^k \otimes \check{e}_j \ \in \mathbf{AL}(n, \mathbb{K}) \cong V \otimes V^T \cong \mathbb{K}^{n^2}, \end{aligned}$$

the dual product can be reformulated by corresponding derivatives:

$$V^T \times V \longrightarrow \mathbb{K},\ \langle \check{e}_j, e^k \rangle = \partial_j e^k = \delta^k_j, \text{ with } \check{e}_j \cong \partial_j = \tfrac{\partial}{\partial e^j},$$

and, correspondingly, also the linear representations,

$$D(g) = D(g)^j_k e^k \partial_j,\ \mathcal{D}(l) = \mathcal{D}(l)^j_k e^k \partial_j,$$

as familiar from the angular momenta with dual positions-momenta:

$$\langle iq_c, x^b \rangle = \delta^b_c,\ \log \mathbf{O}(3) \ni l^a = -\epsilon^{abc} x^b \otimes iq_c \cong -\epsilon^{abc} x^b \partial_c.$$

A derivative formulation can be extended to "appropriate" non-polynomial functions, e.g., to position functions $\vec{x} \longmapsto f(\vec{x})$.

The polynomial algebra $\bigvee V$ for the direct sum $V = V_1 \oplus \dots \oplus V_r$ of vector spaces V_j with the r fundamental representations of a simple rank r Lie algebra contains all its finite-dimensional irreducible representations, e.g., for the Dirac space $V = V_L \oplus V_R \cong \mathbb{C}^4$ with left- and right-handed Weyl spinor polynomials $\mathbb{C}^2[z^1, z^2] \otimes \mathbb{C}^2[\overline{z}^1, \overline{z}^2]$ for the Lorentz group $\mathbf{SL}(2, \mathbb{C})$, or for $V = V_q \oplus V_{\overline{q}} \cong \mathbb{C}^6$ with quark and antiquark triplets $V_{q,\overline{q}} \cong \mathbb{C}^3$ for color $\mathbf{SU}(3)$.

9.17 Harmonic Polynomials for Orthogonal Groups

For nontrivial properties with respect to rotations $\mathbf{SO}(3)$ and Lorentz transformations $\mathbf{SO}_0(1,3)$ one can use representation spaces, spanned by products of space(time) translations, $\vec{x} \in \mathbb{R}^3$ and $x \in \mathbb{R}^{1+3}$, and, dually equivalent, (energy-)momenta $\vec{q} \in \mathbb{R}^3$ and $q \in \mathbb{R}^{1+3}$, with the defining representations — in general, representations of orthogonal groups $\mathbf{SO}_0(t,s)$ on their harmonic polynomials.

The totally symmetric "totally traceless" products of power $L = 0, 1, 2, \ldots$ of the momentum direction give, with appropriate normalization, the spherical harmonics $(\vec{\omega})^L = (\frac{\vec{q}}{|\vec{q}|})^L \sim \mathrm{Y}^L(\varphi, \theta)$ which are bases for the Hilbert spaces $\mathbb{C}^{1+2L} \subset L^2(\Omega^2)$ acted on by the irreducible $\mathbf{SO}(3)$-representations. The corresponding totally symmetric "traceless" momentum products $(\vec{q})^L$ define the *harmonic momentum polynomials for* $\mathbf{SO}(3)$:

$$\begin{aligned}
(\vec{q})^0 &= 1,\\
(\vec{q})^1 &= \vec{q} = \{q_a \mid a = 1,2,3\},\\
(\vec{q})^2 &= \{q_a q_b - \tfrac{\delta_{ab}}{3}\vec{q}^2\},\\
(\vec{q})^3 &= \{q_a q_b q_c - \tfrac{\delta_{ab}q_c + \delta_{ac}q_b + \delta_{bc}q_a}{3}\vec{q}^2\},\\
&\ldots
\end{aligned}$$

They span irreducible $(1+2L)$-dimensional spaces with $\mathbf{SO}(3)$-representations (L). In Schur products, a momentum integration over the harmonic $\mathbf{SO}(3)$-polynomials and spherical harmonics:

$$\begin{aligned}
\int d^3q\, (\vec{q})^L \otimes (\vec{q})^{L'} \mu(\vec{q}^2) &= \delta^{LL'} \tfrac{\Gamma(1+L)\Gamma(\frac{3}{2})}{2^L \Gamma(\frac{3}{2}+L)} \int d^3q\, (\vec{q}^2)^L \mu(\vec{q}^2) (\mathbf{1}_3)^L,\\
\int d^3q\, (\vec{q})^L \otimes (\vec{q})^{L'} 2\delta(\vec{q}^2 - 1) &= \delta^{LL'} \tfrac{2^L \Gamma(1+L)^2}{\Gamma(2+2L)} 4\pi (\mathbf{1}_3)^L,
\end{aligned}$$

with $\frac{\Gamma(\frac{3}{2})}{\Gamma(\frac{3}{2}+L)} = 2^{2L} \frac{\Gamma(1+L)}{\Gamma(2+2L)}$, displays their orthogonality and involves, apart from multiplicity factors, the *harmonic* $\mathbf{SO}(3)$*-units (projectors)*:

$$(\mathbf{1}_3)^L = \frac{(\frac{\partial}{\partial \vec{q}})^L \otimes (\vec{q})^L}{\Gamma(1+L)} \cong \mathbf{1}_{1+2L} : \begin{cases} (\mathbf{1}_3)^0 = 1,\\ (\mathbf{1}_3)^1 \cong \delta_a^c \cong \mathbf{1}_3,\\ (\mathbf{1}_3)^2 \cong \frac{\delta_a^c \delta_b^d + \delta_a^d \delta_b^c}{2} - \frac{\delta_{ab}\delta^{cd}}{3} \cong \mathbf{1}_5,\\ \ldots \end{cases}$$

$$\text{e.g., } (\mathbf{1}_3)^2 \vec{q} \otimes \vec{q} = (\vec{q})^2 : \left(\frac{\delta_a^c \delta_b^d + \delta_a^d \delta_b^c}{2} - \frac{\delta_{ab}\delta^{cd}}{3}\right) q_c q_d = q_a q_b - \frac{\delta_{ab}}{3}\vec{q}^2.$$

The *harmonic energy-momentum polynomials* $(q)^L$ *for the Lorentz group* $\mathbf{SO}_0(1,3)$ are the totally symmetric "traceless" products of the energy-momenta:

$$\begin{aligned}
(q)^0 &= 1,\\
(q)^1 &= q = \{q_j \mid j = 0,1,2,3\},\\
(q)^2 &= \{q_j q_k - \tfrac{\eta_{jk}}{4} q^2\},\\
(q)^3 &= \{q_j q_k q_l - \tfrac{\eta_{jk}q_l + \eta_{jl}q_k + \eta_{kl}q_j}{4} q^2\},\\
&\dots .
\end{aligned}$$

They span irreducible $(1+L)^2$-dimensional spaces with $\mathbf{SO}_0(1,3)$-representations $(\frac{L}{2}|\frac{L}{2})$. In Schur products, an energy-momentum integration over the harmonic $\mathbf{SO}_0(1,3)$-polynomials

$$\int d^4q\,(q)^L \otimes (q)^{L'} \mu(q^2) = \delta^{LL'} \frac{1}{2^L(1+L)} \int d^4q\,(q^2)^L \mu(q^2)(\mathbf{1}_4)^L.$$

displays their orthogonality and involves, apart from multiplicity factors, the *harmonic* $\mathbf{SO}_0(1,3)$*-units (projectors)*:

$$(\mathbf{1}_4)^L = \frac{(\frac{\partial}{\partial q})^L \otimes (q)^L}{\Gamma(1+L)} \cong \mathbf{1}_{(1+L)^2} : \begin{cases} (\mathbf{1}_4)^0 = 1,\\ (\mathbf{1}_4)^1 \cong \delta^l_j \cong \mathbf{1}_4,\\ (\mathbf{1}_4)^2 \cong \frac{\delta^l_j\delta^m_k + \delta^m_j\delta^l_k}{2} - \frac{\eta_{jk}\eta^{lm}}{4} \cong \mathbf{1}_9,\\ \dots \end{cases}$$

The general case yields harmonic $\mathbf{SO}_0(t,s)$-polynomials $(q)^L$ with $(q)^2 = \{q_j q_k - \frac{\eta_{jk}}{n} q^2\}$, $n = t+s$, etc., and

$$\begin{aligned}
(\tfrac{\partial}{\partial q})^L &= (2q)^L (\tfrac{\partial}{\partial q^2})^L,\\
(\tfrac{\partial}{\partial q})^L \otimes (q)^L &= \Gamma(1+L)(\mathbf{1}_n)^L,\\
\int d^nq\,(q)^L \otimes (q)^{L'} \mu(q^2) &= \delta^{LL'} \frac{\Gamma(1+L)\Gamma(\frac{n}{2})}{2^L\Gamma(\frac{n}{2}+L)} \int d^nq\,(q^2)^L \mu(q^2)(\mathbf{1}_n)^L.
\end{aligned}$$

The number of independent monomials of degree L in n variables $q = \{q_j \mid j = 1,\dots,n\}$ is $\binom{n+L-1}{L}$. Those monomials can be used as basis for the totally symmetric products $\bigvee^L \mathbb{C}^n$ of $q \in \mathbb{C}^n$ (homogeneous degree L polynomials) with the action of $\mathbf{SO}_0(t,s) \ni O$ via $q \longmapsto O.q$. Taking into account the invariant square q^2, the degree L polynomials $\bigvee^L \mathbb{C}^n$ decompose into harmonic polynomials up to degree L, e.g., $q \vee q \cong q_j q_k = q_j q_k - \dfrac{\eta_{jk}}{n} q^2 + \dfrac{\eta_{jk}}{n} q^2 \cong (q)^2 + \dfrac{1}{n} q^2 (q)^0$, acted on by irreducible $\mathbf{SO}_0(t,s)$-representations (each of multiplicity one) as follows:

$$\bigvee^L \mathbb{C}^n \cong \begin{cases} \bigoplus_{M=0,2,\dots,L} (q)^M, \text{ even } L = 0,2,\dots,\\ \bigoplus_{M=1,3,\dots,L} (q)^M, \text{ odd } L = 1,,3,\dots . \end{cases}$$

Therefore, the dimensionality of the harmonic $\mathbf{SO}_0(t,s)$-representations is

$$\dim_{\mathbb{C}}(q)^L = \binom{n+L-1}{L} - \binom{n+L-3}{L-2} = \binom{n-2+L}{n-2}\frac{n-2+2L}{n-2+L} = \begin{cases} 2, & n=2, \\ 1+2L, & n=3, \\ (1+L)^2, & n=4, \\ \ldots \end{cases}$$

Harmonic $\mathbf{SO}_0(t,s)$-polynomials $(x)^L$ have a trivial translation invariant:

$$\partial^2 = \vec{\partial}_t^2 - \vec{\partial}_s^2 : \ \partial^2(x)^L = 0.$$

9.18 Harmonic Analysis of Hyperboloids

In analogy to the harmonic analysis of sphere-functions $L^2(\Omega^2)$ and $L^2(\Omega^3)$ with the Peter-Weyl theorem involving spherical harmonics and Legendre polynomials, there is the harmonic analysis of hyperbolic functions, shortly sketched in this section. This is an example for the difficult harmonic analysis of noncompact nonabelian groups which concerns the Iwasawa and Cartan factorized Lorentz groups with maximal noncompact rank 1 motion group $\mathbf{SO}_0(1,1)$ of hyperboloids $\mathcal{Y}^s$:

$$\begin{aligned} s \geq 2: \ \mathbf{SO}_0(1,s) &= \mathbf{SO}(s) \circ \mathbf{SO}_0(1,1) \circ \exp \mathbb{R}^{s-1} \\ &= \mathbf{SO}(s) \circ \mathbf{SO}_0(1,1) \circ \mathbf{SO}(s), \\ \mathcal{Y}^s \cong \mathbf{SO}_0(1,s)/\mathbf{SO}(s) &\cong \mathbf{SO}_0(1,1) \circ \exp \mathbb{R}^{s-1}, \end{aligned}$$

The bounded *Legendre functions* P of the compact group classes, i.e., of the hyperboloids, are positive-type functions for irreducible $\mathbf{SO}_0(1,s)$-representations with complex continuous invariants:

$$\begin{aligned} \mathcal{Y}^s \ni y_s = \begin{pmatrix} \cosh\psi \\ \sinh\psi \cos\varphi \\ \sinh\psi \sin\varphi\, \vec{\omega}_{s-2} \end{pmatrix} &\text{ with } \vec{\omega}_{s-2} \in \Omega^{s-2}, \\ y_s \longmapsto |ip - \tfrac{s-1}{2}\}(\cosh\psi) &= \mathrm{P}^{ip-\frac{s-1}{2}}(\cosh\psi), \ ip \in i\mathbb{R}, \ \mathrm{P}^{ip-\frac{s-1}{2}}(1) = 1. \end{aligned}$$

$s-1$ is the dimension of the nilpotent Lie algebra $\mathbb{R}^{s-1}$.

The positive-type Legendre functions for noncompact $\mathbf{SO}_0(1,s)$ and the hyperboloids $\mathcal{Y}^s$ are the partners of the positive-type Legendre polynomials with integer index for compact $\mathbf{SO}(1+s)$ and the spheres $\Omega^s \cong \mathbf{SO}(1+s)/\mathbf{SO}(s)$,

$$\Omega^s \ni \omega_s = \begin{pmatrix} \cos\theta \\ \sin\theta\cos\varphi \\ \sin\theta\sin\varphi\,\vec{\omega}_{s-2} \end{pmatrix} \longmapsto \mathrm{P}^L(\cos\theta), \ L \in \mathbb{N}, \ \mathrm{P}^L(1) = 1.$$

They are related to each other by the compact-noncompact transition $(i\theta, L) \leftrightarrow (\psi, ip)$:

$$
\begin{aligned}
L \in \mathbb{N}: \mathrm{P}^L(\cos\theta) &= \sum_{k=0}^{L} \frac{(L+k)!}{(L-k)!} \frac{(-\sin^2\frac{\theta}{2})^k}{(k!)^2} &&= \int_{-\pi}^{\pi} \frac{d\varphi}{2\pi} (\cos\theta + i\sin\theta\cos\varphi)^L,\\
iQ \in \mathbb{C}: \mathrm{P}^{iQ}(\cosh\psi) &= \sum_{k=0}^{\infty} \frac{\Gamma(iQ+k+1)}{\Gamma(iQ-k+1)} \frac{(\sinh^2\frac{\psi}{2})^k}{(k!)^2} &&= \int_{-\pi}^{\pi} \frac{d\varphi}{2\pi} (\cosh\psi + \sinh\psi\cos\varphi)^{iQ},\\
\xi \in \mathbb{R},\ iQ \in \mathbb{C}: &\quad \frac{d}{d\xi}(\xi^2-1)\frac{d\mathrm{P}^{iQ}(\xi)}{d\xi} &&= iQ(1+iQ)\mathrm{P}^{iQ}(\xi).
\end{aligned}
$$

The integration goes over the compact subgroup $\mathbf{SO}(2) \subseteq \mathbf{SO}(s)$. The integrands modify the irreducible representations for the abelian groups:

$$
\begin{array}{lll}
(\cos\theta + i\sin\theta\cos\varphi)^L, & \text{e.g., } e^{Li\theta} & \text{for } \mathbf{SO}(1+s),\\
(\cosh\psi + \sinh\psi\cos\varphi)^{ip-\frac{s-1}{2}}, & \text{e.g., } e^{(ip-1)\psi} & \text{for } \mathbf{SO}_0(1,3)
\end{array}
$$

By Inönü–Wigner contraction, e.g., $\mathbf{SO}_0(1,2) \to \mathbf{SO}(2)\ \vec{\times}\,\mathbb{R}^2$, the Legendre functions for the Lorentz groups morph into the Bessel[20] functions $\mathcal{J}_0$ as positive-type function for the translations in the Euclidean groups:

$$
\begin{aligned}
\text{with } p = Pc,\ \psi = \tfrac{r}{c}: \quad & \begin{pmatrix} \cosh\psi & p\sinh\psi \\ \frac{1}{p}\sinh\psi & \cosh\psi \end{pmatrix} = \begin{pmatrix} \cosh\frac{r}{c} & Pc\sinh\frac{r}{c} \\ \frac{1}{Pc}\sinh\frac{r}{c} & \cosh\frac{r}{c} \end{pmatrix}\\
& \xrightarrow{c\to\infty} \begin{pmatrix} 1 & Pr \\ 0 & 1 \end{pmatrix},\\
\text{and } \mathrm{P}^{ip+b}(\cosh\psi) &= \int_{-\pi}^{\pi} \frac{d\varphi}{2\pi} (\cosh\tfrac{r}{c} + \sinh\tfrac{r}{c}\cos\varphi)^{iPc+b}\\
& \xrightarrow{c\to\infty} \int_{-\pi}^{\pi} \frac{d\varphi}{2\pi} (1 + \tfrac{r}{c}\cos\varphi)^{iPc} \to \int_{-\pi}^{\pi} \frac{d\varphi}{2\pi} e^{iPr\cos\varphi}\\
&= \mathcal{J}_0(Pr) = \int \frac{d^2q}{\pi} \delta(\vec{q}^2 - P^2) e^{i\vec{q}\vec{x}},\\
\text{with } & (1 + \tfrac{\alpha}{c})^{\beta c} \xrightarrow{c\to\infty} e^{\alpha\beta},\ \alpha,\beta \in \mathbb{C}.
\end{aligned}
$$

Parallel to the representation-normalized Legendre polynomials, involving normalized Haar and Plancherel measure:

$$
\mathbf{SO}(1+s),\ \zeta = \cos\theta: \begin{cases} \int_{-1}^{1} \frac{d\zeta}{2} \mathrm{P}^{L'}(\zeta)\mathrm{P}^L(\zeta) = \frac{1}{1+2L}\delta^{LL'},\\ \sum_{L=0}^{\infty} (1+2L)\mathrm{P}^L(\zeta)\mathrm{P}^L(\zeta') = \delta(\frac{\zeta-\zeta'}{2}), \end{cases}
$$

the representation-normalized Legendre functions display distributive orthogonality and L^2-completeness,

$$
\mathbf{SO}_0(1,s),\ \kappa = \cosh\psi: \begin{cases} \int_1^{\infty} d\kappa\, \mathrm{P}^{ip-\frac{s-1}{2}}(\kappa)\mathrm{P}^{-ip'-\frac{s-1}{2}}(\kappa) = \frac{1}{\Pi^s(p^2)}\delta(p-p'),\\ \int_0^{\infty} \Pi^s(p^2)dp\, \mathrm{P}^{ip-\frac{s-1}{2}}(\kappa)\mathrm{P}^{-ip-\frac{s-1}{2}}(\kappa') = \delta(\kappa-\kappa'). \end{cases}
$$

They are used for the harmonic analysis (Fock–Mehler[21] transformation) of hyperboloids.

---

[20]Friedrich Bessel (1784–1846).

[21]Gustav Ferdinand Mehler (1835–1895).

The Plancherel measure of the continuous invariant arising in the Schur norm is related to the noncompact $\mathbf{SO}_0(1,1)$-subgroup. It is used for the harmonic analysis of hyperboloid functions and has no discrete series contributions. It has to distinguish between even and odd dimensions:

$$\Pi^s(p^2) = \left|\frac{\Gamma(ip+\frac{s-1}{2})}{\Gamma(ip)}\right|^2 = \begin{cases} \frac{\Gamma(R-\frac{1}{2})^2}{\Gamma(\frac{1}{2})^2} \quad \times \quad p\tanh\pi p \prod_{k=1}^{R-1}\left(1+\frac{p^2}{(k-\frac{1}{2})^2}\right), \\ \qquad \mathcal{Y}^s \text{ with } s=2R=2,4,6,\dots, \text{ in } \mathbf{SO}_0(1,2R), \\ \Gamma(R-1)^2 \quad \times \quad p^2\prod_{k=1}^{R-2}\left(1+\frac{p^2}{k^2}\right), \\ \qquad \mathcal{Y}^s \text{ with } s=2R-1=3,5,7,\dots, \text{ in } \mathbf{SO}_0(1,2R-1), \end{cases}$$
$$\Pi^2(p^2) = p\tanh p,\ \Pi^3(p^2) = p^2.$$

For the minimal nonabelian odd-dimensional case $\mathcal{Y}^3$ with the proper Lorentz group and Plancherel measure $\int_0^\infty p^2 dp$, the principal series $\mathbf{SO}_0(1,3)$-representations are used. The Gamma-factors are related to sphere volumes, $\frac{\Gamma(R-\frac{1}{2})}{\Gamma(\frac{1}{2})} = \frac{2\pi^{R-1}}{|\Omega^{2R-2}|}$ and $\Gamma(R-1) = \frac{2\pi^{R-1}}{|\Omega^{2R-3}|}$ with $\Omega^{s-2} \cong \mathbf{SO}(s-1)/\mathbf{SO}(s-2)$ where $\mathbf{SO}_0(1,s) \supset \mathbf{SO}_0(1,1) \times \mathbf{SO}(s-1)$ for $s = 2R,\ 2R-1$. The functional dependence of the invariant p^2 uses the factorization of the hyperbolic functions (for $z = ip$) with respect to the zeros:

$$\mathbb{C} \ni z \longmapsto \begin{cases} \frac{\pi}{\sin\pi z} = \Gamma(z)\Gamma(1-z),\ \Gamma(\overline{z}) = \overline{\Gamma(z)}, \\ \frac{\sin\pi z}{\pi z} = \prod_{k=1}^{\infty}\left(1-\frac{z^2}{k^2}\right), \\ \cos\pi z = \prod_{k=1}^{\infty}\left(1-\frac{z^2}{(k-\frac{1}{2})^2}\right). \end{cases}$$

9.19 Irreducible Representations of the Lorentz Group

For the Lorentz covering group with the Iwasawa decomposition:

$$\begin{aligned} \mathbf{SO}_0(1,3) &\sim \mathbf{SL}(2,\mathbb{C}) = \mathbf{SU}(2) \circ \mathbf{SO}_0(1,1) \circ \exp\mathbb{R}^2, \\ s &= \begin{pmatrix} \alpha & \beta \\ \gamma & \delta \end{pmatrix} = e^{\frac{i\vec{\theta}}{2}\vec{\sigma}} \circ e^{\frac{\psi}{2}\sigma_3} \circ \begin{pmatrix} 1 & 0 \\ x+iy & 1 \end{pmatrix} \in \mathbf{SL}(2,\mathbb{C}), \\ \hat{s} = s^{*-1} &= \begin{pmatrix} \overline{\delta} & -\overline{\gamma} \\ -\overline{\beta} & \overline{\alpha} \end{pmatrix} \in \mathbf{SL}(2,\mathbb{C}), \end{aligned}$$

the familiar finite-dimensional representations $(L|R)$ are, up to equivalence, the totally symmetric product representations of left- and right-handed Weyl spinor representations

$$2L, 2R \in \mathbb{N}: \ (L|R)(s) = \bigvee^{2L} s \otimes \bigvee^{2R} \hat{s} \in \mathbf{SL}((1+2L)(1+2R),\mathbb{C}).$$

A basis of the $(L|R)$-representation space is given by the product of spinor polynomials of degree $2L$ and $2R$, homogeneous in two complex variables, conjugate to each other $z^{1,2}$ (for left) and $\overline{z}^{1,2}$ (for right)

$$\begin{aligned}
&\mathbb{C}^{2L}[z^1, z^2]\mathbb{C}^{2R}[\overline{z}^1, \overline{z}^2] \in \bigvee^{2L} V_L \otimes \bigvee^{2R} V_R \cong \mathbb{C}^{(1+2L)(1+2R)}, \ V_{L,R} \cong \mathbb{C}^2,\\
&(L|R)(s) \bullet \mathbb{C}^{2L}[z^1, z^2]\mathbb{C}^{2R}[\overline{z}^1, \overline{z}^2],\\
&\text{with } \mathbb{C}^2 \ni z^{\dot{A}} \longmapsto s^{\dot{A}}_{\dot{B}} z^{\dot{B}} \in \mathbb{C}^2, \ \mathbb{C}^2 \ni \overline{z}^A \longmapsto \hat{s}^A_B \overline{z}^B \in \mathbb{C}^2.
\end{aligned}$$

All nontrivial finite-dimensional representations are no Hilbert representations.

The equivalence classes of the *principal series Hilbert representations of the Lorentz group* are characterized by a half-integer and an imaginary number for the Cartan rotation-dilation (chirality) subgroup type (cylinder) $\mathbf{SO}(2) \times \mathbf{SO}_0(1,1) \subset \mathbf{SL}(2,\mathbb{C})$:

$$\begin{aligned}
&\mathbf{irrep}_+^{\text{princ}}\mathbf{SL}(2,\mathbb{C}) \cong \{(j, ip) \,\big|\, 2j \in \mathbb{Z}, \ ip \in i\mathbb{R}\},\\
&\text{unitarily equivalent: } (j, ip) \cong (-j, -ip).
\end{aligned}$$

They act on complex-valued square-integrable functions in one complex variable:

$$\mathbb{C} \ni z \longmapsto f(z) \in \mathbb{C}, \ f \in L^2(\mathbb{C}) \text{ with } \int_{\mathbb{C}} dz\,|f(z)|^2 < \infty, \ dz = dxdy,$$

The coordinates characterize the manifold for the nilpotent classes $\mathbf{SL}(2,\mathbb{C})/[\mathbf{SU}(2) \circ \mathbf{SO}_0(1,1)] \cong e^{\mathbb{R}^2} \ni \begin{pmatrix} 1 & 0\\ z & 1\end{pmatrix}$. The projective variable $z = \frac{z^2}{z^1}$ is a "quotient of Weyl spinors" with Möbius[22] transformation behavior:

$$\begin{aligned}
\begin{pmatrix} z^1\\ z^2\end{pmatrix} \longmapsto s^{-1}\begin{pmatrix} z^1\\ z^2\end{pmatrix} &= \begin{pmatrix} \delta & -\beta\\ -\gamma & \alpha\end{pmatrix}\begin{pmatrix} z^1\\ z^2\end{pmatrix} = \begin{pmatrix} -\beta z^2 + \delta z^1\\ \alpha z^2 - \gamma z^1\end{pmatrix},\\
\text{with } z = \tfrac{z^2}{z^1} : \ s^{-1}\begin{pmatrix} 1\\ z\end{pmatrix} &= \begin{pmatrix} -\beta z + \delta\\ \alpha z - \gamma\end{pmatrix} = (-\beta z + \delta)\begin{pmatrix} 1\\ \frac{\alpha z - \gamma}{-\beta z + \delta}\end{pmatrix},\\
\mathbb{C} \ni z \longmapsto s^{-1} \bullet z &= \tfrac{\alpha z - \gamma}{-\beta z + \delta} \in \mathbb{C}, \ s_1^{-1} \bullet (s_2^{-1} \bullet z) = (s_2 \circ s_1)^{-1} \bullet z,\\
dz &\longmapsto \tfrac{dz}{|-\beta z + \delta|^4}.
\end{aligned}$$

The real six-parametric group with the Möbius (broken linear) transformations is a subgroup of the bijections of the compactified complex numbers $\hat{\mathbb{C}} = \mathbb{C} \cup \{\infty\} \cong \Omega^2$ (Riemannian sphere) and isomorphic to the orthochronous Lorentz group (projective special linear group $\mathbf{PSL}(2,\mathbb{C}) \cong \mathbf{SL}(2,\mathbb{C})/\mathbf{I}(2) \cong \mathbf{SO}_0(1,3)$). It preserves angles and generalized circles.

[22] August Ferdinand Möbius (1790–1868).

With the minimal parabolic subgroup $[\mathbf{SO}(2) \times \mathbf{SO}_0(1,1)] \circ e^{\mathbb{R}^2}$, the $\mathbf{SL}(2,\mathbb{C})$-representations are induced from Cartan subgroup $\mathbf{SO}(2) \times \mathbf{SO}_0(1,1)$-representations:

$$\begin{array}{ccc} \mathbb{C} & \xrightarrow{s\bullet} & \mathbb{C} \\ f\downarrow & & \downarrow (j,ip)(s)\bullet f, \\ \mathbb{C} & \xrightarrow[u^{-2j}(s)\,d^{2(ip-1)}(s)]{} & \mathbb{C} \end{array} \quad \begin{array}{l} s \in \mathbf{SL}(2,\mathbb{C}), \\ u(s) \in \mathbf{U}(1) \cong \mathbf{SO}(2), \\ d(s) \in \mathbf{D}(1) \cong \mathbf{SO}_0(1,1). \end{array}$$

The action on the coordinates comes together with multiplicative factors:

$$\begin{aligned} L^2(\mathbb{C}) \ni f &\longmapsto (j,ip)(s) \bullet f \in L^2(\mathbb{C}), \\ \Big((j,ip)(s) \bullet f\Big)(z) &= u^{-2j}(s)\, d^{2(ip-1)}(s)\, f(s^{-1} \bullet z) \\ &= \left(\tfrac{-\beta z+\delta}{|-\beta z+\delta|}\right)^{-2j} |-\beta z+\delta|^{2(ip-1)} f\left(\tfrac{\alpha z-\gamma}{-\beta z+\delta}\right), \\ \text{group representation: } (j,ip)(s_1) \circ (j,ip)(s_2) &= (j,ip)(s_1 \circ s_2), \end{aligned}$$

which represent the Cartan subgroup with the invariants in the powers:

$$\text{for } s = \begin{pmatrix} e^{\frac{i\theta+\psi}{2}} & 0 \\ 0 & e^{-\frac{i\theta+\psi}{2}} \end{pmatrix} : \Big((j,ip)(s) \bullet f\Big)(z) = e^{ji\theta-(ip-1)\psi} f(e^{i\theta+\psi} z).$$

The nonunitary factor $d^{-2}(s)$ is related to the corresponding factor in the transformation of the measure dz.

The equivalence classes of all *principal series representations of the Lorentz group*, not necessarily Hilbert representations, allow any complex invariant for the noncompact Cartan subgroup $\mathbf{SO}_0(1,1)$:

$$\mathbf{irrep}^{\,\mathrm{princ}}\mathbf{SL}(2,\mathbb{C}) \cong \{(j, iQ) \,\big|\, 2j \in \mathbb{Z}, iQ = ip + b \in i\mathbb{R} + \mathbb{R}\}.$$

As seen in the Cartan subgroup representations:

$$\begin{aligned} \mathbf{U}(1) \times \mathbf{D}(1) \ni u(s) \times d(s) = s_3 &= -\beta z + \delta, \\ \left(\tfrac{-\beta z+\delta}{|-\beta z+\delta|}\right)^{-2j} |-\beta z+\delta|^{2(iQ-1)} &= s_3^{-j+iQ-1} \overline{s}_3^{\,j+iQ-1}, \end{aligned}$$

they contain the unitary representations above for imaginary noncompact invariant:

$$\text{unitary for: } iQ = ip \in i\mathbb{R},$$

and the finite dimensional representations above for real half-integer noncompact invariant with j the difference between right- and left-handed spin:

finite-dimensional for: $s_3^{-j+iQ-1}\overline{s}_3^{j+iQ-1} = s_3^{2L}\overline{s}_3^{2R}$,
$\mathbf{irrep}_{\text{fin}}\mathbf{SL}(2,\mathbb{C}) \cong \{(j, iQ) = (R-L, R+L+1) \mid (2L, 2R) \in \mathbb{N}\times\mathbb{N}\}$.

With the infinitesimal rotations and boosts $(\mathbf{L}^{ik})_{i,k=0}^3 = (\mathbf{J}^a, \mathbf{B}^a)_{a=1,2,3}$ as Lie algebra basis the spectrum of the Killing invariant $-\frac{1}{4}\eta_{in}\eta_{km}\mathbf{L}^{ik}\mathbf{L}^{nm} = \vec{\mathbf{J}}^2 - \vec{\mathbf{B}}^2$ and the chiral invariant $\frac{1}{4}\epsilon_{iknm}\mathbf{L}^{ik}\mathbf{L}^{nm} = \vec{\mathbf{J}}\vec{\mathbf{B}}$ is given as follows:

$$\begin{aligned}
\mathbf{spec}\,(\vec{\mathbf{J}}^2 - \vec{\mathbf{B}}^2) &= \{-j^2 + Q^2 + 1 \mid 2j \in \mathbb{Z},\, iQ \in \mathbb{C}\}\\
&\overset{\text{fin}}{=} \{-2[L(L+1) + R(R+1)] \mid 2L, 2R \in \mathbb{N}\},\\
\mathbf{spec}\,\vec{\mathbf{J}}\vec{\mathbf{B}} &= \{jQ\}\\
&\overset{\text{fin}}{=} \{i[L(L+1) - R(R+1)]\},\\
\mathbf{spec}\,(\vec{\mathbf{J}} \pm i\vec{\mathbf{B}})^2 &= \{-(j \mp iQ)^2 + 1\}\\
&\overset{\text{fin}}{=} \{-4R(R+1)\} \text{ and } \{-4L(L+1)\}.
\end{aligned}$$

The equivalence classes of the *supplementary series Hilbert representations of the Lorentz group* are characterized by a trivial integer for the rotations $\mathbf{SO}(2)$ and a real number for the dilations $\mathbf{SO}_0(1,1)$:

$$\mathbf{irrep}_+^{\text{suppl}}\mathbf{SL}(2,\mathbb{C}) \cong \{(0, b) \mid 0 < b < 1\}.$$

These nonamenable representations (trivial contribution in the Plancherel measure) complete the Hilbert representations. They have a nontrivial Hilbert product inducing function:

$$\langle f | f' \rangle_b = \int_{\mathbb{C}\times\mathbb{C}} dz dz' \, \overline{f(z)} \frac{1}{|z-z'|^{2-2b}} f'(z').$$

9.20 Compact–Noncompact Pairing

A complex Lie algebra $L \cong \mathbb{C}^d$ has compact and noncompact real forms $L^{(p,q)} \cong (i\mathbb{R})^p \oplus \mathbb{R}^q$, $p+q=d$.

An *orthogonal symmetric Lie algebra* $(L_\bullet, \mathtt{I})$ is a real finite-dimensional Lie algebra with a nontrivial involutive automorphism where the fixed elements $L = \{l \in L_\bullet \mid \mathtt{I}(l) = l\}$ constitute a compact Lie algebra. It is called *effective* if the distinguished compact subalgebra does not contain central elements $L \cap \text{centr}\, L_\bullet = \{0\}$.

An orthogonal symmetric Lie algebra has a Killing form κ-orthogonal direct decomposition into involution eigenspaces, the compact Lie subalgebra L, and the $\mathtt{I}$-antisymmetric vector subspace R, compatible with the Lie bracket,

$$L_\bullet = L_+^{\mathtt{I}} \perp L_-^{\mathtt{I}} = L \perp R, \quad \begin{cases} [L, L] \subseteq L,\; [L, R] \subseteq R,\; [R, R] \subseteq L, \\ \kappa(L, R) = \{0\}. \end{cases}$$

An orthogonal symmetric Lie algebra $(L_\bullet, L)$ is *irreducible* if $L_\bullet$ is semisimple and if L does not contain a nontrivial $L_\bullet$-ideal. Then $[R, R] = L$, and $\mathrm{ad}_{L_\bullet} L$ acts irreducibly on R.

There are three types of effective orthogonal symmetric Lie algebras, denoted with subindex $(c, nc, 0)$ and $R = (K, B, P)$,

$$\begin{array}{rl} L_c = L \perp K \text{ semisimple and compact} & \Rightarrow (L_c, \mathtt{I}_c) \text{ of } \mathit{compact\ type}, \\ \left.\begin{array}{r} L_{nc} \text{ semisimple and noncompact,} \\ L_{nc} = L \perp B \text{ is a Cartan decomposition} \end{array}\right\} & \Rightarrow (L_{nc}, \mathtt{I}_{nc}) \text{ of } \mathit{noncompact\ type}, \\ L_0 = L \perp P,\ [P, P] = \{0\},\ P \text{ abelian ideal} & \Rightarrow (L_0, \mathtt{I}_0) \text{ of } \mathit{Euclidean\ type}. \end{array}$$

Therefore, all semisimple real Lie algebras are effective orthogonal symmetric. Every effective orthogonal symmetric Lie algebra has a Killing form orthogonal decomposition into ideals of compact, noncompact, and Euclidean types:

$$(L_\bullet, \mathtt{I}) = (L_c, \mathtt{I}_c) \perp (L_{nc}, \mathtt{I}_{nc}) \perp (L_0, \mathtt{I}_0).$$

Examples are the orthogonal groups:

$$\begin{aligned} L &= \log[\mathbf{SO}(t) \times \mathbf{SO}(s)], \\ (L_c, L_{nc}, L_0) &= (\log \mathbf{SO}(t+s),\ \log \mathbf{SO}_0(t, s),\ \log[\mathbf{SO}(t) \times \mathbf{SO}(s)] \,\vec{\times}\, \mathbb{R}^{t+s}). \end{aligned}$$

There is the abelian case with the phase-, dilation- and translation Lie-algebras:

$$L = \{0\},\ (L_c, L_{nc}, L_0) = (\log \mathbf{U}(1), \log \mathbf{D}(1), \mathbb{R}).$$

In the spherical, hyperbolic, and flat example the antisymmetric subspaces contain the classes of Lie algebras for orthogonal groups $l = -l^T \in \log \mathbf{SO}(s)$. They are the $\mathbb{R}^s$-isomorphic tangent spaces of sphere, hyperboloid, and Euclidean space:

$$\begin{aligned} L \perp K &= \quad \log \mathbf{SO}(1+s) \quad = \log \mathbf{SO}(s) \perp \log \Omega^s \ni \left(\begin{array}{c|c} 0 & \vec{k} \\ \hline -\vec{k}^T & l \end{array}\right), \\ L \perp B &= \quad \log \mathbf{SO}_0(1, s) \quad = \log \mathbf{SO}(s) \perp \log \mathcal{Y}^s \ni \left(\begin{array}{c|c} 0 & \vec{b} \\ \hline \vec{b}^T & l \end{array}\right), \\ L \perp P &= \log[\mathbf{SO}(s) \,\vec{\times}\, \mathbb{R}^s] = \log \mathbf{SO}(s) \,\vec{\oplus}\, \mathbb{R}^s \quad \ni \left(\begin{array}{c|c} 0 & \vec{p} \\ \hline 0 & l \end{array}\right). \end{aligned}$$

The polar decompositions of the complex linear groups give rise to orthogonal symmetric, not effective, Lie algebras:

$$\log \mathbf{GL}(n, \mathbb{C}) = \log \mathbf{U}(n) \perp \mathbb{R}(n),\ \mathbb{R}(n) \cong \mathbb{R}^{n^2}.$$

For an orthogonal symmetric Lie algebra $(L_\bullet, \mathtt{I})$, the complexification $(L_\bullet^*, \mathtt{I}^*)$ with "imaginary" iR is, as a real Lie algebra, orthogonal symmetric:

$$\left(L_\bullet = L\perp R, \mathtt{I}\right) \leftrightarrow \left(L_\bullet^* = L\perp iR, \mathtt{I}^*(l+ir) = l - ir\right).$$

There is a *compact–noncompact duality*: If $(L_\bullet, \mathtt{I})$ is of the compact type with $R = K \cong (i\mathbb{R})^n$ its dual $(L_\bullet^*, \mathtt{I}^*)$ is of the noncompact type $iR = B \cong \mathbb{R}^n$, and vice versa.

A *compact-noncompact Lie algebra pair* (L_c, L_{nc}) with its Euclidean partner L_0 contains direct sums of a Lie algebra L, either trivial or compact, and a compact or noncompact L-invariant subspace $(K, B; P)$, isomorphic as vector spaces, with a relative sign in the Lie brackets and a Lie algebra isomorphism for the complex Lie algebra extension:

$$\left.\begin{array}{rl} L_c = L\perp K & \cong (i\mathbb{R})^s\perp(i\mathbb{R})^n,\\ & (l,k)\\ [l^a, l^b] & = \epsilon_c^{ab} l^c,\\ [l^a, k^j] & = \epsilon_m^{aj} k^m,\\ [k^j, k^m] & = +\epsilon_c^{jm} l^c, \end{array}\right\} \overset{k=\pm ib}{\longleftrightarrow} \left\{\begin{array}{rl} L_{nc} = L\perp B & \cong (i\mathbb{R})^s\perp\mathbb{R}^n,\\ & (l,ib)\\ [l^a, l^b] & = \epsilon_c^{ab} l^c,\\ [l^a, b^j] & = \epsilon_m^{aj} b^m,\\ [b^j, b^m] & = -\epsilon_c^{jm} l^c, \end{array}\right\}$$

$$\underset{c\to\infty}{\overset{b=cp}{\longrightarrow}} \left\{\begin{array}{rl} L_0 = L\perp P & \cong (i\mathbb{R})^s\perp\mathbb{R}^n,\\ & (l,p),\\ [l^a, l^b] & = \epsilon_c^{ab} l^c,\\ [l^a, p^j] & = \epsilon_m^{aj} p^m,\\ [p^j, p^m] & = 0. \end{array}\right.$$

A complex representation of the compact partner by antihermitian matrices for Hilbert representations can be used, after $\pm i$-multiplication for conjugated representations of the noncompact partner with hermitian matrices related to finite-dimensional non-Hilbert representations:

$$\begin{array}{c} (l,k) \longmapsto (\mathcal{D}(l), \mathcal{D}(k) = -\mathcal{D}(k)^\star) \in \log \mathbf{U}(d),\\ \updownarrow\\ (l,b) \longmapsto (\mathcal{D}(l), \pm i\mathcal{D}(k)) \in \mathbf{AL}(d,\mathbb{C}). \end{array}$$

Examples are given by the unitary phase representations and the non-unitary dilation representations, both faithful for $z \neq 0$,

$$\begin{array}{c} \mathbf{U}(1) \ni e^{i\theta} \longmapsto e^{zi\theta} \in \mathbf{U}(1),\ z\in\mathbb{Z},\\ \updownarrow\\ \mathbf{D}(1) \ni e^{\psi} \longmapsto e^{\pm z\psi} \in \mathbf{D}(1), \end{array}$$

and the pairs with the unitary Pauli spinor representation and non-unitary Weyl spinor representations, both two-dimensional and faithful:

$$\begin{array}{c} \mathbf{SU}(2) \times \mathbf{SU}(2) \ni u \longmapsto \bigvee^{2L} e^{i(\vec{\theta}+\vec{\varphi})\frac{\vec{\sigma}}{2}} \otimes \bigvee^{2R} e^{i(\vec{\theta}-\vec{\varphi})\frac{\vec{\sigma}}{2}},\ 2L, 2R \in \mathbb{N}, \\ \updownarrow \\ \mathbf{SL}(2,\mathbb{C}) \ni s \longmapsto \bigvee^{2L} e^{(i\vec{\theta}+\vec{\psi})\frac{\vec{\sigma}}{2}} \otimes \bigvee^{2R} e^{(i\vec{\theta}-\vec{\psi})\frac{\vec{\sigma}}{2}}. \end{array}$$

This partnership picks discrete invariants from the continuous spectrum of the non-compact Lie algebra invariants.

9.21 Special Functions in Basic Physics

Special functions seem to be a mathematical hobby for, well, specialists. It is the merit of Vilenkin[23] to have shown, in explicit detail, that — up to a few unimportant exceptions — all special functions, especially those used in basic physics, can be looked at to be coefficients, i.e., matrix elements $G \ni g \longmapsto f(g) = \langle v|D(g)|w\rangle \in \mathbb{C}$ of group representations $g \longmapsto D(g) \in \mathbf{GL}(V)$. "Behind" (almost) each special function, there is a group acting on a vector space. That applies not only to the "simple" cases, e.g., to spherical and hyperbolic exponentials, sines and cosines, spherical harmonics and Legendre and Laguerre[24] polynomials. The theory of special functions as group functions can be considered to be a part of representation theory which may help to digest their sometimes very complicated explicit expressions and relations.

The infinitesimal structures and the tangent spaces of real Lie groups are given by their Lie algebras $L = \log G$, $\exp L = G$ (covering group), with the operations and irreducible Hilbert representations of the real one-dimensional subgroups by exponentials:

$$\exp\mathbb{R} = \mathbf{D}(1) \cong \mathbb{R},\ \check{\mathbf{D}}(1) \cong i\mathbb{R},\ \exp i\mathbb{R} = \mathbf{U}(1) \cong \mathbb{R}/\mathbb{Z},\ \check{\mathbf{U}}(1) \cong \mathbb{Z}.$$

As invariance groups for the Killing forms of semsimple Lie algebras $L \cong \mathbb{R}^n$ there arise orthogonal groups $\mathbf{O}(t,s)$, $t+s=n$, with $\log \mathbf{O}(t,s) \supseteq L$. $\mathbf{O}(t) \times \mathbf{O}(s)$ is the maximal compact subgroup for positive and negative definiteness.

For the hyperboloid $\mathcal{Y}^s$, the sphere Ω^s and the Euclidean space $\mathbb{E}^s$, the scalar distributions for the definite orthogonal local action group $\mathbf{O}(s)$ with hyperbolic singularities (invariants) $(i\vec{q})^2 = 1$ (positive definite Killing metric — with unit $\vec{q}^2 = -B^2$) on imaginary spheres, and spherical ones $(i\vec{q})^2 = -1$ (negative definite Killing metric — with unit $\vec{q}^2 = P^2$) on real spheres Ω^{s-1} define the hyperbolic *Macdonald functions* $\mathcal{K}$ with, for imaginary argument, the corresponding spherical *Hankel* $\mathcal{H}^{1,2} = \mathcal{J} \pm i\mathcal{N}$ *(Bessel* $\mathcal{J}$ *with Neumann*[25] $\mathcal{N}$*) functions*. Integrals with

[23] Naum Vilenkin (1920–1991).

[24] Edmond Laguerre (1834–1886).

[25] Hector Macdonald (1865–1935), Hermann Hankel (1839–1873), Carl Neumann (1832–1925).

integer poles $\frac{s}{2} - \nu = 1, 2, \ldots$ are important for spacetime representations and will be defined in a form, amenable for an operational interpretation, by their Fourier transforms, i.e., in an analysis with $\mathbb{R}^s \ni \vec{x} \longmapsto e^{i\vec{q}\vec{x}} \in \mathbf{U}(1)$, collected at the invariant singularities $\vec{q}^2 = 0, \mp 1$:

$$\begin{array}{c} \mathbf{O}(s), \\ s = 1, 2, 3, \ldots, \\ r = \sqrt{\vec{x}^2}, \\ \nu \in \mathbb{R}, \\ N = 0, 1, 2, \ldots, \\ a \in \mathbb{R}, \end{array} \left\{ \begin{aligned} \int \frac{d^sq}{\pi^{\frac{s}{2}}} \frac{\Gamma(\frac{s}{2}-\nu)}{(\vec{q}^2)^{\frac{s}{2}-\nu}} e^{i\vec{q}\vec{x}} &= \frac{\Gamma(\nu)}{(\frac{r^2}{4})^\nu}, \\ \int \frac{d^sq}{\pi^{\frac{s}{2}}} \frac{\Gamma(\frac{s}{2}-\nu)}{(\vec{q}^2+1)^{\frac{s}{2}-\nu}} e^{i\vec{q}\vec{x}} &= \frac{2\mathcal{K}_\nu(r)}{(\frac{r}{2})^\nu}, \\ \int \frac{d^sq}{\pi^{\frac{s}{2}}} \frac{\Gamma(\frac{s}{2}-\nu)}{(\vec{q}^2 \pm io-1)^{\frac{s}{2}-\nu}} e^{i\vec{q}\vec{x}} &= \frac{2\mathcal{K}_\nu(\pm ir)}{(\frac{r}{2})^\nu} = \frac{(-\pi\mathcal{N}_\nu \mp i\pi\mathcal{J}_\nu)(r)}{(\frac{r}{2})^\nu}, \\ \int \frac{d^sq}{\pi^{\frac{s}{2}-1}} \delta^{(N)}(1-\vec{q}^2) e^{i\vec{q}\vec{x}} &= \frac{\pi\mathcal{J}_{\frac{s}{2}-1-N}(r)}{(\frac{r}{2})^{\frac{s}{2}-1-N}}, \\ \int \frac{d^sq}{\pi^{\frac{s}{2}}} \frac{\Gamma(\frac{s}{2}+\nu)}{(\vec{q}^2 \pm io+a)^{\frac{s}{2}+\nu}} &= \frac{\Gamma(\nu)}{(a \pm io)^\nu}, \ \nu > 0. \end{aligned} \right.$$

Their meaningful definition depends on the involved Gamma and Beta functions:

$$\begin{aligned} \Gamma(\nu) &= \int_0^\infty dz\, z^{\nu-1} e^{-z} = 2\int_0^\infty dz\, z^{2\nu-1} e^{-z^2}, \text{ for } \nu \in \mathbb{C},\ \nu \neq 0, -1, -2, \ldots, \\ \frac{\Gamma(\nu_1)\Gamma(\nu_2)}{\Gamma(\nu_1+\nu_2)} &= \int_0^1 dz\, z^{\nu_1-1}(1-z)^{\nu_2-1} = \int_1^\infty dz\, \frac{(z-1)^{\nu_2-1}}{z^{\nu_1+\nu_2}}, \text{ for } \nu_{1,2} \in \mathbb{C},\ \Re\,\nu_{1,2} > 0. \end{aligned}$$

For flat spaces, the Fourier transforms give a harmonic analysis with respect to irreducible unitary translation representations. For simple poles, especially for the Dirac contributions $\delta(1 - \vec{q}^2)$, an interpretation with time and position translations and energies and momenta makes sense.

The io-prescription in the "momentum"-distributions computes the integral with a "small" positive $o > 0$ (the Greek omicron o means little o), and takes the result in the limit $o \to 0$. The real-imaginary decomposition involves principal value P and Dirac distributions, related to modulus and phase, respectively:

$$\begin{aligned} a \in \mathbb{R} : \log(a - io) &= \log|a| - i\pi\vartheta(-a), \\ N = 0, 1, 2, \ldots : \quad \frac{\Gamma(1+N)}{(a-io)^{1+N}} &= \frac{\Gamma(1+N)}{a_{\mathrm{P}}^{1+N}} + i\pi\delta^{(N)}(-a) = -(-\frac{\partial}{\partial a})^{1+N} \log(a - io) \\ &= (-\frac{\partial}{\partial a})^k \frac{\Gamma(1+N-k)}{(a-io)^{1+N-k}} \text{ for } k = 0, \ldots, N, \\ \mathbb{R} \ni \nu \neq 1, 2, \ldots : \quad \frac{1}{(a-io)^\nu} &= \frac{\vartheta(a) + \vartheta(-a)e^{i\nu\pi}}{|a|^\nu}. \end{aligned}$$

Via the derivations of the logarithm, the integer order of a pole coincides with the argument of its normalizing Gamma function in the numerator.

The $\mathbf{O}(s)$-scalar Macdonald, Neumann and Bessel functions $\mathcal{K}_\nu, \mathcal{J}_\nu, \mathcal{N}_\nu$ for index ν can be obtained for any dimension s by an appropriate denominator power combination $\frac{s}{2} - \nu$ with matching numerator Gamma function. They have the projection properties for transitions to less dimensions:

$$\begin{aligned} s = n+m \geq 2,\ \mathbf{O}(s) &\supset \mathbf{O}(n),\\ \int \frac{d^m x}{(4\pi)^{\frac{m}{2}}} \int \frac{d^s q}{\pi^{\frac{s}{2}}} \frac{\Gamma(\frac{s}{2}-\nu)}{(\vec{q}^2+io+k)^{\frac{s}{2}-\nu}} e^{i\vec{q}\vec{x}} &= \int \frac{d^n q}{\pi^{\frac{n}{2}}} \frac{\Gamma(\frac{n+m}{2}-\nu)}{(\vec{q}^2+io+k)^{\frac{n+m}{2}-\nu}} e^{i\vec{q}\vec{x}} = \frac{2\mathcal{K}_{\nu-\frac{m}{2}}(\iota r)}{(\frac{r}{2})^{\nu-\frac{m}{2}}},\\ \text{with Killing form-definiteness } k &= (1,-1) = \iota^2\\ \text{for compactness } \iota &= (1,i) \text{ (noncompact, compact).} \end{aligned}$$

For the leading "angular" variable, there is the compact-noncompact partnership, relating sphere and hyperboloid with negative and positive definite metric, respectively, for $s = 1, 2, \ldots$, with $\vec{\omega}_0 = 1$:

$$\begin{gathered} \text{compact } \Omega^s \ni \vec{\omega}_s = \begin{pmatrix} \cos\theta \\ \sin\theta\,\vec{\omega}_{s-1} \end{pmatrix} \leftrightarrow \text{noncompact } \mathcal{Y}^s \ni \mathrm{y}_s = \begin{pmatrix} \cosh\psi \\ \sinh\psi\,\vec{\omega}_{s-1} \end{pmatrix},\\ \text{with } \vec{q}^2 - 1 \leftrightarrow \text{with } \vec{q}^2 + 1,\\ (\theta,\ \cos\theta = \zeta,\ \int_0^{\frac{\pi}{2}} d\theta = \int_0^1 \frac{d\zeta}{\sqrt{1-\zeta^2}}) \leftrightarrow (\psi,\ \cosh\psi = \kappa,\ \int_0^\infty d\psi = \int_1^\infty \frac{d\kappa}{\sqrt{\kappa^2-1}}). \end{gathered}$$

The sphere volume can be obtained with a Dirac distribution, supported by $q_0^2+\vec{q}^2 = +1$ (spherical invariant), or a pole at $\vec{q}^2 = -1$ (hyperbolic invariant):

$$|\Omega^s| = \int d^s\omega = \int d^{1+s}q\, 2\delta(q_0^2+\vec{q}^2-1) = \int \frac{2d^s q}{(\vec{q}^2+1)^{\frac{1+s}{2}}} = \frac{2\pi^{\frac{1+s}{2}}}{\Gamma(\frac{1+s}{2})}.$$

The sphere volume ratio for equal rank action groups $(\mathbf{O}(2R), \mathbf{O}(1+2R))$ is a π-independent line integration, leading to Beta functions (where defined):

$$\begin{aligned} \frac{|\Omega^s|}{2} &= \frac{\pi^{\frac{1+s}{2}}}{\Gamma(\frac{1+s}{2})},\quad \frac{|\Omega^{2R-1}|}{2} = \frac{1}{\Gamma(R)}\pi^R,\quad \frac{|\Omega^{2R}|}{2} = \frac{\Gamma(\frac{1}{2})}{\Gamma(\frac{1}{2}+R)}\pi^R,\\ \frac{|\Omega^s|}{|\Omega^{s-1}|} &= \frac{\Gamma(\frac{1}{2})\Gamma(\frac{s}{2})}{\Gamma(\frac{1+s}{2})} = \int_0^\pi d\theta\,\sin^{s-1}\theta,\quad \frac{|\Omega^{2+s}|}{|\Omega^s|} = \frac{2\pi}{1+s},\quad \frac{\Gamma(\frac{1}{2}+R)}{\Gamma(\frac{1}{2})} = \frac{\Gamma(2R)}{2^{2R-1}\Gamma(R)},\\ \int_0^\pi d\theta\,\sin^\nu\theta &= \int_{-1}^1 d\zeta\,(1-\zeta^2)^{\frac{\nu-1}{2}}\\ &= \begin{cases} \int_0^1 dz\, z^{-\frac{1}{2}}(1-z)^{\frac{\nu-1}{2}} = \frac{\Gamma(\frac{1}{2})\Gamma(\frac{\nu+1}{2})}{\Gamma(\frac{\nu+2}{2})} \text{ with } z = \cos^2\theta,\\ 2^\nu \int_0^1 dz\,[z(1-z)]^{\frac{\nu-1}{2}} = 2^\nu \frac{\Gamma(\frac{\nu+1}{2})^2}{\Gamma(\nu+1)} \text{ with } z = \sin^2\frac{\theta}{2}, \end{cases}\\ \int_0^\infty d\psi\,\sinh^\nu\psi &= \int_1^\infty d\kappa\,(\kappa^2-1)^{\frac{\nu-1}{2}} = \frac{1}{2}\int_0^1 dz\, z^{-\frac{\nu}{2}-1}(1-z)^{\frac{\nu-1}{2}}. \end{aligned}$$

Scalar integrals can be reduced in two-dimensional steps by derivations (*2-sphere spread*) $\frac{d}{d\frac{r^2}{4\pi}} = 2\pi\frac{1}{r}\frac{d}{dr}$ to a lower-dimensional-group dependence. The corresponding orthogonal Lie algebra rank is reduced by one:

$$\begin{aligned} s \geq 3,\ \mathbf{O}(s) \supset \mathbf{O}(s-2):\quad \int d^s q\,\mu(\vec{q}^2)e^{i\vec{q}\vec{x}} &= -\frac{d}{d\frac{r^2}{4\pi}} \int d^{s-2}q\,\mu(\vec{q}^2)e^{i\vec{q}\vec{x}},\\ \text{rank}_{\mathbb{R}} \log \mathbf{O}(s) = R_s:\ R_2 = R_3 = 1 &= R_{s+2} - R_s, \end{aligned}$$

explicitely for $s \geq 4$:

$$\begin{aligned}
\int d^sq\,\mu(\vec{q}^2)e^{i\vec{q}\vec{x}} &= |\Omega^{s-2}|\int_0^\infty q^{s-1}dq\,\mu(q^2)\int_0^\pi d\theta\,\sin^{s-2}\theta\,e^{iqr\cos\theta}\\
&= |\Omega^{s-2}|\int_0^\infty q^{s-1}dq\,\mu(q^2)\int_{-1}^1 d\zeta\,(1-\zeta^2)^{\frac{s-3}{2}}\,e^{iqr\zeta}\\
&= \frac{(s-3)|\Omega^{s-2}|}{2}(-\frac{d}{d\frac{r^2}{4}})\int_0^\infty q^{s-3}dq\,\mu(q^2)\int_{-1}^1 d\zeta\,(1-\zeta^2)^{\frac{s-5}{2}}\,e^{iqr\zeta}\\
&= \frac{(s-3)|\Omega^{s-2}|}{2}(-\frac{d}{d\frac{r^2}{4}})\int_0^\infty q^{s-3}dq\,\mu(q^2)\int_0^\pi d\theta\,\sin^{s-4}\theta\,e^{iqr\cos\theta}\\
&= -\frac{d}{d\frac{r^2}{4\pi}}\int d^{s-2}q\,\mu(\vec{q}^2)e^{i\vec{q}\vec{x}},\text{ with }(s-3)|\Omega^{s-2}| = 2\pi|\Omega^{s-4}|.
\end{aligned}$$

The Fourier transformation of "momentum" distributions is structurally different for even and odd dimensions with respect to the "leading angle"-dependence for polar coordinates on the sphere $\Omega^{s-1}\subset\mathbb{R}^s$, starting with $s=2,3$:

$$\begin{aligned}
\int d^2q\,\mu(\vec{q}^2)e^{i\vec{q}\vec{x}} &= 2\int_0^\infty qdq\,\mu(q^2)\int_0^\pi d\theta\,e^{iqr\cos\theta} &&= 2\int_0^\infty qdq\,\mu(q^2)\int_{-1}^1\frac{d\zeta}{\sqrt{1-\zeta^2}}\,e^{iqr\zeta},\\
\int d^3q\,\mu(\vec{q}^2)e^{i\vec{q}\vec{x}} &= \frac{4\pi}{r}\int_0^\infty qdq\,\mu(q^2)\,\sin qr &&= -\frac{d}{d\frac{r^2}{4\pi}}\int dq\,\mu(q^2)\,e^{iqr}.
\end{aligned}$$

These relations, and similar ones in the following, are somewhat formal in the Fourier integral-form since the radial variable $r=\sqrt{\vec{x}^2}$ is taken from spaces with different dimensions. They have to be taken in the integrated form, e.g., for the Yukawa potential $2\pi\frac{e^{-r}}{r} = -\frac{d}{d\frac{r^2}{4\pi}}e^{-r}$. By 2-sphere spreads, scalar integrals can be reduced to the minimal abelian group-dependence:

$$\begin{aligned}
R=0,1,2,\ldots:\int\frac{d^{1+2R}q}{\pi^R}\,\mu(\vec{q}^2)e^{i\vec{q}\vec{x}} &= (-\frac{d}{d\frac{r^2}{4}})^R\int\frac{dq}{\pi}\,\mu(q^2)e^{iqr}, &&\text{to }\mathbf{O}(1),\\
R=1,2,3,\ldots:\int\frac{d^{2R}q}{\pi^R}\,\mu(\vec{q}^2)e^{i\vec{q}\vec{x}} &= (-\frac{d}{d\frac{r^2}{4}})^{R-1}\int\frac{d^2q}{\pi}\,\mu(\vec{q}^2)e^{i\vec{q}\vec{x}}, &&\text{to }\mathbf{O}(2).
\end{aligned}$$

With a 2-sphere spread for the index increase by one $\nu\longrightarrow\nu+1$:

$$\frac{(2\mathcal{K}_{\nu+1},\pi\mathcal{N}_{\nu+1},\pi\mathcal{J}_{\nu+1})(r)}{(\frac{r}{2})^{\nu+1}} = -\frac{d}{d\frac{r^2}{4}}\frac{(2\mathcal{K}_\nu,\pi\mathcal{N}_\nu,\pi\mathcal{J}_\nu)(r)}{(\frac{r}{2})^\nu},$$

$$\begin{aligned}
\frac{2\mathcal{K}_{\nu+N}(\iota r)}{(\frac{r}{2})^{\nu+N}} &= \int\frac{d^sq}{\pi^{\frac{s}{2}}}\frac{\Gamma(\frac{s}{2}-\nu-N)}{(\vec{q}^2+io+k)^{\frac{s}{2}-\nu-N}}e^{i\vec{q}\vec{x}} = \int\frac{d^{s+2N}q}{\pi^{\frac{s}{2}+N}}\frac{\Gamma(\frac{s}{2}-\nu)}{(\vec{q}^2+io+k)^{\frac{s}{2}-\nu}}e^{i\vec{q}\vec{x}}\\
&= (-\frac{d}{d\frac{r^2}{4}})^N\int\frac{d^sq}{\pi^{\frac{s}{2}}}\frac{\Gamma(\frac{s}{2}-\nu)}{(\vec{q}^2+io+k)^{\frac{s}{2}-\nu}}e^{i\vec{q}\vec{x}} = (-\frac{d}{d\frac{r^2}{4}})^N\frac{2\mathcal{K}_\nu(\iota r)}{(\frac{r}{2})^\nu},
\end{aligned}$$

all (half-)integer index functions above can be obtained as follows:

$$\mathbb{R}_+\ni r\longmapsto\frac{(2\mathcal{K}_\nu,\pi\mathcal{N}_\nu,\pi\mathcal{J}_\nu)(r)}{(\frac{r}{2})^\nu} = \begin{cases}\left(-\frac{d}{d\frac{r^2}{4}}\right)^N\sqrt{\pi}\,(e^{-r},\sin r,\cos r),\\ \qquad\nu=N-\frac{1}{2}=-\frac{1}{2},\frac{1}{2},\frac{3}{2},\ldots,\\ \left(-\frac{d}{d\frac{r^2}{4}}\right)^N(2\mathcal{K}_0,\,\pi\mathcal{N}_0,\,\pi\mathcal{J}_0)\,(r),\\ \qquad\nu=N=0,1,2,\ldots,\end{cases}$$

$$\begin{aligned}
(\mathcal{K}_{-N},\mathcal{N}_{-N},\mathcal{J}_{-N})(r) &= (\mathcal{K}_N,(-1)^N\mathcal{N}_N,(-1)^N\mathcal{J}_N)(r),\\
(\mathcal{K}_{-N-\frac{1}{2}},\mathcal{N}_{-N-\frac{1}{2}},\mathcal{J}_{-N-\frac{1}{2}})(r) &= (\mathcal{K}_{N+\frac{1}{2}},(-1)^N\mathcal{J}_{N+\frac{1}{2}},(-1)^{N+1}\mathcal{N}_{N+\frac{1}{2}})(r).
\end{aligned}$$

The half-integer index functions are relevant for time or 3-position, in general for odd-dimensional manifolds $s = 1 + 2R = 1, 3, \ldots$, ($R$ for rank) with definiteness $k = (1, -1) = \iota^2$ for compactness $\iota = (1, i)$ and hyperbolic and spherical invariant. They start with exponentials and (co)sines as coefficients of the real one-dimensional Lie groups $\mathbf{D}(1) \cong \mathbf{SO}_0(1, 1) \cong \mathbb{R}$ and $\mathbf{U}(1) \cong \mathbf{SO}(2) \cong \mathbb{R}/\mathbb{Z}$:

$$\begin{aligned}
R \geq 0: \int \frac{d^{1+2R}q}{\pi^{\frac{1}{2}+R}} \frac{\Gamma(1+R)}{(\vec{q}^2+io+k)^{1+R}} e^{i\vec{q}\vec{x}} &= \sqrt{\tfrac{r}{2}}\,(2\mathcal{K}_{-\frac{1}{2}}, -\pi\mathcal{N}_{-\frac{1}{2}} - i\pi\mathcal{J}_{-\frac{1}{2}})(r)\\
&= \sqrt{\pi}\,(e^{-r}, -\sin r - i\cos r),\\
R \geq 1: \quad \int \frac{d^{1+2R}q}{\pi^{\frac{1}{2}+R}} \frac{\Gamma(R)}{(\vec{q}^2+io+k)^{R}} e^{i\vec{q}\vec{x}} &= \frac{(2\mathcal{K}_{\frac{1}{2}}, -\pi\mathcal{N}_{\frac{1}{2}} - i\pi\mathcal{J}_{\frac{1}{2}})(r)}{\sqrt{\frac{r}{2}}}\\
&= -\frac{d}{d\frac{r^2}{4}} \sqrt{\pi}\,(e^{-r}, -\sin r - i\cos r)\\
&= \sqrt{\pi}\,\frac{(e^{-r}, \cos r - i\sin r)}{\frac{r}{2}}.
\end{aligned}$$

The hyperbolic Macdonald and spherical Neumann and Bessel functions are defined with half-integer index. They involve the Yukawa potential k_0 and spherical waves j_0:

$$\begin{aligned}
N = 0, 1, 2, \ldots: \; (k_N, n_N, j_N)(r) &= (\tfrac{r}{2})^N (-\tfrac{d}{d\frac{r^2}{4}})^N \frac{(e^{-r}, \cos r, \sin r)}{r}\\
&= \frac{1}{\sqrt{4\pi}} \frac{(2\mathcal{K}_{N+\frac{1}{2}}, -\pi\mathcal{N}_{N+\frac{1}{2}}, \pi\mathcal{J}_{N+\frac{1}{2}})(r)}{\sqrt{\frac{r}{2}}}.
\end{aligned}$$

The integer index functions are relevant for four-dimensional spacetime, in general for even-dimensional manifolds $s = 2R = 2, 4, \ldots$ (R for rank). They integrate exponentials and (co)sines with positive measures for the invariants with definiteness $k = (1, -1)$:

$$\begin{aligned}
R \geq 1: \int \frac{d^{2R}q}{\pi^R} \frac{\Gamma(1+R)}{(\vec{q}^2+io+k)^{1+R}} e^{i\vec{q}\vec{x}} &= \tfrac{r}{2}\,(2\mathcal{K}_{-1}, -\pi\mathcal{N}_{-1} - i\pi\mathcal{J}_{-1})(r)\\
&= \int \frac{d\psi}{\sinh^2\psi} \left(e^{-r\cosh\psi}, \cos(r\cosh\psi) - i\sin(r\cosh\psi)\right)\\
&= \int_1^\infty \frac{d\kappa}{\sqrt{(\kappa^2-1)^3}} \left(e^{-\kappa r}, \cos\kappa r - i\sin\kappa r\right),\\
\int \frac{d^{2R}q}{\pi^R} \frac{\Gamma(R)}{(\vec{q}^2+io+k)^{R}} e^{i\vec{q}\vec{x}} &= (2\mathcal{K}_0, -\pi\mathcal{N}_0 - i\pi\mathcal{J}_0)(r)\\
&= \int d\psi \left(e^{-r\cosh\psi}, \cos(r\cosh\psi) - i\sin(r\cosh\psi)\right)\\
&= 2\int_1^\infty \frac{d\kappa}{\sqrt{\kappa^2-1}} \left(e^{-\kappa r}, \cos\kappa r - i\sin\kappa r\right)\\
&= \int_1^\infty \frac{d\kappa\,\kappa}{\sqrt{(\kappa^2-1)^3}} \left(\frac{e^{-\kappa r}}{\frac{r}{2}}, \frac{\sin\kappa r + i\cos\kappa r}{\frac{r}{2}}\right).
\end{aligned}$$

The even-dimensional integrals integrate singularities of odd-dimensional integrals, e.g.:

$$\begin{aligned}
2\mathcal{K}_0(r) &= \int \frac{d^2q}{\pi} \frac{1}{\vec{q}^2+1} e^{i\vec{q}\vec{x}} = \int dp \int \frac{dq}{\pi} \frac{1}{q^2+p^2+1} e^{iqr}\\
&= \int \frac{dp}{\sqrt{p^2+1}} e^{-\sqrt{p^2+1}r} = 2\int_1^\infty \frac{d\kappa}{\sqrt{\kappa^2-1}} e^{-\kappa r}.
\end{aligned}$$

The integer-index functions have the not so familiar expansions:

$$2\mathcal{K}_0(r) = \sum_{k=0}^{\infty} \frac{(\frac{r^2}{4})^k}{(k!)^2}[2\chi_k - \log \tfrac{r^2}{4}],$$

$$\begin{aligned}
2\mathcal{K}_0(ir) &= [-\pi\mathcal{N}_0 - i\pi\mathcal{J}_0](r),\ \log \tfrac{-r^2+io}{4} = \log \tfrac{r^2}{4} + i\pi,\\
-\pi\mathcal{N}_0(r) &= \sum_{k=0}^{\infty} \frac{(-\frac{r^2}{4})^k}{(k!)^2}[2\chi_k - \log \tfrac{r^2}{4}],\\
\pi\mathcal{J}_0(r) &= \pi\sum_{k=0}^{\infty} \frac{(-\frac{r^2}{4})^k}{(k!)^2} = \int_0^{\pi} d\theta\, \cos(r\cos\theta) = \int_{-1}^{1} \frac{d\zeta}{\sqrt{1-\zeta^2}} \cos\zeta r,
\end{aligned}$$

with the characteristic logarithmic singularities at $r = 0$ for $\mathcal{K}_0$ and $\mathcal{N}_0$ and Euler's constant:

$$\begin{aligned}
\chi_0 &= \Gamma'(1) = \int_0^{\infty} dz\, e^{-z} \log z\\
&= \lim_{k\to\infty}[\log k - (1 + \tfrac{1}{2} + \ldots + \tfrac{1}{k})] = -0.5772\ldots,\\
\chi_k &= \Gamma'(1) + 1 + \tfrac{1}{2} + \ldots + \tfrac{1}{k},\ k = 1, 2, \ldots
\end{aligned}$$

The Bessel-related functions are essentially bounded, i.e., from $L^\infty(\mathbb{R})$:

$$\nu \in \mathbb{R}: \ \frac{\mathcal{J}_\nu(r)}{(\frac{r}{2})^\nu} = \sum_{k=0}^{\infty} \frac{(-\frac{r^2}{4})^k}{\Gamma(k+\nu+1)k!}.$$

The "(energy-)momentum" distributions $\frac{d^s q}{(\vec{q}^2+io+k)^{\frac{s}{2}-\nu}}$, $k = 0, \pm 1$, with positive integer powers,

$$\tfrac{s}{2} - \nu = n = 0, 1, 2, \ldots,$$

i.e., with corresponding *half-integer index ν for odd dimensions $s = 1 + 2R$ and integer index ν for even dimensions $s = 2R$*, are *representation measures*. Their Fourier transforms lead to *representation functions* for $\nu < 0$:

$$\nu < 0,\ \tfrac{s}{2} - \nu = n \in \mathbb{N},\ \text{e.g.,} \begin{cases} \frac{d^2 q}{(\vec{q}^2+io+k)^{(2,3,\ldots)}};\ \mathbf{O}(2),\\ \frac{d^3 q}{(\vec{q}^2+io+k)^{(2,3,\ldots)}};\ \mathbf{O}(3),\\ \frac{d^4 q}{(\vec{q}^2+io+k)^{(3,4,\ldots)}};\ \mathbf{O}(4).\end{cases}$$

The proper *subgroup representation distributions* for $\nu \geq 0$ arise by derivations with $-\vec{\partial}^2 + k \sim \vec{q}^2 + k$ from the representation functions:

$$\nu \geq 0,\ \tfrac{s}{2} - \nu = n \in \mathbb{N},\ \text{e.g.,} \begin{cases} \frac{d^2 q}{\vec{q}^2 + io + k}; & \mathbf{O}(2), \\ \frac{d^3 q}{\vec{q}^2 + io + k}; & \mathbf{O}(2) \subset \mathbf{O}(3), \\ \frac{d^4 q}{(\vec{q}^2 + io + k)^{(1,2)}}; & \mathbf{O}(2) \subset \mathbf{O}(3) \subset \mathbf{O}(4). \end{cases}$$

They distribute embedded subgroup $\mathbf{O}(n) \subset \mathbf{O}(s)$ representation coefficients.

The *characteristic representation measures* for the Fourier-integration on maximally symmetric definite metric odd- and even-dimensional manifolds are with $k = \pm 1$ for the hyperbolic and spherical case, respectively:

$$\mathbf{O}(s): \quad \begin{cases} \frac{d^{1+2R} q}{\pi^R} \frac{\Gamma(1+R)}{(\vec{q}^2 + io + km^2)^{1+R}}, & s = 1 + 2R = 1, 3, \ldots, \\ \frac{d^{2R} q}{\pi^R} \frac{\Gamma(1+R)}{(\vec{q}^2 + io + km^2)^{1+R}}. & s = 2R = 2, 4, \ldots \end{cases}$$

The action with the Laplacian for $k = \iota^2$, $\iota = (1, i)$ relates integrals with different pole orders, $\frac{s}{2} - \nu \neq 1$:

$$\begin{aligned} (-\vec{\partial}^2 + k) \int \frac{d^s q}{\pi^{\frac{s}{2}}} \frac{\Gamma(\frac{s}{2} - \nu)}{(\vec{q}^2 + io + k)^{\frac{s}{2} - \nu}} e^{i\vec{q}\vec{x}} &= \int \frac{d^s q}{\pi^{\frac{s}{2}}} \frac{\Gamma(\frac{s}{2} - \nu)}{(\vec{q}^2 + io + k)^{\frac{s}{2} - \nu - 1}} e^{i\vec{q}\vec{x}} \\ = (-\vec{\partial}^2 + k) \frac{2\mathcal{K}_\nu(\iota r)}{(\frac{r}{2})^\nu} = (\tfrac{s}{2} - \nu - 1) \frac{2\mathcal{K}_{\nu+1}(\iota r)}{(\frac{r}{2})^{\nu+1}} &= (\tfrac{s}{2} - \nu - 1)(-\frac{d}{d\frac{r^2}{4}}) \frac{2\mathcal{K}_\nu(\iota r)}{(\frac{r}{2})^\nu}, \end{aligned}$$

which, with the $\mathbf{O}(s)$-invariant derivative, plays a role as Schrödinger equation:

$$[-\vec{\partial}^2 + (\tfrac{s}{2} - \nu - 1) \frac{d}{d\frac{r^2}{4}}] \frac{2\mathcal{K}_\nu(\iota r)}{(\frac{r}{2})^\nu} = -k \frac{2\mathcal{K}_\nu(\iota r)}{(\frac{r}{2})^\nu}.$$

The first order pole integrals as distributions for the Cartan subgroup $\mathbf{O}(2)$,

$$\int \frac{d^s q}{\pi^{\frac{s}{2}}} \frac{1}{\vec{q}^2} e^{i\vec{q}\vec{x}} = \frac{\Gamma(\frac{s}{2} - 1)}{(\frac{r^2}{4})^{\frac{s}{2} - 1}}, \quad \int \frac{d^s q}{\pi^{\frac{s}{2}}} \frac{1}{\vec{q}^2 + io + k} e^{i\vec{q}\vec{x}} = \frac{2\mathcal{K}_{\frac{s}{2} - 1}(\iota r)}{(\frac{r}{2})^{\frac{s}{2} - 1}} = \frac{(2\mathcal{K}_{\frac{s}{2} - 1}, -\pi\mathcal{N}_{\frac{s}{2} - 1} - i\pi\mathcal{J}_{\frac{s}{2} - 1})(r)}{(\frac{r}{2})^{\frac{s}{2} - 1}},$$

contain Bessel functions as translation representation coefficients, e.g., for Euclidean and Poincaré groups (here the imaginary part) with homogenous equations:

$$\begin{aligned} \frac{1}{\vec{q}^2 + io - 1} = -i\pi\delta(\vec{q}^2 - 1) + \frac{1}{\vec{q}_{\mathrm{P}}^2 - 1}, \quad \int \frac{d^s q}{\pi^{\frac{s}{2} - 1}} \delta(\vec{q}^2 - 1) e^{i\vec{q}\vec{x}} &= \frac{\pi\mathcal{J}_{\frac{s}{2} - 1}(r)}{(\frac{r}{2})^{\frac{s}{2} - 1}}, \\ (-\vec{\partial}^2 - 1) \frac{\pi\mathcal{J}_{\frac{s}{2} - 1}(r)}{(\frac{r}{2})^{\frac{s}{2} - 1}} &= 0. \end{aligned}$$

The real parts are kernels (inverse Laplacians) $(-\vec{\partial}^2 + k)^{-1}$ — for the pole at $\vec{q}^2 = 1$ with the principal value $\vec{q}_{\mathrm{P}}^2$ integration:

$$k = 0, 1, -1: \quad (-\vec{\partial}^2 + k) \int \frac{d^s q}{\pi^{\frac{s}{2}}} \frac{1}{\vec{q}_{\mathrm{P}}^2 + k} e^{i\vec{q}\vec{x}} = (4\pi)^{\frac{s}{2}} \delta(\vec{x}).$$

These distributions are used for interactions.

The different behavior for even and odd dimensions shows up in the rational and logarithmic Laplace functions:

$$\begin{aligned} s=1: &\quad \tfrac{d^2}{dx^2}|x| &&= 2\delta(x),\\ s=2: &-\vec{\partial}^2 \log \tfrac{\vec{x}^2}{4} &&= 2\pi\delta(\vec{x}),\\ s=3: &\quad -\vec{\partial}^2 \tfrac{1}{|\vec{x}|} &&= 4\pi\delta(\vec{x}). \end{aligned}$$

An integral, irreducibly $\mathbf{O}(s)$-dependent via harmonic polynomials $(\vec{x})^L$ or with spherical harmonics $(\frac{\vec{x}}{r})^L = r^{-L}(\vec{x})^L$, can be computed, by derivations, from a scalar one:

$$\int d^s q\,(-2i\vec{q})^L \mu(\vec{q}^2)e^{i\vec{q}\vec{x}} = (-2\vec{\partial})^L \int d^s q\,\mu(\vec{q}^2)e^{i\vec{q}\vec{x}} = (\vec{x})^L(-\tfrac{d}{d\frac{r^2}{4}})^L \int d^s q\,\mu(\vec{q}^2)e^{i\vec{q}\vec{x}},$$
$$L = 0, 1, 2, \ldots$$

This gives for the special functions:

$$\mathbf{O}(s): \begin{cases} \int \frac{d^s q}{\pi^{\frac{s}{2}}} \frac{(-2i\vec{q})^L\Gamma(\frac{s}{2}-\nu+L)}{(\vec{q}^2+io+k)^{\frac{s}{2}-\nu+L}} e^{i\vec{q}\vec{x}} &= (\vec{x})^L(-\frac{d}{d\frac{r^2}{4}})^L \frac{2\mathcal{K}_{\nu-L}(\iota r)}{(\frac{r}{2})^{\nu-L}} = (\vec{x})^L \frac{2\mathcal{K}_\nu(\iota r)}{(\frac{r}{2})^\nu}\\ &= (\vec{x})^L \int \frac{d^s q}{\pi^{\frac{s}{2}}} \frac{\Gamma(\frac{s}{2}-\nu)}{(\vec{q}^2+io+k)^{\frac{s}{2}-\nu}} e^{i\vec{q}\vec{x}}, \end{cases}$$

with the Schrödinger-like equations:

$$(-\vec{\partial}^2+k)\,(\vec{x})^L \frac{2\mathcal{K}_\nu(\iota r)}{(\frac{r}{2})^\nu} = (\tfrac{s}{2}-\nu-1+L)\,(\vec{x})^L \frac{2\mathcal{K}_{\nu+1}(\iota r)}{(\frac{r}{2})^{\nu+1}}.$$

Chapter 10
Free States and Particles

Classical mechanics utilizes the idealized concept of a pointlike particle — a point in the real s-dimensional position space $\mathbb{M}^s$, especially $s = 3$, or in the phase space $\mathbb{P}^{2ns}$ for n particles. With the position-momentum uncertainty of quantum mechanics, and, also, with the detection of spin ("intrinsic angular momentum"), this concept became inappropriate or fuzzy, to say the least. Particle properties are characterized by invariants and eigenvalues for operations: The experimental setup decides about the properties which will be measured with their eigenvalues. For instance, one may measure the position expectation value of a particle. In quantum mechanics, even more in quantum field theory, space and time aquire a new meaning. The "semiquantum" Bohr–Sommerfeld model is not adequate with respect to the corresponding quantum structure — electrons "in" atoms have no orbits and no circulation period. Or, in quantum electrodynamics, a Lagrangian formulated interaction $\overline{\boldsymbol{\Psi}}(x)\gamma_j\boldsymbol{\Psi}(x)\mathbf{A}^j(x)$ with an electron-positron Dirac field $\boldsymbol{\Psi}$ and a field $\mathbf{A}$ for the electromagnetic potential does not imply an electron, a positron and a photon interacting with each other at one spacetime point x. In quantum theory, the classical space-time ontology has to be abandoned. Space and time parametrize operations, implemented by interactions and particles, not arenas for mass point orbits. The local operator products (interaction vertices) signify invariant rearrangements of properties, i.e., scalar representations as products of nonscalar ones, e.g., for the Lorentz group $\mathbf{SL}(2, \mathbb{C})$ or the hyperisospin group $\mathbf{U}(2)$.

With the basic relevance of symmetries and operations, and in a deep abstraction, Wigner defined the "essence" of a *particle* by the Poincaré group — more detailed: a particle is described by an infinite-dimensional Hilbert space acted on by a unitary irreducible representation of the Poincaré group. Positive unitary representations, called Hilbert representations, allow a probability interpretation, here of particle properties. In this impressive radical reduction, all particle properties are explained by a few operations with their eigenvalues and invariants. A particle admits the external translation and rotation operations with the external (spacetimelike) properties mass and spin (polarization), and, as internal operations, electromagnetic windings with

H. Saller, *Operational Symmetries*, DOI 10.1007/978-3-319-58664-9_10

electric charge as internal (chargelike) property — isospin and color symmetry are for interactions only.

With Birkhoff and von Neumann, quantum propositions are Hilbert spaces. A Wigner particle is described by a full vector space, characterized by the "absolute" values for the rotation and translation invariants, spin (polarization) and mass. Measurements give matrix elements for individual "relative" vectors, e.g., for an experiment-determined basis with eigenvectors for a spin direction and position translations. Such a particle definition may be contrasted with the classical idealization of a point particle, characterized by its position and momentum with the time orbit in phase space.

Wigner particles show up in the harmonic analysis of special relativistic fields with respect to the Poincaré group as visible in the on-shell contributions of Feynman propagators. As translation eigenvectors, they have no interactions, they are free. The particle and interaction concepts have some sort of complementarity. Following this definition, *confined* quarks cannot be eigenvectors of spacetime translations, and, therefore, are no particles — how does a quark propagator look like? None of the different types of quark masses can be used as translation invariant. To call confined quarks particles requires a different particle definition. I will stay with Wigner's definition. If one believes in color confinement wherefore, in the context of quantum chromodynamics, there exists no mathematical proof yet, then quarks may be called "hadron-constituting entities (degrees of freedom)" or, together with the confined gluons, "entities (degrees of freedom) for the parametrization of hadrons with their strong interactions." The quark "masses" do not characterize translation representations.

Particle related structures, together with the representations of the nonrelativistic Euclidean group, a Poincaré subgroup, are considered in this chapter, i.e., the representations of interaction free objects.

Since noncompact and compact Lie groups have infinite and finite measures (volumes of parameter spaces), respectively, starting with $\mathbf{D}(1) = \exp \mathbb{R} \cong \mathbb{R}$ as infinite cover group of $\mathbf{U}(1) = \exp i\mathbb{R} \cong \mathbb{R}/\mathbb{Z}$ with parameter space $[0, 2\pi]$, faithful Hilbert representations of noncompact groups are necessarily infinite-dimensional. For the Euclidean and Poincaré group, such representations can be induced by finite-dimensional Hilbert representations of subgroups.

The idea of an induction from subgroups can be visualized in the example of the semidirect Euclidean group $\mathbf{SO}(3) \vec{\times} \mathbb{R}^3$ with the space rotations and space translations as used in the nonrelativistic scattering theory: If a ball is flying on an idealized straight line (translations $\mathbb{R}$ in one direction), the orbit is invariant under the action of axial rotations around this line, i.e., under a subgroup $\mathbf{SO}(2)$ of all space rotations $\mathbf{SO}(3)$. From the full rotation group, there remains as symmetry of the ball line only a polarization subgroup $\mathbf{SO}(2)$. A ball at rest is invariant under all rotations $\mathbf{SO}(3)$.

This example shows, that the different interaction-free motions of the ball with momentum $\vec{q}^2 > 0$ (moving) and $\vec{q}^2 = 0$ (resting) distinguish a fixgroup, called *little group* by Wigner, as the remaining homogeneous symmetry operations for the free particles, in the example the fixgroup $\mathbf{SO}(2)$ in the motion group $\mathbf{SO}(3)$ with 2-sphere manifold $\mathbf{SO}(3)/\mathbf{SO}(2)$ for all flight directions.

Also the Poincaré group $\mathbf{SO}_0(1,3) \vec{\times} \mathbb{R}^4$ with $\mathbb{R}^4$ parametrized by the energy-momenta q has different free motions. The value for the Lorentz invariant square $q^2 = q_0^2 - \vec{q}^2$ characterizes four different types: energy-like $q^2 = m^2 > 0$, light-like $q^2 = 0, q \neq 0$, momentum-like $q^2 < 0$ and trivial $q = 0$. With the exception of the trivial case $q = 0$ (absolute rest), only the causal motion types $q^2 = m^2 \geq 0$ have a real frequency (energy) $q_0 = \sqrt{m^2 + \vec{q}^2}$. It is obvious that for massive particles with $q^2 = m^2 > 0$ the invariance subgroup (fixgroup, little group) in the homogeneous Lorentz group is given by the space rotations $\mathbf{SO}(3) \cong \mathbf{SU}(2)/\mathbf{I}(2)$ around the chosen time axis (rest system) with hyperbolic manifold $\mathbf{SO}_0(1,3)/\mathbf{SO}(3)$ for all time directions. It is somewhat more difficult to understand, that the little group for massless particles, i.e., for light-like motion type $q^2 = m^2 = 0, \ q \neq 0$, is given by the axial rotations $\mathbf{SO}(2) \cong \mathbf{U}(1)$ around the momentum direction. Therefore, Wigner's classification of the unitary irreducible representations of the Poincaré group gives, for the two causal motion types, the experimentally established connection between rotation and translation properties of stable particles:

Massive particles and their fields — electrons, weak bosons, etc. - are acted on by unitary Poincaré group representations induced by $\mathbf{SU}(2) \times \mathbb{R}^4$-representations. The translations $\mathbb{R}^4$ have the energy-momenta as eigenvalues $\mathbb{R}^4 \ni x \longmapsto e^{iqx} \in \mathbf{U}(1)$ with the mass $q^2 = m^2 > 0$ as invariant. The rotation properties are given by the $\mathbf{SU}(2)$-*spin* $J = 0, \frac{1}{2}, 1, \ldots$.

Massless particles and their fields — photons, etc. — are acted on by unitary Poincaré group representations induced by $\mathbf{SO}(2) \times \mathbb{R}^4$. The translations $\mathbb{R}^4 \ni x \longmapsto e^{iqx} \in \mathbf{U}(1)$ have the mass $q^2 = m^2 = 0$ as invariant. The rotation properties are given by the $\mathbf{SO}(2)$-*circularity (polarization, helicity)* $z \in \mathbb{Z}$.

Wigner has classified the Hilbert representations and the spectrum of the characterizing invariants of the Poincaré group. The classification gives no information whatsoever why, in nature, the representations come with the masses and spins actually observed, i.e., it does not explain the experimentally found particle spectrum. In this chapter, only massive particles will be considered.

The last sections with first steps for the operational structure of unstable particles have been worked out with my late friend Walter Blum (1937–2013).

10.1 Irreducible Representations of Translations

Interaction free objects like asymptotic scattering states in nonrelativistic mechanics or free particles in relativistic theories are acted on by translations – by time translations $\mathbb{R}$, by position translations $\mathbb{R}^3$, and by spacetime translations $\mathbb{R}^{1+3}$. Therefore, the mathematical implementation of free objects involves representation structures of the additive groups $\mathbb{R}^n$, $n = 1, 2, \ldots$

The irreducible complex representations of the abelian groups $\mathbb{R}^n$ are, with Schur's lemma, one-dimensional with complex eigenvalues iQ:

$$\mathbb{R}^n \ni x \longmapsto e^{iQx} \in \mathbf{GL}(1, \mathbb{C}), \ \ Qx = Q_1 x_1 + \cdots + Q_n x_n, \ \ iQ \in \mathbb{C}^n \cong \mathbf{irrep}\, \mathbb{R}^n.$$

For *Hilbert representations*, the complex space with the translation weights is reduced to imaginary eigenvalues (weights) $iq \in (i\mathbb{R})^n$ with real (energy-) momenta q:

$$\mathbb{R}^n \ni x \longmapsto e^{iqx} \in \mathbf{U}(1), \quad q \in \mathbb{R}^n \cong \mathbf{irrep}_+\mathbb{R}^n.$$

The simplest case is realized by the harmonic quantum oscillator with — for the product representations — equidistant energy-levels as eigenvalues of the time translations:

$$\mathbb{R} \ni t \longmapsto (e^{iEt})^z = e^{izEt} \in \mathbf{U}(1), \quad E \in \mathbb{R}, \quad z \in \mathbb{Z}.$$

The invariants involve a continuous and discrete structure: The invariant iE characterizes the "basic" unitary representation of the noncompact translation group, the discrete numbers z the product representations $\mathbb{R} \xrightarrow{iE} \mathbf{U}(1) \xrightarrow{z} \mathbf{U}(1)$ with $z = -1$ the dual representation. Those discrete numbers are characteristic examples for the original "quantum numbers" of the "old" Bohr-Sommerfeld quantum theory.

The translations are distinguished by the isomorphy of group $\mathbb{R}^n$ and its irreducible Hilbert representations, characterized by the (energy-)momenta, also an abelian group and, for distinction, if necessary, denoted with a check (for dual space):

$$\begin{aligned}
&\mathbf{irrep}_+\mathbb{R}^n = \{iq \,\big|\, q \in \mathbb{R}^n\} = \check{\mathbb{R}}^n \cong \mathbb{R}^n,\\
&\int d^nx\, e^{iqx}e^{-iq'x} = \delta(\tfrac{q-q'}{2\pi}), \quad \int \tfrac{d^nq}{(2\pi)^n}\, e^{iqx}e^{-iqx'} = \delta(x-x').
\end{aligned}$$

Irreducible dual Hilbert representations have (energy-)momenta which are negative to each other, $e^{ipx} \leftrightarrow e^{-ipx}$. Selfdual translation representations are the sum of dual ones, e.g., of dual irreducible representations:

$$\frac{e^{ipx} + e^{-ipx}}{2} = \cos px.$$

They can be analyzed by positive Dirac distributions:

$$\cos px = \int d^nq\, \frac{\delta(q-p) + \delta(q+p)}{2} e^{iqx}.$$

An example is given by the time behavior of position-momentum $(\mathbf{x}, \mathbf{p})$ of the harmonic oscillator, where the cosine is a diagonal matrix element of the two-dimensional representation of the time translations $\mathbb{R}$ in $\mathbf{SO}(2)$:

$$\mathbb{R} \ni t \longmapsto \cos Et = \int dq_0\, |E|\delta(q_0^2 - E^2)e^{iq_0t},$$

$$\left.\begin{aligned}\tfrac{d\mathbf{x}}{dt} &= E\mathbf{p},\\ \tfrac{d\mathbf{p}}{dt} &= -E\mathbf{x}\end{aligned}\right\} \Rightarrow \begin{pmatrix}\mathbf{x}(t)\\ \mathbf{p}(t)\end{pmatrix} = \begin{pmatrix}\cos Et & \sin Et\\ -\sin Et & \cos Et\end{pmatrix}\begin{pmatrix}\mathbf{x}(0)\\ \mathbf{p}(0)\end{pmatrix}.$$

10.2 Euclidean and Minkowskian Translations

Induced representations of a group start from representations of an "appropriate" subgroup — for a semidirect affine group $G \vec{\times} \mathbb{R}^n$, $G \subseteq \mathbf{GL}(n, \mathbb{R})$, from irreducible representations of the normal subgroup $\mathbb{R}^n$ with the space(-time) translations, characterized by (energy-) momenta, and their fixgroups $H \subseteq G$ with cosets G/H and a G-compatible projection-valued measure thereon.

Adding up, by integration, all unitary irreducible position translation representations:

$$\mathbb{R}^3 \ni \vec{x} \longmapsto e^{i\vec{q}\vec{x}} \in \mathbf{U}(1), \;\; \vec{q} \in \mathbb{R}^3,$$

with their momenta on a 2-sphere $\vec{q}^2 = P^2$, $P > 0$, one obtains spherical Bessel functions as representation coefficients of Euclidean position $\mathbb{E}^3 \cong \mathbf{SO}(3) \vec{\times} \mathbb{R}^3/\mathbf{SO}(3)$ for free (asymptotic) nonrelativistic scattering states:

$$\begin{aligned}\mathbb{E}^3 \ni \vec{x} \longmapsto \int \frac{d^3q}{4\pi P}\, 2\delta(\vec{q}^2 - P^2) e^{i\vec{q}\vec{x}} &= -\frac{1}{Pr}\frac{\partial}{\partial r}\int dq\; \delta(q^2 - P^2) e^{iqr} \\ &= -\frac{1}{P^2 r}\frac{\partial}{\partial r}\cos Pr = \frac{\sin Pr}{Pr} = j_0(Pr) \text{ with } r = |\vec{x}| = \sqrt{\vec{x}^2},\end{aligned}$$

The normalization uses the finite surface $|\Omega^2|$ of the unit 2-sphere $\Omega^2 \cong \mathbf{SO}(3)/\mathbf{SO}(2)$ with fixgroup $\mathbf{SO}(2)$, parametrizing, via the momentum directions, the axial rotation group classes of the compact rotation group:

$$\Omega^2 = \{\vec{\omega} = \frac{\vec{q}}{P} \in \mathbb{R}^3 \,\big|\, \vec{q}^2 = P^2 > 0\}, \;\; |\Omega^2| = \int d^3q\; 2\delta(\vec{q}^2 - 1) = 4\pi.$$

Analogously for spacetime: The addition (integration) of all unitary irreducible translation representations

$$\mathbb{R}^4 \ni x \longmapsto e^{iqx} \in \mathbf{U}(1), \;\; q \in \mathbb{R}^4,$$

with energy-momenta on the mass hyperboloid $q^2 = m^2$, $m > 0$, gives the on-shell part of the Feynman propagator for a free particle. The coefficients of Minkowski spacetime $\mathbb{R}^{(1,3)} \cong \mathbf{SO}_0(1,3) \vec{\times} \mathbb{R}^4/\mathbf{SO}_0(1,3)$ decompose with the characteristic functions $\vartheta(\pm x^2)$ for time- and spacelike translations into Neumann and Macdonald functions, respectively:

$$\begin{aligned}\mathbb{R}^{(1,3)} \ni x \longmapsto\; & \int \tfrac{d^4q}{\mathcal{V}^3 m^2}\, \delta(q^2 - m^2) e^{iqx} \\ &= \tfrac{\pi}{\mathcal{V}^3} \tfrac{\partial}{\partial \frac{m^2x^2}{4}} [-\vartheta(x^2)\pi\mathcal{N}_0(|mx|) + \vartheta(-x^2) 2\mathcal{K}_0(|mx|)] \\ &= \tfrac{\pi}{\mathcal{V}^3} \tfrac{\vartheta(x^2)\pi\mathcal{N}_1(|mx|) + \vartheta(-x^2)2\mathcal{K}_1(|mx|)}{\frac{|mx|}{2}}, \;\text{ with } |x| = \sqrt{|x^2|},\end{aligned}$$

with a timelike oscillating behavior and a spacelike exponential falloff:

$$\begin{aligned}\mathbb{R}^{(1,1)} \ni x \longmapsto \int d^2q\, \delta(q^2-m^2)e^{iqx} &= -\vartheta(x^2)\pi\mathcal{N}_0(|mx|) + \vartheta(-x^2)2\mathcal{K}_0(|mx|)\\ &= \int d\psi\, [\vartheta(x^2)\cos(|mx|\cosh\psi) + \vartheta(-x^2)e^{-|mx|\cosh\psi}],\end{aligned}$$

to be contrasted with the Euclidean two-dimensional free scattering states:

$$\mathbb{E}^2 \ni \vec{x} \longmapsto \int \frac{d^2q}{\pi}\, \delta(\vec{q}^2-P^2)e^{i\vec{q}\vec{x}} = \mathcal{J}_0(|P\vec{x}|) = \int_0^\pi \frac{d\theta}{\pi} \cos(|P\vec{x}|\cos\theta).$$

$\mathcal{V}^3$ is a free normalization factor, e.g., $\mathcal{V}^3m^2 = (2\pi)^3$. The infinities of the Fourier transformed positive Dirac distribution $\delta(q^2-m^2)$ in $\mathcal{K}(0)$ and $\mathcal{N}(0)$ are related to the infinite volume $|\mathcal{Y}^3|$ of the *three-dimensional unit hyperboloid* $\mathcal{Y}^3 \cong \mathbf{SO}_0(1,3)/\mathbf{SO}(3)$ with fixgroup $\mathbf{SO}(3)$, parametrizing, via the momenta, the rotation group classes of the noncompact orthochronous Lorentz group:

$$\mathcal{Y}^3 = \{y = \frac{q}{m} \in \mathbb{R}^4 \,\big|\, q^2 = m^2 > 0,\ \frac{q_0}{m} > 0\},\ |\mathcal{Y}^3| = \int d^4q\, 2\vartheta(q_0)\delta(q^2-1) = \infty.$$

It is possible to characterize the functions for spacetime representations as limits of essentially bounded ones. The mathematically necessary definitions are not discussed here. The hyperboloid volume can be used as a formal normalization factor $\mathcal{V}^3 = |\mathcal{Y}^3|$.

The full Feynman propagators, as used for the expansion of interactions with particle fields, also include the imaginary principal value P off-shell part:

$$\int \frac{d^4q}{\mathcal{V}^3m^2i\pi} \frac{1}{-q^2-io+m^2}e^{iqx}, \text{ with } \frac{1}{i\pi}\frac{1}{-q^2-io+m^2} = \delta(q^2-m^2) + \frac{1}{i\pi}\frac{1}{-q_{\mathrm{P}}^2+m^2}.$$

As used above, the integrals for the momenta with the rotation group and for the energy-momenta with the Lorentz group can be simplified by a *2-sphere spread* related to the introduction of polar coordinates $\mathbb{R}^3 = \mathbb{R}_+ \times \Omega^2$ with the unit 2-sphere Ω^2: A scalar momentum and a scalar energy-momentum integral can be reduced, respectively, to a one-dimensional one with trivial homogeneous group $\mathbf{SO}(1) = \{1\}$ (Fourier-Bessel analysis), and a two-dimensional one with homogeneous abelian Lorentz group $\mathbf{SO}_0(1,1)$ (dilation group):

$$\mathbb{R} \hookrightarrow \mathbf{SO}(3)\vec{\times}\mathbb{R}^3: \begin{cases} \int \frac{d^3q}{4\pi}\, \mu(\vec{q}^2)e^{i\vec{q}\vec{x}} \\ \quad = -\frac{\partial}{\partial r^2}\int dq\, \mu(q^2)\, e^{iqr}|_{r=|\vec{x}|} \\ \quad = \int_0^\infty q^2dq\, \mu(q^2)\, \frac{\sin qr}{qr}.\end{cases}$$

$$\mathbf{SO}_0(1,1)\vec{\times}\mathbb{R}^2 \hookrightarrow \mathbf{SO}_0(1,3)\vec{\times}\mathbb{R}^4: \begin{cases} \int \frac{d^4q}{4\pi}\, \mu(q^2)e^{iqx} \\ \quad = -\frac{\partial}{\partial x^2}\int d^2q\, \mu(q^2)e^{i(q_0x_0-q_3|\vec{x}|)}.\end{cases}$$

10.3 Time and Position Projections

A group representation is a representation of all its subgroups as seen, e.g., in the translation representations integrated over momentum spheres for the Euclidean group and energy-momentum hyperboloids for the Poincaré group.

With a nonrelativistic space-time decomposition, representations of the Poincaré group $\mathbf{SO}_0(1,3) \vec{\times} \mathbb{R}^4$ for $m > 0$ contain representations of time translations $\mathbb{R}$ and of the Euclidean group $\mathbf{SO}(3) \vec{\times} \mathbb{R}^3$. The corresponding projections use the integration over the position translations $\int d^3x$ and over the time translations $\int dx_0$, respectively. The time projection of the Poincaré group coefficients reduces to trivial momenta $\delta(\vec{q})$, i.e., to a rest system:

$$\begin{aligned} \mathbb{R} \ni x_0 \longmapsto \int d^3x \int \frac{d^4q}{(2\pi)^3} \left(\frac{q}{m}\right)^L \delta\left(q^2 - m^2\right) e^{iqx} &= \int dq_0 \left(\frac{q_0}{m}\right)^L \delta\left(q_0^2 - m^2\right) e^{iq_0x_0}, \\ = \left(\frac{1}{im}\frac{\partial}{\partial x_0}\right)^L \frac{\cos mx_0}{m} &= \begin{cases} \frac{\cos mx_0}{m}, & L = 0, 2, \ldots, \\ i\frac{\sin mx_0}{m}, & L = 1, 3, \ldots. \end{cases} \end{aligned}$$

The position projection requires the trivial energy $\delta(q_0)$ which is not supported by the projected Dirac distribution:

$$\mathbb{E}^3 \ni \vec{x} \longmapsto \int dx_0 \int \frac{d^4q}{(2\pi)^3} \left(\frac{q}{m}\right)^L \delta(q^2 - m^2) e^{iqx} = \int \frac{d^3q}{(2\pi)^2} \left(\frac{\vec{q}}{m}\right)^L \delta(\vec{q}^2 + m^2) e^{-i\vec{q}\vec{x}} = 0.$$

Both the time translations coefficients $\cos q_0x_0$ and the position translation coefficients $j_0(qr) = \frac{\sin qr}{qr}$ for free particles arise for energies $q_0^2 > m^2$ over the mass threshold, as seen in a partial integration which reflects the direct product subgroup for the nonrelativistic time-space decompositions:

$$\begin{aligned} &\mathbf{SO}_0(1,3) \vec{\times} \mathbb{R}^4 \supset \mathbb{R} \times [\mathbf{SO}(3) \vec{\times} \mathbb{R}^3], \\ \int \frac{d^4q}{(2\pi)^3} \delta(q^2 - m^2) e^{iqx} &= \int_m^\infty \frac{dq_0}{2\pi^2} \cos q_0x_0 \frac{\sin \sqrt{q_0^2 - m^2}\, r}{r} \\ &= \int_0^\infty \frac{q\,dq}{2\pi^2\sqrt{q^2+m^2}} \cos\sqrt{q^2+m^2}\, x_0 \frac{\sin qr}{r}. \end{aligned}$$

The special cases are obtained as contribution for $q_0 = 0$ (trivial position projection) and $\vec{q} = 0$ (time projection). The projection of nontrivial rotation properties $\left(\frac{q}{m}\right)^L$ gives the corresponding spherical polynomials $\left(\frac{\vec{q}}{m}\right)^L$.

In this connection, and for later use, it is mentioned that the time projection of the full Feynman propagator of a free field, involves a coefficient for time translations:

$$\begin{aligned} \int d^3x \int \frac{d^4q}{i\pi(2\pi)^3} \frac{1}{-q^2 - io + m^2} e^{iqx} &= \int \frac{dq_0}{i\pi} \frac{1}{-q_0^2 - io + m^2} e^{iq_0x_0} \\ &= \frac{e^{-im|x_0|}}{m} = \frac{\cos mx_0 - i\epsilon(x_0)\sin mx_0}{m}. \end{aligned}$$

The position projection gives a Yukawa potential which comes only from the imaginary off-shell part:

$$\begin{aligned}\int dx_0 \int \frac{d^4q}{i\pi(2\pi)^3} \frac{1}{-q^2-io+m^2} e^{iqx} &= \int \frac{d^3q}{i\pi(2\pi)^2} \frac{1}{\vec{q}^2+m^2} e^{-i\vec{q}\vec{x}} \\ &= \frac{i}{2\pi r}\frac{\partial}{\partial r} \int \frac{dq}{\pi} \frac{1}{q^2+m^2} e^{iqr} = \frac{1}{2i\pi} \frac{e^{-mr}}{r}, \\ \text{with } \int \frac{dq}{\pi} \frac{m}{q^2+m^2} e^{iqr} &= e^{-mr}.\end{aligned}$$

The four-dimensional Minkowski-integrals can be computed from two-dimensional Cartan spacetime:

$$\begin{aligned}\int d^4q\, \mu(q^2) e^{iqx} &= -\frac{\partial}{\partial \frac{x^2}{4\pi}} \int d^2q\, \mu(q^2) e^{iqx}, \\ \int \frac{d^2q}{i\pi} \frac{1}{-q^2-io+m^2} e^{iqx} &= \int d\psi\, e^{-\sqrt{-(m\cosh\psi)^2x^2+io}}\end{aligned}$$

10.4 Rotations and Translations

Translations of Euclidean position and Minkowski spacetime are acted on, respectively, by the special rotation group and the orthochronous Lorentz group. They constitute the homogeneous part in *the Euclidean group and the Poincaré group*, respectively, which are both semidirect products:

$$(O, \vec{x}) \in \mathbf{SO}(3) \vec{\times} \mathbb{R}^3, \quad (\Lambda, x) \in \mathbf{SO}_0(1,3) \vec{\times} \mathbb{R}^4.$$

Corresponding dual inhomogeneous groups involve the momenta $\mathbf{SO}(3) \vec{\times} \check{\mathbb{R}}^3$ and the energy-momenta $\mathbf{SO}_0(1,3) \vec{\times} \check{\mathbb{R}}^4$.

The inhomogeneous groups have nonunitary faithful representations with dimensions $3+1$ and $4+1$, respectively, which are convenient for writing down the group multiplication:

$$\begin{aligned}(O, \vec{x}) &\longmapsto \left(\begin{array}{c|c} O & \vec{x} \\ \hline 0 & 1 \end{array}\right) \in \mathbf{GL}(4, \mathbb{R}),\ (O_1, \vec{x}_1) \circ (O_2, \vec{x}_2) = (O_1 O_2, \vec{x}_1 + O_1.\vec{x}_2), \\ (\Lambda, x) &\longmapsto \left(\begin{array}{c|c} \Lambda & x \\ \hline 0 & 1 \end{array}\right) \in \mathbf{GL}(5, \mathbb{R}),\ (\Lambda_1, x_1) \circ (\Lambda_2, x_2) = (\Lambda_1 \Lambda_2, x_1 + \Lambda_1.x_2).\end{aligned}$$

The action of the homogeneous group leads to a decomposition of the space (-time) translations, or of the (energy-)momenta, into disjoint orbits with the general definition of an *orbit* of a translation $x \in \mathbb{R}^n$ under action of an orthogonal group $\mathbf{SO}_0(t, s)$:

$$t + s = n \geq 2, \quad \mathbf{SO}_0(t,s).x = \{R.x \,\big|\, R \in \mathbf{SO}_0(t,s)\} \subset \mathbb{R}^n.$$

Euclidean position has two orbit types — for the trivial and nontrivial translations given, respectively, by the trivial translation itself and by 2-spheres, all isomorphic to the unit sphere Ω^2:

$$\mathbf{SO}(3).0 = \{0\},\ \ \vec{x} \neq 0:\ \mathbf{SO}(3).\vec{x} \cong \Omega^2.$$

Minkowski spacetime has four orbit types: in addition to the trivial orbit, there are timelike one-shell hyperboloids $\mathcal{Y}^3$, spacelike one-shell hyperboloids $\mathcal{Y}^{(1,2)}$ and tip-punctured half-lightcones $\mathcal{C}_0^3$ — all are real three-dimensional manifolds:

$$\mathbf{SO}_0(1,3).0 = \{0\},\quad x \neq 0:\ \mathbf{SO}_0(1,3).x \cong \begin{cases} \mathcal{Y}^3 & \text{for } x^2 > 0, \\ \mathcal{Y}^{(1,2)} & \text{for } x^2 < 0, \\ \mathcal{C}_0^3 & \text{for } x^2 = 0. \end{cases}$$

All orbits have a characteristic *isotropy group*, isomorphic to the fixgroups of all orbit points, with the general definition of a *fixgroup* for a translation $x \in \mathbb{R}^n$:

$$\mathbf{SO}_0(t,s)_x = \{R \in \mathbf{SO}_0(t,s) \mid R.x = x\} \subseteq \mathbf{SO}_0(t,s).$$

For position, the fixgroups are given either by the full group or by the axial rotations around the nontrivial translations:

$$\mathbf{SO}(3)_0 = \mathbf{SO}(3),\ \ \vec{x} \neq 0:\ \ \mathbf{SO}(3)_{\vec{x}} \cong \mathbf{SO}(2).$$

In the Lorentz group for spacetime, there arise real three-dimensional fixgroups for nontrivial translations: rotations for timelike translations, e.g., for $x = \mathbf{1}_2$ (Weyl representation $x = \sigma_j x^j$ of Minkowski spacetime), Lorentz groups in $(1+2)$-dimensions for spacelike translations, e.g., for $x = \sigma_3$, and semidirect groups with axial rotations around the space component for lightlike translations, e.g., for $x = \frac{\mathbf{1}_2+\sigma_3}{2}$ around the third direction σ_3, and two lightlike translations in the polarization-orthogonal directions, here for σ_1 and σ_2:

$$\mathbf{SO}_0(1,3)_0 = \mathbf{SO}_0(1,3),\quad x \neq 0:\ \mathbf{SO}_0(1,3)_x \cong \begin{cases} \mathbf{SO}(3) & \text{for } x^2 > 0, \\ \mathbf{SO}_0(1,2) & \text{for } x^2 < 0, \\ \mathbf{SO}(2) \vec{\times} \mathbb{R}^2 & \text{for } x^2 = 0. \end{cases}$$

The fixgroup for lightlike translations is isomorphic to a subgroup of the nonrelativistic Galilei group:

$$\mathbf{SO}_0(1,3) \supset \mathbf{SO}(2) \vec{\times} \mathbb{R}^2 \subset \mathbf{SO}(3) \vec{\times} \mathbb{R}^3.$$

Orbits are isomorphic to the nontrivially acting group elements, i.e., to the fixgroup classes in the full action group:

$$\mathbf{SO}_0(t,s) \bullet x \cong \mathbf{SO}_0(t,s)/\mathbf{SO}_0(t,s)_x,$$

here for the position and spacetime translation orbits:

$$\{0\} \cong \mathbf{SO}(3)/\mathbf{SO}(3), \qquad \Omega^2 \cong \mathbf{SO}(3)/\mathbf{SO}(2),$$

$$\{0\} \cong \mathbf{SO}_0(1,3)/\mathbf{SO}_0(1,3), \quad \begin{cases} \mathcal{Y}^3 \cong \mathbf{SO}_0(1,3)/\mathbf{SO}(3), \\ \mathcal{Y}^{(1,2)} \cong \mathbf{SO}_0(1,3)/\mathbf{SO}_0(1,2), \\ \mathcal{C}_0^3 \cong \mathbf{SO}_0(1,3)/[\mathbf{SO}(2) \vec{\times} \mathbb{R}^2]. \end{cases}$$

In general, the orthogonal groups have the following fixgroups — where defined, i.e., where the arising dimensions make sense:

$\mathbf{SO}_0(t,s) \vec{\times} \mathbb{R}^{t+s}$:

translation	fixgroup
$x = 0$	$\mathbf{SO}_0(t,s)$
$x^2 > 0$	$\mathbf{SO}_0(t-1,s)$
$x^2 < 0$	$\mathbf{SO}_0(t,s-1)$
$x^2 = 0, x \neq 0$	$\mathbf{SO}_0(t-1,s-1) \vec{\times} \mathbb{R}^{t+s-2}$

For Euclidean position and Minkowski spacetime, the scalar functions above are representation coefficients of the translations with the action of the inhomogeneous groups. For position, they integrate the irreducible translation representation coefficients for the momentum directions $\vec{\omega} \in \Omega^2$ on the unit 2-sphere ($\mathcal{M}$ for Radon measure) where the 2-sphere characterizes the orbits of a nontrivial momentum with "length" P:

$$\mathbf{SO}(3) \vec{\times} \mathbb{R}^3 : \quad \begin{cases} \mathbb{E}^3 \ni \vec{x} \longmapsto |P^2, 0\}(\vec{x}) = \int \frac{d^3q}{2\pi P} \delta(\vec{q}^2 - P^2) e^{i\vec{q}\vec{x}} \\ \quad = \int_0^\infty \frac{|\vec{q}|^2 d|\vec{q}|}{P} \delta(|\vec{q}|^2 - P^2) \int_{\Omega^2} \frac{d^2\omega}{2\pi} e^{i|\vec{q}|\vec{\omega}\vec{x}} \\ \quad = \int_{\Omega^2} \frac{d^2\omega}{|\Omega^2|} e^{iP\vec{\omega}\vec{x}} = \int_{\Omega^2} \frac{d^2\omega}{|\Omega^2|} \cos P\vec{\omega}\vec{x} \\ \text{for } \vec{q} = |\vec{q}|\vec{\omega} \text{ with } \vec{\omega} = \begin{pmatrix} \cos\theta \\ \sin\theta\, \cos\phi \\ \sin\theta\, \sin\phi \end{pmatrix} \in \Omega^2 \subset \mathbb{R}^3, \\ d^2\omega = \sin\theta\, d\theta\, d\phi \in \mathcal{M}(\Omega^2)_+, \ |\Omega^2| = 4\pi. \end{cases}$$

In the following for short $\int_{\Omega^2} d^2\omega = \int d^2\omega$.

For spacetime, the translations use the energy-momentum directions $\mathrm{y} \in \mathcal{Y}^3$ on the unit 3-hyperboloid:

$$\mathbf{SO}_0(1,3) \vec{\times} \mathbb{R}^4 : \quad \begin{cases} \mathbb{R}^{(1,3)} \ni x \longmapsto |m^2, 0\}(x) = \int \frac{d^4q}{V^3 m^2} 2\vartheta(q_0)\delta(q^2 - m^2) e^{iqx} \\ \quad = \int_{\mathcal{Y}^3} \frac{d^3\mathrm{y}}{V^3} e^{im\mathrm{y}x}, \\ \text{for } q = |q|\mathrm{y} \text{ with } \mathrm{y} = \begin{pmatrix} \cosh\psi \\ \sinh\psi\, \vec{\omega} \end{pmatrix} \in \mathcal{Y}^3 \subset \mathbb{R}^4, \\ d^3\mathrm{y} = \sinh^2\psi\, d\psi\, d^2\omega \in \mathcal{M}(\mathcal{Y}^3)_+. \end{cases}$$

In the following for short $\int_{\mathcal{Y}^3} d^3\mathrm{y} = \int d^3\mathrm{y}$. In particle physics, the momentum $\vec{q} \in \mathbb{R}^3$ parametrization with the energy $q_0 = \sqrt{m^2 + \vec{q}^2}$ is more familiar than hyperbolic coordinates with the rapidity $\psi = \operatorname{arsinh} \frac{|\vec{q}|}{m}$, possible with the manifold isomorphy $\mathbb{R}^3 \cong \mathcal{Y}^3$:

$$y = \frac{1}{m}\begin{pmatrix} \sqrt{m^2+\vec{q}^2} \\ \vec{q} \end{pmatrix}, \quad myx = \sqrt{m^2+\vec{q}^2}x_0 - \vec{q}\vec{x}, \quad d^3y = \frac{d^3q}{m^2\sqrt{m^2+\vec{q}^2}}.$$

10.5 Rotation and Translation Invariants

Invariants of Lie algebra operations have to be constructed in the corresponding enveloping algebra (the free algebra with linear combinations of "formal" finite products of the Lie algebra vectors, modulo the equality of Lie bracket and commutator). For semisimple Lie algebras, the number of independent generating invariants coincides with the rank which, for semisimplicity, is the dimension of a maximal abelian Lie subalgebra. The operations in such a Cartan subalgebra are simultaneously diagonalizable or, in physics, simultaneously measurable with vanishing uncertainty if represented by Hilbert space operators. The values of physical invariants (Nöther invariants) determine the representations of the acting operation groups (Lie algebras).

For a vector space representation, both Lie algebra and enveloping algebra come as vector space endomorphisms. The invariant operators $\{\mathbf{I}\}$ from the totally symmetric subalgebra (polynomial algebra) of the enveloping algebra are, for irreducible complex representations, proportional to the unit $\mathbf{I} \longmapsto I\mathbf{1}$ with $I \in \mathbf{spec}\,\mathbf{I} \subseteq \mathbb{C}$. The rotation invariants for polarization and spin, related to compact groups, are from a discrete spectrum, the translation invariants for momentum and mass, related to noncompact groups, from a continuous one.

The real three-dimensional simple rank one Lie algebra of the rotation (spin) group $\mathbf{SU}(2) \sim \mathbf{SO}(3)$ with Cartan subgroup $\mathbf{SO}(2) \cong \mathbf{U}(1)$ is spanned by the angular momenta $\vec{\mathbf{J}} = (\mathbf{J}^a)_{a=1}^3$ with the Lie brackets:

$$[\mathbf{J}^a, \mathbf{J}^b] = -\epsilon^{abc}\mathbf{J}^c.$$

The spin group has one independent *Casimir invariant* — the squared spin (angular momentum) $\vec{\mathbf{J}}^2 = \delta_{ab}\mathbf{J}^a\mathbf{J}^b \longmapsto -J(J+1)\mathbf{1}_{1+2J}$ with the familiar spectrum for the $(1+2J)$-dimensional irreducible representations, starting with the Pauli spinor representation $\vec{\mathbf{J}} \longmapsto \frac{i}{2}\vec{\sigma}$:

$$[\mathbf{J}^a, \vec{\mathbf{J}}^2] = 0, \quad \mathbf{spec}\,\vec{\mathbf{J}}^2 = \{-J(J+1) \,\big|\, 2J \in \mathbb{N} = \{0, 1, 2, \ldots\}\}.$$

The real six-dimensional simple rank two Lie algebra of the Lorentz group $\mathbf{SL}(2,\mathbb{C}) \sim \mathbf{SO}_0(1,3)$ with Cartan subgroup $\mathbf{GL}(1,\mathbb{C}) = \mathbf{D}(1) \times \mathbf{U}(1) \cong \mathbf{SO}_0(1,1) \times \mathbf{SO}(2)$ for special relativity is spanned by $(\mathbf{L}^{ik})_{i,k=0}^3$ with the angular momenta $\mathbf{J}^a = \frac{1}{2}\epsilon^{abc}\mathbf{L}^{bc}$ and the boosts $\mathbf{B}^a = \mathbf{L}^{0a} = -\mathbf{L}^{a0}$, with the Lie brackets:

$$[\mathbf{L}^{ik}, \mathbf{L}^{nm}] = \eta^{in}\mathbf{L}^{km} - \eta^{kn}\mathbf{L}^{im} - \eta^{im}\mathbf{L}^{kn} + \eta^{km}\mathbf{L}^{in} : \quad \begin{cases} [\mathbf{J}^a, \mathbf{J}^b] = -\epsilon^{abc}\mathbf{J}^c, \\ [\mathbf{J}^a, \mathbf{B}^b] = -\epsilon^{abc}\mathbf{B}^c, \\ [\mathbf{B}^a, \mathbf{B}^b] = \epsilon^{abc}\mathbf{J}^c. \end{cases}$$

The noncompact Lorentz group $\mathbf{SL}(2, \mathbb{C}) \sim \mathbf{SO}_0(1,3)$ as partner of the compact doubled spin group $\mathbf{SU}(2) \times \mathbf{SU}(2) \sim \mathbf{SO}(4)$ has two independent invariants, the *Killing and the chiral invariant* — with $\vec{\mathbf{J}}_\pm = \frac{\vec{\mathbf{J}} \pm i\vec{\mathbf{B}}}{2}$ the real and imaginary combination of the related Casimir invariants:

$$\begin{aligned} -\tfrac{1}{2}\eta_{in}\eta_{km}\mathbf{L}^{ik}\mathbf{L}^{nm} &= \vec{\mathbf{J}}^2 - \vec{\mathbf{B}}^2 = 2(\vec{\mathbf{J}}_+^2 + \vec{\mathbf{J}}_-^2), \quad [\mathbf{L}^{ik}, \vec{\mathbf{J}}^2 - \vec{\mathbf{B}}^2] = 0, \\ \tfrac{1}{4}\epsilon_{iknm}\mathbf{L}^{ik}\mathbf{L}^{nm} &= \vec{\mathbf{J}}\vec{\mathbf{B}} \qquad = -i(\vec{\mathbf{J}}_+^2 - \vec{\mathbf{J}}_-^2), \qquad [\mathbf{L}^{ik}, \vec{\mathbf{J}}\vec{\mathbf{B}}] = 0. \end{aligned}$$

The familiar irreducible finite-dimensional representations $(L|R)$ with $2L, 2R \in \mathbb{N}$, starting with the left- and right-handed Weyl spinor representations $(\vec{\mathbf{J}}, \vec{\mathbf{B}}) \longmapsto (\frac{i}{2}\vec{\sigma}, \pm\frac{1}{2}\vec{\sigma})$, are no Hilbert representations for nontrivial dimension $(1+2L)(1+2R)$. The spectrum for the Killing invariant reflects the embedded compact spin structure:

$$\mathbf{spec}_{\,\mathrm{fin}}(\vec{\mathbf{J}}^2 - \vec{\mathbf{B}}^2) = \{-2[L(L+1) + R(R+1)] \,\big|\, 2L, 2R \in \mathbb{N}\}.$$

The spectrum of the chiral invariant (relative orientation of angular momentum and boost) is discrete for the finite-dimensional representations: It is trivial for the real representations which are products of the real four-dimensional Minkowski representation $(\frac{1}{2}|\frac{1}{2})$, and nontrivial in the case $R \neq L$, e.g., conjugated to each other $\pm i\frac{3}{4}\mathbf{1}_2$ for the left- and right-handed Weyl spinor representations $(\frac{1}{2}|0)$ and $(0|\frac{1}{2})$:

$$\mathbf{spec}_{\,\mathrm{fin}}\vec{\mathbf{J}}\vec{\mathbf{B}} = \{i[L(L+1) - R(R+1)] \,\big|\, 2L, 2R \in \mathbb{N}\}.$$

The necessarily infinite-dimensional faithful Hilbert representations of the Lorentz group have continuous chiral spectrum.

The real six-dimensional Lie algebra of the Euclidean group $\mathbf{SU}(2) \vec{\times} \mathbb{R}^3 \sim \mathbf{SO}(3) \vec{\times} \mathbb{R}^3$ for nonrelativistic scattering is spanned by the angular momenta and the momenta $\vec{\mathbf{p}}$ with the Lie brackets:

$$[\mathbf{J}^a, \mathbf{J}^b] = -\epsilon^{abc}\mathbf{J}^c, \quad [\mathbf{J}^a, \mathbf{p}^b] = -\epsilon^{abc}\mathbf{p}^c, \quad [\mathbf{p}^a, \mathbf{p}^b] = 0.$$

The Euclidean Lie algebra as Inönü–Wigner contraction (with the boosts going to the position translations $\vec{\mathbf{B}} = c\vec{\mathbf{p}}$, $c \to \infty$) of the rank two Lorentz Lie algebra $\log \mathbf{O}(1,3)$ is not simple. It has two independent invariants — the squared momentum $-\frac{1}{c^2}(\vec{\mathbf{J}}^2 - \vec{\mathbf{B}}^2) \to \vec{\mathbf{p}}^2$ with positive spectrum for Hilbert representations, in physics used as kinetic energy $\frac{\vec{\mathbf{P}}^2}{2m}$, and the polarization as the momentum projection $\frac{1}{c}\vec{\mathbf{J}}\vec{\mathbf{B}} \to \vec{\mathbf{J}}\vec{\mathbf{p}}$ of the angular momenta with a discrete normalized spectrum:

$$[\mathbf{J}^a, \vec{\mathbf{p}}^2] = 0,\ [\mathbf{p}^a, \vec{\mathbf{p}}^2] = 0, \qquad \mathbf{spec}\,_+\vec{\mathbf{p}}^2 = \{P^2 \geq 0\},$$
$$[\mathbf{J}^a, \vec{\mathbf{J}}\vec{\mathbf{p}}] = 0,\ [\mathbf{p}^a, \vec{\mathbf{J}}\vec{\mathbf{p}}] = 0,\ P^2 > 0: \quad \mathbf{spec}\,\frac{\vec{\mathbf{J}}\vec{\mathbf{p}}}{\sqrt{\vec{\mathbf{p}}^2}} = \{\pm h \mid 2|h| \in \mathbb{N}\}.$$

For the Euclidean group, angular momentum is not translation invariant, $[\mathbf{p}^a, \vec{\mathbf{J}}^2] \neq 0$.

The real one-dimensional abelian Lie algebra $\mathbb{R}$ for the time translations with the Lie group $\mathbf{D}(1) = \exp \mathbb{R} \cong \mathbb{R}$ can use any basis $\mathbf{p}^0$ as invariant, in quantum mechanics represented by Hamiltonians $\mathbf{p}^0 \longmapsto i\mathbf{H}$. The spectrum of the unitary representations is characterized by energies, familiar from $\mathbf{p}^0 \longmapsto iE$ with $t \longmapsto e^{iEt}$ for the group:

$$\mathbf{spec}\,_+\mathbf{p}^0 = \{iE \mid E \in \mathbb{R}\}.$$

The real 10-dimensional Lie algebra of the Poincaré group $\mathbf{SL}(2, \mathbb{C}) \vec{\times} \mathbb{R}^4 \sim \mathbf{SO}_0(1, 3) \vec{\times} \mathbb{R}^4$ for relativistic scattering is spanned by angular momenta and boosts $\mathbf{L}^{ik}$ and the energy-momenta $\mathbf{p}^i$ with the Lie brackets:

$$\begin{aligned}[\mathbf{L}^{ik}, \mathbf{L}^{nm}] &= \eta^{in}\mathbf{L}^{km} - \eta^{kn}\mathbf{L}^{im} - \eta^{im}\mathbf{L}^{kn} + \eta^{km}\mathbf{L}^{in}, \\ [\mathbf{L}^{ik}, \mathbf{p}^n] &= \eta^{in}\mathbf{p}^k - \eta^{kn}\mathbf{p}^i, \quad [\mathbf{p}^i, \mathbf{p}^n] = 0,\end{aligned} \qquad \left\{\begin{aligned}[\mathbf{J}^a, \mathbf{p}^0] &= 0, \\ [\mathbf{J}^a, \mathbf{p}^b] &= -\epsilon^{abc}\mathbf{p}^c, \\ [\mathbf{B}^a, \mathbf{p}^0] &= \mathbf{p}^a, \\ [\mathbf{B}^a, \mathbf{p}^b] &= \delta^{ab}\mathbf{p}^0.\end{aligned}\right.$$

The Poincaré Lie algebra as contraction of the rank two anti-de Sitter Lie algebra $\log \mathbf{O}(2, 3)$ is not simple. It has two independent invariants, first the squared energy-momenta $\mathbf{p}^2 = \eta_{ik}\mathbf{p}^i\mathbf{p}^k$ with continuous spectrum for the Hilbert representations:

$$[\mathbf{L}^{ik}, \mathbf{p}^2] = 0, \quad [\mathbf{p}^i, \mathbf{p}^2] = 0, \quad \mathbf{spec}\,_+\mathbf{p}^2 = \{m^2 \in \mathbb{R}\}.$$

The second invariant uses the Lubanski[1]–Pauli vector $\mathbf{S}^i$, given by the Lorentz group generators, projected on the energy-momenta with the embedded momentum projection of the angular momenta as zeroth component, supplemented by the rest-system spin generator (with $\mathbf{p} \Rightarrow \binom{m}{0}$):

$$\mathbf{S}_m = \frac{1}{2}\epsilon_{iknm}\mathbf{L}^{ik}\mathbf{p}^n \text{ with } \binom{\mathbf{S}_0}{\vec{\mathbf{S}}} \cong \binom{\vec{\mathbf{J}}\vec{\mathbf{p}}}{\vec{\mathbf{J}}\mathbf{p}^0 - \vec{\mathbf{B}} \times \vec{\mathbf{p}}} \Rightarrow m\binom{0}{\vec{\mathbf{J}}}, \ \mathbf{S}_i\mathbf{p}^i = 0,$$
$$[\mathbf{L}^{ik}, \mathbf{S}^n] = \eta^{in}\mathbf{S}^k - \eta^{kn}\mathbf{S}^i, \quad [\mathbf{p}^i, \mathbf{S}^n] = 0, \quad [\mathbf{S}^i, \mathbf{S}^n] = \epsilon^{inkm}\mathbf{S}_k\mathbf{p}_m.$$

The spin invariant is of order four in the Lie algebra elements:

$$\mathbf{S}^2 = \eta^{ik}\mathbf{S}_i\mathbf{S}_k \cong -\vec{\mathbf{J}}^2\mathbf{p}_0^2 + (\vec{\mathbf{J}}\vec{\mathbf{p}})^2 - \vec{\mathbf{B}}^2\vec{\mathbf{p}}^2 + (\vec{\mathbf{B}}\vec{\mathbf{p}})^2 + 2(\vec{\mathbf{J}} \times \vec{\mathbf{B}})\mathbf{p}_0\vec{\mathbf{p}} \Rightarrow -m^2\vec{\mathbf{J}}^2,$$
$$[\mathbf{L}^{ik}, \mathbf{S}^2] = 0, \quad [\mathbf{p}^i, \mathbf{S}^2] = 0.$$

[1] Jozef Lubanski (1914–1946).

For positive translation invariant (nonvanishing mass), its normalized Hilbert representation spectrum is the familiar spin spectrum:

$$m^2 > 0: \quad \mathbf{spec}_+ \left(-\frac{\mathbf{S}^2}{\mathbf{p}^2}\right) = \{-J(J+1) \,\big|\, 2J \in \mathbb{N}\}.$$

It is instructive to schematize the subgroup $\subset$ and contraction $\downarrow$ structures:

$$\begin{array}{lcccc} & & \mathbf{SO}_0(1,3) & & \mathbf{SO}_0(2,3) \\ & \mathbf{SO}(3) \subset & \downarrow & \subset & \downarrow \\ & & \mathbf{SO}(3) \vec{\times} \mathbb{R}^3 & & \mathbf{SO}_0(1,3) \vec{\times} \mathbb{R}^4 \\ \text{number of invariants (order):} & 1 \text{ (two)} & 2 \text{ (two)} & & 2 \text{ (two and four)} \end{array}$$

10.6 Hilbert Metric Inducing Functions

The *cyclic Hilbert representations* of translations, i.e., the unitary ones with a cyclic vector, are characterized by positive (energy-)momentum distributions $\tilde{d} \in \check{\mathcal{M}}(\mathbb{R}^n)_+$ of the irreducible translation representations, leading to an essentially bounded *function of positive type* $d \in L^\infty(\mathbb{R}^n)_+$ as representation coefficient of the translations:

$$\mathbb{R}^n \ni x \longmapsto d(x) = \int \frac{d^n q}{(2\pi)^n} \tilde{d}(q) e^{iqx}.$$

If a positive type function is normalized:

$$d(0) = \int \frac{d^n q}{(2\pi)^n} \tilde{d}(q) = 1,$$

it is called a *state*. The following examples have been used above:

$$\begin{array}{lll} \text{time} & t \in \mathbb{R}: & \int \frac{dq_0}{|\Omega^0|} 2|E|\delta(q_0^2 - E^2) e^{iq_0 t} = \cos Et, \ |\Omega^0| = 2, \\ \text{position} & \vec{x} \in \mathbb{E}^3: & \int \frac{d^3 q}{|\Omega^2| P} 2\delta(\vec{q}^2 - P^2) e^{i\vec{q}\vec{x}} = \frac{\sin Pr}{Pr}, \quad |\Omega^2| = 4\pi, \\ \text{spacetime} & x \in \mathbb{R}^{(1,3)}: & \int \frac{d^4 q}{V^3 m^2} 2\vartheta(q_0)\delta(q^2 - m^2) e^{iqx}. \end{array}$$

A positive type function is almost everywhere a bounded continuous function. It is reflection symmetric (unitary) and bounded by the value for the trivial translation:

$$L^\infty(\mathbb{R}^n)_+ \ni d = \hat{d} \text{ with } \hat{d}(x) = \overline{d(-x)}, \quad |d(x)| \le d(0).$$

The complex Hilbert space (V, d), or short V_d, characterized by the positive type translation function d is one-dimensional $V_d \cong \mathbb{C}$ for irreducible translation representations and necessarily infinite-dimensional for faithful Hilbert representations of the noncompact translations. The explicit construction of the cyclic Hilbert space V_d for a positive type function d is given ahead.

A cyclic Hilbert representation has a cyclic vector $|\mathcal{C}\rangle$, defined by the property that the closure of the finite linear combinations of its orbit vectors gives the representation Hilbert space. For a positive type function there exists a cyclic vector where it is the representation matrix element of. Its normalization is the representation coefficient for the trivial translation:

$$d \in L^\infty(\mathbb{R}^n)_+ : \begin{array}{l} d(x) = \langle \mathcal{C}|x \bullet |\mathcal{C}\rangle_d \text{ with } |\mathcal{C}\rangle \in V_d, \\ d(0) = \langle \mathcal{C}|\mathcal{C}\rangle_d. \end{array}$$

e.g., for the irreducible representations:

$$\langle p|x \bullet |p\rangle = \langle p|e^{ipx}|p\rangle = e^{ipx}, \quad \langle p|p\rangle = 1.$$

The positive type functions constitute a convex cone, i.e.:

$$d_{1,2} \in L^\infty(\mathbb{R}^n)_+, \quad \alpha_{1,2} \geq 0 \Rightarrow \alpha_1 d_1 + \alpha_2 d_2 \in L^\infty(\mathbb{R}^n)_+.$$

The extremal states of the cone characterize the irreducible Hilbert representations with the Dirac (energy-)momentum distributions and the exponentials as positive type functions:

$$\mathbb{R}^n \ni x \longmapsto d_p(x) = e^{ipx}, \quad \tilde{d}_p(q) = \delta\left(\frac{q-p}{2\pi}\right).$$

The trivial representation is characterized by the unit function $d_0(x) = 1$.

A pointwise product of two positive type functions for translations characterizes the product representation:

$$\begin{array}{l} L^\infty(\mathbb{R}^n)_+ \times L^\infty(\mathbb{R}^n)_+ \longrightarrow L^\infty(\mathbb{R}^n)_+, \\ \qquad (d_1, d_2) \longmapsto d = d_1 \cdot d_2 \text{ with } d_1 \cdot d_2(x) = d_1(x) d_2(x), \\ \qquad\qquad \text{and } |\mathcal{C}\rangle = |\mathcal{C}_1\rangle \otimes |\mathcal{C}_2\rangle. \end{array}$$

With the pointwise multiplication and trivial representation d_0 as unit the positive type functions (cyclic representations) constitute a monoid ("group without inversion").

The (energy-)momentum distribution of a product is the convolution product of the factor distributions:

$$d(x) = \int \frac{d^n q}{(2\pi)^n} \tilde{d}(q) e^{iqx} \Rightarrow \tilde{d}_1 * \tilde{d}_2(q) = \widetilde{d_1 \cdot d_2}(q) = \int \frac{d^n p}{(2\pi)^n} \tilde{d}_1(p) \tilde{d}_2(q-p).$$

The slighty incorrect, but convenient notation $\tilde{d}_1 * \tilde{d}_2(q) = \tilde{d}_1(q) * \tilde{d}_2(q)$ will be used. An example is the product of irreducible representations, irreducible again:

$$e^{ip_1x} \cdot e^{ip_2x} = e^{i(p_1+p_2)x} \leftrightarrow \delta\left(\frac{q-p_1}{2\pi}\right) * \delta\left(\frac{q-p_2}{2\pi}\right) = \delta\left(\frac{q-p_1-p_2}{2\pi}\right).$$

Cyclic representations have an involution via conjugation, relating to each other *dual representations*, i.e., representations by linear transformations which are inverse transposed to each other:

$$\text{dual reflection: } L^\infty(\mathbb{R}^n)_+ \leftrightarrow L^\infty(\mathbb{R}^n)_+ : \begin{cases} d \leftrightarrow \overline{d} \text{ with } \overline{d}(x) = \overline{d(x)} \\ \tilde{d}(q) \leftrightarrow \tilde{d}(-q). \end{cases}$$

Selfdual cyclic representations, i.e., with $d = \overline{d}$ and $\tilde{d}(q) = \tilde{d}(-q)$ can be obtained by addition $d + \overline{d}$ of dual partners.

10.7 Hilbert Spaces for Translations

For the irreducible Hilbert representations of the translations $\mathbb{R}^n \ni x \longmapsto e^{ipx} \in \mathbf{U}(1)$, the corresponding Dirac distribution restricts the induced scalar product to one (energy) momentum component $\tilde{f}(p) \in \mathbb{C}$:

$$\begin{aligned} f, f' \in L^1(\mathbb{R}^n) : \ \langle f | f' \rangle_{d^{ip}} &= \int d^n x_1 d^n x_2 \ \overline{f(x_1)} e^{ip(x_2 - x_1)} f'(x_2) \\ &= \int d^n q \ \overline{\tilde{f}(q)} \ \delta(q-p) \ \tilde{f}'(q) = \overline{\tilde{f}(p)} \ \tilde{f}'(p). \end{aligned}$$

The Hilbert space is complex one-dimensional $V_p \cong \mathbb{C}$. A normalized eigenvector $|p\rangle$ with (energy-)momentum p can be used as basis:

$$\overline{|L^1(\mathbb{R})\rangle}_{d^{ip}} = V_p = \mathbb{C}|p\rangle \begin{cases} \text{translation action: } \mathbb{R}^n \times V_p \longrightarrow V_p, \\ \qquad x \bullet |p\rangle = |p(x)\rangle = e^{ipx}|p\rangle, \\ \text{Hilbert product: } \langle p|p\rangle = 1, \\ \qquad \langle p(y)|p(x)\rangle = \langle p|p\rangle(x-y) = e^{ip(x-y)} \\ \text{cyclic vector: } |p;1\rangle = |p\rangle. \end{cases}$$

Here and in the following a translation difference-dependent scalar product matrix element is denoted as

$$\langle a(y)|b(x)\rangle = \langle a|b\rangle(x-y).$$

An example is the time action $\mathbb{R} \ni t \longmapsto e^{iEt} \in \mathbf{U}(1)$ for the harmonic oscillator with invariant energy $E > 0$ where the eigenvectors are related to a creation operator u which spans the complex one-dimensional representation space. The real and imaginary combinations with the annihilation operator, i.e., $\mathrm{u}^*|0\rangle = 0$ for the ground-state vector $|0\rangle$, called Fock vacuum, give position and momentum operators:

$$\begin{aligned}
\text{Hamiltonian: } & \mathbf{H} = E\frac{\mathbf{x}^2+\mathbf{p}^2}{2} = E\frac{\{\mathrm{u},\mathrm{u}^*\}}{2} \text{ with } \begin{cases} \mathbf{x} = \frac{\mathrm{u}+\mathrm{u}^\star}{\sqrt{2}}, \\ -i\mathbf{p} = \frac{\mathrm{u}-\mathrm{u}^\star}{\sqrt{2}}, \end{cases} \\
\text{quantization: } & [i\mathbf{p}, \mathbf{x}] = [\mathrm{u}^*, \mathrm{u}] = \mathbf{1}, \\
\text{Hilbert space basis: } & \{\mathrm{u}|0\rangle = |E\rangle\}, \\
\text{Hilbert product: } & \langle E|E\rangle = \langle 0|\mathrm{u}^\star \mathrm{u}|0\rangle = \langle 0|\{\mathrm{u}^\star, \mathrm{u}\}|0\rangle = 1.
\end{aligned}$$

The irreducible time representation spaces with the creation operator powers $k = 0, 1, \dots$ are complex one-dimensional $\mathbb{C}|kE\rangle$ and orthogonal for $l \neq k$:

$$|kE\rangle = \frac{\mathrm{u}^k}{\sqrt{k!}}|0\rangle, \quad \langle kE| = \langle 0|\frac{\mathrm{u}^{*k}}{\sqrt{k!}}, \quad \langle lE|kE\rangle(t) = \delta^{kl} e^{ikEt}.$$

In analogy to spacetime dependent fields, time dependent position and momentum can be called *quantum "fields" of time translations*: The expectation value of the anticommutators for the commutator-quantized oscillator gives the time translation coefficients with invariant eigenvalue E:

$$\begin{aligned}
&\mathbb{R} \ni t \longmapsto \mathbf{x}(t) = \frac{e^{iEt}\mathrm{u} + e^{-iEt}\mathrm{u}^\star}{\sqrt{2}}, \\
&\mathbb{R} \ni t-s \longmapsto \langle 0|\{\mathbf{x}(s), \mathbf{x}(t)\}|0\rangle = \cos E(t-s) = E \int dq_0\, \delta(q_0^2 - E^2) e^{iq_0(t-s)}.
\end{aligned}$$

The Hilbert space V_d for a cyclic translation representation, characterized by a positive type function $d \in L^\infty(\mathbb{R}^n)_+$, employs all irreducible representations, which support the (energy-)momentum density of this function. Sloppily: The values $\tilde{d}(q)$ of this density normalize individually the "eigenvectors" $|q\rangle$ of the irreducible translation representations — in a first formalization as follows:

$$\mathbb{R}^n: \begin{cases}
\mathbb{R}^n \ni x \longmapsto d(x) = \int \frac{d^nq}{(2\pi)^n}\, \tilde{d}(q) e^{iqx}, \\
\text{distributive Hilbert basis: } \{|q\rangle \,\big|\, q \in \mathbb{R}^n\}, \\
\text{distributive product: } \langle q'|q\rangle = \tilde{d}(q)\delta(\frac{q-q'}{2\pi}), \\
\text{distributive unit decomposition: } \mathrm{id}_{V_d} \cong \int \frac{d^nq}{(2\pi)^n \tilde{d}(q)}\, |q\rangle\langle q|, \\
\text{translation action: } |q\rangle \longmapsto x \bullet |q\rangle = e^{iqx}|q\rangle. \\
\text{distributive translation matrix: } \langle q'|x \bullet |q\rangle = e^{iqx}\tilde{d}(q)\delta(\frac{q-q'}{2\pi}).
\end{cases}$$

This has to be specified in two respects— with respect to the support of the (energy-)momentum density and with respect to the characterization as "distributive basis"of the Hilbert space V_d.

A Dirac density allows the restriction to the supporting (energy-)momenta, e.g., for the harmonic oscillator representations with translation invariants $\pm E$ and $\tilde{d}_E(q_0) = 2\pi|E|\delta(q_0^2 - E^2)$ above. For Euclidean position representations with translation invariant P^2, the momenta are restricted to the 2-sphere with invariant radius $P > 0$:

$$\mathbf{SO}(3)\ \vec{\times}\ \mathbb{R}^3: \begin{cases} \mathbb{E}^3 \ni \vec{x} \longmapsto |P^2, 0\}(\vec{x}) = \int \frac{d^3q}{2\pi P}\, \delta(\vec{q}^2 - P^2) e^{i\vec{q}\vec{x}} = \int \frac{d^2\omega}{|\Omega^2|} e^{iP\vec{\omega}\vec{x}}, \\ \text{distributive Hilbert basis: } \{|P^2, 0; \vec{\omega}\rangle \,\big|\, \vec{\omega} \in \Omega^2\}, \\ \text{distributive product: } \langle P^2, 0; \vec{\omega}'|P^2, 0; \vec{\omega}\rangle = |\Omega^2| \delta(\vec{\omega} - \vec{\omega}'), \\ \qquad\qquad \text{with } \delta(\vec{\omega}) = \frac{1}{\sin\theta}\delta(\theta)\delta(\phi), \\ \text{translation action: } \vec{x} \bullet |P^2, 0; \vec{\omega}\rangle = e^{iP\vec{\omega}\vec{x}} |P^2, 0; \vec{\omega}\rangle. \end{cases}$$

A notational reminder: The representation $|I\}$ is characterized by the invariants, here $|P^2, 0\}$. The distributive "eigenvectors" $\{|I; w\rangle \,\big|\, w\}$ which span the corresponding representation space, display the representation characterizing invariants I before the semicolon, the eigenvalues w after it. This is the analogue, e.g., for spin $\mathbf{SU}(2)$ with the representations $|J\}$ for $J = 0, \frac{1}{2}, 1, \ldots$ and eigenvectors $\{|J; m\rangle \,\big|\, J = -m, \ldots, m\}$ for each invariant J.

A scalar *quantum field of Euclidean position* $\mathbb{E}^3 = \mathbf{SO}(3)\ \vec{\times}\ \mathbb{R}^3/\mathbf{SO}(3)$ parametrizes the scalar Euclidean group $\mathbf{SO}(3)\ \vec{\times}\ \mathbb{R}^3$-coefficients for momentum invariant P by the ground-state value of the anticommutator for a commutator-quantized field:

$$\begin{aligned} \mathbb{E}^3 \ni \vec{x} \longmapsto \mathbf{E}(\vec{x}) &= {}^{\oplus}\!\!\int \frac{d^2\omega}{|\Omega^2|} \left[e^{i\vec{q}\vec{x}} \mathrm{u}(\vec{\omega}) + e^{-i\vec{q}\vec{x}} \mathrm{u}^*(\vec{\omega})\right], \\ &\quad \text{with } |\vec{q}| = P, \\ \text{quantization: } [\mathrm{u}^*(\vec{\omega}'), \mathrm{u}(\vec{\omega})] &= \mathbf{1}|\Omega^2|\, \delta(\vec{\omega} - \vec{\omega}'), \\ \langle 0|\{\mathrm{u}^*(\vec{\omega}'), \mathrm{u}(\vec{\omega})\}|0\rangle &= \langle 0|\mathrm{u}^*(\vec{\omega}')\mathrm{u}(\vec{\omega})|0\rangle \\ &= \langle P^2, 0; \vec{\omega}'|P^2, 0; \vec{\omega}\rangle = |\Omega^2|\, \delta(\vec{\omega} - \vec{\omega}'), \\ \text{distributive Hilbert basis: } \{\mathrm{u}(\vec{\omega})|0\rangle &= |P^2, 0; \vec{\omega}\rangle \,\big|\, \vec{\omega} \in \Omega^2\}, \\ \mathbb{E}^3 \ni \vec{x} - \vec{y} \longmapsto \langle 0|\mathbf{E}(\vec{y})\mathbf{E}(\vec{x})|0\rangle &= |P^2, 0\}(\vec{x} - \vec{y}). \end{aligned}$$

Its harmonic expansion integrates, with the normalized invariant measure $\frac{d^2\omega}{4\pi}$ of the 2-sphere $\Omega^2 \cong \mathbf{SO}(3)/\mathbf{SO}(2)$, translation and axial rotation eigenoperators for momentum direction $\vec{\omega}$ as distributive basis of the representation Hilbert space.

For Minkowski spacetime representations with translation invariant $m^2 > 0$, the energy-momenta are restricted to the 3-hyperboloid for mass m — in the hyperbolic and momentum parametrization:

$$\mathbf{SO}_0(1,3)\ \vec{\times}\ \mathbb{R}^4: \begin{cases} \mathbb{R}^{(1,3)} \ni x \longmapsto |m^2, 0\}(x) = \int \frac{d^4q}{\mathcal{V}^3 m^2}\, 2\vartheta(q_0)\delta(q^2 - m^2) e^{iqx} \\ \qquad = \int \frac{d^3\mathrm{y}}{\mathcal{V}^3} e^{im\mathrm{y}x} = \int \frac{d^3q}{\mathcal{V}^3 m^2 q_0} e^{iqx}, \\ \qquad \text{with } q_0 = \sqrt{m^2 + \vec{q}^2}, \\ \text{distributive basis: } \{|m^2; \mathrm{y}\rangle \,\big|\, \mathrm{y} \in \mathcal{Y}^3\} = \{|m^2; \vec{q}\rangle \,\big|\, \vec{q} \in \mathbb{R}^3\}, \\ \text{distributive product: } \langle m^2; \mathrm{y}'|m^2; \mathrm{y}\rangle = \mathcal{V}^3 \delta(\mathrm{y} - \mathrm{y}'), \\ \qquad \text{with } \delta(\mathrm{y}) = \frac{1}{\sinh^2\psi}\delta(\psi)\delta(\vec{\omega}), \\ \qquad \text{or } \langle m^2; \vec{q}'|m^2; \vec{q}\rangle = q_0 \mathcal{V}^3 m^2\, \delta(\vec{q} - \vec{q}'), \\ \text{translation action: } x \bullet |m^2; \mathrm{y}\rangle = e^{im\mathrm{y}x} |m^2; \mathrm{y}\rangle, \\ \qquad \text{or } x \bullet |m^2; \vec{q}\rangle = e^{iqx} |m^2; \vec{q}\rangle. \end{cases}$$

The harmonic expansion of a free scalar *quantum field of Minkowski spacetime* $\mathbb{R}^{(1,3)} = \mathbf{SO}_0(1,3)\ \vec{\times}\ \mathbb{R}^4/\mathbf{SO}_0(1,3)$, i.e., its particle analysis, e.g., for a stable neutral pion, uses momentum-characterized creation and annihilation eigenoperators $(\mathrm{u}(\vec{q}), \mathrm{u}^\star(\vec{q}))$ of the spacetime translations in the direct integral decomposition with the invariant hyperbolic measure,

$$\begin{aligned}
\mathbb{R}^{(1,3)} \ni x \longmapsto \mathbf{\Phi}(x) &= {}^\oplus\!\!\int \tfrac{d^3q}{\mathcal{V}^3 m^2 q_0}\, [e^{iqx}\mathrm{u}(\vec{q}) + e^{-iqx}\mathrm{u}^\star(\vec{q})],\\
&\text{with } q_0 = \sqrt{\vec{q}^2 + m^2},\\
\text{quantization: } [\mathrm{u}^\star(\vec{p}), \mathrm{u}(\vec{q})] &= \mathbf{1}\mathcal{V}^3 m^2 q_0 \delta(\vec{q} - \vec{p}),\\
\langle 0|\{\mathrm{u}^\star(\vec{p}), \mathrm{u}(\vec{q})\}|0\rangle &= \langle 0|\mathrm{u}^\star(\vec{p})\mathrm{u}(\vec{q})|0\rangle\\
&= \langle m^2; \vec{p}|m^2; \vec{q}\rangle = \mathcal{V}^3 m^2 q_0 \delta(\vec{q} - \vec{p}),\\
\text{distributive Hilbert basis: } &\{\mathrm{u}(\vec{q})|0\rangle = |m^2; \vec{q}\rangle \,\big|\, \vec{q} \in \mathbb{R}^3\}.
\end{aligned}$$

The quantization and Hilbert metric involve the "inverse" of the Lorentz invariant measure $\frac{d^3q}{\mathcal{V}^3 m^2 q_0} \leftrightarrow \mathcal{V}^3 m^2 q_0 \delta(\vec{q} - \vec{p})$ for $q_0 = \sqrt{\vec{q}^2 + m^2}$ of the momentum hyperboloid $\mathcal{Y}^3 \cong \mathbf{SO}_0(1,3)/\mathbf{SO}(3)$. The expectation value $\langle 0| \ldots |0\rangle$ of the (anti-) commutator for the Bose field $\mathbf{\Phi}$ is a representation coefficient of the Poincaré group, for a scalar field representing nontrivially the translations only:

$$\begin{aligned}
\mathbb{R}^{(1,3)} \ni x - y \longmapsto \quad \langle 0|\mathbf{\Phi}(y)\mathbf{\Phi}(x)|0\rangle &= \int \tfrac{d^4q}{\mathcal{V}^3 m^2}\, 2\vartheta(q_0)\delta(q^2 - m^2)e^{iq(x-y)}\\
&= \int \tfrac{d^3q}{\mathcal{V}^3 m^2 q_0} e^{iq(x-y)} = |m^2, 0\}(x - y),\\
\langle 0|\{\mathbf{\Phi}(y), \mathbf{\Phi}(x)\}|0\rangle &= \int \tfrac{d^4q}{\mathcal{V}^3 m^2}\, 2\delta(q^2 - m^2)e^{iq(x-y)}\\
&= \int \tfrac{d^3q}{\mathcal{V}^3 m^2 q_0} 2\cos q(x - y),\\
\langle 0|[\mathbf{\Phi}(y), \mathbf{\Phi}(x)]|0\rangle &= \int \tfrac{d^4q}{\mathcal{V}^3 m^2}\, 2\epsilon(q_0)\delta(q^2 - m^2)e^{iq(x-y)}\\
&= \int \tfrac{d^3q}{\mathcal{V}^3 m^2 q_0} 2i \sin q(x - y).
\end{aligned}$$

With the momentum-characterized creation and annihilation eigenoperators, the concept of a "pointlike" particle has no meaning at all; there are no creation (annihilation) operators for "pointlike" particles.

The basis is called "distributive" since, with the distribution of the irreducible representations $x \longmapsto e^{iqx}$ by $\tilde{d}(q)$, the translation "eigenvectors" involved also have a corresponding Hilbert product distribution and are normalized with distributions. Sometimes, they are also called "improper" vectors. The "proper" vectors of the associated

$$\text{Hilbert space}\, V_d \text{ for } \mathbb{R}^n \ni x \longmapsto d(x) = \int \frac{d^nq}{(2\pi)^n}\, \tilde{d}(q)e^{iqx},$$

are obtained, with Gel'fand and Raikov, starting from the absolute integrable functions $L^1(\mathbb{R}^n)$ of the translations and the Hilbert product inducing positive type function $d \in L^\infty(\mathbb{R}^n)_+$:

$$\begin{aligned}\langle\ |\ \rangle_d : L^1(\mathbb{R}^n)\times L^1(\mathbb{R}^n)\longrightarrow\mathbb{C},\ \langle f|f'\rangle_d &= \int d^nyd^nx\ \overline{f}(y)d(x-y)f'(x)\\ &= \int \frac{d^nq}{(2\pi)^n}\overline{\tilde f}(q)\tilde d(q)\tilde f'(q)\\ \text{for } f(x) &= \int \frac{d^nq}{(2\pi)^n}\tilde f(q)e^{iqx}.\end{aligned}$$

With Lebesgue, the Fourier transformed absolute integrable functions $L^1(\mathbb{R}^n)$ are continuous and vanish at infinity, $L^1(\mathbb{R}^n)\ni f\longmapsto \tilde f\in\mathcal{C}_0(\check{\mathbb{R}}^n)$.

There is a canonical procedure to go over from $L^1(\mathbb{R}^n)$ to the scalar product vector space $|L^1(\mathbb{R}^n)\rangle_d$ of the equivalence classes with a nontrivial norm, and to its Cauchy-completion $V_d=\overline{|L^1(\mathbb{R}^n)\rangle}_d$, a Hilbert space. This is not discussed in more detail. The Hilbert vectors can be spanned by the distributive basis with the (energy-)momentum function values as "continuous coefficients." In general, they are no eigenvectors:

$$\begin{aligned}\overline{|L^1(\mathbb{R}^n)\rangle}_d\ni|f\rangle &= \int\frac{d^nq}{(2\pi)^n}\tilde f(q)|q\rangle,\\ \text{translation action } x\bullet|f\rangle &= \int\frac{d^nq}{(2\pi)^n}\tilde f(q)e^{iqx}|q\rangle.\end{aligned}$$

A more correct notation has to use the direct integral $|f\rangle = {}^{\oplus}\!\int\frac{d^nq}{(2\pi)^n}\tilde f(q)|q\rangle$ for the sum of the "eigenvectors."

For positive type functions with Dirac (energy-)momentum distributions, all this can be simplified as given above: The product for an irreducible Hilbert space with $x\longmapsto e^{ipx}$ and vectors $|p;f\rangle=\tilde f(p)|p\rangle\in\mathbb{C}|p\rangle$, looks in this formulation as follows:

$$\begin{aligned}&\text{Hilbert product for } \mathbb{R}^n\ni x\longmapsto\int d^nq\ \delta(q-p)e^{iqx}=e^{ipx}:\\ \langle p;f|p;f'\rangle &= \int dydx\ \overline{f}(y)e^{ip(x-y)}f'(x)=\int d^nq\ \overline{\tilde f}(q)\delta(q-p)\tilde f'(q)\\ &= \overline{\tilde f}(p)\tilde f'(p).\end{aligned}$$

The Hilbert space for a scalar Euclidean position representation with invariant momentum P^2 is given by the square integrable functions on the 2-sphere with the momentum directions:

$$\begin{aligned}&L^2(\Omega^2)\text{ Hilbert space for } \mathbf{SO}(3)\,\vec\times\,\mathbb{R}^3,\ \vec x\longmapsto j_0(Pr):\\ &\langle P^2,0;f|P^2,0;f'\rangle = \int d^3yd^3x\ \overline{f}(\vec y)j_0(P|\vec x-\vec y|)f'(\vec x)\\ &\qquad = \int\frac{d^3q}{2\pi P}\overline{\tilde f}(\vec q)\delta(\vec q^2-P^2)\tilde f'(\vec q)=\int\frac{d^2\omega}{4\pi}\overline{\tilde f}(P\vec\omega)\tilde f'(P\vec\omega),\end{aligned}$$

and for a scalar free particle representation of Minkowski spacetime with invariant mass $m^2>0$, e.g. for a stable pion, by square integrable functions on the 3-hyperboloid with the energy-momentum directions:

$$\begin{aligned}&L^2(\mathcal{Y}^3)\text{ Hilbert space for } \mathbf{SO}_0(1,3)\,\vec\times\,\mathbb{R}^4,\ x\longmapsto\int\frac{d^4q}{V^3m^2}2\vartheta(q_0)\delta(q^2-m^2)e^{iqx}:\\ &\langle m^2;f|m^2;f'\rangle = \int\frac{d^3y}{V^3}\overline{\tilde f}(my)\tilde f'(my)\\ &\qquad = \int\frac{d^3q}{V^3m^2q_0}\overline{\tilde f}(q)\tilde f'(q),\ \text{with } q_0=\sqrt{m^2+\vec q^2}.\end{aligned}$$

For the Euclidean group, the integral of the distributive eigenvectors over the 2-sphere with the constant function $f(\vec{\omega}) = 1$ is a cyclic vector:

$$|P^2;1\rangle = \int \frac{d^2\omega}{|\Omega^2|}|P^2;\vec{\omega}\rangle \in L^2(\Omega^2): \begin{cases} \langle P^2;1|P^2;1\rangle = \int \frac{d^2\omega d^2\omega'}{|\Omega^2|^2}\langle P^2;\vec{\omega}'|P^2;\vec{\omega}\rangle \\ \qquad = \int \frac{d^2\omega d^2\omega'}{|\Omega^2|}\delta(\vec{\omega}-\vec{\omega}') = 1, \\ \langle P^2;1|\vec{x} \bullet |P^2;1\rangle = \int \frac{d^2\omega d^2\omega'}{|\Omega^2|^2}\langle P^2;\vec{\omega}'|e^{iP\vec{\omega}\vec{x}}|P^2;\vec{\omega}\rangle \\ \qquad = \int \frac{d^2\omega}{|\Omega^2|}e^{iP\vec{\omega}\vec{x}} = \frac{\sin Pr}{Pr} \end{cases}$$

For the Poincaré group, the analogue integral of the distributive eigenvectors over the momentum hyperboloid is no state vector because of the infinite hyperboloid volume:

$$|m^2;1\rangle = \int \frac{d^3y}{\mathcal{Y}^3}|m^2;y\rangle : \begin{cases} \langle m^2;1|m^2;1\rangle = \int \frac{d^3y d^3y'}{(\mathcal{Y}^3)^2}\langle m^2;y'|m^2;y\rangle \\ \qquad = \int \frac{d^3y d^3y'}{\mathcal{Y}^3}\delta(y-y') = \frac{|\mathcal{Y}^3|}{\mathcal{Y}^3} = \infty, \\ \langle m^2;1|x \bullet |m^2;1\rangle = \int \frac{d^3y d^3y'}{(\mathcal{Y}^3)^2}\langle m^2;y'|e^{imyx}|m^2;y\rangle \\ \qquad = \int \frac{d^3y}{\mathcal{Y}^3}e^{imyx} \\ \qquad = \int \frac{d^4q}{\mathcal{Y}^3 m^2}2\vartheta(q_0)\delta(q^2-m^2)e^{iqx}. \end{cases}$$

However, it can be considered as a limit of Hilbert space vectors.

10.8 Orbits and Decomposition-Transmutators

With a nontrivial homogeneous factor $\mathbf{SO}_0(t,s) \vec{\times} \mathbb{R}^{t+s}$ in a semidirect space (time) group, there are translation representations with nontrivial $\mathbf{SO}_0(t,s)$-transformation behavior. Since the fixgroups for the translation representations of position and spacetime are proper subgroups of the homogeneous factors in the Euclidean group, $\mathbf{SO}(2) \subset \mathbf{SU}(2)$, and in the Poincaré group, $\mathbf{SU}(2) \subset \mathbf{SL}(2,\mathbb{C})$, representations with nontrivial homogeneous properties, i.e., scattering states with nontrivial polarization and massive particles with nontrivial spin, have to use embedding transmutators, respectively, for the polarization classes $\mathbf{SU}(2)/\mathbf{SO}(2) \cong \Omega^2$, appropriate for the polarization directions in $d^3q\ \delta(\vec{q}^2 - P^2) \sim d^2\omega$, and the spin classes $\mathbf{SL}(2,\mathbb{C})/\mathbf{SU}(2) \cong \mathcal{Y}^3$, appropriate for the momentum directions in $d^4q\ 2\vartheta(q_0)$ $\delta(q^2 - m^2) \sim d^3y$. The polarization direction orbit and the momentum orbit have corresponding fixgroup orbits (cosets) in the full homogeneous group.

10.8.1 Axial Rotations in Rotations

Vector representation coefficients for the Euclidean group involve the product of spherical Bessel function j_1 and the spherical harmonics $\mathrm{Y}^1(\phi,\theta) \sim \frac{\vec{x}}{r}$:

$$\begin{aligned}
|P^2, \pm\tfrac{1}{2}\}(\vec{x}) &= \int \tfrac{d^3q}{2\pi P}\, \tfrac{1}{2}(\mathbf{1}_2 \pm \tfrac{\vec{\sigma}\vec{q}}{P})\, \delta(\vec{q}^2 - P^2) e^{i\vec{q}\vec{x}} \\
&= \tfrac{1}{2}(\mathbf{1}_2 \mp i\vec{\sigma}\tfrac{\partial}{\partial P\vec{x}}) \int \tfrac{d^3q}{2\pi P}\, \delta(\vec{q}^2 - P^2) e^{i\vec{q}\vec{x}}, \\
\text{projectors: } \mathcal{P}_{\pm\frac{1}{2}}(\tfrac{\vec{q}}{P}) &= \tfrac{1}{2}(\mathbf{1}_2 \pm \tfrac{\vec{\sigma}\vec{q}}{P}) = \mathcal{P}_{\pm\frac{1}{2}}(\tfrac{\vec{q}}{P}) \circ \mathcal{P}_{\pm\frac{1}{2}}(\tfrac{\vec{q}}{P}) \text{ for } \vec{q}^2 = P^2, \\
\tfrac{\partial}{\partial P\vec{x}} \int \tfrac{d^3q}{2\pi P}\, \delta(\vec{q}^2 - P^2) e^{i\vec{q}\vec{x}} &= \tfrac{\vec{x}}{r}\tfrac{\partial}{\partial Pr} j_0(Pr) = -\tfrac{\vec{x}}{r} j_1(Pr), \\
\text{with } j_1(r) &= \tfrac{\sin r - r\cos r}{r^2}, \\
\text{action of } \mathbf{SO}(3) \ni O :\ & \vec{x} \longmapsto O.\vec{x},\ \vec{q} \longmapsto O.\vec{q}.
\end{aligned}$$

The vectors are representations of the homogeneous group orbits: With the Pauli position representation, the 2-sphere $\Omega^2 \subset \mathbb{R}^3$ is representable by nonsingular hermitian (2×2)-matrices with Pauli matrices $\vec{\sigma}$:

$$\begin{aligned}
\vec{q}^2 = P^2 &: \mathbf{SU}(2)/\mathbf{SO}(2) \cong \Omega^2 \longrightarrow \mathbf{GL}(2,\mathbb{C}), \\
\tfrac{\vec{q}}{P} &= \vec{\omega} \longmapsto \tfrac{\vec{\sigma}\vec{q}}{P} = \tfrac{1}{P}\begin{pmatrix} q_3 & q_1 - iq_2 \\ q_1 + iq_2 & -q_3 \end{pmatrix} = \vec{\sigma}\vec{\omega} = \begin{pmatrix} \cos\theta & e^{-i\phi}\sin\theta \\ e^{i\phi}\sin\theta & -\cos\theta \end{pmatrix}, \\
\tfrac{\vec{\sigma}\vec{q}}{P} &\circ \tfrac{\vec{\sigma}\vec{q}}{P} = \vec{\sigma}\vec{\omega} \circ \vec{\sigma}\vec{\omega} = \mathbf{1}_2.
\end{aligned}$$

The rotation group action can be implemented by spin $\mathbf{SU}(2)$-transformations, related to $\mathbf{SO}(3)$ in the familiar way:

$$\mathbf{SU}(2) \ni u \longmapsto O(u) \in \mathbf{SO}(3) \text{ with } O^a_b(u) = \frac{1}{2}\operatorname{tr} u \circ \sigma^b \circ u^* \circ \sigma^a,$$

in the following form:

$$\mathbf{SO}(3) \times \Omega^2 \longrightarrow \Omega^2,\quad u \circ \frac{\vec{\sigma}\vec{q}}{P} \circ u^* = \frac{\vec{\sigma}O(u).\vec{q}}{P}.$$

Thus, the *Pauli representation of the 2-sphere* itself can be factorized with a spin transformation $u(\frac{\vec{q}}{P}) \in \mathbf{SU}(2)$, e.g., to the north pole σ^3 (third momentum axis). This defines the *Pauli transmutator* between spin $\mathbf{SU}(2)$- and axial rotation $\mathbf{SO}(2)$-transformations, determined up to the fixgroup $\mathbf{SO}(2)$ of σ^3:

$$\begin{aligned}
\vec{\sigma}\vec{q} &= u(\tfrac{\vec{q}}{P}) \circ \sigma^3 P \circ u^*(\tfrac{\vec{q}}{P}), \\
u(\tfrac{\vec{q}}{P}) &= \tfrac{1}{\sqrt{2P(q_3+P)}}\begin{pmatrix} q_3 + P & -q_1 + iq_2 \\ q_1 + iq_2 & q_3 + P \end{pmatrix} = \begin{pmatrix} \cos\frac{\theta}{2} & -e^{-i\phi}\sin\frac{\theta}{2} \\ e^{i\phi}\sin\frac{\theta}{2} & \cos\frac{\theta}{2} \end{pmatrix} \in \mathbf{SU}(2), \\
u(0,0,1) &= \mathbf{1}_2.
\end{aligned}$$

The irreducible spin $\mathbf{SU}(2)$-representations (J) with dimensions $1 + 2J$ are decomposable into selfdual axial rotation $\mathbf{SO}(2)$-representations $(\pm h)$ with dimension two $\begin{pmatrix} e^{i|h|\phi} & 0 \\ 0 & e^{-i|h|\phi} \end{pmatrix}$ for $h \neq 0$ and two polarizations $\pm|h|$ (left- and right-circulary polarized), i.e., with two irreducible contributions $e^{\pm i|h|\phi}$:

$$\mathbf{irrep}\,\mathbf{SU}(2) \ni (J) \stackrel{\mathbf{SO}(2)}{\cong} \begin{cases} \bigoplus_{|h|=0,1,\ldots}^{J} (\pm h) \text{ for } J = 0, 1, \ldots, \\ \bigoplus_{|h|=\frac{1}{2},\frac{3}{2},\ldots}^{J} (\pm h) \text{ for } J = \frac{1}{2}, \frac{3}{2}, \ldots, \end{cases}$$

All *transmutators from rotations to axial rotations* arise by the totally symmetric products, denoted by $\overset{2J}{\bigvee}$, of the fundamental Pauli transmutator $u(\frac{\vec{q}}{P}) \in \mathbf{SU}(2)$:

$$\begin{aligned} \mathbf{SU}(2)/\mathbf{SO}(2) \cong \Omega^2 &\longrightarrow \mathbf{SU}(1+2J,) \\ \tfrac{\vec{q}}{P} &\longmapsto (J)(\tfrac{\vec{q}}{P}) = \bigvee^{2J} u(\frac{\vec{q}}{P}),\ \vec{q}^2 = P^2, \\ &\quad (J)(0,0,1) = \mathbf{1}_{1+2J}. \end{aligned}$$

For example, the symmetric square $u(\frac{\vec{q}}{P}) \vee u(\frac{\vec{q}}{P}) \cong O(\frac{\vec{q}}{P})$ of the Pauli transmutator transforms $\begin{pmatrix} 0 \\ 0 \\ P \end{pmatrix} \longmapsto \vec{q}$ and decomposes into polarizations (± 1) and (0):

$$\begin{aligned} (1)(\tfrac{\vec{q}}{P}) \cong O^a_b(\tfrac{\vec{q}}{P}) &= \tfrac{1}{2} \operatorname{tr} u(\tfrac{\vec{q}}{P}) \circ \sigma^b \circ u^\star(\tfrac{\vec{q}}{P}) \circ \sigma^a \cong \tfrac{1}{P} \left(\begin{array}{c|c} \delta^{\alpha\beta} P - \frac{q_\alpha q_\beta}{P+q_3} & q_\alpha \\ -q_\beta & q_3 \end{array} \right) \\ &= \left(\begin{array}{cc|c} e^{i\phi} \cos^2 \frac{\theta}{2} & -e^{i\phi} \sin^2 \frac{\theta}{2} & i e^{i\phi} \frac{\sin\theta}{\sqrt{2}} \\ i \frac{\sin\theta}{\sqrt{2}} & i \frac{\sin\theta}{\sqrt{2}} & \cos\theta \\ -e^{-i\phi} \sin^2 \frac{\theta}{2} & e^{-i\phi} \cos^2 \frac{\theta}{2} & i e^{-i\phi} \frac{\sin\theta}{\sqrt{2}} \end{array} \right) \in \mathbf{SO}(3), \\ &\qquad \text{with } a, b = 1, 2, 3;\ \ \alpha, \beta = 1, 2, \\ (1) &\stackrel{\mathbf{SO}(2)}{\cong} (\pm 1) \oplus (0). \end{aligned}$$

The (3×2)-transmutators $O^{a=1,2,3}_{\alpha=1,2}(\frac{\vec{q}}{P})$ and the (3×1)-transmutators $O^{a=1,2,3}_3(\frac{\vec{q}}{P})$ relate to each other the metrical tensors of $\mathbf{SO}(3)$ and $\mathbf{SO}(2)$:

$$O^a_{\alpha,3}(\frac{\vec{q}}{P}) \delta_{ab} O^b_{\beta,3}(\frac{\vec{q}}{P}) = \left(\begin{array}{c|c} \delta_{\alpha\beta} & 0 \\ \hline 0 & 1 \end{array} \right), \quad \begin{aligned} O^a_\alpha(\tfrac{\vec{q}}{P}) \delta^{\alpha\beta} O^b_\beta(\tfrac{\vec{q}}{P}) &= \delta^{ab} - \tfrac{q^a q^b}{P^2}, \\ O^a_3(\tfrac{\vec{q}}{P}) O^b_3(\tfrac{\vec{q}}{P}) &= \tfrac{q^a q^b}{P^2}. \end{aligned}$$

10.8.2 Rotations in Lorentz Transformations

Lorentz vector representation coefficients occur in the on-shell part of the Feynman propagator of a free Dirac particle with spin $\frac{1}{2}$, e.g., of an electron-positron field:

$$\begin{aligned}
|m^2, \tfrac{1}{2}{}_{\pm}\}(x) &= \int \frac{d^4q}{V^3 m^2}\, \tfrac{1}{2}(\mathbf{1}_4 \pm \tfrac{\gamma q}{m})\, 2\vartheta(q_0)\delta(q^2 - m^2)e^{iqx} \\
&= \tfrac{1}{2}(\mathbf{1}_4 \mp i\gamma\tfrac{\partial}{\partial m x}) \int \frac{d^4q}{V^3 m^2}\, 2\vartheta(q_0)\delta(q^2 - m^2)e^{iqx}, \\
\text{Dirac projectors: } \mathcal{P}_{\frac{1}{2}\pm}(\tfrac{q}{m}) &= \tfrac{1}{2}(\mathbf{1}_4 \pm \tfrac{\gamma q}{m}) = \mathcal{P}_{\frac{1}{2}\pm}(\tfrac{q}{m}) \circ \mathcal{P}_{\frac{1}{2}\pm}(\tfrac{q}{m}) \text{ for } q^2 = m^2, \\
\text{action of } \mathbf{SO}_0(1,3) \ni \Lambda &: x \longmapsto \Lambda.x, \quad q \longmapsto \Lambda.q.
\end{aligned}$$

Starting from the Weyl spacetime representation, the *Dirac representation of the momentum hyperboloid* $\mathcal{Y}^3 \subset \mathbb{R}^{1+3}$ by nonsingular (4×4)-matrices uses Dirac matrices $\{\gamma^j, \gamma^k\} = 2\eta^{jk}\mathbf{1}_4$ which can be chirally constructed $\gamma^j = \begin{pmatrix} 0 & \sigma^j \\ \check\sigma^j & 0 \end{pmatrix}$ by two sets of Weyl matrices $\sigma^j = (\mathbf{1}_2, \vec\sigma)$ (left) and $\check\sigma^j = (\mathbf{1}_2, -\vec\sigma)$ (right):

$$\begin{aligned}
q^2 = m^2 : \mathbf{SL}(2,\mathbb{C})/\mathbf{SU}(2) &\cong \mathcal{Y}^3 \longrightarrow \mathbf{GL}(4,\mathbb{C}), \\
\tfrac{q}{m} \longmapsto \tfrac{\gamma q}{m} &= \tfrac{1}{m}\begin{pmatrix} 0 & \sigma q \\ \check\sigma q & 0 \end{pmatrix} \text{ with } \sigma q, \check\sigma q \in \mathbf{SL}(2,\mathbb{C}), \\
\tfrac{\sigma q}{m} &= \tfrac{1}{m}\begin{pmatrix} q_0 + q_3 & q_1 - iq_2 \\ q_1 + iq_2 & q_0 - q_3 \end{pmatrix} = \sigma y = \cosh\psi \mathbf{1}_2 + \sinh\psi\, \vec\sigma\vec\omega, \\
\tfrac{\sigma q}{m} \circ \tfrac{\check\sigma q}{m} &= \mathbf{1}_2, \quad \tfrac{\gamma q}{m} \circ \tfrac{\gamma q}{m} = \mathbf{1}_4.
\end{aligned}$$

The Lorentz transformations $\mathbf{SO}_0(1,3)$ are implemented by actions of the covering group $\mathbf{SL}(2,\mathbb{C})$:

$$\mathbf{SL}(2,\mathbb{C}) \ni s \longmapsto \Lambda(s) \in \mathbf{SO}_0(1,3) \text{ with } \Lambda^j_k(s) = \tfrac{1}{2}\operatorname{tr} s \circ \sigma_k \circ s^* \circ \check\sigma^j,$$

$$\mathbf{SO}_0(1,3) \times \mathcal{Y}^3 \longrightarrow \mathcal{Y}^3, \quad \begin{cases} s \circ \frac{\sigma q}{m} \circ s^* = \frac{\sigma\Lambda(s).q}{m}, \\ \mathbf{s} \circ \frac{\gamma q}{m} \circ \mathbf{s}^{-1} = \frac{\gamma\Lambda(s).q}{m}, \\ \quad \text{with } \mathbf{s} = \begin{pmatrix} s & 0 \\ 0 & \hat s \end{pmatrix}, \\ \text{where } s, \hat s = s^{-1*} \in \mathbf{SL}(2,\mathbb{C}). \end{cases}$$

This leads to the two Weyl representations $(s, \hat s)$ of the boosts $\mathcal{Y}^3$, called *Weyl transmutators* between Lorentz $\mathbf{SL}(2,\mathbb{C})$- and spin $\mathbf{SU}(2)$-transformations, determined up to spin transformations from the fixgroup in the rest system $\mathbf{1}_2$:

$$\sigma q = s\left(\tfrac{q}{m}\right) \circ \mathbf{1}_2 m \circ s^*\left(\tfrac{q}{m}\right),$$

$$\begin{aligned} s(\tfrac{q}{m}) &= \sqrt{\tfrac{q_0+m}{2m}}\left(\mathbf{1}_2 + \tfrac{\vec{\sigma}\vec{q}}{q_0+m}\right) = \tfrac{1}{\sqrt{2m(q_0+m)}}\begin{pmatrix} q_0+q_3+m & q_1-iq_2 \\ q_1+iq_2 & q_0-q_3+m \end{pmatrix} \\ &= \cosh\tfrac{\psi}{2}\mathbf{1}_2 + \sinh\tfrac{\psi}{2}\vec{\sigma}\vec{\omega} \in \mathbf{SL}(2,\mathbb{C}), \quad s(1,0,0,0) = \mathbf{1}_2, \end{aligned}$$

$$\begin{aligned} \hat{s}(\tfrac{q}{m}) &= \sqrt{\tfrac{q_0+m}{2m}}\left(\mathbf{1}_2 - \tfrac{\vec{\sigma}\vec{q}}{q_0+m}\right) = \tfrac{1}{\sqrt{2m(q_0+m)}}\begin{pmatrix} q_0-q_3+m & -q_1+iq_2 \\ -q_1-iq_2 & q_0+q_3+m \end{pmatrix} \\ &= \cosh\tfrac{\psi}{2}\mathbf{1}_2 - \sinh\tfrac{\psi}{2}\vec{\sigma}\vec{\omega} \in \mathbf{SL}(2,\mathbb{C}), \quad \hat{s}(1,0,0,0) = \mathbf{1}_2. \end{aligned}$$

The decomposable Dirac boost representations $\mathbf{s}(\frac{q}{m})$ contain the solutions of the Dirac equation:

$$\begin{gathered} \gamma q = \mathbf{s}\left(\tfrac{q}{m}\right) \circ \gamma^0 m \circ \mathbf{s}^{-1}\left(\tfrac{q}{m}\right), \quad \mathbf{s}(\tfrac{q}{m}) = \begin{pmatrix} s(\frac{q}{m}) & 0 \\ 0 & s^*(\frac{q}{m}) \end{pmatrix}, \quad \gamma^0 = \begin{pmatrix} 0 & \mathbf{1}_2 \\ \mathbf{1}_2 & 0 \end{pmatrix}, \\ (\gamma q - m\mathbf{1}_4)\mathbf{s}\left(\tfrac{q}{m}\right) = 0. \end{gathered}$$

The finite-dimensional irreducible Lorentz group representations $(L|R)$ with dimensions $(1+2L)(1+2R)$ decomposes into irreducible spin representations:

$$\mathbf{irrep}_{\text{fin}}\mathbf{SL}(2,\mathbb{C}) \ni (L|R) \overset{\mathbf{SU}(2)}{\cong} \bigoplus_{J=|L-R|}^{L+R} (J).$$

All finite-dimensional representations of the 3-hyperboloid (boost representations), i.e., all finite-dimensional *transmutators from Lorentz group to rotation group*, can be built by the totally symmetric products of the two fundamental Weyl transmutators $s(\frac{q}{m}), \hat{s}(\frac{q}{m}) \in \mathbf{SL}(2,\mathbb{C})$:

$$\begin{aligned} \mathbf{SL}(2,\mathbb{C})/\mathbf{SU}(2) \cong \mathcal{Y}^3 &\longrightarrow \mathbf{SL}((1+2L)(1+2R),\mathbb{C}), \\ \tfrac{q}{m} &\longmapsto (L|R)\left(\tfrac{q}{m}\right) = \bigvee^{2L} s\left(\tfrac{q}{m}\right) \otimes \bigvee^{2R} \hat{s}\left(\tfrac{q}{m}\right), \quad q^2 = m^2, \\ & (L|R)(1,0,0,0) = \mathbf{1}_{(1+2L)(1+2R)}. \end{aligned}$$

For example, the vector representation $s \otimes s^\star \cong \Lambda = (\frac{1}{2}|\frac{1}{2})$ transforms $\begin{pmatrix} m \\ 0 \end{pmatrix} \longmapsto q$ and gives two irreducible transmutators from Lorentz group to rotation group, the first column $\Lambda_0^{j=0,1,2,3}(\frac{q}{m})$ a (4×1)-transmutator to a spin 0-representation and the three remaining columns (4×3)-transmutators $\Lambda_{a=1,2,3}^{j=0,1,2,3}(\frac{q}{m})$ to a spin 1-representation:

$$\begin{aligned} (\tfrac{1}{2}|\tfrac{1}{2})\left(\tfrac{q}{m}\right) = \Lambda_k^j\left(\tfrac{q}{m}\right) &= \tfrac{1}{2}\operatorname{tr} s\left(\tfrac{q}{m}\right) \circ \sigma_k \circ s^\star\left(\tfrac{q}{m}\right) \circ \check{\sigma}^j \\ &\cong \tfrac{1}{m}\left(\begin{array}{c|c} q_0 & q_a \\ \hline q_b & \delta_{ab}m + \frac{q_a q_b}{m+q_0} \end{array}\right) \in \mathbf{SO}_0(1,3), \quad \begin{cases} j,k = 0,1,2,3, \\ a,b = 1,2,3, \end{cases} \\ (\tfrac{1}{2}|\tfrac{1}{2}) &\overset{\mathbf{SO}(3)}{\cong} (0) \oplus (1). \end{aligned}$$

The four columns of the transmutators $\Lambda^j_{0,a}(\frac{q}{m})$ relate to each other the metrical tensors of $\mathbf{SO}_0(1,3)$ and $\mathbf{SO}(3)$:

$$\Lambda^j_{0,a}\left(\frac{q}{m}\right)\eta_{jk}\Lambda^k_{0,b}\left(\frac{q}{m}\right) = \left(\begin{array}{c|c} 1 & 0 \\ \hline 0 & -\delta_{ab}\end{array}\right), \quad \begin{aligned}\Lambda^j_a\left(\tfrac{q}{m}\right)\delta^{ab}\Lambda^k_b\left(\tfrac{q}{m}\right) &= -\eta^{jk} + \tfrac{q^j q^k}{m^2},\\ \Lambda^j_0\left(\tfrac{q}{m}\right)\Lambda^k_0\left(\tfrac{q}{m}\right) &= \tfrac{q^j q^k}{m^2}.\end{aligned}$$

The transmutators, parametrized by the (energy-)momenta, are representatives of the fixgroup classes, here of $\mathbf{SU}(2)/\mathbf{SO}(2) \cong \Omega^2$ and $\mathbf{SL}(2,\mathbb{C})/\mathbf{SU}(2) \cong \mathcal{Y}^3$:

$$\begin{aligned} u(\tfrac{\vec{q}}{P}) &\in u(\tfrac{\vec{q}}{P})\mathbf{SO}(2) \text{ since } o \circ \sigma^3 \circ o = \sigma^3 \text{ for } o \in \mathbf{SO}(2),\\ s(\tfrac{q}{m}) &\in s(\tfrac{q}{m})\mathbf{SU}(2) \text{ since } u \circ \mathbf{1}_2 \circ u^* = \mathbf{1}_2 \text{ for } u \in \mathbf{SU}(2).\end{aligned}$$

Therefore, they have a complicated transformation behavior under the full group: A left transformation with the full group gives the transmutator for the transformed (energy-)momenta up to a right transformation with a Wigner (axial) rotation from the fixgroup, dependent on the left acting group element and the (energy-)momentum:

$$\begin{aligned} v \in \mathbf{SU}(2) : v \circ u\left(\tfrac{\vec{q}}{P}\right) &= u\left(\tfrac{O(v).\vec{q}}{P}\right) \circ o\left(\tfrac{\vec{q}}{P}, v\right) \text{ with } o\left(\tfrac{\vec{q}}{P}, v\right) \in \mathbf{SO}(2),\\ w \in \mathbf{SL}(2,\mathbb{C}) : w \circ s\left(\tfrac{q}{m}\right) &= s\left(\tfrac{\Lambda(w).q}{m}\right) \circ u\left(\tfrac{q}{m}, w\right) \text{ with } u\left(\tfrac{q}{m}, w\right) \in \mathbf{SU}(2).\end{aligned}$$

The explicit form of the Wigner (axial) rotations can be obtained from these equations:

$$\begin{aligned} o\left(\tfrac{\vec{q}}{P}, v\right) &= u^*\left(\tfrac{O(v).\vec{q}}{P}\right) \circ v \circ u\left(\tfrac{\vec{q}}{P}\right),\\ u\left(\tfrac{q}{m}, w\right) &= s^{-1}\left(\tfrac{\Lambda(w).q}{m}\right) \circ w \circ s\left(\tfrac{q}{m}\right).\end{aligned}$$

10.9 Polarization and Spin Projectors

The decomposition of representations of the full homogeneous rotation (spin) and Lorentz group into representations of the "little" fixgroups for axial rotations and spin, respectively, is reflected by a decomposition of the corresponding harmonic polynomials into polynomials, related to polarization (axial) and spin projectors, respectively.

10.9.1 Polarized States for Nonrelativistic Scattering

The transmutator from rotation group $\mathbf{SO}(3)$ to axial rotation group $\mathbf{SO}(2)$ in the rectangular (3×2)-submatrix $O^{a=1,2,3}_{\alpha=1,2}(\frac{\vec{q}}{P})$ is used for a scattering state to transmute from spin 1 to polarization (± 1):

$$\begin{aligned}\mathbb{E}^3 \ni \vec{x} \longmapsto |P^2, \pm 1\}(\vec{x}) &= \int \frac{d^3q}{2\pi P}\, \mathcal{P}_{\pm 1}\left(\frac{\vec{q}}{P}\right) \delta(\vec{q}^2 - P^2) e^{i\vec{q}\vec{x}} \\ &\cong \int \frac{d^3q}{2\pi P} \left(\delta^{ab} - \frac{q^a q^b}{P^2}\right) \delta(\vec{q}^2 - P^2) e^{i\vec{q}\vec{x}} \\ &= \int \frac{d^3q}{2\pi P}\, O^a_\alpha\left(\frac{\vec{q}}{P}\right) \delta^{\alpha\beta} O^b_\beta\left(\frac{\vec{q}}{P}\right) \delta(q^2 - m^2) e^{i\vec{q}\vec{x}}.\end{aligned}$$

The irreducible Hilbert representations of the Euclidean group $\mathbf{SU}(2) \vec{\times} \mathbb{R}^3$ are induced by representations of $\mathbf{SO}(2) \times \mathbb{R}^3$ for scattering states. The polarization $h \in \frac{\mathbb{Z}}{2}$ is defined by the axial rotations around the momentum. An *axial system*, defined by a distinguished position direction, e.g., $\vec{\sigma}\vec{q} = \sigma^3 P$ or $\begin{pmatrix} q_1 \\ q_2 \\ q_3 \end{pmatrix} = \begin{pmatrix} 0 \\ 0 \\ P \end{pmatrix}$, allows the definition of polarization properties. All axial systems are parametrized by a 2-sphere.

Hilbert representations of the Euclidean group involve $\mathbf{SO}(2)$-*embedding projectors* $\mathcal{P}_{\pm h}(\frac{\vec{q}}{P})$:

$$\begin{aligned}|h| = 0, \tfrac{1}{2}, 1, \ldots :\ \vec{x} \longmapsto |P^2, \pm h\}(\vec{x}) &= \int \frac{d^3q}{2\pi P}\, \mathcal{P}_{\pm h}\left(\frac{\vec{q}}{P}\right) \delta(\vec{q}^2 - P^2) e^{i\vec{q}\vec{x}} \\ &= \mathcal{P}_{\pm h}\left(-i\frac{\partial}{\partial P\vec{x}}\right) |P^2, 0\}(\vec{x}).\end{aligned}$$

In an axial system with $\mathcal{P}_{\pm h}(\frac{\vec{q}}{P}) \Rightarrow \mathcal{P}_{\pm h}(0, 0, 1)$, the polarization projectors are proportional to units, used for the Hilbert metric of the polarization degrees of freedom:

$$\begin{aligned}\mathcal{P}_0\left(\frac{\vec{q}}{P}\right) &= 1, \text{ or } = \frac{q^a q_b}{P^2} = O\left(\frac{\vec{q}}{P}\right)^a_3 O\left(\frac{\vec{q}}{P}\right)^3_b \Rightarrow \left(\begin{array}{c|c} \mathbf{0}_2 & 0 \\ \hline 0 & 1 \end{array}\right), \\ \mathcal{P}_{\pm\frac{1}{2}}\left(\frac{\vec{q}}{P}\right) &= \tfrac{1}{2}(\mathbf{1}_2 \pm \frac{\vec{\sigma}\vec{q}}{P}) \quad = u\left(\frac{\vec{q}}{P}\right) \frac{\mathbf{1}_2 \pm \sigma^3}{2} u^*\left(\frac{\vec{q}}{P}\right) \Rightarrow \frac{\mathbf{1}_2 \pm \sigma^3}{2} = \begin{pmatrix} 1 & 0 \\ 0 & 0 \end{pmatrix}, \begin{pmatrix} 0 & 0 \\ 0 & 1 \end{pmatrix}, \\ \mathcal{P}_{\pm 1}\left(\frac{\vec{q}}{P}\right) &= \delta^a_b - \frac{q^a q_b}{P^2} \quad = O\left(\frac{\vec{q}}{P}\right)^a_\alpha \delta^\beta_\alpha O\left(\frac{\vec{q}}{P}\right)^\beta_b \Rightarrow \left(\begin{array}{c|c} \mathbf{1}_2 & 0 \\ \hline 0 & 0 \end{array}\right), \ \ldots\end{aligned}$$

With the $\mathbf{SO}(2)$-decomposition of $\mathbf{SU}(2)$-representations, the harmonic $\mathbf{SO}(3)$-polynomials $(\vec{q})^L$ decompose into $\mathbf{SO}(2)$-representations with corresponding projectors $\mathcal{P}_{\pm h}(\frac{\vec{q}}{P})$, and vice versa: The projector $\mathcal{P}_{\pm h}(\frac{\vec{q}}{P})$-related $\mathbf{SO}(2)$-representations are real linear combinations of the polynomials $(\frac{\vec{q}}{P})^L$ with maximal degree $2|h|$, e.g., for the $\mathbf{SO}(2)$-decomposition $(1) = (\pm 1) \oplus (0)$:

$$\begin{aligned}\frac{q^a q^b}{P^2} - \tfrac{1}{3}\delta^{ab} &= -\tfrac{1}{3}\left(\delta^{ab} - \frac{q^a q^b}{P^2}\right) + \tfrac{2}{3}\frac{q^a q^b}{P^2}, \quad \text{i.e., } \left(\frac{\vec{q}}{P}\right)^2 \cong -\tfrac{1}{3}\mathcal{P}_{\pm 1}\left(\frac{\vec{q}}{P}\right) + \tfrac{2}{3}\mathcal{P}_0\left(\frac{\vec{q}}{P}\right), \\ \delta^{ab} - \frac{q^a q^b}{P^2} &= \tfrac{2}{3}\delta^{ab} - \left(\frac{q^a q^b}{P^2} - \tfrac{1}{3}\delta^{ab}\right), \quad \text{i.e., } \mathcal{P}_{\pm 1}\left(\frac{\vec{q}}{P}\right) \cong \tfrac{2}{3}\left(\frac{\vec{q}}{P}\right)^0 - \left(\frac{\vec{q}}{P}\right)^2.\end{aligned}$$

10.9.2 Spinning Particles in Relativistic Fields

The transmutator from Lorentz group $\mathbf{SO}_0(1, 3)$ to rotation group $\mathbf{SO}(3)$ in the rectangular (4×3)-submatrix $\Lambda^{j=0,1,2,3}_{a=1,2,3}(\frac{q}{m})$ is used for embedding a massive spin 1 particle in a Lorentz vector field, e.g., for the on-shell contribution of the Feynman propagator for a stable neutral weak boson $\mathbf{Z}^j$:

$$\begin{aligned}\mathbb{R}^{(1,3)} \ni x \longmapsto |m^2, 1\}(x) &= \int \frac{d^4q}{V^3m^2}\, \mathcal{P}_1\left(\tfrac{q}{m}\right) 2\vartheta(q_0)\delta(q^2 - m^2)e^{iqx} \\ &\cong \int \frac{d^4q}{V^3m^2}\left(-\eta^{jk} + \frac{q^jq^k}{m^2}\right) 2\vartheta(q_0)\delta(q^2 - m^2)e^{iqx} \\ &= \int \frac{d^4q}{V^3m^2}\, \Lambda_a^j\left(\tfrac{q}{m}\right)\delta^{ab}\Lambda_b^k\left(\tfrac{q}{m}\right) 2\vartheta(q_0)\delta(q^2 - m^2)e^{iqx}.\end{aligned}$$

Flat spacetime fields with particles are acted on by irreducible Hilbert representations of the Poincaré cover group $\mathbf{SL}(2,\mathbb{C}) \vec{\times} \mathbb{R}^4$ which are induced by representations of $\mathbf{SU}(2) \times \mathbb{R}^4$ for massive particles. The spin $J \in \frac{\mathbb{N}}{2}$ for a stable massive particle, $m > 0$, is defined in a rest system with a distinguished time direction, e.g., $\sigma^j q_j = \mathbf{1}_2 m$ or $\begin{pmatrix} q_0 \\ \vec{q}\end{pmatrix} = \begin{pmatrix} m \\ 0\end{pmatrix}$. It allows the definition of the rotation properties of the resting particle. All rest systems are parametrized by a 3-hyperboloid. Half-integer spin can be embedded with left and right chirality, denoted by a subsign, e.g., for Weyl spinors by $J_\pm = \frac{1}{2}_\pm$.

Hilbert representations for the Poincaré group with massive particles $m > 0$ lead to $\mathbf{SU}(2)$- *embedding projectors* $\mathcal{P}_J(\frac{q}{m})$:

$$\begin{aligned}J = 0, \tfrac{1}{2}_\pm, 1, \tfrac{3}{2}_\pm, 2, \ldots :\ x \longmapsto |m^2, J\}(x) &= \int \frac{d^4q}{V^3m^2}\, \mathcal{P}_J\left(\tfrac{q}{m}\right) 2\vartheta(q_0)\delta(q^2 - m^2)e^{iqx} \\ &= \mathcal{P}_J\left(-i\frac{\partial}{\partial mx}\right) |m^2, 0\}(x)\end{aligned}$$

In a rest system with $\mathcal{P}_J(\frac{q}{m}) \Rightarrow \mathcal{P}_J(1,0,0,0)$, the spin projectors are proportional to spin $\mathbf{SU}(2)$-units, used for the Hilbert metric of the spin degrees of freedom:

$$\begin{aligned}\mathcal{P}_0\left(\tfrac{q}{m}\right) &= 1, \text{ or } = \frac{q^jq_k}{m^2} &&= \Lambda_0^j\left(\tfrac{q}{m}\right)\Lambda_k^0\left(\tfrac{q}{m}\right) &&\Rightarrow \left(\begin{array}{c|c} 1 & 0 \\ \hline 0 & \mathbf{0}_3\end{array}\right), \\ \mathcal{P}_{\frac{1}{2}_\pm}\left(\tfrac{q}{m}\right) &= \tfrac{1}{2}\left(\mathbf{1}_4 \pm \frac{\gamma q}{m}\right) &&= \mathbf{s}\left(\tfrac{q}{m}\right)\frac{\mathbf{1}_4 \pm \gamma_0}{2}\mathbf{s}^{-1}\left(\tfrac{q}{m}\right) &&\Rightarrow \frac{\mathbf{1}_4 \pm \gamma_0}{2} \cong \begin{pmatrix}\mathbf{1}_2 & 0 \\ 0 & 0\end{pmatrix}, \begin{pmatrix} 0 & 0 \\ 0 & \mathbf{1}_2\end{pmatrix}, \\ \mathcal{P}_1\left(\tfrac{q}{m}\right) &= \delta_k^j - \frac{q^jq_k}{m^2} &&= \Lambda_a^j\left(\tfrac{q}{m}\right)\delta_b^a\Lambda_k^b\left(\tfrac{q}{m}\right) &&\Rightarrow \left(\begin{array}{c|c} 0 & 0 \\ \hline 0 & \mathbf{1}_3\end{array}\right), \quad \ldots\end{aligned}$$

With the $\mathbf{SU}(2)$-decomposition of $\mathbf{SL}(2,\mathbb{C})$-representations, the harmonic $\mathbf{SO}_0(1,3)$-polynomials $(\frac{q}{m})^L$ decompose into $\mathbf{SU}(2)$-representation with corresponding projectors $\mathcal{P}_J(\frac{q}{m})$ and vice versa: The projectors $\mathcal{P}_J(\frac{q}{m})$-related $\mathbf{SU}(2)$-representations are real linear combinations of harmonic polynomials $(\frac{q}{m})^L$ with maximal degree $L = 2J$, e.g., for the spin decomposition $(\frac{1}{2}|\frac{1}{2}) = (0) \oplus (1)$:

$$\begin{aligned}\frac{q^jq^k}{m^2} - \tfrac{1}{4}\eta^{jk} &= \tfrac{1}{4}\left(-\eta^{jk} + \frac{q^jq^k}{m^2}\right) + \tfrac{3}{4}\frac{q^jq^k}{m^2}, \quad \text{i.e., } \left(\tfrac{q}{m}\right)^2 \cong \tfrac{1}{4}\mathcal{P}_1\left(\tfrac{q}{m}\right) + \tfrac{3}{4}\mathcal{P}_0\left(\tfrac{q}{m}\right), \\ -\eta^{jk} + \frac{q^jq^k}{m^2} &= -\tfrac{3}{4}\eta^{jk} + \left(\frac{q^jq^k}{m^2} - \tfrac{1}{4}\eta^{jk}\right), \quad \text{i.e., } \mathcal{P}_1\left(\tfrac{q}{m}\right) \cong -\tfrac{3}{4}\left(\tfrac{q}{m}\right)^0 + \left(\tfrac{q}{m}\right)^2.\end{aligned}$$

10.10 Hilbert Spaces for Euclidean Position and Minkowski Spacetime

Nontrivial homogenous representation properties have to be taken account of in the distributive Hilbert bases. The distributive bases with nontrivial homogeneous group properties are acted on by representations of the fixgroup.

For the Euclidean group the representations of the fixgroup $\mathbf{SO}(2) \ni e^{i\sigma_3\phi}$ are characterized by a polarization h, left and right for $\pm|h|$:

$$\mathbf{SU}(2) \vec{\times} \mathbb{R}^3 : \begin{cases} \mathbb{E}^3 \ni \vec{x} \longmapsto |P^2, \pm h\}(\vec{x}) = \int \frac{d^3q}{2\pi P} \mathcal{P}_{\pm h}(\vec{q})\delta(\vec{q}^2 - P^2)e^{i\vec{q}\vec{x}}, \\ \text{invariants: } P^2 > 0, \quad \pm h = 0, \pm\frac{1}{2}, \pm 1, \ldots, \\ \text{distributive basis : } \{|P^2, h; \vec{\omega}\rangle \,\big|\, \vec{\omega} \in \Omega^2\}, \\ \text{distributive product : } \langle P^2, h; \vec{\omega}'|P^2, h; \vec{\omega}\rangle = |\Omega^2|\delta(\vec{\omega} - \vec{\omega}'), \\ \text{distributive unit : } \mathbf{1}_{L^2(\Omega^2)} \otimes \mathbf{1} \cong \int \frac{d^2\omega}{4\pi} |P^2, h; \vec{\omega}\rangle\langle P^2, h; \vec{\omega}|, \\ \mathbb{R}^3 - \text{action : } \vec{x} \bullet |P^2, h; \vec{\omega}\rangle = e^{iP\vec{x}\vec{\omega}}|P^2, h; \vec{\omega}\rangle \\ \mathbf{SO}(2)\text{-action : } e^{i\sigma_3\phi} \bullet |P^2, h; \vec{\omega}\rangle = e^{ih\phi}|P^2, h; \vec{\omega}\rangle. \end{cases}$$

The representations of the Poincaré group for free massive stable particles have a fixgroup $\mathbf{SU}(2) \ni u$, characterized by a spin J – in the momentum parametrization:

$$\mathbf{SL}(2, \mathbb{C}) \vec{\times} \mathbb{R}^4 : \begin{cases} \mathbb{R}^{(1,3)} \ni x \longmapsto |m^2, J\}(x) = \int \frac{d^4q}{\mathcal{V}^3 m^2} \mathcal{P}_J(q) 2\vartheta(q_0)\delta(q^2 - m^2)e^{iqx}, \\ \text{invariants : } m^2 > 0, \quad J = 0, \frac{1}{2}_\pm, 1, \frac{3}{2}_\pm, \ldots, \\ \text{distributive basis : } \{|m^2, J; \vec{q}, \alpha\rangle \,\big|\, \vec{q} \in \mathbb{R}^3, \quad \alpha = -J, \ldots, J\}, \\ \text{distributive product : } \langle m^2, J; \vec{q}', \alpha'|m^2, J; \vec{q}, \alpha\rangle \\ \qquad = \delta_{\alpha\alpha'} q_0 \mathcal{V}^3 m^2 \, \delta(\vec{q} - \vec{q}'), \\ \text{with } q_0 = \sqrt{m^2 + \vec{q}^2}, \\ \text{distributive unit : } \mathbf{1}_{L^2(\mathcal{Y}^3)} \otimes \mathbf{1}_{1+2J} \\ \qquad \cong \int \frac{d^3q}{q_0 \mathcal{V}^3 m^2} |m^2, J; \vec{q}, \alpha\rangle\langle m^2, J; \vec{q}, \alpha|, \\ \mathbb{R}^4\text{-action : } x \bullet |m^2, J; \vec{q}, \alpha\rangle = e^{iqx}|m^2, J; \vec{q}, \alpha\rangle \\ \mathbf{SU}(2) - \text{action : } u \bullet |m^2, J; \vec{q}, \alpha\rangle = (\bigvee^{2J} u)^{\alpha'}_{\alpha} |m^2, J; \vec{q}, \alpha'\rangle. \end{cases}$$

The proper Hilbert space vectors arise from the distributive eigenvectors by transmutators, transmuting from fixgroup $\mathbf{SO}(2)$ and $\mathbf{SU}(2)$-representations to representations of the full group $\mathbf{SU}(2)$ and $\mathbf{SL}(2, \mathbb{C})$, respectively, and by integration with square integrable (energy-)momentum functions:

$$\begin{aligned} \mathbf{SU}(2) \vec{\times} \mathbb{R}^3 : \quad & |P^2, J; f\rangle = \int \frac{d^2\omega}{|\Omega^2|} f_\alpha(\vec{\omega}) \, (J)^\alpha_h(\vec{\omega}) \, |P^2, |h|; \vec{\omega}, h\rangle, \\ & \text{with } (\pm h) \in (J), \quad f_\alpha \in L^2(\Omega^2), \\ \mathbf{SL}(2, \mathbb{C}) \vec{\times} \mathbb{R}^4 : \; & |m^2, L, R; f\rangle = \int \frac{d^3q}{\mathcal{V}^3 m^2 q_0} f_k(\tfrac{q}{m}) \, (L|R)^k_\alpha(\tfrac{q}{m}) \, |m^2, J; \vec{q}, \alpha\rangle \\ & \text{with } (J) \in (L|R), \quad f_k \in L^2(\mathcal{Y}^3). \end{aligned}$$

Those vectors have the full group transformation properties, involving the Wigner rotation, e.g., for a Lorentz transformation $w \in \mathbf{SL}(2, \mathbb{C})$, acting on the transmutator

$(L|R)(\frac{q}{m})$ and producing the Wigner rotation $(J)(\frac{q}{m}, w)$ of the distributive basis vectors:

$$\begin{aligned} w \in \mathbf{SL}(2,\mathbb{C}) : w \bullet \int \tfrac{d^3q}{V^3 m^2 q_0} f_k\left(\tfrac{q}{m}\right)(L|R)^k_\alpha\left(\tfrac{q}{m}\right)\, |m^2, J; \vec{q}, \alpha\rangle \\ = \int \tfrac{d^3q}{V^3 m^2 q_0} f_k\left(\tfrac{q}{m}\right)(L|R)^k_\gamma\left(\tfrac{\Lambda(w).q}{m}\right)\, (J)^\gamma_\alpha\left(\tfrac{q}{m}, w\right)\, |m^2, J; \vec{q}, \alpha\rangle \end{aligned}$$

In the scalar product, the transmutator products produce the linear representations of the orbit, e.g., for the Euclidean group vectors:

$$\langle P^2, J; f|P^2, J; f'\rangle = \int \frac{d^2\omega}{|\Omega^2|} \overline{f_\gamma(\vec{\omega})}\, [(J)^* \circ (J)]^\alpha_\gamma(\vec{\omega})\, f'_\alpha(\vec{\omega})$$

A polarization nontrivial *quantum field of the Euclidean group* involves a rotation-axial group transmutation, e.g., the (3×2)-transmutator $O^a_h(\frac{\vec{q}}{P})$ from $\mathbf{SO}(3)$-rotation property $L = 1$ to axial $\mathbf{SO}(2)$-rotation helicity $h = \pm 1$. It has a harmonic expansion with translation and axial rotation eigenoperators for momentum direction $\vec{\omega}$ and helicities h:

$$\begin{aligned} \mathbf{E}^a(\vec{x}) = {}^{\oplus}\!\!\int \tfrac{d^2\omega}{|\Omega^2|}\, O^a_h(\tfrac{\vec{q}}{P})[e^{i\vec{q}\vec{x}}\mathrm{u}^h(\vec{\omega}) + e^{-i\vec{q}\vec{x}}\mathrm{u}^*_h(\vec{\omega})], \\ a = 1, 2, 3,\ h = \pm 1,\ \text{with } |\vec{q}| = P, \\ \text{quantization: } [\mathrm{u}^*_h(\vec{\omega}'), \mathrm{u}^{h'}(\vec{\omega})] = \delta^{h'}_h |\Omega^2|\, \delta(\vec{\omega} - \vec{\omega}'), \\ \text{distributive basis: } \{\mathrm{u}^h(\vec{\omega})|0\rangle = |P^2, h; \vec{\omega}\rangle \,\big|\, \vec{\omega} \in \Omega^2,\ h = \pm 1\}, \\ \langle 0|\mathrm{u}^*_h(\vec{\omega}')\mathrm{u}^{h'}(\vec{\omega})|0\rangle = \langle P^2, h; \vec{\omega}'|P^2, h'; \vec{\omega}\rangle = \delta^{hh'}|\Omega^2|\, \delta(\vec{\omega} - \vec{\omega}'), \end{aligned}$$

The ground-state value of the anticommutator of the commutator-quantized field characterizes the Euclidean group representation:

$$\begin{aligned} \langle 0|\{\mathbf{E}^a(\vec{y}), \mathbf{E}^b(\vec{x})\}|0\rangle = \int \tfrac{d^3q}{2\pi P}\, O^a_h(\tfrac{\vec{q}}{P})\delta^{hh'} O^b_{h'}(\tfrac{\vec{q}}{P})\delta(\vec{q}^2 - P^2)e^{i\vec{q}(\vec{x}-\vec{y})} \\ = \int \tfrac{d^3q}{2\pi P}\, (\delta^{ab} - \tfrac{q^a q^b}{P^2})\delta(\vec{q}^2 - P^2)e^{i\vec{q}(\vec{x}-\vec{y})}. \end{aligned}$$

A spin nontrivial *quantum field of the Poincaré group*, e.g., for a stable neutral spin 1 particle, involves a Lorentz-rotation group transmutation, e.g., a spin 1 particle field with the (4×3)-transmutator $\Lambda^k_a(\frac{q}{m})$ from Lorentz vector properties $(\frac{1}{2}|\frac{1}{2})$ to rotation vector properties (1). It has a harmonic expansion with translation and rotation eigenoperators for energy-momenta q and spin directions a:

$$\begin{aligned} \mathbf{Z}^k(x) = {}^{\oplus}\!\!\int \tfrac{d^3q}{V^3 m^2 q_0}\, \Lambda^k_a(\tfrac{q}{m})[e^{iqx}\mathrm{u}^a(\vec{q}) + e^{-iqx}\mathrm{u}^*_a(\vec{q})], \\ k = 0, 1, 2, 3;\ a = 1, 2, 3,\ \text{with } q_0 = \sqrt{m^2 + \vec{q}^2}, \\ \text{quantization: } [\mathrm{u}^*_a(\vec{q}'), \mathrm{u}^b(\vec{q})] = \delta^b_a q_0 V^3 m^2\, \delta(\vec{q} - \vec{q}'), \\ \text{distributive basis: } \{\mathrm{u}^a(\vec{q})|0\rangle = |m^2, 1; \vec{q}, a\rangle \,\big|\, \vec{q} \in \mathbb{R}^3,\ a = 1, 2, 3\}, \\ \langle 0|\mathrm{u}^*_a(\vec{q}')\mathrm{u}^b(\vec{q})|0\rangle = \langle m^2, 1; \vec{q}', a|m^2, 1; \vec{q}, b\rangle = \delta^{ab} q_0 V^3 m^2\, \delta(\vec{q} - \vec{q}'), \end{aligned}$$

The ground-state value of the anticommutator of the commutator-quantized vector field characterizes the Poincaré group representation (on-shell contribution of the Feynman propagator):

$$\begin{aligned}\langle 0|\{\mathbf{Z}^j(y),\mathbf{Z}^k(x)\}|0\rangle &= \int \frac{d^4q}{V^3m^2}\,\Lambda_a^j(\tfrac{q}{m})\delta^{ab}\Lambda_b^k(\tfrac{q}{m})2\delta(q^2-m^2)e^{iq(x-y)}\\ &= \int \frac{d^4q}{V^3m^2}\,(-\eta^{jk}+\tfrac{q^jq^k}{m^2})2\delta(q^2-m^2)e^{iq(x-y)}\end{aligned}$$

The proper vectors have as Hilbert product:

$$\begin{aligned}|m^2,\tfrac{1}{2},\tfrac{1}{2};f\rangle &= \int \frac{d^3q}{V^3m^2q_0}\,f_k(\tfrac{q}{m})\,\Lambda_a^k(\tfrac{q}{m})\,|m^2,1;\vec{q},a\rangle\\ \langle m^2,\tfrac{1}{2},\tfrac{1}{2};f|m^2,\tfrac{1}{2},\tfrac{1}{2};f'\rangle &= \int \frac{d^3q}{V^3m^2q_0}\,\overline{f_j\left(\tfrac{q}{m}\right)}\,(\Lambda\circ\Lambda)^{jk}\left(\tfrac{q}{m}\right)\,f'_k\left(\tfrac{q}{m}\right),\\ (\Lambda\circ\Lambda)^{jk}\left(\tfrac{q}{m}\right) &= \Lambda_a^j\left(\tfrac{q}{m}\right)\delta^{ab}\Lambda_b^k\left(\tfrac{q}{m}\right) = -\eta^{jk}+\tfrac{q^jq^k}{m^2}.\end{aligned}$$

10.11 Harmonic Analysis for Euclidean Position and Minkowski Spacetime

There is not only the Hilbert product $\langle I;w|I;w'\rangle$ for vectors from one representation space, e.g., $\langle J;m|J;m'\rangle$ for spin, there is also the *Schur product* $\{I|I'\}$ for different representations, e.g., $\{J|J'\}$. The Schur product is important for the harmonic analysis, i.e., for the decomposition of a representation into irreducible ones.

The Schur product for particle (translation) representations is defined by integration over the translation group, i.e., as convolution at the trivial translation:

$$f(x) = \int \frac{d^nq}{(2\pi)^n}\,\tilde{f}(q)e^{iqx},\quad f*f'(0) = \int d^nx\, f(-x)f'(x) = \int \frac{d^nq}{(2\pi)^n}\,\tilde{f}(q)\tilde{f}'(q).$$

The Schur product of two cyclic translation representations is given by the integral over the characterizing positive type functions:

$$\{d_2|d_1\} = \int d^nx\,\overline{d_2(x)}d_1(x) = \int \frac{d^nq}{(2\pi)^n}\,\tilde{d}_1(q)\tilde{d}_2(q).$$

The irreducible translation representations with different (energy-)momenta are Schur-orthogonal with the Plancherel distribution:

$$\text{for } \mathbb{R}^n:\ \{p'|p\} = \int d^nx\, e^{-ip'x}e^{ipx} = \delta\left(\frac{p-p'}{2\pi}\right),$$

as well as the selfdual representations, e.g., for the time translations:

$$\text{for } \mathbb{R}:\ \{E'^2|E^2\} = \int dt\, \cos E't\cos Et = 2\pi|E|\delta(E^2-E'^2).$$

Coefficients of Hilbert representations for the Euclidean group have a spectral decomposition into polarization components $(\pm|h|)$ for $\mathbf{SO}(2)$, and, for fixed polarization, a decomposition with respect to the translation invariants P^2 with positive distributions $\rho_{\pm h}(P^2)$:

$$\text{for } \mathbf{SU}(2)\vec{\times}\mathbb{R}^3: \quad \sum_{\pm h=0,\pm\frac{1}{2},\pm 1,\ldots} \int_0^\infty dP^2\, \rho_{\pm h}(P^2) \int \frac{d^3q}{2\pi P}\, \mathcal{P}_{\pm h}\left(\frac{\vec{q}}{P}\right) \delta(\vec{q}^2 - P^2) e^{i\vec{q}\vec{x}}$$
$$= \sum_{\pm h=0,\pm\frac{1}{2},\pm 1,\ldots} \int_0^\infty dP^2\, \rho_{\pm h}(P^2)\, \mathcal{P}_{\pm h}\left(-i\frac{\partial}{\partial P\vec{x}}\right) |P^2, 0\}(\vec{x}).$$

The translation invariant shows up in the *Laplace equation*:

$$(\vec{\partial}^2 + P^2)|P^2, \pm h\} = 0.$$

An example for the harmonic analysis with respect to the Euclidean group is the familiar decomposition of the plane wave $e^{iPz} \in L^\infty(\mathbb{E}^3)_+$ for fixed translation invariant P^2 into positive type functions for irreducible representations of the Euclidean group with different rotation invariants L and Plancherel measure $d_L = 1 + 2L$ as used in the description of nonrelativistic scattering:

$$e^{iPz} = e^{iPr\cos\theta} = \sum_{L=0}^{\infty} (1+2L)\mathrm{P}^L(\cos\theta) i^L j_L(Pr), \quad (\vec{\partial}^2 + P^2)e^{iPz} = 0,$$
$$e^{iPr} = \sum_{L=0}^{\infty} (1+2L) i^L j_L(Pr),$$

$$\text{with} \begin{cases} i^L j_L(r) = \int_{-1}^1 \frac{d\zeta}{2}\, \mathrm{P}^L(\zeta) e^{ir\zeta} = \mathrm{P}^L\left(-i\frac{d}{dr}\right) \int_{-1}^1 \frac{d\zeta}{2}\, e^{ir\zeta} \\ \qquad = (ir)^L \left(-\frac{1}{r}\frac{d}{dr}\right)^L \frac{\sin r}{r} \text{ (spherical Bessel functions)}, \\ \mathrm{P}^L(\cos\theta) = \sqrt{\frac{4\pi}{1+2L}}\, \mathrm{Y}_0^L(\phi,\theta) \sim \left(\frac{z}{r}\right)^L \text{ (Legendre polynomials)}, \\ \mathrm{Y}_m^L(\phi,\theta) i^L j_L(r) = (-2i\vec{\partial})_m^L j_0(r) = \int \frac{d^3q}{2\pi}\, (2\vec{q})_m^L \delta(\vec{q}^2 - 1) e^{i\vec{q}\vec{x}}. \end{cases}$$

For the asymptotic behavior the large distance limit of the spherical Bessel functions is relevant:

$$\text{for } r \to \infty: \ i^L j_L(r) \to \frac{e^{ir} - (-1)^L e^{-ir}}{2ir} = \begin{cases} \frac{\sin r}{r}, & L = 0, 2, \ldots, \\ -i\frac{\cos r}{r}, & L = 1, 3, \ldots. \end{cases}$$

The representations of Euclidean position are Schur-orthogonal for different invariant momenta P:

$$\{P'^2, 0|P^2, 0\} = \int d^3x\, \frac{\sin Pr}{Pr}\frac{\sin P'r}{P'r} = \int \frac{2\pi d^3q}{PP'}\, \delta(\vec{q}^2 - P^2)\delta(\vec{q}^2 - P'^2)$$
$$= \frac{(2\pi)^2}{P}\, \delta(P^2 - P'^2).$$

The $\mathbf{SO}(2)$-embedding projectors are, in general, combinations of harmonic rotation group polynomials. In contrast to different harmonic $\mathbf{SO}(3)$-polynomials, they are in general, *not Schur-orthogonal for different polarization* (see ahead for the Poincaré group).

Coefficients of Hilbert representations for the Poincaré group with massive particles $m > 0$ have a spectral decomposition into irreducible components for $\mathbf{SU}(2)$ with invariant spin J, and, for fixed spin J, a Källen–Lehmann-[2] decomposition for the translation invariant masses m^2, positively distributed by $\rho_J(m^2)$:

$$\begin{aligned}\text{for } \mathbf{SL}(2,\mathbb{C}) \vec{\times} \mathbb{R}^4: \quad &\sum_{J=0,\frac{1}{2}\pm,1,\dots} \int_0^\infty dm^2\, \rho_J(m^2) \int \frac{d^4q}{\mathcal{V}^3 m^2}\, \mathcal{P}_J\left(\frac{q}{m}\right) 2\vartheta(q_0)\delta\left(q^2 - m^2\right) e^{iqx}\\ = &\sum_{J=0,\frac{1}{2}\pm,1,\dots} \int_0^\infty dm^2\, \rho_J(m^2)\mathcal{P}_J\left(-i\frac{\partial}{\partial mx}\right) |m^2, 0\}(x).\end{aligned}$$

The translation invariant occurs in the *Klein–Gordon*[3] *equation*:

$$(\partial^2 + m^2)|m^2, J\} = 0.$$

Poincaré group representations are — up to the infinite volume of the hyperboloid $|\mathcal{Y}^3| = \int d^4q\; 2\vartheta(q_0)\delta(q^2 - 1) = \infty$ — Schur-orthogonal for different masses m:

$$\begin{aligned}\{m'^2, 0|m^2, 0\} &= \int d^4x \int \frac{d^4q d^4q'}{(\mathcal{V}^3 m^2)^2} 2\vartheta(q_0)\delta(q^2 - m^2) 2\vartheta(q_0')\delta(q'^2 - m'^2) e^{i(q-q')x}\\ &= \int \frac{(2\pi)^4 d^4q}{(\mathcal{V}^3 m^2)^2} 4\vartheta(q_0)\delta(q^2 - m^2)\delta(q^2 - m'^2)\\ &= \frac{(2\pi)^4 2|\mathcal{Y}^3|}{(\mathcal{V}^3)^2} \frac{1}{m^2}\delta(m^2 - m'^2).\end{aligned}$$

The $\mathbf{SU}(2)$-projectors which embed spinning particles in relativistic fields are, in general, related to combinations of harmonic Lorentz group polynomials. In contrast to different harmonic Lorentz group polynomials, they are in general, *not Schur-orthogonal for different spins*, e.g., for a scalar and a vector particle:

$$\begin{aligned}&\mathcal{P}_0\left(\frac{q}{m}\right) = 1, \quad \mathcal{P}_1\left(\frac{q}{m}\right) \cong \left(\frac{q}{m}\right)^2 - \frac{3}{4}\left(\frac{q}{m}\right)^0 = \frac{q^j q^k}{m^2} - \eta^{jk},\\ &\int d^4q\, (q^j q^k - \eta^{jk} q^2)\mu(q^2) = -\frac{3}{4}\eta^{jk} \int d^4q\, q^2 \mu(q^2).\end{aligned}$$

10.12 Unstable Particles

Unstable particles are characterized by a *mass and a width (inverse lifetime)*. Their decay distinguishes future and past which are related to each other by time reflection. Their positive type functions for cyclic translation representations are Fourier transformed Breit[4]–Wigner densities of energy–momenta.

---

[2]Gunnar Källén (1926–1968), Harry Lehmann (1924–1998).

[3]Oskar Klein (1894–1977), Walter Gordon (1893–1939).

[4]Gregory Breit (1899–1981).

10.12.1 Reflections of Translations

Time translations together with their reflection $t \leftrightarrow -t$ constitute a nonabelian semi-direct group:

$$\begin{aligned}&\mathbf{I}(2)\,\vec{\times}\,\mathbb{R} \ni (\epsilon, t) \longmapsto \left(\begin{array}{c|c}\epsilon & t\\ \hline 0 & 1\end{array}\right) \in \mathbf{GL}(2,\mathbb{R}),\\ &\epsilon \in \mathbf{I}(2) = \{1, -1\},\quad (\epsilon_1, t_1)\circ(\epsilon_2, t_2) = (\epsilon_1\epsilon_2, t_1 + \epsilon_1 t_2).\end{aligned}$$

The position reflection is a direct factor in the *full rotation group*,

$$\mathbf{O}(3) = \mathbf{I}(2) \times \mathbf{SO}(3),\quad \mathbf{I}(2) = \{\mathbf{1}_3, -\mathbf{1}_3).$$

The *full Lorentz group* contains, as factor for the *orthochronous Lorentz group*, a discrete *Klein group*:

$$\mathbf{O}(1,3) = \mathbf{I}(2)\,\vec{\times}\,\mathbf{SO}(1,3) = \mathbf{I}(2)\,\vec{\times}\,[\mathbf{I}(2) \times \mathbf{SO}_0(1,3)].$$

The Klein group involves three reflections: the Lorentz invariant spacetime reflection $-\mathbf{1}_4$ and, dependent on a space-time decomposition, i.e., on a rest system, an only rotation invariant position reflection $\mathtt{P}\circ\mathtt{P} = \mathbf{1}_4$ with associated time reflection $\mathtt{T}\circ\mathtt{T} = \mathbf{1}_4$,

$$\mathbf{I}(2)\times\mathbf{I}(2) = \{\mathbf{1}_4,\ \mathtt{P} = \left(\begin{array}{c|c}1 & 0\\ \hline 0 & -\mathbf{1}_3\end{array}\right),\ \mathtt{T} = -\mathtt{P} = \left(\begin{array}{c|c}-1 & 0\\ \hline 0 & \mathbf{1}_3\end{array}\right),\ -\mathbf{1}_4 = \mathtt{T}\circ\mathtt{P}\}.$$

For relativistic particles, there are two ways to embed the nonrelativistic time reflection — either as only rotation invariant $\mathtt{T}$, or embedded in the Lorenz invariant spacetime reflection $-\mathbf{1}_4$. In the following the Lorentz invariant reflection is used in the extended inhomogenous group with the *special Lorentz group*,

$$\begin{aligned}&\mathbf{SO}(1,3)\,\vec{\times}\,\mathbb{R}^4 \ni (\epsilon\Lambda, x) \longmapsto \left(\begin{array}{c|c}\epsilon\Lambda & x\\ \hline 0 & 1\end{array}\right) \in \mathbf{GL}(5,\mathbb{R}),\\ &\epsilon \in \mathbf{I}(2) = \{1,-1\},\quad \Lambda \in \mathbf{SO}_0(1,3),\\ &(\epsilon_1\Lambda_1, x_1)\circ(\epsilon_2\Lambda_2, x_2) = (\epsilon_1\epsilon_2\Lambda_1\circ\Lambda_2, x_1 + \epsilon_1\Lambda_1.x_2).\end{aligned}$$

10.12.2 Breit–Wigner Functions for Time

The time reflection group $\mathbf{I}(2)$ is discrete. Therefore, and with Wigner, *time reflection* can be represented in quantum theories by an *antilinear* transformation. For complex energies $iQ = iE - b \in \check{\mathbb{C}}$ as eigenvalues for all irreducible time translation representations $t \longmapsto e^{iQt}$ (not only unitary), the reflection group is implemented by the identity and the complex conjugation:

$$\mathbf{I}(2) = \{1, -1\} \longrightarrow \{\mathrm{id}_{\check{\mathbb{C}}}, *\} : \quad \begin{cases} \mathbf{I}(2) \times \check{\mathbb{C}} \longrightarrow \check{\mathbb{C}}, \\ iE - b \overset{-1}{\leftrightarrow} (iE - b)^* = -iE - b. \end{cases}$$

A complex energy has as two-elementic reflection orbit $\{iE - b, -iE - b\}$ for real energies E and width $b > 0$. The conventional width Γ is obtained by normalization $\Gamma = 2b$.

The characteristic functions for *future and past* have an energy pole in the *upper and lower complex energy plane*, respectively:

$$\begin{aligned} \vartheta(\pm t) &= \pm \int \frac{dq_0}{2i\pi} \frac{1}{q_0 \mp io} e^{iq_0 t} = \begin{cases} 1, \ \pm t > 0, \\ 0, \ \pm t < 0, \end{cases} \\ \vartheta(t) + \vartheta(-t) &= \int \frac{dq_0}{2\pi} \delta(q_0) e^{iq_0 t} = 1, \\ \vartheta(t) - \vartheta(-t) &= \int \frac{dq_0}{2i\pi} \frac{1}{(q_0)_{\mathrm{P}}} e^{iq_0 t} = \epsilon(t) = \frac{t}{|t|} \\ &\text{(principal value integration).} \end{aligned}$$

Summation of time representations on a nontrivial $\mathbf{I}(2)$-energy orbit gives the *Breit–Wigner states* with positive type functions $L^\infty(\mathbb{R})_+$:

$$\begin{aligned} b > 0 : \ \mathbb{R} \ni t \longmapsto & \int \frac{dq_0}{2\pi} \left(\frac{1}{iq_0 - iE + b} + \frac{1}{-iq_0 + iE + b} \right) e^{iq_0 t} \\ &= \int \frac{dq_0}{2i\pi} \left(\frac{1}{q_0 - E - ib} - \frac{1}{q_0 - E + ib} \right) e^{iq_0 t} = \int \frac{dq_0}{\pi} \frac{b}{(q_0 - E)^2 + b^2} e^{iq_0 t} \\ &= \vartheta(t) e^{i(E + ib)t} + \vartheta(-t) e^{i(E - ib)t} \quad = e^{iEt - b|t|}. \end{aligned}$$

Poles with a positive imaginary part, $E + ib$, have future-supported Fourier transforms; they are called *advanced poles* (forwards), in contrast to *retarded poles* (backwards) with negative imaginary part, $E - ib$, and past-supported Fourier transforms.

For nontrivial width, the energy-distribution supporting pair $E \pm ib$ with the representation characterizing two reals (E, b) are eigenvalues of irreducible time translations, however not of unitary ones. The Breit–Wigner states with $b > 0$ characterize cyclic faithful, not irreducible, time translation representations. They are Fourier transformed positive *energy densities*. The Dirac distribution arises for stable states:

$$\begin{aligned} |E, ib\}(t) &= e^{iEt - b|t|} = \int dq_0 \, \delta_b(q_0 - E) e^{iq_0 t} \\ \delta_b(q_0 - E) &= \frac{1}{\pi} \frac{b}{(q_0 - E)^2 + b^2} = \delta\left(\frac{q_0 - E}{2\pi}\right) * \delta_b(q_0), \\ \delta_0(q_0 - E) &= \delta(q_0 - E). \end{aligned}$$

In the language of residues in the complex energy plane, time translation representations for stable particles are obtained in the distributive limit of vanishing width $0 < b = o \to 0$, i.e., in the Breit–Wigner approximation of the Dirac distribution:

$$\begin{aligned} b = 0 : \ \mathbb{R} \ni t \longmapsto & \oint \frac{dq_0}{2i\pi} \frac{1}{q_0 - E} e^{iq_0 t} = \int dq_0 \, \delta(q_0 - E) e^{iq_0 t} = e^{iEt}, \\ \frac{1}{2i\pi} & \left(\frac{1}{q_0 - E - io} - \frac{1}{q_0 - E + io} \right) = \delta(q_0 - E). \end{aligned}$$

The representation normalized Breit–Wigner state has as Schur-norm its Plancherel density, given by the inverse width (related to the mean lifetime, more ahead):

$$|E, ib\}(t) = \int \frac{dq_0}{\pi} \frac{b}{(q_0 - E)^2 + b^2} e^{iq_0 t} = e^{iEt - b|t|}, \quad \{E, ib|E, ib\} = \frac{1}{b}.$$

The Breit–Wigner energy densities for *selfdual time representations* are the normalized sums of those for two pairs $\{E \pm ib\}$ and $\{-E \pm ib\}$ with conjugated energy poles:

$$\begin{aligned}
b \geq 0: \quad \delta_b(|q_0| - |E|) &= \tfrac{1}{4i\pi} \left[\tfrac{1}{q_0 - E - ib} - \tfrac{1}{q_0 - E + ib} + \tfrac{1}{q_0 + E - ib} - \tfrac{1}{q_0 + E + ib} \right] \\
&= \tfrac{1}{2\pi} \left[\tfrac{b}{(q_0 - E)^2 + b^2} + \tfrac{b}{(q_0 + E)^2 + b^2} \right] \\
&= \tfrac{1}{2i\pi} \left[\tfrac{E + ib}{q_0^2 - (E + ib)^2} - \tfrac{E - ib}{q_0^2 - (E - ib)^2} \right], \\
\delta_0(|q_0| - |E|) &= |E| \delta(q_0^2 - E^2).
\end{aligned}$$

Also the selfdual Breit–Wigner states contain the characteristic exponential damping factor:

$$|E^2, ib\}(t) = \int dq_0 \; \delta_b(|q_0| - |E|) e^{iq_0 t} = e^{-b|t|} \cos Et,$$

10.12.3 Hilbert Spaces for Unstable Particles

The Breit–Wigner energy-densities characterize the Hilbert product of infinite-dimensional spaces:

$$\mathbb{R} \ni t \longmapsto e^{iEt - b|t|}: \begin{cases} \text{distributive basis: } \{|E, ib; q_0\rangle \; \big| \; q_0 \in \mathbb{R}\}, \\ \text{distributive product: } \langle E, ib; q_0' | E, ib; q_0\rangle \\ \qquad\qquad = \frac{2b}{(q_0 - E)^2 + b^2} \delta\left(\frac{q_0 - q_0'}{2\pi}\right), \\ \text{translation action: } t \bullet |E, ib; q_0\rangle = e^{iq_0 t} |E, ib; q_0\rangle, \\ \text{time reflection: } \alpha |E, ib; q_0\rangle \overset{-1}{\leftrightarrow} \langle E, ib; q_0 | \overline{\alpha}, \quad \alpha \in \mathbb{C}. \end{cases}$$

The time reflection $-1 \ni \mathbf{I}(2)$ acts antilinearly with the exchange of bra $\langle \ldots |$, standing for past or in-going, and ket $| \ldots \rangle$, standing for future or out-going.

The proper Hilbert vectors use energy functions, square integrable with the Breit–Wigner functions:

$$\begin{aligned}
|E, ib; f\rangle &= \int \tfrac{dq_0}{2\pi} f(q_0) |E, ib; q_0\rangle, \\
\langle E, ib; f | E, ib; f'\rangle &= \int \tfrac{dq_0}{2\pi} \overline{f(q_0)} \tfrac{2b}{(q_0 - E)^2 + b^2} f'(q_0).
\end{aligned}$$

The sum over the distributive basis, i.e., with $f(q_0) = 1$, gives a cyclic vector:

$$|E, ib; 1\rangle = \int \frac{dq_0}{2\pi}|E, ib; q_0\rangle, \quad \langle E, ib; 1|t \bullet |E, ib; 1\rangle = e^{iEt-b|t|}.$$

The one-dimensional irreducible time representation for stable states can be reached from the infinite-dimensional spaces for unstable states as vanishing width limit:

$$\mathbb{R} \ni t \longmapsto e^{iEt}: \begin{cases} \text{distributive basis: } \{|E, 0; q_0\rangle \bigm| q_0 \in \mathbb{R}\}, \\ \text{distributive product: } \langle E, 0; q_0'|E, 0; q_0\rangle = \delta\left(\frac{q_0-E}{2\pi}\right)\delta\left(\frac{q_0-q_0'}{2\pi}\right), \\ \langle E, 0; f|E, 0; f'\rangle = \int \frac{dq_0}{2\pi}\,\overline{f(q_0)}\,\delta\left(\frac{q_0-E}{2\pi}\right) f'(q_0) = \overline{f(E)}f'(E), \end{cases}$$

i.e., all frequency functions use the fixed energy value $f(E) \in \mathbb{C}$ only, and the relevant representation space is complex one-dimensional $f(E)|E\rangle \in \mathbb{C}|E\rangle$.

A Breit–Wigner state is not only the sum, but also the *(convolution) product of its past (in-going) and future (out-going) contributions*, as $d = \xi * \hat{\xi}$ and $\tilde{d} = \tilde{\xi} \cdot \overline{\tilde{\xi}}$:

$$\begin{aligned} e^{iEt-b|t|} &= \int \frac{dq_0}{2\pi}\frac{2b}{(q_0-E)^2+b^2}e^{iq_0t} = \int \frac{dq_0}{2\pi}\frac{\sqrt{2b}}{q_0-(E-ib)} \cdot \frac{\sqrt{2b}}{q_0-(E+ib)}e^{iq_0t} \\ &= [\vartheta(-t)\sqrt{2b}e^{i(E-ib)t}] * [\vartheta(t)\sqrt{2b}e^{i(E+ib)t}]. \end{aligned}$$

This allows a scalar product formulation for $L^2(\mathbb{R}_+)$ with correspondingly modified *energy-functions having future-supported Fourier transforms*:

$$\begin{aligned} \langle E, ib; f|E, ib; f'\rangle &= \int \frac{dq_0}{2\pi}\,\overline{f(q_0)}\,\frac{2b}{(q_0-E)^2+b^2}\,f'(q_0) \\ &= \int \frac{dq_0}{2\pi}\,\overline{F(q_0)} \cdot F'(q_0), \text{ with } F(q_0) = \frac{\sqrt{2b}\,f(q_0)}{q_0-(E+ib)}, \\ \int \frac{dq_0}{2i\pi}\,F(q_0)e^{iq_0t} &= \vartheta(t)\sqrt{2b}\,f(E+ib)e^{i(E+ib)t} \in L^2(\mathbb{R}_+), \\ \vartheta(t)\sqrt{2b}\,e^{i(E+ib)t} &= \int \frac{dq_0}{2i\pi}\,\frac{\sqrt{2b}}{q_0-(E+ib)}e^{iq_0t} \in L^2(\mathbb{R}_+). \end{aligned}$$

Time dependent Breit–Wigner states can be used in analogy to position-dependent Schrödinger wave functions with the expectation values in the Schur-normalized wave functions $\sqrt{b}e^{iE-b|t|}$:

$$\langle a\rangle = \frac{\{E, ib|a|E, ib\}}{\{E, ib|E, ib\}}.$$

The expectation values of the time powers, especially the mean lifetime $\langle |t|\rangle = \frac{1}{2b} = \frac{1}{\Gamma}$, and the expectation values of the energies as conjugated time derivatives are:

$$\begin{aligned}
\langle |t|^n\rangle &= b\int dt\ |t|^n\ e^{-2b|t|} = b\left(-\tfrac{\partial}{\partial 2b}\right)^n \tfrac{1}{b} = \tfrac{n!}{(2b)^n} = 1, \tfrac{1}{2b}, \tfrac{1}{2b^2}, \ldots,\\
\langle t^n\rangle &= \delta^n_{2m}\tfrac{n!}{(2b)^n} = 1, 0, \tfrac{1}{2b^2}, \ldots,\\
\langle (-id_t)^n\rangle &= \begin{cases} 1, & n=0,\\ b\int dt\ [E + ib\epsilon(t)]e^{-2b|t|} = E, & n=1,\\ b\int dt\ [(E+ib\epsilon(t))^2 + 2b\delta(t)]e^{-2b|t|} = E^2 + b^2, & n=2,\\ \ldots, \end{cases}
\end{aligned}$$

This gives the Fourier-transformation induced reciprocity ("Fourier uncertainty") between the deviations of time and time derivative:

$$\Delta a^2 = \langle a^2\rangle - \langle a\rangle^2 : \left\{\begin{array}{c} \Delta t^2 = \frac{1}{2b^2},\\ \Delta(-id_t)^2 = b^2, \end{array}\right\} \Rightarrow \Delta t^2 \Delta(-id_t)^2 = \frac{1}{2}.$$

10.13 Collectives with Unstable Particles

Unstable particles belong to collectives. If there exist conservation laws for electric charge, lepton and baryon number, the collectives are characterized by one fixed value for these invariants. Different spins, either all integer or all half-integer, can occur in one collective. Spacetime collectives contain non-orthogonal members with either integer spins (bosonic collectives) or half-integer spins (fermionic collectives).

10.13.1 Metrical Matrices for Collectives

Two Breit–Wigner states, characterizing representations for stable or unstable particles, are not Schur-orthogonal for a nontrivial width sum $b + b' = 2(\Gamma + \Gamma') > 0$:

$$\begin{aligned}
|E, ib\}(t) &= e^{iEt - b|t|},\quad \{E', ib'|E, ib\} = \tfrac{2(b+b')}{(E-E')^2 + (b+b')^2},\\
\text{e.g., } \{E', 0|E, ib\} &= \tfrac{2b}{(E-E')^2 + b^2},\quad \{E', 0|E, 0\} = 2\pi\delta(E - E').
\end{aligned}$$

For a stable member, the limit $0 > b \to 0$ has to be performed in the distributive sense, i.e., after the energy-integration. Breit–Wigner states for selfdual time representations have as Schur product:

$$|E^2, ib\}(t) = e^{-b|t|}\cos Et,\quad \{E'^2, ib'|E^2, ib\} = \tfrac{2(b+b')}{(E-E')^2 + (b+b')^2} + \tfrac{2(b+b')}{(E+E')^2 + (b+b')^2}.$$

Breit–Wigner collectives are characterized by energy-dependent (2×2) *metrical pair matrices*, with the Schur products:

$$\mathbf{S}_0 = \begin{pmatrix} \{E, ib|E, ib\} & \{E, ib|E', ib'\} \\ \{E', ib'|E, ib\} & \{E', ib'|E', ib'\} \end{pmatrix} = \begin{pmatrix} \frac{1}{b} & \frac{2b_+}{E_-^2+b_+^2} \\ \frac{2b_+}{E_-^2+b_+^2} & \frac{1}{b'} \end{pmatrix},$$
$$\text{with } E_- = E - E', \quad b_+ = b + b'.$$

A metrical pair matrix (scalar product matrix) $\mathbf{S}_0$ is positive, i.e., hermitian with positive diagonal elements and a positive discriminant (determinant):

$$\mathbf{S}_0 = \mathbf{S}_0^* \succeq 0 \text{ since } \begin{cases} b, b' \geq 0, \\ \det \mathbf{S}_0 = \frac{1}{bb'} - \frac{4b_+^2}{(E_-^2+b_+^2)^2} = \frac{E_-^4+2E_-^2 b_+^2+(b^2-b'^2)^2}{bb'(E_-^2+b_+^2)^2} \geq 0. \end{cases}$$

$\mathbf{S}_0$ has not to be unitary: Unitarity is an appropriate concept for a linear transformation, not for a scalar product matrix.

Schur-normalized Breit–Wigner states $\sqrt{b}e^{iE-b|t|}$

$$\mathbf{S} = \begin{pmatrix} \sqrt{b} & 0 \\ 0 & \sqrt{b'} \end{pmatrix} \mathbf{S}_0 \begin{pmatrix} \sqrt{b} & 0 \\ 0 & \sqrt{b'} \end{pmatrix} = \begin{pmatrix} 1 & \sqrt{bb'}\frac{2b_+}{E_-^2+b_+^2} \\ \sqrt{bb'}\frac{2b_+}{E_-^2+b_+^2} & 1 \end{pmatrix},$$

are related by an orthogonal transformation to an orthogonal, in general not normalized basis:

$$\mathbf{S} = u \begin{pmatrix} 1+\sqrt{bb'}\frac{2b_+}{E_-^2+b_+^2} & 0 \\ 0 & 1-\sqrt{bb'}\frac{2b_+}{E_-^2+b_+^2} \end{pmatrix} u^T, \text{ with } u = \frac{1}{\sqrt{2}}\begin{pmatrix} 1 & -1 \\ 1 & 1 \end{pmatrix} \in \mathbf{SO}(2).$$

For $b_+ > 0$, an orthonormal basis requires dilation transformations. The special case of stable particles, $b_+ = 0$, $\mathbf{S} = \mathbf{1}_2$, is decomposable into two orthonormal state vectors.

A distributive basis with "improper" eigenvectors $\{|E, ib; q_0\rangle \mid q_0 \in \mathbb{R}\}$ for the energies q_0 has a distributive Hilbert product:

$$\begin{pmatrix} \langle E, ib; q_0'|E, ib; q_0\rangle & \langle E, ib; q_0'|E', ib'; q_0\rangle \\ \langle E', ib'; q_0'|E, ib; q_0\rangle & \langle E', ib'; q_0'|E', ib'; q_0\rangle \end{pmatrix} = \tilde{\mathbf{d}}(q_0)\, \delta\left(\frac{q_0 - q_0'}{2\pi}\right).$$

Here, the factorization of the Schur products in the energy integration:

$$\begin{aligned} \{E', ib'|E, ib\} &= \int \frac{dq_0}{2\pi} \frac{2b}{(q_0-E)^2+b^2} \frac{2b'}{(q_0-E')^2+b'^2} \\ &= \int \frac{dq_0 dq_0'}{(2\pi)^2} \frac{\sqrt{2b}}{q_0-(E+ib)} \frac{\sqrt{2b'}}{q_0-(E'-ib')}\, \delta\left(\frac{q_0-q_0'}{2\pi}\right) \frac{\sqrt{2b'}}{q_0'-(E'+ib')} \frac{\sqrt{2b}}{q_0'-(E-ib)}. \end{aligned}$$

yields the distributive metrical pair matrix:

$$\tilde{\mathbf{d}}(q_0) = \begin{pmatrix} \frac{2b}{[q_0-(E+ib)][q_0-(E-ib)]} & \frac{2\sqrt{bb'}}{[q_0-(E+ib)][q_0-(E'-ib')]} \\ \frac{2\sqrt{bb'}}{[q_0-(E'+ib')][q_0-(E-ib)]} & \frac{2b'}{[q_0-(E'+ib')][q_0-(E'-ib')]} \end{pmatrix}$$

$$= \begin{pmatrix} \frac{\sqrt{2b}}{q_0-(E+ib)} \\ \frac{\sqrt{2b'}}{q_0-(E'+ib')} \end{pmatrix} \otimes \left(\frac{\sqrt{2b}}{q_0-(E-ib)}, \frac{\sqrt{2b'}}{q_0-(E'-ib')} \right).$$

It is positive semidefinite with the energy densities as positive diagonal elements:

$$(\tilde{d}(q_0),\ \tilde{d}'(q_0)) = \left(\frac{2b}{(q_0-E)^2+b^2}, \frac{2b'}{(q_0-E')^2+b'^2} \right),$$

and a vanishing determinant — therefore one positive and one trivial eigenvalue:

$$0 = \det[\tilde{\mathbf{d}}(q_0) - \xi(q_0)\mathbf{1}_2] \Rightarrow \xi(q_0) = \frac{2b}{(q_0-E)^2+b^2} + \frac{2b'}{(q_0-E')^2+b'^2} \text{ and } \xi(q_0) = 0.$$

The Fourier transformed matrix elements describe the time development, either as convolution product or as sum:

$$\begin{aligned} \int \frac{dq_0}{2\pi} \frac{\sqrt{2b}}{q_0-(E+ib)} \cdot \frac{\sqrt{2b'}}{q_0-(E'-ib')} e^{iq_0 t} &= [\vartheta(t)\sqrt{2b}e^{i(E+ib)t}] * [\vartheta(-t)\sqrt{2b'}e^{i(E'-ib')t}] \\ &= \frac{2\sqrt{bb'}}{-iE_-+b_+}[\vartheta(t)e^{i(E+ib)t} + \vartheta(-t)e^{i(E'-ib')t}]. \end{aligned}$$

This yields the time dependent metric inducing pair matrix, i.e., a matrix with time representation coefficients:

$$\mathbb{R} \ni t \longmapsto \mathbf{d}(t) = \int \frac{dq_0}{2\pi} \tilde{\mathbf{d}}(q_0) e^{iq_0 t},$$

with the diagonal positive type functions:

$$\mathbb{R} \ni t \longmapsto (d(t),\ d'(t)) = (e^{iEt-b|t|},\ e^{iE't-b'|t|}).$$

The full positive-type function Breit–Wigner pair matrix is the sum of future and past contributions:

$$\begin{aligned} \mathbf{d}(t) = \vartheta(t) &\begin{pmatrix} e^{i(E+ib)t} & \frac{2\sqrt{bb'}}{-iE_-+b_+} e^{i(E+ib)t} \\ \frac{2\sqrt{bb'}}{iE_-+b_+} e^{i(E'+ib')t} & e^{i(E'+ib')t} \end{pmatrix} \\ +\vartheta(-t) &\begin{pmatrix} e^{i(E-ib)t} & \frac{2\sqrt{bb'}}{iE_-+b_+} e^{i(E'-ib')t} \\ \frac{2\sqrt{bb'}}{-iE_-+b_+} e^{i(E-ib)t} & e^{i(E'-ib')t} \end{pmatrix}, \end{aligned}$$

or, also, their convolution with tensor product $\circledast = * \otimes$:

$$\mathbf{d}(t) = \vartheta(t)\begin{pmatrix} \sqrt{2b}e^{i(E+ib)t} \\ \sqrt{2b'}e^{i(E'+ib')t} \end{pmatrix} \circledast \vartheta(-t)\left(\sqrt{2b}e^{i(E-ib)t}, \sqrt{2b'}e^{i(E'-ib')t}\right).$$

A hermitian *metrical matrix* for a collective **PC** with finitely many members $N_{\mathbf{PC}}$ contains the elements:

$$\mathbf{d}_{\iota\kappa}(q_0) = \frac{2\sqrt{b_\iota b_\kappa}}{[q_0 - (E_\iota + ib_\iota)][q_0 - (E_\kappa - ib_\kappa)]}, \quad \iota, \kappa \in \mathbf{PC}.$$

It is positive semidefinite with one nontrivial eigenvalue by the trace, i.e., the sum of the diagonal elements: In general, a nontrivial matrix, decomposable into a vector and a 1-form $M = v \otimes \omega$, has rank 1:

$$\mathbf{b} = \vec{\mathbf{b}} \otimes \vec{\mathbf{b}}^* = \begin{pmatrix} b_1 \\ \vdots \\ b_n \end{pmatrix} \otimes \left(\overline{b_1}, \ldots, \overline{b_n}\right) = \begin{pmatrix} b_1\overline{b_1} & \ldots & b_1\overline{b_n} \\ \vdots & \vdots & \vdots \\ b_n\overline{b_1} & \ldots & b_n\overline{b_n} \end{pmatrix},$$

$$\det[\xi \mathbf{1}_n - \mathbf{b}] = (\xi - \sum_{\iota=1}^{n} |b_\iota|^2)\, \xi^{n-1},$$

$$\xi(q_0) = \operatorname{tr} \mathbf{d}(q_0) = \sum_{\iota \in \mathbf{PC}} \frac{2b_\iota}{(q_0 - E_\iota)^2 + b_\iota^2}, \quad \int \frac{dq_0}{2\pi} \operatorname{tr} \mathbf{d}(q_0) = N_{\mathbf{PC}}.$$

The proper vectors contain energy functions as coefficients for the distributive basis:

$$|\mathbf{PC}; f\rangle = \sum_{\iota \in \mathbf{PC}} \int \frac{dq_0}{2\pi} f_\iota(q_0) |E_\iota, ib_\iota; q_0\rangle.$$

Their Hilbert product is induced by the metrical matrix,

$$\langle f | f' \rangle_{\mathbf{PC}} = \sum_{\iota, \kappa \in \mathbf{PC}} \int \frac{dq_0}{2\pi} \overline{f_\iota}(q_0) \mathbf{d}_{\iota\kappa}(q_0) f'_\kappa(q_0).$$

10.13.2 The Neutral Kaon Collective

An example for a collective is given by the neutral spinless kaons, long- and short-lived (K_L, K_S):

$$|E, ib\}(t) = e^{iEt - b|t|}, \qquad |E', ib'\}(t) = e^{iE't - b'|t|},$$

$$\text{normalized: } K_S(t) = \sqrt{\tfrac{\Gamma_S}{2}} e^{im_S t - \frac{\Gamma_L}{2}|t|}, \qquad K_L(t) = \sqrt{\tfrac{\Gamma_L}{2}} e^{im_L t - \frac{\Gamma_S}{2}|t|}.$$

The kaon masses are used for the energies of Breit–Wigner states, the width normalization involves a conventional factor $\frac{1}{2}$:

$$\begin{array}{ll} E = m_S \sim 498\frac{\text{MeV}}{c^2}, & E_- = m_L - m_S \sim 35 \times 10^{-13}\frac{\text{MeV}}{c^2}, \\ 2b = \Gamma_S \sim 72 \times 10^{-13}\frac{\text{MeV}}{c^2}, & 2b' = \Gamma_L \sim 0.13 \times 10^{-13}\frac{\text{MeV}}{c^2}. \end{array}$$

With

$$E_- \sim b \sim b_+, \quad \frac{b'}{b_+} \sim \frac{0.13}{72} \sim 2 \times 10^{-3}, \quad \sqrt{bb'}\frac{2b_+}{E_-^2 + b_+^2} \sim \sqrt{\frac{b'}{b_+}} \sim 4.5 \times 10^{-2},$$

the normalized non-diagonal kaon scalar product matrix is computed — up to phases — as follows:

$$\mathbf{S} = \int dt \begin{pmatrix} \overline{K_L(t)}\, K_L(t) & \overline{K_L(t)}\, K_S(t) \\ \overline{K_S(t)}\, K_L(t) & \overline{K_S(t)}\, K_S(t) \end{pmatrix} \sim \begin{pmatrix} 1 & 4.5 \times 10^{-2} \\ 4.5 \times 10^{-2} & 1 \end{pmatrix}.$$

The orthogonal non-particle vectors are combinations of the particles:

$$\begin{aligned} \begin{pmatrix} K_0 \\ K_{\overline{0}} \end{pmatrix} &= u \begin{pmatrix} K_S \\ K_L \end{pmatrix} = \begin{pmatrix} \frac{K_S - K_L}{\sqrt{2}} \\ \frac{K_S + K_L}{\sqrt{2}} \end{pmatrix}, \\ \int dt \begin{pmatrix} \overline{K_0(t)}\, K_0(t) & \overline{K_0(t)}\, K_{\overline{0}}(t) \\ \overline{K_{\overline{0}}(t)}\, K_0(t) & \overline{K_{\overline{0}}(t)}\, K_{\overline{0}}(t) \end{pmatrix} &\sim \begin{pmatrix} 1 + 4.5 \times 10^{-2} & 0 \\ 0 & 1 - 4.5 \times 10^{-2} \end{pmatrix}. \end{aligned}$$

10.14 Unstable Relativistic Particles

A relativistic embedding of Breit–Wigner energy-densities gives Breit–Wigner energy-momentum-densities with a width for both energies and momenta.

10.14.1 Breit–Wigner Functions for Position

An implementation of the position reflection uses conjugated momentum invariants,

$$\begin{aligned} P, B \geq 0: \quad \delta_B(\vec{q}^2 - P^2) &= \frac{1}{2i\pi}\left[\frac{1}{\vec{q}^2 - (P + iB)^2} - \frac{1}{\vec{q}^2 - (P - iB)^2}\right] = \frac{1}{\pi}\frac{2BP}{(\vec{q}^2 - P^2 + B^2)^2 + 4B^2P^2} \\ \delta_0(\vec{q}^2 - P^2) &= \delta(\vec{q}^2 - P^2). \end{aligned}$$

The *position Breit–Wigner states* with the exponential damping factor are representation coefficients of $\mathbf{O}(3) \vec{\times} \mathbb{R}^3$,

$$\begin{aligned}\mathbb{E}^3 \ni \vec{x} \longmapsto |P^2, iB\}(\vec{x}) &= \int \frac{d^3q}{2\pi P}\delta_B(\vec{q}^2 - P^2)e^{i\vec{q}\vec{x}} \\ &= -\frac{1}{Pr}\frac{\partial}{\partial r}\int \frac{dq}{\pi}\frac{2BP}{(q^2-P^2+B^2)^2+4B^2P^2}e^{iqr} \\ &= e^{-Br}\frac{\sin Pr}{Pr}.\end{aligned}$$

Here, a 2-sphere spread and the following integrals have been used:

$$x \in \mathbb{R},\ r = |x|,\ a \in \mathbb{R},\ b \geq 0: \left\{\begin{aligned} \frac{\partial}{\partial r}\int \frac{dq}{i\pi}\frac{1}{q^2-(a+ib)^2}e^{iqx} &= \frac{\partial}{\partial r}\frac{e^{i(a+ib)r}}{a+ib} = ie^{i(a+ib)r}, \\ \frac{\partial}{\partial r}\int \frac{dq}{\pi}\frac{2ba}{(q^2-a^2+b^2)^2+4b^2a^2}e^{iqx} &= \frac{\partial}{\partial r}\frac{a\cos ar + b\sin ar}{a^2+b^2}e^{-br} \\ &= -\sin ar\, e^{-br}.\end{aligned}\right.$$

10.14.2 Breit–Wigner Densities of Energy-Momenta

For a Lorentz invariant antilinear implementation of the reflection, $-\mathbf{1}_4 \longmapsto *$, the Poincaré group representation coefficients are decomposed into two conjugated contributions with the advanced and retarded poles in the upper and lower complex q^2-plane, respectively,

$$m^2 \geq 0: \ \delta(q^2 - m^2) = \frac{1}{2i\pi}\left[\frac{1}{q^2 - m^2 - io} - \frac{1}{q^2 - m^2 + io}\right].$$

For unstable particles, there is no rest sytem and no sharp mass. The Breit–Wigner densities split the poles with the positive translation invariant m^2 into an advanced-retarded (complex conjugate) pair $m^2 + ib^2 \overset{-\mathbf{1}_4}{\leftrightarrow} m^2 - ib^2$ with a relativistic width b^2:

$$\begin{aligned} b^2 \geq 0: \ \delta_{b^2}(q^2 - m^2) &= \frac{1}{2i\pi}\left[\frac{1}{q^2-m^2-ib^2} - \frac{1}{q^2-m^2+ib^2}\right] \\ &= \frac{1}{\pi}\frac{b^2}{(q^2-m^2)^2+b^4}, \\ \delta_0(q^2 - m^2) &= \delta(q^2 - m^2).\end{aligned}$$

The Fourier transform is a representation coefficient of the Poincaré group with reflection $\mathbf{SO}(1,3) \vec{\times} \mathbb{R}^4$:

$$\mathbb{R}^{(1,3)} \ni x \longmapsto |m^2, ib^2\}(x) = \int \frac{d^4q}{V^3 m^2\pi}\frac{b^2}{(q^2 - m^2)^2 + b^4}e^{iqx}.$$

The Breit–Wigner energy-momentum densities characterize the distribution of the Hilbert product:

distributive basis: $\{|m^2, ib^2; q\rangle \bigm| q \in \mathbb{R}^4\}$,
distributive product: $\langle m^2, ib^2; q'|m^2, ib^2; q\rangle = \frac{V^3 m^2}{\pi}\frac{b^2}{(q^2-m^2)^2+b^4}\delta(q - q')$,
translation action: $x \bullet |m^2, ib^2; q\rangle = e^{iqx}|m^2, ib^2; q\rangle$,
spacetime reflection: $\alpha|m^2, ib^2; q\rangle \overset{-\mathbf{1}_4}{\leftrightarrow} \langle m^2, ib^2; q|\overline{\alpha}, \ \ \alpha \in \mathbb{C}$.

The proper Hilbert vectors use energy-momentum functions, square integrable with the Breit–Wigner densities,

$$\begin{aligned} |m^2, ib^2; f\rangle &= \int \frac{d^4q}{V^3m^2} f(q)|m^2, ib^2; q\rangle, \\ \langle m^2, ib^2; f|m^2, ib^2; f'\rangle &= \int \frac{d^4q}{V^3m^2\pi} \overline{f(q)} \frac{b^2}{(q^2-m^2)^2+b^4} f'(q). \end{aligned}$$

The sum over the distributive basis gives a cyclic vector,

$$|m^2, ib^2; 1\rangle = \int \frac{d^4q}{V^3m^2}|m^2, ib^2; q\rangle, \quad \langle m^2, ib^2; 1|x \bullet |m^2, ib^2; 1\rangle = |m^2, ib^2\}(x).$$

10.14.3 Time-Position Projections for Unstable Particles

In Breit–Wigner energy-momentum densities for spacetime, the Breit–Wigner energy and momentum densities depend on the squared energies and momenta, with projections on energy- and momentum-functions:

$$(q_0^2, \vec{q}^2) \longmapsto \frac{1}{\pi} \frac{b^2}{(q_0^2 - \vec{q}^2 - m^2)^2 + b^4} = \begin{cases} \frac{1}{\pi} \frac{b^2}{(q_0^2-\mathcal{E}^2)^2+b^4}, & \mathcal{E}^2 = \vec{q}^2 + m^2, \\ \frac{1}{\pi} \frac{b^2}{(\vec{q}^2-\mathcal{P}^2)^2+b^4}, & \mathcal{P}^2 = q_0^2 - m^2. \end{cases}$$

The energy-projections for fixed $\vec{q}^2$ give squared-energy densities with maximum at $q_0^2 = \mathcal{E}^2$. The momentum-projections give, for energies $q_0^2 > m^2$, squared-momentum densities with maximum at $\mathcal{P}^2$. For energies $q_0^2 < m^2$, the maximum "becomes imaginary."

In the harmonic decomposition according to the nonrelativistic time and position decomposition,

$$\int \frac{d^4q}{\pi} \frac{b^2}{(q^2-m^2)^2+b^4} e^{iqx} = -\frac{2\pi}{r}\frac{\partial}{\partial r} \int \frac{dq_0 dq}{\pi} \, e^{i(q_0x_0-qr)} \frac{b^2}{(q_0^2-q^2-m^2)^2+b^4},$$

the projective integrations involve complex energies and momenta

$$\begin{aligned} \vec{q}^2 + m^2 + ib^2 &= (E + i\tfrac{\Gamma}{2})^2, \text{ with } E \geq \tfrac{\Gamma}{2} > 0, \\ q_0^2 - m^2 + ib^2 &= (P + iB)^2, \text{ with } P, b > 0. \end{aligned}$$

The momentum (energy) integration yields the position (time) projection: The harmonic decomposition of the representation coefficients with momentum-dependent functions $\vec{q} \longmapsto (E, \frac{\Gamma}{2})(\vec{q})$, and energy-dependent ones $q_0 \longmapsto (P, B)(q_0)$, looks as follows:

$$\begin{aligned} \int \frac{d^4q}{\pi} \frac{b^2}{(q^2-m^2)^2+b^4} e^{iqx} &= 4\pi \int_0^\infty q^2 dq \, e^{-\frac{\Gamma}{2}|x_0|} \frac{E \cos Ex_0 + \frac{\Gamma}{2} \sin E|x_0|}{E^2 + \frac{\Gamma^2}{4}} \frac{\sin qr}{qr} \\ &= 2\pi \int_0^\infty dq_0 \, e^{iq_0x_0} \, e^{-Br} \, \frac{\sin Pr}{r}. \end{aligned}$$

Chapter 11
Operational Position and the Atomic Spectrum

What characterizes objects and particles "in" time and space; what determines the "quantum ladder" for (binding) energies and masses and their characteristic units? In the operational framework, quantum theory of time, position and spacetime is determined by representations of the time-, position- and spacetime-defining operation groups. Free particles and interactions implement time and position operations, abelian and nonabelian ones, respectively.

For quantum mechanics "in" — better "of" or "for" 3-position, the Hilbert space $L^2(\mathbb{R}^3)$ with the square integrable position coefficients is of paramount importance. The $\mathbb{R}^3$-position may be a parametrization of different operational positions, e.g., of the abelian translation group, or for the position operator of the nonabelian Heisenberg Lie group with Lie algebra $[\mathbf{x}^a, \mathbf{p}^b] = \delta^{ab} i\mathbf{1}$, or of a coset manifold with classes of an operation group, interpreted as position, e.g., of hyperbolic position $\mathcal{Y}^3 \cong \mathbb{R}^3$. In the Schrödinger picture of nonrelativistic quantum mechanics, the energy as time invariant is expressed by position representation eigenvalues and invariants — for the Hamiltonian $-i\frac{d}{dt} \cong \mathbf{H} = \frac{\vec{\mathbf{p}}^2}{2} + V(\vec{\mathbf{x}}) \cong -\frac{1}{2}\vec{\partial}^2 + V(\vec{x})$.

Free objects and particles represent abelian translations, i.e., flat spaces, with the invariant $\mathbf{H} = \frac{\vec{\mathbf{p}}^2}{2}$. Composite states and interactions represent nonabelian operations, which, for basic structures, come as motion groups of curved position manifolds as determined by the potential $V(\vec{\mathbf{x}})$, e.g., as rotation groups $\mathbf{SO}(1+s)$ for the spheres Ω^s or as Lorentz groups $\mathbf{SO}_0(1, s)$ for the hyperboloids $\mathcal{Y}^s$, both with $s \geq 2$. In general relativity, the potential V is related to the metrical tensor $\mathbf{g}$ and the curvature $\mathbf{R}$ of the underlying manifold, here of the position.

"Flat" and "curved" can be put in correspondence with "free" and "interacting" and, also, with time and position (space), as given in the following table to indicate a coarse parallelism of concepts:

	Manifold	Group	Diff. eq.	Dynamics	Fields
Time	one-dimensional	abelian	linear	free	free leptons (on shell)
Position	curved	nonabelian	nonlinear	interacting	confined quarks (off shell)

H. Saller, *Operational Symmetries*, DOI 10.1007/978-3-319-58664-9_11

A curvature is measured in inverse areas (squared lengths), i.e., with natural action units $\hbar$ and c, it is proportional to squared energies, momenta or masses.

Particles and interactions implement different kinds of operations: The concept of an interaction is, in a sense, complementary to that of a particle, especially with respect to the translation properties: Interactions are incompatible with (do not "know") translations; in interactions, field-theoretically characterized by nonlinear (non-quadratic) invariant vertices, homogeneous symmetry properties and their representations are rearranged. Apparently, it is possible for a scattering experiment to analyze the initial and final states in terms of particles in probability interpretable Hilbert space representations. However, it is impossible to follow the identity of particles through an interaction — where and when, in a pair annihilation, do the electron and the positron end, where and when starts the photon? To talk about a particle identity through an interaction is similar to talking about an electron jumping from one atomic orbit to another one — both make no sense.

The physically most important problem of nonrelativistic quantum theory is the atomic spectrum with its energies related to the eigenvalues of the Hamiltonian with the Kepler–Coulomb potential. Its nature-realized structure tells us about the position model represented by the Schrödinger wave functions. As first elaborated by Pauli and Fock, the Lenz–Runge invariance (classical perihel conservation) of the Hamiltonian for the time translations involves a noncompact-compact pair of position operation groups with the Lorentz group $\mathbf{SO}_0(1,3)$ and the rotation group $\mathbf{SO}(4)$ as motion groups for their corresponding rotation group classes, i.e., for the 3-hyperboloid $\mathcal{Y}^3$ and the 3-sphere Ω^3, respectively. It will be argued ahead that the wave functions of the nonrelativistic hydrogen atom are coefficients of Hilbert representations of hyperbolic position $\mathcal{Y}^3 \cong \mathbf{SO}_0(1,3)/\mathbf{SO}(3)$.

For a Lie group, acting on time, position, spacetime, $G \times G/H \longrightarrow G/H$, given by cosets, the semidirect tangent group $G \vec{\times} \mathbb{R}^n$ for the interaction free objects contains the Lie algebra classes $\log G/\log H \cong \mathbb{R}^n$ as translations.

11.1 Operational Positions

The *maximally symmetric three-dimensional operational positions* with definite metric parametrize rotation group classes. They are given by compact spherical, and noncompact flat and hyperbolic manifolds:

$$\begin{aligned}
\Omega^3 &\cong \mathbf{SO}(4)/\mathbf{SO}(3) &&\cong [\mathbf{SU}(2)\times\mathbf{SU}(2)]/\mathbf{SU}(2), && k = i^2 = -1,\\
\mathbb{E}^3 &\cong \mathbf{SO}(3)\vec{\times}\mathbb{R}^3/\mathbf{SO}(3) &&\cong \mathbf{SU}(2)\vec{\times}\mathbb{R}^3/\mathbf{SU}(2), && k = 0^2 = 0,\\
\mathcal{Y}^3 &\cong \mathbf{SO}_0(1,3)/\mathbf{SO}(3) &&\cong \mathbf{SL}(2,\mathbb{C})/\mathbf{SU}(2), && k = 1^2 = 1,
\end{aligned}$$

$$\text{with metric } ds_k^2 = \frac{d\vec{x}^2}{(1-k\frac{\vec{x}^2}{4})^2} = \begin{pmatrix} d\theta^2 \\ dr^2 \\ d\psi^2 \end{pmatrix} + \begin{pmatrix} \sin^2\theta \\ r^2 \\ \sinh^2\psi \end{pmatrix} d\omega_2^2,$$

$$\Omega^2 \cong \mathbf{SO}(3)/\mathbf{SO}(2) \text{ with metric } d\omega_2^2 = d\Theta^2 + \sin^2\Theta\, d\varphi^2.$$

The positions $\mathbb{M} \in \{\Omega^3, \mathbb{E}^3, \mathcal{Y}^3\}$ can be taken as bases of coset bundles $\mathbf{SO}(3)(\mathbb{M})$ with $\mathbf{SO}(3) \sim \mathbf{SU}(2)$ as typical fiber.

The conformally Euclidean compact sphere Ω^3 with Killing-form definiteness $k = -1$ — with invariant $P^2 > 0$ for metric $\frac{d\vec{x}^2}{(1+\frac{P^2\vec{x}^2}{4})^2}$ — may be called *Einstein position* since it was used in the first attempt of Einstein for a cosmological model. The two noncompact partners are flat *Euclidean position* $\mathbb{E}^3$ with $k = 0$ and conformally Euclidean hyperbolic *Kepler position* $\mathcal{Y}^3$ with $k = 1$ or invariant $(iB)^2 < 0$ for metric $\frac{d\vec{x}^2}{(1-\frac{B^2\vec{x}^2}{4})^2}$. Euclidean position can be considered as "flattened," i.e., as Inönü–Wigner contracted curved position, where, for hyperbolic position, the boosts are contracted to the position translations. This is mathematically isomorphic to the original Inönü–Wigner expansion of the Galilei velocity transformations into the Lorentz group boosts $\mathbf{SO}(3) \vec{\times} \mathbb{R}^3 \xleftarrow{0 \leftarrow \frac{1}{c}} \mathbf{SO}_0(1,3)$ with the speed of light as the contraction–expansion parameter.

Without time and position connecting transformations, one may define *spherical, flat and hyperbolic non-relativistic space-times* $\mathbb{R} \times \mathbb{M}$ with the Einstein, Euclid and Kepler position, respectively, as direct factor for time $\mathbb{R}$.

With the introduction of half-integer spin structures in complex representation spaces, the real rotation and Lorentz groups can be replaced by their twofold cover groups. Therefore, the maximally symmetric three-dimensional positions can be formulated as classes of the unitary group $\mathbf{SU}(2) \sim \mathbf{SO}(3)$. The isotropy group for the 3-sphere Ω^3 is the "diagonal" group $(u, u) \in \mathbf{SU}(2) \times \mathbf{SU}(2)$. The Lorentz group $\mathbf{SL}(2, \mathbb{C}) \sim \mathbf{SO}_0(1,3)$ is familiar as a special relativistic spacetime group, i.e., acting on the spacetime tangent translations $\mathbb{R}^4$. As the motion group of hyperbolic position $\mathcal{Y}^3$, it does not affect time.

Related to real forms of the simplest simple complex Lie algebra $A_1 \cong \mathbb{C}^3$ and its complex three-dimensional Lie group $\mathbf{SL}(2, \mathbb{C}) \sim \mathbf{O}(3, \mathbb{C})$, the sphere and the hyperboloid and their motion groups constitute compact–noncompact pairs as defined by the imaginary–real transition $(i\mathbb{R})^3 \leftrightarrow \mathbb{R}^3$, which relates their global rank two Lie algebras with the rotation Lie algebra $\log \mathbf{SU}(2) \cong (i\mathbb{R})^3$:

$$\begin{aligned}
\log[\mathbf{SU}(2) \times \mathbf{SU}(2)] &\cong (i\mathbb{R})^3 \oplus (i\mathbb{R})^3 \\
&\Rightarrow \log[\mathbf{SU}(2) \times \mathbf{SU}(2)]/\log \mathbf{SU}(2) \cong (i\mathbb{R})^3, \\
\log \mathbf{SL}(2, \mathbb{C}) &\cong (i\mathbb{R})^3 \oplus \mathbb{R}^3 \\
&\Rightarrow \log \mathbf{SL}(2, \mathbb{C})/\log \mathbf{SU}(2) \cong \mathbb{R}^3.
\end{aligned}$$

The Lie algebras — schematically $\vec{\mathbf{J}} \oplus \vec{\mathbf{R}}$ — of the three real six-dimensional motion groups have two independent invariants — always a discrete one (for compact Cartan subgroup $\mathbf{SO}(2)$) and, in addition, a discrete or a continuous (for noncompact Cartan subgroup $\mathbf{SO}_0(1,1) \cong \mathbb{R}$) one:

$$\begin{aligned}
&[\vec{\mathbf{J}}, \vec{\mathbf{J}}] = -\vec{\mathbf{J}}, \quad [\vec{\mathbf{J}}, \vec{\mathbf{R}}] = -\vec{\mathbf{R}}, \\
&[\vec{\mathbf{R}}, \vec{\mathbf{R}}] = k\vec{\mathbf{J}}, \quad \text{invariants: } -\vec{\mathbf{J}}^2 + \frac{1}{k}\vec{\mathbf{R}}^2, \ \vec{\mathbf{J}}\vec{\mathbf{R}}, \\
&\text{for } k = \iota^2 = (-1, 1, 0), \ \iota = (i, 1, 0),
\end{aligned}$$

and explicitly in bases and with structure constants:

$$\log \mathbf{SO}(4), \ k = -1: \begin{cases} [\mathbf{J}^a, \mathbf{J}^b] = -\epsilon^{abc}\mathbf{J}^c, \\ [\mathbf{J}^a, \mathbf{K}^b] = -\epsilon^{abc}\mathbf{K}^c, \ [\mathbf{K}^a, \mathbf{K}^b] = -\epsilon^{abc}\mathbf{J}^c, \\ \text{invariants: } -\vec{\mathbf{J}}^2 - \vec{\mathbf{K}}^2, \ \vec{\mathbf{J}}\vec{\mathbf{K}}, \end{cases}$$

$$\text{noncompact} \leftrightarrow \text{compact} \updownarrow \ \vec{\mathbf{B}} \leftrightarrow i\vec{\mathbf{K}}$$

$$\log \mathbf{SO}_0(1,3), \ k = 1: \begin{cases} [\mathbf{J}^a, \mathbf{J}^b] = -\epsilon^{abc}\mathbf{J}^c, \\ [\mathbf{J}^a, \mathbf{B}^b] = -\epsilon^{abc}\mathbf{B}^c, \ [\mathbf{B}^a, \mathbf{B}^b] = +\epsilon^{abc}\mathbf{J}^c, \\ \text{invariants: } -\vec{\mathbf{J}}^2 + \vec{\mathbf{B}}^2, \ \vec{\mathbf{J}}\vec{\mathbf{B}}, \end{cases}$$

$$\text{contraction: curved} \xrightarrow{\iota \to 0} \text{flat} \downarrow \ \text{with } \vec{\mathbf{B}} = \frac{1}{\iota}\vec{\mathbf{p}}$$

$$\log \mathbf{SO}(3) \vec{\times} \mathbb{R}^3, \ k = 0: \begin{cases} [\mathbf{J}^a, \mathbf{J}^b] = -\epsilon^{abc}\mathbf{J}^c, \\ [\mathbf{J}^a, \mathbf{p}^b] = -\epsilon^{abc}\mathbf{p}^c, \ [\mathbf{p}^a, \mathbf{p}^b] = 0, \\ \text{invariants: } \vec{\mathbf{p}}^2, \ \vec{\mathbf{J}}\vec{\mathbf{p}}. \end{cases}$$

The groups $\mathbf{SO}(4)$ and $\mathbf{SO}_0(1,3)$ have a compact-noncompact pair of symmetric orthogonal Lie algebras. The flat structures arise by contraction from the curved ones.

The tangent Euclidean group $\mathbf{SO}(3) \vec{\times} \mathbb{R}^3$, common to all three maximally symmetric positions above, is implemented by free scattering states; nonflat hyperbolic position and its motion group $\mathbf{SL}(2,\mathbb{C}) \sim \mathbf{SO}_0(1,3)$ is implemented by the nonrelativistic atomic spectrum (more ahead).

11.2 Special Position Functions

The representation coefficients and distributions of maximally symmetric operational positions with odd dimensions $s = 1, 3, \ldots$, use Macdonald, Bessel and Neumann functions with half-integer indices. Characteristic $\mathbf{O}(3)$-invariant special functions come with a positive (spherical), a trivial (flat), or a negative (hyperbolic) invariant as poles $\vec{q}^2 = (P^2, 0, -B^2)$ for definiteness $k = \iota^2 = (-1, 0, 1)$ in the complex $|\vec{q}|$-plane with the compact-noncompact relation $i|P| \leftrightarrow |B|$:

$$\left.\begin{aligned} \int \frac{d^3q}{2\pi^2}\,\frac{1}{\vec{q}^2+io-P^2}e^{i\vec{q}\vec{x}} &= \frac{e^{-i|P|r}}{r},\\ \int \frac{d^3q}{2\pi^2}\,\frac{1}{\vec{q}^2}e^{i\vec{q}\vec{x}} &= \frac{1}{r},\\ \int \frac{d^3q}{2\pi^2}\,\frac{1}{\vec{q}^2+B^2}e^{i\vec{q}\vec{x}} &= \frac{e^{-|B|r}}{r}, \end{aligned}\right\} \quad \begin{aligned} &\int \frac{d^3q}{2\pi^2}\,\frac{1}{\vec{q}^2+io+k}e^{i\vec{q}\vec{x}} = \frac{e^{-\iota r}}{r},\\ &\iota = (i, 0, 1). \end{aligned}$$

The simple pole distribution with positive momentum invariant $\vec{q}^2 = P^2 > 0$ contains contributions for two different groups via the real-imaginary decomposition $\pm\frac{1}{i\pi}\frac{1}{a\mp io} = \delta(a) \pm \frac{1}{i\pi}\frac{1}{a_{\mathrm{P}}}$ for $a \in \mathbb{R}$ into positive Dirac δ-distribution and principal value P contribution:

$$\text{for } \mathbb{E}^3: \quad \begin{aligned} \int \frac{d^3q}{2\pi}\,\delta(\vec{q}^2 - P^2)e^{i\vec{q}\vec{x}} &= \frac{\sin|P|r}{r},\\ \int \frac{d^3q}{2\pi^2}\,\frac{1}{\vec{q}_{\mathrm{P}}^2 - P^2}e^{i\vec{q}\vec{x}} &= \frac{\cos Pr}{r}. \end{aligned}$$

The Dirac distribution gives an $(r = 0)$-regular spherical Bessel function as scalar translation representation coefficient of the Euclidean group $\mathbf{SO}(3)\ \vec{\times}\ \mathbb{R}^3$ for a free scattering state with the momentum P^2 as translation invariant in the homogeneous Laplace equation:

$$\text{for } \mathbb{E}^3: \quad (-\vec{\partial}^2 - P^2)\frac{\sin|P|r}{r} = 0.$$

The principal value contribution (spherical Neumann function) and the Kepler and Yukawa potentials (hyperbolic Macdonald functions) are representation distributions. The kernels $(-\vec{\partial}^2 + k)^{-1}$ are used for interactions. They have an $(r = 0)$-singularity (distributions) and obey inhomogeneous Laplace equations:

$$\begin{aligned} (-\vec{\partial}^2 - P^2)\frac{\cos Pr}{r} &= 4\pi\delta(\vec{x}),\\ -\vec{\partial}^2\frac{1}{r} &= 4\pi\delta(\vec{x}),\\ (-\vec{\partial}^2 + B^2)\frac{e^{-|B|r}}{r} &= 4\pi\delta(\vec{x}). \end{aligned}$$

The tangent $\mathbb{R}^3$-kernel (Laplacian kernel) as limit of spherical and hyperbolic potential is given by the Kepler potential:

$$\frac{\cos Pr}{r} \xrightarrow{P\to 0} \frac{1}{r} \xleftarrow{B\leftarrow 0} \frac{e^{-|B|r}}{r}.$$

The three-dimensional potentials are distributions of one-dimensional position representation coefficients, i.e., of the compact and noncompact abelian Cartan subgroups:

$$\mathbb{E}^1 \cong \mathbb{R} \cong \mathcal{Y}^1 \cong \mathbf{SO}_0(1,1),\ \ \Omega^1 \cong \mathbf{SO}(2),$$

$$\begin{aligned}
\mathbb{E}^1 \subset \mathbb{E}^3,\ \Omega^1 \subset \Omega^3 : \int \frac{dydz}{2\pi} \int \frac{d^3q}{2\pi^2} \frac{1}{\vec{q}^2+io-P^2} e^{i\vec{q}\vec{x}} &= \int \frac{dq}{\pi} \frac{1}{q^2+io-P^2} e^{iqx} = \frac{e^{-i|Px|}}{i|P|},\\
\mathcal{Y}^1 \subset \mathcal{Y}^3 : \int \frac{dydz}{2\pi} \int \frac{d^3q}{2\pi^2} \frac{1}{\vec{q}^2+B^2} e^{i\vec{q}\vec{x}} &= \int \frac{dq}{\pi} \frac{1}{q^2+B^2} e^{iqx} = \frac{e^{-|Bx|}}{|B|}.
\end{aligned}$$

In contrast to one-dimensional abelian positions with the exponential representation coefficients:

$$\begin{aligned}
\text{for}\ \mathbb{E}^1 \text{ and } \Omega^1 : \int \tfrac{dq}{\pi} \tfrac{i|P|}{q^2+io-P^2} e^{iqx} = e^{-i|Px|},\ \ & (-\partial^2 - 1)e^{-i|x|} = 2i\delta(x),\\
& (-\partial^2 - 1)\cos x = 0,\\
& (-\partial^2 - 1)\sin|x| = -2\delta(x),\\
\text{for}\ \mathcal{Y}^1 : \int \tfrac{dq}{\pi} \tfrac{|B|}{q^2+B^2} e^{iqx} = e^{-|Bx|},\ \ & (-\partial^2 + 1)e^{-|x|} = 2\delta(x),
\end{aligned}$$

the pole order differentiates, for three-dimensional positions with nonabelian rotations, between *subgroup representation distributions for interactions* (potentials, simple momentum poles, exponential position densities), e.g. $\frac{e^{-|B|r}}{r}$, and singularity free *representations functions for negative energy states* (momentum dipoles), e.g. $e^{-|B|r}$:

$$\begin{aligned}
\text{for}\ \Omega^3 : \int \tfrac{d^3q}{\pi^2} \tfrac{i|P|}{(\vec{q}^2+io-P^2)^2} e^{i\vec{q}\vec{x}} &= e^{-i|P|r} = \tfrac{|P|}{2i\pi} \tfrac{e^{-i|P|r}}{r} * \tfrac{e^{-i|P|r}}{r},\\
\text{for}\ \mathcal{Y}^3 : \int \tfrac{d^3q}{\pi^2} \tfrac{|B|}{(\vec{q}^2+B^2)^2} e^{i\vec{q}\vec{x}} &= e^{-|B|r} = \tfrac{|B|}{2\pi} \tfrac{e^{-|B|r}}{r} * \tfrac{e^{-|B|r}}{r}.
\end{aligned}$$

The residues are normalized with π^2. Simple hyperbolic poles at $\vec{q}^2 = -1$ in three position dimensions give the Yukawa potential; dipoles, etc., give $r = 0$-regular representation coefficients.

The derivations of spherical and hyperbolic representation coefficients lead to a spherical and hyperbolic (Yukawa) potential, respectively:

$$\begin{aligned}
&\text{for}\ \mathbb{E}^3 : (-\vec{\partial}^2 - 1)\tfrac{\sin r}{r} = 0,\\
&\text{for}\ \Omega^3 : (-\vec{\partial}^2 - 1)e^{-ir} = \tfrac{2i}{r}e^{-ir},\ (-\vec{\partial}^2 - 1)^2\, e^{-ir} = 8\pi i\delta(\vec{x}),\\
&\text{for}\ \mathcal{Y}^3 : (-\vec{\partial}^2 + 1)e^{-r} = \tfrac{2}{r}e^{-r},\ \ (-\vec{\partial}^2 + 1)^2\, e^{-r} = 8\pi\delta(\vec{x}).
\end{aligned}$$

The invariant "non-flat" derivations for the eigenvalues (invariants) Q^2 involve the nontrivial potential,

$$(-\vec{\partial}^2 + \frac{\partial}{\partial\frac{r^2}{4}})e^{-iQr} = (-\vec{\partial}^2 - \frac{2iQ}{r})e^{-iQr} = Q^2 e^{-iQr},\ iQ = (i|P|, |B|).$$

The identity for the functions with hyperbolic invariant $\vec{q}^2 = -B^2$ is the Schrödinger equation for the rotation invariant ground-state coefficient of the nonrelativistic hydrogen atom with the Coulomb potential in the Hamiltonian $\mathbf{H} \cong -\frac{1}{2}\vec{\partial}^2 - \frac{1}{r}$ (more ahead):

$$\begin{aligned}
\text{for } \mathbb{E}^3: \quad & -\vec{\partial}^2 \frac{\sin|P|r}{r} = P^2 \frac{\sin|P|r}{r},\\
\text{for } \Omega^3: \quad & (-\vec{\partial}^2 - \tfrac{2i|P|}{r})e^{-i|P|r} = P^2 e^{-i|P|r},\\
\text{for } \mathcal{Y}^3: \quad & (-\vec{\partial}^2 - \tfrac{2|B|}{r})e^{-|B|r} = -B^2 e^{-|B|r}.
\end{aligned}$$

11.3 Schrödinger's Position Representations

A dynamics for nonrelativistic quantum mechanics comes with three intrinsic units — in addition to the universal action unit $\hbar$ and the "individual" mass m, there is a "normalization unit" of the specific potential. With such intrinsic units, only dimensionless variables will be used in the following.

For a quantum Hamiltonian $\mathbf{H} = \frac{\mathbf{p}^2}{2} + V(\mathbf{x})$, the mass point equations of motion $\frac{d\mathbf{x}}{dt} = \frac{\partial \mathbf{H}}{\partial \mathbf{p}} = \mathbf{p}$ and $\frac{d\mathbf{p}}{dt} = -\frac{\partial \mathbf{H}}{\partial \mathbf{x}} = -\frac{dV}{d\mathbf{x}}$ for phase space time orbits $t \longmapsto (\mathbf{x}(t), \mathbf{p}(t))$ are inappropriate. In nonrelativistic wave mechanics, one looks for time translation eigenvectors, realized by position-dependent Schrödinger functions:

$$\begin{aligned}
\boldsymbol{\psi}_E(t) = e^{iEt}|E\rangle: \quad & \frac{d}{dt}\boldsymbol{\psi}_E(t) = i\mathbf{H}\boldsymbol{\psi}_E(t) = iE\boldsymbol{\psi}_E(t),\\
|E\rangle \cong \psi_E \in L^2(\mathbb{R}^3): \quad & \mathbf{H}|E\rangle \cong \left(-\frac{1}{2}\vec{\partial}^2 + V(\vec{x})\right)\psi_E(\vec{x}) = E\psi_E(\vec{x}).
\end{aligned}$$

The Schrödinger functions are coefficients of position representations. They lead to dilation factors in the position measures $|\psi(\vec{x})|^2 d^3x$ for probabilities of operators from position motion groups like translations, rotations for spheres, or Lorentz transformations for hyperboloids. In Schrödinger's wave mechanics, the energies as time translation invariants are expressed by position representations.

For a rotation-invariant dynamics with time translation generator (Hamiltonian) $\mathbf{H} = \frac{\vec{\mathbf{p}}^2}{2} + V(\mathbf{r})$, $[\mathbf{H}, \vec{\mathbf{J}}] = 0$, the position translations are decomposed $\mathbb{R}^3 = \mathbb{R}_+ \times \Omega^2$ into radial translations and rotations. In a derivative representation the radial and angular momentum squares act on differentiable functions as follows:

$$\vec{\mathbf{p}}^2 = \mathbf{p}_r^2 + \frac{\vec{\mathbf{J}}^2}{r^2} \cong -\vec{\partial}^2, \ [\vec{\mathbf{p}}^2, \vec{\mathbf{J}}] = 0, \quad \begin{cases} i\mathbf{p}_r \cong \frac{1}{r}d_r r = d_r + \frac{1}{r},\\ \mathbf{p}_r^2 \cong -d_r^2 - \frac{2}{r}d_r,\\ \vec{\mathbf{J}}^2 \cong -\frac{1}{\sin^2\theta}[(\sin\theta\frac{\partial}{\partial\theta})^2 + (\frac{\partial}{\partial\varphi})^2]. \end{cases}$$

In a Schrödinger equation the position wave functions can be decomposed with respect to its radial and rotation representation part.

In contrast to the spherical harmonics, appropriate for asymptotically free scattering waves, with the $r \to 0$ ambiguity in $\frac{\vec{x}}{r}$, the *harmonic* $\mathbf{SO}(3)$*-polynomials* as product with the corresponding radial power are defined also for $\vec{x} \to 0$. They are eigenfunctions for a trivial translation invariant and, therefore, appropriate for negative energy (bound) waves:

$$(\vec{x})^L_m = r^L \mathrm{Y}^L_m(\varphi,\theta): \left.\begin{array}{r}\vec{\mathbf{J}}^2\mathrm{Y}^L_m(\varphi,\theta) = L(1+L)\mathrm{Y}^L_m(\varphi,\theta),\\ \mathbf{p}_r^2 r^L = -L(1+L)r^{L-2}\end{array}\right\} \Rightarrow \vec{\partial}^2(\vec{x})^L_m = 0.$$

The $\vec{x}$-homogeneous harmonic polynomials span the irreducible $\mathbf{SO}(3)$-representation spaces $\mathbb{C}^{1+2L}$. Position polynomials can be decomposed into harmonic polynomials and r-parametrized invariant coefficients.

11.4 Euclidean Position and Free Scattering

The finite-dimensional faithful nondecomposable time representations of the free mass point, acting on the complexified phase space $\mathbb{C}^2$, with $\mathbf{H} = \frac{\mathbf{p}^2}{2} \longmapsto \begin{pmatrix} 0 & 1 \\ 0 & 0 \end{pmatrix}$ and $t \longmapsto e^{t\mathbf{H}} \longmapsto \begin{pmatrix} 1 & t \\ 0 & 1 \end{pmatrix} \in \mathbf{U}(1,1)$, are not definite unitary. They have no probability interpretation.

Nonrelativistic scattering theory is formulated in terms of infinite-dimensional faithful unitary representations of the Euclidean group $\mathbf{SO}(3) \vec{\times} \mathbb{R}^3$ and of its rotation group classes, the Euclidean position $\mathbb{E}^3$. The solutions of the Schrödinger equation,

$$\mathbf{H} = \frac{\vec{\mathbf{p}}^2}{2},\quad 0 \le E = \frac{P^2}{2} \in \mathbf{spec}\,_+\mathbf{H}: \quad (\vec{\partial}^2 + P^2)\psi(\vec{x}) = 0,$$

for trivial translation invariant $P = 0$ are fixgroup $\mathbf{SO}(3)$-representations with harmonic polynomials, $\vec{\partial}^2(\vec{x})^L = 0$. For nontrivial momentum $P > 0$, the solutions are coefficients of the irreducible Hilbert space representations of the Euclidean group, induced by fixgroup $\mathbf{SO}(2)$-representations:

$$P > 0,\ \ h = 0, 1, 2, \ldots: \ \ \{|\vec{q}, \pm h\rangle \,\big|\, \vec{q} \in \mathbb{R}^3,\ \vec{q}^2 = P^2\},$$

with dual $\mathbf{SO}(2)$-eigenvalues $\pm h \in \mathbb{Z}$ (*helicity*) for axial rotations around the momentum direction $\frac{\vec{q}}{P} = \vec{\omega} \in \Omega^2$.

The rotation representations for $\Omega^2 \cong \mathbf{SO}(3)/\mathbf{SO}(2)$ are harmonically analyzable by the spherical harmonics $\{\vec{\omega} \longmapsto \mathrm{Y}^L_m(\vec{\omega}) \,\big|\, L = 0, 1, \ldots; |m| \le L\}$ as orthonormal Hilbert basis of the 2-sphere functions $L^2(\Omega^2)$, decomposable into Hilbert spaces with irreducible action of the rotations:

$$L^2(\Omega^2) = \bigoplus_{L=0}^{\infty} \mathbb{C}^{1+2L}(\Omega^2).$$

With the wave function decomposition:

$$\psi(\vec{x}) = \sum_{L=0}^{\infty} \sum_{m=-L}^{L} \mathrm{Y}^L_m(\varphi, \theta)\psi^m_L(r),$$

a rotation-invariant Hamiltonian is decomposable into the generators $\mathbf{H}_L$ for each angular momentum:

$$\mathbf{H} = \frac{\vec{\mathbf{p}}^2}{2} + V(r) = \bigoplus_{L=0}^{\infty} \mathbf{H}_L,$$

$$\text{with} \begin{cases} \mathbf{H}_L = \frac{\mathbf{p}_r^2}{2} + \frac{L(1+L)}{2r^2} + V(r) \\ \quad = r^n[-\frac{1}{2}d_r^2 - \frac{1+n}{r}d_r + \frac{L(1+L)-n(1+n)}{2r^2} + V(r)]\frac{1}{r^n}, \\ n = 0, \pm 1, \pm 2, \ldots. \end{cases}$$

Attention has to be paid to the small distance $r \to 0$ behavior, prepared with the powers r^n. Compact radial representations are spread with a factor $\frac{1}{r}$, then $\frac{1}{r^2}$ for the probability, to the 2-sphere:

$$\psi(\vec{x}) = \sum_{L=0}^{\infty} \sum_{m=-L}^{L} (\frac{\vec{x}}{r})^L_m \frac{D_L(r)}{r} \Rightarrow [d_r^2 - \frac{L(1+L)}{r^2} + 2(E - V(r))]D_L(r) = 0,$$

$$D_L(r) = r\psi_L(r).$$

An irreducible compact radial position representation requires a constant potential:

$$D_0(r) = e^{\pm iPr} \Rightarrow \mathbf{H} = \frac{\vec{\mathbf{p}}^2}{2} + V_0, \quad E - V_0 = \frac{P^2}{2}.$$

For large radial translations $r \to \infty$ the *centrifugal potential* $\frac{L(1+L)}{r^2}$, and hence the angular momentum dependence, vanishes. For a potential leveling off stronger than $\frac{1}{r}$ the product $D_L(r) = r\psi_L(r)$ of the translation parameter r with the wave function is a coefficient for the representation of the radial translations. The scattering eigenfunctions for trivial angular momentum and, for large distances $r \to \infty$, for any L are:

$$\psi_0(r) = \frac{D_0(r)}{r} = \alpha\frac{\sin Pr}{r} + \beta\frac{\cos Pr}{r}.$$

The radial coefficients of plane waves $e^{iPz} = \sum_{L=0}^{\infty}(1 + 2L)\mathrm{P}^L(\cos\theta)i^L j_L(Pr)$ define the *spherical Bessel functions* $j_L \in L^\infty(\mathbb{R}_+)$ in the expansion of the irreducible translation $\mathbb{R}^3$-representation coefficient with respect to representations of the Euclidean group $\mathbf{SO}(3) \vec{\times} \mathbb{R}^3$. The essentially bounded representation coefficients of the Euclidean group in $L^\infty(\mathbb{R}^3)$ are products $\mathrm{Y}^L_m(\varphi, \theta)j_L(Pr)$ of matching spherical Bessel functions and spherical harmonics:

$$\begin{aligned}
P > 0: \ \mathbb{E}^3 \ni \vec{x} \longmapsto & \int \frac{d^3q}{2\pi P} \begin{pmatrix} 1 \\ i\frac{\vec{q}}{|\vec{q}|} \end{pmatrix} \delta(\vec{q}^2 - P^2) e^{i\vec{q}\vec{x}} \\
&= \begin{pmatrix} j_0(Pr) \\ \frac{\vec{x}}{r} j_1(Pr) \end{pmatrix} = \begin{pmatrix} \frac{\sin Pr}{Pr} \\ \frac{\vec{x}}{r} \frac{\sin Pr - Pr\cos Pr}{P^2r^2} \end{pmatrix}, \\
i^L \mathrm{Y}^L_m(\varphi, \theta) j_L(r) \sim & (i\frac{\vec{x}}{2})^L \frac{\sqrt{\pi}}{2} \frac{\mathcal{J}_{L+\frac{1}{2}}(r)}{(\frac{r}{2})^{L+\frac{1}{2}}} = (i\frac{\vec{x}}{2})^L (-\frac{\partial}{\partial \frac{r^2}{4}})^L \frac{\sin r}{r} \\
&= (-i\vec{\partial})^L \frac{\sin r}{r} = \int \frac{d^3q}{2\pi} (\vec{q})^L \delta(\vec{q}^2 - 1) e^{i\vec{q}\vec{x}}.
\end{aligned}$$

The direct integral Hilbert space:

$$L^2(\mathbb{R}_+) = \{f \bigm| \int_0^\infty \frac{2P^2dP}{\pi} |f(P)|^2 < \infty\} = {}^{\oplus}\!\!\int_0^\infty \frac{P^2dP}{\pi} \mathbb{C}|P\rangle$$

has the spherical Bessel functions for each angular momentum as distributive basis (not Hilbert space vectors) with distributive orthogonality and completeness (technically the same relations) — for the momenta P, not for the angular momenta L:

$$L = 0, 1, 2, \ldots: \ L^2(\mathbb{R}_+) \text{ has distributive basis } \{r \longmapsto j_L(Pr) \bigm| P > 0\},$$
$$\text{with} \begin{cases} \int_0^\infty r^2dr \ j_L(P'r)\ j_L(Pr) = \frac{\pi}{2P^2}\delta(P - P'), \\ \int_0^\infty \frac{2P^2dP}{\pi} \ j_L(Pr)\ j_L(Pr') = \frac{1}{r^2}\delta(r - r'). \end{cases}$$

Via momentum measures (wave packets) one obtains Hilbert space vectors in the *Fourier–Bessel transformation*:

$$L^2_{dP}(\mathbb{R}_+) \cong L^2_{dr}(\mathbb{R}_+) \text{ with } \begin{cases} \psi_L(r) = \int_0^\infty \frac{2P^2dP}{\pi} \ j_L(Pr) f_L(P), \\ \int_0^\infty r^2dr \ \overline{\psi_L(r)}\psi'_L(r) = \int_0^\infty \frac{2P^2dP}{\pi} \ \overline{f_L(P)} f'_L(P). \end{cases}$$

The generalization with respect to position dimensions $s = 2, 3, \ldots$ is obvious: The faithful representations of the Euclidean group $\mathbf{SO}(s) \vec{\times} \mathbb{R}^s$ are characterized by momentum invariant P^2 and irreducible fixgroup $\mathbf{SO}(s-1)$ representations. They start with Bessel functions as scalar coefficients:

$$P > 0: \ \mathbb{E}^s \ni \vec{x} \longmapsto \int \frac{d^sq}{\pi^{\frac{s}{2}-1} P^{s-2}} \delta(\vec{q}^2 - P^2) e^{i\vec{q}\vec{x}} = \frac{\pi \mathcal{J}_{\frac{s}{2}-1}(Pr)}{(\frac{Pr}{2})^{\frac{s}{2}-1}}$$

With the polar decomposition $\mathbb{R}^s = \mathbb{R}_+ \times \Omega^{s-1}$, the Hilbert space for fixed momentum $P > 0$ is given by the square integrable function $L^2(\Omega^{s-1})$ on the $(s-1)$-sphere. The harmonic analysis with respect to the Euclidean group works with the product Hilbert space $L^2(\mathbb{R}_+) \times L^2(\Omega^{s-1}) \cong L^2(\mathbb{R}^s)$.

11.5 Kepler Position and the Atomic Spectrum

Representations of the radial translations $\mathbb{R}_+$ in states for a rotation symmetric potential come after the separation of the harmonic $\mathbf{SO}(3)$-polynomials. This leads to the Schrödinger equations for the $\mathbb{R}_+$-representation coefficients:

$$[\vec{\partial}^2 - 2V(r) + 2E]\psi(\vec{x}) = 0,$$
$$\psi(\vec{x}) = \sum_{L=0}^{\infty} \sum_{m=-L}^{L} (\vec{x})^L_m d_L(r) \Rightarrow \left[d_r^2 + \frac{2(1+L)}{r}d_r - 2V(r) + 2E\right] d_L(r) = 0.$$

The representations can involve an $\mathbb{R}_+$-reparametrization $r \longmapsto \zeta(r)$. The $\mathbb{R}_+$-representation coefficient as a product with an exponential,[1]

$$\frac{\zeta}{2} \longmapsto d_L(r) = F_L(\zeta)\, e^{-\frac{\zeta}{2}},$$

gives the Schrödinger equation for the remaining function:

$$\left[(d_r\zeta)^2 d_\zeta^2 + \left(d_r^2\zeta - (d_r\zeta)^2 + \frac{2(1+L)}{r}d_r\zeta\right) d_\zeta \right.$$
$$\left. + \frac{1}{4}(d_r\zeta)^2 - \frac{1}{2}d_r^2\zeta - \frac{1+L}{r}d_r\zeta - 2(V-E)\right] F_L(\zeta) = 0.$$

After separating an exponential with a complex "momentum" Q as noncompact invariant, there remain the radial equations:

$$d_L(r) = F_L(\zeta)e^{-\frac{\zeta}{2}}, \quad \frac{\zeta}{2} = iQr, \quad Q \in \mathbb{C},$$
$$\Rightarrow \left[\zeta d_\zeta^2 + (2+2L-\zeta)d_\zeta - \left(1 + L + \frac{rV(r)}{iQ}\right) + \frac{\zeta}{4}\left(1 - \frac{2E}{Q^2}\right)\right] F_L(\zeta) = 0.$$

Quantum numbers of negative energy (bound) states, i.e., the position representation invariants, are determined with the condition of square integrability $L^2(\mathbb{R}^3)$ of the position wave functions.

The Hamiltonian for two masses with Kepler interaction is decomposed into center of mass free scattering Hamiltonian for Euclidean position and relative "bound state" Hamiltonian for Kepler position:

$$\mathbf{H}_{12} = \frac{\vec{\mathbf{p}}_1^2}{2m_1} + \frac{\vec{\mathbf{p}}_2^2}{2m_2} + \frac{\epsilon}{|\mathbf{x}_1 - \mathbf{x}_2|} = \mathbf{H}_{\text{scatt}} + \mathbf{H} = \frac{\vec{\mathbf{P}}^2}{2M} + \frac{\vec{\mathbf{p}}^2}{2m} + \frac{\epsilon}{\mathbf{r}}.$$

---

[1] The normalization $\frac{\zeta}{2}$ is chosen with respect to the Laguerre polynomials ahead.

The normalization of the potential $V(r) = \frac{g^2\epsilon}{\underline{r}} = \frac{\epsilon}{r}$ gives an intrinsic velocity unit $[\frac{g^2}{\hbar}] = \frac{\text{m}}{\text{s}}$ with $\frac{g^2}{\hbar} = \alpha_{\text{Som}} c z_1 z_2$, $\alpha_{\text{Som}} \sim \frac{1}{137}$, for a Coulomb potential and charge numbers $z_{1,2} \in \mathbb{Z}$. Especially for negative energy (bound) states, it does not make sense to analyze the Kepler Hamiltonian $\mathbf{H} = \mathbf{H}_0 + \mathbf{H}_{\text{int}}$, in a perturbation expansion, by free scattering states, related to the kinetic term $\frac{\vec{\mathbf{p}}^2}{2m}$ and interacting via $\frac{\epsilon}{\mathbf{r}}$. The Kepler Hamiltonian $\mathbf{H}$ as a whole has to be diagonalized with respect to time translation representations, characterized by energies as invariants which makes the interpretation of the hydrogen atom as a two particle bound state somewhat inappropriate.

The nonrelativistic Kepler dynamics with time translation generator,

$$\mathbf{H} = \frac{\vec{\mathbf{p}}^2}{2} + \frac{\epsilon}{\mathbf{r}}, \quad \epsilon = \pm 1 \text{ repulsive, attractive,}$$

has, in addition to the rotation symmetry, a Lenz–Runge "perihelion" invariance which is the reason for the "periodicity" of the "periodic" system, i.e., of the atomic spectrum. Angular momentum $\vec{\mathbf{J}} = i\vec{\mathbf{x}} \times \vec{\mathbf{p}}$ and Lenz–Runge vector $\vec{\mathbf{R}} = i[\vec{\mathbf{p}}, \frac{\vec{\mathbf{J}}^2}{2} + \epsilon\mathbf{r}]$ are conserved $[\vec{\mathbf{J}}, \mathbf{H}] = 0 = [\vec{\mathbf{R}}, \mathbf{H}]$ and orthogonal to each other $\vec{\mathbf{J}}\vec{\mathbf{R}} = 0$ (trivial invariant). With the sum and difference $\vec{\mathbf{J}}_\pm = \frac{1}{2}(\vec{\mathbf{J}} \pm \frac{\vec{\mathbf{R}}}{\sqrt{-2\mathbf{H}}})$ for negative energies the Hamiltonian can be written as $2(\vec{\mathbf{J}}_+^2 + \vec{\mathbf{J}}_-^2) = 1 + \frac{1}{2\mathbf{H}}$. The orthogonality of angular momentum and Lenz–Runge vectors involves the equality of the invariants $\vec{\mathbf{J}}_+^2 = \vec{\mathbf{J}}_-^2 = -J(J+1)$ in the energy $-4J(J+1) = 1 + \frac{1}{2E}$.

The Kepler dynamics has, for an attraction, both negative energy (imaginary momentum, bound) waves and positive energy (real momentum, scattering) solutions. The noncompact position as represented by the wave functions of the nonrelativistic hydrogen atom is not flat Euclidean position $\mathbb{E}^3$, but maximally symmetric hyperbolic position $\mathcal{Y}^3 \cong \mathbf{SO}_0(1,3)/\mathbf{SO}(3)$, parametrizing the rotation group classes in the Lorentz group. $\mathbb{E}^3$ and $\mathcal{Y}^3 = \exp \mathbb{R}^3$ are $\mathbb{R}^3$-isomorphic as manifolds, not as homogeneous spaces.

In the position representation, the Kepler Hamiltonian is the sum of the Laplacian and the Laplacian kernel:

$$\mathbf{H} = \frac{\vec{\mathbf{p}}^2}{2} + \frac{\epsilon}{\mathbf{r}} \cong -\frac{1}{2}\vec{\partial}^2 - \frac{4\pi\epsilon}{\vec{\partial}^2} = \left(\frac{i\vec{\partial}}{\sqrt{2}} - \frac{\sqrt{-4\pi\epsilon}}{i\vec{\partial}}\right)\left(\frac{i\vec{\partial}}{\sqrt{2}} + \frac{\sqrt{-4\pi\epsilon}}{i\vec{\partial}}\right),$$

$$\text{with} \vec{\mathbf{p}}^2 \cong -\vec{\partial}^2, \quad \frac{1}{r} = \int \frac{d^3q}{2\pi^2}\frac{1}{\vec{q}^2}e^{i\vec{q}\vec{x}} \cong -\frac{4\pi}{\vec{\partial}^2}.$$

For the Kepler potential and a nonrelativistic, i.e., purely "kinetic" energy (time translation invariant), given by half of the squared "momentum" (position invariant) there remains a Laplacian differential equation:

$$V(r) = \frac{\epsilon}{r} \text{ and } E = \frac{Q^2}{2} \Rightarrow [\zeta d_\zeta^2 + (2 + 2L - \zeta)d_\zeta - (1 + L + \frac{\epsilon}{iQ})]F_L(\zeta) = 0.$$

The equations for the Kepler potential have $(r = 0)$-regular solutions with confluent hypergeometric functions ${}_1\mathrm{F}_1(1 + L + \frac{\epsilon}{iQ}; 2 + 2L; \zeta)$:

$$\begin{aligned}
\zeta, \alpha, \gamma \in \mathbb{C} :& [\zeta d_\zeta^2 + (\gamma - \zeta)d_\zeta - \alpha]\, {}_1\mathrm{F}_1(\alpha; \gamma; \zeta) = 0,\\
{}_1\mathrm{F}_1(\alpha; \gamma; \zeta) &= \sum_{k=0}^{\infty} \frac{\Gamma(\alpha + k)}{\Gamma(\alpha)} \frac{\Gamma(\gamma)}{\Gamma(\gamma + k)} \frac{\zeta^k}{k!}, \quad {}_1\mathrm{F}_1(\alpha; \gamma; 0) = 1,\\
{}_1\mathrm{F}_1(\alpha; \alpha; \zeta) &= e^\zeta.
\end{aligned}$$

The confluent hypergeometric functions with negative integer $\alpha = -N$ and real argument are, up to a normalization factor, the Laguerre polynomials:

$$\begin{aligned}
&\xi \in \mathbb{R},\ N = 0, 1, \ldots, && [\xi d_\xi^2 + (1 + \lambda - \xi)d_\xi + N]\mathrm{L}_N^\lambda(\xi) = 0,\\
&\lambda \in \mathbb{R},\ \lambda \neq -1, -2, \ldots : && {}_1\mathrm{F}_1(-N; 1 + \lambda; \xi) \sim \mathrm{L}_N^\lambda(\xi).
\end{aligned}$$

They have degree N (radial quantum number, knot number) and real order λ, which is used for the angular momentum L-dependence. The Rodrigues formula for the Laguerre polynomials contains derivatives of the function $\xi \longmapsto \xi^\lambda e^{-\xi}$:

$$\begin{aligned}
\mathrm{L}_N^\lambda(\xi) &= \left(\xi^{-\lambda} e^\xi \frac{d}{d\xi} \xi^\lambda e^{-\xi}\right)^N \frac{\xi^N}{N!} = \frac{1}{N!} \xi^{-\lambda} e^\xi \frac{d^N}{d\xi^N} \xi^{\lambda+N} e^{-\xi}\\
&= \sum_{k=0}^{N} \frac{\Gamma(1 + \lambda + N)}{\Gamma(1 + N - k)\Gamma(1 + \lambda + k)} \frac{(-\xi)^k}{k!} = \begin{cases} 1, & N = 0,\\ 1 + \lambda - \xi, & N = 1,\\ \frac{(1+\lambda)(2+\lambda)}{2} - (2 + \lambda)\xi + \frac{\xi^2}{2}, & N = 2,\\ \ldots \end{cases}
\end{aligned}$$

For each $\lambda \notin -\mathbb{N}$, the representation coefficients

$$\mathbb{R}_+ \ni \frac{\xi}{2} \longmapsto \mathrm{L}_N^\lambda(\xi)\, e^{-\frac{\xi}{2}},$$

give a basis for a Hilbert space — with the measures $\int_0^\infty \xi^\lambda e^{-\xi} d\xi$ for group and $\sum_{N=0}^{\infty} \frac{N!}{\Gamma(1+\lambda+N)}$ for representations:

$$L^2(\mathbb{R}_+) \text{ has basis } \{\xi \longmapsto \xi^{\frac{\lambda}{2}} \mathrm{L}_N^\lambda(\xi) e^{-\frac{\xi}{2}} \mid N = 0, 1, \ldots\}$$

$$\text{with } \begin{cases} \int_0^\infty \xi^\lambda e^{-\xi} d\xi \, \mathrm{L}_N^\lambda(\xi) \, \mathrm{L}_{N'}^\lambda(\xi) = \frac{\Gamma(1+\lambda+N)}{N!} \delta_{NN'}, \\ \sum_{N=0}^\infty \frac{N!}{\Gamma(1+\lambda+N)} \mathrm{L}_N^\lambda(\xi) \, \mathrm{L}_N^\lambda(\xi') = \xi^{-\lambda} e^\xi \delta(\xi - \xi'). \end{cases}$$

Negative energy (bound) state and positive energy (scattering) solutions are related to each other by the noncompact-compact relation $|B| \leftrightarrow i|P|$ of the hyperbolic $\mathbf{SO}_0(1,1)$-invariant:

$$d_L(r) = {}_1\mathrm{F}_1\left(1 + L + \frac{\epsilon}{iQ}; 2 + 2L; 2iQr\right) e^{-iQr} \text{ for } iQ = |B| \leftrightarrow iP.$$

For the attractive interaction $V(r) = -\frac{1}{r}$ and *states with negative energy* (*binding energy*) and square-integrable hyperbolic wave functions, the radial *imaginary "momentum"* and, therefore, the energy is "quantized" — the related noncompact invariant is determined by the compact ones with the degree of the Laguerre polynomials:

$$\begin{array}{r} \epsilon = -1, \\ Q^2 = -B^2 < 0, \\ E = -\frac{B^2}{2}, \end{array} \Rightarrow \begin{cases} d_L(r) \sim {}_1\mathrm{F}_1(1 + L - \frac{1}{|B|}; 2 + 2L; 2|B|r) e^{-|B|r} \\ \quad \sim \mathrm{L}_N^{1+2L}(2|B|r) e^{-|B|r}, \\ \mathbb{N} \ni \frac{1}{B} = n = 1 + 2J = 1 + L + N, \quad L, N = 0, 1, \ldots, \\ \psi_{Lm}^{2J}(\vec{x}) \sim (\frac{\vec{x}}{n})_m^L \, \mathrm{L}_N^{1+2L}(\frac{2r}{n}) \, e^{-\frac{r}{n}}, \\ \psi_E(t) = e^{iEt}|E\rangle, \ |E\rangle = |J, J; L, m\rangle \cong \psi_{Lm}^{2J} \in L^2(\mathbb{R}^3). \end{cases}$$

The product of spatially decreasing exponentials, characterized by the noncompact "momentum" invariants B, the harmonic $\mathbf{SO}(3)$-polynomials of degree L and the Laguerre polynomials with degree $N = 2J - L$ (knot number) give square integrable wave functions. For fixed principal quantum number $n = 1 + 2J$, i.e., for energy $E = -\frac{1}{2n^2}$, the products of Laguerre and harmonic $\mathbf{SO}(3)$-polynomials give n^2 harmonic $\mathbf{SO}(4)$-polynomials (more ahead).

In contrast to the negative energy (bound)-state exponential $e^{-|B|r}$, the *hyperbolic scattering solutions* in the Kepler potential $V(r) = \frac{\epsilon}{r}$ — not leveling off stronger than $\frac{1}{r}$ — with positive energy and real momentum from a continuous spectrum, are obtained after the separation of a unitary exponential e^{iPr}. The radial equations have the real $(r = 0)$-regular solutions with confluent hypergeometric functions:

$$\begin{array}{r} \epsilon = \pm 1, \\ Q^2 = P^2 > 0, \\ E = \frac{P^2}{2}, \end{array} \Rightarrow \begin{cases} d_L(r) \sim {}_1\mathrm{F}_1(1 + L + \frac{\epsilon}{iP}; 2 + 2L; +2iPr) e^{-iPr} \\ \quad = {}_1\mathrm{F}_1(1 + L - \frac{\epsilon}{iP}; 2 + 2L; -2iPr) e^{+iPr}, \\ P \in \mathbb{R}, \ L = 0, 1, \ldots, \\ \psi_{Lm}^P(\vec{x}) \sim (P\vec{x})_m^L \, {}_1\mathrm{F}_1(1 + L + \frac{\epsilon}{iP}; 2 + 2L; 2iPr) e^{-iPr}. \end{cases}$$

The large distance behavior of the hyperbolic scattering solutions does not lead to spherical Bessel functions for free particles, representing flat Euclidean position $\mathbb{E}^3$. For the curved hyperbolic position $\mathcal{Y}^3$-representations, the asymptotic Bessel functions $j_L(r) \to \frac{\sin(r-\frac{L\pi}{2})}{r}$ are modified by a radial-dependent logarithm and a constant phase α_L:

$$r \to \infty: \ (P\vec{x})^L_m \ {}_1\mathrm{F}_1(1+L+\frac{\epsilon}{iP}; 2+2L; 2iPr)e^{-iPr}$$
$$\to \mathrm{Y}^L_m(\varphi, \theta)\frac{\sin(Pr - \frac{L\pi}{2} - \frac{\epsilon}{P}\log 2Pr + \alpha_L)}{Pr}.$$

The confluent hypergeometric functions contain the factor $\frac{1}{\Gamma(1+L+\frac{\epsilon}{iQ})}$ which, for the negative energy (bound) solutions with negative integer $1+L-\frac{1}{|B|} = -N$, leads to the Laguerre polynomials. The factor for the scattering solutions with

$$L = 0, 1, 2, \ldots: \ \left|\frac{\Gamma(1+L)}{\Gamma(1+L+\frac{\epsilon}{iP})}\right|^2 = \prod_{k=L+1}^{\infty}\left(1+\frac{1}{P^2k^2}\right) = \frac{\sinh\frac{\pi}{P}}{\frac{\pi}{P}}\prod_{k=1}^{L}\frac{1}{1+\frac{1}{P^2k^2}},$$
$$\text{with } \frac{\sinh\frac{\pi}{P}}{\frac{\pi}{P}} = \prod_{k=1}^{\infty}\left(1+\frac{1}{P^2k^2}\right),$$

is related to the Plancherel measures ("hyperbolic R-tails" $\prod_{k=R}^{\infty}\cdots$) for the harmonic analysis of the odd-dimensional position hyperboloids $\mathcal{Y}^{2R-1}$.

11.6 Dipoles for Hyperbolic 3-Position

In contrast to the representations of flat Euclidean space (scattering waves, free particles, simple momentum poles), Hilbert representations of three-dimensional hyperbolic position $\mathcal{Y}^3 = e^{\mathbb{R}^3} \ni e^{\vec{\sigma}\vec{x}} \leftrightarrow \vec{x} \in \mathbb{R}^3$ (isomorphic as manifolds) with nonabelian degrees of freedom use momentum dipoles. This will be shown in the nonrelativistic hydrogen atom negative energy (bound) waves with Hilbert representations of the noncompact nonabelian group $\mathbf{SO}_0(1,3)$. The Hilbert space $L^2(\Omega^2)$ on the momentum orientation sphere $\frac{\vec{q}}{|\vec{q}|} \in \Omega^2$ for nonrelativistic scattering differs from the Hilbert space $L^2(\mathcal{Y}^3)$ on hyperbolic position $\vec{x} \in \mathcal{Y}^3$.

The measure of the unit 3-sphere as the orientation manifold of the rotation group $\mathbf{SO}(3)$ in the invariance group $\mathbf{SO}(4)$ for negative energy (bound) waves has a momentum parametrization by an imaginary dipole $\vec{q}^2 = -1$:

$$\frac{1}{\sqrt{\vec{q}^2+1}}\binom{1}{\vec{q}} \in \Omega^3 \subset \mathbb{R}^4 \Rightarrow |\Omega^3| = \int d^3\omega = \int d^3q \, \frac{2}{(\vec{q}^2+1)^2} = 2\pi^2.$$

A curvature unit with inverse length $|B|$ is implemented by $\int d^3q \, \frac{2}{(\vec{q}^2+B^2)^2} = \frac{2\pi^2}{|B|}$.

The Fourier-transformed Ω^3-measure gives the hydrogen *ground-state coefficient with hyperbolic invariant* as a scalar representation coefficient of 3-position space:

$$\mathbf{SO}_0(1,3)/\mathbf{SO}(3) \cong \mathcal{Y}^3 \cong \mathbb{R}^3 \ni \vec{x} \longmapsto \int \frac{d^3q}{\pi^2} \frac{|B|}{(\vec{q}^2+B^2)^2} e^{i\vec{q}\vec{x}} = e^{-|B|r}.$$

It involves the half-integer index Macdonald function for $N = 0$:

$$\int \frac{d^3q}{\pi^2} \frac{\Gamma(2+N)}{(\vec{q}^2+1)^{2+N}} e^{i\vec{q}\vec{x}} = \frac{1}{\sqrt{\pi}} \left(\frac{r}{2}\right)^{\frac{1}{2}+N} 2\mathcal{K}_{\frac{1}{2}+N}(r) = \left(\frac{r}{2}\right)^{1+2N} \left(-\frac{d}{d\frac{r^2}{4}}\right)^{1+N} e^{-r},$$
$$\text{for } N = -1, 0, 1, 2, \ldots.$$

In the negative energy (bound) waves, position as a noncompact hyperboloid is represented in the form of Fourier-transformed Ω^3-measures with a continuous invariant B^2 for the imaginary "momenta" $\vec{q}^2 = -B^2$ on a 2-sphere Ω^2.

The Kepler negative energy (bound) waves are coefficients of infinite-dimensional cyclic representations of the Lorentz group $\mathbf{SO}_0(1,3) = \mathbf{SO}(3) \circ \mathbf{SO}_0(1,1) \circ \exp \mathbb{R}^2$ (Iwasawa factorization). With the Cartan subgroups $\mathbf{SO}(2) \times \mathbf{SO}_0(1,1)$, the irreducible representations are characterized by one integer and one continuous invariant. In the language of induced representations, the negative energy (bound) waves of the hydrogen atom are rotation $\mathbf{SO}(3)$-intertwiners on the group $\mathbf{SO}_0(1,3)$ ($\mathcal{Y}^3$-coefficient) with values in Hilbert spaces with $\mathbf{SO}(3)$-representations in $(1+2J)^2$-dimensional $\mathbf{SO}(4)$-representations (J, J).

The rotation dependence $\vec{x}$ is effected by momentum derivation $i\frac{\partial}{\partial \vec{q}}$ of the Ω^3-measure:

$$i\vec{x}e^{-r} = \int \frac{d^3q}{\pi^2} \frac{4\vec{q}}{(\vec{q}^2+1)^3} e^{i\vec{q}\vec{x}} \text{ with } \frac{4\vec{q}}{(\vec{q}^2+1)^3} = -\frac{\partial}{\partial \vec{q}} \frac{1}{(\vec{q}^2+1)^2}.$$

The 3-vector factor $\frac{2\vec{q}}{\vec{q}^2+1} = \frac{\vec{q}}{|\vec{q}|} \sin\chi$ is uniquely supplemented to a parametrization of the unit 3-sphere by a normalized 4-vector:

$$\begin{pmatrix} \cos\chi \\ \frac{\vec{q}}{|\vec{q}|} \sin\chi \end{pmatrix} = \frac{1}{\vec{q}^2+1} \begin{pmatrix} 1-\vec{q}^2 \\ 2\vec{q} \end{pmatrix} = \begin{pmatrix} p_0 \\ \vec{p} \end{pmatrix} = p \in \Omega^3 \subset \mathbb{R}^4, \;\; p_0^2 + \vec{p}^2 = 1.$$

The unit 4-vector $\mathrm{Y}^{(\frac{1}{2},\frac{1}{2})}(p) \sim \begin{pmatrix} p_0 \\ \vec{p} \end{pmatrix} \in \Omega^3$ is the analogue to the unit 3-vector $\mathrm{Y}^1(\frac{\vec{q}}{|\vec{q}|}) \sim \frac{\vec{q}}{|\vec{q}|} \in \Omega^2$ used for the buildup of the 2-sphere harmonics $\mathrm{Y}^L(\frac{\vec{q}}{|\vec{q}|}) \sim (\frac{\vec{q}}{|\vec{q}|})^L$. Analoguously, the Ω^3-harmonics are the totally symmetric traceless products $\mathrm{Y}^{(J,J)}(p) \sim (p)^{2J}$, e.g., for $J = 1$, the nine independent components in the (4×4)-matrix and the decomposition $9 \overset{\mathbf{SO}(3)}{=} 1 \oplus 3 \oplus 5$ with the respective components

$p_0^2 - \frac{1}{4}$, $p_0\vec{p}$, and $p_a p_b - \frac{\delta_{ab}}{3}\vec{p}^2$:

$$\mathrm{Y}^{(1,1)}(p) \sim (p)^2_{jk} = p_j p_k - \frac{\delta_{jk}}{4} \cong \left(\begin{array}{c|c} p_0^2 - \frac{1}{4} & p_0 p_a \\ \hline p_0 p_b & p_a p_b - \frac{\delta_{ab}}{4} \end{array}\right),$$
$$\text{with } p_a p_b - \frac{\delta_{ab}}{4} = p_a p_b - \frac{\delta_{ab}}{3}\vec{p}^2 - \frac{\delta_{ab}}{3}\left(p_0^2 - \frac{1}{4}\right) \text{ for } p^2 = 1.$$

The Kepler negative energy (bound) waves in $(1+2J)^2$-multiplets for **SO**(4) come with momentum poles of order $2+2J$:

$$\frac{1}{B} = 1+2J : \mathcal{Y}^3 \ni \vec{x} \longmapsto \int \frac{d^3q}{\pi^2}\frac{1}{(\vec{q}^2+1)^2}(p)^{2J} e^{i\vec{q}B\vec{x}} = \int \frac{d^3q}{\pi^2}\frac{1}{(\vec{q}^2+1)^{2+2J}}\begin{pmatrix} 1-\vec{q}^2 \\ 2\vec{q}\end{pmatrix}^{2J} e^{i\vec{q}B\vec{x}},$$
$$\text{with} p = \frac{1}{\vec{q}^2+1}\begin{pmatrix} 1-\vec{q}^2 \\ 2\vec{q}\end{pmatrix}, \text{ and } E = -\frac{B^2}{2}.$$

The hyperboloid curvature units (binding energies) $-2E_{2J} = \frac{1}{(1+2J)^2}$ involve the Plancherel density $d_{(J,J)} = (1+2J)^2$ for **SO**(4), modified, in the hydrogen atom $\frac{1}{(1+2J)^2 l_{\mathrm{Bohr}}^2}$ by the Bohr length $l_{\mathrm{Bohr}} = \frac{\hbar}{\alpha_{\mathrm{Som}} c m_e} \sim 0.5 \times 10^{-10}$ m as intrinsic unit. The nonrelativistic hydrogen spectrum has both a noncompact and a compact structure: The energy $E_0 = -\frac{B_0^2}{2}$ of the ground-state (0, 0) is taken from an $\mathbf{SO}_0(1,1)$-related continuous spectrum. It is no translation invariant. The **SO**(4)-nontrivial states (J, J) as symmetric product states of $(\frac{1}{2}, \frac{1}{2})$ are characterized by the discrete numbers $(1+2J)$ with the noncompact invariants $\frac{1}{B^2} = (1+2J)^2 \frac{1}{B_0^2}$ for the product representations:

$$\mathcal{Y}^3 \ni \vec{x} \longmapsto \int \frac{d^3q}{\pi^2}\frac{B_0}{(\vec{q}^2+B_0^2)^2}(p)^{2J} e^{\frac{i\vec{q}\vec{x}}{1+2J}}, \text{ with } p = \frac{1}{\vec{q}^2+B_0^2}\begin{pmatrix} B_0^2-\vec{q}^2 \\ 2B_0\vec{q}\end{pmatrix}.$$

The Fourier transformations with the 3-sphere measure

$$\mu(\vec{x})e^{-r} = \int \frac{d^3q}{\pi^2}\frac{1}{(\vec{q}^2+1)^2}\tilde{\mu}(\vec{q})e^{i\vec{q}\vec{x}} :$$

$\mu(\vec{x})$	$\tilde{\mu}(\vec{q})$
1	1
r	$\frac{3-\vec{q}^2}{\vec{q}^2+1}$
$i\frac{\vec{x}}{2}$	$\frac{2\vec{q}}{\vec{q}^2+1}$
$\frac{r^2}{3}$	$\frac{4(1-\vec{q}^2)}{(\vec{q}^2+1)^2}$
$ir\vec{x}$	$\frac{2\vec{q}\,(5-\vec{q}^2)}{(\vec{q}^2+1)^2}$
$\vec{x}\otimes\vec{x} - \mathbf{1}_3\frac{r^2}{3}$	$-\frac{6(\vec{q}\otimes\vec{q}-\mathbf{1}_3\frac{\vec{q}^2}{3})}{(\vec{q}^2+1)^2}$

are used for the ground-state singlet with a dipole,

$$J=0,\ \frac{1}{B}=1:\ e^{-Br}=\int\frac{d^3q}{\pi^2}\frac{B}{(B^2+\vec{q}^2)^2}e^{i\vec{q}\vec{x}}=\int\frac{d^3q}{\pi^2}\frac{1}{(\vec{q}^2+1)^2}e^{i\vec{q}B\vec{x}},$$
$$(-\vec{\partial}^2+B^2)^2\,e^{-Br}=8\pi B\delta(\vec{x}).$$

for the negative energy (bound)-state quartet with $\mathrm{Y}^{(\frac{1}{2},\frac{1}{2})}(p)\sim(p)^1$ and tripole, leading to Laguerre polynomials:

$$J=\frac{1}{2},\ \frac{1}{B}=2:\ \int\frac{d^3q}{\pi^2}\frac{1}{(\vec{q}^2+1)^2}\begin{pmatrix}p_0\\ \vec{p}\end{pmatrix}e^{i\vec{q}B\vec{x}}=\int\frac{d^3q}{\pi^2}\frac{1}{(\vec{q}^2+1)^3}\begin{pmatrix}1-\vec{q}^2\\ 2\vec{q}\end{pmatrix}e^{i\vec{q}B\vec{x}}$$
$$=\begin{pmatrix}\frac{Br-1}{2}\\ i\frac{B\vec{x}}{2}\end{pmatrix}e^{-Br}=\begin{pmatrix}-\frac{1}{4}\mathrm{L}_1^1(2Br)\\ iB\vec{x}\,\frac{1}{2}\,\mathrm{L}_0^3(2Br)\end{pmatrix}e^{-Br},$$

and for the negative energy (bound)-state nonet with $\mathrm{Y}^{(1,1)}(p)\sim(p)^2$ and quadrupole:

$$J=1,\ \frac{1}{B}=3:\ \int\frac{d^3q}{\pi^2}\frac{1}{(\vec{q}^2+1)^2}\begin{pmatrix}p_0^2-\frac{1}{4}\\ p_0\vec{p}\\ \vec{p}\otimes\vec{p}-\frac{\mathbf{1}_3}{3}\vec{p}^2\end{pmatrix}e^{i\vec{q}B\vec{x}}=\int\frac{d^3q}{\pi^2}\frac{4}{(\vec{q}^2+1)^4}\begin{pmatrix}3(\frac{1-\vec{q}^2}{4})^2-\frac{\vec{q}^2}{4}\\ \vec{q}\,\frac{1-\vec{q}^2}{2}\\ \vec{q}\otimes\vec{q}-\frac{\mathbf{1}_3}{3}\vec{q}^2\end{pmatrix}e^{i\vec{q}B\vec{x}}$$
$$=\begin{pmatrix}\frac{1}{4}(1-2Br+\frac{2B^2r^2}{3})\\ iB\vec{x}\frac{Br-2}{6}\\ -\frac{B^2}{6}(\vec{x}\otimes\vec{x}-\mathbf{1}_3\frac{r^2}{3})\end{pmatrix}e^{-Br}=\begin{pmatrix}\frac{1}{12}\mathrm{L}_2^1(2Br)\\ -iB\vec{x}\,\frac{1}{12}\mathrm{L}_1^3(2Br)\\ -B^2(\vec{x}\otimes\vec{x}-\mathbf{1}_3\frac{r^2}{3})\,\frac{1}{6}\mathrm{L}_0^5(2Br)\end{pmatrix}e^{-Br}.$$

The Kepler representation (J,J) of $\mathbf{SO}(4)$ as the highest-dimensional representation in the totally symmetric product of the defining four-dimensional representation $\bigvee^{2J}(\frac{1}{2},\frac{1}{2})\cong(J,J)\oplus\ \ldots$ is decomposable with respect to the position rotation $\mathbf{SO}(3)$-properties into irreducible representations of dimension $(1+2L)$ with integers $L=0,1,\ldots,2J$ for angular momentum $\vec{\mathbf{J}}=\vec{\mathbf{J}}_++\vec{\mathbf{J}}_-$:

$$(J,J)\overset{\mathbf{SO}(3)}{\cong}\bigoplus_{L=0,1,\ldots,2J}(L),\quad(1+2J)^2=\sum_{L=0}^{2J}(1+2L),$$
$$2J=L+N\Rightarrow(L,N)=(2J,0),(2J-1,1),\ldots,(0,2J).$$

With semisimple rank two Lie algebra, the irreducible $\mathbf{SO}(4)$-representations are characterized by two invariants. The difference $2J-L=N$ characterizing the classes $\mathbf{SO}(4)/\mathbf{SO}(3)\cong\Omega^3$ is the radial quantum number or knot number. It is the degree of the Laguerre polynomials in the factorization of the Fourier-transformed harmonic $\mathbf{SO}(4)$-polynomials by harmonic $\mathbf{SO}(3)$-polynomials and $\mathbf{SO}(3)$-invariant Laguerre polynomials:

$$\int\frac{d^3q}{\pi^2}\frac{B}{(\vec{q}^2+B^2)^2}\left(\frac{p}{B}\right)^{2J}e^{i\vec{q}\vec{x}}\Rightarrow\{(B\vec{x})^L\mathrm{L}_N^{1+2L}(2Br)e^{-Br}\,|\,L,N=0,1,\ldots,\ L+N=2J\}.$$

As seen in experiments, there is, in addition to the Lenz–Runge **SO**(4), a twofold degeneracy in the atoms. It originates from an "internal" spin **SU**(2)-property of the electron not contained in the nonrelativistic scheme above. It can be added by an ad hoc doubling leading to doubled multiplicities $2n^2 = 2, 8, 18, \ldots$.

For a nucleus with positive charge number z and the electron with charge number -1 and the reduced mass $\frac{1}{m} = \frac{1}{m_e} + \frac{1}{m_N}$ the energy eigenvalues are, in atomic units with the Rydberg energy:

$$E_{2J} = -\frac{1}{(1+2J)^2}\frac{E_{\text{Ryd}}(m, z^2)}{2}, \quad \frac{E_{\text{Ryd}}(m, z^2)}{2} = z^2\frac{m}{2}(\alpha_{\text{Som}}c)^2 \sim z^2 \times 13.6\,\text{eV}.$$

The energy sum of all energy-degenerated eigenvectors is always one-half of the intrinsic energy unit, $(1+2J)^2 E_{2J} = -\frac{E_{\text{Ryd}}(m,z^2)}{2}$. Induced by the Kepler potential $\frac{1}{r}$, the spherical spread of the energy goes with the radial scaling $\frac{r}{n} = r\sqrt{-2E}$ for the weight "area" $n^2 = (1+2J)^2$ (**SO**(4)-multiplicity).

Obviously, the **SO**(4)-periodicity for the Kepler-potential negative energy (bound) states is strictly valid only for the nonrelativistic hydrogen atom. It is modified by relativistic corrections and can serve as a first order approximation for the other atoms. In the periodic system of the atoms, additional interactions are relevant which lead to more than one not completely filled shell, etc.

11.7 Maximally Symmetric Positions

For general dimensions, the *maximally symmetric operational positions with definite metric* parametrize rotation group classes by compact spherical, noncompact flat and hyperbolic manifolds:

$$\begin{aligned}
\Omega^s &\cong \mathbf{SO}(1+s)/\mathbf{SO}(s), \quad k = i^2 = -1,\\
\mathbb{E}^s &\cong \mathbf{SO}(s) \vec{\times} \mathbb{R}^s/\mathbf{SO}(s), \; k = 0^2 = 0,\\
\mathcal{Y}^s &\cong \mathbf{SO}_0(1,s)/\mathbf{SO}(s). \quad k = 1^2 = 1,
\end{aligned}$$

$\mathbb{M} \in \{\Omega^s, \mathbb{E}^s, \mathcal{Y}^s\}$ can be taken as bases of coset bundles $\mathbf{SO}(s)(\mathbb{M})$ with $\mathbf{SO}(s)$ as typical fiber.

11.7.1 Spherical and Hyperbolic Coefficients

First for the abelian groups $\mathcal{Y}^1 \cong \mathbf{SO}_0(1,1)$, $\mathbb{E}^1 \cong \mathbb{R}$, and $\Omega^1 \cong \mathbf{SO}(2)$:

The positive Radon measures of the $\mathbb{R}$-dual group $iq \in \check{\mathbb{R}}$ with the momenta are characterized by their support. The extremal states for the irreducible translation representations in $\mathbf{U}(1)$ acting on one-dimensional Hilbert spaces are the Dirac measures $\delta_P \in \mathcal{M}(\check{\mathbb{R}})_+$ supported by the real momentum P for the invariant imaginary eigenvalue $iP \in i\mathbb{R}$:

$$\mathbb{R} \ni x \longmapsto e^{iPx} = \int dq\ \delta(q-P)e^{iqx} = \oint \frac{dq}{2i\pi}\frac{1}{q-P}e^{iqx}.$$

Dirac distributions can be expressed by residual integration, sketched as follows:

$$\text{e.g., } \int dq\ \delta(f(q)) \sim \oint \frac{dq}{2i\pi}\frac{1}{f(q)} \sim \sum_{q_0,\, f(q_0)=0} \frac{1}{|f'(q_0)|}\oint \frac{dq}{2i\pi}\frac{1}{q-q_0}.$$

The representation normalization is effected by the residual normalization $\text{Res}\frac{1}{q-P} = 1$ of the energy invariant. The Schur product of the irreducible Hilbert representations gives the Plancherel density $\delta(\frac{P}{2\pi})$, i.e., the dual ("inverse") of the Plancherel measure $\frac{dp}{2\pi}$:

$$\text{Schur product: } \int dx\ e^{-iP'x}e^{iPx} = \delta\left(\frac{P-P'}{2\pi}\right).$$

The self-dual representations on two-dimensional Hilbert spaces are cyclic with positive Radon distributions, invariant under the reflection $P \leftrightarrow -P$, and supported by two reflected eigenvalues $\pm iP \in i\mathbb{R}$ and positive invariant P^2:

$$\mathbb{R} \ni x \longmapsto \cos Px = \int dq\ |q|\delta(q^2-P^2)e^{iqx},$$

$$\begin{aligned}\text{Schur product: } \int dx\ \cos P'x \cos Px &= \frac{1}{2}\left[\delta\left(\frac{P-P'}{2\pi}\right) + \delta\left(\frac{P+P'}{2\pi}\right)\right]\\ &= |P|\delta\left(\frac{P^2-P'^2}{2\pi}\right) = \delta\left(\frac{|P|-|P'|}{\pi}\right).\end{aligned}$$

Irreducible faithful translation representations $x \longmapsto e^{Bx}$ with real eigenvalue $B \in \mathbb{R}$ are not unitary. From the residual form of the *coefficients of the circle* $\mathbf{SO}(2)$ and its two-dimensional representations with positive invariant $P^2 \geq 0$:

$$\mathbb{R} \ni x \longmapsto \cos Px = \oint \frac{dq}{i\pi}\frac{q}{q^2-P^2}e^{iqx},$$

one obtains by the compact-noncompact transition $P \to iB$ the *coefficients for the hyperbola* $\mathbf{SO}_0(1,1) \cong \mathbb{R}$ and its cyclic infinite-dimensional Hilbert representations with negative invariant $-B^2$:

$$\mathbb{R} \ni x \longmapsto e^{-|Bx|} = \int \frac{dq}{\pi} \frac{|B|}{q^2+B^2} e^{iqx},$$
$$\frac{1}{\pi} \frac{|B|}{q^2+B^2} = \frac{1}{2i\pi} \left(\frac{1}{q-i|B|} - \frac{1}{q+i|B|} \right).$$

The singularities of the function in the Radon measure (imaginary momenta $\pm iB$) are not eigenvalues for unitary translation representations. The representation normalizations are effected by the unit normalization of all energy-residues, with $+1$ for positive $\oint$ and -1 for negative integration direction.

In contrast to the spherical coefficients for the irreducible representations, the hyperbolic ones for the cyclic representations of the noncompact group are not orthogonal for different invariants:

$$\text{Schur product: } \int dx \, e^{-|B'x|} e^{-|Bx|} = \frac{2}{|B|+|B'|}.$$

11.7.2 Odd-Dimensional Hyperboloids and Spheres

Hilbert representations of hyperboloids and spheres with nonabelian degrees of freedom use higher-order momentum poles.

Distributions of s-dimensional momenta $\vec{q} \in \mathbb{R}^s$ with the action of the rotation group $\mathbf{SO}(s)$ are used for representations of the hyperboloids $\mathcal{Y}^s$ and spheres Ω^s. Flat spaces and hyberboloids are isomorphic as manifolds. The sphere measures are defined recursively:

$$s = 1, 2, \ldots : \; \vec{\omega}_s = \begin{pmatrix} \cos\theta \\ \sin\theta \, \vec{\omega}_{s-1} \end{pmatrix} = \frac{1}{\sqrt{\vec{q}^2+1}} \begin{pmatrix} 1 \\ \vec{q} \end{pmatrix} \in \Omega^s \subset \mathbb{R}^{1+s},$$
$$d^0\omega = 1, \; d^s\omega = (\sin\theta)^{s-1} d\theta \, d^{s-1}\omega = d^s q \, \frac{2}{(\vec{q}^2+1)^{\frac{1+s}{2}}}.$$

The residual representations of nonabelian noncompact hyperboloids and compact spheres with $s \geq 2$ have to embed the nontrivial representations of the abelian groups with continuous and integer dual invariants, respectively:

$$\mathbf{SO}_0(1,1) \cong \mathcal{Y}^1 \ni x \longmapsto \int \frac{dq}{\pi} \frac{|B|}{q^2+B^2} e^{iqx} = e^{-|Bx|},$$
$$\mathbf{SO}(2) \cong \Omega^1 \ni e^{ix} \longmapsto \begin{cases} \int \frac{dq}{i\pi} \frac{P}{q^2-io-P^2} e^{iqx} = e^{iP|x|}, \\ \int dq \, |q| \delta(P^2-\vec{q}^2) e^{iqx} = \cos Px, \\ \qquad P = 0, 1, 2, \ldots . \end{cases}$$

The invariant poles $\{\pm P\}$ and $\{\pm iB\}$ on a discrete sphere $\Omega^0 = \{\pm 1\}$ are embedded, for the nonabelian case, in real or imaginary singularity spheres $\Omega^{s-1} \cong \mathbf{SO}(s)/\mathbf{SO}(s-1)$, whose motion group is the maximal compact factor in Iwasawa and Cartan factorizations:

$$\begin{aligned}\mathbf{SO}_0(1,s) &= \mathbf{SO}(s) \circ \mathbf{SO}_0(1,1) \circ \exp \mathbb{R}^{s-1} \\ &= \mathbf{SO}(s) \circ \mathbf{SO}_0(1,1) \circ \mathbf{SO}(s).\end{aligned}$$

In the following, odd-dimensional maximally symmetric positions from the noncompact–compact pairs $(\mathcal{Y}^{2R-1}, \Omega^{2R-1})$ with hyperboloids and spheres are considered, e.g.,

$$\begin{aligned}&\mathcal{Y}^{2R-1} \cong \mathbf{SO}_0(1,2R-1)/\mathbf{SO}(2R-1) = \mathbf{SO}_0(1,2R_s+1)/\mathbf{SO}(2R_s+1), \\ &\text{with } R_s = R-1 = 0,1,2\ldots,\end{aligned}$$

where R is the Lie algebra rank for the global group $\mathbf{SO}_0(1,2R-1)$ and $R_s = R-1$ the rank of the local group $\mathbf{SO}(2R_s+1)$. The Lorentz groups for even dimensions $2R$ have a unique Cartan subgroup type,

$$\mathbf{SO}_0(1,2R-1) \supseteq \mathbf{SO}(2)^{R-1} \times \mathbf{SO}_0(1,1),\ s = 2R-1 = 1,3,5,\ldots,$$

in contrast to odd dimensions, e.g., the two Cartan subgroup types $\mathbf{SO}_0(1,1)$ and $\mathbf{SO}(2)$ for $\mathbf{SO}_0(1,2)$. The real rank 1 of the orthogonal groups $\mathbf{SO}_0(1,2R-1)$ gives the real (noncompact) rank 1 for the odd-dimensional hyperboloids, i.e., one continuous invariant.

The representation coefficients and proper distributions (for interactions) for hyperboloids $\mathcal{Y}^{2R-1}$ use the Fourier-transformed measure of the momentum sphere Ω^{2R-1} with singularity sphere $\{\vec{q} \in \mathbb{R}^{2R-1} \mid \vec{q}^2 = -1\} \cong \Omega^{2R-2}$, i.e., for imaginary "momenta" as eigenvalues (where defined):

$$\begin{aligned}&\text{for } \mathcal{Y}^{2R-1}: & \int \frac{d^{2R-1}q}{\pi^R} \frac{\Gamma(R-N)}{(\vec{q}^2+1)^{R-N}} e^{i\vec{q}\vec{x}} &= \left(-\frac{d}{d\frac{r^2}{4}}\right)^N e^{-r} \\ &2R-1 = 1,3,\ldots, & &= \left(1, \tfrac{2}{r}, \tfrac{4(1+r)}{r^3}, \ldots\right) e^{-r}, \\ & & \text{for } N &= 0,1,\ldots,R-1, \\ & & \int \frac{d^{2R-1}q}{\pi^R} \frac{\Gamma(R+N)}{(\vec{q}^2+1)^{R+N}} e^{i\vec{q}\vec{x}} &= \left(\tfrac{r}{2}\right)^{1+2N} \left(-\frac{d}{d\frac{r^2}{4}}\right)^{1+N} e^{-r} \\ & & &= \left(1, \tfrac{1+r}{2}, \ldots\right) e^{-r}, \\ & & \text{for } N &= 0,1,\ldots\end{aligned}$$

The coefficients for $N = 0$ characterize the cyclic ground state. For a continuous unit, the invariant is at the pole $\vec{q}^2 = -B^2$.

$\mathbf{SO}(2R)$-multiplets arise via the parametrization of the compact partner sphere $\frac{1}{\vec{q}^2+B^2}\begin{pmatrix} B^2-\vec{q}^2 \\ 2|B|\vec{q} \end{pmatrix} \in \Omega^{2R-1} \subset \mathbb{R}^{2R}$,

$$\text{for } \mathcal{Y}^{2R-1},\ R = 1, 2, \ldots,$$
$$\vec{x} \longmapsto \begin{cases} \int \frac{d^{2R-1}q}{\pi^R} \frac{|B|\,\Gamma(R)}{(\vec{q}^2+B^2)^R} e^{i\vec{q}\vec{x}} & = e^{-|B|r}, \\ \int \frac{d^{2R-1}q}{\pi^R} \frac{|B|\,\Gamma(1+R)}{(\vec{q}^2+B^2)^{1+R}} \begin{pmatrix} B^2-\vec{q}^2 \\ 2|B|\vec{q} \end{pmatrix} e^{i\vec{q}\vec{x}} & = \begin{pmatrix} 1-R+|B|r \\ i|B|\vec{x} \end{pmatrix} e^{-|B|r}, \end{cases}$$
$$\int d^{2R-1}x\, e^{-|B'\vec{x}|} e^{-|B\vec{x}|} = \frac{(4\pi)^R}{(|B'|+|B|)^{2R-1}}.$$

$\int \frac{d^{2R-1}q}{i\pi^R}$ generalizes the normalization $\int \frac{dq}{i\pi}$ of the one-dimensional residue for a quadratic invariant. The order of the singularity is related to the rank R of the acting group $\mathbf{SO}_0(1, 2R-1)$, i.e., to the dimension of the Cartan subgroups $\mathbf{SO}(2)^{R-1} \times \mathbf{SO}_0(1,1)$.

All representation coefficients for $\mathcal{Y}^{2R-1} \cong \mathbb{R}^{2R-1}$ with different continuous invariants $B^2 > 0$ are elements of the Hilbert space $L^2(\mathbb{R}^{2R-1})$.

The distributions on hyperboloids $\mathcal{Y}^{2R-1}$ with $r = 0$-singularities arise, by derivatives $(-\frac{d}{d\frac{r^2}{4}})^N e^{-r}$, from the $r = 0$ regular representation coefficients. For $R \geq 2$, the Schrödinger equations come with (non)trivial potentials in the invariant (non-)flat position derivatives $-\vec{\partial}^2 + 2V(r) \cong 2\mathbf{H}$ — for Euclidean $\mathbb{E}^{2R-1}$ with $\frac{\pi \mathcal{J}_{R-\frac{3}{2}}(r)}{(\frac{r}{2})^{R-\frac{3}{2}}} = (-\frac{d}{d\frac{r^2}{4}})^{R-1}\sqrt{\pi}\cos r$, spherical Ω^{2R-1} and hyperbolic $\mathcal{Y}^{2R-1}$ positions, for the invariant $iQ = (i|P|, B)$, respectively:

$$R = 1: \quad \begin{cases} \left(-\frac{d^2}{dx^2} - P^2\right)\cos Px = 0, \\ \left(-\frac{d^2}{dx^2} - Q^2\right) e^{-iQ|x|} = 2iQ\delta(x). \end{cases}$$

$$R \geq 2: \quad \begin{cases} (-\vec{\partial}^2 - P^2)\left(\frac{d}{d\frac{P^2r^2}{4}}\right)^{R-1} \cos Pr = 0, \\ (-\vec{\partial}^2 - Q^2)e^{-iQr} = \frac{(R-1)2iQ}{r} e^{-iQr}, \\ \left[-\vec{\partial}^2 + (R-1)\frac{\partial}{\partial\frac{r^2}{4}}\right] e^{-iQr} = \left[-\vec{\partial}^2 - \frac{(R-1)2iQ}{r}\right] e^{-iQr} = Q^2 e^{-iQr}, \\ (-\vec{\partial}^2 - Q^2)^R e^{-iQr} = 2iQ\Gamma(R)(4\pi)^{R-1}\delta(\vec{x}). \end{cases}$$

Only for "our" position dimension $s = 2R-1 = 3$, the Kepler potential $\frac{1}{r}$ is also the inverse of the Laplacian $\vec{\partial}^2$:

$$R \geq 2: \quad \int \frac{d^{2R-1}q}{\pi^R} \frac{\Gamma(R-1)}{\vec{q}^2} e^{i\vec{q}\vec{x}} = \frac{2\Gamma(2R-3)}{r^{2R-3}}.$$

The representation coefficients of odd-dimensional spheres, corresponding to those of the hyperboloids, are obtained by a real–imaginary transition (by noncompact-compact pairing), starting from $\int \frac{d^{2R-1}q}{i\pi^R} \frac{\Gamma(R)}{(\vec{q}^2-io-1)^R} e^{i\vec{q}\vec{x}} = e^{ir}$ — for units with $|B| \leftrightarrow i|P|$. They involve multipole Feynman distributions (derived Dirac distributions) with supporting singularity sphere Ω^{2R-2} for real momenta with, for appropriate normalization, integer invariants $\vec{q}^2 = N^2$, $N = 0, 1, 2, \ldots$,

$$\text{for } \Omega^{2R-1},\ R = 1, 2, \ldots:\ \vec{x} \longmapsto \begin{cases} \int \frac{d^{2R-1}q}{i\pi^R} \frac{N\,\Gamma(R)}{(\vec{q}^2-io-N^2)^R} e^{i\vec{q}\vec{x}} = e^{iNr}, \\ \int \frac{d^{2R-1}q}{\pi^{R-1}} N\delta^{(R-1)}(N^2 - \vec{q}^2) e^{i\vec{q}\vec{x}} = \cos Nr. \end{cases}$$

In contrast to hyperboloids $\mathcal{Y}^s \cong \mathbb{R}^s$, spheres are compact. Therefore, the Fourier-transformed functions of the spheres for $\mathbf{SO}(1+s)$-representations do not have to give positive Radon measures of the momenta.

The irreducible $\mathbf{SO}(1+s)$-representation spaces used for the sphere Ω^s are finite-dimensional. A basis is given by the spherical harmonics $(\vec{\omega}_s)^L$, $L = 0, 1, \ldots$, with $r \sim \theta$:

$$\vec{\omega}_s = \begin{pmatrix} \cos\theta \\ \sin\theta\, \vec{\omega}_{s-1} \end{pmatrix} \in \Omega^s \subset \mathbb{R}^{1+s},\ \ s = 1, 2 \ldots.$$

In the nontrivial case, $L \neq 0$, $\Omega^1 = \mathbf{SO}(2)$ is acted on by the two-dimensional $\mathbf{SO}(2)$-representations $\begin{pmatrix} \cos L\theta & \sin L\theta \\ -\sin L\theta & \cos L\theta \end{pmatrix}$, Ω^2 by the $(1+2L)$-dimensional harmonic $\mathbf{SO}(3)$-representations (L), and Ω^3 by the $(1+L)^2$-dimensional harmonic $\mathbf{SO}(4)$-representations $(\frac{L}{2}, \frac{L}{2})$. Different invariants L characterize finite-dimensional Schur-orthogonal subspaces of the infinite-dimensional Hilbert space $L^2(\Omega^s)$.

11.8 Normalized Position Representations

The invariants for the negative energy (bound) waves of the nonrelativistic hydrogen atom can be determined as eigenvalues for square integrable solutions of the position Schrödinger equation. The noncompact energy and compact angular momentum invariants are not independent: In a *noncompact–compact reciprocity*, the continuous invariants B_{2J}^2 of the unitary $\mathbf{SO}_0(1,3)$-representations for rank-1 position $\mathcal{Y}^3 \supset \mathbf{SO}_0(1,1)$ are determined by the dimensions $(1+2J)^2$ of $\mathbf{SO}(4)$-representation spaces, $1 + \frac{1}{2\mathbf{H}} = 1 + \frac{1}{B_{2J}^2} = 2(\vec{\mathbf{J}}_+^2 + \vec{\mathbf{J}}_-^2) = -4J(J+1)$. These $\mathbf{SO}(4)$-representations (J, J) are the totally symmetric products of the real four-dimensional one $(\frac{1}{2}, \frac{1}{2})$ which is the compact partner of the real four-dimensional Minkowski representation $(\frac{1}{2}|\frac{1}{2})$ of the Lorentz group $\mathbf{SO}_0(1,3)$.

As will be shown in the following, this relation between continuous and integer invariants reflects a *representation normalization* of the hyperboloid coefficients. The energy invariants are given by multipole singularities $2E = \vec{q}^2 = -B^2$ in the Fourier-transformed wave functions with $B_L > 0$:

$$\mathcal{Y}^3 \cong \mathbb{R}^3 \ni \vec{x} \longmapsto \int \frac{d^3q}{\pi^2} \frac{1}{(\vec{q}^2 + B_L^2)^2} \frac{(2\vec{q})^L}{(\vec{q}^2 + B_L^2)^L} e^{i\vec{q}\vec{x}} = \frac{(i\vec{x})^L}{(1+L)!} \frac{e^{-B_L r}}{B_L},$$
$$L = 2J = 0, 1, \ldots .$$

Via π^2, the coefficients are residually normalized: In addition to the hyperboloid $\mathcal{Y}^3$-characteristic measure $\frac{d^3q}{\pi^2} \frac{1}{(\vec{q}^2+B_L^2)^2}$, there are $\mathbf{SO}(4)$-representations spanned, for their $\mathbf{SO}(3)$-components, by normalized harmonic $\mathbf{SO}(3)$-polynomials $(\frac{2\vec{q}}{\vec{q}^2+B_L^2})^L$.

In general, the harmonic representation coefficients of odd-dimensional hyperboloids $\mathcal{Y}^{2R-1} \cong \mathbf{SO}_0(1, 2R-1)/\mathbf{SO}(2R-1)$, $R = 1, 2, \ldots$,

$$\mathcal{Y}^{2R-1} \ni \vec{x} \longmapsto \int \frac{d^{2R-1}q}{\pi^R} \frac{\Gamma(R)}{(\vec{q}^2 + B_L^2)^R} \frac{(2\vec{q})^L}{(\vec{q}^2 + B_L^2)^L} e^{i\vec{q}\vec{x}} = \frac{\Gamma(R)}{\Gamma(R+L)} (i\vec{x})^L \frac{e^{-B_L r}}{B_L},$$

with representation invariants B_L^2, have as dual products with harmonic $\mathbf{O}(2R-1)$-polynomials $(\vec{q})^L$ and harmonic projectors $(\mathbf{1}_{2R-1})^L$:

$$\begin{aligned}
&\int \frac{d^{2R-1}q}{\pi^R} \frac{\Gamma(R)}{(\vec{q}^2 + B_L^2)^R} \frac{(2\vec{q})^L}{(\vec{q}^2 + B_L^2)^L} \otimes (\vec{q})^{L'} \\
&= \delta^{LL'} \frac{\Gamma(1+L)\Gamma\left(R-\frac{1}{2}\right)}{\Gamma(R-\frac{1}{2}+L)} \int \frac{d^{2R-1}q}{\pi^R} \frac{\Gamma(R)}{(\vec{q}^2 + B_L^2)^R} \frac{(\vec{q}^2)^L}{(\vec{q}^2 + B_L^2)^L} (\mathbf{1}_{2R-1})^L \\
&= \delta^{LL'} \frac{\Gamma(1+L)\Gamma(R)}{\Gamma(R+L)} \frac{1}{B_L} (\mathbf{1}_{2R-1})^L .
\end{aligned}$$

The normalization conditions for the dual products, starting with the scalar case, give equations for the invariants in the compact-noncompact reciprocity:

$$\frac{\Gamma(1+L)\Gamma(R)}{\Gamma(R+L)} \frac{1}{B_L} = 1.$$

The noncompact representation invariants for $\mathbf{SO}_0(1, 2R-1)$ are determined by compact rotation $\mathbf{SO}(2R-1)$-invariants via dimensions of symmetric vector space products:

$$\frac{1}{B_L} = \frac{\Gamma(R+L)}{\Gamma(1+L)\Gamma(R)} = \begin{cases} 1, & R = 1, \\ 1+L, & R = 2, \\ \frac{(2+L)(1+L)}{2}, & R = 3, \\ \ldots, \end{cases}$$

$$\binom{R+L-1}{L} = \frac{\Gamma(R+L)}{\Gamma(1+L)\Gamma(R)} = \dim_{\mathbb{C}} \bigvee^L \mathbb{C}^R.$$

The invariants show a dimensional equipartition $B_L \frac{\Gamma(R+L)}{\Gamma(1+L)\Gamma(R)} = B_0 = 1$. For the nonrelativistic hydrogen atom, this is similar to a flux quantization $B_0 = (1+L)B_L$ of the basic invariant with respect to the dimensions of the symmetric products of Pauli spinor spaces $\mathbb{C}^2$:

$$R = 2 : \quad \frac{1}{B_L} = 1 + L = \dim_{\mathbb{C}} \bigvee^{L} \mathbb{C}^2.$$

B_0 can be used as intrinsic unit, $(\vec{x}, \vec{q}) = (B_0 \underline{\vec{x}}, \frac{\underline{\vec{q}}}{B_0})$.

The Schur product for the cyclic representations of the hyperboloid $\mathcal{Y}^{2R-1}$ is orthogonal for different "angular momentum" L:

$$\int \frac{d^{2R-1}q}{\pi^R} \frac{\Gamma(R)(2\vec{q})^L}{(\vec{q}^2 + B_L^2)^{R+L}} \otimes \frac{\Gamma(R)(2\vec{q})^{L'}}{(\vec{q}^2 + B_{L'}^2)^{R+L'}} = \delta^{LL'} \frac{\Gamma(1+L)\Gamma(R)^2}{\Gamma(R+L)} \frac{1}{2^{2R-1} B_L^{2R+2L+1}} (\mathbf{1}_{2R-1})^L.$$

Different noncompact invariants $B \neq B'$ for equal compact invariant L do not involve orthogonality, e.g. for $L = 0$:

$$\int \frac{d^{2R-1}q}{\pi^R} \frac{\Gamma(R)}{(\vec{q}^2 + B^2)^R} \frac{\Gamma(R)}{(\vec{q}^2 + B'^2)^R} = \frac{\Gamma(R)^2}{BB'(B+B')^{2R-1}}.$$

11.9 Harmonic Oscillators

The Hamiltonian as the time translation generator with creation-annihilation operators for s degrees of freedom:

$$s = 1, 2, \ldots : \ \mathbf{H} = \sum_{a=1}^{s} \mathbf{H}_a = \omega_0 \frac{\{\mathrm{u}^a, \mathrm{u}_a^\star\}}{2}, \quad [\mathrm{u}_b^\star, \mathrm{u}^a] = \delta_b^a \mathbf{1},$$

can be related, with the definition of s hermitian position-momentum pairs,

$$\mathbf{x}^a = \frac{\mathrm{u}_a^\star + \mathrm{u}^a}{\sqrt{2}}, \quad i\mathbf{p}^a = \frac{\mathrm{u}_a^\star - \mathrm{u}^a}{\sqrt{2}}, \quad \mathbf{H} = \omega_0 \frac{\mathbf{p}^a\mathbf{p}^a + \mathbf{x}^a\mathbf{x}^a}{2} = \omega_0 \frac{\vec{\mathbf{p}}^2 + \vec{\mathbf{x}}^2}{2},$$

to the s-dimensional isotropic harmonic quantum Bose oscillators, implementing the Heisenberg Lie algebra $\log \mathbf{H}(s)$,

$$\begin{aligned} &\log \mathbf{H}(s) = \mathbb{R}^s \ \vec{\oplus}\ \mathbb{R}^{1+s} \ni \mathbf{x}^a, \mathbf{p}^a, i\mathbf{1}, \ \text{with } [\mathbf{x}^b, \mathbf{p}^a] = \delta_b^a i\mathbf{1}, \\ &\mathbf{H}(s) = e^{\mathbb{R}^s} \ \vec{\times}\ \mathbb{R}^{1+s}, \end{aligned}$$

as Hilbert space operations with position $\log \mathbf{H}(s)/\mathbb{R}^{1+s} \cong \mathbb{R}^s$. The Hamiltonian has no scattering solutions, i.e., no representations of Euclidean translations $\mathbb{E}^s$.

The normalization of the potential $\underline{\mathbf{H}} = \frac{\vec{\mathbf{p}}^2}{2m} + \frac{k_0\vec{\mathbf{x}}^2}{2} = \hbar\omega_0\frac{\vec{\mathbf{p}}^2+\vec{\mathbf{x}}^2}{2}$ yields an intrinsic frequency unit $[\omega_0^2] = [\frac{k_0}{m}] = \frac{1}{s^2}$, in the following intrinsic units are used, formally $\omega_0 = 1$.

The quantum structure of the oscillators uses the Bose quantum algebras $\mathbf{Q}_-(\mathbb{C}^{2s}) \cong \bigotimes^s \mathbf{Q}_-(\mathbb{C}^2)$ with $[\mathbf{H}, \mathrm{u}^a] = \mathrm{u}^a$, $[\mathbf{H}, \mathrm{u}^\star_a] = -\mathrm{u}^\star_a$, and the Fock form related scalar product and Hilbert spaces.

11.9.1 Position Representation

For the one-dimensional oscillator, creation and annihilation operator can be represented in a Rodriguez form with position translation generators (derivatives):

$$\mathbf{x} \cong x,\ i\mathbf{p} \cong \frac{d}{dx} \Rightarrow \begin{cases} \sqrt{2}\mathrm{u} \cong x - \frac{d}{dx} = -e^{\frac{x^2}{2}}\frac{d}{dx}e^{-\frac{x^2}{2}}, \\ \sqrt{2}\mathrm{u}^\star \cong x + \frac{d}{dx} = e^{-\frac{x^2}{2}}\frac{d}{dx}e^{\frac{x^2}{2}}, \end{cases}$$

$$\mathbf{H} = \frac{\{\mathrm{u}, \mathrm{u}^\star\}}{2} = \frac{\mathbf{p}^2 + \mathbf{x}^2}{2} \cong \frac{1}{2}\left(-\frac{d^2}{dx^2} + x^2\right).$$

The Fock state vector $|0\rangle$ is representable by a positive definite rapidly decreasing function:

$$V_-^{\mathrm{Fock}} \ni |0\rangle \cong \psi^0 \in \mathcal{S}(\mathbb{R}) \subset L^2(\mathbb{R}) : \begin{cases} \mathrm{u}^\star|0\rangle = 0 \iff \frac{d}{dx}\left(e^{\frac{x^2}{2}}\psi^0(x)\right) = 0, \\ \langle 0||0\rangle = 1 \iff \int dx|\psi^0(x)|^2 = 1, \\ \Rightarrow \psi^0(x) = \frac{1}{\sqrt{\sqrt{\pi}}}e^{-\frac{x^2}{2}}. \end{cases}$$

From the cyclic ground-state vector a scalar product space basis $\{\psi^k \mid k = 0, 1, \dots\}$ with eigenvectors $\mathbf{H}|k\rangle = k|k\rangle$ can be constructed by the action of the creation operator powers:

$$\frac{\mathrm{u}^k}{\sqrt{k!}}|0\rangle = |k\rangle \cong \psi^k : \quad \begin{cases} \psi^k(x) = \frac{1}{\sqrt{k!}}\left(\frac{x-\frac{d}{dx}}{\sqrt{2}}\right)^k \psi^0(x) = \frac{1}{\sqrt{2^k k!\sqrt{\pi}}}e^{\frac{x^2}{2}}(-\frac{d}{dx})^k e^{-x^2} \\ \quad = \frac{1}{\sqrt{2^k k!\sqrt{\pi}}}e^{-\frac{x^2}{2}}\mathrm{H}^k(x), \end{cases}$$

with the Rodriguez formula for the *Hermite polynomials*:

$$\mathrm{H}^k(x) = e^{x^2}\left(-\frac{d}{dx}\right)^k e^{-x^2} = k!\sum_{n=0}^{[\frac{k}{2}]}\frac{(-1)^n}{n!(k-2n)!}(2x)^{k-2n} = (-1)^k\mathrm{H}^k(-x)$$

$$\text{with } \left[\frac{2n}{2}\right] = \left[\frac{1+2n}{2}\right] = n,\quad \deg \mathrm{H}^k = k,\quad \left(\frac{d^2}{dx^2} - 2x\frac{d}{dx} + 2k\right)\mathrm{H}^k(x) = 0$$

Up to a factor from $\{1, x\}$, the Hermite polynomials are Laguerre polynomials with $\deg \mathrm{L}_N^\lambda(x^2) = N$, depending on the squared length x^2:

$$\mathrm{H}^{2N}(x) = N!(-4)^N \mathrm{L}_N^{-\frac{1}{2}}(x^2),\qquad \text{e.g.,}\begin{cases}\mathrm{H}^0(x) = 1,\\ \mathrm{H}^2(x) = -2(1-2x^2),\\ \mathrm{H}^4(x) = 4(3-12x^2+4x^4),\end{cases}$$

$$\mathrm{H}^{1+2N}(x) = N!(-4)^N 2x\,\mathrm{L}_N^{\frac{1}{2}}(x^2),\qquad \text{e.g.,}\begin{cases}\mathrm{H}^1(x) = 2x,\\ \mathrm{H}^3(x) = -4x(3-2x^2),\\ \mathrm{H}^5(x) = 8x(15-20x^2+4x^4).\end{cases}$$

The Fourier transformation is a scalar product space isomorphism for rapidly decreasing functions and their completions:

$$\mathcal{S}(\mathbb{R}) \cong \mathcal{S}(\check{\mathbb{R}}) \subset L^2_{dx}(\mathbb{R}) \cong L^2_{dp}(\check{\mathbb{R}}).$$

With the position-momentum symmetry of the oscillator Hamiltonian the positive position representation ground-state coefficient is essentially its own Fourier transform:

$$\begin{aligned}\sqrt{2^k k!}\psi^k(x) &= \mathrm{H}^k(x)\psi^0(x) = \int\frac{dp}{2\pi}\,e^{ipx}\mathrm{H}^k(i\frac{d}{dp})\tilde{\psi}^0(p),\\ \mathcal{S}(\mathbb{R}) \ni \psi^0(x) &= \frac{1}{\sqrt{\sqrt{\pi}}}\,e^{-\frac{x^2}{2}} \overset{\mathbf{F}}{\leftrightarrow} \sqrt{2\sqrt{\pi}}\,e^{-\frac{p^2}{2}} = \tilde{\psi}^0(p) \in \mathcal{S}(\check{\mathbb{R}}),\\ 1 &= \int\frac{dp}{\sqrt{2\pi}}e^{\frac{(x+ip)^2}{2}} = \int_C\frac{dz}{\sqrt{2\pi}}e^{-\frac{z^2}{2}},\end{aligned}$$

with the normalized Gaussian $\mathbb{R}$-measure $\int\frac{dp}{\sqrt{2\pi}}e^{-\frac{p^2}{2}} = 1$.

The position wave functions ψ^k solve the Schrödinger equation with the energy eigenvalues k for the harmonic oscillator:

$$\frac{1}{2}\left(-\frac{d^2}{dx^2}+x^2\right)\psi^k(x)=\left(k+\frac{1}{2}\right)\psi^k(x).$$

k gives the degree of the Hermite polynomials.

Each normalized state vector ψ defines a positive measure $|\psi(x)|^2dx$ on position with Lebesgue measure basis. The Hilbert space has the structure of an orthogonal direct integral with the positive function for the ground-state coefficient (normal or Gaussian distribution) and an $\mathbb{C}$-isomorphic Hilbert space at each position:

$$L^2_{dx}(\mathbb{R})\cong{}^{\perp}\!\!\int_{\mathbb{R}}|\psi^0(x)|^2dx\,\mathbb{C}(x),\ \ |\psi^0(x)|^2dx=\frac{e^{-x^2}}{\sqrt{\pi}}dx,\ \ \mathbb{C}(x)\cong\mathbb{C}.$$

Orthonormality and completeness of the harmonic oscillator eigenfunctions as Hilbert basis with the infinite-dimensional representation of the Heisenberg group $\mathbf{H}(1)=\mathbb{R}\,\vec{\times}\,\mathbb{R}^2$ are expressed with the positive ground-state coefficient:

$$\{x\longmapsto e^{-\frac{x^2}{2}}\mathrm{H}^k(x)\,\big|\,k=0,1,\dots\}\text{ basis of }L^2_{dx}(\mathbb{R}),$$

$$\text{with}\ \begin{cases}\int\frac{e^{-x^2}}{\sqrt{\pi}}dx\ \mathrm{H}^k(x)\ \mathrm{H}^{k'}(x)=2^kk!\ \delta_{kk'},\\ \sum_{k=0}^{\infty}\frac{1}{2^kk!}\ \mathrm{H}^k(x)\ \mathrm{H}^k(x')=e^{x^2}\sqrt{\pi}\delta(x-x').\end{cases}$$

The degree-k polynomials multiplying the ground-state exponential span a unital algebra. The degree reflects the tensor powers in the quantum algebra grading.

The position-momentum uncertainty gives the minimal value:

$$\Delta\mathbf{x}^2=\int\frac{dx}{\sqrt{\pi}}e^{-x^2}x^2=\frac{1}{2}\langle0|\mathrm{u}^\star\mathrm{u}|0\rangle=\frac{1}{2}=\Delta\mathbf{p}^2=\Delta\mathbf{x}\Delta\mathbf{p}.$$

For the s-dimensional oscillator with Heisenberg group $\mathbf{H}(s)$ one has product representations:

$$L^2(\mathbb{R}^s)\cong\bigotimes^s L^2(\mathbb{R}),\ \ \psi^0(\vec{x})=\psi^0(x_1)\cdots\psi^0(x_s)=\frac{1}{\pi^{\frac{s}{4}}}e^{-\frac{r^2}{2}}.$$

For the three-dimensional harmonic oscillator and harmonic polynomials $(\vec{x})^L_m$, the radial representation coefficients

$$\psi(\vec{x})=\sum_{L=0}^{\infty}\sum_{m=-L}^{L}(\vec{x})^L_m d_L(r)\Rightarrow\left[d_r^2+\frac{2(1+L)}{r}d_r-2V(r)+2E\right]d_L(r)=0.$$

with squared radial dependence $d_L(r) = F_L(r^2)e^{-\frac{r^2}{2}}$ give as square-integrable solutions products of a Laguerre polynomial of degree N with a quantum number independent exponential:

$$\left[\frac{1}{2}(-d_r^2 + r^2) - \frac{1+L}{r} d_r\right] F_L(r^2) = E F_L(r^2) \Rightarrow \begin{cases} d_L(r) \sim \mathrm{L}_N^{\frac{1+2L}{2}}(r^2)e^{-\frac{r^2}{2}}, \\ E_{LN} = \frac{3}{2} + k = \frac{3}{2} + L + 2N, \\ L, N = 0, 1, \dots, \\ \psi^k_{Lm}(\vec{x}) \sim (\vec{x})^L_m \, \mathrm{L}_N^{\frac{1+2L}{2}}(r^2)\, e^{-\frac{r^2}{2}}. \end{cases}$$

The harmonic oscillator solutions for each angular momentum $L = 0, 1, 2, \dots$ constitute a Hilbert space basis:

$$\begin{aligned} &L^2(\mathbb{R}_+)\text{-basis}: \ \{\xi \longmapsto \xi^{\frac{1+2L}{4}} \mathrm{L}_N^{\frac{1+2L}{2}}(\xi)e^{-\frac{\xi}{2}} \mid N = 0, 1, \dots\}, \\ &\int_0^\infty dr\, \mathrm{L}_N^{\frac{1+2L}{2}}(r^2)\, e^{-r^2} r^{2+2L} \mathrm{L}_{N'}^{\frac{1+2L}{2}}(r^2) = \frac{1}{2}\int_0^\infty \xi^{\frac{1+2L}{2}} e^{-\xi} d\xi\, \mathrm{L}_N^{\frac{1+2L}{2}}(\xi)\, \mathrm{L}_{N'}^{\frac{1+2L}{2}}(\xi) \\ &\qquad = \frac{\Gamma(\frac{3}{2}+N+L)}{2N!}\delta_{NN'}. \end{aligned}$$

The Hamiltonian with creation and annihilation operators $(\mathrm{u}^a, \mathrm{u}^\star_a)$, $a = 1, 2, 3$, generates time orbits of the Hilbert vectors with k quanta. The creation operator polynomials of degree k, acting on the Fock ground-state vector $|0\rangle$, give the Schrödinger wave functions as position representation with degree-k polynomials:

$$\begin{aligned} |k; n_1, n_2, n_3\rangle &= \frac{(\mathrm{u}^1)^{n_1}(\mathrm{u}^2)^{n_2}(\mathrm{u}^3)^{n_3}}{\sqrt{k!}}|0\rangle \\ &\cong \{\vec{x} \longmapsto \mathrm{H}^k(\vec{x})e^{-\frac{r^2}{2}}\} \in L^2(\mathbb{R}^3), \ \ k = n_1 + n_2 + n_3, \\ \mathrm{u}^\star_a|0\rangle &= 0, \\ \mathrm{u}^a(t) &= e^{it}\mathrm{u}^a, \ \ |k; n_1, n_2, n_3\rangle(t) = e^{ikt}|k; n_1, n_2, n_3\rangle, \\ \mathrm{H}^k(\vec{x}) &\sim (\vec{x})^L_m \, \mathrm{L}_N^{\frac{1+2L}{2}}(r^2), \text{ e.g.,} \begin{cases} \mathrm{H}^0(\vec{x}) \sim 1, \\ \mathrm{H}^1(\vec{x}) \sim \vec{x}, \\ \mathrm{H}^2(\vec{x}) \sim (\vec{x})^2, \quad \mathrm{L}_1^{\frac{1}{2}}(r^2) = \frac{3}{2} - r^2, \\ \mathrm{H}^3(\vec{x}) \sim (\vec{x})^3, \ \vec{x}\, \mathrm{L}_1^{\frac{3}{2}}(r^2) = \vec{x}(\frac{5}{2} - r^2). \end{cases} \end{aligned}$$

11.9.2 *"Color"* SU(3) *for 3-Position*

The complex embedding of the three positions and momenta leads, for an isotropic oscillator, to a "color" **SU**(3)-invariance with Gell–Mann matrices $\{\lambda^A\}_{A=1}^8$:

$$\chi_A\lambda^A = \begin{pmatrix} \chi_3 + \frac{\chi_8}{\sqrt{3}} & \chi_1 - i\chi_2 & \chi_4 - i\chi_5 \\ \chi_1 + i\chi_2 & -\chi_3 + \frac{\chi_8}{\sqrt{3}} & \chi_6 - i\chi_7 \\ \chi_4 + i\chi_5 & \chi_6 + i\chi_7 & -2\frac{\chi_8}{\sqrt{3}} \end{pmatrix}, \quad \mathbf{C} = \frac{i}{2}\mathrm{u}^a\lambda_a^b\mathrm{u}_b^\star, \quad [\mathbf{C}, \mathbf{H}] = 0.$$

In general, the isotropic s-dimensional quantum oscillator for s quantized bosonic degrees of freedom displays an $\mathbf{AL}(s, \mathbb{C})$-structure with a unitary $\mathbf{U}(s)$-symmetry where $\mathbf{U}(1)$ is generated by the Hamiltonian $i\mathbf{H}$, and $\mathbf{SU}(s)$ by the Lie algebra elements $\mathbf{C}(s) = \frac{i}{2}\mathrm{u}^a\tau(s)_a^b\mathrm{u}_b^\star$ with the $s^2 - 1$ generalized Pauli matrices $\{\tau^A(s) \mid A = 1, \dots, s^2 - 1\}$.

The irreducible $\mathbf{SU}(3)$-representations are characterized by two integers as generating invariants for the $\mathbf{SU}(3)$-operations with rank 2:

$$\begin{aligned} \mathbf{irrep}\,\mathbf{SU}(3) &= \{[2C_1, 2C_2] \mid 2C_{1,2} \in \mathbb{N}\}, \\ \dim_{\mathbb{C}}[2C_1, 2C_2] &= (1 + 2C_1)(1 + 2C_2)(1 + C_1 + C_2). \end{aligned}$$

The harmonic oscillator representations $[k, 0]$ (singlet, triplet, sextet, etc.) with one trivial $\mathbf{SU}(3)$-invariant $C_2 = 0$ are the totally symmetric products of $\mathbf{SU}(3)$-triplets,

$$\begin{aligned} [k, 0] &= \bigvee^k [1, 0] \in \bigvee^k \mathbb{C}^3 \cong \mathbb{C}^{\binom{2+k}{2}}, \\ \dim_{\mathbb{C}}[k, 0] &= \binom{2+k}{2} = 1, 3, 6, \dots, \quad k = 0, 1, 2, \dots. \end{aligned}$$

The embedded rotation group, generated by the transposition antisymmetric Lie subalgebra $\log \mathbf{SO}(3) = \{\varphi_a\mathbf{J}^a \mid \varphi_a \in \mathbb{R}\}$,

$$\mathbf{SO}(3) \hookrightarrow \mathbf{SU}(3) \text{ with } \begin{cases} \chi_A i\frac{\lambda^A - (\lambda^A)^T}{2} = \begin{pmatrix} 0 & \chi_2 & \chi_5 \\ -\chi_2 & 0 & \chi_7 \\ -\chi_5 & -\chi_7 & 0 \end{pmatrix} = \begin{pmatrix} 0 & \varphi_3 & -\varphi_2 \\ -\varphi_3 & 0 & \varphi_1 \\ \varphi_2 & -\varphi_1 & 0 \end{pmatrix}, \\ \mathbf{J}^a = \epsilon^{abc}\mathrm{u}^b\mathrm{u}_c^\star, \quad (\mathbf{J}^1, \mathbf{J}^2, \mathbf{J}^3) = (\mathbf{C}^7, -\mathbf{C}^5, \mathbf{C}^2), \end{cases}$$

comes with the real five-dimensional orientation manifold, given by the rotation group orbits $\mathbf{SU}(3)/\mathbf{SO}(3)$ in the "color" group, which describes the $\binom{4}{2} - 1$ relative phases of the three angular momenta directions in complex quantum structures.

The energy (suitably normalized) as principal quantum number $E = k = L + 2N$ (polynomial degree) is the power of the product representations $\bigvee^k [1, 0]$ of the time translations $\mathbb{R}$. It is the sum of the angular momentum quantum number L for $\mathbf{SO}(3)$ and twice the radial quantum number (knot number) N for the rotation group classes in the real five-dimensional space $\mathbf{SU}(3)/\mathbf{SO}(3)$. One has with the angular momentum degeneracy $1 + 2L$ the energy degeneracy given by the dimension $\binom{2+k}{2}$ of the $\mathbf{SU}(3)$-representation $[k, 0]$:

$$[k,0] \overset{\mathbf{SO}(3)}{\cong} \begin{cases} \bigoplus_{L=0,2,\dots,k} (L),\ k=0,2,\dots, \\ \bigoplus_{L=1,3,\dots,k} (L),\ k=1,3,\dots, \end{cases} \quad \text{with} \begin{pmatrix} 2+k \\ 2 \end{pmatrix} = \begin{cases} \sum_{\dots}(1+2L), \\ \sum_{\dots}(1+2L), \end{cases}$$

$$E_{LN} - \frac{3}{2} = k = L + 2N \Rightarrow (L,N) = \begin{cases} (k,0),(k-2,1),\dots,(0,\frac{k}{2}), \\ (k,0),(k-2,1),\dots,(1,\frac{k-1}{2}). \end{cases}$$

11.10 Quantum Fields of Positions

With dual bases of dual representation spaces as creation and annihilation operators for quantum algebras, the characteristic position-representation coefficients can be identified with expectation values of quantum operators.

11.10.1 Quantum Fields of Euclidean Position

A *quantum field of Euclidean position* $\mathbb{E}^3 = \mathbf{SO}(3) \vec{\times} \mathbb{R}^3/\mathbf{SO}(3)$, e.g., for momentum invariant $P > 0$ and helicities $h = 0$ and $h = \pm 1$,

$$\mathbb{E}^3 \ni \vec{x} \longmapsto \begin{cases} \mathbf{E}(\vec{x}) = {}^{\oplus}\!\int \frac{d^2\omega}{|\Omega^2|} [e^{i\vec{q}\vec{x}}\mathrm{u}(\vec{\omega}) + e^{-i\vec{q}\vec{x}}\mathrm{u}^\star(\vec{\omega})], \\ \mathbf{E}^a(\vec{x}) = {}^{\oplus}\!\int \frac{d^2\omega}{|\Omega^2|} O^a_h(\frac{\vec{q}}{P})[e^{i\vec{q}\vec{x}}\mathrm{u}^h(\vec{\omega}) + e^{-i\vec{q}\vec{x}}\mathrm{u}^\star_h(\vec{\omega})],\ a = 1,2,3, \end{cases}$$

parametrizes the Euclidean group $\mathbf{SO}(3) \vec{\times} \mathbb{R}^3$-coefficients by the ground-state value of the anticommutator for a commutator-quantized field:

$$\mathbb{E}^3 \ni \vec{x} - \vec{y} \longmapsto \begin{cases} \langle 0|\{\mathbf{E}(\vec{y}), \mathbf{E}(\vec{x})\}|0\rangle = |P^2, 0\}(\vec{x} - \vec{y}) \\ \qquad = \int \frac{2d^3q}{|\Omega^2|P} \delta(\vec{q}^2 - P^2) e^{i\vec{q}(\vec{x}-\vec{y})}, \\ \langle 0|\{\mathbf{E}^a(\vec{y}), \mathbf{E}^b(\vec{x})\}|0\rangle = |P^2, \pm 1\}(\vec{x} - \vec{y}) \\ \qquad = \int \frac{2d^3q}{|\Omega^2|P} (\delta^{ab} - \frac{q^a q^b}{P^2}) \delta(\vec{q}^2 - P^2) e^{i\vec{q}(\vec{x}-\vec{y})}. \end{cases}$$

The harmonic expansion of the $\mathbf{SO}(2) \times \mathbb{R}^3$-induced representation coefficients integrates, with the normalized invariant measure $\frac{d^2\omega}{4\pi}$ of the 2-sphere $\Omega^2 \cong \mathbf{SO}(3)/\mathbf{SO}(2)$, arising from $d^3q\ \delta(\vec{q}^2 - 1)$, and a rotation-axial group transmutation, in the $h = \pm 1$-example the (3×2)-transmutator $O^a_h(\frac{\vec{q}}{P})$, translation and axial rotation eigenoperators for momentum direction $\vec{\omega} = \frac{\vec{q}}{P}$ and helicity h as distributive basis of the representation Hilbert space:

$$\text{quantization: } [\mathrm{u}^\star_{h'}(\vec{\omega}')\mathrm{u}^h(\vec{\omega})] = \mathbf{1}\delta^h_{h'}|\Omega^2|\,\delta(\vec{\omega}-\vec{\omega}'),$$
$$\text{distributive Hilbert basis: } \{\mathrm{u}^h(\vec{\omega})|0\rangle = |P^2,h;\vec{\omega}\rangle \,\big|\, \vec{\omega}\in\Omega^2\},$$
$$\langle P^2,h';\vec{\omega}'|P^2,h;\vec{\omega}\rangle = \langle 0|\mathrm{u}^\star_{h'}(\vec{\omega}')\mathrm{u}^h(\vec{\omega})|0\rangle = \langle 0|\{\mathrm{u}^\star_{h'}(\vec{\omega}'),\mathrm{u}^h(\vec{\omega})\}|0\rangle$$
$$= \delta^h_{h'}|\Omega^2|\,\delta(\vec{\omega}-\vec{\omega}').$$

11.10.2 Quantum Fields of Hyperbolic Position

The positive-type functions for the hyperbola $\mathcal{Y}^1 \cong \mathbf{SO}_0(1,1)$ with invariant $q^2 = -B^2$ characterize a scalar product for an infinite-dimensional Hilbert space:

$$f,f'\in L^1(\mathbb{R}):\ \langle -B^2;f|-B^2;f'\rangle = \int dx_1dx_2\,\overline{f(x_1)}e^{-|B(x_2-x_1)|}f'(x_2)$$
$$= \int \tfrac{dq}{\pi}\,\overline{\tilde{f}(q)}\,\tfrac{|B|}{B^2+q^2}\,\tilde{f}'(q),$$

$$\overline{|L^1(\mathbb{R})\rangle}_{-B^2}: \begin{cases} \text{distributive basis: } \{|-B^2;q\rangle \,\big|\, q\in\mathbb{R}\},\\ \langle -B^2;q'|-B^2;q\rangle = \frac{|B|}{B^2+q^2}\delta(\frac{q-q'}{\pi}),\\ x\bullet|-B^2;q\rangle = e^{iqx}|-B^2;q\rangle,\\ \text{cyclic vector: } |-B^2;1\rangle = {}^{\oplus}\!\int\frac{dq}{\pi}|-B^2;q\rangle,\\ \langle -B^2;1|-B^2;1\rangle = 1.\end{cases}$$

For hyperbolic position $\mathcal{Y}^3 = \mathbf{SO}_0(1,3)/\mathbf{SO}(3) \ni e^{\vec{\sigma}\vec{x}} \leftrightarrow \vec{x}\in\mathbb{R}^3$, the Kepler negative energy (bound) waves in $(1+2J)^2$-multiplets for $\mathbf{SO}(4)$ come with momentum poles of order $2+2J$ and harmonic $\mathbf{SO}(4)$-polynomials $(p)^{2J} \sim \mathrm{Y}^{(J,J)}(p)$:

$$\mathcal{Y}^3 \ni \vec{x} \longmapsto |B^2,J\}(\vec{x}) = \int \frac{d^3q}{\pi^2}\frac{B}{(B^2+\vec{q}^2)^2}(\frac{p}{B})^{2J}e^{i\vec{q}\vec{x}} = \int\frac{d^3q}{\pi^2}\frac{1}{(1+\vec{q}^2)^2}(p)^{2J}e^{i\vec{q}B\vec{x}},$$
$$\text{with}\begin{cases} p\cong p^j = \frac{1}{1+\vec{q}^2}\begin{pmatrix}1-\vec{q}^2\\ 2\vec{q}\end{pmatrix},\ p_0^2+\vec{p}^2 = 1,\\ B^2 = \frac{1}{(1+2J)^2},\ 2J = 0,1,2,\ldots.\end{cases}$$

The scalar product of $L^1(\mathcal{Y}^3)$ involves the positive-type function, with intrinsic unit $B^2=1$ for $J=0$:

$$\langle 0;f|0;f'\rangle = \int d^3x_1d^3x_2\,\overline{f(\vec{x}_1)}\,e^{-|\vec{x}_2-\vec{x}_1|}f'(\vec{x}_2) = \int\frac{d^3q}{\pi^2}\,\overline{\tilde{f}(\vec{q})}\,\frac{1}{(1+\vec{q}^2)^2}\,\tilde{f}'(\vec{q}).$$

It can be put in the form of square-integrability $L^2(\mathcal{Y}^3)$ by absorption of the square-integrable square root of the positive-type function:

$$L^1(\mathcal{Y}^3) \longrightarrow L^2(\mathcal{Y}^3), f(\vec{x}) \longmapsto \psi(\vec{x}) = \xi * f(\vec{x}),$$

$$\tilde{f}(\vec{q}) \longmapsto \tilde{\psi}(\vec{q}) = \frac{\sqrt{8\pi}}{1+\vec{q}^2}\tilde{f}(\vec{q}),$$

$$\text{with} \int \frac{d^3q}{\pi^2} \frac{1}{(1+\vec{q}^2)^2} e^{i\vec{q}\vec{x}} = e^{-r} = d(\vec{x}) = \xi * \xi(\vec{x}),$$

$$\int \frac{d^3q}{2\pi^2} \frac{1}{1+\vec{q}^2} e^{i\vec{q}\vec{x}} = \frac{e^{-r}}{r} = \sqrt{2\pi}\xi(\vec{x})$$

$$\Rightarrow \langle 0; f|0; f'\rangle = \int \frac{d^3q}{\pi^2} \overline{\tilde{f}(\vec{q})} \frac{1}{(1+\vec{q}^2)^2} \tilde{f}'(\vec{q}) = \int \frac{d^3q}{(2\pi)^3} \overline{\tilde{\psi}(\vec{q})}\, \tilde{\psi}'(\vec{q})$$

$$= \int d^3x\, \overline{\psi(\vec{x})}\, \psi'(\vec{x}).$$

Therefore, all infinite-dimensional Hilbert spaces for different continuous invariants $B^2 > 0$ can be transformed to subspaces of one Hilbert space $L^1(\mathcal{Y}^{2R-1}) \longrightarrow L^2(\mathcal{Y}^{2R-1}) \cong L^2(\mathbb{R}^{2R-1})$.

A scalar state $\{\vec{x} \longmapsto e^{-r}\} \in L^\infty(\mathcal{Y}^{2R-1})_+$ characterizes an infinite-dimensional Hilbert space with a faithful cyclic $\mathbf{SO}_0(1, 2R-1)$-representation as familiar for $R = 2$ from the principal series representations of the Lorentz group $\mathbf{SO}_0(1, 3)$. The positive-type function defines the Hilbert product for $L^1(\mathcal{Y}^{2R-1})$:

$$\text{distributive Hilbert basis: } \{|0; \vec{q}\rangle \mid \vec{q} \in \mathbb{R}^{2R-1}\},$$

$$\langle 0; \vec{q}'|0; \vec{q}\rangle = \frac{1}{(1+\vec{q}^2)^R} \frac{|\Omega^{2R-1}|}{2} \delta(\vec{q} - \vec{q}'),$$

$$\text{Hilbert vectors: } |0; f\rangle = {}^{\oplus}\!\!\int \frac{2d^{2R-1}q}{|\Omega^{2R-1}|} \tilde{f}(\vec{q})|0; \vec{q}\rangle,$$

$$\langle 0; f|0; f'\rangle = \int \frac{2d^{2R-1}q}{|\Omega^{2R-1}|} \overline{\tilde{f}(\vec{q})} \frac{1}{(1+\vec{q}^2)^R} \tilde{f}'(\vec{q}),$$

$$\text{cyclic vector: } |0; 1\rangle = {}^{\oplus}\!\!\int \frac{2d^{2R-1}q}{|\Omega^{2R-1}|} |0; \vec{q}\rangle$$

$$\text{with} \int \frac{4d^{2R-1}q\, d^{2R-1}q'}{|\Omega^{2R-1}|^2} \langle 0; \vec{q}'|e^{i\vec{q}\vec{x}}|0; \vec{q}\rangle = e^{-r}.$$

There is a representation of each abelian noncompact subgroup in the Cartan decomposition $\mathcal{Y}^{2R-1} \sim \mathbf{SO}_0(1, 1) \times \Omega^{2R-2}$ with the action on a distributive basis and hence on the Hilbert vectors:

$$\mathbf{SO}_0(1, 1) - \text{representations for all } \vec{\omega} \in \Omega^{2R-2} : e^{\vec{\omega}\vec{x}} \longmapsto e^{i|\vec{q}|\vec{\omega}\vec{x}} = e^{i\vec{q}\vec{x}} \in \mathbf{U}(1),$$

$$\text{action of all } \mathbf{SO}_0(1, 1) : |0; \vec{q}\rangle \longmapsto e^{i\vec{q}\vec{x}}|0; \vec{q}\rangle.$$

Quantum fields of hyperbolic position $\mathcal{Y}^3 = \mathbf{SO}_0(1,3)/\mathbf{SO}(3)$ parametrize the $\mathcal{Y}^3$-representation coefficients for the hyperbolic ground-state $|\mathcal{Y}\rangle$, starting with the scalar field:

$$
\begin{aligned}
J = 0:\ \mathbf{Y}(\vec{x}) &= {}^{\oplus}\!\!\int \frac{d^3q}{\pi^2}\, e^{i\vec{q}\vec{x}}\mathrm{u}(\vec{q}),\ \mathbf{Y}^\star(\vec{x}) = {}^{\oplus}\!\!\int \frac{d^3q}{\pi^2}\, e^{-i\vec{q}\vec{x}}\mathrm{u}^\star(\vec{q}),\\
\text{quantization:}[\mathrm{u}^\star(\vec{q}\,')\mathrm{u}(\vec{q})] &= \mathbf{1}\frac{\pi^2}{(1+\vec{q}^2)^2}\,\delta(\vec{q}-\vec{q}\,'),\\
\text{distributive Hilbert basis: } &\{\mathrm{u}(\vec{q})|\mathcal{Y}\rangle = |0;\vec{q}\rangle \,\big|\, \vec{q}\in\mathbb{R}^3\},\\
\langle 0;\vec{q}\,'|0;\vec{q}\rangle &= \langle\mathcal{Y}|\mathrm{u}^\star(\vec{q}\,')\mathrm{u}(\vec{q})|\mathcal{Y}\rangle = \langle\mathcal{Y}|\{\mathrm{u}^\star(\vec{q}\,'),\mathrm{u}(\vec{q})\}|\mathcal{Y}\rangle\\
&= \frac{\pi^2}{(1+\vec{q}^2)^2}\,\delta(\vec{q}-\vec{q}\,'),\\
\mathcal{Y}^3 \ni \vec{x}-\vec{y} &\longmapsto \langle\mathcal{Y}|\{\mathbf{Y}^\star(\vec{y}),\mathbf{Y}(\vec{x})\}|\mathcal{Y}\rangle = \int \frac{d^3q}{\pi^2}\,\frac{1}{(1+\vec{q}^2)^2}e^{i\vec{q}(\vec{x}-\vec{y})}.
\end{aligned}
$$

The scalar field determines, via the measure $\frac{d^3q}{(1+\vec{q}^2)^2}$, the hyperbolic normalization of the scalar product.

The non-scalar coefficients with the harmonic $\mathbf{SO}(4)$-polynomials $(p)^{2J}$ start with the four-dimensional vector representations where the components of the harmonic polynomial $(p)^1$ can be collected in a special unitary 2×2-matrix:

$$
(J,J) = (\tfrac{1}{2},\tfrac{1}{2}):\ \begin{cases} (p)^1 \cong \mathbf{1}_2 p_0 + i\vec{\sigma}\vec{p} = \begin{pmatrix} p_0+ip_3 & i(p_1-ip_2)\\ i(p_1+ip_2) & p_0-ip_3\end{pmatrix}\\ \quad = \frac{1}{1+\vec{q}^2}\begin{pmatrix} 1-\vec{q}^2+2iq_3 & 2i(q_1-iq_2)\\ 2i(q_1+iq_2) & 1-\vec{q}^2-2iq_3\end{pmatrix} \in \mathbf{SU}(2).\end{cases}
$$

With respect to the induction by $\mathbf{SU}(2)$-representations it can be decomposed by transmutators for $[\mathbf{SU}(2)\times\mathbf{SU}(2)]/\mathbf{SU}(2) \cong \mathbf{SO}(4)/\mathbf{SO}(3)$ in a product of the two spinor representations $e^{(\vec{\theta}\pm\vec{\varphi})i\frac{\vec{\sigma}}{2}}$ for $\vec{\theta}=0$:

$$
\begin{aligned}
(\tfrac{1}{2},\tfrac{1}{2}) &\ \underset{\mathbf{SO}(3)}{=}\ (\tfrac{1}{2},0)\otimes(0,\tfrac{1}{2})\\
&\ \cong\ (0)\oplus(1):
\end{aligned}
\ \begin{cases}
\frac{\vec{\varphi}}{|\vec{\varphi}|} = \frac{\vec{p}}{|\vec{p}|} = \frac{\vec{q}}{|\vec{q}|},\\
\cos|\vec{\varphi}| = p_0,\ \frac{\vec{\varphi}}{|\vec{\varphi}|}\sin|\vec{\varphi}| = \vec{p},\\
(\tfrac{1}{2},0)(p) = (0,\tfrac{1}{2})^\star(p) = e^{\vec{\varphi}i\frac{\vec{\sigma}}{2}}\\
\quad = \sqrt{\frac{1+p_0}{2}}(\mathbf{1}_2 + i\frac{\vec{\sigma}\vec{p}}{1+p_0}) \in \mathbf{SU}(2),\\
(p)^1 \cong (\tfrac{1}{2},\tfrac{1}{2})(p) = (\tfrac{1}{2},0)(p)\circ(0,\tfrac{1}{2})^\star(p)\\
= e^{2\vec{\varphi}i\frac{\vec{\sigma}}{2}} = \mathbf{1}_2 p_0 + i\vec{\sigma}\vec{p} = \frac{\mathbf{1}_2(1-\vec{q}^2)+2i\vec{\sigma}\vec{q}}{1+\vec{q}^2} \in \mathbf{SU}(2).
\end{cases}
$$

Similarily for the nine-dimensional representation as product of three-dimensional $\mathbf{SU}(2)$-representations where the components of the harmonic polynomial $(p)^2$ come in an orthogonal 3×3-matrix:

$$(1,1) \underset{\mathbf{SO}(3)}{=} (1,0) \otimes (0,1) \cong (0) \oplus (1) \oplus (2) \begin{cases} (1,0)(p) = (0,1)^\star(p) = \exp \vec{\varphi}\vec{\mathcal{L}} \\ \qquad \cong \delta_{ab} p_0 + \frac{p_a p_b}{1+p_0} - \epsilon_{abc} p_c \in \mathbf{SO}(3), \\ (1,1)(p) = (1,0)(p) \circ (0,1)^\star(p) = \exp 2\vec{\varphi}\vec{\mathbf{J}} \in \mathbf{SO}(3), \\ (1,1)(p) \cong \delta_{ab}(p_0^2 - \vec{p}^2) - 2\epsilon_{abc} p_0 p_c + 2 p_a p_b \\ \qquad = \frac{4\delta_{ab}}{3}(p_0^2 - \frac{1}{4}) - 2\epsilon_{abc} p_0 p_c + 2(p_a p_b - \frac{\delta_{ab}}{3}\vec{p}^2) \\ \cong (p)^2 \cong p_i p_j - \frac{1}{4}\delta_{ij} \\ \qquad = \left(\begin{array}{c|c} p_0^2 - \frac{1}{4} & p_0 p_a \\ \hline p_0 p_b & -\frac{\delta_{ab}}{3}(p_0^2 - \frac{1}{4}) + p_a p_b - \frac{\delta_{ab}}{3}\vec{p}^2 \end{array} \right). \end{cases}$$

In general, the decomposition of the $(1+2J)^2$-component harmonic $\mathbf{SO}(4)$-polynomial $(p)^{2J}$ involves $\mathbf{SU}(2)$-representations, all in special unitary $(1+2J) \times (1+2J)$-matrices:

$$\begin{aligned} (J,J) = (J,0) \otimes (0,J) &\overset{\mathbf{SO}(3)}{\cong} (0) \oplus (1) \oplus (2) \oplus \cdots \oplus (2J), \\ (J,0)(p) = (0,J)^\star(p) &= \bigvee^{2J} (\tfrac{1}{2}, 0)(p) = e^{\vec{\varphi}\mathcal{D}^J(\vec{\mathbf{I}})} \in \mathbf{SU}(1+2J), \\ (p)^{2J} \cong (J,J)(p) = (J,0)(p) \circ (0,J)^\star(p) &= e^{2\vec{\varphi}\mathcal{D}^J(\vec{\mathbf{I}})} \in \mathbf{SU}(1+2J). \end{aligned}$$

Illustrated by the *quantum spinor field of hyperbolic position* $\mathcal{Y}^3$:

$$\begin{aligned} J = \frac{1}{2}: \quad \mathbf{Y}^B_{\frac{1}{2}}(\vec{x}) &= {}^{\oplus}\!\!\int \frac{d^3q}{\pi^2} [e^{\vec{\varphi}(\vec{q}) i \frac{\vec{\sigma}}{2}}]^B_A \, e^{i\frac{\vec{q}}{2}\vec{x}} \mathrm{u}^A(\vec{q}), \quad A, B = 1, 2, \\ \mathbf{Y}^\star_{\frac{1}{2}A}(\vec{x}) &= {}^{\oplus}\!\!\int \frac{d^3q}{\pi^2} [e^{\vec{\varphi}(\vec{q}) i \frac{\vec{\sigma}}{2}}]^B_A \, e^{-i\frac{\vec{q}}{2}\vec{x}} \mathrm{u}^\star_B(\vec{q}), \\ \text{quantization:} [\mathrm{u}^\star_B(\vec{q}\,')\mathrm{u}^A(\vec{q})] &= \mathbf{1}\delta^A_B \frac{\pi^2}{(1+\vec{q}^2)^2} \delta(\vec{q} - \vec{q}\,'), \\ \text{distributive Hilbert basis:} \{\mathrm{u}^A(\vec{q})|\mathcal{Y}\rangle &= |\tfrac{1}{2}; \vec{q}, A\rangle \,\big|\, \vec{q} \in \mathbb{R}^3, \ A = 1, 2\}, \\ \langle \tfrac{1}{2}; \vec{q}\,', A' | \tfrac{1}{2}; \vec{q}, A\rangle &= \langle \mathcal{Y} | \mathrm{u}^\star_{A'}(\vec{q}\,')\mathrm{u}^A(\vec{q})|\mathcal{Y}\rangle = \langle \mathcal{Y} | \{\mathrm{u}^\star_{A'}(\vec{q}\,'), \mathrm{u}^A(\vec{q})\}|\mathcal{Y}\rangle \\ &= \delta^A_{A'} \frac{\pi^2}{(1+\vec{q}^2)^2} \delta(\vec{q} - \vec{q}\,'), \\ \mathcal{Y}^3 \ni \vec{x} - \vec{y} \longmapsto \langle \mathcal{Y} | \{\mathbf{Y}^\star_{\frac{1}{2}}(\vec{y}), \mathbf{Y}_{\frac{1}{2}}(\vec{x})\} | \mathcal{Y}\rangle &= \int \frac{d^3q}{\pi^2} \frac{(p)^1}{(1+\vec{q}^2)^2} e^{i\frac{\vec{q}}{2}(\vec{x}-\vec{y})}. \end{aligned}$$

the *quantum fields of hyperbolic position* $\mathcal{Y}^3$ *for general principal quantum number* $n = 1 + 2J$ use transmutations of $\mathbf{SU}(2)$-representations for J:

$$n = 1, 2, \ldots : \quad \mathbf{Y}^b_J(\vec{x}) = {}^{\oplus}\!\!\int \frac{d^3q}{\pi^2} \, (J, 0)^b_a(p) e^{i\frac{\vec{q}}{n}\vec{x}} \mathrm{u}^a(\vec{q}), \quad a, b = 0, 1, \ldots, 1 + 2J,$$

$$\mathbf{Y}^\star_{Ja}(\vec{x}) = {}^{\oplus}\!\!\int \frac{d^3q}{\pi^2} \, (J, 0)^b_a(p) e^{-i\frac{\vec{q}}{n}\vec{x}} \mathrm{u}^\star_b(\vec{q}),$$

$$\mathcal{Y}^3 \ni \vec{x} - \vec{y} \longmapsto \langle \mathcal{Y} | \{ \mathbf{Y}^\star_J(\vec{y}), \mathbf{Y}_J(\vec{x}) \} | \mathcal{Y} \rangle = \int \frac{d^3q}{\pi^2} \frac{(p)^{2J}}{(1 + \vec{q}^{\,2})^2} e^{i\frac{\vec{q}}{n}(\vec{x} - \vec{y})}.$$

Chapter 12
Operational Spacetime and the Particle Spectrum

The particle concept (with Wigner's definition) — sufficient for the asymptotics without interaction (flat spacetime approach) — is a spacetime projection on familiar pictures, therefore both useful and restricted. Today's quantum theory is a special relativistic (Minkowskian) quantum field theory. The concept of a field as used in the standard model of electroweak and strong interactions is more basic than that of a particle: All particles are embedded in fields, but not all fields are parametrizable completely by particle structures. This holds especially for mass zero fields as used for the electroweak and gravitational interactions, and for strong interaction parametrizing quarks and gluons, if confined. Apparently, not all interactions are harmonically analyzable by unitary representations of the Poincaré group and its Lorentz group classes, i.e., of Minkowski spacetime.

For relativistic scattering, the perturbative expansion of an interaction by particle embedding Feynman propagators with simple energy-momentum poles encounters "divergencies" which can be sedated, in "renormalizable" theories, by the regularization-renormalization procedure. The price to be paid is the principal indeterminacy of some basic physical properties like mass ratios and coupling constants. To parametrize a propagator, as used in Feynman diagrams for a perturbation expansion of field theory, only by "real" particles is incomplete, as seen, e.g., in the non-particle implemented Coulomb interaction, coming together with the two polarized photonic degrees of freedom in the four-component electromagnetic field. To completize "real" or "on-shell" particles, and to include interactions, the slippery concept of "virtual" or "off-shell" particles is introduced. The causally supported simple pole off-shell contributions cause the "divergencies."

With the requirement of local inertia systems, spacetimes as used in general relativity are postulated to be Riemannian manifolds with a causal signature $(1, 3)$-metric, determined by the energy-momentum tensor. Fields of tangent Minkowski spacetime are defined by particles; the Poincaré group describes, but does not specify the particle spectrum, i.e., their actual mass, spin (polarization) and electromagnetic charge. With the harmonic flat space analysis and a perturbation expansion,

H. Saller, *Operational Symmetries*, DOI 10.1007/978-3-319-58664-9_12

appropriate for scattering theory, there goes a decomposition of a dynamics (Lagrangian) into free particle contribution (kinetic terms, on-shell) and interaction (off-shell). To understand the particle spectrum, i.e., for non-scattering states, such a decomposition, related to classical equations of motion, does not make sense. With Dirac and Heisenberg: Particles cannot explain particles — in quantum theory, the concept of building bricks loses its sense. The bound state picture is of limited value in a quantum world. Spacetimes, in general, cannot be implemented and harmonically analyzed by particles and their Feynman propagators, i.e., by Hilbert spaces with representations of tangent flat Minkowski spacetime, characterized by the two invariants for translations (mass) and rotations (spin, polarization). It is proposed that, in some analogy to the identification of interaction and nonflat geometry (curvature) in general relativity, nonabelian motion and isotropy groups of spacetime — especially with nonflat position — parametrize the basically relevant operations whose representations determine interactions and particle properties. The invariants of the non-Minkowski spacetime representations can determine the particle properties; the tangent Lie algebra structure leads to Minkowski spacetime and free particles. The representation normalizations will be related to coupling constants like Sommerfeld's constant in the residue of the energy-momentum propagator $\frac{\alpha_{\text{Som}}}{q^2}$ with a massless pole for electromagnetism.

A similar situation exists for nonrelativistic position: There, Hilbert representations of flat Euclidean position $\mathbb{E}^3$ with invariant $\vec{\mathbf{p}}^2 = 2\mathbf{H}$ are not appropriate to explain the nonrelativistic atomic spectrum, i.e., the negative energy (bound) states cannot be analyzed by positive energy (scattering) states like $\frac{\sin |P|r}{r} = \int \frac{d^3q}{2\pi}\, \delta(\vec{q}^2 - P^2)e^{i\vec{q}\vec{x}} \in L^\infty(\mathbb{R}^3)_+$ (simple momentum poles). The negative energy (bound) state adequate hyperbolic Kepler position with invariant $\vec{\mathbf{p}}^2 - \frac{2}{\mathbf{r}} = 2\mathbf{H}$ and the related $\mathbf{SL}(2, \mathbb{C})$-representations with the characteristic dipole measure $e^{-|B|r} = \int \frac{d^3q}{\pi^2} \frac{|B|}{(\vec{q}^2+B^2)^2} e^{i\vec{q}\vec{x}} \in L^\infty(\mathcal{Y}^3)_+, \mathcal{Y}^3 \cong \mathbb{R}^3$, $(\vec{\partial}^2 + B^2)^2 e^{-|B|r} = 8\pi|B|\delta(\vec{x})$, for the cyclic hyperbolic ground state is embeddable in spacetime. Its extension by the noncompact abelian causal group $\mathbf{D}(1) \cong \mathbb{R}$ for dilations and (eigen)time translations and representations with energy poles like $\cos Et = \int dq\, \epsilon(q)q\delta(q^2 - E^2)e^{iqt} \in L^\infty(\mathbb{R})_+$, $(-d_t^2 - E^2)\cos Et = 0$, for the cyclic spherical ground state defines electroweak spacetime $\mathbb{D}(2) = \mathbf{D}(1) \times \mathcal{Y}^3$. As unitary relativity $\mathbb{D}(2) \cong \mathbf{GL}(2, \mathbb{C})/\mathbf{U}(2)$, it parametrizes the classes of internal hyper-isospin operations $\mathbf{U}(2)$ in the dilation- and phase-extended Lorentz group $\mathbf{GL}(2, \mathbb{C}) = \mathbf{D}(1) \times \mathbf{U}(1) \circ \mathbf{SL}(2, \mathbb{C})$ (hyper-Lorentz group), familiar as homogeneous action groups in the electroweak standard model. In a fibration, $\mathbf{GL}(2, \mathbb{C})$ can be considered as a coset bundle with noncompact electroweak spacetime $\mathbb{D}(2)$ as base manifold and compact hyperisospin $\mathbf{U}(2)$ as typical fiber. Its representations are proposed to determine the basic interactions and the particle spectrum.

The representations of hyperbolic Kepler position $\mathcal{Y}^3 \cong \mathbf{SL}(2, \mathbb{C})/\mathbf{SU}(2)$ display the noncompact-compact reciprocity $\frac{1}{B^2} = (1 + 2J)^2$ for the (binding) energies $2E = \vec{q}^2 = (iB)^2$ of the atoms, which determines "nonrelativistic kinetic energies with imaginary momenta," i.e., the hyperboloid curvature, related to the noncompact hyperbolic motion group $\mathbf{SL}(2, \mathbb{C}) \supset \mathbf{SO}_0(1, 1)$ with continuous invariants, by the principal quantum number $1 + 2J = 1, 2, \ldots$ ($\mathbf{SU}(2)$-multiplicities as Plancherel measure), related to the compact group $\mathbf{SU}(2) \supset \mathbf{SO}(2)$ with discrete invariants. The expression of the energy as time $\mathbf{D}(1)$-invariant by position representations, $2\mathbf{H} = \mathbf{p}^2 + 2V(\mathbf{x})$, as in nonrelativistic quantum mechanics, is no longer possible for relativistic spacetimes. Correspondingly, there is no naive takeover of the Schrödinger equation. For electroweak spacetime, the time $\mathbf{D}(1)$- and the position $\mathcal{Y}^3$-representation structures are "embedded" in $\mathbb{D}(2)$-representations and a noncompact-compact reciprocity where the two continuous invariants (m^2, M^2) for the maximal noncompact abelian motion group $\mathbf{D}(1) \times \mathbf{SO}_0(1,1) \subset \mathbf{GL}(2, \mathbb{C})$ come in the logarithm of their ratio $\int_{m^2}^{M^2} \frac{d\kappa^2}{\kappa^2} = \log \frac{M^2}{m^2}$.

In contrast to the inverse quadratically spaced energy levels in the atomic spectrum $2E = -\frac{m_R c^2}{(1+2J)^2} \in [-m_R c^2, 0[$ with the Rydberg energy $m_R c^2 \sim 27.2$ eV as unit, the range of experimentally relevant masses for particle physics is rather large, even for logarithmic ratios ($\log 10 \sim 2.3$) which are characteristic for even-dimensional spacetimes:

mass m_1	$\frac{m_1}{m_2}$	decimal	$\log \frac{m_1^2}{m_2^2}$
electron $m_e \sim 0.5$MeV	$\frac{m_e}{m_R} = \frac{1}{\alpha_{\text{Som}}^2}$	1.9×10^4	19.6
proton $m_p \sim 938$MeV	$\frac{m_p}{m_e}$	1.8×10^3	15
weak boson $m_W \sim 80.4$GeV	$\frac{m_W}{m_p}$	0.8×10^2	8.8
Planck mass $m_P \sim 1.2 \times 10^{19}$GeV	$\frac{m_P}{m_W}$	4.5×10^{17}	81

In analogy to the determination of the energy levels in the nonrelativistic atomic spectrum by the normalization of hyperbolic position representations, the basic mass ratios on the level of the elementary particles will be related to the normalization of electroweak spacetime representations. The external spin and internal isospin multiplicities are important ingredients for such a normalization. A first, also quantitative comparison of theory and experiments shows that such a procedure may be – at least – not completely hopeless.

12.1 Operational Spacetimes

Spacetime coordinates parametrize operations. Similar to operational positions with global motion and local isotropy groups, there are distinguished operational spacetimes. Minkowski spacetime as tangent vector space approximation for physical spacetimes is characterized by the Poincaré motion group with the abelian translations and the Lorentz group as local isotropy (equivalence) group, and implemented by free particles. It cannot explain the specific properties of interactions and particles.

The representations and invariants of a distinguished operational spacetime with a nonflat position are candidates for the determination of the particle spectrum.

Einstein's relativity and gravity in $(n = 4)$-dimensional spacetime involves a *dynamical theory of dilation representations*: In addition to maximally $\binom{n}{2} = 6$ independent compact rotation degrees of freedom, the $\binom{n+1}{2} = 10$ components of the symmetric spacetime metrical tensor $\mathbf{g}$ contain maximally $n = 4$ independent dilation degrees of freedom for the lengths (units) of the $n = 4$ principal axes of the metrical hyperboloid. This is visible in the form $\mathbf{g}_{ij} = \mathbf{e}_i^k \eta_{kl} \mathbf{e}_j^l$, diagonalized to an orthonormal $\mathbf{O}(t, s)$-form with $t + s = n$ by an n-bein (tetrad) $\mathbf{e}(x) = \mathbf{O}(x) \circ \mathbf{d}(x) \in \mathbf{GL}(n, \mathbb{R})$, involving the maximal compact subgroup from the Iwasawa and Cartan factorizations of the structure group for the bases, explicitly for $n = 1 + 3$:

$$\begin{aligned} \mathbf{GL}(n, \mathbb{R}) &= \mathbf{O}(n) \circ \mathbf{D}(1) \circ \mathbf{SO}_0(1, 1)^{n-1} \circ \exp \mathbb{R}^{\binom{n}{2}} \\ &= \mathbf{O}(n) \circ \mathbf{D}(1) \circ \mathbf{SO}_0(1, 1)^{n-1} \circ \mathbf{O}(n), \end{aligned}$$

$$\mathbf{g} = \mathbf{O} \circ \operatorname{diag} \mathbf{g} \circ \mathbf{O}^T = \mathbf{O} \circ \mathbf{d} \circ \eta \circ \mathbf{d}^T \circ \mathbf{O}^T = \mathbf{e} \circ \eta \circ \mathbf{e}^T,$$

$$\text{with } \mathbf{O}(x) \in \mathbf{O}(4), \quad \operatorname{diag} \mathbf{g} = \mathbf{d} \circ \eta \circ \mathbf{d}^T,$$

$$\eta = dt^2 - d\vec{x}^2, \ \mathbf{d}(x) \cong \begin{pmatrix} e^{\psi_0(x)} & 0 & 0 & 0 \\ 0 & e^{\psi_1(x)} & 0 & 0 \\ 0 & 0 & e^{\psi_2(x)} & 0 \\ 0 & 0 & 0 & e^{\psi_3(x)} \end{pmatrix} \in \mathbf{D}(1) \times \mathbf{SO}_0(1, 1)^3.$$

The n-bein (moving basis) parametrizes the manifold $\mathbf{GL}(n, \mathbb{R})/\mathbf{O}(t, s)$ for classes of tangent space bases with $n^2 - \binom{n}{2} = \binom{n+1}{2} = \binom{n}{2} + n = 10$ real dimensions. The metric and the tetrad which embed the potentials of non-relativistic theories are dilation representation functions (distributions). In Einstein's equations, their derivations $\partial\partial\mathbf{g} \sim \partial\partial\mathbf{e} \sim (\partial\partial\mathbf{O}, \ \partial\partial\mathbf{d})$, i.e., the representations of the generators of the metrical rotations and dilations in the spacetime curvature are determined by the energy-momentum tensor, which, in flat spacetime, gives the $n = 4$ translation currents. Translations can be considered to be dilation generators, $\mathbb{R} = \log \mathbf{D}(1)$.

A position motion group $\mathbf{SO}(3)$ for spherical symmetry leaves maximally two independent dilations $\begin{pmatrix} e^{\psi_0} & 0 \\ 0 & e^{|\vec{\psi}|}\mathbf{1}_3 \end{pmatrix}$ — for time and position. A famous example is the Schwarzschild metric with the Schwarzschild length $\ell = \frac{GM}{c^2}$ as sole unit for a rotation parabolic position with static dilation properties outside the horizon embedding the Newton potential $-\vec{\partial}^2 \frac{\ell}{r} = \ell 4\pi\delta(\vec{x})$ — in the Cartesian parametrization:

$$\mathbf{e}(x) = \mathbf{d}(r) \cong \begin{pmatrix} \frac{1-\frac{\ell}{r}}{1+\frac{\ell}{r}} & 0 \\ 0 & (1 + \frac{\ell}{r})^2 \mathbf{1}_3 \end{pmatrix} \in \mathbf{D}(1) \times \mathbf{SO}_0(1, 1) \text{ for } |\vec{x}| = r > \ell,$$

$$\begin{aligned} e^{\psi_0(r)} &= \frac{1-\frac{\ell}{r}}{1+\frac{\ell}{r}} = 1 - \frac{2\ell}{r} + \ldots, \quad \tanh \frac{\psi_0(r)}{2} = -\frac{\ell}{r}, \\ e^{|\vec{\psi}|(r)} &= (1 + \tfrac{\ell}{r})^2 = 1 + \frac{2\ell}{r} + \ldots, \quad \tanh \frac{|\vec{\psi}|(r)}{4} = \frac{\frac{\ell}{2r}}{1+\frac{\ell}{2r}}. \end{aligned}$$

Position ("space") is embedded in spacetime. *Friedman*[1] *universes* $\mathbb{M}$ display the time development of the three maximally symmetric (conformally Euclidean) operational positions $(\Omega^3, \mathbb{E}^3, \mathcal{Y}^3) \subset \mathbb{M}$ for the rotation group classes with definite metric $ds^2_{Q^2} = \frac{d\vec{x}^2}{(1-\frac{Q^2\vec{x}^2}{4})^2}$ and position invariants (units) $Q^2 = (-P^2, 0, B^2)$ in the local dilation factors. They are used for cosmological spacetime models with *Robertson–Walker*[2] *metrical tensors*,

$$\mathbf{g} = dt^2 - \rho^2(t) ds^2_{Q^2}, \quad \text{with } \mathbf{e}_{Q^2}(x) = \mathbf{d}_{Q^2}(t, r) \cong \begin{pmatrix} 1 & 0 \\ 0 & \frac{\rho(t)}{1-\frac{Q^2 r^2}{4}} \mathbf{1}_3 \end{pmatrix} \in \mathbf{D}(1),$$

which involve, as position expansion factor ("radius"), a time representation coefficient $t \longmapsto \rho(t)$ with a time invariant (unit). A trivial time representation $\rho = 1$ characterizes non-relativistic space-times $\mathbb{R} \times \mathbb{M}$ which have no time and position connecting transformations, e.g., non-relativistic Kepler space-time $\mathbb{R} \times \mathcal{Y}^3$ as represented by the non-relativistic hydrogen atom. In general, Friedman universes have two invariants — a time and a position unit.

The three *maximally symmetric spacetimes with* $(1, 3)$-*signature metric* parametrize classes of the noncompact Lorentz group (local symmetry or local equivalence classes). They have 10-dimensional motion groups (global symmetry) — the de Sitter group $\mathbf{SO}_0(1, 4)$, the Poincaré group $\mathbf{SO}_0(1, 3) \vec{\times} \mathbb{R}^4$, and the anti-de Sitter group $\mathbf{SO}_0(2, 3)$. Implementing the noncompact and compact subgroups $\mathbf{SO}_0(1, 1)$, $\mathbb{R}$, and $\mathbf{SO}(2)$, the "radius" grows or shrinks exponentially, stays constant, or oscillates, respectively, with time:

$$\begin{aligned} \text{de Sitter spacetime: } \rho(t) &\sim e^{\pm bt}, & \mathbf{SO}_0(1, 4)/\mathbf{SO}_0(1, 3) &\cong \mathcal{Y}^{(1,3)}, \\ \text{Minkowski spacetime: } \rho(t) &\sim 1, & \mathbf{SO}_0(1, 3) \vec{\times} \mathbb{R}^4/\mathbf{SO}_0(1, 3) &\cong \mathbb{R}^{(1,3)}, \\ \text{anti-de Sitter spacetime: } \rho(t) &\sim e^{\pm ipt}, & \mathbf{SO}_0(2, 3)/\mathbf{SO}_0(1, 3) &\cong \mathcal{Y}^{(3,1)}. \end{aligned}$$

Both the de Sitter and anti-de Sitter spacetime are a one-shell hyperboloid, spacelike $\mathcal{Y}^{(1,3)}$ and timelike $\mathcal{Y}^{(3,1)}$. The conformally Minkowskian metrical tensors can be parametrized by a flat metric with one overall dilation factor and one spacetime invariant (unit):

$$\begin{aligned} \mathbf{g} &= \left(\frac{dx^2}{(1+b^2x^2)^2}, \; dx^2, \; \frac{dx^2}{(1-p^2x^2)^2} \right) \cong \frac{dx^2}{(1-kx^2)^2}, \; k = (-1, 0, 1), \\ &\text{with } x^2 = x_0^2 - \vec{x}^2, \; dx^2 = dx_0^2 - d\vec{x}^2. \end{aligned}$$

[1] Alexander Friedman (1888–1925).

[2] Howard Percy Robertson (1903–1961), Arthur Geoffrey Walker (1909–2001).

The three maximally symmetric spacetimes – here for signature $(t, s) = (1, 3)$ – are related to each other by Inönü–Wigner contraction (expansion) with the representation invariants, a compact-noncompact pair $ip \leftrightarrow b$, as the contraction parameter:

$$\begin{array}{ccccc} \frac{\sin pt}{p} & \longrightarrow & t & \longleftarrow & \frac{\sinh bt}{b} \\ & p\to 0 & & 0\leftarrow b & \\ \mathbf{SO}_0(t+1, s) & \longrightarrow & \mathbf{SO}_0(t, s) \vec{\times} \mathbb{R}^{t+s} & \longleftarrow & \mathbf{SO}_0(t, s+1). \end{array}$$

In a complex representation, the two nonflat positions which parametrize $\mathbf{SU}(2)$-classes, i.e., compact spherical Einstein position $\Omega^3 \cong [\mathbf{SU}(2) \times \mathbf{SU}(2)]/\mathbf{SU}(2) \cong \exp(i\mathbb{R})^3$, and noncompact hyperbolic Kepler position $\mathcal{Y}^3 \cong \mathbf{SL}(2, \mathbb{C})/\mathbf{SU}(2) \cong \exp \mathbb{R}^3$, can be supplemented by a corresponding pair with compact time $\mathbf{U}(1) = \exp i\mathbb{R}$ (phase group) and noncompact time $\mathbf{D}(1) = \exp \mathbb{R}$ (dilation group):

$$\begin{array}{l} \mathbb{U}(2) \cong \mathbf{U}(1) \times \Omega^3 \cong [\mathbf{U}(2) \times \mathbf{U}(2)]/\mathbf{U}(2) \cong \exp(i\mathbb{R})^4, \\ \mathbb{D}(2) \cong \mathbf{D}(1) \times \mathcal{Y}^3 \cong \quad \mathbf{GL}(2, \mathbb{C})/\mathbf{U}(2) \qquad \cong \exp \mathbb{R}^4. \end{array}$$

As real forms of $\mathbf{GL}(2, \mathbb{C})$, taken as complex four-dimensional Lie group, the global motion groups $\mathbf{U}(2) \times \mathbf{U}(2) \cong \exp[(i\mathbb{R})^4 \oplus (i\mathbb{R})^4]$ and $\mathbf{GL}(2, \mathbb{C}) \cong \exp[(i\mathbb{R})^4 \oplus \mathbb{R}^4]$, both real eight-dimensional, constitute a compact-noncompact pair. The local compact fixgroup $\mathbf{U}(2) \cong \exp(i\mathbb{R})^4$ is real four-dimensional. It is the maximal compact group in the Iwasawa decomposition of the full linear group $\mathbf{GL}(2, \mathbb{C}) = \mathbf{U}(2) \circ \mathbf{D}(1) \circ \mathbf{SO}_0(1, 1) \circ \exp \mathbb{R}^2$.

Only the noncompact manifold $\mathbb{D}(2) \cong \mathbf{GL}(2, \mathbb{C})/\mathbf{U}(2)$ with a hyperbolic Kepler position submanifold comes with the general relativistic nontrivial $(1, 3)$-causal structure, compatible with the motion group. It is interpreted as the classes of the *hyper-isospin group in the hyper-Lorentz group* and called *electroweak spacetime*. Electroweak spacetime is not maximally symmetric. It can be parametrized by the strictly future (positive) subset of Minkowski spacetime which also supports the interaction contributions ("virtual particles") of Feynman propagators. The definition of electroweak spacetime does not involve the real eight-dimensional color group $\mathbf{SU}(3)$ as a constitutive operation group.

12.2 Special Spacetime Distributions

Scalar Feynman distributions with one invariant for possibly indefinite orthogonal groups, implemented by singularities $q^2 = k = 0, \pm 1$ (with units $q^2 = km^2$) in the complex q^2-plane, give complex argument Macdonald functions $\mathcal{K}$ — where the involved Gamma and Beta functions are defined:

$$\mathbf{O}(t,s): \quad n=t+s=1,2,3,\ldots,\quad x^2=\vec{x}_t^2-\vec{x}_s^2,\quad q^2=\vec{q}_t^2-\vec{q}_s^2,\quad \nu\in\mathbb{R},\quad N=0,1,2,\ldots,\quad a\in\mathbb{R},$$

$$\begin{cases}
\int \frac{d^nq}{i^t\pi^{\frac{n}{2}}}\frac{\Gamma(\frac{n}{2}-\nu)}{(-q^2-io)^{\frac{n}{2}-\nu}}e^{iqx}=\frac{\Gamma(\nu)}{(\frac{-x^2+io}{4})^{\nu}},\\
\int \frac{d^nq}{i^t\pi^{\frac{n}{2}}}\frac{\Gamma(\frac{n}{2}-\nu)}{(-q^2-io+1)^{\frac{n}{2}-\nu}}e^{iqx}=\frac{2\mathcal{K}_\nu(\sqrt{-x^2+io})}{(\frac{-x^2+io}{4})^{\frac{\nu}{2}}}\\
\qquad\qquad -\delta_\nu^{N+1}i\pi\sum_{k=0}^{N}\frac{1}{(N-k)!}\delta^{(k)}(-\frac{x^2}{4}),\\
\int \frac{d^nq}{i^s\pi^{\frac{n}{2}}}\frac{\Gamma(\frac{n}{2}-\nu)}{(q^2-io+1)^{\frac{n}{2}-\nu}}e^{iqx}=\frac{2\mathcal{K}_\nu(\sqrt{x^2+io})}{(\frac{x^2+io}{4})^{\frac{\nu}{2}}}\\
\qquad\qquad -\delta_\nu^{N+1}i\pi\sum_{k=0}^{N}\frac{1}{(N-k)!}\delta^{(k)}(\frac{x^2}{4}),\\
\nu>0: \int \frac{d^nq}{i^t\pi^{\frac{n}{2}}}\frac{\Gamma(\frac{n}{2}+\nu)}{(-q^2-io+a)^{\frac{n}{2}+\nu}}=\int \frac{d^nq}{i^s\pi^{\frac{n}{2}}}\frac{\Gamma(\frac{n}{2}+\nu)}{(q^2-io+a)^{\frac{n}{2}+\nu}}=\frac{\Gamma(\nu)}{(a-io)^{\nu}}.
\end{cases}$$

In addition, there occur, for nontrivial signature $st>0$ and integers $\nu=N+1=1,2,\ldots$, "lightcone"-supported ($x^2=0$) Dirac distributions. The corresponding "anti-Feynman" distributions are obtained by canonical conjugation.

In the case of simple poles, especially for the on-shell contributions $\delta(q^2-km^2)$, e.g., of $\mathbf{O}(1,3)$-Feynman propagators, an interpretation of the Fourier transformation with time and position translations and energies and momenta makes sense. Simple poles $\frac{n}{2}-\nu=1$ are for abelian groups; nonabelian groups use higher order poles $\frac{n}{2}-\nu=2,3,\ldots$ The invariants at higher order singularities $q^2=km^2$ characterize non-translation representations (therefore in the following often the quotation marks in "energies" or "momenta").

The real and imaginary arguments go with "time- and spacelike"-supported contributions, i.e., for opposite definiteness:

$$|x|=\sqrt{|x^2|}: \quad \sqrt{\mp x^2+io}=\vartheta(\pm x^2)i|x|+\vartheta(\mp x^2)|x|.$$

The related exponentials give Macdonald and Hankel (with Neumann $\mathcal{N}$ and Bessel $\mathcal{J}$) functions, which go with exponential, cosine and sine:

$$\begin{aligned}
e^{-\sqrt{\mp x^2+io}}&=\vartheta(\pm x^2)e^{-i|x|}+\vartheta(\mp x^2)e^{-|x|},\\
\text{with } e^{-i|x|}&=\cos|x|-i\sin|x|,\\
\frac{2\mathcal{K}_\nu(\sqrt{\mp x^2+io})}{(\frac{\mp x^2+io}{4})^{\frac{\nu}{2}}}&=\vartheta(\pm x^2)\frac{2\mathcal{K}_\nu(i|x|)}{(i|\frac{x}{2}|)^{\nu}}+\vartheta(\mp x^2)\frac{2\mathcal{K}_\nu(|x|)}{|\frac{x}{2}|^{\nu}},\\
\text{with } 2\mathcal{K}_\nu(i|x|)&=-\pi\mathcal{N}_\nu(|x|)-i\pi\mathcal{J}_\nu(|x|).
\end{aligned}$$

Here, the decomposition into "time- and spacelike" contribution is different from the real-imaginary decomposition.

For scalar integrals, 2-sphere spreads $\pm\frac{\partial}{\partial\frac{x^2}{4\pi}}$ produce, in steps of two for the dimensions and in steps of one for the orthogonal Lie algebra ranks, higher-dimensional integrals, with the somewhat formal Fourier integral-relations:

$$\begin{aligned}
&\mathbf{O}(t,s) \to \mathbf{O}(t-2R_t, s-2R_s) : \ 2R_t < t, \ 2R_s < s, \\
&R_t + R_s = 0, 1, \dots : \int d^n q \, \mu(q^2) e^{iqx} = (-1)^{R_s} \Big(\frac{\partial}{\partial \frac{x^2}{4\pi}}\Big)^{R_t+R_s} \int d^{n-2(R_t+R_s)} q \, \mu(q^2) e^{iqx},
\end{aligned}$$

for example, scalar Euclidean integrals in odd dimensions and scalar Minkowskian integrals in even dimensions:

$$\begin{aligned}
\mathbf{O}(2R_s+1) \to \mathbf{O}(1) : \quad & \int \frac{d^{2R_s+1}q}{\pi^{R_s+1}} \, \mu(\vec{q}^2) e^{i\vec{q}\vec{x}} = (-\frac{\partial}{\partial \frac{\vec{x}^2}{4}})^{R_s} \int \frac{dq}{\pi} \, \mu(q^2) e^{iq|\vec{x}|}, \\
\mathbf{O}(1, 2R-1) \to \mathbf{O}(1,1) : \quad & \int \frac{d^{2R}q}{\pi^R} \, \mu(q^2) e^{iqx} = (-\frac{\partial}{\partial \frac{x^2}{4}})^{R-1} \int \frac{d^2 q}{\pi} \, \mu(q^2) e^{i(q_0 x_0 - q_3 |\vec{x}|)}.
\end{aligned}$$

Here and in the following for spaces with odd dimension $2R_s + 1$: The orthogonal Lie algebra rank is $R_s = R - 1$ with $\mathbf{O}(2R_s + 1) = \mathbf{O}(2R - 1)$.

Nontrivial $\mathbf{O}(t, s)$-properties with $L = 0, 1, 2, \dots$ are effected by harmonic "(energy-)momentum" polynomials and derivatives,

$$\begin{aligned}
&(q)^L, \ (\tfrac{\partial}{\partial q})^L = (2q)^L (\tfrac{\partial}{\partial q^2})^L, \\
&\tfrac{\partial}{\partial q} \otimes q = \mathbf{1}_n + q \otimes q \, 2\tfrac{\partial}{\partial q^2}, \\
&\tfrac{\partial}{\partial q} \otimes \tfrac{\partial}{\partial q} = (\mathbf{1}_n + q \otimes q \, 2\tfrac{\partial}{\partial q^2}) 2\tfrac{\partial}{\partial q^2}, \ \dots.
\end{aligned}$$

The derivatives, acting on multipoles, raise the pole order, $k = 0, \pm 1$,

$$\begin{aligned}
&x \int d^n q \, \mu(q) e^{iqx} \simeq \int d^n q \, i \tfrac{\partial \mu(q)}{\partial q} e^{iqx} \text{ (if defined)}, \\
&\tfrac{\partial}{\partial q} \tfrac{\Gamma(N)}{(-q^2+k)^N} = \tfrac{2q \, \Gamma(N+1)}{(-q^2+k)^{N+1}}, \text{ etc.}, \quad \tfrac{(2q)^L \, \Gamma(N+L)}{(-q^2+k)^{N+L}}.
\end{aligned}$$

12.3 Maximally Symmetric Spacetimes

The three maximally symmetric signature (t, s)-manifolds with dimension $n = t + s = 1, 2, \dots$ are conformally flat. They are characterized by one spacetime invariant (unit) and have, for $n \geq 2$, constant curvature — two general hyperboloids and a flat space:

$$\left.\begin{aligned}
\mathcal{Y}^{(t,s)} &\cong \mathbf{SO}_0(t, 1+s)/\mathbf{SO}_0(t,s), \quad k = -1, \\
\mathbb{R}^{(t,s)} &\cong \mathbf{SO}_0(t,s) \,\vec{\times}\, \mathbb{R}^n/\mathbf{SO}_0(t,s), \ k = 0, \\
\mathcal{Y}^{(s,t)} &\cong \mathbf{SO}_0(1+t, s)/\mathbf{SO}_0(t,s), \quad k = 1,
\end{aligned}\right\} \quad \begin{aligned} \mathbf{g} &= \tfrac{dx^2}{(1+km^2x^2)^2}, \\ x^2 &= \vec{x}_t^2 - \vec{x}_s^2, \end{aligned}$$

with time $\mathbb{R}^{(1,0)}$, the Euclidean spaces $\mathbb{E}^s = \mathbb{R}^{(0,s)}$, Minkowski spacetime $\mathbb{R}^{(1,3)}$, the spheres $\Omega^s = \mathcal{Y}^{(0,s)}$, the causal hyperboloids $\mathcal{Y}^s = \mathcal{Y}^{(s,0)}$, and the (anti-)de Sitter universes $\mathcal{Y}^{(1,3)}$ and $\mathcal{Y}^{(3,1)}$.

The tangent translation groups of the maximally symmetric spacetimes arise by contractions of the nonflat spacetime motion groups, e.g., the position translations from the boosts $\mathbf{SO}_0(1,3)/\mathbf{SO}(3) \to \mathbb{R}^3$. Vice versa, the translations of flat spacetimes are expanded into operations of a simple group as exemplified by the (anti-)de Sitter spacetimes.

The relevant representation coefficients for the flat spaces (translations) use Dirac (energy-)momentum distributions which are related to general Feynman propagators with simple poles, characteristic for abelian groups:

$$\mathbb{R}^{(t,s)}: \quad \int d^n q\, \delta(-q^2+km^2)e^{iqx} \text{ from } \int \frac{d^n q}{i\pi} \frac{1}{-q^2-io+km^2} e^{iqx},$$

e.g., for time $\delta(-q_0^2+E^2)$ with energies, and for Euclidean position $\delta(\vec{q}^2-P^2)$ with momenta. All unitary Poincaré group $\mathbf{SL}(2,\mathbb{C}) \vec{\times} \mathbb{R}^4$-representations with nontrivial translations are induced by unitary representations of the direct product subgroups $\mathbf{SU}(2)\times\mathbb{R}^4$, $\mathbf{SO}(2)\times\mathbb{R}^4$ and $\mathbf{SU}(1,1)\times\mathbb{R}^4$ for $k=1,0,-1$, respectively. Only causal spacetime invariants $k=1,0$, i.e., $km^2\geq 0$, are used as masses for, respectively, massive and massless particles (more below).

For odd dimensions $n=1+2R=1,3,\ldots$, e.g., for one-dimensional time and three-dimensional positions, the relevant scalar coefficients for $\mathcal{Y}^{(t,s)}$ and $\mathcal{Y}^{(s,t)}$ for cyclic (ground) states and — for $n\geq 3$ — interactions involve half-integer index functions. The corresponding functions and distributions have one invariant $m^2>0$ (or analogously with $m^2<0$) in the exponentials — as a spherical or a hyperbolic invariant for "time- and spacelike" support, respectively:

$$\begin{aligned}
\int \frac{d^{1+2R}q}{i^t\pi^{1+R}} \frac{\Gamma(1+R)}{(-q^2-io+m^2)^{1+R}} e^{iqx} &= \frac{e^{-\sqrt{-m^2x^2+io}}}{|m|} = \vartheta(x^2)\frac{e^{-i|mx|}}{|m|} + \vartheta(-x^2)\frac{e^{-|mx|}}{|m|},\\
\int \frac{d^{1+2R}q}{i^t\pi^{1+R}} \frac{\Gamma(1+R)}{(-q^2-io+m^2)^{1+R}} &= \frac{1}{|m|},\\
\int \frac{d^{1+2R}q}{i^t\pi^{1+R}} \frac{\Gamma(R)}{(-q^2-io+m^2)^{R}} e^{iqx} &= \frac{e^{-\sqrt{-m^2x^2+io}}}{\sqrt{\frac{-x^2+io}{4}}} = \vartheta(x^2)\frac{e^{-i|mx|}}{i\frac{|x|}{2}} + \vartheta(-x^2)\frac{e^{-|mx|}}{\frac{|x|}{2}}.
\end{aligned}$$

The relevant scalar integer index functions for even dimensions $n=2R=2,4,\ldots$, e.g., for (anti-)de Sitter spacetimes $\mathcal{Y}^{(1,3)}$ and $\mathcal{Y}^{(3,1)}$, integrate – with a positive measure – the exponentials for odd dimensions over an invariant line, i.e., $\frac{e^{-\iota|\kappa x|}}{|\kappa|}$ for states and $\frac{e^{-\iota|\kappa x|}}{\frac{|x|}{2}}$ for interactions with compactness $\iota=(1,i)$:

$$\begin{aligned}
\int \frac{d^{2R}q}{i^t\pi^{R}} \frac{\Gamma(1+R)}{(-q^2-io+m^2)^{1+R}} e^{iqx} &= \int_{m^2}^{\infty} \frac{d\kappa^2}{2\sqrt{(\kappa^2-m^2)^3}} \frac{e^{-\sqrt{-\kappa^2x^2+io}}}{|\kappa|},\\
\int \frac{d^{2R}q}{i^t\pi^{R}} \frac{\Gamma(1+R)}{(-q^2-io+m^2)^{1+R}} &= \frac{1}{m^2},\\
\int \frac{d^{2R}q}{i^t\pi^{R}} \frac{\Gamma(R)}{(-q^2-io+m^2)^{R}} e^{iqx} &= \int_{m^2}^{\infty} \frac{d\kappa^2}{2\sqrt{(\kappa^2-m^2)^3}} \frac{e^{-\sqrt{-\kappa^2x^2+io}}}{\sqrt{\frac{-x^2+io}{4}}}.
\end{aligned}$$

The cyclic (ground) state values for $x=0$ exhibit the representation invariant at the singularity $q^2=m^2$.

The *residual normalization* takes into account the pairwise rank structure for orthogonal operations (more ahead). For odd-dimensional spaces, the residual integration $\frac{d^{1+2R}q}{\pi^{1+R}}$ is used for a "point-residue," and $\frac{d^{2R}q}{\pi^R}$ for even-dimensional spaces, leaving a dilation κ^2-integration for a "line-residue".

There are the Schrödinger-like equations for translation representation coefficients:

$$\mathbf{O}(t, n-t): \ \partial^2 \int d^n q\, \delta(-q^2+m^2)e^{iqx} = -m^2 \int d^n q\, \delta(-q^2+m^2)e^{iqx},$$

and, for the representation functions of odd-dimensional spaces, $1+2R \geq 3$:

$$\mathbf{O}(t, 1+2R-t): \ (\partial^2+m^2) \int \frac{d^{1+2R}q}{i^t \pi^{1+R}} \frac{\Gamma(1+R)}{(-q^2-io+m^2)^{1+R}} e^{iqx} = \int \frac{d^{1+2R}q}{i^t \pi^{1+R}} \frac{\Gamma(1+R)}{(-q^2-io+m^2)^R} e^{iqx},$$

$$\left(\partial^2 - R\frac{\partial}{\partial\frac{x^2}{4}}\right)e^{-\sqrt{-m^2x^2+io}} = \left(\partial^2 - \frac{2Rm^2}{\sqrt{-m^2x^2+io}}\right)e^{-\sqrt{-m^2x^2+io}} = -m^2 e^{-\sqrt{-m^2x^2+io}},$$

where the $\mathbf{O}(t, s)$-invariant "non-flat" derivatives involve a hyperbolic or a spherical Yukawa potential:

$$\begin{aligned}\mathbf{O}(0, 1+2R): \ &(-\vec{\partial}^2 - \tfrac{2R\iota|m|}{r})e^{-\iota|m|r} = -m^2 e^{-\iota|m|r},\\ &\text{with } \sqrt{m^2} = \iota|m|, \ \iota = (1, \pm i), \ \ \iota^2 = k = (1, -1).\end{aligned}$$

The corresponding Schrödinger-like equations for even-dimensional spaces, $2R \geq 2$, hold for the measure-integrated exponentials:

$$\mathbf{O}(t, 2R-t): \ \ (\partial^2+m^2) \int \frac{d^{2R}q}{i^t \pi^R} \frac{\Gamma(1+R)}{(-q^2-io+m^2)^{1+R}} e^{iqx} = \int \frac{d^{2R}q}{i^t \pi^R} \frac{\Gamma(1+R)}{(-q^2-io+m^2)^R} e^{iqx},$$

$$(\partial^2 - R\frac{\partial}{\partial\frac{x^2}{4}}) \int_{|m|}^{\infty} \frac{d\kappa}{\sqrt{(\kappa^2-m^2)^3}}\, e^{-\sqrt{-\kappa^2x^2+io}} = -m^2 \int_{|m|}^{\infty} \frac{d\kappa}{\sqrt{(\kappa^2-m^2)^3}}\, e^{-\sqrt{-\kappa^2x^2+io}}.$$

12.4 Minkowski Spacetime

Canonical quantum field theory works with particle fields. They represent the Poincaré group $\mathbf{SO}_0(1,3) \vec{\times} \mathbb{R}^4$ as motion group of Minkowski spacetime and are also used for an expansion of interactions.

The *simple energy-momentum pole* Feynman propagators of a neutral scalar field $\mathbf{\Phi}$, or of a Dirac field $\mathbf{\Psi}$ for idealized[3] particles with spin $\frac{1}{2}$ (Dirac matrices γ_j), and isospin $I \in \{0, \frac{1}{2}, 1, \ldots\}$ (unit matrix $\mathbf{1}_{1+2I}$), and normalization factors $\rho(m^2) > 0$,

$$\begin{aligned}\langle 0|\mathbf{\Phi}(y)\mathbf{\Phi}(x)|0\rangle_{\text{Feynman}} &= \rho(m^2) \int \frac{d^4q}{i\pi(2\pi)^3} \frac{1}{-q^2-io+m^2} e^{iq(x-y)},\\ \langle 0|\mathbf{\Psi}(y)\overline{\mathbf{\Psi}}(x)|0\rangle_{\text{Feynman}} &= \rho(m^2) \int \frac{d^4q}{i\pi(2\pi)^3} \frac{(\gamma q+\mathbf{1}_4 m)\times\mathbf{1}_{1+2I}}{-q^2-io+m^2} e^{iq(x-y)},\end{aligned}$$

[3]So far, no mass-degenerate nontrivial isospin multiplet particles have been found — in contrast to a mass-degeneracy for spin.

decompose into free particle and interaction contribution as given by the *io*-distribution – by no means only a mathematical sublety. The real *on-shell* $q^2 = m^2$ *contribution (particles)* from the positive Dirac distribution

$$\frac{1}{i\pi}\frac{1}{-q^2-io+m^2} = \delta(q^2 - m^2) + \frac{1}{i\pi}\frac{1}{-q_{\rm P}^2+m^2},$$

is the expectation value $\langle 0|\ldots|0\rangle$ of the anticommutator for the Bose field $\mathbf{\Phi}$ and of the commutator for the Fermi field $\mathbf{\Psi}$ in the Fock ground-state vector $|0\rangle$ for translation coefficients. It is an *induced representation matrix element of the Poincaré group*, for a scalar field representing nontrivially the translations with the momenta from a hyperboloid $\mathbf{SO}_0(1,3)/\mathbf{SO}(3)$:

$$\begin{aligned}\langle 0|\{\mathbf{\Phi}(y), \mathbf{\Phi}(x)\}|0\rangle &= \rho(m^2)\int \frac{d^4q}{(2\pi)^3}\, 2\delta(q^2 - m^2)e^{iq(x-y)},\\ \mathbb{R}^{(1,3)} \ni x \longmapsto \int \frac{d^4q}{\pi}\, \delta(q^2 - m^2)e^{iqx} &= m^2\frac{\vartheta(x^2)\pi\mathcal{N}_1(|mx|)+\vartheta(-x^2)2\mathcal{K}_1(|mx|)}{|\frac{mx}{2}|}.\end{aligned}$$

The imaginary *off-shell contribution* with the principal value simple pole is not a coefficient of a Poincaré group representation, therefore the name "virtual particles." It is a distribution of an $\mathbf{O}(1,1)$-coefficient, used for $\mathbf{O}(1,3)$-interactions and only causally supported, not by all translations:

$$\begin{aligned}-\epsilon(x_0 - y_0)\langle 0|[\mathbf{\Phi}(y), \mathbf{\Phi}(x)]|0\rangle &= \rho(m^2)\int \frac{d^4q}{i\pi(2\pi)^3}\frac{2}{-q_{\rm P}^2+m^2}e^{iq(x-y)}\\ &= 0 \text{ for } (x-y)^2 < 0,\\ \mathbb{R}^4_+ \cup \mathbb{R}^4_- \ni x \longmapsto \int \frac{d^4q}{i\pi^2}\frac{1}{-q_{\rm P}^2+m^2}e^{iqx} &= -\epsilon(x_0)\int \frac{d^4q}{\pi}\,\epsilon(q_0)\delta(q^2 - m^2)e^{iqx}\\ &= i\big[m^2\vartheta(x^2)\frac{\pi\mathcal{J}_1(|mx|)}{|\frac{mx}{2}|} - \delta(\tfrac{x^2}{4\pi})\big].\end{aligned}$$

It arises by causal ordering, via $\epsilon(x_0 - y_0)$, from the on-shell quantization of the free scalar field via the commutator:

$$\begin{aligned}[\mathbf{\Phi}(y), \mathbf{\Phi}(x)] &= \mathbf{1}\rho(m^2)\int \frac{d^4q}{(2\pi)^3}\, 2\epsilon(q_0)\delta(q^2 - m^2)e^{iq(x-y)},\\ [i\partial^a\mathbf{\Phi}(y), \mathbf{\Phi}(x)]\Big|_{x_0=y_0} &= \mathbf{1}\rho(m^2)\int \frac{d^4q}{(2\pi)^3}\, 2\epsilon(q_0)q^a\delta(q^2 - m^2)e^{-i\vec{q}(\vec{x}-\vec{y})}\\ &= \mathbf{1}\rho(m^2)\, 2\delta^a_0\delta(\vec{x} - \vec{y}),\end{aligned}$$

i.e., the off-shell feature is caused by the $\epsilon(x_0 - y_0)$-multiplication. The *causally supported quantization* characterizes locality and microcausality. For free fields, the factor $\rho(m^2) > 0$, i.e., the residual normalization at the pole $q^2 = m^2$, can be chosen arbitrarily, e.g., $\rho(m^2) = 1$ or $\rho(m^2) = (2\pi)^3$.

A perturbative Feynman propagator expansion of interactions needs the off-shell propagators: For example, in a Feynman graph for the Compton scattering of an electron e and a photon γ, described by the gauge vertex $\overline{\mathbf{\Psi}}(x)\gamma_j\mathbf{\Psi}(x)\mathbf{A}^j(x)$, the intermediate electron e^* cannot be on the mass shell, as seen in the electron's rest system:

$$e\gamma \to e^* \to e\gamma : \left\{ \begin{array}{l} q_e = (m, 0),\ q_\gamma = (p, p),\ p \neq 0, \\ \Rightarrow q_{e^*} = q_e + q_\gamma = (m + p, p), \\ \quad q_{e^*}^2 = m^2 + 2pm \neq m^2. \end{array} \right.$$

Equally, an electron cannot radiate a photon. The virtual electron comes with the principal value part of the electron propagator. The naive off-shell interaction supplements of the on-shell particles are remarkably successful – up to the divergencies.

Perturbation theory calculations for relativistic scattering constitute a whole industry and yield, for quantum electrodynamics, one of the most accurate agreements in physics between theory and experiment. With the Lorentz invariant scattering amplitude $\langle f|T|i\rangle$ from an initial to a final state vector one obtains the transition amplitude per unit time $\Gamma_{fi} = 2\pi\delta(E_f - E_i)|\langle f|T|i\rangle|^2$ by using the expansion with the interaction Hamiltonian $\langle f|T|i\rangle = \langle f|H_{int}|i\rangle + \sum_n \frac{\langle f|H_{int}|n\rangle\langle n|H_{int}|i\rangle}{E_i - E_n} + \ldots$.

For a particle field propagator, the time projection, effected by position integration, gives as real part from the on-shell (free particle) contribution a harmonic oscillator cosine, i.e., a time representation coefficient with the mass as invariant. The position projection, effected by time integration, is nontrivial for the off-shell (interaction) contribution (imaginary part) only. It gives, as position distribution, a Yukawa potential with the inverse mass as its range:

$$\begin{aligned} \int d^3x \int \frac{d^4q}{i\pi(2\pi)^3} \frac{1}{-q^2 - io + m^2} e^{iqx} &= \int \frac{dq_0}{i\pi} \frac{1}{-q_0^2 - io + m^2} e^{iq_0x_0} \\ &= \frac{\cos mx_0 - \epsilon(x_0) i \sin |m| x_0}{|m|} = \frac{e^{-i|mx_0|}}{|m|}, \\ \int d^3x \int \frac{d^4q}{(2\pi)^3} \delta(-q^2 + m^2) e^{iqx} &= \int dq_0\, \delta(-q_0^2 + m^2) e^{iq_0x_0} = \frac{\cos mx_0}{|m|}, \\ \int dx_0 \int \frac{d^4q}{i\pi(2\pi)^3} \frac{1}{-q^2 - io + m^2} e^{iqx} &= \int dx_0 \int \frac{d^4q}{i\pi(2\pi)^3} \frac{1}{-q_{\mathrm{P}}^2 + m^2} e^{iqx} \\ &= \int \frac{d^3q}{4i\pi^3} \frac{1}{\vec{q}^2 + m^2} e^{-i\vec{q}\vec{x}} = \frac{e^{-|m|r}}{2\pi i r}. \end{aligned}$$

The hyperbolic position dependence of the Yukawa potential comes from the position projection of a causally supported off-shell distribution with $\vartheta(x^2)$, not from a spacelike dependence with $\vartheta(-x^2)$: Interactions are spacetime-embedded by causally supported distributions of subgroup representations.

The on-shell functions for "real particles" fulfill a homogeneous Klein–Gordon or Dirac equation for the translation invariant of the Poincaré group representation $\partial^2 = -m^2$, whereas the off-shell distributions for "virtual particles," i.e., for interactions are introduced and motivated as Green's[4] kernels of the corresponding inhomogeneous equations:

$$\begin{aligned} (\partial^2 + m^2) \int \frac{d^4q}{(2\pi)^3} \begin{pmatrix} \delta(q^2 - m^2) \\ \frac{1}{i\pi} \frac{1}{-q_{\mathrm{P}}^2 + m^2} \end{pmatrix} e^{iqx} &= \begin{pmatrix} 0 \\ -2i\delta(x) \end{pmatrix}, \quad \begin{array}{ll} (d_{x_0}^2 + m^2) \cos mx_0 & = 0 \\ (-\vec{\partial}^2 + m^2)\rho \frac{e^{-|m|r}}{4\pi r} & = \rho\delta(\vec{x}) \end{array}, \\ (\partial^2 + m^2)\langle 0|\boldsymbol{\Phi}(0)\boldsymbol{\Phi}(x)|0\rangle_{\text{Feynman}} &= -2i\rho(m^2)\, \delta(x), \\ (i\gamma\partial + \mathbf{1}_4 m)\langle 0|\boldsymbol{\Psi}(0)\overline{\boldsymbol{\Psi}}(x)|0\rangle_{\text{Feynman}} &= -2i\rho(m^2)\, \mathbf{1}_4 \otimes \mathbf{1}_{1+2I}\, \delta(x). \end{aligned}$$

A particle field Feynman propagator has three parts, which can be related to the three maximally symmetric operational positions ($\mathbb{E}^3$, Ω^3, $\mathcal{Y}^3$): In addition to the

---

[4]George Green (1793–1841).

on-shell "flat" contribution with the ($r = 0$)-regular spherical Bessel function $\frac{\sin |P|r}{r}$ for the free Euclidean $\mathbb{E}^3$-scattering states, there are the causally supported off-shell "non-flat" contributions for energies over the mass threshold with the spherical Neumann function $\frac{\cos Pr}{r}$ (spherical Laplacian Ω^3-kernel, $\mathbf{SO}(2)$-distribution), and for energies under the mass threshold (imaginary momenta) with the Yukawa potential, i.e., the hyperbolic Macdonald function $\frac{e^{-|B|r}}{r}$ (hyperbolic Laplacian $\mathcal{Y}^3$-kernel, $\mathbf{SO}_0(1,1)$-distribution), both singular at $r = 0$:

$$\int \frac{d^4q}{(2\pi)^3} \begin{pmatrix} \delta(q^2 - m^2) \\ \frac{1}{i\pi}\frac{1}{-q_{\rm P}^2+m^2} \end{pmatrix} e^{iqx}$$
$$= \int \frac{dq_0}{(2\pi)^2} e^{iq_0x_0} \left[\vartheta(q_0^2 - m^2)\begin{pmatrix} \frac{\sin |P|r}{r} \\ \frac{\cos Pr}{ir} \end{pmatrix} + \vartheta(m^2 - q_0^2)\begin{pmatrix} 0 \\ \frac{e^{-|B|r}}{ir} \end{pmatrix}\right]$$
$$= \int_0^\infty \frac{dP^2}{(2\pi)^2} \frac{\cos\sqrt{m^2+P^2}x_0}{\sqrt{m^2+P^2}} \begin{pmatrix} \frac{\sin |P|r}{r} \\ \frac{\cos Pr}{ir} \end{pmatrix} + \int_0^{m^2} \frac{dB^2}{(2\pi)^2} \frac{\cos\sqrt{m^2-B^2}x_0}{\sqrt{m^2-B^2}} \begin{pmatrix} 0 \\ \frac{e^{-|B|r}}{ir} \end{pmatrix},$$
$$\text{with } |P| = \sqrt{q_0^2 - m^2} \text{ and } |B| = \sqrt{m^2 - q_0^2},$$

$$\int \frac{d^4q}{i\pi(2\pi)^3} \frac{1}{-q^2-io+m^2} e^{iqx} = \int \frac{dq_0}{(2\pi)^2} \frac{e^{iq_0x_0+i\sqrt{q_0^2+io-m^2}\, r}}{ir}.$$

The harmonic expansion of a free scalar field, i.e., its particle analysis, uses distributive momentum-characterized creation and annihilation eigenoperators ($\mathrm{u}(\vec q)$, $\mathrm{u}^\star(\vec q)$) of the spacetime translations with momenta as eigenvalues in the direct integral decomposition with the invariant hyperbolic measure,

$$\begin{aligned} \boldsymbol{\Phi}(x) &= {}^\oplus\!\!\int \frac{d^3q}{q_0(2\pi)^3} \, [e^{iqx}\mathrm{u}(\vec q) + e^{-iqx}\mathrm{u}^\star(\vec q)], \\ d^4q\, 2\vartheta(q_0)\delta(q^2 - m^2) &\to \frac{d^3q}{q_0} \text{ with } q_0 = \sqrt{\vec q^2 + m^2}. \end{aligned}$$

The harmonic analysis of a Dirac spinor field, e.g., of an electron-positron field, with irreducible left and right Weyl spinor contributions involves translation and rotation eigenoperators for particles ($\mathrm{u}^A(\vec q)$, $\mathrm{u}^\star_A(\vec q)$), e.g., negatively charged electrons, $z = -1$, and antiparticles ($\mathrm{a}_A(\vec q)$, $\mathrm{a}^{A\star}(\vec q)$), then positively charged positrons, $z = +1$, with spin direction (eigenvalue) "up" and "down," $A = 1, 2$:

$$\boldsymbol{\Psi}(x) \cong \begin{pmatrix} \mathbf{l}^C \\ \mathbf{r}^{\dot C} \end{pmatrix}(x) = \sqrt{m}\, {}^\oplus\!\!\int \frac{d^3q}{q_0(2\pi)^3} \begin{pmatrix} s_A^C(\frac{q}{m})\, [e^{iqx}\mathrm{u}^A(\vec q) + e^{-iqx}\mathrm{a}^{\star A}(\vec q)] \\ \hat s_A^{\dot C}(\frac{q}{m})\, [e^{iqx}\mathrm{u}^A(\vec q) - e^{-iqx}\mathrm{a}^{\star A}(\vec q)] \end{pmatrix},$$

$$\boldsymbol{\Psi}^*(x) \cong \begin{pmatrix} \mathbf{l}^\star_{\dot C} \\ \mathbf{r}^\star_C \end{pmatrix}(x) = \sqrt{m}\, {}^\oplus\!\!\int \frac{d^3q}{q_0(2\pi)^3} \begin{pmatrix} s^{-1}{}^A_{\dot C}(\frac{q}{m})\, [e^{-iqx}\mathrm{u}^\star_A(\vec q) + e^{iqx}\mathrm{a}_A(\vec q)] \\ \hat s^{-1}{}^A_C(\frac{q}{m})\, [e^{-iqx}\mathrm{u}^\star_A(\vec q) - e^{iqx}\mathrm{a}_A(\vec q)] \end{pmatrix},$$
$$\text{with } q_0 = \sqrt{\vec q^2 + m^2}.$$

$(s,\hat{s})(\frac{q}{m}) = \sqrt{\frac{q_0+m}{2m}}(\mathbf{1}_2 \pm \frac{\vec{\sigma}\vec{q}}{q_0+m})$ are the momentum dependent boost representations (Weyl transmutators), which embed the spin group $\mathbf{SU}(2)$ into the Lorentz group $\mathbf{SL}(2,\mathbb{C})$. The Dirac equation describes this dependence for free massive spin-$\frac{1}{2}$ particles. Originally, the free field Dirac equation $\gamma_j\partial^j\boldsymbol{\Psi} = im\boldsymbol{\Psi}$ and its extension by an interaction was considered as a one-particle equation with $\boldsymbol{\Psi}$ as a Schrödinger-like wave function. This led to problems with negative energy or negative probability solutions, interpreted away with the invention of the spurious Dirac sea. All these problems prove irrelevant in the interpretation of $\boldsymbol{\Psi}$ as a quantum field with creation and annihilation operators for particles and antiparticles. Quantum fields are no Schrödinger functions. The relativistic Dirac–Schrödinger equation, e.g., for the hydrogen atom with the Coulomb interaction, involves an 1-electron matrix element — in the free case $\langle 0|\boldsymbol{\Psi}^*(x)|m^2,\frac{1}{2},-1;\vec{q},A\rangle$ as part of the on-shell two point function $\langle 0|\boldsymbol{\Psi}^*(x)\boldsymbol{\Psi}(y)|0\rangle$.

The quantization of canonical Bose fields is the position distribution (quantization density) of the quantum mechanical commutator $[i\mathbf{p},\mathbf{x}] = \mathbf{1}$. The spacetime dependence of canonically quantized free fields is the position distribution of the time translations, $d_t \hookrightarrow \partial^j$, implemented by harmonic oscillators with

$$\mathbf{x}(t) = \frac{e^{i|E|t}\mathrm{u}+e^{-i|E|t}\mathrm{u}^\star}{\sqrt{2}},\quad \left\{\begin{aligned} [\mathbf{x}(s),\mathbf{x}(t)] &= \mathbf{1}i\sin|E|(t-s)\\ &= |E|\int dq_0\;\epsilon(q_0)\delta(q_0^2-E^2)e^{iq_0(t-s)},\\ \langle 0|\{\mathbf{x}(s),\mathbf{x}(t)\}|0\rangle &= \cos E(t-s)\\ &= |E|\int dq_0\;\delta(q_0^2-E^2)e^{iq_0(t-s)}.\end{aligned}\right.$$

$$\begin{aligned}\langle 0|\mathbf{x}(s)\mathbf{x}(t)|0\rangle_{\text{Feynman}} &= \langle 0|\{\mathbf{x}(s),\mathbf{x}(t)\}|0\rangle - \epsilon(t-s)\langle 0|[\mathbf{x}(s),\mathbf{x}(t)]|0\rangle\\ &= \int\frac{dq_0}{i\pi}\,\frac{|E|}{-q_0^2-io+E^2}e^{iq_0(t-s)} = e^{-i|E(t-s)|}.\end{aligned}$$

The quantization and Hilbert metric involve the "inverse" of the Lorentz invariant measure $(2\pi)^3q_0\delta(\vec{q}-\vec{p}) \leftrightarrow \frac{d^3q}{(2\pi)^3q_0}$ for $q_0 = \sqrt{\vec{q}^2+m^2}$ of the momentum hyperboloid $\mathcal{Y}^3$:

$$\begin{aligned} [\mathrm{u}^\star(\vec{p}),\mathrm{u}(\vec{q})] &= \mathbf{1}\;\rho(m^2)q_0\delta(\tfrac{\vec{q}-\vec{p}}{2\pi}),\\ \langle 0|\{\mathrm{u}^\star(\vec{p}),\mathrm{u}(\vec{q})\}|0\rangle &= \langle 0|\mathrm{u}^\star(\vec{p})\mathrm{u}(\vec{q})|0\rangle = \langle m^2;\vec{p}|m^2;\vec{q}\rangle = \rho(m^2)q_0\delta(\tfrac{\vec{q}-\vec{p}}{2\pi}),\\ &\text{distributive Hilbert space basis: } \{|m^2;\vec{q}\rangle = \mathrm{u}(\vec{q})|0\rangle \mid \vec{q}\in\mathbb{R}^3\}.\end{aligned}$$

The related Hilbert space is constituted by the square integrable functions $L^2(\mathcal{Y}^3)$ of the mass shell (momentum hyperboloid) $q^2 = m^2$. The distributive Hilbert space basis for the Dirac field contains eigenvectors for spin $\frac{1}{2}$ particles and antiparticles, e.g., for electrons and positrons with charge numbers $z = \mp 1$:

$$\text{for } \boldsymbol{\Psi}: \quad \left\{\begin{aligned} |m^2,\tfrac{1}{2},-1;\vec{q},A\rangle &= \mathrm{u}^A(\vec{q})|0\rangle,\\ |m^2,\tfrac{1}{2},+1;\vec{q},A\rangle &= \mathrm{a}_A(\vec{q})|0\rangle\end{aligned}\right. \mid \vec{q}\in\mathbb{R}^3,\ A=1,2\}.$$

In this framework, it is by no way obvious how to define the position for a "pointlike" particle and a particle orbit. The quantum mechanical position-momentum "equivalence" does not exist, there is no Heisenberg Lie-algebra. Also, an implementation of the nonrelativistic position-momentum uncertainty, or of an energy-time uncertainty, is by no way straightforward. A Poincaré group representation is determined by the invariants mass und spin (polarization) with the momentum and spin (polarization) direction for a possible distributive eigenvalue basis. The quantum algebra for the creation and annihilation operators allows the quantum representations of the Poincaré group with the Minkowski translations.

The field action on the Fock ground-state vector gives a cyclic vector $|\mathcal{C}\rangle = \mathbf{\Phi}(0)|0\rangle$, summing up all translation "eigenvectors" with spherical waves over the threshold $q_0 \ge |m|$:

$$\begin{aligned} x \bullet |\mathcal{C}\rangle = \mathbf{\Phi}(x)|0\rangle &= {}^{\oplus}\!\!\int \tfrac{d^3q}{q_0(2\pi)^3}\, e^{iqx}|m^2;\vec{q}\rangle, \text{ with } q_0 = \sqrt{\vec{q}^2 + m^2},\\ \langle\mathcal{C}|x \bullet |\mathcal{C}\rangle = \langle 0|\mathbf{\Phi}(0)\mathbf{\Phi}(x)|0\rangle &= \rho(m^2) \int \tfrac{d^4q}{(2\pi)^3}\, 2\vartheta(q_0)\delta(q^2 - m^2)e^{iqx},\\ \int \tfrac{d^4q}{\pi}\, 2\vartheta(q_0)\delta(q^2 - m^2)e^{iqx} &= m^2 \tfrac{\vartheta(x^2)[\pi\mathcal{N}_1(|mx|) - i\epsilon(x_0)\pi\mathcal{J}_1(|mx|)] + \vartheta(-x^2)2\mathcal{K}_1(|mx|)}{|\frac{mx}{2}|}\\ &\quad + i\epsilon(x_0)\delta(\tfrac{x^2}{4\pi}). \end{aligned}$$

Feynman propagators are expectation values of time-ordered products with an explicitly spacetime-dependent order factor $\epsilon(x_0 - y_0)$:

$$\tfrac{\{\mathbf{\Phi}(y),\mathbf{\Phi}(x)\} - \epsilon(x_0 - y_0)[\mathbf{\Phi}(y),\mathbf{\Phi}(x)]}{2} = \vartheta(x_0 - y_0)\mathbf{\Phi}(x)\mathbf{\Phi}(y) + \vartheta(y_0 - x_0)\mathbf{\Phi}(y)\mathbf{\Phi}(x).$$

The creation-annihilation operator assignment is compatible with the time order: annihilation is possible only after creation. The characteristic functions of the causal bicone are harmonically analyzed by an "energy-momentum" dipole and a derived Dirac distribution for trivial invariant:

$$\vartheta(\pm x_0)\vartheta(x^2) = \int \tfrac{d^4q}{2\pi^3}\, \tfrac{1}{[(q_0 \mp io)^2 - \vec{q}^2]^2} e^{iqx}, \quad \begin{cases} \vartheta(x^2) = \int \tfrac{d^4q}{\pi^3}\, \tfrac{1}{(q_{\mathrm{P}}^2)^2} e^{iqx},\\ \epsilon(x_0)\vartheta(x^2) = -\int \tfrac{d^4q}{i\pi^2}\, \epsilon(q_0)\delta'(q^2)e^{iqx}. \end{cases}$$

The particle induced off-shell energy-momentum distributions of the Feynman propagators for the interactions are no translation coefficients, and not convolutable; their undefined products cause the divergencies in the perturbation expansion of an interaction with canonically quantized particle related fields. The translations incompatible interactions are representation distributions, i.e., linear mappings of representation functions. They can be added, however not multiplied with each other: The pointwise product of two interactions, e.g., of two Yukawa potentials $\frac{e^{-|B_1|r}}{r} \cdot \frac{e^{-|B_2|r}}{r} = \frac{e^{-|B_+|r}}{r^2}$, or of their relativistic embeddings, e.g., $\frac{1}{r} \hookrightarrow \delta(x^2)$ and $\delta(x^2) \cdot \delta(x^2)$, does not make sense as an interaction. The convolution of two scalar Feynman propagators with general real power denominators, $\nu_{1,2} \in \mathbb{R}$ (more ahead):

$$\begin{aligned} \Pi(m_1^2, \nu_1, m_2^2, \nu_2; q^2) &= \int d^4q_1 d^4q_2\, \tfrac{\Gamma(1+\nu_1)}{(-q_1^2 - io + m_1^2)^{1+\nu_1}}\, \tfrac{\delta(q_1 + q_2 - q)}{i\pi^2}\, \tfrac{\Gamma(1+\nu_2)}{(-q_2^2 - io + m_2^2)^{1+\nu_2}}\\ &= \int_0^1 dz\, \tfrac{z^{\nu_1}(1-z)^{\nu_2}\, \Gamma(\nu_1+\nu_2)}{[-z(1-z)q^2 - io + zm_1^2 + (1-z)m_2^2]^{\nu_1+\nu_2}}, \end{aligned}$$

is not defined for non-positive integers $\nu_1 + \nu_2 = 0, -1, \ldots$. The convolution of the propagators of a positively and negatively charged pion from the pointwise product of distributions $\langle \boldsymbol{\Phi\Phi}^\star \rangle(x) \cdot \langle \boldsymbol{\Phi\Phi}^\star \rangle(x)$ occurs in the renormalization of the coupling constant of the electromagnetic field by the vacuum polarization $\Pi(m^2, \nu_1, m^2, \nu_2; 0) = \frac{\Gamma(1+\nu_1)\Gamma(1+\nu_2)}{\Gamma(2+\nu_1+\nu_2)} \frac{\Gamma(\nu_1+\nu_2)}{(m^2-io)^{\nu_1+\nu_2}}$, undefined ("divergent") for particle propagators with $\nu_{1,2} = 0$.

The convolution of the on-shell contributions (relativistic phase space integral) is well defined:

$$\begin{aligned} \int d^4q_1 d^4q_2 \vartheta(\pm q_{10})\delta(q_1^2 - m_1^2) \frac{\delta(q_1+q_2-q)}{4\pi} \vartheta(\pm q_{20})\delta(q_2^2 - m_2^2) \\ = \frac{\sqrt{\Delta(q^2)}}{4q^2} \vartheta(\pm q_0)\vartheta(q^2 - m_+^2). \end{aligned}$$

It contains the *threshold factor* (discriminant) Δ, symmetric in the three translation invariants $\{q^2, m_1^2, m_2^2\}$:

$$\begin{aligned} -z(1-z)q^2 + zm_1^2 + (1-z)m_2^2 = 0 &\Rightarrow z_{1,2} = \frac{q^2 - m_+m_- \pm \sqrt{\Delta(q^2)}}{2q^2}, \\ \Delta(q^2) &= (q^2 - m_+^2)(q^2 - m_-^2) \\ &= (q^2)^2 + m_1^4 + m_2^4 - 2(q^2m_1^2 + m_1^2m_2^2 + m_2^2q^2), \\ \text{with } m_\pm &= |m_1| \pm |m_2|. \end{aligned}$$

12.5 Coulomb and Newton Interactions

With Wigner, particles — here, polarized massless photons for the electromagnetic field and possibly existing polarized massless gravitons for the gravitational field — are defined for Minkowski spacetime only. Massless Wigner particles with nontrivial polarization do not exist for two-dimensional spacetime $\mathbf{SO}_0(1,1) \vec{\times} \mathbb{R}^2$ — they become possible as polarized particles with the additional axial rotations in the Cartan subgroup $\mathbf{SO}_0(1,1) \times \mathbf{SO}(2) \subset \mathbf{SO}_0(1,3)$ of the Lorentz group.

Massless particles with energy-momentum $q \neq 0$ and $q^2 = 0$ have no rest system. They have light systems. To define a light system, one needs a decomposition into time and position, e.g., in a rest system of a massive particle, not provided by a massless particle. With a given position and $q = (|\vec{q}|, \vec{q})$, $\vec{q} \neq 0$, a light system has a flight direction by the momentum of a massless particle.

Projectors like $\mathcal{V} \cong \delta_i^k - \frac{q^k q_i}{q^2}$ for spin 1 and $\mathcal{S} \cong \frac{q^k q_i}{q^2}$ for spin 0 are not defined for lightlike energy-momenta $q^2 = 0$. With the fixgroup $\mathbf{SO}(2) \vec{\times} \mathbb{R}^2 \subset \mathbf{SO}_0(1,3)$ of the nontrivial energy-momenta $q \in \mathbb{R}^4$ with $q^2 = 0$, and the fixgroup in the fixgroup $\mathbf{SO}(2) \subset \mathbf{SO}(2) \vec{\times} \mathbb{R}^2$ for trivial boost $\psi = 0 \in \mathbb{R}^2$, massless particles have no spin $\mathbf{SO}(3)$, they have only polarization $\mathbf{SO}(2)$ around the flight direction, e.g., polarization (± 1) for the photons and polarization (± 2) for gravitons. The four-component Lorentz vector for the electromagnetic field $\mathbf{A}$ decomposes into an $\mathbf{SO}(2)$-dublet (± 1) for the two polarized particle degrees of freedom and two $\mathbf{SO}(2)$-singlets (0) for the

Coulomb and gauge degree of freedom. The 10-component symmetric Lorentz tensor for the tetrad related gravitative field $\mathbf{E}$ in the Minkowski spacetime expansion of the metrical tensor $\mathbf{g} = \mathbf{e} \circ \eta \circ \mathbf{e}^T = \eta + 2\mathbf{E} + \ldots$ contains an irreducible Lorentz group nonet $(1|1)$ and singlet $(0|0)$ and decomposes with respect to polarization $\mathbf{SO}(2)$ into two particle degrees of freedom (± 2) and eight nonparticle degrees of freedom as follows:

$$\begin{aligned}
\mathbf{A}^i : \quad & (\tfrac{1}{2}|\tfrac{1}{2}) \overset{\mathbf{SO}(2)}{\cong} (\pm 1) \oplus 2 \times (0),\\
\mathbf{E}^{ij} : (1|1) \oplus & (0|0) \overset{\mathbf{SO}(2)}{\cong} (\pm 2) \oplus 2 \times (\pm 1) \oplus 4 \times (0).
\end{aligned}$$

The free equation for the electromagnetic vector field contains the ∂^2-multiplied spin-1 projector,

$$\partial_i \mathbf{F}^{ki} = (\delta_i^k \partial^2 - \partial^k \partial_i)\mathbf{A}^i = \mathcal{V}_i^k \partial^2 \mathbf{A}^i = 0,$$

whereas the free equation for the Einstein tensor field does not contain pure projectors, but a combination of the ∂^2-multiplied spin-2 and spin-0 projectors:

$$\begin{aligned}
\check{\mathbf{R}}_\bullet^{kl} &= [\delta_i^k \delta_j^l \partial^2 - \delta_j^l \partial^k \partial_i - \delta_j^k \partial^l \partial_i - \eta^{kl}\eta_{ij}\partial^2 + \eta^{kl}\partial_j\partial_i + \eta_{ij}\partial^k\partial^l]\mathbf{E}^{ij}\\
&= (\tfrac{\mathcal{V}_i^k \mathcal{V}_j^l + \mathcal{V}_i^l \mathcal{V}_j^k}{2} - \mathcal{V}^{kl}\mathcal{V}_{ij})\partial^2 \mathbf{E}^{ij}\\
&= [(\mathcal{P}_2)_{ij}^{kl} - 2(\mathcal{P}_0)_{ij}^{kl}]\partial^2 \mathbf{E}^{ij} = 0,
\end{aligned}$$

$$\text{with } (\mathcal{P}_0)_{ij}^{kl} = \tfrac{1}{3}\mathcal{V}^{kl}\mathcal{V}_{ij}, \quad (\mathcal{P}_2)_{ij}^{kl} = \tfrac{\mathcal{V}_i^k \mathcal{V}_j^l + \mathcal{V}_i^l \mathcal{V}_j^k}{2} - \tfrac{1}{3}\mathcal{V}^{kl}\mathcal{V}_{ij}.$$

The Einstein tensor $\check{\mathbf{R}}_\bullet = \mathbf{R}_\bullet - \frac{\mathbf{g}}{2}\mathbf{R}_\bullet^\bullet$ is the combination of the Ricci tensor $\mathbf{R}_\bullet$ with the metric multiplied curvature scalar $\mathbf{R}_\bullet^\bullet = \mathbf{g}_{li}\mathbf{R}_\bullet^{li} = \mathbf{g}_{li}\mathbf{R}_j^{lij}$.

With the divergenceless antisymmetric field strengths and the divergenceless Einstein tensor,

$$\begin{aligned}
\partial_k \partial_i \mathbf{F}^{ki} &= 0,\\
\partial_k \check{\mathbf{R}}_\bullet^{kl} &= 0,
\end{aligned}$$

the vector and tensor fields for massless particles are determined up to the divergences of a scalar and a vector field, respectively (with the connection $\boldsymbol{\Gamma}$):

$$\mathbf{A}^k \longmapsto \mathbf{A}^k + \partial^k \alpha \Rightarrow \mathbf{F}^{kj} \longmapsto \mathbf{F}^{kj},$$

$$\left.\begin{aligned}
\mathbf{E}^{kl} &\longmapsto \mathbf{E}^{kl} + \partial^k \xi^l + \partial^l \xi^k,\\
\boldsymbol{\Gamma}_j^{kl} &\longmapsto \boldsymbol{\Gamma}_j^{kl} + 2\partial^k \partial^l \xi_j,
\end{aligned}\right\} \Rightarrow \check{\mathbf{R}}_\bullet^{kl} \longmapsto \check{\mathbf{R}}_\bullet^{kl}.$$

The electromagnetic current and the Minkowski spacetime energy-momentum tensor are conserved, $\partial_k \mathbf{J}^k = 0$ and $\partial_k \mathbf{T}^{kl} = 0$.

The freedom in the definition of the polarization (± 1)-embedding Lorentz vector field $\mathbf{A}$ is connected with the $\mathbf{U}(1)$-gauge transformation of the electromagnetic field.

The corresponding freedom in the definition of the polarization (± 2)-embedding Lorentz tensor field $\mathbf{E}$ in the tetrad expansion $\mathbf{e} = \mathbf{1} + \mathbf{E} + \ldots$ can be related to the reparametrizations of the spacetime manifold:

$$x_j \longmapsto \overline{x}_j(x) = x_j + \lambda\xi_j(x), \quad \mathbf{e}_j^k = \partial^k \overline{x}_j = \delta_j^k + \lambda \partial^k \xi_j.$$

These reparametrizations up to order λ are the remainder of the parametrization independence of Einstein's gravity.

The gauge transformation analogy between the electromagnetic potential $\mathbf{A}$ and the metrical tensor $\mathbf{g}$, or the tetrad $\mathbf{e}$, is superficial: With respect to the Lorentz group $\mathbf{SO}_0(1,3)$ for orthogonal frames, not the tetrad $\mathbf{e}$, but the connection $\mathbf{\Gamma}$ is the analogue to the $\mathbf{U}(1)$-connection, the electromagnetic gauge field $\mathbf{A}$, both with inhomogeneous transformation behavior:

$$\begin{array}{ccccc} & & \mathbf{e} & & \\ & & \downarrow \partial & & \\ & \mathbf{A} & \leftrightarrow & \mathbf{\Gamma} & \\ \mathbf{U}(1) & \partial\downarrow & & \downarrow\partial & \mathbf{SO}_0(1,3). \\ & \mathbf{F} & \leftrightarrow & \mathbf{R}. & \end{array}$$

The matter equations for linearized gravity are invariant up to order λ only with a corresponding coordinate reparametrization, e.g., $\mathbf{\Phi}(x) \longmapsto \mathbf{\Phi}(\overline{x})$. A full reparametrization invariance requires the full nonlinear theory.

The particle field equations can be reduced to Klein–Gordon equations $\partial^2 \mathbf{\Phi} = 0$ for the translation-invariant mass $m^2 = 0$, where the Lorentz vector properties of the scalar squared spacetime derivative ∂^2 do not combine with the Lorentz properties of the fields. To obtain such a decoupling for electrodynamics with Sommerfeld's normalization $g^2 = 4\pi\alpha_{\text{Som}} \sim 0.9 \times 10^{-1}$, the divergence of the vector field has to vanish. For gravity with the Planck area $\ell^2 = \frac{8\pi G\hbar}{c^3} \sim (0.8 \times 10^{-34}\text{m})^2$, *harmonic coordinates* are used, where the divergence of the Einstein tensor vanishes:

$$\begin{aligned} &\text{for } \partial_k \mathbf{A}^k = 0 \Rightarrow \partial^2 \mathbf{A}^k = -g^2 \mathbf{J}^k, \\ &\text{for } \partial_k(\mathbf{E}^{kl} - \tfrac{\eta^{kl}}{2}\mathbf{E}_i^i) = 0 \Rightarrow \begin{cases} \partial^2(\mathbf{E}^{kl} - \frac{\eta^{kl}}{2}\mathbf{E}_i^i) = -\ell^2 \mathbf{T}^{kl}, \\ \partial^2 \mathbf{E}^{kl} = -\ell^2(\mathbf{T}^{kl} - \frac{\eta^{kl}}{2}\mathbf{T}_i^i). \end{cases} \end{aligned}$$

The propagators of massless vector and tensor fields in spacetime are, up to gauge terms,

$$\begin{aligned} \langle 0|\mathbf{A}^k(y)\mathbf{A}^i(x)|0\rangle_{\text{Feynman}} &= g^2 \int \frac{d^4q}{i\pi(2\pi)^3} \frac{-\eta^{ki}}{-q^2 - io} e^{iq(x-y)}, \\ \langle 0|\mathbf{E}^{kl}(y)\mathbf{E}^{ij}(x)|0\rangle_{\text{Feynman}} &= \ell^2 \int \frac{d^4q}{i\pi(2\pi)^3} \frac{\frac{\eta^{ki}\eta^{lj} + \eta^{li}\eta^{kj} - \eta^{kl}\eta^{ij}}{2}}{-q^2 - io} e^{iq(x-y)}. \end{aligned}$$

They can be obtained from the projectors for massive spin-0, -1, and -2 fields by omitting the gauge-related contributions $\frac{q\otimes q}{m^2}$ in the vector projector $\mathcal{V}^i_k = \delta^i_k - \frac{q_k q^i}{m^2}$. The projector inverse of the kinetic term in the free equation for the Einstein tensor field,

$$\begin{aligned} &[(\mathcal{P}_2)^{kl}_{ij} - 2(\mathcal{P}_0)^{kl}_{ij}]\partial^2 \mathbf{E}^{ij} = 0,\\ &(\mathcal{P}_2 - 2\mathcal{P}_0)(\mathcal{P}_2 - \tfrac{1}{2}\mathcal{P}_0) = \mathcal{P}_2 + \mathcal{P}_0, \end{aligned}$$

gives the relevant combination:

$$\begin{aligned} (\mathcal{P}_1)^i_k &= \mathcal{V}^i_k = \delta^i_k + \tfrac{q\otimes q}{m^2}\text{-terms},\\ (\mathcal{P}_2 - \tfrac{1}{2}\mathcal{P}_0)^{ij}_{kl} &= \tfrac{\delta^i_k\delta^j_l + \delta^i_l\delta^j_k - \eta_{kl}\eta^{ij}}{2} + (\tfrac{q\otimes q}{m^2} \text{ and } \tfrac{q\otimes q\otimes q\otimes q}{m^4})\text{-terms}. \end{aligned}$$

For interacting vector and tensor fields, the normalizations (pole residues) are the coupling constants, i.e., g^2 is related to the fine-structure constant and ℓ^2 to Newton's constant.

The nonrelativistic Kepler potential $\frac{1}{r}$ can be embedded in the off-shell contributions of massless relativistic fields with any Lorentz group representation $(J|J)$: The Coulomb and Newton potentials are the position projections — via time integration — of the off-shell contributions ("virtual particles") in the vector and tensor field propagators, i.e., of the lightcone-supported Dirac distribution $\frac{1}{r} \hookrightarrow \delta(x^2)$:

$$\begin{aligned} \int \tfrac{d^4q}{i\pi(2\pi)^3}\tfrac{1}{-q^2-io}e^{iqx} &= \int \tfrac{d^4q}{(2\pi)^3}\delta(q^2)e^{iqx} + \int \tfrac{d^4q}{i\pi(2\pi)^3}\tfrac{1}{-q_{\mathrm{P}}^2}e^{iqx}\\ &= \tfrac{1}{2\pi^2(-x^2+io)} = -\tfrac{1}{2\pi^2}\tfrac{1}{x_{\mathrm{P}}^2} + \tfrac{1}{2i\pi}\delta(x^2),\\ \int dx_0 \int \tfrac{d^4q}{i\pi(2\pi)^3}\tfrac{1}{-q^2-io}e^{iqx} &= \int dx_0 \int \tfrac{d^4q}{i\pi(2\pi)^3}\tfrac{1}{-q_{\mathrm{P}}^2}e^{iqx} = \int \tfrac{dx_0}{2i\pi}\delta(x^2)\\ &= \int \tfrac{d^3q}{4i\pi^3}\tfrac{1}{\vec{q}^2}e^{-i\vec{q}\vec{x}} = \tfrac{1}{2i\pi r}, \end{aligned}$$

which is, up to $-2i$, the inverse of the invariant spacetime Laplacian,

$$\mathbf{O}(1,3): \ \partial^2 \int \tfrac{d^4q}{i\pi(2\pi)^3}\tfrac{1}{-q^2-io}e^{iqx} = \tfrac{1}{2i\pi}\partial^2\delta(x^2) = -2i\delta(x),$$

and for $\mathbf{O}(1, 2R-1),\ R \geq 2$: $(\frac{\partial^2}{4\pi})^{R-1}\delta(x^2) = \delta(x)$.

There are no projectors from a four-component Lorentz vector field $\mathbf{A}_k$ to the two polarized massless photons, which can be expressed by their energy-momenta, and, similarily, no projectors from a 10-component symmetric Lorentz tensor field $\mathbf{E}_{kl}$ to the two polarized massless gravitons. The boost representations $\Lambda(\frac{q}{m})$ for a transformation to a rest system are not defined for $m = 0$. The embedding of the two particle degrees of freedom with polarization in Lorentz group-compatible fields with four and 10 components, respectively, involves two and eight additional nonparticle components.

For the four-component electromagnetic field, the propagator involves the four-dimensional unit

$$(\tfrac{1}{2}|\tfrac{1}{2}),\ \mathbf{1}_{(\frac{1}{2}|\frac{1}{2})} = \mathbf{1}_4 \cong \delta^i_k.$$

The indefinite metric $-\eta^{ki}$ contains a definite Hilbert submetric $\mathbf{1}_2$ (lower right corner in the metrical matrix) for the two polarized particle components (± 1) (subindex $\pm$ gives the metric for $-\eta$):

$$\text{metric of } (\tfrac{1}{2}|\tfrac{1}{2}): \begin{cases} -\eta = \begin{pmatrix} -1 & 0 & 0 \\ 0 & 1 & 0 \\ 0 & 0 & \mathbf{1}_2 \end{pmatrix}, \\ (\tfrac{1}{2}|\tfrac{1}{2}) \overset{\mathbf{SO}(2)}{\cong} (0)_- \oplus (0)_+ \oplus (\pm 1)_+. \end{cases}$$

The other two components with trivial polarization constitute a *Witt pair* $\begin{pmatrix} -1 & 0 \\ 0 & 1 \end{pmatrix} \cong \begin{pmatrix} 0 & 1 \\ 1 & 0 \end{pmatrix}$ with neutral signature for the Coulomb and gauge degree of freedom, both nonparticlelike.

For the 10-component gravity field, the 10-dimensional symmetric tensor unit can be decomposed into the two harmonic units for the irreducible Lorentz group representations, $10 = 9 + 1$:

$$\begin{aligned} (\tfrac{1}{2}|\tfrac{1}{2}) \vee (\tfrac{1}{2}|\tfrac{1}{2}) &= (1|1) \oplus (0|0), \\ \mathbf{1}_4 \vee \mathbf{1}_4 &= \mathbf{1}_{(1|1)} + \mathbf{1}_{(0|0)} \cong [\tfrac{\delta_k^i \delta_l^j + \delta_l^i \delta_k^j}{2} - \tfrac{\eta_{kl}\eta^{ij}}{4}] + \tfrac{\eta_{kl}\eta^{ij}}{4}. \end{aligned}$$

The graviton propagator comes with a projector combination:

$$\frac{\delta_k^i \delta_l^j + \delta_l^i \delta_k^j - \eta_{kl}\eta^{ij}}{2} = (\mathbf{1}_{(1|1)} - \mathbf{1}_{(0|0)})_{kl}^{ij}.$$

$\mathbf{1}_{(1|1)}$ gives the pair of polarized gravitons (± 2) with definite metric $\mathbf{1}_2$ (lower right corner in the metrical matrix). There is a triplet of Witt pairs with neutral signature $(\mathbf{1}_3, -\mathbf{1}_3)$. The remaining degree of freedom with metric 1 is paired with the one degree of freedom in $-\mathbf{1}_{(0|0)}$ with metric -1:

$$\text{metric of } (1|1): \begin{cases} (\eta \vee \eta)_{(1|1)} = \begin{pmatrix} -\mathbf{1}_3 & 0 & 0 & 0 \\ 0 & 1 & 0 & 0 \\ 0 & 0 & \mathbf{1}_3 & 0 \\ 0 & 0 & 0 & \mathbf{1}_2 \end{pmatrix}, \\ (1|1) \overset{\mathbf{SO}(2)}{\cong} (0)_- \oplus 2 \times (0)_+ \oplus (\pm 1)_- \\ \qquad \oplus (\pm 1)_+ \oplus (\pm 2)_+, \end{cases}$$

$$\text{metric of } (0|0): \begin{cases} -(\eta \vee \eta)_{(0|0)} = -1, \\ (0|0) \overset{\mathbf{SO}(2)}{\cong} (0)_-. \end{cases}$$

In analogy to the one neutral pair for electrodynamics, the four neutral Witt pairs in gravity $\begin{pmatrix} -\eta & 0 \\ 0 & \eta \end{pmatrix} \cong \begin{pmatrix} 0 & \eta \\ \eta & 0 \end{pmatrix}$ for $\eta = \begin{pmatrix} 1 & 0 \\ 0 & -\mathbf{1}_3 \end{pmatrix}$, with the Newton and gauge degrees of freedom, are nonparticlelike.

The Lorentz group representations for Minkowski spacetime gravity constitute an irreducible real 10-dimensional representation of $\mathbf{SL}_0(4,\mathbb{R}) \sim \mathbf{SO}_0(3,3)$, which is considered with a Lorentz invariant metric of signature (6,4):

$$\begin{array}{rcccc} \textbf{irrep}\,\mathbf{SL}_0(4,\mathbb{R}) \ni [2,0,0] & \overset{\mathbf{SO}_0(1,3)}{\cong} & (0|0) & \oplus & (1|1), \\ \begin{pmatrix} -\mathbf{1}_4 & 0 \\ 0 & \mathbf{1}_6 \end{pmatrix} & \cong & (-1) & \oplus & \begin{pmatrix} -\mathbf{1}_3 & 0 \\ 0 & \mathbf{1}_6 \end{pmatrix}. \end{array}$$

Summarizing: For long-range interactions in Minkowski spacetime, only two Lorentz group representations are used and experimentally relevant — the vector representation $(\frac{1}{2}|\frac{1}{2})$, e.g., for the electromagnetic field, and the tensor with scalar representation $(1|1) \oplus (0|0)$ for the gravitational field. Their Cartan subgroup $\mathbf{SO}(2) \times \mathbf{SO}_0(1,1)$ (polarization, boost) decomposition displays, respectively, one pair of massless photons of polarization $(\pm 1)_+$ and one Witt pair $(0)_\pm$, and one pair of massless gravitons $(\pm 2)_+$, still to be detected, and four Witt pairs $(\pm 1)_\pm$, $2 \times (0)_\pm$. In general, a Lorentz group representation $(J|J)$ includes, in addition to a massless $(\pm 2J)$-polarized particle pair, $(1+2J)^2 - 2$ non-particle modes:

$$\begin{array}{rcl} (J|J) \overset{\mathbf{SO}(3)}{\cong} \bigoplus_{L=0}^{2J} (L) & \overset{\mathbf{SO}(2)}{\cong} & (1+2J) \times (0) \oplus 2J \times (\pm 1) \oplus \ldots \oplus (\pm 2J), \\ \text{with } (L) & \overset{\mathbf{SO}(2)}{\cong} & (0) \oplus (\pm 1) \oplus \ldots \oplus (\pm L). \end{array}$$

$(1+2J)$ modes in $(J|J)$ have trivial polarization ("spin zero"), e.g., two in the electromagnetic field and four $(1+3)$ in the gravitational field.

12.6 General Feynman Propagators

The Feynman propagators for general position dimension $s = 1, 2, \ldots$ (where defined) have simple energy-momentum poles. They involve representation coefficients of the associated Poincaré group $\mathbf{SO}_0(1,s) \vec{\times} \mathbb{R}^{1+s}$ for Minkowski spacetime and free particles, given by the real part with the positive energy-momentum delta distribution (on-shell):

$$\mathbf{O}(1,s): \quad \int \frac{d^{1+s}q}{i\pi^{\frac{1+s}{2}}} \frac{1}{-q^2 - io} e^{iqx} = \frac{\Gamma(\frac{s-1}{2})}{(\frac{-x^2+io}{4})^{\frac{s-1}{2}}},$$

$$\int \frac{d^{1+s}q}{i\pi^{\frac{1+s}{2}}} \frac{1}{-q^2 - io + 1} e^{iqx} = \int \frac{d^{1+s}q}{\pi^{\frac{1+s}{2}}} [\pi\delta(q^2 - 1) - \frac{i}{-q_{\mathrm{P}}^2 + 1}] e^{iqx}$$

$$= \frac{\vartheta(-x^2) 2\mathcal{K}_{\frac{1-s}{2}}(|x|) - \vartheta(x^2)\pi\mathcal{N}_{\frac{1-s}{2}}(|x|)}{|\frac{x}{2}|^{\frac{s-1}{2}}} - i\pi\Big[\vartheta(x^2)\frac{\mathcal{J}_{\frac{1-s}{2}}(|x|)}{|\frac{x}{2}|^{\frac{s-1}{2}}} + \delta^N_{\frac{s-3}{2}} \sum_{k=0}^{N} \frac{1}{(N-k)!}\delta^{(k)}(-\frac{x^2}{4})\Big].$$

The on-shell (free particle) function of a Feynman propagator obeys a homogeneous equation with respect to $\partial^2 + m^2$. The off-shell interaction distribution (here the imaginary part with the principal value energy-momentum distribution) is, up to a scalar factor, the inverse operator $(\partial^2 + m^2)^{-1}$ (Green's function). It is a causally supported distribution of the maximal noncompact abelian Lorentz subgroup $\mathbf{SO}_0(1,1) \subseteq \mathbf{SO}_0(1,s)$,

$$(\partial^2 + m^2) \int \frac{d^{1+s}q}{\pi^{\frac{1+s}{2}}} \begin{pmatrix} \pi\,\delta(-q^2+m^2) \\ -\frac{i}{-q_{\mathrm{P}}^2+m^2} \end{pmatrix} e^{iqx} = \begin{pmatrix} 0 \\ -i\,\delta(\frac{x}{\sqrt{4\pi}}) \end{pmatrix}.$$

The representations for the on-shell and off-shell contributions are characterized by the same invariant m^2.

There are no lightcone-supported Dirac distributions $\delta(x^2)$ for one-dimensional position with the rotation-free abelian group $\mathbf{SO}_0(1,1)$, and for all even-dimensional positions with abelian subgroup $\mathbf{SO}(2)$ as shown in the simplest examples:

$$\mathbf{O}(1,s): \quad \int \frac{d^{1+s}q}{i\pi} \frac{1}{-q^2-io+m^2} e^{iqx} = \left(-\frac{\partial}{\partial \frac{x^2}{4\pi}}\right)^{R-1} \begin{cases} \int \frac{d^2q}{i\pi} \frac{1}{-q^2-io+m^2} e^{iqx}, \\ \qquad s = 2R-1 = 1, 2, \ldots, \\ \int \frac{d^3q}{i\pi} \frac{1}{-q^2-io+m^2} e^{iqx}, \\ \qquad s = 2R = 2, 4, \ldots, \end{cases}$$

$$\mathbf{O}(1,1): \begin{cases} \int \frac{d^2q}{i\pi} \frac{1}{-q^2-io+m^2} e^{iqx} = \int d\psi\, e^{-\sqrt{-m^2x^2+io}\cosh\psi} \\ \qquad = 2\int_{|m|}^{\infty} \frac{d\kappa}{\sqrt{\kappa^2-m^2}} [\vartheta(x^2)e^{-i|\kappa x|} + \vartheta(-x^2)e^{-|\kappa x|}] \\ \qquad = \vartheta(x^2) 2\mathcal{K}_0(i|mx|) + \vartheta(-x^2) 2\mathcal{K}_0(|mx|), \end{cases}$$

$$\mathbf{O}(1,2): \begin{cases} \int \frac{d^3q}{i\pi} \frac{1}{-q^2-io+m^2} e^{iqx} = \frac{\partial}{\partial \frac{x^2}{4\pi}} \frac{e^{-\sqrt{-m^2x^2+io}}}{2|m|} = \pi \frac{e^{-\sqrt{-m^2x^2+io}}}{\sqrt{-\frac{x^2}{4}+io}} \\ \qquad = \vartheta(x^2)\pi \frac{e^{-i|mx|}}{i\frac{|x|}{2}} + \vartheta(-x^2)\pi \frac{e^{-|mx|}}{\frac{|x|}{2}}. \end{cases}$$

The time projections via position integration with the particle representations coincide for all dimensions:

$$\mathbf{O}(1,s) \longrightarrow \mathbf{O}(1,0): \quad \int \frac{d^sx}{2^s\pi^{\frac{1+s}{2}}} \int \frac{d^{1+s}q}{i\pi^{\frac{1+s}{2}}} \frac{1}{-q^2-io+m^2} e^{iqx} = \int \frac{dq_0}{i\pi} \frac{1}{-q_0^2-io+m^2} e^{iq_0x_0}$$
$$= \frac{\cos mx_0 - \epsilon(x_0) i \sin |m| x_0}{|m|} = \frac{e^{-i|mx_0|}}{|m|}.$$

The position distributions, embedded by the simple pole off-shell contributions (interactions), are exhibited by the position projection via time integration:

$$\mathbf{O}(1,s) \longrightarrow \mathbf{O}(0,s): \begin{cases} \int \frac{dx_0}{2\pi^{\frac{1}{2}}} \int \frac{d^{1+s}q}{\pi^{\frac{1+s}{2}}} \frac{1}{-q^2-io} e^{iqx} = \int \frac{d^sq}{\pi^{\frac{s}{2}}} \frac{1}{\vec{q}^2} e^{-i\vec{q}\vec{x}} = \frac{\Gamma(\frac{s-2}{2})}{(\frac{r}{2})^{s-2}}, \\ \int \frac{dx_0}{2\pi^{\frac{1}{2}}} \int \frac{d^{1+s}q}{\pi^{\frac{1+s}{2}}} \frac{1}{-q^2-io+1} e^{iqx} = \int \frac{d^sq}{\pi^{\frac{s}{2}}} \frac{1}{\vec{q}^2+1} e^{-i\vec{q}\vec{x}} = \frac{2\mathcal{K}_{\frac{s-2}{2}}(r)}{(\frac{r}{2})^{\frac{s-2}{2}}}. \end{cases}$$

They are Laplacian kernels for interactions, $\mathbf{SO}(2) \subseteq \mathbf{SO}(s)$-distributions for $s \geq 2$, generalizing the Yukawa potential for $s = 3$:

$$(-\vec{\partial}^2 + m^2) \int \frac{d^s q}{\pi^{\frac{s}{2}}} \frac{1}{\vec{q}^2+m^2} e^{-i\vec{q}\vec{x}} = \delta(\frac{\vec{x}}{\sqrt{4\pi}}),$$

$$\frac{2\mathcal{K}_{\frac{s-2}{2}}(r)}{(\frac{r}{2})^{\frac{s-2}{2}}} = \begin{cases} (-\frac{d}{d\frac{r^2}{4}})^{\frac{s-1}{2}} \sqrt{\pi} e^{-r} = \sqrt{\pi}\,(1,\ \frac{2}{r},\ \frac{4(1+r)}{r^3}, \ldots)\, e^{-r},\ s = 1, 3, \ldots, \\ (-\frac{d}{d\frac{r^2}{4}})^{\frac{s-2}{2}} 2\mathcal{K}_0(r) = \int_1^\infty \frac{2d\kappa}{\sqrt{\kappa^2-1}} (-\frac{d}{d\frac{r^2}{4}})^{\frac{s-2}{2}} e^{-\kappa r}, \quad s = 2, 4, \ldots. \end{cases}$$

They differ for different position dimensions. Only for one-dimensional position, the projection is a negative energy (bound) state exponential without $(r = 0)$-singularity. It is a representation coefficient of the abelian Lorentz group $\mathbf{SO}_0(1, 1)$. The corresponding $\mathbf{SO}_0(1, 3)$-position coefficient comes with a dipole, embedding the ground-state of the nonrelativistic hydrogen atom, in general, with an order R-pole for $2R$-dimensional spacetime:

$$\mathbf{O}(1, 2R-1):\ \int \frac{dx_0}{2\pi} \int \frac{d^{2R}q}{\pi^R} \frac{\Gamma(R)}{(-q^2-io+M^2)^R} e^{iqx} = \int \frac{d^{2R-1}q}{\pi^R} \frac{\Gamma(R)}{(\vec{q}^2+M^2)^R} e^{-i\vec{q}\vec{x}} = \frac{e^{-|M|r}}{|M|}.$$

These higher order pole Feynman measures embed coefficients of spherical Ω^{2R-1} and hyperbolic $\mathcal{Y}^{2R-1}$ position representations with nonabelian motion groups for $R \geq 2$ and without spatial $(r = 0)$-singularities:

$$\begin{aligned} \int \frac{d^{2R}q}{i\pi^R} \frac{\Gamma(R)}{(-q^2-io+M^2)^R} e^{iqx} &= \mathcal{K}_0(\sqrt{-M^2x^2+io}) = \int dq_0 \frac{e^{iq_0x_0+i\sqrt{q_0^2+io-M^2}\,r}}{\sqrt{q_0^2+io-M^2}} \\ &= \int dq_0 e^{iq_0x_0}\Big[\vartheta(q_0^2 - M^2)\frac{e^{i|P|r}}{|P|} + \vartheta(M^2 - q_0^2)\frac{e^{-|B|r}}{i|B|}\Big], \\ \text{with } |P| &= \sqrt{q_0^2 - M^2} \text{ and } |B| = \sqrt{M^2 - q_0^2}. \end{aligned}$$

The imaginary part is the causally supported one:

$$\begin{aligned} \vartheta(x^2)\mathcal{K}_0(\sqrt{-M^2x^2+io}) &= i\int dq_0 e^{iq_0x_0}[\vartheta(q_0^2 - M^2)\frac{\sin|P|r}{|P|} - \vartheta(M^2 - q_0^2)\frac{e^{-|B|r}}{|B|}] \\ &= 2\vartheta(x^2)\int_M^\infty \frac{d\kappa}{\sqrt{\kappa^2-M^2}} e^{-i|\kappa x|}. \end{aligned}$$

12.7 Spherical and Hyperbolic Invariants for Indefinite Orthogonal Groups

Orthogonal groups for real even- and odd-dimensional spaces constitute different Cartan series. They come in pairs with respect to the rank, i.e., the number of independent invariants: The Lie algebras $\log \mathbf{SO}_0(t, s)$, both with even $t + s = 2R = 2, 4, \ldots$ (Cartan series D_R) and with odd $t + s = 1 + 2R = 3, 5, \ldots$ (Cartan series B_R), have rank $R = 1, 2, \ldots$. Two additional real dimensions $t + s + 2$ are necessary to raise the rank by one $R + 1$. This becomes relevant for the Lorentz groups where

$\mathbf{SO}_0(1, 2R-1)$ with odd-dimensional position has a unique Cartan subgroup type — a *noncompact Cartan cylinder* $\mathbf{SO}_0(1, 1) \times \mathbf{SO}(2)^{R-1}$, e.g., the proper Lorentz group $\mathbf{SO}_0(1, 3)$ with self-dual dilations and axial rotations $\begin{pmatrix} \cosh\psi & \sinh\psi & 0 & 0 \\ \sinh\psi & \cosh\psi & 0 & 0 \\ 0 & 0 & \cos\theta & -\sin\theta \\ 0 & 0 & \sin\theta & \cos\theta \end{pmatrix} \in \mathbf{SO}_0(1, 1) \times \mathbf{SO}(2)$, whereas $\mathbf{SO}_0(1, 2R)$ with even-dimensional position has two types of Cartan subgroups, either a *compact Cartan torus* $\mathbf{SO}(2)^R$ or a Cartan cyclinder $\mathbf{SO}_0(1, 1) \times \mathbf{SO}(2)^{R-1}$, e.g., for $\mathbf{SO}_0(1, 2)$ with either $\begin{pmatrix} 1 & 0 & 0 \\ 0 & \cos\theta & -\sin\theta \\ 0 & \sin\theta & \cos\theta \end{pmatrix} \in \mathbf{SO}(2)$ or $\begin{pmatrix} \cosh\psi & \sinh\psi & 0 \\ \sinh\psi & \cosh\psi & 0 \\ 0 & 0 & 1 \end{pmatrix} \in \mathbf{SO}_0(1, 1)$. The difference between odd- and even-dimensional spacetimes shows up also in the implementation of either only imaginary (compact) or only real (noncompact) infinitesimal operations, or of all complex operations, $\mathbb{R} + i\mathbb{R} = \mathbb{C}$, as seen for the "smallest" nonabelian Lorentz groups with their Cartan subgroups:

$$\begin{array}{lll} \mathbf{SL}(2, \mathbb{R}) \sim \mathbf{SO}_0(1, 2) \text{ with} & \left\{ \begin{array}{l} \text{either} \quad \mathbf{SO}(2) \cong \exp i\mathbb{R} \ni e^{i\sigma_3\theta}, \\ \quad \text{or } \mathbf{SO}_0(1, 1) \cong \exp\mathbb{R} \ni e^{\sigma_3\psi}, \end{array} \right. \\ \quad\cap \qquad\qquad \cap & \\ \mathbf{SL}(2, \mathbb{C}) \sim \mathbf{SO}_0(1, 3) \text{ with } \mathbf{SO}(2) \times \mathbf{SO}_0(1, 1) \cong \mathbf{U}(1) \times \mathbf{D}(1) \cong \exp\mathbb{C}. & \end{array}$$

The Lorentz group is locally isomorphic $\mathbf{SL}(2, \mathbb{C}) \sim \mathbf{O}(1, 3) \sim \mathbf{O}(3, \mathbb{C})$ to the rotation group in three complex dimensions, considered as real Lie group. As complex Lie algebra, $\log \mathbf{O}(3, \mathbb{C})$ has dimension three and rank one.

The indefinite orthogonal group for the metric $q^2 = \vec{q}_t^{\,2} - \vec{q}_s^{\,2}$ has an Iwasawa factorization with two compact factor subgroups for opposite definiteness:

$$1 \le t \le s: \ \mathbf{O}(t, s) = [\mathbf{O}(t) \times \mathbf{O}(s)] \circ \mathbf{SO}_0(1, 1)^t \circ \exp\mathbb{R}^{t(s-1)}.$$

Maximally symmetric spacetimes with (t, s)-metric are characterized by a constant curvature and one invariant (unit). In the not maximally symmetric cases with an invariant (unit) for both metrical sectors, one uses the $\mathbf{O}(t, s)$-invariantly embedded $\mathbf{O}(t)$-distribution $\frac{d^tq\,\Gamma(1+R_t)}{(-\vec{q}_t^{\,2}+m_t^2)^{1+R_t}} \hookrightarrow \frac{d^nq\,\Gamma(1+R_t)}{(-q^2+m_t^2)^{1+R_t}}$ (spherical invariant $\vec{q}_t^{\,2} = m_t^2 > 0$ for negative $\mathbf{O}(t)$-definiteness) and $\mathbf{O}(s)$-distribution $\frac{d^sq\,\Gamma(1+R_s)}{(\vec{q}_s^{\,2}+m_s^2)^{1+R_s}} \hookrightarrow \frac{d^nq\,\Gamma(1+R_s)}{(-q^2+m_s^2)^{1+R_s}}$ (hyperbolic invariant $\vec{q}_s^{\,2} = -m_s^2 < 0$ for positive $\mathbf{O}(s)$-definiteness) where the point invariants for the subgroups are distributed on lines: $-\vec{q}_t^{\,2} + m_t^2 \hookrightarrow -\vec{q}_t^{\,2} + (m_t^2 + \vec{q}_s^{\,2})$ and $\vec{q}_s^{\,2} + m_s^2 \hookrightarrow \vec{q}_s^{\,2} + (m_s^2 - \vec{q}_t^{\,2})$. The group product related convolution product of the two subgroup spacetime distributions leads to the pointwise product of the two "energy-momentum" functions in the scalar representation coefficients:

$$\mathbf{O}(t) \times \mathbf{O}(s) \hookrightarrow \mathbf{O}(t, s),$$
$$\int d^nq\, \frac{\Gamma(1+R_t)}{(-q^2+m_t^2)^{1+R_t}} e^{iqx} * \int d^nq\, \frac{\Gamma(1+R_s)}{(-q^2+m_s^2)^{1+R_s}} e^{iqx} \sim \int d^nq\, \frac{\Gamma(1+R_t)\Gamma(1+R_s)}{(-q^2+m_t^2)^{1+R_t}(-q^2+m_s^2)^{1+R_s}} e^{iqx}.$$

$\mathbf{O}(t, s)$-scalar integrals, $1 \le t \le s$, with two invariants are obtained by 2-sphere spreads from $\mathbf{O}(1, 1)$-, $\mathbf{O}(1, 2)$- and $\mathbf{O}(2, 2)$-scalar integrals with odd-odd, odd-even and even-even subdimensions and distributed subgroups $\mathbf{O}(1) \times \mathbf{O}(1), \mathbf{O}(1) \times \mathbf{O}(2)$ and $\mathbf{O}(2) \times \mathbf{O}(2)$, respectively.

For nontrivial signature with odd-odd subdimensions adding up to even dimensions,

$$\mathbf{O}(t, s) = \mathbf{O}(1 + 2R_t, 1 + 2R_s) \text{ with } n = t + s = 2(1 + R_t + R_s), \\ \text{for } R_t, R_s = 0, 1, 2, \ldots,$$

one has the characteristic scalar representation coefficients for cyclic (ground) states:

$$\begin{aligned} &\int \frac{d^{2(1+R_t+R_s)}q}{\pi^{1+R_t+R_s}} \frac{\Gamma(1+R_t)}{(-q^2-io+m_t^2)^{1+R_t}} \frac{\Gamma(1+R_s)}{(-q^2-io+m_s^2)^{1+R_s}} e^{iqx} \\ &= \int_0^1 dz\, z^{R_s}(1-z)^{R_t} \int \frac{d^{2(1+R_t+R_s)}q}{\pi^{1+R_t+R_s}} \frac{\Gamma(2+R_t+R_s)}{(-q^2-io+m_z^2)^{2+R_t+R_s}} e^{iqx} \\ &= \int_{m_s^2}^{m_t^2} \frac{dm^2}{m_t^2-m_s^2} \frac{(m^2-m_s^2)^{R_t}(m_t^2-m^2)^{R_s}}{(m_t^2-m_s^2)^{R_t+R_s}} \int \frac{d^{2R}q}{\pi^R} \frac{\Gamma(1+R)}{(-q^2-io+m^2)^{1+R}} e^{iqx}, \\ &\quad \text{with } R = 1 + R_t + R_s,\ \ m_z^2 = (1-z)m_t^2 + zm_s^2 = m^2. \end{aligned}$$

The denominators are joined with the generalized Beta function relation, adjusting pole orders and Gamma function arguments:

$$\begin{aligned} \frac{\Gamma(\nu_1)}{D_1^{\nu_1}} \frac{\Gamma(\nu_2)}{D_2^{\nu_2}} &= \int_0^1 dz\, (1-z)^{\nu_1-1} z^{\nu_2-1} \frac{\Gamma(\nu_1+\nu_2)}{[(1-z)D_1+zD_2]^{\nu_1+\nu_2}} \\ &= \int_{D_2}^{D_1} dD\, \frac{(D-D_2)^{\nu_1-1}(D_1-D)^{\nu_2-1}}{(D_1-D_2)^{\nu_1+\nu_2-1}} \frac{\Gamma(\nu_1+\nu_2)}{D^{\nu_1+\nu_2}} \end{aligned}$$

It reduces for $D_1 = D_2$ to the definition of the Beta function:

$$\Gamma(\nu_1)\Gamma(\nu_2) = \int_0^1 dz\, (1-z)^{\nu_1-1} z^{\nu_2-1} \Gamma(\nu_1+\nu_2),\ \Re\,\nu_{1,2} > 0.$$

The Lie algebra of $\mathbf{O}(1 + 2R_t, 1 + 2R_s)$ has rank $R = 1 + R_t + R_s$ — one genuine noncompact $\mathbf{O}(1, 1)$-related real rank in addition to the sum of the imaginary ranks $R_{t,s}$ for its compact subgroups $\mathbf{O}(1 + 2R_{t,s})$. The integral $\int_{m_s^2}^{m_t^2} dm^2$ sums up a line singularity for the $\mathbf{O}(t, s)$-invariant $q^2 = m^2$ with the invariant for $x = 0$ in the minimal case showing the characteristic *logarithmic connection of spherical and hyperbolic invariant*:

$$\begin{aligned} \mathbf{O}(1) \times \mathbf{O}(1) \hookrightarrow \mathbf{O}(1, 1):\ \int_{m_s^2}^{m_t^2} \frac{dm^2}{m_t^2-m_s^2} \int \frac{d^2q}{i\pi} \frac{1}{(-q^2-io+m^2)^2} &= \frac{1}{m_t^2-m_s^2} \int_{m_s^2}^{m_t^2} \frac{dm^2}{m^2} \\ &= \frac{1}{m_t^2-m_s^2} \log \frac{m_t^2}{m_s^2}. \end{aligned}$$

For odd-even $\mathbf{O}(1 + 2R_t, 2R_s)$, starting with $\mathbf{O}(1, 2)$, and even-even subdimensions $\mathbf{O}(2R_t, 2R_s)$, starting with $\mathbf{O}(2, 2)$, the subgroups-related rank $R_t + R_s$ stays constant.

Important Lorentz group examples are even-dimensional causal spacetimes embedding odd-dimensional positions, with $q^2 = q_0^2 - \vec{q}^2$:

$$\mathbf{O}(1) : \int \frac{dq_0}{i\pi} \frac{1}{-q_0^2 - io + m_t^2} e^{iq_0 x_0} = \frac{e^{-i|m_t x_0|}}{|m_t|}, \quad R_t = 0,$$

$$\mathbf{O}(1 + 2R_s) : \int \frac{d^{1+2R_s} q}{\pi^{1+R_s}} \frac{\Gamma(1+R_s)}{(\vec{q}^2 + m_s^2)^{1+R_s}} e^{i\vec{q}\vec{x}} = \frac{e^{-|m_s|r}}{|m_s|}, \; R_s = R - 1 = 0, 1, 2, \ldots,$$

$$\begin{aligned} \mathbf{O}(1, 2R - 1) : \int \frac{d^{2R} q}{i\pi^R} & \frac{\Gamma(R)}{(-q^2 - io + m_t^2)(-q^2 - io + m_s^2)^R} e^{iqx} \\ &= \int_0^1 dz\, z^{R-1} \int \frac{d^{2R} q}{i\pi^R} \frac{\Gamma(1+R)}{(-q^2 - io + m_z^2)^{1+R}} e^{iqx} \\ &= \int_{m_s^2}^{m_t^2} \frac{dm^2}{m_t^2 - m_s^2} \left(\frac{m_t^2 - m^2}{m_t^2 - m_s^2}\right)^{R-1} \int_{m^2}^{\infty} \frac{d\kappa^2}{2\sqrt{(\kappa^2 - m^2)^3}} \frac{e^{-\sqrt{-\kappa^2 x^2 + io}}}{|\kappa|}. \end{aligned}$$

The normalization of these representations will be discussed ahead.

12.8 Chirally Degenerate Ground-State

Electrodynamics and gravity, in the Minkowski spacetime approach, are implemented by massless fields which, in a light system with $\mathbf{SO}(2)$-operations (polarization), contain $\mathbf{SO}(2)$-scalar non-particle degreees of freedom. Also, a *degenerate ground-state* ("spontaneous symmetry breakdown") comes with long-range Lorentz scalar interactions, characterized, qualitatively, by the "broken" symmetries of a dynamically relevant Lie group G, i.e., by the degeneracy-effecting operations in the classes G/H (degeneracy manifold) with respect to the remaining "unbroken" symmetry for a distinguished subgroup (fixgroup) $H \subseteq G$, and, quantitatively, by a dilation unit, a "breakdown mass." The long-range interactions and the particles in relativistic field theories come in representations of the "unbroken" local symmetry group H, possibly trivial $H \cong \{1\}$.

Examples are the charge related circle-degeneracy, $\Omega^1 \cong \mathbf{U}(1)$, of the superconductor, the sphere-degeneracy, $\Omega^2 \cong \mathbf{SO}(3)/\mathbf{SO}(2)$, of the ferromagnet, the weak Goldstone manifold-degeneracy, $\mathcal{G}^3 \cong \mathbf{U}(2)/\mathbf{U}(1)$, of the electroweak standard model, or the dilation-translation line-degeneracy, $\mathbf{D}(1) = \exp \mathbb{R}$, for nontrivial masses. As illustrated, e.g., by superfluid helium-3 with spin-angular momentum symmetry $\mathbf{SU}(2) \times \mathbf{SO}(3)$, a dynamics, invariant under a group G, can have different ground-states (phases) with degeneracy orbits G/H_ι for different invariance groups $H_\iota \subseteq G$. This is in some analogy to different boundary conditions for a classical dynamics or different cosmological models.

A "dynamical breakdown" establishes the degeneracy-implementing massless fields as product representations of spacetime. In the model of Nambu and Jona-Lasinio,[5] the massless chiral Goldstone boson for a $\mathbf{U}(1)$-degeneracy ("breakdown") is a Lorentz pseudoscalar bound state of regularized self-interacting massive Dirac fermions.

[5] Yoichiro Nambu (1921–2015), Giovanni Jona-Lasinio (1932–).

12.8.1 Chirality

The relative phase of the $\mathbf{SL}(2,\mathbb{C})$-irreducible left- and right-handed Weyl spinors in a Dirac spinor $(\frac{1}{2}|0) \oplus (0|\frac{1}{2})$ is affected by the *chiral transformations* (external phase) as subgroup of the indefinite special unitary group $\mathbf{U}(1) \cong \mathbf{U}(\mathbf{1}_2)_3 \subset \mathbf{SU}(2,2) \subset \mathbf{GL}(4,\mathbb{C})$,

$$\begin{aligned} \mathbf{U}(1) : \boldsymbol{\Psi} &\longmapsto e^{\frac{\alpha}{2}\gamma_5}\boldsymbol{\Psi}, \quad \overline{\boldsymbol{\Psi}} \longmapsto \overline{\boldsymbol{\Psi}} e^{\frac{\alpha}{2}\gamma_5}, \\ \boldsymbol{\Psi} &= \begin{pmatrix} \frac{\mathbf{1}_4 - i\gamma_5}{2} \\ \frac{\mathbf{1}_4 + i\gamma_5}{2} \end{pmatrix} \boldsymbol{\Psi} = \begin{pmatrix} \mathbf{l} \\ \mathbf{r} \end{pmatrix} \longmapsto \begin{pmatrix} e^{-i\frac{\alpha}{2}\mathbf{1}_2}\mathbf{l} \\ e^{i\frac{\alpha}{2}\mathbf{1}_2}\mathbf{r} \end{pmatrix}. \end{aligned}$$

The chiral invariant $\frac{1}{4}\epsilon_{ijkl}\mathbf{L}^{ij}\mathbf{L}^{kl} = \vec{\mathbf{J}}\vec{\mathbf{B}}$ (spin-boost orientation) with spectrum $i[L(1+L) - R(1+R)]$ for the $(1+2L)(1+2R)$-dimensional representations $(L|R)$ has the reflected values $\pm i\frac{3}{4}$ for left- and right-handed Weyl spinors.

The additional common $\mathbf{U}(1)$-phase (internal phase) $\boldsymbol{\Psi} \longmapsto e^{i\frac{\alpha_0}{2}\mathbf{1}_4}\boldsymbol{\Psi}$ with $\mathbf{U}(1) \cong \mathbf{U}(\mathbf{1}_4) \subset \mathbf{GL}(4,\mathbb{C})$ can be used for (hyper)charge operations. It plays no role in the model of Nambu and Jona-Lasinio. In the electroweak standard model, external and internal phases, i.e., chirality and hypercharge, are connected in the ground-state-induced symmetry rearrangement.

The $2^n = 16 = 1+4+6+4+1 = \sum_{k=0}^{n}\binom{n}{k}$ elements in a basis for the real 16-dimensional Clifford algebra for $(n=4)$-dimensional Minkowski spacetime with $\{\gamma_i, \gamma_j\} = 2\eta_{ij}\mathbf{1}_4$, represented by bilinear Dirac spinors $\overline{\boldsymbol{\Psi}}\Gamma\boldsymbol{\Psi}$, involve five different couplings Γ:

$$\begin{aligned} &\mathbf{1}_4,\ \gamma_i,\ \gamma_{ij} = -\tfrac{1}{4}[\gamma_i, \gamma_j],\ \gamma_5\gamma^i = \tfrac{\epsilon^{ijkl}}{3!}\gamma_j\gamma_k\gamma_l,\ \gamma_5 = \tfrac{\epsilon^{ijkl}}{4!}\gamma_i\gamma_j\gamma_k\gamma_l, \\ &\text{with } \{\gamma_i, \gamma_5\} = 0,\ \ \gamma_5^2 = -\mathbf{1}_4, \end{aligned}$$

with vector γ_i and axial vector $\gamma_5\gamma_i$ as chiral invariants, and two chiral $\mathbf{SO}(2)$-dublets, given by scalar and pseudoscalar $(\mathbf{1}_4, \gamma_5)$, and by the tensor via $(\gamma_{ij}, \gamma_5\gamma_{ij} = \frac{1}{2}\epsilon_{ijkl}\gamma^{kl})$, e.g.,

$$\mathbf{U}(1): \ \begin{pmatrix} \overline{\boldsymbol{\Psi}}\mathbf{1}_4\boldsymbol{\Psi} \\ \overline{\boldsymbol{\Psi}}\gamma_5\boldsymbol{\Psi} \end{pmatrix} \longmapsto \begin{pmatrix} \cos\alpha & \sin\alpha \\ -\sin\alpha & \cos\alpha \end{pmatrix} \begin{pmatrix} \overline{\boldsymbol{\Psi}}\mathbf{1}_4\boldsymbol{\Psi} \\ \overline{\boldsymbol{\Psi}}\gamma_5\boldsymbol{\Psi} \end{pmatrix}, \quad \overline{\boldsymbol{\Psi}}\frac{\mathbf{1}_4 \pm i\gamma_5}{2}\boldsymbol{\Psi} = \begin{pmatrix} \mathbf{l}^*\mathbf{r} \\ \mathbf{r}^*\mathbf{l} \end{pmatrix} \longmapsto \begin{pmatrix} e^{i\alpha}\,\mathbf{l}^*\mathbf{r} \\ e^{-i\alpha}\mathbf{r}^*\mathbf{l} \end{pmatrix}.$$

The five quartic Lorentz scalar couplings,

$$\Gamma \otimes \Gamma = \begin{pmatrix} \mathbf{1}_4 \otimes \mathbf{1}_4 \\ \gamma_i \otimes \gamma^i \\ \gamma_{ij} \otimes \gamma^{ij} \\ \gamma_5\gamma_i \otimes \gamma_5\gamma^i \\ \gamma_5 \otimes \gamma_5 \end{pmatrix} \text{ for } (\overline{\boldsymbol{\Psi}}\Gamma\boldsymbol{\Psi})(\overline{\boldsymbol{\Psi}}\Gamma\boldsymbol{\Psi}) = \begin{pmatrix} s \\ v \\ t \\ a \\ p \end{pmatrix},$$

combine the three chirally invariant 4-fermion couplings:

$$\mathbf{U}(1): \ \ (v, a, s+p) \longmapsto (v, a, s+p),$$

The two Fierz-antisymmetric combinations $\{s+p-\frac{v+a}{2}, v-a\}$, i.e., $\Gamma^A_B\Gamma^C_D = -\Gamma^C_B \Gamma^A_D$, are nontrivial for Fermi Dirac fields. The chirally invariant coupling with a scalar term $s+p-\frac{v+a}{2}$ is used in the nonlinear interaction of the model of Nambu and Jona-Lasinio with the equation of motion:

$$i\gamma\partial\boldsymbol{\Psi} = \tfrac{1}{2}\Gamma\boldsymbol{\Psi}(\overline{\boldsymbol{\Psi}}\Gamma\boldsymbol{\Psi}), \text{ with } \Gamma\otimes\Gamma = \tfrac{1}{4}(\mathbf{1}_4\otimes\mathbf{1}_4 + \gamma_5\otimes\gamma_5 - \tfrac{\gamma_i\otimes\gamma^i+\gamma_5\gamma_i\otimes\gamma_5\gamma^i}{2}).$$

This quartic interaction is distinguished as the square of the radial part $\mathbf{R}^2 = \boldsymbol{\Phi}\boldsymbol{\Phi}^*$ in the scalar and pseudoscalar Dirac field product, which yield the analogue of the $\mathbf{U}(1)\times\mathbf{D}(1)$-factorization of a basic chiral Higgs field $\boldsymbol{\Phi} = \mathbf{l}^*\mathbf{r} = e^{i\alpha}\mathbf{R}$:

$$\overline{\boldsymbol{\Psi}}\frac{\mathbf{1}_4\pm i\gamma_5}{2}\boldsymbol{\Psi} = e^{\pm i\alpha}\mathbf{R} \Rightarrow \begin{cases} \mathbf{R}^2 = (\overline{\boldsymbol{\Psi}}\frac{\mathbf{1}_4+i\gamma_5}{2}\boldsymbol{\Psi})(\overline{\boldsymbol{\Psi}}\frac{\mathbf{1}_4-i\gamma_5}{2}\boldsymbol{\Psi}) = (\mathbf{l}^*\mathbf{r})(\mathbf{r}^*\mathbf{l}) \\ \quad = \frac{1}{2}(\overline{\boldsymbol{\Psi}}\Gamma\boldsymbol{\Psi})(\overline{\boldsymbol{\Psi}}\Gamma\boldsymbol{\Psi}), \\ e^{\pm i\alpha} = \frac{\overline{\boldsymbol{\Psi}}\frac{\mathbf{1}_4\pm i\gamma_5}{2}\boldsymbol{\Psi}}{\mathbf{R}}. \end{cases}$$

It arises in the corresponding classical Lagrangian:

$$\mathbf{L}(\boldsymbol{\Psi}) = i\overline{\boldsymbol{\Psi}}\gamma\partial\boldsymbol{\Psi} - \tfrac{1}{4}(\overline{\boldsymbol{\Psi}}\Gamma\boldsymbol{\Psi})(\overline{\boldsymbol{\Psi}}\Gamma\boldsymbol{\Psi}).$$

12.8.2 Chiral Degeneracy and Representation Normalization

A particle Dirac field has a Fock ground-state vector $|0\rangle$ for its Feynman propagator:

$$\langle 0|\boldsymbol{\Psi}(y)\overline{\boldsymbol{\Psi}}(x)|0\rangle_{\text{Feynman}} = \rho(m^2)\int \frac{d^4q}{i\pi(2\pi)^3}\frac{\gamma q+m\mathbf{1}_4}{-q^2-io+m^2}e^{iq(x-y)} = \rho(m^2)\mathbf{k}^m(x-y).$$

An interaction determines the normalization of the Poincaré group representation: A coupling constant $g_0\Gamma\otimes\Gamma$ can be absorbed into the normalization $\rho(m^2)\longmapsto g_0\rho(m^2)$.

A nontrivial mass term $m\neq 0$ characterizes a chirally $\mathbf{U}(1)$-degenerate ground-state. The local group (unbroken subgroup, fixgroup of the ground-state orbit) is trivial, $G/H = \mathbf{U}(1)/\{1\}$. Particles are chiral singlets. The nonlinear field equation yields, in a first-order approximation, a *self-consistency condition ("gap equation")* for the mass m as the chiral breakdown dilation unit:

$$\begin{aligned} i\gamma\partial\boldsymbol{\Psi} &= -\Gamma\boldsymbol{\Psi}\,\text{tr}\,\Gamma\rho(m^2)\mathbf{k}^m(0) + \tfrac{1}{2} :\Gamma\boldsymbol{\Psi}(\overline{\boldsymbol{\Psi}}\Gamma\boldsymbol{\Psi}): = -m\boldsymbol{\Psi} + \dots \\ &\Rightarrow m = \tfrac{1}{4}\,\text{tr}\,\mathbf{1}_4\rho(m^2)\mathbf{k}^m(0) = \rho(m^2)\int\frac{d^4q}{i\pi(2\pi)^3}\frac{m}{-q^2-io+m^2}. \end{aligned}$$

Here, the distribution $\mathbf{k}^m$ of the spacetime translations, used for a free particle, is taken for the trivial translation $x=0$. With the volume of the mass hyperboloid $\int d^4q\, 2\vartheta(q_0)\delta(q^2-1) = |\mathcal{Y}^3|$, it is "divergent"; $\mathbf{k}^m(0)$ does not make sense. A perturbative approach with particle fields is inappropriate for a bound state problem. The model of Nambu and Jona-Lasinio is nonrenormalizable.

With the extraction of the Dirac field mass term, the chirally invariant Lagrangian is decomposed into a ground-state adapted kinetic (free) term, no longer chirally invariant, and a remaining interaction, also not chirally invariant, to be considered with respect to the less symmetric ground-state:

$$\mathbf{L}(\Psi) = i\overline{\Psi}\gamma\partial\Psi + m\overline{\Psi}\Psi - \tfrac{1}{4} : (\overline{\Psi}\Gamma\Psi)(\overline{\Psi}\Gamma\Psi) :$$

The double-dot prescription $: \ldots :$ is defined by the subtraction of the mass term as the leading bilinear ground-state contribution. The chiral invariance of the full Lagrangian is restored by the interaction with the long range Nambu–Goldstone field (chiral degree of freedom, more ahead).

The translation ground-state vector $|0\rangle$ for particle fields in Minkowski spacetime has to be replaced by a ground-state vector $|\mathcal{U}\rangle$ from a *chiral ground-state* $\mathbf{U}(1)$-*manifold* for fields with an interaction. Such a ground-state will be implemented in the form of an ad hoc regularization, e.g., by a dipole at mass M^2 in the regularized "Feynman propagator":

$$\begin{aligned} x-y \longmapsto \langle\mathcal{U}|\Psi(y)\overline{\Psi}(x)|\mathcal{U}\rangle &= d^{M^2,m}(x-y)\\ &= \rho(m^2)\int \frac{d^4q}{i\pi(2\pi)^3}\frac{\gamma q + m\mathbf{1}_4}{(-q^2-io+m^2)(-q^2-io+M^2)^2}e^{iq(x-y)}. \end{aligned}$$

Other regularizations are possible.

With respect to spacetime translation representations, the dipole regularization involves a ghost pair (Witt pair) with an indefinite $\mathbf{U}(1,1)$-metric. It cannot be interpreted as a Wigner particle with a unitary Poincaré group representation. The modification of the particle field propagator for Minkowski spacetime by the regularization, here by the dipole with the "flattening" (contraction) $M^4 d^{M^2,m} \to \rho(m^2)\mathbf{k}^m$ for $\frac{1}{M^2} \to 0$, will be related to the transition to representations of curved space. This connects a nontrivial curvature and interaction and will be discussed ahead.

The dipole-regularized "Feynman propagator" can be used as a function of the spacetime translations $\mathbb{R}^4 \ni x \longmapsto d^{M^2,m}(x)$, defined for the trivial translation in the modified consistency equation,

$$\begin{aligned} m\mathbf{1}_4 &= \langle\mathcal{U}|\Psi(x)\overline{\Psi}(x)|\mathcal{U}\rangle = d^{M^2,m}(0)\\ &= \rho(m^2)\mathbf{1}_4\int \tfrac{d^4q}{i\pi(2\pi)^3}\tfrac{m}{(-q^2-io+m^2)(-q^2-io+M^2)^2}\\ &= -m\mathbf{1}_4\tfrac{\rho(m^2)}{8\pi^2(m^2-M^2)}(\tfrac{m^2}{m^2-M^2}\log\tfrac{M^2}{m^2}+1). \end{aligned}$$

For a chiral degeneracy, the consistency equation is a *representation normalization*. It determines the normalization factor $\rho(m^2)$ by the ratio $\frac{m^2}{M^2}$ of the chiral breakdown mass and the regularization mass:

$$\begin{aligned} m \neq 0 : \tfrac{1}{m}d^{M^2,m}(0) = \mathbf{1}_4 \Rightarrow \tfrac{8\pi^2}{\rho(m^2)} &= \int_0^1 dz\,\tfrac{z}{(1-z)m^2+zM^2}\\ &= -\tfrac{1}{m^2-M^2}(\tfrac{m^2}{m^2-M^2}\log\tfrac{M^2}{m^2}+1). \end{aligned}$$

The positive definite expression on the right hand side is a typical normalization for four-dimensional spacetime (more ahead).

12.8.3 Massless Chiral Boson

The Green's distribution of the free Dirac equation with mass,

$$(i\gamma\partial + m\mathbf{1}_4)\mathbf{k}^m(x) = -2i\delta(x) \Rightarrow \mathbf{k}^m(x) = \int \frac{d^4q}{i\pi(2\pi)^3} \frac{\gamma q + m\mathbf{1}_4}{-q^2 - io + m^2} e^{iqx},$$

is not a function. Only for a free theory (Minkowski spacetime) can the Feynman propagator $\rho(m^2)\mathbf{k}^m$ be identified, up to a constant factor, with the Green's distribution $\mathbf{k}^m$. In the regularization used, it is related to the "Feynman propagator" of the interacting Dirac field by a convolution with a dipole regulator:

$$\begin{aligned} d^{M^2,m}(x) &= \tfrac{1}{2i}\, y^{M^2} * \mathbf{k}^m(x) = \tfrac{1}{2i} \int d^4x'\, y^{M^2}(x - x')\mathbf{k}^m(x'), \\ y^{M^2}(x) &= \rho(m^2) \int \frac{d^4q}{i\pi(2\pi)^3} \frac{1}{(-q^2 - io + M^2)^2} e^{iqx}. \end{aligned}$$

The Green's distribution inverts the particle field differential operator:

$$\begin{aligned} (i\gamma\partial + m\mathbf{1}_4)\mathbf{\Psi} &= \tfrac{1}{2} : \Gamma\mathbf{\Psi}(\overline{\mathbf{\Psi}}\Gamma\mathbf{\Psi}) :, \\ \mathbf{\Psi}(x) &= -\tfrac{1}{4i} \int d^4x'\, \mathbf{k}^m(x - x') : \Gamma\mathbf{\Psi}(\overline{\mathbf{\Psi}}\Gamma\mathbf{\Psi})(x') : . \end{aligned}$$

The bilinear local products of the Dirac field for bosonic matrix elements with a state vector $|\mathrm{B}\rangle$,

$$\Gamma(x) = \langle \mathcal{U} | \overline{\mathbf{\Psi}}\Gamma\mathbf{\Psi}(x) | \mathrm{B}\rangle \ \text{ for } \Gamma \in \{\mathbf{1}_4, \gamma_i, \gamma_5\gamma_i, \gamma_5\},$$

have the equations of motion:

$$\Gamma_1(x) = -\tfrac{1}{4i} \int d^4x'\, \langle \mathcal{U} | \overline{\mathbf{\Psi}}(x)\Gamma_1\mathbf{k}^m(x - x') : \Gamma\mathbf{\Psi}(\overline{\mathbf{\Psi}}\Gamma\mathbf{\Psi})(y) : |\mathrm{B}\rangle.$$

They can be linearized with the "Feynman propagator" of the interacting Dirac field to yield first-order eigenvalue equations,

$$\Gamma_1(x) = \mathbf{K}_{\Gamma_1}^{\Gamma_2} * \Gamma_2(x) = \int d^4x'\, \mathbf{K}_{\Gamma_1}^{\Gamma_2}(x - x')\, \Gamma_2(x'),$$

with the *tangent kernel matrix* $\mathbf{K}$:

$$\mathbf{K}_{\Gamma_1}^{\Gamma_2}(x) = \tfrac{1}{2i}\, \mathrm{tr}\, (\Gamma_1 \otimes \Gamma_2) \circ (\mathbf{k}^m \otimes d^{M^2,m})(x) = \tfrac{1}{2i}\, \mathrm{tr}\, \Gamma_1\mathbf{k}^m(x)\Gamma_2 d^{M^2,m}(x).$$

The corresponding convolution product for the energy-momenta (translation eigenvalues) distributions gives the distributions for eigenvalues of the product representations:

$$\tilde{\Gamma}_1(q) = \tilde{\mathbf{K}}_{\Gamma_1}^{\Gamma_2}(q)\tilde{\Gamma}_2(q), \ \ [\delta_{\Gamma_1}^{\Gamma_2} - \tilde{\mathbf{K}}_{\Gamma_1}^{\Gamma_2}(q)]\tilde{\Gamma}_2(q) = 0,$$

$$\begin{aligned}\text{with } \tilde{\mathbf{K}}_{\Gamma_1}^{\Gamma_2}(q) &= \tfrac{1}{2i}\,\mathrm{tr}\,(\Gamma_1 \otimes \Gamma_2) \circ (\tilde{\mathbf{k}}^m \circledast \tilde{d}^{M^2,m})(q)\\ &= -\rho(m^2)\,\mathrm{tr}\int \tfrac{d^4p}{i\pi(2\pi)^3}\Gamma_1 \tfrac{\gamma(p-q)+m\mathbf{1}_4}{-(p-q)^2-io+m^2}\Gamma_2 \tfrac{\gamma p+m\mathbf{1}_4}{(-p^2-io+m^2)(-p^2-io+M^2)^2}.\end{aligned}$$

The convolution $\circledast$ contains a tensor product for Clifford algebra elements.

The pointwise product $\rho(m^2)\mathbf{k}^m(x) \otimes \mathbf{k}^m(x)$ of the Green's distribution and the particle field propagator is not defined because of the distributional off-shell contributions. With the regularization for the quantization of the interacting Dirac field, the product makes sense.

The bilinear scalar and pseudoscalar products $(\overline{\mathbf{\Psi}}\mathbf{1}_4\mathbf{\Psi}, \overline{\mathbf{\Psi}}\gamma_5\mathbf{\Psi})$ constitute a chiral dublet. The degenerate ground-state is characterized by the particle mass,

$$\langle \mathcal{U}|\overline{\mathbf{\Psi}}\tfrac{\mathbf{1}_4+i\gamma_5}{2}\mathbf{\Psi}|\mathcal{U}\rangle\langle \mathcal{U}|\overline{\mathbf{\Psi}}\tfrac{\mathbf{1}_4-i\gamma_5}{2}\mathbf{\Psi}|\mathcal{U}\rangle = 4m^2;$$

i.e., the degeneracy manifold for $m^2 > 0$ is a circle $\Omega^1 = \mathbf{SO}(2) \cong \mathbf{U}(1)$ with radius $2\,m$. A scalar ground-state, defined by

$$\begin{aligned}\langle \mathcal{U}|\overline{\mathbf{\Psi}}\mathbf{1}_4\mathbf{\Psi}|\mathcal{U}\rangle &= 4m,\\ \langle \mathcal{U}|\overline{\mathbf{\Psi}}\gamma_5\mathbf{\Psi}|\mathcal{U}\rangle &= 0,\end{aligned} \qquad \langle \mathcal{U}|\mathbf{l}^*\mathbf{r}|\mathcal{U}\rangle = \langle \mathcal{U}|\mathbf{r}^*\mathbf{l}|\mathcal{U}\rangle = 2m,$$

strips the particles of the chiral degree of freedom. In a first-order approximation, the pseudoscalar Dirac field product is the infinitesimal chiral field, i.e., the Goldstone degree of freedom,

$$\begin{aligned} e^{i\alpha} &= 1 + i\alpha + \ldots = \tfrac{\overline{\mathbf{\Psi}}(\mathbf{1}_4+i\gamma_5)\mathbf{\Psi}}{2\mathbf{R}} = 1 + i\tfrac{\overline{\mathbf{\Psi}}\gamma_5\mathbf{\Psi}}{4m} + \ldots\\ \Rightarrow \quad \alpha &= \tfrac{\overline{\mathbf{\Psi}}\gamma_5\mathbf{\Psi}}{4m} + \ldots,\end{aligned}$$

for the degeneracy transformation of the ground-state vector $|\mathcal{U}\rangle$. The *equation of motion (translation analysis) for the composite Nambu–Goldstone field* as the corresponding matrix element $\gamma_5(x) = \langle \mathcal{U}|\overline{\mathbf{\Psi}}\gamma_5\mathbf{\Psi}(x)|\pi\rangle$ with the "pion" state vector $|\pi\rangle$ is given as follows:

$$\begin{aligned}&\tilde{\gamma}_5(q) = \tilde{\mathbf{K}}_{\gamma_5}^{\gamma_5}(q^2)\tilde{\gamma}_5(q), \ \ [1 - \tilde{\mathbf{K}}_{\gamma_5}^{\gamma_5}(q^2)]\tilde{\gamma}_5(q) = 0,\\ &\text{with } \tilde{\mathbf{K}}_{\gamma_5}^{\gamma_5}(q^2) = -\tfrac{\rho(m^2)}{4}\,\mathrm{tr}\int \tfrac{d^4p}{i\pi(2\pi)^3}\gamma_5 \tfrac{\gamma(p-q)+m\mathbf{1}_4}{-(p-q)^2-io+m^2}\gamma_5 \tfrac{\gamma p+m\mathbf{1}_4}{(-p^2-io+m^2)(-p^2-io+M^2)^2}.\end{aligned}$$

With $\gamma_5\gamma p\gamma_5 = \gamma p$ and $\gamma_5\gamma_5 = -\mathbf{1}_4$, thence $\gamma_5(\gamma p + m\mathbf{1}_4)\gamma_5(\gamma p + m\mathbf{1}_4) = (p^2 - m^2)\mathbf{1}_4$, it has a mass zero $q^2 = 0$ solution $\tilde{\mathbf{K}}_{\gamma_5}^{\gamma_5}(0) = 1$ if compared with the consistency equation for the chiral breakdown, $m \neq 0$:

$$1 = \int \tfrac{d^4q}{i\pi(2\pi)^3} \tfrac{\rho(m^2)}{(-q^2-io+m^2)(-q^2-io+M^2)^2} = \begin{cases} \tfrac{1}{4m}\,\mathrm{tr}\int \tfrac{d^4q}{(2\pi)^4}\,\tilde{d}^{M^2,m}(q),\\ \tfrac{1}{4}\,\mathrm{tr}\int \tfrac{d^4q}{i\pi(2\pi)^3}\,\tilde{\mathbf{k}}^m(q)\tilde{d}^{M^2,m}(q).\end{cases}$$

The consistency equation for the fermion mass (degeneracy parameter) as representation normalization leads to the eigenvalue equation for the massless field.

The expansion of the resolvent for the pseudoscalar kernel at the mass zero solution determines the normalization $\rho(0)$ of the corresponding spacetime translation representation as the residue at the massless pole:

$$\frac{1}{1-\tilde{\mathbf{K}}^{\gamma_5}_{\gamma_5}(q^2)} = \frac{1}{q^2}\frac{q^2}{1-\tilde{\mathbf{K}}^{\gamma_5}_{\gamma_5}(q^2)} = \frac{\rho(0)}{q^2} + \dots \text{ with } \frac{1}{\rho(0)} = -\frac{\partial\tilde{\mathbf{K}}^{\gamma_5}_{\gamma_5}}{\partial q^2}(0).$$

The characteristic structures of the chiral model do not depend on a "perturbative" expansion. They are given by the relation between the normalization, in the "gap" equation, of the spacetime representation by the regularized "Feynman propagator" and the ensuing massless solution via the singularity of the resolvent $\frac{1}{1-\tilde{\mathbf{K}}^{\gamma_5}_{\gamma_5}(q^2)}$ with the normalized kernel $\tilde{\mathbf{K}}^{\gamma_5}_{\gamma_5}$.

12.9 Electroweak Spacetime

The complex functions of electroweak spacetime are the coefficients of unitary relativity:

$$\begin{aligned}\mathbf{GL}(2,\mathbb{C}) &= \mathbf{U}(2)\circ\mathbf{D}(1)\circ\mathbf{SO}_0(1,1)\circ\exp\mathbb{R}^2,\\ \mathbf{GL}(2,\mathbb{C})/\mathbf{U}(2) &= \mathbb{D}(2)\ni d\longmapsto\varphi(d)\in\mathbb{C}.\end{aligned}$$

The hyperisospin $\mathbf{U}(2)$-induced representations of the hyper-Lorentz group $\mathbf{GL}(2,\mathbb{C})$ are subrepresentations $\mathbf{GL}(2,\mathbb{C})\times\mathbf{U}(2)\subset\mathbf{GL}(2,\mathbb{C})\times\mathbf{GL}(2,\mathbb{C})$ of the two-sided regular representation. They act on $\mathbf{U}(2)$-intertwiners, i.e., on mappings that connect spacetime points, parametrizing unitary classes, with hyperisospin orbits in a Hilbert space with $\mathbf{U}(2)$-representation. With respect to the homogeneous operations, fields of electroweak spacetime are transmutators φ for the transition $\mathbf{U}(2)\subset\mathbf{GL}(2,\mathbb{C})$ from the internal hyperisospin group to the external hyper-Lorentz group. They associate each hyperisospin orbit in the hyper-Lorentz group, parametrized by four real parameters $g\mathbf{U}(2)\cong d\cong(d_\mu)_{\mu=0}^3\in\mathbb{D}(2)$, with linear mappings $\{V\longrightarrow W\}$ from a vector space V with $\mathbf{U}(2)$-action to a vector space W with $\mathbf{GL}(2,\mathbb{C})$-action:

$$\begin{array}{ccc} \mathbb{D}(2) & \xrightarrow{g\bullet} & \mathbb{D}(2) \\ \varphi\downarrow & & \downarrow {}_g\varphi_u \\ W\otimes V^T & \xrightarrow[D(g)\otimes R(u^{-1})]{} & W\otimes V^T \end{array} \qquad \begin{aligned}\mathbb{D}(2)&\cong\mathbf{GL}(2,\mathbb{C})/\mathbf{U}(2),\\ \mathbf{GL}(2,\mathbb{C})\ni g&\longmapsto D(g)\in\mathbf{GL}(W),\\ \mathbf{U}(2)\ni u&\longmapsto R(u)\in\mathbf{GL}(V),\\ W\cong\mathbb{C}^m,&\ V\cong\mathbb{C}^n,\end{aligned}$$

$$d\longmapsto\varphi(d)\text{ with }{}_g\varphi_u(d) = D(g)\circ\varphi(g^{-1}\bullet d)\circ R(u^{-1}).$$

In the bi-regular representation the subgroup $\mathbf{SU}(2)$ acts twice, i.e., independently for the pair spin-isospin $\mathbf{SU}(2) \times \mathbf{SU}(2)$ with isomorphic groups.

The action group $\mathbf{SL}(2, \mathbb{C}) \times \mathbf{U}(2)$ with external and internal transformations as subgroup of the both-sided regular $\mathbf{GL}(2, \mathbb{C})$-action is realized as homogeneous group in the electroweak standard model: The representations act faithfully for $\mathbf{SL}(2, \mathbb{C}) \times \mathbf{U}(2)$ on the left-handed spinor-isospinor lepton field $\mathbb{R}^{(1,3)} \ni x \longmapsto \mathbf{l}^A_\alpha(x) \in \mathbb{C}^2 \otimes \mathbb{C}^2$ with Feynman propagator $\langle \mathbf{l}\mathbf{l}^\star \rangle(x) \sim \int \frac{d^4q}{i\pi^2} \frac{\sigma^j q_j \times \mathbf{1}_2}{-q^2 - io + m^2} e^{iqx}$, for its adjoint group $\mathbf{SO}_0(1, 3) \times \mathbf{SO}(3)$ on the isotriplet gauge vector field $x \longmapsto \mathbf{A}^j_a(x) \in \mathbb{C}^4 \otimes \mathbb{C}^3$, for $\mathbf{SL}(2, \mathbb{C}) \times \mathbf{U}(1)$ on the right-handed isosinglet lepton field $x \longmapsto \mathbf{r}^{\dot{A}}(x) \in \mathbb{C}^2 \otimes \mathbb{C}$, and for its adoint group $\mathbf{SO}_0(1, 3) \times \{1\}$ on the hypercharge gauge vector field $x \longmapsto \mathbf{A}^j_0(x) \in \mathbb{C}^4 \otimes \mathbb{C}$. The $\mathbf{SL}(2, \mathbb{C})$-Lorentz group representations are induced by the $\mathbf{SU}(2)$-isospin representations: With the notable exception of the Lorentz scalar isodublet Higgs field, all isospin $\mathbf{SU}(2)$-representations of the standard model fields are isomorphic to subrepresentations of their Lorentz group $\mathbf{SL}(2, \mathbb{C})$-representations with $V \cong \mathbb{C}^n \subseteq \mathbb{C}^m \cong W$ which can be related to Frobenius' reciprocity for induced representations.

12.9.1 Dilation and Translation Parametrizations of Causal Spacetimes

In the following, electroweak spacetime $\mathbb{D}(2) = \mathcal{D}^4 \cong \mathbf{GL}(2, \mathbb{C})/\mathbf{U}(2)$ will be considered as the $(R = 2)$-member of the series with the *even-dimensional causal spacetimes* for the classes of the odd-dimensional orthogonal groups:

$$R = 1, 2, \ldots : \quad \mathcal{D}^{2R} \cong \mathbf{D}(1) \times \mathcal{Y}^{2R-1} \cong \exp \mathbb{R}^{(1,2R-1)},$$
$$\mathbf{D}(1) \cong \exp \mathbb{R},$$
$$\mathcal{Y}^{2R-1} \cong \exp \mathbb{R}^{2R-1} \cong \mathbf{SO}_0(1, 2R-1)/\mathbf{SO}(2R-1).$$

Since the maximal noncompact abelian motion group $\mathcal{D}^2 = \mathbf{D}(1) \times \mathbf{SO}_0(1, 1)$, called *Cartan spacetime*, is two-dimensional, $\mathcal{D}^{2R}$ has real rank two.

Electroweak spacetime is also the $(n = 2)$-member in the series of the causal manifolds for the unitary group classes:

$$n = 1, 2, \ldots : \quad \mathbb{D}(n) = \mathbf{GL}(n, \mathbb{C})/\mathbf{U}(n) = \mathbf{D}(1) \times \mathbf{SL}(n, \mathbb{C})/\mathbf{SU}(n).$$

This series starts with the dilation (causal) group $\mathbf{D}(1)$. $\mathbb{D}(n) \cong \exp \mathbb{R}(n)$ has real dimension n^2 and real rank n. $\mathcal{D}^{2R}$ and $\mathbb{D}(n)$ meet only for four dimensions, i.e., $\mathcal{D}^4 \cong \mathbb{D}(2) \cong \mathbf{D}(1) \times \mathcal{Y}^3$.

Two analogue series contain the compact partners with imaginary ranks and rational invariants:

$$\begin{aligned}&\mathcal{D}^{2R} \leftrightarrow \mathcal{U}^{2R} \cong \mathbf{U}(1) \times \mathbf{SO}(2R)/\mathbf{SO}(2R-1) \cong \mathbf{U}(1) \times \Omega^{2R-1},\\&\mathbb{D}(n) \leftrightarrow \mathbb{U}(n) \cong [\mathbf{U}(n) \times \mathbf{U}(n)]/\mathbf{U}(n).\end{aligned}$$

with $\mathcal{U}^4 \cong \mathbb{U}(2) \cong \mathbf{U}(1) \times \Omega^3$.

The additive reals (translations) $\mathbb{R}$ and its multiplicative exponentials (dilations) $\mathbf{D}(1) = \exp \mathbb{R}$ with Lie algebra $\log \mathbf{D}(1) = \mathbb{R}$ are embedded in higher dimensional causal spacetimes $\mathcal{D}^{2R}$ with corresponding Minkowski spacetime $\mathbb{R}^{(1,2R-1)}$ as tangent spaces:

$$\begin{aligned}\mathbb{R}^{(1,2R-1)} &\cong \log[\mathbf{D}(1) \times \mathbf{SO}_0(1, 2R-1)]/\log \mathbf{SO}(2R-1)\\&\cong [\mathbf{SO}_0(1, 2R-1) \,\vec{\times}\, \mathbb{R}^{2R}]/\mathbf{SO}_0(1, 2R-1),\end{aligned}$$

$$\begin{aligned}\mathcal{D}^{2R} &= \{d = \begin{pmatrix} d_0 \\ \vec{d} \end{pmatrix} = e^{\psi_0}\begin{pmatrix} \cosh |\vec{\psi}| \\ \sinh |\vec{\psi}|\, \vec{\omega}_{2R-2} \end{pmatrix} \mid \vec{\omega}_{2R-2} = \frac{\vec{\psi}}{|\vec{\psi}|} \in \Omega^{2R-2},\ \psi_0 \in \mathbb{R},\ |\vec{\psi}| \in \mathbb{R}_+\},\\ \text{with } e^{\psi_0} &= |d| = \sqrt{d_0^2 - \vec{d}^2},\ e^{|\vec{\psi}|} = \sqrt{\frac{d_0+|\vec{d}|}{d_0-|\vec{d}|}}.\end{aligned}$$

An example is the Weyl representation of electroweak spacetime by exponentials of the spacetime translations, given by hermitian (2×2)-matrices:

$$\begin{aligned}&\mathbf{GL}(2, \mathbb{C}) \times \mathbb{D}(2) \longrightarrow \mathbb{D}(2),\ \ \mathbf{GL}(2, \mathbb{C}) \times \mathbb{R}(2) \longrightarrow \mathbb{R}(2),\\&\mathbf{GL}(2, \mathbb{C}) = \exp \mathbf{AL}(2, \mathbb{C}) \supset \mathbb{D}(2) = \exp \mathbb{R}(2),\\&\mathbb{R}^{(1,3)} = \mathbb{R}(2) \ni \psi = \begin{pmatrix} \psi_0 + \psi_3 & \psi_1 - i\psi_2 \\ \psi_1 + i\psi_2 & \psi_0 - \psi_3 \end{pmatrix} = u \circ \begin{pmatrix} \psi_0 + |\vec{\psi}| & 0 \\ 0 & \psi_0 - |\vec{\psi}| \end{pmatrix} \circ u^\star,\\&\mathcal{D}^4 = \mathbb{D}(2) \ni e^{\psi} = e^{\psi_0}(\cosh |\vec{\psi}|\, \mathbf{1}_2 + \frac{\sinh |\vec{\psi}|}{|\vec{\psi}|}\, \vec{\psi}\vec{\sigma})\\&\qquad = d = \begin{pmatrix} d_0 + d_3 & d_1 - id_2 \\ d_1 + id_2 & d_0 - d_3 \end{pmatrix} = u \circ \begin{pmatrix} d_0 + |\vec{d}| & 0 \\ 0 & d_0 - |\vec{d}| \end{pmatrix} \circ u^\star,\ \ d_0 \pm |\vec{d}| > 0,\end{aligned}$$

with the Cartan subgroup and the rotations:

$$\begin{aligned}&\mathbf{D}(1) \times \mathbf{SO}_0(1, 1) \ni \begin{pmatrix} d_0 + |\vec{d}| & 0 \\ 0 & d_0 - |\vec{d}| \end{pmatrix} = \begin{pmatrix} e^{\psi_0+|\vec{\psi}|} & 0 \\ 0 & e^{\psi_0-|\vec{\psi}|} \end{pmatrix},\\&\Omega^2 \ni \vec{\omega}_2 = \frac{\vec{d}}{|\vec{d}|} = \frac{\vec{\psi}}{|\vec{\psi}|} \longmapsto u = \begin{pmatrix} \cos \frac{\theta}{2} & -e^{-i\varphi} \sin \frac{\theta}{2} \\ e^{i\varphi} \sin \frac{\theta}{2} & \cos \frac{\theta}{2} \end{pmatrix}\\&\qquad = \frac{1}{\sqrt{2|\vec{\psi}|(\psi_3+|\vec{\psi}|)}}\begin{pmatrix} \psi_3 + |\vec{\psi}| & -\psi_1 + i\psi_2 \\ \psi_1 + i\psi_2 & \psi_3 + |\vec{\psi}| \end{pmatrix} \in \mathbf{SU}(2).\end{aligned}$$

Ahead, the dimensionless parameters obtain units via the invariants of the group representations.

In contrast to time $\mathbf{D}(1)$, electroweak spacetime $\mathbb{D}(2) = \exp \mathbb{R}(2)$ is no group. The noncompact dilations $\mathbf{D}(1)$ constitute, in a relativistic framework, the group for the Lorentz invariant "eigentime" $e^{\psi_0} \in \mathbf{D}(1)$. The noncompact odd-dimensional hyperboloids are the position submanifolds $\begin{pmatrix} \cosh |\vec{\psi}| \\ \sinh |\vec{\psi}|\, \vec{\omega}_{2R-2} \end{pmatrix} \in \mathcal{Y}^{2R-1}$.

Spacetime $\mathcal{D}^{2R}$ is an orbit of $\mathbf{D}(1) \times \mathbf{SO}_0(1, 2R-1)$. For the future cone, foliated by position hyperboloids $\mathcal{Y}^{2R-1}$, the action of the causal group $\mathbf{D}(1)$ may be called *hyperbolic hopping*, from position hyperboloid to position hyperboloid, and the action of the dilative Lorentz subgroup $\mathbf{SO}_0(1,1) \subseteq \mathbf{SO}_0(1,2R-1)$ *hyperbolic stretching*, inside each position hyperboloid,

$$\mathbf{D}(1): \begin{pmatrix} d_0 \\ d_3 \end{pmatrix} = e^{\psi_0} \begin{pmatrix} \cosh \psi_3 \\ \sinh \psi_3 \end{pmatrix} \longmapsto e^{\chi_0} \begin{pmatrix} d_0 \\ d_3 \end{pmatrix} = e^{\chi_0+\psi_0} \begin{pmatrix} \cosh \psi_3 \\ \sinh \psi_3 \end{pmatrix},$$

$$\mathbf{SO}_0(1,1): \begin{pmatrix} d_0 \\ d_3 \end{pmatrix} = e^{\psi_0} \begin{pmatrix} \cosh \psi_3 \\ \sinh \psi_3 \end{pmatrix} \longmapsto \begin{pmatrix} \cosh \chi_3 & \sinh \chi_3 \\ \sinh \chi_3 & \cosh \chi_3 \end{pmatrix} \begin{pmatrix} d_0 \\ d_3 \end{pmatrix} = e^{\psi_0} \begin{pmatrix} \cosh(\chi_3+\psi_3) \\ \sinh(\chi_3+\psi_3) \end{pmatrix},$$

$$\Rightarrow \begin{pmatrix} d_0 \\ d_3 \end{pmatrix} = e^{\psi_0 \mathbf{1}_2 + \psi_3 \sigma_1} \begin{pmatrix} 1 \\ 0 \end{pmatrix}.$$

The maximal compact motion group in the Iwazawa decomposition

$$\mathbf{SO}_0(1,2R-1) = \mathbf{SO}(2R-1) \circ \mathbf{SO}_0(1,1) \circ \exp \mathbb{R}^{2(R-1)},$$

nontrivial for $2R = 4, 6, \ldots$, rotates each position hyperboloid. Cartan spacetime $\mathcal{D}^2$ does not know the concepts spin or polarization.

The tangent translation parametrization $(x_0, \vec{x})$ is appropriate for a relativistic formulation and an embedding $\mathbf{D}(1) \times \mathcal{Y}^3 \cong \mathbb{R}^4_+ \hookrightarrow \mathbb{R}^4$ in tangent Minkowski spacetime structures with particles and a corresponding harmonic analysis (ahead). The manifold isomorphy of the dilation group with the positive numbers $\mathbf{D}(1) \ni d = e^{\psi_0} \cong \vartheta(t)t \in \mathbb{R}_+$, is embedded in higher dimensions: A parametrization of causal spacetime is possible by the open future cone of the corresponding Minkowski spacetime with the manifold isomorphy:

$$\begin{aligned} \mathcal{D}^{2R} \ni d &\cong \vartheta(x_0)\vartheta(x^2)x \in \mathbb{R}_+^{(1,2R-1)} = \{x \in \mathbb{R}^{(1,2R-1)} \mid x^2 > 0,\ x_0 > 0\}, \\ d = e^{\psi} &= u \circ \begin{pmatrix} e^{\psi_0+|\vec{\psi}|} & 0 \\ 0 & e^{\psi_0-|\vec{\psi}|} \end{pmatrix} \circ u^\star \cong x = u \circ \begin{pmatrix} x_0+|\vec{x}| & 0 \\ 0 & x_0-|\vec{x}| \end{pmatrix} \circ u^\star. \end{aligned}$$

12.9.2 The Causal Cartan Plane

The causal spacetimes $\mathcal{D}^{2R}$ have real rank 2 as dimension of a Cartan plane $\mathcal{D}^2 = \mathbf{D}(1) \times \mathbf{SO}_0(1,1)$ with causal operations and Lorentz dilations as maximal noncompact abelian group.

The characteristic functions of the causal Cartan line $\mathbf{D}(1)$ and plane $\mathcal{D}^2$ (spacetime without rotation degrees of freedom) can be unitarily analyzed (Fourier-transformed) by an advanced "energy(-momentum)" distribution (measure) with a pole at a trivial invariant $q_0 = 0$ and $q^2 = 0$, respectively,

$$\begin{aligned} \text{for } \mathbf{D}(1): &\quad \vartheta(x_0) = \int \frac{dq_0}{2i\pi} \frac{1}{q_0 - io} e^{iq_0x_0}, \\ \text{for } \mathcal{D}^2 = \mathbf{D}(1) \times \mathbf{SO}_0(1,1): &\quad \vartheta(x)\pi = \int \frac{d^2q}{2\pi} \frac{1}{-(q-io)^2} e^{iqx}, \\ \text{with } \vartheta(\pm x) = \vartheta(\pm x_0)\vartheta(x^2) &\text{ for } (q \mp io)^2 = (q_0 \mp io)^2 - q_3^2. \end{aligned}$$

The Fourier transform of an "energy-momentum" function, holomorphic in the lower (upper) complex "energy" q_0 plane, i.e., with $q_0 \mp io$, is valued in the future (past) cone, i.e., supported by causal line and plane with $\vartheta(\pm x_0)$.

In a Lorentz group action-compatible translation parametrization, the harmonic analysis of the Cartan plane uses $\mathbf{SO}_0(1,1)$-invariants $q^2 = m^2$ as "energy-momentum" singularities:

$$\mathbf{D}(1): \begin{cases} \int \frac{dq_0}{2i\pi} \frac{1}{q_0 - io - m} e^{iq_0 x_0} = \vartheta(x_0) e^{imx_0}, \\ \int \frac{dq_0}{2\pi} \frac{iq_0}{-(q_0 - io)^2 + m^2} e^{iq_0 x_0} = \vartheta(x_0) \cos mx_0, \end{cases}$$

$$\mathcal{D}^2 = \mathbf{D}(1) \times \mathbf{SO}_0(1,1): \int \frac{d^2q}{2\pi} \frac{1}{-(q-io)^2 + m^2} e^{iqx} = \vartheta(x)\pi\mathcal{J}_0(|mx|),\ |x| = \sqrt{|x^2|}.$$

The Bessel functions with half-integer index start from cosine oscillations

$$\mathbb{R} \ni x \longmapsto \cos mx = \int dq\ |q|\delta(q^2 - m^2) e^{iqx},$$

with one invariant m^2. They are used for real odd-dimensional spaces, e.g., for one-dimensional time and for three-dimensional position, with energies and momenta as eigenvalues. The Bessel functions with integer index are relevant for real even-dimensional spaces, e.g., for four-dimensional spacetime. $\mathcal{J}_0$ integrates $\mathbb{R}$-representation with a $\mathbf{D}(1)$-measure supported by a line, e.g., $\int_{|m|}^{\infty} d\kappa$ which starts at the invariant m^2:

$$\begin{aligned} \mathbb{R} \ni x \longmapsto \pi\mathcal{J}_0(mx) &= \int d\psi\ \sin(|mx| \cosh \psi) = 2\int_{|m|}^{\infty} \frac{d\kappa}{\sqrt{\kappa^2 - m^2}} \sin|\kappa x| \\ &= \int_0^{\pi} d\theta\ \cos(|mx| \cos\theta) = 2\int_0^{|m|} \frac{d\kappa}{\sqrt{m^2 - \kappa^2}} \cos|\kappa x|. \end{aligned}$$

The amplitudes of the $\mathcal{J}_0$-oscillations are exponentially damped. $\mathcal{J}_0$ embeds both spherical and hyperbolic representation coefficients, disentanglable for causal Cartan $(1,1)$-spacetime: the time projection yields an oscillating coefficient of the causal operations $\mathbf{D}(1)$ whereas the position projection gives an exponentially decreasing coefficient of the Lorentz dilations $\mathbf{SO}_0(1,1)$:

$$\begin{aligned} \mathcal{D}^2 &\longrightarrow \mathbf{D}(1), & \text{with } \vartheta(x_0)x_0 &\longmapsto \int dx_3\ \vartheta(x)\mathcal{J}_0(|mx|) = \vartheta(x_0)\frac{\sin mx_0}{m}, \\ \mathcal{D}^2 &\longrightarrow \mathbf{SO}_0(1,1), & \text{with } x_3 &\longmapsto \int dx_0\ \vartheta(x)\mathcal{J}_0(|mx|) = \frac{e^{-|mx_3|}}{|m|}. \end{aligned}$$

The spherical and hyperbolic invariants for the $\mathbf{D}(1)$-oscillations and the $\mathbf{SO}_0$ $(1,1)$-fall-off must not coincide. Two continuous invariants (units) for the rank 2 causal plane are implementable, in a residual representation, by two poles at $(\pm|m|)$ and $(\pm i|M|)$ in the complex energy and momentum planes with the pointwise product of two functions for hyperbolic non-relativistic space-time:

$$\mathbf{D}(1) \times \mathbf{SO}_0(1,1) \ni (\vartheta(x_0)x_0, x_3) \longmapsto \int \frac{dq_0 dq_3}{\pi^2} \frac{iq_0|M|}{[-(q_0-io)^2+m^2](q_3^2+M^2)} e^{i(q_0x_0-q_3x_3)}$$
$$= \vartheta(x_0)\cos mx_0\, e^{-|Mx_3|}.$$

In a relativistic Lorentz group-compatible form, the two poles show up as the endpoints of an $\mathbf{O}(1,1)$-*invariant line singularity* $q^2 = \kappa^2 \in [M^2, m^2]$, characteristic for even-dimensional causal spacetimes,

$$\int \frac{d^2q}{2\pi} \frac{1}{[-(q-io)^2+m^2][-(q-io)^2+M^2]} e^{iqx} = \int_{M^2}^{m^2} \frac{d\kappa^2}{m^2-M^2} \int \frac{d^2q}{2\pi} \frac{1}{[-(q-io)^2+\kappa^2]^2} e^{iqx}$$
$$= -\vartheta(x) \int_{M^2}^{m^2} \frac{d\kappa^2}{m^2-M^2} \frac{\partial \pi \mathcal{J}_0(|\kappa x|)}{\partial \kappa^2}$$
$$= -\vartheta(x) \frac{\pi \mathcal{J}_0(|mx|) - \pi \mathcal{J}_0(|Mx|)}{m^2-M^2},$$

$$\int \frac{d^2q}{2\pi} \frac{iq}{[-(q-io)^2+m^2][-(q-io)^2+M^2]} e^{iqx} = \frac{\partial}{\partial x} \int_{M^2}^{m^2} \frac{d\kappa^2}{m^2-M^2} \int \frac{d^2q}{2\pi} \frac{1}{[-(q-io)^2+\kappa^2]^2} e^{iqx}$$
$$= \vartheta(x) \frac{x}{|x|} \frac{m\pi \mathcal{J}_1(|mx|) - M\pi \mathcal{J}_1(|Mx|)}{m^2-M^2},$$

with $\frac{\partial}{\partial x} \mathcal{J}_0(|x|) = \frac{x}{2} \frac{\partial}{\partial \frac{x^2}{4}} \mathcal{J}_0(|x|) = -\frac{x}{|x|} \mathcal{J}_1(|x|)$.

12.9.3 Even-Dimensional Causal Spacetimes

For even-dimensional spacetime $\mathcal{D}^{2R}$ in general, the characteristic future functions $\vartheta(x) = \vartheta(x_0)\vartheta(x^2)$ in the translation cone parametrizations have the harmonic analysis with an advanced "energy(-momentum)" distribution $(q-io)^2 = (q_0-io)^2 - \vec{q}^2$:

$$\mathbf{SO}_0(1, 2R-1): \ \vartheta(x)\pi \begin{pmatrix} 1 \\ x \end{pmatrix} = \int \frac{d^{2R}q}{2\pi^R} \frac{\Gamma(R)}{[-(q-io)^2]^R} \begin{pmatrix} 1 \\ \frac{2iqR}{-(q-io)^2} \end{pmatrix} e^{iqx}.$$

The spherical degrees of freedom show up in the order of the pole and in the normalization of the integration $\frac{d^{2R}q}{2\pi^R}$.

The causal spacetime coefficients with one invariant are

$$\mathbf{SO}_0(1, 2R-1): \quad \int \frac{d^{2R}q}{2\pi^R} \frac{\Gamma(R)}{[-(q-io)^2+m^2]^R} \begin{pmatrix} 1 \\ \frac{2iqR}{-(q-io)^2+m^2} \end{pmatrix} e^{iqx} = \vartheta(x) \begin{pmatrix} 1 \\ x \end{pmatrix} \pi \mathcal{J}_0(|mx|).$$

Even-dimensional spacetime $\mathcal{D}^{2R}$ with two continuous invariants (real rank $r = 2$) is represented as the Fourier-transformed product of two "energy-momentum" distributions, one with a simple pole — spherical invariant for dilation (time) $\mathbf{D}(1)$ — and the other one with a pole of order $R = 1, 2, \ldots$ — hyperbolic invariant for position $\mathcal{Y}^{2R-1}$, e.g., in the Lorentz scalar and vector functions:

$$\begin{aligned}
\mathcal{D}^{2R} \ni \vartheta(x)x \longmapsto & \int \frac{d^{2R}q}{2\pi^R} \frac{\Gamma(R)}{[-(q-io)^2+m^2][-(q-io)^2+M^2]^R} e^{iqx} \\
&= \int_0^1 dz\, z^{R-1} \int \frac{d^{2R}q}{2\pi^R} \frac{\Gamma(1+R)}{[-(q-io)^2+(1-z)m^2+zM^2]^{1+R}} e^{iqx} \\
&= \int_{M^2}^{m^2} \frac{d^R\kappa^2}{m^2-M^2} \int \frac{d^{2R}q}{2\pi^R} \frac{\Gamma(1+R)}{[-(q-io)^2+\kappa^2]^{1+R}} e^{iqx} \\
&= -\vartheta(x) \int_{M^2}^{m^2} \frac{d^R\kappa^2}{m^2-M^2} \frac{\partial \pi \mathcal{J}_0(|\kappa x|)}{\partial \kappa^2},
\end{aligned}$$

$$\begin{aligned}
\mathcal{D}^{2R} \ni \vartheta(x)x \longmapsto & \int \frac{d^{2R}q}{2\pi^R} \frac{iq\, \Gamma(R)}{[-(q-io)^2+m^2][-(q-io)^2+M^2]^R} e^{iqx} \\
&= -\vartheta(x) \frac{\partial}{\partial x} \int_{M^2}^{m^2} \frac{d^R\kappa^2}{m^2-M^2} \frac{\partial \pi \mathcal{J}_0(|\kappa x|)}{\partial \kappa^2} \\
&= \int_{M^2}^{m^2} \frac{d^R\kappa^2}{m^2-M^2} \int \frac{d^{2R}q}{2\pi^R} \frac{i}{2} \frac{\partial}{\partial q} \frac{\Gamma(R)}{[-(q-io)^2+\kappa^2]^R} e^{iqx} \\
&= \vartheta(x) \frac{x}{2} \int_{M^2}^{m^2} \frac{d^R\kappa^2}{m^2-M^2} \pi \mathcal{J}_0(|\kappa x|).
\end{aligned}$$

The spacetime coefficients contain the characteristic dimension-dependent integration with a *finite positive spectral* $\mathbf{D}(1)$*-measure over a line with* $\mathbf{O}(1, 2R-1)$*-invariants*:

$$\begin{aligned}
&\int_0^1 dz\, z^{R-1} = \int_{M^2}^{m^2} \frac{d^R\kappa^2}{m^2-M^2} \text{ with } \kappa^2 = (1-z)m^2 + zM^2, \\
&\text{where } d^R\kappa^2 = d\kappa^2 \left(\frac{m^2-\kappa^2}{m^2-M^2}\right)^{R-1}, \quad R = 1, 2, \ldots.
\end{aligned}$$

Lorentz group $\mathbf{SO}_0(1, 2R-1)$-properties, and nontrivial properties with respect to the maximal compact group $\mathbf{SO}(2R-1)$ for $R \geq 2$, are obtained by derivations $x \sim \frac{\partial}{\partial q}$ and harmonic $\mathbf{SO}_0(1, 2R-1)$-polynomials $(x)^L$ and $(q)^L$.

A pointwise product form for hyperbolic non-relativistic space-time, not compatible with the Lorentz group action looks as follows:

$$\int \frac{dq_0\, d^{2R-1}q}{\pi^{1+R}} \frac{iq_0\Gamma(R)}{[-(q_0-io)^2+m^2](\vec{q}^2+M^2)^R} e^{i(q_0x_0-\vec{q}\vec{x})} = \vartheta(x_0) \cos mx_0 \frac{e^{-|M|r}}{|M|}.$$

In the relativistic Lorentz group-compatible form, the representation coefficients for the factor groups arise by projection: The simple pole embeds the representation of the abelian time operations (dilations) $\mathbf{D}(1) \cong \mathbb{R}$ with rank 1, e.g., in a Lorentz vector representation:

$$\begin{aligned}
&\int \frac{d^{2R}q}{2\pi^R} \frac{iq}{-(q-io)^2+m^2} e^{iqx} = \frac{x}{2} \left(\frac{\partial}{\partial \frac{x^2}{4}}\right)^R \vartheta(x)\pi\mathcal{J}_0(|mx|), \\
&\left(\frac{\partial}{\partial \frac{x^2}{4}}\right)^R \vartheta(x^2)\mathcal{J}_0(|mx|) = \vartheta(x^2)\frac{\mathcal{J}_R(|mx|)}{(-\frac{|mx|}{2})^R} + \sum_{k=0}^{R-1} \frac{(m^2)^{R-1-k}}{(R-1-k)!}\delta^{(k)}(-\frac{x^2}{4}).
\end{aligned}$$

It is a position distribution of a time representation coefficient: The time projection by integration over position, i.e., for trivial momenta $\int \frac{d^{2R-1}x}{(2\pi)^{2R-1}} e^{-i\vec{q}\vec{x}} = \delta(\vec{q})$, contains a positive-type function $\cos \in L^\infty(\mathbb{R})_+$ for time representations:

$$\int \frac{d^{2R-1}x}{(2\pi)^{2R}} \int d^{2R}q \, \frac{iq}{-(q-io)^2+m^2} e^{iqx} = \int \frac{dq_0}{2\pi} \frac{iq_0}{-(q_0-io)^2+m^2} e^{iq_0x_0}$$
$$= \vartheta(x_0) \cos mx_0.$$

The representation of hyperbolic position $\mathcal{Y}^{2R-1} \cong \exp \mathbb{R}^{2R-1}$ with the Lorentz group $\mathbf{SO}_0(1, 2R-1)$ as motion group is embedded by a multipole of order R:

$$\int \frac{d^{2R}q}{2\pi^R} \frac{\Gamma(R)}{[-(q-io)^2+M^2]^R} e^{iqx} = \vartheta(x)\pi \mathcal{J}_0(|Mx|).$$

It is a time distribution of a position representation coefficient. The projection on the position hyperboloid with the $\mathbf{SO}_0(1, 2R-1)$-coefficients by integration over time, i.e., for trivial energy $\int \frac{dx_0}{2\pi} e^{iq_0x_0} = \delta(q_0)$, gives a negative energy (bound) state exponential $\exp \in L^2(\mathcal{Y}^{2R-1})$ for hyperbolic position:

$$\int \frac{dx_0}{\pi} \int \frac{d^{2R}q}{2\pi^R} \frac{\Gamma(R)}{[-(q-io)^2+M^2]^R} e^{iqx} = \int \frac{d^{2R-1}q}{\pi^R} \frac{\Gamma(R)}{(\vec{q}^2+M^2)^R} e^{-i\vec{q}\vec{x}} = \frac{e^{-|M|r}}{|M|}$$

With the $r = 2$ (real rank) continuous invariants and the $r_c = R-1$ (imaginary rank) discrete invariants of the orthogonal group $\mathbf{SO}(2R-1)$ with a Cartan torus $\mathbf{SO}(2)^{R-1}$, the acting group $\mathbf{D}(1) \times \mathbf{SO}_0(1, 2R-1)$ has rank $r + r_c = 1 + R$.

Unitary relativity $\mathcal{D}^4 = \mathbf{D}(1) \times \mathcal{Y}^3$ with three position dimensions as the minimal nonabelian case has imaginary rank 1 for the compact $\mathbf{SO}(2) \subset \mathbf{SO}(3)$-rotation degrees of freedom. The $\mathcal{D}^4$-representation coefficients with two continuous invariants (real rank 2 for noncompact $\mathbf{D}(1) \times \mathbf{SO}_0(1, 1)$) involve a position-representing dipole, e.g., with Lorentz group vector properties,

$$\mathbf{GL}(2, \mathbb{C})/\mathbf{U}(2) \cong \mathbb{R}^4_+ \ni \vartheta(x)x \longmapsto \int \frac{d^4q}{2\pi^2} \frac{iq}{[-(q-io)^2+m^2][-(q-io)^2+M^2]^2} e^{iqx}$$
$$\sim \frac{x}{2}\left(\frac{\partial}{\partial \frac{x^2}{4}}\right)^2 \vartheta(x)\pi \mathcal{J}_0(|mx|) \, * \, \vartheta(x)\pi \mathcal{J}_0(|Mx|).$$

With a different motivation and interpretation, a similar "two-point function" for Minkowski spacetime $\mathbb{R}^{(1,3)}$ with Wigner-particle metric and translation ground-state $|0\rangle$ was proposed by Heisenberg in his attempts for a unified field theory of elementary particles (nonlinear spinor theory) with a simple positive metric pole for the baryonic and an ad hoc regularizing indefinite metric dipole related to the leptonic sector. Here, the spacetime function is motivated as coefficient of electroweak spacetime $\mathcal{D}^4$: The simple pole, embedding time (dilation) $\mathbf{D}(1)$-representations e^{imx_0}, may be extended and used for a tangent Poincaré group representation with Wigner particles. The dipole, embedding Hilbert representations of hyperbolic position $\mathcal{Y}^3$ via a negative energy (bound) state exponential $e^{-|M|r}$, has no particle asymptotics by a Minkowski spacetime representation. It will be proposed to describe interactions, not parametrizable by Wigner-particles.

12.9.4 Spacetime Hilbert Spaces

Obviously, the representation coefficients of spacetime $\mathcal{D}^{2R}$ and the related Hilbert spaces differ from those of its tangent Minkowski spacetime $\mathbb{R}^{(1,2R-1)}$, used with $R = 2$ for particles. They are not Feynman propagators. The free particle Hilbert spaces involve (energy-)momentum functions, in contrast to the non-translational Hilbert spaces with space(-time) functions.

The Hilbert spaces $L^2(\Omega^2)$ for Euclidean group representations, i.e., the square integrable functions of the momentum directions on the 2-sphere $\vec{q}^2 = P^2$, are not appropriate for negative energy (bound) state representations with Hilbert space $L^2(\mathcal{Y}^3)$, comprising the square integrable functions of Kepler position. Similarily, the Hilbert spaces for Poincaré group representations, e.g., the square integrable functions $L^2(\mathcal{Y}^3)$ of the momentum hyperboloid for massive particles (mass shell $q^2 = m^2 > 0$), are not appropriate for electroweak spacetime representations with Hilbert space $L^2(\mathcal{D}^4)$, comprising the square integrable functions of causal spacetime. The special relativistic space $L^2(\mathcal{Y}^3)$ for Wigner particles has as nonrelativistic subspace $L^2(\Omega^2)$ for nonrelativistic free scattering states, e.g., $\frac{\sin Pr}{Pr}$. The space $L^2(\mathcal{D}^4)$ for electroweak spacetime has as position related subspace $L^2(\mathcal{Y}^3)$ for hyperbolic position, e.g., $e^{-|B|r}$, as used for the negative energy (bound) states of the Kepler potential:

free (scattering)	interacting (bound)
$L^2(\Omega^2)$ for momenta: $\vec{q} \in \Omega^2,\ \vec{q}^2 = P^2,$ $\delta(\vec{q}^2 - P^2)$	$L^2(\mathcal{Y}^3)$ for position: $e^{\vec{x}\vec{\sigma}} \in \mathcal{Y}^3,$ $\frac{1}{(\vec{q}^2+B^2)^2}$
$\mathbb{E}^3 \supset \Omega^2 \subset \mathcal{Y}^3 \subset \mathbb{R}^{(1,3)}$	$\mathcal{Y}^3 \subset \mathcal{D}^4$
$L^2(\mathcal{Y}^3)$ for energy-momenta: $q \in \mathcal{Y}^3,\ q^2 = m^2,$ $\delta(q^2 - m^2)$	$L^2(\mathcal{D}^4)$ for spacetime: $e^{x_0\mathbf{1}_2+\vec{x}\vec{\sigma}} \in \mathcal{D}^4,$ $\frac{1}{(-q^2+m^2)(-q^2+M^2)^2}$

The Fourier transform of the advanced (retarded) "energy-momentum" integration is, up to the order function, the corresponding principal value integration:

$$\begin{aligned} \int \frac{d^{2R}q}{2\pi} \frac{\Gamma(1+N)}{[-(q\mp io)^2+1]^{1+N}} e^{iqx} &= \vartheta(\pm x_0) \int \frac{d^{2R}q}{\pi} \frac{\Gamma(1+N)}{(-q_{\mathrm{P}}^2+1)^{1+N}} e^{iqx}, \\ \int \frac{d^{2R}q}{\pi} \frac{\Gamma(1+N)}{(-q_{\mathrm{P}}^2+1)^{1+N}} e^{iqx} &= i\epsilon(x_0) \int d^{2R}q\, \epsilon(q_0)\delta^{(N)}(q^2-1)e^{iqx}, \\ &\text{for } N = 0, 1, 2, \ldots \end{aligned}$$

All these spacetime distributions are contained in the Feynman distributions:

$$
\begin{aligned}
\int \frac{d^{2R}q}{i\pi^R} \frac{\Gamma(R-N)}{(-q^2-io+1)^{R-N}} e^{iqx} &= \quad (-\frac{\partial}{\partial\frac{x^2}{4}})^N \; 2\mathcal{K}_0(\sqrt{-x^2+io}),\; N=0,1,\dots R-1,\\
\int \frac{d^{2R}q}{i\pi^R} \frac{\Gamma(R+N)}{(-q^2-io+1)^{R+N}} e^{iqx} &= (\frac{x^2}{4})^N (\frac{\partial}{\partial\frac{x^2}{4}})^N 2\mathcal{K}_0(\sqrt{-x^2+io}),\; N=0,1,2,\dots,\\
\int \frac{d^{2R}q}{i\pi^R} \frac{\Gamma(R)}{(-q^2-io+m^2)^R} e^{iqx} &= \int_{m^2}^\infty \frac{d\kappa^2}{\sqrt{(\kappa^2-m^2)\kappa^2}} \; e^{-\sqrt{-\kappa^2x^2+io}}\\
&= \vartheta(x^2)2\mathcal{K}_0(i|mx|)+\vartheta(-x^2)2\mathcal{K}_0(|mx|),\\
2\mathcal{K}_0(i|mx|) = 2\int_{|m|}^\infty \frac{d\kappa}{\sqrt{\kappa^2-m^2}} \; e^{-i|\kappa x|} &= -\pi\mathcal{N}_0(|mx|)-i\pi\mathcal{J}_0(|mx|).
\end{aligned}
$$

Electroweak spacetime representation coefficients, so far given by causally supported Bessel functions, e.g., $\vartheta(x)x \longmapsto \vartheta(x)\pi\mathcal{J}_0(x)$, arise as causally supported part $\vartheta(x)2\mathcal{K}_0(i|mx|)$ of Fourier transformed Feynman distributions with the unitary contribution, where the exponential is integrated with a positive measure:

$$
\begin{aligned}
&R=1,2,\dots,\quad L=0,1,2,\dots:\\
\mathcal{D}^{2R}:\; \vartheta(x)x \longmapsto\; &\vartheta(x)\int \frac{d^{2R}q}{i\pi^R} \frac{(m_L^2-M_L^2)\;\Gamma(R)}{(-q^2-io+m_L^2)(-q^2-io+M_L^2)^R}(2iq)^L e^{iqx}\\
&= \vartheta(x)(2\partial)^L \int_{M_L^2}^{m_L^2} d^R\kappa^2 \int \frac{d^{2R}q}{i\pi^R} \frac{\Gamma(1+R)}{(-q^2-io+\kappa^2)^{1+R}} e^{iqx}\\
&= \vartheta(x)(x)^L(\frac{\partial}{\partial\frac{x^2}{4}})^L \int_{M_L^2}^{m_L^2} d^R\kappa^2 \int_{|\kappa|}^\infty \frac{d\mu}{\sqrt{(\mu^2-\kappa^2)^3}} \; e^{-i\mu|x|},
\end{aligned}
$$

$$
\begin{aligned}
(\partial^2+m_L^2)(\partial^2+M_L^2)^R \int \frac{d^{2R}q}{\pi^R} &\frac{\Gamma(R)\;(2iq)^L}{(-q^2-io+m_L^2)(-q^2-io+M_L^2)^R} e^{iqx}\\
&= \Gamma(R)(4\pi)^R(2\partial)^L\delta(x).
\end{aligned}
$$

Multipoles in the characteristic measure $\frac{d^{2R-1}q}{\pi^R} \frac{\Gamma(R)}{(\vec{q}^2+M^2)^R}$ with $R \geq 2$ do not serve as cutoff as in Minkowski spacetime particle theories, but implement nonflat hyperbolic position with nonabelian Lorentz group and the related negative energy (bound) state exponential. They cannot be connected with simple pole Wigner-particle propagators; the position related hyperbolic invariant (curvature) M^2 cannot be used as particle mass, i.e., as an invariant for tangent translations $\mathbb{R}^{(1,2R-1)}$. Multipoles parametrize interactions, "off-shell" and causally supported, and, possibly, can be related to strong interactions, usually implemented by $\mathbf{SU}(3)$-colored quark and gluon degrees of freedom, postulated to be confined, i.e., without translation action (no Wigner-particles). The embedded (eigen)time (dilation) representations with $\frac{dq_0}{i\pi} \frac{1}{-q_0^2-io+m^2}$ (simple poles) can be extended to Minkowski translations representation coefficients for time related spherical invariants m^2, i.e. to particle degrees of freedom. Harmonic $\mathbf{O}(1,2R-1)$-polynomials q_0^L, $(\vec{q})^L \hookrightarrow (q)^L$ implement both energy- and rotation $\mathbf{O}(2R-1)$-dependence.

Both factors come, in the product of the $\mathbf{D}(1) \times \mathcal{Y}^{2R-1}$-contributions, with a residual normalization $\frac{d^{2R}q}{i\pi^R}$. There remains a dilation line integration over $\mathbf{O}(1,2R-1)$-invariants with $d^R\kappa^2 = d\kappa^2\left(\frac{m^2-\kappa^2}{m^2-M^2}\right)^{R-1}$. For $(q)^L$ with $L=0,1,2$ the representation coefficients are functions, for $L \geq 3$ one obtains distributions proper. More on the normalization ahead.

12.10 Products of Representation Distributions

The convolution product $(f_1 * f_2)(x)$ of (space-)time coefficients is effected by the pointwise product of "energy(-momentum)" functions $(\tilde{f}_1 \cdot \tilde{f}_2)(q)$. It generalizes representations of *group elements products* like $g_1 g_2 \longmapsto D(g_1 g_2) = D(g_1) \circ D(g_2)$ — for group functions $L^1(G) * L^1(G) \subseteq L^1(G)$. In contrast, the *representations products* like $g \longmapsto (D_1 \otimes D_2)(g) = D_1(g) \otimes D_2(g)$ are characterized by the pointwise product $(f_1 \cdot f_2)(x)$ of (space-)time coefficients — for group functions $L^\infty(G) \cdot L^\infty(G) \subseteq L^\infty(G)$, which is effected by the convolution product $(\tilde{f}_1 * \tilde{f}_2)(q)$ of the Fourier transforms with "energy(-momentum)" distributions.

The familiar Feynman integrals involve convolution products of energy-momentum Feynman propagators whose Fourier transforms contain, from the on-shell contributions, product representation coefficients of the spacetime Poincaré group. The convolutions of the off-shell contributions are not defined (divergent).

Product representations of a compact group with discrete invariants and counting Plancherel measure can be characterized by the integer multiplicities of their irreducible subrepresentations with finite dimensions, e.g., for the product of two octet representations of the color group $\mathbf{SU}(3)$:

$$\begin{aligned} [1,1] \otimes [1,1] &= [0,0] \oplus 2 \times [1,1] \oplus [2,0] \oplus [0,2] \oplus [2,2], \\ 8 \otimes 8 &= 1 \oplus 2 \times 8 \oplus 10 \oplus \overline{10} \oplus 27. \end{aligned}$$

That is more complicated for noncompact groups with their possibly infinite-dimensional representations and continuous invariants, i.e., with continuously supported Plancherel measure and continuous multiplicities (normalizations).

The pointwise product of positive-type functions for translation $\mathbb{R}^n$ representations is related to the positive-type function for the product representation. With a normalization factor, e.g., $N = 1$ or $N = (2\pi)^n$, the pointwise product leads to the convolution product of the Fourier transforms with the "(energy-)momentum" distributions:

$$\begin{aligned} d(x) &= \int \frac{d^n q}{N}\, \tilde{d}(q) e^{iqx}, \quad (d_1 \cdot d_2)(x) = \int \frac{d^n q}{N}\, \widetilde{d_1 \cdot d_2}(q) e^{iqx}, \\ \widetilde{d_1 \cdot d_2}(q) &= \int d^n q_1 d^n q_2\, \tilde{d}_1(q_1) \frac{\delta(q_1 + q_2 - q)}{N} \tilde{d}_2(q_2) = \tilde{d}_1(q) \,\underset{N}{*}\, \tilde{d}_2(q). \end{aligned}$$

The convenient, but with respect to the arguments not quite correct notation $(\tilde{d}_1 * \tilde{d}_2)(q) = \tilde{d}_1(q) * \tilde{d}_2(q)$ is used. The "(energy-)momentum" addition in the convolution product $* \sim \delta(q_1 + q_2 - q)$ is normalized with a complex factor N.

12.10.1 Residual Normalizations

For irreducible translation $\mathbb{R}$ representations, $N = 1$ is the *residual normalization* factor for the Dirac distribution, and $N = 2i\pi = \oint \frac{dq}{q}$ for the $\mathbf{U}(1)$-phase of the simple pole in the complex energy- or momentum-plane:

$$\begin{aligned} e^{imt} &= \int dq\, \delta(q-m)e^{iqt}, \ \delta(q-m_1) * \delta(q-m_2) = \delta(q-m_1-m_2), \\ \vartheta(t)e^{imt} &= \int \frac{dq}{2i\pi} \frac{1}{q-io-m} e^{iqt}, \qquad \frac{1}{q-io-m_1} \frac{*}{2i\pi} \frac{1}{q-io-m_2} = \frac{1}{q-io-m_1-m_2}. \end{aligned}$$

For $\mathbb{R}$- and $\mathbb{R}^3$ and half integer-index Bessel functions with quadratic invariant, the convolution products are characterized by the sum of the invariants:

$$\begin{aligned} e^{i|mx|} &= \int \frac{dq}{i\pi} \frac{|m|}{q^2-io-m^2} e^{iqx}, & \frac{|m_1|}{q^2-io-m_1^2} \frac{*}{i\pi} \frac{|m_2|}{q^2-io-m_2^2} &= \frac{|m_+|}{q^2-io-m_+^2}, \\ e^{-|mx|} &= \int \frac{dq}{\pi} \frac{|m|}{q^2+m^2} e^{iqx}, & \frac{|m_1|}{q^2+m_1^2} \frac{*}{\pi} \frac{|m_2|}{q^2+m_2^2} &= \frac{|m_+|}{q^2+m_+^2}, \\ e^{i|m|r} &= \int \frac{d^3q}{i\pi^2} \frac{|m|}{(\vec{q}^2-io-m^2)^2} e^{i\vec{q}\vec{x}}, & \frac{|m_1|}{(\vec{q}^2-io-m_1^2)^2} \frac{*}{i\pi^2} \frac{|m_2|}{(\vec{q}^2-io-m_2^2)^2} &= \frac{|m_+|}{(\vec{q}^2-io-m_+^2)^2}, \\ e^{-|m|r} &= \int \frac{d^3q}{\pi^2} \frac{|m|}{(\vec{q}^2+m^2)^2} e^{i\vec{q}\vec{x}}, & \frac{|m_1|}{(\vec{q}^2+m_1^2)^2} \frac{*}{\pi^2} \frac{|m_2|}{(\vec{q}^2+m_2^2)^2} &= \frac{|m_+|}{(\vec{q}^2+m_+^2)^2}, \end{aligned}$$
with $m_+ = |m_1| + |m_2|$.

For real odd-dimensional spaces $n = 1 + 2R = 1, 3, \ldots$ with 2-sphere spreads $\mathbf{O}(1) \xrightarrow{(4\pi \frac{\partial}{\partial x^2})^R} \mathbf{O}(1+2R)$ and point singularities, the residual normalization in the complex "(energy-)momentum" space is $\frac{d^{1+2R}q}{\pi^{1+R}}$, starting with $\frac{dq}{\pi}, \frac{d^3q}{\pi^2}, \ldots$.

The pointwise products of integer-index Bessel functions for even-dimensional spaces $\mathbb{R}^{2R}$, and, therefore, the related "(energy-)momentum" convolutions are more complicated. They are characterized by an integration over an invariant line (spectral measure) from the difference of the invariants to their sum, as seen in the following example with the threshold factor $\Delta(\kappa^2)$:

$$\begin{aligned} \pi \mathcal{J}_0(mr) &= \int_0^{m^2} \frac{d\kappa^2}{\sqrt{(m^2-\kappa^2)\kappa^2}} \cos \kappa r, \\ \pi \mathcal{J}_0(m_1 r) \cdot \pi \mathcal{J}_0(m_2 r) &= \int_{m_-^2}^{m_+^2} \frac{d\kappa^2}{\sqrt{(m_+^2-\kappa^2)(\kappa^2-m_-^2)}} \pi \mathcal{J}_0(\kappa r). \end{aligned}$$

For real even-dimensional spaces $n = 2R = 2, 4, \ldots$, with 2-sphere spreads $\mathbf{O}(1,1) \xrightarrow{(4\pi \frac{\partial}{\partial x^2})^{R-1}} \mathbf{O}(1, 2R-1)$ and line singularities, the residual normalization is $\frac{d^{2R}q}{\pi^R}$ in the complex "energy-momentum" space, starting with $\frac{d^2q}{\pi}, \frac{d^4q}{\pi^2}, \ldots$ (more ahead).

12.10.2 Convolution Products of Feynman Distributions

For a convolution of "(energy-)momentum" distributions for space(-time) Feynman distributions, it is convenient to rationalize the Feynman distributions by separating residual normalizations, exhibited in $\frac{*}{N} \sim \frac{1}{N}\delta(q_1 + q_2 - q)$:

$$\begin{array}{c}\mathbf{O}(t,s):\\ n=t+s,\\ n=1,2,\ldots,\\ t=0,1,\ldots,\\ m_{1,2}^2\in\mathbb{R},\end{array}\left\{\begin{array}{l}(\frac{\partial}{\partial q})^{L_1}\frac{\Gamma(\frac{n}{2}+\nu_1)}{(-q^2-io+m_1^2)^{\frac{n}{2}+\nu_1}}\frac{*}{i^t\pi^{\frac{n}{2}}}\otimes(\frac{\partial}{\partial q})^{L_2}\frac{\Gamma(\frac{n}{2}+\nu_2)}{(-q^2-io+m_2^2)^{\frac{n}{2}+\nu_2}}\\ \quad=(\frac{\partial}{\partial q})^{L_1}\otimes(\frac{\partial}{\partial q})^{L_2}\int_0^1 dz\,\frac{z^{\frac{n}{2}+\nu_1-1}(1-z)^{\frac{n}{2}+\nu_2-1}\,\Gamma(\frac{n}{2}+\nu_1+\nu_2)}{[-z(1-z)q^2-io+zm_1^2+(1-z)m_2^2]^{\frac{n}{2}+\nu_1+\nu_2}},\\ \text{with }(\frac{\partial}{\partial q})^L=(2q)^L\,(\frac{\partial}{\partial q^2})^L.\end{array}\right.$$

The products lead to a positive $\mathbf{D}(1)$-measure over an $\mathbf{O}(t,s)$-invariant line:

$$\int_0^1 dz=\frac{1}{m_1^2-m_2^2}\int_{m_2^2}^{m_1^2}d\kappa^2\text{ with }\left\{\begin{array}{l}\kappa^2=zm_1^2+(1-z)m_2^2,\\ z=\frac{\kappa^2-m_2^2}{m_1^2-m_2^2},\ 1-z=\frac{m_1^2-\kappa^2}{m_1^2-m_2^2}.\end{array}\right.$$

In the explicit calculation of the convolution, the denominators are joined with the extended Beta function relation above $\Gamma(\frac{n}{2}+\nu_1)*\Gamma(\frac{n}{2}+\nu_2)\sim\Gamma(n+\nu_1+\nu_2)$. After rearranging the "(energy-)momenta" $(q_1,q_2)\leftrightarrow(p,q)=(q_1-q_2,q_1+q_2)$, the residue is taken $\int\frac{d^np}{i^t\pi^{\frac{n}{2}}}\frac{\Gamma(\frac{n}{2}+\nu)}{(-p^2-io+a)^{\frac{n}{2}+\nu}}=\frac{\Gamma(\nu)}{(a-io)^\nu}$ with respect to the "relative (energy-)momenta" $p=q_1-q_2$. The convolution product depends on the "center of mass (energy-)momenta" $q=q_1+q_2$. Nontrivial $\mathbf{O}(t,s)$-properties with $L=0,1,2,\ldots$ are effected by the convolution-compatible "(energy-)momentum" derivatives.

With a shift $\nu\to\nu+\frac{1}{2}$ and an additional normalization $*\to\frac{*}{\Gamma(\frac{1}{2})}$, the appropriately normalized convolutions for real odd-dimensional spaces are (the $\frac{\partial}{\partial q}$-operations are left out):

$$\begin{array}{c}\mathbf{O}(t,1+2R-t):\\ n=1+2R=1,3,\ldots,\end{array}\left\{\begin{array}{l}\frac{\Gamma(1+R+\nu_1)}{(-q^2-io+m_1^2)^{1+R+\nu_1}}\frac{*}{i^t\pi^{1+R}}\frac{\Gamma(1+R+\nu_2)}{(-q^2-io+m_2^2)^{1+R+\nu_2}}\\ \quad=\frac{1}{\Gamma(\frac{1}{2})}\int_0^1 dz\,\frac{z^{R+\nu_1}(1-z)^{R+\nu_2}\,\Gamma(\frac{3}{2}+R+\nu_1+\nu_2)}{[-z(1-z)q^2-io+zm_1^2+(1-z)m_2^2]^{\frac{3}{2}+R+\nu_1+\nu_2}}\\ \quad=(\frac{\partial}{\partial q^2})^R\frac{1}{\Gamma(\frac{1}{2})}\int_0^1 dz\,\frac{z^{\nu_1}(1-z)^{\nu_2}\,\Gamma(\frac{3}{2}+\nu_1+\nu_2)}{[-z(1-z)q^2-io+zm_1^2+(1-z)m_2^2]^{\frac{3}{2}+\nu_1+\nu_2}},\end{array}\right.$$

e.g., for one-dimensional time and three-dimensional position:

$$\mathbf{O}(1,0):\left\{\begin{array}{l}\frac{\Gamma(1+\nu_1)}{(-q^2-io+m_1^2)^{1+\nu_1}}\frac{*}{i\pi}\frac{\Gamma(1+\nu_2)}{(-q^2-io+m_2^2)^{1+\nu_2}}\\ \quad=\frac{1}{\Gamma(\frac{1}{2})}\int_0^1 dz\,\frac{z^{\nu_1}(1-z)^{\nu_2}\,\Gamma(\frac{3}{2}+\nu_1+\nu_2)}{[-z(1-z)q^2-io+zm_1^2+(1-z)m_2^2]^{\frac{3}{2}+\nu_1+\nu_2}},\end{array}\right.$$

$$\mathbf{O}(0,3):\left\{\begin{array}{l}\frac{\Gamma(2+\nu_1)}{(\vec{q}^2-io+m_1^2)^{2+\nu_1}}\frac{*}{\pi^2}\frac{\Gamma(2+\nu_2)}{(\vec{q}^2-io+m_2^2)^{2+\nu_2}}\\ \quad=\frac{1}{\Gamma(\frac{1}{2})}\int_0^1 dz\,\frac{z^{1+\nu_1}(1-z)^{1+\nu_2}\,\Gamma(\frac{5}{2}+\nu_1+\nu_2)}{[z(1-z)\vec{q}^2-io+zm_1^2+(1-z)m_2^2]^{\frac{5}{2}+\nu_1+\nu_2}}.\end{array}\right.$$

For real even-dimensional spaces, the appropriately normalized convolutions are:

$$\mathbf{O}(t, 2R-t): \quad n = 2R = 2, 4, \ldots, \quad \begin{cases} \frac{\Gamma(R+\nu_1)}{(-q^2-io+m_1^2)^{R+\nu_1}} \underset{i^t\pi^R}{*} \frac{\Gamma(R+\nu_2)}{(-q^2-io+m_2^2)^{R+\nu_2}} \\ = \int_0^1 dz \, \frac{z^{R+\nu_1-1}(1-z)^{R+\nu_2-1}\,\Gamma(R+\nu_1+\nu_2)}{[-z(1-z)q^2-io+zm_1^2+(1-z)m_2^2]^{R+\nu_1+\nu_2}} \\ = (\frac{\partial}{\partial q^2})^{R-1} \int_0^1 dz \, \frac{z^{\nu_1}(1-z)^{\nu_2}\,\Gamma(1+\nu_1+\nu_2)}{[-z(1-z)q^2-io+zm_1^2+(1-z)m_2^2]^{1+\nu_1+\nu_2}} \end{cases}$$

e.g., for Cartan and Minkowski spacetime:

$$\mathbf{O}(1,1): \begin{cases} \frac{\Gamma(1+\nu_1)}{(-q^2-io+m_1^2)^{1+\nu_1}} \underset{i\pi}{*} \frac{\Gamma(1+\nu_2)}{(-q^2-io+m_2^2)^{1+\nu_2}} \\ = \int_0^1 dz \, \frac{z^{\nu_1}(1-z)^{\nu_2}\,\Gamma(1+\nu_1+\nu_2)}{[-z(1-z)q^2-io+zm_1^2+(1-z)m_2^2]^{1+\nu_1+\nu_2}}, \end{cases}$$

$$\mathbf{O}(1,3): \begin{cases} \frac{\Gamma(2+\nu_1)}{(-q^2-io+m_1^2)^{2+\nu_1}} \underset{i\pi^2}{*} \frac{\Gamma(2+\nu_2)}{(-q^2-io+m_2^2)^{2+\nu_2}} \\ = \int_0^1 dz \, \frac{z^{1+\nu_1}(1-z)^{1+\nu_2}\,\Gamma(2+\nu_1+\nu_2)}{[-z(1-z)q^2-io+zm_1^2+(1-z)m_2^2]^{2+\nu_1+\nu_2}}. \end{cases}$$

The convolution of simple poles, $R + \nu_{1,2} = 1$, leads to a singularity in $\Gamma(2-R)$ for $R \geq 2$.

12.10.3 Point and Line Convolutions

The integration for convoluted "(energy-)momentum" distributions is completely different for half-integer powers, relevant for real odd-dimensional spaces, e.g., for time and 3-position, with the addition of two invariants as in $e^{-\iota|m_1 x|}e^{-\iota|m_2 x|} = e^{-\iota|m_+ x|}$ with compactness $\iota \in \{1, \pm i\}$ and $m_+ = |m_1| + |m_2|$, and for integer powers, relevant for real even-dimensional spaces, e.g., for causal spacetimes, with the "addition" of two invariant lines $|m_1|\cosh\psi_1 + |m_2|\cosh\psi_2$, as for $\int d\psi\, e^{-\iota|mx|\cosh\psi}$.

For (half-)integer powers in the denominator, the remaining integrals can be reduced, by derivations with respect to the "(energy-)momentum" invariants, to a ($\frac{3}{2}$-power of a) quadratic polynomial $D(z)$ with the threshold factor $\Delta(q^2)$ as discriminant:

$$\frac{n}{2} + \nu_1 + \nu_2 = \begin{cases} 1 + N \Rightarrow \frac{1}{D(z)} = \frac{1}{-z(1-z)q^2-io+zm_1^2+(1-z)m_2^2}, \\ \frac{3}{2} + N \Rightarrow \frac{1}{D(z)^{\frac{3}{2}}} = \frac{4}{\Delta(q^2+io)} \frac{d^2\sqrt{D(z)}}{dz^2}, \end{cases}$$

$$\text{with } \Delta(q^2) = (q^2 - m_+^2)(q^2 - m_-^2), \quad N = 0, 1, 2, \ldots.$$

For half-integer powers, the z-integration cancels the m_-^2-pole from the discriminant $\Delta(q^2)$, and leaves, for the product distribution (measure), one characterizing invariant singularity at $q^2 = m_+^2$, e.g., for time $\mathbb{R}$ with $q^2 = q_0^2$, and positions $(\mathcal{Y}^{2R-1}, \mathbb{E}^{2R-1}, \Omega^{2R-1})$ with $q^2 = -\vec{q}^2$:

$$\mathbf{O}(t, 1+2R-t): \ \frac{\Gamma(1+R)}{(-q^2-io+m_1^2)^{1+R}} \underset{i^t\pi^{1+R}}{*} \frac{\Gamma(1+R)}{(-q^2-io+m_2^2)^{1+R}} = \left(\frac{\partial}{\partial q^2}\right)^R \int_0^1 \frac{dz}{2D(z)^{\frac{3}{2}}}$$
$$= \frac{|m_+|}{|m_1 m_2|} \frac{\Gamma(1+R)}{(-q^2-io+m_+^2)^{1+R}},$$
$$\text{with } \int_0^1 \frac{dz}{2D(z)^{\frac{3}{2}}} = \frac{2}{(-q^2-io+m_-^2)(-q^2-io+m_+^2)} \left.\frac{d\sqrt{D(z)}}{dz}\right|_0^1 = \frac{|m_+|}{|m_1 m_2|} \frac{1}{-q^2-io+m_+^2},$$

$$\underset{q^2=m_1^2}{\bullet} \quad * \quad \underset{q^2=m_2^2}{\bullet} \quad = \quad \underset{q^2=m_+^2}{\bullet}$$

For integer powers, the remaining line integral $\int_0^1 dz = \int_{m_2^2}^{m_1^2} \frac{d\kappa^2}{m_1^2-m_2^2}$ occurs, for example, for Minkowski and causal spacetimes $(\mathbb{R}^{(1,2R-1)}, \mathcal{D}^{2R})$ with $q^2 = q_0^2 - \vec{q}^2$:

$$\mathbf{O}(t, 2R-t): \ \frac{\Gamma(R)}{(-q^2-io+m_1^2)^R} \underset{i^t\pi^R}{*} \frac{\Gamma(R)}{(-q^2-io+m_2^2)^R} = \left(\frac{\partial}{\partial q^2}\right)^{R-1} \int_0^1 \frac{dz}{D(z)},$$

$$\int_0^1 \frac{dz}{D(z)} = \int_0^1 dz \left[\frac{1}{-z(1-z)q_P^2+zm_1^2+(1-z)m_2^2} + i\pi\delta(-z(1-z)q^2+zm_1^2+(1-z)m_2^2)\right]$$

$$= \frac{2}{\sqrt{|\Delta(q^2)|}} \left[\begin{array}{c} \vartheta(m_-^2-q^2)\log\left|\frac{\Sigma(q^2)-2\sqrt{\Delta(q^2)}}{m_+^2-m_-^2}\right| \\ +\,\vartheta(q^2-m_-^2)\vartheta(m_+^2-q^2)\arctan\frac{2\sqrt{-\Delta(q^2)}}{\Sigma(q^2)} \\ +i\pi\,\vartheta(q^2-m_+^2) \end{array}\right],$$

$$\text{with } \Delta(q^2) = (q^2-m_+^2)(q^2-m_-^2), \quad \Sigma(q^2) = (q^2-m_+^2)+(q^2-m_-^2),$$

$$\underset{q^2=m_1^2}{\bullet} \quad * \quad \underset{q^2=m_2^2}{\bullet} \quad =$$

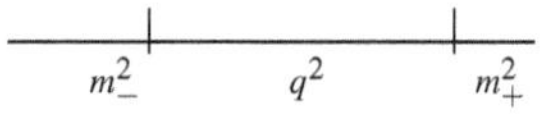

In contrast to the pole structure for $q^2 - m_+^2$ in the half-integer case, the convolution product for the integer case depends on the two variables $\{q^2 - m_+^2, q^2 - m_-^2\}$ with two masses $m_1^2 - m_2^2 = m_+ m_-$. The contribution over the threshold with $\vartheta(q^2 - m_+^2)$ (here imaginary part) is the on-shell convolution for Minkowski spacetime. The three contributions can be seen in parallel to the two real off-shell contributions — for imaginary and real momenta $\vec{q}^2 = q_0^2 - m^2$ — and the one imaginary on-shell contribution in the analysis of a Feynman propagator above, related to the maximally symmetric positions $\mathcal{Y}^{2R-1}$, Ω^{2R-1} and $\mathbb{E}^{2R-1}$, respectively.

In contrast to the characteristic distributions and convolutions in the case of odd-dimensional spaces, e.g., for time and for positions, correspondingly normalized with π and π^2, which contain *invariant pole singularities* (zero-dimensional), the characteristic distributions and convolutions for even-dimensional spaces, e.g. for causal $\mathbf{O}(1, 2R-1)$-spacetime with embedded point singularities, display pole distributions only before an integration over an *invariant line singularity* (one-dimensional). They can be written with spectral functions, for example,

$$\int_0^1 dz \, \frac{1}{-zq^2+m^2} = \int_{m^2}^\infty \frac{d\kappa^2}{\kappa^2} \frac{1}{-q^2+\kappa^2},$$
$$\int_0^1 dz \, \frac{1-z}{-zq^2+m^2} = \int_{m^2}^\infty \frac{d\kappa^2(\kappa^2-m^2)}{(\kappa^2)^2} \frac{1}{-q^2+\kappa^2}.$$

After z-integration logarithms arise. The logarithm and the absence of the 1-sphere factor 2π for the full circle is typical for a "non-circle" integration. In general, one has for a function $\hat{\mathbb{C}} \ni z \longmapsto f(z) \in \hat{\mathbb{C}}$ holomorphic on the integration curve with $\alpha \neq \infty$ the residual results (where applicable):

$$\int_{\alpha_2}^{\alpha_1} dz\, f(z) = \sum \text{res}[f(z) \log \tfrac{z-\alpha_1}{z-\alpha_2}],$$

$$\int_{\alpha}^{\infty} dz\, f(z) = -\sum \text{res}[f(z) \log(z-\alpha)],$$

$$\int_{-\infty}^{\infty} dz\, f(z) = 2i\pi \sum \text{res} f(z),$$

with the sum of all residues in the closed complex plane $\hat{\mathbb{C}} = \mathbb{C} \cup \{\infty\}$, cut along the integration curve, for the examples above,

$$\begin{aligned} \int_0^1 dz\, \tfrac{1}{-zq^2+m^2} &= \text{res}\left[\tfrac{1}{-zq^2+m^2} \log \tfrac{z-1}{z}\right] &= \frac{\log(1-\frac{q^2}{m^2})}{-q^2}, \\ \int_0^1 dz\, \tfrac{1-z}{-zq^2+m^2} &= \sum \text{res}\left[\tfrac{1-z}{-zq^2+m^2} \log \tfrac{z-1}{z}\right] &= \frac{(1-\frac{m^2}{q^2})\log(1-\frac{q^2}{m^2})-1}{-q^2}. \end{aligned}$$

In the second case, there is a nontrivial residue at the holomorphic point $z = \infty$.

12.10.4 Convolution Modules and Algebras

The convolution for the nonrelativistic scattering momentum distributions describes product representations of the Euclidean group with the momentum between the invariants $P_-^2 \leq \vec{q}^2 \leq P_+^2$, $P_\pm = |P_1| \pm |P_2|$, and the volume of the momentum spheres $\Omega^{s-1}(\vec{q}^2) \subset \mathbb{R}^s$ for $s \geq 1$:

$$\begin{aligned} \mathbf{SO}(1+s) \vec{\times} \mathbb{R}^{1+s} :\; & \delta(\vec{q}^2 - P_1^2) * \delta(\vec{q}^2 - P_2^2) \\ & = |\Omega^{s-1}(\vec{q}^2)| \frac{2}{\sqrt{-\Delta(\vec{q}^2)}} \vartheta(P_+^2 - \vec{q}^2)\vartheta(\vec{q}^2 - P_-^2), \\ \text{with } & |\Omega^{s-1}(\vec{q}^2)| = |\Omega^{s-1}| (\tfrac{-\Delta(\vec{q}^2)}{4\vec{q}^2})^{\frac{s-1}{2}},\; |\Omega^{s-1}| = \frac{2\pi^{\frac{s}{2}}}{\Gamma(\frac{s}{2})}, \\ & -\Delta(\vec{q}^2) = (P_+ - \vec{q}^2)(\vec{q}^2 - P_-^2), \end{aligned}$$

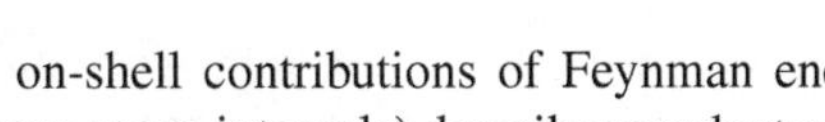

The convolution of the on-shell contributions of Feynman energy-momentum distributions (relativistic phase space integrals) describes product representations of the Poincaré group, e.g., a pair of free particles over the threshold $q^2 \geq m_+^2$ with $m_+ = |m_1| + |m_2|$ and the volume of the momentum spheres $\Omega^{s-1}(q^2) \subset \mathbb{R}^s$ for $s \geq 1$:

$$\begin{aligned}\mathbf{SO}_0(1,s)\ \vec{\times}\ \mathbb{R}^{1+s}:\ \vartheta(\pm q_0)\delta(q^2-m_1^2)\ &* \vartheta(\pm q_0)\delta(q^2-m_2^2)\\ &= |\Omega^{s-1}(q^2)|\frac{1}{\sqrt{\Delta(q^2)}}\vartheta(\pm q_0)\vartheta(q^2-m_+^2),\\ \text{with } |\Omega^{s-1}(q^2)| &= |\Omega^{s-1}|\left(\frac{\Delta(q^2)}{4q^2}\right)^{\frac{s-1}{2}},\quad |\Omega^{s-1}| = \frac{2\pi^{\frac{s}{2}}}{\Gamma(\frac{s}{2})},\\ \Delta(q^2) &= (q^2-m_+^2)(q^2-m_-^2),\end{aligned}$$

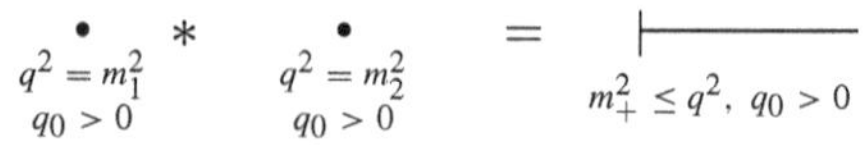

The convolution of Feynman distributions with either $q^2 + io$ or $q^2 - io$ gives the corresponding definite integration prescription via $zio + (1-z)io = io$:

$$\begin{aligned}\mu_1(q^2+io) * \mu_2(q^2+io) &= \mu_{1*2}(q^2+io),\\ \mu_1(q^2-io) * \mu_2(q^2-io) &= \mu_{1*2}(q^2-io).\end{aligned}$$

In general, a convolution of factors with different integration prescriptions $\mu_1(q^2 + io) * \mu_2(q^2 - io)$ does not determine a unique prescription for their product with indefinite $zio - (1-z)io = (2z-1)io$ where $2z - 1 \in [-1, 1]$ for $z \in [0, 1]$.

The Feynman representation distributions $d^n q \frac{\Gamma(\frac{n}{2}+\nu)}{(-q^2-io+m^2)^{\frac{n}{2}+\nu}}$ with positive integer denominator power for $\nu > 0$, i.e.,

$$\left.\begin{array}{ll}\text{for odd } n = 1+2R: & \nu = \frac{1}{2}, \frac{3}{2}, \ldots\\ \text{for even } n = 2R: & \nu = 1, 2, \ldots\end{array}\right\} \Rightarrow \frac{n}{2} + \nu = 1+R, 2+R, \ldots,$$

span a convolution algebra. As spacetime functions, they can be multiplied pointwise. The Feynman distributions with positive integer denominator power $\frac{n}{2} + \nu = 0, 1, 2, \ldots$ span a convolution module for these representation distributions. As spacetime distributions, they can be multiplied pointwise with the representation functions,

$\mathbf{O}(t, n-t)$:

$*$	$\nu > 0$	$\frac{n}{2}+\nu \ge 0$
$\nu > 0$	$\nu > 0$	$\frac{n}{2}+\nu \ge 0$
$\frac{n}{2}+\nu \ge 0$	$\frac{n}{2}+\nu \ge 0$	−

with integer $\frac{n}{2} + \nu \ge 0$,

$$\begin{aligned}\frac{\Gamma(\frac{n}{2}+\nu_1)}{(-q^2-io+m_1^2)^{\frac{n}{2}+\nu_1}}\ \frac{*}{i^t\pi^{\frac{n}{2}}}\ \frac{\Gamma(\frac{n}{2}+\nu_2)}{(-q^2-io+m_2^2)^{\frac{n}{2}+\nu_2}} &= \int_0^1 dz\, \frac{z^{\frac{n}{2}+\nu_1-1}(1-z)^{\frac{n}{2}+\nu_2-1}\ \Gamma(\frac{n}{2}+\nu_1+\nu_2)}{[-z(1-z)q^2-io+zm_1^2+(1-z)m_2^2]^{\frac{n}{2}+\nu_1+\nu_2}},\\ \frac{\Gamma(\frac{n}{2}+\nu_1)}{(-q^2-io+m^2)^{\frac{n}{2}+\nu_1}}\ \frac{*}{i^t\pi^{\frac{n}{2}}}\ \frac{\Gamma(\frac{n}{2}+\nu_2)}{(-q^2-io)^{\frac{n}{2}+\nu_2}} &= \int_0^1 dz\, \frac{z^{-\nu_2-1}(1-z)^{\frac{n}{2}+\nu_2-1}\ \Gamma(\frac{n}{2}+\nu_1+\nu_2)}{[-(1-z)q^2-io+m^2]^{\frac{n}{2}+\nu_1+\nu_2}}.\end{aligned}$$

The representation distributions arise by derivations $(\partial^2 + m^2)^N$ from the representation functions. The proper distributions for $0 \ge \nu \ge -\frac{n}{2}$ are kernels of corresponding differential equations, e.g., $(\partial^2 + m^2)^{\frac{n}{2}+\nu}\mathbf{k}(x) = \delta(x)$, used for interactions. They distribute subgroup representation coefficients. The simple poles (Green's functions) distribute abelian (Cartan) subgroups $\mathbf{SO}(2)$ and $\mathbf{SO}_0(1, 1)$:

$$\mathbf{O}(t, 1+2R-t): \begin{cases} \frac{\Gamma(1+R)}{(-q^2-io+m_1^2)^{1+R}} \frac{*}{i^t\pi^{1+R}} \frac{\Gamma(1+\nu)}{(-q^2-io+m_2^2)^{1+\nu}} \\ = \frac{1}{\Gamma(\frac{1}{2})} \int_0^1 dz \frac{z^R(1-z)^\nu \, \Gamma(\frac{3}{2}+\nu)}{[-z(1-z)q^2-io+zm_1^2+(1-z)m_2^2]^{\frac{3}{2}+\nu}}, \end{cases}$$

$$\text{e.g. } \mathbf{O}(1+2R): \begin{cases} \frac{\Gamma(1+R)|m_1|}{(\vec{q}^2+m_1^2)^{1+R}} \frac{*}{\pi^{1+R}} \frac{\Gamma(1+R)|m_2|}{(\vec{q}^2+m_2^2)^{1+R}} = \frac{\Gamma(1+R)|m_+|}{(\vec{q}^2+m_+^2)^{1+R}}, \\ e^{-|m_1\vec{x}|} \cdot e^{-|m_2\vec{x}|} = e^{-|m_+\vec{x}|}, \\ R \geq 1: \frac{\Gamma(1+R)|m_1|}{(\vec{q}^2+m_1^2)^{1+R}} \frac{*}{\pi^{1+R}} \frac{\Gamma(R)}{(\vec{q}^2+m_2^2)^R} = \frac{\Gamma(R)}{(\vec{q}^2+m_+^2)^R}, \\ e^{-|m_1\vec{x}|} \cdot \frac{e^{-|m_2\vec{x}|}}{|\vec{x}|} = \frac{e^{-|m_+\vec{x}|}}{|\vec{x}|}, \end{cases}$$

$$\mathbf{O}(t, 2R-t): \begin{cases} \frac{\Gamma(1+R)}{(-q^2-io+m_1^2)^{1+R}} \frac{*}{i^t\pi^R} \frac{\Gamma(1+\nu)}{(-q^2-io+m_2^2)^{1+\nu}} \\ = \int_0^1 dz \frac{z^R(1-z)^\nu \, \Gamma(2+\nu)}{[-z(1-z)q^2-io+zm_1^2+(1-z)m_2^2]^{2+\nu}}, \\ \text{e.g., } R=2: \frac{2}{(-q^2-io+m^2)^3} \frac{*}{i^t\pi^2} \frac{1}{-q^2-io} = \frac{1}{m^2(-q^2-io+m^2)}. \end{cases}$$

Space(-time) Feynman functions and distributions as Fourier-transformed "(energy-)momentum" distributions have pointwise multiplication properties in parallel with essentially bounded group functions $L^\infty(G)$ and group Radon distributions $\mathcal{M}(G)$:

$\cdot$	$L^\infty(G)$	$\mathcal{M}(G)$
$L^\infty(G)$	$L^\infty(G)$	$\mathcal{M}(G)$
$\mathcal{M}(G)$	$\mathcal{M}(G)$	–

$*$	$\mathcal{M}(G)$	$L^\infty(G)$
$\mathcal{M}(G)$	$\mathcal{M}(G)$	$L^\infty(G)$
$L^\infty(G)$	$L^\infty(G)$	–

$\mu_1 \cdot \mu_2(g)$ $\mu_1 * \mu_2(g)$

$$G \subseteq \mathcal{M}(G) \subseteq L^\infty(G)'.$$

The positive type functions $L^\infty(G)_+$ characterize the cyclic Hilbert representations. The Radon distributions (measures) contain the group and are dual to the essentially bounded functions.

The convolution of Feynman "energy-momentum" distributions is the pointwise multiplication of spacetime functions — separate for causal and spacelike support with $\vartheta_{t,s} \sim \vartheta(\pm x^2)$:

$$f = \vartheta_t f_t + \vartheta_s f_s : \ f \cdot g = \vartheta_t(f_t \cdot g_t) + \vartheta_s(f_s \cdot g_s),$$

and, therefore, can be used for product representations of electroweak spacetime parametrized by causal spacetime translations $\vartheta(x^2)x \in \mathbb{R}^4_{\text{caus}}$, for example:

$$
\begin{aligned}
e^{-\sqrt{-m^2x^2+io}} &= \vartheta(x^2)e^{-i|mx|} + \vartheta(-x^2)e^{-|mx|},\\
e^{-\sqrt{-m_1^2x^2+io}}e^{-\sqrt{-m_2^2x^2+io}} &= \vartheta(x^2)e^{-i|m_+x|} + \vartheta(-x^2)e^{-|m_+x|},
\end{aligned}
$$

$$
\begin{aligned}
2\mathcal{K}_0(\sqrt{-m^2x^2+io}) &= \vartheta(x^2)2\mathcal{K}_0(i|mx|) + \vartheta(-x^2)2\mathcal{K}_0(|mx|),\\
2\mathcal{K}_0(\sqrt{-m_1^2x^2+io})2\mathcal{K}_0(\sqrt{-m_2^2x^2+io}) &= \vartheta(x^2)2\mathcal{K}_0(i|m_1x|)2\mathcal{K}_0(i|m_2x|)\\
&\quad + \vartheta(-x^2)2\mathcal{K}_0(|m_1x|)2\mathcal{K}_0(|m_2x|).
\end{aligned}
$$

A further projective decomposition is possible for future and past support $\vartheta(\pm x)x \in \mathbb{R}^4_\pm$ according to the decomposition of the unit:

$$
1 = \vartheta(x^2) + \vartheta(-x^2) = [\vartheta(x) + \vartheta(-x)] + \vartheta(-x^2).
$$

12.11 Normalizations of Feynman Measures

The representation normalization of an $\mathbf{O}(t,s)$-scalar Feynman measure, if defined, is its value for the trivial group element $x = 0$ and given by the integral over all "(energy-)momenta," i.e., over the individual q-dependent normalizations of the representations $x \longmapsto e^{iqx} \in \mathbf{U}(1)$:

$$
\mu(x) = \int d^nq\, \tilde\mu(q)e^{iqx},\quad \mu(0) = \int d^nq\, \tilde\mu(q).
$$

It is representation-normalized with $\mu(0) = 1$ as exemplified by the consistency equation for the chiral "breakdown" above. With residual normalization, the remaining representation normalizations for odd and even-dimensional spaces (where defined) involve dimension-related powers of the characteristic invariants:

$$
\begin{aligned}
a\in\mathbb{R},\ \ \mathbf{O}(t,n-t):& \int \frac{d^nq}{i^t\pi^{\frac{n}{2}}}\frac{\Gamma(\frac{n}{2}+\nu)}{(-q^2-io+a)^{\frac{n}{2}+\nu}} &&= \frac{\Gamma(\nu)}{(a-io)^{\nu}}, && \nu > 0,\\
\mathbf{O}(t,1+2R-t):& \int \frac{d^{1+2R}q}{i^t\pi^{1+R}}\frac{\Gamma(1+R+\nu)}{(-q^2-io+a)^{1+R+\nu}} &&= \frac{\Gamma(\frac{1}{2}+\nu)}{\Gamma(\frac{1}{2})(a-io)^{\frac{1}{2}+\nu}}, && \tfrac{1}{2}+\nu > 0,\\
\mathbf{O}(t,2R-t):& \int \frac{d^{2R}q}{i^t\pi^{R}}\frac{\Gamma(1+R+\nu)}{(-q^2-io+a)^{1+R+\nu}} &&= \frac{\Gamma(1+\nu)}{(a-io)^{1+\nu}}, && 1+\nu > 0.
\end{aligned}
$$

The convolution product of two Feynman measures for trivial "center of mass (energy-)momentum," if defined, is their product normalization:

$$
\begin{aligned}
\langle\mu_1,\mu_2\rangle &= \int \frac{d^nx}{(2\pi)^n}\,\mu_1(-x)\mu_2(x) = \mu_1(x) \underset{(2\pi)^n}{*} \mu_2(x)|_{x=0}\\
&= \int d^nq\, \tilde\mu_1(q)\tilde\mu_2(q) = \tilde\mu_1(-q) * \tilde\mu_2(q)|_{q=0}.
\end{aligned}
$$

With respect to such a product, e.g., with test functions and distributions like $\mathcal{M}(G) \subseteq L^\infty(G)'$, one may have orthogonality, $\langle\mu,f\rangle = 0$, or duality, $\langle\mu,f\rangle = 1$, etc., as exemplified by the eigenvalue equation for the massless mode in the chiral model above.

Products of $\mathbf{O}(t,s)$-scalar Feynman measures with real invariants $m^2 \in \mathbb{R}$,

$$\begin{aligned}
\int \frac{d^n q}{i^t \pi^{\frac{n}{2}}} \frac{\Gamma(\frac{n}{2}+\nu_1)}{(-q^2-io+m_1^2)^{\frac{n}{2}+\nu_1}} \frac{\Gamma(\frac{n}{2}+\nu_2)}{(-q^2-io+m_2^2)^{\frac{n}{2}+\nu_2}} &= \int_0^1 dz\, \frac{z^{\frac{n}{2}+\nu_1-1}(1-z)^{\frac{n}{2}+\nu_2-1}\,\Gamma(\frac{n}{2}+\nu_1+\nu_2)}{[zm_1^2+(1-z)m_2^2-io]^{\frac{n}{2}+\nu_1+\nu_2}}\\
&= \frac{1}{m_1^2-m_2^2}\int_{m_2^2}^{m_1^2} \frac{\Gamma(\frac{n}{2}+\nu_1+\nu_2)\,d\kappa^2}{(\kappa^2-io)^{\frac{n}{2}+\nu_1+\nu_2}} \frac{(\kappa^2-m_2^2)^{\frac{n}{2}+\nu_1-1}(m_1^2-\kappa^2)^{\frac{n}{2}+\nu_2-1}}{(m_1^2-m_2^2)^{n+\nu_1+\nu_2-2}}\\
&= \frac{\Gamma(\frac{n}{2}+\nu)^2}{\Gamma(n+2\nu)} \frac{1}{(m^2)^{\frac{n}{2}+2\nu}},\ \text{for } \nu_1=\nu_2=\nu,\ m_1^2=m_2^2=m^2,
\end{aligned}$$

display, in the relevant contributions for (half-)integer powers, important for odd- and even-dimensional spaces, a *rational* and a *logarithmic* structure, respectively:

$$\mathbf{O}(1,0):\ \begin{cases} \left.\frac{|m_1|}{-q^2-io+m_1^2} \overset{*}{\underset{i\pi}{}} \frac{|m_2|}{-q^2-io+m_2^2}\right|_{q=0} = \frac{1}{2}\int_0^1 dz \left.\frac{|m_1 m_2|}{[-z(1-z)q^2-io+zm_1^2+(1-z)m_2^2]^{\frac{3}{2}}}\right|_{q=0}\\
\qquad = \frac{|m_1 m_2|}{m_1^2-m_2^2}\int_{|m_2|}^{|m_1|}\frac{d\kappa}{\kappa^2} = \frac{1}{|m_1|+|m_2|}, \end{cases}$$

$$\mathbf{O}(1,1):\ \begin{cases} \left.\frac{1}{-q^2-io+m_1^2} \overset{*}{\underset{i\pi}{}} \frac{1}{-q^2-io+m_2^2}\right|_{q=0} = \int_0^1 dz \left.\frac{1}{-z(1-z)q^2-io+zm_1^2+(1-z)m_2^2}\right|_{q=0}\\
\qquad = \frac{1}{m_1^2-m_2^2}\int_{m_2^2}^{m_1^2}\frac{d\kappa^2}{\kappa^2} = \frac{1}{m_1^2-m_2^2}\log\frac{m_1^2}{m_2^2}. \end{cases}$$

The orthogonal products with harmonic $\mathbf{O}(t,s)$-polynomials $(q)^L$, $L=0,1,2,\ldots$, are proportional to the harmonic units (projectors) $(\mathbf{1}_n)^L$. They involve characteristic *external multiplicity factors*:

$$\int d^n q\, \mu(q^2)(q)^L \otimes (q)^{L'} = \delta^{LL'}\frac{\Gamma(\frac{n}{2})\Gamma(1+L)}{2^L\,\Gamma(\frac{n}{2}+L)}\int d^n q\, \mu(q^2)(q^2)^L\,(\mathbf{1}_n)^L,$$

with the simplest examples for $L=1,2$ and $\mathbf{O}(t,s)$-metric $\eta_{ij} \cong \begin{pmatrix}\mathbf{1}_t & 0\\ 0 & -\mathbf{1}_s\end{pmatrix}$:

$$\int d^n q\, \mu(q^2) q^i q_j = \frac{1}{n}\int d^n q\, \mu(q^2) q^2\ \delta^i_j,$$

$$\int d^n q\, \mu(q^2)(q^i q^k - \frac{\eta^{ik}}{n}q^2)(q_j q_l - \frac{\eta_{jl}}{n}q^2) = \frac{2}{n(n+2)}\int d^n q\, \mu(q^2)(q^2)^2\left(\frac{\delta^i_j\delta^k_l+\delta^i_l\delta^k_j}{2} - \frac{\eta^{ik}\eta_{jl}}{n}\right).$$

The distributions μ are characterized by invariants of the $\mathbf{O}(t,s)$-representations acting on $(q)^L$. There remain scalar integrals, for example:

$$\begin{aligned}
\int \frac{d^n q}{i^t\pi^{\frac{n}{2}}} \frac{\Gamma(\frac{n}{2}+\nu_1)\,\Gamma(\nu_2)\,(-q^2-io)^\lambda}{(-q^2-io+m_1^2)^{\frac{n}{2}+\nu_1}(-q^2-io+m_2^2)^{\nu_2}} &= \frac{\Gamma(\frac{n}{2}+\lambda)}{\Gamma(\frac{n}{2})}\int_0^1 dz\, \frac{z^{\frac{n}{2}+\nu_1-1}(1-z)^{\nu_2-1}\,\Gamma(\nu_1+\nu_2-\lambda)}{[zm_1^2+(1-z)m_2^2-io]^{\nu_1+\nu_2-\lambda}},\\
\int \frac{d^n q}{i^t\pi^{\frac{n}{2}}} \frac{\Gamma(\frac{n}{2}+\nu)\,(-q^2-io)^\lambda}{(-q^2-io+m^2)^{\frac{n}{2}+\nu}} &= \frac{\Gamma(\frac{n}{2}+\lambda)}{\Gamma(\frac{n}{2})}\frac{\Gamma(\nu-\lambda)}{(m^2-io)^{\nu-\lambda}}.
\end{aligned}$$

12.12 Normalized Spacetime Representations

The representation normalizations for position $\mathcal{Y}^{2R-1}$ and spacetime $\mathcal{D}^{2R}$ with nonabelian motion groups for $R \geq 2$ determine the continuous invariants for noncompact Cartan subgroups $\mathbf{D}(1) \cong \mathbf{SO}_0(1,1)$, e.g., for dilations or translations, by the discrete invariants for compact Cartan subgroups $\mathbf{U}(1) \cong \mathbf{SO}(2)$, e.g., for spin or isospin rotations. For odd-dimensional spaces, the noncompact invariants at point singularities for "energies", "momenta" or masses are related by rational functions to the discrete numbers (Plancherel multiplicities) for the compact invariants, whereas even-dimensional spaces with line singularities employ logarithms for this connection.

The measures of the odd- and even-dimensional maximally symmetric spacetimes are both normalizable by the representation invariant:

$$\begin{array}{rl} \mathbf{O}(t, 1+2R-t): & \int \frac{d^{1+2R}q}{i^t \pi^{1+R}} \frac{\Gamma(1+R)}{(-q^2-io+m^2)^{1+R}} = \frac{1}{|m|}, \\ \mathbf{O}(t, 2R-t): & \int \frac{d^{2R}q}{i^t \pi^R} \frac{\Gamma(1+R)}{(-q^2-io+m^2)^{1+R}} = \int_{|m|}^{\infty} \frac{d\kappa}{\sqrt{(\kappa^2-m^2)^3}} = \frac{1}{m^2}, \end{array}$$

which can be taken as intrinsic unit.

The simplest example is the normalization of the time representation coefficients with the energy as intrinsic unit:

$$\mathbf{D}(1): \ x_0 \longmapsto \int \frac{dq_0}{i\pi} \frac{1}{-q_0^2-io+E^2} e^{iq_0 x_0} = \frac{e^{-i|Ex_0|}}{|E|},$$

For odd-dimensional hyperbolic position $\mathcal{Y}^{2R-1} \cong \mathbf{O}(1, 2R-1)/\mathbf{O}(2R-1)$ with measure $\frac{d^{2R-1}q}{\pi^R} \frac{\Gamma(R)}{(\vec{q}^2+1)^R}$ and harmonic $\mathbf{O}(2R)$-polynomials with $p = \frac{1}{\vec{q}^2+1} \begin{pmatrix} 1-\vec{q}^2 \\ 2\vec{q} \end{pmatrix} \in \Omega^{2R-1}$, e.g., for 3-position as represented by the atomic spectrum, with the representation coefficients

$$\begin{array}{l} R = 1, 2, \ldots, \\ L = 0, 1, \ldots: \end{array} \ \mathcal{Y}^{2R-1}: \vec{x} \longmapsto \int \frac{d^{2R-1}q}{\pi^R} \frac{\Gamma(R)}{(\vec{q}^2+B_L^2)^R} \frac{(2\vec{q})^L}{(\vec{q}^2+B_L^2)^L} e^{i\vec{q}\vec{x}}$$
$$= \frac{\Gamma(R)}{\Gamma(R+L)} (i\vec{x})^L \frac{e^{-|B_L|r}}{|B_L|},$$

compact and noncompact invariants are not independent. The product normalization with the representation distributions for the harmonic $\mathbf{O}(2R-1)$-polynomials:

$$\int \frac{d^{2R-1}q}{(2\pi)^{2R-1}} (\vec{q})^L e^{i\vec{q}\vec{x}} = (-i\vec{\partial})^L \delta(\vec{x}),$$

determines the hyperbolic invariants $|B_L|$ for a Cartan subgroup $\mathbf{SO}_0(1,1) \subseteq \mathcal{Y}^{2R-1}$ in a reciprocity with the compact invariants for a Cartan subgroup $\mathbf{SO}(2) \subset \mathbf{SO}(2R-1)$, $R \geq 2$:

$$\begin{aligned}(\mathbf{1}_{2R-1})^L &= \int \frac{d^{2R-1}q}{\pi^R} \frac{\Gamma(R)}{(\vec{q}^2+B_L^2)^R} \frac{(2\vec{q})^L}{(\vec{q}^2+B_L^2)^L} \otimes (\vec{q})^{L'}\\ &= \delta^{LL'} \frac{\Gamma(R)\Gamma(1+L)}{\Gamma(R+L)} \frac{1}{|B_L|} (\mathbf{1}_{2R-1})^L,\\ &\qquad\qquad\qquad\qquad L\\ R = 1, 2, \ldots, \quad L = 0, 1, \ldots : \quad \frac{1}{|B_L|} &= \binom{R+L-1}{L} = \dim_{\mathbb{C}} \bigvee \mathbb{C}^R,\end{aligned}$$

e.g., $\frac{1}{B_L^2} = (1+L)^2$ in the atomic spectrum for $\mathcal{Y}^3$. The representation invariant $|B_0|$ for the scalar coefficients can be used as intrinsic unit, e.g., the Rydberg energy $\underline{B}_0^2 = m_R c^2$ in $\frac{1}{\underline{B}_L^2} = (1+L)^2 \frac{1}{\underline{B}_0^2}$. For abelian position $\mathcal{Y}^1 = \mathbf{SO}_0(1,1)$ without rotations there is no L-dependence, $|B_L| = |B_0|$.

12.12.1 Logarithmic Mass Ratios for Indefinite Spacetimes

The characteristic measure for an indefinite even-dimensional manifold with odd-odd submanifolds involves, in addition to the residual normalization, an integration over both invariants (units) in the embedded characteristic submanifold measures:

$$\begin{aligned}\mathbf{O}(1+2R_t, 1+2R_s) : \int \frac{d^{2(1+R_t+R_s)}}{\pi^{1+R_t+R_s}} &\frac{\Gamma(1+R_t)}{(-q^2-io+m^2)^{1+R_t}} \frac{\Gamma(1+R_s)}{(-q^2-io+M^2)^{1+R_s}}\\ = \int_0^1 dz\, (1-z)^{R_t} z^{R_s} &\int \frac{d^{2(1+R_t+R_s)}q}{\pi^{1+R_t+R_s}} \frac{\Gamma(2+R_t+R_s)}{(-q^2-io+(1-z)m^2+zM^2)^{2+R_t+R_s}}\\ = \int_0^1 dz\, \frac{(1-z)^{R_t} z^{R_s}}{(1-z)m^2+zM^2} &= \frac{1}{m^2-M^2} \int_{M^2}^{m^2} \frac{d\kappa^2}{\kappa^2} \frac{(\kappa^2-M^2)^{R_t}(m^2-\kappa^2)^{R_s}}{(m^2-M^2)^{R_t+R_s}}.\end{aligned}$$

A spacetime coefficient with two continuous invariants (units), characterizing time and position $(E^2, B^2) \leftrightarrow (\mathbf{D}(1), \mathcal{Y}^{2R-1})$, is, for the nonrelativistic case, in the $\mathbf{O}(1) \times \mathbf{O}(2R-1)$-compatible, but Lorentz group-incompatible form with the two residual integrations $\int \frac{dq_0}{i\pi}$ and $\int \frac{d^{2R-1}q}{\pi^R}$ normalized by the product of the two invariants (point normalizations):

$$\mathbf{O}(1) \times \mathbf{O}(2R-1) : \quad \int \frac{dq_0}{i\pi} \frac{1}{-q_0^2-io+E^2} \int \frac{d^{2R-1}q}{\pi^R} \frac{\Gamma(R)}{(\vec{q}^2+B^2)^R} = \frac{1}{|E||B|}.$$

In the Lorentz group $\mathbf{O}(1, 2R-1)$-compatible form with embedded time and position $(m^2, M^2) \leftrightarrow (\mathbf{D}(1), \mathcal{Y}^{2R-1})$,

$$\begin{aligned}\mathbf{O}(1, 2R-1) : \int \frac{d^{2R}q}{i\pi^R} &\frac{\Gamma(R)}{(-q^2-io+m^2)(-q^2-io+M^2)^R}\\ = \frac{1}{m^2-M^2} \int_{M^2}^{m^2} d^R\kappa^2 &\int \frac{d^{2R}q}{i\pi^R} \frac{\Gamma(1+R)}{(-q^2-io+\kappa^2)^{1+R}}\\ = \int_0^1 dz\, \frac{z^{R-1}}{(1-z)m^2+zM^2} &= \frac{1}{m^2-M^2} \int_{M^2}^{m^2} \frac{d^R\kappa^2}{\kappa^2} = -\frac{1}{m^2-M^2} \mathrm{Log}_R \frac{M^2}{m^2},\end{aligned}$$

the representation normalization connects both invariants by a characteristic line integration $\int_{M^2}^{m^2} \frac{d^R\kappa^2}{\kappa^2}$ for a finite dilation measure (line normalization). In the case of different invariants, $m^2 \neq M^2$, i.e., for not maximally symmetric spacetimes, a multiplication with the difference $\int_{M^2}^{m^2} d\kappa^2 = m^2 - M^2$ of the two invariants in the integration length defines the characteristic dimensionless "energy-momentum" measures

$$\text{for a Fourier analysis of} \mathcal{D}^{2R}: \frac{d^{2R}q}{i\pi^R} \frac{(m^2-M^2)\ \Gamma(R)}{(-q^2-io+m^2)(-q^2-io+M^2)^R}.$$

If the minimal abelian case,

$$\frac{1}{i\pi} \frac{m^2-M^2}{(-q^2-io+m^2)(-q^2-io+M^2)} = \delta(q^2 - M^2) - \delta(q^2 - m^2) + \dots,$$

is taken with an additive decomposition for a flat spacetime $\mathbb{R}^{(1,1)}$-metric, the on-shell contributions of the simple poles have opposite metric (no Hilbert representation of the Poincaré group). Such an additive decomposition is not appropriate for Cartan spacetime $\mathcal{D}^2 = \mathbf{D}(1) \times \mathbf{SO}_0(1, 1)$ and, in general, for $\mathcal{D}^{2R} = \mathbf{D}(1) \times \mathcal{Y}^{2R-1}$ where the coefficients $\frac{1}{-q^2-io+m^2} \cdot \frac{\Gamma(R)}{(-q^2-io+M^2)^R}$ represent a product space.

The characteristic $\mathcal{D}^{2R}$-measures are normalized by the *logarithmic R-tails* with the dimensionless ratio $\mu^2 = \frac{M^2}{m^2} > 0$ of the hyperbolic and spherical invariant:

$$\begin{aligned}
R = 1, 2, \dots : -\text{Log}_R\mu^2 &= \int_0^1 dz\, \frac{(1-\mu^2)z^{R-1}}{1-(1-\mu^2)z} = \int_{\mu^2}^1 \frac{d^R\kappa^2}{\kappa^2} = \int_{\mu^2}^1 \frac{d\kappa^2}{\kappa^2} \left(\frac{1-\kappa^2}{1-\mu^2}\right)^{R-1} \\
&= -\frac{1}{(1-\mu^2)^{R-1}} \left[\log\mu^2 + \sum_{k=1}^{R-1} \frac{(1-\mu^2)^k}{k} \right] = \sum_{k=R}^{\infty} \frac{(1-\mu^2)^{k-(R-1)}}{k} \\
&= - \begin{cases} \log\mu^2, & R = 1, \\ \frac{\log\mu^2}{1-\mu^2} + 1, & R = 2, \\ \dots \end{cases} \\
-\text{Log}_1\mu^2 = -\log\mu^2, &\quad -\text{Log}_{R+1}\mu^2 = -\frac{1}{1-\mu^2}\text{Log}_R\mu^2 - \frac{1}{R}.
\end{aligned}$$

With the factor $m^2 - M^2 \sim 1 - \mu^2$, the logarithmic tails are positive for $\mu^2 < 1$, i.e., for a smaller hyperbolic invariant $M^2 < m^2$, and negative for $\mu^2 > 1$, i.e., for a smaller spherical invariant $M^2 > m^2$, and monotoneously decreasing:

$$-\frac{\text{Log}_R\mu^2}{1-\mu^2} = \int_0^1 dz\, \frac{z^{R-1}}{1-(1-\mu^2)z} > 0, \quad -\frac{\partial}{\partial\mu^2}\text{Log}_R\mu^2 = -\int_0^1 dz\, \frac{z^{R-1}}{[1-(1-\mu^2)z]^2} \le 0,$$

with the following expansions and special values:

$$-\text{Log}_R\mu^2 = \begin{cases} -\log\mu^2 - \sum_{k=1}^{R-1} \frac{1}{k} + \dots, & \mu^2 \ll 1, \\ \frac{1-\mu^2}{R} + \dots \to 0, & \mu^2 \to 1, \\ -\frac{1}{R-1} + \dots, \quad R \neq 1, & \mu^2 \gg 1. \end{cases}$$

12.12.2 Multiplicities for External Degrees of Freedom

For the Lorentz group-compatible representations of even-dimensional spacetime $\mathcal{D}^{2R} = \mathbf{D}(1) \times \mathcal{Y}^{2R-1}$ with harmonic $\mathbf{O}(1, 2R-1)$-polynomials,

$$\begin{array}{l} R = 1, 2, \ldots, \\ L = 0, 1, \ldots : \end{array} \quad \mathcal{D}^{2R} : \vartheta(x)x \longmapsto \vartheta(x)d^L(x),$$
$$\begin{aligned} d^L(x) &= \int \frac{d^{2R}q}{i\pi^R} \frac{(m_L^2 - M_L^2)\,\Gamma(R)}{(-q^2-io+m_L^2)(-q^2-io+M_L^2)^R} (2iq)^L e^{iqx} \\ &= \int_{M_L^2}^{m_L^2} d^R\kappa^2 \int \frac{d^{2R}q}{i\pi^R} \frac{\Gamma(1+R)}{(-q^2-io+\kappa^2)^{1+R}} (2iq)^L e^{iqx} \\ &= (2\partial)^L d^0(x^2) = (x)^L \left(\frac{\partial}{\partial \frac{x^2}{4}}\right)^L d^0(x^2), \end{aligned}$$

the inverse spacetime $\mathcal{D}^{2R}$-derivatives with the dual ("inverse") harmonic dependence, i.e., the spacetime distributions:

$$\frac{1}{(\partial)^L} \sim \int \frac{d^{2R}q}{i\pi^R} \frac{(iq)^L}{(-q^2-io)^L} e^{iqx} = \frac{1}{\Gamma(L)}(\partial)^L \frac{\Gamma(R-L)}{(-\frac{x^2}{4}+io)^{R-L}} = \frac{1}{\Gamma(L)}\left(\frac{x}{2}\right)^L \frac{\Gamma(R)}{(-\frac{x^2}{4}+io)^R},$$

are used for the external multiplicity factors in the product normalizations:

$$\begin{aligned} (\mathbf{1}_{2R})^L &= \int \frac{d^{2R}q}{i\pi^R} \frac{(m_L^2 - M_L^2)\,\Gamma(R)}{(-q^2-io+m_L^2)(-q^2-io+M_L^2)^R} (2iq)^L \otimes \frac{(iq)^{L'}}{(-q^2-io)^{L'}} \\ &= \delta^{LL'} \frac{\Gamma(R)\Gamma(1+L)}{\Gamma(R+L)} \int_{M_L^2}^{m_L^2} d^R\kappa^2 \int \frac{d^{2R}q}{i\pi^R} \frac{\Gamma(1+R)}{(-q^2-io+\kappa^2)^{1+R}} (\mathbf{1}_{2R})^L \\ &= -\delta^{LL'} \frac{\Gamma(R)\Gamma(1+L)}{\Gamma(R+L)} \operatorname{Log}_R \frac{M_L^2}{m_L^2} (\mathbf{1}_{2R})^L. \end{aligned}$$

The line singularity normalizations give a compact-noncompact reciprocity for the logarithmic tails of the ratio of the two noncompact invariants:

$$\begin{array}{l} R = 1, 2, \ldots, \\ L = 0, 1, 2, \ldots : \end{array} \quad -\operatorname{Log}_R \frac{M_L^2}{m_L^2}\Big|_{\text{ext}} \sim d_{\text{ext}} = \binom{R+L-1}{L}.$$

This normalization with the external multiplicities has to be supplemented by the normalization with respect to the internal degrees of freedom.

12.12.3 Multiplicities for Internal Degrees of Freedom

A representation of four-dimensional electroweak spacetime $\mathbb{D}(2) \cong \mathbf{GL}(2, \mathbb{C})/\mathbf{U}(2)$ uses, for external operations with Lorentz $\mathbf{SL}(2, \mathbb{C})$ and internal operations with isospin $\mathbf{SU}(2)$, a subrepresentation of the bi-regular $\mathbf{GL}(2, \mathbb{C})$-representation $\mathbf{GL}(2, \mathbb{C}) \times \mathbf{U}(2) \longrightarrow \mathbf{GL}(W) \times \mathbf{GL}(V)$ in a product group. The nonscalar representations start for $\mathbf{GL}(2, \mathbb{C})$ with the (2×2)-Weyl representation $q \sim \sigma^j q_j = \begin{pmatrix} q_0 + q_3 & q_1 - iq_2 \\ q_1 + iq_2 & q_0 - q_3 \end{pmatrix}$ with unit $\sigma^0 = \mathbf{1}_2$ and its internal $\mathbf{U}(2)$-partner, the Pauli

(2×2)-representation with unit $\mathbf{1}_2$. In general, the irreducible harmonic $\mathbf{SL}(2, \mathbb{C})$-dependence $(q)^L$ from the totally symmetric powers of $W_1 \cong \mathbb{C}^2$ comes with the corresponding unrestricted tensor powers of $V_1 \cong \mathbb{C}^2$ with decomposable $\mathbf{SU}(2)$-representations:

$$\mathcal{D}^4: \ \vartheta(x)x \longmapsto \vartheta(x) \int \frac{d^4q}{i\pi^2} \frac{(m_L^2 - M_L^2)}{(-q^2 - io + m_L^2)(-q^2 - io + M_L^2)^2} (2iq)^L \times \mathbf{1}_{2^L} e^{iqx},$$

$$\text{with} \begin{cases} \text{external (spin):} & W_L \cong \bigvee^L \mathbb{C}^2 \cong \mathbb{C}^{1+L}, \quad (q)^L \in \mathbf{AL}(1+L, \mathbb{C}), \\ \text{internal (isospin):} & V_L \cong \bigotimes^L \mathbb{C}^2 \cong \mathbb{C}^{2^L}, \quad \bigotimes^L \mathbf{1}_2 = \mathbf{1}_{2^L} \in \mathbf{U}(2^L). \end{cases}$$

$I = \frac{L}{2}$ is the maximal isospin in $\mathbb{C}^{2^L}$:

maximal I	$\frac{1}{2}$	1	$\frac{3}{2}$	$\dots$
$\mathbb{C}^{2^{2I}}$	$\mathbb{C}^2$	$\mathbb{C}^4 = \mathbb{C}^3 \oplus \mathbb{C}$	$\mathbb{C}^8 = \mathbb{C}^4 \oplus 2 \times \mathbb{C}^2$	$\dots$

The normalization of electroweak spacetime $\mathbf{GL}(2, \mathbb{C})/\mathbf{U}(2)$ by the $\mathbf{GL}(2, \mathbb{C}) \times \mathbf{U}(2)$-unit takes into account the internal multiplicity $d = 2^L$,

$$\begin{aligned} (\mathbf{1}_4)^L \times (\mathbf{1}_d \otimes \mathbf{1}_d) &= \int \frac{d^4q}{i\pi^2} \frac{(m_L^2 - M_L^2)}{(-q^2 - io + M_L^2)^2(-q^2 - io + m_L^2)} (2iq)^L \times \mathbf{1}_d \otimes \frac{(iq)^L \times \mathbf{1}_d}{(-q^2 - io)^L} \\ &= \frac{\Gamma(1+L)}{\Gamma(2+L)} \int_{M_L^2}^{m_L^2} d^2\kappa^2 \int \frac{d^4q}{i\pi^2} \frac{2}{(-q^2 - io + \kappa^2)^3} (\mathbf{1}_4)^L \times (\mathbf{1}_d \otimes \mathbf{1}_d) \\ &\overset{\text{Fierz}}{=} -\frac{1}{1+L} \operatorname{Log}_2 \frac{M_L^2}{m_L^2} (\mathbf{1}_4)^L \times \frac{\mathbf{1}_d \otimes \mathbf{1}_d}{d} + \dots, \end{aligned}$$

from the Fierz recoupling $(\mathbf{1}_d)_\alpha^\beta (\mathbf{1}_d)_\gamma^\delta = \frac{1}{d} (\mathbf{1}_d)_\alpha^\delta (\mathbf{1}_d)_\gamma^\beta + \dots$, necessary for the decomposition of the product representations. The internal Fierz recoupling is the analogue to the external "energy-momenta" recoupling $(q_1, q_2) \leftrightarrow (q_1 + q_2, q_1 - q_2)$ into relative and center-of-mass contributions,

$$\begin{aligned} &\text{Pauli matrices } (\tau^a)_{a=1}^{d^2-1} \in \mathbf{AL}(d, \mathbb{C}), \ d = 2, 3, \dots, \\ &\begin{pmatrix} \mathbf{1}_d \otimes \mathbf{1}_d \\ \tau^a \otimes \tau^a \end{pmatrix} \overset{\text{Fierz}}{=} F \begin{pmatrix} \mathbf{1}_d \otimes \mathbf{1}_d \\ \tau^a \otimes \tau^a \end{pmatrix}, \ F = \frac{1}{d} \begin{pmatrix} 1 & 1 \\ d^2 - 1 & -1 \end{pmatrix}, \ F \circ F = \mathbf{1}_2. \end{aligned}$$

The rearranged unit $\mathbf{1}_d \otimes \mathbf{1}_d$ is normalized with the internal dimension d.

Altogether one obtains the external-internal normalization conditions with the noncompact-compact reciprocity,

$$\text{electroweak spacetime } \mathcal{D}^4 \text{ with} \begin{cases} L = 2J = 2I & = 0, 1, 2, \dots, \\ d_{\text{ext}} = 1 + 2J & = 1, 2, 3, \dots, \\ d_{\text{int}} = d = 2^{2I} & = 1, 2, 4, \dots, \\ -\operatorname{Log}_2 \frac{M_L^2}{m_L^2} = d_{\text{ext}} d_{\text{int}} & = (1 + 2J)2^{2I} \\ \qquad = (1 + L)2^L = 1, 4, 12, \dots, \end{cases}$$

and the association of spacetime $\mathcal{D}^4$-coefficients and numerical mass values for $L = 0, 1, 2$:

$(J,I) = (\frac{L}{2}, \frac{L}{2})$	$\frac{(2iq)^{2J} \times \mathbf{1}_{2^{2I}}}{(-q^2-io+M_L^2)^2(-q^2-io+m_L^2)}$	$-\mathrm{Log}_2 \frac{M_L^2}{m_L^2} = (1+2J)2^{2I}$	$\frac{M_L^2}{m_L^2}$
$(0,0)$	$\frac{1}{(-q^2-io+M_0^2)^2(-q^2-io+m_0^2)}$	$-\mathrm{Log}_2 \frac{M_0^2}{m_0^2} = 1$	$\frac{M_0^2}{m_0^2} \sim e^{-2} \sim 10^{-1}$
$(\frac{1}{2}, \frac{1}{2})$	$\frac{2iq \times \mathbf{1}_2}{(-q^2-io+M_1^2)^2(-q^2-io+m_1^2)}$	$-\mathrm{Log}_2 \frac{M_1^2}{m_1^2} = 4$	$\frac{M_1^2}{m_1^2} \sim e^{-5} \sim 10^{-2.2}$
$(1,1)$	$\frac{(-4q \otimes q + \eta q^2) \times \mathbf{1}_4}{(-q^2-io+M_2^2)^2(-q^2-io+m_2^2)}$	$-\mathrm{Log}_2 \frac{M_2^2}{m_2^2} = 12$	$\frac{M_2^2}{m_2^2} \sim e^{-13} \sim 10^{-5.6}$

In analogy to the condition for hyperbolic position $\mathcal{Y}^3$, one may call this an equipartion of the logarithmic tail to the external and internal dimensions of the electroweak spacetime $\mathcal{D}^4$-representations.

12.13 Particles and Interactions of Spacetime

An attempt will be given to relate the mathematical representation structures above with experimentally observed physical structures: Electroweak spacetime $\mathcal{D}^4 \cong \mathbf{GL}(2, \mathbb{C})/\mathbf{U}(2)$, parametrizable by strict Minkowski spacetime future, is fundamentally realized by irreducible spinor-isospinor representations for the external-internal group $\mathbf{GL}(2, \mathbb{C}) \times \mathbf{U}(2)$ (without indices $\frac{M_1^2}{m_1^2} = \frac{M^2}{m^2}$) with the representation coefficient:

$$\begin{aligned}(J,I) = (\tfrac{1}{2}, \tfrac{1}{2}) : \mathcal{D}^4 \ni \vartheta(x)x \longmapsto \vartheta(x) \int \frac{d^4q}{i\pi^2} \frac{(m^2-M^2)\; 2iq \times \mathbf{1}_2}{(-q^2-io+m^2)(-q^2-io+M^2)^2} e^{iqx},\\ \text{with } -\mathrm{Log}_2 \frac{M^2}{m^2} = (1+2J)(1+2I) = 4 \Rightarrow \frac{M^2}{m^2} \sim e^{-5} \sim 10^{-2.2}.\end{aligned}$$

This spacetime convolution product can be interpreted as a position $\mathcal{Y}^3$-distribution of a time $\mathbf{D}(1)$-representation by a pole with spherical invariant m^2 and as a time $\mathbf{D}(1)$-distribution of a position $\mathcal{Y}^3$-representation by a dipole with hyperbolic invariant M^2. A spectral decomposition for Poincaré group and Minkowski spacetime representations with a linear combination of simple poles and a dipole would involve positive and negative metric contributions — it is inappropriate here.

In a tangent space extension, the simple pole may give rise to an on-shell Wigner-particle structure for the leptonic sector with a particle field $\mathbf{l}$:

$$\begin{aligned}\mathbf{D}(1) \ni x_0 \longmapsto \int dq_0\, q_0\, \delta(-q_0^2 + m^2) e^{iq_0x_0} \hookrightarrow \int \frac{d^4q}{\pi}\, q \times \mathbf{1}_2\, \delta(-q^2+m^2) e^{iqx}\\ \sim\ \langle \mathbf{l}\, \mathbf{l}^* \rangle(x),\end{aligned}$$

whereas the dipole has no connection with a Wigner-particle structure. It may be related to causally supported strong interactions, usually parametrized by the gluon and the quark fields $\mathbf{q}$, postulated to be confined (never "on-shell", i.e., no translation representations):

$$\begin{aligned}\mathcal{Y}^3 \ni \vec{x} \longmapsto \int \frac{d^3q}{\pi^2} \frac{\vec{q}}{(\vec{q}^2+M^2)^2} e^{-i\vec{q}\vec{x}} \hookrightarrow \int \frac{d^4q}{\pi^2} \frac{q \times \mathbf{1}_2}{(-q_{\mathrm{P}}^2+M^2)^2} e^{iqx} \\ \sim \langle \mathbf{q}\,\mathbf{q}^* \rangle(x).\end{aligned}$$

In a field language, the representation matrix element can be related to a *lepton-quark product field* $\chi = \mathbf{l} \otimes \mathbf{q}$ with $\langle \mathbf{l} \otimes \mathbf{q}\,\mathbf{q}^* \otimes \mathbf{l}^* \rangle(x) \sim \langle \mathbf{l}\mathbf{l}^* \rangle * \langle \mathbf{q}\mathbf{q}^* \rangle(x)$ where the leptonic degrees of freedom $\mathbf{l}$ embed time $\mathbf{D}(1)$ and the quark degrees of freedom $\mathbf{q}$ position $\mathcal{Y}^3$.

The $\mathbf{SU}(3)$-color triplet degrees of freedom $\mathbf{1}_3$ in a conventional quark-particle flat spacetime formulation $\langle \mathbf{q}\,\mathbf{q}^* \rangle(x) \sim \int \frac{d^4q}{\pi^2} \frac{q \times \mathbf{1}_2 \times \mathbf{1}_3}{-q^2-io+M^2} e^{iqx}$ may be introduced via a "custodian" nonabelian $\mathbf{SU}(3)$ to distinguish lepton and quark degrees of freedom in a Minkowski spacetime interpretation. The color singlet spacetime convolution cube comes with a dipole "energy-momentum" function $[\frac{q \times \mathbf{1}_2 \times \mathbf{1}_3}{q^2}]^3 \sim \frac{q \times \mathbf{1}_2}{(q^2)^2}$, more general for $(2R-1)$-dimensional position $\mathcal{Y}^{2R-1} \subset \mathcal{D}^{2R}$ with $\mathbf{SU}(2R-1)$-color degrees of freedom and an order R-pole $[\frac{q \times \mathbf{1}_{2R-1}}{q^2}]^{2R-1} \sim \frac{q}{(q^2)^R}$. In such an interpretation the "custodian" color group $\mathbf{SU}(2R-1)$ for a simple pole (usual quarks for $R=2$) may be looked at to be a totally antisymmetric $(2R-1)$-root of a pole of order R with $\bigwedge^{2R-1} \mathbf{SU}(2R-1) = \{1\}$, generalizing the cyclotomic groups $\mathbf{I}(2R-1) = \sqrt[2R-1]{1}$. The dimensions for position and unitary "custodian" color group are related to each other, e.g., three position dimensions use quark triplets.

Color confinement has the interpretation that a nontrivial nonabelian color representation is incompatible with a nontrivial abelian translation representation.

A lepton-quark association to the electroweak spacetime representation coefficient leads to a ratio $\frac{M}{m} \sim \frac{1}{12}$ of an average quark "mass" M (hyperbolic or interaction invariant) to an average lepton mass m (spherical or particle invariant). The proliferation of lepton and quark degrees of freedom into different families, e.g., $\mathbf{l} \longrightarrow \{\mathbf{e}, \boldsymbol{\mu}, \boldsymbol{\tau}\}$, may be related to an isospin-degeneracy (breakdown) of the ground state, not discussed here.

The complex infinite-dimensional vector space with the $\mathcal{D}^4$-action may be spanned by a basis $\{\chi_A^\alpha(x) \mid A = 1, 2;\ \alpha = 1, 2;\ x \in \mathcal{D}^4\}$ with Weyl spinor-isospinors and the electroweak ground state $|\mathcal{D}\rangle$ characterizing coefficient:

$$\begin{aligned}x, y \in \mathcal{D}^4: \ \langle \mathcal{D}|\chi_A^\alpha(y)\chi^{*B}_{\beta}(x+y)|\mathcal{D}\rangle &\cong \langle \chi\chi^* \rangle(x) \\ &= \vartheta(x) \int \frac{d^4q}{i\pi^2} \frac{(m^2-M^2)\ 2iq \times \mathbf{1}_2}{(-q^2-io+m^2)(-q^2-io+M^2)^2} e^{iqx}.\end{aligned}$$

The causally supported spacetime field χ has no harmonic analysis with respect to the Poincaré group, especially no translations representations and no corresponding equation of motion. The convolution product for $\mathcal{D}^4 = \mathbf{D}(1) \times \mathcal{Y}^3$ has projections to the simple pole particle $\mathbf{D}(1)$- and dipole interaction $\mathcal{Y}^3$-distributions:

$$\begin{aligned}
\langle \chi\chi^*\rangle(x) &\sim \langle \mathbf{l}\,\mathbf{l}^*\rangle * \langle \mathbf{q}\mathbf{q}^*\rangle(x),\\
(\partial^2+M^2)^2\langle \chi\chi^*\rangle(x) &= \vartheta(x)\int \frac{d^4q}{i\pi^2}\frac{(m^2-M^2)\,2iq\times\mathbf{1}_2}{-q^2-io+m^2}e^{iqx} \sim \langle \mathbf{l}\,\mathbf{l}^*\rangle(x),\\
(\partial^2+m^2)\langle \chi\chi^*\rangle(x) &= \vartheta(x)\int \frac{d^4q}{i\pi^2}\frac{(m^2-M^2)\,2iq\times\mathbf{1}_2}{(-q^2-io+M^2)^2}e^{iqx} \sim \langle \mathbf{q}\mathbf{q}^*\rangle(x).
\end{aligned}$$

12.14 Massless Poles for D(1)-Degeneracy

Nontrivial masses for nontrivial dilation representations signify a $\mathbf{D}(1)$-degeneracy of the spacetime ground state. The non-particle $\mathbf{SO}(2)$-scalar (unpolarized) contributions in the vector and tensor representations of electroweak spacetime $\mathcal{D}^4$ are proposed as long ranged degeneracy $\mathbf{D}(1)$-degrees of freedom for electrodynamics and gravity.

The normalization of the $\mathcal{D}^4$-representations involves the L-kernels with the isospin multiplicity $d = 2^L$:

$$\mathbf{K}^L(q, \tfrac{M_L^2}{m_L^2}) = \frac{\Gamma(R)}{(-q^2-io+M_L^2)^R}\frac{(m_L^2-M_L^2)(2iq)^L\times\mathbf{1}_d}{(-q^2-io-m_L^2)} \circledast \frac{(iq)^L\times\mathbf{1}_d}{i\pi^R\,(-q^2-io)^L} \text{ for } R=2.$$

The kernels can be computed from

$$\begin{aligned}
&\frac{\Gamma(R+\nu_1+L)(2iq)^L}{(-q^2-io+\kappa^2)^{R+\nu_1+L}} \circledast \frac{\Gamma(\nu+L)(iq)^L}{i\pi^R\,(-q^2-io)^{\nu+L}} = (\tfrac{\partial}{\partial q})^L\otimes(\tfrac{\partial}{2\partial q})^L\frac{\Gamma(R+\nu_1)}{(-q^2-io+\kappa^2)^{R+\nu_1}} \underset{i\pi^R}{*} \frac{\Gamma(\nu)}{(-q^2-io)^{\nu}}\\
&= (\tfrac{\partial}{\partial q})^L\otimes(\tfrac{\partial}{2\partial q})^L\int_0^1 dz\,\frac{(1-z)^{R+\nu_1-1}z^{\nu-1}\Gamma(\nu_1+\nu)}{[-z(1-z)q^2-io+(1-z)\kappa^2]^{\nu_1+\nu}}\\
&= (\tfrac{\partial}{\partial q})^L\otimes(\tfrac{\partial}{2\partial q})^L\int_0^1 dz\,\frac{(1-z)^{R-\nu-1}z^{\nu-1}\Gamma(\nu_1+\nu)}{(-zq^2-io+\kappa^2)^{\nu_1+\nu}}\\
&= (\tfrac{\partial}{\partial q})^L\otimes(q)^L\int_0^1 dz\,\frac{(1-z)^{R-\nu-1}z^{\nu-1+L}\Gamma(\nu_1+\nu+L)}{(-zq^2-io+\kappa^2)^{\nu_1+\nu+L}},
\end{aligned}$$

with $\mu^2 = \frac{M_L^2}{m_L^2}$ for $\nu_1 + L = 1$:

$$\begin{aligned}
\mathbf{K}^{L+\nu}(q,\mu^2) &= \int_{\mu^2}^1 d^R\kappa^2\,\frac{\Gamma(1+R)(2iq)^L\times\mathbf{1}_d}{(-q^2-io+\kappa^2)^{1+R}} \circledast \frac{\Gamma(L+\nu)(iq)^L\times\mathbf{1}_d}{i\pi^R\,\Gamma(L)(-q^2-io)^{L+\nu}}\\
&= \frac{1}{\Gamma(L)}[(\tfrac{\partial}{\partial q})^L\times\mathbf{1}_d]\otimes[(q)^L\times\mathbf{1}_d]\int_{\mu^2}^1 d^R\kappa^2\int_0^1 dz\,\frac{(1-z)^{R-1-\nu}z^{\nu-1+L}\Gamma(1+\nu)}{(-zq^2-io+\kappa^2)^{1+\nu}},
\end{aligned}$$

in the limit $\nu \to 0$:

$$\mathbf{K}^L(q,\mu^2) = \frac{1}{\Gamma(L)}[(\tfrac{\partial}{\partial q})^L\times\mathbf{1}_d]\otimes[(q)^L\times\mathbf{1}_d]\int_{\mu^2}^1 d^R\kappa^2\int_0^1 dz\,\frac{(1-z)^{R-1}z^{L-1}}{-zq^2-io+\kappa^2}.$$

This gives explicitly for electroweak spacetime ($R = 2$) with the spinor-isospinor representation ($L = 1$):

$$\mathbf{K}^1(q,\mu^2) = (\mathbf{1}_4 + 2q\otimes q\tfrac{\partial}{\partial q^2})\int_{\mu^2}^1 d^2\kappa^2\int_0^1 dz\,\frac{1-z}{-zq^2-io+\kappa^2}\,\mathbf{1}_2\otimes\mathbf{1}_2.$$

The expansion at $q^2 = 0$ with Fierz recoupling in the isospin units $\mathbf{1}_d\otimes\mathbf{1}_d \overset{\text{Fierz}}{=} \frac{1}{d}\mathbf{1}_{d^2} + \ldots$:

$$\begin{aligned}\mathbf{K}^L(q,\mu^2) &= \frac{\Gamma(1+L)}{\Gamma(L)}\int_{\mu^2}^1 d^R\kappa^2 \int_0^1 dz\, \frac{(1-z)^{R-1}z^{L-1}}{-zq^2-io+\kappa^2}\,(\mathbf{1}_{2R})^L \times \frac{1}{d}\mathbf{1}_{d^2}+\dots\\ &= [K_0^L(\mu^2)+(q^2+io)K_1^L(\mu^2)+\dots]\,(\mathbf{1}_{2R})^L\times\mathbf{1}_{d^2}+\dots\end{aligned}$$

gives the absolute coefficient, used above for the normalization :

$$\begin{aligned}K_0^L(\mu^2) &= \frac{\Gamma(R)\Gamma(1+L)}{\Gamma(R+L)d}\int_{\mu^2}^1 \frac{d^R\kappa^2}{\kappa^2} = 1 \text{ for normalization},\\ \text{with } \frac{1}{1-\mu^2}\int_{\mu^2}^1\frac{d^R\kappa^2}{\kappa^2} &= \int_0^1 dz\,\frac{z^{R-1}}{1-(1-\mu^2)z} = -\frac{\mathrm{Log}_R\mu^2}{1-\mu^2} > 0.\end{aligned}$$

The q^2-proportional coefficient for $L \neq 0$:

$$\begin{aligned}K_1^L(\mu^2) &= \frac{\Gamma(R)\Gamma(1+L)L}{\Gamma(1+R+L)d}\int_{\mu^2}^1 \frac{d^R\kappa^2}{(\kappa^2)^2},\\ \text{with } \frac{1}{1-\mu^2}\int_{\mu^2}^1\frac{d^R\kappa^2}{(\kappa^2)^2} &= \int_0^1 dz\,\frac{z^{R-1}}{[1-(1-\mu^2)z]^2} = \frac{\partial}{\partial\mu^2}\mathrm{Log}_R\mu^2 > 0,\end{aligned}$$

has the mass ratio μ^2-dependence:

$$\frac{\partial}{\partial\mu^2}\mathrm{Log}_R\mu^2 = \frac{\partial}{\partial\mu^2}\left\{\begin{array}{l}\log\mu^2, \quad R=1,\\ \frac{\log\mu^2}{1-\mu^2}+1,\ R=2,\\ \dots\end{array}\right\} = \frac{1}{\mu^2}+\dots, \text{ for } \mu^2 \ll 1.$$

The spectrum of the kernels $\mathbf{K}$ is given by the singularities of the corresponding resolvents $\frac{\mathbf{1}}{\mathbf{1}-\mathbf{K}}$. The normalization conditions $K_0^L(\mu^2)=1$ lead to massless poles, $q^2=0$. The geometrical series expansions at these poles give the propagators for the massless interactions with the residues their normalizations g_L^2 (coupling constants):

$$\frac{\mathbf{1}}{\mathbf{1}-\mathbf{K}^L(q,\mu^2)} = \frac{1}{K_1^L(\mu^2)(-q^2-io)}(\mathbf{1}_{2R})^L\times\mathbf{1}_{d^2}+\dots = \frac{g_L^2}{-q^2-io}(\mathbf{1}_{2R})^L\times\mathbf{1}_{d^2}+\dots$$

The normalization of the massless pole is proportional to the mass ratios of hyperbolic (interaction) and spherical (particle) invariant:

$$\begin{aligned}\text{for } R=2,\ L=1,2,\dots:\ g_L^2 &= \frac{1}{K_1^L(\mu^2)} = \frac{(1+L)(2+L)d_{\mathrm{int}}}{L}\mu^2,\\ g_1^2 &= 2\cdot 3\cdot 2\,\frac{M^2}{m^2} \sim 12\,e^{-5} \sim \frac{1}{12}.\end{aligned}$$

This may be compared for $L=1$ with the electroweak standard model coupling constants, e.g., with Sommerfeld's constant $g_e^2 = 4\pi\alpha_{\mathrm{Som}} \sim \frac{1}{11}$ for the massless electromagnetic interaction.

Bibliography

1. V. Bargmann, Irreducible representations of the Lorentz group, *Ann. of Math.*, 48 (1947), 568–640
2. H. Bauer, *Wahrscheinlichkeitstheorie und Grundzüge der Masstheorie* (1964), Sammlung Göschen, Gruyter, Berlin
3. H. Behnke, F. Sommer, *Theorie der analytischen Funktionen einer komplexen Veränderlichen* (1962), Springer-Verlag, Berlin, Göttingen, Heidelberg
4. G. Birkhoff, J.v. Neumann, The Logic of Quantum Mechanics, *Ann. of Mathematics* 37 (1936), 823–843
5. S. Bochner, Monotone Funktionen, Stieltjessche Integrale und harmonische Analyse, *Math. Ann.*, 108 (1933), 378–410
6. H. Boerner, *Darstellungen von Gruppen* (1955), Springer, Berlin, Göttingen, Heidelberg
7. M. Born, *Einstein's Theory of Relativity* (1962), Dover Publications, New York
8. N. Bourbaki, *Théorie des Ensembles, Chapitre 1-4* (1970), Hermann, Paris
9. N. Bourbaki, *Algebra I, II, Chapters 1–7* (1989, 1990), Springer, Berlin, Heidelberg, New York, London, Paris, Tokyo
10. N. Bourbaki, *Algèbre, Chapitre 9* (Formes sesquilineaires et formes quadratiques), Hermann, Paris (1959)
11. N. Bourbaki, *Lie Groups and Lie Algebras, Chapters 1-3* (1989), Springer, Berlin, Heidelberg, New York, London, Paris, Tokyo
12. N. Bourbaki, *Groupes et Algèbres de Lie, Chapitres 4–8* (1968–1975), Hermann, Paris
13. N. Bourbaki, *Groupes et Algèbres de Lie, Chapitre 9* (Groupes de Lie réels compacts), Masson, Paris etc. (1982)
14. N. Bourbaki, *Topological Vector Spaces, Chapter 1–5* (1987), Springer, Berlin, Heidelberg, New York, London, Paris, Tokyo
15. N. Bourbaki, *Intégration, Chapitres 1–8* (1952-1959), Hermann, Paris
16. N. Bourbaki, *Théories Spectrales* (1967), Hermann, Paris
17. R. Carnap, *An Introduction to the Philosophy of Science* (edited by Martin Gardner) (1995), Dover Publications, New York
18. H.S.M. Coxeter, *Regular Polytopes* (1973), Dover, New York
19. P.A.M. Dirac, *The Principles of Quantum Mechanics* (Second Edition, 1935), Oxford At the Clarendon Press
20. J. Dixmier, *C*-Algebras* (1969, English Translation 1977), North-Holland Publishing Company, Amsterdam, New York, Oxford
21. V. Fock, Zur Theorie des Wasserstoffatoms, *Zeitschrift für Physik* 98 (1935), 145–154
22. G.B. Folland, *A Course in Abstract Harmonic Analysis* (1995), CRC Press, Boca Raton, Ann Arbor, London, Tokyo
23. F.G. Frobenius, *Gesammelte Abhandlungen* (1968), Springer Verlag, Berlin

H. Saller, *Operational Symmetries*, DOI 10.1007/978-3-319-58664-9

24. W. Fulton, J. Harris, *Representation Theory* (1991), Springer
25. G. Gamow, *One Two Three ... Infinity* (1947), Viking Press
26. M. Gardner, *The New Ambidextrous Universe* (1990), W.H. Freeman, New York
27. I.M. Gelfand and D.A. Raikov, Irreducible Unitary Representations of locally bicompact groups, *Mat. Sb.* 13 (55) (1942), 301–316
28. I.M. Gelfand, M.A. Neumark, *Unitäre Darstellungen der klassischen Gruppen* (1950, German Translation 1957), Akademie Verlag, Berlin
29. I.M. Gel'fand, G.E. Shilov, *Generalized Functions I (Properties and Operations)* (1958, English translation 1963), Academic Press, New York and London
30. I.M. Gel'fand, G.E. Shilov, *Generalized Functions II (Spaces of fundamental and generalized Functions)* (1958, English translation 1968), Academic Press, New York and London
31. I.M. Gel'fand, M.I. Graev, N.Ya. Vilenkin, *Generalized Functions V (Integral Geometry and Representation Theory)* (1962, English translation 1966), Academic Press, New York and London
32. R. Gilmore, *Lie Groups, Lie Algebras and Some of Their Applications* (1974), John Wiley & Sons, New York, London, Sidney, Toronto
33. K. Gottfried, V.F. Weisskopf, *Concepts of Particle Physics I, II* (1986), Oxford University Press, New York
34. P.R. Halmos, *Measure Theory* (1950), Van Nostrand, New York
35. W. Heisenberg, *Einführung in die einheitliche Feldtheorie der Elementarteilchen* (1967), Hirzel, Stuttgart
36. W. Heisenberg, *Gesammelte Werke Abteilung C: Allgemeinverständliche Schriften, I – V*, Piper, München, Zürich (1984–1989)
37. S. Helgason, *Differential Geometry, Lie Groups and Symmetric Spaces* (1978), Academic Press, New York. London, Sydney, Tokyo, Toronto etc.
38. S. Helgason, *Groups and Geometric Analysis* (1984), Academic Press, New York. London, Sydney, Tokyo, Toronto, etc.
39. J. Hucks, Global Structure of the Standard Model, Anomalies and Charge Quantization, *Phys. Rev.* D 43 (1991), 2709–2717
40. E. Inönü and E.P. Wigner, On the contraction of groups and their representations, *Proceedings of the National Academy of Sciences* 39 (1952), 510–525
41. N. Jacobson, *Lie Algebras* (1961), Dover, New York
42. G. Källen, *Helv. Phys. Acta*, 25 (1952), 417–434
43. A.A. Kirillov, *Elements of the Theory of Representations* (1976), Springer-Verlag, Berlin, Heidelberg, New York
44. A. Knapp, *Representation Theory of Semisimple Groups* (1986), Princeton University Press, Princeton
45. A. Koestler, *The Sleepwalkers* (1959), Penguin Inc., New York
46. T. Kugo, I. Ojima, *Progr. of Theor. Phys.* 60 (1978), 1869–1889
47. H. Lehmann, *Nuovo Cimento* 11 (1954), 342–357
48. G.W. Mackey, On induced representations of groups (1951), Amer. J. Math. 73, 576–592
49. G.W. Mackey, *Induced Representations of Group and Quantum Mechanics* (1968), W.A. Benjamin, New York, Amsterdam
50. A. Messiah, *Quantum Mechanics I, II* (1965), North Holland, Amsterdam
51. Y. Nambu, G. Jona-Lasinio, Dynamical Model of Elementary Particles Based on an Analogy with Superconductivity. I, II *Phys. Rev.* 122 (1961), 345–358, and 124 (1961), 246–254
52. M.A. Neumark, *Lineare Darstellungen der Lorentzgruppe* (1958, German Translation 1963), VEB Deutscher Verlag der Wissenschaften, Berlin
53. J. von Neumann, *Collected Works* (1961), Pergamon, New York
54. A. Pais, *Inward Bound: Of Matter and Forces in the Physical World* (1988), Oxford University Press, New York
55. F. Peter, H. Weyl, Die Vollständigkeit der primitiven Darstellungen einer geschlossenen kontinuierlichen Gruppe, *Math. Ann.* 97 (1927), 737–755
56. H. Reichenbach, *The Philosophy of Space and Time* (1927), Dover, New York

57. B. Russell, *History of Western Philosophy* (1945), Routledge, London
58. B. Russell, *Wisdom of the West* (1970), Macdonald, London
59. H. Saller, *Operational Quantum Theory I - Nonrelativistic Structures* (2006), Springer, New York etc.
60. H. Saller, *Operational Quantum Theory II - Relativistic Structures* (2006), Springer, New York etc.
61. H. Saller, *Operational Spacetime - Interactions and Particles* (2009), Springer, New York etc.
62. H. Saller, Operational Spacetimes, *International Journal for Theoretical Physics*, 56 (2017), 168–220
63. W. Schadewaldt, *Der Gott von Delphi und die Humanitätsidee*, Insel Taschenbuch 1291, Frankfurt am Main (1990)
64. J. Schmidt, *Mengenlehre I* (1966), BI Hochschultaschenbücher 56/56a, Mannheim
65. R.S. Strichartz, Harmonic analysis on hyperboloids, *Journal of Functional Analysis* **12**, 341-383 (1973)
66. F. Treves, *Topological Vector Spaces, Distributions and Kernels* (1967), Academic Press, New York, London
67. N.Ja. Vilenkin, A.U. Klimyk, *Representations of Lie Groups and Special Functions* (1991), Kluwer Academic Publishers, Dordrecht, Boston, London
68. S. Weinberg, A model of leptons, *Phys. Rev. Lett.* 19 (1967), 1264–1266
69. C.F.v. Weizsäcker, *Zum Weltbild der Physik* (1993), Hirzel, Stuttgart
70. H. Weyl, *Sitzungsber. d. Preuß. Akad. d. Wissensch.* (1918), 465–480, *Ann. d. Physik* 59 (1919), 101–133
71. H. Weyl, *Raum, Zeit, Materie* (1923), Wissenschaftliche Buchgesellschaft, Darmstadt
72. H. Weyl, *Z. Physik* 56 (1929), 330–352
73. H. Weyl, *The Theory of Groups and Quantum Mechanics* (1931), Dover, New York
74. H. Weyl, *The Classical Groups* (1946), Princeton U. Press, Princeton, N.J.
75. H. Weyl, *Symmetry* (1983), Princeton U. Press, Princeton, N.J.
76. E.P. Wigner, Über die Operation der Zeitumkehr in der Quantenmechanik, *Göttinger Nachrichten* 31 (1932), 546–549
77. E.P. Wigner, On unitary representations of the inhomogeneous Lorentz group, *Annals of Mathematics* 40 (1939), 149-204
78. E.P. Wigner, The Unreasonable Effectiveness of Mathematics in the Natural Sciences, *Communications in Pure and Applied Mathematics*, 13 (1960), 1–14

Index

H. Saller, *Operational Symmetries*, DOI 10.1007/978-3-319-58664-9

Kolibri 360 GmbH
Ferdinand-Jühlke-Straße 7
99095 Erfurt, Deutschland
produktsicherheit@kolibri360.de